HANDBOOK OF
INDUSTRIAL ROBOTICS

HANDBOOK OF
INDUSTRIAL ROBOTICS

SHIMON Y. NOF, *Editor*
School of Industrial Engineering
Purdue University
West Lafayette, Indiana

With a Foreword by Isaac Asimov

JOHN WILEY & SONS
New York • Chichester • Brisbane • Toronto • Singapore

Library of Congress Cataloging in Publication Data:
Main entry under title:

Handbook of industrial robotics.

  Includes index.
  1. Robots, Industrial—Handbooks, manuals, etc.
I. Nof, Shimon Y., 1946–
TS191.8.H36   1985   629.8'92   84-20969
ISBN 0-471-89684-5

Printed in the United States of America

10 9 8 7 6 5 4 3 2

*This handbook is dedicated
to all of us who believe in
the wonders of human ingenuity
and robot servitude for the
betterment of our life*

CONTRIBUTORS

Gerald J. Agin, Senior Research Scientist, The Robotics Institute, Carnegie-Mellon University, Pittsburgh, Pennsylvania

James S. Albus, Acting Chief, Industrial Systems Division, National Bureau of Standards, Washington, D.C.

James M. Apple, Jr., Senior Vice President, Systecon, Inc., Duluth, Georgia

Robert U. Ayres, Professor, Department of Engineering and Public Policy, Carnegie-Mellon University, Pittsburgh, Pennsylvania

Jean-Louis Barre, Systems Engineer, Cybotech, Indianapolis, Indiana

Antal K. Bejczy, Jet Propulsion Laboratory, Pasadena, California

Christian Blume, Institute for Information III, University of Karlsruhe, Karlsruhe, West Germany

M. C. Bonney, Department of Production Engineering and Production Management, University of Nottingham, Nottingham, United Kingdom

Wayne J. Book, Associate Professor, School of Mechanical Engineering, Georgia Institute of Technology, Atlanta, Georgia

Patrick J. Bowles, Applications Engineer, Advanced Technology Section, Major Appliance Business Group, General Electric Company, Louisville, Kentucky

Rodney A. Brooks, Department of Computer Science, Stanford University, Stanford, California

Timothy J. Bublick, Manager, Application Engineering, The DeVilbiss Company, Toledo, Ohio

Stephen J. Buckley, Staff Programmer, IBM Corporation, Boca Raton, Florida

Charles F. Carter, Jr., Technical Director, Cincinnati Milacron, Inc., Cincinnati, Ohio

Richard (Ben) Cartwright, Project Engineer, Systems Division, Unimation, Inc., Danbury, Connecticut

Michael J. W. Chen, Senior Systems Scientist, Machine Intelligence Corporation, Sunnyvale, California

Fred A. Ciampa, Application Consultant, Robotics and Automation Center, Ford Motor Company, Dearborn, Michigan

Gale F. Collins, Senior Associate Information Developer, IBM Corporation, Boca Raton, Florida

Tibor Csakvary, Cyber Technologies, Inc., Pittsburgh, Pennsylvania

Mark Cutkosky, Research Associate, Department of Mechanical Engineering, Carnegie-Mellon University, Pittsburgh, Pennsylvania

Michael P. Deisenroth, Associate Professor, Department of Mechanical Engineering and Engineering Management, Michigan Technological University, Houghton, Michigan

R. Dillman, Institute for Information III, University of Karlsruhe, Karlsruhe, West Germany

Wallace D. Dreyfoos, Chief Planning Engineer, Manufacturing Research Department, Lockheed-Georgia Company, Marietta, Georgia

Michael E. Duncan, Engineering Department, Cambridge University, Cambridge, England

Joseph F. Engelberger, President, Unimation, Inc., Danbury, Connecticut

L. Wayne Garrett, Manager, Equipment Development, Advanced Technology Section, Major Appliance Business Group, General Electric Company, Louisville, Kentucky

J. A. Gleave, Department of Production Engineering and Production Management, University of Nottingham, Nottingham, United Kingdom

J. L. Green, Department of Production Engineering and Production Management, University of Nottingham, Nottingham, United Kingdom

Yukio Hasegawa, Professor, System Science Institute, Waseda University, Tokyo, Japan

Lane A. Hautau, Accounts Executive, GMF Robotics Corporation, Troy, Michigan

Larry L. Hollingshead, Cincinnati Milacron, Inc., Cincinnati, Ohio

Kenneth R. Honchell, Industrial Robot Division, Cincinnati Milacron, Inc., Lebanon, Ohio

Seiuemon Inaba, President, Fanuc Corporation, Nino-shi, Japan

Kenichi Isoda, Manager, Automation Engineering Department, Production Engineering Research Laboratory, Hitachi Ltd., Tokyo, Japan

Joseph Jablonowski, Senior Editor, American Machinist, New York, New York

Peter G. Jones, Manager of Customer Support, Cybotech, Indianapolis, Indiana

James A. Kaiser, Senior Manufacturing Project Engineer, Fisher Body General Office, General Motors Corporation, Warren, Michigan

Avinash C. Kak, Professor, School of Electrical Engineering, Purdue University, West Lafayette, Indiana

John A. Kallevig, Corporate Production Technology Laboratory, Honeywell, Inc., Golden Valley, Minnesota

Dan Kedrowski, Senior Welding Engineer, Cybotech, Indianapolis, Indiana

M. P. Kelly, BL Technology, Cowley Body Plant, Oxford, England

Keith L. Kerstetter, Manager Application Engineering, IBM Advanced Manufacturing Systems, IBM Corporation, Boca Raton, Florida

Jerry Kirsch, Kirsch Technologies, St. Clair, Michigan

Kerry E. Kirsch, Kirsch Technologies, St. Clair, Michigan

Thomas E. Klotz, Sales Manager, Mazak Corporation, Florence, Kentucky

Kazuhiko Kobayashi, Manager, Industrial Robot Design Department, Narashino Works, Hitachi Ltd., Tokyo, Japan

Yoram Koren, Associate Professor, Faculty of Mechanical Engineering, Technion-Israel Institute of Technology, Haifa, Israel

Sten Larsson, Project Engineer, Industrial Robot Division, Vasteras, Sweden

P. Levi, Institute for Information III, University of Karlsruhe, Karlsruhe, West Germany

Duncan B. Lowe, Technical Director, Taylor Hitec Limited, Lancashire, United Kingdom

Thomás Lozano-Pérez, Artificial Intelligence Laboratory, Massachusetts Institute of Technology, Cambridge, Massachusetts

J. Y. S. Luh, Professor, School of Electrical Engineering, Purdue University, West Lafayette, Indiana

Ralph L. Maiette, Manager, Systems Engineering, UAS Automation Systems, Inc., Bristol, Connecticut

Ann M. Martin, Deputy Director, Division of National Vocational Programs, U.S. Department of Education, Washington, D.C.

József Marton, Senior Researcher, Computer and Automation Institute, Hungarian Academy of Sciences, Budapest, Hungary

William E. McIntosh, Application Engineer, Imaging and Control Systems, Honeywell, Inc., Golden Valley, Minnesota

Ronald D. McCleary, UAS Automation Systems, Inc., Bristol, Connecticut

John D. Meyer, President, Tech Tran Corporation, Naperville, Illinois

Donald Michie, Professor, Machine Intelligence, University of Edinburgh, Edinburgh, Scotland

Steven M. Miller, Assistant Professor, Graduate School of Industrial Administration, Carnegie-Mellon University, Pittsburgh, Pennsylvania

George E. Munson, Senior Vice President, Robot Systems, Inc., Atlanta, Georgia

Yasuo Nakagawa, Senior Researcher, Production Engineering Research Department, Hitachi Ltd., Tokyo, Japan

David Nitzan, Director, Robotics Department, SRI International, Menlo Park, California

Shimon Y. Nof, Associate Professor, School of Industrial Engineering, Purdue University, West Lafayette, Indiana

Jack W. Posey, School of Industrial Engineering, Purdue University, West Lafayette, Indiana

Ronald D. Potter, Vice President, Robot Systems, Inc., Atlanta, Georgia

Ulrich Rembold, Professor, Institute for Information III, University of Karlsruhe, Karlsruhe, West Germany

Charles A. Rosen, Chief Scientist, Machine Intelligence Corporation, Sunnyvale, California

Gavriel Salvendy, Professor, School of Industrial Engineering, Purdue University, West Lafayette, Indiana

Victor Scheinman, Vice President of Advanced Research, Automatix, Inc., Billerica, Massachusetts

Rolf D. Schraft, Director, Fraunhofer Institute for Manufacturing Engineering and Automation, Stuttgart, West Germany

Joachim Schuler, Fellow Scientist, Fraunhofer Institute for Manufacturing Engineering and Automation, Stuttgart, West Germany

Albert M. Sciaky, Associate Director, Manufacturing Technology Center, IIT Research Institute, Chicago, Illinois

Mario Sciaky, President, Sciaky S.A., Vitry-Sur-Seine, France

Warren P. Seering, Associate Professor, Department of Mechanical Engineering, Massachusetts Institute of Technology, Cambridge, Massachusetts

Bruce S. Smith, Application Engineer, ASEA, Inc., Troy, Michigan

Randall C. Smith, Robotics Department, SRI International, Menlo Park, California

Paul F. Stregevsky, Manufacturing Administrative Analyst, Manufacturing Research Department, Lockheed-Georgia Company, Marietta, Georgia

Rajan Suri, Assistant Professor, Division of Applied Sciences, Harvard University, Cambridge, Massachusetts

Ken Susnjara, President, Thermwood Robotics Corporation, Dale, Indiana

Michio Takahashi, Senior Researcher, Production Engineering Research Laboratory, Hitachi Ltd., Yokohama, Japan

Kazuo Tanie, Senior Research Scientist, Mechanical Engineering Laboratory, Ministry of International Trade and Industry, Ibaraki, Japan

William R. Tanner, President, Productivity Systems, Inc., Farmington, Michigan

William E. Uhde, Manager, Systems Consulting, UAS Automation Systems, Inc., Bristol, Connecticut

Tibor Vámos, Director, Computer and Automation Institute, Hungarian Academy of Sciences, Budapest, Hungary

Michael W. Walker, Associate Professor, Department of Electrical and Computer Engineering, Clemson University, Clemson, South Carolina

Martin C. Wanner, Fellow Scientist, Fraunhofer Institute for Manufacturing Engineering and Automation, Stuttgart, West Germany

Hans J. Warnecke, Professor and Head, Fraunhofer Institute for Manufacturing Engineering and Automation, Stuttgart, West Germany

John A. White, Director, Material Handling Research Center, Georgia Institute of Technology, Atlanta, Georgia

Daniel E. Whitney, C.S. Draper Laboratory, Inc., Cambridge, Massachusetts

Paul K. Wright, Professor, Department of Mechanical Engineering, Carnegie-Mellon University, Pittsburgh, Pennsylvania

Y. F. Yong, Department of Production Engineering and Production Management, University of Nottingham, Nottingham, United Kingdom

Joseph P. Ziskovsky, Manager—Robotics, GCA Corporation, PaR Systems, St. Paul, Minnesota

FOREWORD

LOOKING AHEAD

In 1939, when I was 19 years old, I began to write a series of science fiction stories about robots. At the time, the word *robot* had been in existence for only 18 years; Karel Capek's play, *R.U.R.*, in which the word had been coined, having been performed for the first time in Europe in 1921. The concept, however, that of machines that could perform tasks with the apparent "intelligence" of human beings, had been in existence for thousands of years.

Through all those years, however, robots in myth, legend, and literature had been designed only to point a moral. Generally, they were treated as examples of overweening pride on the part of the human designer; an effort to accomplish something that was reserved to God alone. And, inevitably, this overweening pride was overtaken by Nemesis (as it always is in morality tales), so that the designer was destroyed, usually by that which he had created.

I grew tired of these myriad-told tales, and decided I would tell of robots that were carefully designed to perform certain tasks, but with *safeguards built in;* robots that might conceivably be dangerous, as any machine might be, but no more so.

In telling these tales, I worked out, perforce, certain rules of conduct that guided the robots; rules that I dealt with in a more and more refined manner over the next 44 years (my most recent robot novel, *The Robots of Dawn,* was published in October, 1983). These rules were first put into words in a story called "Runaround," which appeared in the March, 1942, issue of *Astounding Science Fiction.*

In that issue, on page 100, one of my characters says, "Now, look, let's start with the three fundamental Rules of Robotics . . ." and he proceeds to recite them. (In later stories, I took to referring to them as "the Three Laws of Robotics" and other people generally say "Asimov's Three Laws of Robotics.")

I am carefully specific about this point because that line on that page in that story was, as far as I know, the very first time and place that the word *robotics* had ever appeared in print.

I did not deliberately make up the word. Since *physics* and most of its subdivisions routinely have the "-ics" suffix, I assumed that "robotics" was the proper scientific term for the systematic study of robots, of their construction, maintenance, and behavior, and that it was used as such. It was only decades later that I became aware of the fact that the word was in no dictionary, general or scientific, and that I had coined it.

Possibly every person has a chance at good fortune in his life, but there can't be very many people who have had the incredible luck to live to see their fantasies begin to turn into reality.

I think sadly, for instance, of a good friend of mine who did not. He was Willy Ley who, for all his adult life was wedded to rocketry and to the dream of reaching the moon; who in his early twenties helped found rocket research in Germany; who, year after year wrote popular books on the subject; who, in 1969, was preparing to witness the launch of the first rocket intended to land on the moon; and who then died six weeks before that launch took place.

Such a tragedy did not overtake me. I lived to see the transistor invented, and solid-state devices undergo rapid development until the microchip became a reality. I lived to see Joseph Engelberger (with his interest sparked by my stories, actually) found Unimation, Inc., and then keep it going, with determination and foresight, until it actually constructed and installed industrial robots and grew enormously profitable. His devices were not quite the humanoid robots of my stories, but in many respects they were far more sophisticated than anything I had ever been equipped to imagine. Nor is there any doubt that the development of robots more like mine, with the capacities to see and to talk, for instance, are very far off.

I lived to see my Three Laws of Robotics taken seriously and routinely referred to in articles on robotics, written by real roboticists, as in a couple of cases in this volume. I lived to see them referred to familiarly, even in the popular press, and identified with my name, so that I can see I have secured for myself (all unknowingly, I must admit) a secure footnote in the history of science.

I even lived to see myself regarded with a certain amount of esteem by legitimate people in the field of robotics, as a kind of grandfather of them all, even though, in actual fact, I am merely a chemist by training and a science-fiction writer by choice—and know virtually nothing about the nuts and bolts of robotics; or of computers, for that matter.

But even after I thought I had grown accustomed to all of this, and had ceased marveling over this amazing turn of the wheel of fortune, and was certain that there was nothing left in this situation that had the capacity to surprise me, I found I was wrong.

Let me explain . . .

In 1950 nine of my stories of robots were put together into a volume entitled *I, Robot* (the volume, as it happens, that was to inspire Mr. Engelberger).

On the page before the table of contents, there are inscribed, in lonely splendor *The Three Laws of Robotics:*

1. *A robot may not injure a human being, or, through inaction, allow a human being to come to harm.*
2. *A robot must obey the orders given it by human beings except where such orders would conflict with the First Law.*
3. *A robot must protect its own existence as long as such protection does not conflict with the First or Second Law.*

And underneath, I give my source. It is *Handbook of Robotics, 56th Edition, 2058 A.D.*

Unbelievable. Never, until it actually happened, did I ever believe that I would *really* live to see robots, *really* live to see my three laws quoted everywhere. And certainly I never actually believed that I would ever *really* live to see the first edition of that handbook published.

To be sure, it is *Handbook of Industrial Robotics,* for that is where the emphasis is now, in the early days of robotics—but I am certain that, with the development of robots for the office and the home, future editions will need the more general title. I also feel that so rapidly does the field develop, there will be new editions at short intervals. And if there are new editions every 15 months on the average, we will have the fifty-sixth edition in 2058 A.D.

But matters don't stop here. Having foreseen so much, let me look still further into the future. I see robots rapidly growing incredibly more complex, versatile, and useful than they are now. I see them taking over all work that is too simple, too repetitive, too stultifying for the human brain to be subjected to. I see robots leaving human beings free to develop creativity, and I see humanity astonished at finding that almost everyone *can* be creative in one way or another. (Just as it turned out, astonishingly, once public education became a matter of course, that reading and writing was not an elite activity but could be engaged in by almost everyone.)

I see the world, and the human outposts on other worlds and in space, filled with cousin-intelligences of two entirely different types. I see silicon-intelligence (robots) that can manipulate numbers with incredible speed and precision and that can perform operations tirelessly and with perfect reproducibility; and I see carbon-intelligence (human beings) that can apply intuition, insight, and imagination to the solution of problems on the basis of what would seem insufficient data to a robot. I see the former building the foundations of a new, and unimaginably better society than any we have ever experienced; and I see the latter building the superstructure, with a creative fantasy we dare not picture now.

I see the two together advancing far more rapidly than either could alone. And though this, alas, I will not live to see, I am confident our children and grandchildren will, and that future editions of this handbook will detail the process.

ISAAC ASIMOV

New York, New York
January 1985

PREFACE

The story of modern industrial robotics unveils over three main periods. In the 1920s, the period of early conceptualization of robots, there prevailed the physical fear of monstrous humanlike machines; in the 1960s, following the installation of pioneering robots in industry, there appeared the skepticism, sometimes mixed with ridicule, as to whether robots are at all practical; in the 1980s, with increasing robot deployment and proven success, the major issue has become whether robots are going to replace us all. Albeit very different from one relatively short period to another, such strong feelings toward a new technology are not surprising. Robots possess the two very crucial properties of life: free motion and built-in intelligence. Moreover, one cannot but admire their ability and promise to humbly take over dangerous, unpleasant, and demeaning chores—to perform indefatigably, with precision and no protest, work by command.

When Isaac Asimov wrote his Three Laws of Robotics in 1940, his purpose was to guide robots in their attitude toward humans. At present, our society is more concerned with our own attitude toward robots. Therefore, for this first edition of the *Handbook of Industrial Robotics,* I offer to add the following laws that, together with future ones, may comprise the "Robotic Codex."

THE THREE LAWS OF ROBOTICS APPLICATIONS

1. Robots must continue to replace people on dangerous jobs. (This benefits all.)
2. Robots must continue to replace people on jobs people do not want to do. (This also benefits all.)
3. Robots should replace people on jobs robots do more economically. (This will initially disadvantage many, but inevitably will benefit all as in the first and second laws.)

The overwhelming and growing amount of information about industrial robotics, and in particular, its multidisciplinary nature, have created the need for this comprehensive handbook. The development of the handbook, which started in mid-1982, was guided by the following five objectives:

1. To combine up-to-date material, prepared by leading authorities on research, development, and applications of industrial robotics, with emphasis on the industrial aspects. This is particularly important in view of the current abundance of diversified mechanical, electrical, and computer engineering robotics activity on one hand, and industrial, practical needs for integrated robotic systems on the other hand.
2. To provide engineers and decision makers in industry with important contemporary overview of the field of industrial robotics.
3. To present techniques that are available, or will shortly be available, for practitioners in this area.
4. To provide in one volume significant material that can be used in courses on robotics by university and continuing education students.
5. To motivate and encourage more investigators to become active in this field, and to further advance its technical level.

To accomplish successfully these challenging objectives, a group of distinguished experts, comprising the Editorial Board listed on page ii, was invited to assist me in deciding the structure and contents of the handbook. From each of them I gained outstanding inputs, and for their invaluable recommendations I am deeply indebted.

The many authors who contributed to this volume represent the multitude of disciplines necessary to fully cover all the important aspects of industrial robotics. They were invited because of their unique expertise in their individual subject area. These authors, from industry, universities, and governments around the world, have written about theories and techniques that in most cases they themselves invented, developed, and proved, or that they personally implemented. Each chapter was carefully reviewed by two independent reviewers from the Editorial Board, and by myself. Recommended revisions, which were intended to assure consistency and high quality, were then incorporated by the authors. I have been most privileged to work with all of them, and thank them for their wisdom and friendship.

With all the above, a creative endeavor of such magnitude is severely constrained by the rapid developments and tremendous innovations in this young and dynamic technology. We have made every effort to eliminate errors and provide the most up-to-date and reliable information available at printing time. However, any shortcomings that remain are my own responsibility. My consolation is that future editions are bound to be better.

The chapters are grouped in thirteen major parts arranged logically by the development of industrial robotics (Part 1); robots as computerized mechanical systems (Parts 2, 3, and 4); robotics implementation (Parts 5, 6, and 7); and the six major industrial robotic application areas (Parts 8 through 13). These are followed by an extensive industrial robotics terminology, which was especially prepared for the handbook and is based mainly on the material in the previous chapters. The handbook concludes with five reference appendixes and a detailed index.

I wish especially to thank the many people who helped me so ably in creating this handbook. From John Wiley and Sons: Thurman R. Poston, Editor of Handbooks, who guided me with his seasoned advice throughout the whole project; Balwan R. Singh, who meticulously supervised the excellent copy editing provided by Nancy Burleson; Valda Aldzeris and Tina Marzocca, who painstakingly coordinated the complex preproduction process; Carolyn Joseph, the designer; and Ed Cantillon, who supervised the book through the production process. From Purdue University: Thanks to Jack W. Posey, who prepared the index for this handbook in cooperation with the chapter authors; Diane Schafer, who assisted me kindly with correspondence and other office work; Ed Fisher, Hannan Lechtman, Oded Maimon, Andy Robinson, Cristy Sellers, and Bob Wilhelm, my graduate students who provided talented assistance in a variety of editorial tasks.

I am also very grateful to my gracious colleagues, who inspired my work on the handbook with numerous suggestions: Moshe Barash, Tibor Csakvary, John DiPonio, Ehud Lenz, Alan Letzt, John Luh, Colin Moodie, Richard (Lou) Paul, Charlie Rosen, Gavriel Salvendy, Jim Solberg, Andy Whinston, and Dan Whitney.

Finally, I would like to express my gratitude to my wife Nava for her untiring help and counsel throughout the duration of this project; my parents, Dr. Jacob and Yaffa Nowomiast; and to our daughters, Moriah and Jasmin, for their cheerful support.

SHIMON Y. NOF (NOWOMIAST)

West Lafayette, Indiana
May 1985

CONTENTS

PART 6   APPLICATION PLANNING: TECHNIQUES

PART 7   APPLICATION PLANNING: INTEGRATION

PART 8   FABRICATION AND PROCESSING

PART 9   WELDING

PART 10   MATERIAL HANDLING AND MACHINE LOADING

PART 11   ASSEMBLY

PART 12   INSPECTION, QUALITY CONTROL, AND REPAIR

PART 13   FINISHING, COATING, AND PAINTING

ROBOTICS TERMINOLOGY, *by J. Jablonowski and J. W. Posey*

APPENDIX   INDUSTRIAL ROBOTICS AROUND THE WORLD

INDEX

PART 1
DEVELOPMENT OF INDUSTRIAL ROBOTICS

CHAPTER 1

HISTORICAL PERSPECTIVE OF INDUSTRIAL ROBOTICS

JOSEPH F. ENGELBERGER

Unimation, Inc.
Danbury, Connecticut

Any historical perspective on robotics should at the outset pay proper homage to science fiction. After all, the very words *robot* and *robotics* were coined by science fiction writers. Karel Capek gave us *robot* in his 1922 play *Rossum's Universal Robots* (RUR), and Isaac Asimov coined the word *robotics* in the early 1940s to describe the art and science in which we robotists are engaged today.

There is an important distinction between these two science fiction writers. Capek decided that robots would ultimately become malevolent and take over the world—Asimov from the outset built circuits into his robots to assure mankind that robots would always be benevolent. A handbook on industrial robotics must surely defend the Asimov view. That defense begins with the history of industrial robotics—a history that overwhelmingly finds benefits exceeding costs and portends ever-rising benefits.

Science fiction aside, a good place to start the history is in 1956. At that time George C. Devol had marshalled his thoughts regarding rote activities in the factory and his understanding of available technology that might be applied to the development of a robot. His patent application for a programmable manipulator was made in 1954, and it issued as patent number 2,988,237 in 1961. This original patent was destined to be followed by a range of others that would flesh out the principles to be used in the first industrial robot.

Also in 1956, Devol and Joseph Engelberger met at a fortuitous cocktail party. Thus began an enduring relationship that saw the formation and growth of Unimation Inc. The first market study for robotics was also started in 1956 with field trips to some 15 automotive assembly plants and some 20 other diverse manufacturing operations. Figure 1.1 is a reproduction of an actual data sheet prepared during this first market study.

Giving a fairly tight specification regarding what was needed to do simple but heavy and distasteful tasks in industry, the original design team set to work. First came appropriate components and then a working robot in 1959. Shortly thereafter Devol and Engelberger celebrated again—we see them in Figure 1.2 being served their cocktails, this time by a prototype Unimate industrial robot.

By 1961 the prototype work had progressed far enough to let an industrial robot venture forth. Figure 1.3 shows the first successful robot installation: a die casting machine is tended in a General Motors plant.

At this juncture it may be well to step back and retrospectively evaluate whether or not robotics should have become a successful innovation. A useful vantage point is provided by a 1968 Air Force sponsored study called Project Hindsight. The objective of the study was to determine what circumstances are necessary for an innovation to become successful. Project Hindsight concluded that there were three essential prerequisites for success:

1. There must be a perceived need.
2. Appropriate technology and competent practitioners must be available.
3. There must be adequate financial support.

For robotics there was a perceived need, certainly in the eyes of Devol and Engelberger, although it would be many years before this perception would be broadly shared. Appropriate technology was available, and very competent practitioners could be drawn from aerospace and electronic industries.

CONSOLIDATED CONTROLS CORPORATION

UNIMATION SURVEY DATA SHEET

DATE: ___5-14-56___

OBSERVER: ___MJD___

LOCATION: _____

TYPE OF WORK PERFORMED:

Press blanking of side panels from sheet stock.

SEQUENCE OF PRESENT OPERATION:

Sheet steel put into die against 3 locating pins. Trim drops through bed, stamped part withdrawn and stacked.

APPROXIMATE CYCLES PER MINUTE: 3

MAXIMUM NO. OF SEQUENCES:

Horizontal	Vertical	Rotary
8	6	4

HORIZONTAL TRAVERSE, ACCURACY, AND MAX. SPEED: 3 FT; $\pm 1/8$"; 6 IN/SEC
VERTICAL   "   "   "   "   "   : 4 FT; $\pm 1/8$"; 6 IN/SEC
ROTARY   "   "   "   "   "   : 270°; ± 12 MIN; 60°/SEC

HAND ACTION REQUIRED:

Suction cup

APPROXIMATE WEIGHT OF PART: 2 LBS

NO. OF OPERATORS: 18 per shift - 2 shifts

PROCESS MODIFICATION REQUIRED:

None

AVAILABLE AREA: 4 FT x 5 FT

/lb
8-23-56

Fig. 1.1. Reproduction of actual data sheet used in first field market study.

Fig. 1.2. Devol and Engelberger being served cocktails by a prototype Unimate robot.

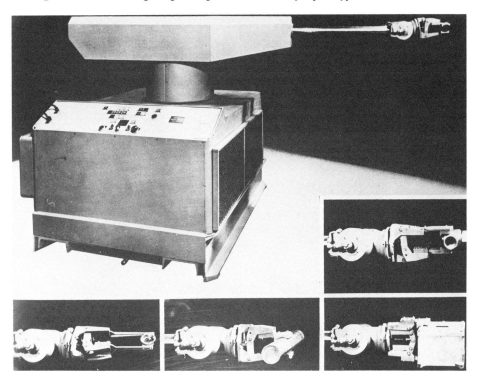

Fig. 1.3. First robot installation (die casting machine in a General Motors plant).

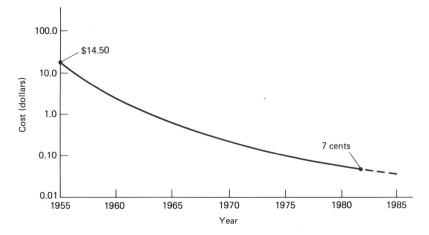

Fig. 1.4. Cost of a typical computation function. (Data according to IBM.)

Finally, venturesome financial support was brought to bear from such companies as Condec Corporation and Pullman Inc.

Back in 1922, and still in 1940, it was quite possible for Capek and Asimov to perceive a need for robots. There were certainly many heinous jobs that created subhuman working conditions. It might also have been possible in those times to gather financial support (witness all of the harebrained schemes that Mark Twain innocently sponsored); however, the technology simply was not at hand.

There are three technologies that were born during or after World War II that are crucial to successful robotics. First, servo mechanisms theory was unknown before World War II. Second, digital computation came into its own after World War II, and, finally, solid state electronics made it all economically feasible.

It is interesting to look at what has happened to the cost of electronic computation since the first tentative steps were made to produce a control system for an industrial robot. Figure 1.4 is a semilog plot of the cost of a typical computation function versus time. What in 1955 might have cost $14.00 by 1982 would cost seven cents. That is a 200-fold reduction in cost. It allows today's roboticist to luxuriate in computational hardware and make his heavy investments in the software. In 1956 one of the design challenges was to accomplish necessary functions with frugal use of electronics hardware. One of Unimation's triumphant decisions was to go solid state in its controller design at a time when vacuum tube controller execution would have been substantially cheaper. At that time a five-axis controller for a numerically controlled machine tool could have been acquired at an Original Equipment

Fig. 1.5. Current robot applications.

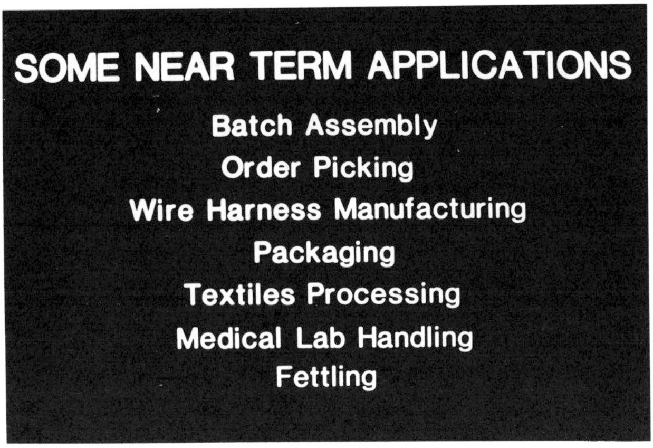

SOME NEAR TERM APPLICATIONS

Batch Assembly

Order Picking

Wire Harness Manufacturing

Packaging

Textiles Processing

Medical Lab Handling

Fettling

Fig. 1.6. Near-term robot applications.

Manufacturer (OEM) discount price of about $35,000.00. Unimation engineers prided themselves on a purpose-built design that could be achieved in 1959 for $7000.00.

For the first robot the cost to manufacture was 75% electronic and 25% hydromechanical. Today that cost ratio is just reversed.

One should note that automation was already flourishing before World War II. There were many high-volume products that were made in large quantities by what today is called "hard automation." Charlie Chaplin in his 1936 movie "Modern Times" was able to decry satirically the role of a human floundering in an automated manufacturing scene. However, all of that automation used mechanics that today we archly call "bang-bang." It is ironic that the word *robot* has become so glamorous that some companies, and even some countries, include mechanisms using this "bang-bang" technology in their categorization of robotics. The Japanese recognize "limited sequence" robots, which are conceptually turn-of-the-century technology, as being the single largest segment of the Japanese robot population (more about the Japanese role in this historic perspective shortly).

In 1961 the first industrial robot went to work, and Unimation's founder and president proved just how clouded his crystal ball was by going from 1961 until 1975 before his company was able to show a profit. The publicity was great; it attracted many abortive competitive efforts. But those who provided that third ingredient, money, were sorely disappointed in the slow progress. Just consider: the first robot worked quite well! It is now in the Smithsonian Institute. Some of its brethren are still functioning today. Many of the earliest robots built have accumulated more than 100,000 hours of field operation and that is more than 50 man-years of working. The concept was viable, the product

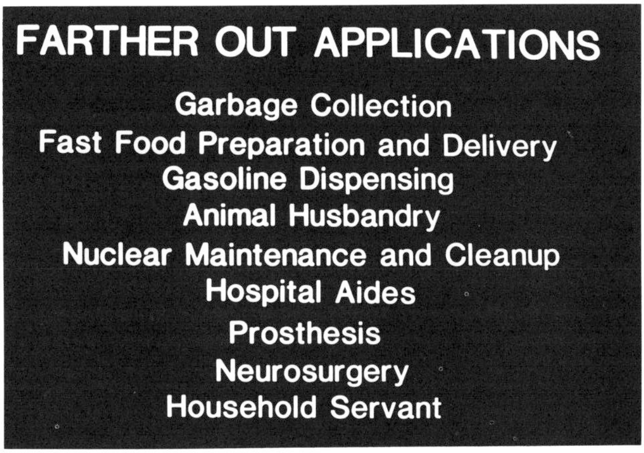

FARTHER OUT APPLICATIONS

Garbage Collection

Fast Food Preparation and Delivery

Gasoline Dispensing

Animal Husbandry

Nuclear Maintenance and Cleanup

Hospital Aides

Prosthesis

Neurosurgery

Household Servant

Fig. 1.7. Long-range applications.

was viable. Why was it taking so long to gain sufficient acceptance to make robot manufacture economically viable?

In retrospect we recognize that manufacturing is an extremely conservative activity. It does not take to change lightly. Worse than that, no one really needs a robot! Anything a robot can do, a human can also do. The only justification for hiring a robot is that a robot will work at a cheaper rate. Even that justification is not convincing if one's competitors are not making extensive use of robots. The institutional load was a formidable one, and at times it seemed almost insurmountable.

Enter the Japanese. In the early 1960s Japanese visitors to Unimation increased in frequency, and by 1967 Engelberger was an invited speaker in Tokyo. In the United States it was difficult to gain the attention of industrial executives, but the Japanese filled the hall with 700 manufacturing and engineering executives who were keenly interested in the robot concept. They followed the formal presentation with three hours of enthusiastic questioning. In 1968 Kawasaki Heavy Industries took a license under all of Unimation Inc.'s technology, and by 1971 the fever had spread and the world's first robot association was formed—not in the United States but in Japan. The Japan Industrial Robot Association (JIRA) started out with an opening membership of 46 companies and with representatives having personal clout in the industrial community. The first president of JIRA was Mr. Ando, the Executive Vice President of Kawasaki Heavy Industries, a three billion dollar company.

Thereafter the rest of the industrial world slowly began to awaken. The Robot Institute of America was founded in 1975, well after the first International Symposium on Industrial Robotics (ISIR) was held in Chicago in 1970. That first ISIR attracted 125 attendees despite coincidence with a crippling snowstorm. Before this handbook is published the thirteenth ISIR will also be history, and all indications are that 1200 will attend the conference itself, and the industrial exhibition will attract some 25,000 visitors.

Perhaps the institutional job has finally been accomplished. Look at the industrial giants who are attempting to stake out claims in the robotics arena. Beyond the biggest in Japan who are already well represented, we have such companies as General Motors, General Electric, Westinghouse, IBM, and United Technologies in the United States, and major European industrialists such as G.E.C. in England, Siemens in Germany, Renault in France, Fiat in Italy. Add to these a legion of smaller companies who fragment the market and make their mark in specialized robots, in robot peripherals, or in consulting and in robotic system design.

The governments of virtually every major industrial country in the world, capitalist or communist, have declared robotics to be an arena of intense national interest worthy of support from public coffers. So obviously robotics has arrived, hasn't it? Or, really, has it? We have a plethora of robot manufacturers, very few of whom are profitable. There is a shakeout under way unique in industrial history. It is occurring before any robot manufacturer has achieved great financial success.

The commercially available technology is not remarkably different from what existed 20 years ago. Moreover, none of the obvious applications is even close to saturation. Figure 1.5 lists applications that have been proven both technically and economically and still represent great robotic opportunities. There is little imagination necessary to go beyond the current level of commercially available technology to the addition of rudimentary vision or modest tactile sensing ability to accomplish another broad spectrum of jobs such as those listed in Figure 1.6. Further, jobs outside of the industrial robot stamping ground are already on the technically visible horizon. Some of these are listed in Figure 1.7.

What wonderful good luck to have founded a company, nay, even an industry, when one is young enough to participate during the industry's adolescence and to speculate on the tremendous technical excitement ahead as robotics reaches its majority. A handbook on industrial robotics will need be a living document for at least the balance of this century to keep up with the inevitable expansion of the technology. From the historical perspective one wishes the editors good health, long life, and a proclivity for conscientious reporting.

CHAPTER 2
THE ROLE OF ROBOTS IN AUTOMATING WORK

CHARLES F. CARTER, JR.

Cincinnati Milacron, Inc.
Cincinnati, Ohio

2.1. SOME PERSPECTIVES

Historically, we have automated work for three basic reasons:

1. The energy required to perform the task or the surrounding environment is beyond human endurance.
2. The skill required to produce a useful output is beyond human capability.
3. The demand for output (product) is so great that there is motivation to seek better methods.

A fourth factor plays an important but secondary role: the availability of a new technology that can be brought to bear on the task in question.

The role of robots in automating work fits these basic reasons as we can see from an examination of typical applications. Further, the robot itself is made possible only because of the availability of computer-related technology.

If we consider robotics as a technology, we can better understand the significance of current applications by looking at the three stages of a technology as described by Naisbitt.[1] During the first stage the technology is related to what may be considered nonthreatening applications that reduce the chance that it will be rejected. Robots are still exploiting this early stage in such applications as welding, handling hot or heavy parts, and working in hot, unsafe ambient conditions. This first stage is also compatible with the first in the hierarchy of reasons given for automating work.

The second stage improves on existing technologies and methods. Here we begin to see some disruption of the status quo and a need to change organizations and systems to take advantage of the new technology. Robot technology is now entering this phase, and, as we show in the discussion of the role of robots in computer-aided manufacturing (CAM), significant system changes or accommodations are required on the part of users. This phase in some respects relates to the second in the list of reasons for automation.

The second phase in the life of a technology is usually of long duration, and robot technology will be no exception. Experts expect that the work going on now to improve sensors with tactile, vision, and force characteristics, plus work on controls to incorporate artificial intelligence, will bring us third- and fourth-generation robots. However, success in this work will still maintain the robot in the second phase of the technology. This is because all of this work is aimed at imparting to the robot human skills and judgment, replacing or enhancing existing technology—a difficult but entirely foreseeable task.

In the third phase of a technology new directions and uses are found that were not predicted at the outset. As an example, rocket engine and space technology have spawned the significant business of satellite communications, which is having a most profound influence on our lives. But early predictions relating to applications dwelled on exploration, manufacturing in space, and military uses. In this context the technology of the robot is too young to foretell applications beyond the enhancement of man's skills.

2.2   A CLOSER LOOK—METALWORKING

Most current robot applications are in the metalworking manufacturing industry. This industry, there-fore, provides the best source for information concerning the nature and growth of applications. As we have mentioned before, the first applications usually involved the handling of parts in a hostile environment, such as that around furnaces or die-casting machines. Improved accuracy and control made applications in spot welding viable, and for several years almost all published colored pictures of robots depicted this application.

Next, it was logical to consider arc welding, but this application is not fully matured because it requires sensing of process variables and related responses that require further development. In parallel with welding, painting was developed as well as more sophisticated applications requiring workpiece manipulation.

Robot technology with respect to metalworking has now arrived at the phase where applications will be considered in the context of what needs to be done on a plantwide basis to improve manufacturing effectiveness. The incorporation of robots will now be evaluated on effectiveness in improving capital utilization, inventory value, and quality. The practitioners must therefore be aware of what changes will have the most impact on these factors. There is not now a broad understanding of or sensitivity to these important factors among those who plan the automation of work.

Studies show that production equipment in metalworking is poorly utilized.[2] This poor utilization is related to the large amount of unscheduled time during the course of a year plus an inability to achieve a good percentage of value-added time even when the machines are scheduled. Ayers and Miller[3] have analyzed this poor utilization of capital with respect to other costs and conclude that the gains possible from improving utilization exceed by an order of magnitude the gains possible by merely reducing labor costs. This conclusion should aid in defining the long-term role of the robot in automating work. The role is obviously not the mere replacement of the human operator.

In addition to capital equipment, another large consumer of financial resource in manufacturing is inventory. It is not unusual for the value of in-process inventory to exceed the value of production equipment for a manufacturer in the metalworking industry. Any consideration of return on investment must take into account this important element. The actions that bring about a reduction in inventory also tend to reduce lead time and improve response to changing demand. Practitioners look for these beneficial results when they consider the introduction of new machines or methods in manufacturing. Robots will be no exception.

The third characteristic of the manufacturing process that acts to stimulate the introduction of new technology is quality. Improved product quality is now almost universally recognized to bring not only greater market acceptance but also reduced manufacturing cost. The cost of improved quality is usually offset by lower costs in assembly, test, rework, scrap, and warranty.

For the purpose of determining the role that robots might play in bringing improvements to the areas just described in broad terms, it may be helpful to examine in more detail some of the factors leading to poor utilization and high inventory. The assumption here is that a perception of underlying factors must be gained before solutions can be effectively applied.

An estimate of manufacturing equipment utilization may be gained by referring to Figure 2.1. Here we see that large blocks of time are unavailable for productive use because our plants are closed. During the available productive time, various conditions reduce utilization so that a productive fraction of only 6% is left. The purpose here is not to suggest that we should change our social structure so that manufacturing plants can be open more hours per year. Indeed, it will be increasingly difficult to find people to work in a manufacturing environment, let alone working more on second and third shifts. This figure points out that the productive fraction must be improved, and must be improved without requiring large numbers of people to work at undesirable times. Even when the typical machine is scheduled to be used, the productive fraction is only slightly more than 25% of that scheduled time. And this productive fraction is not totally consumed in adding value.

Further details with respect to individual machine utilization can also be helpful in developing an overview. A view of grinding machine utilization may be obtained by referring to Table 2.1.

Here we see that 60% of the scheduled time for a grinding machine is consumed in activities other than actual grinding. This does not account for the fact that schedule time is undoubtedly less than the total time available for production.

Machines dedicated to high production present no better picture with respect to utilization or productivity. Data shown in Tables 2.2, 2.3, and 2.4 show the status of productivity in an environment where machines are dedicated and losses related to setup would not be an important factor.

Table 2.2 indicates that the accepted standard for output on a typical high-production machine or system is 80% of the calculated optimum. In actual practice, only 59% of the optimum is achieved.

The result of further analysis of the reasons for loss in productivity is shown in Table 2.3. The percentage for equipment failure represents the total time charged to that element, which encludes the time to have the proper maintenance skill respond to the problem. Work force control relates to

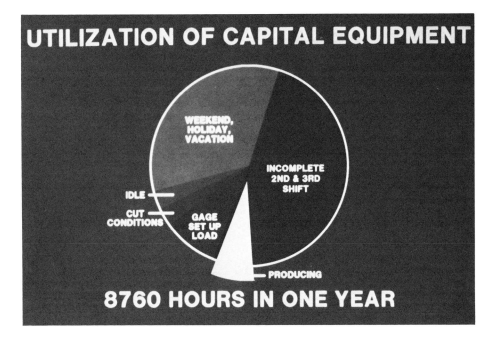

Fig. 2.1. Typical use of available time in batch manufacturing shop. Productive fraction is 6%.

the total problem of operator availability, which includes responsiveness to a particular problem plus such general problems as absenteeism and tardiness.

It is interesting to restructure these losses as has been done in Table 2.4 to show that 64% of total losses may be attributed to the factors relating to management effectiveness.

The situation is not markedly improved, with respect to actual time-in-cut, by the introduction of numerical control. The general productivity gains due to numerical control are well documented but are sometimes misinterpreted. Improvements frequently accrue as a result of the combination of operations in one setting rather than improvements in the cycling of machines.

Table 2.5 shows the results of a study of numerically controlled machining centers. Here we see that the actual metalcutting time consumes less than a quarter of the time a machine is scheduled

TABLE 2.1. TYPICAL GRINDING MACHINE UTILIZATION

Activity	Percent Time
In cycle	40
Gaging, loading	20
Setup	20
Waiting and idle	13
Repair and technical	7

TABLE 2.2. STATUS OF PRODUCTIVITY—HIGH PRODUCTION

Machine optimum	100%
Established standard	80%
Actual output	59%

TABLE 2.3. SOURCES OF PRODUCTIVITY LOSS—HIGH PRODUCTION

Equipment failure	42%
Machine in wait mode	34%
Work force control	16%

TABLE 2.4. MANAGEMENT-RELATED PRODUCTIVITY LOSSES

Skill trades response	14%
Machine in wait mode	34%
Work force control	16%
Percent of total loss	64%

TABLE 2.5. TIME UTILIZATION OF MACHINING CENTERS

Activity	Percent Time
Metalcutting	23
Positioning, tool changing	27
Gaging, loading	18
Setup	5
Waiting and idle	14
Repair and technical	13

and that considerable improvement could be made mainly by management procedures to reduce waiting and idle time. One of the positive features of numerical control, that of reduced setup time, is substantiated by these data.

The data reported in the preceding sections are from actual measurements but depict average values for those measurements. Obviously, situations can be cited that would not correspond to these values. However, they are presented as valid, representative values from which general conclusions can be drawn.

The point to be made in presenting these data is that gains in individual machine productivity depend on improving all aspects of machine usage. Traditionally, emphasis has been given to the reduction of actual cutting time. Significant future improvement will depend on greater emphasis being given to the noncutting elements of machine usage. Some of these can be brought about by machine improvements. Others can be achieved through changes in shop procedures. For instance, the careful selection and scheduling of the mix of parts coming to a machine can drastically reduce setup time. This is one of the promises of group technology.

The complexity of manufacturing is well recognized by all who are close to the function. But many are not aware that the conventional practices in manufacturing relating to schedules and lead time have evolved to cope with complexity. Consider a typical part moving through several metalworking operations to be completed. The schedule is made weeks in advance. The assumption is that at each move machines, tools, and fixtures will be available and ready to work on the part. If the part requires 5–20 moves, what are the chances of everything happening as planned several weeks in advance? Low.

Superimposed on the inherent complexity of the process are many disturbances that cause delays and require schedule changes. Typically disturbances are caused by engineering changes, vendor lateness, material conditions, machine availability, process problems, and emergency orders, and the list goes on.

To cope with complexity and disturbances, manufacturing practice has evolved two procedures. The first is to create buffers of work for each station or machine. This assures that work will always be available for a machine, but this also creates a large in-process inventory. The second procedure involves the changing of priorities to comply with the current status. This is done through the use of daily lists, various colored tags, and people to expedite work through the shop. Of course, work that receives special status acts as a disturbance to regularly scheduled work.

Anyone working in manufacturing knows all of this but may not recognize what action to take to improve the situation. The diagram in Figure 2.2 helps to provide a good image of the makeup of the shop schedule. It shows that the schedule is primarily queue and move time. Increasing the predictability and reliability of the manufacturing process is obviously more important than increasing the speed.

With these details in mind we have a guide for assessing the value of introducing new technology into the manufacturing process. Specifically with respect to robots, future applications will be justified less on the elimination of labor and speedup of the process and more on the impact of improving utilization and reducing part movement, complexity, and lot size requirements. These actions will bring about increased predictability, which in turn paves the way for reduced work-in-process and reduced lead time.

A further understanding of the role of robots in bringing about these improvements in manufacturing may be gained by examining the concept of the flexible manufacturing system (FMS). The physical embodiment of this concept ranges from one machine processing a selected variety of workpieces without operator attention for changing setups and machine cycles to a number of machines or process stations completing the required operations on a selected variety of workpieces.

Figure 2.3 represents the single-machine configuration. Different workpieces are fixtured on the pallets in front of the machine and automatically indexed for machining. Figure 2.4 represents the next step in complexity with a two-machine system serviced by a robot. The robot brings parts into the system, transfers parts between machines, presents finished parts to a gage, and places parts on an exit conveyance. Figure 2.5 depicts what may be termed a classical FMS with multiple stations performing a variety of tasks.

In the figure, parts are stored at the left. Adjacent to the storage, parts are loaded (or unloaded) onto pallets for distribution throughout the system by wire-guided carts. The black lines on this model indicate the paths available to the carts. The lines would not be visible in an actual installation. The machine tools appear in the center of the figure with expansion shown to the right. The computer room is upper center, and chip handling is right of center.

Regardless of the type or complexity, all FMSs display similar requirements for successful operation. The requirements relate to the need for much more detailed advance planning and scheduling. Selection of workpieces must be aimed at accommodating the size, range, accuracy, tooling, and weight constraints

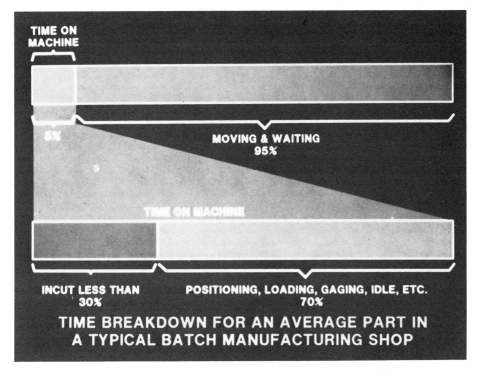

TIME ON MACHINE

5%

MOVING & WAITING
95%

TIME ON MACHINE

INCUT LESS THAN
30%

POSITIONING, LOADING, GAGING, IDLE, ETC.
70%

TIME BREAKDOWN FOR AN AVERAGE PART IN A TYPICAL BATCH MANUFACTURING SHOP

Fig. 2.2. High percentage of wait time leads to large work-in-process inventory.

Fig. 2.3. Single-machine system arranged to work on several different workpieces without operator attention.

of the system. Scheduling of workpieces through the system must be aimed at minimizing disruption caused by tool changes requiring manual intervention and maximizing machine utilization. Of course, manufacturing engineers attempt to do the same things in a traditional manufacturing environment, but optimal results are more difficult to attain because of increased complexity (wider variety of parts) and poorer control (part movement through more functional and organizational entities). In the case of an FMS, the work stations (except for system load/unload) are usually numerically controlled and therefore cycles are predictable. This, coupled with the requirement for planning and scheduling, produces the desired results of short lead time, lower in-process inventory, predictable schedule, and predictable quality level.

The concept of FMS has been applied to a wide variety of parts, and approximately 130 systems exist at the present time. Descriptions of representative systems[4,5] are usually limited to technical features and documented operating data, and benefits are seldom available. Generally the beneficial results mentioned are achieved, but available data is limited. However, a detailed comparison of 119 systems[6] concludes that only six are rated as flexible. Most systems deal with only a very limited part-number mix—frequently as low as two or three—or deal with such a narrow part classification that a stated high part-number application is misleading in terms of required flexibility. This is important to know since it would be incorrect to assume that the state-of-the-art supports 130 systems producing a variety of parts limited only by the size, weight, and accuracy constraints of the system.

The reasons for the actual versus the assumed flexibility of FMS can be explained in part and may suggest a role for robots. The primary reason for restricted flexibility is the requirement for a large number of tools to machine a wide variety of workpieces. Almost all FMSs in use are limited in the capability to handle automatically too few tools for wide part selection. Some additional capabilities are now emerging, but at considerable added expense. Robots may be applied to the problem. The situation is exactly parallel to the use of a pallet changer on a tool-changing machining center. Early models had pallet changers but not enough tool storage capacity to machine very different workpieces. Pallet changers fell in popularity until tool storage capacity increased. Likewise true flexibility in robots will require a fast, reliable means to change a variety of gripping or toollike devices.

In addition to enhancements required in techniques for transporting tools to the machining stations, several other hardware-related improvements are required for true flexibility. One is providing higher accuracy of bored holes and locations of holes without hard tooling. Another is a requirement for machine configurations with greater hole-making efficiency while maintaining versatility. The drillhead changing machine is a current answer to this problem, but further work must be done on this configuration.

Obviously, numerical control allows many different shapes to be cut from a well-defined initial geometry in turning, and numerical control is now being utilized in high production for parts of

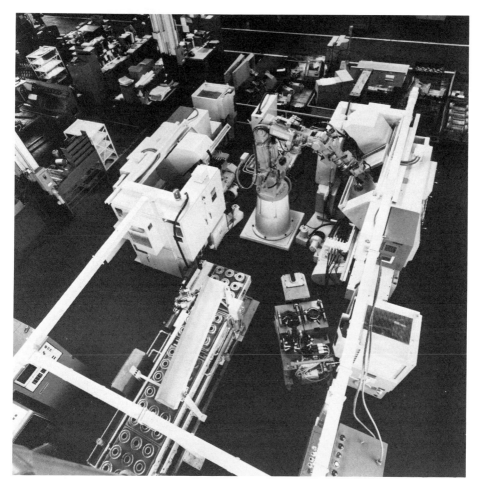

Fig. 2.4. Two-machine system—turning machines loaded by a robot.

rotation. However, in both turning and grinding the initial workpiece geometry that can be machined is limited by the lack of versatility of work-holding devices. Innovation in the area of work holding is one of the major future requirements to be fulfilled to bring more versatility to the manufacture of parts of rotation.

Attention must also be turned to support service improvements in mechanical aids that will be required as machines receive less operator attention. Even though adaptive control will make a machine more immune to in-process variables, there is a strong requirement for improved tool regrinding and reconditioning. The variability that results from these support activities has a direct influence on variations in size and finish. Operator attention is usually required to adjust for these variations.

Chip curl and resulting snarls are still a problem in the machining of steel in spite of developments in chip breaker geometry on tools. Problems associated with the control of chips must be solved if steel is to be machined in an environment with no operator attention.

All of these factors or variables are brought together in one system and act to drive the flexibility to a low level to achieve successful operation. We can expect the increasing use of FMS to depend on the solution of these problems.

What was thought to be the primary barrier to system acceptance has not developed. Ten years ago, it was predicted by many knowledgeable observers that software development would represent the greatest barrier to large-scale flexibility automation. The predictions overlooked several factors. The rapid development of microcomputers has allowed cost-effective distribution of control tasks and with it the logical partitioning of software. The development of computer operating systems designed for real-time environments has eased the burden of the applications programmer. The tremendous

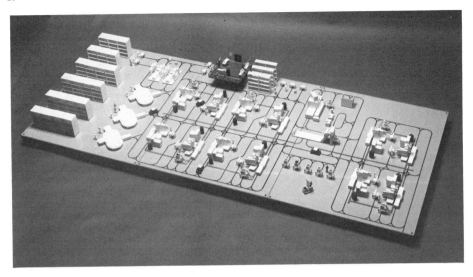

Fig. 2.5. Complete flexible manufacturing system with wire-guided carts for work and tool transport.

worldwide effort aimed at developing more powerful languages and improved programming techniques produces fallout for all disciplines. Universities, institutes, and research firms have applied countless man-hours of effort to software-related problems because this type of investigation requires little capital investment, is suited to multiple approaches and solutions, and can utilize analysts with only limited manufacturing knowledge.

The preceding discussion should not imply that problems in simulation, scheduling, and real-time control of a complex system are trivial. The problems are large and serious. The point is that in comparison to hardware considerations, software problems are well in hand.

With an understanding of the foregoing assessment of the status of flexible automation, we should be in a position to evaluate the conditions for the successful application of robots in similar situations. The first conclusion to be drawn is that successful applications of automation are generally related to situations involving high volumes or a high degree of uniformity. Robots readily fit into this environment and bring a degree of versatility greater than hard automation since they can perform multiple but repetitive tasks at no additional cost. Figure 2.6 depicts such a situation where a robot applied sealant to an automotive body—an application requiring versatility, but a high-volume repetitive task. However, when robots are applied to low-volume or nonuniform situations the organization of the work must be changed to take on some of the characteristics of high-volume production. Parts must be placed in a precise location and orientation prior to handling. Parts must be clamped or held in a specific orientation and position for processing such as welding. Figure 2.7a illustrates a welding fixture and rotary table arrangement for the robot welding of cabinets. Figure 2.7b shows the cabinet in place for welding. This is a degree of work organization not required for manual welding. The processing method must be reasonably repetitive. None of these requirements is present in low volume. It is easy to neglect the fact that low-volume processes are inherently flexible, and any attempt to bring automation to such processes requires a sacrifice of flexibility. Of course, the reduced flexibility is not detrimental if proper planning accompanies the introduction of these more automated methods.

One beneficial result of the change to more uniformity when a robot is introduced is improved quality. Every weld is made, and made under uniform conditions. Every part is properly indexed or positioned or inserted. Every operation is performed. In fact, the justification for some robot applications in low-volume relates to improved quality.

2.3. WHAT TO EXPECT

It is interesting to speculate on how fast the applications of a new technology will grow. Those who report on such matters usually pose questions to experts in the field in a manner suggesting resistance to change. Why isn't technology X growing faster? Why don't more companies use application Y? Of course, the real answer to these questions is that we do not know what the rate of growth should be. Perhaps the rate already exceeds that which is economically justified. Nevertheless, the expert will struggle to suggest several reasons to explain a perceived slow growth.

Forecasts of the growth of robot applications in the United States place dollar volume by 1992 at

Fig. 2.6.   Robot applying sealant to automobile body.

anywhere from 1 billion to 5 billion dollars and units from 20,000 to more than 250,000. Obviously the projections are of limited value beyond the simple fact that there will be continued, significant growth. It will be more helpful to examine the current characteristics of robots and comment on enhancements required to broaden the scope of applications. As these enhancements become commercially viable, the observer can expect a corresponding increase in applications.

The characteristics of range and load capacity should be considered together since the combination has a direct bearing on accuracy. For instance, a robot with a reach of 100 in. and a load capacity of 100 lb may have a repeat accuracy within a band of 0.020 in. Whereas a smaller robot with a reach of 50 in. and a load capacity of 50 lb may have a repeat accuracy within a band of 0.012 in. Since many simple applications requiring the positioning, loading, or assembly of parts demand greater accuracy, improvements are required to broaden applications. These improvements will require greater mechanical rigidity and more precise control. Improving rigidity without sacrificing range is a formidable design task.

Presently most robots are programmed by teaching each move and position required to perform the task. This does not present a serious problem where the task is to be performed continuously or repeated at frequent intervals. However, this method of programming becomes impractical for applications involving low volume combined with nonrepetitive tasks. For robot applications to make serious inroads into this enormous application area, two enhancements are required. There must be the ability to program the robot off-line in a manner similar to numerically controlled machine tools. Such capability is available today but on a very limited basis. In addition, the absolute accuracy capability must be improved so that the robot can be commanded (rather than taught) to go to any position with an accuracy approaching the repeat accuracy. Currently robot absolute positioning errors are of an order of magnitude greater than repeat errors. The required improvement may be achieved through some calibration technique rather than by means of greater rigidity and precision.

Most robot applications depend on the fact that the organization of the task is highly structured and repeatable. Parts to be moved have consistent physical characteristics and are placed in a specific location. Parts to be welded are fixtured and clamped in a repeatable and uniform manner. However, most tasks in the world of work are not uniform or structured. If robots are to be applied to a significant number of work tasks, they must acquire abilities to adjust to changes in the task. At the same time, some degree of uniformity may be added to the work environment to accommodate the robot, but this will not always be justified. To be responsive to variability or lack of structure in the

Fig. 2.7. (*a*) Fixture for robot-aided welding of steel cabinet. (*b*) Cabinet in place on rotary table for welding.

task, the robot will be required to sense and react to touch and required force. For instance, the simple task of assembling a shaft into a hole requires the robot to sense the force and recognize variable force limits with respect to penetration distance. This is a degree of sophistication not required if the parts are pregaged and the task tested to known limits—conditions not present in low-volume manufacturing.

The most widely acknowledged sense required to allow robots to work in an unstructured environment is vision. That feature is now available to a limited extent. Robots can now distinguish and react to part shape and orientation. This allows a degree of nonuniformity. However, the assessment of and reaction to a scene is required to meet most work situations. This total scene assessment is a function handled almost unconsciously by a human operator, but one which requires artificial intelligence techniques and great computational power when performed automatically.

Another characteristic required to perform many unstructured tasks is mobility. Processes currently performed by robots such as welding cannot be performed on buildings, ships, or large pieces of equipment. Robots have been given a degree of mobility by being placed on a track or overhead rail. But the mobility required for tasks like those mentioned is not available.

The foregoing comments are made to emphasize that very sophisticated enhancements will be needed to enable robots to perform tasks that are considered quite simple when performed by human operators. There are literally millions of such tasks, and that is the motivating fact behind continued robot development.

The primary reason that the growth rate of a technical device is difficult to forecast is because breakthroughs in critical hardware elements cause step inputs, and unforeseen applications can become important. On the other hand, the absence of these inputs can become important or can result in disappointing growth. A paper presented at the Twelfth International Symposium on Industrial Robots[7], which cataloged papers at all such meetings from the first, enables us to make some observations about trends in robot technology.

The technique of artificial intelligence was discussed at the first symposium. It is now commercial to a limited extent but not widely used. In retrospect, we can say that the application of this technique has been difficult to achieve and slower than expected 12 years ago. Applications in metalworking continue to receive the most attention but the impending, significant broadening of applications is signalled by a wider variety of application papers at the twelfth symposium, with applications ranging from shearing sheep to decorating chocolates. Figure 2.8 illustrates the use of a robot to decorate chocolate candy. It will be this broadening of applications away from metalworking, but with existing

Fig. 2.8.   Robot decorating chocolate candy.

technology, that will be most important for growth in the next three to five years. Then technical breakthroughs will allow growth in the direction of more sophisticated applications—that is, sophisticated for the robot but not for the human operator.

REFERENCES

1. Naisbitt, John, *Megatrends: Ten New Directions for Transforming our Lives,* Warner Books Inc., New York, 1982.
2. Carter, C. F., Jr. Toward Flexible Automation, *Manufacturing Engineering,* August 1982.
3. Ayres, Robert U., and Steven M. Miller, *Robotics Applications and Social Implications,* Ballinger Publishing Co., Cambridge, MA, 1983.
4. Hutchinson, G. K., Flexible Manufacturing Systems in the United States, *Automation Manufacturing,* 1979.
5. Dupont-Gatelmand, C., A Survey of Flexible Manufacturing Systems, *Journal of Manufacturing Systems,* Vol. 1(1), 1982.
6. Ito, Y., Present Status and Trends in Flexible Manufacturing Systems, *Journal of Japan Society of Mechanical Engineers,* **85**(761), 1982.
7. Fleck, J., The Development and Diffusion of Industrial Robots, Proceedings 12th International Symposium on Industrial Robots, Paris, France, 1982.

CHAPTER 3

ROBOTS AND MACHINE INTELLIGENCE

CHARLES A. ROSEN

Machine Intelligence Corporation
Sunnyvale, California

3.1. INTRODUCTION

The factory of the far future is visualized as composed of a complex array of computer-controlled processes, programmable machine tools, and adaptive, sensor-mediated fixed and mobile industrial robots. These systems will be operated and maintained by a small cadre of skilled technicians, and supervised by a smaller group of highly professional engineers, computer scientists, and business people. Planning, design, production, distribution, and marketing of products will depend critically on computers, used as information and knowledge-processing tools by the staff and as autonomous controllers (in the general sense) for each manufacturing process. Systems of such complexity must of necessity evolve, since, at present, major components of these systems are not yet capable of performing required functions, or are not cost-effective when they can. Even when such subsystems have attained acceptable performance, there will still remain the difficult and laborious problems of standardization, interfacing, and integration into smoothly operating factory systems.

What is the difference between a so-called "intelligent" computer system and all other computer systems? The criteria for "intelligence" vary with time. In a relatively short period of history, less than thirty years, the explosive growth of available computer science and technology has provided us with the means for supplementing and supplanting human intellectual functions far beyond our present capabilities for exploitation. At an early date, arithmetic computation or "number crunching" was considered a function performable only by intelligent natural species. In a remarkably short time (as measured on an evolutionary scale) early pioneers realized the potential of the symbol-processing capabilities of the digital computer, a revolutionary advance in abstraction rivaled by few historical events. The encoding, manipulation, and transformation of symbols, representing objects of the world, actions, induction and deduction processes, natural laws, theories and hypotheses, cause and effect, are intellectual functions that are now being performed with increasing sophistication by computers.

It is now commonplace to consider important computer applications, such as storage and retrieval, data management systems, modeling, word processing, graphics, process controllers, computer games, and many others as merely information-processing techniques devoid of intelligence. Somewhat higher in abstraction, pattern recognition systems, initiated by the development of optical character recognition techniques, have led in theory and practice to explosive growth involving the extraction and classification of relevant information from complex signals of every type. Many of these applications have become commonplace and are no longer considered as "intelligent" systems. At present it is acceptable to label programs as part of "machine intelligence" or "artificial intelligence," when they are concerned with studies of perception and interpretation, natural language understanding, common-sense reasoning and problem solving, learning, and knowledge representation and utilization (expert systems). After 20 years of primarily empirical development, including conceptualization, debugging, and analysis of computer programs, only a few implementations of this technology are now being introduced into industry, and doubtless they are already considered as "mechanistic" rather than "intelligent" systems. In the following sections the current thrust toward implementing robot systems (that progressively become more and more "intelligent") is explored.

3.2. AVAILABLE ROBOT SYSTEMS

3.2.1. First-Generation Robot Systems

The majority of robots in use today are first-generation robots with little (if any) computer power. Their only "intelligent" functions consist of "learning" a sequence of manipulative actions, choreographed by a human operator using a "teach-box." These robots are "deaf, dumb, and blind." The factory world around them must be prearranged to accommodate their actions. Necessary constraints include precise workpiece positioning, care in specifying spatial relationships with other machines, and safety for nearby humans and equipment. In many instances costs incurred by these constraints have been fully warranted by increases in productivity and quality of product and work life. The majority of future applications in material handling, quality control, and assembly will require more "intelligent" behavior for robot systems based on both cost and performance criteria.

3.2.2. Second-Generation Robot Systems

The addition of a relatively inexpensive computer processor to the robot controller led to a second generation of robots with enhanced capabilities. It now became possible to perform, in real time, the calculations required to control the motions of each degree-of-freedom in a cooperative manner to effect smooth motions of the end-effector along predetermined paths, for example, along a straight line in space. Operations by these robots on workpieces in motion along an assembly line could be accommodated. Some simple sensors, such as force, torque, and proximity, could be integrated into the robot system, providing some degree of adaptability to the robot's environment.

Major applications of second-generation robots include spot welding, paint spraying, arc welding and some assembly—all operations that are part of automated manufacturing. Perhaps the most important consequence has been the growing realization that even more adaptability is highly desirable and could be incorporated by full use of available sensors and more computer power.

3.2.3. Third-Generation Robot Systems

Third-generation robot systems have been introduced in the past few years, but their full potential will not be realized and exploited for many years. They are characterized by the incorporation of multiple computer processors, each operating asynchronously to perform specific functions. A typical third-generation robot system includes a separate low-level processor for each degree of freedom, and a master computer supervising and coordinating these processors as well as providing higher-level functions.

Each low-level processor receives internal sensory signals (such as position and velocity) and is part of the servosystem controlling that degree-of-freedom. The master computer coordinates the actions of each degree-of-freedom; can perform coordinate transformation calculations to accommodate different frames of reference; can interface with external sensors, other robots, and machines; store programs; communicate with other computer systems. Although it is possible to perform all the functions listed with a single computer, the major trend in design appears to favor distributed hierarchical processing, the resulting flexibility and ease of modification justifying the acceptably small incremental costs incurred by use of multiple processors.

3.3. INTELLIGENT ROBOT SYSTEMS

3.3.1. Adaptive, Communicating Robot Systems

A third-generation robot equipped with one or more advanced external sensors, interfaced with other machines, and communicating with other computers could be considered to exhibit some important aspects of intelligent behavior. Interfaced with available machine vision, proximity, and other sensor systems (e.g., tactile, force, torque), the robot would acquire randomly positioned and oriented workpieces; inspect them for gross defects; transport them to assigned positions in relation to other workpieces; do insertions or other mating functions, while correcting its actions mediated by signals from force, torque, and proximity sensors; perform fastening operations; and finally verify acceptable completion of these intermediate assembly processes. Its computer would compile statistics of throughput, inspection failures by quantity and type, and would communicate status with neighboring systems and to the master factory system computer. The foregoing scenario is just one of many feasible today. The major functional elements of such an intelligent system are the following:

1. The capability of a robot system to adapt to its immediate environment by sensing changes or differences from some prespecified standard conditions and by computing, in real time, the necessary corrections for trajectories and/or manipulative actions.

2. The capability of interacting and communicating with associated devices (such as feeders and other robots) and with other computers so that a smoothly integrated manufacturing system can be implemented, incorporating fail-safe procedures and alternate courses of actions to maintain production continuity.

Clearly, the degree of intelligence exhibited by such systems depends critically on the complexity of the assigned sequence of operations and how well the system performs without failure. At present the state of the art in available machine vision and other sensory systems requires considerable constraints to be engineered into the system and therefore limits applications to relatively simple manufacturing processes. However, rapid progress in developing far more sophisticated machine vision, tactile, and other sensory systems can be expected, with consequent significant increases in adaptability in the next two to five years. The level of "intelligence," however, will reside primarily in the overall system design, quite dependent on the sophistication of the master program that orchestrates and controls the individual actions of the adaptive robots and other subsystems.

3.3.2. Programming the Adaptive Robot

Programming the adaptive robot consists, roughly, of two parts:

1. The first is a program that controls the sequence(s) of manipulative actions, specifying motions, paths, speed, tool manipulation, and so on. Several different sequences may be "taught" and stored and called up as required by some external sensory input or as a result of a conditional test.

The programming of these sequences has traditionally been implemented by using a "teach box" for first- and most second-generation robot systems. This method of on-line programming is very attractive as it is readily learned by factory personnel who are not trained software specialists.

2. The second is a program that controls the remainder of the adaptive robot's functions, such as sensory data acquisition, coordinate transformations, conditional tests, communications with other devices and computers. Programming this part is off-line and does require a professional programmer.

It is likely that some form of "teach-box" programming will be retained for many years, even for third-generation robots, since the replacement of off-line programming would require the development of a complex computer model of the robot and its total immediate environment, including dynamic as well as static characteristics. The master program that controls adaptive behavior and communications will call up, as subroutines, the manipulative sequences taught on-line as described.

Machine intelligence research has, for many years, included the development of high-level programming languages designed specifically for robotic assembly (see Chapters 22, 23). An appropriate language would permit off-line programming of complex assembly operations, with perhaps some calls to special on-line routines.

Ultimately one may strive for the goal of using a natural language to direct the robot. This may develop at increasing levels of abstraction. For example, an instruction as part of a program controlling an assembly task might be:

PICK UP THE BOLT FROM BIN NUMBER ONE

At this level the "smart" system must interpret the sentence, be able to recognize the object (bolt) using a vision sensor, plan a trajectory to bin number one (the position of which is modeled in its memory), acquire one bolt (after determining its position and orientation using a vision sensor), check (by tactile sensing) that the bolt has been acquired, and then await the next high-level instruction.

In a more advanced system operating at a higher level of abstraction, the instruction might be:

FASTEN PART A TO PART B WITH A QUARTER-INCH BOLT

This presumes that previous robot actions had brought parts A and B together in correct mating positions. The "smarter" system, here hypothesized, would know where all sizes of bolts are kept and carry out the implied command of picking a quarter-inch bolt from its bin, aligning it properly for insertion, inserting it, after visually determining the precise position of the mating holes, and checking it after insertion for proper seating. A high-level *planning* program, having interpreted the instruction, would invoke the proper sequence of subroutines.

It is apparent that one can increase robotic intelligent behavior indefinitely by storing more and more knowledge of the world in computer memory, together with programmed sequences of operations required to make use of the stored knowledge. Such levels of intelligent behavior, while attainable, are still at the research stage but can probably be demonstrated in laboratories within five years.

3.3.3. Future Developments in Intelligent Robotic Systems

The development of intelligent robot systems is truly in its earliest stages. The rapid growth of inexpensive computer hardware and increases in software sophistication are stimulating developments in machine intelligence, especially those to be applied usefully in commerce and industry. General acceptance of third-generation adaptive robot systems will lead to the widespread belief that much more intelligence in our machines is not only possible but also highly desirable. It is equally probable that expectations will be quite unrealistic, and exceed capabilities.

The following sections examine some interesting aspects of machine (or "artificial") intelligence (AI) research relevant to future robotics systems.

Sensors

Sensing and interpreting the environment are key elements in intelligent adaptive robotic behavior (as in human behavior). Physicists, chemists, and engineers have provided us with a treasure of sensing devices, many of which perform only in laboratories. With modern solid-state techniques in packaging, ruggedization, and miniaturization, these sensors can be adapted for robot use in factories.

Extracting relevant information from sensor signals and subsequent interpretation will be the function of inexpensive high-performance computer processors. We can expect that with these advanced sensors, a robot will, in time, have the capability to detect, measure, and analyze data about its environment considerably beyond unaided human capabilities, using both passive and active means for interaction. Sensory data will include many types of signals: signals from the whole electromagnetic spectrum from static magnetic fields to X-rays; acoustic signals ranging from subsonic to ultrasonic; measurements of temperature, pressure, humidity; measurements of physical and chemical properties of materials using many available spectroscopic techniques; detection of low-concentration contaminants; electrical signals derived from testing procedures (including physiological); and many more sensory modalities.

One may expect that such sensors with their integrated computer processors will be made available in modular form with standardized computer interfaces to be selected as optional equipment for robotic systems.

Knowledge-Based (Expert) Systems

The technology of knowledge-based (expert) systems, typified by Stanford's "Dendral," "Mycin," and SRI's "Prospector" systems, has been developed sufficiently for near-term implementation in factories. In such systems, carefully selected facts and relations about a large body of specialized information/ knowledge in a well-defined restricted domain has been encoded with the aid of one or more high-level human experts in that domain. A trained practitioner (but not necessarily an expert in that domain) can access the encoded expertise in an interactive give-and-take interchange with the computer program. The program can include empirical rules, laws of physics, models of processes, tabled values, and data bases of many types.

It is expected that expert systems will be highly useful for the factory-of-the-future at many levels of the production process. In CAD/CAM (computer-aided design and computer-aided manufacturing), expert systems can aid the designer in selection of materials, presentation of available purchased parts, selection of mechanisms, analysis of stress and temperature distributions, methods of assembly and details of many other manufacturing processes. In particular, an expert system could be developed to aid the designer and manufacturing engineer in the design of parts and assemblies destined to be produced by robots, not by humans. Relaxation of requirements for manual dexterity or for visual scene analysis, for example, would greatly enhance the performance of existing robot and sensor systems, with their severely constrained sensory and manipulative capabilities. Design of workpieces that are easy to identify, inspect, handle, mate with other parts, and assemble requires a new form of expertise, which, when acquired, can be transferred to a computer-based expert system.

There are many other uses for expert systems in the total manufacturing process, such as in purchasing, marketing, inventory control, line balancing, quality control, distribution and logistics, and others. It is expected that many proprietary programs will be developed and be offered to the manufacturing community.

Continuous Speech Recognition and Understanding

Word and phrase recognition systems with limited vocabularies are available commercially today. A few systems can handle a few simple sentences. In most instances, prior "training" of the system is required for each user. Research is proceeding to develop a continuous speech recognition system with an extended vocabulary and, if possible, speaker independence. Such a system would depend heavily on concurrent research in natural language processing systems.

Even with only a moderately advanced phrase or sentence recognition system, an attractive on-line programming method for "teaching" a robotic system is suggested. The manipulative parts of the task can be "taught" using a teach-box or joysticks, while the programmer "talks" to the computer with instructions regarding interrupts, sensing, tests, branching, communications, timing, setting accurate positions, and so on. This combination will generate a complete program, which can be edited, modified, and debugged on-line. No computer modeling is required—the real world is its own model. In future developments, special computer programs can be written that will optimize the performance of the robot system, given as input the relatively crude program generated, on-line, by the human.

The programming system outlined may have other attractive uses. Any advanced robot system could be operated in teleoperator mode, that is, under continuous operator control, or under semiautonomous control, in which the operator sets up the robot system for some repetitive task and a subroutine then takes over to complete the assigned task. In this mode a human can be time-shared, using the robot system as a "slave" to do the dangerous or less intellectually demanding parts of a task.

Other Aspects of Machine Intelligence

The subfields of problem solving, planning, automatic programming and verification, learning, and, in general, common-sense reasoning are all in the very early stages of development. They cannot now be considered as viable near-term options for at least the next five years. Incremental advances to enhance the intelligent behavior of our robotic systems will be incorporated at an accelerated pace when a large number of third-generation adaptive systems are in place and functioning cost-effectively. At that time, the conservative manufacturing community will have become accustomed to the notion of machines that can adapt behavior according to conditions that cannot be precisely predetermined. They will then be prepared to accept additional intelligent functions certain to result from accelerating machine intelligence research programs now under way, worldwide.

Hybrid Teleoperator/Robot Systems for Services and Homes

In the foregoing sections developments and applications of robot systems have been considered primarily for the manufacturing industries. In the United States only 20% of the working populace of approximately 100 million people are engaged in manufacturing, about 4% in agriculture, about 5% in the mining and extraction industries, and the remainder in the so-called service industries. The services include military, construction, education, health and social, transportation and distribution, sales, fire fighting, public order and security, financial, recreation, and white-collar support services. It is the author's considered opinion that adaptations and extensions of present robot/teleoperator systems will be developed for service use within the next generation exceeding in number and total value all the installations in factories. Further, one can also predict a mass market for robot/teleoperator systems developed for use in the home. These will serve primarily as servants or aids to the aged and physically disabled who have limited physical capabilities for activities, such as lifting, carrying, cleaning, and other household chores. By the turn of this century it is estimated that well over 15% of the total U.S. populace will be in this class (approximately 35 million individuals).

At present fully autonomous robots cannot cope with the more difficult environmental conditions in the relatively unstructured home or outdoors. Common-sense reasoning capabilities would have to be developed for a "self-acting" robot. However, a partially controlled robot (hybrid teleoperator/robot) can be developed within the present state of the art that would be economically viable for the majority of these service tasks. Most of the intelligence would be supplied by the human operator with the most modern "user friendly" interfaces to control the physical motion and manipulation of the robot, using switches, joysticks, and spoken word and phrase input devices. Subroutines controlling often-used manipulative procedures would be called up by the operator, to be implemented autonomously by the robot/teleoperator, fitted with available sensors. Seldom-occurring tasks would be performed by the operator effecting step-by-step control of the robot system in the same manner that present industrial robots are "trained" using a "teach-box." A specialized procedure could be stored and called up as needed. In short, dangerous, arduous, and repetitive physical manipulation of objects and control of simple manipulative actions would be performed by our new "slaves"; the target of these "slaves" will be progressively to minimize human detailed control as we learn to improve our robot systems.

One can debate the social desirability of making possible the cost-effective elimination of many manual tasks. There is little doubt that in our free-market system the development of such systems cannot be prevented, only slowed down. The thrust toward the implementation of these technologies is worldwide, and international competition will guarantee that these new systems will be made available.

When the market for specialized robot systems approaches the size of the automotive industry, the price for a teleoperator/robot system will be comparable to (or less than) that of a car, which has evolved into a comparatively far more complicated system demonstrating the successful integration of electronic, mechanical, thermodynamic, and many other technologies, together with effective "user-friendly" control.

For the home we can visualize a small mobile vehicle fitted with a relatively slow-moving arm and hand, visual and force/tactile sensors, controlled by joysticks and speech, with a number of accessories specialized for carrying objects, cleaning, and other manipulative tasks. High speed and precision would not be necessary. It would be all-electric, clean, and safe to operate. Its on-board minicomputer could be used for purposes other than for controlling the robot/teleoperator, for example, for recreation, record keeping, and security. It would be particularly useful for the aged and handicapped but would not be limited to these groups. Providing a versatile assistant for those with reduced strength and other physical disabilities appears to be an effective substitute for expensive live-in or visiting household help and care.

There are many opportunities for introducing teleoperator/robot systems for military and commercial service. The U.S. Army has initiated a program for developing material-handling systems for support services and is studying their potential use under battle conditions. Both wheeled and legged mobile robot systems are under development. Teleoperator/robot systems could be effectively used in the construction and agricultural industries. Applications in space, deep seas, mining, and in the Arctic are being explored. A host of other commercial applications appear feasible, including loading and unloading trucks, fire fighting, handling dangerous and noxious chemicals, painting and cleaning structures and buildings (outdoors), road maintenance, and so on. These will require more expensive and sophisticated machines, probably specialized for the particular applications, and ranging widely in size, load-handling capacity, precision, and speed. Finally, there are more imaginative applications that will be addressed: the hobby market, game playing, dynamic shop-window displays, science fiction movies, choreographed robots for modern dance, and others.

3.4. SUMMARY

After 25 years of laboratory research and development, machine intelligence technology is being exploited to a small but rapidly growing degree in manufacturing, primarily applied to improving the adaptability of robots through the use of sensors. Expert systems, planning, and advanced programming languages (including natural language) will provide significant improvements within the next 5–10 years. Hybrid teleoperator/robot systems will be developed for the service industries and ultimately may constitute the largest market for "smart" robots. We are now in a transition between the solid-state revolution and the information age. By early in the twenty-first century we can anticipate enjoying the era of the intelligent/mechanical slave.

PART 2

MECHANICAL DESIGN
OF ROBOTS

CHAPTER 4

MECHANICAL DESIGN OF AN INDUSTRIAL ROBOT

WARREN P. SEERING

Massachusetts Institute of Technology
Cambridge, Massachusetts

VICTOR SCHEINMAN

Automatix, Inc.
Billerica, Massachusetts

4.1. THE DESIGN PROCESS

The mechanical design of an industrial robot requires application of engineering expertise in a variety of areas. Important disciplines include machine design, structures design, and mechanical, control, and electrical engineering. Traditionally, robot design decisions have been based largely on use of simple design specifications relating to number of joints, size, load capacity, and speed. Robots have been designed not to perform specific tasks but to meet general performance criteria. Manipulator bearings, shafts, links, and other structural elements are selected for strength and stiffness to produce a manipulator meeting the work envelope and mechanical accuracy requirements. Motors are sized to meet worst case or average case gravity and acceleration torques or loads. Gearing is specified to meet gear tooth load limits in bending and surface stress. Bearings and shafts are selected and sized for life under estimated loads and to meet structural requirements such as deflection limits and clearance bores for cables or air lines. Links are sized to provide the required work range and to have a loaded deflection well below the accuracy specification. Component selection is generally made by looking through catalogs for appropriate components and by designing and building the manipulator around available and appropriately priced standard hardware. This produces a manipulator with unpredictable dynamic performance and, as a result, with uncertain performance specifications.

Early robots were designed with general motion capability under the assumption that they would find the largest market if they could perform the widest variety of tasks. This flexibility proved to be expensive in both cost and performance. Robots are now beginning to be designed with a specific set of tasks in mind. Overall size, number of degrees of freedom, and basic configuration are determined from task specifications for reach, work envelope, and reorientation requirements. Also considered are types of motion requirements, such as controlled-path motion for arc welding, continuous-path motion for spray painting, absolute positioning for CAM-based assembly, repeatability for materials handling, and fine resolution for precise, real-time sensor-based motions.

This chapter presents a set of considerations for the mechanical designer setting out to design a manipulator. It can also serve as a source of guidelines for use in evaluating an existing or evolving design or a compilation of lists of features to be considered when selecting a commercially available robot.

The first and most important phase in the process of designing a robot is defining the range of tasks for which the robot is to be built. This range of tasks should be specified as carefully as possible so that detailed manipulator properties and feature specifications may be developed. No single robot configuration will perform well on tasks of widely varying description. Therefore a robot should be designed to have only the flexibility it needs to perform the range of tasks for which it is intended. This range of tasks must be selected even though it may be difficult to do so. (See more in Chapter 5.)

All design decisions will be made based on this choice of tasks. Several alternative configurations should be considered in detail before one is chosen. This detailed consideration includes sizing of the most important system components and evaluation of dynamic system performance. On the basis of these evaluations the design configuration best suited to the tasks to be performed should be chosen. Before detailed drawings are started it is useful to check out the design by building a very simple mockup. This allows the designer to discover and solve problems associated with system geometry, structural integrity, cabling, and workspace utilization.

There are many possible paths through the maze of choices encountered during detailed design of an industrial robot. Often design specifications place conflicting demands on system components. The following sections discuss a number of important design considerations. The first two discuss system specifications, and the next four deal with system configuration. The final six are concerned with system performance characteristics. These twelve subsections are followed by a discussion of detailed design consideration of the major mechanical systems components of a robot. The final section presents several algorithms for choosing robot actuators and transmission ratios.

4.2. DESIGN CONSIDERATIONS

4.2.1. System Specifications

Range, Reach, and Work Envelope

Manipulator work envelope layouts must include considerations of regions of limited accessability (not all degrees of freedom will be fully available throughout much, if not all, of the workspace). These constraints arise from limited joint travel range, link lengths, the angles between axes, or a combination of these. Revolute joint manipulators generally work better in the middle of their work envelopes than at extremes. Manipulator links and joint travel should be chosen to leave margins for reorientations required because of changes of end effectors (tool offset angles and tool lengths will usually alter the work envelope).

Load Capacity

Load capacity, a frequent robot specification, is closely coupled with acceleration and velocity performance. In the case of assembly robots, acceleration and structural stiffness are more important design parameters than peak velocity or maximum load capacity, as minimizing small motion times is generally a top priority. In the case of arc welding, where slow-speed controlled-path motion is required, velocity jitter and path-following accuracy are important. Load capacity should be seen as a variable. It is wise to design and specify a manipulator in terms of useful load capacity as a function of performance rather than just in terms of maximum capacity. Choice of load specification must take into account load inertia and effective gravity and oscillation torque loads seen at the grip points. These factors strongly affect wrist and gripper design and drive selection. In general, load capacity is more a function of manipulator acceleration and peak wrist torque than any other factor. The load also affects manipulator static structural deflection, steady-state motor torque, system natural frequency, damping, and the choice of servosystem control gains for stability.

4.2.2. System Configuration

Joint Configuration

Manipulator configuration is determined by motion, control, obstacle avoidance, and structural requirements. Cartesian manipulators (with or without revolute wrist axes) have the simplest transform and control equation solutions. Their prismatic (straight-line motion), orthogonal axes make it easy and quick to compute desired positions of the links for any gripper orientation. Because their major motion axes do not dynamically couple (to a first order), their control equations are also simplified. Manipulators with all revolute joints are generally harder to control, but they feature less physical structure for a given working volume. It is generally easier to design and build a good revolute joint than a long-motion prismatic joint. The workspaces of revolute joint manipulators can easily overlap for coordinated multiarm tasks.

Final selection of the configuration should capitalize on specific kinematic or structural features. For example, a requirement for a very precise vertical straight-line motion may dictate the choice of a simple prismatic vertical axis rather than two or three revolute joints requiring coordinated control.

Number of Degrees of Freedom

Although 6 degrees of freedom (DF) are the minimum required to place the tip of a manipulator at any arbitrary location within its accessible workspace, most simple or preplanned tasks can be performed with less than 6 DF either because they can be set up carefully to eliminate certain axis motions, or because the tool or task does not require full specification of location. Generally, adding DF increases cycle time and reduces load capacity and accuracy for a given manipulator configuration and drive system.

Joint Travel Range

For revolute joint configurations, the shoulder and elbow joints determine the gross volume of the work envelope. The wrist joints generally determine the orientation range about a location within this work envelope. Larger joint travel may increase the number of possible manipulator configurations that will reach a particular location, permitting more alternative access envelopes. This is a useful feature when working in confined spaces or in the presence of obstacles. Wrist joint travel in excess of 360 degrees and up to 720 degrees can be useful for situations requiring controlled-path (e.g., straight-line) motion, or synchronized motion such as conveyor tracking. Continuous last-joint rotation is desirable in certain cases like loading or unloading a rotating machine or mating threaded parts.

Drive Configuration

Typically, a manipulator joint will consist of at least four major components: the mechanical power source, the joint position feedback device, the transmission, and the joint axis structure. Selection of these individual components is discussed in later sections. Depending on the sample rates and bandwidth of the controller, a tachometer may be a useful addition. Typical sample frequencies for a digital joint servo with just an encoder in the feedback loop are 200–2000 samples per second. By adding a tachometer and analog velocity loop, smooth motions may be obtained with position sample rates as low as five per second. Smaller manipulators generally require higher sample rates to be compatible with their higher structural natural frequencies, and shorter electrical and mechanical time constants.

In low-performance manipulators (less than 0.5 g load acceleration), system inertia is not as important as gravity torques. Here compensation for gravity torques through counterbalancing (by mass, springs, or air pressure) can help performance. In high-performance manipulators, system inertia becomes increasingly important. Placement of heavy drives and joints close to the first rotation or motion axis reduces system inertia and can improve performance. Here, the trade-off between drives at the joints with high inertia and high stiffness, and long transmission-link drives with low inertia and low stiffness becomes important. This choice dictates the major physical characteristics of a manipulator design.

4.2.3. System Performance

System Velocity

Maximum joint velocity (angular or linear) is not an independent value. It is usually limited by servo bus voltage or maximum allowable motor speed. For manipulators with high accelerations, even small point-to-point motions may be velocity limited. For low-acceleration arms, only gross motions will be velocity limited. A general design guideline is that most motions should be performed with the system at its velocity limit part of the time. A more detailed discussion of design with velocity limits is presented in a later section.

System Acceleration

In most modern manipulators, because the payload mass is small when compared with the manipulator mass, more power is spent accelerating the manipulator than the load. Acceleration affects gross motion time as well as cycle time (gross motion time plus nulling or settling time). Manipulators capable of greater acceleration tend to be stiffer manipulators. In high-performance arms, acceleration is a more important design parameter than velocity or load capacity.

Repeatability

This specification indicates the ability of the manipulator to return repeatedly to the same position. Depending on the method of teaching or programming the manipulator, most manufacturers intend this figure to indicate the radius of a sphere enclosing the set of locations to which the arm returns when sent from the same location by the same program with the same load and setup conditions. This sphere may not include the target point because calculation round-off errors, simplified calibration, precision limitations, and differences during the teaching and execution modes can cause significantly larger errors than those just due to friction, unresolved joint and drive backlash, fixed servo gain, and structural and mechanical assembly clearances and play. The designer must seriously consider the real meaning of the repeatability specification required. Repeatability is important when performing precisely repetitive tasks such as blind assembly or machine loading. Typical repeatability specifications are from ±2 mm for large spot-welding robots to ±0.005 mm for very precise micropositioners.

Resolution

This specification represents the smallest incremental motion that can be produced by the manipulator. Resolution is important in sensor-controlled robot motion and in fine positioning. Although most manufacturers calculate system resolution from resolution of encoders, resolvers, or analog to digital converters, or from motor step size, this calculation is misleading because system friction, windup, backlash, and kinematic configuration adversely affect system resolution. Typical encoder or resolver resolution is 2^{12} to 2^{20} counts for full axis or joint travel, but actual resolution may vary from one part in 2^{10} to 2^{18} for revolute joints and 0.2 mm to 0.002 mm for prismatic joints. The useful resolution of a multijoint serial-link manipulator is somewhat poorer than that of the individual joints.

Accuracy

This specification covers the ability of a robot to position its end effector at a preprogrammed location in space. Usually a coordinate transformation is assumed between world and joint coordinates. The precision of this positioning is a function of the precision of the arm model in the computer (joint type, link lengths, angles between joints, any accounting for link or joint deflections under load, etc.), the precision of the world, tool, and fixture model, and the completeness and accuracy of the arm solution routine. Although most higher-level robot programming languages support arm solutions, for computation speed they all use simplified solutions and model only ideal kinematic configurations. Thus manipulator accuracy becomes a matter of matching the robot geometry to the robot solution in use by precisely measuring, calibrating, and adjusting link lengths, joint angles, and mounting positions. Robot accuracy is important in the performance of nonrepetitive types of tasks programmed from a data base, or for taught tasks that have been remapped or offset owing to measured changes in the installation. Typical accuracies for manipulators range from ±100 mm for noncalibrated manipulators that have poor computer models to ± 0.01 mm for machine-tool-like manipulators that have simple accurate models and solutions and precisely manufactured and measured kinematic elements.

Component Life and Duty Cycle

The three subassemblies in an electrically powered robot with the greatest failure problems are motor brushes and commutators (wear), gear teeth (scoring and fatigue), and power and signal cables (flex life). Worst-case motion cycles must be assumed as most current robot installations are used in generally repetitive tasks. Small-motion design-cycle life (less than 5% range of joint travel) should be 20–50 million full bidirectional cycles. Large-motion cycle life (greater than 50% of full joint range) should be typically 1 to 20 million cycles. Mean time between failures should be a minimum of 2000 hours on line, and ideally at least 5000 operating hours should pass between major component preventive maintenance replacement schedules.

Most manipulators have individual motion-cycle times which are a small percentage of their motor thermal time constants. Short-term peak performance is frequently limited by maximum gear stress, whereas long-term (continuous) performance is limited by motor heating. Rather than design for equal levels of short- and long-term performance, cost savings and performance improvements can result from designing for an anticipated duty cycle. This allows the use of smaller motors and amplifiers than might be required for a 100% duty cycle robot. Temperature sensors can be used to sense excessive duty cycle conditions.

In the course of operation, unforeseen or unexpected situations may occasionally result in a *crash* involving the manipulator, its tools, the workpiece, or other objects in the workplace. These accidents may result in no, little, or extensive damage, depending in large part on the design of the manipulator. Crash-resistant design options should be considered early in the design process if the time or money cost of such accidents is significant. Typical damage due to accidents include fracture or shear failures of gear teeth or shafts, dented or bent link structures, slipping of gears or pulleys on shafts, cut or severely abraded or deformed wires, cables or hoses, and broken connectors, fittings, or limit stops or switches.

4.3. DETAILED DESIGN OF MAJOR COMPONENTS

4.3.1. Robot Structures

Although all robot structures are flexible to a degree, some are substantially more flexible than others. Only two structural types, flexible and rigid, are considered here. Rigid structures are defined as those for which both the kinematic solution and the control algorithms assume all links to be rigid. Most commercially available robot arms are of this type. Control of these rigid manipulators assumes that there is no structural deflection, whereas in fact, for certain loading conditions, system deflections can be significant and will result in decreased accuracy.

Some robots have a *gravity produced deflection* term in their control algorithms. Others employ strain-sensors to measure end-point loads and deflections. These "semirigid" manipulators assume small structural deflection resulting from gravitational load and provide linear corrections to improve accuracy.

For robots with flexible structures, the control algorithms are designed to control the flexibility as well as the gross motion. In general, flexible arms are only found in laboratory and special application settings. Flexibility of commercially available "rigid" arms can dramatically affect their controllability. Control of manipulators is discussed at length in Part 3 of this handbook.

The most important performance characteristics for robot structures are stiffness in bending and in torsion. The two most common types of structures for robot manipulator arms are monocoque or shell structures and beam structures. Although the monocoque structures have lower weight or higher strength-to-weight ratios, they are more expensive and generally more difficult to manufacture. Cast, extruded, or machined hollow-beam-based structures, though not as structurally efficient as pure monocoque designs, are the more cost-effective.

An important structural design consideration is the choice of method of manufacturing. Typical designs include bolted, welded assemblies, and epoxied assemblies of cast elements. Although bolted assembly is straightforward, inexpensive, and easily maintained, there are associated problems including creep and hysteresis at the bolted connections and dimension changes resulting from assembly and disassembly. Welded and cast structures are much less susceptible to creep and hysteresis deformation; however, in many cases they require secondary manufacturing operations such as thermal stress relieving and finish machining. The typical minimum wall thickness for sand castings is 5.0 mm. This is generally thicker than strength specifications would require. Thinner walls can be obtained through the use of plaster-mold casting, die casting, or investment casting; however, these processes are more expensive for small volumes. Minimal wall thickness of monocoque structures is often specified for resistance to puncture or denting, rather than just for strength.

Today aluminum and steel are the most common materials for robot structures. However, thermoplastics and glass or carbon-fiber reinforced plastics are beginning to be used. For large production runs, plastic structured arms can be significantly less expensive. To decrease weight in aluminum and steel arms, one can either taper the wall thickness or configure the gross dimensions so the links become smaller at the end closest to the payload.

Integration of the structure with the joint mechanisms and power train hardware poses a design challenge. Positioning of bearings for transmission elements is extremely important, as deformation in the joint at the bearing housings can adversely affect precision by reducing preload and allowing backlash or free play. Inadequate structural stiffness can also adversely affect overall manipulator precision by allowing changes in gear center spacing, excessive shaft windup, or binding caused by large drive forces and torques.

A consideration in the design of robot structures is the effect of workplace and drive- and actuator-produced temperature variations. Of the most commonly used materials, steel structures have the best (i.e., least) response to thermal changes. Although change in dimension as a function of temperature is a problem, more serious is the issue of compatibility of thermal expansion coefficients among various elements of the robot. For example, steel structures make better housings for steel bearings, and they will maintain center distances on steel gears where aluminum structures will not. Because the robot is often not mounted to the same base as the workpiece, the dimensional stability of the entire robot and workplace system must be carefully considered. Another important consideration is structural distortion caused by localized heating from motors, transmissions, electronics, and workpieces. The designer must carefully choose the location and mounting method of these elements to minimize the resultant problems. Thermal and loading effects on plastic structures vary significantly, depending on the type of plastic and manufacturing method chosen.

4.3.2. Robot Joints

Robot joints can be catagorized generally as either prismatic or revolute joints. Other types, such as spherical or universal joints, are not discussed separately here as they are generally implemented as combinations of the two primary classes.

There are two basic types of prismatic or linear motion joints: single-stage and multiple-stage or telescoping joints. Single-stage joints are made up of a moving surface that slides linearly along a fixed surface. Multiple-stage joints are actually sets of nested or stacked single-stage joints. Single-stage joints feature simplicity and high stiffness, whereas the primary advantage of telescoping joints is their retracted-state compactness and large extension ratio. Telescoping joints have a lower joint inertia for some motions because part of the joint may remain stationary.

The primary functions of bearings in prismatic joints are to facilitate motion in a single direction and to prevent motion in all other directions, both linear and rotational. Preventing these unwanted motions poses the more challenging design problem. Deformations in the structure can significantly affect bearing surface configuration, which affects performance. In severe cases, roller deflection under load may cause binding, which precludes motion. For high-precision prismatic joints, ways must be made straight over long distances. The required precision grinding on multiple surfaces can be expensive. Expensive and bulky covers are required to shield and seal a prismatic bearing and way.

The primary criterion for evaluating prismatic joints is the stiffness-to-weight ratio. Achieving a good stiffness-to-weight ratio requires the use of hollow structure for the moving elements rather than solid rods. Bearing spacing is extremely important in design for stiffness. If spacing is too short, system stiffness will be inadequate no matter how great the bearing stiffness. Major causes for failure in prismatic joints are foreign particle contamination and Brinelling of the ways caused by excessive ball loading and by shock loads. Excessive preload can also lead to Brinell failure. The large exposed precision surfaces in most prismatic joints make them much more sensitive than revolute joints to improper handling and environmental effects. They are also significantly more difficult to manufacture, properly assemble, and align.

Common types of sliding elements for prismatic motion are bronze or thermoplastic impregnated bushings. These bushings have the advantage of being low in cost, of having relatively high load capacity and of working with nonhardened or superficially hardened (i.e., chrome-plated) surfaces. Because the local or contact stress on the moving element is distributed and low this element may be made of thin tubing. Another type of bushing in common use is the ball bushing. Ball bushings have the advantages of lower friction and greater precision than plain bushings. However, they require that the contacting surface of the joint be through or case hardened (generally to R_c 55 or greater) and of sufficient case and wall thickness to support the point ball loads and resulting high contact stresses.

Ball slides are also commonly used in robot prismatic joints. The distinction between ball bushings and ball slides is that ball bushings operate on cylindrical surfaces whereas ball slides operate on ground ways of various configurations. There are two basic categories of ball slides, recirculating and nonrecirculating. Nonrecirculating ball slides are used primarily for light load or short travel applications. They feature high precision and very low friction at the expense of being quite sensitive to shock and relatively poor at accommodating moment loading. Recirculating ball slides are somewhat less precise but can carry higher loads than nonrecirculating ball slides. They can also be set up to carry relatively large moment loads. For a given way length, the travel is greater for a recirculating ball slide than for a nonrecirculating ball slide.

Another common type of prismatic robot joint is made up of cam followers or ball or roller bearings rolling on extruded, drawn, machined, or ground surfaces. Both needle and roller bearings are in common use as cam followers; the ball bearings are less common. In high-load applications the way surfaces must be hardened before they are finish ground. Cam followers generally appear in prismatic joints in sets of 6 to 16 units. They can be purchased with eccentric mounting shafts which facilitate setup and adjustment.

Two less common types of linear or prismatic joints are flexures and air bearings. Flexures, joints whose motions result from elastic bending deformations of beam support elements, are used primarily for small quasi-linear motions. Air bearings for precise motion require smooth surfaces and close control of tolerances. Less precise air bearings can use machined or even cast surfaces, however such practice results in large clearances which cause low stiffness and result in large air flows. Multiple air pads or separate ball or roller bearings on ways are generally used to handle moment loads.

Revolute (rotary motion) joints are designed to allow pure rotation while minimizing radial and axial motions. There are many design issues to be considered when designing a revolute joint. The most important measure of the quality of a revolute joint is its stiffness or resistance to all undesired motion. Key factors to be considered in design for stiffness are bearing shaft, housing and diameters, clearances and tolerances, mounting configuration of the bearings, and implementation of proper bearing preloading. Bearing size is not always based on load-carrying capacity; rather, the bearing chosen often will be the smallest one that is stiff enough in both bending and torsion to give desired system stiffness that will fit on the shaft. Because joint shafts will frequently be torque-transmitting members, they must be designed both for bending and torsional stiffness. The first axis of the PUMA™ Robot is an example of such a joint.

An important factor in maintaining stiffness in a revolute joint is choice of bearing-mount configuration. The interface between the mount and the structure is as important as the interface between the mount and the bearing. The mount and mounting arrangement must also be designed to accommodate preloading of the bearings. Axial preloading of ball or tapered roller bearings improves system accuracy and stiffness by minimizing bearing radial and axial play. Preloads can be achieved through the use of selective assembly or spring elements, shim spacers, or threaded collars.

4.3.3. Actuators

The three most popular types of robot actuators are hydraulic, pneumatic, and electromagnetic. Hydraulic actuators, chosen as power sources for the earliest industrial robots, offer very large force capability and high power-to-weight ratios. In a hydraulic system the power is provided mechanically from a pump while the solenoid or servo control valve is driven electrically from a lower-power control circuit.

The hydraulic power supply is bulky and energy inefficient, and cost of the proportional, fast-response servo valve is high.

Pneumatic actuators are primarily found in simple manipulators. Typically they provide uncontrolled motion between mechanical limit stops. These actuators provide good performance in point-to-point motion, they are simple to control and are low in cost. Although a few small actuators may be run with typical factory air supplies, extensive use of pneumatic-actuated robots will require the purchase and installation of dedicated compressed air sources which may be expensive and are very energy inefficient.

The most common types of actuators in robots today are electromagnetic actuators, typically DC motors. There are a wide variety of types of DC motors, each with its own advantages and disadvantages. The most common types of electromagnetic actuators for inexpensive robots are stepper motors. These motors provide open loop position and velocity control. They are relatively low in cost and they interface easily to electronic drive cirucits. Recent developments in control systems have permitted each stepper motor "step" to be divided into many incremental microsteps. As many as 10,000 or more microsteps per revolution can be obtained. Motor magnetic stiffness, however, is lower at these microstepping positions. Typically, stepper motors are run in an open loop configuration. In this mode they are underdamped systems and are prone to vibration, which can be damped either mechanically or through application of closed loop control algorithms. Power-to-weight ratios are lower for stepper motors than for other types of electric motors.

The permanent-magnet, direct-current, brush-commutated motor is widely available and comes in many different types and configurations. The lowest-cost permanent-magnet motors are the ceramic (ferrite) magnet motors. Motors with alnico magnets have higher energy product and produce higher motor constants than equivalent sized motors with ceramic magnets. (Motor constant is defined as torque produced divided by square root of power consumed.) Rare-earth (samarium-cobalt) motors have the highest energy product magnets, and in general produce the largest peak torques because they can accept large currents without demagnetization. However, these larger currents cause increased brush wear and more rapid motor heating.

Another subset of DC permanent-magnet brush motors are ironless rotor motors. Typically these motors have rotors made of copper conductors enclosed in epoxy glass cup or disk rotor structures. The advantages of these motors include low inertia and negligible inductance, which reduces arcing, extends brush life, and gives them short electrical and mechanical time constants. Because these motors have no iron in the rotor they have very little residual magnetism and consequently very low cogging torques. Disk-type motors have several advantages. They have short overall lengths, and because their rotors have many commutation segments they produce a smooth output with low torque ripple. A disadvantage of ironless armature motors is that they have a low thermal capacity due to low mass and limited thermal paths to their case. As a result, they have rigid duty cycle limitations or require forced-air cooling when driven at high-torque levels.

The weakest links in most motor designs are the bearings and brushes. Brushless DC motors, also classed as synchronous AC motors, have been developed. They substitute magnetic and optical switches and sensors and electronic switching circuitry for the graphite brushes and copper bar commutators, thus eliminating the friction, sparking, and wear of commutating parts. Brushless DC motors generally have good performance at low cost because of the decreased complexity of the motor. However, the controllers for these motors are more expensive because they must include all the switching circuitry. There is a strong trend toward brushless DC motors because of their increased reliability and improved thermal capacity. This improved thermal capacity occurs because in brushless motors the rotor is a passive magnet and the wire windings are in the stator, giving them good thermal conductivity to the motor case.

4.3.4. Transmissions

Many types of transmission elements are in use in robot design. The purpose of the transmission is to transmit mechanical power from a source to a load. Choice of transmission elements depends on power requirements, the nature of the desired motion, and the placement of the power source with respect to the joint. The primary considerations in transmission design are stiffness, efficiency, and cost.

Gears are the most common transmission elements in robots today. Factors to consider in gear design are material choice, choice of material surface treatment, and manufacturing precision. Considerations in designing geared transmissions are gear ratio, type of gear, gear shaft support, control of center distances, and lubrication.

Spur gears are most commonly used for parallel axis transmissions or for prismatic motions in the rack and pinion configuration. Spur gears have the advantage of producing minimum axial forces, which minimizes the need for controlling play in the gear mount.

Helical gears are also used in robot transmissions. They have several specific advantages. Because gear reductions are often quite large in robot transmissions, lack of adequate gear tooth contact ratio can be a problem. For given gear ratios and gear sizes, helical gears have higher contact ratios and as a result produce smoother output. They also tend to be quieter. The primary disadvantage to helical gears is that they produce axial gear loads that must be constrained to maintain drive stiffness.

The limiting factor in gear transmission stiffness is the stiffness of the gear teeth; each tooth acts as an elastic cantilever during the time that it is loaded. To maximize stiffness, the largest possible gear diameters should be chosen. Choice of gear ratios is discussed in the next section.

Rack and pinion transmissions are in common use in robots, particularly for long linear motions in which the rack can be mounted to the structure so that the structure carries the loads applied by the pinion. Another common linear motion transmission element in robot design is the ball screw. Ball screws feature high efficiency, moderate stiffness, and short leads which offer large mechanical advantages. Screws can be purchased both in precision (ground) and commercial (rolled) grades. Precision ball screws are purchased with ball nuts as matching pairs. They typically have lead accuracies of better than one part in 50,000, whereas commercial grade screws have lead precisions of approximately one part in 2,000. To obtain best possible accuracy and zero backlash, ball nuts are used in preloaded pairs. Precision screws are preloaded with shim spacers, and nonuniformities in the system are absorbed through elastic deformation of system components. Commercial grade screws use elastic members such as spring washers located between preloaded nuts to take up small relative motions between the nuts. Vibration problems can result within this ball-nut-spring system. Problems can also occur as a result of torsional vibration or windup in the screw, particularly where long screw lengths are used.

A common revolute joint transmission element in robot design is the *Harmonic Drive*, a patented unit (USM Corp.). These drives feature in-line parallel shafts and very high transmission ratios in compact packages. With selective assembly procedures, near zero backlash harmonic drives can be produced. Static friction in these drives is high, and manufacturing tolerances often result in cyclic friction torque variation called cogging.

Power is often transmitted in robots through torsion shafts or weight-saving torque tubes. Transmitting power at high angular velocities also minimizes required shaft diameter, wall thickness, and weight. Fatigue life is an important consideration, particularly if aluminum shafts are used.

Several robot manufacturers use toothed positive drive belts as transmission elements. They are used primarily when low-cost power transmission is required over large distances, or as a simple interface between the motor and the first stage of gear reduction. Transmission ratios are limited because there is generally a minimum pulley size based on belt fatigue life. Drive stiffness in a belt transmission is a function of the belt material and belt tensioning system. Belts containing fine fibers of materials such as Kevlar ™, which have high stiffness modulus to weight ratios, can be driven around smaller pulleys because the Kevlar reinforcing bands themselves consist of flexible microscopic fibers.

A common transmission element in low-cost robots is the stranded cable or flat alloy steel band. These elements are easy to configure and repair and are relatively efficient. Stiffness in cables and bands, as with stiffness in belts, is primarily a function of the choice of material. Chains are another common transmission element. They are relatively low in cost and are good for high-load applications. Problems with chains are weight and wear. High preloads are required on chains to overcome the effects of chain droop caused by weight. This droop creates a reduction in system stiffness. To minimize droop in long sections of chains, solid or tubular push-pull rods are often substituted for lengths of the chain. The Unimation Unimate 2000 and 4000 and the Hitachi "Process" robot both use push-pull rods that connect chain segments that engage the sprockets.

Hydraulic lines are in fact transmission elements. Heavy-wall rigid tubing provides relatively high stiffness. Flexible tubing is much more versatile but exacts a high stiffness penalty because the elastic modulus of this tubing is small compared to that of the fluid. Hydraulic systems can be used to transmit power over long distances. System performance is limited by fluid viscosity, pressure drop, and time constant of the fluid lines.

Linkages or linkage structures may be considered as transmission elements, although they are often structural elements as well. The key advantage to linkage elements is that they can be configured to offer variable transmission ratios at different positions throughout their operating range. Though the links in linkage drives are usually quite stiff, the limitation of stiffness in linkage systems is in the bearings and shafts that connect the various links.

The second important characteristic in evaluating transmission system performance is efficiency. Most robot transmission elements have good efficiencies when they are transmitting at or near their rated power levels. Transmissions with high static friction such as harmonic drives with the low backlash option or belt drives with heavily preloaded bearings and high belt pretension are not very efficient at power transmission levels that are a small percentage of their rated limits. Other important considerations in choosing transmission elements are those of system geometry, compactness, and simplicity. Planetary spur gears and harmonic drives are among the most compact forms of transmission elements. Positive-drive belts, chains, and cable drives are among the simplest and the easiest to manufacture.

In choosing transmission elements, one must consider the time required for adjustment and set-up procedures. Proper backlash control of gears requires adjustments of distance between gear centers. Ball screws purchased with preloaded ball nuts require careful alignment at setup. Setup of belts, bands, cables, and chains consists of adjustment of tension idlers or of center distances. Backlash in harmonic drives is controlled through selective assembly and adjustment of component spacing. The setup of linkages involves length adjustment alignment and preloading of the pivot bearings.

The lubricant of choice in most robots is grease. Because configurations and orientations of the joint vary, and because motion is intermittent and of relatively low speed, other lubricants would be difficult to use, and grease serves adequately in most situations where transmission heating is not severe.

Smoothness of the power transmission is another design consideration, especially where low-speed motion is required such as in arc welding. Bands and linkages provide the smoothest drive. Ball screws also provide smooth drives if they are clean and in good condition. Gears are not as smooth as screws and belts, but in general they are smoother than chains and Harmonic Drives. Proper setup is critical for smoothness of operation of gears.

In general, when large reduction ratios are required, transmission elements of choice will be Harmonic Drives, gears, or ball screws. For a low transmission ratio, rack and pinion drives, single-stage gearing, belts, cables, or chains may be used. Direct-drive systems, in which the power source is directly attached to the joint, are also becoming increasingly attractive solutions.

4.3.5. Wiring and Routing of Cables and Hoses

Internal wiring of base and shoulder joints has never been a big problem because joints and structures are large and access through the robot base is simple. Wiring of wrist joints is more difficult, but in most cases careful planning can permit internal routing of cables. Selection of thin section bearings and larger-diameter tubular shafts for joints provides stiffness, light weight, and room to pass cables and hoses. Adequate flex lengths and coiling at joints as well as the use of high-strand and flat-section wire or flex circuits is necessary. Supplying internal electrical power and signals or air lines to end effectors is difficult on manipulators having three-axis wrists because their complex and compact design leaves little flex and coil room for these lines. Many manufacturers choose to run these lines outside the structure for ease in replacement and simplicity in design at the expense of the extra length and of their getting in the way. Until end effector interface standards are developed, users will probably end up adding their own extra external cables anyway.

For noise immunity most signal and power wires are shielded. A typical joint or servoed end effector might have three or four shielded cable sets associated with it: encoder wires, motor wires, tachometer wires, and travel limit and initialization switch wires. These are grouped and interfaced to the controller through one to four multipin connectors mounted in the robot base.

4.4. ALGORITHMS FOR CHOOSING ACTUATORS AND TRANSMISSION RATIOS

The choice of actuators and transmissions in high-performance machines and manipulators has a very significant effect on the time required for a specified machine task. For systems operating with mechanical constraints, such as limits on motor speed, designing for greatest achievable acceleration will not always minimize move time. This section evaluates effects on performance of choices of actuators and transmission mechanisms.

4.4.1. Maximum Acceleration in a System for a Given Amplitude of Actuator Effort

Proper choice of system transmission ratio is necessary to produce maximum system acceleration. A typical 1 DF mechanical system consists of a torque or force source, a transmission, and an inertial load. One such system, a DC motor ball-screw system, is presented schematically in Figure 4.1. In general the inertial load M can represent a mass as in a prismatic joint or a rotary inertia as in a rotating element. For systems with varying geometry, such as the PUMA™ robot, rotary inertia about a given joint varies as a function of system orientation. Such systems should be designed to give maximum acceleration at a position near the center of the workspace. For the model of Figure 4.1, the torque source is assumed to have a rotary inertia J and to produce a torque T. The transmission ratio r is defined as the ratio of transmission output velocity to input velocity. For a gear transmission system, r is the gear ratio defined as the ratio of output angular velocity to input angular velocity. For the case of a ball-screw system, r is the lead or the ratio of output translational velocity to input angular velocity in radians per second. For a rack and pinion system, r is also the ratio of the output translational velocity to the input angular velocity and is equal in magnitude to the radius of the pinion pitch circle.

For many high-performance applications, it is desirable to maximize system output acceleration a. The equation of motion for the system in Figure 4.1 is

$$T = \left(\frac{J}{r} + Mr\right)a \tag{4.1}$$

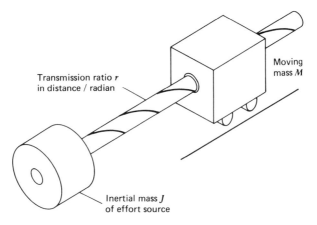

Transmission ratio r
in distance / radian

Moving
mass M

Inertial mass J
of effort source

Fig. 4.1. Simple model of a servo system.

A value can be found for r that minimizes the quantity in brackets, thus maximizing acceleration for a given actuator effort.

$$\frac{\partial}{\partial r}\left(\frac{J}{r} + Mr\right) = -\frac{J}{r^2} + M = 0$$

or

$$r' = \sqrt{\frac{J}{M}} \tag{4.2}$$

In this equation r' is the transmission ratio that will yield the greatest system acceleration for a given actuator effort. Note that substitution of r' for r in Eq. (4.1) produces equal "impedance" contributions for the rotary inertia and the moving mass. This choice of r then gives an "impedance-matched" system. Figure 4.2 is a plot of a^* versus r^*. The terms a^* and r^* are dimensionless parameters defined as

$$r^* = \frac{r}{r'} \tag{4.3}$$

$$a^* = \frac{a}{a_{max}} \tag{4.4}$$

where from Eq. (4.1) with $r = r'$

$$a_{max} = \frac{T/\sqrt{J}}{2\sqrt{M}} \tag{4.5}$$

In Figure 4.2 a^* is seen to be insensitive to small changes in r^* near $r = r'$. However, a^* decreases markedly with larger decreases in r^*. From Eq. (4.5), peak system acceleration a_{max} is seen to be proportional to $T/\sqrt{J}$. Hence, assuming optimal transmission ratio selection, for a bigger motor to yield a greater acceleration, the percentage increase in torque must be greater than the percentage increase in $\sqrt{J}$.

Many systems, such as DC servo motors with current amplifiers, can produce constant torque output independent of angular velocity (within the maximum speed limit). For other systems, torque is a function of angular velocity ω. In these latter cases the transmission ratio calculated in Eq. (4.2) will give maximum achievable acceleration at all values of ω. Peak acceleration will of course occur at the value of ω for which T is greatest.

4.4.2. Effects of Changes in Driven Mass on System Performance

Let $M = M_c + M_p$, the sum of the constant and payload masses. Figure 4.3 shows the effect on system performance of changes in M_p over a range of values of the transmission ratio r^*. Also plotted is a line of peak acceleration for variations in M_p/M_c. This plot illustrates that optimal value for the

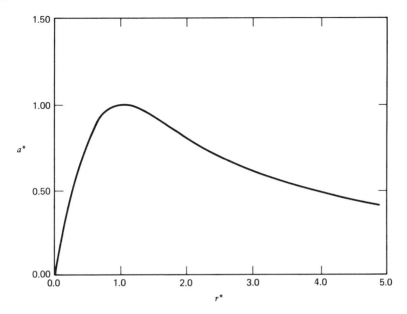

Fig. 4.2. Achievable system acceleration as a function of lead choice; a^* is normalized acceleration, and r^* is transmission ratio normalized about the optimum.

transmission ratio does not vary significantly through a range of small values for M_p/M_c. This is fortunate as the transmission ratio for most systems is difficult to change during operation. Figure 4.3 also shows that for values of r^* less than 0.4, system acceleration at a given torque is virtually independent of M_p/M_c. This fact has led many system designers to choose low values of r^* to simplify the problem of system control. As can be seen from the figure, this choice imposes a substantial performance penalty.

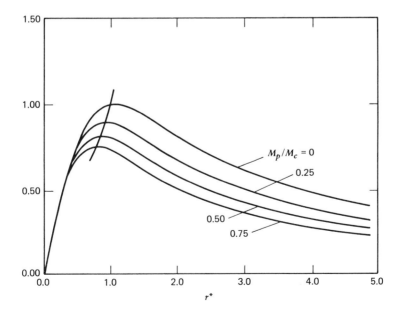

Fig. 4.3. The effect on system performance of increase in moving mass.

4.4.3. Motor Selection for Maximum Acceleration

Equation (4.5) quantifies the dependence of system acceleration on system torque and inertia. Assume for the moment that J_s, the inertia of the transmission element, is small as compared with J_m, the inertia of the motor. Equation (4.5) can be rewritten

$$\frac{T}{\sqrt{J_m}} \sim 2a_{max}\sqrt{M} \qquad (4.6)$$

for the case in which the optimal transmission ratio has been chosen.

In Figure 4.4 are plotted values of rotor inertia versus motor torque for several families of motors ranging in size (and cost) over two orders of magnitude. Values for T and J_m were obtained from manufacturers' catalogs. From Eq. (4.6), lines of constant $a_{max}\sqrt{M}$ can be plotted on the figure. If the moving mass of the system is known, the system acceleration produced by each of the motors can be read directly from the chart. For Figure 4.4 a moving mass of 50 lb (23 kg) was chosen, and constant acceleration lines were plotted for accelerations up to 6 G.

The plot in Figure 4.4 reveals that for most families of motors, larger and more expensive motors do not produce significantly greater acceleration. In fact, for brushless motors, the largest motor included produces the lowest acceleration. There is, however, an advantage to using larger motors. Because they have larger values of J, larger values of r' [see Eq. (4.2)] will be chosen for maximum acceleration. And for a larger chosen value of r', the peak velocity of the moving mass will be larger for a given peak motor angular velocity.

Once a transmission element for a system has been chosen and its rotary inertia J_s determined, a corrected value for system acceleration can be obtained from Figure 4.4. The addition of the inertia J_s simply shifts the location of the point representing the chosen motor upward to account for the fact that the ordinate J is now taken to represent the sum of the inertias J_s and J_m. For a given value of J_s, system performance will be affected less for larger values of J_m.

Correction of the value of J by inclusion of J_s will increase the value of r'. In certain cases, this change in r', because of limitations on transmission element design, will necessitate a change in dimension of a transmission element, resulting in an increase in J_s. When this occurs, an iteration process must be employed to establish optimal choices of J_s and r'. It should be noted, however, that for transmission elements such as ball screws, transmission ratios r' are only available in discrete increments. This is not a serious disadvantage, as from Figure 4.2 we again note that system performance is relatively insensitive to small variations in r about the optimal value $r = r'$.

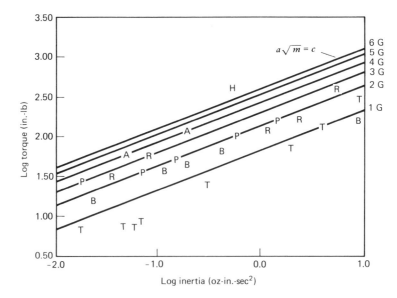

Fig. 4.4. Motor performance for various families of motors in systems with optimized lead. Each letter on the chart represents a commercially available motor: T = torque; B = brushless; P = pancake; A = air; H = hydraulic; R = rare earth.

4.4.4. Motor Selection for a Given System

Many times the design engineer cannot specify all system parameters. Rather, the task may be to specify a motor to drive an existing system. In this case J_s, r, and M are known, and the governing system equation can be written

$$T = ac_1 J_M + ac_2 \tag{4.7}$$

where

$$c_1 = \frac{1}{r} \qquad c_2 = \frac{J_s}{r} + Mr$$

This equation shows torque required to produce a given acceleration to be a linear function of motor inertia.

Figure 4.5 gives constant acceleration lines on a plot of torque versus motor inertia. As can be seen from Eq. (4.7), for a motor change to produce greater acceleration of the translating mass, increase in torque must be greater than (a/r) times the increase in J_m. Hence, as indicated in Figure 4.5, the greater the initial value of a, the greater must be the change in T for a given change in J_m if system acceleration is to be increased.

4.4.5. Determining Optimal Transmission Ratio for Velocity Limited Systems

Maximizing acceleration does not guarantee that a system will be time optimal for all moves. Top speed (r times ω_{max}) may be reached in the minimum time, but choice of transmission ratio might unnecessarily limit this top speed for a given allowable peak motor or transmission angular velocity.

Consider the velocity profile for one of the incremental moves shown in Figure 4.6. This move could be produced by a DC servo motor that has constant torque but is speed limited to top speed ω_{max}. Note that for $r = r'$ the slopes at the beginning and end of the move are steepest. The area under the curve represents the distance traveled during the move. For the case in which the system does not dwell at peak velocity, the transmission ratio that minimizes total move time is $r = r'$. If the system dwells at top speed, $r = r'$ is no longer the optimal ratio. Notice that for r less than r', the slope of the velocity versus time line is decreased as is the top speed. Thus as r decreases from r', the area under the curve will always decrease for a move of fixed time duration. For r slightly

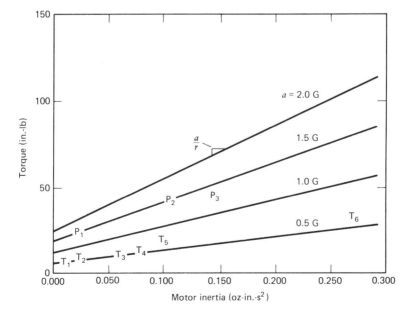

Fig. 4.5. Performance curves for motors in systems with fixed transmission ratios. Each letter on the chart represents a commercially available motor.

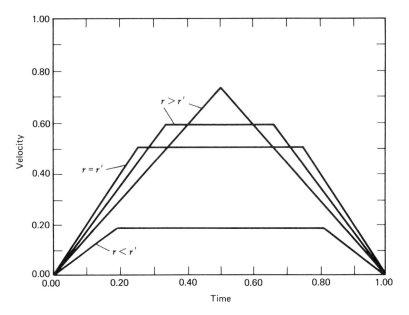

Fig. 4.6. Typical incremental move velocity profiles.

larger than r', the area under the curve increases with r. At some value of increasing r, the area starts to decrease.

For a typical move in which the system reaches peak velocity it can be shown that travel time t_t is

$$t_t = \frac{\omega_{max}(J + Mr^2)}{T} + \frac{d}{r\,\omega_{max}} \tag{4.8}$$

where d is the total move distance. Taking the partial derivative of this equation with respect to transmission ratio gives

$$\frac{\partial t_t}{\partial r} = 0 = \frac{2rM\omega_{max}}{T} - \frac{d}{\omega_{max}r^2}$$

or

$$r_{opt} = \sqrt[3]{\frac{Td}{2M\omega_{max}^2}} \tag{4.9}$$

where r_{opt} is the transmission ratio that minimizes time required to perform a specified move d for the case in which the system dwells at $\omega = \omega_{max}$.

Generally, computer-controlled robots do not have a fixed move distance. Rather, they move through a range of distances. The transmission ratio of choice for a given system design is that which does the best job of minimizing move time throughout the range of interest of moves. Because optimal lead is proportional to the cube root of the move distance, a lead can be chosen that is close to optimal for a fairly large range of move distances.

4.5. SUMMARY

The mechanical design of a manipulator is an iterative process involving evaluation and choice among a large number of engineering and technical considerations in several disciplines. The final design should be based on a specific set of task requirements rather than on broad specification. Properly identifying and understanding these requirements is a key to meeting the design goals. Design and choice of specific components also involves trade-offs. A purely static, rigid-body approach to manipulator design is often used but is not always sufficient. Mechanical system stiffness, natural frequencies, control system compatibility, and workpiece properties must be considered. Although certain detailed design decisions can be made through the application of straightforward algorithms, the multitude of factors that must be considered transform the problem into one of good engineering judgment as well.

BIBLIOGRAPHY

D.C. Motors, Speed Controls, Servo Systems, third edition, Electro-Craft Corporation, Hopkins, Minnesota.

Design and Application of Small Standardized Components, Stock Drive Products, New Hyde Park, New York, 1983.

Glegg, Gordon L., *The Design of Design,* Cambridge University Press, 1979.

Metals Handbook, ninth edition, American Society for Metals, Metals Park, Ohio, 1978.

Paul, R. P., *Robot Manipulators: Mathematics, Programming and Control,* The M.I.T. Press, Cambridge, MA, 1981.

Pieper, D. L., and Roth, B., The Kinematics of Manipulators under Computer Control, Proceedings of the 2nd International Congress on the Theory of Machines and Mechanism, Vol. 2, 1969, pp. 159–168.

Roth, B., Performance Evaluation of Manipulators from a Kinematic Viewpoint, *Performance Evaluation of Programmable Robots and Manipulators,* National Bureau of Standards Special Publication 459, pp. 39–62.

Seering, W. P., Directions in Robot Design, *Journal of Mechanisms, Transmissions, and Automation in Design,* Vol. 1, March 1983.

Shigley, Joseph E., and Mitchell, Larry D., *Mechanical Engineering Design,* fourth edition, McGraw-Hill, 1983.

Source Book on Gear Design Technology and Performance, American Society for Metals, Metals Park, Ohio, 1983.

Spotts, M. F., *Design of Machine Elements,* fifth edition, Prentice-Hall, 1978.

Sunada, W., and Dubowsky, S., On the Dynamic Analysis and Behavior of Industrial Robotic Manipulators with Elastic Members, *Journal of Mechanisms, Transmissions, and Automation in Design,* Vol. 1, March 1983.

Yang, D. C. H., and Lee, T. W., On the Workspace of Mechanical Manipulators, *Journal of Mechanisms, Transmissions, and Automation in Design,* Vol. 1, March 1983.

CHAPTER 5

MECHANICAL DESIGN OF THE ROBOT SYSTEM

HANS J. WARNECKE

ROLF D. SCHRAFT

MARTIN C. WANNER

Fraunhofer Institute for Manufacturing
Engineering and Automation
Stuttgart, West Germany

5.1. THE STRUCTURE OF INDUSTRIAL ROBOTS

5.1.1. Kinematics

The task of an industrial robot in general is to move a body (workpiece or tool) with a maximum of 6 degrees of freedom (DF) (three translations, three rotations) into another point and orientation within the workspace. The complexity of the task determines the required kinematic construction.

Industrial robots according to VDI 2861[1] are kinematic chains with several links and joints. The number of DFs of the system determines how many independently driven and controlled axes are needed to move a body in a defined way in space. In the kinematic description of a robot we distinguish:

Arm. An interconnected set of links and powered joints that support or move a wrist and hand or end effector. With the arm we have a one-dimensional movement per axis. One axis constitutes a path, two axes a surface, three axes and more, a working space.

Wrist. A set of joints between the arm and the hand that allow the hand to be oriented to the workpiece. The wrist is for orientation and small changes in position.

Figure 5.1 shows the following definitions:

The reference system defines the base of the robot and also, in most cases, the zero position of the axes and the wrist.

The tool system describes the position and orientation of a workpiece or tool with 6 DF (X, Y, Z and A, B, C).

The robot (arm and wrist) is the link between reference and tool system.

As far as axes are concerned we distinguish the following:

A rotatory axis is an assembly connecting two rigid members that enables one to rotate in relation to the other around a fixed axis.

A translatory axis is an assembly between two rigid members enabling one to have a linear motion in contact with the other.

A complex joint is an assembly between two closely related rigid members enabling one to rotate in relation to the other about a mobile axis.

Figure 5.2 gives an overview of the symbols used in VDI 2861 and in this chapter. The kinematic chain can be combined by translatory and rotatory axes. Complex joints are also possible. Figure 5.3 shows an example of the preparation of a kinematic chain using the symbols and terms of VDI 2861.[1]

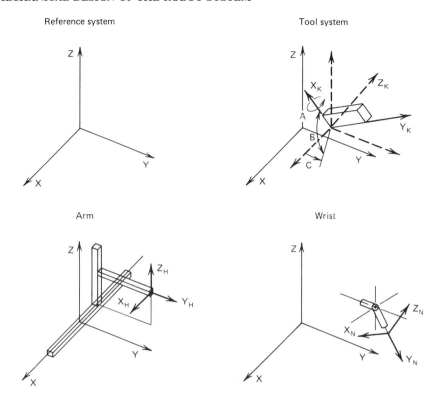

Fig. 5.1. Definition of coordinate systems for the handling task and the robot.

According to Reference 2 the number of possible variations of an industrial robot structure can be determined as follows.

$$V = 6^{\mathrm{DF}} \qquad \begin{aligned} &V = \text{number of variations} \\ &\mathrm{DF} = \text{number of degrees of freedom} \end{aligned}$$

These considerations show that a very large number of different chains can be built; for example, for six axes 46656 chains are possible. However, a large number are inappropriate for kinematic reasons. Figure 5.4 shows criteria and examples of inappropriate kinematic chains.

Further restrictions are given by the following facts:

Positioning accuracy decreases with the number of axes.

With the number of axes the computing time for continuous path control increases considerably.

Every additional axis produces additional costs (power train, brake, motor, measuring system, control of the drive, memory).

Power transmission becomes more difficult as the number of axes increases.

Today the number of axes is limited; industrial robots normally have up to four principal arm axes and three wrist axes. Figure 5.5 shows the most important chains of today. With new developments in the field of mechanical and controller design it can be expected that further kinematic chains will enter the market with success.

With the selection of the kinematic chain important decisions are made regarding the kind of robot control (CP, PTP) and the number of programmable axes.

Following Spur,[3] Table 5.1 shows the functional relationship between the task, number of programmable axes, and robot control as function of the arrangement of the arm. This table points out some fundamental problems in robot design: with translatory axes we have a complex mechanical hardware (see Section 5.10) against a simple controller design. The opposite is true for robots with rotatory

System	Name	Symbol
Translatory axis telescopic	X , Y , Z	
Translatory axis Transverse	U , V , W	
Rotatory axis Pivot	A , B , C	
Rotatory axis Hinge	D , E , P	
Gripper		
Tool		
Separation of arm and wrist	/	
More than one independent chain in the robot system	Beginning (End)	

Fig. 5.2.   VDI-symbols for industrial robots.

axes. Therefore the selection of the kinematic chain is one of the most important decisions in the mechanical and controller design process.

5.1.2. Type of Installation

Industrial robots can also be distinguished by type of installation:

1. With the floor installation the base plate is mounted on a foundation. Depending on the construction, the basic unit contains cable distributor, energy supply, oil container, and so on.
2. With the console installation the robot and working machine form one unit.
3. Gantry installation is used when working space must remain accessible or certain operations (e.g., gripping into a high box, difficult welding operations) cannot be realized by the floor installations. Most common is the conversion of a floor installation robot into a gantry type with additional translatory axes.

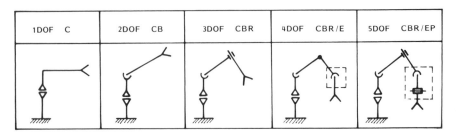

Fig. 5.3.   Creation of kinematic chains.

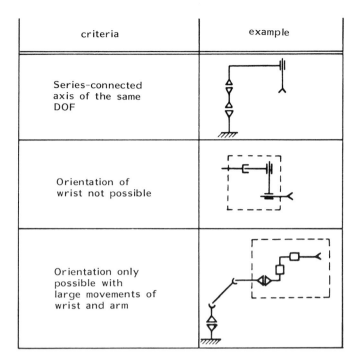

criteria	example
Series-connected axis of the same DOF	
Orientation of wrist not possible	
Orientation only possible with large movements of wrist and arm	

Fig. 5.4. Undesirable kinematic chains: criteria and examples.

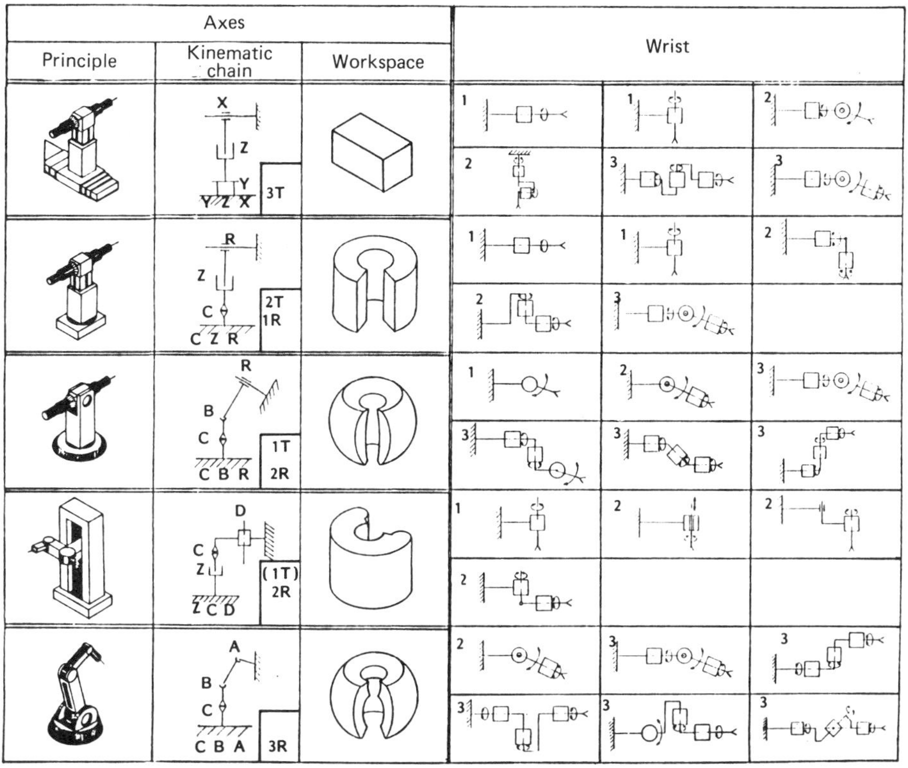

Fig. 5.5. Typical arm and wrist configurations of industrial robots.

TABLE 5.1. FUNCTIONAL RELATIONSHIP BETWEEN TASK, NUMBER OF PROGRAMMABLE AXES, AND ROBOT CONTROL AS A FUNCTION OF ARM ARRANGEMENT

Task / Kinematic chain		3T		2T 1R		1T 2R		1T 2R		3R		>3 axes >3R	
Movement A and B	Orientation to surface	PTP	CP	PTP	CP	PTP	CP	PTP	CP	PTP	CP	PTP	CP
Z (A, B)	None —	1		1		2	2	1	2		2		2
	dx	2		2		3	3	2	3		3		3
	dx dy	3		3		4	4	3	4		4		4
	dx dy dz	4		4		5	5	4	5		5		5
Z (A, B)	None	[2]	2		2	3	2	3	2		3		3
	dx	[3]	3		3	4	3	4	3		4		4
	dx, dy	[4]	4		4	5	4	5	4		5		5
	dx, dy, dz	[5]	5		5	6	5	6	5		6		6
Z (A, B)	None		3		3		3		3		3		3
	dx		4		4		4		4		4		4
	dx, dy		5		5		5		5		5		5
	dx, dy, dy		6		6		6		6		6		6

Not necessary   [ ] Possible under certain circumstances

No solution/without practical value

4. A new development is an industrial robot freely movable in one plane on the floor. One such development goes in the direction of a guided vehicle including robot, system control, and magazine.

From the foregoing systems many variants can be derived. Figure 5.6 gives an overview of the different designs. In any case the type of installation is related to the kinematic chain.

5.2. TASK-RELATED DESIGN

In many cases production technology determines the design of the robot system. The most important fields of application and their influence on the design are discussed in this section.

5.2.1. Spray Painting

In spray painting the following criteria are to be considered:[4]

1. In most cases the surface of the workpiece is complicated. Spray painting must be carried out at a certain angle to the surface. Therefore a high degree of mobility of the kinematic chain is needed.
2. Errors in the path corners lead to overlaps (thickness of the layer). The same is true for large variations in velocity.
3. Since the robot is usually programmed by manual tracing of the desired path, it is essential that only a low guiding power be needed within the whole working space.
4. Conveyor tracking may be required.

Spray-painting robots should have an easy "teach in" procedure and in most cases continuous path control. For safety reasons and to achieve high accelerations and decelerations, hydraulic drive systems are common. An example is shown in Figure 5.7.

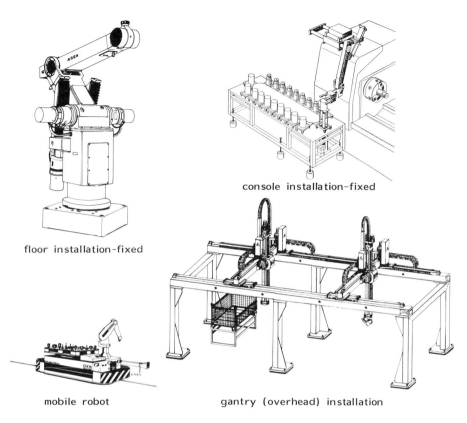

console installation-fixed

floor installation-fixed

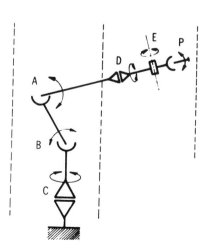

mobile robot

gantry (overhead) installation

Fig. 5.6. Types of installations of industrial robots.

Fig. 5.7. Example of a spray-painting robot with six axes.
(Photo courtesy of AOIP-Kremlin.)

5.2.2. Spot Welding

In the design of robots for spot welding the following requirements must be considered:

1. High acceleration and decelerations.
2. Mechanical mobility (as a rule at least five programmable axes).
3. Extreme reliability of the components.
4. Spot welding is usually performed in a large working space and with high loads.
5. It is difficult to run the power supply to the working tool.
6. Repeatability $\leqq$ 1 mm at visible seams, $\leqq$ 3 mm at invisible seams.
7. Often long and tall forearm owing to collision problems.

Robots for spot welding should be designed for floor or gantry installation or both. Figure 5.8 shows a typical spot-welding robot.

5.2.3. Arc Welding

The essential characteristics for arc welding robots are the following:

1. Processing of external sensor data is necessary.
2. Forearm should swing in positive and negative direction (often difficult accessibility to welding seams).
3. In most cases continuous path control is required. Options like circular interpolation are useful.
4. High welding velocities are to be realized.
5. Often a swivelling workpiece positioner is needed.
6. Programming comfort is required in the computation of dependences of welding current and welding voltage, wire feeding, and welding velocity.

In arc welding smaller units also have a high market share. An example is shown in Figure 5.9. A new trend proceeds in the direction of a transportable unit.

5.2.4. Assembly

The requirements in assembly are essentially different from those of the previous technologies.

1. The weights of workpieces in assembly are normally low ($<$ 1 kg).
2. Very short cycle times.
3. High positioning accuracy ($<$ 0.1 mm).
4. In most cases movements are parallel to X, Y, Z, and there is small working space

Especially in assembly a clear direction in the development of industrial robots cannot yet be determined. Even in the long run, inexpensive "pick and place" units (pneumatic with one or two programmable

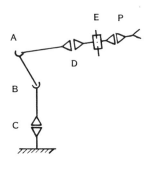

Fig. 5.8. Example of a spot-welding robot with six axes. (Photo courtesy of ASEA.)

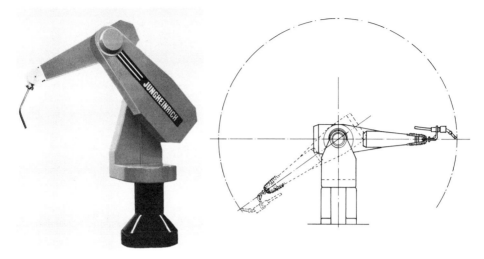

Fig. 5.9. Example of an arc-welding robot. (Photo courtesy of Jungheinrich.)

axes) may keep an important share of the market. Robots with translatory axes are often used in assembly as shown in Figure 5.10.

5.2.5. Workpiece Handling

The task of industrial robots for workpiece handling is to move workpieces from one point to another with defined positions and orientations. The following kinds of applications can be distinguished:

1. Handling on presses.
   (a) Very short cycle times.
   (b) Special design, often cartesian coordinate robot.
   (c) Conditions of installation and accessibility are to be considered in particular.
2. Handling on forging presses.
   (a) High speed and heavy workpieces.
   (b) Robot must be resistant to dirt, heat, and shock.
   (c) Floor installation recommended.
3. Handling on die casting and injection molding machines.
   (a) Often gantry or console type installation.

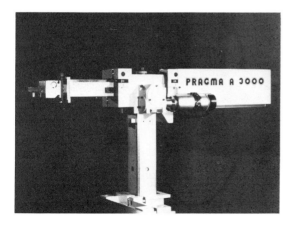

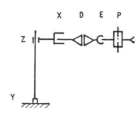

Fig. 5.10. Example of an assembly robot with three to six axes. (Photo courtesy of DEA-PRAGMA.)

(b) Cycle times with injection molding machines are shorter than with die casting machines.
(c) Very often simple movements lead to a simple programming language. Changeover frequency is low.
(d) High positioning accuracy.
(e) Insensitive to heat and dirt.
4. Handling on machine tools.
(a) High positioning accuracy, short cycle times, and also quite often heavy workpieces.
(b) Simple programming language owing to simple movements.

Gantry-type or console-installed robots are often used in workpiece handling. An example is shown in Figure 5.11.

5.3. ORGANIZATION AND STEPS IN DESIGN PROCEDURE

This section presents a general overview of the necessary steps in the design procedure.[5]

5.3.1. Organization

It is useful to separate exactly the different fields of engineering and to define the links between the different engineering groups as early as possible. A common solution is as follows:

1. **Group 1.** Design of the mechanical system.
    (a) Selection of the kinematic chain, calculation of forces and moments.
    (b) Drive system, power train; measuring systems (path, speed) and brakes.
    (c) Construction of the axes.
    (d) Passive sensor systems.
2. **Group 2.** Robot control and language.
    (a) Development of the robot control, control of the drives, information processing.
    (b) Operating systems and program interpreter; language development.
    (c) Interface to the mechanical system, external PC, and sensors.

3. **Group 3.** Development of sensors and external data processing.

5.3.2. Steps in the Design Procedure

The conception, development, and testing of the robot system require not only new ideas and experience in the scientific field but also the ability to transform these into a methodical procedure. At the starting

Fig. 5.11. Example of a gantry type workpiece-handling robot. (Photo courtesy of DÜRR.)

point of such a project it is useful to formulate clearly the sequence of all steps. The essential steps for the mechanical system design are the following:

1. Analysis of the state of the art, the products of competitors, and possible gaps in the market.
2. Study of the market and setting up the requirements by worksite analysis and overall determination of important parameters of the production technology.
3. Setup of the performance specifications. Based on this step preparation of a net plan including timetable.
4. Preparation of a sketch of overall designs. Selection by means of value analysis.
5. Splitting the design into the components. Selection of possible solutions.
6. Combination of the components to an overall design. Calculation of the costs for alternatives. Selection of the basic design and start of detailed design.
7. Order of long-lead items and, parallel to that, testing important or newly introduced components.
8. Performance testing of the complete system on a test stand. Correction of the faults. (See also Chapter 10.)
9. Working out a system for documentation and maintenance.
10. Start for production.

The following sections discuss these essential steps in the design procedure.

5.4. PROFILE OF THE REQUIREMENTS BY WORKSITE ANALYSIS

Usually the foundation for an analysis is the present worksite. The human worker together with the already existing manufacturing system performs operations. The subject for the investigation is the kind and frequency of these operations. A most useful approach was developed by Herrmann.[6.2] This procedure was tested in practical applications by some German companies and the IPA with success. The principal steps are:

1. Detailed worksite analysis of the present system.
2. Formulation and evaluation of alternative system solutions.
3. Analysis of requirements for the robot system, including peripheral devices.

Independent of the application the following data should be presented to the robot designers:

1. Number and weight of the workpieces to be handled.
2. Required movements in the working space.
3. Required working range, considering peripheral units.
4. Required positioning accuracy.
5. Changeover frequency of the production system.
6. Sensor function for recognition and quality control.
7. Gripper and tool options.
8. Kind of machining functions (e.g., drilling).
9. Peripheral units needed (e.g., vibratory hopper).
10. Careful examination of the production technology (see Section 5.2).

Experience has shown that more than 100 worksites should be investigated. Unfortunately this very expensive and time-consuming research is completely ignored by many robot manufacturers.

5.5. PERFORMANCE SPECIFICATIONS FOR THE ROBOT SYSTEM

The setup of performance specifications is the next logical step. These can be classified as follows:

1. Essential requirements which must be fulfilled in any case.
2. Minimum requirements (limiting values that must not be exceeded).
3. Desirable requests.

The following requirements, at least, should be established in the performance specifications before the start of the design study:

1. The kind of motions and kinematic chain.
2. Geometrical dimensions.
3. Velocities and accelerations referring to the axes.
4. Drive system and control.
5. Positioning accuracy.

With the increasing number of specifications the number of possible solutions decreases rapidly. From the engineering point of view, this aspect is welcome in most cases.

5.6.   PREPARATION OF A SKETCH OVERALL DESIGN

It is not useful to approach the overall design in too great detail because the large number of possible solutions may lead in different directions. The purpose of this section is to draw the mechanical designer's attention to specific problems in this step.

5.6.1.   Geometrical Dexterity

Within a given workspace the geometric dexterity describes the ability of the robot to achieve a wide range of orientations of the hand with the tool center point in a specified position, as described in Jou and Waldron.[7] Figure 5.12 shows as an example (after Jou and Waldron[7]) the workspace (outer curve) of a three-link robot. The hatched region is that reachable with the hand horizontally oriented. The geometric dexterity is very important in practical application. The designer has the choice to compare different chains, including their arm lengths, and to check the performance specifications.

5.6.2.   Kinematic Chain Suitable for Arithmetic Processor

With the determination of intersection points of the kinematic chain the designer can give useful contributions to the controller design team. The axes should intersect in one plane, and the axes of the wrist even in one point, to avoid unnecessary calculations in the forward and reverse calculation process. Other problems are singularity fields or ambiguous solutions in the coordinate transformation.[8]

5.6.3.   Forces and Moments on the Robot Structure

For the designer of the robot structure it is most important to determine the forces and moments in the joints as a function of the various kinematic chains and possible arm movements. Two calculation procedures are most common:

1. Lagrange equations for systems of less than four joints.[9,10]
2. Newton-Euler formulation for more general models.[11]

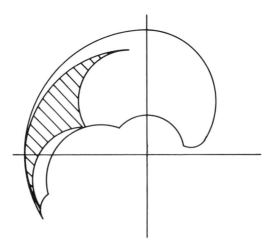

Fig. 5.12.   Example of a geometric dexterity. (After Jou and Waldron.[7])

Under certain assumptions it is possible to include friction in these models. These data must be provided by measuring procedures. Depending on the kinematic chain and the type of bearing, the friction may become a factor well worth considering.

Another factor concerns the forces and moments as functions of the kinematic chain. Robots with translatory axes have nearly independent movements of the single axes, which is not true for rotatory axes where coriolis, centripetal, and gravity terms lead to nonlinear couplings. The influence of these parameters on the trailing error is described in Duelen and Wendt.[12]

5.6.4. Selection of Material

The design of the arm includes an early decision about the materials used. The construction must be stiff and light. At the moment, steel and aluminium alloy are challenged by fibrous materials like CFRP (carbon fiber).[14] Positive properties of these new materials are:

High tensile strength.

High damping.

Low weight (one-fourth compared with steel, two-thirds compared with aluminium alloy).

On the negative side we should mention:

Reduction in strength due to temperature and humidity effects.

Considerable problems in molding complicated surfaces.

Difficult connection of the links.

Very high costs.

Taking these facts under review we must consider that with the use of the finite element method considerable weight savings can be achieved; furthermore, only 30–50% of the total arm weight can be affected by any of these weight-saving measures. We still need drives, gears, power transmission. The use of fibrous materials should not be completely ruled out if the cost factor is declining considerably.

5.7. SPLITTING THE DESIGN INTO COMPONENTS

The selection of available components is an important step in the overall mechanical design.

5.7.1. Selection of the Drive System

The kind of drive system is a characteristic feature of the robot system including controller. We distinguish:

1. **Pneumatic Drive Systems.** These are inexpensive, with a simple, robust construction and low weight for fast movements. However, a closed positional control is difficult to realize. The application range is for "pick and place" machines positioned by mechanical stops and for short-stroke axes in assembly robots.

2. **Hydraulic Drive Systems.** These allow a high concentration of power within small dimensions and weights, which is particularly advantageous for the wrist. Power train and transmission are simple; furthermore, high accelerations can be realized. Major drawbacks are leakage losses, the price for the hydraulic pump, friction, temperature changes, the high-input power, and, under certain circumstances, the layout of the controller.

3. **Electric Drive Systems.** The most important advantages are the wide range of possible options for speed control, the high reliability, and the simple reset after breakdown of energy.

The most important types of electric motors with their characteristics are the following:

1. **DC-motor.** A gear is necessary for high moments at low speed, also a measuring system and position control. This drive system is often used for industrial robots. One disadvantage is caused by the backlash in the power train. For a direct drive without gear, high-torque motors can be used. These systems are still in development, mainly for wrist axes.

2. **Stepping Motor.** This is controlled by providing in advance a certain number of path increments for the desired position. It is inexpensive because no measuring system is needed. At high

moments and collisions with obstacles the stepping motor may lose steps. Then feedback is needed, and the price increases considerably.

3. **Three-Phase Motor.** Speed is controlled by varying the frequency. This is an expensive technology, but there is no need of speed reduction.

Figure 5.13 shows the control structure of the most commonly used systems, the DC-motor with gear and the stepping motor. Electrical drive systems can be selected by the following criteria:

1. Reliability of the total system.
2. High starting torque.
3. Low moment of inertia.
4. Reduction of costs due to simplification of the mechanical design.
5. Low heat generation and reasonable dissipation of heat.
6. Speed control.

The arm drives can be equipped with brakes for safety reasons.

5.7.2. Selection of Path-Measuring System

Path-measuring systems are part of the mechanical construction and have the task to provide permanently for each axis its position coordinate as a reference coordinate during the programming cycle and as the actual coordinate for the position-control loop during the working cycle.

Positioning accuracy is determined also by the resolution of the path-measuring systems. These are in close interaction with features of the mechanical construction: the rigidity of the construction, backlash, and reverse error should be related to the resolution and linearity of the path-measuring system. We distinguish translatory (potentiometer, ultrasound) and rotatory systems (resolver, absolute optical encoder, and incremental encoder). The possibilities of the construction with the drive systems are as follows:

1. Directly on the robot axes. Translatory movements of axes are often transformed to rotatory movements of the path-measuring system.
2. Indirectly, as backup of the drive system (example in Figure 5.13). Backlash and reversal error may be compensated.

Common path-measuring systems are explained in Table 5.2.

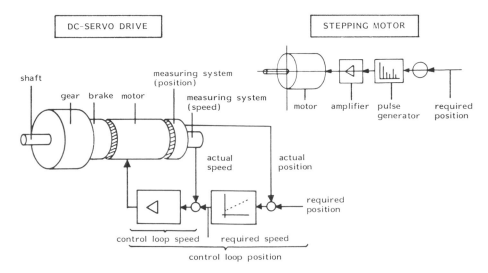

Fig. 5.13.   DC-Motor drive systems and control.

TABLE 5.2. COMMON PATH-MEASURING SYSTEMS

System	Data Register	Principle	Reference Point	Resolution	Advantage	Disadvantage
			Analog			
Potentiometer	Absolute / Signal / Distance	Brush / Resistive track	Fixed	$\dfrac{\Delta L}{L} \sim 10^{-3}$	Inexpensive Small dimensions	Linearity Contact measuring Mechanical uncertainties
Ultrasound	Absolute / Signal / Distance	Impulse Magnet / Reciever / Magnetostrictive wire / Ultrasound impulse / Transmitter	Fixed	$\dfrac{\Delta L}{L} \sim 2 \cdot 10^{-4}$	Long translatory measuring range Direct measuring method	Expensive Difficult data processing
Resolver	Incremental / Signal / Distance	S4 S2 S1 S3 / Stator winding 1 / Stator winding 2 / R2 Rotor R4	Ambiguous	$\dfrac{\Delta \varphi}{\varphi} \sim 7 \cdot 10^{-3} \div 10^{-5}$	Robust Simple design Resolver inexpensive	Digitization of signals expensive Data ambiguous
			Digital			
Absolute encoder	Cycle—absolute / Signal / Distance	Light sources / Light detectors / Interface (electronic) / Absolute position output / Absolute coded disks / Encoder shaft	Fixed	$\dfrac{\Delta \varphi}{\varphi} \sim 10^{-3} \div 10^{-5}$	Unambiguous representation of shaft angle High resolution No addition of errors	Limited measuring range or very expensive
Incremental encoder	Absolute / Signal / Distance	Light source / Lens / Index-plate / Light cell detector / Incrementally coded disks / Encoder shaft	Ambiguous	$\dfrac{\Delta \varphi}{\varphi} \sim 4 \cdot 10^{-3}$	Unlimited measuring range Inexpensive Simple design	Addition of errors Revolution marker necessary Disturbance pulse control

TABLE 5.3. MOST COMMONLY USED POWER TRANSMISSION SYSTEMS

System	Principle	Characteristics	Transmission of Movement	Speed Reduction	Transmission at Distance	Application
Spur gearing		First rotatory arm axis High moments	$\frac{R}{R}$	Yes	No	Arm
Bevel gearing		Special case for flange installation	$\frac{R}{R}$	Yes	No	Arm
Worm gear		High transmission ratio Heavy in weight Heat problems	$\frac{R}{R}$	Yes, high	No	Arm Wrist Gripper
Planetary gear		Expensive Heavy in weight	$\frac{R}{R}$	Yes, high	No	Arm
Harmonic drive		Very high transmission ratio Small dimensions Light in weight	$\frac{R}{R}$	Yes, very high	No	Arm Wrist
Chain drive		No backlash Heavy in weight No vibration	$\frac{R}{R}, \frac{T}{R}, \frac{R}{T}$	Possible	Yes	Slide module Wrist
Toothed belt drive		Backlash and vibration problematic Very light in weight	$\frac{R}{R}, \frac{T}{R}, \frac{R}{T}$	Possible	Yes	Wrist Gripper

Type	Characteristics	Motion conversion		Application
Bowden wire	Very good for transmission on distance. Axial extension	$\frac{R}{R}, \frac{T}{R}, \frac{R}{T}$	No	Yes	Wrist
Four bar-linkage	Very good for precise power transmission on distance	$\frac{R}{R}$	Possible	Yes	Arm
Slider crank chain	Special case application	$\frac{R}{T}, \frac{T}{R}$	No	Yes	Arm, Wrist
Screw drive spindle	High transmission. Friction (lubrication) problem	$\frac{R}{T}$	Yes, high	Yes	Wrist
Recirculating ball nut and screw	High transmission. Very high precision and reliability. Expensive	$\frac{R}{T}$	Yes, very high	Yes	Arm, Wrist
Toothed rack	Precise. Inexpensive	$\frac{R}{T}, \frac{T}{R}$	Yes	Yes	Arm, Wrist, Gripper
Cylinder	A lot of variations possible for hydraulic and pneumatic systems	$\frac{T}{T}$	Yes	Yes	Arm, Wrist, Gripper

5.7.3. Power Transmission Systems

The following are typical tasks for power transmission systems:

1. Transmission of movement: translation to rotation (T/R) or rotation to translation (R/T).
2. Speed reduction: Rotation to rotation (R/R), translation to translation (T/T), (R/T) and (T/R).

Depending on the kind of application, the following criteria should be considered:

1. High transmission ratio with low moments of inertia and size.
2. Low bearing clearance and reversal error.
3. Weight.
4. Long useful life and easy maintenance.
5. Power transmission at distance, movability.

Table 5.3 shows the most commonly used power transmission systems with their essential characteristics and fields of application.[15]

5.7.4. Bearings

In general bearings and guideways are under high loading. For rotatory axes tapered roller bearings or shoulder bearings are used prestressed in axial direction. Warping should be carried out by means of screws during the assembly.

The aim of low bearing clearance is to keep positioning, reversal, and trailing error as low as possible. The influence and measuring procedure of these errors are described in Chapter 10.

5.7.5. Couplings

Unsuited couplings can create serious problems. In the selection of possible components we should review the following requirements:

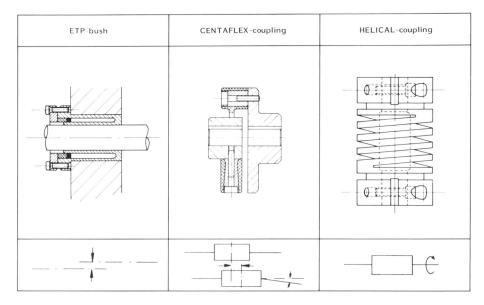

Fig. 5.14. Power transmission components with minimum backlash

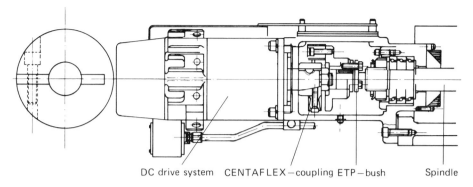

DC drive system   CENTAFLEX—coupling ETP—bush   Spindle

Fig. 5.15. Application of ETP-bush and CENTAFLEX coupling in a drive system. (After T. Hashimoto.)

1. Keep the reversal error as low as possible.
2. Construct with low weight and restrict the technical expenditure.
3. Reduce vibration and overload in the drives.
4. Keep the system free of maintenance.

According to Hashimoto[16] the following components are in common use:

1. **ETB-Bush.** These are applied for smaller forces with a possible misalignment between shaft and hub. The bush is put onto the shaft. By means of a flange the medium of pressure lies against the hub with constant pressure. A high torsional moment is possible.
2. **CENTAFLEX-Coupling.** It is possible that thermal expansion and deformation of the bearings may cause canting in the connection of two shafts. This leads to increased friction in the

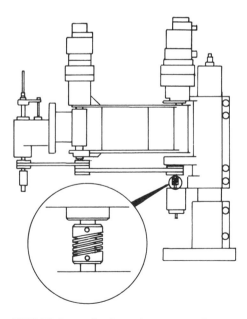

Fig. 5.16. Application of HELICAL coupling in a drive system with stepping motor. (After T. Hashimoto.)

bearings. One solution is an interconnection with a CENTAFLEX-coupling, which is compliant in axial and radial direction and very rigid in torsion.

3. **HELICAL-Coupling.** With changing moments and oscillating vibrations in the system it is useful to interconnect a torsion spring between the drive and the path-measuring systems. The HELICAL-coupling is able to transmit a regular rotation at a very high speed without reversal error.

Figure 5.14 shows the described components of power transmission with their most important characteristics. In Figure 5.15 an application of an ETB-bush and a CENTAFLEX-coupling is presented connecting a DC-motor drive with a spindle for a translatory movement in the hand axes of a robot. The requirements are interruption of heat transfer, absorption of vibration and push loadings, as well as easy assembly and disassembly. Figure 5.16 shows an application of the HELICAL-coupling with an assembly robot of the SCARA family. It connects the stepping motor with the toothed belt for orientation of the hand axis. The main problem here is the transfer function of the belt drive.

5.8. COMBINATIONS OF THE COMPONENTS TO AN OVERALL DESIGN

After preparation of sketch overall designs and evaluation of the components we must calculate the hardware costs for the different solutions which are technically on the same level. After the selection of the most promising alternative for the performance specifications we start the detailed overall design.

Here it is possible to split the procedure into design for wrist and for arm. For the wrist the following parameters should be taken into account:

1. Number of degrees of freedom, possibility of additional axes, gripper options.
2. Points of intersection for the wrist. Avoidance of unnecessary transformation for the robot control.
3. Movability of the wrist and gripper. Coordination with the arm.
4. Type of power transmission and guidance through the axes.
5. Arrangement of the drive system.
6. Guidance of the energy supply and signal lines.
7. Accessibility for maintenance.

A wrist with one or two axes is far less prone to problems than a three-axes wrist.
For the arm we must consider the following:

1. Relation of bearing clearance, backlash, friction, and wear in the drive systems to the stiffness of the axes.

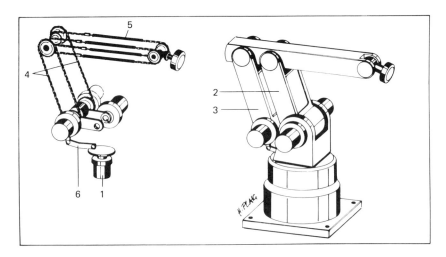

Fig. 5.17. General view of the Hitachi "Process Robot" with the power transmission system.

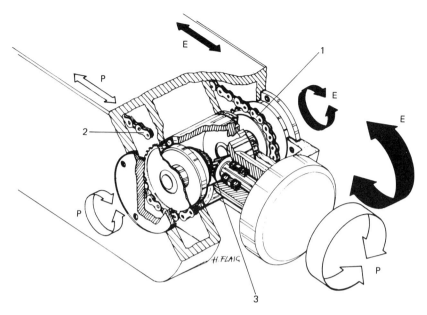

Fig. 5.18. Two degrees of freedom wrist axes. Hitachi "Process-Robot."

2. Actual utilizable working space for the arm, including wrist and gripper.

3. The checking of the possibilities of active and passive counterbalance.

4. Arrangement of the drive systems.

5. Guidance of the energy supply.

As a case study we discuss in detail the overall design of the Hitachi "Process Robot." One major requirement was the need to keep the orientation of the wrist in all positions in the workspace by mechanical means. The aim was to simplify the robot control and increase the speed in CP-mode.

Figure 5.17 shows the design concept. Here the rotatory axis C (1) is located in the base. The two other arm axes (2 and 3) are coupled in a parallelogram. Axis 3 is supported by a lever arm (6) for parallel guidance. The wrist is moved by the chains (4). This parallelogram solution of the arm and the attached chain drive keeps the orientation of the axis E (5).

Another simplification of the controller was achieved by the symmetric mechanical design. All axes of the arm intersect in one plane and the wrist in one point.

The use of elastic couplings in the drive system and shock absorbers in the chain drives leads to a structure with low reversal error. To make the design insensitive to vibrations, this case (Figure 5.17) featured shock-absorbers (5) in the wrist drive system and the fork-bearing layout of the arm axes. With all the important drive systems near the base it was possible to keep the influence of the moving masses within controllable limits. Sometimes the weight of the drive systems is used as counterbalance (as in the example of the KUKA IR 100) (Fig. 5–33).

The wrist was easy to realize for the requirement of 2 DF. Figure 5.18 shows the overall concept. Here the axis E is transformed over a chain (1) from a translatory to a rotatory movement, whereas axis P is moved rectangular to axis E over chain (2) and bevel gear (3).

5.9. WEAK POINTS IN THE MECHANICAL DESIGN

Robots are often used in lines linked with timed sequence, for example, spot welding and assembly. As a rule 80% of all malfunctions can be associated with the peripheral devices, including controller.[14] If the complete system must run with an availability of 99.5%, the mechanical system of the robot must be designed for 99.9% availability. For this we must consider the potential weak points with great care (Table 5.4 gives an overview):

TABLE 5.4. POTENTIAL WEAK POINTS IN MECHANICAL DESIGN

weak points	system	measures to be taken	calculation method	measuring method
deformation of the total structure and single components		- increase stiffness - reduce weight of the structure - counterbalance	- finite element method	refer 10.4.1
dynamic deformation		- increase stiffness - reduction of the moved masses - distribution of masses	- frequency and time related methods	refer 10.4.5
backlash		- reduce backlash in gear - use of stiff power transmission systems	- experience of the component manufacturer	refer 10.4.1
bearing clearance		- prestressed bearings	- construction of arm by FEM - experience of component manufacturer	refer 10.4.1
friction		- bearing clearance - improve lubrication	- experience of the component manufacturer	refer 10.4.1
thermal effects		- isolate heat source	- finite element method	refer 10.4.1 and 10.4.3
path measuring system		- depending on system location and connection with the mech. system	- component manufacturer	

1. The deformation of the total structure and components can be compensated by the controller with input of correction factors. The designer has options like increase of stiffness, weight reduction, counterbalance, and layout of the bearings.

2. Dynamic deformation is hard to control. Two methods have been investigated[17]:
   (a) Linear control applied in velocity control loop.
   (b) Bang-bang control applied in the position control loop. From the mechanical point of view we can increase the stiffness, reduce the moving masses, and think about the mass distribution of the robot. A useful tool in the evaluation process is the experimental modal analysis described in Chapter 10.

3. Backlash, bearing clearance, and friction are nonlinear characteristics causing inaccuracy and instability of the servomechanism. With a proper mechanical design many problems can be solved as outlined in Table 5.4.

4. Thermal effects are often compensated by the controller. Sometimes the isolation of the heat source may be possible.

5. A source of considerable trouble could be path-measuring systems, owing to bad connection with the mechanical system.

6. If the mechanical workspace is not identical with the workspace set by the controller, serious malfunctions in practical application are common. (See also Section 10.4.)

5.10. DESIGN EXAMPLES

This section introduces industrial robots based on essentially different principles of design and application.

5.10.1. Modular Design of an Assembly Robot—Bosch FMS

The Bosch system is a family of pneumatic and electrically driven handling modules that are compatible among each other. Such a system gives a fair chance to design tailor-made solutions for each application.

Modular systems are very popular in workpiece handling, handling on machine, and assembly. Figure 5.19 gives an overview of the total Bosch system. It should be noted that a large number of gripper options are also available.

Horizontal and Vertical Slide Module: 1 and 3

Figure 5.20 shows the slide module consisting of the Bosch profile (1) and the slide unit (2), which is guided by prestressed ball bearings (3). Power transmission is carried out by a toothed rack drive (4). The pneumatic and electric energy supply is protected by the cable suspension (5) parallel to the moving direction. The vertical slide modules are equipped with pneumatic weight balance (6) and brakes.

Rotatory Arm Module: 2

The rotatory arm module can be assembled between the slide modules to create a cylindrical coordinate robot.

Linear Module: 4

The electric linear module (Figure 5.21) has three major components: the block unit (1), guidance (2), and air transmission (3). The drive system includes the motor (4), gear (5), and path-measuring system (6). The rotation of the motor is transformed by a toothed rack drive (7) to a translatory movement. The drive block is moved in a guide rod (8).

A special problem is energy supply—electric power is carried by a spiral cable (9), air through special tubes.

Pneumatic Short-Stroke Module: 5

In assembly a quick vertical short-stroke axis is needed quite often. This requirement is accommodated by the short-stroke sledge (Figure 5.22), which can be combined with the linear module and the gripper rotation module.

Gripper Rotation Module: 6

The gripper rotation module (Figure 5.23) is used for rotary movements of the gripper in one axis. Basic elements include the block unit and turn table (5). The block unit includes the motor (1), toothed belt (2), worm gear (3), and the path-measuring system (4). The hardware interface is standardized for possible combinations with pneumatic units.

For the modular design of industrial robots the following general conclusions are possible:

1. To a large extent they are suitable only for translatory axes.
2. With increasing number of axes (more than five) the modular system becomes too complicated.
3. The design can be tailored to a specific task.
4. Optimal coordination with peripheral devices is possible if the total system includes modular subunits.
5. Layout planning and construction can be simplified if CAD/CAM techniques are available.
6. The modular design approach is well suited for workpiece handling and assembly.

Figure 5.24 shows a CAD drawing of the modular Bosch system for a gantry-type construction. In a philosophy similar to the mechanical design Bosch has developed a modular controller system running from programmable controllers (PCs) for peripheral devices to a robot control.

5.10.2. Robot with Horizontal Rotatory Axes—Dainichi Kiko PT 300 H

The Dainichi Kiko company has developed a family of industrial robots with the following characteristics:

1. The mechanical construction of the robot is very similar whether built with vertical or horizontal rotatory axes. Figure 5.25 shows the principle. The frame system of both robots includes:
   (a) V-version: B, A, and D axes.
   (b) H-version: C, D, and P axes.
   For the V-version we have an additional C-axis and a second wrist axis E; the H-version has a translatory short-stroke hand axis within the basic frame system.

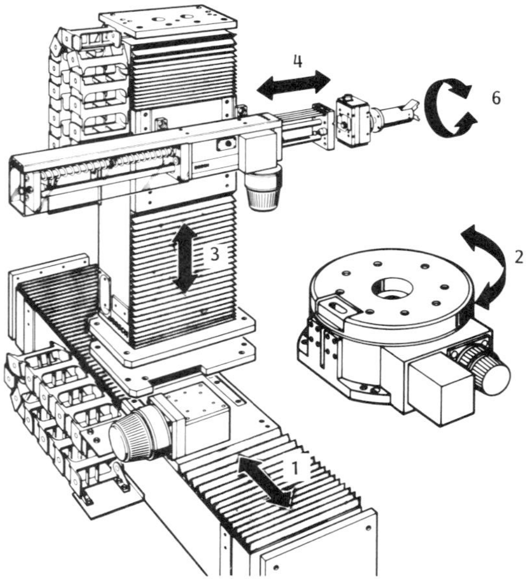

System	Components	
	1. slide module horizontal option : pneumatic	stroke : 240/ 400/ 800 1200/ 1600/ 2000
	2. rotatory arm module	diameter : 400/ 630
	3. slide module horizontal option : pneumatic	stroke : 240/ 400/ 800 1200/ 1600/ 2000
	4. linear module option : pneumatic	stroke : 160/ 240/ 400/ 560
	5. short stroke slide module–pneumatic	stroke : 35/ 50
	6. gripper rotation module option : pneumatic	

Fig. 5.19. Modular construction of an assembly robot. (Illustration courtesy of Bosch.)

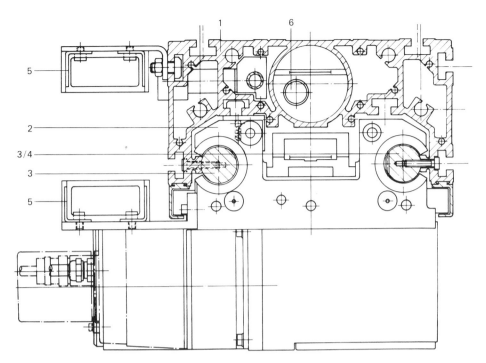

Fig. 5.20. Horizontal and vertical slide module. (Illustration courtesy of Bosch.)

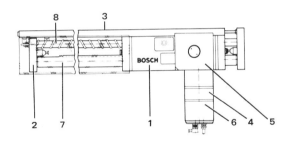

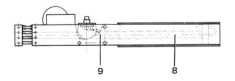

Fig. 5.21. Linear module. (Illustration courtesy of Bosch.)

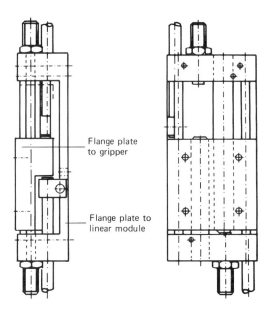

Fig. 5.22. Pneumatic short-stroke slide module. (Illustration courtesy of Bosch.)

2. Series of identical kinematic chains and different workloads, armlengths, and number of program-mable axes:

	Workload	Armlength C and D (H-Version) B and A (V-Version)	Number of Axes
PT 200 series	2 kg	450 mm	H: 3–4/ V: 5
PT 300 series	5 kg	700 mm	H: 3–4/ V: 5
PT 800 series	25 kg	1500 mm	V: 5

For the PT 800 series only the V-version is available.

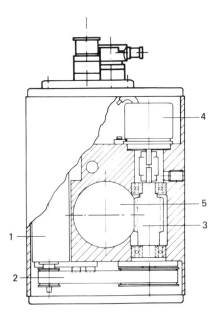

Fig. 5.23. Gripper rotation module. (Illustration courtesy of Bosch.)

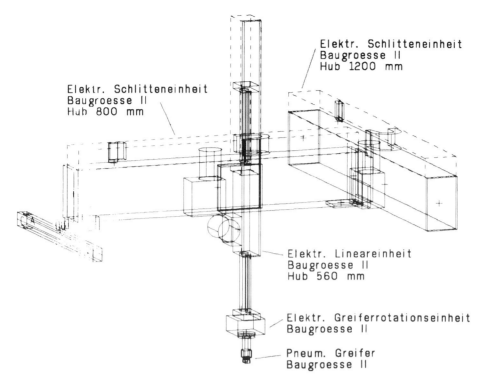

Elektr. Schlitteneinheit
Baugroesse II
Hub 1200 mm

Elektr. Schlitteneinheit
Baugroesse II
Hub 800 mm

Elektr. Lineareinheit
Baugroesse II
Hub 560 mm

Elektr. Greiferrotationseinheit
Baugroesse II

Pneum. Greifer
Baugroesse II

Fig. 5.24. CAD drawing of the modular BOSCH system. Modules are in gantry-type construction. Notes are in the German original. (Illustration courtesy of Bosch.)

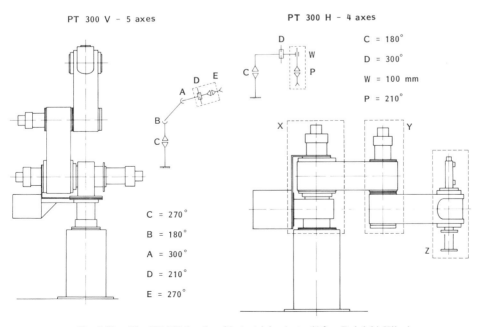

PT 300 V – 5 axes

PT 300 H – 4 axes

C = 180°
D = 300°
W = 100 mm
P = 210°

C = 270°
B = 180°
A = 300°
D = 210°
E = 270°

Fig. 5.25. The PT 300 family of industrial robots. (After Dainichi Kiko.)

3. For all robots the same type of controller is available in different versions:

	Position Control	Number of Axes	Simultaneously Controllable Axes
R 510	PTP	5	1
A 200	PTP	6	3 + 3
A 250	PTP + CP	6	3 + 3
A 300	CP	6	6

The mechanical design of the PT 300 H is described in more detail in the following sections.

Frame, C-Axis, Drive System of the C and P Axes

Detail X of Figure 5.25 is further shown by Figure 5.26. Here the drive of the arm axis C is performed by a DC motor with an incremental path-measuring system (1). The motor is supported by frame 4. Speed reduction is achieved by the use of an harmonic-drive unit (2) over the flexspline (2b) on the drive shaft (3). The drive shaft (3) and the arm of axis C (6) are connected axially over a conical clamping element (5) and radially over a flange. The wrist drive P (orientation) (7) is assembled face to face on axis C and secured by the frame (8). The drive of the wrist P (7) includes a DC servomotor (7), an incremental path-measuring system, and an harmonic-drive gear (8) engaging a toothed-belt drive (9).

Power Transmission of Axis P, Connection of the Axes C and D, and Drive System of Axis D

Figure 5.27 shows detail Y of Figure 5.25. Here the toothed-belt drive (9) coming from axis C (6) is transmitted over a flying mounted hollow shaft (11) to the toothed-belt drive (10) in axis D (16). Axis D (16) is driven over a DC-motor (12), harmonic drive (13), and drive shaft (14). Axis D is radially connected by an ETP-bush (15).

Wrist Axes W and P

Figure 5.28 shows detail Z of Figure 5.25. For the vertical W-axis the following options are offered:

1. Programmable with DC motor (17) and spindle (19/20) for a translatory movement.
2. Pneumatic drive with fixed stop. Here the cylinder (22) is performing the translatory movement.

For both options the P-axis remains identical. The P-axis is driven over the toothed-belt drive (10) while the vertical axis W (18) rotates as well. The P-axis is a parallelogram solution, as already described in the example of the Hitachi "Process Robot." Electric as well as pneumatically driven grippers or tools can be joined with the flange plate (21).

The concept of the PT 300 H was derived from the SCARA robot family. That means two axes that cover a surface, a short vertical axis W, and an axis P for orientation. The concept was carefully tailored for specific tasks in assembly, palletizing, and machining, with considerable success on the market.

In summarizing the facts the following should be noted:

1. All members of the robot family are designed for certain tasks. The requirements related to the various tasks are covered by the total family of robots with different workloads, numbers of axes, the kind of robot control, and so on.
2. There is no expensive surplus, which is not needed for the task.
3. Changes of the mechanical components are possible.
4. Costs for the development of the robot control and programming language can be covered by the total family.
5. Sometimes there are very low batch numbers for certain types (risk for the producer).

5.10.3. Hydraulic-Driven Robot with Cylindrical Coordinates—ZF T III L

The robot model T III L was designed by Zahnradfabrik Friedrichshafen for special requirements in workpiece handling. Figure 5.29 shows the overall concept. The frame system consists of four servohydraulic axes: C, Z, R (arm), and D (wrist). The following options are possible:

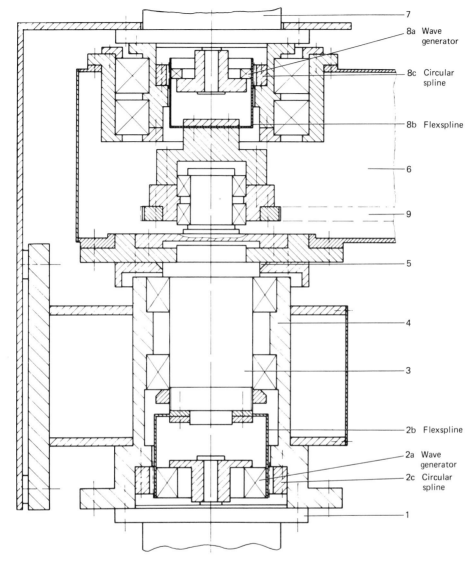

Fig. 5.26. Installation and drive system of the C and P axis of the Dainichi Kiko PT 300 H. (After Dainichi Kiko.)

1. Another wrist axis P.
2. Translatory Y-axis.
3. Several gripper solutions.
4. Controller for four to six axes.

The requirements in workpiece handling of heavy parts were carefully considered, leading to the following overall balanced design:

1. Minimum space requirements constituted the principal reason for the telescopic R-axis.
2. The long, tall R-axis was designed to meet the requirements for a very small workpiece input channel.
3. Solutions for high temperatures and dirt had been introduced.

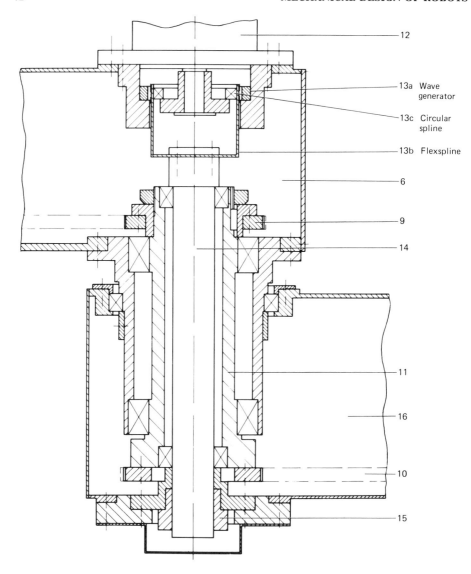

Fig. 5.27. Connection C and D axis with drive system for D axis in the Dainichi Kiko PT 300 H. (After Dainichi Kiko.)

4. The requirement for a high load at high speed was the principal reason for the selection of the hydraulic drive system.

Figure 5.30 shows the T III L in process of assembly at the manufacturer. For the definition of the axes, refer to Figure 5.29.

R-Axis

The principle of the transmission of movement and the design of the bearings is shown in Figure 5.31 (detail X of Figure 5.29). Here the total stroke consists of a substroke X of the lafette housing (3) and of a substroke Y of the tube (1). The tube (1) is moved over a double gear (7) with one gear engaging the toothed rack (4), which is fixed in the lafette housing, and the other gear the rack (2) fixed with the tube. The clearance is adjusted by a screw (6), and the pressure is produced over roll (13) borne on a rocker. The rotation of the double gear is produced by the movement of the servo

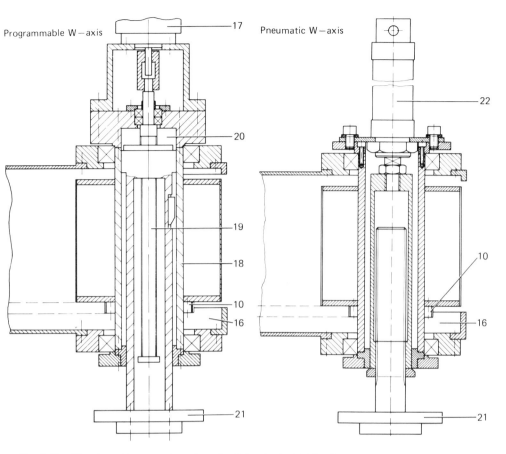

Programmable W—axis   Pneumatic W—axis

Fig. 5.28. Electric and pneumatic wrist axis W in the Dainichi Kiko PT 300 H. (After Dainichi Kiko.)

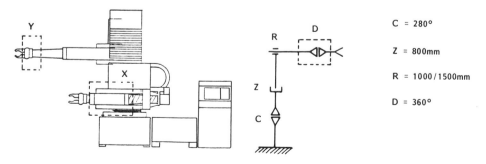

C = 280°

Z = 800mm

R = 1000/1500mm

D = 360°

Fig. 5.29. Robot with cylindrical workspace and hydraulic drive system, the ZF T III L. (Illustration courtesy of Zahnradfabrik Friedrichshafen.)

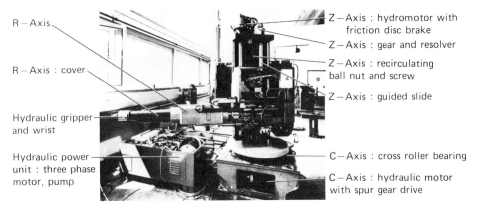

R — Axis

R — Axis : cover

Hydraulic gripper
and wrist

Hydraulic power
unit : three phase
motor, pump

Z — Axis : hydromotor with
friction disc brake

Z — Axis : gear and resolver

Z — Axis : recirculating
ball nut and screw

Z — Axis : guided slide

C — Axis : cross roller bearing

C — Axis : hydraulic motor
with spur gear drive

Fig. 5.30. ZF T III L under assembly at the manufacturer. (Photo courtesy of Zahnradfabrik Friedrichshafen.)

cylinder (12) and the support on the toothed rack (4); at the same time, path measuring is performed by belt drive (9) on the resolver (10).

Adjustable roller bearings (11) are part of the slide construction. The total stroke can be changed by the selection of different wheel sets of the double gear (7/8).

Wrist Axes D, P, and Double Gripper

Figure 5.32 shows the layout with two hydraulic wrist axes. Six main lines (1) supply the rotatory drives of axes D (2) and P (4) and the cylinder for the gripper (8). Path measuring is performed by the resolvers (3 and 5). The double gripper (7) is opened and closed by the cylindrial slide valve (6).

This case study gives an indication of how technological requirements lead to a special design. On the market the T III L is quite successful for applications in difficult environments, like bath hardening, handling at forging presses, and milling machines. It should be noted that this robot was developed by a potential user.

5.10.4. Multipurpose Robot with Six Axes—KUKA IR 100

The design principle of multipurpose robots is to cover certain different tasks at a reasonable price. One example is the KUKA IR 100/160 series, which was designed for the following tasks:

1. Spot and arc welding.
2. Assembly of heavy workpieces.
3. Machining with industrial robots.

Figure 5.33 shows the KUKA IR 100/160 with six programmable axes. For the design the following specifications were most important:

1. A high degre of mobility, different installations (for example, gantry type).
2. A high degree of stiffness of the axes and no bearing clearance; therefore very good repeatability and low reversal error.
3. Good long-term behavior and easy maintenance.

The following more or less new design principles were introduced:

1. All arm axes are equipped with prestressed roller bearings.
2. Closed hydraulic counterbalance of the second axis.
3. All motors of the wrist are located at the end of the third axis for static counterbalance. Universal joints are used between motor and gear of the wrist.

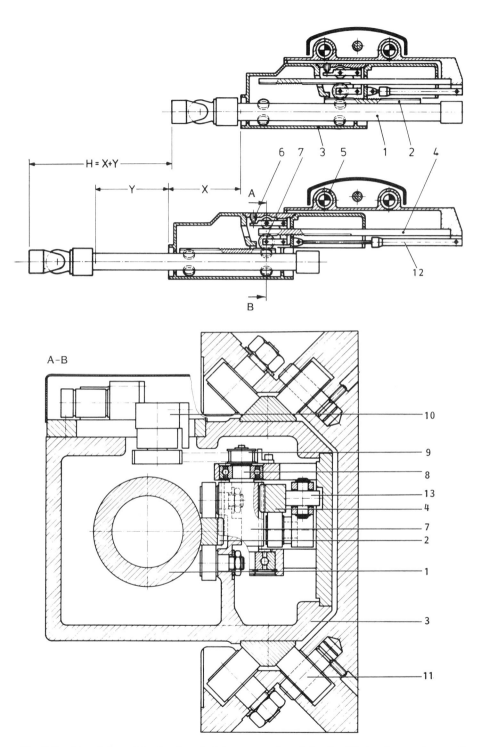

Fig. 5.31. Translation of the R-axis: principle, construction of the the slide-bearings, and connection with resolver. (Illustration courtesy of Zahnradfabrik Friedrichshafen.)

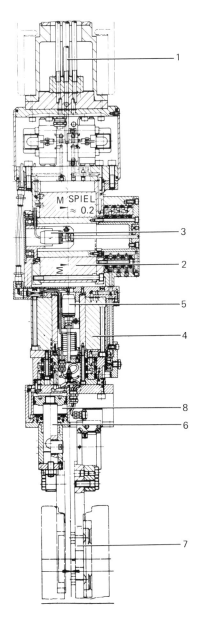

Fig. 5.32. Two degrees of freedom hydraulic wrist axes with gripper. (Illustration courtesy of Zahnradfabrik Friedrichshafen.)

Wrist D, E, P

Illustrated in Figure 5.34, the most important features of this design are the following:

1. Compact construction, all three axes move in one point; low moments because of the small dimensions.
2. Harmonic-drive gear integrated into wrist.

The wrist is explained by the power transmission of the different axes:

1. Axis D: universal joint (1) on hollow shaft (2), harmonic-drive gear (3) on axis D (4).
2. Axis E: universal joint (5) on hollow shaft (6), gear (7), toothed-belt drive (8), harmonic-drive gear (9) on axis E (10).

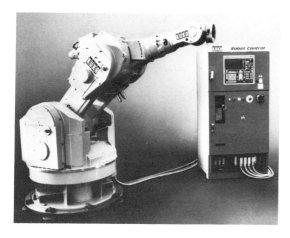
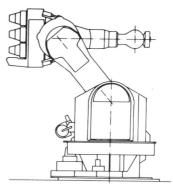

Fig. 5.33. KUKA IR 100/160 universal robot. (Photo courtesy of KUKA Schweissanlagen and Roboter GmbH.)

3. Axis P: Universal joint (11) on shaft (12), gear (13), toothed-belt drive (14), gear (15), harmonic-drive gear (16) on axis P (17).

For this case study the following conclusions are possible:

1. A multipurpose design like the KUKA IR 100 may lead to very high production numbers if the price can be kept within certain limits.
2. KUKA, as many other successful robot manufacturers, entered the market from a specific production technology know-how (for KUKA welding), integrating the robot into the whole production system.
3. The multipurpose robot, in general, can be used in different departments of the same company with considerable advantages in maintenance and programming. The same approach is more or less true for modular systems or robot families.
4. A major and important drawback is the fact that multipurpose designs are generally more expensive because, in any case, some functions are not needed.

5.11. SUMMARY

The following conclusions should be made regarding the mechanical design of industrial robots:

1. The requirements for industrial robots are closely related to the task. A decision about the design principle (universal robot, modular robot, or special design robot) should be made according to the market volume.

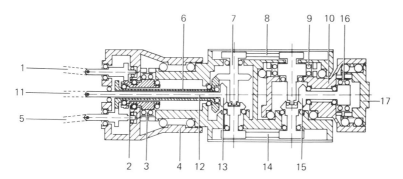

Fig. 5.34. Wrist axis with 3 DF. (Illustration courtesy of KUKA Schweissanlagen and Roboter GmbH.)

2. To get a good overall solution the designer should think about the practical applications. He or she should evaluate the possible flexibility of the robot, grippers, tools, and peripheral units and integrate all components to one system. In most cases peripheral units are the bottleneck in flexible automation.

3. Only a balanced combination of mechanical design, robot control, programming language, and peripheral units makes a highly valuable solution.

4. Robots must be measured according to international definitions, classifications, and test methods.

Future development in the field of the mechanical design may go into the following directions:

1. Simplification for the mechanical system.

2. Introduction of new drive and path-measuring systems.

3. Reduction on research and development time. Application of CAD/CAM for design, layout, planning, performance testing, and manufacturing.

4. New techniques in coordinate transformation, robot control, and path planning.

5. Considerable improvements in movability.

Essential steps forward can always be expected when industrial robots emerge into new fields of application.

ACKNOWLEDGMENTS

The authors wish to thank Mr. Drexel and Mr. Kaufmann (Bosch), Mr. Zimmer and Dr. Wörn (KUKA), Mr. Christen (Stiefelmayer), and Mr. Sauter and Mr. Manogg (ZF) for their most useful contributions and suggestions to the text of this chapter.

REFERENCES

1. VDI 2861, Blatt 1 und 2, Entwurf, Bezeichnungen und Kenngrößen von Handhabungseinrichtungen, *VDI-Verlag.*

2. Warnecke, H. J., and Schraft, R. D., *Industrial Robots,* IFS Publications Ltd., Bedford, 1982.

3. Spur, G., *Industrieroboter,* Hanser-Verlag, Munich, 1979.

4. Schraft, R. D., and Schiele, G., Industrieroboter zum Lackieren, *I-Lack,* February 1982, pp. 65–73.

5. Wanner, M. C., and Weiss, K., Systematische Vorgehensweise bei der Konzeption, der Entwicklung und Ausarbeitung von Handhabungseinrichtungen, *Technische Rundschau,* February 1982, pp. 16–17.

6. Herrmann, G., *Analyse von Handhabungsvorgängen im Hinblick auf deren Anforderungen an programmierbare Handhabungsgeräte in der Teilefertigung,* Dr.-Ing. dissertation, University of Stuttgart, 1976.

7. Jou, T. M., and Waldron, K. J., Geometric design of manipulators using interactive computer graphics, *6th IFToMM Congress on Theory of Machines and Mechanisms,* New Delhi, 1983.

8. Schmieder, L., Kinematik und Betriebsprogramme von rechnergesteuerten Manipulatoren, *Lehrgang R1.08,* March 1979, DFVLR Oberpfaffenhofen.

9. Horn, K. P., Kinematics, statics and dynamics of Two-D manipulators, *MIT Working paper 99,* June 1975.

10. Horn, K. P., Hirokawa, K., and Vizirani, V., Dynamics of a three degree of freedom kinematic chain, *MIT A.I. Memo 478,* October 1977.

11. Takano, M., Yashima, K., and Yada, S., Development of computer simulation system of kinematics and dynamics of robots, *Journal of the Faculty of Engineering, University of Tokyo (B),* no. 4, 1982.

12. Duelen, G., and Wendt, W., Ein Regelungsverfahren zur Verminderung von Bahnabweichungen bei Handhabungsgeräten, *ZwF,* October 1982, pp. 441–445.

13. Hopfengärtner, H., Lageregelung schwingungsfähiger Servosysteme am Beispiel eines Industrieroboters, *Regelungstechnik,* January 1981, pp. 3–10.

14. Zimmer, E., Industrieroboter—mechanische Konstruktion, *Konstruktion,* June 1983, pp. 221–227.

15. *Handbook of Small Standardized Components,* Master catalog 757, Stock Drive Products, New York, 1982.

16. Hashimoto, T., Power transmission equipments of the industrial robot, *Robot,* October 1981, pp. 75–82.

17. Futami, S., Kyura, N., and Nanai, S., Intelligent servo system: an approach to control-configured robot, *12th ISIR,* Paris, June 1982, pp. 381–390.

18. Manogg, H., Industrieroboter, konzipiert für die Handhabung von Werkstücken, *Konstruktion,* June 1983, pp. 239–245.

BIBLIOGRAPHY

Blume, C., and Dillmann, R., *Freiprogrammierbare Manipulatoren,* Vogel-Verlag, Würzburg, 1981.

Blume, C., and Jakob, W., *Programmiersprachen für Industrieroboter,* Vogel-Verlag, Würzburg, 1983.

Engel, G., *Konzipierung und Auslegung modular aufgebauter Handhabungssyteme,* VDI-Verlag, Düsseldorf, 1980.

ISO/TC 97/SC 8, *US Contribution for Discussion of Robots: Glossary of Terms for Robotics,* National Bureau of Standards, Washington, DC, 1981.

Makino, H., A kinematical classification of robot manipulators, 6th International Symposium on Industrial Robots, *IFS Publications Ltd.,* Bedford, 1976.

Paul, R. P., *Robot Manipulators: Mathematics, Programming, and Control,* MIT Press, Cambridge, 1982.

Schraft, R. D., *Systematisches Auswählen und Konzipieren von programmierbaren Handhabungsgeräten,* Dr.-Ing. dissertation, University of Stuttgart, 1977.

Steusloff, H., *Methods for Very Advanced Industrial Robots,* Springer-Verlag, 1980.

Volmer, J., *Industrieroboter,* VEB-Technik Verlag, Berlin, 1981.

CHAPTER **6**

KINEMATICS AND DYNAMICS

MICHAEL W. WALKER

Clemson University
Clemson, South Carolina

6.1. INTRODUCTION

The objective of this chapter is to provide a systematic methodology for the kinematic and dynamic analysis of manipulators. It is divided into four sections: Introduction, Kinematics, Dynamics, and Conclusion.

In the kinematics section two coordinate systems are used to describe the position of the manipulator: joint coordinates and link coordinates. The study of manipulator kinematics is concerned with the mapping of joint coordinates to link coordinates—and the inverse mapping of link coordinates to joint coordinates. Since this is not a one-to-one transformation, some difficulties arise and are discussed. In addition to position, the mapping of velocity and acceleration from joint coordinates to link coordinates and the inverse mapping are presented.

The dynamics section presents the equation of motion for a manipulator. It is shown how the equations in the kinematics section for position, velocity, and acceleration of the link coordinates can be used with the Newton-Euler equations of motion of a free rigid body to obtain the model for an open-chain manipulator. The dynamics of a manipulator containing closed kinematic chains is more complex, and references are given for further study.

The conclusion summarizes the chapter and proposes the development of computer procedures that are suggested throughout the chapter. The collection of these procedures into a library has proven to be a valuable tool in the design, modeling, and control of a manipulator. The purpose of describing these procedures is not only to provide guidelines from which the reader can create a useful set of routines, but also to emphasize the relationships between the various concepts described in this chapter.

TERMINOLOGY

m = number of degrees of freedom of the manipulator

$\mathbf{q}$ = $m \times 1$ vector of joint variable positions

$\dot{\mathbf{q}}$ = $m \times 1$ vector of joint variable velocities

$\ddot{\mathbf{q}}$ = $m \times 1$ vector of joint variable accelerations

$\mathbf{A}(j,i)$ = 4×4 homogeneous transform matrix for the link i coordinate referenced to the link j coordinate

$\mathbf{A}_i(\mathbf{q})$ = $\mathbf{A}(0,i)$ evaluated with joint variables equal to $\mathbf{q}$

$$= \begin{bmatrix} \mathbf{x}_i & \mathbf{y}_i & \mathbf{z}_i & \mathbf{p}_i \\ 0 & 0 & 0 & 1 \end{bmatrix}$$

$\mathbf{J}(\mathbf{q})$ = 6×6 Jacobian matrix

$\mathbf{w}_i$ = 3×1 vector, angular velocity of link i coordinates

$\dot{\mathbf{w}}_i$ = 3×1 vector, angular acceleration of link i coordinates

$\mathbf{v}_i$ = 3×1 vector, linear velocity of link i coordinates

$\dot{\mathbf{v}}_i$ = 3×1 vector, linear acceleration of link i coordinates

$\dot{\mathbf{v}}_i$ $= 3 \times 1$ vector, linear acceleration of link i center of mass

$\mathbf{f}_i$ $= 3 \times 1$ vector, force exerted on link i by link $i - 1$

$\mathbf{n}_i$ $= 3 \times 1$ vector, moment exerted on link i by link $i - 1$

$\mathbf{F}_i$ $= 3 \times 1$ vector, total force exerted on link i

$\mathbf{N}_i$ $= 3 \times 1$ vector, total moment exerted on link i

$\mathbf{H}(\mathbf{q})$ $= m \times m$ symmetrix, nonsingular generalized moment of inertia matrix

$\mathbf{C}(\mathbf{q},\dot{\mathbf{q}})$ $= m \times 1$ vector specifying centrifugal and coriolis effects in the manipulator dynamics model. Note: $\mathbf{C}(\mathbf{q}, \mathbf{0}) = \mathbf{0}$

$\mathbf{g}(\mathbf{q})$ $= m \times 1$ vector specifying the effects due to gravity in the manipulator dynamics model

$\mathbf{k}$ $= 6 \times 1$ vector of external forces and moments on link m acting through link m coordinates. The first three elements are comprised of the 3×1 moment vector, and the last elements are comprised of the 3×1 force vector.

$\mathbf{u}$ $= m \times 1$ vector of torques (forces) of each joint actuator

mass_i $=$ mass of link i

$\mathbf{r}_i$ $=$ position of link i center of mass with respect to link i coordinates expressed in the base link coordinates

$\mathbf{I}_i$ $= 3 \times 3$ moment of inertia matrix of link $\mathbf{i}$ about the center of mass of link i expressed in the base link coordinates

$\hat{\mathbf{r}}_i$ $=$ position of link i center of mass with respect to link i coordinates expressed in link i coordinates

$\hat{\mathbf{I}}_i$ $= 3 \times 1$ moment of inertia matrix of link i about the center of mass of link i expressed in link i coordinates.

6.2. KINEMATICS

The purpose of a manipulator is to manipulate its end effector. Some other names for end effector are *hand, gripper,* and *tool.* It is that part of the manipulator which physically interfaces with its environment. To perform a task the robot must know where the object to be worked on is located and what the location of the end effector should be with respect to that object. For this purpose one needs a kinematic model of the manipulator. This section presents this model and shows how it is used to define the position, velocity, and acceleration of each link coordinate and, hence, the end effector.

6.2.1. Homogeneous Transforms

Figure 6.1 shows three coordinate systems. The position and orientation of system c is known with respect to system b, and the position and orientation of system b is known with respect to system a. The problem is to determine the position and orientation of system c with respect to system a.

Orientation can be parameterized in many ways.[20] Here orientation is defined in terms of the direction cosine matrix. This is a 3×3 matrix whose columns are three unit vectors that represent the x, y, and z axes of the right-handed orthogonal coordinate system.

A convenient way of transforming both the orientation and position from one coordinate system to another is with the use of homogeneous transforms. A homogeneous transform is a 4×4 matrix of the form:

$$\mathbf{A}(a,b) = \begin{bmatrix} \mathbf{x} & \mathbf{y} & \mathbf{z} & \mathbf{p} \\ 0 & 0 & 0 & 1 \end{bmatrix}$$

$$= \begin{bmatrix} & \mathbf{D} & & \mathbf{p} \\ 0 & 0 & 0 & 1 \end{bmatrix}$$

where $\mathbf{A}(a,b)$ is the transform of coordinate system b with respect to a; $\mathbf{D}$ is the 3×3 direction cosine matrix of coordinate system b expressed in coordinate system a; and $\mathbf{p}$ is the 3×1 vector denoting the position of coordinate system b expressed in a. The transform for c expressed in b is denoted by $\mathbf{A}(b,c)$. Given $\mathbf{A}(a,b)$ and $\mathbf{A}(b,c)$, the transform $\mathbf{A}(a,c)$ is computed by simply multiplying these two transforms together:

$$\mathbf{A}(a,c) = \mathbf{A}(a,b)\mathbf{A}(b,c)$$

Using this technique one has a simple notation and also an easy technique of computing the position and orientation of any coordinate system with respect to any other coordinate system. The routine

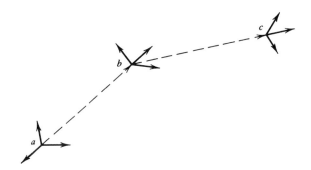

Fig. 6.1. Series of three coordinate systems.

MULMAT (output = **A**, input = **B,C**) is used to multiply homogeneous transforms **B** and **C** together and store the result in **A**.

$$\mathbf{A} = \mathbf{BC}$$

The inverse of the transform is easily computed:

$$\mathbf{A}(a,b)^{-1} = \mathbf{A}(b,a) = \begin{bmatrix} \mathbf{D'} & -\mathbf{D'p} \\ 0 \quad 0 \quad 0 & 1 \end{bmatrix}$$

where ′ denotes transpose.

Any homogeneous transform can be defined in terms of the product of six special transforms. This is because there are only three independent elements in the direction cosine matrix and three for the position. These special transforms are as follows:

1. A rotation about the x axis at angle β

$$\mathbf{Rot}(x,\ \beta) = \begin{bmatrix} 1 & 0 & 0 & 0 \\ 0 & c(\beta) & -s(\beta) & 0 \\ 0 & s(\beta) & c(\beta) & 0 \\ 0 & 0 & 0 & 1 \end{bmatrix}$$

2. A rotation about the y axis at angle β

$$\mathbf{Rot}(y,\ \beta) = \begin{bmatrix} c(\beta) & 0 & s(\beta) & 0 \\ 0 & 1 & 0 & 0 \\ -s(\beta) & 0 & c(\beta) & 0 \\ 0 & 0 & 0 & 1 \end{bmatrix}$$

3. A rotation about the z axis at angle β

$$\mathbf{Rot}(z,\ \beta) = \begin{bmatrix} c(\beta) & -s(\beta) & 0 & 0 \\ s(\beta) & c(\beta) & 0 & 0 \\ 0 & 0 & 1 & 0 \\ 0 & 0 & 0 & 1 \end{bmatrix}$$

4. A translation along the x axis a distance δ

$$\mathbf{Trans}(x,\ \delta) = \begin{bmatrix} 1 & 0 & 0 & \delta \\ 0 & 1 & 0 & 0 \\ 0 & 0 & 1 & 0 \\ 0 & 0 & 0 & 1 \end{bmatrix}$$

5. A translation along the y axis a distance δ

$$\mathbf{Trans}(y,\ \delta) = \begin{bmatrix} 1 & 0 & 0 & 0 \\ 0 & 1 & 0 & \delta \\ 0 & 0 & 1 & 0 \\ 0 & 0 & 0 & 1 \end{bmatrix}$$

6. A translation along the z axis a distance δ

$$\textbf{Trans}(z, \delta) = \begin{bmatrix} 1 & 0 & 0 & 0 \\ 0 & 1 & 0 & 0 \\ 0 & 0 & 1 & \delta \\ 0 & 0 & 0 & 1 \end{bmatrix}$$

6.2.2. Forward Kinematics

Forward kinematics refers to the computation of the position or motion of each link as a function of the joint variables. The next section presents a method for determining the position, velocity, and acceleration of each link given the position, velocity, and acceleration of the joint variables.

6.2.3. Link Coordinate Position and Orientation

The joint coordinate is the position of the joint variable. If a joint is rotational, its position is measured in radians. If a joint is translational, it is measured in units of length. To each link of the manipulator is attached a right-handed coordinate system composed of three orthogonal unit vectors. These coordinate systems are called link coordinates, and their position and orientation are defined in terms of homogeneous transformation matrices.

One particularly suitable method for assigning link coordinates is attributed to Hartenberg and Denavit.[1] In this method four parameters are used to describe the position of successive link coordinates, Figure 6.2. The parameters are a, α, d, and θ. The definitions of these parameters are:

a_i = the shortest distance between z_i and z_{i-1}
α_i = the angle between z_i and z_{i-1}
d_i = the shortest distance between x_i and x_{i-1}
θ_i = the angle between x_i and x_{i-1}

Only one of these four parameters is variable and is denoted by q_i throughout the rest of this chapter. If joint i is rotational, then θ_i is the joint variable and d_i, a_i, and α_i are constants. If joint i is translational, then d_i is the joint variable and θ_i, a_i, and α_i are constants.

When determining the kinematic parameters of a manipulator, it is helpful to follow a systematic method. The following algorithm works well.

ALGORITHM 6.1. DETERMINATION OF KINEMATIC PARAMETERS

Step 1. Draw a sketch of the manipulator.

Step 2. Number each link from 0 to m, starting at the base of the manipulator as link 0 out to the last link, m.

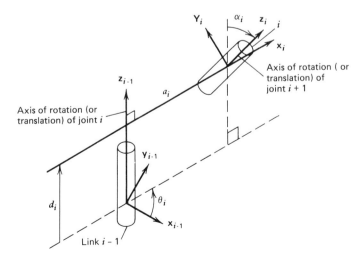

Fig. 6.2. Parameters relating adjacent link coordinate systems.

Step 3. Number the joint between link i and $i - 1$ as joint i.

Step 4. Draw in the unit vector z_{i-1} for the joint i, $i = 1$, m.
z_{i-1} = axis of rotation if joint i is rotational.
z_{i-1} = axis of translation if joint i is translational.

Step 5. Determine a_i, $i = 1$, m, the minimum distance between z_{i-1} and z_i.

Step 6. Draw in the unit vector x_i in the direction of z_{i-1} to z_i such that x_i is perpendicular to both z_{i-1} and z_i.

Step 7. Determine d_i, $i = 1$, m, the minimum distance between x_{i-1} and x_i.

Step 8. Determine θ_i, $i = 1$, m, the angle of rotation, positive in the right-hand sense, about the z_{i-1} axis between x_{i-1} and x_i.

Step 9. Determine α_i, $i = 1$, m, the angle of rotation, positive in the right-hand sense, about the x_i axis between z_{i-1} and z_i.

Since there is no link $m + 1$ coordinate system, d_m, α_m, and a_m can be set to zero if joint m is rotational. If joint 1 is rotational, d_0 can be set to zero.

One can describe the position and orientation of link i coordinates with respect to link $i - 1$ coordinates, $A(i - 1, i)$, as:

$$
A(i - 1, i) = \text{Trans}(z, d_i)\text{Rot}(z, \theta_i)\text{Trans}(x, a_i)\text{Rot}(x, \alpha_i)
$$

$$
= \begin{bmatrix}
c(\theta_i) & -s(\theta_i)c(\alpha_i) & s(\theta_i)s(\alpha_i) & c(\theta_i)a_i \\
s(\theta_i) & c(\theta_i)c(\alpha_i) & -c(\theta_i)s(\alpha) & s(\theta_i)a_i \\
0 & s(\alpha_i) & c(\alpha_i) & d_i \\
0 & 0 & 0 & 1
\end{bmatrix} \tag{6.1}
$$

The routine HOMOJ (output = E, input = i) is used to compute the transform $A(i - 1, i)$ and store the results in the array E.

To find the position and orientation of link i coordinates with respect to link $i - 2$ coordinates, $A(i - 2, i)$, one simply multiplies successive link coordinate transforms together:

$$
A(i - 2, i) = A(i - 2, i - 1)A(i - 1, i)
$$

By using the routines MULMAT and HOMOJ, one can obtain the position and orientation of any link coordinate system with respect to any other link coordinate. The position and orientation of link m coordinates with respect to the base coordinates is

$$
A(0, m) = A(0, 1)A(1, 2) \cdots A(m - 2, m - 1)A(m - 1, m) \tag{6.2}
$$

Finally, the manipulator usually has some sort of tool attached to the last link as part of the end effector. This tool also has a coordinate system affixed to it. The position and orientation of the tool coordinates with respect to link m coordinates are denoted by **TOOLM**. The position and orientation of the tool coordinates with respect to the manipulator base coordinates are denoted by **TOOLB**. Note that **TOOLM** is a constant matrix whereas **TOOLB** is a function of the joint variables. It is

$$
\text{TOOLB} = A(0, m)\text{TOOLM} \tag{6.3}
$$

6.2.4. Velocity of the End Effector

The velocity of the end effector is a function of the velocity of the joint variables. The procedure is similar to computing the position of the end effector. The angular and linear velocities of each link coordinate are computed one link at a time starting at the base of the manipulator. If the angular and linear velocities of the ith link coordinates are known, the velocity of the $i + 1$ link coordinate is easily computed based upon whether joint $i + 1$ is rotational or translational.

For joint i rotational

$$
\begin{aligned}
w_{i+1} &= w_i + z_i\dot{q}_{i+1} \\
v_{i+1} &= v_i + w_{i+1} \times p_{i+1}^*
\end{aligned} \tag{6.4}
$$

For joint i translational

$$
\begin{aligned}
w_{i+1} &= w_i \\
v_{i+1} &= v_i + z_i\dot{q}_{i+1} + w_{i+1} \times p_{i+1}^*
\end{aligned} \tag{6.5}
$$

where $p_{i+1}^* = p_{i+1} - p_i$ and $\times$ denotes vector cross product.

The routine is VEL (output $= \mathbf{w}_{i+1}$, $\mathbf{v}_{i+1}$, input $= \mathbf{w}_i$, $\mathbf{v}_i$, $\mathbf{z}_i$, $\mathbf{p}^*_{i+1}$, $\dot{q}_{i+1}$, type$_{i+1}$). Note that $\mathbf{z}_i$ is the first three elements of the third column of $\mathbf{A}(0, i)$ and $\mathbf{p}_i$ is the first three elements of the fourth column of $\mathbf{A}(0, i)$. Therefore, routines MULMAT and HOMOJ can be used to compute $\mathbf{z}_i$ and $\mathbf{p}^*_{i+1}$. The variable type$_{i+1}$ is used to inform the procedure whether joint $i + 1$ is rotational or translational and, hence, whether to use Eq. (6.4) or (6.5) to compute the link velocities.

These equations can be expanded and the components due to each joint velocity can be obtained.[16,17] For a 6-DF manipulator, the equation for the angular and linear velocity of link 6 coordinates is

$$V_6 = J(q)\dot{q} \qquad (6.7)$$

where V_6 is a 6×1 vector composed of the angular and linear velocities of link 6 coordinates:

$$V_6 = \begin{bmatrix} \mathbf{w}_6 \\ \mathbf{v}_6 \end{bmatrix}$$

$J(q)$ is the Jacobian matrix:

$$J(q) = [\mathbf{s}_1 \ \mathbf{s}_2 \ \mathbf{s}_3 \ . \ . \ . \ \mathbf{s}_6]$$

The columns of $J(q)$ are of the form:

$$\mathbf{s}_j = \begin{bmatrix} \mathbf{z}_{j-1} \\ \mathbf{z}_{j-1} \times (\mathbf{p}_6 - \mathbf{p}_{j-1}) \end{bmatrix}$$

if joint j is rotational;

$$\mathbf{s}_j = \begin{bmatrix} \mathbf{0} \\ \mathbf{z}_{j-1} \end{bmatrix}$$

if joint j is translational. $\mathbf{0}$ is the 3×1 null vector.

The Jacobian matrix is very useful for solving the inverse problem in kinematics. The routine JACOBI (output $= \mathbf{J}$, input $= \mathbf{q}$) is used to compute the Jacobian matrix and store the result in $\mathbf{J}$. Note that the elements of $\mathbf{s}_j$, the jth column of $\mathbf{J}$, can be computed using the routines MULMAT and HOMOJ.

6.2.5. Link Acceleration

Link accelerations are computed in a way similar to the way velocities are computed. Given the acceleration of the ith link coordinate, the acceleration of the $i + 1$ link coordinate is for joint i rotational

$$\dot{\mathbf{w}}_{i+1} = \dot{\mathbf{w}}_i + \mathbf{z}_i \ddot{q}_{i+1} + \mathbf{w}_i \times \mathbf{z}_i \dot{q}_{i+1}$$
$$\dot{\mathbf{v}}_{i+1} = \dot{\mathbf{v}}_i + \dot{\mathbf{w}}_{i+1} \times \mathbf{p}^*_{i+1} + \mathbf{w}_{i+1} \times (\mathbf{w}_{i+1} \times \mathbf{p}^*_{i+1}) \qquad (6.8)$$

For joint i translational

$$\dot{\mathbf{w}}_{i+1} = \dot{\mathbf{w}}_i$$
$$\dot{\mathbf{v}}_{i+1} = \dot{\mathbf{v}}_i + \mathbf{z}_i \ddot{q}_{i+1} + \dot{\mathbf{w}}_{i+1} \times \mathbf{p}^*_{i+1} + 2\mathbf{w}_{i+1} \times (\mathbf{z}_i \dot{q}_{i+1}) + \mathbf{w}_{i+1} \times (\mathbf{w}_{i+1} \times \mathbf{p}^*_{i+1}) \qquad (6.9)$$

These equations can be expanded, and the components attributable to each joint acceleration can be obtained. For a 6-DF manipulator the equation for the angular and linear acceleration of link 6 coordinates is

$$\dot{V}_6 = J(q)\ddot{q} + \dot{J}(q)\dot{q} \qquad (6.10)$$

where V_6 and $J(q)$ are defined in the previous section.

To compute $\dot{\mathbf{w}}_{i+1}$ and $\dot{\mathbf{v}}_{i+1}$ the routine ACCEL (output $= \dot{\mathbf{w}}_{i+1}$, $\dot{\mathbf{v}}_{i+1}$, input $= \dot{\mathbf{w}}_{i+1}$, $\dot{\mathbf{v}}_i$, $\mathbf{w}_{i+1}$, $\mathbf{z}_i$, $\mathbf{p}^*_{i+1}$, $\dot{q}_{i+1}$, $\ddot{q}_{i+1}$, type$_{i+1}$) is used. The inputs to this routine can be computed with the use of routines MULMAT, HOMOJ, and VEL. The variable type$_{i+1}$ plays the same role in this procedure as it does in the procedure VEL().

6.2.6. Inverse Kinematics

Trajectories can be planned in joint space, but they can also be planned in terms of the position and orientation of the end effector. If so done, the position and orientation of the end effector must be transformed into joint coordinate motions to drive the joint servo controllers.

6.2.7 Joint Positions

This section describes a technique for obtaining the joint position for each joint in order to obtain a specified position and orientation of the end effector. A manipulator with at least 6 DF is required to position the end effector in any position and orientation. Therefore we let m, the number of degrees of freedom of the manipulator, equal 6 in the following.

The desired homogeneous transform for the end effector will be denoted by the 4×4 matrix **Tr**. This matrix is input to the inverse kinematics routine. The objective of this routine is to determine the values of the joint variables so that the end effector will have the same position and orientation as denoted by **Tr**. Therefore the equation to solve is

$$\textbf{TOOLB} = \textbf{Tr}$$

Substituting in the expression for **TOOLB** from Eq. (6.3) and solving for **A**(0, 6) gives

$$\textbf{A}(0, 6) = \textbf{Tr}(\textbf{TOOLM})^{-1} = \textbf{A}_d$$

where $\textbf{A}_d$ is the desired position of link 6 coordinate system.

It is unfortunate that there does not exist an algorithm that can be applied for any manipulator to find a closed-form solution of this equation. However, there does exist a solution for most existing manipulators. The criterion is that all wrist axes should intersect at a common point. The position of this point is then only a function of the first three joint variables. If this position can be obtained from the input **Tr** matrix, then the first three joint variables can be determined. Once the first three joint variables are obtained, the last three joint variables are determined such that the end effector has the correct orientation. This is best illustrated by an example.

Consider the Stanford model manipulator, Figure 6.3. This is a manipulator with six joints, five rotational and one translational. The kinematic parameters for this manipulator are given in Table 6.1. Using these values in Eq. (6.1) gives the $\textbf{A}(i - 1, i)$ matrices given in Table 6.2.

It can be seen from Figure 6.3 that the position where the wrist axes intersect is $\textbf{p}_3$, the position of link 3 coordinates. This can be obtained from $\textbf{A}_d$.

$$\textbf{p}_3 = \textbf{p}_d - d_6 \textbf{z}_d$$

The vector $\textbf{p}_3$ is the first three elements of the fourth column of **A**(0, 3). Using Eq. (6.2) gives

$$A(0, 3) = \textbf{A}(0, 1)\textbf{A}(1, 2)\textbf{A}(2, 3)$$

$$= \begin{bmatrix} s_1 & c_1 c_2 & c_1 s_2 & -s_1 d_2 + d_3 c_1 s_2 \\ c_1 & s_1 c_2 & s_1 s_2 & c_1 d_2 + d_3 s_1 s_2 \\ 0 & -s_2 & c_2 & d_3 c_2 \\ 0 & 0 & 0 & 1 \end{bmatrix}$$

Therefore θ_1, θ_2, and d_3 must satisfy the equation

$$\begin{bmatrix} -s_1 d_2 + d_3 c_1 s_2 \\ c_1 d_2 + d_3 s_1 s_2 \\ d_3 c_2 \end{bmatrix} = \textbf{p}_d - d_6 \textbf{z}_d \tag{6.11}$$

Referring to Figure 6.3, the only variable that influences the length of $\textbf{p}_3$ is d_3. Therefore the length of $\textbf{p}_3$ squared should only be a function of the joint variable d_3.

$$(\textbf{p}_3)^2 = (d_2)^2 + (d_3)^2$$
$$= (\textbf{p}_d - d_6 \textbf{z}_d)^2$$

Therefore

$$d_3 = \{(\textbf{p}_d - d_6 \textbf{z}_d)^2 - (d_2)^2\}^{1/2}$$

Note that the argument to the square root function being negative is an unrealizable position. If this ever occurs, then something has gone wrong with the path-planning routine, that is, the routine that computes the desired homogeneous transform of the end effector **Tr**.

The third component of Eq. (6.10) is only a function of θ_2 and d_3. Therefore θ_2 is given by

$$\theta_2 = \pm \arccos \frac{(p_{dz} - d_6 z_{dz})}{d_3}$$

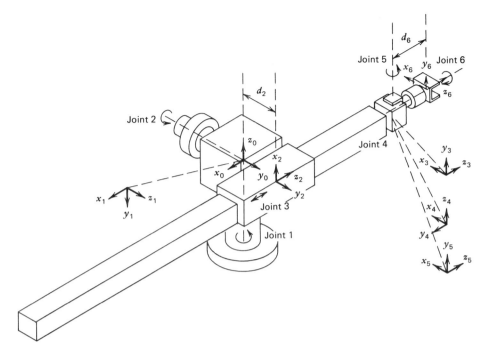

Fig. 6.3. Link coordinates for Stanford model manipulator.

where p_{dz} and z_{dz} denote the z components of $\mathbf{p}_d$ and $\mathbf{z}_d$. The $\pm$ indicates that two solutions are possible. They could be called the right-handed and left-handed solutions.

Finally, since θ_2 and d_3 have already been solved, either the first element, the second element, or both the first and second elements of Eq. (6.10) may be used to compute θ_1. Using both elements gives:

$$\begin{bmatrix} -d_2 & d_3 s_2 \\ d_3 s_2 & d_2 \end{bmatrix} \begin{bmatrix} s_1 \\ c_1 \end{bmatrix} = \begin{bmatrix} p_{dx} - d_6 z_{dx} \\ p_{dy} - d_6 z_{dy} \end{bmatrix}$$

Solving for s_1 and c_1 gives

$$s_1 = \frac{d_3 s_2 (p_{dy} - d_6 z_{dy}) - d_2 (p_{dx} - d_6 z_{dx})}{(d_2)^2 + (d_3 s_2)^2}$$

$$c_1 = \frac{d_3 s_2 (p_{dx} - d_6 z_{dx}) + d_2 (p_{dy} - d_6 z_{dy})}{(d_2)^2 + (d_3 s_2)^2}$$

and $\theta_1 = \text{atan2}(s_1, c_1)$, where atan2 is a standard Fortran arctangent function whose arguments are sine and cosine of the angle.

TABLE 6.1. KINEMATIC PARAMETERS FOR THE STANFORD MODEL MANIPULATOR

Joint	θ	d	α	a
1	Variable	0	−90 deg	0
2	Variable	16.2 cm	90 deg	0
3	−90 deg	Variable	0	0
4	Variable	0	−90 deg	0
5	Variable	0	90 deg	0
6	Variable	24.7 cm	0	0

TABLE 6.2. $A(i-1, i)$ MATRICES FOR THE STAN-FORD MODEL MANIPULATOR

$$A(0, 1) = \begin{bmatrix} c_1 & 0 & -s_1 & 0 \\ s_1 & 0 & c_1 & 0 \\ 0 & -1 & 0 & 0 \\ 0 & 0 & 0 & 1 \end{bmatrix} \quad A(1, 2) = \begin{bmatrix} c_2 & 0 & s_2 & 0 \\ s_2 & 0 & -c_2 & 0 \\ 0 & 1 & 0 & d_2 \\ 0 & 0 & 0 & 1 \end{bmatrix}$$

$$A(2, 3) = \begin{bmatrix} 0 & 1 & 0 & 0 \\ -1 & 0 & 0 & 0 \\ 0 & 0 & 1 & d_3 \\ 0 & 0 & 0 & 1 \end{bmatrix} \quad A(3, 4) = \begin{bmatrix} c_4 & 0 & -s_4 & 0 \\ s_4 & 0 & c_4 & 0 \\ 0 & -1 & 0 & 0 \\ 0 & 0 & 0 & 1 \end{bmatrix}$$

$$A(4, 5) = \begin{bmatrix} c_5 & 0 & s_5 & 0 \\ s_5 & 0 & -c_5 & 0 \\ 0 & 1 & 0 & 0 \\ 0 & 0 & 0 & 1 \end{bmatrix} \quad A(5, 6) = \begin{bmatrix} c_6 & -s_6 & 0 & 0 \\ s_6 & c_6 & 0 & 0 \\ 0 & 0 & 1 & d_6 \\ 0 & 0 & 0 & 1 \end{bmatrix}$$

c_i and s_i denote cosine (θ_i) and sin (θ_i), respectively

The reason for solving for s_1 and c_1 is that they uniquely specify which quadrant θ_1 is located in. If only one element is used, for instance p_{dx}, then the equation would be

$$-s_1 d_2 + d_3 c_1 s_2 = p_{dx} - d_6 z_{dx}$$

This equation is of the form

$$a s_1 + b c_1 = c$$

This equation can be solved by substituting for s_1 and c_1 their values as a function of t_1, where t_1 is the tangent of one-half of θ_1. The result is

$$\theta_1 = 2 \arctan(t_1)$$

where

$$t_1 = \frac{a \pm \{a^2 + b^2 - c^2\}^{1/2}}{b + c}$$

The sign to choose is dependent on the sign and relative magnitude of a, b, and c.

The first three joint variables are used to control position, and the last three are used to control orientation. Note that since the last three axes intersect, rotating any wrist axis does not move the position of the point where they intersect.

To solve for the last three axes, the homogeneous transform of link 6 coordinates with respect to link 3 coordinates must first be computed. It is

$$\begin{aligned} A(3, 6) &= A(3, 4)A(4, 5)A(5, 6) \\ &= \begin{bmatrix} c_4 c_5 c_6 - s_4 s_6 & -c_4 c_5 s_6 - s_4 c_6 & c_4 s_5 & d_6 c_4 s_5 \\ s_4 c_5 c_6 + c_4 s_6 & -s_4 c_5 s_6 + c_4 c_6 & s_4 s_5 & d_6 s_4 s_5 \\ -s_5 c_6 & s_5 s_6 & c_5 & d_6 c_5 \\ 0 & 0 & 0 & 1 \end{bmatrix} \\ &= A(0, 3)^{-1} A_d \\ &= \begin{bmatrix} a_{11} & a_{12} & a_{13} & a_{14} \\ a_{21} & a_{22} & a_{23} & a_{24} \\ a_{31} & a_{32} & a_{33} & a_{34} \\ 0 & 0 & 0 & 1 \end{bmatrix} \end{aligned}$$

Note that the a_{ij} are functions of the input transformation matrix and the first three joint variables, which are known at this stage. The a_{ij} can therefore be evaluated now. The third row, third column gives the value of θ_5.

$$\theta_5 = \pm \arccos(a_{33})$$

Again two possible solutions exist. Given θ_5, s_4 and c_4 can be obtained from column three, rows one and two.

$$s_4 = \frac{a_{23}}{s_5}$$

$$c_4 = \frac{a_{13}}{s_5}$$

and

$$\theta_4 = \text{atan2}(s_4, c_4)$$

Given θ_5, s_6 and c_6 can be obtained from row three, columns one and two.

$$s_6 = \frac{a_{32}}{s_5}$$

$$c_6 = \frac{-a_{31}}{s_5}$$

and

$$\theta_6 = \text{atan2}(s_6, c_6)$$

If θ_5 is equal to zero, then the foregoing equations cannot be used. In this case there are an infinite number of possible solutions, and some sort of choice must be made to pick one of the solutions. When θ_5 is equal to zero, $A(3, 6)$ becomes

$$A(3, 6) = \begin{bmatrix} c_{46} & -s_{46} & 0 & 0 \\ s_{46} & c_{46} & 0 & 0 \\ 0 & 0 & 1 & d_6 \\ 0 & 0 & 0 & 1 \end{bmatrix}$$

where

$$c_{46} = \cos(\theta_4 + \theta_6)$$
$$s_{46} = \sin(\theta_4 + \theta_6)$$

Therefore,

$$c_{46} = a_{11}$$
$$s_{46} = a_{21}$$

and

$$\theta_4 + \theta_6 = \text{atan2}(a_{21}, a_{11}) \tag{6.12}$$

Thus there are an infinite number of solutions. A common technique is to pick one value, for instance θ_6, and then compute the other from Eq. (6.12).

For every different manipulator the problem of solving the inverse kinematics problem will be different, but the basic technique will always be the same. Solve for the first three joint variables to get the correct position where the wrist axes intersect; then solve for the last three joint variables to get the correct orientation of the end effector. For additional information on closed-form solutions, refer to References 2 and 3.

For control, the speed of solution of the inverse kinematics problem is very critical. For this reason a procedure as described previously is recommended. However, for modeling purposes it is useful to have a routine that can be used for any manipulator. One popular method is to use Newton's method for solving the nonlinear kinematic equations. Let q_1 denote the position of the joint variables at time t_1 and q_2 denote the position of the joint variables at time t_2. Let $t_2 = t_1 + \Delta t$, where Δt is a small increment in time. The columns $A_6(q_1)$ are denoted by $x(q_1)$, $y(q_1)$, $z(q_1)$, and $p(q_1)$ to represent their values at time t_1. Similarly $x(q_2)$, $y(q_2)$, $z(q_2)$, and $p(q_2)$ denote the columns of $A_6(q_2)$. The small change in the orientation and position of the link 6 coordinates can be obtained from Eq. (6.7). Multiplying both sides by Δt gives

$$J(q_1)\Delta t \dot{q} = \Delta t V_6$$

Note that

$$\Delta t \dot{q} \simeq \Delta q = q_2 - q_1$$

$$\Delta t V_6 = \begin{bmatrix} \Delta t w_6 \\ \Delta t v_6 \end{bmatrix}$$

$$\Delta t w_6 \simeq \frac{x(q_1) \times x(q_2) + y(q_1) \times y(q_2) + z(q_1) \times z(q_2)}{2}$$

$$= e_{ro}$$

$$\Delta t v_6 \simeq p(q_2) - p(q_1)$$

$$= e_{rp}$$

where e_{ro} is the orientation error vector and e_{rp} is the position error vector. Define

$$e_r = \begin{bmatrix} e_{ro} \\ e_{rp} \end{bmatrix}$$

then

$$J(q_1)\Delta q \simeq e_r \qquad (6.13)$$

Solving for Δq gives:

$$\Delta q \simeq J(q_1)^{-1} e_r$$

Two additional procedures are needed to solve for Δq: ERROR [(output $= e_r$, input $= A_6(q_1)$, $A_6(q_2)$]
to compute e_r, and RATE (output $= \Delta q$, input $= q_1$, e_r) to compute Δq using Eq. (6.13). Therefore
we now have a method of determining the difference in the position of the manipulator in joint space
based on the difference in position and orientation of link 6 coordinates. To apply this to the problem
of inverse kinematics, we let the matrix $A_6(q_2)$ be equal to A_d, the desired homogenous transformation
matrix of link 6 coordinates. The solution obtained for Δq will then be an approximation to the
difference in the current estimate of the joint variables and the desired values of the joint variables.
Since this is only an approximate solution, an iterative procedure must be used to solve for the joint
variable positions. The following algorithm works well.

ALGORITHM 6.2. INVERSE KINEMATICS

Step 1.   Let $i = 0$ and guess the solution q_i for the joint variable positions.
Step 2.   Compute e_r, the error vector, using the procedure ERROR [output $= e_r$, input $= A_6(q_i)$,
          A_d.]
Step 3.   Compute the estimate of the difference between q_i and the desired solution by using the
          procedure RATE (output $= \Delta q_i$, input $= q_i$, e_r).
Step 4.   Set $q_{i+1} = q_i + \Delta q_i$.
Step 5.   If the magnitude of Δq_i is less than some predefined stopping criterion, then STOP, else
          continue.
Step 6.   Set $i = i + 1$ and go to Step 2.

The routine REVKIN (output $= q$, input $= A_d$) is used to solve the inverse kinematics problem
by implementing Algorithm 6.2. Note that the only additional routines needed are ERROR and RATE.
The procedure RATE needs a linear equation solution routine. The library routine LIN (output $= y$,
input $= B$, x) is used to solve the linear equation

$$y = Bx$$

The solution of Eq. (6.13) suffers from the same problems as the solution of any other linear equation.
If the Jacobian matrix is nonsingular, a unique solution can be obtained. If the Jacobian matrix is
singular, then either there exists no exact solution or an infinity of solutions exist. The linear equation
routine LIN must consider the singular positions and attempt to obtain a solution in some sense.
For example, the routine might minimize the sum of squares of residuals.[11] It is also important to
note that even if the Jacobian matrix is nonsingular, it may still be ill-conditioned. Ill-conditioning is
related to the fact that the Jacobian matrix is nearly singular. When this occurs, the solution of the
equation becomes very sensitive to the values of the elements of the Jacobian matrix. Small errors in
the coefficients in the Jacobian matrix result in large errors in the answers that are computed.

6.2.8. Joint Velocities

As described previously, the velocity of the end effector can be computed using Eqs. (6.4) and (6.5). Given the velocity of the end effector, the velocity of the joint variables can be obtained by solving the linear equation 6.6.[16,17]

$$\dot{q} = J(q)^{-1} V_6$$

Note that the routine RATE (output $= \dot{q}$, input $= q$, V_6) defined in the preceding section can be used to solve for the joint rates.

6.2.9. Joint Accelerations

Joint accelerations can be obtained in a manner analogous to obtaining joint velocities by using Eq. (6.6).

$$\ddot{q} = J(q)^{-1} [\dot{V}_6 - \dot{J}(q)\dot{q}]$$

The routine RESACC (output $= \ddot{q}$, input $= \dot{V}_6$, q, $\dot{q}$) is used to solve for the joint accelerations. Note that $\dot{J}(q)\dot{q}$ can be computed using ACCEL( ) by setting all joint acceleration inputs to the procedure equal to zero.

6.3. DYNAMICS OF MANIPULATORS

This section presents the model of the manipulator, a necessary tool in the design of a manipulator control system.

6.3.1. Open-Chain Mechanism

All manipulators can be classified into one of two categories: those that contain closed kinematic loops and those that do not (open-chain mechanisms). Nearly all manipulators contain closed kinematic loops. Even electric manipulators that appear to be open-chain mechanisms usually contain on the output shaft of each motor a gearing system that is a system of closed kinematic loops. The model for manipulators of the open-chain type is presented here because of its simplicity and the fact that useful approximations for many other types of manipulators can be obtained from this model.

The equations of motion for the manipulator can be written in the form

$$H(q)\ddot{q} + C(q, \dot{q}) + g(q) + J(q)'k = u \tag{6.14}$$

This equation is used in simulations and in the development of control algorithms. Associated with each of these problems are the problems of inverse dynamics and forward dynamics, which are discussed in the following sections. But first, a few observations are made associated with the dynamics of the manipulator.

If the angular and linear acceleration of each link is known, then the linear acceleration of each link center of mass is obtained by using the following equation:

$$\dot{\hat{v}}_i = \dot{v}_i + \dot{w}_i \times r_i + w_i \times (w_i \times r_i) \tag{6.15}$$

Note that the angular velocity and acceleration of the center of mass of link i are the same as for the link i coordinate system. Routine ACCNT (output $= \hat{v}_i$, input $= \dot{v}_i$, $\dot{w}_i$, r_i) is used to compute the linear acceleration of the center of mass of link i.

If the angular velocity, the angular acceleration, and the linear acceleration of a link's center of mass is known, then the total force and moment exerted on that link can be obtained from the Newton-Euler equations[19]

$$\begin{aligned} F_i &= \text{mass}_i \dot{\hat{v}}_i \\ N_i &= I_i \dot{w}_i + w_i \times (I_i w_i) \end{aligned} \tag{6.16}$$

Routine TOTFOR (output $= F_i$, N_i, input $= \text{mass}_i$, I_i, w_i, $\dot{w}_i$, $\dot{\hat{v}}_i$) is used to compute F_i and N_i for each link of the manipulator using Eq. 6.16.

Note that I_i and r_i are functions of the position and orientation of link i, whereas $\hat{I}_i$ and $\hat{r}_i$ are constants. Their relationship is given by the following equations.

$$\begin{aligned} I_i &= D_i \hat{I}_i (D_i)' \\ r_i &= D_i \hat{r}_i \end{aligned}$$

where $\mathbf{D}_i$ is the upper left 3×3 submatrix of $\mathbf{A}_i(\mathbf{q})$. The procedure TRFIN (output $= \mathbf{I}_i, \mathbf{r}_i$, input $= \mathbf{A}_i(\mathbf{q}), \hat{\mathbf{I}}_i, \hat{\mathbf{r}}_i$) is used to compute $\mathbf{I}_i$ and $\mathbf{r}_i$ using this equation.

The total force and moment exerted on link i is composed of the force and moment exerted by link $i - 1$ and link $i + 1$ plus the force and moment due to gravity. If i is equal to m, there is no $\mathbf{f}_{m+1}$ or $\mathbf{n}_{m+1}$, since there is no link $m + 1$. Therefore $\mathbf{f}_{m+1}$ and $\mathbf{n}_{m+1}$ are zero. In the inverse dynamics algorithm, it is convenient to set them equal to any external force and moment exerted on link $\mathbf{m}$.

The effects of gravity can be included in the model by setting the linear acceleration of the base, $\dot{\mathbf{v}}_0$ equal to the acceleration due to gravity and $\dot{\mathbf{w}}_0$ equal to zero in Eqs. (6.8) and (6.9). The effect of gravity will automatically be contained in the $\mathbf{F}_i$ and $\mathbf{N}_i$. $\mathbf{w}_0$ and $\mathbf{v}_0$ are set to zero in Eqs. (6.4) and (6.5).

The total force and moment exerted on link i is

$$\mathbf{F}_i = \mathbf{f}_i - \mathbf{f}_{i+1}$$
$$\mathbf{N}_i = \mathbf{n}_i - \mathbf{n}_{i+1} - (\mathbf{p}_i^* + \mathbf{r}_i) \times \mathbf{f}_i + \mathbf{r}_i \times \mathbf{f}_{i+1}$$

Solving for $\mathbf{f}_i$ and $\mathbf{n}_i$ gives

$$\mathbf{f}_i = \mathbf{F}_i + \mathbf{f}_{i+1}$$
$$\mathbf{n}_i = \mathbf{N}_i + \mathbf{n}_{i+1} + (\mathbf{p}_i^* + \mathbf{r}_i) \times \mathbf{F}_i + \mathbf{p}_i^* \times \mathbf{f}_{i+1} \tag{6.17}$$

Routine LINKFOR (output $= \mathbf{f}_i, \mathbf{n}_i$, input $= \mathbf{F}_i, \mathbf{N}_i, \mathbf{f}_{i+1}, \mathbf{n}_{i+1}, \mathbf{p}_i^*, \mathbf{r}_i$) is used to compute $\mathbf{f}_i$ and $\mathbf{n}_i$ using Eq. (6.17).

Finally, $\mathbf{u}_i$ is the projection of $\mathbf{f}_i$ or $\mathbf{n}_i$ along the axis of motion of joint $\mathbf{i}$.

$$\mathbf{u}_i = \begin{cases} \mathbf{z}_{i-1}'\mathbf{n}_i & \text{if joint } i \text{ is rotational} \\[2mm] \mathbf{z}_{i-1}'\mathbf{f}_i & \text{if joint } i \text{ is translational} \end{cases} \tag{6.18}$$

Routine JOTFOR (output $= \mathbf{u}_i$, input $= \mathbf{z}_{i-1}, \mathbf{n}_i, \mathbf{f}_i$, type$_i$) is used to compute $\mathbf{u}_i$ using Eq. (6.18).

Thus in a roundabout way the function for $\mathbf{u}$ in Eq. (6.14) has been derived. A more efficient computational procedure can be obtained by letting all of the quantities associated with link i in the foregoing equations be referenced to link i coordinates.[6] This simplifies the computational procedure since $\mathbf{r}_i$ and $\mathbf{I}_i$ need not be computed.

6.3.2. Inverse Dynamics

For control purposes it is important to be able to compute the required torque for specific trajectories very quickly. Depending on the power of the processor being used, it may be necessary to use approximations for these terms.[3,18] However, if the purpose of having the model is for simulation, or if the processor being used for control is of sufficient power, the foregoing equations can be used directly. This problem is referred to as the inverse dynamics problem. Given the position, velocity, and acceleration of the joint variables, the torque needed to move the manipulator can be computed by the following algorithm.

ALGORITHM 6.3.   INVERSE DYNAMICS

Step 1. Set $i = 0$, $\mathbf{A}(0, 0) = \mathbf{I} =$ the identity matrix, $\mathbf{w}_0 = \dot{\mathbf{w}}_0 = \mathbf{v}_0 = 0$.

Step 2. Compute $\mathbf{A}(0, i + 1)$ by computing $\mathbf{A}(i, i + 1)$ with the procedure HOMOJ [output $= \mathbf{A}(i, i + 1)$, input $= i + 1$] and multiplying it times $\mathbf{A}(0, i)$ using procedure MULMAT [output $= \mathbf{A}(0, i + 1)$, input $= \mathbf{A}(0, i), \mathbf{A}(i, i + 1)$].

Step 3. Compute $\mathbf{I}_{i+1}$ and $\mathbf{r}_{i+1}$ by calling procedure TRNINT [output $= \mathbf{I}_{i+1}, \mathbf{r}_{i+1}$, input $= \mathbf{A}(0, i + 1), \hat{\mathbf{I}}_{i+1}, \hat{\mathbf{r}}_{i+1}$].

Step 4. Compute $\mathbf{w}_{i+1}$ and $\mathbf{v}_{i+1}$ by calling procedure VEL (output $= \mathbf{w}_{i+1}, \mathbf{v}_{i+1}$, input $= \mathbf{w}_i, \mathbf{v}_i, \mathbf{z}_i, \mathbf{p}_{i+1}^*, \dot{q}_{i+1}$).

Step 5. Compute $\dot{\mathbf{w}}_{i+1}$ and $\dot{\mathbf{v}}_{i+1}$ by calling procedure ACCEL (output $= \dot{\mathbf{w}}_{i+1}, \dot{\mathbf{v}}_{i+1}$, input $= \dot{\mathbf{w}}_i, \dot{\mathbf{v}}_i, \mathbf{w}_i, \mathbf{z}_i, \mathbf{p}_{i+1}^*, \dot{q}_{i+1}, \ddot{q}_{i+1}$, type$_{i+1}$).

Step 6. Compute $\hat{\mathbf{v}}_{i+1}$ by calling procedure ACCNT (output $= \hat{\mathbf{v}}_i$, input $= \dot{\mathbf{v}}_i, \dot{\mathbf{w}}_i, \mathbf{r}_i$).

Step 7. Compute $\mathbf{F}_{i+1}$ and $\mathbf{N}_{i+1}$ by calling procedure TOTFOR (output $= \mathbf{F}_{i+1}, \mathbf{N}_{i+1}$, input $= \mathbf{I}_{i+1}, \mathbf{w}_{i+1}, \dot{\mathbf{w}}_{i+1}, \hat{\mathbf{v}}_{i+1}$).

Step 8. Set $i = i + 1$.

Step 9. If $i < m$, go to step 2, else continue.

Step 10. Set $\mathbf{f}_{m+1}$ and $\mathbf{n}_{m+1}$ equal to the force and moment acting through link m coordinates which are equivalent to all of the external forces and moments exerted on link m. $[\mathbf{n}_{m+1}' \ \mathbf{f}_{m+1}'] = \mathbf{k}'$.

Step 11. Compute $\mathbf{f}_i$ and $\mathbf{n}_i$ by calling procedure LINKFOR (output $= \mathbf{f}_i$, $\mathbf{n}_i$, input $= \mathbf{F}_i$, $\mathbf{N}_i$, $\mathbf{f}_{i+1}$,
$\mathbf{n}_{i+1}$, $\mathbf{p}_i^*$, $\mathbf{r}_i$).
Step 12. Compute $\mathbf{u}_i$ by calling procedure JOTFOR (output $= \mathbf{u}_i$, input $= \mathbf{z}_{i-1}$, $\mathbf{n}_i$, $\mathbf{f}_i$, type$_i$).
Step 13. Set $i = i - 1$.
Step 14. If $i = 0$ STOP, else go to Step 10.

The procedure INVDYN (output $= \mathbf{u}$, input $= \mathbf{q}$, $\dot{\mathbf{q}}$, $\ddot{\mathbf{q}}$, $\dot{\mathbf{v}}_0$, $\mathbf{k}$) uses the preceding algorithm to compute
the actuator force or moment given the position, velocity, and acceleration of the joint variables, the
force and moment exerted on the end effector, and the acceleration of the base link 0 coordinates.

6.3.3. Forward Dynamics

The opposite problem is referred to as the forward dynamics problem. To simulate the manipulator,
one needs the state of the manipulator and the derivative of the state. The state variables are position
and velocity of the joint variables. The derivative of the position is velocity, which is a state variable
and therefore known to the simulation. The derivative of the velocity is acceleration, and this must
be obtained from Eq. (6.14). If the vector $\mathbf{b}(\mathbf{q}, \dot{\mathbf{q}}, \mathbf{k})$ is defined as

$$\mathbf{b}(\mathbf{q}, \dot{\mathbf{q}}, \mathbf{k}) = \mathbf{C}(\mathbf{q}, \dot{\mathbf{q}}) + \mathbf{g}(\mathbf{q}) + \mathbf{J}(\mathbf{q})'\mathbf{k}$$

then the acceleration of the joint variables can be obtained by solving the linear equation

$$\mathbf{H}(\mathbf{q})\ddot{\mathbf{q}} = \mathbf{u} - \mathbf{b}(\mathbf{q}, \dot{\mathbf{q}}, \mathbf{k}) \qquad (6.19)$$

From Eq. (6.13), the numerical value of $\mathbf{b}(\mathbf{q}, \dot{\mathbf{q}}, \mathbf{k})$ can be computed by calling the routine INVDYN
(output $= \mathbf{b}$, input $= \mathbf{q}$, $\dot{\mathbf{q}}$, $\mathbf{0}$, $\dot{\mathbf{v}}_0$, $\mathbf{k}$) with the joint acceleration set equal to zero. The resultant value
of $\mathbf{u}$ will then be equal to the vector $\mathbf{b}(\mathbf{q}, \dot{\mathbf{q}}, \mathbf{k})$.
 All that is needed now is the matrix $\mathbf{H}(\mathbf{q})$. To obtain this matrix, routine INVDYN (output $=$
$\mathbf{h}_i$, input $= \mathbf{q}$, $\mathbf{0}$, $\mathbf{e}_i$, $\mathbf{0}$, $\mathbf{0}$) is called. Now the input velocity of the joint variables is zero, the acceleration
due to gravity is zero, the external forces and moments are zero, and the acceleration of all of the
joint variables is zero except q_j, which is set equal to 1. Thus $\mathbf{e}_i$ is a vector with all values equal to
zero except for the ith element being equal to 1. Referring to Eq. (6.13), the $\mathbf{u}$ computed from this
routine is the $\mathbf{h}_i$, the ith column of $\mathbf{H}(\mathbf{q})$. Repeating this procedure for each column, the entire matrix
$\mathbf{H}(\mathbf{q})$ is obtained, which can then be used in Eq. (6.19) to obtain the joint variable accelerations.
 Given the joint forces, the state of the manipulator, the external forces and moments exerted on
the end effector, and the acceleration due to gravity, the following algorithm can be used to obtain
the joint accelerations.

ALGORITHM 6.4. FORWARD DYNAMICS

Step 1. Compute the vector $\mathbf{b}(\mathbf{q}, \dot{\mathbf{q}}, \mathbf{k})$ by calling the procedure INVDYN (output $= \mathbf{b}$, input $= \mathbf{q}$,
$\dot{\mathbf{q}}$, $\mathbf{0}$, $\dot{\mathbf{v}}_0$, $\mathbf{k}$).
Step 2. Set $i = 1$.
Step 3. Compute $\mathbf{h}_i$, the ith column, of the moment of inertia matrix $\mathbf{H}(\mathbf{q})$ by calling the procedure
INVDYN (output $= \mathbf{h}_i$, input $= \mathbf{q}$, $\mathbf{0}$, $\mathbf{e}_i$, $\mathbf{0}$).
Step 4.
 Set $i = i + 1$.
Step 5. If $i \leq m$, go to Step 3, else continue.
Step 6. Compute the joint accelerations using the procedure LIN (output $= \mathbf{q}$, input $= \mathbf{H}(\mathbf{q})$,
$\mathbf{u} - \mathbf{b}$).
Step 7. Stop.

The procedure FORDYN (output $= \ddot{\mathbf{q}}$, input $= \mathbf{q}$, $\dot{\mathbf{q}}$, $\dot{\mathbf{v}}_0$, $\mathbf{k}$, $\mathbf{u}$) is used to compute the joint accelerations
using the foregoing algorithm at each step of time in the simulation. The acceleration of the joint
variables can be integrated to obtain the joint velocities and positions. For additional information on
the simulation of open-chain manipulators using the Newton-Euler formulation of the equations of
motion, see References 4, 5, and 8.

6.3.4. General-Purpose Mechanism Programs

Most industrial manipulators contain a large number of closed kinematic loops. For example, joint
actuator drive mechanisms, which are composed of a DC motor with a gearing mechanism, are systems
of closed kinematic loops. As an approximation, one can include the mass of the motor armature
and the gears into the mass of the link and use the model for open-chain mechanism and ignore the
effect of the momentum of the rotating armature and gears. If this is small compared to the momentum

of the manipulator links, a reasonably close response to the model and the real system can be expected.

For a more complex mechanism, the use of more powerful analysis packages is called for. There exist several general-purpose computer programs such as ADAMS, IMP, DRAM, MEDUSA, and DYMAC for the simulation of such mechanisms.[9,10,14,21-26]

6.4. CONCLUSION

This chapter has presented some of the theory and common practice used in solving basic problems in manipulator kinematics and dynamics. The development of the kinematics was based upon the use of homogeneous transforms to describe the position and orientation of the link coordinates. The mapping of joint coordinates to link coordinates and the inverse mapping was presented in the forward and inverse kinematics sections for position, velocity, and acceleration. In the dynamics section the equations of motion for an open-chain manipulator were presented. It was shown how the results of the kinematics section could be used to obtain the equations of motion of an open-chain manipulator. References were made to other methods for formulating the dynamics of more complex manipulators.

With this background the reader can assemble a collection of procedures that can be used in the design and modeling of a manipulator. Throughout this chapter several procedures have been described which are associated with various equations or algorithms. The collection of these routines has proven useful in the design, modeling, and control of manipulators. Their development could logically proceed in the following order: forward kinematic routines, inverse kinematic routines, inverse dynamics routines, forward dynamics routines. It is hoped that the reader will attempt to build a collection of these procedures. In this way one can create some useful tools as well as obtain a better understanding of the basic problems in manipulator kinematics and dynamics.

REFERENCES

1. Denavit, J., and Hartenberg, R. S., A Kinematic Notation for Lower-Pair Mechanisms Based on Matrices, *ASME Journal of Applied Mechanics,* June 1965, pp. 215–221.

2. Mayer, G. E., A Systematic Approach for Obtaining Solutions to the Kinematic Equations of Simple Manipulators, Master of Science Thesis, Purdue University, July 1979.

3. Paul, R. P., *Robot Manipulators: Mathematics, Programming, and Control,* MIT Press, Cambridge, MA, 1981.

4. Orin, D. E., McGhee, R. B., Vukobratovic, M., and Hartoch, G., Kinematic and Kinetic Analysis of Open-Chain Linkages Utilizing Newton-Euler Methods, *Mathematical Biosciences,* Vol. 43, No. 1/2, February 1979, pp. 107–130.

5. Stepanenko, Y., and Vukobratovic M., Dynamics of Articulated Open-Chain Active Mechanisms, *Mathematical Biosciences,* Vol. 28, No. 1/2, 1976.

6. Luh, J. Y. S., Walker, M. W., and Paul, R. P. C., On-Line Computational Scheme for Mechanical Manipulators, *ASME Journal of Dynamic Systems, Measurement, and Control,* Vol. 102, June 1980, pp. 69–76.

7. Hollerbach, J. M., A Recursive Lagrangian Formulation of Manipulator Dynamics and a Comparative Study of Dynamics and a Comparative Study of Dynamics Formulation Complexity, *IEEE Transactions on Systems, Man, and Cybernetics,* Vol. SMC-10, No. 11, Nov. 1980, pp. 730–736.

8. Walter, M. W., and Orin, D. E., Efficient Dynamic Computer Simulation of Robotic Mechanisms, *ASME Journal of Dynamic Systems, Measurement, and Control,* September 1982, Vol. 104, pp. 205–211.

9. Paul, B., Analytical Dynamics of Mechanisms—A Computer Oriented Overview, *Mechanism and Machine Theory,* 1975, Vol. 10, Pergamon Press, New York, pp. 481–507.

10. Kaufman, R. E., Mechanism Design by Computer, *Machine Design,* October 28, 1978, pp. 94–100.

11. Nobel, B., *Applied Linear Algebra,* Prentice-Hall, Englewood Cliffs, New Jersey, 1969.

12. Renaud, M., Coordinated Control of Robots-Manipulators: Determination of the Singularities of the Jacobian Matrix, *Proceedings First Yugoslav Symposium on Industrial Robotics and Artificial Intelligence,* Dubrovnik, September 13–15, 1979, pp. 153–165.

13. Uicker, J. J. Jr., Denavit, J., and Hartenberg, R. S., An Iterative Method for the Displacement Analysis of Spatial Mechanisms, *ASME Journal of Applied Mechanics,* June 1966, pp. 309–314.

14. Uicker, J. J. Jr., Dynamic Behavior of Spatial Linkages, Part 1: Exact Equations of Motion; Part 2: Small Oscillations About Equilibrium, *ASME Journal of Engineering for Industry,* February 1969, pp. 251–265.

15. Paul, R. P., Modeling, Trajectory Calculation and Servoing of a Computer Controlled Arm, Stanford Artificial Intelligence Laboratory, Stanford University, AIM 177, 1972.

16. Whitney, D. E., Resolved Motion Rate Control of Manipulators and Human Prosthesis, *IEEE Transactions on Man-Machine Systems,* Vol. MMS-10, No. 2, June 1969, pp. 47–53.

17. Whitney, D. E., The Mathematics of Coordinated Control of Prosthetic Arms and Manipulators, *ASME Journal of Dynamic Systems, Measurement, and Control,* December 1972, pp. 303–309.

18. Bejczy, A. K., Robot Arm Dynamics and Control, JPL Technical Memorandum 33–669, February 1974.

19. Symon, K. R., *Mechanics,* Addison-Wesley, New York, 1961.

20. Wertz, J. R., *Spacecraft Attitude Determination and Control,* D. Reidel Publishing Company, Dordrecht, Holland, 1978.

21. Sheth, P. N., and Uicker, J. J. Jr., IMP (Integrated Mechanism Program): A Computer-Aided Design Analysis System for Mechanisms and Linkages, *Transactions ASME, Journal of Engineering in Industry, 94 Ser. B,* 1972, pp. 454–466.

22. Sheth, P. N., A Digital Computer Based Simulation Procedure for Multiple Degree of Freedom Mechanical Systems with Geometric Constraints, Ph.D. Thesis, University of Wisconsin, Madison, 1972.

23. Uicker, J. J. Jr., User's Guide for IMP (Integrated Mechanism Program): A Problem Oriented Language for the Computer-Aided Design and Analysis of Mechanisms, NSF Rep. Res. Grant GK-4552, University of Wisconsin, Madison, 1973.

24. Chace, M. A., and Sheth, P. N., Adaptation of Computer Techniques to the Design of Mechanical Dynamic Machinery, ASME paper 73-DET-58, 1973.

25. Dix, R. C., and Lehman, T. J., Simulation of the Dynamics of Machinery, *Transactions ASME, Journal of Engineering in Industry, 94 Ser. B,* 1972, pp. 433–438.

26. Orlandea, N., Node-Analogous, Sparsity-Oriented Methods for Simulation of Mechanical Dynamic Systems, Ph.D. Dissertation, University of Michigan, 1973.

CHAPTER 7

DESIGN OF GRIPPERS

PAUL K. WRIGHT

MARK R. CUTKOSKY

Carnegie-Mellon University
Pittsburgh, Pennsylvania

7.1. INTRODUCTION

John Milton's famous statement, "No man is an island," could also be applied to industrial robots. The function of a robot is to interact with its surroundings. The robot does this by manipulating objects and tools to fulfill a given task. The robot *gripper* or *end-of-arm tooling* becomes a bridge between the computer-controlled arm and the world around it. The design of the gripper should reflect this role, matching the capabilities of the robot to the requirements of the task. The ideal gripper design should be synthesized from independent solutions to the three considerations shown in Figure 7.1.

The first industrial robots were nearly islands. They were used primarily as stand-alone machines for painting, spot welding, or pick-and-place work in which parts were moved from one location to another without much attention paid to how the parts were picked up or put down. For pick-and-place work, simple beak-like grippers were used, and the ability of the robot to grasp and manipulate parts was at best equal to that of a person using fireplace tongs.

Since then, robots have been put to work in more challenging applications. The objects they grasp may have complicated shapes and they may be fragile. The tasks the robots perform may involve assembling parts or fitting them into clamps and fixtures. These tasks place greater demands on the accuracy of the arm and also the gripper. Once a part has been picked up it must be held securely and in such a way that the position and orientation remain accurately known with respect to the robot arm. While the object is being manipulated during, say, an assembly task, forces arise between the object held by the robot and the mating parts. The robot is constrained by this contact, and the actions of the robot and the gripper determine whether the assembly will go smoothly or whether the parts will become damaged in the process. For assembly, the "fingers" should not slip, and the gripper, as a whole, should be compliant enough to prevent contact forces from doing any harm.

The actions required of the robot and the gripper will vary, depending on the tasks being done. The same effect is seen in the way that a human task determines the choice of grip and the actions of the hand. If one picks up a pencil to hand it to somebody, the way one holds the pencil is entirely different from the way it is held for writing. The remarkable thing about the human hand is that it can be adapted to so many different tasks: writing with a pencil, kneading dough, groping in the dark, and playing the guitar, to name a few. In addition to being a gripper, the human hand is a sensory organ and an organ of communication. The grippers presented in Section 7.3 are all very crude in comparison, although the last few examples begin to show some flexibility. It would be a mistake, however, to assume that the human hand would be the ideal gripper for an industrial robot. Manufacturing represents a world much more restricted than the one for which the human hand is designed. In fact, for many manufacturing tasks the human hand is not the best gripper, which is why workers use pliers, wrenches, tweezers, work gloves, and numerous other tools to help them.

7.2. PREHENSION AND GRIP

Human prehension combines the choice of a grip, the act of grasping or picking up an object, and the control of an object by using the hand. A desire, internally or externally generated, triggers responses in the eye and the mind. An instant later the hand takes a position over the object, forms a grip, and begins with the task. As shown in Figure 7.2, the process is controlled by feedback loops which include eyesight for approximate positioning of the hand and touch for adjusting the force and position of the fingers.

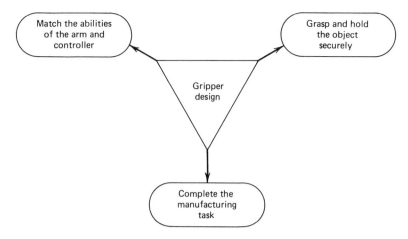

Fig. 7.1. Requirements of a gripper.

Prehension is task oriented, as the pencil example in the last section illustrates. Choosing a grip is an important step in the process of prehension. The block diagram in Figure 7.3 is greatly simplified, but it illustrates the relationships between the different activities included in prehension. In recent years, a few papers have been published that deal with the subcomponents of Figure 7.3. Asada,[1,2] Cutkosky,[3] and Salisbury[4,5] consider the force balance for an object held by a gripper with several fingers. For a given three-dimensional object, Asada[1] and Cutkosky[3] have considered the problem of choosing the optimum finger position around the object. For example, if the finger strengths and friction conditions are known, it is possible to find the grip that is stiffest or that will resist the greatest force or torque without slipping. For example, if the industrial three-fingered hand (shown

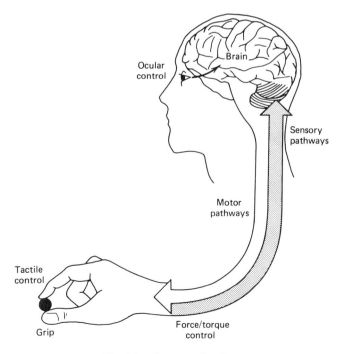

Fig. 7.2. Human prehension.

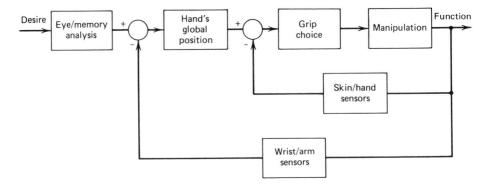

Fig. 7.3. Feedback diagram of the actions involved in prehension.

in Figure 7.17) were used to pick up an irregularly shaped ashtray from a table, the analyses in Asada[1,2] and Cutkosky[3] could predict the most secure finger orientations.

Control laws for multijointed, three-finger grippers have also been presented.[4-7] Small hands have been built on this basis that can, for example, screw a nut onto a bolt. For such movements, a kinematic analysis of the rolling between the finger tips and the object is needed. These are the few examples of robot hands with active, closed-loop control of the fingers. In the research community a number of similar hands are currently being built. For the next five years, however, closed-loop control of the fingers is impractical for industrial use. This is because the hands are mechanically complex, it is awkward to coordinate and control the fingers, and it is difficult to transmit information between the fingers and the robot.

Grippers for current industrial use have passive fingers. In fact, a good design allows a single set of passive fingers to conform to several different object shapes and to be more flexible than one might expect. Before exploring such designs, let us consider some of the common grips used by human hands.

A review of pertinent medical literature[8] reveals as many as eight basic categories of grip, but, for manufacturing work, only two of these are of primary importance: the "three-fingered" grip shown in Figure 7.4 and the "wrap-around" grip used to hold a large screwdriver or a hammer. The three-fingered grip is used for 90% of light domestic and manufacturing tasks; it is adaptable to many object sizes (from a pea to a softball); and it makes use of the stable structure of ligaments in the back of the hand, as illustrated in Figure 7.4. However, it is not as strong as the wrap-around grip, which involves friction between the faces of all the fingers and the tool and uses the power of all five digits, including the very strong muscle running from the small finger along the outside of the hand and into the arm.

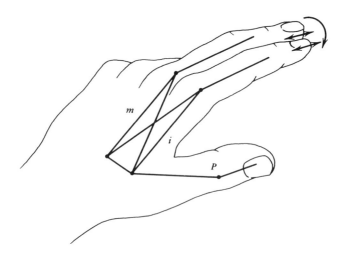

Fig. 7.4. The human three-fingered grip.

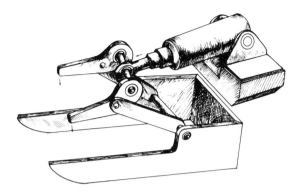

Fig. 7.5. Flexible gripper for turbine blades.

The same principles can be applied to industrial grippers, such as those shown in Figures 7.5 and 7.6. Figure 7.5 is essentially a three-fingered gripper. The lower face is a very wide thumb. The upper two fingers, by virtue of the ball-joint linkage between them, can conform to the object being gripped. They have the degrees of freedom shown in Figure 7.4. The gripper in Figure 7.5 also has some of the attributes of the wrap-around grip, since the spacing between the index and middle fingers is expanded (in comparison to the human hand), and the thumb is widened to resemble a platform or a palm with some scooping ability. As a result, the grip is stronger and more stable. This amalgamation of two primary grip types leads to a design specifically suited to heavy manufacturing.

7.3. INDUSTRIAL GRIPPER EXAMPLES

The quickest way of becoming acquainted with the design of current robot grippers is to look at some examples. Figures 7.5–7.18 have been selected to give some indication of the wide variety of gripping techniques and gripper materials in use. The designs are briefly discussed below.

Figure 7.7 shows a simple, compact design of the most common gripper style used today. Grippers resembling this one are sold in many sizes made from steel, plastic, or aluminum. The actuator for closing the gripper may be a hydraulic or pneumatic cylinder or an electric motor. Only a short motion of the actuator is required to completely open and close the jaws, and, because the linkage works as a toggle mechanism, the gripping force is high at the end of the actuator stroke. One drawback is that the fingers close by rotating through an angle so that the finger surfaces are not parallel to each other unless the gripper is closed.

Figure 7.8 shows a gripper in which the fingers move toward each other without rotating so that the gripping surfaces always remain parallel. For gripping flat or rectangular objects this design is much more secure than the angular-motion gripper in Figure 7.7. The driving mechanism consists of two gear sectors turned either by a central rack connected to a cylinder or by a central worm screw connected to an electric motor. The gripping force of this design is limited by the maximum practical gear size and gear tooth loads. There are many other ways of accomplishing parallel motion, such as mounting the fingers on a threaded shaft that has a right-hand screw at one end and a left-hand screw at the other. The combination of an electric servomotor and a threaded shaft is particularly suited for light, compact grippers used in electronics assembly, although miniature pneumatic cylinders have also been used.[9]

Figure 7.9 shows a simple gripper that uses a conical punch as a cam for closing the fingers. Springs are used to keep the fingers open when the punch retracts. The fingers cannot open far, but a high gripping force is available without using a large cylinder or motor.

For gripping tubes, jars, and other items with holes in them, an internal gripper is often most practical. Figure 7.10 shows a simple design that requires no external actuators. The gripper consists

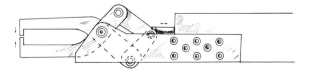

Fig. 7.6. Gripper for irregular forgings.

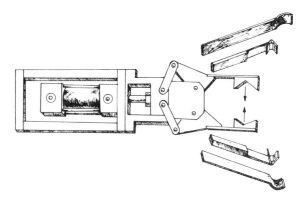

Fig. 7.7. Two-fingered gripper with changeable fingertips (adapted from designs in Reference 22.)

of a cylindrical rubber membrane surrounding a hydraulic cylinder and a piston. This is inserted into a hole in the object to be gripped, and, as the robot starts to lift, the piston rises inside the cylinder, forcing fluid into the membrane. The membrane expands, gripping the insides of the hole. When the robot puts the object down, the piston is lowered inside the cylinder, deflating the membrane and releasing the gripped part. A light spring keeps the piston normally in the lowered position. The gripper could easily be modified to grip the cylindrical objects from the outside.

For objects with complex or irregular surfaces, a gripper with more flexibility is required. The gripper shown in Figure 7.5 is designed to hold turbine blades. The blades come in a variety of shapes and sizes, but they all taper and twist between the root and the tip of the blade. As discussed in Section 7.2, this gripper combines some of the attributes of the human three-fingered grip and wrap-around grip. The upper fingers, driven by a single pneumatic cylinder, pull the blade backward until the rear edge of the blade rests against two teeth mounted in the lower fingers. This ensures that the blade is correctly aligned within the gripper. Microswitches (see Section 7.6, Sensors in Grippers) are used to detect whether the upper fingers are fully closed and whether the rear edge of the blade is pushed against the teeth in the lower fingers. The gripper shown in Figure 7.6 is designed to handle rough forgings. Like the gripper in Figure 7.9, it uses two upper fingers that are connected by a ball-joint linkage. This gripper is used as an example in the discussion in Section 7.5 on flexibility and compliance.

Figure 7.11 shows a simple gripper used for handling large, lightweight items such as cardboard cartons. The gripper is an angular-motion gripper, like the design in Figure 7.7.

Figure 7.12 shows a gripper designed for gripping fragile objects. It could even be used for handling fresh fruit. The two inflated fingers are made of an elastomeric material. When the pneumatic pressure in them is released, they curl inward, wrapping around the object between them and pressing it gently against the small "palm" that is mounted between the fingers.[10] The fingers have a high coefficient of friction and consequently are able to hold a wide variety of object sizes and shapes without slipping.

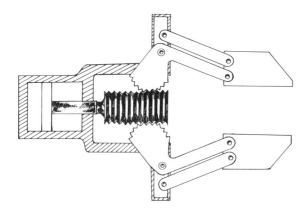

Fig. 7.8. Rack and pinion parallel-jaw gripper (adapted from designs in References 22 and 25).

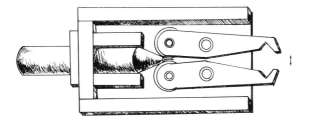

Fig. 7.9. Two-fingered gripper with high gripping force (adapted from designs in Reference 22).

The gripper is not precise but is compliant and forgiving of minor robot errors. The advantages of compliance are discussed in Section 7.5.

Large flat objects are often difficult to grasp. Figure 7.13 shows one solution in which vacuum cups are used for picking up metal plates, panes of glass, or large, lightweight boxes. Since the vacuum cups are usually made of an elastomeric material, this gripper, like that in Figure 7.12, is compliant. The gripper is tolerant of errors in the orientation of the part and is especially suited for pick-and-place work. If the part must be located accurately with respect to the gripper, guides can be mounted to meet the edges or corners of the plate or box being grasped.

A variation on the vacuum cup gripper is shown in Figure 7.14. It has been used extensively for lifting fragile silicon wafers. A compressed air supply (which is often easier to obtain than a vacuum pump) and a venturi are used to create a gentle vacuum that lifts the wafer and holds it in place. Another variation on this gripper uses a stream of air directed onto the center of the silicon wafer so that the air flows uniformly toward the periphery of the wafer, creating aerodynamic lift. Guides are used to center the wafer while it floats, suspended just beneath the gripper.

Magnetic grippers are used extensively on ferrous materials. They range from tiny electromagnets for picking up wire brads or nails to the giant electromagnets used on junkyard cranes. If the workpiece must be held without slipping, it becomes important to achieve good contact between the magnet and the workpiece surface. If the workpieces have flat surfaces, this is easily achieved by mounting the magnets in a slightly compliant structure (see Section 7.5). For rough workpieces the method shown in Figure 7.15 is useful. In this arrangement a bag or membrane filled with small iron particles is attached to the underside of the electromagnet. As the robot sets the electromagnet down upon the workpiece the bag of particles conforms to local surface variations, ensuring good contact. When the electromagnet is energized the particles become a rigid mass and the workpiece is rigidly held.

Flexible materials such as fabric, wire, or string pose special handling problems. They are poorly constrained, and it is difficult to extract just one string or sheet from a bundle or a pile. Figure 7.16 shows a unique gripper for picking up pieces of cloth, one by one, from a stack. A jet of air is first blown on the stack. This causes the top piece of cloth to ruffle and lift slightly from the stack so that the thin lower blade of the gripper can be slipped beneath it. The gripper then closes. A light source on the lower blade of the gripper and a detector on the upper blade are used to ensure that the gripper has trapped just a single layer of cloth.[11]

For robots that are expected to perform a variety of tasks, interchangeable grippers or fingers can be used. For example, Figure 7.7 shows a variety of finger styles. Entire grippers can also be changed if necessary. At McDonnell Douglas Corp. bayonet fittings between the end of the arm and the grippers allow quick changes to suit different tasks during a cycle.[12] The disadvantages of interchangeable grippers are the cost of maintaining an inventory of grippers, the time required to switch between them, and the difficulty of making connections for sensors and actuators.

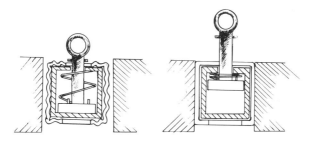

Fig. 7.10. Self-expanding internal gripper (adapted from Reference 22).

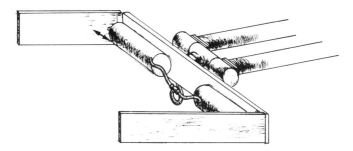

Fig. 7.11. Gripper for boxes and cartons (adapted from Reference 22).

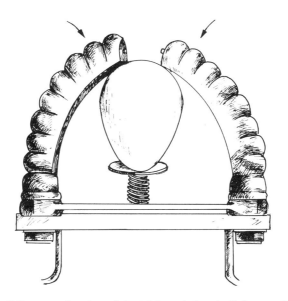

Fig. 7.12. Soft pneumatic gripper (adapted from designs in References 26, 27, 10).

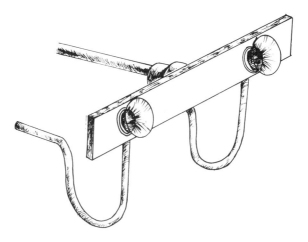

Fig. 7.13. Vacuum gripper with suction cups (adapted from designs in Reference 22).

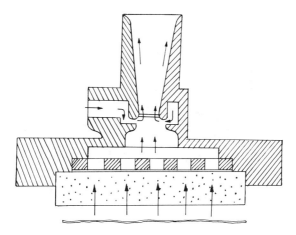

Fig. 7.14. Venturi vacuum gripper for flat plates (adapted from designs in Reference 22).

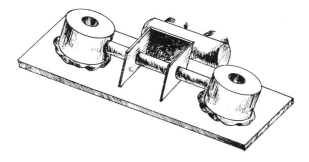

Fig. 7.15. Magnetic gripper (adapted from designs in Reference 22).

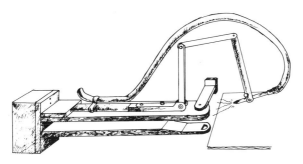

Fig. 7.16. Gripper for fabric.

These disadvantages are avoided by using more flexible grippers. Figure 7.17 shows one such gripper in which two of the three fingers are closed in unison by a single motor. In addition, two of the fingers can be swiveled, using a second motor, so that the gripper can adopt different gripping orientations. To grip a long rod or plate the two fingers can be rotated so that they are parallel to each other and facing the third finger. For gripping a small cylindrical object the two fingers can be rotated so that all three fingers move radially inward, meeting at the center.[13]

Figure 7.18 shows a flexible gripper in which the fingers wrap around objects, exerting a gentle pressure against them.[14,15] Like the gripper shown in Figure 7.12, it is suited for fragile objects and objects with irregular shapes. In fact, a large version of this gripper was tested as a device for rescuing

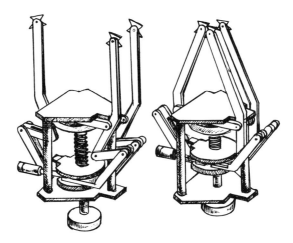

Fig. 7.17. Industrial three-fingered hand (adapted from design in Reference 13).

people. The gripper consists of many links joined together, like a section of bicycle chain. The joints between the links have pulleys, with thin wires wrapped around them. Pulling on the wires causes the chains to change their curvature, wrapping completely about the object to be gripped. The gripping pressure is uniformly distributed about the part.

7.4. FORCE ANALYSIS

The intent here is not to review classical mechanics but to emphasize the considerations that are most important in modeling grippers. A more extensive discussion of the static analysis of conventional grippers can be found in Chapter 8, which contains a number of worked examples and provides a convenient, systematic way of determining loads and actuator forces.

The actuator, acting by way of the fingers, applies forces to the object or tool being held. Such forces must resist the externally applied forces and moments arising from gravity, robot accelerations, and contact with fixtures. The external force acting on the grasped part is, in general, a six-element vector with three translational and three rotational components. Free-body diagrams drawn on a sheet of paper are adequate for analyzing pairs of force components, provided that the other force components acting simultaneously in and out of the paper are not forgotten. If elastic deflections of the gripper and its fingers must be considered, a matrix approach (similar to the approach used in modeling elastic trusses and space frames) is preferable.[3,5]

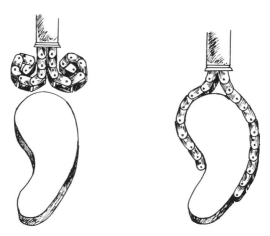

Fig. 7.18. Conformal gripper (adapted from design in References 14, 15).

Estimating dynamic loads is less straightforward than the calculation of gripping forces and gravity loading. Chen[16] recommends an additional safety factor of 1.2 –2.0 for dynamic loads, which is consistent with the safety factors commonly used in moving industrial machinery. The maximum working acceleration of the robot is usually an available number and for a typical robot will be at least 4.2 m/sec² (200 in./sec²), or about one-half the acceleration of gravity. These accelerations, however, are likely to be small compared with the dynamic accelerations that occur when a robot comes in contact with a fixture (up to four times the acceleration of gravity). Contact forces are one of the main reasons for considering compliance in the gripper design as discussed in Section 7.5.

Once the forces acting upon the part have been estimated, it is possible to determine the required gripping forces. These will depend on the coefficient of friction between the gripper and the part and on how well the gripper can enclose or entrap the part. For a gripper with steel fingers, grasping a steel part, the coefficient of friction will usually be 0.3 or greater. For a gripper with rubber fingers the coefficient is generally 1.0 or greater, although this will depend greatly on whether the surfaces become dirty or greasy.[17,18] If the required gripping force seems excessive for the part, or for the available actuators, it will be necessary to look at compliant materials for the gripper fingers and to consider very flexible gripper designs that conform to the shape of the part, increasing the gripping surface area (see Section 7.5).

When the gripping force is established, the required actuator force or torque can be computed for a given gripper design. One way to do this is to start at the fingers and work backward to the actuator, balancing forces on each link. However, if the gripper is complex, a more convenient approach is to find the ratio of the velocity of the actuator V_a to the velocity of the fingertips V_f. This is easily done from a scale drawing of the gripper, or on a computer-aided design system, by moving the actuator a small amount and determining how far the fingertips move. For example, if the fingertips move 2 mm for 1 mm of actuator travel, the velocity ratio is ½. The actuator force P is then found by equating the amount of work done by the actuator and by the fingertips

$$V_a \cdot P = 2 V_f \cdot F$$

where P is the actuator force, and F is the force at the fingertips.

In Table 7.1 this method is used to determine typical gripping forces and actuator forces for the gripper styles of Figures 7.7–7.9. For many grippers, the velocity ratio will vary as a function of how far the gripper is closed. If the gripper will be picking up objects of different sizes, this ratio must be determined for each size.

7.5. FLEXIBILITY AND COMPLIANCE

Flexibility is defined as the ability of a gripper to conform to parts that have irregular shapes and to adapt to parts that are inaccurately oriented with respect to the gripper. Flexible grippers are therefore especially useful for acquiring parts positioned imprecisely on a table or in a fixture and for grasping rough workpieces, such as castings and forgings.

Compliance is defined as the ability of a mechanism or structure to elastically deform in response to forces and torques. When the forces are removed, the structure returns to an equilibrium position.

Flexibility of the gripper makes a great contribution to how firmly the parts are held. Figure 7.19 shows a schematic of a long bar held by the upper and lower jaws of a gripper. In the upper view, the bar is smooth and regular and there are four points of contact. In the lower view there is an exaggerated irregularity and only three-point contact. The part is shown subject to a force and a moment. In general, the applied force F and moment M will vary in sign and magnitude as the robot rotates the gripper and bar (perhaps turning them upside-down). Under these conditions, for a given gripping force, the grip with four-point contact is substantially less likely to slip than the grip with three-point contact. The gripper in Figure 7.19 could be designed to have four-point contact with the irregular bar by mounting the upper jaw so that it could rock or pivot slightly. This is one common way of increasing the flexibility of a gripper.

Figure 7.6 is an example of a flexible gripper designed to grasp rough forgings. Two factors influencing the design of the gripper were the temperature of the forgings (538°C or 1000°F) and the need to grasp the forgings without disturbing their initial orientation. As discussed in Section 7.3, the upper fingers are connected by a ball-joint linkage, allowing the upper fingers to settle independently against curved or irregular surfaces. The upper and lower fingers are driven by a single hydraulic cylinder, and the entire assembly consisting of the cylinder and the fingers can pivot slightly with respect to the chassis of the gripper. As the hydraulic cylinder retracts, the fingers settle, one by one, against the forging. The cylinder rod continues to travel until all fingers are pressing firmly against the part and any play in the linkage has been taken up. At this point, a hydraulic brake is actuated so that the gripper/part assembly becomes rigid, and the orientation of the part is preserved.

Compliance represents an ability to absorb minor impacts by allowing elastic deformations to occur within the gripper structure. It keeps contact forces from becoming excessive if the gripper interacts with guides or fixtures.

TABLE 7.1. COMMON GRIPPING FORCE AND ACTUATOR REQUIREMENTS FOR GRIPPER STYLES IN FIGURES 7.7–7.9

Gripper Style	Velocity Ratio $(V_a/V_f)^a$	Friction Coefficient $(\mu)^b$	Grip Force $(2F)^c$	Actuator Force $(P)^d$	Pneumatic Cylinder Diameter $(D)^e$
Gripper in Figure 7.7, with steel fingers	3.0	0.3	333 N (75 lbf)	111 N (25 lbf)	2.0 cm (0.8 in.)
With rubber-surfaced fingers	3.0	1.0	100 N (22.5 lbf)	33 N (7.5 lbf)	1.1 cm (0.4 in.)
Gripper in Figure 7.8, with steel fingers	0.3	0.3	333 N (75 lbf)	1110 N (250 lbf)	6.4 cm. (2.5 in.)
With rubber-surfaced fingers	0.3	1.0	100 N (22.5 lbf)	333 N (75 lbf)	3.5 cm. (1.4 in.)
Gripper in Figure 7.9, with steel fingers	3.8	0.3	333 N (75 lbf)	88 N (20 lbf)	1.8 cm (0.7 in.)
With rubber-surfaced fingers	3.8	1.0	100 N (22.5 lbf)	26 N (5.9 lbf)	1.0 cm. (0.4 in.)

^{a} As discussed in Section 7.4, the ratio V_a/V_f is equal to the velocity of the actuator divided by the velocity at which the fingertips move together. The ratios in the table are typical values for when the grippers are nearly closed. The ratio for the design in Figure 7.7 becomes considerably smaller when the gripper is only partly closed (as in gripping an oversized object).

^{b} The coefficient of friction μ will depend on the materials used and on how clean and smooth the surfaces are. The numbers used in the table are conservative values for steel against steel and rubber against steel.

^{c} The grip force $2F$ is calculated for a compact metal object weighing approximately 25 N (5.6 lbf). The object is picked up with a high-speed robot, capable of accelerating 9.8 m/sec² (386 in./sec²), about equal to the acceleration of gravity. The maximum acceleration the part could experience is therefore $2g$, corresponding to a force of 50 N. We use a dynamic safety factor of 2 so the required grip force is given by $2\mu F \geq 100$ N.

^{d} The actuator force P is a function of the grip force and the velocity ratio: $2F = P(V_a/V_f)$.

^{e} The pneumatic cylinder diameter is chosen so that the required actuator force will be achieved as long as the air supply pressure remains above 35 N/cm² (50 psi).

    As an alternative, a compliant wrist unit can be mounted between the gripper and the robot arm. Such wrists, called remote-center-of-compliance (RCC) devices, are commercially available. Typically they consist of two metal discs, 50–150 mm (2–6 in.) in diameter, separated by angled springs. One disc is fitted to the arm and the other to the gripper so that the gripper can float with respect to the arm.* As a result, minor "bumps" are absorbed. In assembly operations these might occur, for example, when fitting a peg into a chamfered hole. By choosing an appropriate angle and stiffness for the springs, it is possible to project the "center" of compliance out to the tip of the peg in such a way that initial contact between the peg and the hole produces no tilting (and consequent jamming). Instead, compliant deflections caused by the contact forces will orient the axis of the peg in the right direction for smooth assembly.[19,20]

    An additional advantage of compliance is increased safety. A compliant structure has the ability to absorb impact forces. Errors resulting from programming mistakes, robot inaccuracies, and misaligned fixtures may arise, and, without compliance, damage may result. In fact, it is usually desirable to go a step beyond compliance and to make the gripper capable of breaking away or collapsing in the event that collisions occur.

7.6. SENSORS IN GRIPPERS

Sensors are increasingly used in industrial grippers and can be divided into three categories of increasing cost and complexity.

---

* Refer to Chapter 64, Part Mating in Assembly.

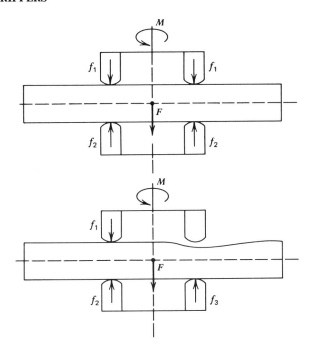

Fig. 7.19. Holding a bar with three and four points of contact.

Binary sensors include microswitches, optical and magnetic switches, and bimetallic thermal switches. These are generally inexpensive, rugged, and easy to interface with robot controllers. They are used to sense the presence or absence of a part, to check whether a variable such as pressure or temperature is within a permissible range, and as limit-switches.

Analog sensors include thermocouples, linear-variable-differential-transformers (LVDTs), strain gages, and piezo-electric sensors. These are more expensive than the preceding sensors and generally require some signal instrumentation and analog-to-digital conversion. They are used when the magnitude of a quantity is required.

Sensor arrays and sensors requiring low-level signal processing include pressure-sensitive arrays used on the fingers and palm of a gripper, optical arrays, and actively stimulated piezo-electric devices. For the most part, sensors of this type are confined to research laboratories, although arrays of pressure-sensitive rubber are beginning to be marketed for industrial use. All of these devices require a considerable amount of signal processing, and this is usually accomplished with a dedicated microprocessor that communicates over serial or parallel lines with a higher level computer such as the robot controller.

7.7. DESIGN GUIDES FOR PASSIVE GRIPPERS

At this point it is possible to summarize the important considerations in designing a gripper. The design Guides 1–6, which follow, are for current industrial grippers with passive fingers. In Section 7.8 Guides 7–9 are given for the next generation of industrial grippers.

We suggest that the designer consider Guides 1–6 (and Guides 7–9 in Section 7.8) sequentially, while reviewing the typical gripper styles in Figures 7.5–7.18. After this process, the guides should be reconsidered once or twice until the design is finalized.

GUIDE 1:   Study the parts to be grasped and the tasks to be performed.

Try to express the gripping requirements as abstractly as possible. This will require a force analysis of the kind discussed in Section 7.4. It is also possible at this stage to decide whether contact forces must be distributed over large areas or not. Are the parts fragile? compliant? slippery? irregularly shaped? The flexibility and compliance required of the gripper will depend on these characteristics and on task-related criteria as discussed in Section 7.5. Finally, what sort of sensory information is required from the gripper? Simple on/off devices will be adequate for most current applications.

GUIDE 2:   Determine additional requirements, not directly related to the act of acquiring and gripping parts.

It may be necessary for the gripper(s) to be automatically disengaged from the wrist of the robot. Environmental conditions including high temperatures or abrasive dirt should also be considered at this time. Other factors may include stringent weight allowances or a cramped working space for the gripper. Basic design decisions will be made in response to these requirements. For example, it may be necessary to locate the actuators remotely from the fingers of the gripper, to make some parts of the gripper from compliant materials, or to design the gripper so that it will break away in the event of a crash.

GUIDE 3:   Determine specific solutions to the requirements in Guides 1 and 2.

The idea here is to develop independent solutions to the individual design requirements. At this stage, no single combination of sensors, mechanisms or actuators should be considered. This modular approach keeps the design flexible and open to innovation and makes it easier, at a later stage, to evaluate how well competing designs satisfy each of the requirements.[21]

For example, if there is a requirement that power be transmitted from remote actuators, any sort of flexible transmission devices including cables, gear trains, chains, hydraulic lines, or rotary shafts may do the job. Similarly, if the task requires force information, foil strain gages, piezo-electric load cells, or piezo-resistive device will work. At this point it is usually not necessary to specify details such as whether the sensors should be mounted on the gripper fingers or on intermediate links.

As Figures 7.5–7.18 show, there is a variety of solutions to gripping requirements. More examples can be found in books such as that by Lundstrom, Glemme, and Rooks[22] which contain numerous drawings of unusual industrial grippers. Another common source of inspiration is nature. The human hand is the most obvious example, but, as discussed in Section 7.2, it is unnecessarily complex for most manufacturing tasks. Simpler designs from nature include a bird's beak, a dog's mouth, a lobster's claw, an elephant's trunk, and the tentacles of an octopus.

GUIDE 4:   Begin to develop designs combining the foregoing modular solutions.

Experiment with different combinations of the solutions determined in Guide 3. Additional concerns invoke the kinds of questions habitually asked by designers: Is the design serviceable? robust? economical? How could it be made with even fewer moving parts?

GUIDE 5:   Consider designs with two or three grippers mounted together at the end of the arm.

Doing this makes it unnecessary to build a single gripper that will perform all of the robot tasks. Another advantage is that the robot becomes more productive since it does not have to move back and forth as often. For example, a robot can use one gripper to carry a rough part over to a finishing station, use a second gripper to pick up the finished part, load the rough part with the first gripper, and return with the finished part.

GUIDE 6:   Redesign the part and/or the task.

At this stage it should be possible to see how changes in the part or in the robot task could simplify the design of the gripper and improve the ability of the robot to accomplish the task. As pointed out in Section 7.1, the gripper is a bridge between the robot and its environment. Therefore the design of the gripper should not be isolated from the design of the fixtures and parts that it interacts with, but should be part of a joint design. For example, if a gripper is required to pick up castings, it is often useful to design the castings with tabs or other identifying marks to make it easier to establish their orientation. The tabs or marks can be removed later when the casting is machined into a finished part. As another example, it has been shown that parts become easier for robots to assemble when particular chamfers are chosen for the mating surfaces.[23]

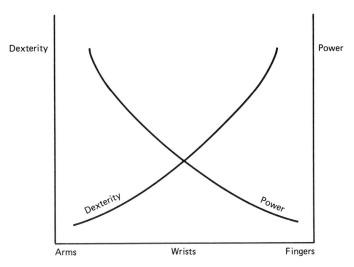

Fig. 7.20. Trade-off between arm, wrist, and finger manipulation.

7.8. FUTURE DEVELOPMENTS: DESIGN GUIDES FOR ACTIVE GRIPPER/WRIST UNITS

In the passive industrial grippers shown in Figures 7.5–7.18 there is neither the sense of feeling nor the possibility for active reorientation that we take for granted as humans going about our daily lives. A passive gripper can only open and close, and any manipulation must be done by the adjoining robot arm. Passive grippers can be viewed as part of the open loop in Figure 7.3. In the future, as industrial robot controllers and sensors improve, it will become practical to use feedback for the gripper to close the loops shown in Figure 7.3. The advantages of active finger sets and active wrists are now briefly reviewed.

In Section 7.5 on compliance the concept of a passive, compliant wrist was introduced. Such wrists can be modified to measure the magnitude and direction of the "bumps" or "tilts" experienced by the part during an assembly operation. (There are a variety of ways of obtaining this quantitative information using analog sensors such as those mentioned in Section 7.6.) In addition, the sensory information can be used to control or adjust the wrist, at which point the wrist can be defined as *active*. There are a number of tasks for which the ability to measure forces or deflections and to control the wrist is useful. For example, during a routing operation the deflections of the routing tool relative to the arm can be monitored for smoother contour following. (Routing, similar to edge milling, is used to trim flat plates to the correct size.)

Further in the future it will be practical to build industrial robots with "feeling" fingers. A recent sensor developed by Hackwood et al.[24] is effectively a skin that can sense normal and shear forces and that could gather data for an active fingertip. As discussed in Section 7.2, some of the control laws for active fingers have been formulated, and prototype hands have been constructed that could be used with skin sensors.

From an industrial design viewpoint it should be emphasized that there are many industrial tasks that can be performed with a current industrial arm, an *active* wrist, and a passive set of fingers. For heavy manufacturing tasks, an active wrist with passive fingers may actually be able to do the job better than a gripper with active fingers. This point can be clarified in two ways. First, Figure 7.20 shows the trade-off that occurs when one adds progressively more manipulative ability, and more complexity, to the robot. Going across the schematic figure, an arm with a passive finger set (gripper) could be successively improved with a passive wrist, then an active wrist, and finally with active fingers. However, the price for this added dexterity is a loss of power as might be measured by the gripping force or tool/work interaction force. Second, we do not use our own fingers in an active way when employing the wrap-around grip on a heavy wrench or a hammer. When we need power, the fingers adopt a passive, strong grip, and we rely on muscles in our wrist and arm. As industry moves into the next generation of "feeling" robots, if the designer can pick those tasks that need the wrap-around grip with an active wrist, initial applications will be successful. These ideas follow, recapitulated in a set of design guides for future work:

GUIDE 7: Heavier manufacturing tasks that are currently done by humans using a wrap-around grip holding hand tools are the first candidates for "feeling" robots since such tasks are best done by an active wrist and passive fingers.

There is a gray line between those tasks involving large hand tools and lighter operations that would best be done with active fingers using the three-fingered grip of Figure 7.4. Using a light screwdriver or a pair of tweezers for electronic assembly are examples. However, Guide 7 can be coupled with the following.

GUIDE 8: Manufacturing tasks that involve forces or moments greater than 9–22 N (2–5 lbf) lead to the use of a wrap-around grip on hand tools and an active wrist with passive fingers, rather than an active three-fingered grip.

The 9–22 N figure comes from preliminary experiments that identify the force level at which the three-fingered grip slips during a task, suggesting reorienting the fingers for a wrap-around grip. Once this level is established, the designer can consider the maximum interaction forces that are likely to occur for a given task and then proceed to Guide 9.

GUIDE 9: Manufacturing tasks can be categorized according to the trade-off among "arms, wrists, and fingers," and practical applications can be selected in the factory.

At present, manipulators equipped with the passive fingers shown throughout this chapter are most appropriate for industrial application. However, for future sensor-based applications the last three design guides have been included to show which manufacturing tasks are the first candidates for active wrist/gripper designs.

REFERENCES

1. Asada, H., Studies on Prehension and Handling by Robot Hands with Elastic Fingers, Ph.D. thesis, Kyoto University, April 1979.

2. Hanafusa, H. and Asada, H., Stable Prehension by a Robot Hand with Elastic Fingers, *Seventh International Symposium on Industrial Robots,* Tokyo, Japan, October 1977, pp. 361–368.

3. Cutkosky, M. R., A Cybernetic Approach to Grasping and Prehension for Automated Manufacturing, Ph.D. thesis, Carnegie-Mellon University, 1984 (expected).

4. Salisbury, J. K., Kinematic and Force Analysis of Articulated Hands, Ph.D. thesis, Stanford University, July 1982.

5. Salisbury, J. K. and Craig, J. J., Articulated Hands: Force Control and Kinematic Issues, *Robotics Research,* Vol. 1, No. 1, 1982, pp. 4–17.

6. Okada, T. and Tsuchiya, S., On a Versatile Finger System, *Seventh International Symposium on Industrial Robots,* October 1977, pp. 345–352.

7. Okada, T., Computer Control of Multijointed Finger System for Precise Handling, *IEEE Transactions on Systems, Man and Cybernetics,* Vol. SMC-12, No. 3, May 1982, pp. 289–299.

8. Tubiana, R., *The Hand,* W. B. Saunders Co., Philadelphia, 1981.

9. Sanderson, A. C. and Perry, G., Sensor-Based Robotic Assembly Systems: Research and Applications in Electronic Manufacturing, *Proceedings of the IEEE,* Vol. 71, No. 7, July 1983, pp. 856–871.

10. Inside Japan, *Assembly Automation (GB),* February 1982, pp. 58.

11. Taylor, P. et al., The Application of Robots in the Garment Manufacturing Industry, *Robotics Initiative, 2nd Annual Grantees Conference,* SRC, London, September 1983, pp. 37G.

12. Eastwood, M. A. and Ennis, G. E., ICAM Robotic System for Aerospace Batch Manufacturing, Fourth Quarterly Interim Technical Report IR-812-8 (IV), McDonnell Douglas Corp., August 1979, U.S. Air Force Contract F33615-78-C-5189, Project 812-8, Task B.

13. Skinner, F., Designing a Multiple Prehension Manipulator, *Journal of Mechanical Engineering,* Vol. 97, No. 9, September 1975, pp. 30–37.

14. Hirose, S. and Umetani, Y., The Development of Soft Gripper for the Versatile Robot Hand, *Proceedings, Seventh International Symposium on Industrial Robots,* 1977, pp. 353–360.

15. Hirose, S. and Umetani, Y., The Development of Soft Gripper for the Versatile Robot Hand, *Mechanism and Machine Theory (GB),* 1978, pp. 351–358.

16. Chen, F. Y., Force Analysis and Design Considerations of Grippers, *Industrial Robot (UK)*, Vol. 9, No. 4, December 1982, pp. 243–249.

17. Bowden, F. P. and Tabor, D., *The Friction and Lubrication of Solids*, Oxford University Press, London, 1950.

18. Fuller, D. D., Friction, in Baumeister, T., Ed., *Marks' Standard Handbook for Mechanical Engineers*, McGraw-Hill, New York, 1978, pp. 25–32.

19. Whitney, D. E. et al., Part Mating Theory for Compliant Parts, First Report, The Charles Stark Draper Laboratory Inc., August 1980, NSF Grant No. DAR79-10341.

20. Cutkosky, M. R. and Wright, P. K., Position Sensing Wrists for Industrial Manipulators, *Proceedings, Twelfth International Symposium on Industrial Robots*, Paris, France, June 1982, pp. 427–438.

21. Rinderle, J. R., Measures of Functional Coupling in Design, Ph.D. thesis, Massachusetts Institute of Technology, June 1982.

22. Lundstrom, G., Glemme, B., and Rooks, B. W., *Industrial Robots—Gripper Review*, International Fluidics Services Ltd., 35–39 High Street, Kempston, Bedford, England, 1977.

23. Whitney, D. E., Gustavson, R. E., and Hennessey, M. P., Designing Chamfers, *The International Journal of Robotics Research*, Vol. 2, No. 4, 1983, pp. 3–18.

24. Hackwood, S. et al., A Torque Sensitive Array for Robotics, *The International Journal of Robotics Research*, Vol. 2, No. 2, 1983, pp. 46–50.

25. Sheldon, O. L. et al., Robots and Remote Handling Methods for Radioactive Materials, *Second International Symposium on Industrial Robots*, Chicago, May 1972, pp. 235–256.

26. Warnecke, H. J. and Schmidt, I. I., Flexible Grippers For Handling Systems—Design Possibilities and Experiences, *Fifth I.C.P.R.*, Chicago, August 1979, pp. 320–324.

27. Robots, an American Breed by A.T.I., Air Technical Industries, Mentor, Ohio, Catalog No. R-21-C.

CHAPTER 8

DESIGN OF ROBOT HANDS

KAZUO TANIE

Ministry of International Trade and Industry
Ibaraki, Japan

8.1 INTRODUCTION

Generally the gripper for industrial robots is used for special purposes—a device to handle limited shapes of objects and limited functions. This kind of gripper makes designing easy and also keeps machinery costs relatively inexpensive, but versatility and dexterity are reduced. In some applications the simplification of gripper function may be more important than versatility and dexterity from the point of view of economics. In others, however, the gripper will be required to handle and manipulate many different objects of varying weights, shapes, and materials. The universal grippers, actually robot hands, will be suitable in such case.

Currently, the development of a universal gripper and the investigation of manipulation using it are under way. There are no practical universal grippers or hands at present. Therefore, the mechanical design of special-purpose grippers is mainly discussed in the following sections to complement Chapter 7. Only an outline of recent developments on universal grippers is described, with gripper functions and related design factors.

8.2. FUNCTIONS OF GRIPPERS AND RELATED FACTORS

Human-hand grasping is divided into six different types of prehension: palmar, lateral, cylindrical, spherical, tip, and hook, which were identified by Schlesinger (Figure 8.1).[1] Crossley classified manipulation functions by human hand into nine types: trigger grip, flipping a switch, transfer pipe to grip, use cutters, pen screw, cigarette roll, pen transfer, typewrite, and pen write.[2]

There are several factors relating to these variations of function. Important factors are the number of fingers, the number of joints for each finger, and the number of degrees of freedom of a hand.

A human arm including a hand has five fingers: the thumb, the index finger, the middle finger, the third finger, and the little finger. The whole arm structure has 27 DF, 20 of which are for the hand.[3] Each finger except the thumb has three joints, and each can produce 4 DF motion. The thumb has two joints with 3 DF. There is 1 DF in the palm. To approximate a subset of the human grasp and hand manipulation, the relation between gripper structure and its function must be considered.

To achieve minimum gripping function, a gripper needs two fingers connected to each other using a joint with 1 DF for its open-close motion. If the gripper has two rigid fingers, it has only the capability of grasping objects of limited shapes and is not able to enclose objects of various shapes. Also, this type of gripper cannot have manipulation function because all degrees of freedom are used to maintain prehension.

There are two ways to improve the capability to accommodate the change of object shapes. One solution is to put joints on each finger. The other is to increase the number of fingers up to a maximum of five. The manipulation function will also emerge from this. To manipulate objects it is usually necessary that the gripper have more fingers and joints driven externally and independently than does the gripper used only for grasping objects. The more fingers, joints, and degrees of freedom a gripper has, the more versatile and dexterous it can become. Table 8.1 shows approximate relations between the numbers of fingers and joints and the functions of grippers.

8.3. GRIPPER CLASSIFICATION

A gripper can be designed to have several fingers, joints, and degrees of freedom, as mentioned before. Any combination of these factors gives different grasping modalities to a gripper. Also, a gripper can

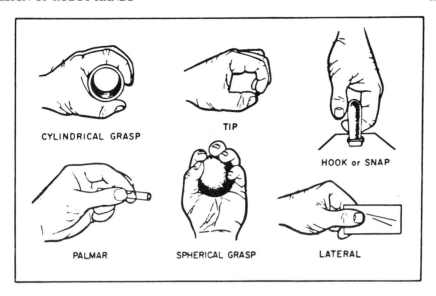

CYLINDRICAL GRASP

TIP

HOOK or SNAP

PALMAR

SPHERICAL GRASP

LATERAL

Fig. 8.1. The various types of hand prehension. (Schlesinger.)

be designed to include several kinds of drive methods. A discussion of classifications according to grasping modalities follows later. The drive system is described in next section.

In general, grippers can be classified according to type of grasping modality as follows:

1. Mechanical finger.
2. Special tool.
3. Universal finger.

Modality 1 corresponds to grippers with fingers designed for a special purpose. This category includes less versatile and less dexterous finger grippers with fewer numbers of joints compared with modality 3. However, they economize in the device cost.

This type of finger can be subject to finer classifications, for example, the number of fingers, typically

TABLE 8.1.   APPROXIMATE RELATIONS BETWEEN THE NUMBERS OF FINGERS AND JOINTS AND THE FUNCTIONS OF GRIPPER.[a]

Type of Finger	Functions		
	Grasping	Shape Accommodation	Manipulation
2fG-Rf[b]	O	X	X
2fG-Af[c]	O	O	X
3fG-Rf	O	O	X
3fG-Af	O	O	O
5fG-Rf	O	O	X
5fG-Af	O	O	O

[a] O: Some of the grippers can involve the function.
X: None of the grippers can involve the function.

[b] nfG-Rf ($n = 2$, 3, and 5) means n-finger-type gripper with rigid fingers.

[c] nfG-Af ($N = 2$, 3, and 5) means n-finger-type gripper with articulate fingers.

two-, three-, and five-finger types. For industrial applications, the two-finger gripper is the most popular. The three- and five-finger grippers, with some exceptions, are customarily used for prosthetic hands for amputees.

Another classification is the number of grippers, single or multiple, mounted on the wrist of robot arm. Multigripper systems (Figure 8.2) enable effective simultaneous execution of more than two different jobs. Design methods for each individual gripper in a multigripper system are subject to those of single grippers.

Classification of the mode of grabbing results in external and internal systems. The external gripper (Figure 8.3) is used to grasp the exterior surface of objects with closed fingers, whereas the internal gripper (Figure 8.4) grips the internal surface of objects with open fingers. There are two finger-movement classifications: translational finger grippers and swinging finger grippers. The translational gripper can move its own fingers, keeping them parallel. The swinging gripper involves a swinging motion of fingers.

Another classification may be possible according to the number of degrees of freedom included by gripper structures. Typical mechanical grippers belong to the classification of 1 DF. A few grippers can be found with more than 2 DF.

Modality 2 is a special-purpose device for holding objects. Vacuum cups and electromagnets are typical devices in this class. In some applications the objects to be handled may be too large or too thin for finger grippers to grasp them. Here this gripper has a great advantage over the others.

Modality 3 is comprised of multipurpose grippers of usually more than three fingers and/or more than one joint on each finger which provides the capability to perform a wide variety of grasping and manipulative assignments. Almost all grippers in this category are under development, as mentioned in Section 8.1. The following sections concentrate on finger grippers.

8.4. DRIVE SYSTEM FOR GRIPPERS

In typical robot systems there are three kinds of drive methods: electric, pneumatic, and hydraulic. Pneumatic drive can be found in gripper systems of almost all industrial robots. The main actuator systems in pneumatic drive are the cylinder and motor. They are usually connected to on-off solenoid valves which control their directions of movement by electric signal. For adjusting the speed of actuator motion, air flow regulation valves are needed. A compressor is used to supply air (maximum working pressure, 10 kg/cm²) to actuators through valves.

The pneumatic system has the merit of being less expensive than other methods, which is the main reason that many industrial robots use it. Another advantage of the pneumatic system relates to the low degree of stiffness of the air-drive system. This feature of the pneumatic system can be used effectively to achieve compliant grasping, which is necessary to one of the most important functions of grippers: to grasp objects with delicate surfaces carefully. On the other hand, the relatively limited stiffness of the system makes precise position control difficult. Air servo valves are being developed for this purpose but are not practical enough for widespread use.

The electric-drive system is also popular. There are typically two kinds of actuators, DC motors and step motors. In general, each motor requires appropriate reduction gear systems to provide proper output force or torque. Direct-drive torque motors (DDM) are commercially available[4] but are too expensive to be used in normal industrial application. There are few examples of robot grippers using DDM. In the electric system a servo power amplifier is also needed to provide a complete actuation system. Electric drive has a lot of merit for actuating robot articulation. First, a wide variety of

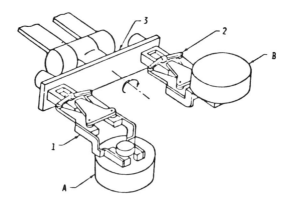

Fig. 8.2. Multigripper system.

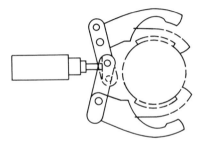

Fig. 8.3. External gripper.

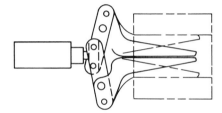

Fig. 8.4. Internal gripper.

products are commercially available. Second, constructing flexible signal-processing and control systems becomes very easy because they can be controlled by electric signals, and this enables the use of computer systems as control devices. Third, actuators in electric systems, especially those using DC motors, can be used for both force and position control. However, there are some drawbacks of the electric system as follows:

1. It is a little expensive compared with the pneumatic system.
2. The transient response is lower than those in pneumatic and hydraulic systems.
3. It is less stiff than hydraulic systems.
4. It can not be used in an explosive environment because of its spark and heating.

Hydraulic drives used in robot systems are electrohydraulic drive systems. They have almost the same system configuration as pneumatic systems, though their features are different from each other. A typical hydraulic drive system consists of actuators, control valves, and power units. There are three kinds of actuators in the system: piston cylinder, swing motor, and hydraulic motor. To achieve position control using electric signals electrohydraulic conversion devices are available. For this purpose electromagnetic or electrohydraulic servo valves are used. The former provides on-off motion control, and the latter is used to get continuous position control. Hydraulic drive gives accurate position control and load-invariant control because of the high degree of stiffness of the system. On the other hand, it makes force control difficult because high stiffness causes high pressure gain, which has a tendency to make the force control system unstable. Another claimed advantage of hydraulic systems is that the ratio of the output power per unit weight can be lower than in other systems if high pressure is supplied. Facts show this drive system can provide an effective way to construct a compact high-power system.

Outside of the foregoing three types of drive, there are a few other drive methods. One method uses a springlike elastic element. A spring is commonly used to guarantee automatic release action of grippers driven by pneumatic or hydraulic systems. Figure 8.5 shows an example of a spring-loaded

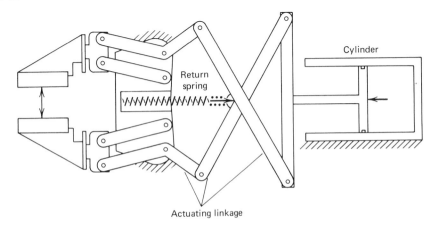

Fig. 8.5. Spring-loaded linkage gripper. (Courtesy of Dow Chemical Co., Ltd.)

linkage gripper using a pneumatic cylinder.[5] Gripping action is performed by means of one-directional pneumatic action, while the spring force is used for automatic release of the fingers. This method considerably simplifies the design of the pneumatic or hydraulic network and its associated control system.

The spring force can also be used for grasping action. In this case, the grasping force is obviously influenced by the spring force. To produce a strong grasping force, it is necessary to use a spring with a high degree of stiffness. This usually causes the undesirable requirement for high-power actuators for the release action of the fingers. Therefore the use of spring force for grasping action is limited in low-grasping-force grippers for handling small machine parts such as pins, nuts, and bolts.

The reason the spring force can be used for a one-directional motion of the pneumatic and the hydraulic actuator is that the piston can be moved easily by the force applied to the output axis (piston rod). The combination of a spring and electric motor is not viable because normal electric motors include a gear reduction system which makes it difficult to transmit the force inversely from the output axis.

Another interesting method uses electromagnets. The electromagnet actuator consists of a magnetic head constructed with a ferromagnetic core, conducting coil, and actuator rod made of ferrous materials. When the coil is activated, the magnetic head attracts the actuator rod, and the actuator displacement is locked at a specified position. When the coil is not activated, the actuator rod can be moved freely. This type of actuator is usually employed with a spring and produces two output control positions.

Figure 8.6 shows a gripper using the electromagnetic drive. The electromagnetic actuator (1) produces the linear motion to the left along the $L–L$ line. The motion is converted to grasping action through the cam (2). The releasing action is performed by the spring (3).

The actuator displacement that this kind of actuator can make is commonly limited to a small range because the force produced by the magnetic head decreases according to the increase of the actuator displacement. Therefore this drive method can be effectively used only for gripping small workpieces.

In the design of effective gripper systems, the selection of drive system is a very important problem. Selection depends on the kinds of jobs required of the robot. Briefly, if a gripper has some joints that need positional control, an electric or hydraulic system is a better choice. If not, a pneumatic system is better. For robots required to work in a combustible environment, for instance a spray-painting environment, pneumatic or hydraulic systems are suitable. If force-control function is needed at some joints, for example, to control grasping force, electric or pneumatic systems are recommended.

8.5. MECHANICAL GRIPPERS

Several kinds of gripper functions described in Section 8.3 can be realized using various mechanisms. From observation of the usable pair elements in gripping devices, as Chen found,[6] the following kinds are identified: (1) linkage, (2) gear-and-rack, (3) cam, (4) screw, (5) cable and pulley, and so on. The selection of these mechanisms is affected by the kind of actuators to be employed and the kind of grasping modality to be used. In the past, many gripper mechanisms have been proposed. However,

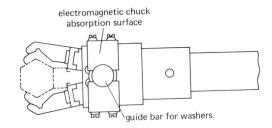

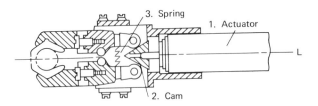

Fig. 8.6. Gripper using an electromagnetic drive. (Courtesy of Seiko-seiki Co., Ltd.)

fewer mechanisms have been put into practical use. The following sections explain the practical gripper mechanisms. A number of other possible gripper mechanisms can be found in Chen's list,[6] if needed.

8.5.1. Mechanical Grippers with Two Fingers

Swinging Gripper Mechanisms

This is the most popular mechanical gripper for industrial robots. It can be designed for limited shapes of an object, especially cylindrical workpieces. Figure 8.7 shows a typical example. Mechanisms to be adopted depend on the type of actuators used. If actuators are used that produce linear movement, like pneumatic or hydraulic piston cylinders, the device contains a pair of slider-crank mechanisms.

Figure 8.8 shows an example of a pair of slider-crank mechanisms commonly adopted by grippers using pneumatic or hydraulic piston cylinder drive. When the piston 1 is pushed by hydraulic or pneumatic pressure to the right, the elements in the cranks, 2 and 3, rotate counterclockwise with the fulcrum A1 and clockwise with the fulcrum A2, respectively, when γ is less than 180°. These rotations make the grasping action at the extended end of the crank elements 2 and 3. The releasing action can be obtained by moving the piston to the left. The motion of this mechanism obviously has a dwell position at $\gamma = 180°$. To obtain effective grasping action, $\gamma = 180°$ must be avoided. An angle γ ranging from 160 to 170° is commonly used.

Figure 8.9 shows another example of a mechanism for swinging grippers that use the piston cylinder, the swing-block mechanism. The sliding rod 1, actuated by a pneumatic or a hydraulic piston, transmits motion by way of the two symmetrically arranged swinging-block linkages 1–2–3–4 and 1–2–3′–4′ to grasp or release the object by means of the subsequent swinging motions of links 4 and 4′ at their pivots A1 and A2.

Figure 8.10 is a typical example of a gripper using a rotary actuator in which the actuator is placed at the cross point of the two fingers. Each finger is connected to the rotor and the housing of the actuator, respectively. The actuator movement directly produces grasping and releasing actions. This gripper mechanism includes a revolute pair, which can be schematically illustrated as shown in Figure 8.11a. For proper grasping and releasing action, another drive method can be adopted for this mechanism. Figure 8.11b shows the drive method that uses the motion of a cylinder piston instead of a rotary actuator. Figure 8.11c shows the piston movement converted to the grasping-releasing action through the use of a cam.

Figure 8.12a shows a cross-four-bar link mechanism with two fulcrums, A and B. This mechanism is sometimes used to make a finger-bending motion. Figure 8.12b shows a typical example of a finger constructed with the cross-four-bar link. There are two ways for activating this mechanism. First, a rotary actuator can be used at point A or B to rotate the element AD or BC. The actuator movement

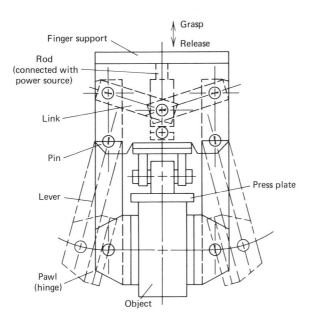

Fig. 8.7. An example of swing gripper. (Courtesy of Yasukawa Electric Mfg. Co., Ltd.)

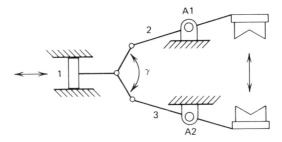

Fig. 8.8. Schematic of a pair of slider-crank mechanisms.

produces the rotation of the link element *CD*, which produces a bending motion of the finger. Second, a slider-crank mechanism activated by a cylinder piston can be used to rotate the element *AD* or *BC*. The lower illustration in Figure 8.12*b* depicts this type of drive. The finger-bending motion can be obtained in the same way as with rotary actuators. The use of a cross-four-bar link offers the capability of enclosing the object with the finger. This mechanism can be used with grippers of more than three fingers as well as in two-finger grippers.

Translational Gripper Mechanisms

Translational mechanisms are used widely in grippers of industrial robots. The mechanism is a little complex compared to the swinging type.

The simplest translational gripper uses the direct motion of the piston cylinder. Figure 8.13 shows this type of gripper using a hydraulic piston cylinder. As depicted in the figure, the finger motion corresponds to the piston movement without any connecting mechanisms between them. The drawback to this method is that the actuator size decides the gripper size. This can sometimes make it difficult to design the desired size of gripper. The method is suitable for the design of wide-opening translational grippers.

Figure 8.14 shows a translational gripper using a pneumatic or hydraulic piston cylinder, which includes a dual-rack gear mechanism and two pairs of the symmetrically arranged parallel-closing linkages. This is a widely used translational gripper mechanism. The pinion gears are connected to the elements *A* and *A'*, respectively. When a piston rod moves toward the left, the translation of the rack causes the two pinions to rotate clockwise and counterclockwise, respectively, and produces the release action, keeping each finger direction constant. The grasping action occurs when the piston rod moves to the right in the same way. There is another way to rotate the two pinions. Figure 8.15 shows the mechanism using a rotary actuator and gears in lieu of the piston cylinder and rack.

Figure 8.16 shows two examples of translational gripper mechanism using rotary actuators. Figure 8.16*a* consists of an actuator and rack-pinion mechanism. The advantage of this kind of gripper is

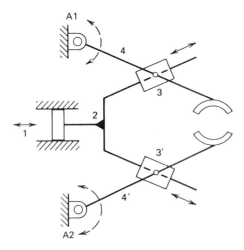

Fig. 8.9. Schematic of swing-block mechanism.

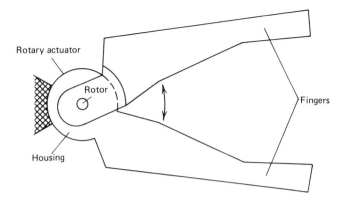

Fig. 8.10. Gripper using a rotary actuator.

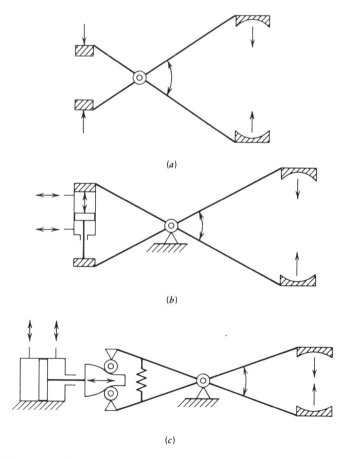

(a)

(b)

(c)

Fig. 8.11. (a) Schematic of mechanism that includes a revolute pair. (b) Gripper using a revolute pair and a piston cyliner. (c) Gripper using a cam and a piston cylinder.

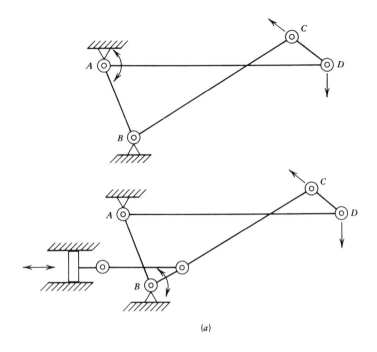

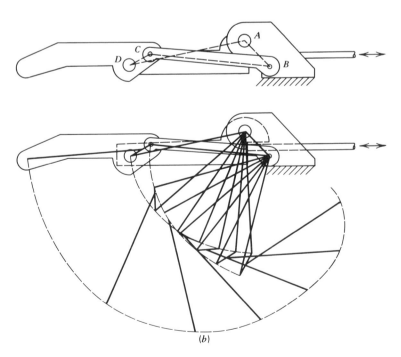

Fig. 8.12. (*a*) Schematic of cross-four-bar link mechanism. (*b*) A finger using cross-four-bar link mechanism.

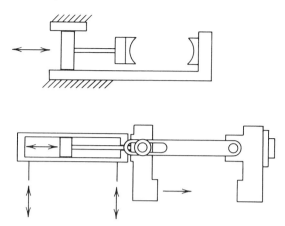

Fig. 8.13. Translational gripper using a cylinder piston.

that it can accommodate a wide range of dimensional variations. Figure 8.16*b* includes two sets of ball-screw mechanisms and an actuator. This type of gripper enables accurate control of finger positions.

Consideration of Finger Configuration

When using rigid fingers for mechanical grippers, the finger configuration must be contrived to accommodate the shape of the object to be handled. To grasp the object tightly, it is effective to make object-shaped cavities on the contact surface of the finger as shown in Figure 8.17*a*. A cavity is designed to conform to the periphery of the object of a specified shape. For example, if cylindrical workpieces are handled, a cylindrical cavity is made. The finger with this type of cavity has the merit of grasping single-size workpieces more tightly with wider contact surface. The establishment of wide contact surface is important to prevent the grasping force from localizing.

Structure will limit the capability of the gripper to accommodate change in the dimension of an object to be handled. Versatility of the gripper can be slightly improved by the use of a finger with multiple cavities for objects of differing size and shape. Figure 8.17*b* shows examples of fingers with multiple cavities.

In manufacturing, many tasks involve cylindrical workpieces. For handling cylindrical objects, the finger with a V-shaped cavity may be adopted instead of the object-shaped cavity. Each finger

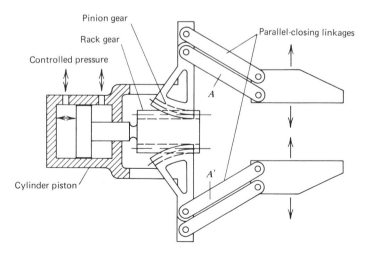

Fig. 8.14. Translational gripper including parallel-closing linkages driven by a cylinder piston and a dual-rack gear.

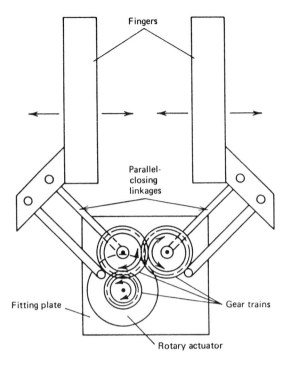

Fig. 8.15. Translational gripper including parallel-closing linkages driven by a rotary actuator and gears.

contacts the object at two spots on the contact surface of the cavity during the gripping operation. The two-spot contact applies a larger grasping force to a limited surface of the grasped object and may sometimes distort or scratch the object. However, there are many tasks where this problem is not significant, and the device has a great advantage over the gripper with object-shaped cavities. One advantage is that it can accommodate a wide range of diameter variations in the cylindrical workpiece, allowing the shape of the cavity to be designed independent of the dimensions of the cylindrical object. Another advantage is that it is easy to make, resulting in reduced machinery costs. Figure 8.18a gives a typical geometrical configuration of a grasping system for a cylindrical object using the gripper including two fingers, each with a V-shaped cavity. There is some relation between the configuration parameters of the V-shaped cavity and the diameter of possible cylindrical workpieces to be grasped. Suppose that parameters of the grasping system, γ, β, R, R', L, l, a, and b; symbols B, C, Q, B', C', Q', and O; and the coordinate system O–xy are defined as shown in the figure.

From the geometrical conditions, the cylindrical workpiece grasped and the gripper construction cannot intersect. This leads to the following inequality

$$x - R < \Delta s \tag{8.1}$$

where Δs is the width of the gripper element as shown in Figure 8.18. The distance between the center of cylindrical workpiece and origin O is x, which can be expressed as the following equations.

$$x = \sqrt{\left[L^2 + \left(\frac{R}{\sin \gamma} + a \right)^2 - 2L \left(\frac{R}{\sin \gamma} + a \right) \cos \beta \right] - l^2} \tag{8.2}$$

In the swinging-type gripper, β keeps a constant value during the grasping operation. If $\beta = 90°$, which is often used, Eq. (8.2) becomes

$$x = \sqrt{L^2 + \left(\frac{R}{\sin \gamma} + a \right)^2 - l^2} \tag{8.3}$$

Using the translational gripper in which each cavity block is kept parallel to every other, the following equation can be obtained:

$$\cos \beta = \frac{(R' - 1)}{L} \qquad \left(R' = \frac{R}{\sin \gamma} + a\right) \tag{8.4}$$

After substitution of Eq. (8.4) into Eq. (8.2), x can be expressed by the following:

$$x = \sqrt{L^2 - \left(\frac{R}{\sin \gamma} + a - l\right)^2} \tag{8.5}$$

For the object to contact each finger at the two spots on the surface of the cavity, D, γ, and b must satisfy the following inequality.

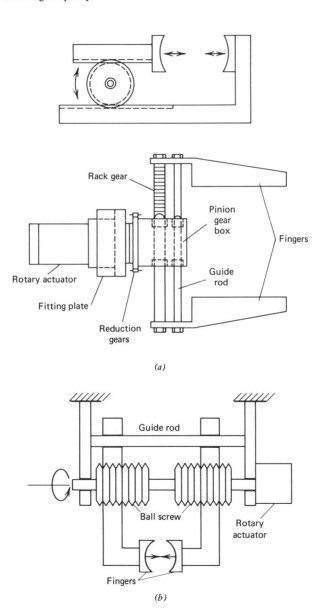

(a)

(b)

Fig. 8.16. (a) A translational gripper operated by a rotary actuator with rack-pinion mechanism and an example of this type of gripper. (Courtesy of Tokiko Co., Ltd.) (b) A translational gripper using a rotary actuator and ball-screw mechanism.

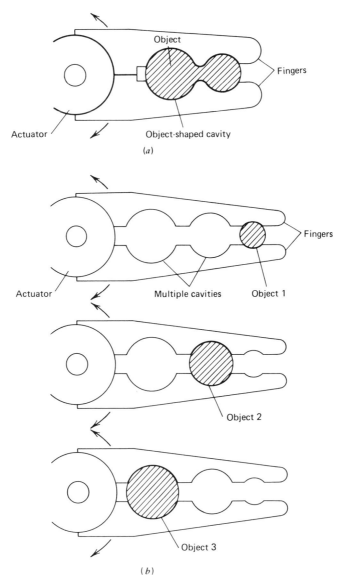

Fig. 8.17. (a) Finger with an object-shaped cavity. (b) Finger with multiple object-shaped cavities.

$$D < 2b \tan \gamma \qquad (8.6)$$

where D is the diameter of the cylindrical object and equals $2R$.

If the swinging gripper is assumed, another condition must be considered because the longitudinal direction of each finger varies with objects of differing size. Obviously, to hold an object safely in the gripper without slippage, the extended line QC must cross the extended line $Q'C'$ at point P in front of the gripper. The upper diameter limit of an object that can be grasped by the gripper with cavities of given structural parameters results in the following inequality:

$$D < 2 \sin \gamma \cdot \left[L \cdot \tan\left(\frac{\pi}{2} - \gamma\right) + \frac{l}{\tan(\pi/2 - \gamma)} \right] \qquad (8.7)$$

For the special case l equal to 0, which corresponds to Figure 8.10, the preceding inequality becomes

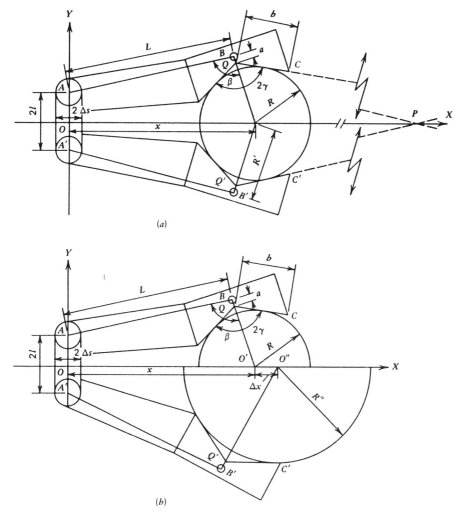

Fig. 8.18. (a) Geometrical configuration of grasping system for a cylindrical object using two-finger gripper with a V-shaped cavity. (b) The deviation of center position of grasped objects.

$$D < 2 \sin \gamma \cdot L \tan \left(\frac{\pi}{2} - \gamma \right) \tag{8.8}$$

If the translational gripper is used, the inequality (8.7) or (8.8) can be ignored because the attitudes of the finger and the V-shaped cavity are kept constant.

Equations (8.1), (8.6), (8.7), and (8.8) must be considered to find the maximum possible diameter of an object to be grasped by the gripper with specific cavities.

From Eqs. (8.2), (8.3), and (8.5), it can be recognized that the x coordinate of the center of the cylindrical object varies with the diameter. Figure 8.18b explains the deviation ($O'O'' = \Delta x$) of the center positions of two different-sized objects grasped by a gripper with $l = 0$. The deviation will be undesirable for some tasks, where the reduction must be considered.[7]

Figure 8.19 shows the case when using swinging-type gripper with $2\gamma = 140°$, the relation between the diameter of the grasped object D and the value Dev (see Figure 8.18b) where Dev corresponds to the deviation of the center position of the object, which can be calculated using Eq. (8.2). The dotted curves in the figure represent the limitations of diameter of the object determined by the inequality of Eq. (8.7). The figure reveals, obviously, that the gripper with longer fingers gives smaller deviation of the center position. If translational grippers are used, the deviation is smaller

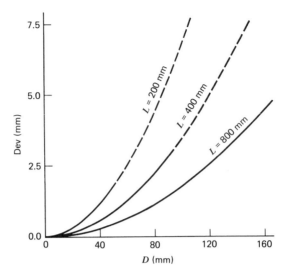

Fig. 8.19. The relation between the diameter of the grasped object D and the value Dev $= x - L$ ($\gamma = 140°$).

than that of swinging grippers when fingers of the same size are employed for each type of gripper. A well-designed translational gripper using a rack-pinion or ball-screw mechanism, as shown in Figure 8.16, can make the deviation almost zero. The gripper shown in Figure 8.13 also gives the same effect.

To provide the capability to completely conform to the periphery of objects of any shape, a soft-gripper mechanism has been proposed, which is shown in Figure 8.20.[8] The segmental mechanism is schematically illustrated in Figure 8.21. The adjacent links and pulleys are connected with a spindle and are free to rotate around it. This mechanism is manipulated by a pair of wires, each of which is driven by an electric motor with gear reduction and clutch. One wire is called a *grip wire*, which produces the gripping movement. The other is a *release wire*, which pulls antagonistically and produces the release movement from the gripping position. When the grip wire is pulled against the release wire, the finger makes a bending moment from the base segment. During this process of wire traction the disposition of each gripper's link is determined by the mechanical contact with an object. When the link i makes contact with an object and further movement is hindered, the next link, $(i + 1)$, begins to rotate toward the object until it makes contact with the object. This results in a finger motion conforming to the peripheral shape of the object. In this system it is reported that the proper selection of pulleys enables the finger to grasp the object with uniform grasping pressure.

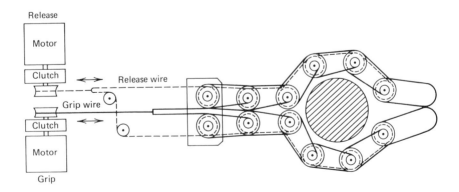

Fig. 8.20. The soft gripper. (Courtesy of Hirose et al., Reference 8.)

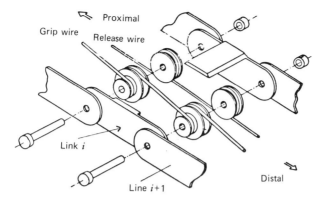

Proximal

Grip wire

Release wire

Link i

Line $i+1$

Distal

Fig. 8.21. Segmental mechanism of the soft gripper (Courtesy of Hirose et al., Reference 8.)

Calculation of Grasping Force or Torque

The maximum grasping force or grasping torque is as important a specification for gripper design as the geometrical configuration. The actuator output force or torque must be designed to satisfy the specification of the maximum force. How hard the robot must grasp the object depends on the weight of the object, the friction between the object and the fingers, how fast the robot is to move, and the relation between the direction of movement and the fingers' position on the object.

The worst case is when the lines of action of the gravity and the acceleration force to the grasped object are parallel to the contact surface of the fingers. Then friction alone has to hold the object. Therefore this situation is assumed in evaluating the maximum grasping force or grasping torque.

There are several ways to grasp objects. Figure 8.22 shows two examples of typical grasping modalities that require different consideration in the calculation of the maximum grasping force. Figure 8.22a illustrates grasping the cylindrical object on two spots using a swing gripper and keeping the center of gravity of the object inside the line between the two fingers. Figure 8.22b illustrates grasping the object with flat surface on two spots using a translational gripper and keeping the center of gravity of the object outside the line between the two fingers.

For Figure 8.22a, normal forces at two contact points, NA, and NB, are related to the grasping torque as the following, which can be derived from the sum of moments at hinge P.

$$NA = \frac{\tau}{\cos(\gamma - \theta)[a \tan \theta + l \cos \theta] [\tan(\gamma - \theta) + \tan(\theta + \gamma)]}$$

$$NB = \frac{\tau}{\cos(\gamma + \theta)[a \tan \theta + l \cos \theta] [\tan(\gamma - \theta) + \tan(\theta + \gamma)]}$$

(8.9)

where each parameter is defined in Figure 8.22. Under the grasping force action, friction forces, which enable the object to be held at the contact points of fingers, are expressed by μNA, μNB, where μ is a friction coefficient. The forces at the contact points caused by acceleration and gravity must be smaller than the friction forces. Assuming acceleration force caused by nG (n times the earth's gravity) applies to the center of gravity of the grasped object, forces FA and FB at the contact points A and B are calculated, respectively, from the sum of moments at the center of gravity of the object as follows. For point A,

$$FA = \frac{n\overline{W} \cos(\gamma + \theta)}{4 \cos \gamma \cos \theta}$$

(8.10)

For point B,

$$FB = \frac{n\overline{W} \cos(\gamma - \theta)}{4 \cos \gamma \cos \theta}$$

(8.11)

Grasping torque must be determined from the condition of grasping without slippage to satisfy the following inequalities simultaneously. For point A,

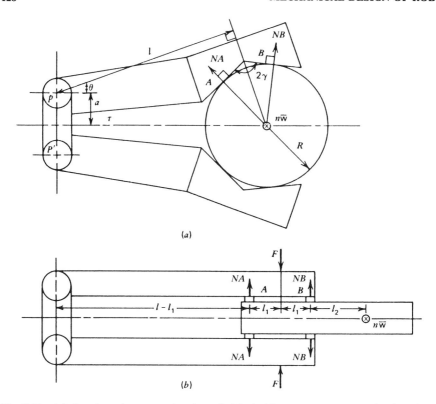

Fig. 8.22. (a) A swing gripper grasping the cylindrical object on two spots and keeping the center of gravity of the object inside the line between the two fingers. (b) A translational gripper grasping the object with a flat surface on two spots and keeping the center of gravity of the object outside the line between the two fingers. $\bigotimes$: Center of gravity of object; $\overline{W}$: weight of object.

$$FA < \mu NA \tag{8.12}$$

For point B,

$$FB < \mu NB \tag{8.13}$$

The grasping forces for Figure 8.22b can be evaluated in the same way. Friction and acceleration forces at contact points and the inequalities of grasping force are given as follows.

$$NA = \frac{F}{2}, \qquad NB = \frac{F}{2} \tag{8.14}$$

$$FA = \frac{l_2}{2l_1} n\overline{W}, \qquad FB = \left(1 + \frac{l_2}{2l_1}\right) n\overline{W} \tag{8.15}$$

$$FA < \mu NA, \qquad FB < \mu NA \tag{8.16}$$

After the maximum grasping force or torque has been determined, the force or torque that the actuator must generate can be considered. The calculation of those values requires the conversion of the actuator output force or torque to the grasping force or torque, which depends on the kind of actuator used and the kind of mechanism employed. Table 8.2 shows the relation between the actuator output force or torque and the grasping force in grippers that include the various kinds of mechanisms and actuators described above.

TABLE 8.2. RELATIONS BETWEEN THE ACTUATOR OUTPUT FORCE OR TORQUE AND THE GRASPING FORCE

Type of Gripper Mechanism and Configuration Parameters	Relations between Actuator Output and Grasping force
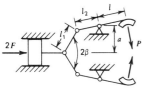	$lP = l_2 \left[\tan \beta \sqrt{1 - \left(\dfrac{l_1 \sin \beta - a}{l} \right)^2} - \dfrac{l_1 \sin \beta - a}{l} \right] \cdot F$
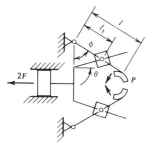	$lP = l_1 \dfrac{\cos (\phi - \theta)}{\cos \theta} \cdot F$
	$lP = \tau$
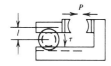	$P = \dfrac{\tau}{l}$
	$P = \dfrac{\tau}{r \tan \alpha}$
r: the pitch radius of the thread α: pitch angle of the thread (rectangular)	
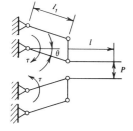	$P = \dfrac{\tau}{l_1 \cos \theta}$

8.5.2. Mechanical Hands with Three or Five Fingers

Three-finger Hand

The increase of the number of fingers and degrees of freedom will greatly aid the improvement of the versatility of grippers. However, this also complicates the design process. Although design methods for this type of gripper have still not been established there are a few examples that have been put into practical use.

The simplest example is a gripper with three fingers and one joint driven by an appropriate actuator system. The main reason for using the three-finger gripper is its capability of grasping the object in three spots, enabling both a tighter grip and the holding of a spherical object of differing size keeping the center of the object at a specified position. Three-point chuck mechanisms are typically used for this purpose. Figure 8.23 gives an example of this gripper. Each finger motion is performed using a ball-screw mechanism. Electric motor output is transmitted to screws attached to each finger through bevel gear trains which rotate the screws. When each screw is rotated clockwise or counterclockwise the translational motion of each finger will be produced, which results in the grasping-releasing action.

The configuration of the grasping-mode switching system using three fingers[9] is shown by Figure 8.24. This includes four electric motors and three fingers and can have four grasping modes, as shown in Figure 8.25, each of which can be achieved by the finger-turning mechanism. All fingers can be bent by motor-driven cross-four-bar link mechanisms, and each finger has one motor.

The finger-turning mechanism is called a *double-dwell mechanism,* which is shown in Figure 8.26. Gears that rotate the fingers are shown and double-headed arrows indicate the top edge of the finger's bending planes for each prehensile mode. This mechanism transfers the state of gripper progressively from three-jaw, to wrap, to spread, and to tip prehension. The gears for fingers 2 and 3 are connected to the motor-driven gear directly, whereas the gear for finger 1 is connected to the motor-driven gear through a coupler link. Rotating the motor-driven gear in three-jaw position, finger 1 rotates, passes through a dwell position, and then counterrotates to reach the wrap position. Similarly, finger 1 is rotated out of its spread position but is returned as the mechanism assumes tip prehension. Finger 2 is rotated counterclockwise 60° from its three-jaw position to the wrap position, then counterclockwise 120° into the spread position, then counterclockwise 150° into the tip position. Finger 3 rotates through identical angles but in a clockwise direction. A multiprehension system of this type is effective for picking up various-shaped objects.

Five-Finger Hand

A small number of five-finger hands have been developed in the world, with only a few for industrial use. Almost all of them are prosthetic hands for amputees. In the development of prosthetic arms, cosmetic aspects are more important to the mental state of the handicapped than functions. This

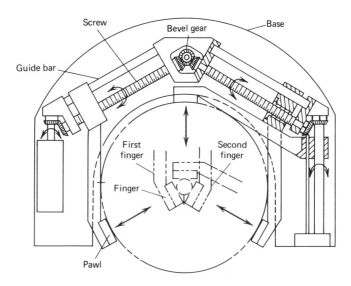

Fig. 8.23.   Gripper using three-point chuck mechanism. (Courtesy of Yamatake Honeywell Co., Ltd.)

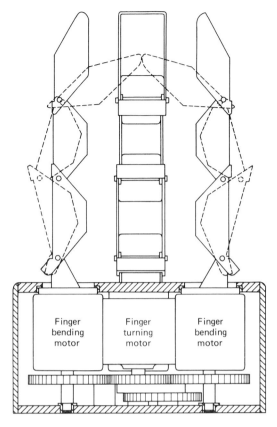

Fig. 8.24. Multiple prehension gripper system. (Courtesy of Skinner, Reference 9.)

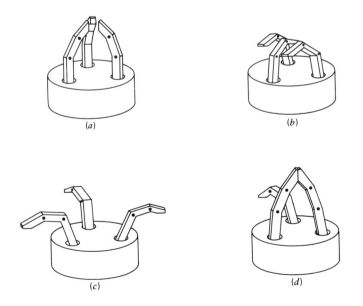

Fig. 8.25. Mechanical equivalent prehensile modes. (*a*) Three-jaw position; (*b*) wrap position; (*c*) spread position; (*d*) tip position. (Courtesy of Skinner, Reference 9.)

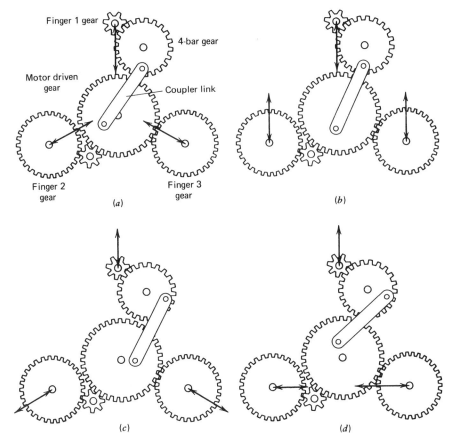

Fig. 8.26. The double-dwell finger-turning mechanism. (Courtesy of Skinner, Reference 9.)

requires anthropomorphism in the design of prosthetic hands. For industrial use, the function is more important than cosmetic aspects of the gripper. Therefore anthropomorphism is beyond consideration in the design of industrial grippers. This is why there are only a few five-finger industrial grippers. Nevertheless, five-finger grippers that have been developed so far for prosthetic use are described in the following part of this section because they include many mechanisms that will be effective in the design of industrial grippers.

A handicapped person must produce control signals for operation of a prosthetic arm. The number of independent control signals available determines how many degrees of freedom the prosthetic device can have. Typical models of a five-finger gripper for prostheses have only one degree of freedom. Each finger is connected to a motor by appropriate mechanisms.

Figure 8.27 shows an example, called the WIME Hand.[10] Each finger is constructed using a cross-four-bar link mechanism which gives the finger proper bending motion. One element of each of the five sets of cross-four-bar links includes a crank rod. All crank rods are connected to the spring-loaded plate (1), which is moved translationally by a electric motor drive-screw mechanism (2). When the motor rotates clockwise or counterclockwise, the plate (1) moves toward the left or the right, respectively, and activates the cross-four-bar link of each finger to bend the finger and to produce the grasping operation. To ensure that the gripper holds the object with the equilibrium of the forces between the fingers and the object, the arrangement of fingers must be carefully considered. In typical five-finger hands, the thumb faces the other four fingers and is placed equidistant from the index finger and middle finger so the tips of the fingers can meet at a point when each finger is bent (see Figure 8.28).

If each finger connects to the drive system rigidly, finger movements are decided by the motion of the drive system. The finger configuration can not accommodate the shape change of grasped objects. To remedy this problem, the motor output can be transmitted to each finger through flexible elements.

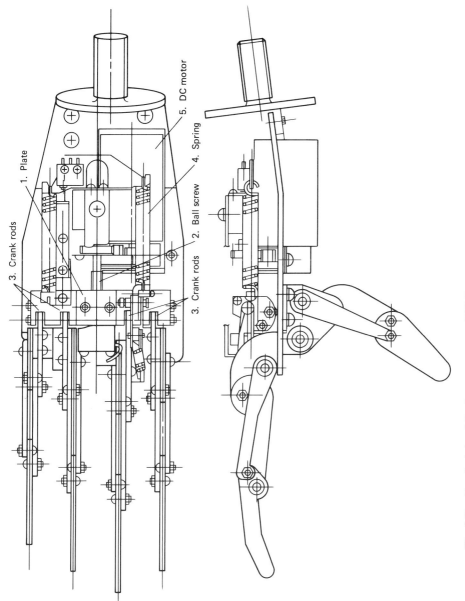

Fig. 8.27. Prosthetic hand for forearm amputee, WIME Hand (Courtesy of Imasen Engineering Co., Ltd.)

1. Plate
2. Ball screw
3. Crank rods
4. Spring
5. DC motor

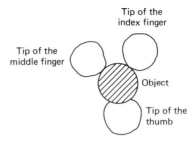

Fig. 8.28. Three-point pinch.

Figure 8.29 shows examples of this type of gripper.[11] There are two pairs of fingers: the index and the middle fingers, and the ring and the little fingers. Each pair is connected to lever 1 through a pivot. Also, lever 1 is connected to lever 2 through lever 3 at joints *A* and *B*. Lever 2 has a fulcrum point at *D* and is supported by the spring *S*. The grasping operation is executed by pulling lever 2 at the end *C*. Assuming that the index and middle fingers touch the object, movements of these fingers will stop, but the third and little fingers can still continue to move till those fingers touch the object because lever 2 can rotate at joint *A*. This causes the fingers to accommodate the shape of the object.

It is possible to move each finger independently, if the cross-four-bar link for each finger is driven by a different motor. Usually this type of gripper requires small motors to be installed in the finger.

8.6. UNIVERSAL GRIPPERS

Universal grippers with many degrees of freedom like the human hand have been researched by several investigators. Increasing the degrees of freedom causes several problems. One difficult problem is how to install in the gripper the actuators necessary to activate all degrees of freedom. This requires miniature actuators that can produce enough power to drive the gripper joint. Commercially available actuators are too large to attach at each joint of the finger. The use of SME (shape memory effect) actuators is one solution, but it is not practical at present. The most frequent solution to this problem is to use cable and pulley mechanisms that enable a motor to be placed at an appropriate position away from the joint.

There are two examples of universal grippers. Each uses cable pulley mechanisms and DC motor drive. Figure 8.30 shows an example of the gripper, which includes three fingers: a thumb, an index finger, and a middle finger.[12] Usually three fingers provide enough functions for universal grippers. Each finger contains two or three segments made of 17-mm brass rods bored to be cylindrical. The tip of each segment is truncated at a slope of 30° so that the finger can be bent at the maximum

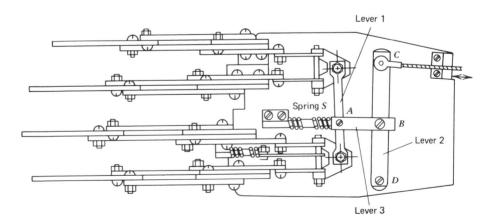

Fig. 8.29. Prosthetic hand equipped with accommodation mechanism. (Courtesy of Mechanical Engineering Lab., MITI, Japan.)

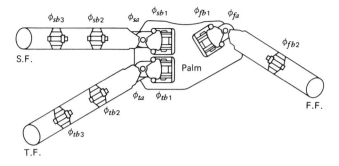

Fig. 8.30. Versatile finger system (Courtesy of Okada, Reference 12.)

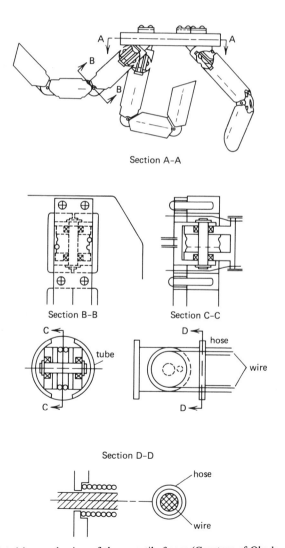

Section A-A

Section B-B   Section C-C

Section D-D

Fig. 8.31. Joint drive mechanism of the versatile finger (Courtesy of Okada, Reference 12.)

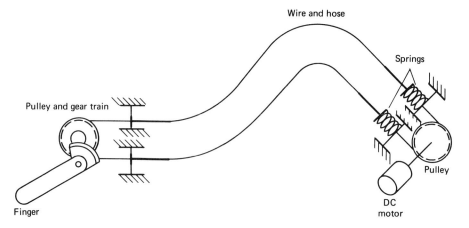

Fig. 8.32. Wire-pulley drive system with wire-guiding hoses supported by springs. (Courtesy of Sugano et al., Reference 13.)

angle of 45° at each joint—not only inward but also outward. This makes the workspace of the finger more extensive than that of the human finger.

The thumb has three joints. Each of the index and the middle fingers has four joints. Each joint includes 1 DF, which is driven using a wire-pulley mechanism and electric motors. A pulley is placed at each joint, around which two wires are wound after an end of each wire is fixed on the pulley. The wire is guided through coillike hoses so that it cannot interfere with the complicated finger motion. Using coillike hoses is effective in protecting the wire and also in making it possible to eliminate relaying points for guiding wire. To make the motions of the fingers flexible and to make the gripper system more compact, the wires and hoses are installed through the finger tubes. The section drawing of the gripper system in Figure 8.31 explains the joint drive mechanisms. Motors for driving respective joints are located together within a trunk separated from the gripper system.

Fig. 8.33. Five-finger hand with 14 DF. (Courtesy of Sugano et al., Reference 13.)

Using coillike hoses reduces the complexity of the wire-guidance mechanism, but it raises the problem of friction between hoses and wires. To resolve this problem it is effective to support the hoses with flexible elements like springs.[13] Figure 8.32 shows the construction of a wire-pulley drive system with wire-guiding hoses supported by springs. If a hose whose ends are rigidly fixed is bent, the hose will be extended. This reduces the cross-section area of the hose, which increases the friction. In the system of Figure 8.32, the spring, instead of the hose, will be extended when the hose is bent. Therefore the cross-section area of the hose will not be affected by the change of hose configuration, which prevents friction from increasing. Figure 8.33 shows the five-finger gripper system using the drive mechanism for each joint shown in Figure 8.32. The system has 14 total DF independently capable of being driven by electric motors (DC). All fingers except the thumb have 3 DF, and the thumb has 2 DF. Each joint axis arrangement is designed so that the tip of each finger can move to specified locations in the finger workspace.

The system in Figure 8.30 is reported to perform some manipulative tasks, screwing a bolt, bar- and sphere-turning tasks through the use of the playback control method, whereas the system of Figure 8.33 is capable of playing the piano under preprogrammed control. To achieve stable prehension using universal grippers, each joint for the finger must be moved cooperatively. For this purpose a stable prehension control algorithm has been proposed based on potential method.[14]

REFERENCES

1. Schlesinger, G., Der Mechanische Aufbau der kunstlichen Glieder, *Ersatzglieder und Arbeitshilfen,* Part II, Springer, Berlin, Germany, 1919.

2. Crossley, F. R. E. and Umholts, F. G., Design for a Three-Fingered Hand, in Heer, E., Ed., *Robot and Manipulator Systems,* Pergamon Press, Oxford, 1977, pp. 85–93.

3. Morecki, A., Ekiel, J., and Fidelus, K., Some Problems of Controlling a Live Upper Extremity and Bioprosthesis by Myopotential External Control of Human Extremities, *Proceedings of the Second Symposium on External Control of Human Extremities,* Belgrade, Yugoslavia, 1967.

4. Asada, H. and Kanade, T., Design of Direct-Drive Mechanical Arms, Robotics Institute, Carnegie-Mellon University, Pittsburgh, Pennsylvania, 1981.

5. Sheldon, O. L., Robots and Remote Handling Methods for Radioactive Materials, *Proceedings of the Second International Symposium on Industrial Robots,* Chicago, Illinois, pp. 235–256.

6. Chen, F. Y., *Gripping Mechanisms for Industrial Robots, Mechanism and Machine Theory,* Vol. 17, No. 5, Pergamon Press, Oxford, 1982, pp. 299–311.

7. Osaki, S. and Kuroiwa, Y., Machine Hand for Bar Works, *Journal of Mechanical Engineering Laboratory,* Vol. 23, No. 4, Mechanical Engineering Laboratory, MITI, Tsukuba, Japan, 1969 (in Japanese).

8. Hirose, S. and Umetani, Y., The Development of a Soft Gripper for the Versatile Robot Hand, *Proceedings of the Seventh International Symposium on Industrial Robots,* Tokyo, Japan, 1977, pp. 353–360.

9. Skinner, F., Design of a Multiple Prehension Manipulator System, ASME Paper, 74-det-25, 1974.

10. Kato, I., Yamakawa, S., Ichikawa, K., and Sano, M., Multi-functional Myoelectric Hand Prosthesis with Pressure Sensory Feedback System—Waseda Hand 4P, *Proceedings of the Third International Symposium on External Control of Human Extremities,* Belgrade, Yugoslavia, 1969.

11. Maeda, Y., Fujikawa, A., Tanie, K., Abe, M., Ohno, T., Tani, K., Honda, F., Inanaga, T., Yamanaka, T., and Kato, I., A Hydraulically Powered Prosthetic Arm with Seven Degrees of Freedom (Prototype-I), *Bulletin of Mechanical Engineering Laboratory,* No. 27, Tsukuba, Japan, 1977.

12. T. Okada, On a Versatile Finger System, *Proceedings of the 4th International Symposium on Industrial Robots,* Tokyo, Japan, 1977, pp. 345–352.

13. Sugano, S., Nakagawa, J., Tanaka, Y., and Kato, I., The Keyboard Playing by an Anthropomorphic Robot, *Preprints of Fifth CISM-IFTomm Symposium on Theory and Practice of Robots and Manipulators,* Udine, Italy, 1984, pp. 113–123.

14. Hanafusa, H., and Asada, H., Stable Prehension by a Robot Hand with Elastic Fingers, in Brady, M. et al., Eds., *Robot Motion,* MIT Press, Cambridge, Massachusetts, 1983, pp. 337–359.

BIBLIOGRAPHY

Engelberger, J. F., *Robotics in Practice,* Kogan Page Ltd., London, 1980.

Kato, I., Ed., *Mechanical Hand Illustrated,* Survey Japan, Tokyo, Japan, 1982.

Timoshenko, S. and Young, D. H., *Engineering Mechanics,* McGraw-Hill, New York, 1956.

CHAPTER 9
TELEOPERATOR ARM DESIGN

WAYNE J. BOOK

Georgia Institute of Technology
Atlanta, Georgia

A teleoperator is a manipulator that requires the command or supervision of a human operator. The manipulator is remote from the operator, as is implied by the name. The manipulator arm for a teleoperator has many design problems in common with the arm for an autonomous robot. Unlike the robot, however, the teleoperator has a human involved with each execution of the task. As a consequence, the human interface of the teleoperator is more critical than for most autonomous robots. The operator can exercise his judgement and skill in completing the task, even in the face of unforeseen circumstances. The distinction between the robot and the teleoperator is blurred when the operator only supervises the operation of the teleoperator or when a robot is being lead through a motion by its human programmer.

Industrial applications of teleoperators are numerous and typically involve work conditions inappropriate for the human. The environment may be hazardous or unpleasant, or the forces and reach may be greater than the human can directly provide. If the task is predictable and repetitious, an autonomous robot is appropriate. If the judgment and skill of a human are needed, or if the task is one of a kind, use of a teleoperator should be considered. Examples include the nuclear and munitions industries, foundries, and resource exploration and extraction. Teleoperator technology applies to cranes, backhoes, and other material-handling equipment. As the industrialization of space becomes a reality, teleoperators such as the remote manipulator system on the space shuttle and smaller arms will be essential. Figure 9.1 shows a range of applications. As autonomous robots are applied to tasks with smaller batch size, thus requiring more frequent reprogramming, the behavior of these systems during teaching will be more critical. Programming of the robot raises many of the same issues as teleoperation.

The intent of this chapter is to present those issues and alternatives in the design of mechanical arms that are unique to their use as teleoperators. The material is presented in a general way to apply to the diverse range in configuration, size, type of human interface, and purpose found in teleoperators. Design trade-offs in teleoperator arm design can be made if performance of the arm can be predicted as a function of the design parameters and the task to be performed. The difficulty in predicting the performance of teleoperators is due to the variability of that unique and essential component of the system: the human operator. As a consequence of this rather poorly understood component, one should not expect to predict performance of teleoperators with the certainty possible for autonomous robots that have only mechanical and electrical components. It is important, however, to struggle with the often incomplete results available at this time to build a qualitative and quantitative model of the effects of design decisions on the effectiveness of the teleoperator man-machine system. The designer should also have reference to a suitable handbook on human factors, as such information is not duplicated here. One example of such is the *Human Engineering Guide to Equipment Design.*[1]

9.1. TELEOPERATOR SUBSYSTEMS AND TERMINOLOGY

The terminology used in this chapter is introduced in this section and in Figure 9.2. A teleoperator system consists of a *remote unit* which carries out the remote manipulation, a *control unit* for input of the operator's commands, and a *communications channel* for linking the control and remote unit.

The remote unit generally consists of the *manipulator arm* with an *end effector* for grasping or otherwise engaging the workpiece and special tools. Each articulation or *joint* of the manipulator provides a *degree of freedom* (DF) to the manipulator. Commonly used joints are *rotational* or hinge joints providing one axis of rotation and *prismatic* or sliding joints with one direction of translation. The motion of the joints determines the motion of the end effector. A minimum of 6 DF are necessary for the end effector to be arbitrarily positioned in the workspace with arbitrary orientation. Usually

138

one *joint actuator* is provided per joint. Common actuator types are electromagnetic, hydraulic, and pneumatic. The motion of the joint actuator determines the position of the joint and is often *servo controlled* based on the error between the desired position or velocity and the measured position or velocity of the joint. Servo control is also called *feedback* or *closed-loop control*. *Open-loop control* requires no measurement of the actual position for feedback to the controller. Electric stepping motors and hydraulic actuators can provide holding torque without feedback and can be controlled open loop.

The control unit must contain some way for the operator to input the desired activities of the remote unit. For symbolic input this may be a button box or keyboard. For analog input this may be a *hand controller, joy stick,* or *master* unit kinematically equivalent to the manipulator or *slave* unit. The control unit must also provide some way to transform the operator inputs to commands compatible for transmission to the remote unit and a transmission interface.

Fig. 9.1. Examples of teleoperator systems. (*a*) A Manmate 1600 teleoperator in a forging operation. The Manmate is an entirely hydraulic master–slave arm that scales up the operator's motions. (Courtesy of Canadian General Electric.) (*b*) A Diver Equivalent Manipulator System (DEMS), unilateral electric master. (Courtesy of Canadian General Electric.) (*c*) The DEMS teleoperator system hydraulic slave. (Courtesy of Canadian General Electric.) (*d*) The Model M2 servo manipulator by Sargent Industries' Central Research Laboratories (CRL) provides two master and two slave arms (force reflecting) and a remotely positionable camera. It is shown here attached to a transporter for additional coverage. (Courtesy of Oak Ridge National Laboratory.) (*e*) The remote manipulator system (RMS) on board the space shuttle is shown from the operator's viewpoint. It is designed to remove up to 29,500 kg (65,000 lb) from the cargo bay in zero gravity. (Courtesy of the National Aeronautics and Space Administration.) (*f*) The RMS uses a hand controller shown in the upper left in various selectible modes. (Courtesy of the National Aeronautics and Space Administration.)

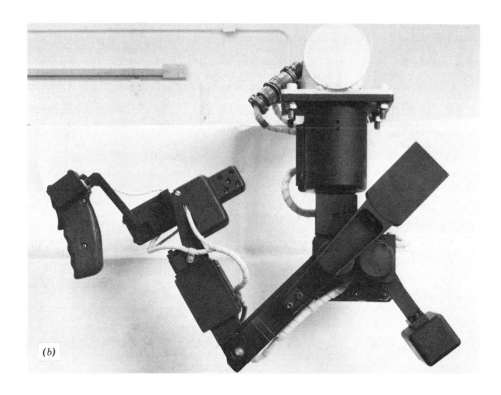

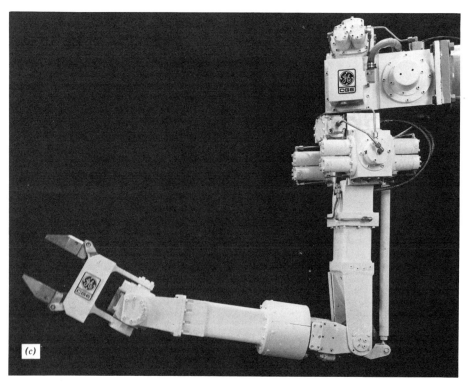

Fig. 9.1. (*Continued*)

Fig. 9.1. (*Continued*)

(e)

(f)

Fig. 9.1. (*Continued*)

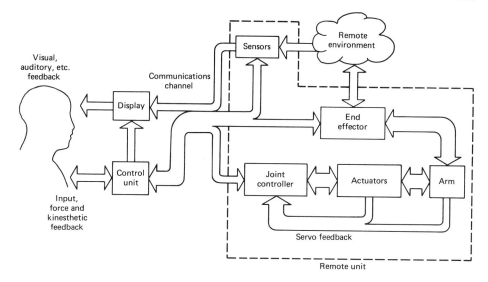

Fig. 9.2. Teleoperator terminology displayed in a conceptual block diagram.

The communications channel may consist of individual wires carrying analog signals to command each joint, or the signals may be digitized and multiplexed for efficient use of the channel capacity. The physical medium of the communications channel could be a coaxial cable, a radio link, or a sonar link, depending on the environment through which one must communicate.

9.2. MAJOR CATEGORIES OF TELEOPERATORS

One way to categorize teleoperators is by their source of power. The earliest teleoperators were powered by the operator through a direct mechanical connection, either metal cables, tapes, or rods. The separation distance of the remote unit from the operator was limited, as were the speeds and forces of operation. The operator directly supplied and controlled the power to the various degrees of freedom. An advantage of the direct drive was the direct-force feedback to the operator.

Externally powered teleoperators opened up a new range of opportunities and problems. The operator now inputs signals to a control system that modulates the power to the remote unit. The power the operator must provide is greatly reduced, but his or her effectiveness may be diminished unless the forces experienced by the remote unit can be displayed to the operator in a natural manner.

Another means of categorizing teleoperators is by the human interface, arranged, as follows, roughly in order of increasing sophistication:

1. *On-off control* permits joint actuators to be turned on or off in each direction at fixed velocity. This is the most primitive interface available and the simplest to implement. Simultaneous joint motion may be possible, but true joint coordination is not.

2. *Joint-rate control* requires the operator to specify the velocity of each separate joint, and thus mentally transforms the coordinates of the task into arm-joint coordinates. This is usually not an easy task, especially for the novice operator. The difficulty varies with the joint arrangement and the path of motion. Three orthogonal prismatic joints are easier to move in straight-line motion than three rotational joints capable of the same motion. Many industrial devices (backhoes, cranes, forklift trucks) use joint-rate control because of the simple hardware required to implement it.

3. *Master-slave control* allows the operator to specify the end position of the slave (remote) end effector by specifying the position of a master unit. Commands are resolved into the separate joint actuators either by the kinematic similarity of the master and slave units or mathematically by a control unit performing a transformation of coordinates. A scaling in size from the master to the slave unit may multiply the reach of the operator. For extreme scaling, the limited position resolution of the operator may render this interface at a disadvantage.

4. *Master-slave control with force reflection* incorporates the features of simple master-slave control and in addition provides to the operator resistance to motions of the master unit that correspond to the resistance experienced by the slave unit. Sometimes the master and slave unit are identical

except for the end effector. In other cases the master unit experiences reduced resistance to amplify the operator's strength and reduce his or her fatigue. This interface is also called bilateral master-slave control.

5. *Resolved motion rate control* allows the operator to specify the velocity of the end effector. The commands are resolved into the remote axes mathematically. The interface for the human may be a button box or joy stick. A button box still requires one operator action per coordinate axis, but the coordinates of the input can be made suitable for the human and the task being performed. A joystick allows one operator action to affect several degrees of freedom in a natural way. (As many as six have been implemented.) Another approach to controlling 6 DF is to use two joy sticks or hand controllers, one for translation and one for rotation.

6. *Supervisory control* takes many forms, but in general allows the operator to specify some of the desired motion symbolically instead of by analog. The computer interpreting the symbolic commands then issues the joint commands. This mode of operator interface results in a hybrid between teleoperator and robot. It is discussed in more detail in other chapters of this handbook.

9.3. THE DESIGN CONTEXT

The design of a teleoperator ultimately involves the specification of components and parameters. These components and parameters result in characteristics of the manipulator behavior. Knowing the characteristics of the teleoperator and the characteristics of the task one can determine the relative performance of that teleoperator by some performance index. Optimization of a design requires that the penalties associated with the cost and reliability of the components be considered as well as their performance. This chapter considers only the relationship between performance and characteristics unique to teleoperators. For a wide range of opinions on performance evaluation the reader is referred to the report of a workshop on the subject sponsored by the National Bureau of Standards.[2]

9.3.1. Performance Measures for Teleoperators

The most relevant quantifiable measures of performance for teleoperators are based on task-completion time. Measures that are considered here are the task total time, time-effectiveness ratio (time relative to the unencumbered hand), and unit-task time (time for elemental task components). Operator fatigue, success ratio, and satisfaction are hard to quantify but nonetheless important. Quantifying these performance measures requires that the task or a range of tasks be specified. They are task-dependent measures to some extent. The most relevant tasks to be specified are the tasks for which the teleoperator will be used. Unfortunately, the tasks are not often known in advance with great certainty owing to the general-purpose nature of teleoperators. The performance measures have been combined in various ways in an attempt to explain experimental results better.

Information-Based Performance Measures

One successful measure is the information transmission rate achieved by the teleoperator. This is not totally task independent but has been correlated with simple characterizations of the task. The information transmitted by the operator is equal to the reduction in uncertainty in the relative position of the end effector and the target, usually measured in bits. The time required to transmit the information determines the information transfer rate. The experimental determination of these correlations is based on measuring the task completion time, calculating an index of difficulty, and then normalizing the result. The index of difficulty I_d proposed by Fitts[3] for use in direct manual positioning tasks (unencumbered hand) is:

$$I_d = \log_2 \frac{2A}{B}$$

where $A =$ the distance between targets in a repetitive motion
$B =$ the width of the target.

This index of difficulty and its variations have been applied to teleoperators by Book and Hannema[4] for a simple manipulator of programmable dynamics and by McGovern[5] for more complex manipulators with fixed dynamics. Hill[6] combined this measure with the unit-task concept, described later, to predict task times. He and his co-workers document claims that only the fine-motion or final-positioning phase is governed by information transmission. The gross-motion phase of manipulation is governed by arm dynamics, both human and manipulator.

Time-Effectiveness Ratio

One popular and easily understood measure of performance is the task time multiplier or time-effectiveness ratio. When multiplied by the task time for the unencumbered hand it yields the task time for the teleoperator. This is a useful rough estimate of performance, but varies significantly with the task type. There is some indication that it is approximately constant for similar manipulators and for narrow variation of the task difficulty.

Unit-Task Times

Perhaps the most practical method for predicting the performance of teleoperators at a specific task is based on completion times for component subtasks or unit tasks. The time required for units such as "move," "grasp," "apply pressure," and "release," some of which have parameters, can be approximately predicted for a given existing manipulator. Unfortunately, they have not been related to the manipulator characteristics or the design parameters in any methodical way.

9.3.2. Task Characterization

Since all tasks that a teleoperator might be called on to perform cannot be included in a test of performance of a manipulator, important features of tasks must be characterized to evaluate its performance. More important, for design where no teleoperator yet exists, a general, if approximate, characterization is needed on which to base predictions. The simple Fitts law type of characterization requires only the ratio of distance moved to tolerance of the target position, that is, A/B. More complex measures of the task consider the degrees of freedom constrained when the end effector is at the final position. Positioning within a given tolerance in one dimension allows the five remaining dimensions free, hence one degree of constraint. Inserting a sphere in a circular hole leaves the remaining four dimensions free, hence a two degrees of constraint task. A rectangular peg placed in a rectangular hole requires three positions, and two orientations be constrained, leaving only one degree of freedom. The tolerance in each degree of constraint can be used to specify the difficulty of the final positioning for insertions. The self-guiding feature of insertions with chamfers, for example, is difficult to account for and depends on manipulator characteristics such as compliance and force reflection.

As mentioned in the section on performance measures for teleoperators, a task can also be described by its component subtasks. The unit-task concept for industrial workers was developed in the 1880s by Taylor[7] and later refined by Gilbreth.[8] Hill[9] proposed a modified list of unit tasks for manipulators as shown in Table 9.1. The limited experimentation that has been done with this system indicates promise. For design optimization one would like unit-task times as a function of the manipulator characteristics. These are not available, but one can predict which characteristics will most affect the various units. These effects have been qualitatively or quantitatively reported by various researchers.

TABLE 9.1. UNIT TASK DESCRIPTIONS FOR MANIPULATORS

Unit	Description
MOVE (d)	Transport end effector a distance d mm. Enhanced by greater arm speed. Hill[15] documents a linear dependence on d.
TURN (a)	Rotate end effector about the long axis of the forearm a degrees. Enhanced by force feedback (50–100%) or passive compliance. Degraded by friction and backlash.
APPLY PRESSURE	Apply force to overcome resistance with negligible motion. Enhanced by force feedback. Degraded by friction.
GRASP	Close end effector to secure object. Enhanced by gripper force feedback and gripper speed.
RELEASE	Open end effector to release object. Enhanced by gripper force feedback and speed.
PRE-POSITION (t)	Align and orient object in end effector within a tolerance of t mm. Degraded by friction, backlash, and low bandwidth.
INSERT (t)	Engage two objects along a trajectory with tolerance t mm. Enhanced by force feedback. Degraded by friction, backlash, and low bandwidth. Nonlinear with tolerance.[a]
DISENGAGE (t)	Reverse INSERT. Greatly enhanced by force feedback at small t (factor of 2). Degraded by friction, backlash, and low bandwidth.
CRANK	Follow a constrained circular path, pivoting at the "elbow." Greatly enhanced by force feedback (50–100%) or passive compliance.
CONTACT	Insure contact with a surface, switch, etc. Enhanced by force feedback.

[a] See Reference 15.

9.3.3. Teleoperator Characteristics

The important characteristics of a teleoperator in predicting performance should be traceable to the design decisions that resulted in those characteristics. With the wide variety of teleoperators of interest it is more productive to work with abstract characteristics that will be relevant to most if not all designs. Relating these characteristics to parameters and components for teleoperators is similar for robots and other systems. Some of the characteristics can be presented quantitatively whereas others can only be discussed.

Characteristics considered by various researchers, designers, and authors include:

Reach and workspace shape and volume.

Space occupied by the remote unit and command unit.

Dexterity.

Degrees of freedom.

Velocity, acceleration, and force obtainable with and without payload.

Accuracy, repeatability, and resolution of position, force, and velocity.

Backlash, dead band, or free motion between the input and response.

Coulomb or dry friction and viscous friction of command and remote unit.

Bandwidth or frequency response of the remote unit to small amplitude inputs.

Time delay between issuance of a command and initiation of the resulting action by the remote unit.

Rigidity or compliance of the remote unit to externally applied forces.

Inertia of the remote unit and of the command unit.

Static loads that must be counteracted by the actuators of the remote unit or the operator.

The relative significance of each of these characteristics is subject to interpretation. Their effects on performance are discussed later with the design decisions commonly affecting the characteristic. Some of the effects were already presented in Table 9.1.

9.4. PERFORMANCE PREDICTIONS

Many researchers have predicted performance of an existing manipulator for hypothetical tasks to be remotely performed. For design purposes one must predict the performance of a manipulator that does not yet exist so that design decisions can be made to achieve specified or desired performance. The limited existing results are presented in the context of design, and qualitative discussion of other characteristics is given. Also, the evaluation of existing manipulators is summarized and referenced.

9.4.1. Task Time Based on Index of Difficulty for Varying Bandwidth, Backlash, and Coulomb Friction

Perhaps the most methodical results obtained applicable to design are attributed to Hannema and Book.[4] The results are based on experiments using a simple, 2-DF manipulator with programmable characteristics. The characteristics considered are among the most important for determining manipulator performance: arm servo bandwidth, backlash (lost motion), and coulomb or dry friction. The task considered was a simple repetitive positioning task that involved moving to and tapping within a tolerance band. The layout of the experiment is shown schematically in Figure 9.3. The relative task times for various values of task and manipulator parameters were obtained. The following caveats should be observed when applying the results:

Only positioning tasks were considered, and the task had only one degree of constraint since positioning anywhere within a linear band was permitted. Tasks involving the application of force are not addressed in these experiments.

Simple master-slave control was used, with no force reflecting or bilateral capabilities.

Combinations of manipulator characteristics were not considered. Variations were made in only one characteristic while holding all other characteristics constant. This allowed the effects of that characteristic to be isolated.

The Information Transmission Model

An extended information transmission model similar to the one proposed by Welford[10] as an extension of Fitts' model was used. It allows for different information transmission rates for the gross-movement or travel phase and the fine-movement or positioning phase reflecting different channel capacities.

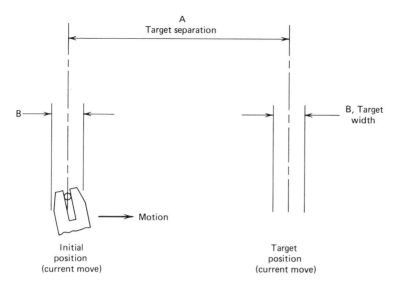

Fig. 9.3. The experiment used to determine the effect of several arm characteristics.

$$t = b_1 \log_2 \frac{2A'}{W_0} + b_2 \log_2 \frac{W_0}{B} + a$$

or

$$t = b_1 \log_2 A' - b_2 \log_2 B + t_0$$
$$t_0 = a + b_1 + (b_2 - b_1) \log_2 W_0$$

where t = task completion time
$A' = A + B/2$
A = distance between centers of the targets
B = width of the targets
a = a reaction delay time
b_1, b_2 = 1/(information transmission rate for the subtasks)
W_0 = the distance from the initial location to the point that separates the two parts of the task.

Welford suggested that two separate control processes were involved. A motor-control process governs gross motion and contributes to the first term in the foregoing expression for t. The fine motion contributes to the second term in t, and the final term is a constant related to the reaction time delay. See also Hill and McGovern[11] for application of the information approach to manipulators.

Correlation with Experiments

In the experiments a controller was used that decoupled the joint motions, providing each with a transfer function (in the absence of other simulated effects) of

$$\frac{\text{slave joint angle}}{\text{master joint angle}} = \frac{w^2}{(s + w)^2}$$

Backlash within both joints of the master arm was simulated. The maximum angle error due to backlash was 0, 10.5, 21, and 31°. Coulomb friction was simulated as if it occurred within each of the joints of the slave arm corresponding to torques of approximately 0, 25, 50, and 75% of the maximum torque of the joint actuator. Values of the travel distance used were 8, 16, 32, and 64 cm. Values of the target width used were 1, 2, and 4 cm.

A multiple regression analysis was performed on the experimental data obtained. A convenient plot of the regression results can be obtained by expressing t in the form

$$t = t_1 + t_2$$

where

$$t_1 = t_0 + (b_1 - b_2)\log_2 W + b_1 \log_2 \frac{A'}{W}$$

$$t_2 = b_2 \log_2 \frac{W}{B}$$

Choosing $W = 8$ cm as a reference distance is arbitrary but convenient since all values of B will then contribute a positive time to t. Figures 9.4 through 9.6 plot t_1 and t_2 separately versus the two task parameters $\log_2 A'$ and $\log_2 B$. The total time is represented as the vertical distance between two curves of the same value of manipulator characteristic. For example, a natural frequency of 14 rad/sec can be found on Figure 9.4a. Find on the two ordinate scales the values of interest for A' and B, for example, $A' = 32$ cm, $B = 2$ cm. The value of t_1 is read from the t_1 axis where the upper $\omega = 14$ line crosses $A' = 16$ cm. The value of t_2 is read from the t_2 axis where the lower $\omega = 14$ line crosses $B' = 2$ cm. The total predicted time for 30 repetitions is $t_1 + t_2$.

The manipulator design problem poses a question that is better answered by Figures 9.4b, 9.5b, and 9.6b. In these figures times t_1 and t_2 are plotted versus the manipulator characteristics, natural frequency, coulomb friction, or backlash. The task parameters are constant along the lines shown. This information can be coupled with a design strategy to optimize the design.

Other values could be chosen for W that would result in a different intercept but the same slope. The constant t_0 has been included in t_1. The data have been used to estimate the transition distance from gross to fine motion, W_{0e}. This value and the regression coefficients are shown in Table 9.2.

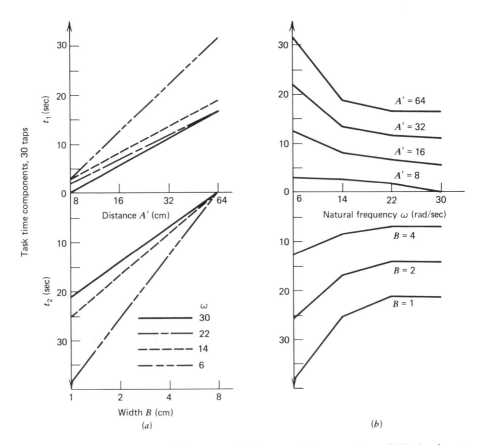

Fig. 9.4. Regression results: natural frequency. (a) Gross t_1 and fine t_2 motions. (b) Design format. (*Source:* Book and Hannema.[4])

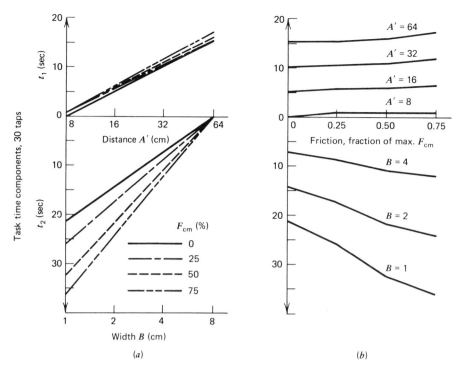

Fig. 9.5. Regression results: coulomb friction. (*a*) Gross t_1 and fine t_2 motions. (*b*) Design format. (*Source:* Book and Hannema.[4])

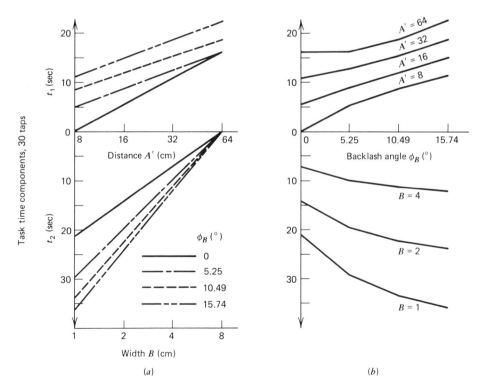

Fig. 9.6. Regression results: backlash. (*a*) Gross t_1 and fine t_2 motions. (*b*) Design format. (*Source:* Book and Hannema.[4])

TABLE 9.2.   REGRESSION COEFFICIENTS FOR MANIPULATOR CHARACTERISTICS

Natural Frequency (rad/sec)	Coulomb Friction (%)	Back-lash (°)	Regression Coefficient			Correlation Coefficient	Transfer Point
			b_1	b_2	t_0		
6	0	0	9.47	12.89	13.14	.987	14.35
14	0	0	5.39	8.48	11.81	.980	14.17
22	0	0	4.82	7.03	8.47	.987	14.23
30	0	0	5.45	7.09	4.96	.979	8.07
30	25	0	4.83	8.64	12.25	.976	9.30
30	50	0	5.03	10.83	18.16	.979	8.77
30	75	0	5.43	12.08	20.75	.983	8.67
30	0	5.25	3.72	9.90	23.67	.961	14.27
30	0	10.49	3.42	11.29	32.10	.968	16.90
30	0	15.74	3.66	12.09	36.52	.971	20.17

For changes in natural frequency (Figure 9.4) the information rates b_1 and b_2 remain in roughly the same proportion, indicating equal transmission rates in gross- and fine-motion phases, as would be assumed by Fitts' model. Increasing ω beyond 30 rad/sec brings little improvement based on the trends observed.

For coulomb friction variations (Figure 9.5) the gross-motion times and W_{0e} are almost constant, but the fine-motion times are noticeably influenced. If the simulated friction had reduced the maximum torque as real friction does, the gross-motion time would be expected to suffer as well.

Backlash (Figure 9.6) greatly increases fine-motion time and W_{0e} as if the visual feedback available during fine motion were needed. The amount of backlash simulated is large, greater than most teleoperators would contain. Hannema[12] discusses some aspects of this in more detail.

Examples of how the experimental information could be used in design follow but are not intended to be all-inclusive.

EXAMPLE 1:   How much increase in task time should be expected when a teleoperator end effector is changed to a heavier one, increasing the arm inertia and thus reducing the joint natural frequencies from 22 to 14 rad/sec? The design task involves moves of 32 cm to a target width of 2 cm.

The experimental data presented showed an increase from 26 to 30 sec for a similar repetitive task. The designer should expect about 15% increase in task time.

EXAMPLE 2:   If a backlash of 5° can be eliminated, what is the expected reduction in task time for a short gross motion (8 cm) with 1 cm tolerance?

The value of t_1 was reduced in the experiments from 5 to 0 sec, and t_2 was reduced from 30 to 21 sec, for a total percentage reduction on the order of 40%.

EXAMPLE 3:   What is the most likely payoff for a precision teleoperator design: improving joint natural frequency beyond 30 rad/sec or reducing a substantial coulomb friction?

Little if any improvement is expected from improving the bandwidth beyond 30 rad/sec, whereas lowering friction shows considerable sensitivity.

Results for Multiple Degrees of Constraint

By using the concept of degree of constraint it is possible to extend the index of difficulty to more complex tasks. This has been done for two manipulators with quite different characteristics and the results presented.[13] The total task time was assumed to be the sum of travel, positioning, and insertion times. The positioning and insertion times were related to the degree of constraint, with the index of difficulty being the sum of the indices for each constraint taken separately.

One of the useful concepts of this work was that the positioning and insertion task may involve one, two, or three phases. If the index of difficulty is low enough (less than 5 bits), it may be accomplished with only the open-loop travel phase, dependent on only gross-motion characteristics. If the index of difficulty is between roughly 5 and 10 bits, it will be completed within a fine-motion phase. Fine-motion characteristics influence both the time required and the upper limit on bits for completion in this phase. If the task tolerances are very high relative to the fine-motion capabilities of the manipulator, a third phase may be required. In this phase a random variation of end position occurs about the nominal target. The probability of completing the task in any given time interval is constant and depends on the standard deviation of the end-point error. Only the second phase is limited by information channel capacity.

The fine-motion phase begins with an uncertainty that depends on the accuracy of the open-loop move of the first phase. If the open-loop move results in a Gaussian distribution of final positions with standard deviation s, the reduction in uncertainty (information transmitted in the fine-motion phase) results from truncating the tails of that Gaussian distribution to lie within plus or minus $\Delta x/2$. This is displayed schematically in Figure 9.7. Precise tasks with low tolerance relative to the open-loop accuracy (represented by s) require a large reduction of uncertainty in the fine-motion phase. Coarse tasks require little reduction in uncertainty after the open-loop move and, in fact, probably are completed without a fine-motion phase. When $s/\Delta x > 0.5$ the information transmitted is well approximated as

$$H = 2.05 + \log_2 \frac{s}{\Delta x}$$

This approximation was used by Hill to obtain the index of difficulty for the fine-motion phase. He found that using this index of difficulty and normalizing with respect to the zero degree of freedom task time (essentially an open-loop move), three quite different manipulators were reduced to a single curve as shown in Figure 9.8. The manipulators used were the unencumbered hand, the Ames manipulator (an accurate, low backlash unit based on a hard space suit), and the Rancho arm (a much less precise unit).

9.4.2. Time-Effectiveness Ratio versus Manipulator and Task Type

Work by Vertut[14] complements the previously described work in the following ways:

Commercial, multidegree-of-freedom manipulators were used. As a consequence, however, the effects of separate manipulator characteristics could not be isolated.

Bilateral (force reflecting) and unilateral master-slave, joint rate control, resolved rate control, and on-off control were represented.

More complicated tasks were performed. As a consequence, however, the effects of the various task characteristics could not be isolated.

Operator fatigue as well as task time were considered.

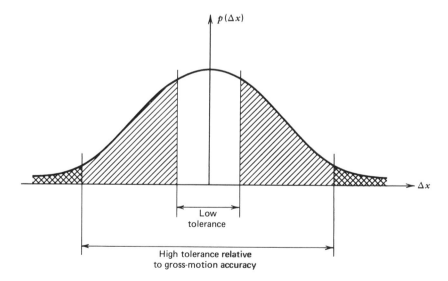

Fig. 9.7. The probability density function of position after an open-loop move determines the remaining uncertainty in position. Shaded areas show large reduction of uncertainty needed in time-motion phase when tolerance is low. Cross-hatched areas show low reduction of uncertainty needed in time-motion phase when tolerance is large.

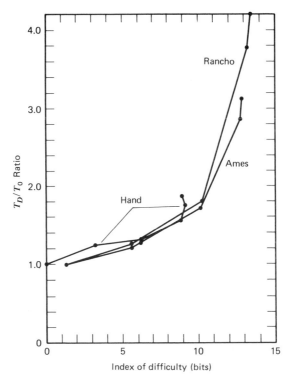

Fig. 9.8. Normalized time for multidegree of constraint assembly. T_0 = zero degree of constraint time, T_D = task time for constrained assembly. Index of difficulty based on an initial open-loop move. (*Source:* Hill.[6.13])

Vertut presented his results in terms of time-effectiveness ratio, that is, compared to the times of the unencumbered human hand. The general time-effectiveness ratio predictions are shown in Figure 9.9 for manipulators of six types. The types are generally the same as listed under categories previously with the abbreviations as follows:

Light-duty master slave—bilateral (LD)
Heavy-duty master slave—bilateral (HD)
Position control—unilateral master slave (PC)
Resolved motion rate control (RMRC)
Rate control (RC)
On-off control (OOC)

The task types are described as follows:

Pick and Place (PP). A positioning task plus a grasp and release task. It consists of simply picking up an object and placing it in a new, specified location.
Simple Assembly (SA). A removal and insertion task. A simple insertion of a peg in a hole is an example.
Normal Assembly (NA). Involves insertion and turning, for example.

Figure 9.9 shows the significance of the human interface in determining the overall task time and the relative difficulty of the various task types. Time-effectiveness ratios vary by a factor of more than 100 for the different types of interfaces. As is always the case with comparison of complete systems, one cannot attribute with certainty all the effects to one characteristic.

McGovern[5] also considered the time-effectiveness ratio as a means for performance prediction. His results are in rough agreement with Figure 9.9. He also breaks the time-effectiveness ratio into

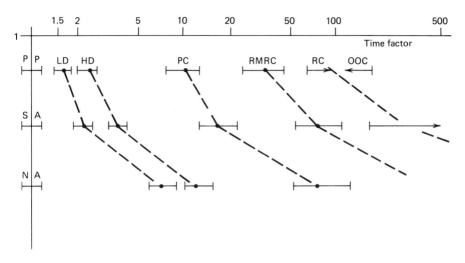

Fig. 9.9. General chart for time-effectiveness ratio. See text for nomenclature. (*Source:* Vertut.[14])

components for the gross motion (reach phase) and fine motion (positioning phase) for a positioning task not considered by Vertut. He shows that the effectiveness in the reach phase (1.25–1.5) is much better for teleoperators he considered (type PC) than for the positioning phase (2.5–11). The variability between manipulators is also much less for the reach times.

9.4.3. Unit-Task Times for Teleoperators

Prediction of total task times from the sum of unit-task times has been effectively demonstrated for existing manipulators. Experiments[9] performed to determine the unit-task times for the specific combination of operator and manipulator have then been applied to other tasks with predictions within 20%. The variability of the operators and the limited data for correlation account for a large fraction of this variation.

The results for two manipulators, one force-reflecting master slave and one unilateral (nonforce-reflecting) master slave, were obtained. The force-reflecting manipulator was a Model H cable-driven manipulator located at the Lawrence Berkeley Laboratory. The unilateral manipulator was the Ames Arm at SRI International. Both arms are reported to be high-quality, low-backlash, high-stiffness arms. The operator sat 2 m from the task viewing it directly. The results appearing in Figures 9.10

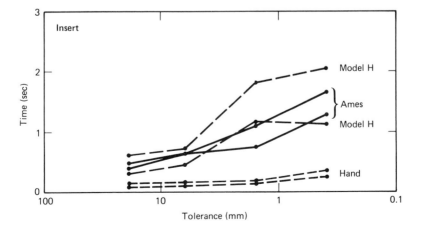

Fig. 9.10. *Insert* unit task versus tolerance. Sample includes three manipulators (including hand) and two subjects. (*Source:* Hill.[9])

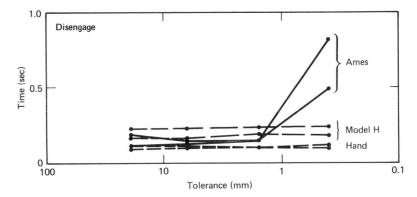

Fig. 9.11. *Disengage* unit task time versus tolerance. Sample includes three manipulators (including hand) and two subjects. (*Source:* Hill.[9])

through 9.15 present the results for two subjects. Although a small sample size was used, the results present the designer with some concept of the range of values of unit-task time and how they vary with the unit parameter. Units not having a parameter appear in Table 9.3.

An example of breaking a knob-turning task into motion elements is given by Hill.[15] In the task the operator touches a plate (signaling the start of the task), grasps the vertical handle of a rotary switch, turns it 90° clockwise, then 180° counterclockwise, and finally 90° clockwise, returning it to vertical. He releases the knob and touches the plate, signaling the end of the task. The motion elements are as follows with unit parameters in parentheses:

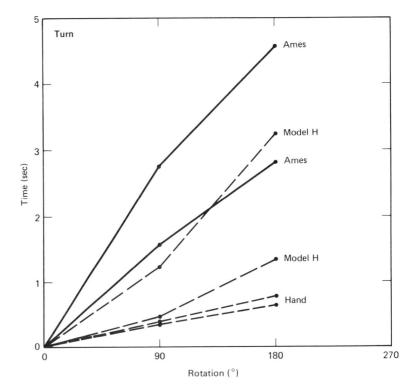

Fig. 9.12. *Turn* unit task time versus angle. Sample includes three manipulators (including hand) and two subjects. (*Source:* Hill.[9])

Contact
Move (250 mm)
Grasp knob
Turn (+90°)
Dwell time
Turn (−180°)
Dwell time
Turn (+90°)
Release
Move (250 mm)
Pre-position (76.2)

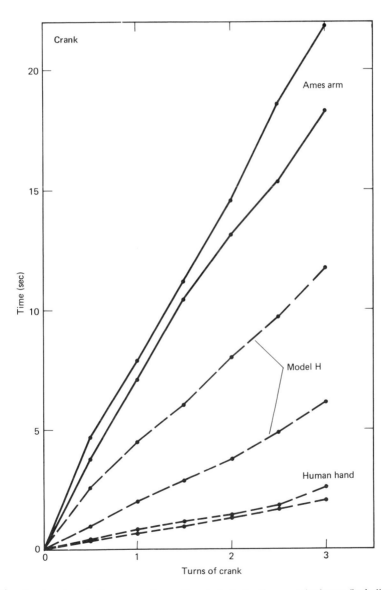

Fig. 9.13. *Crank* unit task time versus turns. Sample includes three manipulators (including hand) and two subjects. (*Source:* Hill.[9])

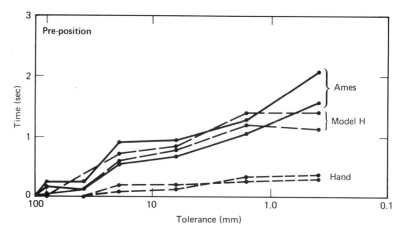

Fig. 9.14. *Pre-position* unit task time versus tolerance. Sample includes three manipulators (including hand) and two subjects. (*Source:* Hill.[9])

To estimate the effect of manipulator characteristics backlash, coulomb friction, and bandwidth on the unit-task times the designer can draw from the results of this and the previous section. The gross-motion time of the previous section corresponds to the move-unit time, and variations with manipulator characteristics should be similar. Fine-motion time variations should be similar to variations in pre-position time and roughly similar to the other final-positioning and assembly-unit times.

9.5. SUMMARY

In this chapter the unique aspects of designing an arm for use in a teleoperator have been addressed. The focus has been on the human and his interface to the manipulator control. The important classes of interfaces have been described. Three measures of teleoperator performance have been presented: information-transmission rate, times relative to the unencumbered human, and unit-task times. Good design decisions require a prediction of the performance of the teleoperator in terms of the alternatives and the application. The available results have been surveyed and presented. Much work remains before a unified approach can be applied to teleoperator design. Progress is being made in that direction, and the existing work presented gives a suitable framework for design considerations.

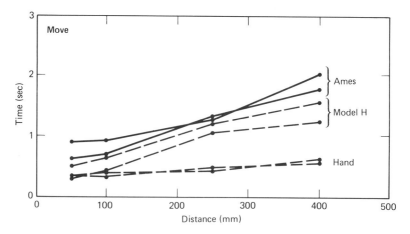

Fig. 9.15. *Move* unit task time versus distance. Sample includes three manipulators (including hand) and two subjects. (*Source:* Hill.[9])

TABLE 9.3.   UNIT-TASK TIMES[a]

Manipulator	Subject		Subject		Subject	
	1	2	1	2	1	2
Hand	0.20	0.20	0.40	0.35	0.35	0.28
Model H	2.56	3.04	1.92	1.88	0.40	0.83
Ames	4.60	7.36	2.88	2.76	0.39	.0.85

[a] Units without parameters. Times in seconds.

REFERENCES

1. Morgan, Clifford T., Chapanis, A., Cook, J. S., III, and Lund, M. W., Eds., *Human Engineering Guide to Equipment Design,* 2nd ed., McGraw-Hill, New York, 1963.

2. Sheridan, Thomas B., Ed., *Performance Evaluation of Robots and Manipulators,* U.S. Department of Commerce, National Bureau of Standards Special Publication 459, October 1976.

3. Fitts, P. M., The Information Capacity of Human Motor System in Controlling the Amplitude of Movement, *Journal of Experimental Psychology,* Vol. 47, June 1954, pp. 381–391.

4. Book, Wayne J., and Hannema, Dirk, Master-Slave Manipulator Performance for Various Dynamic Characteristics and Positioning Task Parameters, *IEEE Transactions on Systems Man and Cybernetics,* Vol. SMC-10, No. 11, November 1980, pp. 764–771.

5. McGovern, Douglas E., Factors Affecting Control Allocation for Augmented Remote Manipulation, Ph.D. Thesis, Stanford University, November 1974.

6. Hill, J. W., Study to Design and Develop Remote Manipulator Systems, Stanford Research Institute contract report on contract NAS2–8652, July 1976.

7. Taylor, F. W., *Scientific Management,* Harper and Brothers, New York, 1947.

8. Gilbreth, F. B., *Motion Study, a Method for Increasing the Efficiency of the Workman,* Van Nostrand, New York, 1911.

9. Hill, J. W., Study of Modeling and Evaluation of Remote Manipulation Tasks with Force Feedback, Final Report, SRI International Project 7696, March 1979.

10. Welford, A. T., Norris, A. H., and Schock, N. W., Speed and Accuracy of Movement and Their Changes with Age, *Acta Psychologica,* Vol. 30, 1969, pp. 3–15.

11. Hill, J. W., McGovern, D. E., and Sword, A. J., Study to Design and Develop Remote Manipulator Systems, final report, National Aeronautics and Space Administration contract NAS2-7507, SRI, 1974.

12. Hannema, Dirk P., Implementation and Performance Evaluation of a Computer Controlled Master Slave Manipulator with Variable Characteristics, M.S. Thesis, Georgia Institute of Technology, School of Mechanical Engineering, June 1979.

13. Hill, John W., and Matthews, Stephen J., Modeling a Manipulation Task of Variable Difficulty, Twelfth Annual Conference on Manual Control, University of Illinois, May 1976.

14. Vertut, J., Experience and Remarks on Manipulator Evaluation, in Sheridan, T. B., Ed., *Performance Evaluation of Programmable Robots and Manipulators,* U.S. Department of Commerce, NBS SP-459, October 1975, pp. 97–112.

15. Hill, J. W., Two Measures of Performance in a Peg-in-Hole Manipulation Task with Force Feedback, Thirteenth Annual Conference on Manual Control, Massachusetts Institute of Technology, June 1977.

CHAPTER **10**
PERFORMANCE TESTING

HANS J. WARNECKE

ROLF D. SCHRAFT

MARTIN C. WANNER

Fraunhofer Institute for Manufacturing
Engineering and Automation
Stuttgart, West Germany

10.1. GENERAL

The purpose of measuring and testing the performance of industrial robots is the urgent need to compare the characteristics of different robots according to a standardized test program. It is useful to make the following distinctions:

1. **Measuring for the Robot User.** Only those characteristics needed to solve the user's specific problem are evaluated. This procedure may be described as task-oriented performance testing.
2. **Measuring for the Robot Producer.** This evaluation involves determination of weak points in prototypes leading to structural and control redesign and expanding the range of application for already existing robots.
3. **Measuring over a Long Operation Period.** This gives a determination of the long-term behavior of the components.

Unfortunately, the standardization of characteristics and test methods for industrial robots is very slow, largely because industrial robots are still in an early stage of development.

10.2. TEST STAND FOR INDUSTRIAL ROBOTS

To fill this gap for the German industry a test stand and test procedures were developed at the Fraunhofer Institute for Production Automation (IPA)[2] with unified measuring programs. Of paramount importance in the test-stand design is an accurate geometrical coordination between robot and the measuring devices—essential for measuring with the necessary accuracy in the entire workspace.

Figure 10.1 shows the test stand at the IPA-Stuttgart including the platform and the three-dimensional measuring machine with some parts of the signal flow for the processing of geometrical, power, and thermal values. With such a test stand it is possible to get the required data in a reasonable time, which is impossible without the aid of sophisticated test-data processing and software. For shop-floor measurings tape recorders with several input channels are useful if the software from the test stand can be used afterwards.

10.3. MEASURING SENSORS

For the determination of geometrical values a distinction can be made between contacting and noncontacting measuring heads. Contacting methods like a touch-trigger probe and a three-dimensional measuring head with a sphere are used for static measurements and by the producers for adjusting components during robot assembly. Inductive, noncontacting measuring heads are in most cases superior to such devices as ultrasound, laser, and photogrammetry if we consider data processing, distance, resolution linearity, and price for the item.

In selecting measuring systems, the repeatability of modern industrial robots should be remembered.

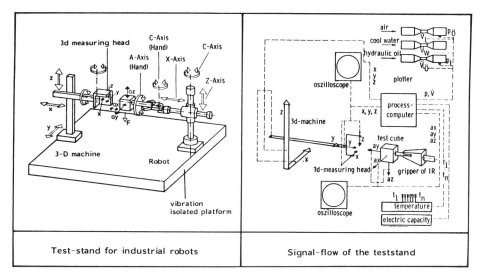

Fig. 10.1. Test stand for industrial robots at the IPA-Stuttgart.

At the IPA two- and three-dimensional measuring heads (see Figure 10.2) performed satisfactorily in measuring the distance to a reference body.

Sensors for nongeometrical values are thermoelements, light barriers, light cells, and accelerometers (piezoceramic, piezoresistive, and inductive).

10.4. TEST DATA

This section describes the following essential test data for industrial robots:

1. Geometrical values: workspace, static behavior, position accuracy (repeatability, reversal error), path accuracy, overshoot, reproduction of the smallest steps, synchronous travel accuracy, long-term behavior.
2. Kinematic values: cycle time, speed, acceleration.
3. Power and noise values.
4. Thermal values.
5. Dynamic values: force (gripping, motion, programming), dynamic compliance (frequency, damping, amplitude, phase), dynamic behavior of the moving structure, data for system modeling and optimization.

10.4.1. Geometrical Values

Workspace

Working space is the envelope reached by the center of the interface between the wrist and the tool, using all available axis motions. The manufacturer's specifications are tested.

1. Workspace with clear separation of arm, wrist, and gripper axes as shown in the example of Figure 10.3 of a painting robot.
2. Workspace that is not utilizable owing to the possible collision of the axes.
3. Difference between the *mechanical* and *control* workspace (structural tolerances, range of the path-measuring systems).

Static Behavior

Static behavior gives an indication of the deformation of a fixed robot structure under different load cases. Two measuring methods are common:

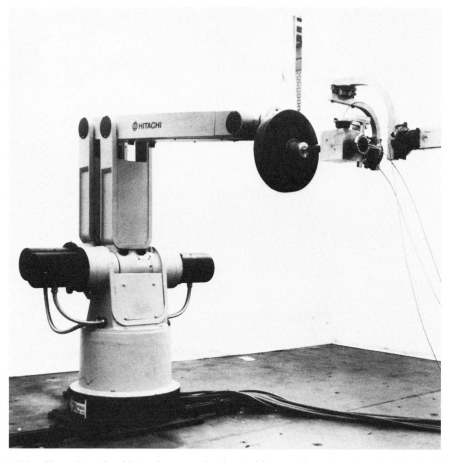

Fig. 10.2. Three-dimensional inductive measuring head with measuring cube and payload fixed with the robot.

1. Measuring at the gripper, which provides general information about the elastic behavior of the structure.
2. Measuring at single axes.

These methods are used for error compensations of the controller and to find potential weak points referring to the axes. Figure 10.4 shows a typical example of some test results. In this case the last axis shows a nonlinear behavior and should be regarded as weak point of the design.

Position Accuracy

Position accuracy is defined as the repeatable accuracy that can be achieved at nominal load and normal operating temperature. In Europe we distinguish between the following:

1. *Repeatability:* deviation between the positions and orientations reached at the end of several similar cycles (three times standard deviation of errors $3s_y$, see Figure 10.5).
2. *Reversal error:* deviation between the positions and orientations reached at the ends of several different paths (numerical value plus or minus the standard deviation errors U, see Figure 10.5).

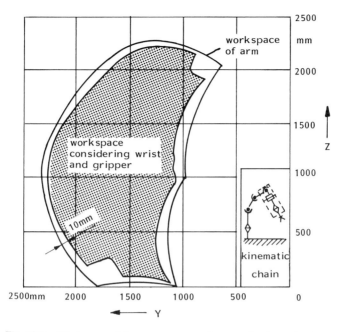

Fig. 10.3. Workspace of the arm versus workspace of wrist and gripper.

The test method includes the following steps:

1. Define a point in the workspace in X, Y, Z with the three-dimensional measuring machine and the three-dimensional noncontacting sensor.
2. Calibrate the system with the test cube attached to the robot hand (see Figure 10.2). This starting point includes an error already.
3. Teach a test cycle to the robot for similar (repeatability) and different paths (reversal error) to the calibrated point.
4. Measure the deviations in X, Y, Z (if possible also the orientations) and repeat the cycle at least 10 times for one point in the space.

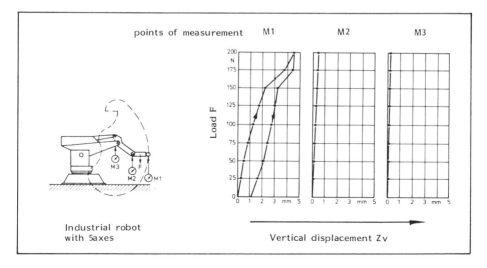

Fig. 10.4. Static behavior of different axes of an industrial robot as a function of the payload.

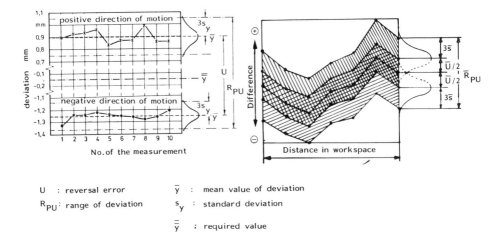

U   : reversal error $\bar{y}$:   mean value of deviation

R_{PU}: range of deviation s_y :   standard deviation

                                $\bar{\bar{y}}$:   required value

Fig. 10.5. Measuring the position accuracy of industrial robots and test results.

Repeatability and reversal error are measured for robots with PTP-control or robots with CP-control and PTP-tasks. Both errors depend on the position in the working space. For the user it is important to know both errors. These errors also influence the task and procedure of the "teach-in."

Path Accuracy

The path accuracy of a path-controlled robot indicates at what level of accuracy programmed path curves can be followed at nominal load. The following measuring methods are common:

1. Photogrammetry. The movement is documented by the use of two or more cameras. Exact, multi-purpose method, but evaluation of data is very complicated.
2. External measuring by laser system.[5] With this multipurpose system accuracy is still a problem.
3. Scale in space, inductive measuring system. Restricted to straight-path movements.

As the last procedure is still important (inexpensive, very accurate, easy evaluation, and most CP-tasks have straight-path movements), we describe it in more detail. A steel scale is arranged in the workspace; the starting and the end point of a straight line is programmed at a certain distance between measuring head (at the robot) and the scale. To determine path accuracy, the robot is run in automatic mode between these two points. In this way it is possible to measure the position and orientation of the robot relative to the scale. The measuring is repeated in different areas of the working space as a function of workload and speed.

Figure 10.6 shows the following typical errors in path accuracy for a robot:

1. Path accuracy or mean path-dispersion error describes the effect of random deviations from the reference straight line (deviation = 0). The dispersion error is stated as the difference between the largest and smallest deviation of the actual path and reference straight line.
2. Trailing error or mean path deviation describes the effect of control circuit settings associated with the drives to the relevant axes of the robot on the actual path. This distance is stated as the difference between the mean actual path and the reference straight line.
3. Overshoot during acceleration and deceleration of the robot.

All errors are functions of the velocity, workload, and number of axes involved in the movement related to different paths in the workspace.

Overshoot

Overshoot of robot structures occurs with violent changes in direction and mass and during acceleration and deceleration. It is measured by feeding a cube (at the robot) into a noncontact three-dimensional measuring head (3-D machine). The data are memorized, and the logarithmic decrement is computed. The measuring must be repeated at different points in the working space. Generally the user is interested

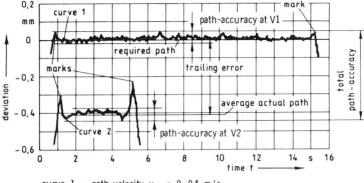

curve 1 : path-velocity v_1 = 0,04 m/s

curve 2 : path-velocity v_2 = 0,15 m/s

Fig. 10.6. Path accuracy—typical errors.

in the maximum amplitude and the decay time, whereas the manufacturer is interested in all parameters and the interactions responsible for the overshoot.

Figure 10.7 shows the overshoot of a robot with Cartesian axes. The performance test was carried out under the same conditions of direction of motion and load. The overshoot to the axes is shown separately.

Reproduction of the Smallest Steps

With very low velocities the so-called slip-stick effect (static friction into sliding friction) may become serious. This phenomenon is hard to control. Measuring can be by the same method as in path accuracy or, at very short movements, with touch trigger. This test may become important for machining tasks with industrial robots. Figure 10.8 graphs the typical slip-stick behavior of a given industrial robot.

Synchronous Travel Accuracy

In some cases the robot must perform tasks that are synchronous to a moving conveyor, for example, spray painting and assembly. In this case the synchronous travel accuracy is very important. Figure

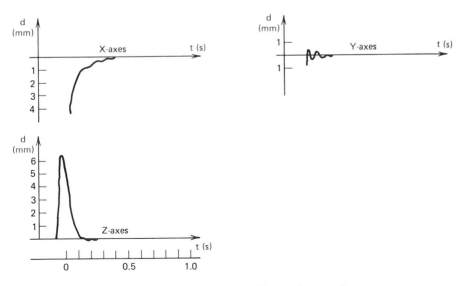

Fig. 10.7. Overshoot of a robot with Cartesian coordinates.

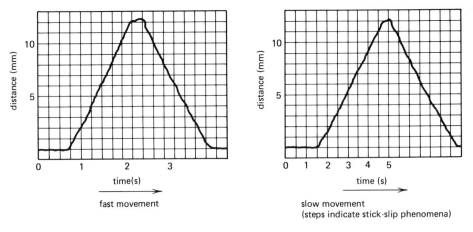

Fig. 10.8. Stick-slip effect as function of the speed.

10.9 shows the measuring procedure for this accuracy and the test results. Here the test cube is fixed on a belt with a known speed. The robot with a two- or three-dimensional measuring head is taught to follow the cube at the same speed and path. The distance S_a is measured.

Long-Term Behavior

The test of long-term behavior gives information on the time required to achieve thermal stability (temperature-dependent deviations). This test is very importrant for hydraulic units. The measuring cycle should last at least four hours with standard test cycles. For comparison the following parameters must be equal: movement sequence and range, load, speed, and measuring conditions. Temperature measuring at different points and the use of infrared cameras supports the search for thermal weak points.

10.4.2. Kinematic Values

The following data should be determined and measured:

1. Attainable cycle times for a defined sequence in different areas of the working space. In most cases the robot supplier gives information about speed and acceleration of the axes. For robots

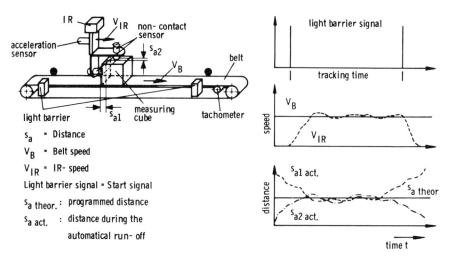

Fig. 10.9. Measuring conveyor tracking: method and typical results.

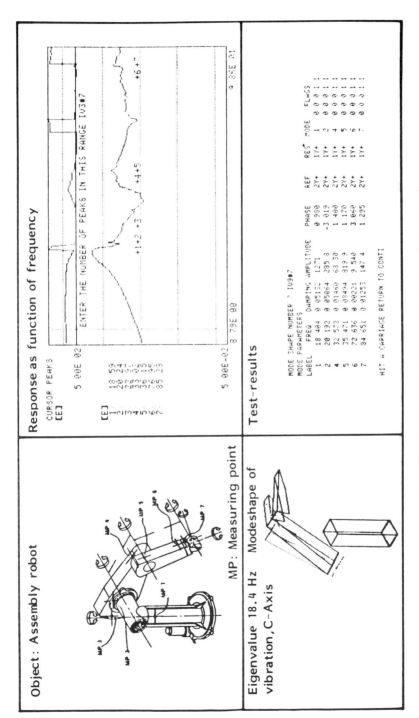

Fig. 10.10. Modal analysis of an assembly robot: method and result.

with more than one rotatory axis the speed in X, Y, Z direction must be related to the area in the workspace.

2. The path of velocity and maximum and average velocity measured by integration of the acceleration. The path of acceleration gives an idea of how to improve the robot control. The maximum acceleration is used to determine the gripping force; in many cases the tangential accelerations can reach high values. Measurement is by three-dimensional devices with inductive and piezoresistive accelerometers.

The kinematic values are closely linked to the dynamic values.

10.4.3. Power and Noise Values

Peak and mean values are determined in continuous operation. With pneumatic robots, for example, the volumetric flow is an important cost factor.

The measuring of the sound pressure is performed at a 1-m distance from the working space. The number of measuring points is equivalent to the difference between the maximum and minimum sound pressure in decibels.

10.4.4. Thermal Values of the Components and Media

Changes in temperature affect deviation of the structure, which is always important in the design of hydraulic units.

10.4.5. Dynamic Values

In addition to measuring at the gripper it is important for the designer to know the dynamic behavior of simple components and of the total structure. Here experimental modal analysis is a useful tool for robot design. With this method we start with the unit forced-response data and extract the mode of vibration without any assumptions about the mass and stiffness distribution. The result is a set of modes including frequency, damping, mode shape, and residues. An example of an assembly robot measured according to this method at the IPA is shown in Figure 10.10.

It is possible to elicit response of the robot structure by the following excitation methods:

1. Shaker (sinus, random).
2. Hammer (impact).
3. Snapback (impact).
4. Drives (sinus, random).

With these data it is possible to create a mathematical model for further design improvements. Another method involves the measuring of data from the path-measuring systems and certain places on the robot arm. Both methods can also be used for tests over a long period and as tools for preventive maintenance.[7,8]

REFERENCES

1. Warnecke, H. J. and Schraft, R. D., *Industrial Robots*, IFS-Publications Ltd., Bedford, 1982.

2. Brodbeck, B. and Schiele, G., *Prüfstand für Industrieroboter*, Forschungsbericht HA 80-032, Humanisierung des Arbeitslebens, Teil 8, Karlsruhe, 1980.

3. Brodbeck, B. and Schiele, G., *Ergebnisse von Messungen von Industrierobotern*, Technische Rundschau, January 1980, pp. 5–7.

4. VDI 2861, Blatt 1 and 2, Bezeichnungen und Kenngrößen von Handhabungseinrichtungen, VDI-Verlag.

5. Robot Check, Three-Dimensional Noncontact Dynamic Robot Measuring, *Selspine AB*, 1983.

6. SDRC/GENRAD, Modal-User Manual, March 1982.

7. Dagalakis, N., Analysis of Robot Performance Operation, *13th International Symposium on Industrial Robots*, Vol. 1, 1983, pp. 7.73–7.95.

8. Warnecke, H. J., Schraft, R. D., and Wanner, M. C., Application of the Experimental Modalanalysis in the Performance Testing Procedure of Industrial Robots, *Proceedings of Robotics Europe Conference*, 1984.

PART 3
ROBOT CONTROL

Chapter 11

DESIGN OF CONTROL SYSTEMS FOR INDUSTRIAL ROBOTS

J. Y. S. LUH

Purdue University
West Lafayette, Indiana

11.1. ROBOT CONTROL

Industrial robots are computer-controlled mechanical manipulators used in industrial applications. The number of joints of commercially available robots varies from three to seven. Typically they have six joints, giving 6 degrees of freedom (DF), with a gripper which is referred to as a hand or an end effector. Each joint of the robot is positionally controlled with a feedback loop.[1]

In reality a robot task is naturally specified in terms of its hand in Cartesian coordinates. The measured variables for feedback purposes are joint displacements and velocities. Thus the goal and the measured quantities are in different coordinate systems as depicted in Figure 11.1. Hence the control of the position and orientation of the hand by the actuators at the joints requires some knowledge of transformations between the hand and joints. The transformations, however, need not be known if the so-called "teaching by doing" procedure is adopted, which is Case 1 shown in Figure 11.1. To implement this procedure, the hand is led by an operator through a number of points on a prescribed Cartesian path along which the hand is required to travel. Thus the robot can be viewed as an analog device through which the points in Cartesian coordinates are transformed into corresponding points in joint coordinates by means of the "teaching by doing" procedure. These joint coordinate points form the corresponding joint trajectories and are then used as the reference input points for the closed-loop positional controller of each joint. Hence the control is done entirely at the joint level, although the goal is the position and orientation of the hand in Cartesian coordinates.

Under some circumstances the "teaching by doing" procedure is not adopted because, for instance, the hand of the robot is too heavy and bulky, or the desired Cartesian path of the hand is generated from the computer and fed directly to the controller. In this situation, the need for the transformation is unavoidable. To determine the transformation, one observes the hand in Cartesian coordinates. It consists of position, described by a position vector $\mathbf{p}(t)$, and orientation, described by three orthonormal vectors $\mathbf{n}(t)$, $\mathbf{s}(t)$, and $\mathbf{a}(t)$ called unit normal, unit slide, and unit approach vectors, respectively. All these vectors are defined with reference to the base coordinates and are indicated in Figure 11.2a. Thus the state of the hand at time t in Cartesian coordinates can be represented by a 4×4 hand matrix

$$\mathbf{H}(t) = \begin{bmatrix} \mathbf{n}(t)\ \mathbf{s}(t)\ \mathbf{a}(t)\ \mathbf{p}(t) \\ 0 \quad 0 \quad 0 \quad 1 \end{bmatrix} \tag{11.1}$$

The last row in $\mathbf{H}(t)$ is added for the convenience of future computation using homogeneous transformation,[2] discussed in detail in Section 11.4.3 on multiple joint controller. The orientation may also be defined in terms of Euler angles with reference to the base coordinates. Initially at $t = t_0$, let $\mathbf{n}(t_0)$, $\mathbf{s}(t_0)$, and $\mathbf{a}(t_0)$ align with $\mathbf{x}_0$, $\mathbf{y}_0$, and $\mathbf{z}_0$, respectively, as shown in Figure 11.2b. Any orientation $[\mathbf{n}(t), \mathbf{s}(t), \mathbf{a}(t)]$ may be obtained by a rotation of γ radians about $\mathbf{z}_0$ so that $\mathbf{s}(t_0)$ aligns with $\mathbf{s}(t_1)$; then a rotation of β radians above $\mathbf{s}(t_1)$ so that $\mathbf{a}(t_0) = \mathbf{a}(t_1)$ aligns with $\mathbf{a}(t)$, and finally a rotation of α radians about $\mathbf{a}(t)$ to obtain the required $\mathbf{n}(t)$ and $\mathbf{s}(t)$. This is equivalent to rotating the $[\mathbf{n}(t_0)$, $\mathbf{s}(t_0)$, $\mathbf{a}(t_0)]$ coordinate, which aligns with $[\mathbf{x}_0, \mathbf{y}_0, \mathbf{z}_0]$ originally, α radians about $\mathbf{z}_0$, then β radians about $\mathbf{y}_0$, and finally γ radians about $\mathbf{z}_0$ again. The relationship is therefore

169

$$\mathbf{n}(t) = \begin{bmatrix} \cos \alpha \cos \beta \cos \gamma - \sin \alpha \sin \gamma \\ \cos \alpha \cos \beta \sin \gamma + \sin \alpha \cos \gamma \\ -\cos \alpha \sin \beta \end{bmatrix} \tag{11.2}$$

$$\mathbf{s}(t) = \begin{bmatrix} -\sin \alpha \cos \beta \cos \gamma - \cos \alpha \sin \gamma \\ -\sin \alpha \cos \beta \sin \gamma + \cos \alpha \cos \gamma \\ \sin \alpha \sin \beta \end{bmatrix} \tag{11.3}$$

$$\mathbf{a}(t) = \begin{bmatrix} \sin \beta \cos \gamma \\ \sin \beta \sin \gamma \\ \cos \beta \end{bmatrix} \tag{11.4}$$

Consequently the state of the hand at time t in Cartesian coordinates with reference to the base coordinates may also be represented by a six-dimensional vector $[\mathbf{p}(t)'\ \boldsymbol{\theta}(t)']$ where $[\boldsymbol{\theta}(t)'] = [\alpha\ \beta\ \gamma]$ and $()' = $ transpose of $()$.

The hand, however, is driven by the actuators at the joints. Intuitively, if all the joint displacements are known, the position and orientation of the hand are determined. Let n be the number of joints. For $i = 1, 2, \ldots, n$, let q_i be the displacement of the ith joint with respect to its own reference point. Then, for any given robot with known geometrical dimensions, there is a relation

$$[\mathbf{p}(t)'\ \boldsymbol{\theta}(t)']' = \mathbf{f}[q_1, q_2, \cdots, q_n] \tag{11.5}$$

where $\mathbf{f}(\cdot)$ is a 6×1 vector-valued function. This relation is known, but almost always nonlinear, which complicates the problem.[3] Since, in reality, $[\mathbf{p}(t)'\ \boldsymbol{\theta}(t)']$ in Cartesian coordinates is specified, but the corresponding $[q_1, \cdots, q_n]$ in joint coordinates is actually needed, one may command the joint actuators to comply with the specification in Cartesian coordinates. The solution requires the inverse vector function $\mathbf{f}^{-1}(\cdot)$ of n dimension. This solution, if it can be found, may not be unique. For the commercially available robots in operation, n is usually either 5 or 6. The geometrical configuration of these robots with proper definitions and ranges of q_i enables one to obtain a unique solution of equation (11.5).[3]

With the knowledge of the transformation between the position and orientation in Cartesian and joint coordinates, it is possible to control the hand, which travels along a desired Cartesian path, in joint coordinates. This implies the control at the joint level and corresponds to Case 2 in Figure 11.1. As an example, consider the Stanford manipulator,[4] which has one prismatic and five revolute joints as shown in Figure 11.3. A block diagram for a joint control of the Stanford manipulator, which has a permanent magnet motor drive, is shown in Figure 11.4. It has an optical encoder for positional feedback with a tachometer feedback for damping. Thus an industrial robot is a positioning device in that each of its joints has a positional control system. Now the question is how does one control the joint to accomplish the goal? Before one arrives at an answer, one must examine the following possible specification: Must the hand follow a specified path? If the answer is no, then one has a simple point-to-point positional control problem. Otherwise, the controller must keep up with path tracking. These two problems are analyzed in the following sections.

11.2. POSITIONAL CONTROL OF A SINGLE JOINT

If there are no path constraints, the controllers have only to make sure that the hand passes through all the specified corner points of the path. The input to the control system is the desired Cartesian

	Cartesian Coordinates	Joint Coordinates
Desired position and orientation	Case 1. Teaching by doing KNOWN $\longrightarrow$ Case 2. $\mathbf{f}^{-1}[\mathbf{p}(t),\boldsymbol{\theta}(t)]$ Case 4. Approximate functions	
Measured actual displacement and velocity	(Case 3. $\mathbf{f}(q_1, q_2, \ldots, q_n)$) $\longleftarrow$	KNOWN
Controller	HAND-LEVEL	JOINT-LEVEL

Fig. 11.1. Knowns and unknowns in different coordinates.

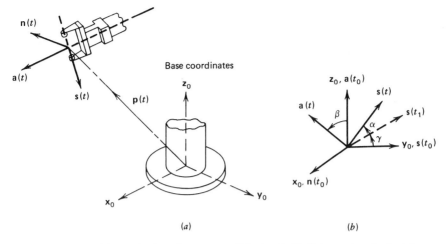

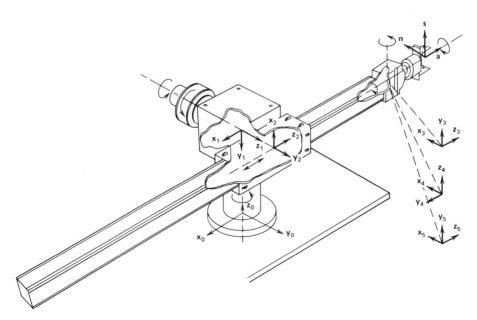

Fig. 11.2. (*a*) Position and orientation vectors of the hand. (*b*) Euler angles of orientation.

corner points of the path, which may be (1) numerically fed into the system, or (2) furnished through so-called *teaching by doing,* that is, corner points are recorded while the hand of the robot is led through these points manually by an operator. Then the coordinates transformation takes place, which computes the corresponding joint coordinates $[q_1, \cdots, q_n]$ of the specified corner points in Cartesian coordinates by means of $f^{-1}(\cdot)$, digitally for case (1), or analogously for case (2), and then positionally controls the robot in joint coordinates from point to point. In practice, a positional servo is used for each joint. To start the discussion, all the joints are considered to be independently controlled. Then each of the joint controllers is very simple. In reality, force interactions among the joints in motion create couplings that complicate the control system. Discussion of these topics follows.

11.2.1. Single Joint Controller

In the following discussion, the robot is considered to have a rigid body structure. Refer to Figure 11.5, the schematic representation of an actuator-gear-load assembly for a single joint, in which

Fig. 11.3. The Stanford manipulator.

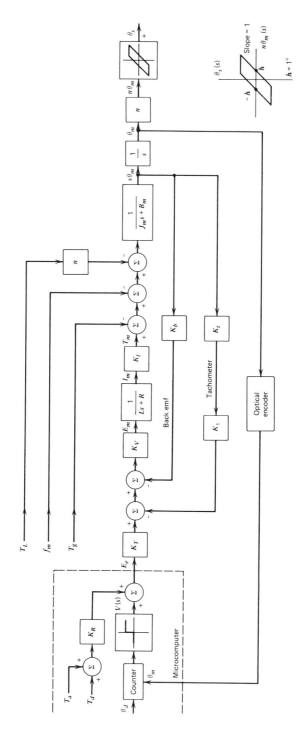

Fig. 11.4. Block diagram of a positional control system.

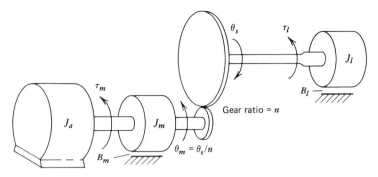

Fig. 11.5. Schematic representation of an actuator-gear-load assembly for one joint.

J_a = actuator inertia of one joint (oz-in-sec²/rad)
J_m = manipulator (robot) inertia of the joint fixtures at actuator side
J_l = inertia of the manipulator link
B_m = damping coefficient at actuator side (oz-in-sec/rad)
B_l = damping coefficient at load side
f_m = average friction torque (oz-in)
τ_g = gravitational torque
τ_m = generated torque at actuator shaft
τ_l = internal load torque
θ_m = angular displacement at actuator shaft (radians)
θ_s = angular displacement at load side

Let N_m, N_s = number of teeth of the gears at the actuator shaft and load shaft, respectively
   r_m, r_s = pitch radii of the gears at the actuator shaft and load shaft, respectively
then

$$n = \frac{r_m}{r_s} = \frac{N_m}{N_s} \leq 1 \tag{11.6}$$

is the gear ratio, so that

$$\theta_s = n\theta_m \tag{11.7}$$

Using D'Alembert's principle, one obtains

$$\tau_l - B_l\dot{\theta}_s = J_l\ddot{\theta}_s \tag{11.8}$$

Apply the same principle at the actuator shaft to yield

$$\tau_m - n\tau_l - B_m\dot{\theta}_m = (J_a + J_m)\ddot{\theta}_m \tag{11.9}$$

Combining Eqs. (11.7), (11.8), and (11.9) yields

$$\tau_m = (J_a + J_m + n^2J_l)\ddot{\theta}_m + (B_m + n^2B_l)\dot{\theta}_m \tag{11.10}$$

where $J_{\text{eff}} = (J_a + J_m + n^2J_l)$ is the effective inertia and $B_{\text{eff}} = (B_m + n^2B_l)$ is the effective damping coefficient at the actuator shaft.

The actuators used in the industrial robots are either hydraulic, pneumatic, or electrical. As an example, the Unimation PUMA and the Stanford manipulators have electrical systems using permanent magnet DC motors. They are armature controlled and their schematic diagram is shown in Figure 11.6. In this figure $v_b(t)$ is the back emf in volts in the armature winding which can be represented by

$$v_b(t) = K_b\dot{\theta}_m(t) \tag{11.11}$$

where K_b is the back emf constant in volts-second per radian. Let L and R be the inductance in henries and resistance in ohms of the motor armature winding, respectively. Since L is in the order

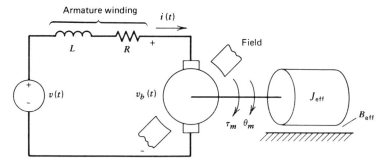

Fig. 11.6. Schematic diagram for an electrical drive system.

of tenths of millihenries, it is practically zero. Then, by applying Kirchhoff's voltage law to the armature circuit, one obtains a frequency domain relation

$$V(s) - K_b s\ \Theta_m(s) = (Ls + R)I(s) \cong RI(s) \tag{11.12}$$

where s is the complex frequency in radians per second. The DC motor is operated in its linear range so that the generated torque is proportional to the armature current. The relation in the frequency domain is

$$T_m(s) = K_I I(s) \tag{11.13}$$

where K_I is the torque constant in ounce-inches per ampere. The motor shaft is mechanically connected to an actuator-gear-load assembly, as indicated in Figure 11.6, with an effective inertia J_{eff} and effective damping coefficient B_{eff} at the actuator shaft. The relations among the mechanical components are described by Eq. (11.10), which has a Laplace transform equivalence

$$T_m(s) = (J_{\text{eff}} s^2 + B_{\text{eff}} s)\ \Theta_m(s) \tag{11.14}$$

Eliminating $T_m(s)$ and $I(s)$ among Eqs. (11.12), (11.13), and (11.14) yields

$$\frac{\Theta_m(s)}{V(s)} = \frac{K_I}{s[RJ_{\text{eff}} s + (RB_{\text{eff}} + K_I K_b)]} \tag{11.15}$$

which is the transfer function, or the feedforward gain, from the applied voltage to the DC motor (input), to the angular displacement of the motor shaft (output).

To construct a positional controller for the angular displacement of the load shaft, it is necessary to convert the displacement into electrical voltage to actuate the DC motor. For a feedback (or closed-loop) controller the actuating signal is the error at time t between the desired and the actual displacements:

$$e(t) = \theta_d(t) - \theta_s(t) \tag{11.16}$$

By means of a potentiometer or an optical encoder/counter assembly, the displacement error is converted into voltage as

$$v(t) = K_\theta e(t) \tag{11.17}$$

which has a transformed equivalence

$$V(s) = K_\theta E(s) = K_\theta[\Theta_d(s) - \Theta_s(s)] \tag{11.18}$$

where K_θ is the conversion constant in volts per radian. Combining all the physical apparatus together, one may construct a block diagram as shown in Figure 11.7a. The feedforward gain, or the open-loop transfer function, is

$$\frac{\Theta_s(s)}{E(s)} = \frac{n K_\theta K_I}{s[RJ_{\text{eff}} s + (RB_{\text{eff}} + K_I K_b)]} \tag{11.19}$$

which is obtained either from the block diagram, Figure 11.7a, or by combining Eqs. (11.15) and (11.18) and the relation $\Theta_s(s) = n\Theta_m(s)$.

To improve the settling time, the signal of the motor shaft velocity is fed back through a tachometer, or the computation of the difference in angular displacements of the shaft during a fixed time interval. The block diagram of the resulting controller is shown as in Figure 11.7b in which K_t is the tachometer constant in volt-seconds per radian, and K_1 is the gain of amplifier in volts per volt. Since the feedback voltage at the motor armature circuit is now $K_b\dot\theta_m(t) + K_1K_t\dot\theta_m(t)$ instead of $K_b\dot\theta_m(t)$ alone, the Laplace transform of the circuit Eq. (11.12) is modified as

$$V(s) - (K_b + K_1K_t)s\,\Theta_m(s) = RI(s) \tag{11.20}$$

Thus the revised open-loop and closed-loop transfer functions can be obtained simply by replacing K_b by $(K_b + K_1K_t)$ in Eq. (11.19). Consequently,

$$\frac{\Theta_s(s)}{E(s)} = \frac{nK_\theta K_I}{RJ_{\text{eff}}s^2 + [RB_{\text{eff}} + K_I(K_b + K_1K_t)]s} \tag{11.21}$$

$$\frac{\Theta_s(s)}{\Theta_d(s)} = \frac{nK_\theta K_I}{RJ_{\text{eff}}s^2 + [RB_{\text{eff}} + K_I(K_b + K_1K_t)]s + nK_\theta K_I} \tag{11.22}$$

For a specific robot, the numerical values of the parameters n, K_I, K_t, K_b, R, J_{eff}, and B_{eff} are either specified (by the component manufacturer) or determined by experiments. As an example, assemblies for joints 1 and 2 of the Stanford manipulator contain respectively a U9M4T and a U12M4T DC motor with an integral tachometer 030/105 by Photocircuits Corporation. The parametric data for the motor-tachometer unit are listed in Table 11.1.[5]

The second to the last line in Table 11.1 shows the average friction torque f_m that exists in each assembly and must be overcome. The effective inertia of each joint of the Stanford Jet Propulsion Laboratory (JPL) manipulator is listed in Table 11.2.[6]

Note that the conversion constant K_θ and amplifier gain K_1, however, must be determined from the parameters corresponding to the structural resonant frequency and the damping ratio of the robot, discussed in the following section.

As mentioned, an average friction torque f_m of the motor-tachometer assembly must be overcome by the motor. Of course, the motor must also compensate the external load torque τ_L, gravitational torque τ_g, and the centrifugal contribution $\tau_c(t)$. These quantities represent the reaction from the physical burden to the robot. Schematically they are inserted in the block diagram of the positional controller, Figure 11.7b, at the point where the torque is generated from the motor. Figure 11.7c shows the revised block diagram of the positional controller in which $F_m(s)$, $T_L(s)$, and $T_g(s)$ are the Laplace transformed variables of f_m, τ_L, and τ_g, respectively. The centrifugal term is a function of $[\theta_s(t)]^2$ whose Laplace transform involves the convolution integrals. Since the mathematical model of the system is linear, the principle of superposition applies. Hence the centrifugal contribution is treated separately in the following discussions.

11.2.2. Determination of K_θ and K_1

From Eq. (11.22), the characteristic equation for the closed-loop controller is

$$s^2 + \frac{[RB_{\text{eff}} + K_I(K_b + K_1K_t)]s}{(RJ_{\text{eff}})} + \frac{nK_\theta K_I}{RJ_{\text{eff}}} = 0 \tag{11.23}$$

which is conventionally expressed as

$$s^2 + 2\zeta\omega_n s + \omega_n^2 = 0 \tag{11.24}$$

where ζ is the damping ratio and ω_n the undamped natural frequency. From Eqs. (11.23) and (11.24), one obtains

$$\omega_n = \sqrt{\frac{nK_\theta K_I}{(RJ_{\text{eff}})}} > 0 \tag{11.25}$$

and

$$\zeta = \frac{RB_{\text{eff}} + K_I(K_b + K_1K_t)}{2\sqrt{nK_\theta K_I R J_{\text{eff}}}} \tag{11.26}$$

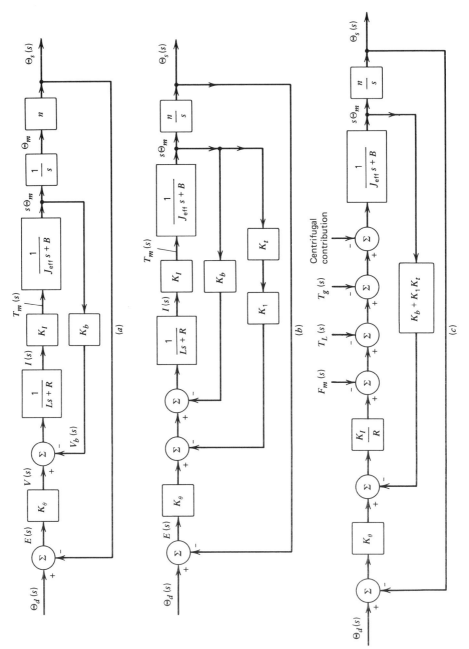

Fig. 11.7. Block diagram of a positional controller.

TABLE 11.1. PARAMETRIC DATA FOR
MOTOR-TACHOMETER UNIT

Model	U9M4T	U12M4T
K_I (oz-in/amp)	6.1	14.4
J_a (oz-in-sec²/rad)	0.008	0.033
B_m (oz-in-sec/rad)	0.01146	0.04297
K_b (volts-sec/rad)	0.04297	0.10123
L (μH)	100.0	100.0
R (ohms)	1.025	0.91
K_t (volts-sec/rad)	0.0149	0.05062
f_m (oz-in)	6.0	6.0
n	0.01	0.01

TABLE 11.2. EFFECTIVE INERTIA OF STANFORD-JPL MANIPULATOR

Joint Number	Minimum Value/No Load (kg-m²)	Maximum Value/No Load (kg-m²)	Maximum Value/Full Load (kg-m²)
1	1.417	6.176	9.570
2	3.590	6.950	10.300
3	7.257	7.257	9.057
4	0.108	0.123	0.234
5	0.114	0.114	0.225
6	0.040	0.040	0.040

Let k_{eff} be the effective stiffness (in oz-in/rad) of the joint of the robot. The restoring torque due to stiffness is $-k_{\text{eff}}\theta$. Thus, by D'Alembert's principle,

$$-k_{\text{eff}}\theta = J_{\text{eff}}\ddot{\theta} \tag{11.27}$$

so that the structural resonant frequency in radians per second is

$$\omega_r = \sqrt{\frac{k_{\text{eff}}}{J_{\text{eff}}}} \tag{11.28}$$

Although k_{eff} for the joint is fixed, J_{eff} varies as the load varies so that ω_r changes accordingly. Let ω be the measured structural resonant frequency of the same joint corresponding to the effective inertia J. Then

$$\omega = \sqrt{\frac{k_{\text{eff}}}{J}} \tag{11.29}$$

Thus, by Eqs. (11.28) and (11.29),

$$\omega_r = \omega\sqrt{\frac{J}{J_{\text{eff}}}} \tag{11.30}$$

The measured ω and its corresponding J for the Stanford manipulator are listed in Table 11.3.[4]
It has been shown[7] that a safety factor of 200% should be included in a conservative design, that is, one sets the undamped natural frequency ω_n no more than one-half of the structural resonant frequency ω_r. Thus by Eqs. (11.25) and (11.30), one obtains

$$\sqrt{\frac{nK_\theta K_I}{RJ_{\text{eff}}}} \leq \omega\frac{\sqrt{J/J_{\text{eff}}}}{2} \tag{11.31}$$

which reduces to

$$K_\theta \leq \frac{(J\omega^2)R}{4nK_I} \tag{11.32}$$

TABLE 11.3. MEASURED FREQUENCY AND JOINT INERTIA OF STANFORD MANIPULATOR

Joint Number	J (kg-m^2)	f(Hz)	ω (=$2\pi f$) (rad/sec)
1	5	4	25.1327
2	5	6	37.6991
3	7	20	125.6636
4	0.1	15	94.2477
5	0.1	15	94.2477
6	0.04	20	125.6636

Relation (11.32) establishes the upper bound of K_θ. It remains to determine the bound on K_1. For practical reasons one avoids the underdamped positional controller for the robot. Thus $\zeta \geq 1$, and from Eq. (11.26) one obtains

$$RB_{\text{eff}} + K_I(K_b + K_1K_t) \geq 2\sqrt{nK_\theta K_I RJ_{\text{eff}}} > 0 \tag{11.33}$$

Again for conservative design, K_θ at the right side of Eq. (11.33) is replaced by its upper bound, which is given by Eq. (11.32). Thus Eq. (11.33) reduces to

$$K_1 \geq \frac{R(\omega\sqrt{JJ_{\text{eff}}} - B_{\text{eff}})}{K_I K_t} - \frac{K_b}{K_t} \tag{11.34}$$

Since J_{eff} varies as the load changes, the lower bound on K_1 changes accordingly. If the load is known ahead of time, J_{eff} can be precomputed to establish the lower bound. On the other hand, if one wishes to simplify the design of the controller by choosing a fixed amplifier gain, then the maximum value of J_{eff} should be used in Eq. (11.34) to avoid any possibility of resulting in an underdamping system.

11.2.3. Steady-State Error for Joint Controller

In the preceding section, the block diagram of the positional controller for an independent single joint of a robot was presented in Figure 11.7c. Because of an addition of the physical burden f_m, τ_L, τ_g, and τ_c to the motor, the closed-loop transfer function of the controller is not the same as given by Eq. (11.22), and it must be modified to include the additions. From Figure 11.7c, it is seen that

$$(J_{\text{eff}}s^2 + B_{\text{eff}}s)\Theta_m = T_m(s) - F_m(s) - T_g(s) - nT_L(s) \tag{11.35}$$

In Eq. (11.35) the centrifugal contribution is not included, but will be treated separately. Now

$$T_m(s) = \frac{K_I[V(s) - s(K_b + K_1K_t)\Theta_s(s)/n]}{R} \tag{11.36}$$

and

$$V(s) = K_\theta[\Theta_d(s) - \Theta_s(s)] \tag{11.37}$$

Thus, after some algebraic manipulation,

$$\Theta_s(S) = \frac{nK_\theta K_I\Theta_d(s) - nR[F_m(s) + T_g(s) + nT_L(s)]}{\Omega(s)} \tag{11.38}$$

where

$$\Omega(s) = RJ_{\text{eff}}s^2 + [RB_{\text{eff}} + K_I(K_b + K_1K_t)]s + nK_\theta K_I \tag{11.39}$$

Whenever $F_m(s)$, $T_g(s)$, and $T_L(s)$ vanish, Eq. (11.38) reduces to Eq. (11.22). Since the position error $e(t)$ is defined as

$$e(t) = \theta_d(t) - \theta_s(t) \tag{11.40}$$

then by Eq. (11.38), Eq. (11.40) can be written as

$$E(s) = \frac{\{RJ_{eff}s^2 + [RB_{eff} + K_I(K_b + K_1K_t)]s\}\Theta_d(s) + nR[F_m(s) + T_g(s) + nT_L(s)]}{\Omega(s)} \quad (11.41)$$

where $E(s)$ is the Laplace transform of $e(t)$. For a constant load, $\tau_L = C_L$. Since $f_m = C_f$ and $\tau_g = C_g$ are also constant, then $T_L(s) = C_L/s$, $F_m(s) = C_f/s$, and $T_g(s) = C_g/s$. Consequently Eq. (11.41) becomes

$$E(s) = \frac{\{RJ_{eff}s^2 + [RB_{eff} + K_I(K_b + K_1K_t)]s\}X(s) + nR[C_f + C_g + nC_L]/s}{\Omega(s)} \quad (11.42)$$

where $X(s)$ replaces $\Theta_d(s)$ to represent a generalized input command.

The steady-state error e_{ss} may be determined by the use of the final value theorem, which states that

$$e_{ss} = \lim_{t \to \infty} e(t) = \lim_{s \to 0} sE(s) \quad (11.43)$$

provided the limits exist.

11.2.4. Steady-State Position Error and Compensation

If the input is a constant displacement C_θ, then

$$X(s) = \Theta_d(s) = \frac{C_\theta}{s} \quad (11.44)$$

By Eq. (11.43) one obtains a steady-state position error

$$e_{ssp} = \frac{R(C_f + C_g + nC_L)}{(K_\theta K_I)} \quad (11.45)$$

Since K_θ has an upper bound given by Eq. (11.32), the error may not be reduced to an arbitrary small value by merely adjusting the parameter K_θ. However, if one knows the value of τ_L, f_m, and τ_g in advance, it is possible to feed forward these quantities into the controller to anticipate the burden. Based on this idea, an anticipated gravitational torque signal τ_a and a desired compensating torque signal τ_d are fed to the controller as an additional input, as shown in Figure 11.8a where $T_a(s)$ and $T_d(s)$ are Laplace transforms of τ_a and τ_d, respectively. With this arrangement, the error given by (11.41) is modified to become

$$E(s) = \frac{\begin{array}{c}\{RJ_{eff}s^2 + [RB_{eff} + K_I(K_b + K_1K_t)]s\}X(s) + nR\{F_m(s) \\ + T_g(s) + nT_L(s) - K_IK_R[T_a(s + T_d(s)]/R\}\end{array}}{\Omega(s)} \quad (11.46)$$

where $X(s)$ replaces $\Theta_d(s)$ to represent a generalized input signal. For an input of a constant displacement $X(s) = \Theta_d(s) = C_\theta/s$, the steady-state position error now becomes

$$e_{ssp} = \frac{\lim_{s \to 0} s\{R[F_m(s) + T_g(s) + nT_L(s)] - K_IK_R[T_a(s) + T_d(s)]\}}{(K_\theta K_I)} \quad (11.47)$$

If $T_a(s) = RT_g(s)/(K_IK_R)$ and $T_d(s) = R[F_m(s) + nT_L(s)]/(K_IK_R)$, then the steady-state error would be zero. In practice, one may set

$$\tau_a = \left(\frac{R}{K_IK_R}\right)\hat{\tau}_g \quad (11.48)$$

and

$$\tau_d = \frac{R}{K_IK_R}(\hat{f}_m + n\hat{\tau}_L) \quad (11.49)$$

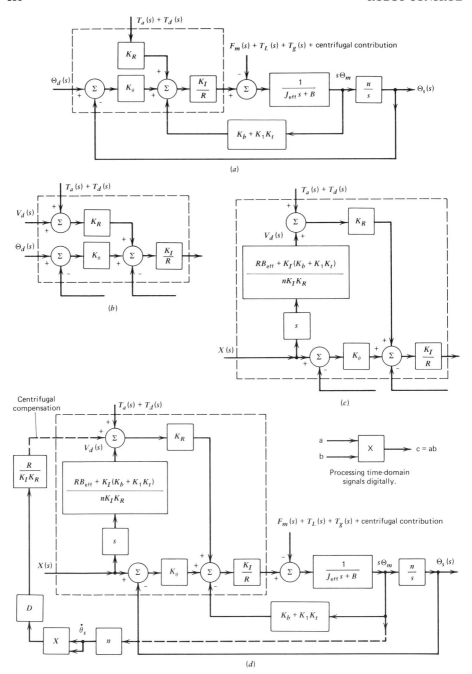

Fig. 11.8. Controller with anticipated burden and feedforward compensation.

to reduce the error, where $\hat{\tau}_g$, $\hat{f}_m$, and $\hat{\tau}_L$ are the estimates of τ_g, f_m, and τ_L, respectively. For a given task, the value τ_L, which includes the compliant torque, is usually known. Thus $\hat{\tau}_L$ can be estimated directly. The measured value of f_m through experimentation is ordinarily used for $\hat{f}_m$. The value of $\hat{\tau}_g$ is normally computed, which is discussed in the section on multiple joint controller.

What is the contribution to the positional steady-state error e_{ssp} from the centrifugal term $\tau_c(t)$? Since $\tau_c(t) = D[\dot{\theta}_s(t)]^2$ where D is a proportional constant, and since $\dot{\theta}_s(t) \to 0$ as $t \to \infty$ for a stable positional controller, the contribution is therefore none. Thus no feedforward compensation is required for the centrifugal term as far as the positional steady-state error is concerned.

For the commercially available industrial robots, unfortunately the friction torque is dominant. Since the static friction varies and is not always known in advance, it is difficult to compensate. Because of the dominance, its inaccurate compensation affects the merit of the feedforward compensation of other terms. It is anticipated, however, that the frictional effect will be reduced to a minimum in the future design of industrial robots, as in the direct-drive arms produced at the Carnegie-Mellon University[8] and MIT.[9] When the friction torque is small, it has been shown[8] that the response follows the input command closely when all the compensations are fed forward.

11.3. CONVEYOR FOLLOWING WITH SINGLE-JOINT CONTROLLERS

Quite often the robot is required to follow a moving conveyor with a constant speed and perform a certain task on the conveyor. In this case the input signal of the desired position $\theta_d(t)$ must be updated very frequently, say, every $\frac{1}{60}$ sec, to synchronize the moving conveyor. In essence the input is a ramp signal $C_v t$ with a constant slope so that $X(s) = C_v/s^2$.

11.3.1. Velocity Error and Compensation

Applying the final value theorem to Eq. (11.46), with $X(s) = C_v/s^2$ one obtains the steady-state velocity error

$$e_{ssv} = \frac{[RB_{eff} + K_I(K_b + K_1 K_t)]C_v}{(nK_\theta K_I)} + e_{ssp} \tag{11.50}$$

Note that the contribution from the centrifugal term τ_c is not included in Eq. (11.50). Its effect, however, does exist since $\dot{\theta}_s(t) \neq 0$ when the robot is in motion. In fact, the effect becomes significant when the robot moves at a high speed, which is intuitively true from the view point of physics. Essentially, the centrifugal term affects the path along which the robot travels. Its feedforward compensation, however, is discussed separately later. Now, since the controller requires $\zeta \geq 1$ to avoid underdamping, Eq. (11.33) holds so that Eq. (11.50) becomes

$$e_{ssv} \geq 2C_v \sqrt{\frac{RJ_{eff}}{(nK_\theta K_I)}} + e_{ssp} \tag{11.51}$$

If the controller is designed conservatively to require $\omega_n \leq \omega_r/2$, then Eq. (11.32) holds and Eq. (11.51) reduces to

$$e_{ssv} \geq 4\left(\frac{C_v}{\omega}\right)\sqrt{\frac{J_{eff}}{J}} + e_{ssp} \tag{11.52}$$

which gives a lower bound for the steady-state velocity error.

To reduce the steady-state velocity error, additional feedforward signal v_d, corresponding to the desired constant slope of the ramp signal, is fed to the controller at the same place where τ_a and τ_d are injected. This is indicated in Figure 11.8b where $V_d(s)$ is the Laplace transform of v_d. As a result, $E(s)$ given by Eq. (11.46) is modified to become

$$E(s) = \frac{\begin{aligned}&\{RJ_{eff}s^2 + [RB_{eff} + K_I(K_b + K_1 K_t)]s\}X(s) - nK_I K_R V_d(s)\\ &\quad + nR\{F_m(s) + T_g() + nT_L(s) - K_I K_R[T_a(s) + T_d(s)]/R\}\end{aligned}}{\Omega(s)} \tag{11.53}$$

so that e_{ssv} in Eq. (11.50) now becomes

$$e_{ssv} = \frac{\{[RB_{eff} + K_I(K_b + K_1 K_t)]C_v - nK_I K_R \lim_{s \to 0} sV_d(s)\}}{(nK_\theta K_I)} + e_{ssp} \tag{11.54}$$

Consequently the first term of e_{ssv} vanishes if

$$V_d(s) = \frac{s(C_v/s^2)[RB_{eff} + K_I(K_b + K_1 K_t)]}{(nK_I K_R)} \tag{11.55}$$

But $C_v/s^2 = X(s)$ represents the ramp input signal. Hence

$$v_d = \frac{(dx/dt)[RB_{\text{eff}} + K_t(K_b + K_1K_t)]}{(nK_1K_R)} \tag{11.56}$$

which can be obtained directly from the input terminal of the controller[10] as indicated in Figure 11.8c. Since $x(t)$ is a ramp input $C_v t$, or $X(s) = C_v/s^2$, then dx/dt is the constant slope C_v, or $sX(s) = C_v/s$. Of course one may obtain dx/dt by computing the quotient $[x(t_i) - x(t_{i-1})]/(t_i - t_{i-1})$, where $x(t_i)$ and $x(t_{i-1})$ are values of two consecutive input signals. These arrangements will automatically take care of the steady-state error in the original positional-control mode. When the controller leaves the conveyor-following mode and enters the positional-control mode, $x(t)$ is a step input C_θ, or $X(s) = \Theta_d(s) = C_\theta/s$. Then $dx/dt = C_\theta\delta(t)$, or $sX(s) = C_\theta$ (an impulse that is absorbed by the energy-storing elements of the system), so that the compensation for the velocity error vanishes.

11.3.2. Compensation for the Centrifugal Term

The centrifugal contribution $\tau_c(t)$ can be computed from $D[\dot{\theta}_s(t)]^2$ where D is a proportional constant and $\dot{\theta}_s$ is the velocity of the robot link. The value of the velocity can be measured at the output shaft by means of a tachometer. The value of parameter D depends on the geometrical configuration of the robot and is discussed in detail in the section on the multiple-joint controller. Once D is determined $\tau_c(t)$ can be obtained at the output terminal. The resulting value is then used for compensation. To feed forward this compensating term to the system at the same point that T_a and T_d enter, as indicated in Figure 11.8c, a gain factor of $R/(K_1K_R)$ must be included to cancel the existing gains in the path. Figure 11.8d shows the schematic arrangement of the feedforward compensation for the centrifugal contribution.

11.4. CONTROLLER FOR ROBOT WITH MULTIPLE JOINTS

Intuitively, the motion of each joint of a robot is not independent of other joints. There are force and moment interactions among the moving joints that cause inadequacy in the use of the preceding positional controller for each joint. Thus an additional compensation is needed to overcome the interaction. To determine the compensation for interaction it is necessary to analyze the dynamic behavior of the robot.

11.4.1. Lagrangian Formulation of Dynamic Equation

A discussion of the Lagrangian equation can be found in most of physics textbooks. It represents the dynamic behavior of a system of rigid bodies, and it has the form

$$\frac{d}{dt}\left(\frac{\partial L}{\partial \dot{q}_i}\right) - \frac{\partial L}{\partial q_i} = \tau_i, \qquad i = 1, 2, \cdots, n \tag{11.57}$$

where q_i = generalized coordinates
$L = L(q_1, \cdots, q_n, \dot{q}_1, \cdots, \dot{q}_n)$ = Lagrangian
τ_i = generalized forcing function

The generalized coordinate q_i represents the displacement θ_s of joint i. The Lagrangian is also defined as

$$L = (\text{kinetic energy of the system}) - (\text{potential energy of the system})$$

By applying the Lagrangian equation to a robot with n joints (or n links), one obtains[6,10]

$$\tau_i = \sum_{j=1}^{n} D_{ij}\ddot{q}_j + J_{ai}\ddot{q}_i + \sum_{j=1}^{n} D_{ijj}(\dot{q}_j)^2 + \sum_{\substack{j=1 \\ j \neq k}}^{n} \sum_{k=1}^{n} D_{ijk}\dot{q}_j\dot{q}_k + D_i \tag{11.58}$$

where

$$D_{ij} = \sum_{p=\max(i,j)}^{n} \text{Tr}[\mathbf{U}_{pj}\mathbf{J}_p(\mathbf{U}_{pi})'] \tag{11.59}$$

$$D_{ijk} = \sum_{p=\max(i,j,k)}^{n} \text{Tr}[\mathbf{U}_{pjk}\mathbf{J}_p(\mathbf{U}_{pi})'] \tag{11.60}$$

$$D_i = -\sum_{p=i}^{n} m_p \hat{\mathbf{g}}' \mathbf{U}_{pi} \hat{\mathbf{r}}_p \tag{11.61}$$

where $\text{Tr} = $ trace operator
  $(\ )' = $ transpose of $(\ )$
  $\tau_i = $ input generalized force for joint i
  $m_p = $ mass of link p
  $\hat{\mathbf{r}}_p = $ a vector describing the center of mass of link p with respect to pth coordinate system
  $\hat{\mathbf{g}}' = [0, 0, 9.8, 0 \text{ m/sec}^2]$ is a gravitational acceleration vector at a sea level base
  $\mathbf{J}_p = $ inertia matrix for link p

$$\mathbf{U}_{pj} = \frac{\partial T_o^p}{\partial q_j} = \begin{cases} (\mathbf{T}_o^{j-1})\mathbf{Q}_j(\mathbf{T}_{j-1}^p) & \text{for } p \geq j \\ 0 & \text{otherwise} \end{cases} \tag{11.62}$$

$$\mathbf{U}_{pjk} = \frac{\partial^2 T_o^p}{\partial q_j \partial q_k} = \begin{cases} (\mathbf{T}_o^{j-1})\mathbf{Q}_j(\mathbf{T}_{j-1}^{k-1})\mathbf{Q}_k(\mathbf{T}_{k-1}^p) & \text{for } p \geq k \geq j \\ (\mathbf{T}_o^{k-1})\mathbf{Q}_k(\mathbf{T}_{k-1}^{j-1})\mathbf{Q}_j(\mathbf{T}_{j-1}^p) & \text{for } p \geq j \geq k \\ 0 & \text{otherwise} \end{cases} \tag{11.63}$$

$$\mathbf{Q}_j = \begin{cases} \begin{bmatrix} 0 & -1 & 0 & 0 \\ 1 & 0 & 0 & 0 \\ 0 & 0 & 0 & 0 \\ 0 & 0 & 0 & 0 \end{bmatrix} & \text{if joint } j \text{ is rotational} \\[4em] \begin{bmatrix} 0 & 0 & 0 & 0 \\ 0 & 0 & 0 & 0 \\ 0 & 0 & 0 & 1 \\ 0 & 0 & 0 & 0 \end{bmatrix} & \text{if joint } j \text{ is translational} \end{cases} \tag{11.64}$$

$\mathbf{T}_j^k = 4 \times 4$ matrix that transforms any vector expressed in kth coordinate system to the same vector expressed in jth coordinate system
$q_k = $ generalized coordinate (i.e., joint displacement)

11.4.2 Coupling between Joints and Compensation

For each joint i, the required torque or force is divided into five groups as shown in Eq. (11.58). The first group represents the contribution from inertias of all the joints. Unlike the single-joint case in which all the joints are considered to move independently with no interaction, now there are contributions from coupling inertias between joints. These torque terms $\sum_{\substack{j-1 \\ j \neq i}}^{n} D_{ij}\ddot{q}_j$ must be fed forward in the controller for joint i, as shown in Figure 11.9, to compensate the interaction between joints. The second term in Eq. (11.58) represents the inertia torque of the actuator of joint i which has already been included in J_{eff} term as outlined during the discussion on the single-joint controller. The last term results from the gravitational acceleration, which has also been compensated by the feedforward term τ_a. This is the anticipated gravitational torque signal which must be computed by Eq. (11.48), that is, $\tau_a = (R/K_I K_R)/\hat{\tau}_g$ where $\hat{\tau}_g$ is the estimate of gravitational torque τ_g. Intuitively, one uses D_i for the best estimate for τ_g for joint i controller. Thus by Eq. (11.61), one sets

$$\hat{\tau}_g = D_i = -\sum_{p=i}^{n} m_p \hat{\mathbf{g}}' \mathbf{U}_{pi} \hat{\mathbf{r}}_p \tag{11.65}$$

for joint i.

The third and fourth groups in Eq. (11.58) represent the contributions from, respectively, the centrifugal term and the Coriolis force. Again, these torque terms must be fed forward in the controller for joint i to compensate the physical interactions between joints as shown in Figure 11.9. This figure depicts the complete block diagram of the controller for joint i of an industrial robot, $i = 1, 2, \cdots,$ n. To implement these n controllers, the values of the feedforward elements D_{ij}, D_{ijk}, and D_i must be computed for the specific robot, which is discussed in the following sections.

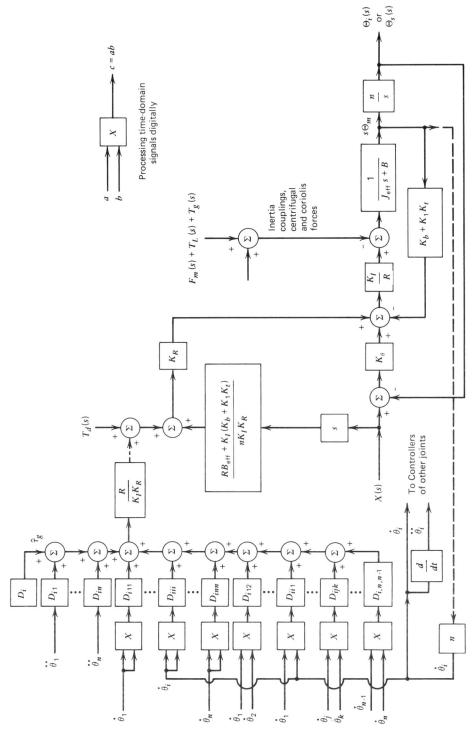

Fig. 11.9. Block diagram of a complete controller for joint i of a robot having n joints.

11.4.3. Computation of Compensation for Coupling Inertia

The computation of term D_{ij} is, unfortunately, very complicated and time-consuming. To illustrate the difficulty, Eq. (11.58) is expanded for a six-joint robot, $n = 6$, as follows:

$$
\begin{aligned}
\tau_i = {}& D_{i1}\ddot{q}_1 + D_{i2}\ddot{q}_2 + \cdots + D_{i6}\ddot{q}_6 + J_{ai}\ddot{q}_i \\
& + D_{i11}\dot{q}_1^2 + D_{i22}\dot{q}_2^2 + \cdots + D_{i66}\dot{q}_6^2 \\
& + D_{i12}\dot{q}_1\dot{q}_2 + D_{i13}\dot{q}_1\dot{q}_3 + \cdots + D_{i16}\dot{q}_1\dot{q}_6 \\
& + \cdots \\
& + D_{i45}\dot{q}_4\dot{q}_5 + \cdots + D_{i56}\dot{q}_5\dot{q}_6 + D_i, \qquad i = 1, 2, \cdots, 6
\end{aligned}
\tag{11.66}
$$

For $i = 1$, the term $D_{i1} = D_{11}$ is further expanded as shown in Figure 11.10 in which $\theta_i = q_i$, $i = 1, 2, \cdots, 6$. Obviously it is not a simple computational task, especially when the position-dependent and orientation-dependent parameters change as the robot moves. Therefore it warrants the effort of searching for methods of simplifying the computation. There are three known approaches of simplification, namely, geometric/numeric, composite, and differential transformation. Bejczy's geometric/numeric evaluation[11] deals with the nature of joints whether revolute or prismatic. Thus the $\mathbf{T}_j^k$ matrices in Eqs. (11.62), (11.63), and (11.64) can be simplified in advance. Since many elements in the 4×4 matrices are zeros, the resulting expressions for D_i, D_{ij}, and D_{ijk} are less complicated.[11] The composite technique by Luh and Lin[12] involves the comparison of all the terms in Newton-Euler formulation of the dynamic equation[13] in a computer. Some of the terms may be eliminated under various criteria. The remaining terms are then rearranged in a Lagrangian formulation. The upshot is a computer output of a simplified equation in symbolic form. Paul's differential transformation[10] converts the partial derivatives of the matrix transformation $\partial \mathbf{T}_a^p / \partial q_j$ into the matrix product of the transformation and a differential matrix which reduces to a much simpler form. Discussion of the third approach follows. To facilitate the discussion, it is necessary to introduce the homogeneous transformation, which is a 4×4 matrix that represents the rotation and translation of vectors in some coordinate systems.

Refer to Figure 11.11a, which shows two aligned coordinate systems $(\mathbf{x}, \mathbf{y}, \mathbf{z})$ and $(\mathbf{x}', \mathbf{y}', \mathbf{z}')$. A point P is fastened upon $(\mathbf{x}', \mathbf{y}', \mathbf{z}')$, that is, when $(\mathbf{x}', \mathbf{y}', \mathbf{z}')$ moves with respect to $(\mathbf{x}, \mathbf{y}, \mathbf{z})$, point P moves with it. Suppose $(\mathbf{x}', \mathbf{y}', \mathbf{z}')$ is rotated γ radians about z axis as shown in Figure 11.11b. Since P moves together with $(\mathbf{x}', \mathbf{y}', \mathbf{z}')$, its location in that coordinate system remains unchanged. However, the location of P in $(\mathbf{x}, \mathbf{y}, \mathbf{z})$ coordinates changes. Refer to Figure 11c,

$$
\begin{cases}
a_1 = a \cos \gamma - b \sin \gamma \\
b_1 = a \sin \gamma + b \cos \gamma \\
c_1 = c
\end{cases}
\tag{11.67}
$$

which can be written as

$$
\begin{bmatrix} a_1 \\ b_1 \\ c_1 \\ 1 \end{bmatrix} =
\begin{bmatrix}
\cos \gamma & -\sin \gamma & 0 & 0 \\
\sin \gamma & \cos \gamma & 0 & 0 \\
0 & 0 & 1 & 0 \\
0 & 0 & 0 & 1
\end{bmatrix}
\begin{bmatrix} a \\ b \\ c \\ 1 \end{bmatrix}
\tag{11.68}
$$

$$
\begin{aligned}
D_{11} = {}& m_1 k_{122}^2 \\
& + m_2 [k_{211}^2 s^2\theta_2 + k_{233}^2 c^2\theta_2 + r_2(2\bar{y}_2 + r_2)] \\
& + m_3 [k_{322}^2 s^2\theta_2 + k_{333}^2 c^2\theta_2 + r_3(2\bar{z}_3 + r_3)s^2\theta_2 + r_2^2] \\
& + m_4 \{\tfrac{1}{2}k_{411}^2 [s^2\theta_2(2s^2\theta_4 - 1) + s^2\theta_4] + \tfrac{1}{2}k_{422}^2(1 + c^2\theta_2 + s^2\theta_4) \\
& \qquad + \tfrac{1}{2}k_{433}^2 [s^2\theta_2(1 - 2s^2\theta_4) - s^2\theta_4] + r_3^2 s^2\theta_2 + r_2^2 - 2\bar{y}_4 r_3 s^2\theta_2 + 2\bar{z}_4(r_2 s\theta_4 + r_3 s\theta_2 c\theta_2 c\theta_4)\} \\
& + m_5 \{\tfrac{1}{2}(-k_{511}^2 + k_{522}^2 + k_{533}^2) [(s\theta_2 s\theta_5 - c\theta_2 s\theta_4 c\theta_5)^2 + c^2\theta_4 c^2\theta_5] \\
& \qquad + \tfrac{1}{2}(k_{511}^2 - k_{522}^2 + k_{533}^2)(s^2\theta_4 + c^2\theta_2 c^2\theta_4) \\
& \qquad + \tfrac{1}{2}(k_{511}^2 + k_{522}^2 - k_{533}^2) [(s\theta_2 c\theta_5 + c\theta_2 s\theta_4 s\theta_5)^2 + c^2\theta_4 s^2\theta_5] + r_3^2 s^2\theta_2 + r_2^2 \\
& \qquad + 2\bar{z}_5 [r_3(s^2\theta_2 c\theta_5 + s\theta_2 s\theta_4 c\theta_4 s\theta_5) - r_2 c\theta_4 s\theta_5]\} \\
& + m_6 \{\tfrac{1}{2}(-k_{611}^2 + k_{622}^2 + k_{633}^2) [(s\theta_2 s\theta_5 c\theta_6 - c\theta_2 s\theta_4 c\theta_5 c\theta_6 - c\theta_2 c\theta_4 s\theta_6)^2 + (c\theta_4 c\theta_5 c\theta_6 - s\theta_4 s\theta_6)^2] \\
& \qquad + \tfrac{1}{2}(k_{611}^2 - k_{622}^2 + k_{633}^2) [(c\theta_2 s\theta_4 c\theta_5 s\theta_6 - s\theta_2 s\theta_5 s\theta_6 - c\theta_2 c\theta_4 c\theta_6)^2 + (c\theta_4 c\theta_5 s\theta_6 + s\theta_4 c\theta_6)^2] \\
& \qquad + \tfrac{1}{2}(k_{611}^2 + k_{622}^2 - k_{633}^2) [(c\theta_2 s\theta_4 s\theta_5 + s\theta_2 c\theta_5)^2 + c^2\theta_4 s^2\theta_5] \\
& \qquad + [r_6 c\theta_2 s\theta_4 s\theta_5 + (r_6 c\theta_5 + r_3)s\theta_2]^2 + (r_6 c\theta_4 s\theta_5 - r_2)^2 \\
& \qquad + 2\bar{z}_6 [r_6(s^2\theta_2 c^2\theta_5 + c^2\theta_4 s^2\theta_5 + c^2\theta_2 s^2\theta_4 s^2\theta_5 + 2s\theta_2 c\theta_2 s\theta_4 s\theta_5 c\theta_5) \\
& \qquad\qquad + r_3(s\theta_2 c\theta_2 s\theta_4 s\theta_5 + s^2\theta_2 c\theta_5) - r_2 c\theta_4 s\theta_5]\}
\end{aligned}
$$

Fig. 11.10. Coefficient of inertia term for joint 1.

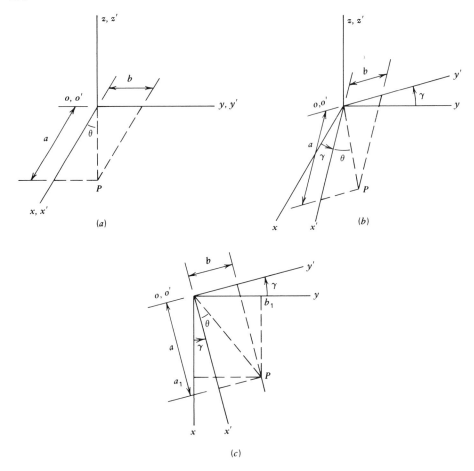

Fig. 11.11. Rotation of coordinates with reference to base coordinates.

The reason for the addition of the fourth row and fourth column in the matrix of Eq. (11.68) will be clear when the translation of the position is introduced. In the meantime the matrix in Eq. (11.68) is the homogeneous transformation that rotates the vector or point P, and hence the coordinates (x', y', z'), γ radians about z axis. For convenience, the matrix is denoted by $\mathbf{R}(\mathbf{z}, \gamma)$ so that Eq. (11.68) may be written as

$$\begin{bmatrix} a_1 \\ b_1 \\ c_1 \\ 1 \end{bmatrix} = \mathbf{R}(\mathbf{z}, \gamma) \begin{bmatrix} a \\ b \\ c \\ 1 \end{bmatrix} \tag{11.69}$$

Likewise,

$$\mathbf{R}(\mathbf{x}, \alpha) = \begin{bmatrix} 1 & 0 & 0 & 0 \\ 0 & \cos\alpha & -\sin\alpha & 0 \\ 0 & \sin\alpha & \cos\alpha & 0 \\ 0 & 0 & 0 & 1 \end{bmatrix}, \quad \mathbf{R}(\mathbf{y}, \beta) = \begin{bmatrix} \cos\beta & 0 & \sin\beta & 0 \\ 0 & 1 & 0 & 0 \\ -\sin\beta & 0 & \cos\beta & 0 \\ 0 & 0 & 0 & 1 \end{bmatrix} \tag{11.70}$$

Now suppose (x', y', z') is rotated β radians about y axis and then α radians about x axis. Again, the location of point P in (x', y', z') does not change but in (x, y, z) changes from (a_1, b_1, c_1) to (a_2, b_2, c_2) as

$$\begin{bmatrix} a_2 \\ b_2 \\ c_2 \\ 1 \end{bmatrix} = \mathbf{R}(\mathbf{x}, \alpha)\mathbf{R}(\mathbf{y}, \beta)\mathbf{R}(\mathbf{z}, \gamma) \begin{bmatrix} a \\ b \\ c \\ 1 \end{bmatrix} \tag{11.71}$$

or

$$\begin{bmatrix} a_2 \\ b_2 \\ c_2 \\ 1 \end{bmatrix} = \begin{bmatrix} R_{11} & R_{12} & R_{13} & 0 \\ R_{21} & R_{22} & R_{23} & 0 \\ R_{31} & R_{32} & R_{33} & 0 \\ 0 & 0 & 0 & 1 \end{bmatrix} \begin{bmatrix} a \\ b \\ c \\ 1 \end{bmatrix} \tag{11.72}$$

where

$$\begin{cases} R_{11} = \cos \beta \cos \gamma \\ R_{12} = -\cos \beta \sin \gamma \\ R_{13} = \sin \beta \\ R_{21} = \cos \alpha \sin \gamma + \sin \alpha \sin \beta \cos \gamma \\ R_{22} = \cos \alpha \cos \gamma - \sin \alpha \sin \beta \sin \gamma \\ R_{23} = -\sin \alpha \cos \beta \\ R_{31} = \sin \alpha \sin \gamma - \cos \alpha \sin \beta \cos \gamma \\ R_{32} = \sin \alpha \cos \gamma + \cos \alpha \sin \beta \sin \gamma \\ R_{33} = \cos \alpha \cos \beta \end{cases} \tag{11.73}$$

Now suppose $(\mathbf{x}', \mathbf{y}', \mathbf{z}')$ is translated t_x, t_y, and t_z units, respectively, along x, y, and z axes. Then

$$\begin{bmatrix} a_3 \\ b_3 \\ c_3 \\ 1 \end{bmatrix} = \begin{bmatrix} a_2 \\ b_2 \\ c_2 \\ 1 \end{bmatrix} + \begin{bmatrix} t_x \\ t_y \\ t_z \\ 0 \end{bmatrix} = \begin{bmatrix} R_{11} & R_{12} & R_{13} & t_x \\ R_{21} & R_{22} & R_{23} & t_y \\ R_{31} & R_{32} & R_{33} & t_z \\ 0 & 0 & 0 & 1 \end{bmatrix} \begin{bmatrix} a \\ b \\ c \\ 1 \end{bmatrix} \tag{11.74}$$

Let

$$\mathbf{L}(t_x, t_y, t_z) = \begin{bmatrix} 1 & 0 & 0 & t_x \\ 0 & 1 & 0 & t_y \\ 0 & 0 & 1 & t_z \\ 0 & 0 & 0 & 1 \end{bmatrix} \tag{11.75}$$

denote the aforementioned linear translation. Then the 4×4 matrix in (11.74) may be written as

$$\mathbf{L}(t_x, t_y, t_z)\mathbf{R}(\mathbf{x}, \alpha)\mathbf{R}(\mathbf{y}, \beta)\mathbf{R}(\mathbf{z}, \gamma) = \begin{bmatrix} R_{11} & R_{12} & R_{13} & t_x \\ R_{21} & R_{22} & R_{23} & t_y \\ R_{31} & R_{32} & R_{33} & t_z \\ 0 & 0 & 0 & 1 \end{bmatrix} \tag{11.76}$$

which represents a rotation of γ radians about z axis, then a rotation of β radians about y axis followed by a rotation of α radians about x axis, and finally a translation of t_x, t_y, t_z, respectively, along the x, y, and z axes. Thus the 4×4 matrix, which is called the homogeneous transformation, includes rotation as well as translation of the coordinates $(\mathbf{x}', \mathbf{y}', \mathbf{z}')$. Since the matrix multiplications do not commute, the order of multiplying the matrices $\mathbf{L}$ and $\mathbf{R}$ in Eq. (11.76) cannot be interchanged. Should, for example, $\mathbf{R}(\mathbf{x}, \alpha)$ and $\mathbf{R}(\mathbf{y}, \beta)$, or $\mathbf{L}(t_x, t_y, t_z)$ and $\mathbf{R}(\mathbf{z}, \gamma)$ interchange their places in Eq. (11.76), the resulting matrix, and hence the physical order of rotations and translation, would be different.

Refer to Figure 11.12 and suppose that the pth joint of the robot is originally at the point P. Then

$$\mathbf{T}_o^p = \begin{bmatrix} \mathbf{x}_p & \mathbf{y}_p & \mathbf{z}_p & \mathbf{l}_p \\ 0 & 0 & 0 & 1 \end{bmatrix} \tag{11.77}$$

represents the coordinate frame with respect to the base coordinates. Now rotate the joint in the following order: γ radians about z_o axis, β radians about y_o axis, α radians about x_o axis; and finally translate $\mathbf{t}$ with respect to $(\mathbf{x}_o, \mathbf{y}_o, \mathbf{z}_o)$ coordinates. Suppose the resulting orientation and position of the joint is, respectively, $(\mathbf{x}_p^1, \mathbf{y}_p^1, \mathbf{z}_p^1)$ and $\mathbf{l}_p^1$ with reference to $(\mathbf{x}_o, \mathbf{y}_o, \mathbf{z}_o)$. Then the current state of the hand with reference to base coordinates can be represented by

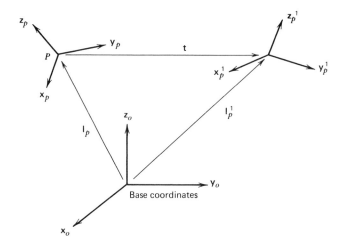

Fig. 11.12. Rotation and translation of coordinates with reference to base coordinates.

$$\begin{bmatrix} \mathbf{x}_p^1 & \mathbf{y}_p^1 & \mathbf{z}_p^1 & \mathbf{l}_p^1 \\ 0 & 0 & 0 & 1 \end{bmatrix} = \begin{bmatrix} R_{11} & R_{12} & R_{13} & t_x \\ R_{21} & R_{22} & R_{23} & t_y \\ R_{31} & R_{32} & R_{33} & t_z \\ 0 & 0 & 0 & 1 \end{bmatrix} \mathbf{T}_o^p \tag{11.78}$$

Now if $\mathbf{T}_o^p$ is perturbed by a small translation and rotation with respect to the base coordinates, then $t_x \leftarrow \delta_o x$, $t_y \leftarrow \delta_o y$, $t_z \leftarrow \delta_o z$, $\alpha \leftarrow \delta_o \alpha$, $\beta \leftarrow \delta_o \beta$, $\gamma \leftarrow \delta_o \gamma$. But $\cos(\delta_o \alpha) \simeq 1$, $\sin(\delta_o \alpha) \simeq \delta_o \alpha$, $\cdots$, $\delta_o \alpha \delta_o \beta \simeq 0$, $\cdots$, and so on. Hence by Eq. (11.73)

$$\begin{bmatrix} R_{11} & R_{12} & R_{13} & t_x \\ R_{21} & R_{22} & R_{23} & t_y \\ R_{31} & R_{32} & R_{33} & t_z \\ 0 & 0 & 1 & 1 \end{bmatrix} = \begin{bmatrix} 1 & -\delta_o \gamma & \delta_o \beta & \delta_o x \\ \delta_o \gamma & 1 & -\delta_o \alpha & \delta_o y \\ -\delta_o \beta & \delta_o \alpha & 1 & \delta_o z \\ 0 & 0 & 0 & 1 \end{bmatrix} \tag{11.79}$$

On the other hand, one may express

$$\begin{bmatrix} \mathbf{x}_p^1 & \mathbf{y}_p^1 & \mathbf{z}_p^1 & \mathbf{l}_p^1 \\ 0 & 0 & 0 & 1 \end{bmatrix} = \mathbf{T}_o^p + \delta_o \mathbf{T}_o^p \tag{11.80}$$

Combining Eqs. (11.78), (11.79), and (11.80) yields the perturbation

$$\delta_o \mathbf{T}_o^p = \begin{bmatrix} 0 & -\delta_o \gamma & \delta_o \beta & \delta_o x \\ \delta_o \gamma & 0 & -\delta_o \alpha & \delta_o y \\ -\delta_o \beta & \delta_o \alpha & 0 & \delta_o z \\ 0 & 0 & 0 & 0 \end{bmatrix} \mathbf{T}_o^p \tag{11.81}$$

The matrix in Eq. (11.81) is the variational operator with respect to the base coordinates. If the variation is referred to pth joint's own coordinates, then it must be premultiplied by a 4×4 transformation matrix

$$\mathbf{T}_p^o = (\mathbf{T}_o^p)^{-1} \tag{11.82}$$

which transforms any vectors or coordinate frames with reference to base coordinates ($\mathbf{x}_o$, $\mathbf{y}_o$, $\mathbf{z}_o$) to the pth joint coordinates ($\mathbf{x}_p$, $\mathbf{y}_p$, $\mathbf{z}_p$). Thus the perturbation on the pth joint coordinate frame with reference to its own coordinates is

$$\delta_p \mathbf{T}_o^p = (\mathbf{T}_o^p)^{-1} \begin{bmatrix} 0 & -\delta_o \gamma_p & \delta_o \beta_p & \delta_o x_p \\ \delta_o \gamma_p & 0 & -\delta_o \alpha_p & \delta_o y_p \\ -\delta_o \beta_p & \delta_o \alpha_p & 0 & \delta_o y_p \\ 0 & 0 & 0 & 0 \end{bmatrix} \mathbf{T}_o^p \tag{11.83}$$

Since

$$(\mathbf{T}_o^p)^{-1} = \begin{bmatrix} x_{px} & x_{py} & x_{pz} & -\mathbf{x}_p'\mathbf{l}_p \\ y_{px} & y_{py} & y_{pz} & -\mathbf{y}_p'\mathbf{l}_p \\ z_{px} & z_{py} & z_{pz} & -\mathbf{z}_p'\mathbf{l}_p \\ 0 & 0 & 0 & 1 \end{bmatrix} \tag{11.84}$$

Then by some algebraic manipulation as shown by Paul,[10]

$$\delta_p \mathbf{T}_o^p = \begin{bmatrix} 0 & -\mathbf{z}_p'\boldsymbol{\Delta}_p & \mathbf{y}_p'\boldsymbol{\Delta}_p & \mathbf{x}_p'(\boldsymbol{\Delta}_p \times \mathbf{l}_p + \mathbf{d}_p) \\ \mathbf{z}_p'\boldsymbol{\Delta}_p & 0 & -\mathbf{x}_p'\boldsymbol{\Delta}_p & \mathbf{y}_p'(\boldsymbol{\Delta}_p \times \mathbf{l}_p + \mathbf{d}_p) \\ -\mathbf{y}_p'\boldsymbol{\Delta}_p & \mathbf{x}_p'\boldsymbol{\Delta}_p & 0 & \mathbf{z}_p'(\boldsymbol{\Delta}_p \times \mathbf{l}_p + \mathbf{d}_p) \\ 0 & 0 & 0 & 0 \end{bmatrix} \tag{11.85}$$

Since $\delta_p \mathbf{T}_o^p$ is the perturbation on the pth joint coordinate frame with reference to its own coordinates, then, based on the results in Eq. (11.81), it can be written as

$$\delta_p \mathbf{T}_o^p = \begin{bmatrix} 0 & -\delta_p\gamma & \delta_o\beta & \delta_p x \\ \delta_p\gamma & 0 & -\delta_p\alpha & \delta_p y \\ -\delta_p\beta & \delta_p\alpha & 0 & \delta_p z \\ 0 & 0 & 0 & 0 \end{bmatrix} \mathbf{T}_p^p \tag{11.86}$$

where $\mathbf{T}_p^p$ is an identity matrix, so that

$$\begin{aligned} \delta_p\alpha &= \mathbf{x}_p'\boldsymbol{\Delta}_p \\ \delta_p\beta &= \mathbf{y}_p'\boldsymbol{\Delta}_p \\ \delta_p\gamma &= \mathbf{z}_p'\boldsymbol{\Delta}_p \\ \delta_p x &= \mathbf{x}_p'(\boldsymbol{\Delta}_p \times \mathbf{l}_p + \mathbf{d}_p) \\ \delta_p y &= \mathbf{y}_p'(\boldsymbol{\Delta}_p \times \mathbf{l}_p + \mathbf{d}_p) \\ \delta_p z &= \mathbf{z}_p'(\boldsymbol{\Delta}_p \times \mathbf{l}_p + \mathbf{d}_p) \end{aligned} \tag{11.87}$$

Equations (11.86) and (11.87) serve as the basis for numerical computation because of their reduced form. Now, if the perturbation on the pth joint coordinate frame is expressed with reference to jth joint coordinates $(\mathbf{x}_j, \mathbf{y}_j, \mathbf{z}_j)$, then it has a form

$$\delta_j \mathbf{T}_o^p = \mathbf{T}_j^p(\delta_p \mathbf{T}_o^p) \tag{11.88}$$

As a limit,

$$\frac{\partial \mathbf{T}_o^p}{\partial q_j} dq_j = \mathbf{T}_j^p(\delta_p \mathbf{T}_o^p) \tag{11.89}$$

Through a lengthy algebraic manipulation, Reference 10 shows that (with the condition that all the cross-inertia terms are ignored since they are relatively insignificant by experiments[6,10].)

$$\begin{aligned} D_{ij} = \sum_{p=\max(i,j)}^{n} m_p\{[&(\delta_p\alpha_i)k_{pxx}^2(\delta_p\alpha_j) + (\delta_p\beta_i)k_{pyy}^2(\delta_p\beta_j) \\ &+ (\delta_p\gamma_i)k_{pzz}^2(\delta_p\gamma_j)] + [(\mathbf{T}_i^o\mathbf{d}_p)'(\mathbf{T}_j^o\mathbf{d}_p)] \\ &+ [\hat{r}_p'((\mathbf{T}_i^o\mathbf{d}_p) \times (\mathbf{T}_j^o\boldsymbol{\Delta}_p) + (\mathbf{T}_j^o\mathbf{d}_p) \times (\mathbf{T}_i^o\boldsymbol{\Delta}_p))]\} \end{aligned} \tag{11.90}$$

where $\delta_p\alpha_i$ is a small rotation of coordinates frame of joint p about x axis with respect to ith joint's coordinates, and k_{pxx} is the radius of gyration xx of joint (or link) p about the origin of pth joint's coordinates, and so on.

11.4.4. Computation of Compensation for Gravity, Centrifugal, and Coriolis Terms

From Eqs. (11.61) and (11.62) one obtains

$$D_i = -\sum_{p=i}^{n} m_p \hat{\mathbf{g}}' \frac{\partial \mathbf{T}_o^p}{\partial q_i} \hat{\mathbf{r}}_p \tag{11.91}$$

By some algebraic manipulation, Reference 10 also shows that

$$D_i = \Gamma' \sum_{p=i}^{n} m_p(\hat{\mathbf{r}}_p^{i-1}) \tag{11.92}$$

where $\hat{\mathbf{r}}_p^{i-1}$ is a vector describing the center of mass of link p with respect to $(i-1)$th coordinates, and

$$\Gamma' = \begin{cases} [-\hat{\mathbf{g}}'\mathbf{y}_{i-1} \ \ \hat{\mathbf{g}}'\mathbf{x}_{i-1} \ \ 0 \ \ 0] & \text{if joint p is revolute} \\ [0 \ \ 0 \ \ 0 \ \ -\hat{\mathbf{g}}'\mathbf{z}_{i-1}] & \text{if joint p is prismatic} \end{cases} \tag{11.93}$$

Since the term D_{ijk} contains a second partial derivative $\partial^2 \mathbf{T}_o^p/(\partial q_j \partial q_k)$, it is not able to simplify Eq. (11.60) for computation. Conventionally, one often ignores the centrifugal and Coriolis terms. The justification is that these two terms are velocity dependent. When the robot starts to move from its initial location and approaches its goal location, the velocities are usually low, and hence the contributions from these two terms are insignificant. Once it picks up the velocity, the robot is traveling in the space, and normally the traveled path is not of importance. Should the path be important, such as in avoiding collision with obstacles, then these two terms may not be ignored. They must be computed either by Eq. (11.60) or by using the Newton-Euler formulation approach,[13] which is a computational scheme. This scheme has been proven to be computationally efficient.[14-16] Also, Bejczy[6,11] used the geometric/numeric approach to show that for the last four joints of the Stanford-JPL manipulator, which has six joints ($n = 6$), the following terms are identically zero:

$D_{333} \ D_{334} \ D_{335} \ D_{336} \ D_{344} \ D_{345} \ D_{346} \ D_{356} \ D_{366}$

$D_{433} \ D_{434} \ D_{435} \ D_{436} \ D_{444} \ D_{446} \ D_{455} \ D_{466}$

$D_{533} \ D_{534} \ D_{535} \ D_{536} \ D_{555} \ D_{556} \ D_{566}$

$D_{633} \ D_{634} \ D_{635} \ D_{636} \ D_{644} \ D_{666}$

Thus it is possible to reduce the computational task if the geometrical configuration of the robot is known and if an analysis is carried out.

As mentioned before, the Newton-Euler formulation yields a computationally efficient scheme. Further shortening of computing time is possible by means of parallel computations using a computer with multiple central-processing units (CPU). A variable branch-and-bound method, which determines an optimum ordered schedule for each of the CPUs, was developed by Luh and Lin.[17] When the computational task is reduced and the computational time is shortened, it is then feasible to have a real-time controller for the robot.

11.5. PATH TRACKING BY ROBOT WITH MULTIPLE JOINTS

If the robot is required to travel along a prescribed path, the controller must keep up with path tracking. There are two alternatives to achieve the control of the desired path along which the hand of the robot travels: control at the hand level or at the joint level. In either case the transformation between the Cartesian and joint coordinates is required. With positional controllers the path tracking can be accomplished by dividing the Cartesian path into a number of segments. Each end point of the segments is transformed into joint coordinates, and then the positional control is applied from point to point in joint coordinates. This approach corresponds to Case 2 in Figure 11.1, which was briefly mentioned previously in Section 11.3 on conveyor following. A number of facts related to this approach should be mentioned. By transforming all the end points of segments of the Cartesian path, one essentially constructs n corresponding trajectories in joint coordinates, one for each of the n joints. If these segments, $[d\mathbf{p}(t)' \ d\boldsymbol{\theta}(t)']$, are very short, the increments of joint displacement dq_i between adjacent points are very small so that $\sin dq_i \simeq dq_i$ and $\cos dq_i \simeq 1$. Thus the transformation $\mathbf{f}(\cdot)$ defined by Eq. (11.5) becomes a differential transformation, which is usually linear. This transformation is the Jacobian matrix of the displacement, which contains trigonometric functions of the joint displacement with respect to the joint coordinates before the differential increment takes place.[18] Analytically, the solution dq_i in terms of $d\mathbf{p}$ and $d\boldsymbol{\theta}$ can be obtained simply by inverting the Jacobian matrix. Although it is sometimes possible, it is usually difficult since the Jacobian is quite complicated. Numerical solution is also possible but usually requires long computing time. Moreover, the Jacobian matrix becomes singular when the robot reaches a degenerate position at which the solution dq_i is not unique (i.e., more than one value of dq_i yields a same $d\mathbf{p}$ and $d\boldsymbol{\theta}$.) An alternative method proven to work successfully is to differentiate the solution of Eq. (11.5) directly[18] so that matrix inversion is avoided. This is possible since for a given robot with fixed dimensions the transformation $\mathbf{f}$ is known. Using this approach, one must set dq_i to zero if it is physically impossible due to constraints, or if it is undetermined so that the solution is forced to be unique. It usually results in a simpler expression.[18]

Conventionally, a sequence of sampling points is specified on the desired Cartesian path, such as the seam in arc welding, through which the hand must pass. If the control is planned at the hand level that corresponds to Case 3 in Figure 11.1, the hand is controlled to travel along straight-line segments between adjacent sampling points, although the desired path is not necessarily a straight line in some practical applications. Thus the piecewise linear path along which the hand actually travels inscribes the desired path. Intuitively the error between these two paths depends on the number of specified sampling points. To achieve the straight-line motion in the Cartesian coordinates, Whitney[19] applied the Jacobian and its inverse repeatedly to constrain the joint velocity. Paul[20] took a different approach by selecting refined sampling points on the straight-line segments at fixed, short time durations while the hand is in motion. These refined points are transformed into joint coordinates during motion execution for the purpose of straight-line-segment path control. Unless Eq. (11.58) is in its reduced form, as discussed in the preceding section, both methods require a considerable amount of real time computation, which lowers the upper bound of the sampling rate and deteriorates the accuracy of path control.

To avoid the on-line coordinate transformation of the refined sampling points, the control may be planned again at the joint level. Taylor[21] preselected enough intermediate points on the Cartesian straight-line segment and transformed them into joint coordinates during the planning stage. Every pair of the adjacent, transformed points is then connected by a straight-line segment along which the joints are controlled to move. Thus points between Cartesian intermediate points need not be computed for transformation. However, a straight-line segment in joint coordinates does not necessarily correspond to a straight line in Cartesian coordinates. Thus an error results between the Cartesian straight-line path and the actually traveled path corresponding to the straight-line segment in joint coordinates. To achieve better accuracy, the error is not allowed to exceed a prespecified amount by adding enough numbers of Cartesian intermediate points. In doing so, although the error described is forced to stay within its bound, the original error associated with the desired Cartesian path still exists and is not changed. As described earlier, this error is caused by the fact that the Cartesian straight-line segments only inscribe the desired path.

To eliminate the inscribing error, the joint trajectories may be constructed directly from the Cartesian path so that the intermediate step of inscribing straight-line segments is avoided. That is, the desired Cartesian path may be transformed into n corresponding joint trajectories, one for each joint, to serve as reference inputs to the joint-level controllers. Intuitively one is tempted to apply the inverse Jacobian. However, the transformation is defined on a point-to-point basis. Thus to represent the Cartesian path by n joint trajectories, enough points on the Cartesian path must be selected and transformed into joint coordinates first. Then one may apply some curve-fitting procedure to the n sets of joint points. The resulting n functions of approximation, with each function approximating the desired trajectory of one joint, are the necessary inputs for the joint-level controllers. This approach corresponds to Case 4 in Figure 11.1.

To construct n approximate functions for the joints, one selects enough points (or knots) on the originally given path in Cartesian coordinates and then transforms them into angular displacements of n joints. The function of approximation for the joint trajectory passes through points corresponding to those that are selected from the Cartesian path. A possible solution for this purpose is to spline lower-degree polynomials together.[22] Paul[23] and Finkel[24] have investigated the calculation of trajectories by interpolating selected knots. Both of their methods require solving a system of $3(N - 1)$ or $4(N - 1)$ linear equations where N is the number of selected points. A simpler computational method, which requires solving only a system of $N - 2$ equations to spline $N - 1$ polynomials together for each joint, is given in Reference 25, which is summarized as follows.

11.5.1. Formulation of Cubic Polynomial Joint Trajectories

The position and orientation of the hand of a manipulator can be represented by a 4×4 matrix $\mathbf{H}(t)$ as described by Eq. (11.1) in Section 11.1 on robot control. Joint values corresponding to $\mathbf{H}(t)$ can be solved depending on the structure of the manipulator. One example of the joint solution for the PUMA manipulator is given in Reference 3.

Let $\mathbf{H}_i = \mathbf{H}(t_i)$. The hand is required to pass a sequence of N hand matrices $\{\mathbf{H}_1, \mathbf{H}_2, \ldots, \mathbf{H}_N\}$, referred to as N knots. To construct the joint trajectories, N knots are first transformed into joint vectors $[q_{11}, q_{21}, \ldots, q_{n1}], [q_{12}, q_{22}, \ldots, q_{n2}], \ldots, [q_{1N}, q_{2N}, \ldots, q_{nN}]$, where q_{ji} is the displacement of joint j at knot i corresponding to $\mathbf{H}_i$. The cubic polynomial trajectory is then constructed for each joint to fit the joint sequence $q_{j1}, q_{j2}, \ldots, q_{jN}$. In this section the procedure of constructing joint trajectories deals with one joint at a time. It is not necessary to specify the joint number j, and hence q_{ji} is replaced by q_i for simplicity.

The main objective is to construct a joint trajectory that fits a number of joint displacements at a sequence of time instants by using cubic polynomial functions. Let $t_1 < t_2 < t_3 < t_4 < \cdots < t_{N-2} < t_{N-1} < t_N$ be an ordered time sequence. At the initial time $t = t_1$, the joint displacement q_1, joint velocity v_1, and joint acceleration a_1 are specified; in a like manner q_N, v_N, and a_N at the terminal time $t = t_N$. In addition, joint displacements q_k at $t = t_k$ for $k = 3, 4, \ldots, N - 2$ are also specified for the joint trajectory to pass through. However, q_2 and q_{N-1} are not fixed, which are the two extra knots required to give enough freedom for solving the problem under the constraints. Let $Q_i(t)$ be a cubic polynomial function defined on the time interval $[t_i, t_{i+1}]$. The problem of trajectory interpolation is to spline $Q_i(t)$, for $i = 1, 2, \ldots, N - 1$, together such that the required displacement, velocity, and acceleration are satisfied and the displacement, velocity, and acceleration are continuous on the entire time interval $[t_1, t_N]$.

Because $Q_i(t)$ is cubic, the second time derivative $Q_i''(t)$ must be a linear function of t. Hence, $Q_i''(t)$ can be expressed as

$$Q_i''(t) = \frac{t_{i+1} - t}{h_i} Q_i''(t_i) + \frac{(t - t_i)}{h_i} Q_i''(t_{i+1}), \qquad i = 1, 2, \ldots, N - 1 \tag{11.94}$$

where $h_i = t_{i+1} - t_i$. Integrating $Q_i''(t)$ twice and imposing the conditions $Q_i(t_i) = q_i$ and $Q_i(t_{i+1}) = q_{i+1}$ leads to the following interpolating functions:

$$Q_i(t) = \frac{Q_i''(t_i)}{6h_i} (t_{i+1} - t)^3 + \frac{Q_i''(t_{i+1})}{6h_i} (t - t_i)^3$$

$$+ \left[\frac{q_{i+1}}{h_i} - \frac{h_i Q_i''(t_{i+1})}{6} \right] (t - t_i) + \left[\frac{q_i}{h_i} - \frac{h_i Q_i''(t_i)}{6} \right] (t_{i+1} - t), \quad i = 1, 2, \ldots, N - 1 \tag{11.95}$$

The continuity condition for velocities gives

$$Q_i'(t_i) = Q_{i-1}''(t_i) \qquad i = 2, 3, \ldots, N - 1 \tag{11.96}$$

which leads to the following equations:

$$\frac{h_{i-1}}{h_i} Q_{i-1}''(t_{i-1}) + \frac{2(h_i + h_{i-1})}{h_i} Q_i''(t_i) + Q_i''(t_{i+1})$$

$$= \frac{6}{h_i} \left[\frac{q_{i+1} - q_i}{h_i} - \frac{q_i - q_{i-1}}{h_{i-1}} \right], \qquad i = 2, 3, \ldots, N - 1 \tag{11.97}$$

The unspecified joint displacements of the two extra knots can be expressed in terms of boundary values at the beginning and end knots together with $Q_1''(t_2)$, $Q_{N-2}''(t_{N-1})$. Consequently,

$$q_2 = q_1 + h_1 v_1 + \frac{h_1^2}{3} a_1 + \frac{h_1^2}{6} Q_1''(t_2) \tag{11.98}$$

$$q_{N-1} = q_N - h_{N-1} v_N + \frac{h_{N-1}^2}{3} a_N + \frac{h_{N-1}^2}{6} Q_{N-2}''(t_{N-1}) \tag{11.99}$$

Substituting Eqs. (11.98) and (11.99) into Eq. (11.97) yields a system of $(N - 2)$ linear equations with $(N - 2)$ unknowns $Q_i''(t_i)$ for $i = 2, 3, \ldots, N - 1$:

$$A \begin{bmatrix} Q_2''(t_2) \\ Q_3''(t_3) \\ \vdots \\ Q_{N-1}''(t_{N-1}) \end{bmatrix} = \begin{bmatrix} (N-2)\text{--dimensional constant} \\ \text{vector} \end{bmatrix} \tag{11.100}$$

where

$$A = \begin{bmatrix} a_{11} & a_{12} & & & & 0 \\ a_{21} & a_{22} & a_{23} & & & \\ & a_{32} & a_{33} & a_{34} & & \\ & & & \vdots & & \\ 0 & & & & a_{N-2,N-3} & a_{N-2,N-2} \end{bmatrix}$$

$$= \begin{bmatrix} 3h_1 + 2h_2 + \dfrac{h_1^2}{h_2} & h_2 & & & & & 0 \\[2mm] h_2 - \dfrac{h_1^2}{h_2} & 2(h_2 + h_3) & h_3 & & & & \\[2mm] & h_3 & 2(h_3 + h_4) & h_4 & & & \\ & & & \ddots & & & \\ & & & & 2(h_{N-3} + h_{N-2}) & h_{N-2} - \dfrac{h_{N-1}^2}{h_{N-2}} & \\[2mm] 0 & & & & h_{N-2} & 3h_{N-1} + 2h_{N-2} + \dfrac{h_{N-1}^2}{h_{N-2}} \end{bmatrix}$$

Note that $Q_i''(t_{i+1}) = Q_{i+1}''(t_{i+1})$ because of continuity conditions on accelerations. The banded structure of matrix $\mathbf{A}$ makes it easy to solve Eq. (11.100) for the unknowns $Q_i''(t_i)$. Note that the resulting solution is in terms of time intervals h_is and the given values of joint displacements.

It is noticed that the trajectory problem just described has a unique solution, that is, the system matrix $\mathbf{A}$ of equation (11.100) is nonsingular. To prove the assertion, observe that h_i terms are time intervals and must be positive. Thus, all rows of the system matrix in Eq. (11.100) except row 2 and row $N - 3$ satisfy the inequality

$$|a_{ii}| > \sum_{j \neq i} |a_{ij}|, \qquad \text{for row } i \tag{11.101}$$

(1) If $h_2 \geq h_1$ and $h_{N-2} \geq h_{N-1}$, row 2 and row $N - 3$ will also satisfy the inequality of Eq. (11.101). Thus the system matrix $\mathbf{A}$ becomes strictly diagonal-dominant and nonsingular. (2) If $h_1 > h_2$, one may perform the row operation by subtracting (row 1) $\times (h_2 - h_1^2/h_2)/(3h_1 + 2h_2 + h_2^2/h_2)$ from row 2 to eliminate a_{21}. Consequently,

$$a_{22}' = \frac{h_1^2 - h_2^2}{3h_1 + 2h_2 + h_1^2/h_2} + 2(h_2 + h_3) \quad \text{and} \tag{11.102}$$

$$a_{23}' = h_3$$

Because $h_1 > h_2$, it is apparent that $|a_{22}'| > |a_{23}'|$. The similar result exists for $h_{N-1} > h_{N-2}$. Thus the system matrix $\mathbf{A}$ is equivalent to a strictly diagonal-dominant matrix. Consequently Eq. (11.100) has a unique solution.

11.5.2. Joint Level Path Controller

Once the desired Cartesian hand path is transformed into a corresponding set of joint trajectories, the equivalent path control may be achieved at the joint level by tracking the joint trajectories. A simple scheme for the joint trajectories tracking is shown in Figure 11.13. In the figure $\mathbf{q}$ and $\mathbf{q}_d$ are actually measured and desired displacements for n joints, respectively. To obtain a stably computed $\ddot{\mathbf{q}}$, the constants k_1 and k_2 must be so chosen that the roots of the characteristic equation $\ddot{\epsilon} + k_1 \dot{\epsilon} + k_2 \epsilon = 0$ have negative real parts. The block labeled with "Input torque/force program" represents a process that computes the required torque vector $\boldsymbol{\tau}$ for all the points using Eq. (11.58). For better accuracy the measured $\mathbf{q}$ and $\dot{\mathbf{q}}$ should be updated no less than, say, once every $\frac{1}{60}$ sec for commercially available robots. (That is, the sampling frequency is no less than roughly five times the resonant frequency of the robot.) This implies that $\ddot{\mathbf{q}}$ as well as $\boldsymbol{\tau}$ must be computed once every $\frac{1}{60}$ sec. Consequently the speed of on-line computation becomes a major factor in the controller. This problem has been discussed in Section 11.4 on multiple-joints controller.

11.6. ADAPTIVE CONTROLLER

To improve the performance such as smoothing the robot motion, reducing the deviation of actually traveled path from the planned path, minimizing the execution time, and so on, it is usually required to compute the dynamical behavior of the robot using Lagrangian or Newton-Euler equations. As shown before, the on-line computation of the desired joint torque has been a bottleneck of the control scheme. An alternative approach is the use of an adaptive controller.

Young[26] proposed a scheme using variable structure theory for the manipulator. To implement the scheme the motion trajectory must lie on the switching surface, which, itself, is difficult to implement as one has experienced in the minimum-time control theory. Dubowsky and DesForges[27] simplified the problem by introducing the model-referenced adaptive control. The performance at the robot's

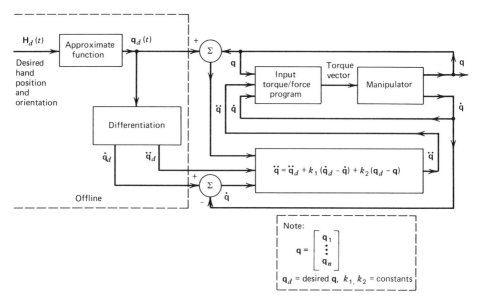

Fig. 11.13. Block diagram of a joint-level controller.

high speed is, however, not verified. Koivo and Guo[28] applied Borison's self-tuning regulator[29] to the robot control using a model of linear difference equation with white Gaussian noise. A similar approach using a pole placement regulator has been investigated by Leininger and Wang,[30] and Walters and Bayoumi.[31] The parameters were iteratively obtained from the minimum variance estimator. The control variable (input torque) was then computed for the minimum average deviation of the joint trajectories from the nominal trajectories. Although the computation of either Lagrangian or Newton-Euler equations is completely avoided, the iterative computing time is still too long for an on-line operation, even based on a two-stage observation. Also, the use of the linear model ignores the nonlinear characteristics of the robot with coupling among its joints and thus introduces undesirable errors. Lee and Chung[32] adopt a similar idea of parameter estimation, and perturb the motion of the manipulator in the vicinity of its desired path. The requirement of computing the manipulator dynamics still using Newton-Euler equations does not eliminate the computation bottleneck.

In this section, quadratic terms are added in the difference equation to include centrifugal force terms as an improvement on the model. The discrete-time system provides a natural formulation for processing iterative estimations and computations on a digital computer. All the parameters are estimated from the iterative minimum variance estimator.[33] To speed up the computation and to ensure the convergence of the iteration, the stochastic approximation formulation proposed by Astrom and Eykhoff[34] is adopted. The input torques are then computed for the minimum average deviation of the trajectory traveled.

11.6.1. Discrete-Time Model of Manipulator

Refer to Eq. (11.58), the Lagrangian formulation of robot dynamics. Let $q_i(k)$ and $q_{im}(k)$ be the actual displacement of joint i, $i = 1, 2, \ldots, n$, and its model, respectively, at time k. At that time, $q_i(k - \alpha)$ for $\alpha = 1, 2, \ldots, N$ are assumed known through measurement while the joint torques $\tau_i(k - \alpha - 1)$ have been determined. The main purpose is to determine $\tau_i(k - 1)$. To include the nonlinear effect caused by the centrifugal torque/force, the model is chosen to be

$$q_{im}(k) = a_i^o + \sum_{\alpha=1}^{N} [a_i^\alpha q_i(k - \alpha) + b_i^\alpha \tau_i(k - \alpha)] + \sum_{\beta=1}^{N} \sum_{\gamma=1}^{N} c_i^{\beta\gamma} \, q_i(k - \beta) q_i(k - \gamma) \qquad (11.103)$$

where a_i^o, a_i^α, b_i^α, and $c_i^{\beta\gamma} = c_i^{\gamma\beta}$ are parameters yet to be determined. In Eq. (11.58), parameters D_i, D_{ij}, and D_{ijk} vary as the robot moves. It is assumed that they vary slowly so that they are practically constant on every time interval $[k, k + 1]$. For this reason, the parameters a_is, b_is, and c_is are assumed constant. Although the model given in Eq. (11.103) is nonlinear, it still does not include the interactions between joints. Even so, the estimation of the parameters is quite involved.

Before $\tau_i(k-1)$ is determined, $q_i(k)$ is not known and cannot be measured. If $q_{im}(k)$ is used as its model, then the modeling error is

$$\epsilon_i(k) = q_i(k) - q_{im}(k) \tag{11.104}$$

Since $q_i(k)$ is unknown at time k, so is $\epsilon_i(k)$. Assume it is Gaussianly distributed and uncorrelated with zero mean and variance σ_i^2. Then the unknown parameters a_i^o, a_i^α, b_i^α, and $c_i^{\beta\gamma}$ may be estimated by the minimum variance estimator, that is, by minimizing the mean-square error

$$J_i(k-1) = \sum_{\alpha=0}^{k-1} E[\epsilon_i(\alpha)^2] \tag{11.105}$$

where E denotes the expected value.

11.6.2. Estimation of Parameters

Since the manipulator is modeled by Eq. (11.103), which does not include interactions among its joints, all the estimation of parameters and the computation of controls are for one specific joint. Thus the subscript i is dropped in this section to simplify the notation.

For the estimation problem outlined in the preceding section, there is more than one possible way to approach the problem.

Kalman Filter

Let $\mathbf{Q}(k-1)$ be a $2N+1 + N(N+1)/2$ dimensional vector defined by

$$\mathbf{Q}'(k-1) = [1\ q(k-1)\ \cdots\ q(k-N)\ \tau(k-1)\ \cdots\ \tau(k-N)\ q(k-1)^2\ q(k-1)q(k-2)$$
$$\cdots\ q(k-1)q(k-N)\ q(k-2)^2\ q(k-2)q(k-3)\ \cdots\ q(k-N)^2] \tag{11.106}$$

where $(\cdot)' =$ transpose of $(\cdot)$. Let $\mathbf{p}$ be a parameter vector with the same dimension defined by

$$\mathbf{p}'(k) = [a^o a^1\ \cdots\ a^N b^1\ \cdots\ b^N c^{11} c^{12}\ \cdots\ c^{1N} c^{22} c^{23}\ \cdots\ c^{NN}] \tag{11.107}$$

Since the parameters are assumed to be constant, then

$$\mathbf{p}(k) = \mathbf{p}(k-1) = \mathbf{p} \tag{11.108}$$

This is a noiseless system equation for the parameters. By Eqs. (11.106) and (11.107), the combination of Eqs. (11.103) and (11.104) yields

$$q(k) = \mathbf{Q}'(k-1)\mathbf{p} + \epsilon(k) \tag{11.109}$$

which is the linear measurement equation in $\mathbf{p}(k)$ with additive, Gaussianly distributed white noise. The system consisting of Eqs. (11.108) and (11.109) has a discrete Kalman filter (Reference 33, p. 107). Updated parameter estimate:

$$\hat{\mathbf{p}}(k_+) = \hat{\mathbf{p}}(k_-) + \mathbf{K}(k)[q(k) - \mathbf{Q}'(k-1)\hat{\mathbf{p}}(k_-)] \tag{11.110}$$

Parameter estimate extrapolation:

$$\hat{\mathbf{p}}(k_-) = \hat{\mathbf{p}}(k-1)_+) \tag{11.111}$$

Kalman gain vector:

$$\mathbf{K}(k) = \frac{\mathbf{P}(k_-)\mathbf{Q}(k-1)}{\mathbf{Q}'(k-1)\mathbf{P}(k_-)\mathbf{Q}(k-1) + \sigma^2} \tag{11.112}$$

Error covariance matrix extrapolation:

$$\mathbf{P}(k_-) = \mathbf{P}[(k-1)_+] \tag{11.113}$$

Error covariance matrix update:

$$\mathbf{P}(k_+) = [\mathbf{I} - \mathbf{K}(k)\mathbf{Q}'(k-1)]\mathbf{P}(k_+) \tag{11.114}$$

with $\mathbf{I}$ = identity matrix, and initial conditions

$$E[\mathbf{p}(0)] = \hat{\mathbf{p}}(0) = E[\mathbf{p}] \tag{11.115}$$

$$E[\{\mathbf{p}(0) - \hat{\mathbf{p}}(0)\}\{\mathbf{p}(0) - \hat{\mathbf{p}}(0)\}'] = \mathbf{P}(0) \tag{11.116}$$

An alternative representation of the noiseless system equation for the parameters and its linear measurement equation is as follows. Let

$$\mathbf{\rho}'(k - 1) = [1 \; q(k - 1) \; \cdots \; q(k - N) \; \tau(k - 1) \; \cdots \; \tau(k - N)] \tag{11.117}$$

be a $(2n + 1)$ dimensional vector, and

$$\Phi = \begin{bmatrix} a^0 & \dfrac{a^1}{2} & \cdots & \dfrac{a^n}{2} & \dfrac{b^1}{2} & \cdots & \dfrac{b^n}{2} \\[2mm] \dfrac{a^1}{2} & c^{11} & \cdots & c^{1n} & 0 & \cdots & 0 \\[1mm] \cdot & \cdot & & \cdot & \cdot & & \cdot \\ \cdot & \cdot & & \cdot & \cdot & & \cdot \\ \cdot & \cdot & & \cdot & \cdot & & \cdot \\ \dfrac{a^n}{2} & c^{n1} & \cdots & c^{nn} & 0 & \cdots & 0 \\[1mm] \dfrac{b^1}{2} & 0 & \cdots & 0 & 0 & \cdots & 0 \\[1mm] \cdot & \cdot & & \cdot & \cdot & & \cdot \\ \cdot & \cdot & & \cdot & \cdot & & \cdot \\ \cdot & \cdot & & \cdot & \cdot & & \cdot \\ \dfrac{b^n}{2} & 0 & \cdots & 0 & 0 & \cdots & 0 \end{bmatrix} \tag{11.118}$$

be a $(2n + 1)$ by $(2n + 1)$ constant parameter matrix. Combining Eqs. (11.103), (11.104), (11.117), and (11.118) yields

$$q(k) = \mathbf{\rho}'(k - 1)\Phi\mathbf{\rho}(k - 1) + \epsilon(k) \tag{11.119}$$

This is a scalar equation. Let $\mathbf{\rho}^+(k - 1) = [\mathbf{\rho}'(k - 1)\mathbf{\rho}(k - 1)]^{-1}\mathbf{\rho}'(k - 1)$ be the Moore-Penrose pseudo inverse[35] of $\mathbf{\rho}(k - 1)$. Multiply Eq. (11.119) by $\mathbf{\rho}^+(k - 1)$ to yield

$$q(k)\mathbf{\rho}^+(k - 1) = \mathbf{\rho}'(k - 1)\Phi\mathbf{\rho}(k - 1)\mathbf{\rho}^+(k - 1) + \epsilon(k)\mathbf{\rho}^+(k - 1) \tag{11.120}$$

Now $\mathbf{\rho}^+\mathbf{\rho} = [\mathbf{\rho}'\mathbf{\rho}]^{-1}\mathbf{\rho}'\mathbf{\rho} = 1$ and $\mathbf{\rho}\mathbf{\rho}^+ = [\mathbf{\rho}'\mathbf{\rho}]^{-1}\mathbf{\rho}\mathbf{\rho}' \neq I$. Since for an arbitrary vector $\mathbf{c}$, one has $\|\mathbf{\rho}\mathbf{\rho}^+\mathbf{c}\| \leq \|\mathbf{c}\|$, then the solution to

$$q(k)\mathbf{\rho}^+(k - 1) = \mathbf{\rho}'(k - 1)\Phi + \epsilon(k)\mathbf{\rho}^+(k - 1) \tag{11.121}$$

is a minimum norm solution[36] to Eq. (11.120). To solve the original problem, use Eq. (11.121), which is a linear measurement equation in parameter matrix Φ. The noiseless system equation for the parameter is

$$\Phi(k) = \Phi(k - 1) = \Phi \tag{11.122}$$

Equations (11.121) and (11.122) represent the same system described by Eqs. (11.108) and (11.109), and the resulting Kalman filters are equivalent.

Least-Squares Estimation

It has been shown (Reference 34, p. 149) that, by an algebraic manipulation, the recursive equations of the least-squares estimate can be obtained directly from the Kalman filter as:

$$\hat{\mathbf{p}}(k) = \hat{\mathbf{p}}(k - 1) + \Gamma(k - 1)[q(k) - \mathbf{Q}'(k - 1)\hat{\mathbf{p}}(k - 1)] \tag{11.123}$$

where the vector

$$\Gamma(k-1) = \frac{\mathbf{P}(k-1)\mathbf{Q}(k-1)}{[\sigma^2 + \mathbf{Q}'(k-1)\mathbf{P}(k-1)\mathbf{Q}(k-1)]} \tag{11.124}$$

and the covariance matrix

$$\mathbf{P}(k) = [\mathbf{I} - \Gamma(k-1)\mathbf{Q}'(k-1)]\mathbf{P}(k-1) \tag{11.125}$$

Stochastic Approximation

Although the Kalman filter and the least-squares estimation are applicable to estimate the parameters in this problem, the computation is quite involved. To reduce the computational burden, one may adopt the technique of stochastic approximation by replacing Eq. (11.124) by that from Reference 34, (p. 148).

$$\Gamma(k-1) = \frac{\mathbf{AQ}(k-1)}{(k-1)} \tag{11.126}$$

where $\mathbf{A}$ is a positive definite matrix and could be $\mathbf{I}$ for simplicity. It has been shown[37] that the choice of Eq. (11.126) for $\Gamma(k-1)$ results in the convergence of the estimation of $\mathbf{p}(k)$. Although the algorithm of stochastic approximation always reduces computational effort and converges, the resulting estimates, however, have larger variances than those of the least-squares estimate.

11.6.3. Path-Tracking Controller

Let $q_{id}(t)$ be the desired displacement of joint i for all t. Ideally one wishes that the deviation of the actual displacement $q_i(t)$ of joint i from $q_{id}(t)$ be zero for all t. This is a difficult problem to solve. Since, in our discussion, the dynamic model for the manipulator is of discrete-time type with a Gaussianly distributed modeling error, and the parameters of the model are estimated based on the minimum modeling error, the trajectory-tracking control input is determined in the sense of minimum average deviation. Thus the trajectory constraint is not strict.

At each time instant k, let $q_{id}(k)$ be the desired displacement of joint i. Whenever the actual displacement at time $k-1$, $q_i(k-1)$, is fixed, $q_i(k)$ is determined by the torque of the joint i actuator $\tau_i(k-1)$. Again omit subscript i to simplify the notation. Thus $\tau(k-1)$ can be obtained by minimizing the conditional expectation

$$J[\tau(k-1)] = E\{[q_d(k) - q(k)]^2 | q(k-1)\} \tag{11.127}$$

By Eq. (11.109) and noting that $\epsilon_i(k)$ has zero mean, one obtains

$$\begin{aligned} J[\tau(k-1)] &= E\{[q_d(k) - \mathbf{Q}'(k-1)\mathbf{p} + \epsilon(k)]^2 | q(k-1)\} \\ &= [q_d(k) - \mathbf{Q}'(k-1)\mathbf{p}]^2 + \sigma^2 \end{aligned} \tag{11.128}$$

which has a derivative

$$\frac{dJ[\tau(k-1)]}{d\tau(k-1)} = 2b^1[q_d(k) - \mathbf{Q}'(k-1)\mathbf{p}] \tag{11.129}$$

Since $b_i^1 \neq 0$, the zero derivative leads to

$$\mathbf{Q}'(k-1)\mathbf{p} = q_d(k) \tag{11.130}$$

or

$$\tau(k-1) = \frac{\left\{ \begin{aligned} q_d(k) - a^0 - a^1 q(k-1) - \sum_{\alpha=2}^{N}[a^\alpha q(k-\alpha) + b^\alpha \tau(k-\alpha)] \\ - \sum_{\beta=1}^{N}\sum_{\gamma=1}^{N} c^{\beta\gamma} q(k-\beta)q(k-\gamma) \end{aligned} \right\}}{b^1} \tag{11.131}$$

In Eq. (11.131) the values of parameters a^α, b^α, and $c^{\beta\gamma}$ are the estimates obtained in Section 11.6.2 on estimation of parameters. The computed control is for the immediately following sampling period to ensure the stability of the system.

11.6.4. Desired Joint Trajectory

In reality, the desired path traveled by the hand or end effector of the manipulator is specified in Cartesian coordinates. Intuitively one may determine the equivalent joint displacement $q_{id}(k)$ by means of inverse Jacobian transformation. Whenever the sampling frequency is changed, the point of q_i between the time instants k and $k + 1$ may be obtained by interpolation. To store the values of $q_{id}(k)$ for all k in a computer requires a large memory. This problem can be avoided by pretransforming the three-dimensional Cartesian path into n-dimensional joint trajectories in terms of spline-function approximations. The discussion on this subject was presented previously in Section 11.5.1.

The block diagram for the overall process is shown as in Figure 11.14. It shows the information flow for one joint during one sampling period. It is seen that both the estimation of model parameters and the computation of the control must be done on-line and completed during one sampling period. The speeds of estimation and computation are crucial since they determine the upper bound of the sampling frequency; the higher the frequency, the more accuracy the process will have.

In the preceding sections, estimations by means of Kalman filter, least-squares estimate and stochastic approximation have been presented. Among them there is a trade-off between speed and accuracy. The method of stochastic approximation is preferable since it requires the least amount of computing time and its iterations always converge. But the resulting estimate has the largest variance. The model of the mechanical manipulator includes a quadratic term representing the centrifugal forces. Its purpose is to reduce the modeling error at the expense of more computation. Because of the trade-off, designers must use personal judgment on their assigned tasks and then decide which alternative should be used.

11.7 CONTROLLER WITH TORQUE/FORCE FEEDBACK

It has been recognized that in the design of teleoperators (i.e., remote manipulators) for speed, accuracy, dexterity, and load capacity, force-reflecting capability must be included in the control system.[38] An illustrative example involves fitting a part together with another part in assembly. As the tolerances of the mating parts become tight, it is impractical to require the manipulator and its controller to furnish such high-positional accuracy. Alternatively, the controller must combine the positional control and the force control in compliance with the positional constraint imposed by the task geometry. Goertz and Bevilacqua[39] in 1952 suggested a design of a force-reflecting positional controller for a teleoperator. It required a position sensor and a torque sensor on the input shaft as well as on the output shaft. The resulting double-loop control system yielded a complicated stability problem.

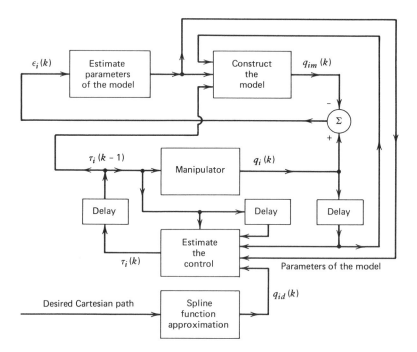

Fig. 11.14. Overall process for joint i during one sampling period.

A number of force-control schemes for manipulators requiring no direct measurement of forces by sensors have appeared in literature. In 1969 Whitney[19] developed a resolved-rate-motion scheme. It enabled one to generate the desired joint torques for the manipulator to reach the next desired position, which was computed from differentially approximated motions. To generate the desired torques, the Jacobian matrix must be inverted to complete the closed-loop control system, which introduced computational difficulties. Luh, Walker, and Paul[40] extended Whitney's idea to include resolved accelerations for smooth transitions to different velocities. The disadvantage of inverting the Jacobian matrix still existed. Inoue's method[41] of 1971 assumed that no information about the load was available before the manipulator moved. His programming software allowed the user to choose some joints to have force control capabilities. For these joints, he assumed that the external torque was proportional to the settling time of the measured positional error. With this information, the input torque was then adjusted to reduce the error. The disadvantage was that the user had to decide which joints were to be force controlled. Paul[23] in 1972 suggested that the required torque for each joint could be precalculated by Lagrangian equations. The computed torque value became the reference to which the actual joint torque had to be controlled. When the manipulator started to move, however, its geometrical configuration and hence the actually required torque for each joint changed. Since the precomputed torque requirement was not updated, errors could result. Nevins and Whitney[42] in 1973 suggested the idea of using the force vector assembler. Their discussion was based on a sensing wrist with three extensional strain gages and three strain gage shear bridges.[43] In 1976 Paul and Shimano[44] developed a compliance scheme with the assumption that a sensing wrist was available so that the load forces and moments with respect to the axes of the Cartesian coordinates could be measured on-line. The discussions on the compliance and on the Scheinman sensing wrist with a Maltese cross configuration were amplified in details later by Shimano[45] in 1978. The extraction of joint torque information from the sensor was done by computing a 6×8 matrix.

At the Draper Laboratory, an experiment on a single-axis hydraulic test apparatus was performed with force, position, and velocity feedbacks[46] in 1974. The controller was modeled as a linear system except that the sensitivity of the force sensor was represented by an on-off nonlinear element. Six-joint torque-controlled assembly was also demonstrated. The dynamic models and their stability were discussed by Whitney in 1977.[47]

Actual force controllers employing the Scheinman sensing wrist were implemented in 1979–1981 by Craig and Raibert[48,49] and Salisbury.[50] Craig and Raibert's scheme was based on Mason's theoretical framework[51] of natural and artificial constraints. It contained two separate closed loops: one for positional control and the other for force control. Salisbury added a deadband-and-limiting nonlinear device to reduce the limit cycle and the excessive loop gain due to large force errors caused by impact transients. Since the signal from the sensing wrist was referenced to Cartesian coordinates, the Jacobian transform of the coordinate systems had to be performed every sampling period.

Most recently, in 1980, Wu and Paul[52] completed an experiment on a single-joint manipulator with a joint sensor. A completely linear model was adopted. It was concluded that the joint-torque control system had high gain and wide bandwidth. Because the feedback loop was closed around the joint and all the hardware was of the analog type, the scheme avoided the computational difficulties, contained no differential approximation, and provided fast response. Based on Wu and Paul's experiment,[52] the first two joints of the Stanford manipulator, Figure 11.3, have been redesigned and fabricated to include torque sensors and the associated electronics by Luh, Fisher, and Paul.[5] The results of experiments performed on the modified manipulator and a stability analysis are described in Reference 5. Specifically, the reference gives the following report: The boom of joint 3 was stretched all the way out to yield a distance of 33.5 in. from the tip of the boom to the center line of the output shaft of joint 2. At the tip of the boom an average force of 32 oz, or 1072 oz-in., was required to overcome the pure mechanical, frictional torque for each of the two joints. With the joint torque feedback in the control loop, the average force is reduced from 32 to 1 oz. Thus the effective frictional torque for each joint was reduced to $1 \times 33.5 = 33.5$ oz-in.

11.8. SUMMARY

For industrial robots whose assignments are limited to usual tasks involving mainly synchronization, controllers can be designed by conventional methods based on two factors: damping and structural resonant frequency. It is possible to reduce the steady-state position error and to eliminate the velocity error and error due to centrifugal term by feedforward compensations for individual joints. For multiple-joint controllers, further feedforward compensation is necessary to encounter the force interactions between joints. However, it introduces the computational difficulties so that methods of approximation or simplification are needed.

For the problem of path tracking, the control can be done at either the hand level or the joint level. In either case, the approximation is necessary and the transformation between Cartesian and joint coordinates is required. To avoid the transformation in real time, joint trajectories may be approximated by spline functions in advance. The trade-off is between the accuracy and the computing time.

To circumvent the computation of robot dynamics by means of Lagrangian or Newton-Euler formulations, the discrete-time system may be used to model the robot, and then the parameters of the robot system are estimated by minimum variance estimator or stochastic approximation. Again there is a trade-off between the accuracy of the control and the computing time.

For the assignment of product assembly, task interactions are the main functions. Industrial robots do not have precise position control. The torque/force sensing ability allows robots to perform many labor-intensive tasks for which positionally controlled automation is basically unsuited. Torque/force sensing also reduces the apparent mechanical friction that is a dominant factor of torque consumption in the currently available industrial robots.

REFERENCES

1. Luh, J. Y. S., An Anatomy of Industrial Robots and Their Controls, *IEEE Transactions on Automatic Control,* Vol. 28, No. 2, February 1983, pp. 133–153.

2. Roberts, L. G., *Homogeneous Matrix Representation and Manipulation of N-Dimensional Constructs,* Lincoln Laboratory Document, No. MS 1045, MIT, 1965.

3. Paul, R. P., Shimano, B., and Mayer, G. E., Kinematic Control Equations for Simple Manipulators, *IEEE Transactions on Systems, Man and Cybernetics,* Vol. 11, No. 6, June 1981, pp. 449–455.

4. Scheinman, V. D., *Design of a Computer Controlled Manipulator,* A. I. Memo 92, Artificial Intelligence Laboratory, Stanford University, June 1969.

5. Luh, J. Y. S., Fisher, W. D., and Paul, R. P. C., Joint Torque Control by a Direct Feedback for Industrial Robots, *IEEE Transactions on Automatic Control,* Vol. 28, No. 2, February 1983, pp. 153–161.

6. Bejczy, A. K., *Robot Arm Dynamics and Control,* Technical Memorandum 33–669, Jet Propulsion Laboratory, February 1974.

7. Book, W. J., Maizza-Neto, O., and Whitney, D. E., Feedback Control of Two Beam, Two Joint Systems with Distributed Flexibility, *ASME Transactions, Journal of Dynamic Systems, Measurement and Control,* Vol. 97, No. 4, December 1975, pp. 424–431.

8. Asada, H. T. and Takeyama I., *Control of a Direct-Drive Arm,* Report CMU-RI-TR-82-4, Robot Institute, Carnegie-Mellon University, March 9, 1982.

9. Asada, H. and Youcef-Toumi, K., Analysis and Design of Semi-direct-drive Robot Arms, Proc. 1983 American Control Conference, Vol. 2, June 22–24, 1983, San Francisco, pp. 757–764.

10. Paul, R. P., *Robot Manipulators: Mathematics, Programming and Control,* MIT Press, Cambridge, Massachusetts, 1981.

11. Bejczy, A. K. and Paul, R. P., Simplified Robot Arm Dynamics for Control, *Proceedings of 20th IEEE Conference on Decision and Control,* December 16–18, 1981, San Diego, California, pp. 261–262.

12. Luh, J. Y. S. and Lin, C. S., Automatic Generation of Dynamic Equations for Mechanical Manipulators, *Proceedings of Joint Automatic Control Conference,* June 17–19, 1981, Charlottesville, Virginia, pp. TA-2D.

13. Luh, J. Y. S., Walker, M. W., and Paul, R. P. C., On-line Computational Scheme for Mechanical Manipulators, *ASME Transactions, Journal of Dynamic Systems, Measurement and Control,* Vol. 102, No. 2, June 1980, pp. 69–76.

14. Hollerbach, J. M., A Recursive Lagrangian Formulation of Manipulator Dynamics and a Cooperative Study of Dynamics Formulation Complexity, *IEEE Transactions on Systems, Man and Cybernetics,* Vol. 10, No. 11, November 1980, pp. 730–736.

15. Silver, W. M., On the Equivalence of Lagrangian and Newton-Euler Dynamics for Manipulators, *International Journal of Robotics Research,* Vol. 1, No. 2, Summer 1982, pp. 60–70.

16. Turney, J. L., Mudge, T. N. and Lee, C. S. G., *Connection between Formulations of Robot Arm Dynamics with Applications to Simulation and Control,* Report RSD-TR-4-82, Robot System Division, University of Michigan, November 1981.

17. Luh, J. Y. S. and Lin, C. S., Scheduling of Parallel Computation for a Computer-Controlled Mechanical Manipulator, *IEEE Transactions on Systems, Man and Cybernetics,* Vol. 12, No. 2, March/April 1982, pp. 214–234.

18. Paul, R. P., Shimano, B., and Mayer, G. E., Differential Kinematic Control Equations for Simple Manipulators, *IEEE Transactions on Systems, Man and Cybernetics,* Vol. 11, No. 6, June 1981, pp. 456–460.

19. Whitney, D. E., Resolved Motion Rate Control of Manipulators and Human Prostheses, *IEEE Transactions on Man-Machine Systems,* Vol. 10, No. 2, June 1969, pp. 47–53.

20. Paul, R., Manipulator Cartesian Path Control, *IEEE Transactions on Systems, Man and Cybernetics,* Vol. 9, No. 11, November 1979, pp. 702–711.

21. Taylor, R. H., Planning and Execution of Straight Line Manipulator Trajectories, *IBM Journal of Research and Development,* Vol. 23, No. 4, July 1979, pp. 424–436.

22. Ahlberg, H. H. et al. *The Theory of Splines and Their Applications,* Academic Press, New York, 1967.

23. Paul, R. C., *Modeling, Trajectory, Calculation and Servoing of a Computer Controlled Arm,* A. I. Memo 177, Artificial Intelligence Laboratory, Stanford University, September 1972.

24. Finkel, R. A., *Constructing and Debugging Manipulator Programs,* A. I. Memo 284, Artificial Intelligence Laboratory, Stanford University, August 1976.

25. Lin, C. S., Chang, P. R., and Luh, J. Y. S., Formulation and Optimization of Cubic Polynomial Joint Trajectories for Mechanical Manipulators, *IEEE Transactions on Automatic Control,* 1983, Vol. 28, No. 12, December 1983, pp. 1066–1074.

26. Young, K. K. D., Controller Design for a Manipulator Using Theory of Variable Structure Systems, *IEEE Transactions on Systems, Man and Cybernetics,* Vol. 8, No. 2, February 1978, pp. 101–109.

27. Dubowsky, S. and DesForges, D. T., The Application of Model-Referenced Adaptive Control to Robotic Manipulators, *Journal of Dynamic Systems, Measurement, and Control, Transactions of ASME,* September 1979, pp. 191–200.

28. Koivo, A. J. and Guo, T. H., Control of Robotic Manipulator with Adaptive Controller, *Proceedings 20th IEEE Conference on Decision and Control,* San Diego, California, December 16–18, 1981, pp. 271–276.

29. Borison, U., Self-Tuning Regulators for a Class of Multivariable Systems, *Automatica,* Vol. 15, 1979, pp. 209–215.

30. Leininger, G. G. and Wang, S. P., Pole Placement Self-Tuning Control of Manipulators, Preprints of *IFAC Symposium on Computer-Aided Design of Multivariable Technology Systems,* September 15–17, 1982, West Lafayette, Indiana, pp. 27–29.

31. Walters, R. G. and Bayoumi, M. M., Application of a Self-Tuning Pole-Placement Regulator to an Industrial Manipulator, *Proceedings of 21st IEEE Conference on Decision and Control,* Vol. 1, December 8–10, 1982, Orlando, Florida, pp. 323–329.

32. Lee, C. S. G. and Chung, M. J., An Adaptive Control Strategy for Computer-Based Manipulators, *Proceedings of 21st IEEE Conference on Decision and Control,* Vol. 1, December 8–10, 1982, Orlando, Florida, pp. 95–100.

33. Gelb, A., Kasper, J. F., et al. *Applied Optimal Estimation,* MIT Press, Cambridge, Massachusetts, 1974.

34. Astrom, K. J. and Eykhoff, P., System Identification—A Survey, *Automatica,* Vol. 7, 1971, pp. 123–162.

35. Albert, A., *Regression and the Moore-Penrose Pseudoinverse,* Academic Press, New York, 1972.

36. Rao, C. R. and Mitra, S. K., *Generalized Inverse of Matrices and Its Applications,* Wiley, New York, 1971.

37. Albert, A. E. and Gardner L. A. *Stochastic Approximation and Nonlinear Regression,* MIT Press, Cambridge, Massachusetts, 1967.

38. Goertz, R. C., Fundamentals of General Purpose Remote Manipulators, *Nucleonics,* Vol. 10, No. 11, November 1952, pp. 36–42.

39. Goertz, R. C. and Bevilacqua, F., A Force-Reflecting Positional Servomechanism, *Nucleonics,* Vol. 10, No. 11, November 1952, pp. 43–55.

40. Luh, J. Y. S., Walker, M. W., and Paul, R. P. C., Resolved-Acceleration Control of Mechanical Manipulators, *IEEE Transactions on Automatic Control,* Vol. 25, No. 3, June 1980, pp. 468–474.

41. Inoue, H., Computer Controlled Bilateral Manipulator, *Bulletin of Japan Society of Mechanical Engineers,* Vol. 14, No. 69, March 1971, pp. 199–207.

42. Nevins, J. L. and Whitney, D. E., The Force Vector Assembler Concept, *Proceedings of 1st International Conference on Robots and Manipulator Systems,* Udine, Italy, September 1973.

43. Nevins, J. L., Whitney, D. E., et al., *Exploratory Research in Industrial Modular Assembly,* Sixth Report, Charles Stark Draper Laboratory, MIT, September 1977 to August 1978, pp. 112–120.

44. Paul, R. and Shimano, B., Compliance and Control, *Proceedings of Joint Automatic Control,* 1976, San Francisco, pp. 694–699.

45. Shimano, B. E., *The Kinematic Design and Force Control of Computer Controlled Manipulators,* A.I. Memo 313, Artificial Intelligence Laboratory, Stanford University, March 1978.

46. Jilani, M. A., Force Feedback Hydraulic Servo for Advance Automation Machines, SM Thesis, Mechanical Engineering Department, MIT, November 1974.

47. Whitney, D. E., Force Feedback Control of Manipulator Fine Motions, *ASME Transactions: Journal of Dynamic Systems, Measurement, and Control,* Vol. 99, No. 2, June 1977, pp. 91–97.

48. Craig, J. J. and Raibert, M. H., A Systematic Method of Hybrid Position/Force Control of a Manipulator, *Proceedings 3rd International Computer Software and Applications Conference,* Chicago, November 6–8, 1979, pp. 446–451.

49. Raibert, M. H. and Craig, J. J., Hybrid Position/Force Control of Manipulators, *ASME Transactions: Journal of Dynamic Systems, Measurement, and Control,* Vol. 103, No. 2, June 1981, pp. 126–133.

50. Salisbury, J. K., Active Stiffness Control of a Manipulator in Cartesian Coordinates, *Proceedings 19th IEEE Conference on Decision and Control,* Vol. 1, December 10–12, 1980, Albuquerque, New Mexico, pp. 95–100.

51. Mason, M. T., Compliance and Force Control for Computer Controlled Manipulators, *IEEE Transactions on Systems, Man, and Cybernetics,* Vol. 11, No. 6, June 1981, MIT, April 1979, pp. 418–432.

52. Wu, C. H. and Paul, R. P., Manipulator Compliance Based on Joint Torque Control, *Proceedings 19th IEEE Conference on Decision and Control,* Vol. 1, December 10–12, 1980, Albuquerque, New Mexico, pp. 88–94.

BIBLIOGRAPHY ON CONTROL IMPLEMENTATION

Cristafulli, D. M., Loh, H. H., and Murphy, J. F., The Evolution of a New Industrial Robot Controller from User Specifications to Commercial Product, *SME Paper* No. MR76-607, 1976.

Folin, J., Leary, R., and Plonsky, B., Implementation of a Smart Minicomputer Based Robotic control system, *Proceedings of the Robot 8 Conference,* Detroit, Michigan, June 1984.

Girod, G. F., Utilization of Microprocessors in Robot Automation Systems, *Proceedings Robot VI,* Detroit, Michigan, March 1982, pp. 565–589.

Hohn, R. R., Robot Control Systems and Applications, *International Automatic Control Conference,* Denver, Colorado, June 1979, pp. 750–753.

Koren, Y. and Ulsoy, G., Control of DC Servo-Motor Driven Robots, *Proceedings Robot VI,* Detroit, Michigan, March 1982, pp. 590–602.

Kreinin, G. V., Pneumatic Drives for Industrial Robots, *Machines and Tooling,* Vol. 49, No. 7, 1978, pp. 31–35.

Lundstrom, G., Industrial Robots and Fluid Control Systems, *Industrial Robot,* Vol. 1, No. 6, December 1974, pp. 264–270.

Martin, M. and Menche, H., Electronic Control and Electrical Servomechanism for Industrial Robots, *Proceedings 4th I.S.I.R.,* Tokyo, Japan, November 1974, pp. 339–347.

Resnick, B. J., Robot Interface: Switch Closure and Beyond, *Proceedings Robot II,* November 1978.

Skidmore, M. P., Computer Techniques Used in Industrial Robots, *Industrial Robot,* Vol. 6, No. 4, December 1979, pp. 183–187.

Snyder, W. E. and Mian, M., Microcomputer Control of Manipulators, *Proceedings 9th I.S.I.R.,* Washington, D.C., March 1979, pp. 423–435.

CHAPTER 12

NUMERICAL CONTROL AND ROBOTICS

YORAM KOREN

Technion-Israel Institute of Technology
Haifa, Israel

12.1. INTRODUCTION

Numerically controlled (NC) machine tool systems are in some ways related to industrial robot systems. In both the axes of a mechanical device are controlled to guide a tool that performs a manufacturing process task as shown in Figure 12.1. In NC, the process may be drilling, milling, grinding, welding, and so on, and in robotics—painting, welding, assembly, handling, and so on. In NC, the mechanical handling device is the machine tool, and in robotics the manipulator. In both systems each axis of motion is equipped with a separate driving device and a separate control loop. The driving device may be a DC servomotor, a hydraulic actuator, or a stepping motor. The type selected is determined mainly by the power requirements of the machine, or the robot.[1] (See Part 2 of this Handbook.)

The combination of the individual axes of motion generates, in both systems, the required path of the tool. In NC, the tool is the cutting tool, such as a milling cutter or a drill; in robotics, the tool is the instrument at the far end of the manipulator, which might be a gripper, a welding gun, or a paint-spraying gun. An axis of motion in NC means an axis in which the cutting tool moves relative to the workpiece. This movement is achieved by the motion of the machine tool slides. The main three axes of motion are referred to in NC as X, Y, and Z axes. For example, in the machining center shown in Figure 12.2, X is the longitudinal axis, Y is the vertical axis, and Z is the depth axis.

In a typical NC system the numerical data required for producing a part is maintained on a punched tape and is called the *part program*. The part program plays the same role in NC that task program plays in robotics. The part program is arranged in the form of blocks of information where each block contains the numerical data required to produce one segment of the workpiece. The block contains, in coded form, all the information needed for processing a segment of the workpiece: the segment length, its cutting speed, feedrate, and so on. Dimensional information (length, width, and radii of circles) and the contour form (linear, circular, or other) are taken from an engineering drawing. Dimensions are given separately for each axis of motion (X, Y, etc.). Cutting speed, feedrate, and auxiliary functions (coolant on and off, spindle direction, clamp, gear changes, etc.) are programmed according to surface finish and tolerance requirements.[1-3]

Compared with a conventional machine tool, the NC system replaces the manual actions of the operator. In conventional machining a part is produced by moving a cutting tool along a workpiece by means of handwheels guided by an operator. Contour cuttings are performed by an expert operator by sight. On the other hand, the operator of NC machine tools need not be a skilled machinist. He only has to monitor the operations of the machine, operate the tape reader, and place and remove the workpiece. All intellectual operations that were formerly done by the operator are now contained in the part program. However, since the operator works with a sophisticated and expensive system, intelligence, clear thinking, and good judgment are essential qualifications of a good NC operator.

Preparing the part program for an NC machine tool requires a *part programmer*. The part programmer must possess knowledge and experience in mechanical engineering. Knowledge of tools, cutting fluids, fixture design techniques, use of machinability data, and process engineering are all of considerable importance. The part programmer must be familiar with the function of NC machine tools and machining processes and has to decide on the optimal sequence of operations. He writes the part program manually or by using a computer-assisted language, such as APT.

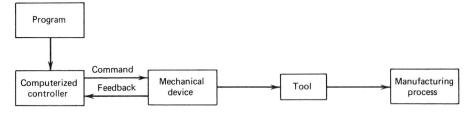

Fig. 12.1. Schematic diagram of a manufacturing system.

The controller of modern NC systems includes a dedicated mini- or microcomputer which performs the functions of data processing and control. These systems are referred to as *computerized numerical control* (CNC) systems and include hardware similar to that of robot systems. Robot systems, however, are more complicated than machine tool CNC systems for reasons discussed later.

12.2. NUMBER OF AXES AND COORDINATE SYSTEMS

Machine tools require the control of the position of the cutting edge of the tool in space. In principle, the control of *three* axes is adequate. Robots require the control of both the position of the tool center point and the orientation of the tool, which is achieved by controlling *six* axes of motion (or degrees of freedom).

The tool center point (TCP) is a point that lies along the last wrist axis at a user-specified distance from the wrist. It can be, for example, the edge of a welding gun or the center of a gripped object. The TCP plays in robotics the same role that the cutting edge plays in machining.

To emphasize the important role of orientation in robotics, let us assume that both the top and the bottom sides of the beam in Figure 12.3 are to be welded. Since the beam is narrow, the required positions, in terms of the *X, Y,* and *Z* coordinates of the TCP, are almost identical in both cases. Nevertheless, to weld the bottom side, the arm must reach the beam in a completely different orientation than required to weld the top side. Note that this drastically changes the position of each joint of the manipulator.

There are robot systems that use more than six axes—as well as CNC machines which use more than three axes of motion. A typical example is the addition of a rotary table to a three-axis milling machine. On the other hand, many robot systems use less than six axes, and there are CNC machines with only two axes of motion. For turning parts on a lathe, for example, only two numerically controlled axes are required. The reason is that the part is symmetrical. Similarly, since a welding gun is a symmetrical tool about one axis, many arc-welding applications require only a five-axis robot manipulator.[4]

12.2.1. Coordinate System

The main three axes of motion in machine tools are referred to as the *X, Y,* and *Z* axes. For example, in a vertical drilling machine, a $+X$ command moves the work table from left to right, a $+Y$ command moves it from front to back, and a $+Z$ command moves the drill toward the top away from the workpiece. In NC and CNC machine tools the *X, Y,* and *Z* axes are always assigned in order to create a right-hand Cartesian coordinate system.

In robotics, several coordinate systems are in use: the world coordinate system (WCS), the tool coordinate system (TCS), the joint coordinate system (JCS), and in intelligent robots also the sensor coordinate system (SCS).

The WCS is a Cartesian coordinate system with the origin at the manipulator base. The *X* and *Y* axes are horizontal, and the *Z* axis is perpendicular to both *X* and *Y*. The WCS is similar to the coordinate system used in CNC machine tools, although in non-Cartesian robots the definition of the *X* and *Y* directions is not natural as with machine tools.

The TCS is a coordinate system assigned to the end effector, or tool. The TCS is very useful in manual teaching of robots. When using the TCS, the teaching is done from the tool's viewpoint, namely, as if the operator were "riding" on the tool and driving it. All displacement and rotation commands refer to the current position of the tool and the direction toward which it is pointing; thus they are the most understandable by the operator.

The term *JCS* refers to the set of all joint position values, and in non-Cartesian robots it is actually not a coordinate system. Coordinates are stored in JCS in most point-to-point robots. Finally, the SCS is a coordinate system assigned to a sensor mounted above the working space of the robot, and is sometimes used with intelligent robots. To conclude, all these four coordinate systems are used in robotics, whereas only the Cartesian coordinate system is used in CNC machine tools.

MILWAUKEE-MATIC® 180

Horizontal (X) travel — 508mm (20")
Vertical (Y) travel — 508mm (20")

Fig. 12.2. CNC machine tool system. (Courtesy of Kearney and Trecker.)

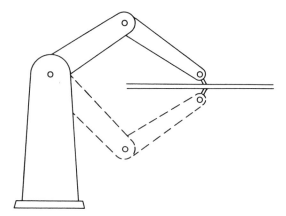

Fig. 12.3. Different robot orientations in beam welding.

12.3. SYSTEM STRUCTURE

As was shown in Figure 12.1, both CNC machines and robot systems include a mechanical device controlled by a computerized controller, which is fed by a task program (in robotics) or part program (in CNC) that dictates the path or trajectory of the tool.

Both CNC and robot systems can be divided into point-to-point (PTP) and continuous-path or contouring systems.[1-3] A typical PTP system is encountered in a spot-welding robot or in a CNC drilling machine. In a spot-welding operation the robot moves until the point to be welded is exactly between the two electrodes of the welding gun, and then the weld is applied. The robot then moves to a new point, and another spot weld is performed. This process is repeated until all the required points on the part are welded. The welding gun is then brought to the starting point, and the system is ready for the next part.

In more general terms, the description of the PTP operation is the following: The robot or the machine tool moves to a numerically defined position, and then the motion is stopped. The tool performs the required task with the robot or the machine stationary. Upon completion of the task, the robot or the machine tool moves to the next point and the cycle is repeated.

In a PTP system, the path and the velocity, while traveling from one point to the next, are without any significance. Therefore, a basic PTP system would require only position counters for controlling the final position of the robot tool to bring it to the target point. The coordinate values for each desired position are loaded into the counters with a resolution that depends on the system's basic resolution unit (BRU). During the motion of the arm the encoder at each joint transmits pulses that represent the position of the joint. Each axis of motion is equipped with a counter to which the corresponding encoder pulses are transmitted. At the beginning of a motion from a point, each axial counter is loaded by the corresponding required axial incremental distance to the next point in BRUs. During the motion of the arm, the contents of each counter are gradually decremented by the pulses arriving from the corresponding encoder. When all counters are at zero the robot is in its new desired position.

In continuous-path robots and CNC machine tools the tool performs the task while the axes of motion are moving, as, for example, in an arc-welding robot or a milling machine. The task of the robot in arc welding is to guide the welding gun along the preprogrammed path. In continuous-path systems all axes of motion may move simultaneously, each at a different velocity. These velocities, however, are coordinated under computer control to trace the required path, or trajectory.

In a continuous-path operation, the position of the machine or the robot tool at the end of each segment, together with the ratio of axes velocities, determines the generated trajectory (e.g., the weld path in arc welding), and at the same time the resultant velocity also affects the quality of the work. For example, variations in the velocity of the welding gun in arc welding result in a nonuniform weld seam thickness (i.e., an unnecessary metal buildup or even holes).

A block diagram of a continuous-path robot system is shown in Figure 12.4. The coordinates of the end points in the task program are stored in the WCS. Based on these points, the robot computer performs trajectory planning and transforms the calculated coordinates to six desired joint position values (i.e., to JCS). The desired position is sent every T_a seconds (typically $T_a = 30$ ms) to the control loops. In robotics, the loop controller usually contains a dedicated microprocessor, whereas in most CNCs the control loops are closed through the computer itself. Each loop controls a corresponding drive unit which actuates one axis of motion of the manipulator.

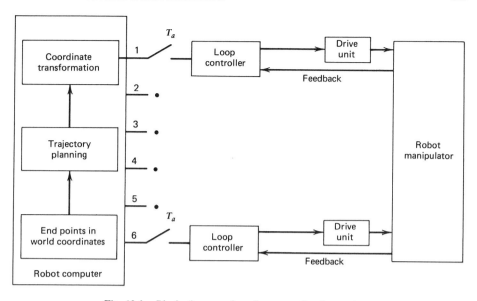

Fig. 12.4. Block diagram of continuous-path robot system.

The axes of motions in many robot systems are coupled, especially in the articulated structure. That means that a load on one axis affects the position accuracy of another axis. The effect of this coupling in machine tools is negligible. Therefore many control loops in robotics are designed to remedy the axial coupling and consequently become more complicated than their machine tool counterparts.

Another major difference is in the structure of the mechanical device. The structure of the robot manipulator is less rigid than that of a machine tool, and therefore in robotics it is more difficult to achieve precise motions with a given level of accuracy.

12.4. INTERPOLATORS

Continuous-path manufacturing systems (e.g., CNC milling machines, arc-welding robots, etc.) require interpolators to determine the path between given end points. Part programs in CNC and task programs in robotics provide only the end points of the segments along the contour, and the system computer interpolates the path between the points and generates in real time the command to the individual axes of motion.[5] Typically, NC and CNC machines are capable of linear and circular interpolation, whereas most robot systems permit only linear motions between end points.

The principle of operation for linear interpolators of CNC machine tools is relatively simple. The axes of motion in machine tools are perpendicular to each other, thus creating a Cartesian coordinate system. A motion along a straight path employs the following relationships:

$$V_x = \left(\frac{x}{l}\right) V \tag{12.1}$$

$$V_y = \left(\frac{y}{l}\right) V \tag{12.2}$$

and

$$V_z = \left(\frac{z}{l}\right) V \tag{12.3}$$

where

$$l = \sqrt{x^2 + y^2 + z^2}$$

The distances x, y, and z are the components of l in the X, Y, and Z directions, respectively, and V is the required velocity along the path (referred to as *feedrate* in machining). Equations (12.1) through

(12.3) provide the basis of linear interpolator algorithms for machine tools. Circular interpolator algorithms are more complicated and can be found in the literature.[6,7]

The interpolators applied in CNC machine tool systems can be used in Cartesian coordinate robots in which the rotating axes of the wrist intersect at one point, as, for example, in the commercial robot shown in Figure 12.5. In these cases the CNC-type interpolator determines the trajectory of the wrist intersection point, and generates velocity commands according to Eqs. (12.1) through (12.3) for the three Cartesian axes of the arm. Another algorithm is applied to determine the desired orientation of the wrist.[5]

However, most robot systems, unlike machine tools, include rotary axes, and therefore the determination of the trajectory between end points is much more complex. In continuous-path robots that include rotary axes, the interpolation process is divided into two stages: trajectory planning and coordinate transformation (see Figure 12.4).

The first stage consists of breaking down the path between the end points into small sections along the same straight line as shown Figure 12.6. This stage is denoted as trajectory planning. Subsequently the motion from the beginning to the end of each small section is obtained by solving the inverse kinematic problem, or, in other words, transforming the points' Cartesian coordinates to corresponding joint commands. For the trajectory in Figure 12.6 the robot computer sends commands to each joint at points 1, 2, 3, and 4, and the manipulator moves successively from point to point. The path between successive points is not predicted and is usually not an exact straight line. By contrast, in NC and CNC machine tool systems linear interpolators always produce straight lines, and the path error does not exceed the resolution unit of the system.[1,2,3]

The deviations from a straight line in robot systems depend also on the moving speed of the manipulator. The spacing of interpolated points between the end points is based on equal time intervals. A typical time interval can be on the order of 25 msec. If the arm moves at 10 mm/sec, then the distance between successive points is 0.25 mm (0.01 in.), but if the required speed is 100 mm/sec,

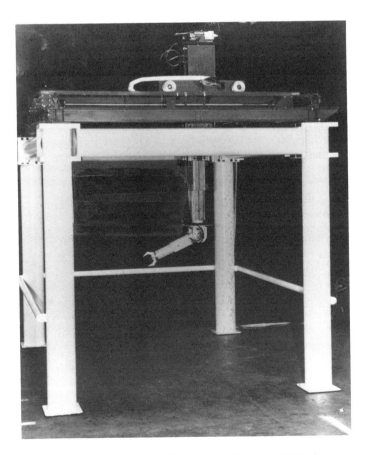

Fig. 12.5. Cartesian coordinate robot. (Courtesy of GCA.)

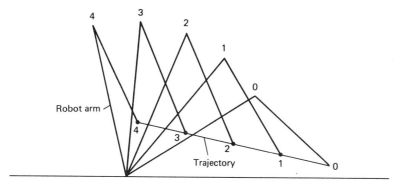

Fig. 12.6. A robot arm follows straight-line trajectory.

then the distance between successive points is 2.5 mm (0.1 in.). Since the path between successive points is not a straight line, the interpolation errors in the latter case are larger.

The distance between successive points in Figure 12.6 is equal. However, in most cases, the trajectory planning stage also includes the calculation of the desired acceleration and deceleration. In these cases, successive points at the beginning and at the end of the trajectory are not equally spaced, as shown in Figure 12.7. Since the distance between points is proportional to velocity, acceleration and deceleration are achieved.

In NC and CNC systems acceleration and deceleration are achieved by varying the V in Eqs. (12.1) through (12.3) according to a predetermined formula. Also notice that the machining feedrates (i.e., axial velocities) in machine tools are smaller by one or two orders of magnitude than in robotics (a typical feedrate for cutting aluminum is 5 mm/sec = 12 in./min, and is much smaller for steel). Therefore, paths that include acceleration and deceleration periods are not so frequently found in machine tools as in robotics.

12.5. CONTROL LOOPS

Most small- to medium-size CNC machine tools and robots utilize DC servomotor actuators to drive the axes of motion. Two alternative approaches exist to control the motions of a mechanical device driven by DC motors. One approach commonly used by robot manufacturers and researchers is to control the torque of the robot arm by manipulating the motor current. Another approach used in CNC machine tools[1,8-10] and by robot manufacturers is to control the motor rotational speed by manipulation of the motor voltage. The first approach, based on manipulation of current, treats the torque produced by the motor as an input to the robot joint, as shown in Figure 12.8. The second approach, based on manipulation of voltage, treats the robot arm as a load disturbance acting on the motor's shaft, as shown in Figure 12.9. This basic distinction is not merely a philosophical one, and has important practical consequences for the final control system design.

A straightforward approach to the control of robot arm motion is to apply at each joint the necessary torque to move the manipulated object and to overcome friction, gravity forces, and dynamic torques due to the moment of inertia. Torque control, based on manipulation of DC motor current, utilizes a *current amplifier* in the motor's drive unit. The problem with this type of system is the need to have an accurate estimate of the moment of inertia at each joint of the robot arm to obtain the desired velocity and trajectory.[11] If the actual value of the inertia is smaller than expected, then the torque applied is larger than required. This torque is translated to higher acceleration and consequently higher velocity. This can have disastrous consequences; for example, a part can be struck and broken since the velocity is not zero as desired at the target position. On the other hand, if the inertia is larger than expected there is a loss of time, since the arm decelerates a long distance before the target point and "creeps" toward it very slowly.

Fig. 12.7. Trajectory planning, including acceleration and deceleration.

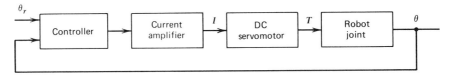

Fig. 12.8.   Control loop utilizing current amplifier.

An important advantage of the torque control approach is that we can maintain a desired torque or force. This is useful in some robotics applications, such as screwing or assembly of mating parts. Another advantage is that when the robot arm encounters resistance (e.g., the gripper touches a rigid obstacle) it maintains a constant torque and does not try to draw additional power from the electrical source.

The alternative approach is to control the velocity of the robot arm by manipulation of the DC motor voltage, utilizing a *voltage amplifier* in the motor's drive unit. A similar approach is also usually employed in CNC machine tools and in hydraulically driven robots.[1,5] The main advantage of this approach in robotics is that variations in the moment of inertia affect only the time constant of the response but do not result in any disastrous consequences and do not affect the time required to reach the target position. The arm always approaches the target smoothly with a very low speed. The problem with this approach is that the torque is not controlled, and the motor will draw from the voltage amplifier whatever current is required to overcome the disturbance torque. In robotics this can lead to burning of the amplifier's fuse when the robot arm encounters a rigid obstacle. Another disadvantage is that this system is not suitable for certain assembly tasks, such as press fitting and screwing, which require a constant torque or force.

The selected control approach to a robot arm should be dependent on the application and the environment in which the robot arm operates. When the arm is free to move along some coordinate (e.g., spray painting robots), the specification of velocity is appropriate. When the robot's end effector might be in contact with another object in such a way as to prevent motion along a coordinate, the specification of torque is appropriate. Note that either velocity or torque may be specified, but not both.

12.6.   PROGRAMMING

NC and CNC machine tools use off-line programming methods, which can be either manual or computer assisted, such as programming with the aid of the APT language. During off-line programming the machine remains in operation while a new part program is being written. Typically, when a part program is ready it is stored on a punched tape (recent systems use floppy disks). The punched tape is taken to the machine shop and loaded onto the tape reader of the NC or CNC machine tool, and the part is subsequently produced.

By contrast, with most robot systems the robot itself is used for the programming stage. At least three programming methods are used in robotics: manual teaching, lead-through teaching, and using a task-programming language.[5,12,13]

Manual teaching is most frequently used in point-to-point robotic systems. With this method a control box (called the teach pendant) is used by an operator during the programming or teaching stage. The operator moves each axis of the robot manually, until the combination of all axial positions yields the desired position of the robot. The operator then stores the coordinates of this position into the computer memory. This process is repeated for each required position until the task program is completed.

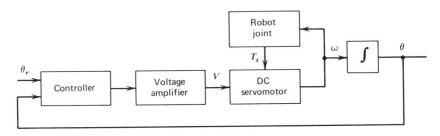

Fig. 12.9.   Control loop utilizing voltage amplifier.

Lead-through teaching is applied in continuous-path applications. It is done by grasping the robot's end effector and leading it through the desired path while simultaneously recording the sequence of positions of each axis. The manipulator actuators are disengaged during the teaching phase. With this method, again, the robot itself is tied up during the programming stage.

The third method is to use robotic programming languages similar to the APT language which is used with NC and CNC machine tools. The main problem, however, in off-line programming of robot systems is the positional errors that are caused by the deflection of the manipulator links. The deflections depend on the payload and the position of the end effector and are not taken into account by the language processor. Such a problem does not exist in NC, where the machine tool structure is rigid and is not affected by the cutting load.

Therefore, in many robot systems (such as the PUMA robot with the VAL language of Unimation Inc.) the programming is divided into two stages. In the first stage the task is programmed by using identifying names for end points (e.g., PNT1). In the second stage the loaded manipulator is brought to each end point and the actual point coordinates are recorded for each identifying name. Thus, we see that the manipulator itself is occupied during programming even when using programming languages, which is not the case in NC systems.

In the future, robot programming will be performed entirely off-line and will use in addition to the language processor a postprocessor, similar to the use of postprocessors in machine tool programming. Postprocessors of machine tools accept as input the output of the general processor (e.g., APT) and generate as output either the punched tape for a particular NC system or information suitable for easily preparing the tape. For each model of NC system a special postprocessor must be available if off-line computer-assisted programming is required. The postprocessor performs computations related to the specific NC system, such as accelerations, and takes into account machine constraints (e.g., the work table size), machining constraints (e.g., maximum feedrate), and special functions (e.g., coolant).

Similarly, each robot needs a postprocessor program. However, robot postprocessors are more complicated and must include a complete kinematic and dynamic simulation of the arm. The robot language processor will provide the postprocessor with identifying names and the estimated payload during each motion, and the postprocessor, in turn, will generate a complete task program. This is the only method by which the robot itself is not tied up during the programming stage.

12.7 ADAPTIVE SYSTEM

Most robots and CNC machines in production are programmable systems that can repeat a sequence of programmed operations as long as necessary. However, these systems are unable to sense and respond to any change in their working environment. For example, assume that a CNC lathe is used to turn a batch of cylindrical workpieces to 200-mm radius from raw material ranging from 201 to 204 mm. The machining feedrate in the part program is calculated as the maximum allowable to remove a load of 4 mm without cutter breakage, although with the smaller load the feedrate can be increased without risking a cutter breakage. In robotics, if a robot had been programmed to grip a part at a certain point, it will always close its gripper jaws at that point, whether a part is there or not. These situations can be remedied only if the systems can adapt themselves to the changing conditions in the environment, sensed with suitable sensors.

Adaptive systems, such as adaptive control (AC) of machine tools and intelligent robots, consist of a conventional CNC or robot system (including the axial position and velocity transducers) and two additional components: (1) at least one sensor able to measure the working conditions, and (2) a computer routine that processes the sensor information and sends suitable signals to correct the operation of the conventional system. In robotics this routine is based on artificial intelligence (AI) algorithms, and in machine tools on simpler strategies.

A typical AC system for a lathe is shown in Figure 12.10. The AC is basically a feedback loop, where the machining feedrate adapts itself to the actual cutting force and varies according to changes in work conditions as cutting proceeds. The CNC controller acts as an internal control loop and executes the original part program from the punched tape. The linkage between the CNC and the AC is through the feedrate correction commands, which are dictated by the strategy of the AC routine.

The AC loop functions in a sampled-data mode. The actual cutting force is sampled every T seconds (typically $T = 0.1$ sec), then converted to a digital signal, and sent to the AC computer. In the computer the actual force is compared with a predetermined allowable reference force, and generates the force error that is used to calculate the feedrate corrections according to an AC strategy. Discussion on various AC strategies for machine tools can be found in the literature,[14] and are beyond the scope of this text.

The sensors used with AC of machine tools are always of a contact type, such as force, torque, and power sensors. In robotics, both contact and noncontact sensors are used. The noncontact sensors include proximity, laser, visual, acoustic, and range sensors.[15] Intelligent or adaptive robots use at least one sensor the data of which is processed by an AI routine. The latter makes real-time decisions that might change the programmed sequence of operation of the robot. A block diagram of an intelligent robot is shown in Figure 12.11. It consists of a two-level hierarchical closed-loop control system. At

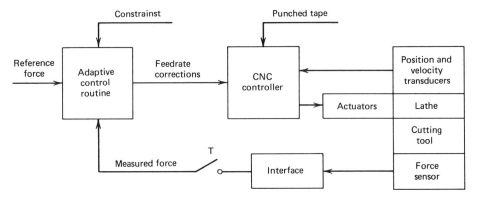

Fig. 12.10. Adaptive control system of a CNC lathe.

the lower level there are position control loops of the individual joints; each includes a position feedback device such as an encoder. In many robots a velocity control loop (with a tachometer as the velocity-measuring device) is contained at the low-control level as well.

The higher level in the hierarchy contains a sensor that is able to sense the robot environment, its associated interface, and the AI algorithm. The loop is closed through the interpolator algorithm, which responds to the original task program instructions, with corrections obtained from the AI algorithm.

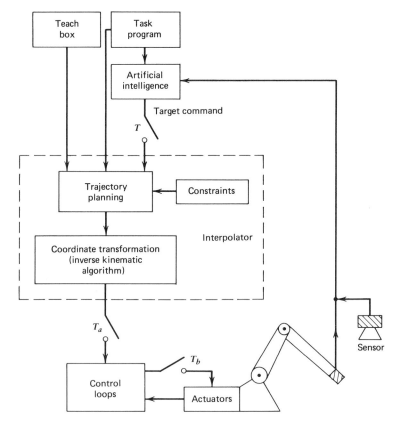

Fig. 12.11. Block diagram of an intelligent robot.

Most intelligent robots operate with three different sampling intervals T, T_a, and T_b. The sensor output is sampled every T seconds (typically 0.05–2 sec) by the AI routine, which processes the data and makes decisions regarding the trajectory. Trajectory information (position and speed) is then sent by the AI routine to the trajectory-planning portion of the interpolator. The coordinate transformation (i.e., the inverse kinematics) subroutine of the interpolator calculates the joint commands and sends them as references to the control loops every T_a seconds (typically $T_a = 0.03$ sec). Each control loop usually includes a microprocessor in which a comparison between the joint references and the actual position takes place. The sample of joint positions as well as the transmission of commands to the actuators is executed every T_b seconds (typically $T_b = 1$ msec). In all robots the relationship $T_b \leq T_a < T$ exists.

REFERENCES

1. Koren, Y., *Computer Control of Manufacturing Systems*, McGraw-Hill, New York, 1983.

2. Olesten, N. O., *Numerical Control*, Wiley-Interscience, New York, 1970.

3. Groover, M. P., *Automation, Production Systems, and Computer-Aided Manufacturing*, Prentice-Hall, Englewood Cliffs, New Jersey, 1980.

4. Engelberger, J., *Robotics in Practice*, AmaCom, A Division of American Management Association, 1980.

5. Koren, Y., *Robotics for Engineers*, McGraw-Hill, New York, 1985.

6. Jordan, B. W., Lenon, W. J., and Holm, B. D., An Improved Algorithm for Generation of Nonparametric Curves, *IEEE Transactions, on Computers*, Vol. C-22, No. 12, December 1973, pp. 1052–1060.

7. Masory, O. and Koren, Y., Reference-Word Circular Interpolators for CNC Systems, *Transactions ASME, Journal of Engineering in Industry*, Vol. 104, November 1982.

8. Beckett, J. T. and Mergler, H. W., Analysis of and Incremental Digital Positioning Servosystem with Digital Rate Feedback, *Transactions ASME, Journal of Dynamic Systems Measuring Control*, Vol. 87, March 1965.

9. Ertell, G., *Numerical Control*, Wiley-Interscience, New York, 1969.

10. Poo, A. N. and Bollinger, J. G., Dynamic Errors in Type I Contouring Systems, *IEEE Transactions Industrial Applications*, Vol. IA-8, No. 4, July 1972, pp. 477–484.

11. Paul, R., *Robot Manipulators: Mathematics, Programming, and Control*, MIT Press, Cambridge, Massachusetts, 1981.

12. Ardayfio, D. D. and Pottinger, H. J., Computer Control of Robotic Manipulators, *Mechanical Engineering*, Vol. 104, No. 8, August 1982, pp. 40–45.

13. Grossman, D. D., Robotic Software, *Mechanical Engineering*, Vol. 104, No. 8, August 1982 pp. 46–47.

14. Ulsoy, G., Koren, Y., and Rasmussen, F., Principal Developments in Adaptive Control of Machine Tools, *Transactions of ASME, Journal of Dynamic Systems Measuring Control*, Vol. 105, No. 2, June 1983, pp. 107–112.

15. Nitzan, D., Assessment of Robotic Sensors, *Proceedings of the 1st International Conference on Robot Vision and Sensory Control*, Stratford on Avon, England, April 1981, pp. 1–8.

CHAPTER 13

SENSORS FOR INTELLIGENT ROBOTS

AVINASH C. KAK

Purdue University
West Lafayette, Indiana

JAMES S. ALBUS

National Bureau of Standards
Washington, D.C.

13.1 INTRODUCTION

Most industrial robots today have little or no sensory capability. Feedback is limited to information about joint positions, combined with a few interlock and timing signals. These robots can function only in environments where the objects to be manipulated are precisely located in the proper position for the robot to grasp. For many industrial applications, this level of performance has been adequate. Until recently most robot applications consisted of taking parts out of die-casting and injection-molding machines. In this task the parts produced are always in exactly the same position in the mold so that the robot needs no sensory capability to find the part or compensate for misalignments. Another principal application has been the spot welding of automobile bodies. Here, the car bodies are positioned and clamped so that each body is always exactly in the same place as the one before. Thus the robot needs no sensory capability to find where to place the welds. Even in those cases where a robot places welds on a moving car body, an optical encoder is attached to the conveyor line to tell the robot how fast the car body is moving. Also, an optical sensor indicates when each car moves into the work area so that the robot can begin its programmed routine. The robot's computer then transforms the coordinate system of the program to follow the conveyor line.

Sensory capabilities are necessary for a robot to function with intelligence, by which is meant the ability to interact with a flexible environment. Such intelligence, for example, permits adaptive motion control in which sensory information is used to modify the commands to a programmable manipulator. In this handbook the major sensor types are discussed in two chapters: in Chapter 14, Vision Systems, Agin[1] discusses the sensors and algorithms used for robot vision today in industry for parts inspection. In Chapter 16, Depth Perception for Robots, Kak discusses various sensors, such as stereo vision, structured light, lasers, and ultrasound, that can be used for range mapping of scenes.

The aim of this chapter is to present an overview and to discuss, sometimes briefly, those areas of robot sensors that are not covered in the more specialized treatments elsewhere in the handbook. In this and the other chapters only those sensors are included that are necessary for a robot to monitor its immediate, proximal, and distal environments. Sensors required by a robot to monitor its own internal state have not been treated in this chapter; usually these are the same as those required for general automation, and have been reviewed in many other publications.[4,8,14,19]

Unfortunately, robot intelligence requires a good deal more than simple acquisition of sensory data. In addition to sensors, what is needed is the ability to organize the sensory data into task-specific models or components. Model representation is a more challenging problem than the design of devices and systems for sensory input. With the rapid progress that has recently been made in all types of sensors, the achievable level of robot intelligence seems constrained primarily by this limitation on the processing of sensory information.

214

To illustrate the problem of model (knowledge) representation, consider the simple scene depicted in Figure 13.1 for vision sensing. That the scene consists of a cube behind a cylinder is readily apparent to human vision. It is also obvious that if a robot used a single camera, it might be difficult to interpret the scene correctly because all the depth information, which is important to human perception, would be lost. But now let us consider a more advanced robot vision system that is capable of depth perception, so that the computer now has a range map of the entire scene. To interpret the scene, the computer could first cluster neighborhood range points and calculate the local surface normal for each cluster; these clusters could be combined into surface patches depending on the permissible variation of surface normals within the same patch; and then the computer could join these patches into surfaces using task-specific constraints. This low-level processing would be followed by the generation of a symbolic description of each surface and a comparison of this description with *stored knowledge* within the computer. Sometimes it might suffice to create a statistical profile of the surfaces, which would then be classified by the computer by drawing on its stored knowledge. Given the extensive nature of computing involved, it is not surprising that the limiting factor in achievable robot intelligence today is the degree of sensory data organization permitted by economic and other factors.

The model (knowledge) organization of the state-of-the-art robots is at a low level, which is consistent with their knowledge base being limited to a small number of readily distinguishable objects or environmental situations. (In vision, for example, this low-level processing may consist of representing an isolated object by its binary silhouette and extracting the associated numerical parameters.) This level of sensory data organization is adequate for functions such as pick-and-place, simple inspection, determining the location and orientation of an isolated component, and elementary assembly tasks like inserting light bulbs in sockets. Some factory systems already exist utilizing low-level organization of the sensory input and using this as a feedback for manipulator control. In most such cases only one or two sensors are used for feedback, and the environment is severely constrained with respect to the other sensors. Examples are the feedback obtained from tactile sensing and a compliant wrist as used by the Hitachi system for assembly,[12] eddy-current sensory feedback used by another Hitachi system for arc-welding,[2] and force sensing used by Olivetti for assembly.[6,7] More complex inspection and assembly operations will require higher-level processing of information from a multiplicity of sensors for the correct interpretation of geometrical relationships among different components in three-dimensional space.

Sensors for robots are usually divided into contact and noncontact categories. Contact sensors may be further subdivided into tactile, touch, and force/torque sensors, whereas noncontact sensors may be divided into proximity and vision sensors. Contact and noncontact sensors play complementary roles for sensory feedback. For example, to insert a peg into a hole, vision sensing with its coarse resolution may be used to find the hole and position the peg close to the hole (even partially inserting it if possible). Feedback from force and torque sensors could then be used for a more precise alignment of the peg with the hole. An optimum algorithm for final insertion would move the peg in such a manner as to minimize the binding force and torque.

As another example of the complementary nature of contact and noncontact sensing, consider the task of picking a workpiece from a table. Again, wide-field-of-view vision sensing with its attendant coarse resolution would identify and locate the workpiece and also determine the optimum gripping locations (holdsites) for the given orientation of the workpiece. The robot could then position the fingers on its end effector near the holdsites. The fingers could subsequently be closed until the force-sensory feedback indicates the required level of force to prevent slippage.

Noncontacting sensing, such as provided by vision, does not always have to be relegated to coarse-resolution functions. In vision, for example, resolution can be enhanced by reducing the field of view or using high-resolution, but more expensive, devices such as image dissector cameras and high-density solid-state linear cameras that can resolve a single image line of illumination into 2048 elements.

Although many sensory interactive control systems are necessary for a robot to accomplish goals

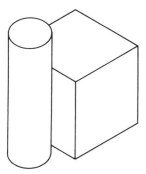

Fig. 13.1. "A cube behind a cylinder" constitutes a simple yet illustrative example for discussing the difficulties associated with knowledge representation for three dimensional scenes.

and execute skilled tasks in an unstructured, uncooperative, and sometimes hostile environment, it must be said that from the standpoint of factory automation, many assembly tasks can be performed with little or no sensing of the environment. For example, engineers at the Kawasaki laboratories in Japan have shown that robots can put together complex assemblies of motors and gearboxes with no more than high-precision position feedback, cleverly designed grippers, and fixtures for holding parts that flex by a slight amount when the parts are brought together. Other experiments have shown that a small amount of vibration or jiggling, together with properly designed tapers and bevels, can accommodate for slight misalignments and prevent jamming when two pieces with close tolerances are assembled.

Assembly of close-fitting parts may also be accomplished without any sensory feedback by using a unique device called the Remote Center Compliance (RCC).[9] As is illustrated later, by projecting the center of compliance into the part that, for example, is undergoing insertion into a hole, this device provides the necessary "give" to prevent jamming, galling, and the like. The center of compliance is the point through which forces act on an object while it is being manipulated for assembly. It is also the point at which lateral and rotational compliances are decoupled. Assembly forces are minimized when this center is located near the point where parts come in initial contact during assembly.

The mystery of the RCC device is perhaps best explained with the help of the following illustrations, which are based on those found in the product literature from Lord Industrial Products, a manufacturer of such devices. First consider the unaided insertion of a shaft into a mating hole as shown in Figure 13.2a. The arrow A is the initial direction of the force applied to the shaft. If there is a lateral error in aligning exactly the axis of the shaft with the axis of the hole, there will come into play a horizontal force on the leading end of the shaft as it makes contact with the chamfer (Figure 13.2b). More likely than not the end result would be a jam as shown in Figure 13.2c, and any further application of the force would only exacerbate the situation. The principal source of difficulty is that the shaft is being *pushed* into the hole.

In contrast, an RCC device causes the shaft to be *pulled* into the hole. Again consider the case when the axes of the shaft and the hole are not exactly lined up. As shown in Figure 13.3a, when the leading end of the shaft now held by an RCC device makes contact with the chamfer, the resulting horizontal force causes the shaft to *translate laterally*, permitting easy insertion (Figure 13.3b). A similar result is obtained when the axes of the shaft and the hole are not parallel to each other as shown in Figure 13.4a. Positioning itself laterally by the mechanism described previously, the shaft will enter the hole. Again because of the give in the RCC, the resulting moments will rotate the shaft about the compliant center and ease further insertion (Figure 13.4b).

Before concluding this introduction, we would like to draw the attention of the reader to earlier surveys of sensors for robots.[3,21,25,29]

13.2. VISION

In Chapter 14, Vision Systems, Agin presents many algorithms for robot vision. Here we add a few introductory comments and then discuss the relative merits of static overhead camera-type vision systems versus the eye-in-hand type.

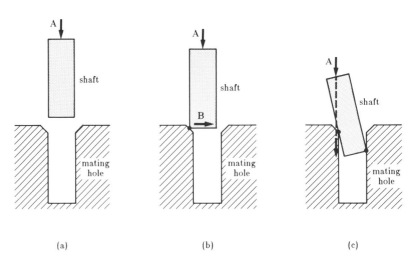

(a)   (b)   (c)

Fig. 13.2. Insertion of a shaft into a mating hole is depicted here. A slight lateral error in aligning the two components results in forces and moments that jam the shaft after only a partial insertion.

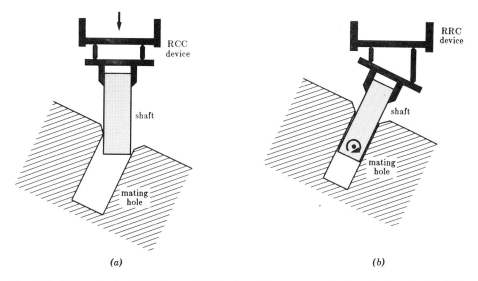

(a)   *(b)*

Fig. 13.3. When the shaft held by an RCC device makes contact with the chamfer, the "give" in the RCC allows a lateral translation that eases further insertion.

Two major applications of vision sensing are for the control of manipulators and for automated inspection. To control manipulation, a vision system must be able to identify and locate workpieces, and in addition it must be able to determine their orientation. That this ability would permit a robot to deal with imprecisely positioned workpieces with random orientations is obvious and is one element in the economic justification of vision for automation. Current systems are able to fulfill these functions, provided the workpieces have a small number of stable positions on a platform (a conveyor belt), and provided for each stable position the object can be distinguished from the other workpieces involved in a given task.

In a recent survey, Rosen[26] has presented an exhaustive listing of the manipulation tasks that require vision. Major categories in this list are the following:

1. Manipulation of separated workpieces on conveyors for the purposes of sorting, packing in a container, feeding into another machine in a prescribed position, and orientation, and so on.
2. Bin picking for the same purposes as listed previously. Bins are commonly used for transporting and buffer-storing workpieces in factories.
3. Manipulation in manufacturing processes for finishing, sealing, deburring, cutting, and so forth.
4. Manipulations required for assembly. These include fitting that involves parts presentation and mating, and fastening that involves spot welding, riveting, bolting, screwing, nailing, gluing, and stapling. For assembly tasks a manipulator with vision is capable of *active accommodation,* which implies that it can compensate for errors in the positioning and orientation of workpieces. With most image-sensing devices used for vision, this compensation is relatively coarse. Finer position control of a workpiece or a tool with respect to another workpiece can be accomplished by *passive accommodation* in which reaction forces and torques as sensed by a compliant wrist are used for the correction of residual positioning errors.

Rosen[28] has also given a detailed listing of applications that could use robot vision for inspection. The broad categories in this list are as follows:

1. Applications requiring qualitative and semi-quantitative mensuration. Inspection tasks here include determination of surface finish properties such as burrs, cracks, voids, stains, and other blemishes; checking for integrity and completeness of a workpiece; label reading and sorting; and so on.
2. Applications requiring highly quantitative mensuration for critical dimensions of key features of a workpiece and the measurement of tool wear.

There does not exist a general-purpose vision system that can cater to the requirements of all (or even most) of the applications listed. Agin has discussed some of the major algorithms that represent a small yet impressive beginning in that direction. The algorithms presented by him can locate and

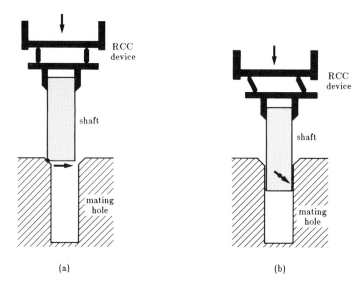

(a)   (b)

Fig. 13.4. In the case shown here the axes of the shaft and the hole are not parallel. Again the give in the RCC facilitates insertion.

identify a workpiece, and determine its orientation, when the workpiece is presented to the vision system in one of its stable positions. These algorithms can also perform qualitative and semi-quantitative visual inspection by using simple features like the area and the perimeter of a binary silhouette of the workpiece and of any holes in that silhouette.

Most current vision systems use a static overhead camera placed above the robot working area, which possesses the advantage that while the vision data are being acquired and processed, the robot can attend to some other part of the industrial process. Its disadvantage, however, is that the vision system becomes ineffective if the robot arm is blocking the workpiece, as might happen during the attempted retrieval of a part. It has been shown by Loughlin[20] that if a camera is mounted on the robot gripper (the result being called a eye-in-hand system), in addition to eliminating the blind-spot-caused problems, it is also possible to employ much lower-resolution imagery without sacrificing accuracy in the calculation of either the location or the orientation of the object.

Generally, algorithms for object recognition require higher-resolution imagery than those for the calculation of object location and orientation. However, more than 70% of vision applications in automated manufacturing do not require objects to be recognized, since they come presorted down the line. In most such applications determining accurately the orientation and location of an object within the cycle time of a robot is important.

Loughlin[20] has shown that owing to parallax errors (an example of which is shown in Figure 13.5) that are associated with the static overhead types, an eye-in-hand vision system can better compute the location and orientation of an object. In Figure 13.5 the parallax error causes an erroneous computation of the center of area when the object is off-center in the field of view of the camera. Such errors are nonexistent for eye-in-hand systems, since each object is examined from a position directly above.

The effective image resolution at close range obtained with the eye-in-hand vision can be comparable to the static case. Also, because of the small matrix size of the image, the computing times can be much shorter. For a 32 × 32 pixel resolution eye-in-hand vision, the image frame acquisition time was 20 msec; and since the processing time on such a small matrix can be kept below 8 msec, the total adds up to less than 28 msec, which is the cycle time of a PUMA robot. Owing to their lower resolution, the eye-in-hand cameras possess an additional advantage of longer depth of fields, which often eliminates the need for automatic focusing, contributing to a reduced cost, size, and weight of the camera.

13.3. DEPTH PERCEPTION FOR ROBOTS

In Chapter 16, "Depth Perception for Robots," the principles of operation and algorithms applied in the major sensory types used for this purpose are discussed. Here we illustrate a simple yet elegant device developed at the National Bureau of Standards, based on the principles of structured light discussed in Chapter 16. This system can also be used to determine the orientation of the parts on a table.

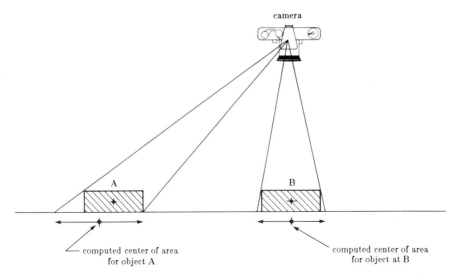

Fig. 13.5. Shown here is the parallax error associated with static overhead camera systems for robot vision. Parallax causes errors in the computation of the center of area when the object is off-center in the field of view of the camera.

As shown in Figure 13.6, light from a slit projector mounted on the robot's wrist is projected into the region in front of the fingertips. When this light strikes an object it makes a bright line that is observed by the television camera also mounted on the robot's wrist. The geometrical relationship between the projector, the camera, and the bright line allows the vision system to compute the position of the object relative to the fingertips. Figure 13.7 shows the calibration chart that the computer uses to transform from row and pixel number of a spot on the bright line to x and y position in the coordinate system of the fingertips. The shape of the line is a depth profile of the object. The slope of any segment of this line indicates the orientation of the corresponding portion of the front surface of the part relative to the fingers. This information is then used by the control system to move the

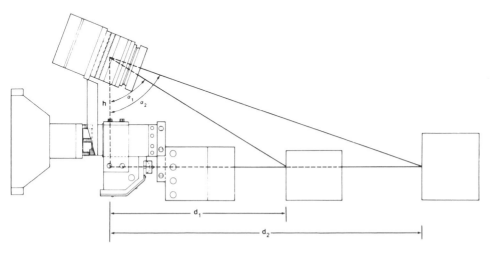

Fig. 13.6. A strobographic flash unit projects a plane of light into the region in front of the robot fingertips. A camera mounted on the robot wrist measures the apparent position of the light reflected from an object and computes the position and orientation of the reflecting surface. If the camera sees a bright mark at angle α_1, the reflecting object must be located at distance d_1. Similarly, if a bright mark is seen at angle α_2, the reflecting object is at distance d_2. The known value of h makes the distance calculation a simple problem in trigonometry (Albus[1]).

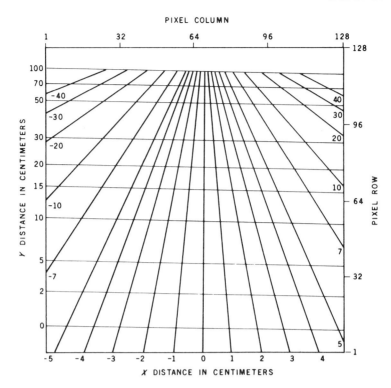

Fig. 13.7. A calibration chart for the vision system shown in Figure 13.6. The pixel row and column of any illuminated point in the television image can be immediately converted to x, y position in a coordinate system defined in the robot fingertips. The x axis passes through the two fingertips, and the y axis points in the same direction as the fingers. The plane of the projected light is coincident with the (x, y) plane so that the z coordinate on every illuminated point is zero (Albus[1]).

robot hand to reach out and grasp an object. In this manner the robot can operate on a random pile of blocks and cylinders and sort them into a regular array. It can also measure the shape of a casting, find the edge of a window frame, detect the crack between a pair of bricks, or measure the angle between two pieces of steel. This is the most basic type of sensory information required to perform tasks in the factory and on the construction site. Further details on this device can be found in Reference 2.

13.4. TACTILE SENSORS

Tactile sensing in the human hand enables the detection of touch, force, pattern, slip, and movement. (After force on the skin exceeds the proportional limit of tactile sensing, stretch sensors in the tendons and muscles of the hand, wrist, and the arm detect the larger forces and transmit signals to the brain.) Recent research has focused on the development of tactile sensors for robots with similar skinlike properties.

As mentioned in the introduction, the roles of vision and tactile sensing are often complementary to each other. With vision sensing, which often spans a wide field of view, a robot can identify workpieces and locate their position, whereas tactile sensing can be used for determining the local shape, orientation, and resistance to gripping pressure once the workpiece is grasped. With taction the mechanical hand of the future should be able to work with soft, delicate, or limp goods.

Harmon[13] has conducted a highly informative survey of researchers and manufacturers in robotics to determine the present and future tactile-sensing requirements and potentials. Of the many applications that Harmon has listed for tactile sensing are included arc-welding tracking, bin-picking, orienting parts, adaptive grasping, batch assembly, detection of jamming (for example, screw cross-threading), fitting of close tolerance parts (down to 0.0005 in tolerances), electronic-component insertion, handling of delicate parts such as light bulbs and soft materials, the ability to test parts for freedom and integrity. As Harmon has pointed out, machines that could automatically grasp and sensibly use small tools designed for human use (e.g., wrench, hammer, pliers, screwdriver) would be revolutionary. On the basis of the responses to his survey questionnaire, Harmon has presented the following specifications as "an oversimplified but illuminating picture of the more-or-less average impression" of a tactile transducer:

1. An array consisting of 10×10 force-sensing elements on a 1-in.[2] flexible surface, much like a human fingertip.

2. Each element should have a response time of 1–10 msec, preferably 1 msec.

3. Threshold sensitivity for the elements ought to be 1 g, with the upper limit of force range at 1000 g.

4. The elements need not be linear, but they *must* have low hysteresis.

5. The skinlike sensing material must be robust, standing up well to harsh industrial environments.

Historically, the earliest tactile sensors were simple microswitches mounted on the inner sides of the fingers of a robot hand. A microswitch does a binary detection of the applied pressure. Ernst[9] was the first to use such binary touch detection. Also, Paul ans Shimano[23] used such switches to determine if a part was present or not, and then to center the hand over the workpiece. Then came pressure-sensitive conductive rubber sensors as used by Goto et al.[12] Their robot hand had four such sensors on the inner side of each of the two gripper fingers; each finger also came equipped with 14 outer contact sensors. The touch information acquired in this manner was used to pick blocks located randomly on a table and to pack them tightly on a pallet. Pneumatic-touch sensing was first used by Garrison and Wang[11]; they built a gripper with 100 pneumatic snap-action touch sensors located on a grid with 0.1×0.1-in. centers.

The information supplied by binary touch sensors is too limited if the aim is to measure gripping forces and to extract information about the object between the fingers. Analog sensors, which get around these limitations, were first used by Hill and Sword.[15] The tactile sensors mounted on the inner sides of the gripper of their manipulator consisted of 6×3 arrays of sensory elements. The force on each sensory element acted against a compliant washer, which displaced a vane that controlled the amount of light received by a phototransistor from a light-emitting diode. Takeda[28] has also built a tactile array that is capable of analyzing the shape of an object. The array consists of an 8×10 matrix of needles that are free to move normal to the face of the gripper. A potentiometer is used to measure the distance between the opposite needles of the two arrays, one mounted on the inner side of each gripper finger. Analog touch sensors can also be designed with carbon fibers.[18]

In the following we present in greater detail two distinctly different approaches to tactile sensing. Although both approaches possess high potential as candidates for commercial tactile sensors of the future, the materials that will be used for pressure transduction may, however, be different from those described. At this time it appears that polyvinylidene (PVF_2), which is piezoelectric and can therefore be used for converting pressure signals into electrical waveforms, may possess many advantages over conductive rubbers. The advantages of PVF_2 include light weight, ruggedness, and low cost. Furthermore, it conforms easily to complex surfaces and possesses a large bandwidth. However, PVF_2 does suffer from one problem, which consists of unwanted voltages generated by its pyroelectric property. It is hoped that this difficulty can be circumvented by insulating the sensor from the environment with a material having a low thermal conductivity. In contrast with PVF_2, conductive rubbers are characterized by nonlinearity, low sensitivity, long time constants, drift, and are prone to hysteresis.

13.4.1. Tactile Sensing Using Compliant and Anisotropically Conducting Materials

A tactile sensor designed by Hillis[16] consists of a monolithic array of 256 individual sensory elements. Each element is of any area less than 0.01 cm[2] and gives an independent analog indication of the force over its surface over the range of 1 to 100 g. The sensor consists of two sheets of wires running perpendicular to each other and separated by a thin elastic medium like a nylon woven mesh or a fine mist of spray paint, as shown in Figure 13.8, the choice being a trade-off between sensitivity and range. Elastic woven meshes allow a large pressure range, whereas mists of paint result in high sensitivity. The contact points at the intersections of the conductors form the sensory elements for measuring local pressure.

In the physical construction of the device, the top sheet of conductors is supplied in the form of a compliant anisotropically conductive silicone rubber (ACS), and the bottom sheet a printed circuit board etched into fine parallel lines. The ACS rubber has the property of being electrically conductive in only one direction owing to the nature of its fabrication: it is constructed of 250-μ layers of silicone rubber impregnated with either graphite or silver, alternating with similar nonconductive layers. The impregnated layers form the conducting paths corresponding to the conductors shown in the top sheet of Figure 13.8. In Hillis's device, the graphite-impregnated layers were gold plated to reduce their linear resistance from a few kilohms per centimeter, which was considered too high for building a sensor, to about 100 ohm/cm. A mechanical drawing of the sensor is shown in Figure 13.9. PC1

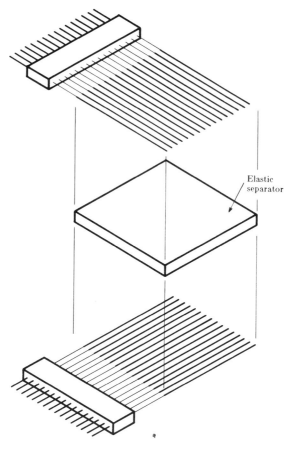

Fig. 13.8. The tactile sensor designed by Hillis[16] consists of a monolithic array of 256 individual sensory elements. The sensor may consist of two sheets of wires running perpendicular to each other separated by a thin elastic medium. The intersection points of the wires form the individual sensory elements for pressure measurement.

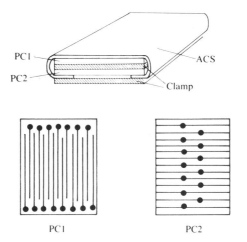

Fig. 13.9. A mechanical drawing of the sensor. PC1 is a printed circuit board etched into fine parallel lines to correspond to the conducting wires of the lower sheet in Figure 13.8. The ACS rubber is wrapped around at the edges as shown, where it makes contact with the etched lines on another printed circuit board PC2 (Hillis[16]).

is a printed circuit board that is etched into fine parallel lines to correspond to the conducting wires of the lower sheet in Figure 13.8. The ACS rubber is wrapped around at the edges as shown, where it makes contact with the etched lines on another printed circuit board PC2. The function of PC2 is merely to give access to the conducting paths of the ACS.

As shown in Figure 13.10, the measured electrical resistance at each sensory element is proportional to the pressure applied to the sensor. The relationship depicted in that figure is nonlinear, and the underlying phenomenon not clearly understood. Hillis has postulated the process depicted in Figure 13.11, which shows that the pressure on the elastomeric ACS deforms the material around the separator, allowing it to contact the metal below.

A tactile image is formed by scanning the array, which is accomplished by applying a voltage to one *column* at a time and measuring the current flowing in each *row* (*columns* and *rows* refer to the two perpendicular sets of wires as depicted in Figure 13.8). (For low-resolution devices, the tactile pattern can be measured directly by attaching separate wires to each sensory element.[5,27]) For example, in Figure 13.12, if the contact A is closed, that fact will emerge when during the scan a voltage is applied to column 2 and a current is measured when row 2 is grounded. A potential problem with this method is the introduction of "phantom" tactile images caused by the crosspoint problem, which is illustrated in Figure 13.12 for the case of a 3 × 3 array. In that figure if contacts A, B, and C are closed, the contact D will appear to be closed also owing to the existence of a conducting path from column 3 to row 3 through contact C, from row 3 to column 2 through B, and finally from row 2 to column 2 through A.

The voltage-mirror approach that Hillis used to get around the crosspoint problem consists of applying a fixed voltage to the column of interest while all the other columns are grounded to eliminate any alternate paths. The rows are also held to a ground potential by injecting whatever current is necessary to cancel the current injected by the active column. The value of the resistance of a crosspoint is inversely proportional to the current that is necessary to pull the corresponding row to ground potential. This method is valid only when the crosspoint resistance is high compared to the linear resistance of the row and column lines. Sample tactile images obtained with such a device using a 16 × 16 array with a sprayed separator (approximately 10^4 dots per square centimeter) are shown in Figure 13.13.

13.4.2. A VLSI Architecture for Tactile Sensing

In the sensor of the preceding section, either a large number of wires must connect the sensor to the processing hardware or bulky multiplexing electronics must be placed close to the sensor where there is little room. Also, a large amount of data must be transmitted to the host computer if there is no on-site preprocessing. In this section, we discuss a VLSI-based tactile sensor designed by Raibert and Tanner[24] that gets around these problems by using the inherently parallel nature of the transduced pressure data to do on-site processing of the tactile image.

The physical structure of the sensor is illustrated in Figure 13.14. A 1-mm-thick sheet of pressure-sensitive conductive plastic (Dynacon B) is placed in contact with a VLSI wafer, which is an *n*MOS integrated circuit. The wafer consists of a two-dimensional array of cells, each of dimension 1.6 × 0.9 mm. The cells measure the pressure by recording the resistivity of the conductive plastic directly

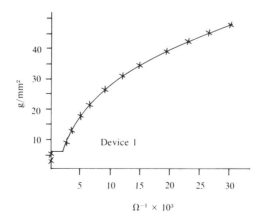

Fig. 13.10. The relationship between the pressure applied and the electrical resistance measured by the tactile sensor (Hillis[16]).

Side view

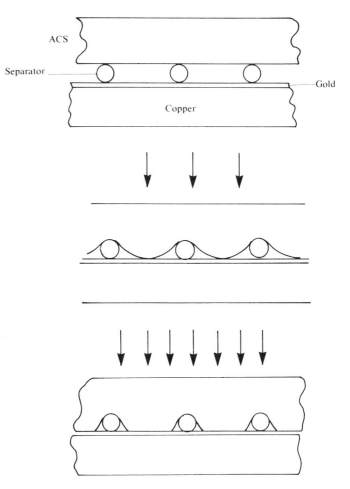

Fig. 13.11. A possible explanation for why the electrical resistance of the ACS decreases with increased applied pressure. (Hillis[16]).

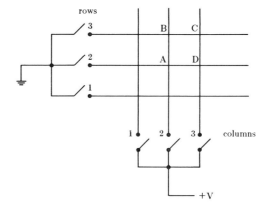

Fig. 13.12. A tactile image is formed by scanning the array, which is accomplished by applying a voltage to one column at a time and measuring the current flowing in each row. For example, if the contact A is closed, that fact will emerge when during the scan a voltage is applied to column 2 and a current is measured when row 2 is grounded. A potential problem with this method is the introduction of "phantom" tactile images caused by the crosspoint problem.

1/8-in. ring ⚪

Cotter pin

Spade lug

Top of screw with slot ⊘

Fig. 13.13. Sample tactile images obtained by Hillis[16].

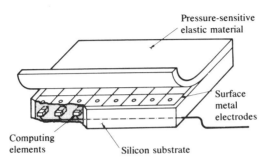

Pressure-sensitive
elastic material

Surface
metal
electrodes

Computing
elements

Silicon substrate

Fig. 13.14. The physical structure of the sensor designed by Raibert and Tanner. A 1-mm thick sheet of pressure-sensitive conductive plastic (Dynacon B) is placed in contact with a VLSI wafer, which is an *n*MOS integrated circuit. The wafer consists of a two-dimensional array of cells, each of dimension 1.6 × 0.9 mm (Raibert and Tanner[24]).

above. The resistivity is measured by sending a test current between a pair of electrodes within the cell, the electrodes being in electrical contact with the plastic. A comparison of the voltage drop across the electrodes and an externally supplied reference voltage yields a signal proportional to the applied force. The electrodes within a cell may be patterned as shown in Figure 13.15. That figure also illustrates the measurement of resistivity of the overlying plastic.

A schematic of the computing structure for processing the transduced pressure signals is shown in Figure 13.16. In addition to the electrodes for measuring the local pressure, each cell contains a processor capable of performing local computations and communicating with its immediate neighbors; such communications are required for computing convolutions of the tactile image. The communications between the cells are also used for shifting the processed tactile data from one cell to another until it reaches the periphery of the array. The data are then serialized for transmission to the host computer.

A block diagram of the digital processor in each cell is shown in Figure 13.17. The voltage V_{press} obtained by passing a test current through the overlying plastic is compared to the reference V_{ref}, and the result given a 1-bit representation. (The pressure at each point may be obtained with more than 1 bit of precision by applying different values of V_{ref} and computing the results for each bit thus obtained.) This 1-bit representation of the pressure is made available to a 1-bit latch for storage through a multiplexer (MUX). (The bits I_1 and I_2 determine which of the four inputs to the MUX will be latched.) Note that as soon as any of the inputs to the MUX are latched, it is also made available to the cell's neighbors to the south and the east. It is through this communication that it is possible to compute convolutions of the type

$$C(m, n) = \sum_{i=0}^{I-1} \sum_{j=0}^{J-1} p(m - i, n - j)h(i, j) \tag{13.1}$$

where $p(m, n)$ represents the transduced pressure values, and $h(i, j)$ the unit sample response of the convolution. Since the pressure data $p(m, n)$ are binary, the multiplications in the foregoing expression can be implemented as conditional additions. In Figure 13.17 the coefficients of convolution, which are also binary, are represented by bit I_6 supplied from the instruction register. For example, assume that $h(0, 0)$ is 1. The first step in implementing the convolution would be to send $I_6 = 1$ to all cells over the global communication line (assuming that the measured data have already been latched). This, with a proper selection for the bit I_4, would cause the latched measured data to be transferred into the accumulator (after passing through the adder, which at this time has nothing to add).

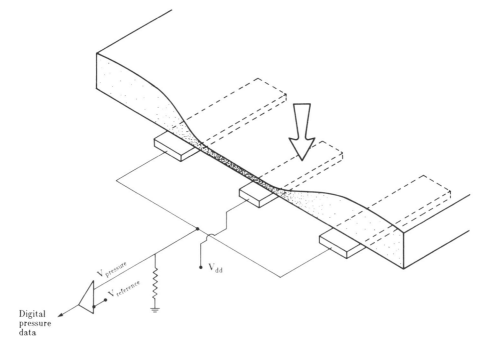

Fig. 13.15. The electrodes within a cell may be patterned as shown here. Also illustrated is the measurement of resistivity of the overlying plastic.

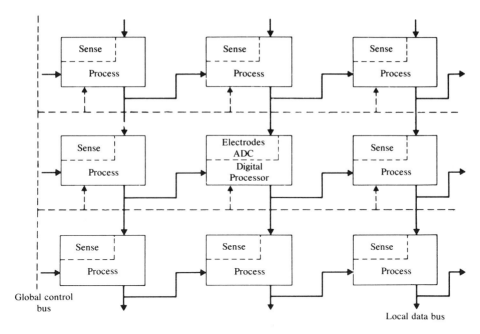

Fig. 13.16. A schematic of the computing structure for processing the transduced pressure signals. In addition to the electrodes for measuring the local pressure, each cell contains a processor capable of performing local computations and communicating with its immediate neighbors; such communications are required for computing convolutions of the tactile image (Raibert and Tanner[24]).

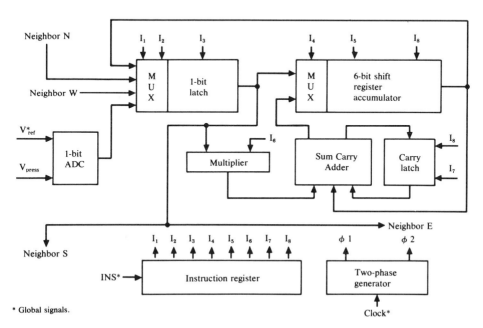

Fig. 13.17. A block diagram of the digital processor in each cell. The voltage V_{press} obtained by passing a test current through the overlying plastic is compared to the reference V_{ref}, and the result given a 1-bit representation. (The pressure at each point may be obtained with more than 1 bit of precision by applying different values of V_{ref} (Raibert and Tanner[26]).

227

At this stage, the latched measured data have moved into the accumulator, and also have been made available to the south and east neighbors. Now, by sending a global instruction with proper values for bits I_1 and I_2, this information can be latched into, let us say, the east neighbor. By the same token, the cell under question will grab (latch) the information from its west neighbor. Now let us assume that $h(0, 1)$ is also 1. By again setting the I_6th bit in the next global instruction to 1, this newly latched information will be made available to the adder, where it will be summed with the previously stored information in the accumulator. A repeated application of this procedure would carry out the convolution in Eq. (13.1).

Raibert and Tanner[24] have fabricated and tested a 6×3 sensing array and proposed a defect-tolerant approach to designing larger arrays. The larger the area of active silicon, the greater the risk of fabricating defective circuitry. For example, for a 25×25 sensor, assuming a cell size of 1×1 mm, the active silicon area would be around 2.5×2.5 cm, which is entirely too large compared to conventional integrated circuits. A large array can be made defect tolerant by incorporating redundancy. In the approach suggested by Raibert and Tanner this is accomplished by duplicating the computing element within each cell and by providing a selector mechanism for choosing between them when one fails.

13.5. FORCE AND TORQUE SENSORS

The forces and torques encountered by a robot arm during assembly can be measured directly by using a wrist force sensor, which basically consists of a structure with some compliant sections and transducers that measure the deflections of the compliant sections. The most common transducer used for this purpose is the strain gage, others being piezoelectric, magnetostrictive, magnetic, and so on.[22] Figure 13.18 shows a strain-gage-type wrist force sensor built at SRI.[25] It is built from a milled 3-in.-diameter aluminum tube having eight narrow elastic beams with no hysteresis. The neck at one end of each beam transmits a bending torque that increases the strain at the other end where it is measured by two foil strain gages. A potentiometer circuit connected to the two strain gages produces

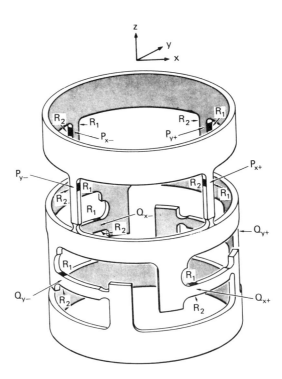

Fig. 13.18. A strain-gage wrist force sensor built at SRI. It is built from a milled 3-in.-diameter aluminum tube having eight narrow elastic beams with no hysteresis. The neck at one end of each beam transmits a bending torque, which increases the strain at the other end where it is measured by two foil strain gages. A potentiometer circuit connected to the two strain gages produces an output that is proportional to the force component normal to the strain-gage plates. (Rosen and Nitzan[25]).

an output that is proportional to the force component normal to the strain-gage planes. This arrangement also automatically compensates for variations in the temperature. This sensor is capable of measuring all three components of force and the three components of torque. Other wrist sensors have been designed by Watson and Drake[30] and Goto et al.[12] Forces and torques can also be measured with pedestal-mounted sensors, as demonstrated by Watson and Drake.[30]

Forces and torques can also be sensed indirectly by measuring the forces acting on the joints of a manipulator. For joints driven by DC electric motors, the force is directly proportional to the armature current; for joints driven by hydraulic motors, it is proportional to back pressure. Inoue[17] and Paul and Shimano[23] have demonstrated assembly using such indirect force sensing. Whereas Inoue programmed a manipulator to insert a shaft into a hole, Paul's work accomplished the assembly of a water pump consisting of a base, a gasket, a top, and six screws. Paul computed the joint forces by measuring the motor current, and his program included compensation for gravity and inertial forces.

13.6. PROXIMITY SENSORS

The basic purpose of a proximity sensor is to indicate without making contact the presence or absence of an object in its vicinity. Utilization of proximity sensors spans applications such as high-speed counting, protection of workers, indication of motion, sensing of ferrous materials (through magnetic effects), level control, reading of coding marks, noncontact and limit switches.

Although all the vision and depth-perception sensors discussed in Chapters 14 and 16 could also be used for proximity indications, it is usually possible to employ much simpler and low-cost devices to meet the objectives here. Some of these devices are based on the same principles. For example, the infrared device in Figure 16.12 and the ultrasonic echo device illustrated in Figure 16.35 make good proximity sensors, although one can now of course take liberties with the accuracy of range calculation electronics. Both these devices also make movement detectors: the infrared device by measuring change in reflected light, and the ultrasonic unit by incorporating Doppler principles. Commercially available proximity sensors also employ radio frequency, magnetic bridges, permanent-magnet hybrids, and Hall effect.

13.7. SUMMARY

In this chapter we have presented an overview on the subject of sensory capabilities for robots. Where available, we listed the tasks that a robot would be able to accomplish through such capabilities and the desired specifications for different sensors. We also touched upon certain sensor types that are not covered elsewhere in this handbook. In particular, we discussed different approaches to tactile sensor design.

REFERENCES

1. Albus, J. S., *Brains, Behavior and Robotics,* McGraw-Hill, New York, 1981.

2. Ando, S., Kusumoto, S., Enomot, K., Tsuchihashi, A., and Kogawa, T., Arc welding robot with sensor, *Proceedings of 7th International Symposium on Industrial Robots,* 1977, pp. 623–630.

3. Binford, T. D., Sensor system for manipulation, Heer, E., Ed., *Proceedings of 1st Conference on Remotely Manned Systems for Exploration and Operations in Space,* 1973, pp. 283–291.

4. Chironis, N. P., *Machine Devices and Instrumentations,* McGraw-Hill, New York, 1966.

5. Broit, M., The utilization of an "artificial skin" sensor for the identification of solid objects, *Proceedings of 9th International Symposium on Industrial Robotics,* March 1979.

6. d'Auria, A. and Salmon, M., Sigma: An integrated general purpose system for automatic manipulation, *Proceedings of 5th International Symposium on Industrial Robots,* 1975, pp. 185–202.

7. d'Auria, A. and Salmon, M., Examples of applications of the Sigma assembly robot, *Proceedings of 6th International Symposium on Industrial Robots,* 1976, pp. G5.37–G5.48.

8. Doeblin, E. O., *Measurement System: Application and Design,* McGraw-Hill, New York, 1966.

9. Drake, S. N., Watson, P. C. and Simunovic, S. N., High speed assembly of precision parts using compliance instead of sensory feedback, *Proceedings of 7th International Symposium of Industrial Robots,* 1977, pp. 87–98.

10. Ernst, H. A., MH-1, A computer-operated mechanical hand, *AFIPS Conference Proceedings, SJCC,* 1962, pp. 39–51.

11. Garrison, R. L. and Wang, S. S. M., Pneumatic touch sensor, *IBM Technical Disclosure Bulletin,* Vol. 16, No. 6, November 1973.

12. Goto, T., Inoyama, T. and Takeyasu, K., Precise insert operation by tactile controlled robot HI-T-HAND Expert-2, *Proceedings of 4th International Symposium on Industrial Robots,* 1974, pp. 209–218.

13. Harmon, L. D., Automated tactile sensing, *International Journal of Robotics,* Vol. 1, No. 2, 1982, pp. 3–32.

14. Harris, C. M. and Crede, C. E., *Shock and Vibration Handbook,* McGraw-Hill, New York, 1961.

15. Hill, J. W. and Sword, A. J., Manipulation based on sensor-directed control: An integrated end effector and touch sensing system, *Proceedings of 17th Annual Human Factor Society Convention,* Washington, D.C., October 1973.

16. Hillis, W. D., A high-resolution imaging touch sensor, *International Journal of Robotics Research,* Vol. 1, No. 2, 1982, pp. 33–44.

17. Inoue, H., Computer controlled bilateral manipulator, *Bulletin of the Japanese Society of Mechanical Engineering,* Vol. 14, pp. 199–207, 1971.

18. Larcombe, M. H. E., Carbon fibre tactile sensors, *Proceedings of 1st International Conference on Robot Vision and Sensory Controls,* 1981, pp. 273–276.

19. Lion, K. S., *Instrumentation in Scientific Research,* McGraw-Hill, New York, 1959.

20. Loughlin, C., Eye-in-hand robot vision scores over fixed camera, *Sensor Review,* Vol. 3, No. 1, 1983, pp. 23–26.

21. Nitzan, D., Assessment of robotic sensors, *Proceedings of 1st International Conference on Robot Vision and Sensory Controls,* 1981, pp. 1–11.

22. Norton, H. N., *Handbook of Transducers for Electronic Measuring Systems,* Prentice-Hall, New Jersey, 1969.

23. Paul, R. and Shimano, B., Compliance and control, in Brady M. et al., Eds., *Robot Motion,* MIT Press, Cambridge, Massachusetts, 1982, pp. 405–417.

24. Raibert, M. H. and Tanner, J. E., Design and Implementation of a VLSI Tactile Sensing Computer, *International Journal of Robotics Research,* Vol. 1, No. 3, 1982, pp. 3–18.

25. Rosen, C. A. and Nitzan, D., Use of sensors in programmable automation, *Computer,* December 1977, pp. 12–23.

26. Rosen, C. A., Machine vision and robotics: Industrial requirements, in Dodd, G. G. and Rossol, L., Eds., *Computer Vision and Sensor-Based Robots,* Plenum Press, New York, 1979.

27. Stojiljkovic, Z. and Clot, J., Integrated behavior of artificial skin, *IEEE Transactions in Biomedical Engineering,* Vol. 24, 1979, pp. 396–399.

28. Takeda, S., Study of artificial tactile sensors for shape recognition algorithm for tactile data input, *Proceedings of 4th International Symposium on Industrial Robots,* 1974, pp. 199–208.

29. Wang, S. S. W. and Will, P. M., Sensors for computer controlled mechanical assembly, *The Industrial Robot,* Vol. 5, No. 1, March 1978, pp. 9–18.

30. Watson, P. C. and Drake, S. H., Pedestal and wrist force sensors for automatic assembly, *Proceedings of 5th International Symposium on Industrial Robots,* 1975, pp. 501–511.

CHAPTER **14**

VISION SYSTEMS

GERALD J. AGIN

Carnegie-Mellon University
Pittsburgh, Pennsylvania

14.1. COMPUTER VISION

Computer vision means the use of computers or other electronic hardware to analyze visual information. It involves the use of visual sensors to create an electronic or numerical analog of a visual scene and some processing to extract intelligence from this representation. The sensors usually are television cameras or similar devices, but the same processing techniques can be applied to scenes derived from sonar, X rays, laser scanners, and so on. Image processing can be used to derive the identity, position, orientation, or condition of objects in the scene. Dimensional measurements can also be made to a limited extent using available two-dimensional cameras; however, high-resolution line cameras together with controlled X-Y-θ stages enable precise measurements to be made.

Computer vision can be useful in a number of industrial applications including inspection, locating, counting, measurement, and control of industrial manipulators. Although the actual and potential applications are quite varied, the underlying principles are the same. Accordingly, the following discussion focuses on techniques rather than applications.

Some important parameters of computer vision systems are cost, speed, adaptability, and capability. For any given function, cost may be expected to decrease and speed to increase over the next 5–10 years, owing mainly to developments in general-purpose computer hardware. Improvements in adaptability and capability may be slower, since they depend on progress in basic understanding of visual processes.

A series of technical reports details progress in industrial computer vision at SRI International (previously Stanford Research Institute) over the last ten years.[1,2]

14.1.1. Products Available

At least a dozen manufacturers offer complete vision systems including cameras, computer, processing software, training procedures, and interfaces to other systems. One of the first on the market was the Machine Intelligence Corporation's VS-100 Vision Module, shown in Figure 14.1. It is a binary vision system, based on the SRI Vision Module.[3] It can analyze a scene in terms of connected components, derive shape features for each of these components independent of position and orientation, and identify the shapes based on these features. Control may be accomplished manually using a light pen on the display screen or through an external computer-to-computer interface.

There are wide differences among commercial vision systems. The field is young, many new ideas are being tried, competition is stiff, and quality is variable. This makes it especially important for the purchaser of vision equipment to understand its principles of operation thoroughly. Since visual problems are so different from each other, success at one visual task does not mean a given system will perform well on another task.

Some pertinent questions for a user of commercial vision systems to ask follow:

1. What kinds of cameras can be supported? Will the system accept a standard RS-170 video signal? Will it work with a linear array camera? What is the limiting spatial resolution?

2. Is the system binary or gray-scale? If both binary and gray-scale capability are claimed, what processing algorithms are available for use with each mode?

3. What processing algorithms are used? Be sure to understand the strengths and limitations of the method.

Fig. 14.1. A binary vision system programmable with a light pen. (Photograph courtesy of Machine Intelligence Corporation.)

4. What is the processing speed? How does the speed depend on the complexity of the image?

5. How much control is readily available over the processing algorithms, parameters, and options? Is there a high-level language available in which to write application programs? Are the lower-level processing routines modifiable by the user?

6. What are the operator controls? Once a system has been set up and tuned by a factory engineer, how much knowledge is needed by the production personnel who use the equipment on a daily basis?

7. What kind of technical support services are available? Is there a clearly written manual that explains the principles of operation as well as the steps needed to set up the system? Are engineers accessible to answer technical questions? How is field maintenance?

Even the best computer vision systems suffer severe limitations on the domains of application and the kinds of visual processing they will do. They tend to treat images in a two-dimensional fashion—in general they cannot deal with three-dimensional objects viewed from an arbitrary perspective. Most of them require special lighting arrangements. Objects must be presented individually. (Some systems can handle multiple objects, but these may not overlap in the image.) They do not deal well with situations not expressly provided for.

Research is under way in a large number of academic and industrial institutions to overcome these limitations. This includes research on the nature of visual processes in humans and animals, as well as on image-processing methods, computer graphics, man-machine communications, pattern recognition, computer architectures, and sensor design. Progress will come from better understanding of the methods available and from the development of special-purpose Very Large Scale Integration (VLSI) architectures to do visual processing rapidly.

14.1.2. Definitions

An *image* is an array

$$\{p_{yx}, \qquad y = 1, \cdots, H; x = 1, \cdots, W\}$$

Each element p_{yx} is called a *pixel* or picture cell. The parameter y corresponds to rows of the image, and x corresponds to columns. x and y form an inverted, left-handed coordinate system with its origin near the upper left corner of the image, with the x axis increasing to the right and the y axis increasing downward. H is the image height or the number of rows in the image, and W is the width or the number of columns. *Spatial resolution* refers to the image dimensions expressed in a form such as 512 × 512.

The *brightness resolution* is the number of discrete values a pixel may take. For example, if $0 \leq p_{yx} \leq 63$, we may refer to the image as having either 64 levels, or 6 bits of brightness resolution. When the brightness resolution is two levels, or 1 bit, we refer to the image as *binary*. Nonbinary images are called *gray-scale*.

There are several distinct purposes for which computer vision may be used. By far the largest use of computer vision today is for *inspection*. Such a system must tell good product from bad. Usually an inspection procedure is developed specifically for a given part, using conventional programming techniques to call on previously written subroutine packages to do specific visual processing steps.

A computer vision system can be made to perform *recognition* or *classification*. Such a system must be capable of telling that a particular part corresponds to one of several alternatives, or perhaps none at all. Generally the system will be provided in advance with the visual characteristics of several different *prototype* objects so that it can classify an *unknown* object in its field of view by comparing it with each of the different prototypes. When the prototypes are provided by using the system to view actual examples of the objects, this is called *training by showing*. Usually the prototypes are stored as a collection of *feature values* that represent numerical quantities or measurements. Those feature values that are independent of position and orientation can be used as a basis for comparison between two images.

A computer vision system can provide *position* and *orientation* of an object in its field of view. Generally the system must be told what the *zero position* of each object is so that it can report displacement from the zero position.

Another possible use for computer vision is in *measurement*. This refers to finding one or more dimensions of a part. The accuracy of a visual measurement depends crucially on the resolution of the image. Because most imaging devices used for computer vision today provide enough resolution for only crude measurements, measurement is not covered in depth in this chapter.

14.2. IMAGING

Setting up a scene for computer imaging involves several considerations: presentation, lighting, camera, optics, and interfaces to the computer.

14.2.1. Presentation and Lighting

Careful attention to imaging considerations is crucial to the success of any computer vision application. Although the details depend on the specific application, the fundamental principle is that the information desired should be as prominent in the picture as possible.

The part should be presented in the center of the the camera's field of view. The spatial resolution available in any given system is limited; therefore the wider the field of view that must be searched for a part, the fewer pixels available for shape analysis. The most usual arrangement is with the camera pointing straight down with the object directly below (*plan view*), but a camera aimed horizontally is also sometimes useful (*elevation view*).

High contrast is required for binary vision, but it is also helpful for gray-scale vision. *Backlighting* is a good technique for creating silhouettes. A backlight can be made with fluorescent bulbs behind a translucent diffusing screen. Such units may also be purchased from drafting-supply houses. The uniformity of illumination of purchased units may be improved by adding a spacer to increase the distance from the bulbs to the diffuser.

Since for many applications a backlight will not be feasible, it will be necessary to consider the background against which the part will be presented. There are only two options here—a dark or a light background. Generally, a light-colored part should get a dark background and a dark part a light one. Sometimes use can be made of colored lights or colored filters to enhance contrast.

To minimize shadows, frontlighting should be diffuse. When a dark object lies on a light-colored background, shadowing can distort the perceived silhouette. With light-colored objects, shadowing can cause gaps and holes in the object's silhouette. This can be useful when internal details of an object are desired.

14.2.2. Cameras

Although there are many different types of television cameras, only two types are generally used for robot vision, the *vidicon* and the *solid-state array*. Although a vidicon can usually be obtained very inexpensively, in most applications the extra expense of a solid-state array camera is justified.

A vidicon is a vacuum tube. An image is focused on a charged photosensitive target, allowing electrons to leak away and discharge the target locally where the light is brightest. An electron beam scans the target, reads the amount of charge remaining, and reestablishes a charge for the next image cycle. Usually there are 480 scans across the active image area, and the output is *interlaced*, so that every other scan line (the odd lines) is read out, followed by the even scan lines. Each scan line is represented at the camera output by a time-varying voltage that represents the brightness profile across that image line. Vidicons are plagued by drift, geometrical distortion, persistence, burn-in, and short lifetimes.

Solid-state array cameras contain evenly spaced arrays of identical photosensitive elements. Light falling on these elements creates a charge, which is later scanned out sequentially cell by cell. The camera output is a sequence of voltage levels representing brightness at the individual cells. Array cameras are available in resolutions ranging from 40×40 to 512×512 and higher. The output may be interlaced for compatibility with conventional video equipment but it need not be. *Linear arrays* are a special class of solid-state arrays that provide a single row of pixels. Linear arrays with resolutions of several thousand elements per line are available. Solid-state cameras tend to be rugged, reliable, and compact. They operate at voltages compatible with most other computer hardware. A typical solid-state television camera is shown in Figure 14.2.

The usual scanning rate is 30 complete frames per second. Sometimes the rate can be varied with the use of external clock signals. With either vidicons or solid-state cameras, the output voltage is proportional to the time integral of light intensity during the frame time. Scanning rates can be increased only when there is sufficient light available. Conversely, if there is insufficient illumination falling on the camera target, slowing down the scanning rate can make more effective use of the available light.

Solid-state array cameras, as well as those vidicons that use silicon for the target material, have a broad spectral sensitivity curve that peaks near 800 nm, in the near infrared region of the spectrum. This means that the brightness of objects may not be what a human observer would see. This is less likely to be a problem when illumination is provided by fluorescent lighting; incandescent lighting emits strongly in the infrared region. Certain lasers and light emitting diodes (LEDs) emit light at 820 nm, which is invisible to a human observer but well matched to the silicon response. Spectral filters are available that can block either the infrared or the visible portion of the spectrum.

The *image aspect ratio* is defined as the ratio of width to height of an image. In the entertainment industry (which represents a much larger market for cameras than does automation) an image aspect ratio of 4:3 is standard. *Pixel aspect ratio* is defined as the ratio of the horizontal spacing to the vertical spacing of individual pixels. Pixel aspect ratio is independent of image aspect ratio and can vary from camera to camera. The number of columns in an image divided by the number of rows is equal to the image aspect ratio divided by the pixel aspect ratio. The pixel aspect ratio must be taken into account when computing the scale factors S_x and S_y mentioned in Section 14.6.2.

Fig. 14.2. VS-100 vision module. Solid-state array camera. Resolution is 244×248 elements. (Photograph courtesy of General Electric.)

14.2.3. Optics

A *lens* is an important part of the overall imaging system. The two most important parameters of a lens are its *focal length* and its *aperture*.

For a given camera, the main effect of changing to a lens of a different focal length is to change the field of view. Short focal lengths give wide-angle views and long focal lengths give "telephoto" views. To cover a given area, a longer focal-length lens requires placing the camera farther away from the area to be viewed. A faraway point of view reduces foreshortening and parallax and keeps the camera out of the way. But it also requires a more substantial mount for the camera, since small vibrations will cause proportionally more image motion.

Aperture is a measure of the light-gathering power of a lens. It is usually specified numerically as *f-number:* the focal length of the lens divided by its effective diameter. The smaller the *f*-number, the wider the lens opening and the more light the lens will admit. Usually a lens contains an adjustable diaphragm that can reduce the width of the aperture (numerically increasing the *f*-number). Aperture also affects the *depth of field* of a camera: the wider the aperture, the shorter the range of distances from the camera at which an object will stay in focus.

Lenses that are specified as "television" lenses are generally lower in quality than lenses produced for other types of cameras. Because television cameras have lower resolution than film, the difference in lens quality is generally not noticeable in the image.

It has been pointed out that most lenses are color corrected for the visible spectrum. When one uses a camera that derives most of its light energy from infrared light, the lens is being used in a manner for which it was not designed. For low-resolution cameras, the effect on image quality is not large enough to be noticed. As resolution improves, however, manufacturers of cameras and lenses will have to confront the problem and redesign lens systems for the appropriate region of the spectrum.

14.2.4. Interfaces

The television camera must be connected to a computing machine that will do the processing. The interface hardware must perform the functions of *timing, digitization,* and *buffering.*

Timing circuitry is necessary to govern the integration time of the camera, to clock the signal out, and to synchronize the digitization hardware. The timing circuitry can also handle functions such as firing strobe lamps at the appropriate instant or synchronizing digitization with the 60-Hz line frequency to minimize image flicker.

Digitization hardware may consist of an analog-to-digital (A-D) converter (for gray-scale systems) or a thresholding circuit (for binary systems). A-D converters for television signals must be fast; to digitize a frame of television input to a 512×480 resolution in $\frac{1}{30}$ sec requires a digitization rate in excess of 7 MHz. Six bits of gray-scale is usual for most applications—8 bits may be used if the television camera is of very good quality.

A thresholding circuit compares its analog input against a threshold level and delivers a 1 if the input is high and a 0 if the input is low. The threshold level should be adjustable, either with a potentiometer or under computer control using a DAC (digital-to-analog converter).

Unless visual processing is done by special-purpose hardware that matches video data rates, it will be necessary to store the image for later processing. The interface can transfer the image directly to computer memory using direct memory access (DMA), or the image may be stored in a separate *frame buffer.* The DMA option is attractive when low cost is desired and when resolution and data rates are low. An image buffer requires additional hardware for transferring its data to the main computer, but it frees valuable memory address space, allows display generation from the image, and facilitates adding further hardware to do fast image-processing functions.

14.3. ONE-DIMENSIONAL SCANNING

For many applications full two-dimensional image analysis is unnecessary—only a single horizontal scan across a scene of interest is needed to extract useful information. Use of a single scan line reduces the amount of data that must be processed. The actual processing can be very much simpler. However, the kinds of situations to which it can be applied are rather restricted, and only limited kinds of information can be extracted from the signal.

One-dimensional processing is frequently useful for looking for spots or blemishes in a product such as paper, textiles, or glass. Usually the product is transported or rotated past a fixed viewing station so that repeated scans are made. The scanning motion causes the one-dimensional scan to cover a two-dimensional area. But as long as each scan line is considered independently of all other scan lines we still refer to the processing as one-dimensional.

Although scanning can be done with a linear array camera, sometimes other devices are used. A laser beam impinging on a rotating mirror can be made to sweep across the area of interest, and detectors can pick up transmitted or reflected light from various angles. Parabolic mirrors can keep the beam always perpendicular to the material surface. Such laser scanners are widely used in the manufacture of synthetic textiles.

The output of a linear array or laser scanner is a time-varying analog waveform. A significant amount of processing can be done using analog hardware. A partial list of useful functions includes:

1. Low-pass filtering to smooth noise.
2. High-pass filtering to remove slow variations due to nonuniform lighting, and so on.
3. Differentiation to emphasize discontinuities in the signal.
4. Integration to obtain an average brightness level.
5. Logical masking to ignore signals outside the region of interest.

The final step in a chain of such processing will usually be thresholding. Any signal over the threshold presumably corresponds to a defect. For simple applications it may be sufficient to couple the threshold circuit output directly to an alarm signal, but most practical systems will analyze the signal further to determine the type or cause of the blemish, to maintain counts of defects, or to perform further checks.

Sometimes a one-dimensional scan can be used for a kind of mask matching.[4] A scan across a known object at a known location can be expected to show certain regions of dark and light. Some simple processing can detect these features. Inspection can be based on the presence of all such expected features. Classification or measurement can be based on the width or spacing of these features.

14.4. DIRECT MATCHING

Direct matching is a class of techniques that match images or portions of images directly with each other. Usually one image is a model or *template,* and the other image is an unknown in which an example of the model is sought. There are two types of situation in which a direct match can be useful in an industrial situation. It can be used for inspection purposes to verify that the entire image (or a major portion of it) corresponds to what is expected. Or it can be used to detect and locate instances of a particular small feature. These feature indications and their locations should be analyzed at a higher level of control. See Section 14.9.2.

Direct matching makes use of some measure of difference between two images. Given two images $\{p_{yx}\}$ and $\{q_{yx}\}$, the difference is $d_{yx} = p_{yx} - q_{yx}$. The individual pixel differences can be combined in a number of ways to yield an overall measure for the image:

The count of pixel differences that exceed a threshold T

$$D_0 = \sum_{x,y} \tau(p_{yx} - q_{yx}) \quad \text{where} \quad \tau(x) = \begin{cases} 1 \text{ if } |x| > T \\ 0 \text{ otherwise} \end{cases}$$

The sum of the absolute values of the pixel differences

$$D_1 = \sum_{x,y} |p_{yx} - q_{yx}|$$

The sum of the squares of the pixel differences

$$D_2 = \sum_{x,y} (p_{yx} - q_{yx})^2$$

A single numerical measure of the pixel similarity of images may be useful in a situation where pieces are fixtured so that there are no positional uncertainties. Such can be the case when inspecting the output of a (usually) highly repeatable process such as printing, punching, labeling, or stamping. If the difference between a part image and a previously trained prototype exceeds a threshold, then the part should probably be rejected.

When images are binary, a small amount of positional uncertainty can be tolerated with the use of "don't-care regions." Although this could take many forms, the most straightforward implementation is as a separate array of bits to indicate whether or not individual pixel differences are to be calculated and summed for the corresponding pixels. Sometimes special knowledge of the inspection situation can be used to specify these don't-care regions. Growing and shrinking operations on the prototype image (see Section 14.7) can be used to create a narrow don't-care region that includes the boundary.

However, in most situations the precise image location is not predictable to within the order of a single pixel. Then the use of direct matching must make use of a variable shift of one image with respect to the other and a search procedure to discover the value of shift that minimizes the difference between images. We will shift the template image q by a vector (u, v), making the pixel difference a function of the shift: $d = p_{yx} - q_{y+v,x+u}$. The overall difference will be one of the forms

$$D_0(u, v) = \sum_{x,y} \tau(p_{yx} - q_{y+v,x+u})$$

$$D_1(u, v) = \sum_{x,y} |p_{yx} - q_{y+v,x+u}|$$

$$D_2(u, v) = \sum_{x,y} (p_{yx} - q_{y+v,x+u})^2$$

The summation is understood to be over the portion where both images overlap.

The third form, D_2, is the square of the *Euclidian difference* between the images. Although it may be harder to compute directly than the other two forms, it has a number of interesting properties.[5] The summation can be expanded as follows:

$$D_2(u, v = \sum_{x,y} p_{yx}^2 + 2 \sum_{x,y} p_{yx}q_{y+v,x+u} + \sum_{x,y} q_{y+v,x+u}{}^2$$

The first term is called the *picture energy,* the second is the *cross-correlation,* and the third is the *template energy.* As long as the template is small and is entirely contained within the picture, the template energy will be independent of u and v. If we can assume that the picture energy is nearly constant in the region of interest, then the cross-correlation is directly related to the difference $D(u, v)$. If we cannot assume the picture energy to be constant, we can divide the cross-correlation by the square root of the picture energy, to arrive at the *normalized cross-correlation.*

Calculation of image differences is computationally intensive. One must choose some value of u and v, evaluate the difference, and repeat for other values of u and v until the range of possible shifts is covered. There is a trade-off between size of the template and size of the search space. When templates are large, as when they are used for inspection, each difference is expensive to compute, so the search space should be kept small. A small amount of preprocessing can be useful in locating a fiducial mark in some predefined location to constrain the search. When templates are small, it becomes conceivable to build specialized hardware to evaluate the difference for all values of u and v in one pass through the image.

Coherent optics can also be used to locate instances of templates. A discussion of these techniques is beyond this chapter. For further information, see Reference 6.

Direct matching can be used to find the shift between two images or between a template and an image. But image rotation is much more difficult. The only way to handle rotation is to store separate templates for the image in numerous orientations.

14.5. CONNECTIVITY ANALYSIS

Binary images are well suited for application to robotic vision. When conditions are suited to the production of binary images, processing can be done quickly and reliably. This stems from the compactness of the data and from the robustness of the processing algorithms.

Compactness is a direct consequence of one-bit brightness resolution. Image buffer memory requires only a single bit plane. Images can be stored in computer memory as packed bytes or words of individual pixel bits and require only one-eighth the storage that would be required for a gray-scale image with one byte per pixel. Further economies can be obtained by the use of *run-length coding.* However, the main benefit to be obtained from compactness is not storage efficiency, but processing efficiency.

Robustness stems from the topological properties of binary images. The images can be segmented in a unique manner into connected components. There is no ambiguity as to what constitutes a single silhouette. Concepts such as enclosure are well defined.

Of course, when imaging conditions are poor, these strengths turn into weaknesses. Poor contrast, random noise, and objects that overlap are poorly handled by binary methods.

14.5.1. Run-Length Coding

Run-length coding of an image takes advantage of spatial coherence. Any given row of a binary image is likely to contain long runs of consecutive 1's and 0's. Run-length coding achieves its economy by storing the column numbers at which transitions take place.

In more formal terms R_y, the run-length representation for row y of a binary image, is the set

$$R_y = \{x \mid p_{y(x-1)} \neq p_{yx}\}$$

We define p_{yx} to be 0 whenever x is outside the range $1 \leq x \leq W$.

The set R_y will always have an even number of elements. The number of elements in the set will be twice the number of consecutive runs of 1's. A line consisting of all 0's will be represented as the empty set $\{ \}$. A line consisting of all 1's will be represented as the two-member set $\{1 \ W + 1\}$.

Note that there is no explicit assumption made here as to whether 1 represents black and 0 white or vice versa. The choice depends on the relative colors of object and background. The convention we observe here is that 0 represents the color of the background.

Whether or not run-length coding will result in a net compression of image data depends on the image. For purposes of illustration, assume an image 256 × 256 pixels square for which a column number can be packed into a single byte. Assume that one additional byte per row will be needed either to serve as an end-of-row indicator or to count the number of transitions on that row. The image of a circle with a diameter of 128 pixels will require only 512 bytes to represent in run-length representation, whereas the bit map would require 8192 bytes. On the other hand, the run-length code could require as much as 65,792 bytes if the worst-case pattern of alternating 1 and 0 were presented.

The main reason we are interested in data compression has little to do with the amount of storage required. A compression will pay bigger dividends in processing time—for the less data there are, the less time will be needed to digest them. The efficiency of the connectivity and feature-extraction algorithms to be described depend to a great extent on the fact that only the transitions are significant, and only the transitions need to be processed.

14.5.2. Connectivity in General

The purpose of connectivity analysis is to separate multiple objects in a scene from each other and from areas of the picture that represent noise or extraneous things. The connected components are called *blobs*. Blobs may represent contiguous areas of 1's or of 0's; they may represent objects, holes in objects, or the background. Figure 14.3 illustrates the intuitive concepts: There are a total of five black and four white blobs in the picture, plus the background. Each blob except the background is totally surrounded by another blob of contrasting color. Two large black blobs represent a connecting rod and a hex nut. Three medium-size white blobs represent holes in these figures. The remaining four small blobs are noise pixels, presumably due to dirt or reflections. One of these noise blobs is a hole in the connecting rod, another one is a black inclusion inside a white hole.

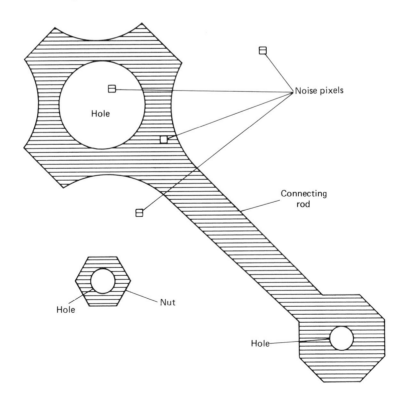

Fig. 14.3. Concepts of connection: There are five black blobs and four white blobs, plus the background. Each blob is totally surrounded by another of contrasting color.

Two pixels, $p_{y_1 x_1}$ and $p_{y_n x_n}$, are part of the same blob if they are the same color and there exists a "path" or a sequence of pixels, $p_{y_1 x_1}$, $p_{y_2 x_2}$, . . . , $p_{y_n x_n}$, such that all are the same color and each pair of adjacent pixels, $p_{y_k x_k}$ and $p_{y_{k+1} x_{k+1}}$, are *neighbors*. Much depends on the definition we choose for neighborhood. If we use the definition that only pixels that are horizontally or vertically adjacent are considered neighbors, then we call the resulting blobs *4-connected*. Figure 14.4 shows all the 4-connected neighbors of a pixel. If neighborhood also includes diagonal connections, then we refer to the blobs as *8-connected*, and Figure 14.5 applies.

Connectivity can apply to both the 1's in an image and the 0's. We want connectivity to have the property that each blob (except the background) will be totally surrounded by another blob of the opposite color. We can achieve this by treating the 1's as 8-connected and the 0's as 4-connected (or vice versa). Another way to achieve this is to use the *6-connected* definition of Figure 14.6, in which pixels are connected along the upper-right-to-lower-left diagonal, but not along the upper-left-to-lower-right one. The advantage of this definition is that it does not depend on the color of the pixels. The algorithm explained later uses the 6-connected convention.

Connectivity analysis involves assigning a unique identifier to each blob. Each pixel in the array then can be labeled with the identifier corresponding to its blob. Intuitively, we may think of a process that starts with an array of 1 and 0 pixels that can each be marked with a blob number. The process must locate one unmarked pixel and mark it with a fresh blob number. It would then propagate that blob number to neighboring pixels and neighbors of the neighbors, and so on, until the number can be propagated no further. It would then go back to find another unmarked pixel, and repeat the process until all pixels are marked.

But there are faster methods available. Connectivity can be determined and each pixel assigned to a blob in a single top-to-bottom pass through the image data. This process examines a row of the picture at a time, comparing each new row to the previous row of the image. At any given point in the analysis a data structure exists that represents the image processed so far. Each run-length segment is added to this data structure by looking for overlaps with corresponding runs in the previous row of the image. Three cases must be recognized. If new segment overlaps a segment in the previous row, then the new segment *continues* a blob already detected, as shown in Figure 14.7. If the new segment lies to the left of the old segment, then the new segment represents the top of a previously undiscovered blob: it *originates* a blob (Figure 14.8). If an unmatched old segment lies to the left of the new segment, the old segment must belong to a blob that does not continue to the new row and must be *terminated* (Figure 14.9). Termination also implies that the oppositely colored blobs on either side of the old segment must be *merged*.

14.5.3. The Connectivity Algorithm

The following data types* are used by the connectivity algorithm. A *boundary descriptor* is a dynamically allocated block of storage that contains two items: a column number and a blob identifier. A *blob descriptor* is a dynamically allocated block of storage that contains at least one item, a beginning row number. Other fields in the blob descriptor will be used to contain various kinds of information related to the size, shape, position, and so on, of the blob. A *blob identifier* is a pointer to a blob descriptor.

The following items of fixed storage are used: The *processed blob list* is the header of a list of blob identifiers. The *active line* is the header of an ordered list of boundary descriptors. The *current boundary pointer* is a pointer to a boundary on this list. The *preceding boundary* and *following boundary* refer to the boundary descriptors that precede and follow the current boundary on the active line.

The result of connectivity analysis is not just a list of blob numbers. Concurrent with connectivity analysis, a great deal of information related to the size, shape, position, and so on, of the blob can be derived and stored in each blob descriptor. This preliminary sort of feature extraction is explained in subsequent sections. There are five standard places during the connectivity analysis for additional processing to take place, called initialization processing, continuation processing, origination processing, termination processing, and merge processing, and these are identified in the following sections.

To Analyze an Image for Connectivity

1. Initialize the processed blob list to the empty list.
2. Allocate a blob descriptor to represent the background. Set its beginning row number to a negative number.

---

* The connectivity algorithm makes extensive use of linked lists and dynamic storage allocation. If the implementation language does not allow pointer variables and dynamic allocation, a sufficient number of each data type must be preallocated in one or more arrays, and indexes into these arrays used as pointers. Lists can be handled as arrays of pointers; insertion or deletion of elements will require shifting data in the array.

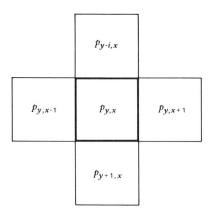

Fig. 14.4. Pixel neighborhoods: pixels that are adjacent horizontally and vertically are 4 connected neighbors.

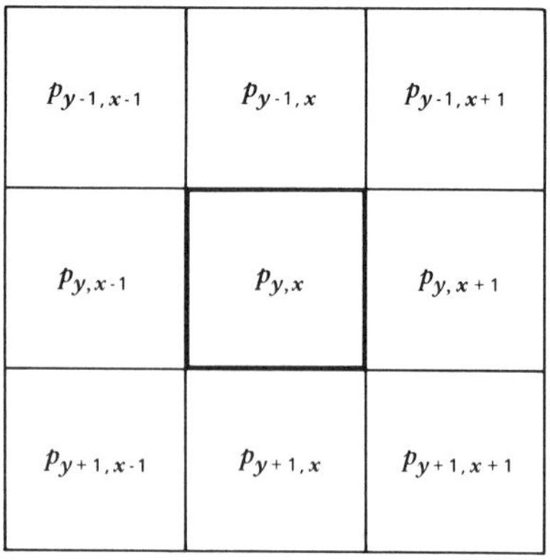

Fig. 14.5. Pixel neighborhoods: pixels that are adjacent horizontally, vertically, and diagonally are 8-connected neighbors.

3. Initialize the active line to contain two boundary descriptors. The first boundary descriptor should have a column number smaller than zero and a blob identifier pointing to the background. The column number of the second descriptor should be a large positive number. Its blob identifier is immaterial.

4. Perform additional *initialization processing,* as needed.

5. Process each row of the image, in top-to-bottom order, as described later.

6. Finish by processing an extra row consisting of all zeros.

To Process a Row

1. Initialize the current boundary pointer to point to the first boundary descriptor in the active line.

2. Obtain the run-length representation of the row, as described in Section 14.5.1. Prefix this list with a negative number and add a large positive number to the end of the list.

3. For every pair of adjacent numbers in the run-length data, in left-to-right order, perform the segment-processing operation, as defined in the next section. [For example, if the run-length data consists of the list $(-1\ 1\ 2\ 3\ 4\ 99)$, then the segment processor will be called five times with the arguments $(-1\ 1)$, $(1\ 2)$, $(2\ 3)$, $(3\ 4)$, and $(4\ 99)$.]

4. If the current boundary pointer does not point to the last boundary in the active line, perform the deletion operation, as defined later, repeatedly until it does.

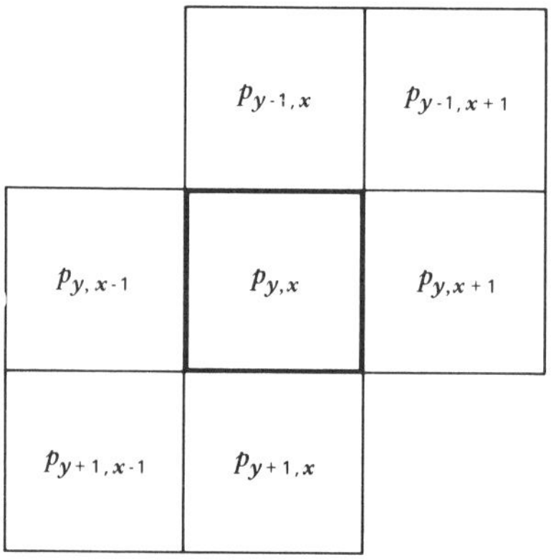

Fig. 14.6. Pixel neighborhoods: pixels that are adjacent horizontally, vertically, and along one diagonal are 6 connected neighbors.

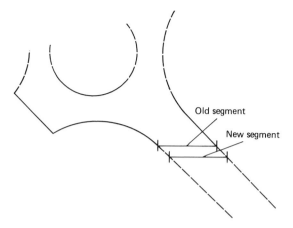

Fig. 14.7. Overlaps for connectivity: The new segment overlaps the old, so it must continue an existing blob.

To Process a Segment, Given a Starting Column Number and an Ending Column Number

1. If the starting column number is greater than or equal to the column number of the following boundary, do the deletion operation, as defined later, and repeat this step.
2. If the ending column number is less than the column number of the current boundary, perform the insertion operation, as defined next, giving the starting and ending column numbers.
3. Perform additional *continuation processing* as required.
4. Copy the starting column number to the current boundary column number.
5. Advance the current boundary pointer to point to the next boundary descriptor in the active line.

To Perform the Insertion Operation, Given a Starting Column Number and an Ending Column Number

1. Obtain the blob identifier contained in the preceding boundary descriptor. Call that blob the *surrounding blob*.
2. Allocate a new blob descriptor, calling it the *new blob*. Set its beginning row number to the row number of the line being processed.
3. Allocate two new boundary descriptions and insert them in the active line immediately before the current boundary. The first of these new boundaries receives the new blob identifier and the starting column number. The second boundary receives the surrounding blob descriptor and the ending column number.
4. Perform additional *origination processing* as needed.

To Perform the Deletion Operation

1. Obtain the blob identifiers from the preceding boundary descriptor, the current boundary, and the following boundary. Call them the *left blob, terminated blob,* and *right blob,* respectively.
2. If the right blob and left blob identifiers point to the same blob, then link the terminated blob identifier onto the processed blob list, and perform additional *termination processing* as needed. Otherwise, perform the merge operation, as defined in the next section.
3. Delete the current boundary and the following boundary from the active list. Let the current boundary pointer point to the first boundary after the deleted ones.

To Perform the Merge Operation

1. Compare the beginning row numbers of the right and left blobs. Call the one with the lower row number the *older blob* and the other one the *younger blob*.

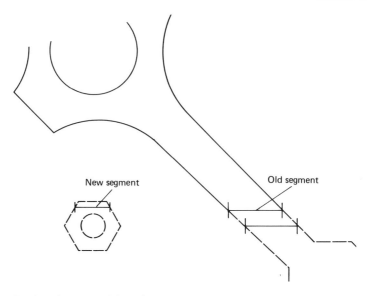

Fig. 14.8. Overlaps for connectivity: The new segment lies to the left of the old one, so it must originate a blob.

2. Perform additional *merge processing* as needed.
3. Search the entire active line to change all instances of the younger blob identifier to the older blob identifier.
4. Since the younger blob is no longer pointed to by any part of the data structure, it may now be deleted, returned to free storage, or otherwise disposed of.

14.5.4. A Simpler Connectivity Method

A simplified version of the foregoing can be used to label the pixels of an array with blob numbers.[7] The only data structure required is an array to hold the blob numbers. Any subsequent feature extraction must operate by rescanning this array.

Rather than give a complete description of the simpler algorithm, the reader is referred to the

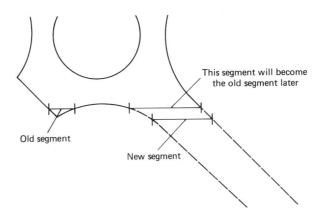

Fig. 14.9. Overlaps for connectivity: The new segment lies to the right of the old one. The blob represented by the old segment must be terminated. The two contrasting blobs on either side of the old segment must be merged.

description in the previous section for the general principles. The differences and simplifications are summarized as follows:

The previous row of the blob number array constitutes the active line. Insertion involves only choosing a small integer for a new blob identifier and inserting it in the appropriate pixels in the new row of the blob number array. The continuation operation requires copying the blob number to the appropriate pixels. Merging can be handled either by going back through the portion of the array already processed to replace the old blob identifier with the new, or through an additional data structure, an *equivalence table*.

14.6. FEATURES OF BINARY SILHOUETTES

A powerful motivation for using binary vision and connectivity analysis is the ability to measure *features* of each blob. The features that we find useful may be divided into two categories: those that are independent of position and orientation and those that are not. Examples from the first category are area, perimeter, elongation, or the count of the number of holes in the shape. These features are useful for object recognition. Examples from the second category include the centroid and various measures of angular orientation.

Many useful features can be derived as a by-product of connectivity analysis. Some are obtainable "instantaneously," but most require a two-stage process of gathering statistics during connectivity, followed by some sort of analysis or postprocessing. Statistics are stored in blob descriptors. Descriptions of statistics gathering in this section are keyed to the description of the connectivity algorithm in the preceding section.

14.6.1. Elementary Features

The simplest feature to calculate is *color*. It requires one bit of storage in the blob descriptor. During initialization processing, set the color field of the background blob to 0. During origination processing, set the color of the new blob to the opposite of the color of the surrounding blob.

Inclusion relationships are also simple to calculate. Each blob descriptor will have a "parent" field that points to the surrounding blob. Additional fields could be allocated for "child" and "sibling" links, but as long as the number of blobs in a scene is not large, these kinds of relationships can be obtained by searching through the processed blob list. During origination processing, set the parent field of the new blob to point to the surrounding blob. During merge processing, the entire active line and the processed blob list must be searched so that if any blob points to the younger blob as a parent, its parent should be changed to the older blob.

The *bounding rectangle* requires four fields in the blob descriptor: minimum x, maximum x, minimum y, and maximum y. During origination processing, initialize minimum y to the current row number, minimum x to a large positive number, and maximum x to a negative number. During continuation processing, compare the starting and ending column numbers with the saved maximum and minimum x, and update if necessary. During merge processing, the maximum and minimum x's of the older blob and the younger blob should be compared, and the correct values written to the older blob. During termination processing, copy the row number to the maximum y field.

The maximum x and maximum y actually point to one column or row beyond the last pixel beyond the blob, so these numbers should be decremented by one before they are used. The *center of the bounding rectangle* is a useful descriptor of blob position. This is the average of maximum and minimum x and of maximum and minimum y.

14.6.2. Accumulating Moments of Area

Moments of area are defined as summations over all the pixels in a blob of various products of x and y. *Area* is the zeroth moment of area, and may be denoted $\Sigma 1$. There are two *first moments* designated Σx and Σy, and three *second moments* designated Σx^2, Σxy, and Σy^2. Space must be allocated in the blob descriptors for as many of these summations as will be used.

Note that the formulas presented here can be applied to the image as a whole as well as to individual blobs. This can be used for computing parameters of aggregations of blobs. For calculating moments of area, connectivity can be bypassed altogether if the image is known to consist only of "interesting" pixels.

When the image is run-length coded, the following formulas can be used to add an entire segment to the accumulations, rather than building them one pixel at a time.

$$\sum_{x=m}^{n-1} 1 = n - m$$

$$\sum_{x=m}^{n-1} x = \frac{n^2 - n}{2} - \frac{m^2 - m}{2} = \tfrac{1}{2}(n - m)(m + n - 1)$$

$$\sum_{x=m}^{n-1} y = y \sum_{x=m}^{n-1} 1 = y(n-m)$$

$$\sum_{x=m}^{n-1} x^2 = \frac{2n^3 + 3n^2 + n}{6} - \frac{2m^3 + 3m^2 + m}{6}$$

$$= \tfrac{1}{12}[3(n-m)(n+m-1)^2 + (n-m)^3 - (n-m)]$$

$$\sum_{x=m}^{n-1} xy = y \sum_{x=m}^{n-1} x = y(n-m)(m+n-1)$$

$$\sum_{x=m}^{n-1} y^2 = y^2 \sum_{x=m}^{n-1} 1 = y^2(n-m)$$

To compute first and second moments, set all six accumulations to zero during origination processing. During continuation processing, increment the moment accumulations according to the preceding formulas. During merge processing, add together the moment accumulations of the older blob and the younger blob and store them in the older blob.

When fixed-point integer arithmetic is used for moment accumulations, there must be sufficient precision in the arithmetic and data storage to accommodate the large sums generated. For an image of dimensions $n \times n$, the maximum value of $\Sigma 1$ is n^2, the maximum Σx or Σy is $\frac{1}{2}n^3$, the maximum Σx^2 or Σy^2 is $\frac{1}{3}n^4$, and the maximum Σxy is $\frac{1}{4}n^4$.

14.6.3. Using Moments of Area: Centroid, and Approximating Ellipse

Assume the existence of scaling parameters S_x and S_y that give the pixel horizontal and vertical pixel spacing. These may be given in millimeters, inches, or other convenient units and represent the distance a point would have to move in the scene so that it moves one pixel in the image. Using these parameters, the area A and the centroid (C_x, C_y) are given by

$$A = S_x S_y \Sigma 1$$

$$C_x = S_x \frac{\Sigma x}{\Sigma 1}$$

$$C_y = S_y \frac{\Sigma y}{\Sigma 1}$$

An ideal ellipse may be specified by five parameters. For our purposes, it is most useful to specify two parameters of size and shape (major axis length and minor axis length), two parameters of position (centroid in x and y), and one parameter of orientation (rotation). It is convenient to add a sixth parameter, a density. Given a blob with its six moments of area, we may solve for the parameters of the ideal ellipse that has the identical moments. This is called *recovering the approximating ellipse*. The centroid formulas given still apply, and the following formulas give the other parameters:

$$A = \frac{4}{\pi} S_x^3 S_y \left(\Sigma x^2 - \frac{(\Sigma x)^2}{\Sigma 1} \right)$$

$$B = \frac{4}{\pi} S_x S_y^3 \left(\Sigma y^2 - \frac{(\Sigma y)^2}{\Sigma 1} \right)$$

$$C = \frac{4}{\pi} S_x^2 S_y^2 \left(\Sigma xy - \frac{\Sigma x \Sigma y}{\Sigma 1} \right)$$

$$E = \sqrt{(A-B)^2 + \frac{4}{\pi} C^2}$$

$$F = \sqrt[4]{AB - C^2}$$

$$\text{Major axis} = \sqrt{\frac{A+B+E}{2F}}$$

$$\text{Minor axis} = \sqrt{\frac{A+B-E}{2F}}$$

$$\text{Rotation} = \tfrac{1}{2} \tan^{-1} \frac{2C}{A - B}$$

$$\text{Density} = \frac{4}{\pi} \frac{S_x S_y \Sigma 1}{2F}$$

When a blob contains holes, the contribution of the area of the holes is not taken into account. If moments are desired of the figure that would result if the holes were filled in, then the summations $\Sigma 1$, Σx, and so on, of the holes should be added to those of the enclosing blob before the formulas of this subsection are evaluated.

14.6.4. Perimeter Length

The *perimeter* of a blob is the boundary between the pixels of the blob and the pixels of its surrounding blob. If we think of pixels as rectangular black and white tiles that cover the plane, then the boundary corresponds to the cracks between tiles of different colors.

The *perimeter length* of a blob is the length of the boundary. To calculate perimeter length we will need the space to accumulate three sums in each blob descriptor: the number of horizontal perimeter segments N_h, the number of vertical segments N_v, and the number of corners in the perimeter N_c. In addition, we will need to know the beginning row number, or minimum y, of each blob. Minimum y may already be available if the bounding rectangle is also being computed.

During origination processing set N_v of the new blob to zero, set N_c to 2, and set N_h to the ending column number minus the starting column number. If it is not being computed elsewhere, set minimum y to the current row number.

During continuation processing, compare the minimum y's of the blobs pointed to by the current boundary descriptor and the following boundary. The blob with the larger y must be updated. Add one to N_v of that blob. If the starting column number of the segment being processed is different from the column number of the current boundary, then add 2 to N_c and add the absolute value of the difference between the column numbers to N_h.

During termination processing add 2 to N_c of the terminated blob, and add the absolute value of the difference between the column numbers of the following boundary and the current boundary to N_h.

During merge processing compare the minimum y's of the terminated blob and the older blob. The blob with the larger y must be updated. Add 2 to N_c of that blob, and add the absolute value of the difference between the column numbers of the following boundary and the current boundary to N_h. Add N_h, N_v, and N_c of the younger blob to the corresponding numbers of the blob being updated.

The perimeter length of a blob is a useful measure for shape recognition. However, when the length is measured along the horizontal and vertical pixel boundary, the measured length may vary considerably from the ideal. Consider a square 10 pixels across as shown in Figure 14.10. Its perimeter will be 40 pixels. But if the square is rotated 45° as shown in Figure 14.11, its perimeter length becomes 54 pixels.

The dependence on orientation can be mitigated by "cutting corners." Each time the boundary makes a right-angle bend, we pretend the perimeter took a "short cut" as shown in Figure 14.12. With this refinement, lines that run horizontally, vertically, or at a 45° diagonal will have a measured perimeter length that agrees with the ideal. However, lines at other orientations will have some error. The worst case is at an angle of 26.5°, at which a line will have a measured perimeter length some

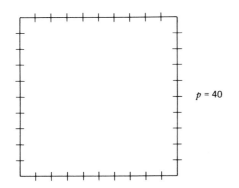

$p = 40$

Fig. 14.10. Perimeter of a square. The perimeter length of a rectangular shape depends on its orientation. A 10-pixel square will have a perimeter of 40 pixels when presented in an upright position.

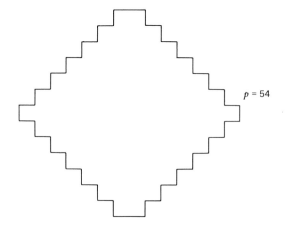

Fig. 14.11. Perimeter of a square. The same square will have an apparent perimeter of 54 pixels when it is oriented diagonally.

9.2% too high. An ideal circle may be expected to have a measured perimeter that is too high by a factor of $(8/\pi)$ $(\sqrt{2} \bullet - 1)$, or 5.48%.

Taking all of the factors mentioned into account (the cutting of corners, the expected 5.48% error, and different horizontal and vertical scaling parameters) yields the following formula:

$$P = 0.948059 \, (S_x N_h + S_y N_v - \tfrac{1}{2}(S_x + S_y - \sqrt{S_x^2 + S_y^2})N_c)$$

14.6.5. Perimeter Lists

Sometimes it is useful to have *lists of perimeter points*—the x- and y-coordinates of points on the perimeter of a blob—in the order they are encountered as the perimeter is followed around the blob. Many features for shape description can be derived using such a list. Or it can be used for display purposes, to encircle a blob of interest on a display screen. The list is extracted during connectivity

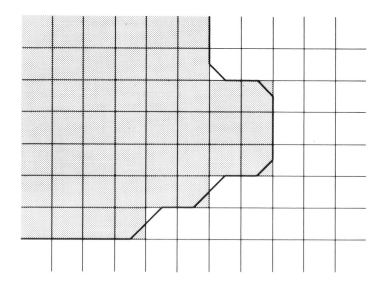

Fig. 14.12. Cutting corners on a perimeter. The diagonal perimeter segments are a better approximation to the perimeter of the actual figure than are the strictly horizontal and vertical segments.

analysis. The complete list will not be available for examination or processing until connectivity has been completed for the blob in question.

The format of perimeter lists may vary depending on the programming language and free-storage allocation facilities available. A singly linked list of storage blocks, each having an x-coordinate, a y-coordinate, and a pointer to the next block in the chain, is easy to implement, but places a considerable demand on the free-storage allocator. A variation on this theme is to have the blocks doubly linked, so that each block contains pointers to the blocks preceding and following it in the chain. Schemes using fixed-length arrays for storage of multiple-perimeter points require care in implementation to avoid excessive waste of memory space. Each block must be large enough to accommodate the longest expected perimeter chain, yet most chains turn out to be quite short.

The center of a given pixel is generally regarded to be at exact integer coordinates. A perimeter segment, lying in the "cracks" between pixels, will have coordinates that are an integer plus one-half. As long as the convention is clearly understood by all routines that process perimeter data, the one-half may be simply dropped and only integers stored in the lists.

For consistency, perimeters should be traced in such a direction that if you "walk" along the perimeter, 1's will be on your right and 0's on your left. Blobs of 1's will be circled in a clockwise direction, and 0's in a counterclockwise direction.

It is relatively simple to devise a routine that will extract a perimeter list from a binary image stored in a fixed pixel array. This can be useful in conjunction with the simplified version of connectivity that generates an array of blob labels, mentioned in Section 14.5.4, "A Simpler Connectivity Method." A starting point must be supplied, and the routine should keep track of the current location and direction along the perimeter. By examining the 4×4 neighborhood of the current location, the next location and direction can be determined. The perimeters extracted by this method will have $(N_v + N_h)$ entries, that is, one point for every pixel of the boundary length.

Extraction of perimeter lists during connectivity requires addition of two fields to the boundary descriptors: a direction bit to tell whether the perimeter at that boundary goes from top to bottom or bottom to top, and a pointer to either the beginning or the end of the perimeter chain associated with that boundary, depending on the direction bit. Each blob descriptor will require a pointer to the header of the perimeter list. (Since the perimeter list, when it is completed, will be circular, any element can serve as the header.)

It is most efficient to store only the corners of the perimeter, where the perimeter changes direction. The resulting list will contain N_c points. Routines that process perimeter lists must take into account the implicit points between corners.

During origination processing, space for two perimeter points must be allocated from free storage, to contain the starting and ending column of the run-length segment being processed and the current row number. A pointer to one or the other of these perimeter points should be stored in the new blob descriptor. A pointer to each point must also be stored in the perimeter chain word of the corresponding boundary descriptor. During continuation processing, if the starting column number of the segment being processed is different from the column number of the current boundary, then allocate two new points. These will each receive the current row number, and the column numbers will come from the current boundary and the new run-length segment. Link the two points into the perimeter chain pointed to by the current boundary descriptor. During termination processing and merge processing two new points must be allocated, to contain the current row number and the column numbers from the current boundary and the following boundary. These two points must be linked onto the perimeter chains pointed to by the respective boundary descriptors, then the two chains can be linked together.

The points of the perimeter list can be used for display purposes. Encircling the largest blob in a scene, or any other blob of interest, is useful for interactive operation of a binary vision system.

The perimeter list can be used for deriving shape features for recognition. One useful class of features is based on *radius statistics*. For every point on the perimeter we may compute the length of the radius vector from that point to the centroid of the blob (or to the center located by any other criterion). The average, minimum, maximum, and standard deviation of this vector length are useful shape features. The angle at which the maximum radius occurs is a useful feature for determining the orientation of a figure. The relationship between the maximum radius vector and the minimum radius vector is sometimes useful for distinguishing between shapes and their mirror images. The count of relative maxima in the radius function corresponds to the number of corners in a blob outline.

It may sometimes be useful to approximate a perimeter list by a sequence of straight-line segments. Any portion of a polygon outline may be fit with one or more straight lines by attempting to fit the entire portion with a single straight line. If the error of the fit exceeds a threshold, the line must be subdivided and each half recursively* fit with the same procedure. Subdivision should be at an extremum, such as the point farthest from the (poorly fitting) straight line.

---

* If the programming language in use does not allow recursive calling of the line-fitting routine, an array of descriptors may hold portions of the overall curve not yet fit, and a single routine will operate one at a time on the descriptors in this array.

When straight lines are represented by the familiar equation $y = Mx + B$, lines that are vertical, or nearly so, are poorly represented, and the distance from any point to that line is hard to compute. A better representation is the equation $Ax + By + C = 0$, with the constraint that $A^2 + B^2 = 1$. The distance from any point (x, y) to the line is then $Ax + By + C$.

The equation of a straight line may easily be calculated from the coordinates of any two points, usually the end points. A more robust procedure is to use a least-squares fit based on an eigenvalue solution,[8] as contained in the following formulas. To fit a sequence of points x_i, y_i, calculate

$$N = \Sigma 1$$

$$r = \Sigma x^2 - \frac{(\Sigma x)^2}{N}$$

$$s = \Sigma xy - \frac{\Sigma x \Sigma y}{N}$$

$$t = \Sigma y^2 - \frac{(\Sigma y)^2}{N}$$

$$\lambda = \tfrac{1}{2}(r + t - \sqrt{(r - t)^2 + 4s^2})$$

$$a = \begin{cases} -s & \text{if } r > t \\ t - \lambda & \text{otherwise} \end{cases}$$

$$b = \begin{cases} r - \lambda & \text{if } r > t \\ -s & \text{otherwise} \end{cases}$$

$$A = \frac{a}{\sqrt{a^2 + b^2}}$$

$$B = \frac{b}{\sqrt{a^2 + b^2}}$$

$$C = \frac{-A\Sigma x - B\Sigma y}{n}$$

The mean-square error of the fit is λ/N.

14.7. GROWING, SHRINKING, AND SKELETONIZING

Growing, shrinking, and skeletonizing represent a class of operations that transform binary images: an *input image* may be transformed by one or more such operations into an *output image*. Some further analysis of the output image will be still necessary to extract meaningful information from the data. Growing and shrinking are well suited to implementation in fast hardware. They are best described using arrays of binary pixels, but they can easily be adapted to run-length representation for a modest gain in speed.

Growing and *shrinking* cause expansion or contraction of the blobs in an image. They are reciprocal operations in the sense that growing the 1's in an image is entirely equivalent to shrinking the 0's. However, growing followed by shrinking will not recover the original image. Instead, repeated growing and shrinking by equal small amounts are useful for smoothing, noise elimination, and detecting blobs based on approximate size. Small irregularities in the image data can be eliminated by this technique. Smoothing can be used to locate irregularities by comparing the smoothed image to the original one. Such techniques can be very useful for inspecting printed-circuit boards and microelectronic components.[9] Growing and shrinking are also useful for creating "don't care regions" for direct image-to-image comparisons.

Growing the 1's in an image is entirely equivalent to shrinking the 0's. Suppose we wish to grow the 1's in an input image $p_{y,x}$, resulting in an output image $q_{y,x}$. $q_{y,x}$ will be 1 if either $p_{y,x}$ is 1 or any neighbor of $p_{y,x}$ is 1, and 0 otherwise. (To shrink the 1's, $q_{y,x}$ should be 0 if either $p_{y,x}$ is 0 or any neighbor of $p_{y,x}$ is 0, and 1 otherwise.) This operation may be repeated an arbitrary number of times to grow or shrink by arbitrary amounts. If each pixel is considered to have four neighbors (as in Figure 14.4), growing and shrinking will proceed more slowly than if they have eight neighbors (Figure 14.5). Small regions tend to grow in diamond shapes using four neighbors, and in squares using eight neighbors.

Skeletonization is similar to shrinking except that guarantees are made that blobs will never shrink so far that they entirely disappear. When blobs that are long and thin are skeletonized, the residue represents a medial axis, which can be used to characterize the original blob. After each shrinking step, if $p_{y,x}$ is 1 and $q_{y,x}$ is 0 and all the neighbors of $q_{y,x}$ are also 0, then change $q_{y,x}$ from 0 back to 1.

The following two-pass algorithm[7] is faster than repeated shrinking; it requires an array in which

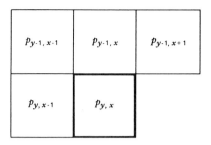

Fig. 14.13. Neighborhoods for distance calculation. Neighborhoods to be used for the forward pass of the skeletonizing algorithm.

to store distances, initialized to the 1's and 0's of the original binary array. A forward pass calculates the shortest distance upward and leftward from each pixel to the nearest boundary, and a backward pass takes into account the distance downward and rightward.

In pass 1, pixels are considered in normal roster order: in row order from top to bottom, and from left to right across each row. As each pixel $p_{y,x}$ is processed, the four pixels above and to the left of this pixel, shown in Figure 14.13, are considered—these are the pixels that have already been processed in this first forward pass. If $p_{y,x}$ is nonzero, it is set to one plus the smallest of $p_{y-1,x-1}$, $p_{y-1,x}$, $p_{y-1,x+1}$, and $P_{y,x-1}$. Figure 14.14 shows the results of the first pass on a portion of an image. Pass 2 is in reverse order, in row order from bottom to top and from right to left back across each row. The five pixels shown in Figure 14.15 are examined, and if $p_{y,x}$ is nonzero, it is set to the smallest among itself and one plus $p_{y+1,x+1}$, $p_{y+1,x}$, $p_{y+1,x-1}$, or $p_{y,x+1}$. Figure 14.16 shows the results of the second pass. Once the distance array has been computed, the skeleton is the set of pixels in the distance array that have no neighbors with higher values. These points are indicated in Figure 14.17.

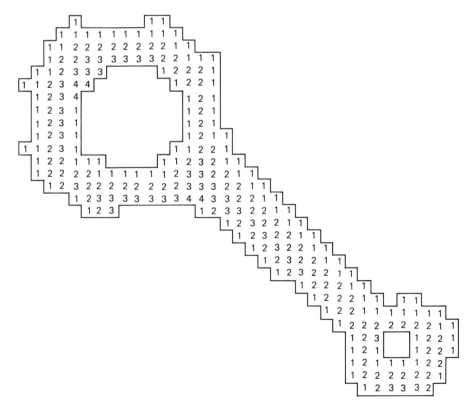

Fig. 14.14. Skeletonizing an image. Results of forward pass.

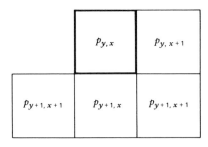

Fig. 14.15. Neighborhoods for distance calculation. Neighborhoods to be used for the backward pass of the skeletonizing algorithm.

14.8. OPERATIONS ON GRAY-SCALE IMAGES

The use of gray-scale information in picture processing for robotic applications gives the potential for greater generality and flexibility than is possible with binary images. The scenes to be viewed can be less constrained; the need to control lighting is reduced (but not by any means eliminated!); and overlapping parts can be less of a problem. However, this is purchased by requiring more of the picture processing. The gross amount of data is larger, and it must be subjected to more sophisticated and more time-consuming operations. The various processing techniques tend to produce results that are ambiguous, and this imposes a requirement for fine-tuning decision thresholds as a function of the application. Fundamental research into the nature of human and computer vision has produced many interesting demonstrations of apparent visual sophistication in the past several years, but most of these techniques today lack either the robustness or the computational efficiency for direct adaptation in a manufacturing environment.

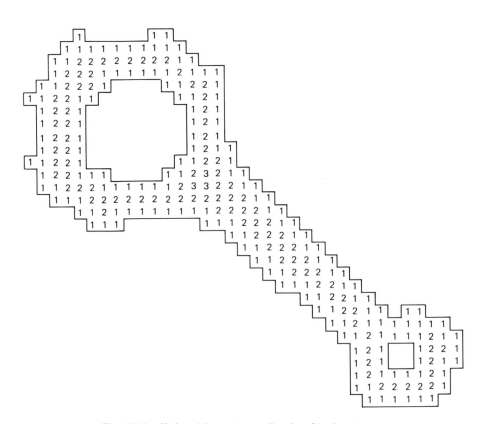

Fig. 14.16. Skeletonizing an image. Results of backward pass.

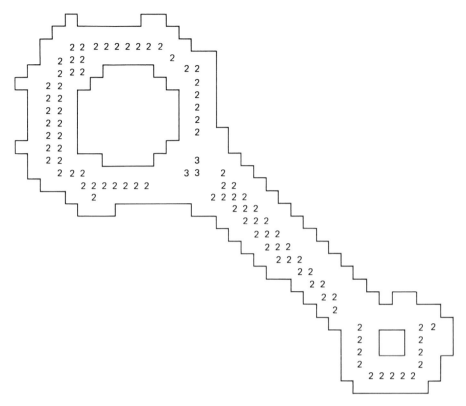

Fig. 14.17. Skeletonizing an image. This shows the skeleton: the set of points with no larger neighbors.

14.8.1. Threshold Setting

Binary picture-processing operations can be applied to gray-level data if there is a way of classifying each pixel as 1 or 0. The major difference between this mode and regular binary image processing is that the threshold criterion may be made a function of the image rather than having to be specified beforehand. Some of the algorithms that follow may be implementable in hardware without buffering an entire gray-scale image. Threshold-setting techniques occupy an intermediate place between binary and gray-scale image processing.

A *histogram* of any given image is a plot of $n(t)$, the number of pixels n at each gray level t. It is usually drawn as a bar graph, as in Figure 14.18. The horizontal axis is labeled with gray values from 0 to the largest possible gray value, and the height of each bar is the total number of pixels of that image having that particular value.

Some useful algorithms can be applied if the gray levels of the image can be assumed to have a *bimodal distribution*. Such a distribution results when the background of a scene is one uniform color and the object being perceived has a different, also uniform, color. The histogram of such a scene will contain two peaks, corresponding to the two colors. To separate the object from the background the threshold should be situated in the valley between the two peaks. For Figure 14.18 this would be in the neighborhood of 36 to 40. Placing the threshold in this valley is best not only from a statistical decision theory standpoint, but also because the image is least sensitive to small changes in lighting or threshold setting at this setting.

A histogram generated from a real gray-scale image may be expected to be ragged rather than smooth—random variations will cause the plot to have many local maxima and minima. Finding the major peaks and valleys must therefore consider a range of brightness levels at one time. An effective way to find the peaks and valleys in a histogram is to find the values of n that are higher than those for plus or minus k adjacent values of t. k should be about one-half the expected distance between peaks and valleys of the histogram, or about 15 in the case of Figure 14.18.

Small changes in an image threshold can cause the apparent sizes of objects in the binary image to grow larger and smaller. If the expected sizes of certain objects are known a priori, this information

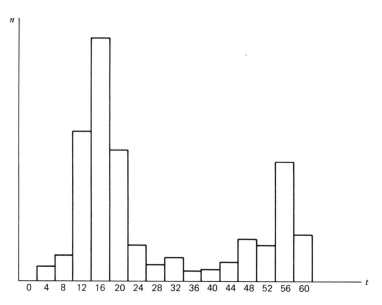

Fig. 14.18. A bimodal histogram. The number of pixels n at each gray level is plotted as a function of the gray value t. The plot is called bimodal because it has two major peaks.

can be used to adjust the threshold by repeated reanalysis of the image with different thresholds until the size of the known object comes out correct.

When a threshold is improperly adjusted the resulting binary image becomes cluttered and ragged. Thus the total number of blobs found by connectivity or the total perimeter of all blobs in the scene can be used as a fairly sensitive indicator of the appropriateness of a threshold. Repeated processing of the same image with different thresholds can be used to find the best threshold.

Spatially varying thresholds can be used, as, for instance, when the illumination field is known to be nonuniform. A spatially variable threshold can be computed as some algebraic function of x and y, or it can be stored in a separate image buffer and subtracted from the "live" image. In either case, the spatially varying component may be combined with the kind of settable threshold just described.

14.8.2. Edge Finding

The purpose of edge finding is to locate the boundaries between regions of an image. If we assume that images consist of regions of nearly uniform intensity, we can define an edge as the boundary between such regions. Most actual images have a more complicated structure than that, but the edges found based on this principle will still be useful.

Edge finding consists of two steps: locating pixels in the image that are likely to be on an edge, and linking candidate edge points together into a coherent edge. The first step is straightforward; the second can be tricky.

Candidate edge points are those where a discontinuity in brightness occurs. They are generally identified by estimating the image *gradient*, or spatial derivative, in two dimensions. The gradient can be characterized by an x-component and a y-component, or by a magnitude and direction. Generally, it is fruitful to estimate the derivatives directly in two orthogonal directions and to convert them into magnitude and direction. For many reasons the spatial derivatives are usually computed over some *window* of several points. The greater the window size, the less the sensitivity to pixel noise and digitization error. But a larger window requires more computation and data buffering and tends to spread out the effects of a sharply defined edge over several pixels.

Within a 2×2 window, the most commonly used gradient estimate is *Roberts' cross*[10]: the two orthogonal derivative estimates are the diagonals of the window as shown in Figure 14.19: $(p_{y+1,x} - p_{y,x+1})$ and $(p_{y+1,x+1} - p_{y,x})$. The gradient magnitude is the square root of the sum of their squares but it is computationally easier to simply add their absolute values. The gradient direction may be crudely quantized into four directions according to the signs and relative magnitudes of the two directional derivatives.

A 3×3 window can be expected to give better results. Although other configurations are possible, the weights most commonly used to estimate the directional derivatives are as shown in Figure 14.20.

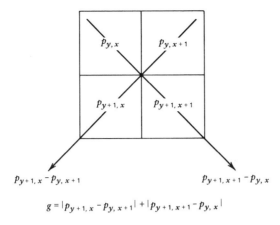

$$g = |p_{y+1,\,x} - p_{y,\,x+1}| + |p_{y+1,\,x+1} - p_{y,\,x}|$$

Fig. 14.19. Roberts' cross. Image gradient is derived from directional derivatives from diagonally adjacent pixels in a 2 × 2 window.

$$d_x = (p_{y-1,x+1} + 2p_{y,x+1} + p_{y+1,x+1}) - (p_{y-1,x-1} + 2p_{y,x-1} + p_{y+1,x-1})$$
$$d_y = (p_{y+1,x-1} + 2p_{y+1,x} + p_{y+1,x+1}) - (p_{y-1,x-1} + 2p_{y-1,x} + p_{y-1,x+1})$$

This has been called the *Sobel gradient*.[11] ("Sobel" rhymes with "noble.") As with Roberts' cross, the magnitude can be calculated exactly as $\sqrt{d_x^2 + d_y^2}$ • or estimated roughly as $|d_x| + |d_y|$. The gradient direction is $\tan^{-1}(d_y/d_x)$.

Gradients such as these are relatively easy to compute with fast hardware. The output is a *gradient image* of magnitude (and direction, if desired). An additional thresholding step is necessary to distinguish candidate edge points from nonedge points. The threshold should not be so high as to exclude real edges or so low that an excessive number of noise points pass as edge candidates. It is usually not possible to find the ideal threshold, so we must make use of higher-level techniques to separate the signal from the noise by finding sequences of adjacent candidate points that form an edge.

One approach is to choose some candidate point with a high gradient magnitude, then use the gradient direction to search for a nearby point with which to extend the edge, and follow this chain from point to point until an end is reached. Such a simple strategy can work in simple cases, but as the situation becomes more complex, methods based on graph theory[12] become advisable. However, they quickly get out of hand in execution time.

-1	0	1
-2	0	2
-1	0	1

Horizontal

-1	-2	-1
0	0	0
1	2	1

Vertical

$$\left\{ \begin{array}{l} h = (p_{y-1,x+1} + 2p_{y,x+1} + p_{y+1,x+1}) - (p_{y-1,x-1} + 2p_{y,x-1} + p_{y+1,x-1}) \\ v = (p_{y+1,x-1} + 2p_{y+1,x} + p_{y+1,x+1}) - (p_{y-1,x-1} + 2p_{y-1,x} + p_{y-1,x+1}) \\ \text{Gradient magnitude} = \sqrt{h^2 + v^2} \\ \text{Gradient direction} = \tan^{-1}\dfrac{v}{h} \end{array} \right\}$$

Fig. 14.20. Sobel gradient operator. Image gradient is derived from directional derivatives across a 3 × 3 window.

A considerable advantage can be obtained if something is known a priori about the nature of the edges being sought. In particular, if the edges are known to be straight lines, several different techniques apply.

Gradient directions will be similar at all points on a straight line or on a group of parallel straight lines. Histogramming the gradient directions of all candidate edge points can identify the prominent edge directions. Given the edge direction, all candidate points can be projected onto the axis of a transformed coordinate system perpendicular to the gradient direction, as shown in Figure 14.21. The peaks of a histogram of the x-values of the projected points will give the location of the edges.

A somewhat more sensitive technique that does not depend on an accurate estimate of the gradient direction is based on the *Hough transform*.[5] ("Hough" rhymes with "rough.") If a straight line is represented by an equation of the form

$$x \sin \theta + y \cos \theta - \rho = 0$$

then all of the lines that pass through the point x_i, y_i are represented by the set of values of θ and ρ that satisfy the equation

$$\rho = x_i \sin \theta + y_i \cos \theta$$

These two equations represent the transformation of a point in $x - y$ space into a curve in $\theta - \rho$ space. Two different points will transform to two different curves, and the intersection of these curves represents the straight line that passes through the two points. If n points lie on a straight line in $x - y$ space, then all n curves will intersect at the point in $\theta - \rho$ space corresponding to the straight line.

Given a number of thresholded gradient points in which we want to find straight lines, allocate a two-dimensional array where one dimension represents equal increments of ρ from zero to the maximum radius of the image, and the other dimension represents equal increments of θ from 0 to 2π. All elements of this array should initially be zero. For each gradient point, calculate ρ as a function of θ for each discrete θ-value, and add 1 to the corresponding array element. (If the gradient direction is known, then only values of θ within some error tolerance of the gradient direction need be incremented.) An array element that ends up with a high count represents a group of collinear points.

The number of array elements needs to be only large enough that multiple lines in the same image will generate separate peaks in the array. Once collinear points have been identified by this technique, the equations at the end of Section 14.6 can be used to refine the estimate of the line parameters.

The Hough transform technique can be generalized to locate any curve that may be characterized by two parameters. (In theory, more than two parameters can be accommodated by use of a higher-dimensional array, but in practice they are unwieldy.) For example, circles of radius R and center (x_c, y_c) are represented by the equation

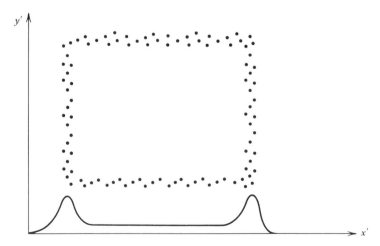

Fig. 14.21. Projecting candidate edge points. Peaks in the projected histogram identify lines parallel to the y direction.

$$\frac{(x - x_c)^2}{R^2} + \frac{(y - y_c)^2}{R^2} = 1$$

If R is known in advance and x and y are the coordinates of a point through which the circle must pass, then the same equation transforms a point in (x, y) space into a circle in (x_c, y_c) space. The remainder of the analysis is similar to the case of straight lines.

14.9. RECOGNITION

The preceding sections describe methods of extracting elemental information from an image. If the domain of image processing is sufficiently constrained, this information may be all that is required. A check for presence or absence of an item, a measurement of area, or a determination of position may suffice for the application, but these all depend on advance knowledge of precisely what is expected in the scene.

A machine vision system can deal with a less constraining environment if it has some higher-level strategy for making sense of what it sees. This includes objects in random position and orientation, multiple objects per scene, and even overlapping objects.

Given a specific object to recognize, many different *ad hoc* methods can be conceived. However, we want methods that can be applied to wide classes of objects without the need to invent new solutions. One very desirable characteristic is *training by showing*. The system should learn about the types of objects it can distinguish from actual examples. This implies the use of a data structure in which to represent *prototypes* and a *matching procedure* to compare prototypes with features extracted from a new image.

14.9.1. Classification Based on Feature Values

The most straightforward kind of recognition is based on numerical feature values, such as a blob's area and perimeter, that are invariant with respect to position and orientation. The area and perimeter of an unknown object can be compared to the areas and perimeters of several prototypes, and the prototype to which the unknown is most similar can be chosen. The idea of "most similar" can be visualized by thinking of the unknown and each of the prototypes as points in a space whose coordinates are perimeter and area. The *nearest neighbor rule* specifies that we choose the prototype nearest in the space to the unknown. The dimensionality of the space is equal to the number of features being considered.

Let there be m prototypes for each of which we have measured n features f_{ij}, $i = 1, \cdots, m$, $j = 1, \cdots, n$, and let the unknown be designated by the feature values f_j, $j = 1, \cdots, n$. Then the *unnormalized feature distance* from the unknown to prototype i may be defined by the formula

$$D_i = \sum_{j=1}^{n} (f_j - f_{ij})^2$$

Then the nearest neighbor is the prototype i that minimizes D_i.

The foregoing formula is sufficient if all the features being measured are of the same order of magnitude. But when features are as different as, say, the number of holes and the area in pixels, then differences in the larger features will swamp out those of the smaller ones unless a normalizing factor is used. There will be n normalizing factors N_j, $j = 1, \cdots, n$, one for each feature. The *normalized feature distance* is defined

$$D_i = \sum_{j=1}^{n} \left(\frac{f_j - f_{ij}}{N_j}\right)^2 \tag{14.1}$$

The N_j's should be related to the expected amount of variation in the feature within a single prototype—that is, if multiple instances of the prototype were measured at different times, the amount we would expect the measured feature value to vary. In the absence of any other information, the magnitude of a "typical" feature value may be used as the normalizing factor.

The amount of variation in feature values can be measured when training takes place. Suppose there are p_i samples of prototype i on which training has taken place, which produced the set of feature values f_{ijk}, $j = 1, \cdots, n$, $k = 1, \cdots, p_i$. The mean feature values f_{ij} are calculated as

$$f_{ij} = \frac{1}{p_i} \sum_{k=1}^{p_i} f_{ijk}$$

The conventional estimate of the variance is given by the formula

$$v_{ij} = \frac{1}{p_i - 1} \sum_{k=1}^{p_i} (f_{ijk} - f_{ij})^2 = \frac{1}{p_i - 1} \left(\sum_{k=1}^{p_i} f_{ijk}^2 - p_i f_{ij}^2\right) \tag{14.2}$$

but we prefer the formula

$$v_{ij} = \frac{1}{p_i} \left[v_{0ij} + \sum_{k=1}^{p_i} (f_{ijk} - f_{ij})^2 \right] = \frac{1}{p_i} \left(v_{0ij} + \sum_{k=1}^{p_i} f_{ijk}^2 - p_i f_{ij}^2 \right) \qquad (14.3)$$

which uses a *prior estimate of the variance* v_{0ij}. The form of Eq. (14.3) avoids a division by zero when the number of samples is one. Additionally, it helps to alleviate the problem when the number of samples is small of greatly underestimated variances given by form (14.2) giving undue weight to their respective features and causing an abnormally high reject rate. In the absence of any better information, a reasonable prior estimate of the variance is the square of 1% of a "typical" feature value.

The normalizing factors N_j of Eq. (14.1) should be the weighted averages of the measured variances

$$N_j = \frac{\sum_{i=1}^{m} p_i v_{ij}}{\sum_{i=1}^{m} p_i}$$

If it may be expected that different prototypes will exhibit different variances, then we can go one step further and use the individual variances themselves as a separate normalizing factor for each prototype and feature.

If our vision system is presented with an object that is not part of the training set, then its normalized feature distance from all of the prototypes may be expected to be rather high. We should be able to tell that the part does not belong by examining the minimum feature distance D_i. What is a reasonable threshold to use for a rejection criterion?

With any system based on a decision threshold, there are two kinds of error we can make: a *false negative* occurs if we mistakenly reject an object that was genuine, and a *false positive* is when we mistakenly accept a bogus object. If we place a high value on the threshold D_i, we are not likely to get many false negatives, but we may allow some false positives. Alternatively, if we tighten up the criterion to keep out spurious parts, we may end up rejecting some good material.

Statistical theory tells us that, under certain idealized assumptions, the distribution of values obtained by calculating D_i for many samples of the same prototype will follow the *chi-squared distribution* (χ^2).[13] This mathematical function depends on the value of n, the number of features used to calculate D_i. A table of relevant values of χ^2/n (chi-squared divided by n) is given in Table 14.1. The table gives the value of a threshold below which D_i/n will fall either 95 or 99% of the time. For example, if four features are being used, then 95% of the time that we measure prototype i, $D_i/4$ will fall below 2.37 (i.e., D_i will be 9.48 or less), and 99% of the time $D_i/4$ will fall below 3.32 (D_i will be below 13.28).

Table 14.1 can be used to choose a threshold rejection value that will correspond to an approximate rate of false negatives. (This rate is only approximate because several of the idealizing assumptions necessary to calculate χ^2 do not hold.) The rate of false positives is not as easy to quantify; it depends on how similar the extraneous objects are to objects in the training set, and on how frequently they occur. In practice the best strategy is to choose an appropriate threshold for D_i/n and to see how it performs. Errors either way can be used to fine-tune the threshold for best performance.

The choice of a feature set to use for recognition is something of an art. There are hundreds of potentially useful measures that can be made of a blob or an image. We would like to find a small

TABLE 14.1. VALUES OF χ^2/n

n	95% Confidence	99% Confidence
1	3.84	6.64
2	3.00	4.61
3	2.61	3.87
4	2.37	3.32
5	2.21	3.02
10	1.83	2.32
15	1.67	2.04
20	1.57	1.88
30	1.46	1.70

subset that is useful. A 1974 version of the SRI vision module[1] used the following seven features for starters:

Perimeter.

Area.

Total area of all holes.

Minimum radius vector length.

Maximum radius vector length.

RMS average radius vector length.

Compactness (perimeter² divided by area).

These seemed to work well on a variety of shapes, although no effort was made to determine their optimality.

To assess the adequacy of a given set of features, it is useful to treat each prototype in turn as an unknown object and attempt to match it to the other $n - 1$ prototypes. The minimum D_i/n should be well above the rejection threshold. If not, then there is the possibility that an unknown object might match two different prototypes. To prevent this, additional features should be added to the recognition set.

14.9.2. Relational Methods

The nearest-neighbor rule and similar methods of pattern matching assume that global feature values represent the whole object and nothing but the object. But we cannot always rely on the objects to be entirely inside the field of view and not overlapping, or depend on our vision algorithms to always give complete and correct results. The methods to be discussed have the potential of dealing with extraneous or incomplete information. However, they are less well understood from a theoretical point of view, and little experience has been gained with them from a practical standpoint. They seem likely to produce satisfactory results for selected problems, but a general-purpose system using relational methods seems farther away.

Relational methods depend on *local features,* or small areas of the scene that are in some way characteristic. A variety of methods can be used to find these local features (see list given later). A prototype consists of a set of specifications of the local features to be found in an object and the spatial relationships among them. Recognition and pose estimation (determining position and orientation) involve a search for some set or group of local features that matches the information in the prototype. Only the portions of the image that actually match some portion of the prototype enter into the matching procedure, so missing or extraneous data are easily accommodated, provided there is still a sufficient amount of correct information. Once a tentative identification has been made, a check can be made, if appropriate, of the information that does not participate in the match.

The list of things that can be used for local features includes, but is not limited to, the following:

Small blobs in binary images. They are especially useful in dealing with parts that have numerous drilled or punched holes.

The perimeter of a blob, after fitting with straight-line segments, which can be used either for its line segments or for its corners.

Edges in gray-scale pictures.

Patches found with correlation-based feature matching.

Special operators that have been reported in the literature, such as an interest operator[14] or a diameter-limited gradient direction histogram.[15]

Local features are always characterized by their position in x and y. In addition, they may have other numbers associated with them such as orientation or size.

A number of *ad hoc* methods can be used for special situations. For example, one might find one or more key features that are unique in the image or are otherwise distinguishing. Then look for other particular features that are in an expected position with respect to the key feature. The distance between two features is useful in this regard.

But *ad hoc* programming is limiting. It requires manual programming for each new part type. As the number of local features to be considered increases, the program complexity increases tremendously. The method to be described is reasonably general, and prototype data can be derived from "training by showing" methods.

Relational matching requires prototype descriptors that describe, for a given part, not only the local features that may be expected to be found in that part's image, but also the spatial relationships between these local features. For example, consider the right triangle of Figure 14.22. We assume

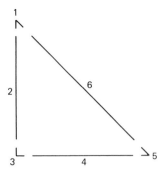

Fig. 14.22. Method of maximal cliques: A prototype right triangle. Three edges and three angles are stored.

our system is capable of using straight-edge segments and corners as local features. The triangle's prototype then consists of the three edges, 2, 4, and 6, together with their lengths and orientations, and three corners, 1, 3, and 5, together with their angles and sizes. Additionally, we store relationship information: the adjacencies of corners to edges (1 is adjacent to 2 and 6, 6 is adjacent to 1 and 5, etc.).

Now suppose the camera sees a new image from which local features are extracted as in Figure 14.23. This is the same right triangle, rotated so that its hypotenuse is horizontal. Poor viewing conditions have obliterated a corner and part of an edge. A relational match will identify the parts of the unknown to corresponding parts of the prototype.

An *assignment* is a tentative match between a feature in the unknown and a feature in the prototype. In our example, we can make the assignment c-4, because edge c in the unknown and edge 4 in the prototype have similar lengths. We also make the assignment c-2, but not 6-c, because edge c in the unknown and edge 6 in the prototype are of different lengths. Figure 14.24 shows all the assignments for the edges and corners of the unknown of Figure 14.23 to those of the prototype of Figure 14.22. Some additional assignments such as b-1 and b-5 have been added on the basis of a liberal error tolerance and noisy data.

Next we must assess the compatibility of the assignments. Two assignments are incompatible if any of the following apply:

1. A contradiction is implied. Assignments b-1 and b-3 are incompatible because they assign the same corner in the unknown to different corners in the prototype.

2. Adjacency relationships are incorrect. Corner 5 and edge 2 are not adjacent in the prototype, while corner b and edge c are adjacent in the unknown; therefore assignments b-5 and c-2 are incompatible.

3. Distances between features are incorrect. The distance between corners 1 and 5 in the prototype is much greater than the distance between corners b and d in the unknown; therefore assignments b-5 and d-1 are incompatible.

The compatibility among all assignments is shown symbolically as a graph structure in Figure 14.25. Each node (circle) represents one assignment. We draw an arc (a line) between two nodes if the two assignments are compatible. The lack of an arc between two nodes indicates that the two assignments are incompatible.

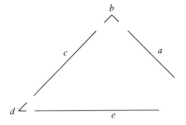

Fig. 14.23. Method of maximal cliques: An unknown figure containing three edges and two corners.

$b - 1$	$c - 2$	$d - 5$
$b - 3$	$c = 4$	$e - 6$
$b - 5$	$d - 1$	

Fig. 14.24. Method of maximal cliques: Tentative assignments showing possible matches between elements of the unknown and the prototype. Match is based only on size, but a liberal error tolerance causes some additional matches to be included.

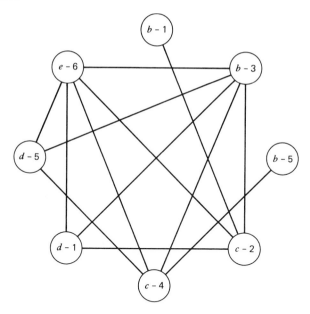

Fig. 14.25. Method of maximal cliques: Compatibility of assignments. Each assignment is represented by a circle. A line between two circles indicates that the assignments are compatible.

A *clique* is a set of assignments that are mutually compatible. For example, the set of assignments {c-2, d-1, e-6} constitutes a clique, but the set {c-2, b-1, e-6} does not because b-1 and e-6 are incompatible. A *maximal clique* is a clique that is not a subset of any larger clique. The clique {c-2, d-1, e-6} is not maximal, because it is included in the larger clique {b-3, c-2, d-1, e-6}. In any given compatibility graph, the largest maximal clique represents the most likely correspondence between the unknown and the prototype.

There are many algorithms for finding maximal cliques in a graph.[16,17] The one that follows, as stated by Bolles,[18] is not the most efficient, but it provides a reasonable trade-off between simplicity and efficiency. The algorithm is stated as a recursive procedure:

```
MaxCs ( C, P, S ) =
    BEGIN
        Setelement Y;                 COMMENT   A local variable;
        If P = EmptySet  THEN
            If S = EmptySet  THEN
                List C as a Maximal Clique
            ELSE RETURN:          COMMENT Dead end;
        Y ← Choose ( P );
        MaxCs ( C∪{Y}, P∩Neighbors(Y), S∩Neighbors(Y) );
        MaxCs ( C, P − {Y}, S ∪ {Y} );
    END;
```

C, P, and S are sets of nodes. The symbols ∪ and ∩ represent the set operations of union and intersection, respectively. "Choose" is a function that arbitrarily returns one element of a set. "Neighbors" is a function that returns the set of all nodes that are compatible with its node argument. To find all the maximal cliques of a graph, make the call MaxCs ({ }, ⟨set of all nodes⟩, { }).

A set is most efficiently represented internal to the computer as a bit-string. Each node in the graph corresponds to a position in the bit-string—a one in that position indicates the node is included in the set. Union and intersection are done very efficiently by logical OR and logical AND.

There are four maximal cliques in the graph of Figure 14.25; these are listed in Figure 14.26. There are two maximal cliques of length 4—one corresponds to the mapping one would expect, and the other corresponds to a mirror-image mapping. If symmetry may be expected in either prototypes or unknowns, then an after-the-fact check can be made to see if the local features are in the correct or the reverse order. Sometimes we include a symmetry breaker in the rules for making assignments

$$\begin{array}{llll}
\{b1 & c2\} & & \\
\{b5 & c4\} & & \\
\{b3 & c2 & d1 & e6\} \\
\{b3 & c4 & d5 & e6\}
\end{array}$$

Fig. 14.26. Method of maximal cliques: Maximal cliques of compatible assignments. Of the two larger maximal cliques, one is the expected mapping and the other is a mirror image.

or compatibilities, for example, the relative orientations of edges. The angular difference between edge 6 and edge 2 of the prototype is $+135°$ or $-45°$, whereas the angular difference between edges e and c in the unknown is $+45°$ or $-135°$. Thus assignments e-6 and c-2 become incompatible when angular orientations are considered.

The foregoing illustrates a fundamental principle when using relational matching techniques—that the stronger the constraints that can be applied, the easier the matching technique and the more reliable the result. The time required to find all maximal cliques in a graph grows rapidly with the increase in the number of nodes and arcs in the graphs; therefore a small gain in limiting the possible interconnections can give a large improvement in run-time efficiency. Naturally, the kinds of constraints that can be applied are very much a function of the particular local features being used. It is up to the designer of a relational-matching system to exploit these constraints as much as possible for the best results.

REFERENCES

1. Rosen, C., et al., Exploratory Research in Advanced Automation, Stanford Research Institute, First Report, December 1973; Second Report, August 1974; Third Report, December 1974; Fourth Report, June 1975; Fifth Report, January 1976; Machine Intelligence Research Applied to Industrial Automation, Sixth Report, November 1976; Seventh Report, August 1977; Eighth Report, August 1978.

2. Nitzan, D., et al., Machine Intelligence Research Applied to Industrial Automation, SRI International, Ninth Report, August 1979; Tenth Report, November 1980; Eleventh Report, January 1982; Twelfth Report, January 1983.

3. Gleason, G. J. and Agin, G. J., A Modular Vision System for Sensor-Controlled Manipulation and Inspection, *Ninth International Symposium on Industrial Robots,* Society of Manufacturing Engineers, Washington, D.C., March 1979, pp. 57–70.

4. Jarvis, J. F., Automatic Visual Inspection of Glass-Metal Seals, *Fourth International Joint Conference on Pattern Recognition,* IEEE Computer Society, Kyoto, Japan, November 1978, pp. 961–965.

5. Duda, R. O. and Hart, P. E., *Pattern Classification and Scene Analysis,* Wiley-Interscience, New York, 1973.

6. Casasent, D., Ed., *Optical Data Processing,* Springer-Verlag, 1978.

7. Rosenfeld, A., Sequential Operations in Digital Picture Processing, *Journal of the ACM,* Vol. 13, No. 4, October 1966, pp. 471–494.

8. Ballard, D. H. and Brown, C. M., *Computer Vision,* Prentice-Hall, Englewood Cliffs, New Jersey, 1982.

9. Uno, T., Mese, M., and Ejiri, M., Defect Detection in Complicated Patterns, *Electrical Engineering in Japan,* Vol. 95, No. 2, March–Apr. 1973, pp. 90–97.

10. Roberts, L. G., Machine Perception of Three-Dimensional Solids, in Tippett, J. T. et al., Eds., *Optical and Electro-Optical Information Processing,* M.I.T. Press, Cambridge, Massachusetts, 1965, pp. 159–197.

11. Sobel, I., Camera Models and Machine Perception, Technical Report AIM-121, Stanford Artificial Intelligence Project, May 1970.

12. Martelli, A., Edge Detection Using Heuristic Search Methods, *Computer Graphics and Image Processing,* Vol. 1, No. 2, August 1972, pp. 169–182.

13. Brownlee, K. A., *Statistical Theory and Methodology in Science and Engineering,* Wiley, New York, 1965.

14. Moravec, H., Towards Automatic Visual Obstacle Avoidance, *Proceedings of the Fifth International Joint Conference on Artificial Intelligence,* Cambridge, Massachusetts, August 1977.

15. Birk, J., et al., Image Feature Extraction using Diameter-Limited Gradient Direction Histograms,

IEEE Transactions on Pattern Analysis and Machine Intelligence, Vol. PAMI-1, No. 2, April 1979, pp. 228–235.

16. Augustson, J. G. and Minker, J., An Analysis of Some Graph-Theoretical Cluster Techniques, *Journal of the ACM,* Vol. 17, No. 4, October 1970, pp. 571–588.

17. Bron, C. and Kerbosch, J., Finding All Cliques of an Undirected Graph, *Communications of the ACM,* Vol. 16, No. 9, September 1973, pp. 575–577.

18. Bolles, R. C., Robust Feature Matching through Maximal Cliques, *Imaging Applications for Automated Industrial Inspection and Assembly,* Society of Photo-Optical Instrumentation Engineers, Washington, D.C., April 1979, pp. 140–149.

CHAPTER 15
INTERFACING A VISION SYSTEM WITH A ROBOT

ULRICH REMBOLD

CHRISTIAN BLUME

University of Karlsruhe
Karlsruhe, West Germany

15.1. INTRODUCTION

It may be necessary for material handling and assembly work to use a vision system. The system must locate the part, identify it, direct the gripper to a suitable grasping position, pick up the part, and bring it to the work area. Frequently this work must be done on a moving conveyor. Both the camera and the robot have unique coordinate systems. The camera identifies the object with reference to its own coordinate system. However, the robot, for grasping, must know where the object lies in reference to its own coordinate system. This requires the transfer of information about the object's location from the coordinate system of the camera to that of the robot. The following two options can be used for coordinates transformation:

1. With the use of a high-level programming language for the robot and/or for the vision system.
2. Without the use of a robot-programming language.

15.2. ROBOT CALIBRATION WITH THE USE OF A HIGH-LEVEL PROGRAMMING LANGUAGE

Most vision systems have the three following properties that facilitate the integration with a robot programming system:

1. Parts to be recognized are described by user-defined symbolic names which are represented by ASCII (American Standard Code for Information Interchange) strings.
2. The position and orientation of the workpiece are determined relative to a Cartesian coordinate system.
3. Parts to be recognized are taught to the system by "showing." In addition, the symbolic name is entered.

Identification of parts is done with the aid of a feature vector including area, number of holes, minimum and maximum diameter, perimeter, and so on. Normally, the user need not enter these parameters. The system automatically generates the feature vector for a part during teach-in. When the program is executed, the system compares the stored feature vector with that produced from the actual image.

The VAL (programming language for Unimation's robots) extension, VAL-IIV, includes several commands for a service program that calibrates the camera, learns the identity of the part, and stores the feature vector under a symbolic object identifier. The user program addresses the vision system with the two additional VAL instructions, VPICTURE and VLOCATE. The result of VPICTURE is that of taking the picture and storing of its pixels in the image buffer. VLOCATE results first in a search for a given object in the image buffer. If the search is unsuccessful, the program flow branches to a special label, or the program execution stops and an error message is printed. In case of a successful search, the vision system stores the position and orientation under a frame identifier that is the same as the name of the object.

The coordinate values of the object frame refer to the Cartesian coordinate system of the camera and to the base coordinate system of the robot. Figure 15.1 shows schematically the interaction between the vision system and the robot program written in VAL IIV.

The user can easily calculate with the aid of relative frames the origin of the camera coordinate system in reference to that of the robot.

With the help of a vector and a rotation matrix a frame describes the position and orientation of the gripper or of the tool of the robot. Simply by looking at the position vector, a relative frame defines this position is obtained by adding the position vector of the frame to the vector of the base frame. The calculation of the relative orientation is done similarily by a matrix multiplication.

For calibration, the robot is directed to take a small disk or ring and to place it into the field of view of the camera. Then the gripper is moved under the teach-in mode to the center of the disk. Here the position and orientation are recorded and stored under the frame with the name calib_robot. Now, the robot is moved out of the field of view of the camera. The following simplified program shows the calculation of the origin and orientation of the camera coordinate system with respect to the robot coordinate system.

```
HERE calib_robot
MOVE home_position (*robot out of field of view*)
VPICTURE
VLOCATE calib_camera, 150
INV inv_calib_camera = calib_camera
SET camera_system = calib_robot : inv_calib_camera
        .
        .                    (*normal program*)
        .
        .
        .

        .
VPICTURE
VLOCATE object, 150
MOVE camera_system : object
```

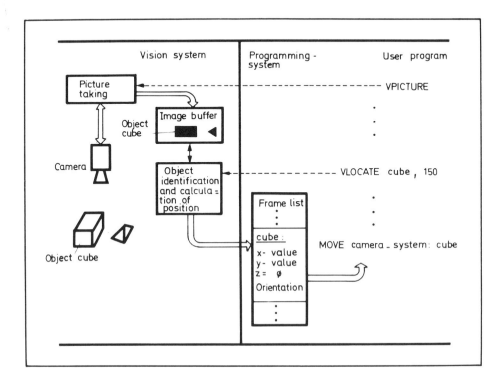

Fig. 15.1. The interaction between a vision system and a robot programming system (VAL IIv).

Figure 15.2 shows the geometric representation of the relative frames necessary for the calculation. For simplification only the position vectors are shown. A similar calculation is done for the orientation of the frame.

The HERE instruction defines the center of the disk as the frame calib_robot with respect to the robot coordinate system. VLOCATE defines the center of the disk as the frame calib_camera with respect to the camera coordinate system, which means that the two vectors point at the same point in space. The frame calib_camera is inverted. The result is the position vector inv_calib_camera which now points from the center of the disk to the origin of the camera coordinate system. The relative frame calib_robot : inv_calib_camera indicates the position and orientation of the camera coordinate system with respect to that of the robot. The vector of the frame points from the origin of the robot coordinate system to that of the camera. To simplify the notation, the result is assigned to the frame camera system. After these calculations have been performed, all following object frames that are determined by VLOCATE are referenced by the MOVE statements relative to the frame camera_ system. This example emphasizes the necessity of the frame concept and of geometrical operations for vision system applications.

15.3. INTERFACING A VISION SYSTEM WITH A ROBOT WITHOUT THE USE OF A PROGRAMMING LANGUAGE

To explain the transposition of the workpiece parameters from the coordinate system of the camera to that of the robot we use Figures 15.3 and 15.4.

A typical vision system is shown in Figure 15.3. To simplify, a photodiode matrix consisting of 64×64 pixel units was selected. The length of one pixel is denoted by one PIX. Present cameras have a resolution of 256×256 pixels or more. Generally the vision system has its own computer which determines from the two-dimensional image of the workpiece its identity, its location, and orienta-

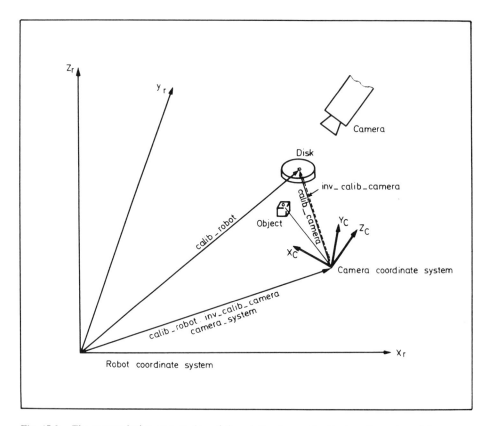

Fig. 15.2. The geometrical representation of the relative frames for the transformation of the camera coordinates into robot coordinates.

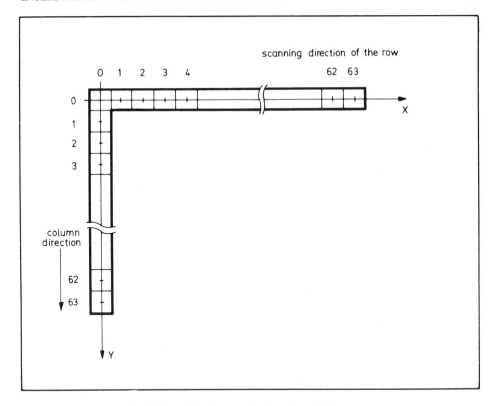

Fig. 15.3.   Cartesian coordinate system of the sensor.

tion. The last two parameters must be transferred to the control computer of the robot. It, in turn, will instruct the control circuit to grasp the object and to bring it to a predetermined location.

We select for our example a simple robot with two translational and two rotational axes, Figure 15.4. These axes determine the operating range of the robot. The movements in direction of the lift and transfer axes of the robot can be represented in terms of number of unit increments of the measuring scales of these axes. The movements of the swivel and rotational axes about their centers can be expressed in units of angular increments of their decoders.

The transformation of the workpiece coordinates from the camera coordinate system to that of the vision system requires trigonometric calculations. They can be quite involved with a complex multiaxis robot. For our sample robot, these calculations are straightforward.

Three aspects must be considered when a vision sensor is interfaced with the industrial robot: the physical interface, the data transfer protocol, and the coordinate transformations. To date there is no standard practice available for the solution of this problem. The designer of such a system relies therefore on his own intuition. He is limited by the capability of the sensor system and by that of the control unit of the manipulator. A physical interface typically can be realized by an RS 232 interface.

For transformation of the workpiece coordinates from the sensor to the robot system, the following concept may be used. For each workpiece known to the sensor system, an action program is assigned by means of a program number. The grasping position of the workpiece, which usually does not correspond to its centroid, is calculated. This is done by bringing the gripper under control to the grasping position and by transferring the coordinates of this position to the sensor computer. From this coordinate axis the grasping position is calculated in reference to the sensor system. With this information the positional values that are needed for the calculation of the grasping position in the work mode are determined.

During the work mode the control unit of the robot waits for the transfer of the grasp coordinates. After a workpiece is identified by the sensor the grasp position must first be calculated in reference to the sensor system, and then this grasp position is transferred to the robot's coordinate system. The coordinate axis obtained by this transformation is transferred to the robot control unit together

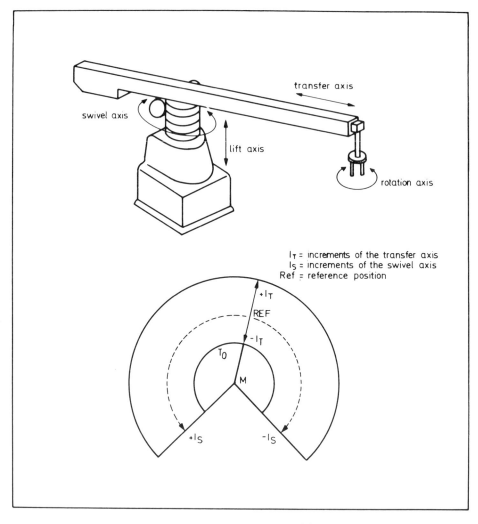

Fig. 15.4. Mechanical components of the robot.

with the program number of the action program assigned to the workpiece in the teach-in phase. The action program, modified by the new coordinate axis, can then be processed by the robot control. The geometric setup of the robot-vision system is shown in Figure 15.5.

15.4. COORDINATE TRANSFORMATION FOR THE TEACH-IN PHASE

The teach-in phase of the coupled sensor system uses the following procedure:

1. Teach the workpiece to the sensor system.
2. Determine the grasp position with the help of the robot.
3. Plan an action program for handling of the workpiece.

In the following paragraphs only the calculation of a grasp position is described. The values that relate to the learning phase are marked with the subscript L. To do this, we start with the setting of the robot-vision system shown in Figure 15.5. The location of the corresponding Cartesian sensor coordinate system with regard to the robot coordinate system is shown in Figure 15.6. The sensor coordinates are described by X and Y, and the robot coordinates by T and S. In the robot system the grasp position G is described by the position of the transfer axis T_G and the position of the

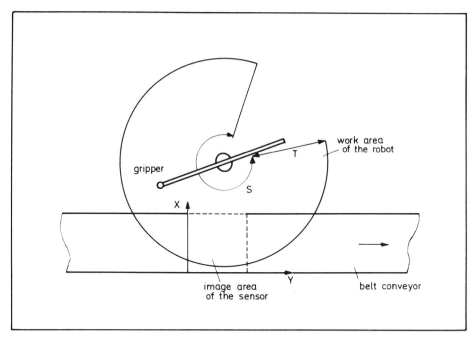

Fig. 15.5. Physical setup of the robot-vision system.

swivel axis S_G when the gripper is above the grasp position. The distance H between G and M can be calculated with regard to the robot system:

$$H_R = T_G + T_O \qquad \text{increment units of the transfer axis } (I_T)$$

And in reference to the sensor system:

$$H_S = \frac{H_R}{F_T}$$

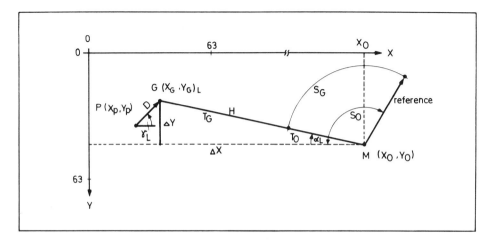

Fig. 15.6. The coordinate system of the robot.

The constant value T_O describes the distance between the gripper and the center of rotation of the swivel axis when the axes are at the reference point. It is measured in the number of increment units of the transfer axis. The constant value F_T represents the ratio of the transfer increment units divided by the length of one pixel (I_T/PIX).

The constant value S_O describes the position of the swivel axis when it is parallel to the sensor-X-axis. The constant value F_S shows the number of increments of the swivel axis per degree of angle (I_S/deg). The coordinates of the point G in the sensor-coordinate system can be calculated from

$$\Delta X = H_S \cos(\alpha)$$
$$\Delta Y = H_S \sin(\alpha)$$

15.5.   COORDINATE TRANSFORMATION DURING THE OPERATING PHASE

During the operating phase the following tasks are carrried out:

1. Calculation of the center of area P_A (X_P, Y_P).
2. Identification of the workpiece.
3. Calculation of the rotational position of the workpiece.
4. Calculation of the grasp position with regard to the sensor and $G_A(X_G, Y_G)$.
5. Transformation of the coordinates of the grasp position to the coordinates of the robot system.

The values obtained during the operating phase are marked by the subscript A. Figure 15.7 shows the interrelation of the different coordinates and the important system parameters during the operating phase. After identification of the workpiece, and after calculation of its rotational angle with regard to the reference position, the grasp position of the workpiece must be calculated with regard to the sensor position.

For the workpiece orientation γ_A one obtains

$$\gamma_A = \gamma_L + \delta$$

in which γ_L is the workpiece orientation and δ, the rotational angle of the workpiece, both in regard to the reference position.

The grasp position $G_A(X_G, Y_G)$ is then calculated from

$$\Delta X = D \cdot \cos(\gamma_A)$$
$$\Delta Y = D \cdot \sin(\gamma_A)$$
$$X_G = X_p + \Delta X$$
$$Y_G = Y_P + \Delta Y$$

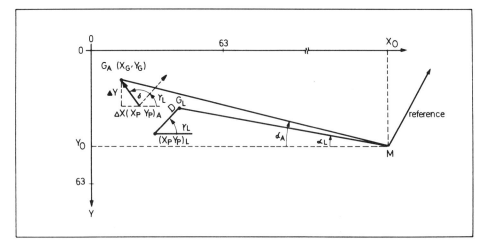

Fig. 15.7.   Coordinate transformation during the work phase.

with X_P, Y_P denoting the center of area of the object obtained during the operating phase and D the distance between the center of area and the grasp position.

Now the grasp position must be transferred to the robot coordinate system. For this the following calculation is made: position of the swivel axis,

$$S_A = S_0 + \frac{F_S \ \arctan(Y_G - Y_0)}{(X_0 - X_G)}$$

Thus the following results are obtained:

$$X_G = X_0 - \Delta X$$
$$Y_G = Y_0 - \Delta Y$$

in which the coordinates X_0, Y_0 describe the center of rotation of the swivel axis in reference to the sensor system. Thus the grasp position of the reference position of the workpiece is known. During the work phase the different positions of each workpiece must be calculated on the basis of the reference position. For this task the distance D between the grasping position $G(X_G, Y_G)$ and the centroid $P(X_P, Y_P)$, and also the workpiece orientation, are calculated during the teach-in phase. The distance D is obtained from

$$D^2 = (X_P - X_G)^2 + (Y_P - Y_G)^2 \qquad \text{(BPKT)}$$

and the workpiece orientation is

$$\gamma = \frac{\arctan \ (Y_P - Y_G)}{(X_G - X_P)} \qquad \text{(degrees)}$$

To calculate the grasp position, and for its transformation to the robot coordinate system, the following additional values must be stored:

1. Distance between the grasp position and the center of area (BPKT).
2. Orientation of the workpiece γ_L (degrees)
3. Angle α_L (degrees).
4. Number of the action program N.
5. Position of the vertical lift axis H_L (I_H).
6. Position of the rotation axis R_L (I_R).

The position of the transfer axis is calculated with the help of the formula

$$T_A = F_T \ [(X_0 - X_G)^2 + (Y_0 - Y_G)^2]^{1/2} - T_0 \qquad (I_T)$$

and the position of the rotational axis is obtained from

$$R_A = R_L + F_R \ \cdot \ (\delta + \alpha_A - \alpha_L) \qquad (I_R)$$
$$\alpha_A = \frac{(S_A - S_0)}{F_S} \qquad \text{(degrees)}$$

The position of the vertical lift axis is

$$H_A = H_L$$

and the number of the action program is

$$N_A = N_L$$

In addition to these parameters, R_L is the position of rotational axis obtained during teach-in; F_R is the number of rotational increments per degree of angle (I_R/degree); H_L is the position of the lift axis obtained from the teach-in phase.

The values S_A, T_A, H_A, R_A, and N_A are then passed over to the robot control unit. Now the robot is able to move to the calculated grasp position.

15.6. MEASURING THE ROBOT POSITION IN REGARD TO THE COORDINATE SYSTEM OF THE SENSOR

In the two preceding sections constant values (which must be measured or calculated beforehand) are used for the calculation and transformation of the grasp position. These constant values can be defined as robot constant values and constant values of the interfaced sensor-robot system.

1. T_O the distance between the center of the gripper in the reference position of the robot and the center of the swivel axis. T_O is measured in increments of the transfer axis units I_T.

   F_S the number of swivel axis increment units per degree of angle (I_S/deg).

   F_R the number of rotational increment units of the gripper per degree of angle (I_R/deg).

2. F_T the number of horizontal transfer axis increment units per pixel units (I_T/PIX).

   S_O the angular offset of the swivel axis when it is located parallel to the sensor X-axis.

   X_O, Y_O the coordinates of the center of rotation of the swivel axis in reference to the sensor system.

The constant values mentioned in group 1 are measured directly at the robot. To obtain T_O, all axes must be moved to the reference point. Thereafter the distance between the center of the gripper and the center of rotation of the swivel axis is measured and multiplied by the number of increments of measurement units.

The constants F_S and F_R are obtained as follows: The robot is brought to a reference position. At this position the value of the controlled variable of the swivel control loop is determined. Then the arm is rotated 180°, and the new controlled variable is read. Thereafter the difference is taken between these two values and divided by 180. The same procedure is used to determine F_R for the rotational axis.

The second group constant values are obtained from the sensor-robot system. For this purpose both the sensor and the robot are brought to a fixed position. With the help of the robot, a disk representing an object is moved between the three points P_1, P_2, and P_3, all of which are in the field of view of the camera, Figure 15.8. A disk is chosen because it has the following criteria:

1. The possibility of an error occurring when calculating the centroid is small.

2. There are no geometric changes when rotated.

3. The grasp position can easily be defined as the centroid of the disk.

The coordinate values (S_1, T_1), (S_2, T_2), and (S_3, T_3) are determined for each point. A prerequisite for this calibration procedure is that the position of the robot between P_1 and P_2 can only be altered by moving the transfer axis. In addition, between the points P_2 and P_3 the robot is only allowed to move about its swivel axis. Because both the centroid and the grasp position are identical with the

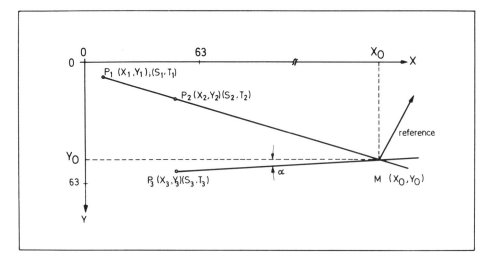

Fig. 15.8. Geometric setup of the robot-vision system.

disk, the grasp position with regard to the sensor system can be obtained by calculating the centroid at the points P_1 to P_3. Thus a fixed assignment between sensor and robot coordinates is obtained.

The constant values can be calculated as follows:

$$F = \frac{(T_1 - T_2)}{[(X_1 - X_2)^2 + (Y_1 - Y_2)^2]^{1/2}} \quad \left(\frac{I_T}{\text{PIX}}\right)$$

where the denominator is the equation of the circle

$$(X_O - X_2)^2 + (Y_O - Y_2)^2 = r^2$$

and

$$(X_O - X_3)^2 + (Y_O - Y_3)^2 = r^2$$

with

$$r = \frac{(T_2 + T_O)}{\text{FT}}$$

Thus the center of rotation of the swivel axis X_O, Y_O can be calculated with regard to the vision system.

REFERENCES

1. Blume, C. and Jakokb, W., Programmiersprachen für Roboter, Vogel-Buchverlag, Würzburg, 1983.
2. Nehr, G. and Martini, P., The Coupling of a Workpiece Recognition System with a Robot, in Pugh, A., Ed., *Robot Vision,* Springer-Verlag, 1983.

CHAPTER 16
DEPTH PERCEPTION FOR ROBOTS

AVINASH C. KAK

Purdue University
West Lafayette, Indiana

16.1. INTRODUCTION

A lack of adequate sensory feedback remains a major limitation of programmable manipulators for computerized manufacturing. Of the four categories of sensors, touch, tactile, proximity, and vision, the latter two are needed not only for automated inspection and recognition of parts and their orientation, but also for controlling a moving manipulator from an initial high speed to a slow approach just before contact, when touch and/or tactile sensors can take over.

Depth information is important for both vision and proximity sensing. Although, for obvious reasons, it is critical for the speed control of a manipulator approaching a workpiece, it can also be useful for the analysis and recognition of a three-dimensional scene around the manipulator. It can serve as an additional and important feature for scene segmentation. After all, the surface of an object can be almost totally characterized by the relative depths of the points on the surface from the viewing system. Depth information may also be used to build three-dimensional moments for shape analysis and for determining object orientation.

Depth information may be extracted by a number of competing technologies. The most fascinating of these, and also the one that until recently has suffered the most from program complexity and high computational cost, uses a pair of cameras for stereo perception. The computational cost of the traditional stereo algorithms is caused by the many correlations required to solve the correspondence problem, which consists of first selecting "candidate pixels" in one image and then locating their corresponding partners in the other. The more recent approaches suggested for stereo can be implemented much more efficiently because the candidate pixels are located by using fast Fourier transform (FFT) algorithms, which with array processing hardware can be implemented rapidly, and the correspondence problem is solved by comparing the locations of zero-crossings in the filtered stereo images. Such comparisons of the locations of zero-crossings can again be carried out rapidly by using specially designed parallel hardware.

Other approaches to depth perception use lasers, ultrasound, infrared, and coded apertures. One advantage shared by most of these techniques is that they do not suffer from the occlusion-caused problems inherent to stereo; by occlusion we mean that near a range discontinuity some parts of a scene might be invisible to one of the two cameras. But then they have their own limitations. For example, when lasers are used at power levels safe for humans to be around, it can take a long time to build up a full-range map for a scene spanning a visual angle of around 45°. With ultrasound, if used in the conventional pulse-echo mode, the lateral resolution is poor. For simple objects, the lateral resolution can be considerably improved (by a factor approaching diffraction limits) by coherent detection and back-propagation. Ultrasound does possess the virtue that it can be used even in the hostile environment usually associated with a manufacturing operation. With direct triangulation using infrared (although triangulation also suffers from occlusion-caused problems) one usually obtains one range point at a time; the total time to build up a full-range map can be considerable. However, infrared triangulation is a rugged and inexpensive approach to depth extraction.

The accuracy of a depth perception technique must be consistent with the task at hand and with the other sensory capabilities of the robot. For example, if the task is to grasp and remove a specified part from a tray of mixed and overlapping objects whose dimensions are on the order of inches, for most applications the depth measurement accuracy would have to be no better than a quarter of an inch. In this case, the depth information would be used for initially homing in toward the object, and the final contact would be made with precision using force or touch sensing. It appears that, for purposes of industrial manufacturing, accuracies of this order can be obtained in near real time by using the technique of structured light discussed in Section 16.3.1. This approach is not feasible for general-purpose robot vision; hence the active interest in all the other approaches discussed in this chapter.

16.2. PERSPECTIVE TRANSFORM AND CAMERA CALIBRATION FOR STEREO RANGE MAPPING

The mathematical relationship between the points in the object space and the corresponding points in a camera image is called a *perspective transform*. The perspective transform is a function of the location of the camera in a fixed coordinate system, its orientation as determined by its pan and tilt angles, and its focal length (which will be variable for a zoom system). Although the transform uniquely determines the image point for any given point in the object space, for a given image point it can give us only the direction of the object point. However, such directions obtained from more than one camera can be combined to pinpoint the location of an object point in a three-dimensional space. This concept forms the basis of depth extraction by stereo imaging.

To use the perspective transform for range mapping, we must therefore first determine the coordinates of the camera, the parameters of its orientation, and its focal length. A couple of options are available for this purpose. The simplest approach is to use electronic hardware that generates the focus, pan, and tilt signals, which can then be fed directly into the computer. This would leave unknown only the coordinates of the location of the camera. Without special hardware, a special interactive calibration process can be designed, which consists of recording the image coordinates of a dozen or so object points whose locations in the three-dimensional space are known and then processing this information for desired parameters. If the camera is mounted on a pan/tilt platform so that it can be aimed, this calibration procedure must be carried out with the platform set in one position. Later the azimuth and elevation angles of the platform can be used to modify the perspective transform equations appropriately.

To be able to write down the perspective transform for a camera, we need to describe how its lens system can be modeled. From this standpoint the following section highlights some of the important characteristics of lenses.

16.2.1. Some Important Things to Know about Camera Lenses

A camera lens can be either of the fixed-focus type, or the variable-focus type if it possesses zooming capability. Fortunately for fixed-focus cameras, image formation can be modeled by an equivalent pinhole system, which facilitates the derivation of relationships between the object points and image points. For example, if a camera is equipped with a lens of focal length f, the image formation may be modeled as shown in Figure 16.1a, where the lens has been replaced by a pinhole a distance of f from the image plane. The size $A \times B$ of the image formation area is about the same for many of the cameras; A and B are, respectively, about 2 and 1.5 in. For a given size for the image plane, it is clear from Figure 16.1b that the view angle ϕ should be directly a function of the focal length. To illustrate this dependence, Table 16.1 shows the relationship between the focal length and the horizontal view angle for a range of lenses made by Canon, Inc. With the eyes fixed straight ahead, the human eye sees clearly within a horizontal view angle of about 45°. From the table, a lens of 50 mm corresponds to this view angle. This lens is therefore called a standard lens. Lenses of shorter focal lengths and therefore wider view angles are called wide-angle lenses, whereas lenses of longer focal length and therefore smaller view angles are called telephoto lenses.

The reader might wonder that since a camera lens can be represented by a pinhole (at least for fixed focal length cameras), why not use a pinhole itself for image formation. Although the advantage of using a pinhole itself for imaging would be an infinite depth of field, it is usually not possible to do so because of the extremely small amount of light that would be permitted through a pinhole to illuminate the detector or the film. (By depth of field we mean the range over which objects are reproduced on the image plane with sharp detail.) Use of a pinhole for imaging would therefore imply an extremely long exposure time. A lens is used to increase the amount of light coming back from an object point. For example, in Figure 16.2 the object point P backscatters the illumination over a wide angle, and the part of it that is within an angle α is intercepted by the camera. The angle α clearly depends on the size of the lens aperture. A camera is characterized by the maximum aperture determined either by the diameter of the lens, or more restrictively by the requirement of keeping the optical distortions below acceptable levels. The diameter of the maximum permissible aperture is expressed as a fraction of the focal length. For example, a lens may be characterized as 50 mm and $f/1.4$, meaning its focal length is 50 mm and that its maximum aperture corresponds to a diameter of $f/1.4$ (= 35.7) mm.

The price paid for using a lens (as opposed to a pinhole) is the decrease in the depth of field of a camera system. However, with cameras of high enough sensitivity and intense lighting (such as might

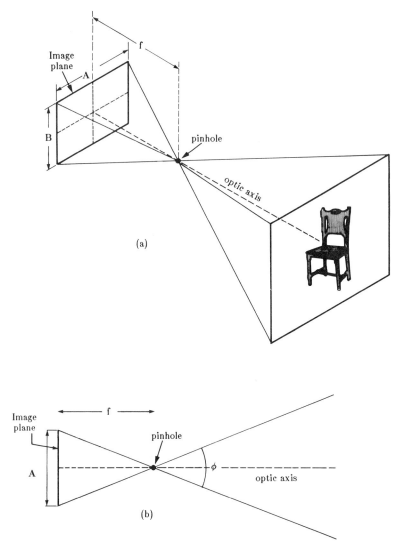

Fig. 16.1. (*a*) If a camera is equipped with a lens of a fixed focal length (i.e., the camera does not possess zooming capability), the process of image formation can be modeled by an equivalent pinhole system shown here. (*b*) For a given size of the image frame (represented by A here), the view angle ϕ is a function of the focal length.

be generated by high power flashes) it is possible to use very small apertures, approaching pinholes. In some applications the resulting increased depth of field renders focusing unnecessary.

To explain the notion of depth of field, note that in Figure 16.3 there is only one point on the optic axis that is brought into focus on the image plane; the object point at P is brought into focus at P'. The distance u to this point is given by

$$\frac{1}{u} + \frac{1}{v} = \frac{1}{f} \tag{16.1}$$

where v is the distance between the lens plane and the image plane. Every other point on the optic axis will be brought into focus at a point other than the image plane. In Figure 16.4, the object points at P_1 and P_2 will create blurry images on the image plane. The diameter of this blurry image

TABLE 16.1. THE RELATIONSHIP BE-
TWEEN THE FOCAL LENGTH f AND THE
VIEW ANGLE ϕ FOR A RANGE OF LENSES
MADE BY CANON, INC.

Focal Length f	View Angle ϕ	
1200 mm	2.1 °	
500 mm	5 °	
200 mm	12 °	
85 mm	29 °	
50 mm	46 °	
28 mm	75 °	
7.5 mm	180 °	

is called the circle of confusion; its values for P_1 and P_2 are denoted by C_1 and C_2 in the figure. The depth of field of a lens is the distance between two range points on two sides of P, whose images on the image plane are spots each of diameter C (Figure 16.5).

For normal photography, depth of field is defined with C equal to $f/1000$. Depth of field is usually shallower in the foreground and deeper in the background. Besides the focal length of the lens, it also depends on the distance u at which the camera is focused and the size of the aperture. These dependencies are contained in the following formula:

$$R = \frac{2CN(m+1)}{m^2} \tag{16.2}$$

where

R = depth of field
C = permissible diameter of the circle of confusion
$N = f$ number of the lens

$\quad = \dfrac{f}{D}$ where D is the diameter of the aperture

m = magnification

$\quad = \dfrac{v}{u}$

The notion of depth of field is particularly important for eye-in-hand camera systems for robot vision. If such a camera system has a large depth of field, no adjustments in the camera focus will be required as the camera is positioned at different distances from the object. Elimination of automatic focusing reduces the cost, size, and weight of the camera; all features that might otherwise make it unsuitable for eye-in-hand use.[21] In such camera systems the depth of field is increased by using coarse sampling for the images, since the constant C can now be set to the large sampling interval. This does not necessarily increase the errors in the computation of object properties because, by bringing the camera closer to the object, the effective sampling of the object image can be comparable to (or even finer than) that of static cameras.

16.2.2. The Perspective Transform

Our derivation of the transform equations is along the lines given by Yakimovsky and Cunningham[45] and as further explained in a tutorial article by Thompson.[42] In this derivation the physical location of the camera is given by the coordinates of its focal center as shown in Figure 16.6. The coordinates

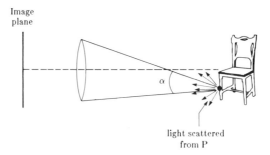

Fig. 16.2. For most imaging situations, every point on an object will scatter the incident illumination over a wide angle. In this figure, the lens of the camera will capture the reflected light that is within an angle α.

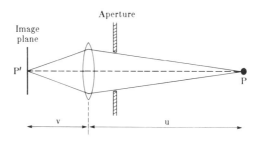

Fig. 16.3. If we confine our attention to object points that lie on the optic axis, then strictly speaking there is only one point (at P) that is brought into focus in the image plane (at P').

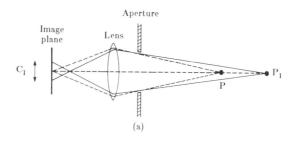

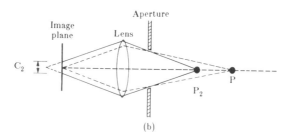

Fig. 16.4. Only one object point P on the optic axis is brought into correct focus on the image plane. All other points, such as those at P_1 and P_2 shown in (a) and (b), respectively, will create blurry images.

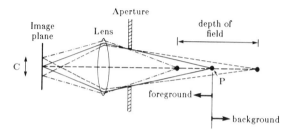

Fig. 16.5. The depth of field depends on the acceptable diameter C for the circle of confusion in the image plane.

of the focal center are the Cartesian components of vector **C**. The focal center is at a distance f from the image plane along the optic axis shown by heavy dashed line in the figure.

The orientation of the camera is denoted by the unit vector $\hat{\mathbf{a}}$, which is perpendicular to the image plane. We use two more unit vectors, $\hat{\mathbf{h}}$ and $\hat{\mathbf{v}}$, orthogonal to each other and also orthogonal to the aiming unit vector $\hat{\mathbf{a}}$. In terms of $\hat{\mathbf{h}}$, $\hat{\mathbf{v}}$, and $\hat{\mathbf{a}}$, all the points in the image plane are described by

$$\mathbf{C} - f\hat{\mathbf{a}} + u\hat{\mathbf{h}} + v\hat{\mathbf{v}} \tag{16.3}$$

for different values of the scalar parameters u and v. The ordered pair (u, v) could be considered to be the coordinates of a point in the image plane. Note that $\mathbf{C} - f\hat{\mathbf{a}}$ is the position vector for the center of the image plane.

The perspective transformation equations must provide us with the image coordinates (u, v) for any given object point **P**. To derive these equations, we simply compare the appropriate similar triangles in Figure 16.6. The result is

$$\frac{u}{f} = \frac{\mathbf{D} \cdot \hat{\mathbf{h}}}{\mathbf{D} \cdot \hat{\mathbf{a}}} \qquad \frac{v}{f} = \frac{\mathbf{D} \cdot \hat{\mathbf{v}}}{\mathbf{D} \cdot \hat{\mathbf{a}}} \tag{16.4}$$

where, as shown in the figure, $\mathbf{D} = \mathbf{P} - \mathbf{C}$ is a vector from the camera focal center to the object point. These two equations are sufficient to determine the image point for a given object point. We now recast these equations in a form suitable for sampled imagery. In other words, instead of using

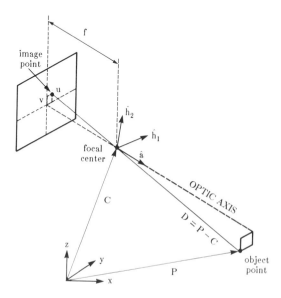

Fig. 16.6. Some parameters of a camera system: the location of the camera is represented by the vector **C** from the origin of the (x, y, z)-coordinate system to the focal center.

the analog coordinates u and v, we show the relationships for their digitized versions I and J. The two versions are tied by

$$u = (I - I_0)\Delta u \quad \text{and} \quad v = (J - J_0)\Delta v \tag{16.5}$$

This relationship is also illustrated in Figure 16.7, where we have shown I_0 and J_0 to represent the indexes for the center of the image frame. The horizontal index I and the vertical index J are measured from the lower left-hand corner of the image frame. Note that we have assumed that the pixel with indexes I_0, J_0 coincides with the center of the (u, v) coordinate plane. If results with subpixel accuracy are desired, and if this assumption is not satisfied, the relationship between the analog and the digital coordinates must be suitably modified by the addition of offsets on the right-hand side.

Substituting Eq. (16.5) in (16.4), we get

$$I - I_0 = \frac{f}{\Delta u} \frac{\mathbf{D} \cdot \hat{\mathbf{h}}}{\mathbf{D} \cdot \hat{\mathbf{a}}} \tag{16.6}$$

$$J - J_0 = \frac{f}{\Delta v} \frac{\mathbf{D} \cdot \hat{\mathbf{v}}}{\mathbf{D} \cdot \hat{\mathbf{a}}} \tag{16.7}$$

These equations are more conveniently expressed as

$$I = \frac{\mathbf{D} \cdot \mathbf{H}}{\mathbf{D} \cdot \hat{\mathbf{a}}} \quad \text{and} \quad J = \frac{\mathbf{D} \cdot \mathbf{V}}{\mathbf{D} \cdot \hat{\mathbf{a}}} \tag{16.8}$$

where

$$\mathbf{H} = \frac{f}{\Delta u} \hat{\mathbf{h}} + I_0 \hat{\mathbf{a}} \tag{16.9}$$

and

$$\mathbf{V} = \frac{f}{\Delta v} \hat{\mathbf{v}} + J_0 \hat{\mathbf{a}} \tag{16.10}$$

The important thing to note here is that $\mathbf{H}$ and $\mathbf{V}$ are neither unit vectors, nor are they perpendicular to the aiming vector $\hat{\mathbf{a}}$.

For camera systems used in practice, the sampling intervals Δu and Δv may not be equal; although when they are equal many pattern-matching processes are simplified. With unequal sampling intervals, for example, some geometrical properties of an object silhouette might change as the object is rotated to a different orientation. For vidicon cameras, the scanning characteristics can be altered to make Δu and Δv the same. However, the same cannot be said of the solid-state cameras, since for these devices the sampling intervals are fixed permanently by the geometrical layout of the light-integrating cells.

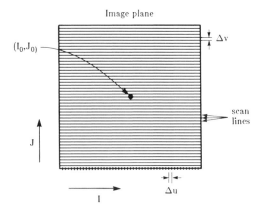

Fig. 16.7. For sampled imagery, the position of an image point is represented by indices I and J measured from the lower left-hand corner; Δu and Δv are the sampling intervals.

Note that the unit vectors $\hat{\mathbf{h}}$ and $\hat{\mathbf{v}}$ are supposed to be along the rows and columns of the digitized image, which follows from the way the indexes I and J were defined. For a vidicon, the rows and the columns are not exactly orthogonal to each other because of the slight diagonal slope of the horizontal scan line, which is caused by the continuous vertical motion of the beam.

Since we are interested in dealing directly with sampled imagery, we drop the unit vectors $\hat{\mathbf{h}}$ and $\hat{\mathbf{v}}$ from any further consideration, and in their place consider the vectors $\mathbf{H}$ and $\mathbf{V}$.

16.2.3. Camera Calibration

The aim of a camera-calibration procedure is to calculate the vectors $\mathbf{C}$, $\hat{\mathbf{a}}$, $\mathbf{H}$, and $\mathbf{V}$. Since the magnitude of $\hat{\mathbf{a}}$ is unity, these represent 11 unknowns. In principle, the calibration can be accomplished by showing to the camera at least six object points whose three-dimensional coordinates are known and then setting up at least 12 equations for the unknowns [each image point generates two equations for the unknowns, these corresponding to the equations of the perspective transform in Eq. (16.8)]. In practice, the computations for the calibration parameters are greatly simplified, provided we show the camera more than 10 noncoplanar points that then lead to more than 20 equations for the unknowns. We now show how the equations are set up for this purpose.

Let the known object points be at locations denoted by position vectors $\mathbf{P}_1, \mathbf{P}_2, \ldots, \mathbf{P}_N$. The corresponding vectors from the camera focal center will be denoted by $\mathbf{D}_m$ (Figure 16.8). These are equal to

$$\mathbf{D}_m = \mathbf{P}_m - \mathbf{C} \tag{16.11}$$

for $m = 1, 2, \ldots, N$. Let (I_m, J_m) denote the image coordinates of the mth object point. Using the perspective transform equations in Eq. (16.8), we have the following relationships:

$$I_m = \frac{\mathbf{D}_m \cdot \mathbf{H}}{\mathbf{D}_m \cdot \hat{\mathbf{a}}} \quad \text{and} \quad J_m = \frac{\mathbf{D}_m \cdot \mathbf{V}}{\mathbf{D}_m \cdot \hat{\mathbf{a}}} \tag{16.12}$$

for $m = 1, 2, \ldots N$. Consider for a moment only the equations generated by the image indexes I_m. These equations may also be expressed as

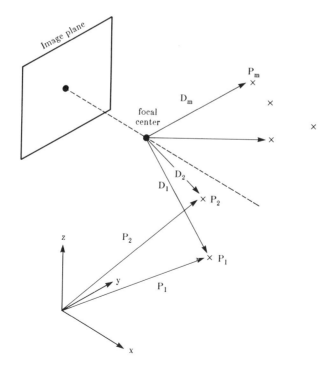

Fig. 16.8. Calibration is accomplished by showing to the camera N object points located at P_1, $P_2, \ldots, P_N$, whose three-dimensional coordinates are known.

$$(\mathbf{D}_m \cdot \mathbf{H} - I_m \mathbf{D}_m \cdot \hat{\mathbf{a}}) = 0 \quad \text{for } m = 1, 2, \ldots, N \tag{16.13}$$

or equivalently as

$$\mathbf{P}_m \cdot \mathbf{H} - \mathbf{C} \cdot \mathbf{H} - I_m \mathbf{P}_m \cdot \hat{\mathbf{a}} + I_m \mathbf{C} \cdot \hat{\mathbf{a}} = 0 \tag{16.14}$$

where we have used Eq. (16.11). Note that these N equations are nonlinear as a function of the unknowns since we have products of the unknowns appearing on the left-hand side. The following "trick" linearizes these equations: we declare two more unknowns C_H and C_a as follows:

$$C_H = \mathbf{C} \cdot \mathbf{H} \quad \text{and} \quad C_a = \mathbf{C} \cdot \hat{\mathbf{a}} \tag{16.15}$$

We may now express the nonlinear Eqs. (16.14) in the following linear form:

$$x_m H_x + y_m H_y + z_m H_z - I_m x_m a_x - I_m y_m a_y - I_m z_m a_z - C_H + I_m C_a = 0 \tag{16.16}$$

for $m = 1, 2, \ldots, N$. This set of equations constitutes N linear equations for the eight unknowns H_x, H_y, H_z, a_x, a_y, a_z, C_H, and C_a. Because the right-hand sides are all zero, these form a homogeneous set of equations. To generate a solution out of these equations, we make a harmless assumption that one of the unknown a_x is equal to 1. Clearly, this violates the requirement that $\hat{\mathbf{a}}$ be a unit vector. However, we can satisfy this requirement by first computing a_y and a_z, and all the other unknowns. All the unknowns will be linearly dependent on a_x. Therefore if we then alter the value of a_x such that $|\hat{\mathbf{a}}|$ is equal to unity, we can also appropriately scale the unknowns. By setting a_x equal to 1, the linear equations given can be recast into the following form:

$$x_m H_x + y_m H_y + z_m H_z - I_m y_m a_y - I_m z_m a_z - C_H + I_m C_a = I_m x_m \tag{16.17}$$

for $m = 1, 2, \ldots, N$. For better visualization, we now express these equations in vector-matrix form by defining a vector $\mathbf{U}$ of unknowns as follows:

$$\mathbf{U} = \begin{bmatrix} H_x \\ H_y \\ H_z \\ a_y \\ a_z \\ C_H \\ C_a \end{bmatrix} \tag{16.18}$$

and then if we use W to denote the matrix of coefficients in Eq. (16.17), those equations may be expressed in the following vector-matrix form

$$[W]\mathbf{U} = \mathbf{B} \tag{16.19}$$

where the vector $\mathbf{B}$ denotes the right-hand side in Eq. (16.17)

$$\mathbf{B} = \begin{bmatrix} I_1 x_1 \\ I_2 x_2 \\ \cdot \\ \cdot \\ \cdot \\ \cdot \\ \cdot \\ I_N x_N \end{bmatrix} \tag{16.20}$$

To protect ourselves against the numerical problems introduced by round-off and digitization errors and the inconsistencies among the equations caused by camera nonlinearities, and so on, we insist on redundancy by making N greater than the number of unknowns. We will set N arbitrarily equal to 10. When N is greater than 7, strictly speaking, there is no solution to the set of equations because of perturbations in $\mathbf{P}_m$'s caused by measurement errors. Consequently instead of finding an exact solution to the system of equations, the problem becomes one of finding a "best possible solution." Such a solution to Eq. (16.19) is taken to mean a set of values for the unknown parameters which minimize the expression

$$\{[W]\mathbf{U} - \mathbf{B}\}^T \{[W]\mathbf{U} - \mathbf{B}\} \tag{16.21}$$

where T denotes the transpose. (Note that an exact solution, if it existed, would make this expression a zero.) It can be shown that the minimizing solution must be a solution of the following *normal* equations[40]:

$$[W]^T[W]\mathbf{U} = [W]^T\mathbf{B} \tag{16.22}$$

In contrast with Eq. (16.19), the system of equations here does possess a unique solution because the matrix $[W]^T[W]$ is 7×7. So we have 7 equations in 7 unknowns. Although these equations could be solved directly by techniques such as Gaussian elimination or the Gauss-Jordan algorithm to yield the following solution

$$\mathbf{U} = |[W]^T[W]|^{-1}[W]^T\mathbf{B} \tag{16.23}$$

in practice it is numerically much more stable to use the Householder or Gram-Schmidt orthogonalization procedure to arrive at this solution.

After we have calculated the unknowns in Eq. (16.18), we compute the magnitude of the resulting vector $\hat{\mathbf{a}}$. We then divide all the unknowns by the ratio of a_x and $|\hat{\mathbf{a}}|$. This normalization is necessary to ensure that $|\hat{\mathbf{a}}|$ is equal to unity.

This procedure gives us the best possible values of the unknowns listed in the column vector in Eq. (16.18). To solve for the rest of the unknowns, we consider the unused set of equations in Eq. (16.12) for the J coordinates of the image points. When these equations are written in an expanded form, we get

$$\mathbf{P}_m \cdot \mathbf{V} - \mathbf{C} \cdot \mathbf{V} - I_m\mathbf{P}_m \cdot \hat{\mathbf{a}} + I_m\mathbf{C} \cdot \hat{\mathbf{a}} = 0 \tag{16.24}$$

Again, to eliminate the nonlinearity we introduce a new variable

$$C_V = \mathbf{C} \cdot \mathbf{V} \tag{16.25}$$

In terms of this variable and the other new variables defined in Eq. (16.15), a linearized version of (16.24) follows

$$x_m V_x + y_m V_y + z_m V_z - I_m x_m a_x - I_m y_m a_y - I_m z_m a_z - C_V + I_m C_a = 0 \tag{16.26}$$

for $m = 1, 2, \ldots, N$. This set of equations may be solved exactly as before to yield an optimum solution for the three components of $\mathbf{V}$ and C_V.

After we have calculated C_H, C_V, and C_a, we can determine the location of the camera (given by the components of the vector $\mathbf{C}$) by solving

$$\begin{aligned} \mathbf{C} \cdot \hat{\mathbf{a}} &= C_a \\ \mathbf{C} \cdot \mathbf{H} &= C_H \\ \mathbf{C} \cdot \mathbf{V} &= C_V \end{aligned} \tag{16.27}$$

which then completes the calibration procedure.

16.3. ACTIVE AND PASSIVE STEREO

As is clear from the preceding section, the perspective transform uniquely determines the image pixel coordinates (I, J) for any given point in the object space. However, for a given pixel in the image it can only give us the direction of the object point. Theoretically such directions obtained from two[*] cameras could be combined to pinpoint the location of an object point in a three-dimensional space. Although essentially this forms the principle behind stereo range mapping, in practice a direct implementation of this triangulation procedure may not work owing to a lack of precise intersection of the computed direction vectors due to round-off and digitization errors. We now present a computational

---

[*] It is possible to do approximate range mapping using a single camera. As the lens of a camera possessing a short depth of field is moved in and out, the image will come into focus at only a particular distance of the lens from the film plane. So with an automatic procedure to determine if the image is in focus, one can use the position of the lens to estimate the range. Note however from Eq. (16.2), the depth of field is inversely proportional to the diameter of the lens aperture. To generate accurate range maps with this procedure would require a lens of impractically large diameter. The reader can himself verify this by using Eq. (16.2) to find the required diameter of a lens for, let us say, a half-centimeter depth of field.

procedure that employs the perspective transform for essentially the same thing, but without any explicit calculation of the intersection of direction vectors.

We assume that the calibration vectors $\mathbf{C}_L$, $\mathbf{H}_L$, $\mathbf{V}_L$, and $\hat{\mathbf{a}}_L$ are for the left camera of a stereo pair, the corresponding vectors for the right camera being represented by $\mathbf{C}_R$, $\mathbf{H}_R$, $\mathbf{V}_R$, and $\hat{\mathbf{a}}_R$. Let the left-image and right-image coordinates of an object point located at $\mathbf{P} = (x, y, z)$ be denoted by (I_L, J_L) and (I_R, J_R), respectively (Figure 16.9). From Eq. (16.8)

$$I_L = \frac{\mathbf{D}_L \cdot \mathbf{V}_L}{\mathbf{D}_L \cdot \hat{\mathbf{a}}_L} \quad \text{and} \quad J_L = \frac{\mathbf{D}_L \cdot \mathbf{H}_L}{\mathbf{D}_L \cdot \hat{\mathbf{a}}_L} \tag{16.28}$$

where $\mathbf{D}_L$ is the vector from the focal center of the left camera to the object point at $\mathbf{P}$; it is related to $\mathbf{C}_L$ by

$$\mathbf{D}_L = \mathbf{P} - \mathbf{C}_L \tag{16.29}$$

From these equations for the coordinates of the image pixel in the left image we compute only the direction of $\mathbf{D}_L$. Since at this time we are not interested in the magnitude of $\mathbf{D}_L$, we can safely assume

$$\mathbf{D}_L \cdot \hat{\mathbf{a}} = 1 \tag{16.30}$$

As a consequence, the expressions for I_L and J_L take the form

$$\mathbf{D}_L \cdot \mathbf{V}_L = I_L \quad \text{and} \quad \mathbf{D}_L \cdot \mathbf{H}_L = J_L \tag{16.31}$$

Equations (16.30) and (16.31) can be written in the following form for the three components of $\mathbf{D}_L$ denoted by $D_{L,x}$, $D_{L,y}$, and $D_{L,z}$:

$$\begin{aligned} D_{L,x}a_{L,x} + D_{L,y}a_{L,y} + D_{L,z}a_{L,z} &= 1 \\ D_{L,x}V_{L,x} + D_{L,y}V_{L,y} + D_{L,z}V_{L,z} &= I_L \\ D_{L,x}H_{L,x} + D_{L,y}H_{L,y} + D_{L,z}H_{L,z} &= J_L \end{aligned} \tag{16.32}$$

These three equations can be solved for the three unknown components of $\mathbf{D}_L$. But note that because of the assumption represented by Eq. (16.30), the computed $\mathbf{D}_L$ at this point only represents the direction information. Although not equal to unity, its magnitude contains no useful information. We might therefore say that the true vector from the focal center of the left camera to the object point is $r\mathbf{D}_L$, where r is an unknown scalar constant.

We use the information in the right image to determine r. For this purpose we again invoke the perspective transform for the pixel coordinates I_R and J_R

$$I_R = \frac{\mathbf{D}_R \cdot \mathbf{V}_R}{\mathbf{D}_R \cdot \hat{\mathbf{a}}_R} \quad \text{and} \quad J_R = \frac{\mathbf{D}_R \cdot \mathbf{H}_R}{\mathbf{D}_R \cdot \hat{\mathbf{a}}_R} \tag{16.33}$$

From Figure 16.9, the vector $\mathbf{D}_R$ may be expressed as

$$\begin{aligned} \mathbf{D}_R &= \mathbf{P} - \mathbf{C}_R \\ &= r\mathbf{D}_L + \mathbf{C}_L - \mathbf{C}_R \end{aligned} \tag{16.34}$$

By substituting Eq. (16.34) in the two equations in Eq. (16.33), we get the following two possibilities for computing r

$$r = \frac{I_R \Delta \cdot \hat{\mathbf{a}}_R - \Delta \cdot \mathbf{V}_R}{\mathbf{D}_L \cdot \mathbf{V}_R - I_R \mathbf{D}_L \cdot \hat{\mathbf{a}}_R} \tag{16.35}$$

or

$$r = \frac{J_R \Delta \cdot \hat{\mathbf{a}}_R - \Delta \cdot \mathbf{H}_R}{\mathbf{D}_L \cdot \mathbf{H}_R - J_R \mathbf{D}_L \cdot \hat{\mathbf{a}}_R} \tag{16.36}$$

where Δ represents

$$\Delta = \mathbf{C}_R - \mathbf{C}_L \tag{16.37}$$

The new vector Δ, called the baseline vector, tells us how the cameras are separated in space regarding the direction and the distance.

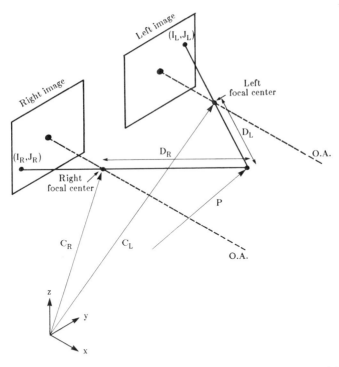

Fig. 16.9. The physical locations of the left and the right cameras are represented by vectors $\mathbf{C}_L$ and $\mathbf{C}_R$. The image elements for the object point $\mathbf{P}$ are at (I_L, J_L) and (I_R, J_R) in the two cameras.

Consider the special case when the baseline vector is parallel to the horizontal plane, the aiming vectors $\hat{\mathbf{a}}_L$ and $\hat{\mathbf{a}}_R$ are parallel to each other (no vergence), and both are perpendicular to the baseline vector. With these assumptions, the following equalities hold

$$\boldsymbol{\Delta} \cdot \mathbf{V}_R = 0 \quad \text{and} \quad \boldsymbol{\Delta} \cdot \hat{\mathbf{a}}_R = 0 \tag{16.38}$$

Therefore, the expression in Eq. (16.35) for r ceases to be applicable, and the expression in Eq. (16.36) reduces to

$$r = \frac{-\boldsymbol{\Delta} \cdot \mathbf{H}_R}{\mathbf{D}_L \cdot \mathbf{H}_R - J_R \mathbf{D}_L \cdot \hat{\mathbf{a}}_R} \tag{16.39}$$

16.3.1. Use of Structured Light

The procedure just presented can work only if we know the left- and the right-image coordinates of an object point. Determination of the pixel pairs in the two images that correspond to the same object point is called the *correspondence* problem. The solution to this problem is made difficult because the image of a small surface on an object may be different in the two images of a stereo pair owing to changes in perspective and the dependence of the surface backscattered light on the viewing angle. Besides, for some of the candidate pixels chosen in one image, it may not be possible to find the corresponding pixels in the other image owing to occlusion. Even before solving the correspondence problem, choosing the candidate pixels in the images also requires some care. For example, if we select a pixel on a smooth surface in one image, it will be impossible to find its corresponding point in the other image, since the gray levels nearly everywhere will be uniform.

The task of selecting candidate pixels in one image and then finding their corresponding pixels in the other is made easier if we project a pattern of light dots or bars on the object, and use the dots themselves or the intersections of light bars as candidate points. Figure 16.10 shows three possible patterns for this purpose. The simplest case is that of a single light dot illuminating the object (Figure 16.10a). This choice resolves immediately all difficulties with both the candidate point selection, and also with solving the correspondence problem. By scanning the light dot, we can build up a range map for the entire scene. The time required to scan the scene is the primary disadvantage of this

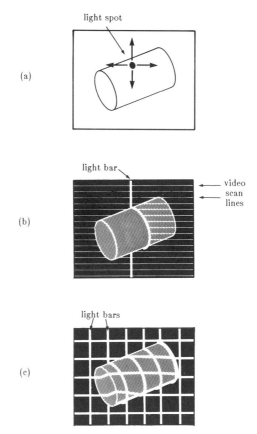

Fig. 16.10. For stereo range mapping, the task of selecting candidate points in one image, and finding their matches in the other, is made easier if we project a pattern of light dots or bars on the object. Three possible patterns are shown: (*a*) A single light dot scans the object. (*b*) A vertical bar scans the object; the candidate points for stereo matching are the intersections of the light bar and the horizontal scan lines. (*c*) A rectangular grid is projected on the object.

method. When the illumination dot is produced by a well collimated beam, it is possible to do away with some of the stereo calculations discussed in the preceding section. We showed in the previous section that with a single camera we can determine the direction of a point in the object space, although not its distance. By using only one camera and finding the direction of the illuminated spot, and then finding the intersection of this direction with the direction of the collimated beam, one can determine the three-dimensional location of the illuminated point.

Concerning the use of a single illumination dot, we must mention another simple procedure that also uses only one camera and with minimal computing yields the range value. In this procedure, first put forward by Sweeney and Hudelson,[41] the camera is used in conjunction with a Fresnel lens as shown in Figure 16.11. By recording the data with a vidicon or a solid-state camera, and doing a simple analysis of the data for the spatial dependence of frequencies along two perpendicular axes in the image plane, one can obtain all the coordinates of the illumination dot.

Faster range maps may be obtained by using a bar of light to illuminate the scene as shown in Figure 16.10*b*. The candidate points for stereo matching are now defined by the intersections of the scan lines and the illuminated line on the object. Matching is again trivial because on each scan line we have only one illuminated point. The pixel coordinates of the corresponding points can be fed into the computing procedure described in the preceding section to determine the range map along the illuminated strip on the object. A complete range map may be obtained by scanning the object with the illuminated strip. Again, if one uses a well-collimated bar for this purpose and knows a priori the direction of the illumination, one can obtain the range map with a single camera (the reasons being the same as for the case of a single illumination dot).

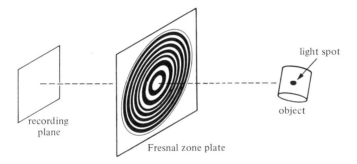

Fig. 16.11. If the object is illuminated with a single dot at a time, the range to the lighted spot can be obtained with minimal computing by looking through a Fresnel lens.

The technique of using a single bar of light may yield poor results if the object surface is nearly parallel to the illuminating direction. This difficulty may be overcome by using two orthogonal bars of light. A scanned version of this generates the projected light pattern shown in Figure 16.10c. Range-mapping techniques that combine direction calculations from a single camera and the a priori known direction of the projected light beams were first demonstrated by Agin and Binford[1] and Shirai.[39] These methods also go by the name of *triangulation ranging*.

Also included in the category of triangulation techniques is another simple procedure that is used for autofocusing in some commercial cameras. The principles of this method are illustrated in Figure 16.12. A rocking light-emitting diode (LED) illuminates the scene with a narrow beam of light. The receiver photodiode looks straight ahead and responds to the diffuse backscatter from the object. For an object at a given range, the receiver will record a large signal only for one value of the angle of transmission of the LED. For computer readout of the range value, the angle of transmission of the LED can be encoded; the angle at which the receiver picks up a maximum signal being directly translatable into the range value. (In autofocus cameras the lens moves with the rocking motion of the transmitting LED, and the lens motion is stopped just as the receiver photodiode peaks out.)

16.3.2. Marr-Poggio-Grimson Algorithm*

If active control of illumination is not allowed, then one must resort to methods discussed here and in the next section for the selection of the candidate points and for solving the correspondence problem. In this section we describe the Marr-Poggio-Grimson (MPG) algorithm for this purpose.[12,13,23,24] This algorithm is a result of attempts at simulating, at least at a computational theoretic level, the human visual system for its ability to perceive depth with binocular fusion.

A few definitions are in order to explain this algorithm. In Figure 16.13a we have shown two cameras fixating at the indicated point.† (The optic axes of the two cameras intersect at the fixation point.) The shown object point casts image elements at locations A and B in the left and the right cameras. Associated with each image element is its angular displacement from the optic axis as shown in the figure. If α_L and α_R are the angular displacements of the two image elements corresponding to the same object point—the positive senses for the displacements being as shown—the *disparity d* of the object point is defined as

$$d = \alpha_L + \alpha_R \tag{16.40}$$

An object point is considered to possess *convergent* or *crossed disparity* if it is in front of the fixation point, as happens in Figure 16.13a. (The reason for this name is that the optic axes of the two cameras will have to cross or converge to fixate at the object point.) Objects behind the fixation point are said to have *divergent* or *uncrossed disparity*, as happens with the object point in Figure 16.13b. Note that for object points that possess convergent disparity, the right-camera image element will always be to the right of the left-camera image element. On the other hand, for object points possessing

---

* This algorithm is based on the Marr-Poggio paradigm of human stereopsis[23,24] and its computer implementation developed by Grimson.[12,13]

† In this and some other figures to follow where we have shown the location of the object point with respect to the fixation point, we have used "eyes" to depict the left and the right cameras of a stereo pair. This was done because the pupil orientations immediately show the location of the fixation point.

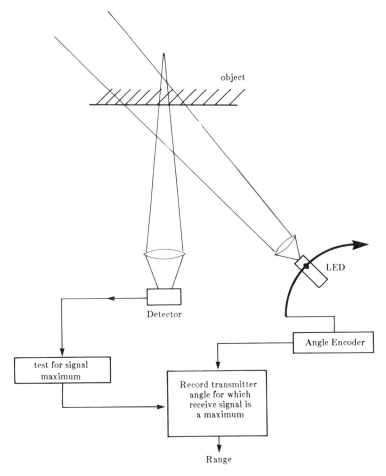

Fig. 16.12. One range value at a time can also be obtained by using the scheme shown here, which consists of illuminating the object with a rocking light emitting diode (LED). The receiver photodiode looks straight ahead. The range value is given by the angle of the LED for which the received signal is a maximum.

divergent disparity, the right-camera image element will always be to the left of the left-camera image element.

Selection of Candidate Points

The following rationale is used for selecting the candidate points in this algorithm:

1. Only those image elements where the gray level sharply changes can be used as candidate points for stereo matching, since if a candidate point is located in a region of uniform gray levels, its corresponding image element in the other image will be difficult to identify.

2. Sharp changes in gray levels correspond to the zero-crossings of the Laplacian of the image. Mathematically, therefore the candidate points will be at the locations of the zero-crossings of

$$\nabla^2 f(x, y) \qquad (16.41)$$

where

$$\nabla^2 \equiv \frac{\partial^2}{\partial x^2} + \frac{\partial^2}{\partial y^2} \qquad (16.42)$$

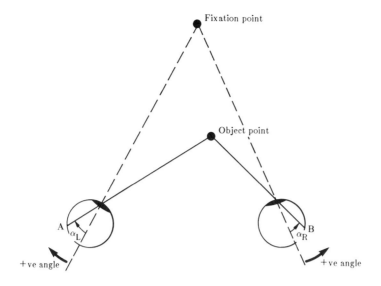

Disparity: $\alpha_L + \alpha_R$

(a)

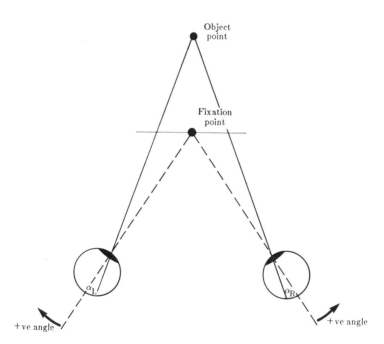

Disparity $= -\alpha_L - \alpha_R$

$$= -(\alpha_L + \alpha_R)$$

(b)

Fig. 16.13. (a) An object point possesses convergent disparity when it lies in front of the fixation point; otherwise, as shown in (b), the disparity is divergent.

3. Before this operation, the image is smoothed by convolution with a Gaussian function of the form

$$G(x, y) = {}^2 \exp - \left(\frac{x^2 + y^2}{2\sigma^2} \right) \tag{16.43}$$

4. The operations in steps 2 and 3 can be combined as a single convolution of the image with the following function

$$\nabla^2 G(x, y) \equiv \frac{r^2 - 2\sigma^2}{\sigma^4} \exp(-(r^2/2\sigma^2)) \tag{16.44}$$

where

$$r^2 = x^2 + y^2 \tag{16.45}$$

In Figure 16.14a we have shown a three-dimensional plot of this operator, and in Figure 16.14b there is a section of this function in the plane $x = 0$. Following Grimson,[12] the width of this function is represented by the distance between the first zeros on either side of the origin. This width, denoted by w_{2D}, is related to σ by

$$w_{2D} = 2\sqrt{2}\sigma \tag{16.46}$$

$\nabla^2 G$ is called the *primal sketch operator*. We can refer to the result obtained by applying the primal sketch operator to an image as the convolved sketch.

Attributes of Candidate Points

Each candidate point is characterized by both the sign change that is represented by the zero-crossing and also by the orientation of the local zero-crossing contour. A positive-to-negative sign change in the convolved sketch implies a low-to-high transition in the gray levels in the original image, and vice versa. To illustrate, in Figure 16.15 we have shown hypothetical gray levels in the vicinity of a pixel A. If we apply the primal sketch operator to these gray levels, the output will contain positive values on the high side of A_1A_2 and negative values on the other, implying the presence of zero-crossings between the two. (For subpixel accuracy work, the locations of zero-crossings can be estimated by interpolation between the positive and the negative values on each row of the primal sketch.) For our discussion here, as each row is scanned left to right, we associate a zero-crossing with a pixel—provided in the convolved result the next value is of the opposite sign. If the next value is zero in the convolved sketch, then the value after that should be of opposite sign. Therefore, in the row marked R in Figure 16.15, we will associate a zero-crossing with the pixel at A, which makes A one of our candidate points. The zero-crossing at this candidate point is characterized by a high-to-low transition in image gray levels; the direction of A_1A_2 is the orientation of the local zero-crossing contour.

An elementary way to estimate the orientation of the local zero-crossing contour is to examine the locations of zero-crossings in one or two lines above and below the candidate point in question. Grimson, although not explaining precisely how he computes the direction of the local zero-crossing contour, uses six quantization levels to represent them (Figure 16.16a). In our implementation, we have combined the sign change and the orientation into a single representation by quantizing the local gradient over 360°, as shown in Figure 16.16b. For this purpose we first compute the digital gradient map for all image points by using the Sobel operator.* This is done by first computing at each pixel the following two as estimates of the x- and y-components of the local gray-level gradient

$$X = (A_3 + 2A_4 + A_5) - (A_1 + 2A_8 + A_7)$$

and

$$Y = (A_1 + 2A_2 + A_3) - (A_7 + 2A_6 + A_5)$$

---

* Since the gradient always points in the direction of maximum change, its direction is always perpendicular to the orientation of the local zero-crossing contour. For example, in Figure 16.15, the direction of the gradient is from B_1 to B_2. A variety of operators are available for computing the gradient of images. We have used the Sobel operator.[36] It is equivalent to use either the local gradient or the local zero-crossing orientation as a measure of the direction of the local edge.

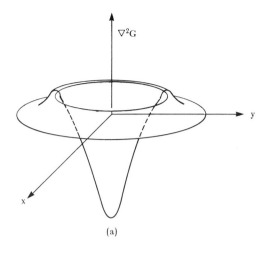

(a)

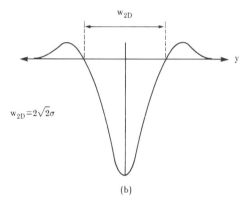

(b)

Fig. 16.14. (a) The primal sketch operator $\nabla^2 G$. (b) A section of the primal sketch operator: its values of the y-axis in (a). The parameter w_{2D} determines the extent of smoothing introduced by the operator.

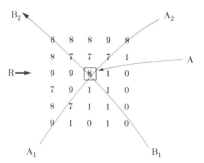

Fig. 16.15. Hypothetical gray levels in the vicinity of a pixel marked A. If we ignore subpixel effects, the local zero-crossing contour will be along the line A_1A_2; the direction of the local gradient at A will be from B_1 to B_2.

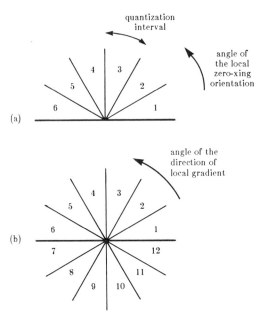

Fig. 16.16. (*a*) The orientation of the local zero-crossing contour is quantized into six angular intervals as shown here. (*b*) To automatically account for the polarity of the zero-crossing as well, the orientation of the local zero-crossing contour (or, equivalently, the direction of the local gradient) is quantized into 12 angular intervals.

where A_m's are the neighbors, numbered as indicated in Figure 16.17, of the pixel. The direction θ of the gradient is then determined from the following relationship

$$\theta = \tan^{-1}\frac{Y}{X}$$

At the location of a verified zero-crossing we compute the orientation by local averaging of the gradient map over 3×3 neighborhoods. This orientation is then classified into the 12 angular intervals, that is, 30° intervals shown in Figure 16.16.

Choice of w_{2D} (*The critical concept*)

How one should go about choosing a value for the width of the primal sketch operator could perhaps be called a critical concept of the Marr-Poggio paradigm, because it is through this concept that *one can explain most easily the need for multichannel implementation.* In a multichannel implementation, the candidate points are first found by using a large w_{2D}. Matching these candidate points in the left and the right images yields a coarsely sampled range map of the scene. This information controls the matching of candidate points from finer channels (i.e., channels characterized by smaller values of w_{2D}). This idea is explained in greater detail later, using cartoon images. We first examine the output of a stereo matcher when given the candidate points produced by a large w_{2D}.

A_1	A_2	A_3
A_8	$f(m,n)$	A_4
A_7	A_6	A_5

Fig. 16.17. Pixel numbering scheme for the Sobel operator.

Assume that for a given choice of w_{2D} the candidate points in the left and the right images are as shown in Figure 16.18. Further assume that we know a priori the maximum disparity, and let us denote it by d_{max}. We have selected a candidate point, marked as A, in the left image for matching; assume that its true match is the candidate point B in the right image. To find this true match, we first transfer the coordinates of A into the right image and mark the resulting point by X. We then construct a search neighborhood around X, the size of this neighborhood being d_{max} on either side of X. Concerning the matching process, there are three important things to bear in mind:

1. All the candidate points in the search neighborhood that represent the same sign change and the same zero-crossing orientation are considered to be potential matches for A. Of all the potential matches, we assume that only one is a true match. All others will be considered to be *false targets*. (For most candidate points in the left image, there will only be one matching element in the right image. However, near large-range discontinuities that cause occlusions in only one camera, one candidate point of one image may correspond to several in the other. For lack of procedures to handle such points adequately, we will ignore them. The number of such points will be greater when the camera separation is longer. Camera separation is a compromise between the desire to enhance range-estimation accuracy by using a large distance between the cameras, and the need to minimize, by reducing camera separation, those regions of the scene that appear occluded in one of the two cameras.)

2. The number of false targets depends on both the size of the search neighborhood and w_{2D}. Since the zero-crossings for the most part cannot be closer than w_{2D}, the larger the w_{2D}, the fewer the candidate points in the search neighborhood. Since there should be only one match within the search neighborhood, and since the size of the search neighborhood is approximately the maximum expected disparity, to minimize the problem of false targets w_{2D} should be approximately equal to d_{max}.

3. By setting w_{2D} equal to d_{max} we will get only one range value per d_{max} distance. Usually this will represent a coarsely sampled range map of the scene.

An Illustration with Cartoon Figures

We now develop several more points about the algorithm with the help of cartoon figures. Let us assume that the scene consists of two trees as shown in Figure 16.19. We will assume that the largest diameter of each tree is much less than $D_3 - D_1$, which is the distance between them. We further assume that the fixation point is at the midway between the two trees, that is, at $D_2 = (D_1 + D_3)/2$. The maximum disparity d_{max} here corresponds approximately to $D_3 - D_2$ (which is the same as $D_2 - D_1$).

Let the left and the right images look as illustrated in Figure 16.20a. We now smooth the images with a Gaussian blur for which w_{2D} is equal to d_{max}. The result is depicted in Figure 16.20b. After smoothing, the gray-level variations within each tree *cannot* occur at a scale smaller than d_{max}. Taking

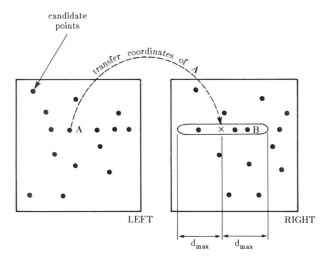

Fig. 16.18. To find a match for the candidate point A in the left image, we construct a search neighborhood around its transferred coordinates (marked X) in the right image and accept all the candidate points therein as potential matches.

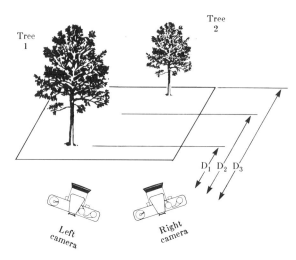

Fig. 16.19. A cartoon scene of two trees separated in depth explains some aspects of the algorithm.

the Laplacian yields the result in Figure 16.20c, where we have shown the zero-crossing contours. *By convolving the image with a primal sketch operator of width w_{2D}, we have ensured that most zero-crossing contours will be at least w_{2D} apart from one another.* We now concentrate our attention on the zero-crossings that are labeled in Figure 16.20c. To find the match for A_1, we transfer its coordinates to the right image and construct a search neighborhood around this location. Since in this search neighborhood only B_1 will be found (idealizing, of course), that is unambiguously the match for A_1. Similarly, for A_2 we will find a match at B_2, and so on. Because of our assumption that the diameter of each tree is much smaller than the distance between them, the following will hold:

$$A_1 \sim B_1 \simeq A_2 \sim B_2 \simeq A_3 \sim B_3$$
$$\simeq \text{disparity that corresponds to the depth of Tree 1 from the fixation point} \quad (16.47)$$

and

$$S_1 \sim T_1 \simeq S_2 \sim T_2 \simeq S_3 \sim T_3$$
$$\simeq \text{disparity that corresponds to the depth of Tree 2 from the fixation point} \quad (16.48)$$

where $A_1 \sim B_1$ is the disparity obtained by matching A_1 with B_1, and so on. That is, the disparity at A_1, A_2, and A_3 is approximately the same, and it corresponds to the depth of Tree 1 from the fixation point. And that the disparity at S_1, S_2, and S_3 is also approximately the same, and it corresponds to the depth of Tree 2 from the fixation point.

So by setting w_{2D} equal to $d_{\max}$, we eliminated the false target problem, and we are now able to determine the depth between the trees. However, our range map is coarse since we only have a half-dozen points on each tree. We do not have enough points to determine the shape of each tree. And also, A_1, A_2, A_3, and so on, may not correspond to any specific points on the trees because they represent points at which the gray level changes in a highly defocused image.

Vergence and the Concept of Multichannel Stereo

To increase the number of range points on Tree 1, we do the following:

1. We use the calculated average disparity for Tree 1 to bring it into vergence. This means that in the computer we change the fixation point from its original site to somewhere near Tree 1. In software this may be accomplished by simply adding an offset to the horizontal coordinates of the right image, the offset being equal to the average of the disparities at A_1, A_2, and A_3.
2. We now convolve both the left and the right images with a new primal sketch operator, the width w_{2D} of which roughly corresponds to the largest expected diameter of the trees. This new w_{2D} is much smaller than that used before. Therefore we should now have a much denser set of candidate points in each image. This is illustrated in Figure 16.20d.

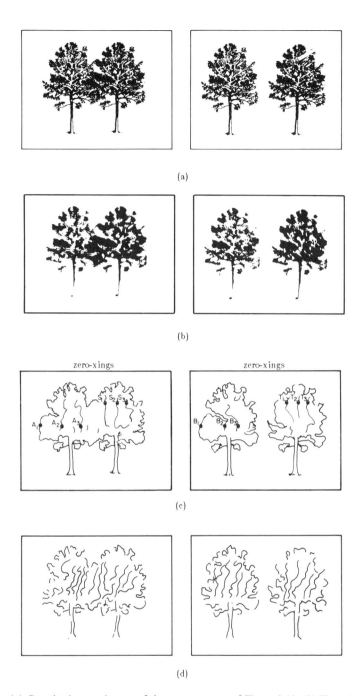

(a)

(b)

(c)

(d)

Fig. 16.20. (*a*) Contrived stereo images of the cartoon scene of Figure 3.19. (*b*) These images depict the results that might be obtained be convolving those shown in (*a*) with a Gaussian blur for which w_{2D} is equal to the maximum expected disparity. (*c*) The zero-crossings shown here are obtained from the Laplacian of the images in (*b*). (*d*) These zero-crossings are obtained by using a Gaussian blur of much smaller w_{2D}.

3. To eliminate the false target problem, the size of the search window is limited to the new w_{2D}.

4. As a result we get a dense set of range values on Tree 1. We should not be able to make any matches in the area of Tree 2.

To increase the number of range points on Tree 2, we can adopt an identical procedure and bring Tree 2 into vergence.

The foregoing procedure constitutes a two-channel stereo for range-mapping the scene of Figure 16.19. The first channel helps us separate Tree 1 from Tree 2, and the second channel gives us additional range values on each tree and therefore defines them better. If we wanted, we could add a third channel, with an even smaller w_{2D}, and separately bring into vergence specific parts of each tree for an even superior definition.

The human visual system is known to possess five different channels for disparity calculation.[13] The values of w_{2D} for these channels are approximately 63, 35, 17, 9, and 4 pixels. By a pixel here is meant the size of a foveal receptor; one such receptor corresponds roughly to an angular interval of 0.4 minutes of the arc. Therefore if we digitize a visual angle of 4° on the side into a 650 × 650 matrix, we will match the sampling capability of the fovea of the human eye. The filter sizes given apply to these sampling rates. (For a robot vision system, a more practical visual angle will be 45°. For such a wide angle it is not possible to sample the images at rates used by the human eye. Using images perceived by the humans as a comparison, the images perceived by a robot will always be undersampled. Such undersampling leads to aliasing artifacts in the representation of edges. This can lead to distortion in edge-based stereo algorithms, of which the current algorithm is an example.)

Vergence for Complex Scenes

The procedure just described would be implementable if the scene consists of a few nonoverlapping (in range) objects. For more complex scenes, the procedure must be somewhat modified. In his computer implementations for simulating human stereopsis, Grimson uses the disparities from the coarse channels to bring image regions within the range of fusion of the finer channels. However, this is not done on an object-to-object basis as discussed before, but on a region-by-region basis. In Grimson's implementation, suppose we are matching the zero-crossings of the $w_{2D} = 9$ channel, each 25 × 25 region of the image is analyzed, and if less than 70% of the zero-crossings in a given region are matched, that region is considered to be out of range of fusion for the channel. In that event no disparities for that region as calculated by the $w_{2D} = 9$ channel are accepted. *Disparity values from the coarser channels are now used for vergence control to bring this region within the range of the finer channel.* As before, vergence in the computer program consists of adding an offset to the coordinates of the pixels of the region; the offset being equal to disparity as determined by the coarse channels.

In our implementation where, for robot vision, images are deliberately undersampled (128 × 128 or 256 × 256 representations for a visual angle of 45°), for a two-channel stereo implementation for a total depth of field of 3 m (10 ft), we have obtained best results with 10 × 10 windows for vergence control. And although the probabilistic considerations dictate a 70% threshold for accepting a region to be within the fusion range of a channel, we have obtained superior results with a 50% threshold.

Note that the concept of vergence is similar to when humans examine a scene without focusing at any particular region, get a rough idea of the relative locations of the major objects in the scene, and then fixate on each object of interest. In the computer, vergence consists merely of shifting one image with respect to another.

Resolution of Ambiguous Matches

In the cartoon illustration just presented, we idealized by assuming that if in the right image we limited the search neighborhood to an interval of w_{2D} on either side of the coordinates of the left-image candidate point, we would eliminate the false target problem. For real images, using Rice's formulation[35] for the probability distribution of separation between zero-crossings, and taking into account the orientation of zero-crossing contours used in the matching process, it may be shown[12] that when the search neighborhood is limited to a distance of w_{2D} on either side of X in Figure 16.18, we have a 20% chance of encountering two potential matches; that is, two zero-crossings of roughly the same orientation.

If there is more than one potential match, the following procedure is used for disambiguating between them. All the potential matches within the search neighborhood are divided into three pools: (1) the divergent disparity pool, which consists of all the potential matches that are to the left of X in Figure 16.18 by more than 1 pixel; (2) the convergent disparity pool, which consists of all the potential matches to the right of X by more than 1 pixel in the same figure; and (3) the zero-disparity pool, which consists of all the potential matches that are within a pixel of X.

If there is only one potential match in all three pools, then that match is accepted and the disparity associated with this match computed and assigned to the candidate point in question. If more than

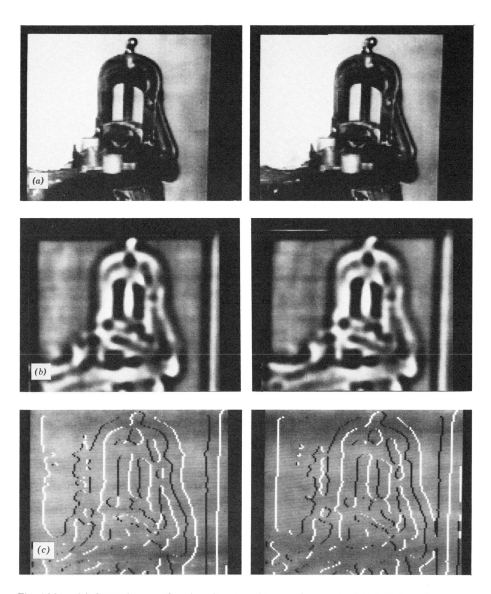

Fig. 16.21. (a) Stereo images of a pipe vise at a distance of approximately 6 ft from the camera baseline. (b) Images obtained by convolving those in (a) with a primal sketch operator characterized by $w_{2D} = 8$. (c) Zero-crossings obtained from the images in (b).

one potential match is found in any of the three pools, then *no* match is assigned to the candidate point; that is, the candidate point in question is left unmatched. If more than one pool contains a potential match, then the candidate pixel is considered to have ambiguous matches. The ambiguity is resolved by using what is known as the *pulling effect,* which consists of examining the unambiguous disparities within a neighborhood (10×10 in our implementation) of the candidate point in question and of the potential matches available, choosing one that is dominant within the neighborhood. To illustrate, if there are two potential matches for a candidate point, one from the convergent pool and one from the zero-disparity pool, and if convergent disparities are in a majority in a 10×10 neighborhood around the candidate point, we select the choice from the convergent pool. Of course, if none among

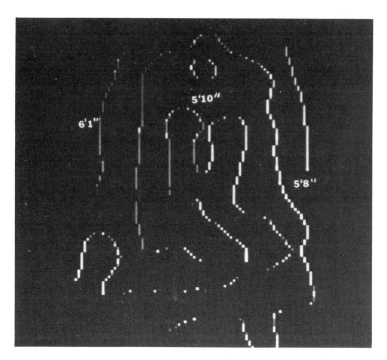

Fig. 16.22. The range map obtained from the stereo images of Figure 16.21. The computed range values for some of the points are shown in the figure. The lighter a pixel, the closer it is to the camera baseline.

the choices available corresponds to the dominant type within the neighborhood, the candidate point is left unmatched. The pulling effect rule reflects the property that, for the most part, disparity can only vary smoothly. It changes abruptly only at range discontinuities.

Results Obtained with a Two-Channel Stereo Algorithm

We first show the results obtained with a single-channel implementation of the foregoing algorithm. Shown in Figure 16.21a is a stereo pair of images (128 × 128) of a pipe vise at a distance of approximately 1.8 m (6 ft) from the camera baseline. When these images are convolved with the primal sketch operator with $w_{2D} = 8$ pixels, we get the convolved sketches of Figure 16.21b. In Figure 16.21c we have shown the zero-crossings obtained from the convolved sketches. The zero-crossings are gray-scale encoded according to the orientation of the local zero-crossing contours. The matching part of the algorithm yields the range map shown in Figure 16.22.

We now illustrate the algorithm with a two-channel implementation for a 45° visual angle and a depth range of approximately 3 m (10 ft). In Figure 16.23a, we have shown a stereo pair of images, which consists of 128 × 128 matrices. The camera fixation was midway between the two stools, and the furthest stool was approximately 2 m (7 ft) from the cameras. The maximum depth translated into a maximum disparity of 12 pixels. To extract the zero-crossings, the images were convolved with the primal sketch operator of Eq. (16.44) with w_{2D} equal to 12 pixels. The resulting images are shown in Figure 16.23b. Figure 16.23c illustrates the zero-crossings obtained from Figure 16.23b. Again, the local orientation of the zero-crossing contour has been encoded into the gray-scale depiction of the zero-crossings. Figure 16.24a shows the disparity map obtained from the coarse-channel zero-crossings of Figure 16.23c. The gray levels in this map are proportional to the range estimated by the coarse channel; the darker the pixel, the greater its distance from the camera baseline. Figure 16.24b shows the histogram of the coarse-channel disparities shown in Figure 16.24a. One might associate objects at different depths with different lobes of the histogram. The histogram in Figure 16.24b may therefore be interpreted as indicating the presence of two objects, one with disparities around 2 and 3, and the other with disparities around −4 and −5. Positive disparities correspond to object points behind the fixation point, and vice versa. These coarse disparities are used for vergence needed to match the outputs of the finer primal sketch operator.

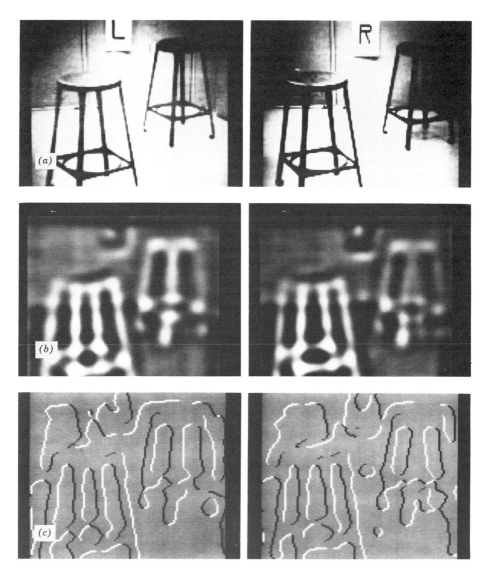

Fig. 16.23. (*a*) Stereo images of a scene consisting of two stools. (*b*) Images obtained by convolving those in (*a*) with a primal sketch operator characterized by $w_{2D} = 12$. (*c*) Zero-crossings obtained from the images in (*b*).

In Figure 16.25*a* we have shown the results obtained by filtering the images of Figure 16.23*a* with a primal sketch operator characterized by $w_{2D} = 6$ pixels. The zero-crossings extracted from these images are illustrated in Figure 16.25*b*. Before these zero-crossings can be matched we must exercise vergence. How this is done is illustrated in Figure 16.26, where we have shown two corresponding pixels A_1 and A_2 that supposedly were matched by the coarse channel. We now construct 12×12 neighborhoods around these pixels. The fine-channel candidate points within these neighborhoods are then matched as before. To the disparity values obtained by matching the fine-channel candidate points we add the coarse-channel disparity obtained from A_1 and A_2. The histogram of the final disparities thus obtained is shown in Figure 16.27. In Figure 16.28 we have shown the final depth map obtained; the darker the pixel, the greater its distance from the camera baseline. The adequacy of this depth map for the purpose of automatic scene interpretation would, of course, depend on how narrowly we constrict the totality of scenes to be understood by the computer.

16.3.3. Barnard-Thompson Algorithm

The MPG (Marr-Poggio-Grimson) algorithms presented in the preceding section can be characterized by the following three features:

1. **Discreteness:** The candidate points selected for matching are distinctive in the sense that they represent locations of transitions in the image gray levels at a scale determined by the width of the primal sketch operator.

2. **Similarity:** This is a measure of how similar the candidate points are. In the MPG algorithm, the similarity within a search neighborhood is measured by comparing the orientation of the local zero-crossing contour.

3. **Consistency:** This is a measure of the continuity of the computed disparities. Since disparities can only be discontinuous at edges that cause occlusions in one camera, this property can be used to suppress those matches that go against the local evidence. In the MPG algorithm, consistency was used for resolving between multiple potential matches for a candidate point (disambiguation).

The Barnard-Thompson algorithm,[5] although neatly fitting into the discreteness-similarity-consistency framework, uses different criteria for each of these properties. Only those candidate points are selected that represent a high gray-level variance in all directions. Selection is made by first surrounding each pixel by a 5×5 neighborhood, computing the gray-level variances in the neighborhood in four directions (horizontal, vertical, and two diagonal), and retaining only the minimum of these variances. The retained value of the variance is called the *initial interest value* of the pixel in question. Clearly a high initial interest value implies that the pixel has a large variance in all four directions. The *final interest values* are obtained by setting to zero the initial values of all but the local maxima. In this manner only those pixels are assigned interest values that are not only the centers of locally large gray-level variations, but also represent larger variations than any of their neighboring points. In Figure 16.29 we have shown the initial interest values assigned to the pixels of the images in Figure 16.23a. For the assignment of the final interest values, we have zeroed out all but the maximas. Of all the candidate points thus obtained in all the rows, only those with final interest values above a predetermined threshold are

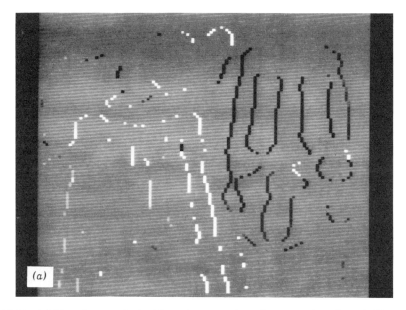

Fig. 16.24. (*a*) Disparity map obtained from the coarse-channel zero-crossings shown in Figure 16.23*c*. The darker a pixel, the greater its range from the camera baseline. (*b*) Histogram of the coarse-channel disparities shown in (*a*).

retained. The value of the threshold depends on the number of candidate points desired for the purpose of matching. Figure 16.30 shows an example of the retained candidate points obtained in this manner.

Perhaps the most distinctive feature of the Barnard-Thompson algorithm is the manner in which the candidate points from the left and the right images are matched. To find a match for a candidate point in the left image, we transfer its coordinates to the right image and then construct a $d_{max} \times d_{max}$ neighborhood around this point as illustrated in Figure 16.31, where d_{max} is the maximum expected disparity. All candidate points in this neighborhood are considered potential matches for the left-image candidate point in question. Let (x_m, y_m) denote the coordinates of the mth candidate point in the left image; and let (x_i, y_i) be the coordinates of the ith possible match within the $d_{max} \times d_{max}$ window (Figure 16.31). With the ith possible match we associate a disparity of $\mathbf{d} \equiv (x_i - x_m, y_i - y_m)$.

The essence of the algorithm lies in assigning a probability $P_m(\mathbf{d})$ to each possible disparity $\mathbf{d}$ for the mth candidate point, then successively updating these probabilities by using the consistency property previously mentioned, and finally retaining the match with the highest probability. Only one final match is accepted for each candidate point, and, of course, some candidate points may have no matches

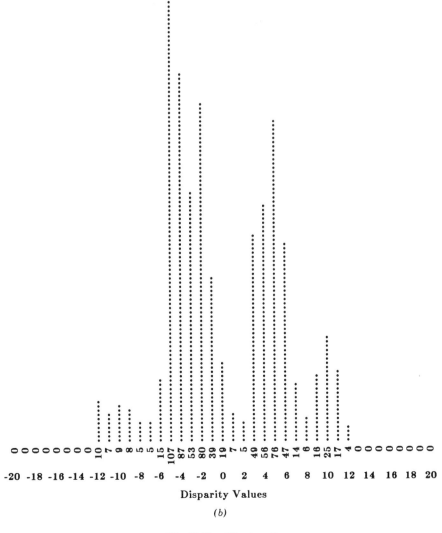

Disparity Values

(b)

Fig. 16.24.   (Continued)

Fig. 16.25. (*a*) The results of the images of Figure 16.23*a* convolved with a primal sketch operator with $w_{2D} = 6$. (*b*) Zero-crossings extracted from (*a*).

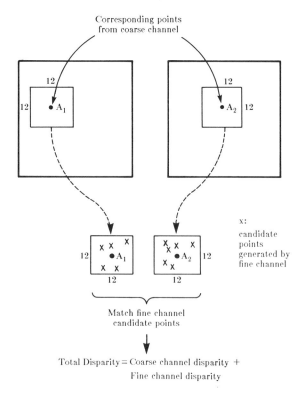

Corresponding points
from coarse channel

x:
candidate
points
generated by
fine channel

Match fine channel
candidate points

Total Disparity = Coarse channel disparity +
Fine channel disparity

Fig. 16.26. The procedure used for implementing vergence.

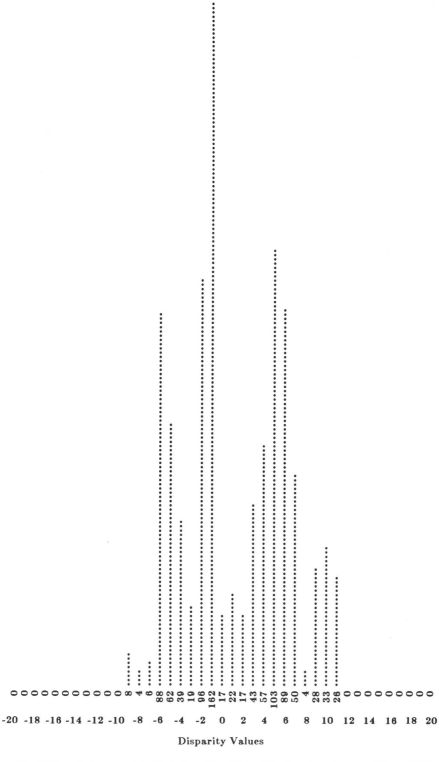

Fig. 16.27. Histogram of the final disparities obtained for the stereo images of Figure 16.23*a*.

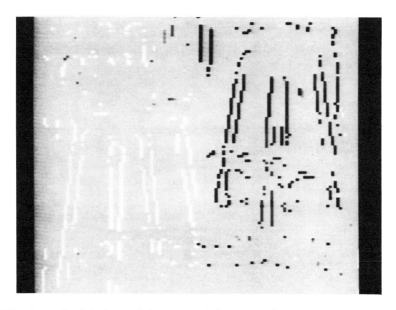

Fig. 16.28. A gray-level depiction of the computed depth map; the darker a pixel, the greatest its distance from the camera baseline.

at all. For example, left-image candidate points that lie in regions that are occluded to the right camera will not have any matches.

The following procedure is used for constructing the initial probabilities for the disparities. We first determine how similar the candidate point (x_m, y_m) is to each of the possible matches within the window in Figure 16.31. Similarity is tested by summing the squares of the differences between a 5×5 window around the (x_m, y_m) point in the left image, and a similar window around the potential match in the right image. Let $S_m(\mathbf{d})$ be this measure for the disparity $\mathbf{d}$. Note that the *smaller* the $S_m(\mathbf{d})$, the more similar the potential match is to the candidate point. Therefore the smaller the $S_m(\mathbf{d})$, the larger should be the associated probability $P_m(\mathbf{d})$. To construct the probability function, we first generate the following set of weights at each (x_m, y_m):

$$W_m(\mathbf{d}) = \frac{1}{1 + C \cdot S_m(\mathbf{d})} \tag{16.49}$$

for some positive constant C (Barnard and Thompson used a value of 10). The advantage of these weights is that regardless of the choice of C they always lie in the interval $[0, 1]$, and a large weight means a more similar potential match. These weights cannot be used directly as probabilities because,

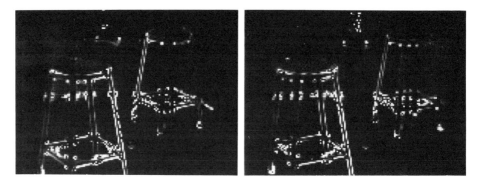

Fig. 16.29. Gray-scale depiction of the initial interest values assigned to the pixels of the stereo imaging of Figure 16.23*a*.

Fig. 16.30. Candidate points retained for the stereo image of Figure 16.29.

first, their sum does not equal unity and, second, they do not directly give us a value for probability of no match.

To convert these weights into probabilities, we first estimate the probability of no match. Let the distinguished label $\mathbf{d^*}$ symbolize the existence of a null disparity (no match), and let $P_m(\mathbf{d^*})$ be the corresponding probability. If we use the superscript 0 to denote the initial value (before using the consistency property) of this (and other) probabilities, we set

$$P_m^0(\mathbf{d^*}) = 1 - \max_{\mathbf{d}} \left[W_m(\mathbf{d}) \right] \tag{16.50}$$

The rationale here is that the largest value of $W_m(\mathbf{d})$ probably corresponds to the true match, and if we treat it as such for a moment, then the right-hand side above must be the probability of a no match. (There is clearly an element of hand-waving here. It must do for lack of a better approach to constructing P_m^0.)

Given $P_m^0(\mathbf{d^*})$, the initial probabilities for other disparities can be estimated by using the Bayes' rule

$$P_m^0(\mathbf{d}) = P_m(\mathbf{d}|m) \left[1 - P_m^0(\mathbf{d^*}) \right], \qquad \mathbf{d} \neq \mathbf{d^*} \tag{16.51}$$

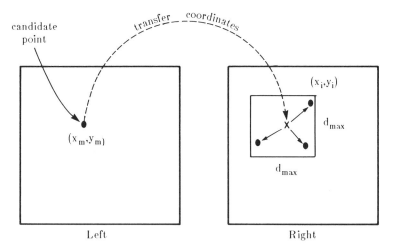

Fig. 16.31. To find a match for a candidate point in the left image, we construct a search neighborhood around its transferred coordinates in the right image.

where $P_m(\mathbf{d}|m)$ is the conditional probability of the disparity at point (x_m, y_m) being equal to $\mathbf{d}$ given that this candidate point is matchable; and $[1 - P_m^0(\mathbf{d}^*)]$ is the probability that this (x_m, y_m) is matchable. The conditional probability is estimated from

$$P_m(\mathbf{d}|m) = \frac{W_m(\mathbf{d})}{\sum_{\mathbf{d}' \neq \mathbf{d}^*} W_m(\mathbf{d}')} \qquad (16.52)$$

An important consequence of this definition is that if we plug this estimate for the conditional densities in Eq. (16.51), we can easily show that the sum of the probabilities for all disparities at (x_m, y_m) (including the null disparity) is equal to unity. In Figure 16.32a, we have displayed the initial probabilities for the candidate points shown in Figure 16.30. The thickness of each line is proportional to the probability of the corresponding match. The darkness of each point is proportional to the probability of the candidate point being matchable.

The probabilities are updated by using the consistency property. The initial probabilities presented depend on only the local neighborhoods around the pixels being paired together. We now increase the probabilities of those disparities that occur often in a region, and decrease the probabilities of those that do not. To account for sampling effects, we must first define what we mean by the disparities being similar. Two disparities, $\mathbf{d}$ and $\mathbf{d}'$, will be considered to be similar provided

$$\max\left(|\mathbf{d}_x - \mathbf{d}'_x|, |\mathbf{d}_y - \mathbf{d}'_y|\right) \leq 1 \qquad (16.53)$$

which says that if either the x- or the y-components of the two disparities differ by no more than one pixel, then they are essentially the same. To compute the new probabilities for the disparities at

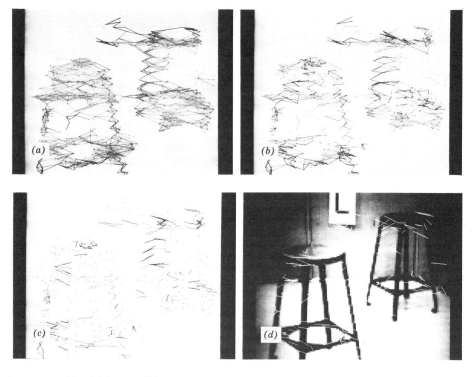

Fig. 16.32. (a) Initial probabilities assigned to possible disparities at the candidate points in the left image of Figure 16.30b. The darkness of each point is proportional to the probability of the candidate's being matchable. The thickness of each line emanating from a bright dot is proportional to the probability of the corresponding disparity. (b) Probabilities updated by using the consistency measure after two iterations. (c) Probabilities after five iterations. (d) Final disparities are shown superimposed on the left image of the stereo pair.

(x_m, y_m), we examine the disparities at neighboring candidate points in the left image and reinforce those probabilities that correspond to frequently occurring disparities while diminishing the others. To explain how this is accomplished, let us assume that we have applied the consistency property k times to yield the probabilities $P_m^k(\mathbf{d})$. With another application of this property we wish to further update the probabilities to yield $P_m^{k+1}(\mathbf{d})$. To obtain this new set of probabilities, we check the disparities of all left-image candidate points in an $R \times R$ neighborhood of (x_m, y_m). (Barnard and Thompson have used $R = 15$ pixels.[5]) For each candidate point (x_n, y_n) in this neighborhood, we compute the following sum

$$Q_m^k(\mathbf{d}) = \sum_{\text{all } n \text{ in } R \times R \text{ and } n \neq m} P_n^k(\mathbf{d}) \qquad \mathbf{d} \neq \mathbf{d*} \qquad (16.54)$$

where by "all n in $R \times R$" we mean the left-image candidate points in an $R \times R$ neighborhood around (x_m, y_m); $n \neq m$ means: do not include the point (x_m, y_m). In actual implementation, since all disparities satisfying the criterion in Eq. (16.53) are considered similar, the argument in the summation is replaced by a sum over all similar disparities at each candidate point (x_n, y_n).

The quantity $Q_m^k(\mathbf{d})$ is a measure of how consistent the disparity $\mathbf{d}$ is with other disparities in a local neighborhood around (x_m, y_m). If $Q_m^k(\mathbf{d})$ is zero, this means that the disparity $\mathbf{d}$ does not exist at any other candidate point in the neighborhood. A large value means that there are several other pairings within the neighborhood that possess the same disparity.

We now use the consistency measure $Q_m^k(\mathbf{d})$ to upgrade the set of probabilities, $P_m^k(\mathbf{d})$, into a new set $P_m^{k+1}(\mathbf{d})$ by first creating a non-normalized set of numbers

$$\hat{P}_m^{k+1}(\mathbf{d}) = \left(A + B \cdot Q_m^k(\mathbf{d}) \right) P_m^k(\mathbf{d}) \qquad \mathbf{d} \neq \mathbf{d*} \qquad (16.55)$$

and

$$\hat{P}_m^{k+1}(\mathbf{d*}) = P_m^k(\mathbf{d*}) \qquad (16.56)$$

The new probabilities are generated by normalizing these numbers

$$P_m^{k+1}(\mathbf{d}) = \frac{\hat{P}_m^{k+1}(\mathbf{d})}{\sum_{\mathbf{d}} \hat{P}_m^{k+1}(\mathbf{d})} \qquad (16.57)$$

The parameters A and B are of great conceptual importance because they determine how the consistency measure at (x_m, y_m) will change the probabilities assigned to the disparities at that point. If A is nonzero, a zero value for the consistency measure for a particular disparity will not force its new probability to go to zero. This is desirable because, with successive updates, information may propagate to (x_m, y_m) from more distant candidate points that would cause the consistency measure to be nonzero for this particular disparity. On the other hand, the constant B determines at what rate a given value of the consistency measure will affect the disparity probabilities. Note that the probability of (x_m, y_m) being unmatchable (i.e., the probability of $\mathbf{d*}$) is affected only by the normalization in Eq. (16.57).

To make the implementation of the algorithm efficient, whenever the probability of a disparity drops below 0.1, it is simply purged. If through this process a candidate point (x_m, y_m) loses all its potential matches, it is declared unmatchable, that is, we set $P_m(\mathbf{d*}) = 1$. Although this procedure should be repeated until the probabilities reach steady state, it is arbitrarily stopped at 10 iterations; those candidate points possessing a disparity of probability of 0.7 or greater are considered to be matched. The matching at other candidate points is considered to be unresolved.

Using these criteria for purging and final acceptance of probabilities, we have shown in Figure 16.32b-d the probability updates obtained by repeated application of the consistency property to the initial values of Figure 16.32a. After 10 iterations, candidate points that still possess multiple matches are considered unmatchable, and updating is stopped. The final disparity values thus obtained are illustrated pictorially in the last of the sequence in Figure 16.32.

16.3.4. Baker-Binford Algorithm

The previous two algorithms are both examples of *edge-based stereo*, since the candidate points for matching represent changes in image intensity. This concept is *not* used in the most common application of stereo algorithms of the past, which is the mapping of terrain elevation contours from aerial images. The algorithms for that application are *area based*, in which for a succession of windows (typically 10×10) in the left image, one locates by cross-correlation the corresponding windows in the right image.[15,37] The disparities estimated from the coordinates of the centers of the corresponding windows are then assigned to the central pixels of the left-image windows. (Instead of cross-correlation, it is computationally less demanding to measure similarity from the sum of the squares of the differences; however, the penalty paid is the sensitivity to contrast differences between the left and the right images.)

Area-based stereos are based on the premise that the corresponding areas in the left and the right images exhibit similar gray-level variations, a property known as *photometric invariance.* This assumption is reasonably justified in photogrammetry where, for the most part, images consist of smoothly varying terrains. However, in the vicinity of sharp range discontinuities, which occur often in robot vision, photometric invariance does not hold, since in the vicinity of such discontinuities some areas of the left image are occluded in the right image, and vice versa. On the other hand, edge-based stereo is based on the premise that stereo images are *geometrically invariant,* which can be explained by saying that along any given scan direction the edges appear in the same order in the left and the right images (although owing to occlusion, some edges that appear in the left image may not appear in the right image, and vice versa). Because of the different underlying premises, edge-based stereo is strong where area correlation is weak, and conversely.

Although an edge-based stereo is incapable of generating a range map for all points in a scene, its advantage is that an edge can be located with subpixel accuracy. A change in image intensity at the site of an edge corresponds to a zero-crossing in the Laplacian of the image. A zero-crossing implies a sign change from one pixel to the next; interpolation between two such adjacent pixels in the Laplacian of the image leads to subpixel accuracy location of an edge. (This is also the reason for the hyperacuity of the human visual system, which refers to our ability to compute the position of a feature in an image to about 5 seconds of arc, whereas the resolution of a foveal receptor is approximately 0.4 minutes of arc. Marr, Poggio, and Hildreth[13] argue that to account for such hyperacuity, one must postulate an interpolation process to isolate the zero-crossings to an accuracy finer than that of the receptor spacing.) Subpixel accuracy is not possible with area-based stereo algorithms. The precision with which the disparity can be calculated here is much poorer and is inversely proportional to the size of the windows used for matching. However, area-based methods do yield depth information at a denser set of points than the edge-based methods.

Baker-Binford algorithm[3] combines the notions of edge-based and area-based techniques to yield disparity values practically everywhere. The first part of the algorithm establishes correspondences between the edges in the two images. The edge descriptions basically highlight the structure of the scene, providing rather sparse disparity measures. The second part of the algorithm then provides fuller stereo detail by correlating image intensities, using for local vergence the information supplied by edge-based correlations. To ease the problem of matching intensity patterns, the unpaired edges are matched on a segment-by-segment basis, each segment of the scan line lying between the edges already paired by the edge-based part of the algorithm. These pairings serve to fill in the gaps of the primary edge-based correlation.

In the interest of robustness and efficiency, the edge-based part of the algorithm analyses the images at two different levels of detail. This *coarse-to-fine analysis* approach, similar in spirit to the multichannel feature of the MPG algorithm, consists of, first, low-pass filtering of the images, matching the edges in the reduced resolution versions thus obtained; and then using the local disparities thus calculated to bring finer edge detail into rough correspondence. For matching, edge points on each line of the reduced resolution images are characterized by the gray levels on two sides of the edge, the contrast sign, and the slope of the edge segment. In the matching procedure, which is based on the *Viterbi algorithm,* these attributes are combined with linear weighting. For full-resolution edge detail, although the same procedure is used for matching, the edge points are further characterized by the local edge angle and, of course, the local disparity as calculated from the reduced resolution images.

The algorithm clearly possesses the discreteness and similarity properties mentioned in the preceding section. The reader might now wonder how the consistency is brought into play. Note that the correlations for the edges are done separately for each line in the images. In other words, only the information contained on one image line from the left image and the corresponding line from the right image is used to correlate the edge points contained on those lines. (When the camera baseline is parallel to the horizontal, these image lines are the horizontal scan lines. In general, the lines in the two images of a stereo pair that contain the corresponding points, fall on the *epipolar line,* which is a projection of the camera baseline on a joint image plane containing both the left and the right images.) The information on the adjacent lines is used to establish *edge connectivity constraint* for weeding out the edge correspondences that violate object surface continuity. The edge connectivity constraint says that a connected sequence of edges in one image should be seen as a connected sequence in the other and that the underlying object surface may be inferred to be a continuous surface detail or a continuous surface contour.

16.4. LASER SENSORS

As Figure 16.33*a* shows, a laser sensor illuminates the object with a collimated beam, and the backscattered light, approximately coaxial with the transmitted beam, is picked up by the receiver. The range is estimated from the time it takes the light to travel from the sensor to the object and back. A

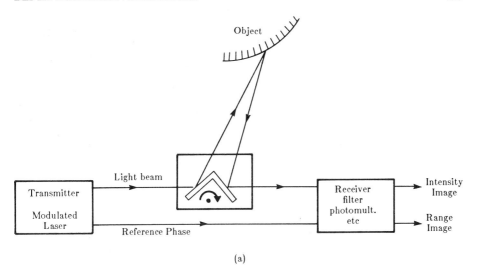

(a)

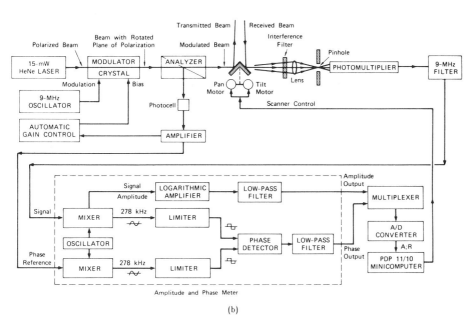

(b)

Fig. 16.33. (a) Principles of operation of a laser sensor. (b) The system used by Nitzan et al. for laser range mapping.

range map is built up by scanning the light beam over the scene. The reflectance at each object point is estimated from the intensity of the backscattered light.

One can, of course, measure the time delays directly for a pulse of light to travel from the sensor to the object and back.[9,20] Since light travels at approximately 30 cm/ns (1 ft/ns), the instrumentation for the direct measurement of time delays must possess a resolution of 50 psec for range accuracy in the vicinity of 6.4 mm (0.25 in.). In the system described,[20] the authors achieved an accuracy of 2 cm in the 1–3-m range with a scan time of 3 min for a 128×128 set of range values.

The time delays can also be measured by modulating a long burst of laser light with a low-frequency sinusoid and measuring the phase difference between the modulations on the backscattered and the transmitted signals.[30] In this section, to bring out the fundamental concepts on which laser sensors are based and the limitations caused by their dynamic range, we mainly describe this latter system.

A detailed diagram of the system used by Nitzan et al. is shown in Figure 16.33b. The plane-polarized output of a 15-mW HeNe laser (of wavelength 632.8 nm) is amplitude modulated at a frequency of 9 MHz by passing it through an ammonium dihydrogen phosphate (ADP) crystal; the modulation is effected by rotating the plane of polarization. The light beam is then split into two parts by passing it through an analyzer. The off-axis part is used as a phase reference and for providing automatic gain control. The on-axis part, which now carries a time-average power of about 6 mW, is deflected by the scanning mirror and illuminates the object. The light reflected from the object passes through a 632.8-nm interference filter with a 2-nm passband to minimize the effect of ambient light. The output of the interference filter goes to a photomultiplier tube, whose output signal passes through a 9-MHz filter. The output of the filter is used for both the amplitude and phase measurements.

An often-cited difficulty with laser depth perception is the length of time it takes to build up a range map. A primary source of this difficulty is the small amount of light that is returned by the object. One could reduce this time by using a higher-power laser; however, that might compromise the safety issue. To explain the considerations that dictate long measurement times, we present some introductory comments about the reflection of light from mirrorlike and mattelike surfaces. We then present the expressions derived by Nitzan et al.[30] for the expected standard deviation in a range measurement and show its dependence on the available signal-to-noise ratio.

As Figure 16.34a shows, consider a collimated beam of laser light that is incident on a surface and that is subtending an angle of θ_i with the normal. Some of this light will be absorbed by the object, while the rest is reflected. Two different components obeying different laws of physics can be identified in the reflected light. One is a mirrorlike specular component obeying the familiar law that θ_r, the angle of reflection, is equal to θ_i, the angle of incidence. The second mechanism, more important for robot vision applications, is called the diffuse reflection: the ratio of the power transported by the specular and the diffuse components is determined by the roughness of the object surface (at a scale determined by the wavelength of illumination used, which for the case of HeNe lasers is around 0.6 microns). As Figure 16.33a shows, for range-mapping applications the returned light intercepted by the sensor is nearly coaxial with the illuminating beam. Therefore, unless the object surface is perpendicular to the laser beam, the received light by the sensor will be totally a diffuse reflection, as opposed to being a specular return. How does the diffuse reflection depend on the orientation of the surface with respect to the illumination beam? This question is answered by *Lambert's cosine law*, which we state after a few definitions.

The intensity of a continuous light beam can be expressed as a function of the time-averaged power per unit area (in watts per square centimeter), a unit area being perpendicular to the direction of propagation of the beam. The strength of a well-collimated beam can also be expressed by the total time-averaged power (in watts) transported by the beam through any of its cross sections. The intensity of the diffuse reflection, by definition noncollimated, is best expressed as a time-averaged power radiated outward per unit solid angle in different directions. Figure 16.34b illustrates a solid angle $d\Omega$ in a direction θ from the normal. If the total radiant flux measured as time-averaged power flowing outward through this solid angle is dP_d, then the time-averaged radiant intensity (in watts per steradian) at the point marked X is

$$I_d = \frac{dP_d}{d\Omega} \tag{16.58}$$

Lambert's cosine law states the following for the angular dependence of diffuse reflection:

$$\bar{I}_d = \frac{1}{\pi} \bar{F}_T \rho_d \cos \theta \tag{16.59}$$

where ρ_d is called the *diffuse reflectance* of the surface and is equal to the ratio of the total power transported by the mechanism of diffuse reflection over all solid angles to F_T, the total power in the incident illumination. The value of ρ_d varies between 0 for ideally black surface and 1 for an ideally white surface; also it is dependent on the wavelength of light used.

Note that although by Lambert's law the diffuse reflection is independent of the angle of the illumination, for the situation depicted in Figure 16.33a we are only interested in the diffuse *backscatter*, that is, the return nearly coaxial with the illumination. Therefore if θ_i is the angle between the illumination and the surface normal, the backscatter will exhibit the cos θ_i dependence on the surface orientation.

In addition to its dependence on the nature of the surface finish and the orientation of the surface, the amount of reflected light picked up by the sensor also depends on how far away the surface is. Let A_R be the effective area of the sensor for capturing the backscattered light, and let r be the range of the backscattering site. The solid angle subtended by this capture area at the object point in question is A_R/r^2. The total returned light picked up by the sensor is then given by

$$\bar{F}_p = \frac{\alpha A_R \bar{F}_T \rho_d \cos \theta_i}{\pi r^2} \tag{16.60}$$

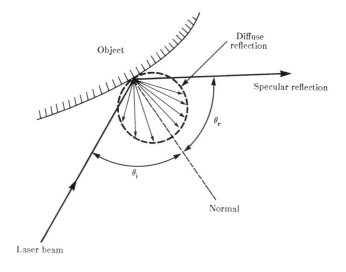

(a)

(b)

Fig. 16.34. (a) When a collimated beam of laser light is incident upon a surface, some of the incident energy is absorbed, the rest is either specularly reflected or diffusely scattered. (b) An elemental solid angle is in a direction θ from the surface normal.

The factor α accounts for the phenomenon that to reduce as much as possible the contributions made by the ambient light, the received backscatter from the object must be filtered through a narrow-band interference filter. To a certain extent this filter will also attenuate the desired signal.

Since the illuminating beam is modulated for the measurement, the time-average power in the transmitted beam can be written as

$$F_T(t) = \bar{F}_T(1 + m \cos \omega_m t) \tag{16.61}$$

In the presence of modulation the expression for the returned light takes the form

$$F_p(t) = \bar{F}_p[1 + m \cos (\omega_m t - \phi)] \tag{16.62}$$

where the time-average radiant flux $\bar{F}_p$ is given by Eq. (16.60), and the phase shift ϕ is

$$\phi = \frac{\omega_m}{c}(2r + r_0) \tag{16.63}$$

where the distance r_0 corresponds to the phase shift introduced by paths internal to the sensor. The extra phase shift is introduced by the light beam traveling a distance r to the object and back. Therefore a measurement of the phase yields directly a value for the depth r.

The accuracy of depth measurement depends on how precisely the phase ϕ can be measured. That leads to the question: What is the smallest phase change that the sensor should be able to measure? From the foregoing formula, a 1-cm change in depth corresponds to a phase change of $0.2°$. Although it is possible to measure phase changes that are as small as $0.1°$,[25] the dynamic range of the received signal makes that difficult. Note that the received signal is a function of $\rho_d \cos \theta / r^2$. If we assume a variation of 0.2–1 for ρ_d, 1–5 m for range, and 0–87° for the angle of incidence, we come up with a variation of 25,000–1 (or 88 dB) in $\rho_d \cos \theta / r^2$. When the received signal is at the bottom end of this dynamic range, and therefore becomes comparable to or buried in noise, it is difficult to measure phase with precision. A few comments follow about the sources of noise and the available signal-to-noise ratios.

When a dark distant object is viewed obliquely, $\rho_d \cos \theta_i / r^2$ takes a small value, and therefore the returned power is also small. At such power levels, quantum noise associated with the emission of electrons in the photomultiplier dominates. The resulting signal-to-noise ratios can be estimated as follows. The electrons in the photomultiplier are emitted by absorption of the incoming light photons. If we use T sec to denote the observation interval, the time-average received power given by Eq. (16.60) translates into an average number of photons denoted by $\bar{n}$, where

$$\bar{n} = \frac{\bar{F}_p T}{hc/\lambda} \tag{16.64}$$

The quantity hc/λ is a unit of energy in one photon, where h is the Planck's constant, c the speed of light, and λ the wavelength of illumination. If the quantum efficiency of the photomultiplier is denoted by η, the number of emitted electrons is

$$\bar{n}_e = \eta \bar{n} \tag{16.65}$$

Since the emission of electrons is a Poisson process, the standard deviation σ_e of the number of emitted electrons is simply $\sqrt{\bar{n}_e}$. Therefore the signal-to-noise ratio (SNR) at the output of the photomultiplier is

$$\text{SNR} = \frac{\bar{n}_e}{\sigma_e} = \sqrt{\bar{n}_e} \tag{16.66}$$

$$= \left[\frac{\alpha \eta \lambda A_R \bar{F}_T T}{\pi hc} \frac{\rho_d \cos \theta_i}{r^2}\right]^{1/2} \tag{16.67}$$

which was obtained by using Eq. (16.60). For weak signals, the phase must be measured in the presence of quantum noise. It has been shown[30] that the noise-induced phase error translates into a range error in the following form:

$$\sigma_r \simeq \frac{c}{\sqrt{2}m\omega_m \text{SNR}} = \frac{1}{2\sqrt{2}\pi} \frac{\lambda_m}{m\text{SNR}} \tag{16.68}$$

where σ_r is the standard deviation of the error in range measurement, and where λ_m is given by

$$\lambda_m = \frac{2\pi c}{\omega_m} \tag{16.69}$$

Substituting Eq. (16.67) in Eq. (16.68), we get for the standard deviation of error in range measurement

$$\sigma_r \simeq \frac{\lambda_m}{m} \left[\frac{hc}{8\pi\alpha\eta\lambda A_R \bar{F}_T T} \frac{r^2}{\rho_d \cos\theta_i} \right]^{1/2} \tag{16.70}$$

This equation shows how the range error depends on system parameters. In Table 16.2 we have shown numerical results obtained by Nitzan et al.[30] by using the expressions presented for low-signal and intermediate-signal cases. (Signal strengths are controlled by the values of ρ_d, θ_i, and r.) The table shows that even at a 40-dB signal-to-noise ratio there is a range uncertainty of 4 cm. Although the range accuracy does indeed improve markedly with increased SNR, the results predicted by the formulas presented are a bit on the optimistic side. That is because, at large signal levels, quantum noise is not the dominant limitation on range resolution. (Other sources of noise include laser noise, ambient noise, thermionic or dark-current noise, etc.) Nitzan et al.[30] have shown how, by adaptive smoothing, the range resolution everywhere can be brought within acceptable levels, the price paid being the increased observation time. Since it can take a long time to build up a complete range map, drift corrections may also become necessary. Such corrections are implemented by pointing the laser beam at fixed-distance reference markers.

16.5. ULTRASONIC SENSORS

Ultrasonic sensors, made popular by the Polaroid range-finder camera, work on the same time-of-flight principle as the laser sensors discussed in the preceding section. A narrow-band pulse of sound consisting of 5–10 cycles of oscillations is transmitted toward the object by a transducer that can act both as a transmitter and a receiver. The time it takes for the pulse to travel to the object and back is proportional to the range. Unlike lasers, since sound travels at a much slower speed, is is easy to measure the time-of-flight directly by employing narrow acoustic pulses, as opposed to using the indirect procedure of amplitude modulation. As depicted in Figure 16.35a, the method of sending forth a pulse of sound and measuring the time it takes for the echo to return from the object is usually referred to as the *pulse-echo* mode. In the pulse-echo mode a range map can be constructed by scanning the transducer, as shown in Figure 16.35a, and recording the range value for each position.

TABLE 16.2. THE INSTRUMENT AND THE COMPUTED SIGNAL PARAMETERS OF THE NITZAN ET AL. SYSTEM [a]

Instrument Parameters	
$\bar{F}_T = 6 \times 10^{-3}$ W	$A_R = 1.5 \times 10^{-4}$ m^2
$\lambda = 0.6328 \times 10^{-6}$ m	$\alpha = 0.25$
$\lambda_m = 33.3$ m	$\eta = 0.1$
$m = 1.0$	$T = 0.01$ sec

Signal Parameters		
Parameters	Low-Signal Strength	Intermediate-Signal Strength
ρ_d	0.02	0.3
θ (degrees)	87	45
r (meters)	5	3
$\bar{F}_p$ (watts)	3×10^{-12}	1.7×10^{-9}
$\bar{n}_{pe}$ (photoelectrons)	10^4	5×10^6
SNR	100 (40 dB)	2300 (67 dB)
σ_r (centimeters)	4	0.2

Source. Nitzan, D. et al., Reference 30.

[a] The last two rows of the table illustrate how the range uncertainty depends on the signal-to-noise ratio.

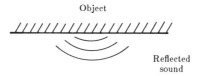

Object

Reflected
sound

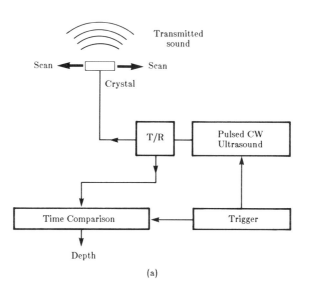

Transmitted
sound

Scan ←———→ Scan

Crystal

T/R → Pulsed CW Ultrasound

Time Comparison ← Trigger

Depth

(a)

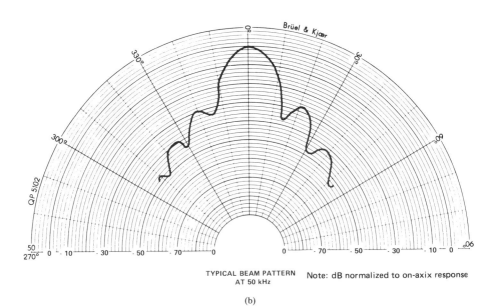

Brüel & Kjær

QP 5102

TYPICAL BEAM PATTERN
AT 50 kHz

Note: dB normalized to on-axix response

(b)

Figure 16.35. (*a*) The pulse-echo technique for measuring range with ultrasound. (*b*) The directivity pattern of the Polaroid ultrasonic range sensor at 50 kHz.

Ultrasonic sensors are operated in the frequency range of 50–200 kHz. To choose the right frequency for a given application, one must be aware of two fundamental characteristics of these sensors: (1) The lower the frequency, the wider the beam produced by a transducer. Figure 16.35b shows the beam pattern for the Polaroid range sensor at 50 kHz; note that the main lobe has a width of 30°. If we assume constant pressure across the face of a circular transducer of radius a, the relationship between the width of the main lobe and the frequency of operation is given by

$$\text{angular width of the main lobe} = 2\sin^{-1}\frac{0.61\lambda}{a} \qquad (16.71)$$

where λ is the wavelength of sound in air and is equal to V_s/f. V_s is the speed of sound in air; it is approximately equal to 343 m/sec at 20°C. The frequency of operation is denoted by f. And (2) the higher the frequency, the greater the attenuation suffered by the acoustic pulse as a result of propagation through air. The attenuation suffered by a plane acoustic wave propagating in x-direction through air is given by

$$e^{-\alpha x}$$

where the attenuation coefficient α is given by

$$\alpha = 1.37 \times 10^{-11} \times f^2 \quad \text{nepers/m} \qquad (16.72)$$

Note the dependence on the square of the frequency. At a frequency of 50 kHz, this translates into a negligible attenuation of 34.2×10^{-3} Np/m, which is equivalent to 0.29 dB/m (using a conversion factor of 8.7 to go from nepers to decibels). However, at a frequency of 200 kHz, the attenuation is at a more noticeable level of 0.548 Np/m, or 4.8 dB/m; and at a frequency of 1 MHz it is over 100 dB/m. (For pulse-echo applications these numbers should be doubled for a given range owing to the round trip involved.)

At the low end of the frequency range, where the attenuation owing to propagation is not a factor, the detectability of the object echo depends considerably on the diffraction losses involved. By diffraction loss, we mean the attenuation of the acoustic signal due to the spreading of the beam as implied by the directivity pattern shown in Figure 16.35b. More important, at the low end of the frequency range the beam width becomes too large, causing a great deterioration in the lateral resolution of the sensor. By *lateral resolution* is meant the ability of the sensor to distinguish between details in the direction of the scan in Figure 16.35a. Therefore the selection of the frequency for this sensor is a trade-off between the poor lateral resolution at the low end of the spectrum and increased propagation attenuation at the upper end.

To illustrate the lateral resolution problem with the pulse-echo implementation, in Figure 16.36 we have shown the range map of a 0.25-in.-diameter steel ball at a distance of 30 cm (1 ft) from the scan line. This experimental result was obtained with the Polaroid transducers operated at 50 kHz.

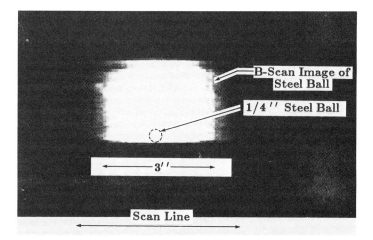

Figure 16.36. A 50-kHz ultrasonic pulse-echo image of a 0.25-in.-diameter steel ball at a distance of 1 ft from the scan line.

Because of the high accuracy with which the arrival times can be measured, the range to an object point can be estimated with relatively high precision. However, in the lateral direction the resolution is poor; here we have a 75-mm (3-in.) wide image for a 6.4-mm (0.25-in.) diameter object.

Lateral Resolution Improvement by Coherent Detection and Back-Propagation

Lateral resolution can be considerably improved (it can even approach diffraction limits for simple objects) by using a two-transducer scheme depicted in Figure 16.37. One transducer used as a transmitter illuminates the object with sound, and the other transducer used as a receiver measures the amplitude and the phase of the object-dispersed sound on a scanning plane. This recorded data constitutes a wavefront of the field as scattered by the object and measured on the scanning plane.

This resolution-improvement technique is based on the concept of back-propagating the measured wavefront back toward the object. To explain the concept of back-propagation, consider a single-point object shown in Figure 16.38, which is illuminated by a plane wave. The scattered field from the object is a spherical wave. If we measured this scattered field in different planes, both the real and the imaginary parts would exhibit oscillatory patterns as depicted in the figure. Figure 16.38 depicts *forward* propagation outward from the point object. The idea in *back*-propagation is to start with the field pattern in, say, the plane marked *X* and then to retrace the field backward toward the point object. And if that could be done, the point object would be reconstructed and would be at the location of the maximum in the back-propagated field pattern. At least theoretically, a more complex object can be considered as a superposition of point objects, and since back-propagation is a linear process, reconstruction should be equally achievable. In practice, it is not possible to reconstruct the entire three-dimensional structure of an object (assuming that the illuminating sound penetrates into

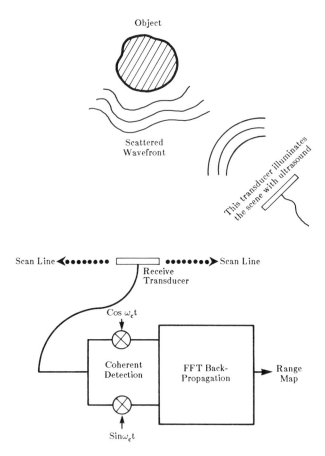

Figure 16.37. Lateral resolution can be considerably improved by using a two-transducer scheme shown here. One transducer illuminates the object with sound, and the other coherently measures the amplitude and the phase of the sound scattered by the object.

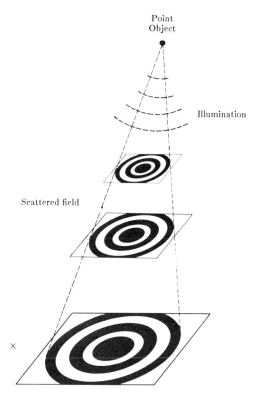

Figure 16.38. An illustration of the concept of back-propagation. Shown here is a spherical wave scattered outward by a single-point object that is illuminated with a plane wave.

the interior), because of difficulty with accounting for material attenuation and interaction between different scattering sites. *However, it should be possible to reconstruct surfaces, although the level of complexity in shape at which that can be done is not known as yet.*

On a computer, back-propagation can be accomplished rather rapidly with fast Fourier transform (FFT) routines. The fields measured in the scan plane are Fourier transformed to yield what is called the *plane wave spectrum* decomposition. In other words, we represent the measured fields as a sum of plane waves. (For a plane wave, surfaces of constant phase are planes. To help with conceptualization, roughly speaking, a giant piston moving back and forth would launch a plane acoustic wave into the air.) The reason for this decomposition is that we know precisely how a plane wave propagates forward and backward; we simply multiply the plane-wave amplitude by a phase factor, called the *transfer function* of propagation. Therefore, to back-propagate the measured wavefront by a distance, we multiply its plane-wave decomposition by the transfer function appropriate to that distance. Let $g(x, y; z_0)$ denote the fields measured as a function of (x, y) in the plane $z = z_0$. Let $G(u, v)$ denote the two-dimensional transform of this field:

$$G(u, v) = \int\int g(x, y, z_0)e^{-j2\pi(ux + vy)}dx\,dy \tag{16.73}$$

where u and v are the spatial frequencies along the x- and the y-directions, respectively. In back-propagating from z_0 to z_1, we multiply $G(u, v)$ by the following transfer function

$$\exp\left(-j2\pi\frac{(z_1 - z_0)}{\lambda}\sqrt{1 - (\lambda u)^2 - (\lambda v)^2}\right) \tag{16.74}$$

where λ is the wavelength of the acoustic field. To reconstruct the field in the plane $z = z_1$, we take the inverse transform of the product. Of course, all the Fourier transforms are computed by using the FFT programs. By the procedure outlined here the measured fields are back-propagated to different depths in the vicinity of the expected range values. The shape of the object surface is obtained by thresholding these fields.

The 0.25-in. steel ball used before for pulse-echo experiments was also imaged with a one-dimensional version of the back-propagation approach presented here. The data were collected on a scan line, as opposed to a plane. A one-dimensional Fourier transform of this data given by the following relationship

$$G(u) = \int g(x; z) e^{-j2\pi ux} dx \qquad (16.75)$$

yields the equivalent of a plane-wave decomposition, which is then multiplied by the transfer function for back propagation

$$\exp\left(j2\pi \frac{(z_1 - z_0)}{\lambda} \sqrt{1 - (\lambda u)^2}\right) \qquad (16.76)$$

The inverse transform of the product gives the back-propagated field. The image obtained is shown in Figure 16.39, which shows a near-perfect reconstruction. The back-propagated fields were thresholded at the half-peak power level for this reconstruction.

In general, the quality of reconstruction by this method depends on the size of the scan area for data collection. This can best be explained by going back to the example of a point object. The fidelity with which the point object is reconstructed depends on the number of cycles of oscillations incorporated in the measurement plane. (The ring pattern shown in Figure 16.38 to represent oscillations is called the Fresnel pattern in optics.) Ideally one should have about 40 rings in the Fresnel pattern to obtain good "focusing" by back-propagation. In practice, acceptable results can be obtained even when this condition is not satisfied. Since the scan line was only 25 cm (10 in.) long, the measured data for the image in Figure 16.39 were equivalent to only about 10 rings in the Fresnel pattern. Since the wavelength at 50 kHz is about 6.4 mm (0.25 in.), the reconstruction in Figure 16.39 has the best resolution that can be obtained at this frequency.

16.6. BIBLIOGRAPHICAL NOTES

The focus of this chapter has been on range-data acquisition. We have not discussed an associated important topic, which is the automatic interpretation of range data. By automatic interpretation we mean a robot looking at a multicomponent scene, using range (and photometric) information to recognize each component individually, and then determining the positional relationships between them. Although space limitation was a primary factor for this deletion, another consideration that influenced our decision was the infancy of this subject. The reader is referred to References 11, 27, 31, and 39 for a representative set of discussions of this subject.

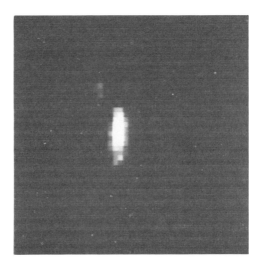

Fig. 16.39. Image by back-propagation of the same 0.25-in. steel ball that was used in Figure 16.36.

The algorithms presented in this chapter for solving the correspondence problem for stereo depth perception constitute, in my opinion, a representative set from what has appeared to date in the literature. The reader is also referred to Burr and Chien[8] and Arnold[2]; these authors have used features such as edges and edge segments and their properties such as length and contrast for solving the correspondence problem. Along similar lines Underwood and Coates[43] have used invariant properties of polyhedral scenes for constructing the matches.

The discussed modifications to the original Marr-Poggio-Grimson algorithm and the depth perception results obtained thereby were first presented by the author in Reference 18. Some preliminary considerations with regard to the architectures for real-time implementation of this algorithm are presented in Reference 38. The Marr-Poggio-Grimson algorithm presented in this chapter uses the principle of coarse-to-fine matching, in which the matching between a stereo pair is first conducted at reduced resolution and subsequently refined with higher-resolution versions of the image. This coarse-to-fine matching has also been used by Moravec[28] and Dev.[10] In this chapter, we only discussed how to obtain depth maps from images of a scene taken from two viewpoints. As mentioned before, the range accuracy is greater the longer the separation between the cameras. However, the price paid for the enhanced accuracy obtained in this manner is the greater incidence of occlusions suffered by one or the other camera, which creates regions of space where no matches can be found. Also, the increased separation between the cameras leads to larger values of disparities for a given depth, which implies larger search regions for finding the corresponding pixels; this in turn increases the search complexity and time. To get around some of these difficulties, one can use more than two views, as proposed by Moravec,[28] Nevatia,[29] and Tsai.[46] In the procedure described by Moravec, disparities computed from different pairs of images are discarded if they are not consistent. In Nevatia's algorithm, multiple views are used to reduce the size of the search neighborhood for finding the correspondence, and Tsai has proposed jointly correlating the data from all the available frames for superior accuracy.

Our treatment of laser sensors considered only those that measure time delays indirectly with amplitude modulation of the laser. However, regarding the discussion of the large dynamic range problems caused by surface obliquity, variations in the diffuse reflectance of surfaces, and the fourth-power dependence on the range, one must note that these problems also occur in the direct methods.[20] It does appear at this time that the direct methods may possess a speed advantage over the modulation-based methods. Jarvis[17] has mentioned a pulsed-laser system using direct measurement of time delays that yields a 64×64 low-resolution range image in 4 sec.

Will et al.[44] and Pennington et al.[32,33] have shown that if a scene contains only polyhedral solids, the locations and the orientations of planar surfaces can be extracted by illuminating the scene with a rectangular grid of lines and two-dimensional filtering of the recorded pattern. The pattern generated by the intersection of the illumination grid and a planar surface consists of a parallel set of bars whose orientation depends on the direction of the surface normal. By performing a Fourier analysis of the recorded pattern, the photometric image can be segmented into different regions, each corresponding to a planar surface of a polyhedron and characterized by a particular surface normal.

Finally, for overviews of varying generality that include the subjects of range mapping and/or range data interpretation for scene analysis, the reader is referred to References 7, 14, and 17. We would also like to draw the attention of the reader to the technique of photometric stereo for the construction of depth maps.[47,48] In this approach, a scene is illuminated by a number of strategically placed light sources, and the resulting shading information obtained with each source combined to yield the final depth map. If one is only interested in relative depth maps, optical flows can be utilized for that purpose[49,50] which are generated when a camera and scene are in relative motion with respect to each other.

ACKNOWLEDGMENTS

My foray into robotics-related research areas is primarily owing to my fascination with anything that smacks of artificial intelligence. Unfortunately, many graduate students have suffered during the course of my developing expertise in these areas. I am particularly grateful to the students who took my course on computer techniques for stereo depth perception and who allowed themselves to be coerced into doing an arduous project on the implementation of the Marr-Poggio paradigm. All these students tested their programs on stereo photos of real scenes that they had themselves taken. Some stereo range-mapping results shown in this chapter came out of that "contest."

My interest in stereo range mapping began with a seminar given at Purdue University by Eric Grimson. I am grateful to him for the initial discussions that propelled me in this direction.

My interest in ultrasonic range mapping is a consequence of the many years that I have spent on both the experimental and theoretical aspects of computed imaging with X rays, ultrasound, and microwaves.

Robert Safranek, currently working with me on a doctoral dissertation on stereo for robot vision, supplied me with most of the results shown on that subject; many thanks go to him. I am also grateful to S. X. Pan, currently a visiting scholar at Purdue, and Malcolm Slaney, Hyun Yang, Mike O'Boyle, and Mike Anderson, who are also working with me on their doctoral dissertations, for many stimulating discussions. They are responsible for supplying me with some of the other results shown in this chapter.

REFERENCES

1. Agin, G. J. and Binford, T. O., Computer description of curved objects, *Proceedings of Third International Joint Conference on Artificial Intelligence*, 1973, pp. 629–640.

2. Arnold, R. D., Local context in matching edges for stereo vision, *Proceedings of Image Understanding Workshop*, 1978, pp. 65–72.

3. Baker, H. H. and Binford, T. O., Depth from edge and intensity based stereo, *International 1981 Joint Conference on Artificial Intelligence*, Vol. 6, pp. 631–636.

4. Binford, T. O., Inferring surfaces from images, *Artificial Intelligence*, Vol. 17, 1981, pp. 205–245.

5. Barnard, S. T. and Thompson, W. B., Disparity analysis of images, *IEEE Transactions PAMI*, PAMI-2, 1981, pp. 333–340.

6. Bajcsy, R. and Lieberman, L., Texture gradient as a depth cue, *Computer Graphics and Image Processing*, Vol. 5, 1976, pp. 52–76.

7. Brady, M., Computational Approaches to Image Understanding, *Computing Surveys*, Vol. 14, 1982, pp. 3–71.

8. Burr, D. J. and Chien, R. T., A system for stereo computer vision with geometric models, *Proceedings of Fifth International Joint Conference on Artificial Intelligence*, 1977, p. 583.

9. Caulfield, H. J., Hirschfeld, T., Weinberg, J. M., and Herron, R. E., Laser Stereometry, *Proceedings of IEEE*, Vol. 65, 1977, pp. 84–88.

10. Dev, P., Perception of depth surfaces in random-dot stereograms: A neural model, *International Journal of Man-Machine Studies*, Vol. 7, 1975, pp. 511–528.

11. Duda, R. O., Nitzan, D., and Barrett, P., Use of range and reflectance data to find planar surface regions, *IEEE Transactions PAMI*, PAMI-1, 1979, pp. 259–271.

12. Grimson, W. E. L., A computer implementation of a theory of human stereo vision, *Philosophical Transactions Royal Society London*, B. 292, 1981, pp. 217–253.

13. Grimson, W. E. L., *From images to surfaces: A computational study of the human early visual system*, M.I.T. Press, Cambridge, Massachusetts, 1981.

14. Hall, E. L., Tio, J. B. K., McPherson, C. A., Sadjadi, F. A., Measuring curved surfaces for robot vision, *Computer*, Vol. 15, December 1982, pp. 42–54.

15. Helava, U. V., Digital Correlation in Photogrammetry Instruments, *Photogrammetria*, Vol. 34, 1978, pp. 19–41.

16. Henderson, R. L., Miller, W. J., and Grosch, C. B., A flexible approach to digital stereo mapping, *Photogrammetry Engineering of Remote Sensing*, Vol. 44, 1978, pp. 1499–1512.

17. Jarvis, R. A., A perspective on range finding techniques for computer vision, *IEEE Transactions PAMI*, Vol. PAMI-5, 1983, pp. 122–139.

18. Kak, A. C., Implementation of the Marr-Poggio paradigm for stereo vision: A perspective for engineering applications, presented at the Workshop on Sensors and Algorithms for 3-D Vision (held in conjunction with the *Conference of the American Association of Artificial Intelligence*), Washington, D.C., 1983. (Being prepared for publication; request copies from the author.)

19. Levine, M. D., O'Handley, D. A., and Yagi, G. M., Computer determination of depth maps, *Computer Graphics and Image Processing*, Vol. 2, 1973, pp. 131–150.

20. Lewis, R. A. and Johnston, A. R., A scanning laser rangefinder for a robotic vehicle, *Proceedings of the Fifth International Joint Conference on Artificial Intelligence*, 1977, pp. 762–768.

21. Loughlin, C., Eye-in-hand robot vision scores over fixed camera, *Sensor Review*, Vol. 3, 1983, pp. 23–26.

22. Lucas, B. D. and Kanade, T., An iterative image registration technique with an application to stereo vision, *Proceedings of the Seventh International Joint Conference on Artificial Intelligence*, August 1981, pp. 674–679.

23. Marr, D. and Poggio, T., A theory of human stereo vision, *Proceedings of the Royal Society London*, B., Vol. 204, 1979, pp. 301–328.

24. Marr, D., *VISION: A Computational Investigation in the Human Representation and Processing of Visual Information*, W. H. Freeman, San Francisco, 1981.

25. Maxwell, D. E., A 5 to 50 MHz direct reading phase meter with hundredth-degree precision, *IEEE Transactions on Instrument Measures*, Vol. IM-15, 1966, pp. 304–310.

26. McVey, E. S. and Lee, J. W., Some accuracy and resolution aspects of computer vision distance measurements, *IEEE Transactions PAMI*, Vol. PAMI-4, 1982, pp. 646–649.

27. Mitiche, A. and Aggarwal, J. K., Detection of edges using range information, *IEEE Transactions PAMI,* Vol. PAMI-5, 1983, pp. 174–178.

28. Moravec, H. P., Rover visual obstacle avoidance, *Proceedings of the Seventh International Joint Conference on Artificial Intelligence,* 1981.

29. Nevatia, R., Depth measurement by motion stereo, *Computer Graphics and Image Processing,* Vol. 5, 1976, pp. 203–214.

30. Nitzan, D., Brain, A. E., and Duda, R. O., The measurement and use of registered reflectance and range data in scene analysis, *Proceedings of IEEE,* Vol. 65, 1977, pp. 206–220.

31. Oshima, M. and Shirai, Y., Object recognition using three-dimensional information, *IEEE Transactions PAMI,* Vol. PAMI-5, 1983, pp. 353–361.

32. Pennington, K. S., Will, P. M., Shelton, G. L., Grid coding: A technique for extraction of differences from scenes, *Optics Communications,* Vol. 2, 1970, pp. 113–119.

33. Pennington, K. S., and Will, P. M., A grid-coded technique for recording 3-dimensional scenes illuminated with ambient light, *Optics Communications,* Vol. 2, 1970, pp. 167–169.

34. Posdamer, J. L. and Altschuler, M. D., Surface measurement by space-encoded projected beam systems, *Computer Graphics and Image Processing,* Vol. 18, 1982, pp. 1–17.

35. Rice, S. O., Mathematical analysis of random noise, *Bell System Technical Journal,* Vol. 24, 1945, pp. 46–156.

36. Rosenfeld, A. and Kak, A. C., *Digital Picture Processing,* Vols. 1 and 2, 2nd ed., Academic Press, 1982.

37. Ryan, T. W., Gray, R. T., and Hunt, B. R., Prediction of correlation errors in stereo-pair images, *Optical Engineering,* Vol. 19, 1980, pp. 312–322.

38. Safranek, R. J. and Kak, A. C., Stereoscopic depth perception for robot vision: Algorithms and Architectures, *Proceedings of the IEEE International Conference on Computer Design: VLSI in Computers,* 1983, pp. 76–79.

39. Shirai, Y., Recognition of polyhedra with a range finder, *Pattern Recognition,* Vol. 4, 1972, pp. 243–250.

40. Stoer, J. and Bulirsch, R. *Introduction to Numerical Analysis,* Springer-Verlag, 1980, pp. 198–208.

41. Sweeney, D. W. and Hudelson, G. D., Optical tracking with Fresnel zone plate coded aperture imaging, unpublished.

42. Thompson, A. M., Camera geometry for robot vision, *Robotics Age,* March/April 1981, pp. 20–27.

43. Underwood, S. A. and Coates, C. L., Visual learning from multiple views, *IEEE Transaction on Computers,* Vol. C-24, 1975, pp. 651–661.

44. Will, P. M. and Pennington, K. S., Grid coding: A preprocessing technique for robot and machine vision, *Artificial Intelligence,* Vol. 2, 1971, pp. 319–329.

45. Yakimovsky, Y. and Cunningham, R., A system for extracting three-dimensional measurements from a stereo pair of TV cameras, *Computer Graphics and Image Processing,* Vol. 7, 1978, pp. 195–210.

46. Tsai, R. Y., Multiframe image point matching and 3-D surface reconstruction, *IEEE Transaction Pattern Analysis and Machine Intelligence,* Vol. PAMI-5, March 1983, pp. 159–174.

47. Ikeuchi, K. and Horn, B. K. P., Numerical shape from shading and occluding boundaries, *Artificial Intelligence,* Vol. 17, 1981, pp. 141–185.

48. Ikeuchi, K., Determination of surface orientations of specular surfaces by using the photometric stereo, *Proceedings of the IEEE,* 1981.

49. Horn, B. K. P. and Schunck, B. G., Determining optical flow, *Artificial Intelligence,* Vol. 17, 1981, pp. 185–204.

50. Lawton, D. T., Processing translational motion sequences, *Computer Vision, Graphics and Image Processing,* Vol. 22, 1983, pp. 116–144.

CHAPTER 17

CONTROL OF REMOTE MANIPULATORS

ANTAL K. BEJCZY

Jet Propulsion Laboratory
California Institute of Technology
Pasadena, California

17.1. REMOTE APPLICATIONS OF ROBOTS

The term *remote applications of robots* refers to teleoperation, which, in turn, means the use of robotic devices having mobility, manipulative and some sensing capabilities, and remotely controlled by a human operator. The remote control can be manual or automatic, or a combination of both. Historically, robot arms and hands are the most important teleoperator devices.

Industrial robots typically *replace* workers performing well-structured, repetitive, and often tedious manual work. Remotely applied robots or teleoperators, on the other hand, *augment* the human manual, sensing, and perceptive/cognitive capabilities and *extend* them to remote, difficult, and dangerous places undesirable or inaccessible by humans. Teleoperators typically perform singular, nonrepetitive and semi-structured work. Industrial robots are essentially machine systems. Teleoperators are essentially man-machine systems.

Remotely applied robots or teleoperators are widely used in the nuclear industry to handle radioactive materials. New technological endeavors in space, deep sea, mining, and the like, will also involve an extensive use of teleoperators. The latest newcomer to the field of teleoperator systems is the 16-m (50-ft) robot arm attached to the Space Shuttle and remotely operated by an astronaut from the Shuttle cockpit. The bibliography at the end of this chapter contains a selected list of references on a variety of recent work related to the control of remote manipulators.

Remote applications of robots raise several issues. The issues can be subdivided into two major groups: (1) the development of proper robot manipulator or locomotion machines capable of coping with the particular environmental and task constraints under which the robot machines will operate; (2) the development of control, information, and man-machine interface devices and techniques required for an efficient and safe operation of robots at remote places. The first group of issues is covered in Chapter 9, "Teleoperator Arm Design."

In this chapter we consider the second group of issues: control, information, and man-machine interface focused at remote application of robot manipulators. Within the frame of this chapter, we consider primarily the devices and techniques which (1) enhance or supplement the visual information for remote robot arm control, (2) facilitate the operator's on-line interaction with computer control of remote robot arms, and (3) promote the use of data-driven automation for remote robot arm control.

General considerations, including human factors, are presented in Section 17.2. Sensors are treated in Section 17.3. Section 17.4 is devoted to the control problem, including a few man-machine interface examples. The information display problem is discussed in Section 17.5.

17.2. GENERAL CONTROL CONSIDERATIONS

Control of robot manipulators is very demanding in any control mode. It requires the coordinated control of several (typically six) manipulator joints while observing a multitude of kinematic, dynamic, and environmental constraints. Then, to follow the specifics of a given task, different sensor signals must be interpreted in real time. Furthermore, manipulation tasks can often be performed in different ways. Consequently robot arm control implies a multilevel decision and monitoring process at both the information feedback and control input channels of the controller.

The human operator is a key control element in remote applications of robots in all (manual and computer) control modes. It is the human presence in the control that provides a versatile response capability to the remotely operated robot in highly variable or unpredictable situations. The human's functional role is bidirectional: the operator receives information from the remotely operated system and sends control commands to it. The operator's manual skill and perceptive, cognitive, and decision-making abilities are important determining factors in the overall task performance.

It is recognized that the human operator's input and output channel capacities are limited. In this sense, the human operator represents a limiting factor in the information and control environment of a remotely operated robot. For example, the operator's information-processing capacity can be easily overloaded. Furthermore, the human input and output channel capacities are not only limited but also asymmetric: the human has much more information receiving (input) channels than information conveying (control output) channels.

Following an operator-centered view (Figure 17.1), the general objectives of control, information, and man-machine interface development for advanced teleoperators are (1) development of devices and techniques that enable the operator to convey control commands to the remote robot in comprehensive and task-related terms, and (2) development of devices and techniques that condense and display the control information in terms and formats compatible to human perception—and attention allocation in the environment of a control station designed for remote robot arm control.

The control and information paths between task description and task execution in remote manipulation pass through several major functional transformation blocks. The main functional elements of a multilevel manipulator control/information organization can be summarized as shown in Figure 17.2. The first or bottom level—represented by block D of Figure 17.2—is embedded in the time continuum of the physical world. The actual motion of the manipulator, the interaction between terminal device and environment, and all sensor signals are generated at this level. This level outputs the greatest amount of information and involves both the control and processing of continuous variables. The second level—represented by blocks B and C of Figure 17.2—can be called the algorithmic level. At

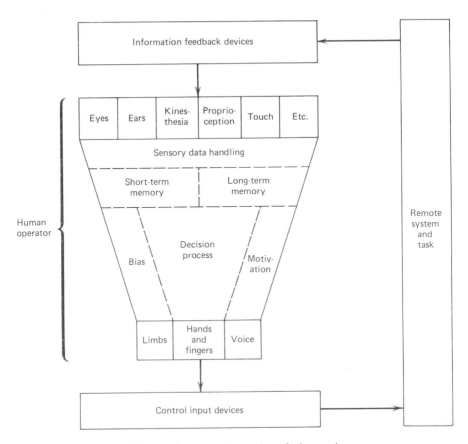

Fig. 17.1. Operator-centered view of teleoperation.

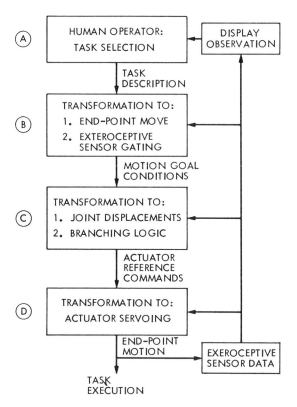

Fig. 17.2. Multilevel manipulator control/information flow paths.

this second level, control and information variables are handled by a *finite* number of computer algorithms, and messages are transmitted to the lower and higher levels of a finite number of time instants. Transformation of the operator's task description both to actuator reference commands and to control context of exteroceptive sensor data are generated by computer algorithms at this second control level. The third and highest control level—represented by block A of Fig. 17.2—is the human operator. In the context of Fig. 17.2, the main function of the human operator is task description and "supervisory" monitoring of task execution. The operator selects, quantifies, or modifies the algorithms for the lower control levels. Of course, this highest control level is based on the human attributes of thinking, learning, judging, and setting "goals" for a machine with known characteristics.

Advanced automation for robot control is data driven. It is inherently flexible since it is programmable. It contrasts the mechanically fixtured rigid automation. The data sources for data-driven automation can be subdivided into two major groups: models and sensors. Data derived from models typically provide a priori information about robot machines and tasks. Data derived from sensors typically provide on-line information about task performance of robots. Broad utilization of both data sources for robot arm control typically requires digital computers.

Remote applications of robots require flexibility in control and in information management to cope with varying and unpredictable task conditions. The use of data-driven automation in teleoperation offers significant new possibilities to enhance overall task performance by providing efficient means for task-related controls and displays.

17.3. SENSORS

The acquisition and use of both visual and nonvisual sensor information in remote robot arm control is of critical importance. Visual information is obtained directly or through stereo or mono television, and can be supplemented with information from ranging devices. Visual information for robot arm control is of geometric nature. It relates to the gross transfer motion of the mechanical arm in the environment and to the position/orientation of the mechanical hand relative to environmental or object coordinates.

Nonvisual sensor information supplements the visual information and is needed in controlling the physical contact or near-contact of the mechanical arm/hand with objects in the environment. It is obtained from proximity, force-torque, and touch-slip sensors integrated with the mechanical hand.[2] These sensors provide the information needed to perform terminal position/orientation and dynamic compliance control with fine robot arm motions. The control information from these sensors is directly referenced to the coordinates of the mechanical hand. The aquisition and use of nonvisual sensor information represents a major challenge in advanced remote manipulator control technology development.

17.3.1. Proximity Sensors

These sensors measure short distances in known directions between mechanical hand and objects. Proximity sensors can utilize electro-optical, electromagnetic, or acoustic measurement principles. Electro-optical proximity sensing requires a light source (typically a light-emitting diode or a low-power laser) and a photodetector. For electromagnetic proximity sensing, capacitance, eddy current, and Hall-effect devices can be employed.[3] Acoustic sensors utilize the sonar principle. Each proximity-sensing technique indicated has advantages and disadvantages as briefly summarized in References 2 and 3.

Several types of electro-optical proximity sensors have been developed at the Jet Propulsion Laboratory (JPL) for different types of mechanical hands and for different uses. In the latest implementation[2], the light source and light detector have been removed from the optical head located on the mechanical hand and have been integrated into the electronics instrumentation package located near the computer interface electronics. Fiber optic cables of low attenuation connect the light source and light detector to the optic head. Application of fiber optics can considerably improve signal quality and simplify instrumentation.

Figure 17.3 shows a proximity sensor system developed at JPL for possible use on a 16-m robot arm aboard NASA's Space Shuttle to aid the operator during payload handling or satellite servicing operations. This sensor system provides simultaneous measurements of short-range depth, pitch, and yaw errors of the mechanical hand relative to a plane target, and supplies the guidance and control information necessary for smooth and successful grasp near the grasp envelope, where visual perception of depth, pitch, and yaw errors is poor. The error information is shown to the operator on a computer-driven "smart" display in an integrated format. The display also uses audio tone and color for event indication. More on the experiments can be found in Reference 4.

17.3.2. Force-Torque Sensing

Force-torque sensors measure the amount of force and torque exerted by the mechanical hand along three hand-referenced orthogonal directions. These sensors also measure forces and torques applied about a point ahead and away from the sensors. These sensors are mounted at the base of the mechanical hand.

Force-torque sensors utilize mechanical force-summing elements that convert applied force into a small mechanical displacement. The mechanical force-summing elements are linked to electrical transduction elements. Several force-summing/transduction element combinations are possible. Strain gages (in particular, semiconductor strain gages) are preferred transducers for six-dimensional force-torque sensors integrated with robot hands.

Several force-torque sensors are applied at JPL to investigate new remote robot-control techniques. Recently, a large force-torque sensor system has been developed at JPL for possible use on the 16-m robot of the Space Shuttle.[5] The overall system configuration is shown in Figure 17.4. The mechanical structure of this sensor is of the Maltese Cross configuration machined from one piece of aluminum to reduce hysteresis; see Figure 17.5. The sensing elements consist of silicon-based semiconductor strain gages bonded to the four deflection bars of the Maltese Cross near the hub. There are two gages on each side of each of the four deflection bars, for a total of 32 gages. The pairs of two gages on the opposite sides of a deflection bar form a computer-matched quad of gages wired as a full bridge. Each full bridge provides a single reading that reflects the differences in the strain levels on the opposite sides of a bar. The full sensor provides eight output readings which can be resolved into three orthogonal force and three orthogonal torque components referenced to a sensor-based coordinate frame through a 6×8 transformation/calibration matrix as indicated in Figure 17.5. Under ideal conditions, only 16 elements of the 6×8 matrix are of significance.

The experimental sensor-claw-display system has been integrated with the simulated full-scale Space Shuttle robot arm at the Johnson Space Center for performance experiments. The sensor data are shown to the operator on a graphic display. The operator's response to the sensor data is through resolved-rate manual control of the Shuttle robot arm. The experiments have established the utility of the sensor-claw-display system for geometrically and dynamically constrained task control. The most interesting result was that all test operators consistently performed a payload-berthing task with success *without* any visual feedback from the work scene; they relied only on graphics display of

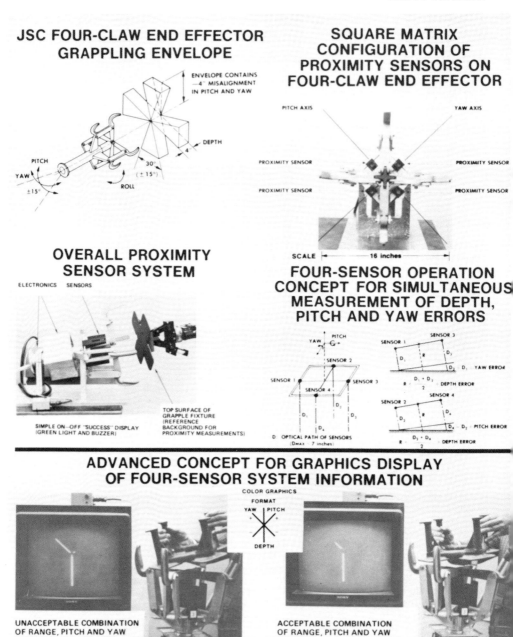

Fig. 17.3. Proximity sensors for space shuttle robot arm applications.

force-torque sensor information during the terminal phase of berthing when the payload guide pins were inside the V-shaped guides of the latch assembly. More on the experiments can be found in Reference 6.

17.3.3. Touch and Slip Sensing

Touch and slip sensors measure the distribution and amount of contact area pressure between hand and objects perpendicular or tangential to the hand, respectively. These sensors can be single-point,

Fig. 17.4. Overall configuration of force-torque sensor-claw-display system.

multiple-point (array), simple binary (yes-no), or proportional sensors. For transducers in these sensors several possibilities exist: strain gages, pressure-conductive elastomers, capacitive and piezoelectric elements, and so on. More on touch-sensing technology can be found in Reference 7. New entries into this field of research include the use of PVF_2 highpolymers[8] and the use of infrared light with fiber optics as demonstrated at JPL.[9] The simplest and least expensive form of tactile sensing is by microswitches.

A "box-sandwich" touch sensor array has been developed at JPL.[10] The construction concept of the sensor is shown in Figure 17.6. The main point in this construction concept is that it forms a closed "box" structure suited to become a claw of a gripper and can be treated as a finished mechanical element in hand design. As shown, this sensor can be used for (1) on-off (binary) indication of tactile pressure or (2) proportional indication of tactile pressure as far as the variable resistance element (e.g., a pressure-conductive plastic) proportionally responds to pressure. There are two more important features associated with the "box-sandwich" construction concept: (1) the sensing points or cells are mechanically and electronically well separated, and (2) the transducer elements are well protected since they are at the bottom layer of the "sandwich" and do not contact the external world.

17.4.   CONTROLS

Data-driven automation techniques provide means for the development of two specific modes of control enhancing the applicability of robots in remote operations: generalized bilateral manual control, and interactive manual-computer control.

17.4.1.   Generalized Bilateral Manual Control

Bilateral manual control permits the operator to feel the forces and torques acting on the robot arm/hand while he manually controls the motion of the robot arm. In this control mode, the operator is kinematically and dynamically "coupled" to the remote robot arm and can command with "feel" and control with a "sense of touch." This control mode can be viewed as a combination of "body language" and "reflective feedback" in control with basic motion primitives corresponding to the three translation and three rotation commands in the task space. This type of man-machine coupling is an important element of "integrated operator control" in teleoperation. (See Fig. 17.7.)

Bilateral, force-reflecting control is a key capability of thousands of master-slave manipulators widely used in the nuclear industry for many years. In these force-reflecting master-slave manipulator systems the master arm is mechanically coupled to the slave arm. In a few cases where the separation between master and slave arms is prohibitively large, the coupling between master and slave is electromechanically implemented through bilateral servo control.

A limiting factor for broadening the application of bilateral, force-reflecting manipulator control technology is the nature of the master arm. In the existing state-of-art implementations, the master arm is a one-to-one size kinematic replica of the slave arm, and each slave arm must have its own master arm.

A new form of bilateral, force-reflecting manual control has been implemented at JPL. It utilizes

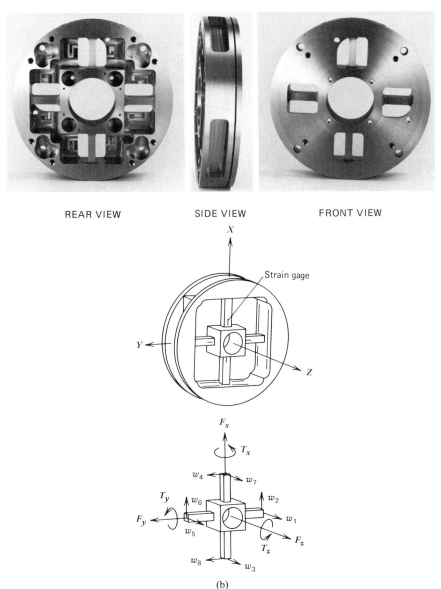

REAR VIEW   SIDE VIEW   FRONT VIEW

(b)

Fig. 17.5. Six-dimensional force-torque sensor. (*a*) Mechanical frame. (*b*) Reference axes.

a general-purpose force-reflecting hand controller.[11] The hand controller is a 6 DF control input device that can be backdriven by forces and torques sensed at the base of the end effector of a remote robot arm. This hand controller is general purpose in the sense that it does not have any geometric and dynamic similarity to the slave arm it controls; it is *not* a replica of any slave arm, but it can be coupled to and used for the control of any remote slave arm.

The positional control relation between the general-purpose hand controller and a remote robot arm is established through mathematical transformation of joint variables measured at both the hand controller and robot arm. Likewise, the forces and torques sensed at the base of the remote robot hand are resolved into appropriate hand-controller motor-drive commands through mathematical transformations to give to the operator's hand the same "feeling" that is "felt" by the remote robot hand.[12] The complex bilateral mathematical transformations are performed by a dedicated minicomputer in real time. These transformations also affect motion synchronization between hand controller and slave arm, referenced to the slave hand, by backdriving the hand controller. Overall system implementation

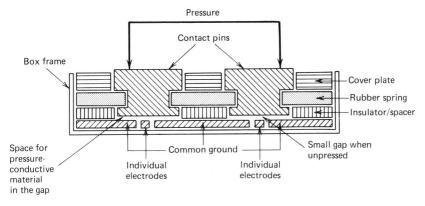

Fig. 17.6. Electromechanical construction concept of "Box-sandwich" touch sensor array.

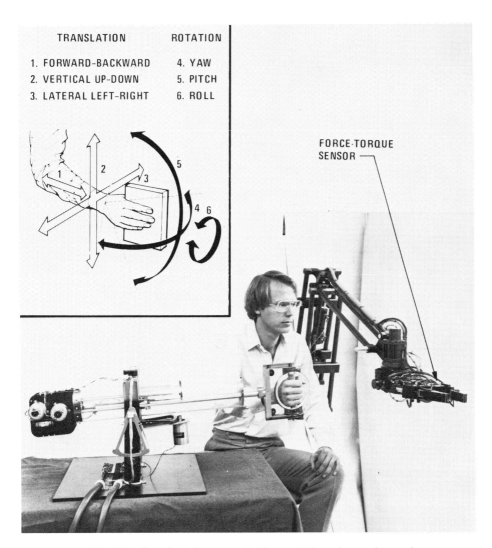

Fig. 17.7. Some body language primitives and bilateral manual control.

is shown in Figure 17.8. A preliminary control system analysis and synthesis of this system can be found in Reference 13. Some experimental results are presented in Reference 14.

The new form of bilateral manual control of remote robot arms described here generalizes the bilateral, force-reflecting manipulator control technique. The main objective is to overcome the limitations and inconveniences inherent to the existing master arms as control input and force feedback devices.

17.4.2. Interactive Manual-Automatic Control Using Sensors

In this mode of control, data from sensors integrated with the remote robot are used to adapt the real-time control actions to changes or variances in task conditions automatically through computer-control algorithms. The function algorithms can be selected by the operator from a preprogrammed menu.

A pilot computer-control system has been developed at JPL for a 6 DF robot arm equipped with proximity and force-torque sensors. This development is aimed to study and evaluate the hardware and software performance implications of man-computer interactive control in teleoperation. Interactive control signifies here a hybrid control capability which allows that some motions of the remote robot in the workspace coordinates are under manual control whereas the remaining motions in the same workspace reference coordinates are under automatic computer control referenced to proximity and force-torque sensor data. A preprogrammed control menu is available to the operator who decides on-line when and which automatic control function should be activated or deactivated. Each automatic control function selection can be accomplished by turning a simple on-off switch addressed directly to the control computer. Some parameters of the automatic control menu can be changed on-line. Note that, in this hybrid control system, the operator has a dual (analog/continuous and digital/discrete) communication with the control computer. Note also that, in extreme cases, all control can be either fully manual control or fully automatic control referenced to sensors.

The structure of the interactive control system software is built on a design concept which states that particular manipulator tasks can be considered as arrangements of interconnected actions which are enforced directly by the operator's continuous manual inputs or by automatic computer control algorithms. In order to synthesize the automatic control of interconnected complex actions, three action categories—primitive, composite, and complex actions—have been introduced. Primitive actions include elementary motions, (e.g., one-step shifts of the mechanical hand) in a given task frame. Composite actions are composed of several primitive actions which are executed sequentially or in parallel (e.g., follow a moving object). Execution of a complex action is determined by precedence rules that define the order of execution of the corresponding composite actions. These rules also specify the

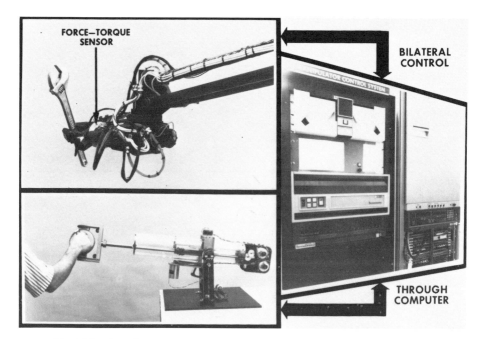

Fig. 17.8. Overall system implementation for generalized bilateral manual control.

conditions that must be satisfied to start or to terminate the execution of the actions. These rules can be expressed graphically by diagrams called Action Precedence Graphs.

The implementation of this pilot system is described in Reference 15. The general system configuration together with the manual and computer control panels are shown in Fig. 17.9. The manual control is normally in resolved-rate or resolved-position mode using the appropriate computer-control algorithms. The capability of executing both manual and computer controls within the same task, function, and action formulation frame facilitates the operational integration of human and computer decisions in remote robot arm control.

17.4.3. Voice Control

Conventional communication between man and machine in a teleoperator control station is indirect. Typically, all control commands the operator inputs to a machine (even the computer commands) require the operator's hand, and all outputs from machine to man typically require the operator's visual attention. This restricted man-machine communication often renders the whole operation inflexible and inefficient. Voice communication with machines offers a new communication channel which is open and within reach most of the time and does not require manual or some specific visual contact between man and machine.

Advancements in computer-based discrete word-voice recognition systems make the direct use of human speech feasible for control applications in a teleoperator station. Several such applications have been developed at JPL.[16] An application system was developed for the control of the Space

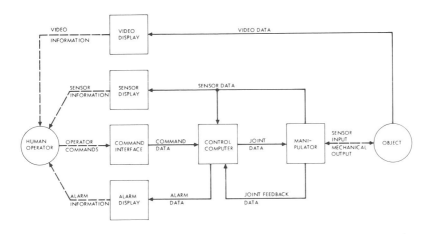

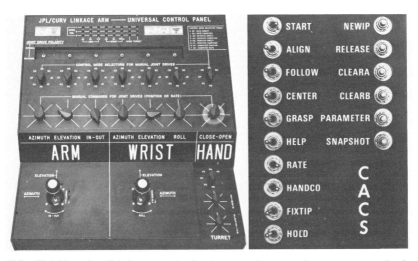

Fig. 17.9. Hybrid analog-digital communication in interactive manual-computer control referenced to sensors.

Shuttle television cameras and monitors while the operator manually controls the Shuttle robot arm. Some Shuttle robot arm tasks are visually very demanding, and can require 50–70 commands to four television cameras and two television monitors within 15–20 min to assure sufficient visual feedback to the operator. The ground-control tests at the Johnson Space Center[17] have shown 96–100% voice-recognition accuracy for the best test runs and resulted in the following major conclusions: (1) the application concept is realistic and acceptable; (2) the use of voice command indeed contributes to a better man-machine interface integration; (3) individual human acoustic characteristics and training have a major impact on system performance.

17.5. INFORMATION DISPLAYS

The stream of data generated by sensors on a "smart hand" (proximity, touch, and force-torque sensors) provides multidimensional information and requires quick (sometimes split-second) control response. In general, the control decision required to respond to the data is also multidimensional. This represents a demanding task and heavy workload for the human operator. It is also recognized that the use of information from sensors on a "smart hand" often requires coordination with visual information (see Fig. 17.10).

17.5.1. Event-Driven Displays

By definition, event-driven displays map a control goal or a set of subgoals into a multidimensional data space based on the fact that control goals or subgoals always can be expressed as a fixed combination of multidimensional sensory data. Event-driven displays can be implemented by real-time computer algorithms which (1) coordinate and evaluate the sensory data in terms of predefined events and (2) drive the graphics display. Flexible display drive algorithms require an open set of task-oriented parameters specifiable by the operator to match the specific needs of a given control task.

Several event-driven graphics displays of proximity, touch-slip, and force-torque sensor data have been developed at JPL.[18] Some displays are in black and white utilizing blinkers for event indications, and some are in color utilizing changes in the color to indicate an event.

Event-driven displays can considerably sharpen the information content of multidimensional sensor data and thereby aid the operator's perceptive task.

INTEGRATED OPERATOR CONTROL

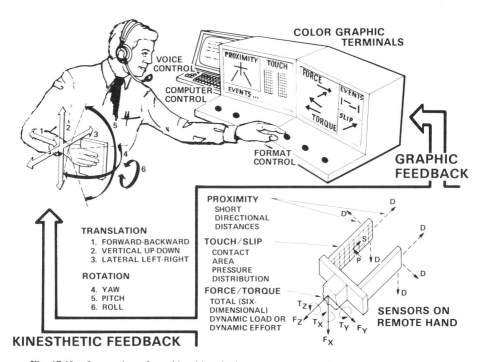

Fig. 17.10.   Integration of graphics, kinesthetic, computer, and voice man-machine interfaces.

17.5.2. Event-Controlled Displays

The need for different types of sensor data displays or for different formats of data displays typically arises in a logical sequence in remote robot-control tasks. For example, when proximity sensor data are needed, then normally there is no need for touch or force-torque sensor data, or vice versa. This sequential logic in the need of sensor information can be utilized to switch automatically between different data displays or formats. Following this concept, event-controlled displays have been implemented at JPL.[19] In the implemented examples, predefined changes in sensor data automatically effect changes in display modes, formats, and parameters, matching the need for a particular information required for remote robot-arm control to different phases of the task. Event-controlled displays require the implementation of state transition nets in real-time computer programs based on even-detection logic.

17.6. CONCLUSIONS

Application of teleoperators as man-extension systems requires flexibility in handling both control and information variables. The use of sensing- and computer-based automation offers significant possibilities for flexibility by providing programmable devices and techniques which permit task-level and "intelligent" two-way communication between human operator and remote robot. The development of these devices and techniques is a multidisciplinary activity and requires careful system design and integration taking also account of human factors requirements.

ACKNOWLEDGMENT

This work has been carried out at the Jet Propulsion Laboratory, California Institute of Technology, under contract with the National Aeronautics and Space Administration.

REFERENCES

1. Sheridan, T. B. and Ferrell, W. R., *Man-Machine Systems,* The MIT Press, Cambridge, Massachusetts, 1974.
2. Bejczy, A. K., Smart Sensors for Smart Hands, in *Progresses in Astronautics and Aeronautics,* Vol. 67, Publ. AIAA, New York, 1979.
3. McDermott, J., Sensors and Transducers, A Special Report, *Electronic Data News,* March 20, 1980.
4. Bejczy, A. K., Brown, J. W., and Lewis, J. L., Evaluation of Smart Sensor Displays for Multidimensional Precision Control of Space Shuttle Remote Manipulator, *Proceedings of the 16th Annual Conference on Manual Control,* Massachusetts Institute of Technology, Cambridge, Massachusetts, May 5–7, 1980.
5. Bejczy, A. K., and Dotson, R. S., A Force-Torque Sensing and Display System for Large Robot Arms, *Proceedings of the IEEE Southeastcon '82,* Destin, Florida, April 4–7, 1982.
6. Bejczy, A. K., Dotson, R. S., Brown, J. W., and Lewis, J. L., Manual Control of Manipulator Forces and Torques using Graphic Display, *Proceedings of the IEEE International Conference on System, Man and Cybernetics,* Seattle, Washington, October 28–30, 1982.
7. Harmon, L. D., Automated Tactile Sensing, *The International Journal of Robotics Research,* Vol. 1, No. 2, MIT Press, Cambridge, Massachusetts, 1982.
8. Dario, P., et al., Touch-Sensitive Skin Uses Piezoelectric Properties to Recognize Orientation of Objects, *Sensor Review,* October 1982, pp. 194–198.
9. Bejczy, A. K., Application of Fiber Optics to Robotics, *International Fiber Optics and Communications,* Vol. 1, No. 6, November 1980.
10. Bejczy, A. K., "Smart Hand"—Manipulator Control Through Sensory Feedback, *Report, JPL D-107,* January 15, 1983.
11. The mechanism of the force-reflecting hand controller was designed by J. K. Salisbury, Jr., Design Division, Mechanical Engineering Department, Stanford University, Stanford, California 1980.
12. Bejczy, A. K., and Salisbury, J. K., Jr., Kinesthetic Coupling Between Operator and Remote Manipulator, *Proceedings of ASME Computer Technology Conference,* Vol. 1, San Francisco, California, August 12–15, 1980; and Controlling Remote Manipulators Through Kinesthetic Coupling, *Computers in Mechanical Engineering,* Vol. 1, No. 1, July 1983, pp. 48–60.
13. Handlykken, M. and Turner, T., Control System Analysis and Synthesis for a Six-Degree-of-Freedom Universal Force-Reflecting Hand Controller, *Proceedings of the 19th IEEE Conference on Decision and Control,* Albuquerque, New Mexico, December 10–12, 1980.
14. Bejczy, A. K. and Handlykken, M., Experimental Results with a Six-Degree-of-Freedom Force-

Reflecting Hand Controller, *Proceedings of the 17th Annual Conference on Manual Control,* UCLA, Los Angeles, California, June 16–18, 1981.

15. Bejczy, A. K. and Vuskovič, M., An Interactive Manipulator Control System, *Proceedings of the 2nd International Symposium on Mini- and Microcomputers in Control,* ACTA Publ., Anaheim, California, 1980.

16. Bejczy, A. K., Dotson, R. S., and Mathur, F. P., Man-Machine Speech Interaction in a Teleoperator Environment, *Proceedings of Symposium on Voice Interactive Systems,* DOD Human Factors Group, Dallas, Texas, May 13–15, 1980.

17. Bejczy, A. K., Dotson, R. S., Brown, J. W., and Lewis, J. L., Voice Control of the Space Shuttle Video System, *Proceedings of the 17th Annual Conference on Manual Control,* UCLA, Los Angeles, California, June 16–18, 1981.

18. Bejczy, A. K. and Paine, G., Event-Driven Displays for Manipulator Control, *Proceedings of the 14th Annual Conference on Manual Control,* University of Southern California, Los Angeles, California, April 25–27, 1978.

19. Paine, G. and Bejczy, A. K., Extended Event-Driven Displays for Manipulator Control, *Proceedings of the 15th Annual Conference on Manual Control,* Wright State University, Dayton, Ohio, March 20–22, 1979.

BIBLIOGRAPHY

Baron, S. and Kleinman, D. C., The Human as an Optimal Controller and Information Processor, *IEEE Transactions on Man-Machine Systems, MMS-10,* 1, 1969, pp. 9–17.

Bejczy, A. K., Distribution of Man-Machine Controls in Space Teleoperation, SAE Aerospace Congress, *Behavioral Objectives in Aviation Automated Systems Symposium Proceedings,* Paper P-114-821496, October 25–28, 1982, Anaheim, California, pp. 15–27.

Bejczy, A. K., Kinesthetic and Graphic Feedback for Integrated Operator Control, *Proceedings of the Sixth Annual Advanced Control Conference,* April 28–30, 1980, Purdue University, West Lafayette, Indiana, pp. 137–147.

Bejczy, A. K., Performance Evaluation of Computer Aided Manipulator Control, *Proceedings of the IEEE International Conference on System, Man and Cybernetics,* November 1976, Washington, D.C.

Bejczy, A., Sensors, Controls, and Man-Machine Interface for Advanced Teleoperation, *Science,* **208,** 1980, pp. 1327–1335.

Bejczy, A. K., Brooks, T. L., and Mathur, F. P., *Servomanipulator Man-Machine Interface Conceptual Design,* JPL Report No. 5030-507, U.S. Dept. of Energy, August 1981.

Bejczy, A. K., Effect of Hand-Based Sensors on Manipulator Control Performance, *Mechanism and Machine Theory,* Vol. 12, 1977, pp. 547–567, Pergamon Press.

Brooks, T., Superman: A System for Supervisory Manipulation and the Study of Human-Computer Interactions, *Master's Thesis,* Man-Machine Systems Laboratory, Massachusetts Institute of Technology, May 1979.

Chu, Y., Chen, K., Clark, C., and Freedy, A., Analysis and Modeling of Information Handling Tasks in Supervisory Control of Advanced Aircraft, *Final Technical Report No. PFTR-1080-82-5,* Perceptronics, Inc., Woodland Hills, California, May 1982.

Estabrook, N., Wheeler, H., Uhler, D., and Hackman D., Development of Deep-Ocean Work System, *IEEE Ocean '65,* 1979, pp. 573–577.

Ferrell, W. R. and Sheridan, T. B., Supervisory Control of Remote Manipulation, *IEEE Spectrum,* October 1967, **4,** 81–88.

Ferrell, W. R., Command Language for Supervisory Control of Remote Manipulation, in E. Heer (Ed.), *Remotely Manned Systems,* California Institute of Technology, Pasadena, California, 1973, pp. 369–373.

Fyler, D. C., Computer Graphic Representation of Remote Environment Using Position Tactile Sensors, M. I. T., *Man-Machine Systems Laboratory Report,* Cambridge, Massachusetts, August 1981.

Groome, R. C., Force Feedback Steering of Teleoperator System, *MS Thesis,* Massachusetts Institute of Technology, Department of Aeronautics and Astronautics, Cambridge, Massachusetts, August 1977.

Handlykken, M., and Turner, T., Control Systems Analysis and Synthesis for a Six Degree-of-Freedom Universal Force-Reflecting Hand Controller, *Proceedings of the Nineteenth IEEE Conference on Decision and Control,* December 10–12, 1980, Albuquerque, New Mexico, pp. 1197–1205.

Hill, J. W. and Sword, A., Manipulators Based on Sensor Directed Control: An Integrated End Effector and Touch Sensing System, *Proceedings of the Seventeenth Annual Conference in Human Factors,* October 1979, Washington, D.C.

Hill, J. W., Study of Modeling and Evaluation of Remote Manipulation Tasks with Force Feedback, *Final Report For JPL, SRI Project 7696 JPL Contract 95-5170*, March 1979.

Jagacinski, R. J., and Miller, R. A., Describing the Human Operator's Internal Model of a Dynamic System, *Human Factors*, **20** (4), 1978, pp. 425–433.

Jelatis, D. C., Characteristics and Evaluation of Master-Slave Manipulators, *Performance Evaluation of Programmable Robots and Manipulators, NBS Special Publication 459*, October 1975, pp. 141–145.

Johnson, E. G. and Corliss, W. R., *Human Factors Applications in Teleoperator Design and Operation*, John Wiley & Sons, Inc., 1971.

Kohler, G. W., *Manipulator Type Book*, Verlag Karl Thiemig, Munchen, FRG, 1981.

Leifer, L., Sun, R., and Van der Loos, H. F. M., Terminal Device Centered Control of Manipulation for a Rehabilitative Robot, *Proceedings of Joint Automatic Control Conference*, 1980.

McCoy Winey III, C., Computer Simulated Visual and Tactile Feedback as an Aid to Manipulator and Vehicle Control, M. I. T., *Man-Machine Systems Laboratory Report*, Cambridge, Massachusetts, May 1981.

Nevins, J. L., Sheridan, T. B., Whitney, D. E., and Woodin, A. E., The Multi-Moded Remote Manipulator System, in E. Heer (Ed.), *Remotely Manned Systems*, California Institute of Technology, Pasadena, California, 1973, pp. 173–187.

Okada, T., Object-Handling System for Manual Industry, *IEEE Transactions on Systems, Man, and Cybernetics*, **SMC-9** (2), February 1979, pp. 79–89.

Person, L. H. and Steinmetz, G. G., The Integration of Control and Display Concepts for Improved Pilot Situational Awareness, *Proceedings of the 34th International Air Safety Seminar*, November 9–15, 1981, Acapulco, Mexico, pp. 122–130.

Setzer, W. and Vossius, G., On the Stability Problem of Human Arm and Hand Movements Controlling External Load Systems, *Proceedings of the First Annual European Conference on Manual Control*, May 25–27, 1981, Delft University of Technology, Delft, The Netherlands, pp. 243–253.

Sheridan, T. B., Supervisory Control: Problems, Theory and Experiment for Application to Human-Computer Interaction in Undersea Remote Systems, *M.I.T. Man-Machines Systems Laboratory Report*, Cambridge, Massachusetts, March 1982.

Sheridan, T. B., Supervisory Control of Remote Manipulators, Vehicles and Dynamic Processes: Experiments in Command and Display Aiding, *M.I.T. Man-Machine Systems Laboratory Report*, Cambridge, Massachusetts, March 1983.

Shultz, R. E., Tesar, D., and Doty, K. L., Computer Augmented Manual Control of Remote Manipulator, *Proceedings of the 1978 IEEE Conference on Decision and Control*, San Diego, California, January 1979.

Spiger, R. J., Farrell, R. J., and Tonkin, M. H., Survey of Multi-Function Display and Control Technology, *NASA Report No. CR-167510*, Boeing Co., 1982.

Stark, L. and Ellis, S. S., Revisited: Cognitive Models in Direct Active Looking, in *Eye Movement, Cognition and Visual Perception*, Fisher, Monty, and Senders (Eds.), Erlbaum Press, New Jersey, 1981, pp. 193–226.

Starr, G. P., Supervisory Control of Remote Manipulation: A Preliminary Evaluation, *Proceedings of the Seventeenth Annual Conference on Manual Control*, June 16–18, 1981, University of California, Los Angeles, California, pp. 95–107.

Starr, G. P., A Comparison of Control Modes for Time-Delayed Remote Manipulation, *IEEE Transactions on Systems, Man, and Cybernetics*, **SMC-9** (4), 1979, pp. 241–246.

Stassen, H. G., Man as Controller, *Introduction to Human Engineering*, Koln: Verlag TUV, Rheinland, 1976.

Vertut, J., Experience and Remarks on Manipulator Evaluation, *Performance Evaluation of Programmable Robots and Manipulators*, NBS Special Publication 459, October 1975, pp. 97–112.

Vykukal, H. C., King, R. F., and Vallotton, W. C., An Anthropomorphic Master-Slave Manipulator System, in E. Heer (Ed.), *Remotely Manned Systems*, California Institute of Technology, Pasadena, California, 1973, pp. 199–205.

Wagner, E. and Hanett, A., MINIMAC—the Remote-Controlled Manipulator with Stereo TV Viewing at the SIN Accelerator Facility, *Transactions of the American Nuclear Society*, Vol. 30, pp. 759–760, 1978.

White, T. N., Modeling the Human Operator's Supervisory Behavior, *Proceedings of the First European Annual Conference on Human Decision Making and Manual Control*, May 25–27, 1981, Delft University of Technology, Delft, The Netherlands, pp. 203–217.

Yoerger, D. R., Supervisory Control of Underwater Telemanipulators: Design and Experiment, M.I.T., *Man-Machine Systems, Laboratory Report*, Cambridge, Massachusetts, August 1982.

PART 4
ROBOT INTELLIGENCE

CHAPTER 18

ELEMENTS OF INDUSTRIAL ROBOT SOFTWARE

LARRY L. HOLLINGSHEAD

Cincinnati Milacron, Inc.
Cincinnati, Ohio

18.1. INTRODUCTION

Robot software is the collection of computer programs and related information that is developed, marketed, manufactured, and sustained for industrial robots.

The purpose of this chapter is to present an overview of the software in modern industrial robots. The discussion begins by examining its role in robot manufacture, application, and operation. The general characteristics of contemporary robot software systems are then described. Next, the fundamental capabilities of modern systems are analyzed and illustrated with specific examples. The final section describes the advanced work in off-line programming sponsored by the U.S. Air Force ICAM project.

The reader of this chapter will benefit from having general familiarity with computer technology and robotics principles. This handbook contains a wealth of information about basic robotics fundamentals, should questions arise. Another excellent source is Joseph Engeleberger's book, *Robotics in Practice*.[8]

18.2. THE ROLE OF SOFTWARE IN INDUSTRIAL ROBOTS

The role of software in industrial robots is multifaceted. Characteristics required at one stage or by a certain group of people are often different from those required by others. For example, the engineer developing a robot application needs to have good task-generation and management tools so that he can do his job quickly and effectively. In production, on the other hand, there is little reason to be concerned with the application process. Goals include performance, ease of operation, reliability, safety, and so on. The robot manufacturer has a third set of needs centered around manufacturing, configurability, and product support issues. So the "role" of software in industrial robots is best seen from at least three points of view:

Operation

Application

Manufacture

The contemporary industrial robot exists to assist in the production of manufactured goods. Virtually all of its life is spent at work. Therefore, the dominant role of software is to provide equipment that consistently delivers quality work with high productivity.

In addition, the robot system must be reliable and easy to operate. Operational ease is provided through software-managed control procedures, through concise and meaningful status reporting and error messages, and with simple, yet thorough, diagnostic functions. System reliability is a composite of both hardware and software factors. The role that software plays in assuring overall system reliability is to detect hardware failures where possible, but, most important, to be reliable itself. There are great strides to be made in understanding and producing reliable software systems in general, and with robots failures can be expensive and/or dangerous.

A different set of features becomes important when the robot is applied to a specific task. The role that software plays at this time is to facilitate the creation of an application program that defines the calculations and actions that the robot is to perform in production. One important factor is the ease with which the system is taught or programmed. The application engineer must be given access to all basic system components in a simple, straightforward manner. For robot-programming language systems it is also necessary to provide data structures, computational power, and appropriate sensor interfaces. Powerful debugging facilities are essential to the task-generation process. Finally, the system software usually includes provisions for displaying, listing, and storing the application task externally.

The production and application aspects of industrial robot software are quite visible. They are important to the robot manufacturer because they must be done well if the product is to be successful. Success is equally dependent on a number of other, less visible, software features and characteristics. The robot supplier relies on software to make his systems as general and as flexible as possible. Doing so allows a greater range of applications for the same product, making the cost per unit lower. The same manufacturing, training, documentation, and support functions serve many customers. Table 18.1 shows the variety of products sharing a common software base for selected manufacturers.

Microprocessors (and therefore software) are commonly used by robot manufacturers to reduce manufacturing cost. A good example of this technique is the display screen found in modern controls. Operator interaction that formerly used expensive push buttons and panel lights is conducted through a single mechanism. Many functions of the typical interpolation and servo components are also implemented with microprocessors, thereby eliminating the cost of expensive electronic hardware.

Finally, the manufacturing process itself can be made more efficient by incorporating software that assists by automating diagnosis during run-off and by providing tools for adjustments made prior to customer shipment.

18.3. CONTEMPORARY ROBOT SOFTWARE SYSTEMS

Robot software systems are frequently classified according to the manner in which the user programs them. Bonner and Shin[3] have published a comparative study of robot languages that defines five classes of robot software and surveys 14 systems. Tomas Lozano-Perez[11] provides an extensive review of requirements for and developments in robot programming systems of three types: guiding, robot level, and task level.

This chapter presents robot software in classes corresponding to the first two of Lozano-Perez. The shop floor guiding machines include those sold by ASEA, Cincinnati Milacron, Cybotech, General Electric, Prab, and Unimation. They are represented in this chapter by the Cincinnati Milacron T3 family.[6] The Automatix RAIL system[1] and the IBM AML approach[10] are used to characterize robot-programming language systems. Some other products in this class are the Unimation robots with VAL, the General Electric machines based on the HELP language, and the International Machine Intelligence arms, which use BASIC.

18.3.1. Shop Floor Machines

The shop floor machines are the most frequently used robots. Their hallmark is simplicity and ease of application. They are programmed with an interactive guiding process, sometimes called *teach by showing*. The user interface is designed to be natural and friendly to trained shop floor personnel.

Teaching and operation are both performed with simple operator interaction. Usually a menu-driven scheme is supplied along with dedicated push buttons and panel lights. All shop floor systems include a hand-held control that the user carries as he moves around the work area while teaching.

The robot task description consists of one or more segments which each represent a path through space. These segments can be grouped together (usually by concatenation) to form a *cycle*. The cycle is the top-level logic in the task. The individual steps that comprise a segment are called *points*. A point consists of geometric information about the location and orientation of the tool attached to the

TABLE 18.1. INDUSTRIAL ROBOT SOFTWARE

Manufacturer	Software	Products	Application
Automatix	AI32/RAIL	Autovision Robovision Cybervision	Inspection Arc welding Assembly
Cincinnati Milacron	Acramatic V4	T3-726 T3-746/756 T3-566/586	General, light-duty Process General, heavy-duty
Cybotech	RC-6	V15 P-15 V80, G80, H80	General, light-duty Painting General, heavy-duty
IBM	RS/1/AML	RS/1	Assembly

robot arm, as well as other data that describe the step. Each point has information about how fast that location is to be approached, whether the arm is to stop at that location or continue smoothly through it, and what, if any, control function is to be performed there. In addition, some systems allow the specification of the method of interpolation (straight-line, circular, etc.) used during the approach to each point.

Shop floor systems are distinguished by the close coupling between the geometric description of the task and the control-sequence specification.

18.3.2. Robot-Programming Language Systems

The robot-programming language systems include the robot arm as an extension of a general-purpose automation element. These devices can sense their environment, perform analysis and computation, store and use shop floor information, and direct the activity of the arm and other manufacturing equipment. In one sense they augment the capabilities of the robot arm to make it "intelligent." In another, they place the arm in the role of end effector in a higher element of automation.

The robot task description consists of data items and a collection of software modules written in the robot-programming language. The data items describe locations in the robot's workspace, parameters such as arm velocity, or welding controller settings, and so on, and other values used during task execution. The software modules are written by the user in a high-level programming language and describe the overall flow of the robot task along with computation, sensing, and data management operations to be done.

Computer science concepts are integral to the robot-programming language systems. Among them, the most significant is the fundamental separation between control logic and data. While the shop floor machines keep a one-to-one relationship between path and the task logic, the robot-programming language method makes a clear distinction. Geometric information about the work or workspace is represented as data that are manipulated by the program. The actual path followed by the arm is determined by the sequence in which task statements are executed as well as by the data that give geometric locations and tool orientation. The flow of control is specified with modern structured programming constructs.

Computer science influences can be seen in many other areas. Simple data files are supported for either communication or information storage and retrieval. Traditional software development tools are also available for documenting and managing task generation.

As capable and flexible as they are, the computer science nature of the robot-programming language systems has been a substantial barrier to their acceptance in industry. Like many powerful tools, they require training and a new way of looking at the job to be done. They are often rejected by shop floor users as being difficult to understand and apply.

18.3.3. Product Foundations

The internal characteristics of robot software are intimately related to the programming tools and electronic hardware chosen by the manufacturer. Nearly all modern controls have a distributed microprocessor hardware architecture, and most robot software is written in a high-level system programming language. Assembly language is found, though, in sections of code where speed is essential (particularly the arm motion element) or where the host microprocessor's primitive data types are inadequate. Table 18.2 lists the computers and programming languages behind many of today's robots.

Commercial real-time operating systems are also becoming part of the robot software base. For example, the Westinghouse MCS-60 robot control system[14] uses Intel's RMX-86 for its core software.

TABLE 18.2. SOFTWARE AND COMPUTER FOUNDATIONS

Manufacturer	Control	Programming Language	Computer
Automatix	AI32	Pascal, Assembly	Motorola 68000
Cincinnati Milacron	Acramatic V4	PL/M, Assembly	Intel iAPX 85, 86
Cybotech	RC-6	LPR, Assembly	Intel iAPX 86
General Electric	Mark C 2000	Pascal, Assembly	Intel iAPX 86
General Electric	Allegro A12	Pascal, Assembly	DEC LSI-11 Intel iAPX 85
IBM	RS/1	Proprietary	IBM Series 1
Westinghouse	MCS-60	PL/M, Assembly	Intel iAPX, 85, 86
Westinghouse	VAL	Assembly	DEC LSI-11

This strategy offers the robot manufacturer advantages of standardized internal interfaces, faster software development, and greater hardware independence. In the past, commercially available real-time operating systems for microprocessors were too slow and filled too much memory. However, as microprocessor speeds increase and memory prices drop, many manufacturers of industrial control equipment can be expected to adopt this approach.

18.4. COMMON ROBOT SOFTWARE ELEMENTS

To describe the functions that robot software typically provides, they are grouped into six common elements as shown in Figure 18.1. Each element is explained in this section with examples taken from contemporary robot systems.

18.4.1. Supervisory Control

The supervisory control element of industrial robot software is responsible for the overall control and coordination of the robot system. These studies usually include both the internal sections of the system and synchronization with the external environment including the operator, associated manufacturing devices, and possibly a higher-level control computer.

The supervisory control element contains a command interpreter that receives and is responsible for the execution of directions from the operator or central control computer. Operator communications are conducted through a combination of push buttons and panel lights and possibly a display or through a keyboard/display screen. In addition, all robot systems provide a hand-held device with buttons/lights that can be carried about the work area. Any of the commands directed to the task generation and management section or to the task interpreter are handled through the supervisory control element. Likewise, error reporting and resolution is handled here.

In addition, there are usually a number of auxiliary operations supplied at this level. One of these is the system generation procedure which configures the manufacturer's generic software to the user's specific configuration (see Figure 18.2). Another is found in the diagnostic tests executed to detect hardware or other system failures. The compilation and reporting of robot performance or maintenance data represents a third example of an auxiliary supervisory control function.

Supervisory control functions are found in the shop floor machines to handle operator interaction in production, during teaching, and when defining system parameters at installation time. These machines typically have simple operator-control panels, and interaction is menu oriented.

Cincinnati Milacron robots operate in one of three modes: *manual, teach,* or *auto.* Auto mode is for production, teach is for teaching, and manual is for direct, axis-by-axis control of arm motion from the pendant. The control panel, shown in Figure 18.3, is used to start or stop the robot, to switch between operating modes, to reset errors, and so on. Teaching and status interrogation are done with a portable CRT terminal whose keyboard is shown in Figure 18.4. These operations can also be performed with the hand-held teach pendant (Figure 18.5), which controls axis motion and task execution in addition.

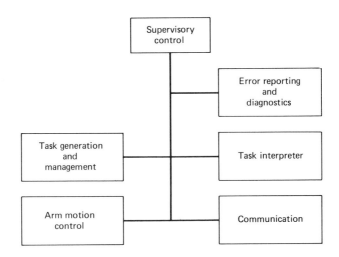

Fig. 18.1. Common robot software elements.

Input/output locations
Serial communication characteristics
Tool transformation
Base transformation
Axis calibration values
English or metric units
Timing limits
Memory sizes
Welding parameters
Start-up file name

Fig. 18.2. Typical configuration parameters.

The robot-programming language supervisory control functions are relatively simple as most commands are handled by the task interpreter as explained in Section 18.4.3.

18.4.2. Task Generation and Management

The task generation and management element is the collection of tools supplied for creating, debugging, and otherwise managing the procedures and data associated with each robot task.

Task creation is currently accomplished in two ways: the interactive guiding method (also called *teach by showing* or *lead-through*), or by writing programs in a *robot-programming language*. Computer-aided design, or CAD, systems are likely to assume more and more responsibility for task creation

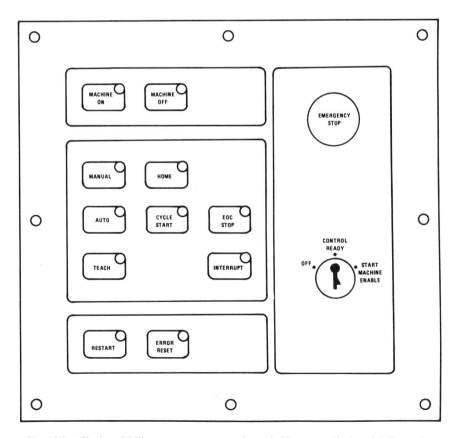

Fig. 18.3. Cincinnati Milacron operator control panel. (Courtesy, Cincinnati Milacron.)

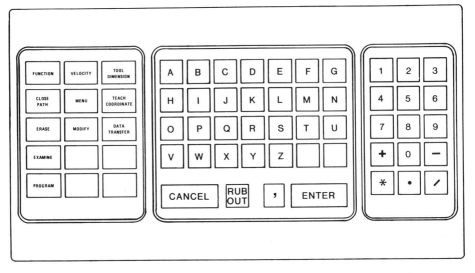

Fig. 18.4. Cincinnati Milacron teach station. (Courtesy, Cincinnati Milacron.)

as they develop. Just as process planning techniques are used to help generate machine tool part programs, it is reasonable to assume that the necessary robot task description could also be generated automatically. In fact, the U.S. Air Force ICAM project (see Section 18.5) has already taken steps in this direction.

Debugging tools, however, are still an essential part of the task generation and management element; they are certain to remain as long as humans write robot task programs (or programs that generate robot tasks!) and until the manufacturing environment becomes completely manageable. Debugging tools serve three major purposes. They permit the user to control and track the execution of the task. The constituent program and data can be displayed and modified. And finally, the robot and associated equipment can be manually operated to prepare the work cell for execution of the specific segment of the task being debugged.

Task management tools also include functions for listing and saving the task program and data. Off-line storage on magnetic tape or disk is the most common choice for saving tasks. As communication networks and central computer control achieve wider use, remote task storage will also be provided. There is usually a means for printing hard-copy listings of the robot task, too.

Tasks are created in the shop floor machines by moving the arm through each path segment, entering control functions and other parameters along the way. Additional information is entered at the console. The Cincinnati Milacron teach station or pendant, for example, can be used to examine the task or to examine/change internal variables and tables of parameters.

Pendant-directed arm motion is done in either an axis-by-axis fashion or in coordinated movement. Movement can typically be in joint or world (rectangular) coordinates. Cylindrical and tool-oriented motion is often provided as well.

Tasks are altered by moving the arm and entering commands to insert, modify, or delete points. Changes can also be made from the operator's console without physically moving the arm.

Debugging is accomplished by selective execution of indiviual steps in the task. Shop floor systems let the operator go through the task one step at a time in either the forward or reverse direction.

Tasks can be displayed at the console or listed on a hard-copy device. Off-line storage is provided on magnetic tape cassettes in most cases, although bubble memory is available in some systems.

Tasks are created in a robot-programming language system by first entering program statements with a text editor, then using the pendant as required to define the location of points. After the initial definition of the software modules in a task, debugging is done by running the task on the robot, one step at a time if necessary, while using the CRT/keyboard to display and/or modify statements or variables. Since these systems are interpretive, iteration in the program development cycle can occur quickly and efficiently. Completed tasks are stored on tape cassette, diskette, or disk.

The text editors are basic, line-oriented tools. Operations such as print, insert, or delete a line are standard. In addition, there are commands for searching the text for a particular word or phrase and replacing it with a new one. The IBM editing facility features *editing system subroutines,* which can be used to construct programs that edit other programs automatically.

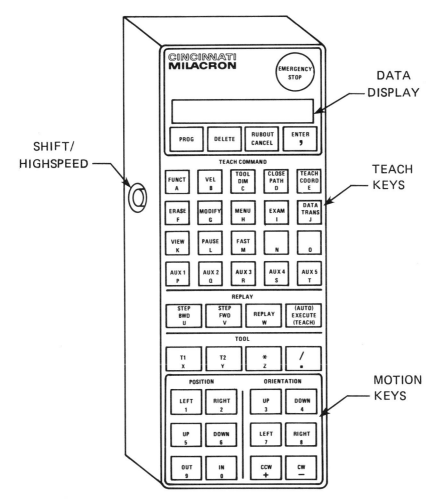

Fig. 18.5. Cincinnati Milacron teach pendant. (Courtesy, Cincinnati Milacron.)

There are special commands for teaching geometric information with the pendant. The Automatix LEARN command allows the operator to define points, sequences of points called *paths,* or a reference frame. To do this, the LEARN or LEARN FRAME command is entered; when the arm is positioned to the desired location, a pendant button is used to save its coordinates. The equivalent AML command is a system subroutine called GUIDE. When GUIDE is executed, the operator moves the arm as desired; pressing the pendant END button causes the GUIDE routine to terminate, returning the current arm position coordinates to the calling program.

Because the robot-programming language systems treat geometric information uniformly and separately from task logic, it is also practical to generate locations computationally or to load them from another source.

The debugging tools range from the usual single-step and breakpoint features to extensive tracing facilities offered by IBM. The single-step mode of operation causes the task interpreter to halt after each statement. The operator may then inspect or modify the task or the workplace before the next statement is executed. The Automatix SINGLESTEP command applies to every statement encountered. The IBM SINGLESTEP system subroutine is executed selectively only above an operator-specified subroutine level. Either system can have operator-initiated breakpoints, and the BREAK subroutine can be used in any AML program.

There are two other debugging system subroutines available in AML. TRACE writes extensive information about the execution of each statement to the display, the printer, or a disk file. WALKBACK shows the statements executed before arriving at the current statement.

18.4.3. The Task Interpreter

The task interpreter controls the step-by-step execution of the robot task. It is responsible for fetching, analyzing, and initiating each step that is performed. The most visible effect of this process is the motion of the arm. The task interpreter must also monitor real-time events and use that information to direct task execution. Communication with the operator and collection of production statistics are also performed.

Most robot systems use an interpreter rather than compiling the task description. The interpretive approach for shop floor machines follows quite naturally from the need to represent, then retrieve and repeat actions performed by the operator. An interpretive approach has many advantages to the robot supplier. It facilitates the creation of a simple, safe user interface that contains items and operations that are natural to robotics. Task-debugging tools can be interactive and give the developer greater control over when and how much of the task is executed. In addition, task representations are compact and are less sensitive to changes in the underlying system hardware. The task interpreters for shop floor robots control the system in production and while replaying various segments of the program during the teaching process.

The interpretation process is closely tied to the geometry of the task; that is, the interpreter causes the arm to move along the path from one point to the next. There is essentially no distinction made between a step and a move to the next point. A decision-making, communication, or other control function can be performed at each point, if necessary. In some cases the system is flexible in allowing the control action to "anticipate" arrival of the tool at the programmed point. A positive or negative anticipation time specifies that the action be performed before or after the point is reached.

The path in space can be defined to be in fixed, absolute locations, or in locations relative to a variable starting position. The path can also be expressed with respect to a moving reference frame for tracking an object traveling along a conveyor.

To increase flexibility, shop floor task interpreters can recognize more than one place along the path to begin motion. The Cybotech RC-6,[7] for example, uses a designated *looping point* to separate a segment into an approach path, which is followed once (the first time), and a working path, which can be repeated many times. Cincinnati Milacron segments can include index points that subdivide them into several, shorter paths.

Decision-making capabilities provided by these interpreters are simple, yet sufficient for traditional robot applications. These are usually tests on digital process signals and internal binary flags. The Cincinnati Milacron T3 family has been extended to include integer variables. Primitive arithmetic expressions and relational operators are supplied for counting and decision making (see Figure 18.6).

Interrupts, system errors, and other asynchronous events can be handled to a limited degree. However, the shop floor machines do not have the sophisticated monitoring and event-handling capability of the IBM AML language discussed in Chapter 21.

The heart of a robot programming language system is its interpreter. While the shop floor machines tend to distinguish between teaching commands and task execution, the robot programming language systems are homogeneous. Commands are executed immediately as received, and there is no inherent distinction between teaching and day-to-day production operation. Furthermore, sequences of commands can come from the operator keyboard, a local disk or tape cassette, or a remote computer.

Automatix and IBM differ in their approach to providing a robot-programming environment. Automatix defined a simple set of functions for robot assembly, seam welding, and computer vision. The selected functions were then implemented as a language much like the popular Pascal. The result is a simple, yet powerful tool for controlling automation equipment. Figure 18.7 shows an example RAIL program segment.

The IBM approach, on the other hand, is to provide the base AML language and a large collection of general-purpose *system subroutines*. There are five classes:

1. Control and error handling.
2. Motion.
3. Sensor and process monitoring.
4. Calculation.
5. Data processing.

Specific robot operations and data structures are then built by combining these subroutines with user-defined AML programs. See Figure 18.8.

The difference between the two programming approaches is indicated by the following example. Suppose the arm is to move quickly to the vicinity of the workpiece before moving to its exact location more slowly. The RAIL language APPROACH command performs this move with one statement, a predefined operation of the interpreter. The AML implementation requires that the user write an AML subroutine that first computes a location and tool orientation for the approach, then issues a

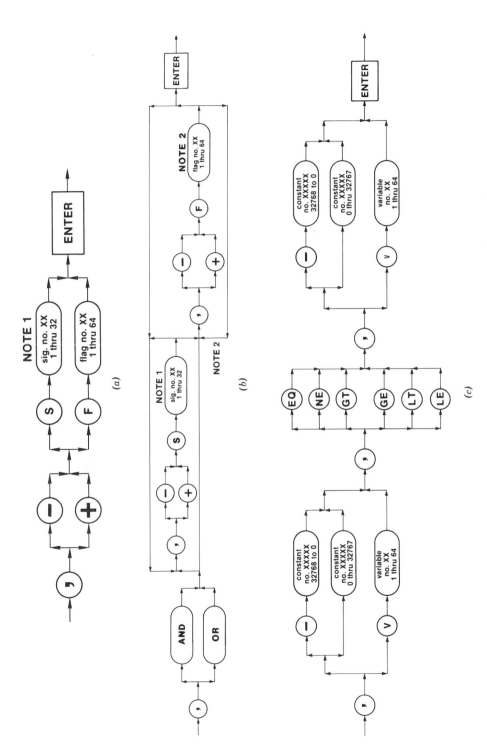

Fig. 18.6. Typical shop floor robot decision conditions. (*a*) Single, (*b*) multiple, and (*c*) numeric comparison. (Courtesy, Cincinnati Milacron.)

A RAIL Program

```
OUTPUT PORT CLAMP 3
INPUT PORT PART_READY 4

FUNCTION SEAM
  GLOBAL FIXTURE, SEAM1
  BEGIN
;
; SEAM is a function in which the robot welds
; a seam that is defined relative to a fixture.
;
    REPEAT
      MOVE SLEW HOME
      WAIT UNTIL PART_READY == ON
      CLAMP = ON
      APPROACH 50 FROM FIXTURE : SEAM1
      WELD FIXTURE : SEAM1 WITH SPEEDSCHED[2],
                     WELDSCHED[2]
      DEPART 50
      CLAMP = OFF
    UNTIL CYCLESTOP == ON
  END
```

Fig. 18.7. An example RAIL program. (Courtesy, Automatix.)

MOVE command. The RAIL command is simple to use, whereas the AML subroutine offers flexibility in specifying exactly how the workpiece is approached.

Both languages feature modern structured programming control constructs such as IF . . . THEN, WHILE, REPEAT. There are multiple data types provided for computation, data collection, operator communication, and sensor interaction. RAIL supports integer, real, string, and logical scalars and arrays. AML also has integer, real, and string scalars; full data structures, including arrays, are provided with *aggregates*.

18.4.4. Arm Motion Control

The arm motion element is the real-time control software that moves the arm as dictated by the task interpreter. The most common form of arm motion is expressed in terms of position, that is, moving from one point in space to another. There are several methods for controlling motion between points. The simplest form is point-to-point where all axes are moved independently to the next location. A second method, commonly used by spray-painting robots, is called continuous path control. Continuous path control software records a stream of closely spaced sets of joint angles during teaching and

An AML Program

```
GOALS: STATIC << · · · >,< · · · >, · · · >; -- Aggregate of 10 goals
  MOVEIT: SUBR;        -- Move arm to desired goal
    G: NEW O;          -- Will contain goal number
    CODE: NEW O;       -- Will contain completion code
    DISPLAY ('ENTER GOAL NUMBER →');-- Prompt
    G = CVTSN ( READ ( O , STRING( 5 ) ) );
                       -- Read the string just entered and convert
                       -- it to numeric; this is the goal number
    CODE = DOIT ( G ); -- Call DOIT to move the arm according
                       -- to the goal number; get completion code
    IF CODE NE O
      THEN RETURN ('NUMBER MUST BE BETWEEEN 1 and 10');
                       -- If DOIT detected an error,
                       -- quit with a message
  END;
```

Fig. 18.8. An example AML program. (Courtesy, IBM.)

reproduces them during production. The most sophisticated and flexible control techniques, called *controlled path,* coordinate axis motion. Coordination can range from simply guaranteeing that all axes arrive at the end point together to performing straight-line or curved motion between points.

Seam-welding applications are now done with specialized control of the path between two points: a sinusoidal "weaving" motion. There are also adaptive techniques that use vision, the welding gun arc current, or other sensors to follow the seam being welded.

Although position control is most common, research is being conducted for control by force, which is expected to be particularly important for assembly operations (see Paul[14]).

Arm motion software is the most critical element in an industrial robot because it affects system performance. The two key factors are the speed at which the arm's servo hardware can be updated and the accuracy with which commands are calculated. In addition, extra consideration must be paid when the programmed path passes through the geometric degeneracy for the arm under control.

The shop floor systems offer state-of-the-art arm motion control features. Velocity is specified on a point-by-point basis, and there is a global velocity multiplier. Arm motion can stop at each point along a segment, or it can flow smoothly through the points (subject to certain geometric and other constraints). In addition, the Cincinnati Milacron control can be configured to optimize the speed of all motions and perform automatic velocity clipping.

The path taken between points is complex in the case of uncoordinated axis-by-axis motion. Two forms of coordinated, or interpolated, motion are also possible. Linear interpolation is used for straight-line moves. Circular interpolation is also supported and is especially convenient for teaching moves along curved paths. Figure 18.9 shows a complex path that can be defined with only seven points using circular interpolation. Many shop floor systems now support welding motion and adaptive seam tracking.

The robot-programming language systems have arm motion control software whose capability is similar to the shop floor machines. However, the difference between the Automatix prepackaged approach and IBM's "roll-your-own" interface is quite evident.

The Automatix motion software supports straight-line, coordinated axis-by-axis ("slew"), and welding paths between points. There is a global velocity factor, and each point or series of points has an associated speed. The RAIL interpreter provides a simple, flexible method for specifying these parameters.

The IBM motion software, on the other hand, is presented in building-block form. There are four parameters that can be specified for a move. In addition to naming which of the axes to be moved and the end point, AML extends to the application designer control over acceleration time, the "travel" time of fixed velocity, the deceleration time, and the settling time (used to compensate for servo lag). Although most robot motion software computes these terms internally, the AML programmer has the option/responsibility for considering these factors.

The IBM RS/1 arm motion is coordinated to the extent that each axis will arrive at the end point at the same time. However, straight-line, circular, or any other control between points is the user's job. The system does supply very powerful monitoring capabilities that can be activated during motion phases.

18.4.5. Communication

Communication software links all other software elements with operator interface devices, with associated shop floor equipment, and, when utilized, with external supervisory computers.

Operator communication is necessary for issuing instructions or requesting information, for reporting system or process status, and for resolving operating errors. Typical devices include a CRT/keyboard,

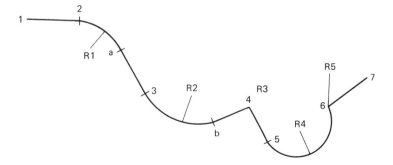

Fig. 18.9. Cybotech interpolation flexibility. (Courtesy, Cybotech.)

a hand-held teaching pendant, and a hard-copy printer. Each robot also has a small number of push buttons and indicators on an operator control panel.

The shop floor equipment present in most robot installations also requires communication support. The most common method used is to place binary input/output interfaces in the robot control itself. When a significant amount of asynchronous operation is required by the application, or a large number of digital signals are used, a separate programmable controller can be added. In any case, the communication software must provide a means for the task interpreter and other software elements to set/clear signals or to read their current state. One other function of the communication software is to monitor for interrupts or other asynchronous events that can be initiated externally. Some other equipment for which communication software is needed includes mass storage devices for saving programs or production data, sensors, and other controllers, for example, welding and vision.

As the level of integration in factory automation increases, robot communication with supervisory computers will become a necessity. Thomas[15] lists 12 features necessary for hierarchical computer control of robots. These features, shown in Figure 18.10, characterize needs for communication software, but they also imply changes to most other software elements in the modern robot. Fussell[9] describes functional specifications for a robot controller designed to be one component in a manufacturing cell.

The communication capabilities of the Cincinnati Milacron T3 are typical of shop floor robots. These machines have an operator control panel, a display screen/keyboard, a teaching pendant, and a magnetic tape cassette (see Figures 18.3–18.5). A simple panel consisting of 12 push buttons and a keyswitch provides basic control functions. The main operator interface device is a 24-line by 80-character CRT with full alphanumeric keyboard. Since Cincinnati Milacron puts full teaching capability at the pendant, this device has 42 keys and a 12-character alphanumeric display. The tape cassette stores user-defined robot tasks as well as the system software. The digital process signal subsystem supports input and output to contacts available in groups of eight inputs and eight outputs.

Some shop floor systems support robot-to-supervisory-computer communications. Full use of this basic communication ability has not been spread through the entire software package, however.

Like the shop floor machines, the robot-programming language systems have communications software for access to operator interface devices, digital process I/O, and related peripheral equipment. However, there is more flexibility in the area of peripheral communication.

The Automatix AI32 controller recognizes six devices: the CRT/keyboard, the teach pendant, each of the two tape drives, a general-purpose RS-232C serial port, and a modem port. Communication can be directed to any of these devices. In addition, normal system I/O can be reassigned from the CRT/keyboard to another device by using the ATTACH command. This technique can be used to drive the AI32 from a remote terminal connected to the modem port or to download programs from a supervisory computer.

AML programs can communicate with the CRT/keyboard *display station,* the printer, a factory communications line, and either the diskette or a fixed disk. All such I/O is file oriented with fixed-size records of user-defined length. There are read, write, and print operations (print converts the data to character format). Communication is directed to one device or another through the use of a *channel* designator.

18.4.6. Error Reporting and Diagnostics

Good error reporting and diagnostic capabilities are a nonobvious, but essential, element in any robot software system. These features serve four functions. Diagnostic and set-up software assists during the manufacture and installation of the robot. Self-diagnosis is usually performed when electrical power is applied to the control. Its purpose is to ensure that the system is ready to operate and, if not, to

High speed, at least 9600 baud

Standard protocol

Security checking

Immediate message initiation

Host up/down loading

Remote task modification during operation

Remote cycle control

Remote variable and I/O modification

Status reporting

Position reporting

Error notification to host

Error response by host

Fig. 18.10. Robot communication features (Courtesy, Cincinnati Milacron)

identify where problems exist. Monitoring software runs constantly when the machine is running to detect operating errors and stop arm motion in dangerous situations. Troubleshooting procedures also frequently involve software written to speed problem isolation and repair.

Safety, that is, the prevention of human injury or equipment damage, is the primary concern addressed by error reporting and diagnostic software. The explanation and resolution of programming or operator errors is also handled in this element. Good diagnostic tools can reduce manufacturing cost. They also speed maintenance and equipment troubleshooting procedures, thereby maximizing up-time and minimizing the robot supplier's field service costs.

Assistance for error reporting and diagnostics is a significant and integral part of the shop floor systems. Diagnostics are resident in the control systems and are activated automatically when electrical power is turned on. The objective for these sets of programs is to isolate a fault to a single, replaceable module—usually an electronic circuit board. When an error is detected, an indication is given to the operator. These diagnostics are useful during manufacture and throughout the normal life of the robot system.

The Cincinnati Milacron "Wake Up" diagnostics[5] illustrate how software can ensure the integrity of a distributed microprocessor robot control system. The purposes of the Wake Up diagnostics are to prevent incorrect system operation and to minimize the time required to make corrections. They are activated when the control is started or reset and occur in three phases, called *levels*. Level 1 testing consists of internal diagnosis by each microprocessor-based board and some cross-checking to guarantee that all boards have passed this level. The communication paths between boards are then tested by the Level 2 diagnostics. Both Level 1 and Level 2 software are stored in ROM on the various circuit boards in the system. Level 3 diagnostics, however, require that the robot operating system be loaded as they check external equipment that can vary from one installation to another.

The results of these tests are reported on four LEDs mounted on the edge of each board. As each diagnostic level succeeds, the corresponding LED is turned on. Note that the internal control hierarchy is observed. A board is not considered to have passed its test until all subordinate boards have reported passing at the same level.

Error-reporting functions that monitor the teaching and production operation represent a major part of the total software effort. For example, the Cincinnati Milacron *Operating/Teach Manual*[6] lists nearly 650 error codes in eight classes; more than 200 of those are programming errors, and there are almost 100 run-time errors detected. Each error condition requires an average of at least one programmed software test; hence, error-checking code is a significant part of the total software package.

The error detection and diagnostic element in robot-programming language systems is comparable to that of the shop floor robots. However, the AML facility for problem reporting and resolution can be considerably more flexible. *Error trap* subroutines can be written to handle errors in any way the user desires (except that the RS/1 arm will always be stopped in dangerous situations). When an error is detected, the currently selected error trap subroutine is activated and an error code and other relevant data are sent to it. With this information, and an IBM-supplied error message file on the system disk, the user can respond to errors in the most appropriate way. Since it is unlikely that a single error subroutine could cover all errors, an AML program can switch from one to another freely. In fact, there can be a designated error trap subroutine associated with each nesting level during task execution.

18.5. THE U.S. AIR FORCE ICAM OFF-LINE APPROACH

The U.S. Air Force is conducting a number of research projects aimed at producing integrated, computer-aided batch-manufacturing systems for the aerospace industry. This program (ICAM), started in 1978, includes the investigation of off-line programming languages for a robot manufacturing cell. The first major language effort, called MCL (Manufacturing Control Language), was developed by McDonnell Douglas Automation Company in the United States.[12]

Off-line programming in general, and the MCL approach in particular, proposed to answer several disadvantages of other robot-programming methods in the batch-manufacturing environment. When there are a large number of parts, therefore many robot tasks are required, programming costs become significant. With off-line programming, the full power of large computer-aided design computers and software tools can be applied to the task generation process. Task definition can occur away from the production area, keeping robot and related manufacturing equipment utilization high. The MCL approach also minimizes the need for using a physical example or model of the manufactured part during teaching. This permits more parallelism in the overall automation process.

Although the robot-programming languages described in Section 18.3 can be applied off-line, the ICAM work is significant because it was designed for a fully integrated CAD/CAM environment. Furthermore, the language was designed to accommodate any type of robot as well as the nonrobotic devices in a manufacturing cell. In anticipation of future widespread use of computer vision systems, the MCL developers introduced *model-based* concepts[2] as a first step toward CAD-driven industrial vision applications.

The MCL system shown in Figures 18.11 and 18.12 consists of three major components. The

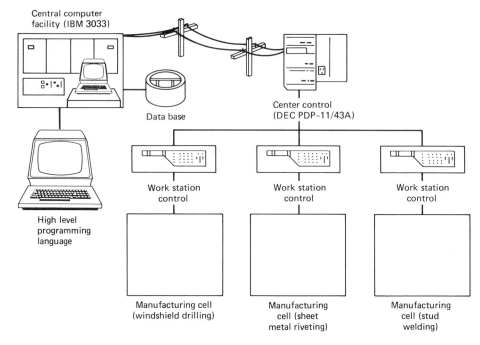

Central computer
facility (IBM 3033)

Data base

Center control
(DEC PDP-11/43A)

High level
programming
language

Work station
control

Work station
control

Work station
control

Manufacturing cell
(windshield drilling)

Manufacturing
cell (sheet
metal riveting)

Manufacturing
cell (stud
welding)

Fig. 18.11. MCL system structure. (Courtesy, U.S. Air Force.)

central computer facility has a large mainframe computer which executes the MCL compiler and contains CAD, vision, and manufacturing data bases. An MCL program is created by compiling source statements and interactively generating geometric descriptions. The result of this process is an APT "CLDATA" file which contains a full description of the task to be performed by the manufacturing cell.

The task description is then transmitted to a center control at the manufacturing site. The center control preprocesses data for, and controls, work stations that perform related manufacturing operations. Preprocessing consists of consistency and resource validation checks followed by conversion of the task description to a set of *workstation interpretable instructions*. For production, the center control transmits these instructions to individual workstation controls and monitors their operation.

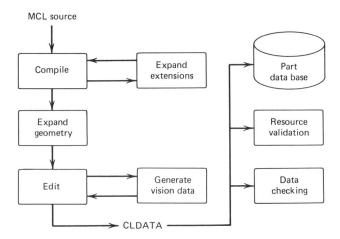

MCL source

Compile

Expand
extensions

Part
data base

Expand
geometry

Resource
validation

Edit

Generate
vision data

Data
checking

CLDATA

Fig. 18.12. MCL organization. (Courtesy, U.S. Air Force.)

The MCL language was based on APT,[13] a standard language used in the aerospace industry and in many other industries where metal cutting is done. APT has established capabilities for defining both geometry and motion. Extensions made to MCL address the specification of geometric information for vision processing, general-purpose decision making, support for specifying locations using multiple reference frames, and process coordination. The language features standard control constructs; supports multiple, asynchronous processes; and has floating point, logical, and string data. There is also a macro-type facility for language extension.

MCL was the initial ICAM research effort addressing off-line programming techniques. Further Air Force contracts have been granted for study in this area, and many other efforts are in progress (see Chapter 20, "Off-Line Programming"). As computer-integrated manufacturing technology advances, many approaches will be developed for highly automated robot task creation. And consequently industrial robot software characteristics will evolve to meet the requirements of operation in an integrated manufacturing environment.

REFERENCES

1. Automatix, Inc., RAIL Software Reference Manual: ROBOVISION and CYBERVISION, Automatix, Inc., Billerica, Massachusetts, 1982.

2. Baumann, Erwin W., Model Based Vision and the MCL Language, McDonnell Douglas Corp., Saint Louis, Missouri, 1981.

3. Bonner, S. and Shin, K. G., A Comparative Study of Robot Languages, *IEEE Computer*, December 1982, pp. 82–96.

4. Carter, W. C., Modular Multiprocessor Design Meets Complex Demands of Robot Control, *Control Engineering*, March 1983.

5. Cincinnati Milacron, Inc., *ACRAMATIC Version 4.0 Control Manual*, Cincinnati Milacron, Inc., Lebanon, Ohio, 1982.

6. Cincinnati Milacron, Inc., *Operating/Teach Manual for T3 Robots*, Cincinnati Milacron, Inc., Lebanon, Ohio, 1982.

7. Cybotech, Inc., *Cybotech V80 Robot System Operation Manual*, Cybotech, Inc., Indianapolis, Indiana, 1981.

8. Engeleberger, Joseph F., *Robotics in Practice*, AMACOM, 1980.

9. Fussell, Paul et al., *A Design of a Controller as a Component of a Robotic Manufacturing System*, Carnegie-Mellon University Robotics Institute, Pittsburgh, Pennsylvania, 1982.

10. IBM Corporation, *IBM Robot System/1 AML Concepts and User's Guide*, IBM Corp., Boca Raton, Florida, 1981.

11. Lozano-Perez, Tomas, *Robot Programming*, Artificial Intelligence Laboratory, Massachusetts Institute of Technology, AI Memo 698a, February 1983.

12. McDonnell Douglas Corporation, *Robotic System for Aerospace Batch Manufacturing, Task B High Level Language User Manual*, Saint Louis, Missouri, 1981.

13. Oldroyd, L. Andrew, MCL: An APT Approach to Robotic Manufacturing, *SHARE 56*, March 1981.

14. Paul, Richard P., *Robot Manipulators: Mathematics, Programming, and Control*, M.I.T. Press, Cambridge, Massachusetts, 1981.

15. Thomas, A. D., *Flexible Automated Manufacturing with Computer-Controlled Robots under Hierarchical Control: Requirements and Benefits*, Cincinnati Milacron, Inc., Lebanon, Ohio, 1982.

CHAPTER 19

ROBOT TEACHING

MICHAEL P. DEISENROTH

Michigan Technological University
Houghton, Michigan

19.1. BASIC CONCEPTS OF TEACH PROGRAMMING

The industrial robot of today is basically an automated mechanism designed to move parts or tools through some desired sequence of motions or operations. As the robot proceeds from one cycle of a work task to the next, the sequence of robot operations may vary to allow the robot to perform other tasks based on changes in external conditions. Additionally, the same type of robot, or even the same robot, may be required to perform a completely different set of motions or operations if the work cell is revised or the desired tasks changed. The robot control program must be able to accommodate a variety of application tasks, and it must also be flexible within a given task to permit a dynamic sequence of operations. The flexibility of the robot is then, to some extent, governed by the types of motions and operations that can be programmed into the control unit and the ease with which that program can be entered and/or modified.

Teach programming is a means of entering a desired control program into the robot controller. In teach programming the robot is manually led through a desired sequence of motions by an operator who is observing the robot and robot motions as well as other equipment within the work cell. The teach process involves the teaching, editing, and replay of the desired path. The movement information, as well as other necessary data, is recorded by the robot controller as the robot is guided through the desired path during the teach process. At specific points in the motion path the operator may also position or sequence related equipment within the work envelope of the robot. Program editing may be used to add supplemental data to the motion control program for automatic operation of the robot or the associated production equipment. Additionally, teach-program editing provides a means of correcting or modifying an existing control program to change an incorrect point or to compensate for a change in the task to be performed. During the teach process the operator may desire to replay various segments of the program for visual verification of the motion or operations. Teach replay features may include both forward and backward relay, single-step operations, and operator-selectable replay motion speeds.

The approach taken in teach programming is somewhat dependent on the control algorithm used to move the robot through a desired path. Therefore we review three basic algorithms before returning to the main topic of this section.

Robots with *point-to-point* control move from one position to the next with no consideration of the path taken by the manipulator. Generally, each axis runs at its maximum or limited rate until it reaches the desired position. Although all axes will begin motion simultaneously, they will not necessarily complete their movements together. Figure 19.1*a* illustrates the trajectory taken by a robot moving with point-to-point motion. *Continuous path* control involves the replay of closely spaced points that were recorded as the robot was guided along a desired path. The position of each axis was recorded by the control unit on a constant-time basis by scanning axes encoders during the robot motion.

The replay algorithm attempts to duplicate that motion. Figure 19.1*b* illustrates a continuous path motion. *Controlled path* motion involves the coordinated control of all joint motion to achieve a desired path between two programmed points. In this method of control, each axis moves smoothly and proportionally to provide a predictable, controlled path motion. In Figure 19.1*c* the path of the end effector can be seen to follow a straight line between the two points of the program.

In teach programming there are two basic approaches taken to guiding the robot through a desired path:

1.   *Teach-pendant programming* involves the use of a portable, hand-held programming unit referred to as a teach pendant. The teach pendant contains a number of buttons or switches that are used to direct the controller in positioning the robot. Teach-pendant programming is normally associated with point-to-point motion and controlled path motion robots.

2.   Continuous path robots utilize *lead-through programming* for teaching a desired path. The operator grasps a handle secured to the arm and guides the robot through the task or motions. Lead-through programming is frequently used for operations such as spray painting or arc welding. Lead-through programming can also be utilized for point-to-point motion programming.

Whether the robot is taught by the teach-pendant method or by the lead-through method, the programming task involves the integration of the following three basic factors:

1.   The coordinates of the points of motion must be identified and stored in the control unit. The points may be stored as individual joint axis coordinates, or geometric coordinates of the tip of the robot may be stored.

2.   The functions to be performed at specific points must be identified and recorded. Functional data can be path oriented, that is, path speed or seam weaving, or point oriented, for example, paint spray on, or wait for signal.

3.   The organization of the point and functional data into logical path sequences and subsequences. This includes establishing what paths should be taken under specific conditions and when various status checks should be made.

The foregoing three factors are integrated into the teaching process and do not exist as separate programming steps. During teach-pendant programming the operator steps through the program, point by point, recording point coordinate and functional data for each point. Path sequence is a direct consequence of the order in which the points are taught.

Teach programming is the most natural way to program an industrial robot. It is on-line programming in a truly interactive environment. During the programming task the operator is free to move around the work cell to obtain the best position to view the operation. The robot controller will limit the speed of the robot, making programming safer. The operator has the ability to coordinate the robot motion with the other pieces of equipment with which the robot must interact. The operator can edit a previously taught program replay; the operator can test the path sequence and then continue to program until the complete program is stored in the memory of the control unit.

It is important to realize that teach programming does not place a significant burden on the operator. Teach programming is easy to learn and requires no special technical skill or education. While the operator is learning to program the robot, he/she is not learning a programming language in the traditional sense. This allows the employment of operators who are most familiar with the operations with which the robot will interact. Painting and welding robots are best taught by individuals skilled in painting and welding. Parts handling and other pick-and-place tasks can be taught by material handlers or other semi-skilled labor.

Another advantage of teach programming is the ease and speed with which programs can be entered and edited. Advanced programming features permit easy entry of some common tasks such as parts stacking, weaving, or line tracking. Stored programs can be reloaded and modified to accommodate changes in part geometry, operating sequence, or equipment location. Commonly used program sequences can be stored and recalled for future applications when needed.

The disadvantage of teach programming stems from the on-line nature and the point-by-point mode of programming. Since the robot is programmed in the actual work cell, valuable production time is lost during the teaching process. If the time is short this may be of little consequence; however,

A) POINT-TO-POINT
PATH MOTION

B) CONTINUOUS PATH
MOTION

C)CONTROLLED PATH
MOTION

Fig. 19.1.   Motion control algorithms determine the actual path taken when the program is replayed in the automatic mode.

teach time can be quite long. If the operations to be programmed involve many points and are complex in nature, have a number of branches and decision points, the time to program on-line may lead to significant equipment downtime. Off-line programming techniques may be more appropriate. (See Chapters 20, "Off-line Programming of Robots," and 21, "A Structured Programming Robot Language.")

Although teach programming has been a basic form of on-line programming for a number of years, it is still receiving a great deal of attention from the robot vendors. It is the most widely used method of programming industrial robots today and will continue to play an important role in robot programming for many years to come. The increased sophistication being added to the taught program capabilities has created a low-cost, effective means of generating a robot program.

19.2. TEACH-PENDANT PROGRAMMING

As discussed earlier, teach-pendant programming involves the use of a portable, hand-held programming unit to direct the robot to desired points within the work envelope. This is best illustrated by the example shown in Figure 19.2. In this example a robot is required to pick up incoming parts from the conveyor on the left, place them into the machining center, and then carry them to the finished parts conveyor on the right. Twin grippers on the end of the arm allow unloading of a finished part followed immediately by loading of a new part, thus reducing wasted motion and overall cycle time. The robot is interfaced to the machining center and to both part conveyors.

An operator will lead the robot, step by step, through one cycle of the operation and record each move in the robot controller. Additionally, functional data and motion parameters will be entered as the points are programmed. The teach pendant is used to position the robot, whereas the controller keyboard may be required for specific data entry. The operator must insure that all interface connections have been made before beginning the teach process. Once the setup is completed the programming process can begin as follows.

19.2.1. Teaching Point Coordinates

1. Move the robot arm until the left gripper is just above the part at the end of the input conveyor, and open the left gripper.
2. Align the gripper axes with the part to be picked up.
3. Store this program by pressing the *record* or *program* button on the teach pendant.
4. Lower the arm until the left gripper is centered on the part to be grasped.
5. Store this point.
6. Close the left gripper so that the part can be lifted.
7. Store this point.
8. Raise the arm so that the part is clear of the conveyor and at a desired level to rotate toward the machining center.
9. Store this point.

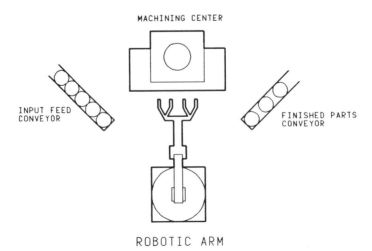

Fig. 19.2. Teach-pendant programming example of a robotic work cell.

10. Move the arm so that the right gripper is in front of the finished part on the machining center, and open the right gripper.

11. Align the gripper axes with the finished part.

12. Store this point.

13. Extend the arm until the right gripper is centered on the finished part.

14. Store this point.

15. Close the right gripper so that the part can be retracted.

16. Store this point.

17. Raise the arm so that the part clears the machining center table slightly.

18. Store this point.

19. Retract the arm so that the finished part is clear of the machine and the unfinished part can be positioned at the desired approach point.

20. Store this point.

21. Rotate the arm so that the unfinished part is positioned in front of the machine and slightly above the table surface.

22. Store this point.

23. Extend the arm so that the unfinished part is in the desired position in the machining center.

24. Store this point.

25. Lower the arm until the unfinished part is on the table.

26. Store this point.

27. Open the left gripper to release the unfinished part.

28. Store this point.

29. Retract the arm so that it is clear of the machining center and ready to rotate to the finished parts conveyor.

30. Store this point.

31. Move the arm until the finished part is above the finished parts conveyor.

32. Store this point.

33. Lower the finished part onto the conveyor.

34. Store this point.

35. Open the right gripper and release the finished part.

36. Store this point.

37. Raise the arm so that it is clear of the conveyor and the finished part.

38. Store this point.

39. Return the robot to the initial position and indicate end of cycle.

The preceding 39 steps provide the controller with the desired program point coordinates. The operator would also enter functional data associated with a number of the programmed points. For example, the robot must be made to wait at the points programmed in steps 3 and 12 until external signals indicate that it is clear to continue; an output signal must be activated at the point programmed at step 30 so that the machining center can begin operations. Additionally, the robot must be taught how to process the first and the last part of the whole batch. The program sequence for these parts differs slightly from the sequence given.

19.2.2. Teaching Functional Data

Functional data can be programmed during the normal teach process. The operator first positions the robot by depressing the appropriate buttons on the teach pendant. Functional data, for instance, desired velocity, can then be specified by keyboard entry or on the pendant. The record button is then pressed, and all relevant data is stored in the robot control memory. The operator then continues until the sequence is complete. If a program branch sequence is to be programmed, the operator must play through the mainline to the point at which the new sequence is to be entered. The control unit can then be altered and the branch sequence programmed point by point.

19.2.3. Robot Control Unit and Teach Pendant

The keyboard and display of the robot control unit is often used in conjunction with the standard teach pendant while in the teach mode. Figure 19.3 illustrates a robot teach pendant that is connected to a control unit with the associated keyboard and display. The control unit contains a number of control push buttons used to control the robot during all of its operations. Additionally, a keyboard panel is present which is used primarily during programming to enter functional data and to facilitate the editing process. The teach pendant is connected to the control unit by a long cable which permits the operator to program from a position that provides good visibility of the actions being performed. It is used to direct robot motion and to cause points to be stored. The teach pendant given in Figure 19.3 has axes buttons (for a six-axis robot) on the face of the lower half of the unit. The left six buttons control robot positioning, and the right six buttons control tool orientation. Each robot axis has two buttons associated with it: one button causes motion in one direction, and the second moves the axis the other direction. These buttons are then used to perform all manual robot positioning.

The ASEA SII Robot Controller shown in Figure 19.4 illustrates recent innovations in pendant design. The unit employs a joystick with a control, two keypads for function and data entry, and a display for operator messages. The joystick can be programmed by keypad entries, to control robot positioning or tool orientation. Moving the joystick to the right causes the robot to move in that direction; moving it forward extends the robot. Raising and lowering of the arm is controlled by the knob at the top of the joystick. Certain keys on the keypad have fixed meaning; others have meanings defined by the operating system software. The operations controlled by these *soft* keys are displayed to the operator in the display screen immediately above the keys. This menu-driven prompting of the operator facilitates rapid learning of the system and eliminates costly errors. An example of this operation would be to consider the subfunctions available when the program position POS key is

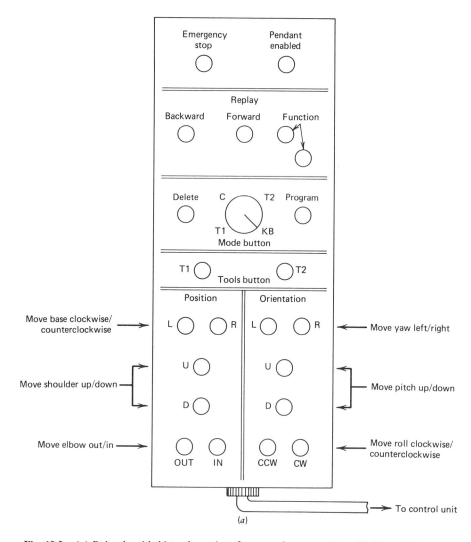

Fig. 19.3. (*a*) Robot hand-held teach pendant for manual programming. (*b*) Control diagram.

depressed. In all, 11 subfunctions are provided at present. The operator is presented with the first four subfunctions given over the first four soft keys and a SCAN subfunction presented over the fifth soft key. If one of these subfunctions is desired, the operator presses the appropriate key. Otherwise the SCAN key is pressed, and the next set of four subfunctions is displayed along with SCAN. The operator can continue to scan through the subfunctions until the desired operation is found. The system operations restrict the operator to the correct set of logical subfunctions.

Since normally the purpose of a robot is to transport some workpiece or tool through a desired path or to a specific point, the operator is usually more concerned with the location of the tool or gripper than any other point of the arm. Many robot control systems track the *tool center point* (TCP) instead of specific joint angles. As points are stored into the memory of the control, the coordinates of the TCP with respect to the robot origin are calculated and stored. This includes both positional data and orientation data. When the robot is in the automatic mode, these TCP data values must be translated into joint position values so that the control algorithms can be executed. This added computational burden increases the complexity of the control program, and many robot controllers simply store the joint angles. This specific factor is of great importance when considering the type of motions available to the operator during the teach process and the range of functions available within the program.

While robot motion involves individual or simultaneous motion of the joints and linkages, the robot programmer is primarily concerned with the motion of the tool or gripper. Early teach pendants restricted robot motion during teaching to axis-by-axis moves. This still exists in a number of controllers that utilize joint axis data to define the position of programmed points. For example, the earlier version of the ASEA pendant provided for coordinated joint motion of the arm during horizontal and vertical teach moves, but restricted all other motion during the teach process to individual axis motion. The new feature of teaching the desired motion of the TCP greatly simplifies teach programming of many parts-handling operations.

19.2.4. Teaching in Different Coordinate Systems

To further enhance the teach process, more modern controllers, like the Cincinnati Milacron Teach Pendant in Figure 19.3 and the ASEA SII Robot Controller in Figure 19.4, permit a number of different programming coordinate systems. Depending on the robot control unit, the operator may be able to position the arm by selecting rectangular, spherical, or cylindrical motion. In the rectangular system the motion of right and left are in a Y plane, in and out are in the X plane, and up and down in the Z plane. In the spherical system the motions are rotation about the base, angular elevation of a ray beginning at the base, and in and out motion along the ray. The cylindrical motion is defined by rotation of the base, in and out motion in a horizontal, radial direction, and up and down motion

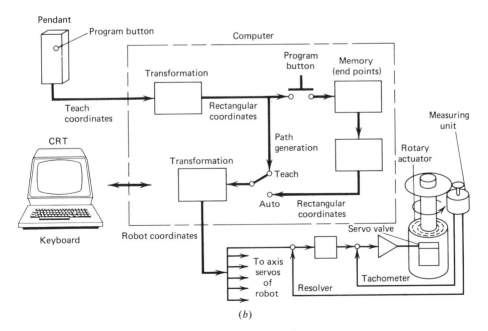

(b)

Fig. 19.3. (*Continued*)

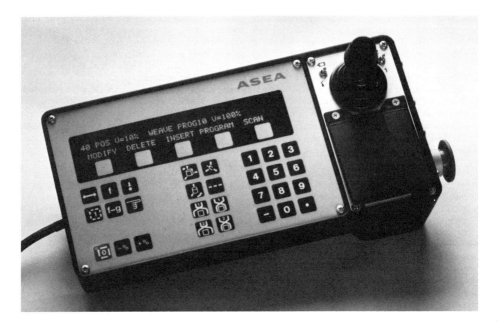

Fig. 19.4.   ASEA SII Robot Controller with joystick and soft function keys.

in a vertical plane. A fourth programming positioning motion that is extremely useful is a wrist-oriented rectangular coordinate system. Here the robot positioning motion is taken in a Cartesian coordinate system that is aligned with the wrist orientation.

Positioning of the TCP during programming does not affect tool orientation. Pitch, roll, and yaw rotations are normally controlled by a separate set of axis buttons or by redefining the function of the joystick. When positioning the TCP by one of the coordinate motions described earlier, the TCP is translated through space. When the orientation of the TCP is changed, the TCP remains fixed, and the robot actuates the motion of the axes to rotate around the specified point. The path taken under program control in the automatic mode is independent of the path taken while teaching, since only end-point data are stored.

Axis-by-axis positioning is acceptable when the end-point data are all that is desirable. If the application requires precise path trajectories, both during programming and replay, coordinated joint motion is highly desirable to maximize operator productivity. This added complexity in the control of the robot during programming requires that the TCP positional data be stored and that the control algorithm be of the continuous path type.

19.2.5.   Teach versus Replay Motion

The relationship between teach path motion control and the real-time replay motion control algorithms is not always fully understood. A robot can be programmed with a joint coordinated motion scheme and replayed in a point-to-point mode. This may be done to minimize the demands placed on the axis servo systems, or to reduce computational burden in the control unit. Robots programmed by axis-by-axis motion may be replayed point-to-point, or by controlled path motion, depending on the results desired. Many robots are limited to point-to-point path motion or a course-controlled path motion. Robots that offer sophisticated path control during programming normally have, and are often limited to, controlled path motion with straight-line motion between the programmed points. Only recently have sophisticated control units begun to offer the point-to-point replay motion after teaching a joint coordinated motion.

19.2.6.   Teaching Interface Signals

Few, if any, industrial robots perform production operations that require no interconnections to the surrounding equipment. In the example presented in Figure 19.2, the robot must be interfaced to both conveyors and the machining center. Input signals are required from the feed conveyor to indicate that a part is ready for pickup, and from the machining center to indicate that the machine is done and a finished part is ready for pickup. It is also desirable to arrange some sort of sensor to insure that the finished-parts conveyor is clear before unloading and placing the next finished part on the unit. The robot must output a signal to the machining center to initiate the start of the machining cycle after the robot has loaded a new part and withdrawn a safe distance. Additional input sensors could be located in the gripper to signal that a part is present, or sensors could be attached to the conveyors to indicate when they are in motion.

Input and output signals are included in the path program through the use of specific functional data associated with specific programmed points as indicated in the following list:

1. Wait for a specific input to go high.
2. Initiate an external process by outputting a low signal on a specific line.
3. Branch to a utility routine if the signal is not active.
4. Output a desired feedrate voltage to the workfeed mechanism.

These input and output signals may be digital in nature, that is, limit switches, relay contacts, photo detectors, and so on, or they may be analog signals, that is, furnace temperature, arc voltage, and so on. Many industrial robot controllers include standard programmable controller interfaces for process interconnection.

It should also be noted that robot control units may include optional servo-controlled systems that can be used to closely coordinate external equipment with path motion. This is specifically true in the case of weld-positioning units that are used to locate the workpiece at some desired attitude. (See Chapter 49, "Robots in Arc Welding.") Although typical weld position units are controlled by cam-driven limit switches, the use of servo-controlled positioning motion greatly enhances weld quality while reducing production cycle time. These external control signals are part of the basic data stored with each path point.

19.2.7. Teaching Program Branching

Decision making allows a robot system to deviate from its normal path program and perform other tasks based on changes in external events. Teach-pendant programming systems permit a variety of capabilities in robot decision making and path modification. The specific features available on a given robot are a function of the sophistication of the control system and the imagination of the control system's programmers. For discussion purposes, basic robot decision making is subdivided into branching and interrupt processing, each of which is further subdivided. Advanced decision-making concepts are discussed in Section 19.3.

Program branch capabilities can be subdivided into standard branching, conditional branching, and offset branching. The first two forms involve branching to a path segment that is defined in absolute space, whereas the latter scheme involves branching to a path that is relative to the point from which the branch was entered.

Standard Branching

A standard branch exists when the robot is programmed to check a single input signal at a given path point and to choose the next destination based on that signal. It must be realized that the robot control unit is not constantly monitoring the status of the input signal, but that the signal is checked only when the robot reaches the branch point. An external condition that is momentary in nature will not cause branching if it goes undetected. The path sequence associated with a branch may return to the point where it was initiated, or it may close on some other motion point. In an actual application, a number of standard branches may be included in the path program. It should also be noted that a number of branches may be initiated from the same physical location. This would permit the controller to examine a number of input signals one at a time and branch when the first signal is found at the desired level.

Conditional Branching

Conditional branching is simply an extension of the standard branch concept. Instead of examining a single input signal to test for a desired state, a number of input signals are examined at the same time to see if they are all in their desired state. These two concepts have been presented separately, however, to emphasize the difference in their nature. If a robot controller is limited to standard branching, the path program may become excessively long or the number of input signals required may be large. Figure 19.5a illustrates a standard or conditional branch off a given mainline program. Functionally they are externally equivalent and have the same appearance.

Offset Branching

Offset branching provides the programmer with the means of establishing a path sequence that is relative to the physical location of the robot arm when the branch is requested. When the branch is

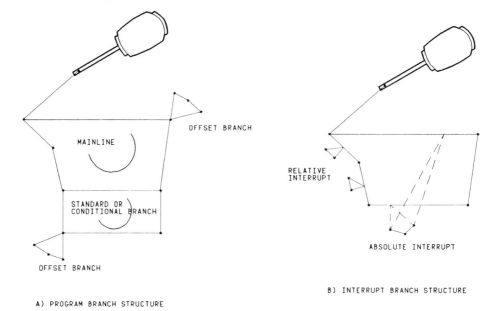

A) PROGRAM BRANCH STRUCTURE

B) INTERRUPT BRANCH STRUCTURE

Fig. 19.5. Program branch and interrupt structures permit deviation from the mainline program.

initiated, each point in the path sequence is offset or translated relative to the current robot position and/or the wrist orientation. Offset branch programming is primarily associated with robot control systems that utilize TCP coordinates as stored data. Offset branches are used when a specific sequence of motions is repeated at a number of points in a path program. An example of offset programming would be the insertion of screws or fasteners during an assembly operation.

19.2.8. Teaching Interrupt Service Routines

Whereas program branches are initiated at the preprogrammed points in the program path, interrupt service routines can be activated by an external input signal anywhere in the cycle. Interrupt signals are constantly monitored by the robot controller. When an interrupt occurs, the robot immediately abandons the current operation and begins the interrupt routine. Interrupt service routines can be subdivided into relative and absolute routines as indicated in Figure 19.5b. A relative interrupt service routine is like an offset branch in that the sequence of points associated with the routine are considered to be relative to the point at which the interrupt occurred. A specific use of this type of routine is found in spot-welding applications. When the robot controller senses that the weld tips have welded themselves to the workpiece, the interrupt service routine causes a series of twisting motions to be executed to free the spot-weld gun tips. Absolute interrupt service routines contain path point descriptions that are fixed in space with respect to the robot. If an absolute interrupt service routine is requested, the robot withdraws from the current operation and begins to follow the path dictated by the routine. If the robot is required to perform a series of operations associated with a specific machine when a malfunction is detected, the service routine would be an absolute interrupt.

19.3. ADVANCED TEACH-PENDANT PROGRAMMING CONCEPTS

A number of special features or options in teach-pendant programming have been created that increase the ease of programming or add flexibility and versatility to a path program. Some of these features are presented in the following sections to illustrate the extent to which teach-pendant programming has developed over the years.

Branch and interrupt structures have been called subroutines by a number of robot manufacturers and could have been presented as such in the preceding section. They somewhat lack all of the features normally associated with high-level language subprograms but do serve a similar purpose. One robot manufacturer has recently announced a programming feature that allows subroutines, or subprogram blocks, to be stored on a floppy disk in the form of a common library. These subroutines are then available for future use when the same or similar sequence of motions is desired. As control path programs are developed, frequently used routines can be added to the library. Such a library will continue to grow in size as well as increase in value. This programming feature promises greatly reduced programming time and cost if the library can be properly maintained and managed.

19.3.1. Teaching and Sensing Functions

Another advanced concept offered to ease the programming task and increase the flexibility of the path motion is the use of sensing functions. These functions include search, contour following, and speed control. The type of sensors used with these functions varies from simple on/off digital sensors to fully analog sensors. As many as three sensors can be used simultaneously during the search and contour functions. This allows the functions to be performed in multiple dimensions.

The search function (see Figure 19.6a) is normally associated with a stacking operation. The robot is manually taught an approach point, a stack empty point, and an exit point. The robot proceeds from the approach point toward the stack empty point at some predefined velocity. When the robot control unit detects the presence of a signal indicating that the top of the stack has been reached, the robot starts a motion toward the exit point. Since the robot will travel slightly beyond the point at which the top of stack is sensed, approach speed and robot compliance are important factors to be considered. The search function can also be used to locate or pick up an object where the exact size or location is not known.

Contour following permits the robot to move along a desired surface that is not defined completely. The robot path program will position the robot gripper with associated sensors at the beginning of the contour to be tracked. As the gripper is moved along the contour, the sensor feeds data back to the control unit to insure that constant contact is maintained, as shown in Figure 19.6b. Contour following can be used to guide the robot motion during automatic replay or it can be used to teach the robot a specific part-oriented path for subsequent operations.

The speed control function (see Figure 19.6c) can be used to adjust the velocity of the robot as it moves along a given path. If little resistance is given to the robot motion, the desired or programmed speed is maintained. As resistance increases the robot decreases motion speed to decrease the resistance level. The speed control function has proven to be useful in routing and deburring operations. The robot moves the tool through the programmed path at the desired speed until cutting forces increase. If more material is being removed, the robot decreases travel speed to maintain cutting parameters.

19.3.2. Teaching and Supervisory Control

Manufacturers of robots programmed by a teach pendant have for some time now provided a communications link between the robot controller and some external computer system. This communications link can be used to store taught programs and down load them for use at some future point in time. A more interesting use of this link can be found, however, in branch modification. The data comprising a complete branch sequence in the robot controller can be transmitted to the external system. This system can then modify the data values and transmit the branch back to the robot controller. When a branch is executed, new data points will be the points of motion of the robot and not the points taught earlier. A higher level of robot decision making is therefore possible by coupling the robot

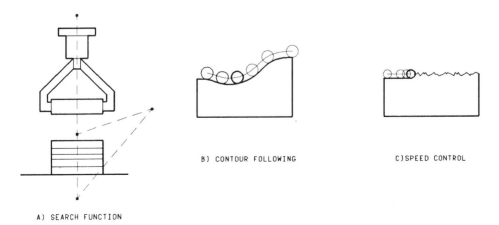

B) CONTOUR FOLLOWING   C)SPEED CONTROL

A) SEARCH FUNCTION

Fig. 19.6. Sensing or adaptive functions permit the taught path to vary as a result of external inputs.

with a supervisory system or some more complex sensor systems employing vision, force, or tactile transducers.

The supervisory control feature has recently been used in two widely diversified applications. First, the branch modification scheme has been mainly responsible for the application success of vision systems used for part location tasks with respect to taught robots. A dummy branch is created to accommodate part pickup, and then the branch is transmitted to the vision control system. This system establishes the robot path that must be taken to pick the part up and revises the point data in the dummy path. The branch is then down loaded, and the robot proceeds according to the new data. Seam tracking for arc-welding operations has been another application area for branch modification. A vision system or laser-scanning device is used to locate the seam to be welded. Positional data and weld parameters can then be down loaded into the control unit to execute the desired weld.

19.3.3. Teaching and Line Tracking

A number of robots provide a tracking option that represents another form of motion adaptation. Many manufacturing operations are carried out on continuously moving production lines. Line tracking involves the ability of the robot to work in conjunction with such lines. There are two basic ways to accomplish line tracking with a teach-pendant robot. Moving-base line tracking requires that the robot be mounted on some form of transport system that will cause the robot to move parallel to the production line. Robot traverse speed is synchronized to the line speed while the robot performs the desired motion. Stationary-base line tracking combines the sophistication of the control unit with sensor data to modify dynamically the robot motion commands to permit work to be performed on the moving line by a stationary robot.

Both moving-base line tracking and stationary-base line tracking permit the user to program the robot in a stationary position. The teach pendant is used to position the robot at desired points which are programmed in the normal manner. During automatic replay a synchronization signal insures that the stationary-base line-tracking robot origin is in the same position relative to the moving line during program replay. Since the points are programmed while the two objects are in the same relative position, the path will be performed as desired. The stationary-base line-tracking system utilizes a position sensory device that sends signals to the control unit indicating the position of the part relative to a fixed origin at all times. This signal is then used to produce a shift in the zero position value of the taught points. Since the joint axis motions must be calculated dynamically during replay anyway, for robots that store TCP, the tracking feature only slightly complicates the computational task.

Moving-base line-tracking robots are easily programmed but require expensive installation of the transport system and may create interference between adjacent stations. Stationary-base line-tracking robots, however, present some interesting programming problems. First, the robot may be able to work on the front, middle, and back of the work as it flows through the work envelope. The fact that the work does move relative to the robot extends the possible working surfaces. The operator must concern himself with the work window from both time and space viewpoints. Although the robot may have expanded the possible working surfaces, each surface is available only during a fraction of the overall design cycle time. Work to be performed on the front of the part must be performed first, while that work surface is accessible.

19.3.4. Abort and Utility Branches

Finally, the operator must consider abort and utility branches. Abort branches represent pretaught safe paths by which the robot can exit a work area if the taught points are no longer feasible. Utility branches are similar to the interrupt service routines discussed before; however, they must be performed as the tool is in motion.

19.3.5. Teaching Arc Welding Functions

A number of programming features are available that adapt general-purpose industrial robots to the task of arc welding. The repeatability, lift capacity, and speed of these robots, coupled with proper weld parameters, can produce welds that are consistent and of high quality. There are added programming features that are required, however, to insure the ease of programming the robot.

A weld schedule function allows the operator to enter the wire feed, speed, and the arc voltage into a table called the "weld schedule." As the operator teaches a given weld path, an index to this schedule is attached to program points to set the desired process parameters. When the path is repeated in the automatic replay mode, the robot control unit will assess these functional values and output analog control setpoints to the weld controller. Since the weld parameters are part of the stored path program, weld quality is reproducible from part to part.

A standard velocity function is important to the welding task. Various weld segments can be programmed at different velocities to produce different weld depths. If the operator desires to change the speed of welding over a given set of points, the teach pendant can be used to edit the necessary stored data. Another standard feature that can be used to ease weld programming is offset branching. A root path of a multipass weld can be programmed once, and then offset as desired to obtain multiple passes. Since the offset feature causes path motion relative to the point the path is entered, a multipass weld could be considered as a single-offset path with a number of entry points.

One programming feature that has been added to robot controllers specifically to accommodate arc welding is a weaving function. The weaving function can be programmed between any two points. The operator teaches the beginning and the ending points of the weave and keys in the following five values:

1. Number of weave cycles.
2. Magnitude of the weave to the right of the line connecting the end points.
3. Magnitude of the weave to the left.
4. Percent dwell on the right.
5. Percent dwell on the left.

19.4. LEAD-THROUGH TEACH PROGRAMMING

Lead-through teach programming is used primarily for continuous path motion-controlled robots. During the lead-through teach process, an operator grasps a removable teach handle, which is attached directly to the arm, and leads the robot through a desired sequence of operations to define the path and relative velocity of the arm. While the operator is teaching the robot, position transducers measure the joint angles associated with each robot axis, and the robot control unit records this information for use during production replay. Automatic replay of the digitized data provides the continuous path motion necessary for precise duplication of the motion and technique utilized by the human operator. Lead-through teach programming and continuous path motion-controlled robots are well suited for operations such as spray painting, sealant application, or arc welding.

Lead-through teach programming can also be used to program a robot for point-to-point motion control. In this mode of lead-through teach programming, only the end points of each path segment are recorded for replay. Path velocity is normally recorded in this mode as a function entry associated with each path segment. Additionally, path replay patterns can be incorporated for use in specific applications such as seam weaving for arc welding.

DeVilbiss/Trallfa arc welding robots offer both continuous path lead-through teaching with the arc on and point-to-point teaching with the arc off. The Thermwood Series Seven robot is a lead-through teach material-handling robot that utilizes the point-to-point mode to program part loading and transfer operations.

The major advantage of lead-through teach programming over teach-pendant programming is directly associated with programming ease. Lead-through teach programming is easy to learn and can be performed by an operator who is directly associated with the production task—painting or welding. To further facilitate programming operations, the robotic arm is kinematically balanced to allow easy movement by the operator during lead-through programming. The counterbalanced arm permits the operator to manipulate the end-of-arm tooling manually in a way similar to manual operations. Lead-through teach-programmed robots often have dual handles. One handle can be used to support the arm while the second is used to guide the tooling.

19.4.1. Lead-Through with a Programming Arm

In spite of design efforts to counterbalance the mechanism, the lead-through robot is often difficult to maneuver. To compensate for this problem, a number of robot vendors market lightweight programming arms that are kinematically identical to the actual manipulator. The operator manipulates the programming arm and joint position data are recorded. The stored data path is then used for replay with the actual production robot. Since the programming arm is used in parallel to the actual manipulator, a second robot control unit will be necessary to digitize joint position data, or the programming arm must be connected to the control unit of the actual manipulator.

Another concept found in lead-through teach programming is a teaching system to program a number of different robots that are all of the same type. The programming facility can be either a lightweight programming arm or a production robot. The teaching system is led through the desired sequence of operations, and the path program is created. This program is then passed to the control unit on which the program is to be automatically replayed. Program transfer can be accomplished in a number of different ways, for example, cassette tape, floppy disc, or a direct communications link. The nature of the application areas associated with lead-through robots can normally tolerate the minor differences between different arms and the resulting differences in replay paths.

19.4.2. Software for Lead-Through Programming

Lead-through teach-programming software systems incorporate basic program-editing features. Often the program is sufficiently short that the entire program can be deleted and a new program recorded. Alternately, the desired path sequence can be broken down into a number of path segments, and each segment can be taught separately. The segments can then be merged into the desired sequence, and the path program is then created. Since the robot controller maintains each segment as a separate entity, the user can delete an undesired segment and reprogram the proper path. The new segment can then be merged into the path program for the desired result. If the program has been created in the continuous path mode, a program segment is the smallest path unit that can be deleted or replaced.

Robot vendors use a number of different schemes in recording the joint-position data associated with lead-through programming. The simplest and most direct scheme is to digitize the data at some present rate and record the value as obtained. The robot controller will then attempt to drive the robot through these points during replay. A second scheme involves digitizing the joint-position data at some multiple of the desired data recording rate. The digitized values can then be mathematically smoothed by a moving-average technique to obtain the desired positional data for recording. This technique can be helpful in removing random fluctuations in the programmed path caused by operator actions. One robot vendor mathematically determines a path that approximates the data gathered and then redistributes points along the mathematical function to insure a more uniform velocity. Irrespective of the scheme chosen for storing joint-position data, velocity control is normally limited to some fractional multiple of the programmed speed. The replay velocity of each program segment can be edited to fine-tune the program. However, the replay motion will contain the acceleration and deceleration patterns exhibited by the programmer.

19.4.3. Lead-Through Robot Examples

An example of a typical lead-through robot is the Graco OM5000 robot. The basic application of this unit is in conventional or electrostatic finishing operations. The robot control system incorporates both point-to-point and continuous path programming modes. A standard control unit can save up to 28 min of continuous path points that can be segmented for each program editing. Program storage options include floppy disc, solid state, or bubble memory. The removable programming handle (see Figure 19.7) has switches for signalling the start and stop of path storage, paint-spray flow or color changes, or triggering of external functions. A remote pendant is available for entering functional data values and program editing. Programs can be created on one OM5000 robot and then transferred to another unit, allowing a single unit to act as the master programming system if desired.

The robotic paint-spraying booth system marketed by GMF Robotics is an example of a complex

Fig. 19.7.   The lead-through teach handle is attached directly to the robot tooling.

system of robots that employs lead-through teach programming. A typical installation might utilize eight or more production robots with one separate, but identical, robot in a booth dedicated to off-line programming. An operator leads the teach-booth robot through the desired path sequence. The path points are then mathematically smoothed to insure a constant paint velocity where the midpoint of the spray fan intersects the surface being painted. The teach booth provides for program verification through actual painting. A supervisory minicomputer can then down load the program to the control unit of a particular desired production robot for operation. Teaching can be done while the system is on-line, and program modifications can be incorporated into the next production cycle. The system was designed to trace a path accurately and repeatedly at a prespecificed constant velocity, regardless of which robot is chosen to execute the stored program.

19.5. CURRENT AND FUTURE TRENDS IN TEACH PROGRAMMING

Recent announcements from robot manufacturers have shown an increased emphasis on the man-machine interface between the operator and the robot system. The introduction of a hand-held, teach pendant with joystick controls and soft function keys is one example. The constant efforts toward achieving better arm balance in lead-through robots is but another. From an operational viewpoint, robot manufacturers are beginning to accept the concept of tool center point motion and the utilization of multiple coordinate systems during programming, which will help to reduce programming time and lead to higher work cell productivity.

The sophistication of the function package supplied by a single manufacturer is increasing. As more and more of the concepts and algorithms become public knowledge, the capability range of the minimum acceptable marketing package will increase. On the other hand, it is obvious that many robot vendors are convinced of the strength of teach programming, as opposed to off-line programming, and are committing resources to increase the capabilities and features offered. The introduction of program variables and the movement toward integrating the robot control program with other work cell control parameters are examples of increased capability.

The flexibility and versatility attributes of teach programming are not always obvious to an observer of the teaching process. At first glance it appears to be somewhat crude and unsophisticated. The ease with which it is accomplished, however, is apparent. The tremendous success of teach programming in practice is really a function of both attributes. It is easily learned and applied, and there is sufficient power in the language to accomplish a variety of needed robot control tasks. Although the future is quite bright for developments in off-line programming areas, the teach programming method has not yet even begun to fade. At present it is, and will remain for some time, the most widely used programming method of computer-controlled industrial robots.

BIBLIOGRAPHY

Akeel, Hadi A., Expanding the Capabilities of Spray Painting Robots, *Robots Today,* Vol. 4, No. 2, April 1982, pp. 50–53.

Colleen, Hans, Giving Robots the Power to Cope, *Robots Today,* Spring 1980, pp. 32–34.

Dawson, Bryan L., Moving Line Applications with a Computer Controlled Robot, Society of Manufacturing Engineers, Paper No. MS77–742, November 1977.

Dorf, Richard C., *Robotics and Automated Manufacturing,* Reston Publishing Co., Reston, Virginia, 1984.

Engelberger, Joseph F., *Robotics in Practice,* American Management Association, 1980.

Green, Robert H., Welding Auto Bodies with Traversing Line-Tracking Robots, *Robotics Today,* Spring 1980, pp. 23–29.

Hohn, Richard E., Application Flexibility of a Computer-Controlled Industrial Robot, Society of Manufacturing Engineers, September 1976, MR76–603.

Holt, H. Randolph, Robot Decision Making, Society of Manufacturing Engineers, November 1977, MS77–751.

Holmes, John G. and Resnick, Bryan J., A Flexible Robot Arc Welding System, *Proceedings of Robots IV,* November 1979.

Warnecke, H. J. and Schraft, R. D., *Industrial Robots: Application Experience,* IFS Publications Ltd., Kempston, Bedford, Great Britain, 1982.

CHAPTER 20

OFF-LINE PROGRAMMING
OF ROBOTS

Y. F. YONG
J. A. GLEAVE
J. L. GREEN
M. C. BONNEY

University of Nottingham
University Park, United Kingdom

20.1. INTRODUCTION

20.1.1. What Is Off-Line Programming?

Present *teach* methods of programming industrial robots have proved to be satisfactory where the proportion of teaching time to production time is small, and also when the complexity of the application is not too demanding. They involve either driving a robot to required positions with a teach pendant or physically positioning the robot, usually by means of a teach arm. Teach methods as such necessitate the use of the actual robot for programming.

Off-line programming may be considered as the process by which robot programs are developed, partially or completely, without requiring the use of the robot itself. This includes generating point coordinate data, function data, and cycle logic. Developments in robot technology, both hardware and software, are making off-line programming techniques more feasible. These developments include greater sophistication in robot controllers, improved positional accuracy, and the adoption of sensor technology. There is currently considerable research activity in off-line programming methods, and it is expected that these techniques will be employed in manufacturing industries within a few years.

20.1.2. Why Should Off-Line Programming Be Used?

Programming a robot by *teaching* can be time-consuming—the time taken quite often rises disproportionately with increasing complexity of the task. As the robot remains out of production, teach programming can substantially reduce the utility of the robot, sometimes to the extent that the economic viability of its introduction is questioned.

Many early robot applications involved mass production processes, for example, spot welding in automobile lines, where the reprogramming time required was either absent or minimal. However, for robot applications to be feasible in the field of small and medium batch production, where the programming times can be substantial, an off-line programming system is essential. The increasing complexity of robot applications, particularly with regard to assembly work, makes the advantages associated with off-line programming even more attractive. These may be summarized as follows:

1. Reduction of robot downtime. The robot can still be in production while its next task is being programmed. This enables the flexibility of the robot to be utilized more effectively.

2. Removal of programmer from potentially hazardous environments. As more of the program development is done away from the robot this reduces the time during which the programmer is at risk from aberrant robot behavior.

3. Single programming system. The off-line system can be used to program a variety of robots without the need to know the idiosyncracies of each robot controller. These are taken care of

by appropriate postprocessors, minimizing the amount of retraining necessary for robot programmers.

4. Integration with existing CAD/CAM systems. This enables interfaces to access standard part data bases, limiting the amount of data capture needed by the off-line system. Centralization of the robot programs within the CAD/CAM system enables them to be accessed by other manufacturing functions such as planning and control.

5. Simplification of complex tasks. The utilization of a high-level computer programming language for the off-line system facilitates the robot programming of more complex tasks.

6. Verification of robot programs. Existing CAD/CAM systems, or the off-line system itself, can be used to produce a solid world model of the robot and installation. Suitable simulation software can then be used to prove out collision-free tasks prior to generation of the robot program.

20.2. DEVELOPMENT OF OFF-LINE PROGRAMMING

20.2.1. Some Parallels with NC

The technique of off-line programming has been employed in the field of numerical controlled (NC) machining for some considerable time, and a significant body of knowledge has been built up in this area. Although there are fundamental differences between the programming/controlling of NC machines and industrial robots, there are similarities in the problems and phases of development for off-line systems. It is therefore instructive to draw some parallels with NC in the course of charting the developments of industrial robot programming methods.

Early NC controllers were programmed directly using codes G, F, S, . . . and so on, with appropriate X, Y, and Z positioning coordinates. These specified certain parameters such as tool movements, feed rates, and spindle speeds, and were specific to individual controllers.

The majority of present-day industrial robots have controllers more or less equivalent to this, that is, they are programmed directly using robot-specific functions. The programming is done entirely on-line, and coordinate positions are taught manually and recorded in the computer memory of the controller. Program sequences can then be replayed.

The next development phase for programming NC controllers utilized high-level computer languages to provide greater facilities for the NC programmer. Programming became truly off-line, with program development being performed on computers remote from the shop floor. High-level textual NC programming languages such as APT and EXAPT developed. Tools and parts could be described geometrically using points, lines, surfaces, and so on, and these were used to generate cutter path data. Appropriate interfaces were required between the languages and specific NC controllers.

This is the corresponding stage for industrial robot programming development today. Textually based systems such as RAPT and ROBEX incorporate high-level programming features similar to APT. Graphically driven systems (CATIA, GRASP) utilize interactive CAD techniques. All the systems generally possess the capability of modeling the kinematics of a 6-DF manipulator.

Just as NC programming progressed toward integration with CAD systems, off-line programming of industrial robots will develop in a similar fashion. Ultimately, systems incorporating direct robot control will be developed to cater to multirobot installations.

20.2.2. Levels of Programming

A useful indicator of the sophistication of a robot-programming system is the level of control of which it is capable. Four levels can be classified as follows:[1]

1. **Joint Level.** This requires the individual programming of each joint of the robot structure to achieve the required overall positions.

2. **Manipulator Level.** This involves specifying the robot movements in terms of world positions of the manipulator attached to the robot structure. Mathematical techniques are used to determine the individual joint values for these positions.

3. **Object Level.** Requires specification of the task in terms of the movements and positioning of objects within the robot installation. This implies the existence of a world model of the installation from which information can be extracted to determine the necessary manipulator positions.

4. **Objective Level.** Specifies the task in the most general form, for example, "spray interior of car door." This requires a comprehensive data base containing not only a world model but also knowledge of the application techniques. In the case of this example, data on optimum spraying conditions and methods would be necessary. Algorithms with what might be termed "intelligence" would be required to interpret the instructions and apply them to the knowledge base to produce optimized, collision-free robot programs.

Programming at object, and particularly objective, levels requires the incorporation of programming constructs to cater to sensor inputs. This is necessary to generate cycle logic. For example, the "IF, THEN, ELSE" construct could be employed in the following manner:

$$\text{IF (SENSOR} = \text{value)} \quad \text{THEN} \quad \text{action 1}$$
$$\text{ELSE} \quad \text{action 2}$$

Most present-day systems, on-line and off-line, provide manipulator-level control. Language systems currently under development are aiming toward the object level of programming, with objective level being a future goal.

20.3. GENERAL REQUIREMENTS FOR AN OFF-LINE SYSTEM

In essence, off-line programming provides an essential link for CAD/CAM. Success in its development would result in a more widespread use of multiaxis robots and also accelerate the implementation of flexible manufacturing systems (FMS) in industry.

As indicated in the preceding section, off-line programming can be affected at different levels of control. Different systems employ different approaches to the programming method. A more detailed discussion of these systems is given in Section 20.5.

Despite their differences, they contain certain common features essential for off-line programming. This following list gives the requirements that have been identified to be important for a successful off-line programming system:

1. Knowledge of the process or task to be programmed.
2. Three-dimensional world model, that is, data on the geometric descriptions of components and their relationships within the workplace.
3. Knowledge of robot geometry, kinematics (including joint constraints and velocity profiles), and dynamics.
4. A computer-based system or method for programming the robots utilizing data from (1), (2), and (3). Such a system could be graphically or textually based.
5. Verifications of programs produced by 4. For example, checking for robot joint constraint violations and collision detection within the workplace.
6. Appropriate interfacing to allow communication of control data from the off-line system to various robot controllers. The choice of a robot with a suitable controller (i.e., one that is able to accept data generated off-line) will facilitate interfacing.
7. Effective man/machine interface. Implicit in off-line programming is the removal of the programmer from the robot. To allow the effective transfer of his skills to a computer-based off-line system, it is crucial that a user-friendly programming interface is incorporated.

20.4. PROBLEMS IN OFF-LINE PROGRAMMING

20.4.1. Overview

$$\text{THEORETICAL} \underline{\quad\quad} \text{ROBOT} \underline{\quad\quad} \text{REAL}$$
$$\text{MODEL} \quad\quad\quad \text{PROGRAM} \quad\quad \text{WORLD}$$

Off-line programming requires the existence of a theoretical model of the robot and its environment; the objective is to use this model to simulate the way in which the robot would behave in real life. Using the model, programs can be constructed that, after suitable interfacing, are used to drive the robot.

The implementation of off-line programming encounters problems in three major areas. First, there are difficulties in developing a generalized programming system that is independent of both robots and robot applications. Second, to reduce incompatibility between robots and programming systems, standards need to be defined for interfaces. Third, off-line programs must account for errors and inaccuracies that exist in the real world. The following sections provide a more detailed discussion of these problem areas.

20.4.2. Problems in Modeling and Programming

The modeling and programming system for off-line work can be categorized into three areas: the geometric modeler, the programming system, and the programming method. Each has its own inherent difficulties, but the major problems arise when attempts are made to generalize functional features. Although generalization (to cater to different types of robots and applications) is necessary to make the system more effective, it is also important to ensure that corresponding increases in complexity do not inhibit the functional use of the system.

Geometric Modeler

A problem with any geometric modeler is the input of geometrical data to allow models to be constructed. In a manual mode this process is time-consuming and error prone. Sufficient attention must be given to improving methods of data capture. One way is to utilize data stored in existing CAD systems. Even so, this necessitates the writing of appropriate interfaces.

The data structure used must be capable not only of representing relationships between objects in the world but also of updating these relationships to reflect any subsequent changes. It must also allow for the incorporation of algorithms used by the robot modeler. To accomplish these requirements efficiently can be a difficult task.

Robot Modeler

An off-line programming system must be able to model the properties of jointed mechanisms. There are several levels at which this may be attempted.

The first level is to develop a robot-specific system, that is, for use with only one type of robot. While this greatly simplifies the implementation, it also limits the scope of application of the system.

A second-level approach is to generalize to a limited class of structures. For example, most commercial robots consist of a hierarchical arrangement of independently controlled joints. These joints usually allow only either rotational or translational motion. Standard techniques exist for modeling such manipulators.[2] Even at this level, controlling the robot in terms of the path to be followed by the mounted tool is not easy. There is no general solution that covers all possibilities. The structures must be subclassified into groups for which appropriate control algorithms can be developed.

The third level is to attempt to handle complex manipulator structures. Some robots have interconnected joints that move as a group and are therefore not mathematically independent. The mathematics of complex mechanisms such as these is not understood in any general sense, although particular examples may be analyzed.

In the near future the kinematics of generalized off-line systems will probably cope with a restricted set of mechanisms as described in the second level. The incorporation of dynamic modeling (to simulate real motion effects such as overshoot and oscillation) is considered to be too complex to accomplish in a generalized fashion.

Programming Method

The geometric and robot modelers provide the capability of controlling robot structures within a world model. A programming method is required to enable control sequences, that is, robot movement sequences, to be defined and stored in a logical manner. The method should allow robot commands, robot functions, and cycle logic to be incorporated within these sequences to enable complete robot programs to be specified.

The latter requirements cause complications when the system is to be applied to different application areas. For example, the functional requirements and robot techniques for arc welding are significantly different from those involved in spray painting. Modularization of the programming method into application areas should ease these complications and produce a more efficient overall system.

The implementation of programming can be at one of several levels, joint, manipulator, object, or objective level, as discussed in Section 20.2.2, "Levels of Programming." Most systems under development are aimed at object level, which requires programmer interaction to specify collision-free, optimized paths. This interaction is considerably simplified if graphics are used within the programming system. This enables the programmer to visualize the problem areas and to obtain immediate feedback from any programmed moves.

Generalization of the programming method to cater to multirobot installations creates many difficulties. The incorporation of time-based programming and methods of communication between separate programs is necessary. Future systems will have greater need for these enhanced facilities as the trend toward FMS increases.

20.4.3. Interfacing

An off-line programming system defines and stores the description of a robot program in a specific internal format. In general, this is significantly different from the format employed by a robot controller for the equivalent program. Hence it is necessary to have a form of interfacing to convert the program description from an off-line system format to a controller format.

One of the major problems lies in the existence of a wide range of different robot controllers together with a variety of programming systems, each employing a different format of program description. To avoid a multiplicity of interfaces between specific systems and controllers, standards need to be defined and adopted.

Standardization could be employed in one or more of the following areas:

1. Programming System. The adoption of a standard system for off-line programming would considerably reduce interfacing efforts. There are presently two main working parties investigating the implications and requirements of this approach. One is a European tripartite effort with members coming from the United Kingdom, France, and West Germany, and the other is based on a set of Japanese proposals working within the CAM-I framework. It is encouraging to note that an initial report[3] from the European group indicates close agreement on ideas.

2. Control System. A standardized robot control system would have similar beneficial effects to standardizing the programming system. Commercial and practical considerations make this approach an unlikely occurrence.

3. Program Format. The definition of a standard format for robot program descriptions would also reduce interfacing problems. Such a format would, of necessity, be independent of the programming systems and controllers. The adoption of CLDATA[4] in numerical control (NC) gives a useful precedent for this approach. Programming systems could incorporate software to generate the standard format from their internal data. Postprocessors would then convert from the standard format to controller-specific formats. IRDATA,[5] based on NC CLDATA, is the most advanced development to date. This proposed standard has the advantage of being based on existing NC terminology, which is familiar to manufacturing engineers. However, the many differences that exist between machine tools and robots make the expansion of CLDATA a formidable task. Nevertheless, it is quite likely that standardization in this area will be the first to occur.

20.4.4. Real-World Errors and Inaccuracies

Owing to implicit differences between an idealized theoretical model and the inherent variabilities of the real world, simulated sequences generally cannot achieve the objective of driving the robot without errors. In practice, the robot does not go to the place predicted by the model, or the workpiece is not precisely at the location as defined in the model. These discrepancies can be attributed to the following components:

1. **The Robot**
    (a) Insufficiently tight tolerances used in the manufacture of robot linkages, giving rise to variations in joint offsets. Small errors in the structure can compound to produce quite large errors at the tool.
    (b) Lack of rigidity of the robot structure. This can cause serious errors under heavy loading conditions.
    (c) Incompatibility between robots. No two robots of identical make and model will perform the same off-line program without small deviations. This is caused by a combination of control system calibration and the tolerancing problems outlined.

2. **The Robot Controller**
    (a) Insufficient resolution of the controller. The resolution specifies the smallest increment of motion achievable by the controller.
    (b) Numerical accuracy of the controller. This is affected by both the word length of the microprocessor (a larger word length results in greater accuracy) and the efficiency of the algorithms used for control purposes.

3. **The Workplace**
    (a) The difficulty in determining precise locations of objects (robots, machines, workpiece) with reference to a datum within the workplace.
    (b) Environmental effects, such as temperature, can adversely affect the performance of the robot.

4. **The Modeling and Programming System**
    (a) Numerical accuracy of the programming system computer—effects such as outlined under 2b.
    (b) The quality of the real-world model data. This determines the final accuracy of the off-line program.

The compounding effects of these errors across the whole off-line programming system can lead to discrepancies of a significant magnitude. For off-line programming to become a practical tool this magnitude must be reduced to a level whereby final positioning adjustments can be accomplished automatically.

To achieve this a combination of efforts will be required. First, the positional accuracy* of the robot must be improved. Positional accuracy is affected by factors such as the accuracy of the arm, the resolution of the controller, and the numeric accuracy of the microprocessor. Second, more reliable

---

* Positional accuracy of the robot is its ability to achieve a commanded world position. This is distinct from *repeatability,* which relates to the variation in position when it repeats a taught move.

methods for determining locations of objects within a workplace must be applied. Third, the incorporation of sensor technology should cater to remaining discrepancies within a system. Improvements in tolerances of the components on which the robot has to work will aid the overall performance of the system.

20.5.   REVIEW OF SOME CURRENT SYSTEMS

There is currently no general-purpose off-line programming system available commercially. Although systems of a limited form, such as VAL[6] and SIGLA,[7] are in practical use, the majority are in the stages of research and development. Some of the major systems, together with their principal features, are identified in this section. The systems considered are GEOMAP,[8] ROBEX,[9] RAPT,[10] GRASP,[11] CATIA,[12] AUTOPASS,[13] ANIMATE/PLACE,[14] and AL.[15]

Apart from this list, mainly European in origin, other systems and development activities exist in off-line programming. These are predominantly within the United States and Japan and include, for reference purposes, ROBOT PERFORMANCE SIMULATOR,[16] ANIMATOR,[17] LAMA,[18] LM,[19] and PLAW.[20] AML, the textual programming system developed by IBM, is described in detail in Chapter 21.

Generally systems lean toward either a textual or graphical approach to the programming method. The textual approach involves programming the robot using a high-level, textual programming language akin to computer programming languages. This provides powerful data manipulation and arithmetic capabilities. Program constructs for the implementation of parallel processes or sensor input exist. The user/programmer specifies the geometric entities, their relationships, and the required task in terms of English-like statements and commands. In the ROBEX system, for example, the definition of a load/unloading task utilizes statements such as the following:[21]

```
START) ONSIG/EVENT, 2, JMP, WZM1
              .
              .
              .
WZM1) GOTO/P2
       CLOSGR
       PUT/PAL, TEACH
       GOTO/P1
       CLOSGR
       GOTO/P2
       OPENGR
       GOTO/SAFP
```

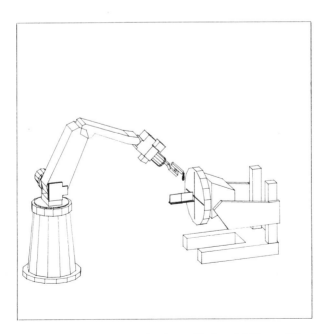

Fig. 20.1.   Display from the GRASP system showing a Cincinnati Milacron T3 robot and a simple welding component mounted on a rotary table.

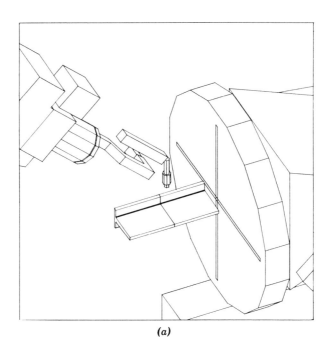

(a)

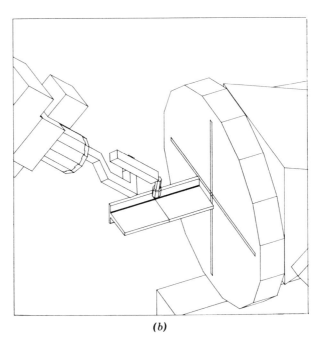

(b)

Fig. 20.2. Task program of a robot performing a simple seam weld. Display output from the GRASP system. (a) Approach point. (b) Start of weld. (c) End of weld. (d) Retract to safe position.

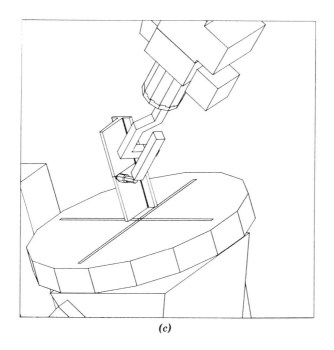

(c)

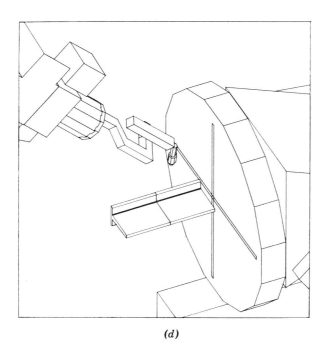

(d)

Fig. 20.2. (Continued)

For this example, a robot responds to a signal from a machine tool (P2), moves to it, grips the machined part, stores it on a pallet according to a predefined position, proceeds to a conveyor belt (P1), grips a raw part, returns to the machine tool, loads it, and returns to a safe position (SAFP).

Some textual systems incorporate a graphics simulator. This is normally used to verify the textual programs in terms of reach constraints and collisions between objects. The programs can then be modified to eliminate such problems. Verified programs are converted to a suitable intermediate format. This is then postprocessed for particular robot controllers.

The graphical approach involves the use of interactive CAD techniques to specify the robot task. The robot and the world model are displayed on a graphics screen. An example of such a display from the GRASP system is seen in Figure 20.1. Programming is done interactively by menu commands, utilizing light pens, tablets, or keyboards. One big advantage with graphics is that it provides immediate visual feedback. Typically, a task program would consist of a series of pictures as shown in Figure 20.2a to d. These represent the move steps. A complete task program would also include textual commands to represent robot functions such as weld, grip, and release. The system stores the program description in a nontextual internal format. This information can be postprocessed as for textual systems.

Table 20.1 is a summary of the general features available within a selection of current systems. This is provided as a guide to the systems' approaches and capabilities and is by no means definitive. Information included is based on published data where available, with systems being given the benefit of the doubt where any exists. For more detailed descriptions of the capabilities of the individual systems reference should be made to the papers cited previously.

The features listed in the table are subdivided into four main sections: the geometric modeler, robot modeler, programming system, and "other features." These have been selected for comparison as they have been identified to be important features for successful off-line systems (see Section 20.3).

20.5.1. Geometric Modeler

All of the systems incorporate some form of geometric modeler. This is used to generate a world model of the installation, so that objects may be referenced during programming. Solid-body modeling is required for a complete three-dimensional description of objects, in turn, necessary for collision checking and avoidance. In graphic simulation, hidden line removal can only be performed with solid models.

20.5.2. Robot Modeler

A robot modeler enables a robot to be defined and controlled by the system. The joint structure, constraints, and velocity data are entered into the system to give a kinematic representation of the robot. Control is effected by specifying the position and orientation of the robot tool and then determining mathematically the individual joint values needed to achieve this.

AL, AUTOPASS, and RAPT define the positions a manipulator must achieve but have no robot modelers. Thus they provide no indication of a robot's ability to perform the task. The systems that include robot modelers have them subclassified into *kinematic, path control,* and *generalized. Kinematic* indicates a modeler that provides point-to-point positioning control. *Path control* indicates the ability to control the robot movement at all points along a particular trajectory. *Generalized* implies that the system is designed to model a range of robots.

20.5.3. Programming System

This section indicates the programming level of the system as discussed in Section 20.2. The tendency toward a textual or graphical approach is also shown, together with the ability to incorporate sensors and hence cycle logic.

20.5.4. Other Features

This last division provides a list of important features that are not highlighted in the other sections.

Reach testing indicates that joint constraints are incorporated within the robot modeler and are checked for any violations during robot movement. *Control output* means that the system is capable of generating robot control data. This may be low-level, robot-specific data, *standard* robot independent data, or in a higher form such as a VAL program.

Dynamic collision indicates that the system is capable of detecting any collisions, not only at programmed points but also during movement between points.

Cycle times require the incorporation of joint velocities within the robot modeler. The time taken for robot moves can be calculated from these, provided that path control is implemented.

Multirobot means that the system is capable of programming and controlling more than one robot simultaneously. That is, it is able to cope with parallel processes.

TABLE 20.1. SUMMARY OF PROGRAMMING SYSTEM [a]

FEATURES	AL (Stanford)	ANIMATE/PLACE (McDonnell Douglas)	AUTOPASS (IBM)	CATIA (Montpellier)	GRASP (Nottingham)	RAPT (Edinburgh)	ROBEX (Aachen)	GEOMAP (Tokyo)
Geometric Modeler								
Nonsolid	✓	✓						✓
Solid			✓	✓	✓	✓	✓	
Robot Modeler								
Kinematic		✓		R	✓		✓	✓
Path control		✓		✓	R		✓	✓
Generalized		✓		✓	✓		✓	
Programming System								
Manipulator level	✓	✓		✓	✓		✓	✓
Object level	✓		✓			✓	✓	✓
Textual	✓		✓			✓		
Graphical		✓		✓	✓		R	
Sensor input	✓		✓		D			
Other Features								
Reach testing		✓		✓	✓		✓	✓
Control output	✓		✓	D	D	✓	✓	✓
Dynamic collision			R	D	D		D	R
Cycle times		✓			✓			✓
Multirobot	✓				✓			
Graphic simulation	✓	✓		✓	✓		✓	✓

[a] ✓ = Available; R = restricted; D = under development.

20.6.   A GRAPHICAL OFF-LINE SYSTEM

20.6.1.   Introduction

An important aspect of off-line programming is the need not only to produce a program in some form for a robot, but also to verify off-line that the robot will be able to execute the program. As far as possible, problems of reach, accessibility, collision, timing, and so on, should be eliminated in the planning stage. Graphical computer-body modeling packages extended to simulate robot behavior have great potential in this area, and several such systems are under development at various centers. In this way, the generation of one or more robot programs can be combined with their off-line verification, and it is possible to produce a visual presentation of the robot performing its task. We describe one such system, known as GRASP (graphical robot applications simulation package). This is a generalized system, under development at Nottingham University, England, aimed at dealing with a wide variety of robots working in many different situations.

20.6.2.   Solid-Body Modeling

The basis of GRASP is a simple body-modeling package. A representation of any object is constructed using as building blocks a set of simple shapes known as *primitives*. A primitive may be a cuboid, a regular *n*-sided prism, an irregular prism, or any general, closed solid. All solids are represented using planar faces, but cylinders and solids of revolution may be approximated using the basic primitives (see Figure 20.3).

The primitives are grouped together into a hierarchical tree structure, so that several of them arranged in appropriate positions may be manipulated as a single entity. Each entity has associated with it a local Cartesian coordinate axis system. The location and orientation of an entity at any level in a model tree may be modified under user control by reference to its local axis system, or to that of another entity.

The model is displayed on a vector refresh or storage tube graphics terminal in wire frame mode. The line of sight and scale may be adjusted so that the model may be viewed from any aspect. It is possible to "zoom in" on particular areas of interest. Hidden line pictures may also be produced, so that edges that would be obscured by true solids are not displayed.

20.6.3.   Robot Modeling

Since GRASP is a robot-modeling system, it is necessary to be able to model the particular properties of robots in some detail. The English-like language used to define basic body models also incorporates

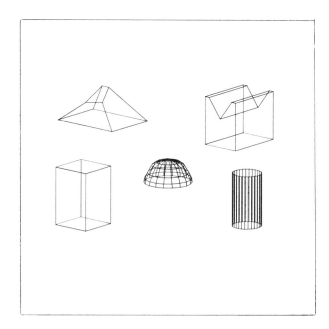

Fig. 20.3.   Some geometric primitives used in the GRASP modeller: cuboid, solid of revolution, cylinder, polyprism, general module.

a high-level statement by which the joint structure, constraints, and other data associated with a robot may be defined. Separate entities modeling the "flesh" of each joint may then be incorporated into the robot model. No special mathematical knowledge is required to construct such a model. However, the robot structure is represented internally in a rational manner using the method of Denavit and Hartenberg.[2] The user interacts with the system in terms of his own definition and is unaware of the underlying mathematics.

Any entity within a model may be nominated as a tool and mounted on a robot. It is then associated in a special way with the robot, and the local axis system of some entity below the tool in the tree is selected as a tool center point (TCP). The TCP is used as a reference system for defining a desired tool position.

20.6.4. Robot Control

There are three basic methods of controlling the simulated position of a robot. The first allows the user to move individual joints through a number of degrees or a specified distance. This is not particularly useful, and it is quite difficult to achieve a desired tool position using this method. It is also possible to record a particular configuration as a park position, which may be invoked at any time.

The most useful method of control is to define the desired position of the tool and allow GRASP to compute the required attitude of each joint in order to achieve that position. This is done in the following manner: The robot joint structure is analyzed when the model is created and classified according to the type of its structure. The classification code is then used to select the appropriate algorithm when it is necessary to compute the joint angles required to move the tool to a particular position.

There are several alternative methods available to the user for defining a tool position. This is usually done by reference to the objects being manipulated (i.e., programming at object level), although coordinate data may be entered. All the methods ultimately determine the location and orientation of the tool center point. One useful method is to create a copy of the tool structure and move it to the desired position using the manipulation facilities available within GRASP. The new position may be verified in a number of different views. This location of the tool copy then defines the desired position of the tool mounted on the robot.

Any number of tools may be mounted on a robot. At any one time one of them is selected as the reference for control.

20.6.5. Defining Robot Programs

The first stage in creating a robot program is to generate a sequence of discrete robot positions or other operations, such as grips and releases. Such a sequence is known in GRASP as a *track*. Each robot position is defined as a location and orientation of the tool center point. These data are stored, not in absolute world coordinates, but relative to the axis system of some useful entity, such as a workstation or workpiece. Thus it is possible to replay the track not only with the workstation in its original position but also to test alternative arrangements of the workplace. Each step along the track essentially refers to a tool position or operation and is thus robot independent. Hence the robot and other items in the workplace may be resited without redefining the track. It is also possible, in the same way, to modify the tool mounting or to test the operation using a different kind of robot. A track may also be edited by inserting, deleting, or modifying steps.

Attempts to move a robot to a position that it cannot attain are reported to the user with helpful diagnostic information. Thus verification is obtained of the ability of the robot to attain discrete positions along a programmed sequence. If problems are found, these can be corrected by altering the workplace layout, modifying the tool mounting, or using a different kind of robot. At this stage, however, nothing has been learned about the motion between the defined positions or the time required to make the move.

20.6.6. Path-Controlled Motion

The next stage in program creation is to define the manner in which the robot moves between positions. For example, it may be constrained to move the tool center point in a straight line at a constant speed between two track steps. A maximum velocity attainable by each joint is stored with the robot model. Another kind of motion determines that each joint moves at its maximum velocity until the required change in attitude has taken place. Other possibilities exist.

Where the motion is defined in terms of the path to be followed by the tool, the motion required of each joint may be rather complex. In this case the joint behavior is modeled by use of cubic splines, computed after sampling the robot position at intervals along the path. Hence, for different kinds of motion, the joint value may be determined along the entire path. It is therefore possible to check whether the required motion violates any joint constraints. For example, a straight-line path may pass out of a robot's working volume, even though both end points lie within range.

20.6.7. Time-Based Motion

Finally, the robot behavior is determined as a function of time. This stage of the program creation is carried out automatically using the information saved in the track and the velocity data stored with the robot. The result is called a *process*. In this way estimates of cycle times are produced, and a check is made that the defined motion does not require any joint to exceed its maximum velocity.

Up to this point, although a particular model may contain several robot models, only one robot at a time has been allowed to move. However, if these robots are required to work simultaneously, processes generated for each robot may be merged and run together. Other objects, such as conveyor belts, may also be caused to perform timed motion. In this manner an entire population system may be simulated, allowing an assessment to be made of the interaction between several moving robots and objects. It is also possible at this stage to produce a good-quality film depicting the operation.

20.6.8. Collision Detection

At present the best method of detecting potential problems of collision and access to work areas is by use of the viewing facilities available in GRASP. Potential problem areas may be viewed in close-up from any angle. If the model is halted at a suspect position, an automatic check may also be carried out to determine whether two objects are interfering with each other. This can be useful since the interpretation of wire frame drawings may be difficult. It is projected that fully automatic collision checking will be incorporated into GRASP.

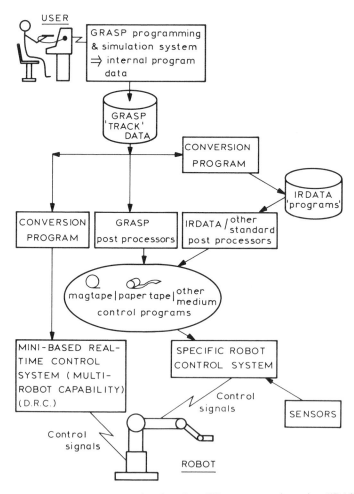

Fig. 20.4. Schematic diagram of options for off-line programming using GRASP.

20.6.9. Off-Line Programming

The tremendous value of off-line simulation has been illustrated. Problems of reach, accessibility, collision, and timing can be detected at the planning stage. Possible solutions may be quickly determined and tested with the minimum of effort. Furthermore, all the information necessary to define an on-line robot program will have been entered into the model during the simulation. The extension of GRASP to incorporate a postprocessor capable of outputting a robot program in some usable form is a natural one.

Three possible approaches are depicted in Figure 20.4. All of them would create some form of output based on the TRACK data stored within the modeler.

1. The most general method would be to develop a conversion program capable of creating a file containing the robot program as defined by some international standard. One possibility is the format specified in IRDATA, currently under development at Aachen in West Germany. It is unlikely, based on past experience in such matters, that a single standard will become universally accepted. It may be necessary, therefore, to have the capability of producing programs in one of several widely used standards. Robot manufacturers will no doubt develop postprocessors capable of converting standard programs into forms acceptable to their machines' specific requirements.

2. Another approach would be to incorporate a set of postprocessors within GRASP, which would produce output in a form directly acceptable to particular makes of robot. The program could be transferred by way of some medium such as magnetic tape or paper tape. Unless standards are evolved, the usefulness of each postprocessor would be limited to a single manufacturer's range of products.

3. A final alternative would be to implement direct robot control (DRC) of one or more on-line robots from the GRASP system. This would be accomplished by means of a real-time control system based in a mini- or microcomputer. Some form of interface program would also be required within GRASP to convert the data to a form acceptable to the control system. This has advantages in terms of shortening the time required for final debugging of programs, enabling necessary changes to be made at the highest level of control within GRASP. It also allows the centralization of control for multirobot installations.

Preliminary experiments have been conducted using coordinate data obtained from GRASP to drive a robot performing welding operations. The parts tested so far have been simple in design, and an accuracy of ± 3 mm has been achieved. Some problems were encountered. It was found that the weld gun was not mounted precisely according to drawings. This resulted in the actual weld point being offset from the point stored in the robot controller as the tool was, therefore, moved to slightly incorrect positions under numerical control. This problem was overcome by building a makeshift jig, which was used to correctly align the gun. Difficulty was also encountered in rotating a worktable to world positions predicted by the simulation because the table controller would not allow any form of numerical control. Only taught positions could be recorded. However, the results are considered to be encouraging. It is envisaged that with a second-generation robot of greater positional accuracy and incorporating sensors the need for on-line fine-tuning could be drastically reduced.

20.7. SUMMARY

Off-line programming is a necessary step toward increasing the use of industrial robots in industry. Although most current off-line systems are in the stages of research and development, some limited systems are in practical use. It is foreseen that, parallel with other second-generation robotic developments, off-line systems will become practical tools within a few years.

Several factors will help to accelerate this. Improvements in robot design will enable the control loop to be closed. Agreement on standards for interfacing programming systems with robots will increase their acceptability in industry. Off-line systems, themselves, will continue to be developed and refined as their requirements become more closely defined.

In many respects the need for off-line programming is necessitated by the unsatisfactory nature of the present teach mode of programming robots. Off-line programming will increase the flexibility and utility of the robot and make it safer by removing the man from the programming environment. Ultimately, the extent of the use and implementation of off-line programming will be determined by the benefits it can provide to manufacturing industry.

ACKNOWLEDGMENT

The authors wish to acknowledge the Science and Engineering Research Council of the United Kingdom for funding the GRASP project.

REFERENCES

1. Latombe, J. C., Une Analyse Structuree d'Outils de Programmation pour la Robotique Industrielle, *Proceedings of the International Seminar on Programming Methods and Languages for Industrial Robots,* IRIA, Rocquencourt, France, June 1979.

2. Denavit, J., Hartenberg, R. S., A Kinematic Notation for Lower-Pair Mechanisms Based on Matrices, *Journal of Applied Mechanics,* June 1955, pp. 215–221.

3. UK–France–Germany, Working Party on Robot Assembly Languages, *A Programme of Work to achieve a Common Robot Assembly Language System,* Science and Engineering Research Council, U.K., September 1982.

4. British Standard 5110—Specification for Programming Languages for the Numerical Control of Machines (Equivalent to ISO 3592 and 4343), 1979.

5. IRDATA—General Structure and Types of Records, Draft VDI 2863, West Germany, 1982.

6. *Users Guide To VAL-II,* The Unimation Robot Programming and Control System, Danbury, Connecticut, September 1982.

7. Salmon, M., Sigla. The Olivetti SIGMA Robot Programming Languages, *Proceedings of the 8th ISIR,* Stuttgart, IFS Ltd., Kempston, Bedford, England, 1978.

8. Sata, T., Kimura, F., and Amano, A., Robot Simulation System as a Task Programming Tool, *Proceedings of the 11th ISIR,* Tokyo, 1981.

9. Weck, M., Eversheim, W., and Zuehlke, D., ROBEX—An Off-line Programming System for Industrial Robots, *Proceedings of the 11th ISIR,* Tokyo, 1981.

10. Popplestone, R. I. and Ambler, A. P., A Language for Specifying Manipulations, *DAI Research Paper 161,* Department of Artificial Intelligence, University of Edinburgh, U.K., 1981.

11. Yong, Y. F. and Bonney, M. C., Simulation—Preventing Some Nasty Snarl-Ups, *A Decade of Robotics—A Special 10th Anniversary Issue,* IFS Publications Ltd., Kempston, Bedford, England, 1983.

12. Borrel, P., Liegeois, A., and Dombre, E., The Robotics Facilities in the CAD-CAM CATIA System, *Developments in Robotics 1983,* IFS Publications Ltd., Kempston, Bedford, England.

13. Lieberman, L. I. and Wesley, M. A., AUTOPASS: An Automatic Programming System for Computer Controlled Mechanical Assembly, *IBM Journal of Research and Development,* Vol. 21, No. 4, July 1977, pp. 321–323.

14. Kretch, S. J., Robotic Animation, *Mechanical Engineering,* August 1982.

15. Mujitaba, M. S., Current Status of the AL Manipulation Programming Language, *Proceedings of the 10th ISIR,* March 1980.

16. Soroka, B. I., Debugging Robot Programs with a Simulator, *Autofact West Conference,* Society of Manufacturing Engineers, Anaheim, California, 1980.

17. Bathor, M. and Siegler, A., "Graphic Simulation for Robot Programming," *4th British Robot Association Annual Conference,* Brighton, U.K., May 1981.

18. Lozano-Perez, T. and Winston, P. H., LAMA: A Language for Automatic Mechanical Assembly, *Proceedings of the 5th IJCAI,* Boston, Massachusetts, 1977, pp. 710–716.

19. Latombe, J. C. and Mazer, E., LM: A High-Level Programming Language for Controlling Assembly Robots, Tripartite Robotics Seminar, Stuttgart, February 1981.

20. Abe R., Ueno, S., Tsujikado, S., and Tagaki, H., The Programming Language for Welding Robot, *Proceedings of the 11th ISIR,* Tokyo, 1981.

21. Eversheim, W., Weck, M., Scholing, H., Zuehlke, D., and Muller, W., Off-line Programming of Numerically Controlled Industrial Robots Using the ROBEX-Programming-Systems, *Annals of the CIRP,* January 1981.

CHAPTER 21

A STRUCTURED PROGRAMMING ROBOT LANGUAGE

STEPHEN J. BUCKLEY
GALE F. COLLINS

IBM
Boca Raton, Florida

21.1. INTRODUCTION

AML is a high-level interactive structured computer programming language designed by IBM for use with a robot system. It provides the function of a general-purpose computer language, offering a variety of data types and operators, language control structures, display station and input/output control, and system identifiers for program readability. AML also provides additional basic resources, such as motion, sensing, and communications, needed to accomplish a robot task.

The current state of robot programming offers a number of robot programming languages as described in Chapter 20, "Off-Line Programming of Robots." However, none of these languages is at *task level*. Task-level languages (those that determine how the task is to be accomplished) are still in the research state (see Chapter 22, "Task-Level Manipulator Programming"). To serve the needs of current users, attention must be paid to the general-purpose computer functions of lower-level languages. These basic functions allow a programmer to build a structure around the robot control features to maximize the application programming capabilities and minimize the programming effort. AML supplies a rich set of program control capabilities so that the programmer has the necessary tools to gain the desired results. Data processing system subroutines are provided to transform the data used with the robotic functions. Program development and debugging facilities are supplied to ease the programming effort. Storage management system subroutines are available to manage the complex systems software efficiently. These features allow the programmer to generate user subroutines and use them with, or instead of, the system-supplied AML subroutines. User subroutines can be nested and used recursively. The programmer has the flexibility to create a tailored environment in which he can write application programs.

A recent five-level classification scheme compares a number of robot languages on the basis of their problem-solving methodologies. (See Reference 2.) Figure 21.1 illustrates this classification scheme.

Level 1. The microcomputer level; requires the user to specify robot tasks in microcomputer language and to perform explicitly all data calculations and conversions necessary to the task.

Level 2. The point-to-point level; requires the user to teach the robot a task by physically guiding it through a series of motions by pressing push buttons on a pendant or moving a joystick. Level 2 is the most common level of robot programming on the market today.

Level 3. The primitive motion level; allows the user to specify point-to-point paths without having to teach each one. Usually a simple, BASIC-like language is provided with one or more motion commands.

Level 4. The structured-programming level; provides greater flexibility in programming. Block-structured control and complex data structures allow a higher degree of calculation and decision making.

Level 5. The task-oriented level; allows the user to specify tasks in English-like terms, concealing low-level aids such as sensors and coordinate transformations. The system must determine how to accomplish the task, using a world model of the objects that are involved. Level 5 is not commercially available today; it is still a research topic.

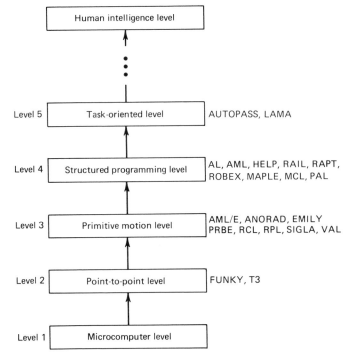

Fig. 21.1. Classification of robot language.

For the user, level 5 is the ideal level; it virtually eliminates concern for the details of the task. However, because level 5 is not available, commercial users must use a level-4, or below, language. AML, a level-4 language, is a good example of a full-bodied, structured programming language.

21.2. FUNCTIONS PROVIDED BY A STRUCTURED PROGRAMMING ROBOT LANGUAGE

This section describes some functions that are useful at the structured programming level, level 4. The majority are provided by AML. Examples and concepts are given to support the fact that AML is truly a structured-programming robot language.

Figure 21.2 shows some of the functions used to program a robot at the structured-programming level. These include basic functions of a general-purpose computer language (such as data processing, program development, and debugging), along with an extension for functions needed for robot control (such as motion, sensing, communication, and vision). The 7565 AML provides all the functions shown in the figure except vision. A research version of AML provides vision functions.[10]

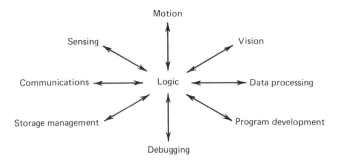

Fig. 21.2. Functions of a structured-programming robotic language.

At the heart of this circle of function is logic in the form of program control. The combination of program control with the various functions creates the impression of intelligence in a robot. In a level-4 language the user, by writing a computer program, provides each robot task with a control algorithm that the robot controller can use to carry out the task. If a task is not properly completed, it is usually because the programmer did not give the robot enough "intelligence" to deal with the problem. This would not occur with a level-5 language, for it is the job of the system to determine how to accomplish a task properly. Therefore it is extremely important for a level-4 language to provide program control capabilities that give the programmer the necessary tools to gain the desired results.

21.2.1. Logic

This section surveys some of the program-control structures of current computer languages that provide a robot with preprogrammed ways to make decisions. There are three kinds of program-control structures: (1) sequential, (2) parallel, and (3) asynchronous. AML provides sequential and asynchronous structures.

Sequential Control

Sequential control allows the execution of a program in ordered steps. Standard structured-programming language constructs for sequential control are branches, IF tests, and WHILE loops.

   In AML the smallest executable entity is an *expression*. An expression is a correctly formed series of constants, variables, subroutine calls, and operators. The following are examples of expressions:

$$12$$
$$36.2$$
$$I+3$$
$$MOVE(JX,10.0)$$

The first two expressions in this list are constants, one an integer and one a real. The third is the addition operator applied to a variable and a constant. The fourth is a call to the system subroutine MOVE. MOVE accepts two parameters in this case, a joint number JX and an absolute goal position for the motion 10.0.

   An expression followed by a semicolon is a *statement*. The following are examples of statements:

$$12;$$
$$36.2;$$
$$I+3;$$
$$MOVE(JX,10.0);$$

Statements can be grouped and defined as a single callable entity. This entity is a *subroutine*.

   Subroutines group common coding sequences together. AML contains a set of system subroutines that are built into the system and available for application use once the initial program load (IPL) is completed. These system subroutines can be used instead of, with, or in addition to user-written subroutines. For example, the following is a user-written subroutine:

```
USERMOVE: SUBR(J,G);
    MOVE(J,BASE+G);
END;
```

This subroutine is similar to the MOVE system subroutine in execution except that the goal position specified in USERMOVE is an offset from a global calibration position named BASE, whereas the goal position specified in MOVE is an absolute position in the workspace. (In this example, USERMOVE, J, and G are user-defined names. SUBR and END are keywords required for subroutine definition.)

   The syntax for user subroutine calls is identical to the syntax for system subroutine calls. This means that without some prior knowledge, a programmer would not automatically know that

```
MOVE(ARM,0);
```

is a call to a system subroutine, and that

```
USERMOVE(ARM,0);
```

is a call to a user-written subroutine. The identical syntax reduces the entry knowledge required to program in AML, since useful subroutines can be written by one user and passed along to another user as if they were system subroutines.

Nested and recursive subroutines are supported by AML. The following example solves the Towers of Hanoi puzzle, where a stack of different-diameter disks must be transferred from one post to another (using an intermediate post) without once placing a larger-diameter disk on top of a smaller-diameter disk.

```
--  PURPOSE: SOLVES THE TOWERS OF HANOI PUZZLE
--
--  PARAMETERS:
--      N,  INTEGER, NUMBER OF DISKS TO MOVE
--      Q,  INTEGER, POST NUMBER TO MOVE FROM
--      R,  INTEGER, POST NUMBER TO MOVE TO
--      S,  INTEGER, POST NUMBER OF AVAILABLE POST
--      AT,  INTEGER, CURRENT DISK NUMBER
--
--  RETURNS:  < >
--
--
TOWER: SUBR( N, Q, R, S, AT );
    IF N EQ 1 THEN RETURN( MOVEIT(AT,Q,R) );
    TOWER( N-1, Q, S, R, N-1 );
    MOVEIT( N, Q, R );
    TOWER( N-1, S, R, Q, N-1 );
    END;
```

In this example, assume that MOVEIT is a user-written subroutine that performs the physical transfer of a disk.

Parameters can be passed in AML either by value or by reference. For example, referencing the following subroutine:

```
S: SUBR(!A,B);
            .
            .
            .
    A=2;
    B=3;
    END;
```

the call:

```
S(X,Y);
```

would change the value of X but not Y. (An exclamation mark preceding a formal parameter declares that the parameter is to be passed by reference.) Parameters are weakly typed, that is, there is no data type checking when a program is entered into the system; data types are resolved at run time. This allows subroutines to be written that do different things depending on the input data types, for example,

```
S: SUBR(P);
    IF ?P EQ 1 THEN . . .
    ELSE IF ?P EQ 2 THEN . . .
```

The ? operator returns the data type of the variable that follows it. Parameters are optional if a default is specified for them. For example, in the following subroutine:

```
S: SUBR(J,G DEFAULT QGOAL(J));
```

the second parameter, if not explicitly specified in a call to S, defaults to the result of the QGOAL system subroutine applied to the first parameter.

All AML subroutines return a value. The RETURN system subroutine can be called to return a value from a subroutine call. Even if a subroutine does not call RETURN, the value < > (a null aggregate, to be explained later) is returned. This allows a subroutine to be called using the results of another subroutine, reducing code size and the need for temporary values.

All AML program control constructs are evaluated as expressions so that their value is available for use in a containing expression. This is different from many other languages in which program control constructs are statements. Certain classic restrictions are relaxed in AML. For example, in the following statement:

$$C = IF \; TEST \; THEN \; A \; ELSE \; B;$$

the result of the IF expression, either A or B, is assigned to C. In this example

$$(IF \; TEST \; THEN \; A \; ELSE \; B) = C;$$

the result of the IF expression, either A or B, is used as the target of the assignment. The programmer need not remember whether an IF expression can be used in one place or another—it can.

The expressions and system subroutines providing sequential program control in AML include the following:

BEGIN . . . END. The keywords BEGIN and END are used to delineate a group of statements which are then treated as a single expression.

IF . . . THEN . . . ELSE. The IF expression is used for basic decision making.

WHILE . . . DO. The WHILE expression allows an expression to loop while a condition is met.

REPEAT . . . UNTIL. The REPEAT expression allows an expression to loop until a condition is met.

BRANCH. The BRANCH system subroutine explicitly switches control to a specified label.

RETURN. The RETURN system subroutine returns to the previous subroutine level. If a parameter is specified to RETURN, it is returned to the previous level as the value of the call.

QUIT. The QUIT system subroutine returns to any specified subroutine level and continues execution at the next statement at that level.

CLEANUP. The CLEANUP system subroutine specifies a subroutine to execute at the termination of the current subroutine level for the purpose of closing open files, returning dynamically allocated data, and other functions. CLEANUP is useful when a subroutine is terminated unexpectedly from a numerically higher subroutine level, for example, an error followed by a QUIT. In the following example an I/O channel is allocated by the subroutine on entry and then returned by its CLEANUP routine. (I/O channels can be used to retrieve stored data in data-driven applications. This topic is examined in more detail later in this section.)

```
APPLIC: SUBR;
   CH: NEW 0;
   APPLCLEAN: SUBR;
      IF CH NE 0 THEN CLOSE(CH);
      END;
   CLEANUP('APPLCLEAN2');
   CH = OPEN('DATAFILE.USER01');

      · (data-driven application code)
      ·
   END;
```

AML is an applicative language; it allows subroutine calls within subroutine parameters. This provides for concise control sequences such as the following:

```
BIN2HEX:SUBR(VAL);
   ALPHA: NEW <'0','1','2','3','4','5','6','7',
               '8','9','A','B','C','D','E','F'>;

   RETURN(APPLY('|',ALPHA(((VAL ROTL<-12,-8,-4,0>) AND 15)+1)));
   END;
```

This subroutine converts a decimal number to hexadecimal printout form. The subroutine can be used to debug digital inputs from external devices connected to a robot.

Parallel Control

Parallel control allows the sequential execution of independent programs. For example, *multitasking* operates under parallel control. In a multitasking programming language, the programmer specifies a number of program segments called *tasks* that execute independently of each other. When the tasks must synchronize execution, special flags called *semaphores* allow them to wait for each other. The semaphores also provide exclusive access to system resources. Data can be shared or special messages can be sent to pass information between tasks. AML does not provide parallel control.

Asynchronous Control

Asynchronous control allows the execution of event-driven steps of a program. Events can be hardware errors, program function key interrupts, device interrupts, or sensors exceeding specified ranges (tripping). In a multitasking environment, asynchronous control is provided by allowing tasks to wait for events. In a single-task environment, special entry points are defined in the program to handle events.

In AML the MONITOR system subroutine can be called to specify a subroutine to obtain control when a sensor trips. The called subroutine is sometimes called a *handler* for sensor trips. Several types of sensor monitors can be specified in a call to MONITOR: tripping within an absolute range, tripping outside an absolute range, and tripping outside a range specified relative to the current reading of a sensor.

AML supplies a system subroutine that can be called to specify a handler for unexpected hardware errors and also a system subroutine that can be called to specify a handler for program function keys.

21.2.2. Motion

There are several ways that motion can be specified in a structured programming language. These ways are summarized in the following section. As noted, all but one way are supported by AML.

Position Teaching

The simplest way to program a robot is to teach a sequence of goal positions and then play them back. Typically, a teach session involves the following:

1. Physically guiding the robot through a path, as is done for a painting or welding robot.
2. Using a teaching device such as a joystick or push-button box to define the robot goals.

It is necessary at the structured-programming language level to store the results of a position-teaching session as a program segment. In this way, the program and position data can be integrated into the rest of the program and then modified in subsequent program refinements if required.

Position teaching in AML is accomplished by pressing push buttons on a hand-held pendant, producing joint-level motion in the positive and negative directions of each joint axis. Program control of the pendant is provided by the GUIDE system subroutine, allowing one position to be defined with each call. GUIDE returns a joint-level position value. Complete paths can be taught by simply writing a subroutine that loops, calling GUIDE, and storing the returned positions.

An AML subroutine can be used for teaching in a tool frame of reference. This allows motion to be taught along the tool frame coordinate axes.

Joint-Level Motion

It is relatively easy to program a Cartesian-joint robot (such as the IBM RS-1 or 7565; see Figure 21.3) at the X-, Y-, and Z-joint level. A set of joint goal positions are specified as a program step, and then the joints are coordinated so they all reach their goal positions at the same time. This coordination allows the Cartesian joints to move in a straight line between goal positions. The X-, Y-, and Z-axes of the IBM 7565 are Cartesian axes and are often programmed at the joint level.

The following examples of calls to the MOVE system subroutine move the specified joints to the indicated absolute positions in a coordinated motion. The absolute positions of joints are defined such that 0 is exactly in the center of physical travel.

```
MOVE(JZ,3.0);
MOVE(<JR,JP,JW>,<-45.0,0.0,0.0>);
MOVE(<JX,JY>,<5.0,8.5>);
MOVE(JG,3.0);
MOVE(JZ,-1.5);
MOVE(JG,1.5);
MOVE(ARM,0.0);
```

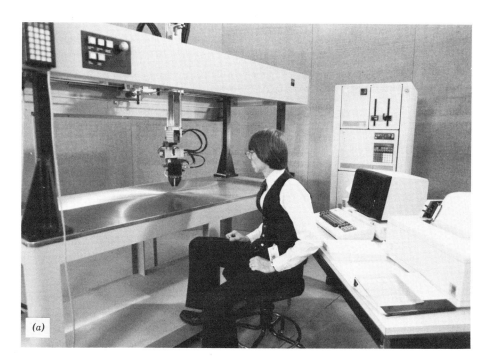

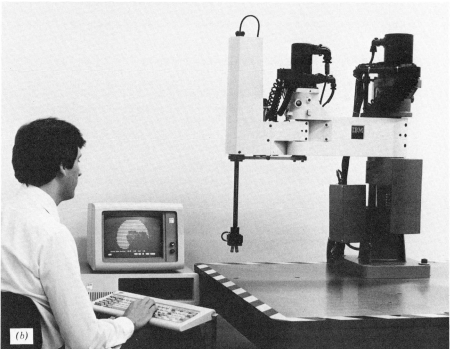

Fig. 21.3. (*a*) Components of the IBM 7565 Manufacturing System (from left to right): manipulator and teach-box; controller; computer terminal and display; printer. (*b*) IBM 7545 Manufacturing System, which is programmed with a subset of AML on a IBM PC computer.

JX, JY, and JZ are the names of the X, Y, and Z joints, respectively; JR, JP, and JW are the names of the roll, pitch, and yaw joints, respectively; JG is the name of the gripper joint; and ARM refers to all joints that are configured on the manipulator.

The programmer has control over the acceleration, top speed, deceleration, and settle characteristics of a motion. For example, in the following call:

MOVE(ARM,P, ,<0.5,1.0,1.0,0>);

the arm is moved to a position specified by the variable P, with the top speed set to half of full speed, the acceleration and deceleration set to full acceleration and deceleration, and no settle checking (the system does not wait for the arm to come to a stop before proceeding with the next motion).

The AMOVE system subroutine provides motion in parallel with program execution. In the following example:

AMOVE(ARM,P);

· (parallel calculations)

WAITMOVE;

calculations are performed in parallel with the move. The two parallel actions are then synchronized by the WAITMOVE call.

Cartesian Motion

In a non-Cartesian robot, programming joint-level motion is very difficult. However, with these robots, users are more concerned with the location of the manipulator tool tip than with the location of the individual joints. This is where Cartesian motion becomes essential.

Cartesian motion control allows a programmer to specify the location of the manipulator tool tip as a position and orientation in space, independent of the positions of the individual joints. A Cartesian position, or *frame*, can refer to the location and orientation of the tool tip. In AML, frames are represented as transformations from the center of the box frame. Three representations for transformations are supported as follows:

1. **Rotation-Matrix Transformations.** A rotation-matrix transformation consists of a three-vector of real numbers representing the offset from the center of another frame and a 3×3 rotation matrix representing the orientation.
2. **Euler-Angle Transformations.** An Euler-angle transformation consists of a three-vector of real numbers representing the offset from the center of another frame and a three-vector of Euler angles representing the orientation.
3. **Joint-Angle Transformations.** A joint-angle transformation consists of a vector of joint positions, along with a three-vector of real numbers representing the offset of the tool tip from the base of the wrist assembly.

System subroutines are provided that convert between the three frame representations; user subroutines can be written in addition to these system subroutines to provide basic Cartesian motion. For example, the following subroutine

```
FMOVE: SUBR(TRANS,COORD DEFAULT <0.,0.,0.>,
            TTOFFSET DEFAULT <0.,0.,0.>,CTLSET);

    GOALSET: NEW TRANSBOX(TRANS,ARM,QGOAL(ARM),TTOFFSET-COORD);
    MOVE(ARM,GOALSET, ,CTLSET);
    END;
```

moves the tool tip to an offset, COORD, in a specified frame, TRANS. The third parameter in the subroutine is the offset of the tool tip from the wrist, and the fourth parameter is the set of motion parameters (speed, acceleration, deceleration, and settle).

Continuous-Path Motion

When a robot must perform a set of coordinated motions in sequence, the resulting motion is often jerky because the robot tries to stop after each goal position. This undesirable jerkiness can be corrected by continuous-path motion control, causing the motion to stop only upon reaching the last point. Intermediate points are passed through as accurately and rapidly as possible.

Accurate Motion

Industrial applications often contain critical positions requiring the robot to maintain a certain tolerance. Unfortunately, today's industrial robots are not always as accurate and repeatable as we would like them to be. There are several reasons for this:

Mechanical changes are made to the robot, fixturing, or tooling.

A gradual shift of fixturing and tooling occurs over time.

Thermal impacts on robot position occur over time.

Robot wear occurs over time.

Parts do not meet tolerance specifications.

Many of these problems can be overcome by a software procedure called *calibration*. Calibration is the periodic locating, by the manipulator, of key positions in the workspace from which other positions have been measured as small offsets or coordinate transformations. The system locates the calibration positions by employing sensing (described in the next section). Motion goals are stored as small offsets or coordinate transformations from the calibration positions. Since small offsets are less likely to be affected by changes in the workspace, they do not have to be corrected during periodic recalibration. Calibration can be performed on a time-scheduled basis or can be performed whenever a discrepancy is detected. In either case, calibration can be performed automatically, thus minimizing manipulator downtime. Calibration can also be performed dynamically, using sensing during the execution of an application.

There are several subroutines in AML that can be used to locate base positions from which offsets can be measured. The FINDPOST subroutine, using a LED sensor in the gripper fingers, can locate a post. The FINDEDGE subroutine, using the LED, can locate an opaque object. The CGRASP subroutine can perform a center-and-grasp action on an object of unknown width positioned between the gripper fingers. There are two AML subroutines that compute the error in a rotary joint at a specified position, CALROLL for the roll joint and CALYP for the yaw and pitch joints.

21.2.3 Sensing

Accurate motion demands that a robot language support sensing, because periodic calibration involves the finding of a number of key positions in the workspace. For example, a post may be used to define a calibration position. The ways to find this or other calibration positions include LED detection sensors, force-sensitive tactile sensors such as strain gages, and proximity sensors such as linear voltage differential transducers (LVDTs).

These sensors can be used not only for error calibration, but for general part mating in situations where part positioning is variable, as in the classic example of the insertion of a peg into a hole. When inserting a peg into a hole, a robot program can use information obtained from tactile sensors to determine where the next motion should be directed. Another, simpler example is the picking up of a part that has an unknown width. The program closes the gripper until a force is detected on one of the fingers. The gripper is then centered on the object using force feedback from the other finger, and closed to a reasonable force level for grasping.

User Sensors

Robot system users must add their own sensors to the workspace to customize their application. A typical user environment requires sensing and signaling devices peculiar to the application. Examples of user sensors are

Safety devices that cause a program interrupt if an operator enters a "danger" zone.

Operator control push buttons.

Parts feeders that cause a program interrupt when empty or broken.

Signal-actuated tools.

Signal-actuated fixtures.

Inspection sensors such as special machinery for measuring part tolerances.

Instrumentation such as microcomputer-controlled gages.

Sensors can be accessed by the program control logic either synchronously or asynchronously. Synchronous accessing of a sensor involves input or output to the sensing device as a sequential step in the program. Asynchronous accessing is event driven. A sensor tripping unexpectedly is a typical example of asynchronous use of an input sensor. Another example is pulsed digital output.

In AML, the pinch, side, and tip strain gages on each gripper finger provide force-sensitive feedback. The LED detector between the gripper fingers provides rudimentary vision in the form of object detection.

User sensors can be defined using the DEFIO system subroutine. User sensors may be accessed as either digital inputs or outputs. Sensors can be used adaptively during a motion in that a motion can be specified to terminate if a sensor trips. This concept is generally referred to as a *guarded move*. In the example

$$M = MONITOR(LED,3,0,0);$$
$$MOVE(ARM,P,M);$$

the LED sensor is monitored so that an object between the gripper fingers trips the sensor. The arm is moved until either the goal position P is reached or the monitor M trips.

The MONITOR system subroutine can also be called to specify a subroutine that is to be called asynchronously if a sensor trips. In the example

$$FDRMON = MONITOR(EMPTY,1,0,0,1.0,'FILLFEEDER');$$

subroutine FILLFEEDER is called if the user-defined sensor EMPTY trips. Synchronous use of a sensor is provided by the SENSIO system subroutine, which performs digital input or output to a sensor.

21.2.4. Vision

Vision provides a more extensive means of sensing the workspace than standard tactile sensing or instrumentation. We summarize some applications that use industrial machine vision systems. The 7565 version of AML does not support vision, although a research version does. See Reference 10 for more information.

Presence Sensing

Presence sensing determines whether an object is at a given point. This is accomplished in many cases by simple LED sensing.

Counting

Counting involves the determination of separate objects in a given view.

Inspection and Measurement

Inspection and measurement applications measure objects in view for adherence to tolerance specifications.

Position and Orientation Finding

Objects are located so that they can be grasped by the manipulator or mated with other parts. Position and orientation finding is currently one of the most important applications of industrial vision. Without vision, special feeders are required to present parts to the manipulator.

Recognition

Recognition allows distinguishable objects to be recognized so that different actions can be taken for different objects. An example of this is keybutton recognition. If the manipulator is programmed to pick up keys marked with the letter R, it detects an error if a key with the letter S is presented in a feeder track. Another example is the sorting of a stream of various parts into specified bins.

21.2.5. Communications

Because of its versatility and intelligence, a robot should not be treated as a stand-alone machine. It must interact with operators and other devices if it is to be integrated into a manufacturing floor. AML supports communication with an operator, another robot, and another computer. These functions are described in the following sections.

Communication with an Operator

A typical industry description of a machine operator is someone with a high school education and little, if any, knowledge of computer programming. Therefore, in many cases, an operator must be shielded from the internal details of a robot program. Communication with an operator generally takes one of the following forms:

1. **Menus.** Presentation of a menu on a display terminal is a way of giving an operator a number of simple choices.

   An AML subroutine provides complete screen and menu processing.

2. **Prompts.** A simple prompt for a Yes/No response may be sufficient in many cases.

   AML provides lower-level access to the display screen with the DISPLAY and READ system subroutines. The PREFILL system subroutine can be used with READ to provide a predefined response to a question. If desired, the operator can type over this response.

3. **Program Function Keys.** Program function keys allow an operator to interrupt a running application to respond to a prompt.

   AML provides a system subroutine that can be called to specify a handler for program function keys.

4. **Pendant Push Buttons and LEDs.** If a pendant is used as a path-teaching device, additional push buttons and LED indicators can be implemented on it for application-dependent interfaces.

   The pendant on the 7565 is equipped with a 10-character display. AML provides a system subroutine to access the display.

5. **User Digital Devices.** Application designers can custom design their own interface hardware, such as switch and push-button panels.

   AML supplies a system subroutine to access pendant push buttons, LEDs, and user digital devices.

6. **Error Trapping.** Trapping of unexpected errors by the application program is an effective way of preventing the operator from seeing the potentially confusing internal details of the program. If the error is a simple one, it may be possible to correct the problem automatically; if not, the operator can be instructed with menus or prompts or to call a programmer.

   AML supplies the ERRTRAP system subroutine to trap errors. Here is an example of a subroutine that traps a floating point underflow error on a division and returns a value of 0.0 for the operation:

```
COMPUTE: SUBR(A,B,C,D);
--  FUNCTION: FOR EACH ELEMENT OF AGGREGATES A,B,C,D COMPUTES
--            (A*B)/(C*D) AND PRINTS THE RESULT
  I: NEW 0;
  N: NEW AGGSIZE(A);
  RESULT: NEW REAL;
  TRAP: SUBR(ERRNUM,ERRDATA);
     --  IF FLOATING UNDERFLOW THEN SET RESULT TO 0.0 AND CONTINUE
     IF ERRNUM EQ 3055 THEN BEGIN
       RESULT = 0.0; QUIT(SUBRLEVEL-1);
       END
     --  IF NOT, LET ERROR FALL THROUGH
     ELSE RETURN(PARMS);
     .END;
  ERRTRAP('TRAP');
  WHILE (I=I+1) LE N DO BEGIN
    RESULT = (A(I)*B(I))/(C(I)*D(I));
    DISPLAY(RESULT,EOL);
    END;
END;
```

Communication with a Host Computer and Other Robots

Factory controls such as production planning, production control, and material handling are generally performed on computers other than the robot system controller. The robot must communicate with these general-purpose computers to receive instructions and send accumulated statistics back for use in subsequent planning. Another important function provided by the communications link is the development of robot programs on host computers and the sending of them to the robot for testing.

A cell of coordinated robots must exchange information and synchronize with each other much like tasks in a multitasking system. However, robots often do not share the same processor and thus must use a communication link for their synchronization. For the 7565, communication with other robots and with host computers is provided by the TP3780 system subroutine. TP3780 is a batch transmission subroutine that writes and reads files to and from any host system that supports Remote Job Entry (RJE) 3780 workstation protocol. For example, in the following call:

TP3780('CTLFILE','DATFLE',,,0);

a control file CTLFILE and a data file DATFLE are sent to a host computer. The control file acts as routing information for the data file.

21.2.6. Data Processing

The previous sections describe the need for robot languages to address the functions of motion, sensing, vision, and communications. Logic, in the form of program control, is needed to control the flow of information to and from these functional areas. The following sections discuss in more detail what forms of information are processed.

Data Types

A robot language must provide the types of data needed for robot applications.

Real numbers for individual joint goal positions.

Arrays for grouping data to store a complete goal position for coordinated moves.

Character strings to identify system resources, such as files.

Integers to identify other system resources, such as joints.

Data in AML are stored as variables. Variables can be either static, dynamic, or global. Static variables are allocated the first time the containing subroutine is called, and they always retain their most recent value. Dynamic variables are reinitialized each time the containing subroutine is called. Global variables are variables that are declared outside of a subroutine.

Here are two examples of variable declarations:

I: STATIC INT;
P: NEW QGOAL(ARM);

The initial value of a variable is given by an arbitrary expression following the keyword STATIC or NEW. (NEW indicates a dynamic declaration.) The first example declares an integer initialized to 0, and the second example declares a vector representing the current position of the arm.

Name binding in AML is dynamic. This means that the most recent declaration of a name in execution is used rather than the lexical nesting rules used by most programming languages. The data types supported include:

Integer

Real

String

Pointer

Aggregate

Aggregate data are data of various types (possibly aggregate) logically grouped together. Aggregates can be nonhomogeneous (of mixed types). Aggregates are used in AML to represent arrays, vectors, robot goal positions, and rotation matrices, among other things. An aggregate with nothing in it (<>) is called a *null aggregate*. The following are examples of aggregate values:

<1,'HELLO',46.7>
<<1,2,3>,<4,5,6>>

Operations

Standard arithmetic operations (such as addition, subtraction, multiplication, and division) must be performed on the given data types to compute joint-level goal positions. Other operations become apparent when beginning to program a robot system: trigonometric functions are needed to convert between joint-level goal positions and Cartesian goal positions; relational operations are necessary to make decisions from sensory feedback; logical, bit-level operations are needed to process input from digital user sensors.

Data operations provided by AML include the following:

Aggregate Indexing. Aggregate indexing has the same syntax as subroutine calls, but instead of calling a subroutine, it indexes a value from an aggregate. In the following example

```
ARRAY: NEW M OF N OF REAL;  --  M X N MATRIX
    .
    .
    .
ARRAY(I,J) = 0.0;                --  ZERO OUT ELEMENT (I,J)
ARRAY(I) = 0.0;                  --  ZERO OUT ROW I
ARRAY(IOTA(M),J) = 0.0;          --  ZERO OUT COLUMN J
```

the last statement indexes several values from the aggregate in one index expression. (The IOTA system subroutine creates the aggregate $<1,2, \ldots, M>$.)

Assignment (=). The assignment operation moves the value of an arbitrary expression into a variable.

Arithmetic (+, −, *, /, IDIV). These are the standard arithmetic operations. The / operator provides real division; IDIV provides integer division.

Relational (EQ, NE, GE, GT, LE, LT). The relational operators return true and false values represented by the integers −1 and 0, respectively, to be used then in subsequent arithmetic operations.

Logical (AND, OR, NOT, XOR). The logical operators perform bit-level operations on integer operands. Integers in AML are assumed to be in two's complement representation.

Trigonometric Functions. System subroutines that provide trigonometric functions include SIN, COS, TAN, ASIN, ACOS, and ATAN.

Vector Functions. System subroutines that provide vector functions include CROSS, DOT, and MAG.

Matrix Functions. System subroutines that provide matrix functions include DOT, INVERSE, and TRANSPOSE.

Secondary Storage

Standard file I/O is needed for a variety of reasons, including the following:

Storing Programs. The volatility of main storage may create the need to store robot programs in secondary nonvolatile storage.

Storing Data for Generic Applications. Generic applications that perform the same basic functions on different configurations of a product, such as component insertion, can be implemented by storing the product-dependent information in a separate data base. (This information can be stored on another computer system. See *Communications*, Section 21.2.5.) Different configurations of a product can be assembled by loading different versions of the data base.

Keeping Statistics. Part and product statistics must be maintained and reported to the production-planning component of the factory network on request. These statistics must be maintained in nonvolatile storage to prevent loss on power-down.

Issuing Messages. Operator messages can be stored on a file to conserve main storage. Direct access to these files is necessary for reasonably fast retrieval of these messages.

Generating Reports. Workstation activity reports can be generated in readable form and stored on a file for subsequent retrieval and printout.

AML provides standard contiguous file support for disk and diskette files. A system subroutine provides direct access to fixed-length records of up to 256 bytes.

Real-time Functions

Robot programs are real-time applications that often require precisely timed behavior. Access to time-of-day clocks and interval timers facilitates real-time I/O precision and execution synchronization. AML provides a system subroutine to access a time-of-day clock and one to access an interval timer.

21.2.7. Storage Management

These discussions are starting to show that robot programs can become quite complex. Often, the programming and data for a complex robot application require a large amount of main storage. If storage is not abundant, storage management is required. Demand paging performed by the underlying programming system is one solution to this problem, but it often leads to unacceptably slow real-time performance. To prevent this, a robot language may provide user-specified storage management functions, some of which are described in the next sections.

Storage Compression

AML provides a system subroutine that causes program storage to be compressed. This operation minimizes fragmentation and creates more contiguous storage for allocations.

Overlays

Overlays are user routines that are called and reside in storage only when they are needed. The storage used by such routines becomes available when the routine is completed. AML does overlays.

Dynamic Storage Allocation and Deallocation

Data storage used by a program can be allocated and deallocated on demand, eliminating maximum requirements and waste caused by static allocation of data areas. AML provides a system subroutine to allocate storage dynamically for a value, and a system subroutine to free the storage used by a stored value.

21.2.8. Program Development

In some programming languages, program development facilities are provided by an external system such as a supporting operating system or an attached computer. This is especially true in compiled languages. If a supporting system is not provided, such as in a closed robot system, the robot language itself must provide program development aids.

Text editing is the development and modification of program text on a file. The file is loaded into storage for execution when editing is complete. As needed, the programmer makes calls to the editor to refine the program and then reloads it into storage after each refinement.

In some systems a program is developed in storage. Once the program is satisfactory, it is saved in a file and subsequently loaded back into storage each time the system is powered on. This method of program development allows a greater degree of interaction between the programmer and the running program. In-storage or "hot" editing is further discussed in the next section.

Both text and storage editing are provided by AML and have the same interface. A screen editor is provided for entry of editing commands in a simple, non-AML syntax. This interface is not specifically tailored for editing AML programs and provides many standard functions. In addition, editing commands can be called from AML programs, in AML syntax. In this way (1) programming function keys can be coded to perform editing functions; (2) subroutines that perform complex editing functions can be coded; (3) the programmer can code a unique editor interface. AML statements can be entered while editing, allowing real-time experimentation and point definition while coding a robot program. Hot editing is provided with certain restrictions, and returning to any point in the active program is allowed.

21.2.9. Debugging

Debugging a robot program is a complex task. It is not unusual for the debugging stage of a robotic application to take more than twice as long as the development stage. Therefore it is critically important for a robotic system to provide powerful debugging aids. If these aids are not provided by a supporting operating system function, the programmer is usually unable to debug the program in the symbolic representation in which he typed it; a machine or assembler-language version of the program must be used. This is why many robot-programming languages supply debugging functions as part of the language. Examples of debugging functions are the following:

Singlestep. Stopping at regular intervals during the execution of the program allows the programmer to examine each step.

Trace. Tracing the execution of a program causes statements to be printed as they are executed.

Stop. A specified stop in a program causes an interrupt to occur at that point in the execution.

Hot Editing. When a program is interrupted by one of the preceding debugging functions or by a system-detected error, program execution is suspended, and a debugging environment is entered. In this environment, the programmer can attempt to determine the cause of the interrupt or problem, if one exists. *Hot editing* is the function that allows the examination of current values of variables and the changing of them if desired. Insertions, deletions, or modifications to statements in the suspended program can be made to refine the program logic. If such changes were made on the text version of the program on a file, the program would have to be reloaded and restarted. Also, fixtures would have to be electronically reset and feeders reloaded; half-assembled parts would have to be disassembled. If the application cycle contains a welding or gluing step, undoing that step could be highly impractical. Hot editing, allowing a program to be modified while it is suspended in execution, provides the ability to tune and test a robot program without costly restart.

21.3. EXAMPLE AML APPLICATION

This section presents an example design of a simple AML application. The example program listing is included. The application consists of assembling a shock assembly. Figure 21.4 shows the shock assembly, with the parts arranged in order of assembly. The design of the application involves the following basic steps:

1. Lay out the work area.
2. Choose the key points.
3. Write the program.
4. Define the points.
5. Test the program.

21.3.1. Lay Out the Work Area

The first step in designing an application is to lay out the work area, taking into consideration the feeders and fixtures required and where they would best be located. The diagram in Figure 21.5 is an example layout of the shock assembly area. The parts needed are: endcaps, springs, bolts, cups, and rubber boots. Figure 21.6 is a representation of what the actual work area looks like for the example program.

The steps involved in assembling the shock are as follows:

1. Remove a bolt from the pallet and place it in the fixture.
2. Place a boot on the bolt.
3. Place a cup on the bolt.
4. Place a spring on the bolt.
5. Place an endcap on the bolt.
6. Place the assembly back in the pallet.

21.3.2. Choose the Key Points

The next step is to establish names for the points on the layout. A point is a location in the work area where the gripper is to pick up or release a part, or perform a function. It may also be a location in space above a part that is going to be accessed. By positioning the gripper to a point above the

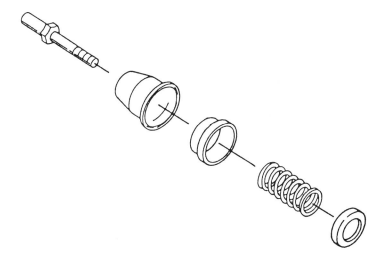

Fig. 21.4. Components of shock assembly, with the parts arranged in order of assembly.

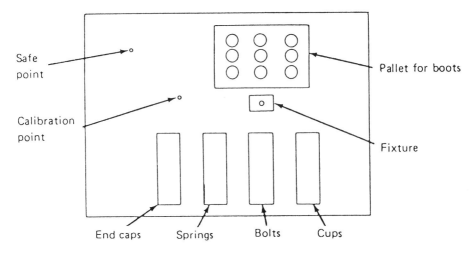

Fig. 21.5. Layout of work area for shock assembly example.

part, there is more freedom in approaching and withdrawing the part. The points for the shock assembly are as follows:

Post1 (the calibration post used as the base point)
Safept
Overbolt
Overhole
Overboot (an array)
Overcup
Overspr
Overcap

Figure 21.7 shows the actual location planned for these points. The base point, Post1, is a calibration point from which the other points are stored as offsets. Periodically during the execution of the application program, this post is recalibrated by calling the AML subroutine FINDPOST. Since all other points in the application are stored as offsets from the calibration post, they never have to be recalibrated.

21.3.3. Write the Program

The next step in designing the application is to write an AML program to perform the assembly, using the chosen point names. The AML editor can be used to prepare this program. This task is made easier by the fact that many utility subroutines have already been written, allowing a higher level of specification than provided in the base level of AML. An example of this is an IBM program product called Program Robot by Example (PRBE), which supplies a number of high-level subroutines that combine sensing and motion.

The actual values of the points are not required at this point. The points can be assigned "dummy" locations for now (such as 0.0). The following is a listing of the program:

```
AMLSHOCK: SUBR(CALIBRATE DEFAULT TRUE,TOPSEED,ITERATIONS);
--
--   FROM PROGRAM: AMLSHOCK VOLUME: RCP002
--   ******************************************
--   *****            PARAMETERS            *****
--   ******************************************

TOPSPEED: NEW IF ?TOPSEED     THEN IF      ?TOPSPEED  GT 2 THEN .2
                              ELSE IF TOPSPEED  GT 1 THEN .2
                              ELSE IF TOPSPEED  LE 0 THEN .2
                              ELSE   TOPSPEED  ELSE .2;
```

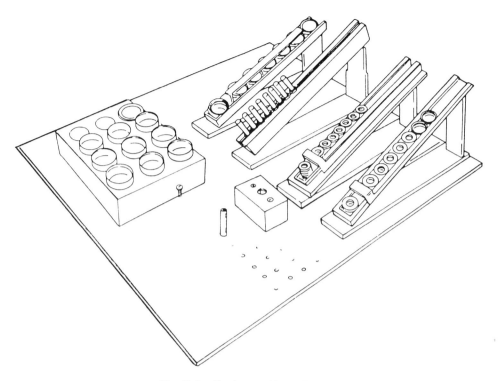

Fig. 21.6. Shock assembly work area.

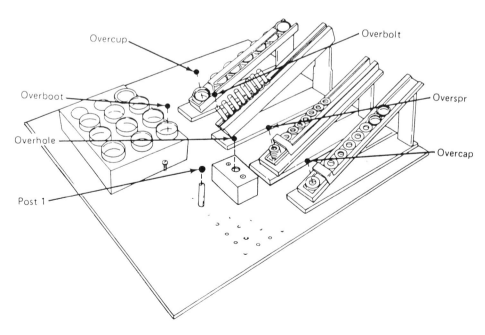

Fig. 21.7. Point locations in work area.

397

```
ITERATIONS: NEW IF ?ITERATIONS THEN IF      ?ITERATIONS GT 2 THEN 1
                                     ELSE IF ITERATIONS GT 99 THEN 1
                                     ELSE IF ITERATIONS LE 0 THEN 1
                                     ELSE    ITERATIONS ELSE        1;
- - - - - - - - - - - - - - - - - - - - - - - - - TRFPOST - - - - - - - - - - - - - - - - - - - - - - - - - -
TRFPOST: SUBR(PN);     -- POSTNAME.  RETURNS: POINT-VALUE
                PNT: NEW 7 OF 0.0;                -- RETURN POSITION.
                SP: NEW SPEED(.2);                -- SET & SAVE SPEED.
     CLEANUP($UND) ;
          UND: SUBR;
               SPEED(SP) ;
               END;
     DISPLAY(EOL,STRING(32),'***FINDPOST***',EOL,STRING(9),
          'USE THE PENDANT TO POSITION THE ARM OVER THE POST: ',PN,'.',
          EOL,STRING(12),
          'THE FINGERS SHOULD BE OPEN AND HANG STRAIGHT DOWN SUCH',EOL,
          STRING(12),'THAT THE "LED" BEAM IS BLOCKED BY THE POST.',
          EOL,EOL,STRING(9),'PRESS THE "END" BUTTON WHEN READY.',
          EOL) ;
RTY:
     PRINT(2,<'ENABLED',EOL>);
     PNT(<JX,JY,JZ>)=GUIDE(ARM)(<JX,JY,JZ>) ;
     PRINT(2,<' ',EOL>);
     IF ?JG EQ 0 THEN RETURN(PNT);-- NO GRIPPER, NO FINDPOST
                                  --  GUIDE WILL HAVE TO SUFFICE
--
     IF ?JR THEN                        -- IF JR AND JG THEN ALL ROTARIES
          MOVE(<JR,JP,JW,JG>,<45,0,90,2.5>) -- ALIGN GRIPPER TO Y.
          ELSE MOVE(<JW,JG>,<135,2.5>); -- ELSE JW DOES JG ALIGNMENT
     IF SENSIO(LED,0) EQ 0 THEN            -- IF LED IS ON, (CTR ON POST)
               CTRMOVE(JX) ;              -- FIND & MOVE TO X-CENTER.
--
     SEARCH(JZ,2) ;              -- SEARCH FOR TOP EDGE OF POST.
     DMOVE(JZ,-.5) ;            -- DOWN OVER POST.
--
     DMOVE(JX,1) ;              -- MOVE TO +X-SIDE.
     CTRMOVE(JX) ;              -- FIND & MOVE TO X-CENTER.
--
     DMOVE(JW,-90) ;           -- ALIGN WITH X-AXIS.
--
     DMOVE(JY,1) ;              -- MOVE TO +Y-SIDE.
     CTRMOVE(JY) ;              -- FIND & MOVE TO Y-CENTER.
--
     PNT(JZ) = SEARCH(JZ,2) ;   -- SEARCH FOR TOP EDGE OF POST.
     RETURN(PNT) ;              -- REPORT RESULTS.
ERR:
     DISPLAY(EOL,STRING(20),'*****POST NOT FOUND AS EXPECTED. *****',
          EOL,STRING(9),
          'USE THE PENDANT TO POSITION THE ARM OVER THE POST: ',
          PN,'.',EOL);
     BRANCH(RTY) ;
- - - - - - - - - - - - - - - - - - - - - - - - - - - - - - - - - - - - - - - - - -
          CTRMOVE: SUBR(J) ;
                    PNT(J) = SEARCH(J,-2) ;  -- FIND +EDGE OF POST.
                    DMOVE(J,-1) ;            -- MOVE TO -SIDE.
                                             -- FIND -EDGE OF POST,
                                             -- *AVERAGE, SAVE RESULT,
                                             -- *AND MOVE TO CENTER.
                    MOVE(J, PNT(J)=(PNT(J)+SEARCH(J,2))/2);
          END;
- - - - - - - - - - - - - - - - - - - - - - - - - - - - - - - - - - - - - - - - - -
          SEARCH: SUBR(J,D) ;
                    MID: NEW MONITOR(LED,3,0,0) ; -- MONITOR ID.
                 CLEANUP($UND) ;
                    UND: SUBR;
```

```
                          ENDMONITOR(MID);
                          END;
                    DMOVE(J,D,MID,<.01>); -- MAKE MOTION SLOW & MONITOR.
                    IF QMONITOR(MID) EQ -1 THEN -- IF WE TRIP,
                          RETURN(QGOAL(J))          -- RTN CURRENT POSITION.
                    ELSE BRANCH(ERR);               -- ELSE, GOTO ERROR LABEL.
          END;
END;
- - - - - - - - - - - - - - - - - - - - - - - - - - - - - - - - - - - - - - - - - -
--  **************************************************
--  *****              BASE POINTS            *****
--  **************************************************
POST1__:    NEW
<0.,0.,0.,0.,0.,0.,0.>;
POST1__:    NEW BEGIN
                    IF CALIBRATE THEN
                        POST1__(ARM)=TRFPOST('POST1__');
                    POST1__
             END;
--  **************************************************
--  *****                ARRAYS               *****
--  **************************************************
OVERBOOT__ARRAY: NEW <
<-6.04377,-6.9217,-1.82102,-45.1941,
-1.23076,18.0848,2.5> + POST1__,
    3,1,
<0.,0.,0.,0.,0.,0.,0.>,
    4,1,
<0.,0.,0.,0.,0.,0.,0.>
              >;
--  **************************************************
--  *****     POINTS RELATIVE TO POST1__      *****
--  **************************************************
SAFEPT__:       NEW
<0.,0.,0.,0.,0.,0.,0.>
    + POST1__;
OVERBOLT__:     NEW
<0.,0.,0.,0.,0.,0.,0.>
    + POST1__;
OVERCUP__:      NEW
>0.,0.,0.,0.,0.,0.,0.>
    + POST1__;
OVERSPR__:      NEW
<0.,0.,0.,0.,0.,0.,0.>
    + POST1__;
OVERCAP__:      NEW
<0.,0.,0.,0.,0.,0.,0.>
    + POST1__;
OVERHOLE__:     NEW
<0.,0.,0.,0.,0.,0.,0.>
    + POST1__;
OVERBOOT__:     NEW
OVERBOOT__ARRAY(1)
    ;
I: NEW INT;             -- LOOP INDICES
J: NEW INT;
--  **************************************************
--  *****          MAIN LINE PROGRAM          *****
--  **************************************************
  DISPLAY(EOP,'USE THE PENDANT TO POSITION THE',
          'ARM TO A SAFE POSITION.',EOL,
          'HIT THE END-BUTTON TO START',
          'PROGRAM: AMLSHOCK EXECUTION.',EOL);
  PRINT(2,<'ENABLED',EOL>); — MESSAGE TO PENDANT
  GUIDE(ARM);
```

```
    PRINT(2,<' ',EOL>);
    SPEED(TOPSPEED);
    WHILE (ITERATIONS = ITERATIONS − 1) GE 0 DO BEGIN
REPEAT BEGIN                                      --START OF OUTER LOOP
  REPEAT BEGIN                                    --START OF INNER LOOP
    REACH(SAFEPT__);
    REACH(OVERBOLT__);
    APPROACH(3.00000);
    GRASP(1500.00);
    WITHDRAW(3.00000);
    POSITION(OVERHOLE__);
    APPROACH(3.00000);
    SETGRIPPER(1.50000);
    WITHDRAW(3.00000);
    REACH(OVERBOOT__);
    APPROACH(3.00000);
    GRASP(1500.00);
    WITHDRAW(3.00000);
    POSITION(OVERHOLE__);
    APPROACH(3.00000);
    SETGRIPPER(1.50000);
    WITHDRAW(3.00000);
    REACH(OVERCUP__);
    APPROACH(3.00000);
    GRASP(1500.00);
    WITHDRAW(3.00000);
    POSITION(OVERHOLE__);
    APPROACH(3.00000);
    SETGRIPPER(1.50000);
    WITHDRAW(3.00000);
    REACH(OVERSPR__);
    APPROACH(3.00000);
    GRASP(1500.00);
    WITHDRAW(3.00000);
    REACH(OVERHOLE__);
    APPROACH(3.00000);
    SETGRIPPER(1.50000);
    WITHDRAW(3.00000);
    REACH(OVERCAP__);
    APPROACH(3.00000);
    GRASP(1500.00);
    WITHDRAW(3.00000);
    REACH(OVERHOLE__);
    APPROACH(3.00000);
    SETGRIPPER(1.50000);
    APPROACH(.500000);
    GRASP(1500.00);
    WITHDRAW(3.00000);
    TRANSPORT(OVERBOOT__);
    APPROACH(3.00000);
    SETGRIPPER(1.50000);
    WITHDRAW(3.00000);
    OVERBOOT=OVERBOOT__ARRAY(1)+I*OVERBOOT__ARRAY(3)
                                    + J*OVERBOOT__ARRAY(5)
    END                                           --END OF INNER LOOP
    UNTIL (J=J+1) EQ OVERBOOT__ARRAY(4);
    J=0
    END                                           --END OF OUTER LOOP
    UNTIL (I=I+1) EQ OVERBOOT__ARRAY(2);
    I=0
    END;                                          --END OF ITERATIONS LOOP
    RETURN;                                       --RETURN TO CALLER
- - - - - - - - - - - - - - - - - - - TROPS - - - - - - - - - - - - - - - - - - - -
```

```
-- **************************************************************************
-- *                        VERB SUBROUTINES                              *
-- *                                                                      *
-- *   THIS SECTION CONTAINS A COLLECTION OF SERVICE                      *
-- *      SUBROUTINES (ONE FOR EACH VERB IN THE BODY OF THE               *
-- *      LOOP).                                                          *
-- *                                                                      *
-- **************************************************************************
--
-- -----------------------------------------------------------------------
   GRASP: SUBR(HOWHARD) ;
                MID: NEW <INT,INT> ;
          IF ?JG EQ 0 THEN RETURN;         -- NO-OP IF NO GRIPPER
          MID = MONITOR(<SLP,SRP>,3,-50,HOWHARD) ;
          CLEANUP($UND);         -- INSURE ENVIRONMENT RESTORED.
                UND: SUBR;
                     ENDMONITOR(MID);
                     END;
          MOVE(JG,0,MID,<.08>); -- MAKE MOTION SLOW & MONITOR.
          END;
-- -----------------------------------------------------------------------
   SETGRIPPER: SUBR(INCHESOPEN);   -- 0.0 TO 3.25
               IF ?JG EQ 0 THEN RETURN;   -- NO-OP IF NO GRIPPER
               TRMOVE(JG,INCHESOPEN);
               END;
-- -----------------------------------------------------------------------
   WITHDRAW:    SUBR(DISTANCE);
                WAITMOVE;
                TRMOVE(<JX,JY,JZ>,QGOAL(<JX,JY,JZ>)+DISTANCE*BOXTRANS()(2,3));
                END;
-- -----------------------------------------------------------------------
   APPROACH:    SUBR(DISTANCE);
                WITHDRAW(-DISTANCE);
                END;
-- -----------------------------------------------------------------------
   TRANSPORT:   SUBR(POINTVALUE);
                   J: NEW <JX,JY,JZ> ;
                   TRMOVE(J,POINTVALUE(J)) ;
                   END;
-- -----------------------------------------------------------------------
   POSITION:    SUBR(POINTVALUE);
        J: NEW SELECT(<1,1,1,?JR,?JP,?JW>,<1,2,3,4,5,6>);
           -- J IS ALL JOINTS BUT GRIPPER AVAILABLE ON SYSTEM
           TRMOVE(J,POINTVALUE(J)) ;
           END;
-- -----------------------------------------------------------------------
   REACH:       SUBR(POINTVALUE);
                TRMOVE(ARM,POINTVALUE(ARM));
                END;
-- -----------------------------------------------------------------------
   TRMOVE:      SUBR(JS,GS);
                AMOVE(JS,GS);
                END;
-- -----------------------------------------------------------------------
   SET_SPEED:  SUBR(FRACTION) ;
        SPEED(FRACTION*TOPSPEED);
        END;
END;-AMLSHOCK: SUBR;
```

21.3.4. Define the Points

Once the program is written, the points must be defined. There are a number of ways to proceed with this task. The most primitive way is as follows:

1. Call the GUIDE subroutine from the terminal to activate the pendant and use the pendant to position the manipulator over the calibration point.

2. Call the FINDPOST subroutine to locate the top of the post and record the position.

3. For each of the other points, call the GUIDE system subroutine and position the manipulator to the appropriate location. GUIDE returns the joint location for each point. Subtract the location of the calibration post from each point location and record this.

4. Replace the "dummy" point definitions in the program with the recorded values.

This is not an automatic way to define application points. Recording of point locations on the part of the programmer can be a tedious task. In practice, programmers tend to write special subroutines that call the GUIDE system subroutine and automatically insert the point location into their program. Software tools of this nature can save an enormous amount of time in point definition and refinement. In fact, the PRBE AML program allows points to be named and defined from a menu-driven screen interface in combination with the pendant. PRBE also allows simple AML programs to be generated automatically by choosing appropriate options in the menus.

21.3.5. Test the Program

As mentioned in Section 21.2.9 on debugging, testing the program is usually the most time-consuming step. There are several reasons for this.

Timing. Justification for robots is largely based on how fast they can accomplish a task. Therefore programmers often spend much of their time working on the speed of the program.

Point Accuracy. In the course of debugging a robot program, a series of problems may arise with the accuracy of the point locations. Some are solved by adjusting the point locations. Some can be solved by slowing down the manipulator. Still others may require changes in the algorithm to make more use of compliance or sensory feedback.

Error Handling. It is difficult for a programmer to anticipate all of the errors that can occur in a complex hardware situation such as a robotic assembly. Adding code to handle unanticipated errors and variability in parts during a debugging session is common.

The application presented in this section was simple enough that these areas did not present serious problems. In more complicated applications, the functions provided by AML for debugging are heavily used.

When a problem arises during testing that interrupts an application, a breakpoint occurs. An error message is displayed on the screen, and a command prompt is issued. To debug the program, the programmer can examine the contents of key variables, query system status, or examine the program logic by calling the editor. If the cause of the problem is determined immediately, it can often be repaired with a change in the program. The AML subroutine editor allows a suspended program to be edited in main storage. Then the programmer can issue a return to the program to continue the application. If the cause of the problem cannot be determined immediately, traces and breakpoints can be set at appropriate places in the program, and a return to the program can be issued.

21.4. CONCLUSIONS

AML in different forms has been in use since 1978. A variety of robotic applications have been implemented with it. AML provides a high-function, structured-programming environment for the development of robotic applications with interfaces that are consistent and easy to understand. Most of the structured-programming robotic language functions are provided by AML. The programming development and debugging functions of AML create a high degree of interaction between the robotic programmer and the application. Until task-oriented languages are commercially available, a general-purpose computer language with robotic extensions is the best tool available for the implementation of complex robotic applications. AML is a good example of this type of language.

ACKNOWLEDGMENTS

AML was conceived as a research probject at IBM Research in Yorktown Heights by Dave Grossman, Jeanine Meyer, Phil Summers, and Russell Taylor. The product level of AML was developed by the IBM Industrial Automation group in Boca Raton, Florida. At the risk of overlooking someone, we would like to acknowledge some of the key individuals in this group: first, Mike Condon, Neil Millett, and Paul VanDyke, for their expertise in the management of quality software development; second, Russell Taylor for all the help that he provided while we were making his invention into a product; last, Yair Gabrieli for his excellent help with the language interpreter, and Glenn Faurot, Dave Heikkinen, Steve Hutchinson, Dave Lasdon, and Jack Sisk for developing the real-time, data processing, and utility components of the software system. This list does not even begin to describe the efforts of dedicated individuals who tested and documented the software, used it and told us what it needed, and provided system support for the development effort. Thanks to Carol Dolan for providing the example code for Section 21.3.

REFERENCES

1. *AML Reference Manual,* IBM Manual SC34–0410.
2. Bonner, Susan and Shin, Kang G., A Comparative Study of Robot Languages, *IEEE Computer,* December 1982, pp. 82–95.
3. Buckley, S. J., An Experimental Interactive Editor, IBM technical report TR54.216.
4. *IBM Robot System/1 General Information Manual and User's Guide,* IBM Manual GA34–0180.
5. *IBM 7535 Manufacturing System User's Guide,* IBM Manual 8508963.
6. *IBM 7565 Manufacturing System Program Robotics by Example (PRBE) User's Guide,* IBM Manual 8509021.
7. *IBM 7565 Manufacturing System AML Reference,* IBM Manual 8509015.
8. *IBM 7565 Manufacturing System AML Screen Editor,* IBM Manual 8509016.
9. *IBM 7565 Manufacturing System General Information,* IBM Manual 8508982.
10. Lavin, M. A. and Lieberman, L. I., AML/V: An Industrial Machine Vision Programming System, *The International Journal of Robotic Research,* Vol. 1, No. 3, Fall 1982.
11. Meyer, J. M., Summers, P. D., and Taylor, R. H., AML: A Manufacturing Language, *The International Journal of Robotic Research,* Vol. 1, No. 3, Fall 1982.

CHAPTER 22

TASK-LEVEL MANIPULATOR PROGRAMMING

TOMÁS LOZANO-PÉREZ

Massachusetts Institute of Technology
Cambridge, Massachusetts

RODNEY A. BROOKS

Stanford University
Stanford, California

22.1. INTRODUCTION

Robots are useful in industrial applications primarily because they can be applied to a large variety of tasks. The robot's versatility derives from the generality of its physical structure and sensory capabilities. However, this generality can be exploited only if the robot's controller can be easily, and hence cost-effectively, programmed.

Three methods of robot programming can be identified; the following list reflects their order of development:

1. Programming by guiding.
2. Programming in an explicit robot-level computer language.
3. Programming by specifying a task-level sequence of states or operations.

22.1.1. Programming by Guiding

The earliest and most widespread method of programming robots involves manually moving the robot to each desired position and recording the internal joint coordinates corresponding to that position. In addition, operations such as closing the gripper or activating a welding gun are specified at some of these positions. The resulting **program** is a sequence of vectors of joint coordinates plus activation signals for external equipment. Such a program is executed by moving the robot through the specified sequence of joint coordinates and issuing the indicated signals. This method of robot programming is known as *teaching by showing* or *guiding* and is explained in Chapter 19, "Robot Teaching."

Robot guiding is a programming method that is simple to use and to implement. Because guiding can be implemented without a general-purpose computer, it was in widespread use for many years before it was cost-effective to incorporate computers into industrial robots. Programming by guiding has some important limitations, however, particularly regarding the use of sensors. During guiding the programmer specifies a single execution sequence for the robot; there are no loops, conditionals, or computations. This is adequate for some applications such as spot welding, painting, and simple materials handling. In other applications, however, such as mechanical assembly and inspection, one needs to specify the desired action of the robot in response to sensory input, data retrieval, or computation. In these cases robot programming requires the capabilities of a general-purpose computer programming language.

22.1.2. Robot-Level Computer Languages

Some robot systems provide computer programming languages with commands to access sensors and to specify robot motions (see Chapters 20, 21). The key advantage of these *explicit* or *robot-level* languages is that they enable the data from external sensors, such as vision and force, to be used in modifying the robot's motions. Through sensing, robots can cope with a greater degree of uncertainty in the position of external objects, thereby increasing their range of application.

Commercially available robot-level languages include VAL,[1] RAIL,[2] and AML.[3] Influential experimental robot-level languages have included WAVE[4] and AL.[5] For a survey of existing robot-level languages, see References 6 and 7.

The key drawback of robot-level programming languages, relative to guiding, is that they require the robot programmer to be expert in computer programming and in the design of sensor-based motion strategies. Robot-level languages are certainly not accessible to the typical worker on the factory floor. Programming at this level can be extremely difficult, especially for tasks requiring complex three-dimensional motions coordinated by sensory feedback. Even when the tasks are relatively simple, as are today's industrial robot tasks, the cost of programming a single robot application may be comparable to the cost of the robot itself. This is consistent with trends in the cost of software development versus the cost of computer hardware. Faced with this situation, it is natural to look for ways of simplifying programming.

22.1.3. Task-Level Programming

Many recent approaches to robot programming seek to provide the power of robot-level languages without requiring programming expertise. One approach is to extend the basic philosophy of guiding to include decision making based on sensing. Another approach, known as *task-level* programming, requires specifying goals for the positions of objects, rather than the motions of the robot needed to achieve those goals. In particular, a task-level specification is meant to be completely robot independent; no positions or paths that depend on the robot geometry or kinematics are specified by the user. Task-level programming systems require complete geometric models of the environment and of the robot as input; for this reason, they are also referred to as *world-modeling* systems. These approaches are not as developed as the guiding and robot-level programming approaches, however.

In a task-level language robot actions are specified only by their effects on objects. For example, users would specify that a pin should be placed in a hole rather than specifying the sequence of manipulator motions needed to perform the insertion. A *task planner* would transform the task-level specifications into robot-level specifications. To do this transformation, the task planner must have a description of the objects being manipulated, the task environment, the robot carrying out the task, the initial state of the environment, and the desired final state. The output of the task planner would be a robot program to achieve the desired final state when executed in the specified initial state. If the synthesized program is to achieve its goal reliably, the planner must take advantage of any capabilities for compliant motion, guarded motion, and error checking. We assume that the planner will not be available when the program is executed. Hence the task planner must synthesize a robot-level program that includes commands to access and use sensory information.

No task-level programming systems are currently commercially available. Indeed, even in research laboratories there is no extant comprehensive prototype. The most notable early designs and partial implementations were AUTOPASS,[8] LAMA,[9] AL (in part),[10] and RAPT.[11] Today there is active research on many of the subproblems of building a task-level system, and active plans to start integrating the resulting modules. One can expect laboratory prototypes to start appearing in the next two to three years, and relatively primitive commercial systems within the next five.

22.1.4. Components of Task-Level Programming

The world model for a task must contain the following:

Geometric descriptions of all objects and robots in the task environment.

Physical descriptions of all objects, for example, mass and angular moments and surface texture.

Kinematic descriptions of all linkages.

Descriptions of the robot system characteristics, for example, joint limits, acceleration bounds, control errors, and sensor error behavior.

A world model for a practical task-level planner also must include explicit specification of the amount of uncertainty there is in model parameters, such as object sizes, positions, and orientations.

Task planning includes the following:

Gross-motion planning.

Grasping planning.

Fine-motion planning.

Gross-Motion Planning

The center of focus in a task-level program is the object being manipulated. However, to move it from one place to another the whole multilinked robot manipulator must move. Thus it is necessary to plan the global motion of the robot and the object to ensure that no collisions will occur with objects in the workspace, whose shape and position are known from the world model.

Grasping Planning

Grasping is a key operation in manipulator programs since it affects all subsequent motions. The grasp planner must choose where to grasp objects so that no collisions will result when grasping or moving them. In addition, the grasp planner must choose grasp configurations so that the grasped objects are stable in the gripper.

Fine-Motion Planning

The presence of uncertainty in the world model affects the kind of motions that a robot may safely execute. In particular, positioning motions are not sufficient for all tasks; guarded motions are required when approaching a surface, and compliant motions are required when in contact with a surface. A task planner must therefore be able to synthesize specifications for these motions on the basis of task descriptions.

The preceding three types of planning produce many interactions. Constraints forced by one aspect of the plan must be propagated throughout the plan. Many of these constraints are related to uncertainties, both initial uncertainties in the world model and the uncertainties propagated by actions of the robot. Recent work has suggested that interactions of the planners and propagation of constraints can be handled by constraint planners relying on symbolic algebraic computations to propagate uncertainties both forward and backward through modeled actions.

22.1.5. Toward Fully Automated Robot Programming

In the near future people will write task-level programs, sequences of relatively simple high-level instructions describing what needs to be done to carry out an assembly task.

In the longer term this task can also be automated. A system called an *assembly planner* will be given an even higher-level command concerning production requirements. It will examine the CAD data base and produce a task-level program. The task planner described will then produce a robot-level program as before.

Automatic planning has long been a domain of interest in artificial intelligence. Typically, the domain has been simplified worlds consisting of blocks to be rearranged. Examples are the HACKER program,[12] which generalized task sequences when it found bugs in them, and NOAH,[13] which generated partially ordered plans for tasks. The techniques of these types of programs are more generally applicable in domains with richer geometry. The only notable, and indeed impressive, planner that has been applied to a realistic domain is the BUILD program.[14] It too used the blocks world, but it had geometric models of blocks and considered the effects of gravity and frictional forces between blocks. It did not consider the magnitudes or effects of uncertainties, however. In the language of this chapter it was a hybrid assembly-planner and task-planner.

22.2. MODELS

The role of models in the synthesis of robot programs is discussed in the remainder of this chapter. First, however, we explore the nature of each of the models needed for a task and how they may be obtained.

22.2.1. The World Model

The geometric description of objects is the principal component of the world model. The major sources of geometric models are computer-aided design (CAD) systems, although computer vision may eventually become a major source of models.[15] There are three major types of commercial CAD systems, differing on their representations of solid objects as follows:

1. Line: objects are represented as lines, and curves are needed to draw them.
2. Surface: objects are represented as a set of surfaces.
3. Solid: objects are represented as combinations of primitive solids.

Line systems and some surface systems do not represent all the geometric information needed for task planning. A list of edge descriptions (known as a *wire frame*), for example, is not sufficient to describe a unique polyhedron.[16] We assume instead that a system based on solid modeling is used.

In these systems, models are constructed by performing set operations on a few types of primitive volumes. Figure 22.1 shows an object constructed by taking the union of three volumes and then subtracting a fourth. Besides the volumetric primitives that make a particular object, it is necessary to specify their spatial interrelationships—usually in terms of coordinate transforms between the local coordinate systems of the primitives.

The descriptions of the primitive and compound objects vary greatly among existing systems. Some of the representational methods currently in use are: polyhedra,[17] generalized cylinders,[18,19] and constructive solid geometry (CSG).[20] For surveys of geometric modeling systems, see References 20 and 21.

The legal motions of an object are constrained by the presence of other objects in the environment, and the form of the constraints depends in detail on the shapes of the objects. This is the fundamental reason that a task planner needs geometric descriptions of objects. There are additional constraints on motion imposed by the kinematic structure of the robot itself. If the robot is turning a crank or opening a valve, then the kinematics of the crank and the valve impose additional restrictions on the robot's motion. The kinematic models provide the task planner with the information required to plan manipulator motions that are consistent with external constraints.

The major part of the information in the world model remains unchanged throughout the execution of the task. The kinematic descriptions of linkages are an exception, however. As a result of the robot's operation, new linkages may be created and old linkages destroyed. For example, inserting a pin into a hole creates a new linkage with one rotational and one translational degree of freedom. Similarly, the effect of inserting the pin might be to restrict the motion of one plate relative to another, thus removing one degree of freedom from a previously existing linkage. The task planner must be appraised of these changes, either by having the user specify linkage changes with each new task state or by having the planner deduce the new linkages from the task state or operations descriptions.

In the planning of robot operations, many of the physical characteristics of objects play important roles. The mass and inertia of parts, for example, determine how fast they can be moved or how much force can be applied to them before they fall over. Similarly, the coefficient of friction between a peg and a hole affects the jamming conditions during insertion. Likewise, the physical constants of the robot links are used in the dynamics computation and in the control of the robot.

The feasible operations of a robot are not sufficiently characterized by its geometrical, kinematical, and physical descriptions. One important additional aspect of a robot system is its sensing capabilities: touch, force, and vision sensing. For task-planning purposes, vision enables obtaining the configuration of an object to some specified accuracy at execution time; force sensing allows the use of compliant motions; touch information could serve in both capacities, but its use remains largely unexplored.[22] In addition to sensing, there are many individual characteristics of manipulators that must be described; velocity and acceleration bounds, positioning accuracy of each of the joints, and workspace bounds are examples.

Much of the complexity in the world model arises from modeling the robot. Fortunately, it needs to be done only once.

22.2.2. The Task Model

A model state is given by the configurations of all the objects in the environment; tasks are actually defined by sequences of states of the world model or transformations of the states. The level of detail in the sequence needed to specify a task fully depends on the capabilities of the task planner.

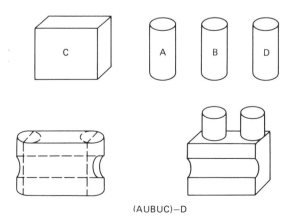

(AUBUC)−D

Fig. 22.1. The two objects in the bottom row are obtained from set operations on primitive volumes A, B, C, D. The different results are obtained from different relative positions of the primitives.

The configurations of objects needed to specify a model state can be provided explicitly, for example, as offsets and Euler angles of rigid bodies and as joint parameters for linkages; but this type of specification is cumbersome and error prone. The following three alternative methods for specifying configurations have been developed:

1. Use a CAD system to position models of the objects at the desired configurations.
2. Use the robot itself to specify robot configurations and to locate features of the objects.[23]
3. Use symbolic spatial relationships among object features to constrain the configurations of objects, for example, $Face_1$ *AGAINST* $Face_2$.[11]

The most common method of positioning object models in CAD systems is by indicating with a light pen the motion they should undergo. With this method, it is difficult to position an object in an arbitrarily oriented corner, since this requires specifying motions parallel to the corner planes, not the global axes. Pointing with the robot is much easier since the physical constraints of objects can be used to guide the motion. A drawback of using the robot is its limited accuracy. Both methods also produce numerical configurations, which are difficult to interpret and modify. In the third method, a configuration is described by a set of symbolic spatial relationships that are required to hold between objects in that configuration. For example, in Figure 22.2 the position of block 1 relative to block 2 is specified by the relations f_3 *AGAINST* f_1 and f_4 *AGAINST* f_2.

Several methods for obtaining configuration constraints from symbolic spatial relationships have been described.[9,10,11,24] The basic procedure for all these methods is to first define a coordinate system on objects and object features, then define equations on object configuration parameters for each of the spatial relationships among features, combine the equations for each object, and finally solve the equations for the configuration parameters of each object. Recent work[25] suggests that inequalities should also be handled and that the solution methods should be able to handle underconstrained situations, producing symbolic constraints.

One advantage of using symbolic spatial relationships is that the configurations they denote are not limited to the accuracy of a light pen or of a manipulator. Another advantage of this method is that families of configurations such as those on a surface or along an edge can be expressed. The relationships, furthermore, are easy to interpret by a human and therefore easy to specify and modify. The principal disadvantage of using symbolic spatial relationships is that they do not specify configurations directly; they must be converted into numbers or equations before they can be used.

Model states are simply sets of configurations. If task specifications were simply sequences of models, then, given a method such as symbolic spatial relationships for specifying configurations, we should be able to specify tasks. This approach has several important limitations, however. One is that a set of configurations may overspecify a state. A classic example[26] of this difficulty arises with symmetric objects, for example, a round peg in a round hole. The specific orientation of the peg around its axis given in a model is irrelevant to the task. This problem can be solved by treating the symbolic spatial relationships themselves as specifying the state, since these relationships can express families of configurations. A more fundamental limitation is that geometric and kinematic models of an operation's final state are not always a complete specification of the desired operation. One example of this is the need to specify how hard to tighten a bolt during an assembly. In general, a complete description of a task may need to include parameters of the operations used to reach one task state from another.

The alternative to task specification by a sequence of model states is specification by a sequence of operations, or more abstractly, transformations on model states. Thus, instead of building a model

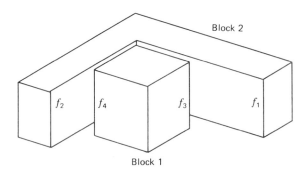

Fig. 22.2. The position of block 1 relative to block 2 can be specified symbolically by $f3$ *against* $f1$ and $f4$ *against* $f2$.

of an object in its desired configuration, we can describe the operation by which it can be achieved. The description should still be object oriented, not robot oriented; for example, the target torque for tightening a bolt should be specified relative to the bolt and not the manipulator. Most operations also include a goal statement involving spatial relationships between objects. The spatial relationships given in the goal not only specify configurations, but also indicate the physical relationships between objects that should be achieved by the operation. Specifying that two surfaces are *against* each other, for example, should produce a compliant motion that moves until the contact is actually detected, not a motion to the configuration where contact is supposed to occur. For these reasons, existing proposals for task-level programming languages have adopted an operation-centered approach to task specification.[8,9]

22.3. GROSS MOTION

Gross robot motions are transfer movements for which the only constraint is that the robot and whatever it is carrying should not collide with objects in the environment. Therefore an ability to plan motions that avoid obstacles is essential to a task planner. Several obstacle avoidance algorithms have been proposed in different domains. In this section we briefly review those algorithms that deal with robot manipulator obstacle avoidance in three dimensions, which can be grouped into the following classes:

1. Hypothesize and test.[39,52,53]
2. Penalty function.[40]
3. Explicit free space.[27-37]

The hypothesize and test method was the earliest proposal for robot obstacle avoidance. The basic method consists of three steps: first, hypothesize a candidate path between the initial and final configuration of the manipulator; second, test a selected set of configurations along the path for possible collisions; third, if a possible collision is found, propose an avoidance motion by examining the obstacle(s) causing the collision. The entire process is repeated for the modified motion.

The main advantage of hypothesize and test is its simplicity. The method's basic computational operations are detecting potential collisions and modifying proposed paths to avoid collisions. The first operation, detecting potential collisions, amounts to the ability to detect non-null geometric intersections between the manipulator and obstacle models.[38] This capability is part of the repertoire of most geometric modeling systems. However, the second operation, modifying a proposed path, can be very difficult. Typical proposals for path modification rely on drastic approximations of the obstacles, such as enclosing spheres. These methods work fairly well when the obstacles are sparsely distributed so that they can be dealt with one at a time. When the space is cluttered, however, attempts to avoid a collision with one obstacle will typically lead to another collision with a different obstacle. This has been observed in practice.[39]

The second class of proposals for obstacle avoidance is based on defining a penalty function on manipulator configurations that encodes the presence of objects. In general, the penalty is infinite for configurations that cause collisions and drops off sharply with distance from obstacles. The total penalty function is computed by adding the penalties from individual obstacles and, possibly, adding a penalty term for deviations from the shortest path. At any configuration, we can compute the value of the penalty function and estimate its partial derivatives with respect to the configuration parameters. On the basis of this local information, the path search function must decide which sequence of configurations to follow. The decision can be made so as to follow local minima in the penalty function. These minima represent a compromise between increasing path length and approaching too close to obstacles.

The penalty function methods are attractive because they seem to provide a simple way of combining the constraints from multiple objects. This simplicity, however, is achieved only by assuming a circular or spherical robot; only in this case will the penalty function be a simple transformation of the obstacle shape. For more realistic robots the penalty function must be much more complex. Otherwise, motions of the robot that reduce the value of the penalty function will not necessarily be safe.

An approach proposed by Khatib[40] is intermediate between these extremes. The method uses a penalty function which satisfies the definition of a potential field[41]; the gradient of this field at a point on the robot is interpreted as a repelling force acting on that point. In addition, an attractive force from the destination is added. The motion of the robot results from the interaction of these forces, subject to kinematic constraints. The key drawback of using penalty functions to plan safe paths is the strictly local information that they provide for path searching. Pursuing the local minima of the penalty function can lead to situations where no further progress can be made. In these cases the algorithm must choose a previous configuration where the search is to be resumed, but in a different direction from the previous time. These backup points are difficult to identify from local information. This suggests that the penalty function method might be combined profitably with a more global method of hypothesizing paths. The free-space methods discussed next might serve this function. Penalty functions are more suitable for applications that require only small modifications to a known path. In these applications, search is not as central as it is in the synthesis of robot programs.

The third class of obstacle avoidance algorithms builds explicit representations of subsets of robot configurations that are free of collisions, the *free space*. Obstacle avoidance is then the problem of finding a path, within these subsets, that connects the initial and final configurations. The proposals differ primarily on the basis of the particular subsets of free space that they represent and in the representation of these subsets.

Widdoes, in an unpublished paper,[37] describes a free-space method for the Stanford Arm.[42] In this method, free space is approximated by grids of regions in the configuration space for the first two joints. Each grid region stores a (possibly null) range of legal values for the third joint. The basic computation used to derive the legal values is to obtain the ranges of the first three joints (*the boom*) that avoid a collision with a point on an obstacle surface. The contents of each of the regions that intersect the range of the first two joint angles are replaced by the intersection of its contents with the new range of the third-joint angle. This computation is carried out on a grid of points covering the surface of all the obstacles. The resulting regions represent an approximation to the free space for the boom.

The foregoing description ignores the last three joints of the manipulator, (the *forearm* or *wrist*). Widdoes' method treats the forearm as a solid attached to the boom; the free space is computed for the modified boom. Changes to the orientation of the forearm are accommodated by computing the free space for three booms, each with a different model for the forearm: one with the forearm joints in their initial configuration, one with the forearm joints in their final configuration, and one approximating the volume swept out by the forearm between the initial and final configurations. The free-space regions resulting from each of these computations are linked into a graph where nodes are regions and where links connect adjacent grid regions whose legal ranges intersect. Links also may be placed between regions obtained from the initial configuration of the forearm to the matching region obtained from the swept volume of the forearm and from there to the matching region obtained from the final configuration of the forearm. The link will be placed only when the range of angles for each region intersect. All paths through this graph represent collision-free paths for the manipulator. Traversing links within a grid involves changing the first three joint angles; traversing links across grids involves changing the last three joint angles.

Udupa[35,36] introduced a new way of computing the free space for the Stanford Arm, approximated by two cylinders, one for the boom and one for the forearm. The next step is to compute a representation of the free space for the boom cylinder, ignoring the forearm cylinder. The basic representation consists of rectangular regions in the configuration space of the boom (formed by the first three joint angles). This representation is similar to Widdoes'; Udupa's representation, however, does not assume a fixed grid size. Instead, Udupa allows the representation of any region to be refined by subdividing it, in a way analogous to the Warnock hidden-line algorithm.[43] The legal range of the third-joint parameter (boom extension) within a region (range of the first two-joint parameters) is computed from the single-line model of the boom and the enlarged obstacles; this is a significant improvement over Widdoes' use of a grid of points on object surfaces. Udupa then locates a safe path for the boom by recursively modifying the straight-line path between the initial and final configurations until it lies completely within the free space. This path is then heuristically modified to allow the forearm to change orientation.

The two methods discussed differ primarily in their approach to computing the free space for the manipulator. The methods were designed for the Stanford Arm, and both exploit some of its special characteristics: the forearm is very small compared to the boom, the links are well approximated by cylinders, and all the joints are rotary except for the third, which is prismatic.

Lozano-Pérez[30] describes an algorithm for computing the free space for a Cartesian manipulator. It is based on computing the free configurations for the manipulator. Figure 22.3 illustrates the method in two dimensions for finding collision-free paths for a polygon without rotation. The moving object and the fixed obstacles are decomposed into unions of convex polygons. A reference point is chosen on the moving object (V_A in the figure). Then each obstacle is grown to compensate precisely for shrinking the moving polygons to the point V_A. In the new *configuration space* a collision-free path for a point corresponds to a collision-free path for the original object in the original space. When rotation is allowed, a third dimension must be added to the configuration space for two-dimensional problems. Three-dimensional problems with rotation result in a six-dimensional configuration space.

Lozano-Pérez[30,31] describes a method for computing the exact configuration space obstacles for a Cartesian manipulator under translation. The free space is represented as a tree of polyhedral cells at varying resolutions. The rotational motion of the manipulator is handled by defining several free-space representations, each using a manipulator model that represents the volume swept out by the rotational links over some range of joint angles (a generalization of Widdoes' use of three grids). Path searching is done by searching a graph whose nodes are cells in the free-space representation and whose links denote overlap between the cells. A related method is described in Lozano-Pérez and Wesley.[32]

Brooks[27] introduced a free-space method based on high-level descriptions (based on the use of generalized cones[18,19]) of the empty corridors between obstacles. These descriptions can then be used

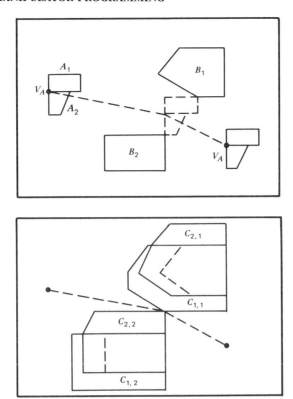

Fig. 22.3. The problem of moving $A = A_1 \cup A_2$ between the obstacles B_1 and B_2 is equivalent to the problem of moving V_A among the modified obstacles $C_{i,j}$ (assuming A cannot rotate).

to compute motion constraints for convex objects moving along the center of the corridors. More recent work[28] has shown how to generalize this approach for a robot with six revolute joints engaged in pick-and-place motions with four Cartesian degrees of freedom (three translation and one rotation of the end effector). Both of these algorithms have been implemented and are very efficient.

The advantage of free-space methods is that by explicitly characterizing the free space they can guarantee to find a path if one exists within the known subset of free space. This does not guarantee, however, that the methods will always find a path when one exists because they only compute subsets of the free space. Moreover, it is feasible to search for short paths, rather than simply finding the first path that is safe. The disadvantage is that the computation of the free space may be expensive. In particular, the other methods may be more efficient for uncluttered spaces. However, in relatively cluttered spaces the other methods will either fail or expend an undue amount of effort in path searching.

Most of the proposed obstacle avoidance methods are fundamentally tied to the use of object approximations. The underlying assumption is that when avoiding collisions object details cannot possibly be important. This assumption does not hold in most applications where the goal is to bring objects into contact. If the obstacle avoidance algorithm fails when the robot is near objects, how is it to grasp objects or approach the site of an assembly operation? The algorithm described in Lozano-Pérez,[30] in contrast, allows motion near to obstacles and is therefore applicable both to planning transfer motions and in choosing grasp configurations. However, the algorithm is limited to Cartesian robots and to approximations of the constraints on rotations. An efficient obstacle avoidance algorithm for general robots with revolute joints engaged in general motions remains to be developed, although a theoretical formulation for such an algorithm exists.[34]

22.4. GRASPING

A typical robot operation begins with the robot grasping an object; the rest of the operation is deeply influenced by choices made during grasping. Several proposals for choosing collision-free grasp configurations on objects exist, but other aspects of the general problem of planning grasp motions have received little attention.

In this section, *target object* refers to the object to be grasped. The surfaces on the robot used for grasping, such as the inside of the fingers, are *gripping surfaces*. The manipulator configuration that grasps the target object at that object's initial configuration is the *initial grasp configuration*. The manipulator configuration that places the target object at its destination is the *final grasp configuration*.

There are three principal considerations in choosing a grasp configuration for objects whose configuration is known.

1. **Safety.** The robot must be safe at the initial and final grasp configurations.
2. **Reachability.** The robot must be able to reach the initial grasp configuration and, with the object in the hand, to find a collision-free path to the final grasp configuration.
3. **Stability.** The grasp should be stable in the presence of forces exerted on the grasped object during transfer motions and parts-mating operations.

If the initial configuration of the target object is subject to substantial uncertainty, an additional consideration in grasping is *certainty:* the grasp motion should reduce the uncertainty in the target object's configuration.

Choosing grasp configurations that are safe and reachable is related to obstacle avoidance; there are significant differences, however. First, the goal of grasp planning is to identify a single configuration, not a path. Second, grasp planning must consider the detailed interaction of the manipulator's shape and that of the target object. Note that candidate grasp configurations are those having the gripping surfaces in contact with the target object while avoiding collisions between the manipulator and other objects. Third, grasp planning must deal with the interaction of the choice of grasp configuration and the constraints imposed by subsequent operations involving the grasped object. Because of these differences, most existing proposals for grasp planning treat it independently of obstacle avoidance.

Most approaches to choosing safe grasps consist of three steps: choose a set of candidate grasp configurations, prune those that are not reachable by the robot or that lead to collisions, then choose the optimal, in some sense, grasp among those that remain.

The initial choice of candidate grasp configurations can be based on considerations of object geometry,[9,15,30,44,45] stability,[46,47] or uncertainty reduction.[48] For parallel-jaw grippers, a common choice is grasp configurations that place the grippers in contact with a pair of parallel surfaces of the target object. An additional consideration in choosing the surfaces is to minimize the torques about the axis between the grippers.

Paul[46] and Taylor[10] consider pruning grasps that are not reachable. Other approaches prune grasps that lead to geometric constraint violations. The collision-avoidance constraints considered have included the following:

1. Potential collisions of gripper and neighboring objects at initial grasp configuration.[9,30,45,47]
2. P-convexity (indicates that all the matter near a geometric entity lies to one side of a specified plane).[44]
3. Existence of collision-free path to initial grasp configuration.[9,30,45]
4. Potential collisions of any part of the manipulator and neighboring objects at initial-grasp configuration, potential collisions of gripper and neighboring objects at final-grasp configuration, potential collisions of any part of the manipulator and neighboring objects at final-grasp configuration, and existence of collision-free path from initial- to final-grasp configuration.[9,30]

A final choice of optimal grasp must be made. One possibility is choosing the configuration that leads to the most stable grasp;[15,49] another is choosing the one least likely to cause a collision in the presence of position error or uncertainty.

Current proposals for grasp planning typically focus on only one aspect of the problem; most focus on finding safe grasp configurations. Even within the aspect of safety, most proposed methods consider only a subset of the constraints needed to guarantee a safe and reachable grasp configuration. In particular, most methods consider only the constraints on gripper configuration in the initial configuration of the target object. None of the methods adequately handles the constraints on grasping imposed by planned motions while grasping the target object. In addition, all of the proposed methods assume a limited class of object models, usually combinations of polyhedra and cylinders, and also a simple type of gripper, usually parallel-jaw grippers. Often, the methods are not readily generalizable to more general object models or more general grippers.

The other aspects of the grasping problem, stability and uncertainty, have received even less attention. The stability condition used in several of the proposals[46,47] amounts to checking that the center of mass of the target object is on or near the axis between the gripper jaws. This condition minimizes the torques on the grip surfaces. Hanafusa and Asada[49] describe an elegant analysis of stable grasping in the absence of friction, but this analysis ignores safety and reachability. Also, the generalization of this method to include friction has not been carried out.

We have heretofore considered grasping when the configurations of all objects in the environment are known. Grasping in the presence of uncertainty in the configuration of the target object and in the configuration of the gripper has received little study. In many applications enough uncertainty is present that a grasping strategy must be designed to reduce the uncertainty in the configuration of the target object relative to the gripper. Often such a strategy must involve sensing. Vision or touch may be used to identify the configuration of an object on a conveyor belt. Many tricks for grasping under uncertainty have been developed for particular applications, but a general theory for synthesizing grasping strategies does not exist. One common class of grasping strategies relies on touch sensing to achieve compliance of the robot gripper to the configuration of the target object.[46] Mason[48] has studied alternative grasping strategies where the target object complies to the gripper motion while under the influence of friction; the goal is to synthesize strategies that grasp the target object at a known configuration in the presence of initial uncertainty.

22.5. FINE MOTION

The presence of uncertainty in the world model and the inherent inaccuracy of the robot affects the kind of motions that a robot may safely execute. In particular, positioning motions are not sufficient for all tasks; guarded motions are required when approaching a surface, and compliant motions are required when in contact with a surface. A task planner must therefore be able to synthesize specifications for these motions on the basis of task descriptions.

Mason[50] describes a method for specifying compliant motions based on kinematic models of the manipulator and the task. The method requires as input a nominal path for the manipulator. Planning the nominal path is akin to the obstacle avoidance problem discussed before: achieving the specified task while avoiding collisions with nearby objects. There is a significant difference, however. In the compliant task, the legal configurations of the robot are constrained by the kinematics of the task as well as the kinematics of the robot. The task kinematics constrain the legal robot configurations to lie on a surface in the robot's configuration space. This type of surface is called a C-surface. The obstacle avoidance algorithm used in planning the nominal path must incorporate this additional constraint. Lozano-Pérez[30] uses a similar constraint to characterize legal grasp configurations: the robot configurations for which the gripper is in contact with the grasped surfaces define a C-surface; legal grasp configurations are defined to be on this surface outside of the configuration-space obstacles for the manipulator. Similarly, a collision-free path for a compliant task is a path on the task's C-surface that remains outside of the configuration-space obstacles. In practice, both the C-surface and the configuration-space obstacles may be difficult to compute exactly.

The foregoing discussion suggests that task planning, for tasks requiring compliant motions, may be done by first finding a collision-free path, on the model C-surface for the task, from the initial to the goal configuration,[30] and then deriving a force-control strategy that guarantees that the path of the robot stays on the actual C-surface, close to the desired path.[50] This two-step procedure assumes that the robot is already on the desired C-surface, that is, in contact with some object in the task. There remains the problem of achieving this contact; this is the role played by guarded motions.

The goal of a guarded motion is achieving a manipulator configuration on an actual C-surface while avoiding excessive forces. The applied forces must be limited to avoid damaging the robot or the objects in the task and to avoid moving objects in known configurations. Because the exact configurations of the objects and the robot are not known, it is necessary to rely on sensors, force or touch, to guarantee that contact is achieved and no excessive forces are exerted. In addition, the speed of the approach must be relatively low so that the robot can be stopped before large forces are generated.

The task planner must specify the approach path to the C-surface and the maximum force that may be exerted for each guarded motion. The approach path may be computed by finding a path that avoids collisions with nearby objects while guaranteeing a collision with the desired C-surface. This is a special case of the obstacle avoidance problem discussed before, one requiring motion close to obstacles. Computing the force threshold for a guarded move is a very difficult problem since the threshold depends on the strength and stability of the objects giving rise to the C-surface. One can always use the smallest detectable force as a threshold, but this choice constrains the manipulator to approach the C-surface very slowly, thus wasting valuable time. It is possible that bounds on these thresholds may be obtained from simple models of the task.[14,48]

When the robot's configuration relative to the task is uncertain, it may not be possible to identify which C-surface in the world model corresponds to the C-surface located by a guarded move. In such situations it is not enough to plan the nominal path; the task planner must generate a sensor-based fine-motion strategy that will guarantee that the desired C-surface is reached. Four examples of such strategies for the problem of bringing a peg partway into a hole are discussed in the literature.

Tilting the Peg

Inoue[51] suggested that by tilting the peg we can increase the likelihood that the initial approach condition will have the peg partway in the hole. Looking at the configuration-space diagram for this case (see Figure 22.4) illustrates why this is so. The uncertainty region now only contacts one of two surfaces, but, more important, by pressing down hard enough we can overcome the frictional forces exerted by the surfaces and guarantee that the peg will slide to the intersection point of the two surfaces. From this point the insertion can proceed as in the case of no uncertainty by maintaining contact with the sides of the hole, that is, a compliant motion on a known C-surface.

Chamfers

Adding chamfers to the hole has the same effect, with respect to approach conditions, as tilting the peg (see Figure 22.4). The chamfers avoid the need to straighten out the peg in the hole, a time-consuming operation.

Search

By moving along the positive or negative x direction after first contact, we can disambiguate which of the C-surfaces the peg is on. The search must be planned as a compliant motion to search for the edge of the hole. If no edge is found after a maximum distance, then the motion is reversed. The maximum length of the motion should be the maximum size of the intersection of each C-surface with the uncertainty region (see Figure 22.4).

Biased Search

The search strategy may require a "wasted" motion to identify which model C-surface corresponds to the actual C-surface found during approach. Inoue[51] suggested that, by introducing a deliberate bias in the initial x position of the peg, we can reduce the search to, at most, one motion. The initial bias needed to guarantee this strategy is the size of the maximum motion for the search (see Figure 22.4).

If faced with a task such as the insertion just described, the task planner must first choose a strategy, and then select the parameters for the motions in the strategy. In this case, the choice of strategy depends on physical characteristics of the task, such as on the presence of chamfers, on the magnitude of the uncertainty region, and on estimates of the time required to execute the strategy. Time estimates can be either worst case, based on maximum motions, or weighted by probability of motions. The last requires estimating the probability distribution for configurations within the uncertainty region and a method for projecting this distribution on C-surfaces. Note that some of the strategies have a limit on the size of the uncertainty for which they are valid. In particular, the first two are valid only when the uncertainty region falls within the slanted C-surface (see Figure 22.4). If the uncertainty is greater, then one of the search strategies must be used.

When the actual state of the world differs from the planner's model in an unexpected way, execution

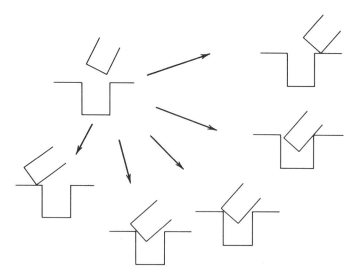

Fig. 22.4. In the presence of uncertainty in the relative position of the peg and the hole, several geometric conditions may result from an insertion motion.

of any of the strategies will have unpredictable effects. It is, therefore, important to include tests in the synthesized program that guarantee that the operation is proceeding as expected. Thus, for example, the locations of C-surface contacts should be checked to test that they occur within the bounds predicted by the model and the expected uncertainty. A simple form of this approach has been used to pick displacement bounds for guarded moves.[9] Brooks[25] suggests more general methods for generating sensory checkpoints.

Two proposals have been advanced for synthesizing sensor-based robot programs to achieve task-level goals.[9,10] The basic idea in both proposals is that motion strategies for particular tasks can be represented as parameterized robot programs, known as *procedure skeletons*. A skeleton has all the motions, error tests, and computations needed to carry out a task, but many of the parameters needed to specify motions and tests remain to be specified. The applicability of a particular skeleton to a task depends, as before, on the presence of certain features in the model and the values of parameters such as clearances and uncertainties. In Taylor's proposal,[10] strategies are chosen by first computing the values of a set of parameters specific to the task, such as the magnitude of uncertainty region for the peg in peg-in-hole insertion, and then using these parameters to choose from among a fixed set of strategies. Having chosen a strategy, the planner computes the additional parameters needed to specify the strategy motions, such as grasp configurations and approach configurations. The planner uses expected execution time as a measure to choose between alternative values of parameters. A program is produced by inserting these parameters into the procedure skeleton that implements the chosen strategy.

The approach to strategy synthesis based on procedure skeletons assumes that task geometry for common tasks is predictable and can be divided into a few classes, each requiring a different skeleton. This assumption is needed because the sequence of motions in the skeleton will only be consistent with a particular class of geometries. More obstacles can always invalidate a given strategy for any particular task as they can prevent key motions of the skeleton.

The synthesis of robust programs must take into account the specific geometric environment of the task. In addition, the use of predefined strategies seems to play an important role in the synthesis process. A definitive framework for the interaction between these two kinds of information in task planning has not emerged. The configuration-space analysis we have used may provide such a framework. Configuration space and C-surfaces have already proved useful in grasp planning, obstacle avoidance, and compliant motion synthesis.[30,31,50]

22.6. UNCERTAINTY PLANNING

All the planning modules described before have had to deal with uncertainty in the world model. Uncertainties in a model of the initial state of the world may be amplified as each action takes place.

22.6.1 Physical Uncertainty

The next sections discuss the three main sources of physical uncertainty.

Mechanical Complexity

Robot manipulators are complex mechanical devices. There are upper bounds on speed and payload and limits to accuracy and repeatability. The absolute positional accuracy of a manipulator is the error that results when it is instructed to position its end effector at a specified position and orientation in space. This may depend on temperature, load, speed of movement, and the particular position within its work area. Furthermore, in a situation where these parameters are all fixed, a manipulator will not in general return to precisely the same location and orientation when commanded to repeat an operation over and over. The error measures the positional repeatability. There can be contributions from stochastic effects and from long-term drift effects which can be corrected by calibration. The positional and repeatability errors of current manipulators are sufficiently large to cause problems in carrying out a large class of planned tasks in the absence of feedback during plan execution. Manipulators can be made more accurate by machining their parts more accurately and increasing their structural stiffness. There is, however, a trade-off between cost and performance of manipulators, so that there is a point of diminishing returns in trying to build ever more accurate mechanisms.

Parts Variability

To make matters worse for the task planner, multiple copies of a mechanical part are never identical in all their dimensions. It is impossible to manufacture parts with exact specifications. Instead, designers specify tolerances for lengths, diameters, and angles. Parts made from the design can take on any physical values for the parameters that fall within the designed tolerances. The effects of these variations might be large enough by themselves to be a significant factor in the success or failure of a planned robot manipulator task. When many parts are assembled into a whole, the individual small variations can combine and become large.

Uncertainty of Initial Position and Orientation

Often the most significant source of uncertainty is the position and orientation of a workpiece when it is first introduced into the task. Mechanical feeders sometimes deliver parts with large uncertainties in position and orientation, sometimes on the order of 50% of the size of the part. Conveyor belts deliver parts with even larger uncertainties. A task planner often includes actions in the plan that are aimed at significantly reducing these initial uncertainties—for instance, the grasp strategies and guarded moves described previously.

Besides physical uncertainty, there will always be uncertainty in the runtime system's knowledge of the state of the world, as all sensors are inaccurate. Usually the maximum inaccuracy can be characterized as a function of sensor reading.

The effects of actions on uncertainties can be modeled in the world model, and so the task planner can propagate uncertainties throughout the sequence of tasks. There have been two approaches to error propagation.

22.6.2. Error Propagation

Numeric Error Propagation[10]

Numeric bounds are estimated for initial errors, the propagation functions are linearized, and linear programming techniques are used to estimate the resultant errors. If the errors are too large for the next task to handle, then deliberate error reduction strategies are introduced, such as sensing or one of the C-surface methods described before.

Symbolic Error Propagation[25]

Unforced decisions, such as workplace location, tool parameters, and compliant motion travel, are represented as symbolic variables. Uncertainties are functions of these variables. Errors are propagated symbolically, and the resultant uncertainties are also symbolic expressions. Constraints on uncertainties necessary for the success of subsequent tasks provide symbolic constraints on the as yet unforced decisions. If the constraints are too severe, sensors must be introduced to the task plan. They too are analyzed symbolically, and constraints on sensor choice are generated. Relaxation methods over the task sequence are applied to satisfying the constraints. The result is that the requirements of task late in the sequence can generate preparatory actions early in the sequence.

22.7. CONCLUSION

Task-level programming offers the potential of significant productivity gains for robot programming. Successful implementation of task-level systems has raised a number of important theoretical and practical issues that are receiving increased attention. Although no complete task-level programming systems have been implemented, a great deal of progress has been made on the basic problems such as collision-free path planning, automatic grasping, sensory planning, and fine-motion planning. The prospects for practical task-level planning systems are currently quite good.

REFERENCES

1. Shimano, B., et al., VAL: a robot programming and control system, Unimation, Danbury, Connecticut, 1977.

2. Franklin, J. W. and Vanderbrug, G. J., Programming Vision and Robotics Systems with RAIL, *SME Robots VI,* March 1982, 392–406.

3. Taylor, R. H., Summers, P. D., and Meyer, J. M., AML: A Manufacturing Language, *International Journal of Robotics Research,* Vol. 1, No. 3, 1982, pp. 19–41.

4. Paul, R. P., WAVE: A model-based language for manipulator control, *Industrial Robot,* March 1977.

5. Goldman, R. and Mujtaba, M. S., *AL User's Manual,* Artificial Intelligence Laboratory, Stanford University, AIM 344, December 1981.

6. Gruver, W. A., Soroka, B. I., Craig, J. J., and Turner, T. L., Evaluation of Commercially Available Robot Programming Languages, *13th International Symposium on Industrial Robots and Robots 7,* Chicago, April 1983, 12-58–12-68.

7. Lozano-Pérez, T., Robot Programming, *Proceedings of the IEEE,* July 1983.

8. Lieberman, L. I. and Wesley, M. A., AUTOPASS: an automatic programming system for computer

controlled mechanical assembly, *IBM Journal of Research and Development,* Vol. 21, No. 4, 1977, pp. 321–333.

9. Lozano-Pérez, T., The design of a mechanical assembly system, Artificial Intelligence Laboratory, Massachusetts Institute of Technology, AI TR 397, 1976.

10. Taylor, R. H., The synthesis of manipulator control programs from task-level specifications, Artificial Intelligence Laboratory, Stanford University, AIM-282, July 1976.

11. Popplestone, R. J., Ambler, A. P., and Bellos, I., An interpreter for a language for describing assemblies, *Artificial Intelligence,* Vol. 14, No. 1, 1980, pp. 79–107.

12. Sussman, G. J., *A Computer Model of Skill Aquisition,* American Elsevier, New York, 1975.

13. Sacerdoti, E. D., *A Structure for Plans and Behavior,* American Elsevier, New York, 1977.

14. Fahlman, S. E., A planning system for robot construction tasks, *Artificial Intelligence,* Vol. 5, No. 1, 1974.

15. Brady, J. M., Parts Description and Acquisition Using Vision, *Proceedings SPIE,* May 1982.

16. Markowsky, G. and Wesley, M. A., Fleshing Out Wire Frames, *IBM Journal of Research and Development,* Vol. 24, No. 5, September 1980.

17. Wesley, M. A., et al., A Geometric Modeling System for Automated Mechanical Assembly, *IBM Journal of Research and Development,* Vol. 24, No. 1, January 1980, pp. 64–74.

18. Binford, T. O., Visual perception by computer, *Proceedings of IEEE Conference on Systems and Control,* 1971.

19. Brooks, R. A., Symbolic Reasoning Among 3-D Models and 2-D Images, *Artificial Intelligence,* Vol. 17, 1981, pp. 285–348.

20. Requicha, A. A. G., Representation of Rigid Solids: Theory, Methods, and Systems, *Computing Surveys,* Vol. 12, No. 4, December 1980, pp. 437–464.

21. Baer, A., Eastman, C., and Henrion, M., Geometric Modeling: A Survey, *Computer Aided Design,* Vol. 11, No. 5, September 1979, pp. 253–272.

22. Harmon, L. D., Automated Tactile Sensing, *International Journal of Robotics Research,* Vol. 1, No. 2, Summer 1982, pp. 3–32.

23. Grossman, D. D. and Taylor, R. H., Interactive Generation of Object Models with a Manipulator, *IEEE Transactions on Systems, Man, and Cybernetics,* SMC-8, 9, September 1978, pp. 667–679.

24. Ambler, A. P. and Popplestone, R. J., Inferring the Positions of Bodies from Specified Spatial Relationships, *Artificial Intelligence,* Vol. 6, 1975, pp. 175–208.

25. Brooks, R. A., Symbolic error analysis and robot planning, *International Journal of Robotics Research,* Vol. 1, No. 4, December 1982.

26. Finkel, R. A., Constructing and debugging manipulator programs, Artificial Intelligence Laboratory, Stanford University, AIM 284, August 1976.

27. Brooks, R. A., Solving the find-path problem by representing free space as generalized cones, Artificial Intelligence Laboratory, Massachusetts Institute of Technology, AI Memo 674, May 1982.

28. Brooks, R. A., Find-Path for a PUMA-class Robot, *American Association of Artificial Intelligence Conference,* Washington, D.C., August 1983.

29. Brooks, R. A. and Lozano-Pérez, T., A Subdivision Algorithm in Configuration Space for Findpath with Rotation, Artificial Intelligence Laboratory, Massachusetts Institute of Technology, AI Memo 684, December 1982 (also IJCAI-83 Proceedings).

30. Lozano-Pérez, T., Automatic planning of manipulator transfer movements, *IEEE Transactions on Systems, Man, and Cybernetics,* SMC-11, No. 10, 1981, pp. 681–689.

31. Lozano-Pérez, T., Spatial planning: a configuration space approach, *IEEE Transactions on Computers,* C-32, No. 2, February 1983.

32. Lozano-Pérez, T. and Wesley, M. A., An algorithm for planning collision-free paths among polyhedral obstacles, *Communications of the ACM,* Vol. 22, No. 10, October 1979, pp. 560–570.

33. Schwartz, J. T. and Sharir, M., On the Piano Movers Problem I: The Case of a Two-Dimensional Rigid Polygonal Body Moving Amidst Polygonal Barriers, Department of Computer Science, Courant Institute of Mathematical Sciences, New York University, Report 39, October 1981.

34. Schwartz, J. T. and Sharir, M., On the Piano Movers Problem II: General Properties for Computing Topological Properties of Real Algebraic Manifolds, Department of Computer Science, Courant Institute of Mathematical Sciences, New York University, Report 41, February 1982.

35. Udupa, S. M., Collision detection and avoidance in computer controller manipulators, *5th International Joint Conference on Artificial Intelligence,* Massachusetts Institute of Technology, 1977.

36. Udupa, S. M., Collision Detection and Avoidance in Computer Controller Manipulators, Ph.D. Thesis, Department of Electrical Engineering, California Institute of Technology, 1977.

37. Widdoes, C., A heuristic collision avoider for the Stanford robot arm (unpublished), Artificial Intelligence Laboratory, Stanford University, 1974.

38. Boyse, J. W., Interference detection among solids and surfaces, *Communications of the ACM,* Vol. 22, No. 1, 1979, pp. 3–9.

39. Pieper, D. L., The Kinematics of Manipulators under Computer Control, Ph.D. Thesis, Department of Computer Science, Stanford University, 1968.

40. Khatib, O., Commande dynamique dans l'espace operationnel des robots manipulateurs en presence d'obstacles, Docteur Ingenieur Thesis, L'Ecole Nationale Superieure de l'Aeronautique et de l'Espace, Toulouse, France, 1980.

41. Symon, K. R., *Mechanics,* Addison Wesley, Reading, Massachusetts, 1971.

42. Scheinman, V. C., Design of a Computer Controlled Manipulator, Stanford Artificial Intelligence Laboratory, AIM 92, June 1969.

43. Warnock, J. E., A Hidden-Surface Algorithm for Computer Generated Halftone Pictures, Computer Science Department, University of Utah, 1969, TR4–15.

44. Laugier, C., *Proceedings of the 11th International Symposium on Industrial Robots,* Tokyo, Japan, October 1981.

45. Wingham, M., Planning how to grasp objects in a cluttered environment, M.Ph. Thesis, Edinburgh, 1977.

46. Paul, R. P., Modelling, trajectory calculation, and servoing of a computer controlled arm, Stanford University, Artificial Intelligence Laboratory, AIM 177, November 1972.

47. Brou, P., Implementation of High-Level Commands for Robots, M.S. Thesis, Department of Electrical Engineering and Computer Science, Massachusetts Institute of Technology, Cambridge, December 1980.

48. Mason, M. T., Manipulator Grasping and Pushing Operations, Technical Report, Artificial Intelligence Laboratory, Massachusetts Institute of Technology, Cambridge, 1982.

49. Hanafusa, H. and Asada, H., Stable prehension of objects by the robot hand with elastic fingers, *Proceedings of the 7th International Symposium on Industrial Robots,* Tokyo, October 1977, pp. 361–368.

50. Mason, M. T., Compliance and force control for computer controlled manipulators, *IEEE Transactions on Systems, Man, and Cybernetics,* SMC-11, No. 6, 1981, pp. 418–432.

51. Inoue, H., Force feedback in precise assembly tasks, Artificial Intelligence Laboratory, Massachusetts Institute of Technology, AIM-308, Cambridge, August 1974.

52. Lewis, R. A., Autonomous manipulation on a robot: summary of manipulator software functions, Jet Propulsion Laboratory, California Institute of Technology, TM 33–679, March 1974.

53. Myers, J. K. and Agin, G. J., A Supervisory Collision-Avoidance System for Robot Controllers, in *Robotics Research and Advanced Applications,* ASME, 1982.

CHAPTER 23

EXPERT SYSTEMS AND ROBOTICS

DONALD MICHIE

University of Edinburgh
Edinburgh, Scotland

23.1. INTRODUCTION

Over the last 10 years industrial robots have progressed along a single track—the track of the deaf, dumb, blind, programmed manipulator. Applicability of expert systems techniques now awaits the arrival on the factory floor of an adequate variety of inexpensive sensory devices. But as soon as robots are equipped, as is now starting in industrial practice, with television cameras, laser rangers, tactile sensors, microphones, and other means of interrogating the environment, the relevance of the new software technology of rule-based programming will become pervasive. Ten years from now the state-of-the-art robot will be the expert robot, skilled, resourceful, responsive, and able to cooperate with human supervisors and with fellow robots to execute tasks inconceivable today. This chapter examines some of the problems and possibilities.

Following this introductory section, Section 23.2 briefly reviews the *expert system* concept as currently understood. Section 23.3 sets forth specimen hand-eye tasks for cooperating robots. The tasks have been carefully chosen to illustrate the competence and teachability in hand-eye skills that we take for granted in humans but have yet to implant in robot control programs. Yet the knowledge engineering branch of artificial intelligence (AI) has begun to demonstrate that skill and teachability *can* be introduced into software.

Section 23.4 conducts a tour of some historical aspects of the AI approach in robotics, including illustrations from the author's laboratory at Edinburgh. Section 23.5 introduces the theme of two-way voice communication between the robot and its supervisor, and also the technique of interrobot voice communication. Section 23.6 outlines a possible robot application of *object-oriented* AI programming languages. Different parts of the robot, each equipped with its own separate microprocessor, are coordinated toward a goal by the exchange of messages between the microprocessors.

Section 23.7 addresses the problem of training students and technical staff in AI programming. A minimal menu of facilities is proposed as a basis for the robotic tutorial stations required. Section 23.8 discusses machine teachability by examples and reports an experimental demonstration of inductive learning of movement strategies.

Section 23.9 considers in greater detail the human-factors benefits and economic payoffs of robot-human and robot-robot voice communication, with an illustrative worked example. Section 23.10 relates the next state in robotics to certain dangers of current software approaches. It has become urgent to improve the user friendliness of information technology's increasingly sophisticated devices, and this will demand a radical break with the past.

23.2. WHAT IS AN EXPERT SYSTEM?

An expert system embodies in a computer the knowledge-based component of an expert skill in such a form that the system can generate intelligent actions and advice and can also on demand justify to the user its line of reasoning. A significant side effect of an expert system is that an improved codification of the given skill can be obtained by translating back into humanly readable terms the system's accumulated knowledge base of facts and rules. The use of expert systems in this way to improve existing knowledge sources is spoken of as *knowledge refining*.

An expert system of nearly 3000 rules called R1 is used at Digital Equipment Corporation for configuring systems to customers' needs.[1] It already outperforms their best technical salesmen. But in knowledge refining, the focus is not on the product that expert systems were originally designed to deliver, namely, interactive advice in conversation with a client, but on the unexpected by-product, the finished knowledge base itself. Expert system languages can be used for the following:

To get knowledge into the machine.

To test it.

To fill gaps.

To extend it.

To modify it.

Finally, the knowledge can be put back into the human world *in unrecognizably improved shape.* This observation was first made in the following manner:

Ivan Bratko, a Yugoslav computer scientist and chess master, during a visit to Edinburgh, chose for study an elementary piece of chess knowledge, namely, how to mate with king and rook against king, from any legal starting position. Although an elementary mate, it is not easy to program, taking (unless artificial intelligence methods are used) approximately two to three months of a graduate student's time. Furthermore, the resulting program is seldom completely correct in its play.

Using one of the Edinburgh advice languages, Bratko was able to complete the programming in two weeks. Subsequently he proved, both exhaustively by computation and also by symbolic proof, that the program was complete and correct. He then translated the program rules back into English. This produced six rules only: they were complete and correct, unlike grandmaster codifications, which turn out to contain errors in those cases tested by exhaustive computation[2-4] and moreover to consist mainly of gaps.

The improvement in the knowledge representation when back-translated from an advice program was unexpected. However, when the possibility was drawn to the attention of workers in other laboratories, they confirmed that they too were able to see indications of the same phenomenon. For example, MYCIN, developed in Feigenbaum's laboratory, was the first expert system of the modern type.[5] Although its knowledge is not broad enough to make the program of significant utility to clinicians, the system is heavily used by medical students, who find the knowledge base more complete and easier to follow than a textbook. Notes follow on some other cases where back-translation from an advice program has produced an improved text for the human learner.

23.2.1. Internal Medicine

The internal medicine project is a collaboration between Dr. Jack Myers, the equivalent of a grandmaster in internal medicine, and a computer scientist, Harry Pople.[6] The aim is to codify Myers' clinical knowledge in the form of an expert system. Clinicians are already using printouts of parts of the knowledge base as reference material.

23.2.2. Organic Chemical Synthesis

In organic chemical synthesis at the University of California, organic chemists use the knowledge base from Wipke's program, SECS.[7] In the domains covered by the program, they regard it as superior to conventional sources.

More striking cases occur where the system has inbuilt learning capabilities. When using these programs, the expert conveys his concepts to the system by means of examples and not through direct programming. The following two examples come into this category.

23.2.3. Soybean Pathology

Until recently, the best classification of soybean diseases was that by the plant pathologist, Dr. Jacobsen, who collaborated in a study directed by R. S. Michalski in the University of Illinois.[8] Michalski's program, AQ11, is capable of generalizing over examples: thus, in this case the machine was part author of the resulting classificatory system, which gave 99% correct allocation of diseased plants as against approximately 83% for the old Jacobsen system. With great difficulty, using the machine to test, recycle, and debug his theory, Jacobsen refined his own system so that it achieved a 93% success rate. Unable to improve it further, he then decided to accept the machine-generated taxonomy.

23.2.4. Mass Spectroscopy

This is the domain of the DENDRAL project at Stanford. The inductive learning module of the program, Meta-DENDRAL, produced rules (previously lacking) for the interpretation of mass spectra of the family of organic chemicals known as the mono- and polyketoandrostanes. These rules formed the basis of a paper subsequently published in the chemical literature.[9]

The question arises: does a different program have to be written for each of these application areas? The knowledge put in by the expert must of course be domain specific, but an induction program which assists his/her input of knowledge should typically show generality, as in the following example.

23.2.5. Lymphatic Cancers

In the differential diagnosis of lymphatic cancers Ivan Bratko successfully used an Edinburgh learning program derived from Ross Quinlan's ID3, which was in turn based on Earl Hunt's CLS (Concept Learning System) developed in the 1960s. The Edinburgh program, Interactive ID3, was developed using the domain of chess. Strategies for chess end games were conveyed by examples. The resulting correct and complete theory of king and pawn against king is itself of interest to chess masters. Bratko then used the same program, in collaboration with a clinician, to construct a diagnostic scheme for the lymphatic cancers. In this experiment, with help only in extracting data from medical case histories, Bratko was able to perform as if he were a clinical expert, although knowing nothing about medicine.[10]

To envisage the principles just illustrated in the context of robotics, one must be prepared to accept that the planning and execution of, for example, the sequence of actions interspersed with perceptual tests by which a product is assembled from parts is the outward sign of a stored body of hand-eye expertise of which the human practitioner may be quite unaware. To discover the rule-based structure of this hidden expertise and to transfer it to computer memory in a form capable of driving expert assembly behavior in a robot is a typical task of expert system engineering.

23.3. EXPERTISE-INTENSIVE HAND-EYE TASKS

As indicated earlier, there is a widespread mental block against the notion that the performance of an everyday manipulative task may involve expertise of the same order as is exemplified when a world expert in internal medicine diagnoses an inflamed gall bladder or a grandmaster classifies a rooks-and-pawns ending. In man, manipulative skill, as other skills including mental, arises from intensive and prolonged experience. The experience necessary for adult-level hand-eye capability is ordinarily acquired during the first few years of life. Hence, it is easy for the adult, who continually plays the hand-eye game with the same intuitive assurance as a grandmaster plays lightning chess, to be forgetful of that early half-decade of daily struggle and practice.

As a corrective to this notion, we have deliberately chosen the following two graded task specifications:

1. Within the range of 7-year-olds to master fairly quickly.
2. Beyond the reach of contemporary robotic accomplishment.
3. Attainable by robots during the 1980s if a decision were taken to apply and further develop standard knowledge-based programming techniques. As a caution against supposing that the attainment would be easy, we have added rough estimates of the scale of effort involved.

To be Demonstrated. An inexpensive micro-based robot system that can be told what to do in words and pictures, and can then work out and test a way to do it, where appropriate, in cooperation with other robots and with the human supervisor.

Specimen Task. Sorting and placing isolated objects on a bounded surface to satisfy specified relations, with error detection and replacing where necessary, as in setting a table or arranging a kit of tools on a bench.

Harder Task. Use of an engineer's prototyping kit to build and test models from pictorial illustrations, again on a multirobot cooperative basis.

Estimated Effort. For task 1, 10 man-years spread over three years. For task 2, 70 man-years spread over five years.

Although substantial by academic standards, such investments are small relative to research and development budgets of major engineering corporations.

23.4. SOME MACHINE INTELLIGENCE HISTORY

The heroic age of experimental AI in robotics (as contrasted with the manufacturing technology school) occupied the late 1960s and early 1970s.

23.4.1. The SRI SHAKEY Project

A distinguished and influential contribution was made to the robot-planning problem by the "SHAKEY" project initiated at Stanford Research Institute by Charles Rosen. SHAKEY was a man-sized mobile robot on wheels equipped with a ranger, camera, and other sensors (See Figure 23.1). The tasks proposed

ANTENNA FOR
RADIO LINK

TELEVISION
CAMERA

RANGE
FINDER

ON-BOARD
LOGIC

CAMERA
CONTROL
UNIT

BUMP
DETECTOR

CASTER
WHEEL

DRIVE
MOTOR

DRIVE
WHEEL

Fig. 23.1. Stanford Research Institute's experimental robot (SHAKEY).

were those involving plan formation. An instruction such as "Go to Room B; find a box; push it until it touches a wall; come back to here" can be converted into effective action only by computational reference to a world model in which are stored the logical and topographic constraints that define the task environment and its causal laws. The forms of representation chosen by the SRI group were those of first-order logic, in which a state of the world was modeled by a list of clauses of the form "AT (ROBOT, A), PUSHABLE (BOX1)" and so forth. Planning was done by heuristic search of a problem graph in which the nodes of the graph correspond with logical descriptions of the foregoing kind and the arcs correspond with actions, such as PUSH, GO, and HALT. With each action name was stored an *add and delete list* detailing the logic clauses to be added to and deleted from the current description to construct a new world description from trial application of the given action. Thus associated with GO (X, Y) would be a list prescribing deletion of the clause AT (ROBOT, X) and addition of the clause AT (ROBOT, Y). Tests for preconditions were also associated with actions: to compute the effect of PUSH (X, Y) prior test on PUSHABLE (X) is indicated.

The importance of the SHAKEY demonstrations for today's tasks remains evergreen, and can be summed up:

1. The use of the predicate logic formalism pointed the way for the subsequent development of the *logic programming* movement, of which the first fruit has been the programming language PROLOG.

2. The use of heuristically guided search through a space of possible worlds gave a solidly based paradigm adequate for routine needs of robot plan formation.

23.4.2. The Edinburgh FREDDY Project

In parallel with SHAKEY the FREDDY project at Edinburgh was addressing complementary issues concerned with the acquisition of perceptual descriptions.[11] FREDDY's mission was to be instructible at short notice (a few days' work by the knowledge engineer) in the names of the parts of new assemblies with their visual appearances, and in the manner of putting them together to form specified simple wooden toys. The robot (see Figure 23.2), using two television cameras, a touch-sensitive manipulator, and a motor-controlled mobile viewing platform, was initially confronted with a jumbled heap of parts and as its first task had to sort them out, identify, classify, and arrange them in standard positions on the platform. The second phase, the actual assembly process, was executed blind. Hence, most of the AI content was concentrated in the earlier, visually driven, phase. The main contributions to technique were as follows:

1. The large-scale use of the *memo function* programming device for speeding up computations through endowing key procedures with rote-learning capability.

2. The introduction and elaboration of relational graphs (today usually called *semantic nets*) as the representational form both for description of images of individual objects and for description of whole platform-top scenes.

3. The development of efficient algorithms for the manipulation and matching of description graphs.

4. As with SHAKEY, the demonstration of an approach to robotics, lacking from the first generation still typical of the industrial scene, that targets the main issue as being that of *integrating* a variety of sensorimotor devices. The same focusing theme motivated analogous work in progress concurrently at MIT and Stanford University.

Fig. 23.2. The Edinburgh hand-eye robot FREDDY-2 photographed in 1972. In addition to the obliquely mounted camera (left of center in the picture) an overhead camera, not visible here, permitted detailed scrutiny of individual objects on the mobile platform. Left to right: H. G. Barrow (vision), S. H. Salter (optical and mechanical engineering), G. Crawford (electronics,), and R. J. Popplestone (manipulator routines).

It is of interest that in the early 1970s SHAKEY and FREDDY were demonstrating levels of behavior that in the 1980s no laboratory in the world can begin to match, although the technological need to do so becomes increasingly pressing. The basis for this state of affairs can be found in the story of the goose that laid the golden eggs. The reader will recall that the farmer and his wife in their impatience killed the magic goose in hopes of finding a boundless store of riches. So in 1973 the relevant funding agencies of Britain and America, impatient with the exploratory motivation of the AI work, in effect issued writs to these experimental robot projects: "Technologize or die!" The SHAKEY and FREDDY projects died. Electrical, control, and production engineers associated with them technologized. The AI scientists largely turned their attention to other domains of problem solving for the study of these same issues of representation, knowledge, planning, and learning.

23.4.3. Subsequent Activity

Hence, since 1973, little specifically AI work has been conducted in the robotics area, though there has been much work on "supporting technologies." Consequently the historical background is concerned mainly with activities of a very few individuals during this intervening period. The AI approach to robotics had been brought to first base in the early 1970s at SRI, MIT, Stanford University, and Edinburgh, as we have seen. My Edinburgh laboratory subsequently concentrated on the study of parallel array-processing architectures for image analysis and the use of inductive learning algorithms for machine acquisition of recognition rules. More recently a number of experimental excursions have been made into the use of manipulators integrated with simple vision for an extension of the inductive learning theme into *scene manipulation,* as described in the next paragraph.

For image analysis, we have investigated the cellular logic CLIP machine (see References 12–22). Results of this work show speedups in vision of at least three orders of magnitude. More recently, preliminary experiments have indicated a family of robot tasks that are essentially doable but that at the same time challenge a full range of knowledge-processing techniques. We call tasks of this kind *scene manipulation.* Early human experience of such tasks occurs in the nursery with cut-out shapes representing objects able to be arranged and rearranged on a viewing surface to represent scenes. An example from later life is setting the table. An industrial counterpart would be the packing by robot of objects into trays. Such a robot must be instructible, including by examples, so as to recognize constituent objects visually and to interpret the scene compounded from them—in the first place into "acceptable" and "unacceptable" categories, and secondarily into better and worse among the acceptable packings. Following interpretation, specific relations responsible for any unacceptability must be identified, and remedial action planned and executed by use of the manipulator. This last is like the diagnosis-and-therapy tasks for which expert systems have been used in various domains.

For factual instruction of the robot, clean-cut and machine-efficient formalisms exist within the general logic programming framework first developed at Edinburgh by Robert Kowalski.[23] For practical demonstration of cost-effective hand-eye systems good PROLOG implementations have been achieved on micros including the IBM PC by Clark and McCabe[38] at Imperial College, London. Adequate facilities for numerical work are also available in state-of-the-art PROLOG systems. The author's own group has demonstrated the feasibility of inductively generating robot control programs using PASCAL-coded elaborations of Quinlan's ID3 program.[24-26] When PROLOG is chosen as the output language of the induction system, both deductive and inductive facilities can be provided in a single package. In Section 23.8 an example is also given of the application of rule induction to shape recognition.

23.5. HUMAN-ROBOT AND ROBOT-ROBOT COMMUNICATION

A new theme is coming into view, namely, the problem of *multirobot* planning, communication, and control. We attach much importance to the solution of the AI and software design problems of cooperative task-execution among shop-floor robotic devices, not least because the first industrial nation successfully to enter this new area will for a while enjoy a critical competitive edge. In the spirit of *human window* considerations,[25,26] the successful approach will model human cognition to a certain degree, so that the robot-human as well as the robot-robot link can be maintained. To give concreteness to this point, first we discuss speech and visual I/O, which impact on this problem at the surface level. We then outline one possible approach to the planning problem based on a class of software facilities known as *object-oriented languages.*

Work at Edinburgh has shown that single-command speech input to a robot, as an optional bypass of the standard key-pad programming of movements, almost doubles the speed at which new trajectories can be taught. The main reason appears to be that the key-pad operator must continually look first at the key pad when keying the command, then at the robot to observe the effects of execution, then back to the key pad, and so on. Using a microphone instead, he/she is free to watch the robot all the time.

It is not generally appreciated that in the coming era of multirobot coordinated signals, speech input will need to be combined with speech output, and that *interrobot communication* may be largely confined to message passing by this human-interceptable channel. The rationale has to do with the avoidance of the cost and inconvenience of building hardware-software interfaces for connecting incompatible devices and also with the cognitive load otherwise imposed on the instructors and supervisors of robot teams if the latter intercommunicate silently.

The following description gives an outline sketch of the experimental facilities developed in the author's laboratory using inexpensive hobbyist equipment, sufficient for student exploration of the points of principle. The user has the option to deliver any command provided by the "teach" key pad by the spoken word used as an alternative channel. Cancellation and correction of a given command can be effected freely both within and between channels. Thus a key-pad command can be overwritten by a spoken correction, and a spoken command can be overwritten from the key pad, in both cases after an intervening key stroke on the computer keyboard. To set up a command vocabulary, the user must speak a number of repetitions of the given command word into the microphone. The default vocabulary of command words is that supplied by the given robot manufacturer for the purpose of key-pad *teaching* of manipulator trajectories. The user is free to create and store alternative vocabularies of synonyms. He/she is also free to include more than one spoken equivalent for any given command. Thus he/she might wish to introduce both "up" and "higher" as equivalents for the command to raise the gripper. The computer echoes, both on the display screen and also by use of its speech synthesizer, its interpretation of the spoken signal. Voice-echoing can be switched on and off by the user at will. In "fail-safe" mode the command will not be executed until confirmation or disconfirmation is signalled by an appropriate key stroke or by an equivalent spoken word.

Under quiet conditions, and subject to the previously stated proviso concerning choice of vocabularies, one can expect from the commercial recognition packages that we have tested about 14 correct recognitions on average out of every 15 words spoken. Recognition is specific to the individual user's voice. A new user must train the device for himself/herself on his/her chosen vocabularies and file these for his/her future use. It is pointless to use speaker A to test the recognizer if it has been trained by speaker B.

The seed of a future development of great importance can be seen in the fact that even a recognition package with performance substantially inferior to the accuracy cited can be relied on to identify spoken commands *with unfailing accuracy* when these are spoken by a robot equipped with a commercial speech synthesizer. The reason for the difference lies in the near exact replication of the same speech waveform on each utterance of a given word. By contrast, a human repeating the same word generates a considerable variation in the corresponding waveforms, over which the recognition device must generalize as far as possible.

The idea of using speech as the basis of interrobot communication seems at first sight perplexing, even bizarre. Yet the benefits as compared with traditional ways of interfacing devices are immediately apparent to the user.

23.5.1. Elimination of Interfacing Costs

First, the costs and complexities of constructing conventional hardware and software interfaces are eliminated. In the case that coordination is desired between initially incompatible devices, the benefits on this count alone may seem overwhelming. For voice interfacing, each robot requires only to *know* the key-pad command vocabulary of every other robot. Such vocabularies typically contain a dozen or two command words. The objection that noisy environmental conditions might jam the voice interfaces is without substance. It is straightforward to have robot A's speaker and robot B's microphone encased together in a soundproof box, with a slave speaker outside for the supervisor's benefit. Likewise, the speaker of robot B and the microphone of robot A are encased together to provide a return communication channel in the B-to-A direction.

23.5.2. Relief of Cognitive Strain

The second striking benefit arises from the complexity of the robot supervisor's task. Even with today's generation of insentient devices with their stereotyped behaviors, the strain of supervising real-time operation of a number of robots can be substantial. Let us now consider sensor-driven behaviors, complicated by message passing between interacting robots. To keep track of what is going on, and to judge whether malfunction is occurring and why, becomes rapidly intractable if the coordinating signals are allowed to pass along silent channels of wire or radio, secret from the would-be supervisor. Voice signals, on the other hand, can be monitored with ease, and a mental picture maintained of what stage has been reached in a task, whose court the ball is in, and what is the cause of a hang-up in execution. Moreover, in suitable cases the supervisor may be able to clear and correct a snarled situation by injecting his/her own spoken commands.

Later in this chapter (Section 23.9) a worked example is given from experiments by D. Croft at Edinburgh, whereby this new mode of operation was first established.

23.6. OBJECT-ORIENTED ROBOT PLANNING

A form of programming stemming from the *actor formalism* of Carl Hewitt[27] has led to a number of languages that depart from the hierarchy of procedure calls that most people regard as the normal flow of control. Instead of procedures we have *actors* or *objects* with behaviors specified in terms of possible input messages received. A behavior includes the emission of new messages addressed to other objects. We exemplify with the following outline sketch of robot plan-formation conceived in terms of the object-oriented language ROSS.

Consider a hypothesize-and-test paradigm according to which each proposed act is first tested in a simulated world to observe its effects, consequences, and feasibilities. This paradigm is in a way very similar to human planning. In many situations humans tend first to perform a mental simulation of a postulated act before actually performing it. Problem solving in the past followed two main routes:

1. That of the Graph Traverser best-next graph search.[28]
2. That of the STRIPS-type theorem proving.[29]

Both routes are fundamental to our present problem-solving techniques. They, however, offer low human comprehensibility and little cognitive significance. In many areas, problem solving is more effectively performed using distributed processing hardware.[33] One such area is in robotics, where multimanipulator cooperation forms the trend in present researches. The following scenario was evolved in collaboration with Mr. Andrew Chun at Illinois University.

We approach problem solving using a hypothesize-and-test paradigm in a distributed environment, which consists of multimanipulators cooperating in the task of simple assembly using polyhedral objects. Assembly consists of complex tasks which will also include the ability to produce structures similar to those handled by BUILD.[30]

The given task is first presented to a scheduler (see Figure 23.3) that hypothesizes a collection of basic acts to be performed. This collection is then distributed to manipulators in a contract negotiation manner.[31] Once the task is received by a manipulator, the manipulator processor will propose certain actions. These actions are then communicated in a conversation-like style to the world model. These actions are simulated in this model to observe the effect, consequence, and feasibility. Necessary modification is then made by the manipulator. Each appropriate action will then be broadcast to all other manipulators. This type of planning and control structure is much more comprehensible to the human user than the graph search of Graph Traverser programs or the logical theorem-proving of STRIPS. In many ways this seems analogous to human plan generation.

The world model consists of a collection of knowledge sources (actors), each of which sustains knowledge of one particular object. Knowledge includes static or dynamic knowledge in either datum or procedural form. Constraints or preconditions can be considered as built-in behaviors of these objects, thus freeing the manipulator processors from constraint considerations. This world model can be classified as what Doran has termed a "perceptual cause and effect model."[32]

23.6.1. Advantages

First, the system will be able to explore problems that may occur in a parallel system, such as deadlocks. The proposed system will also be flexible and modular, since an addition of a manipulator or object to the robot world is simply a matter of adding a new knowledge source or actor.

Owing to the conversation-like control scheme, human comprehensibility is high. The importance of human comprehensibility has been stressed. Acceptance and utilization of any complex decision system depends heavily on the ability of the human user to understand how the system result came about.

At present the world model is centralized to form a simulator. However, distribution of the world model into individual manipulators is not difficult. Each manipulator would need only partial knowledge of the world model at any one time to perform a certain task. This leads to the consideration of routing sensory data only to processors that would be likely to use them. Additional knowledge can be obtained by sending a message to the central world model.[33] Isolation of control and domain knowledge is also achieved in this system, allowing for domain independence, thus enabling the system to be applied to different domains by changing the world model.

23.6.2. Language

The system, being distributed, is most suitably written in an object-oriented language such as ROSS.[34] All processing in ROSS is done in terms of message passing[27] among a collection of *actors* or *objects*. It is useful for modeling and understanding dynamic real-world systems whose complexity makes more analytical tools inappropriate.

23.6.3. Possible Scenario

Figure 23.4 depicts a possible scenario according to the following order for the planning stage:

USER:	tell SCHEDULER to switch block B and block C.
SCHEDULER:	tell HAND 1 to move block B to location of C.
	tell HAND 2 to move block C to location of B.
HAND 1:	tell SIMULATOR plan to pick up block B.
SIMULATOR:	tell HAND 1 ok.
HAND 1:	tell SCHEDULER remember plan (HAND 1 pick up block B).
HAND 2:	tell SIMULATOR plan to pick up block C.
SIMULATOR:	tell HAND 2 block D is on top of C.
HAND 2:	tell SIMULATOR plan to pick up block D.
	tell SCHEDULER remember to replace block D.
SIMULATOR:	tell HAND 2 ok.
HAND 2:	tell SCHEDULER remember plan (HAND 2 pick up block D).
HAND 1:	tell SIMULATOR plan to place block B on location of C.
SIMULATOR:	tell HAND 1 location is occupied.
HAND 1:	ask OTHER HANDS if block C will be removed.
HAND 2:	tell HAND 1 yes.
HAND 1:	tell SCHEDULER remember plan (HAND 1 wait for HAND 2 remove C).
HAND 2:	tell SIMULATOR plan to place block D on useless space.
SIMULATOR:	tell HAND 2 ok.
HAND 2:	tell SCHEDULER remember plan (HAND 2 place block D down).
HAND 2:	tell SIMULATOR plan to pick up block C.
SIMULATOR:	tell HAND 2 ok.
HAND 2:	tell SCHEDULER remember plan (HAND 2 pick up block C)
	tell HAND 1 block C has been removed.
HAND 1:	tell SIMULATOR plan to place block B on location of C.
SIMULATOR:	tell HAND 1 ok.
HAND 1:	tell SCHEDULER remember plan (HAND 1 place block B down on previous location of C).
HAND 2:	tell SIMULATOR plan to place block C on previous location of B.
SIMULATOR:	tell HAND 2 ok.
HAND 2:	tell SCHEDULER remember plan (HAND 2 place block C down on previous location of B).
SCHEDULER:	tell HAND 1 pick up block D.
HAND 1:	tell SCHEDULER ok.
SCHEDULER:	tell myself to remember plan (HAND 1 pick up block D).
SCHEDULER:	tell HAND 1 place block D on block B.
HAND 1:	tell SCHEDULER ok.
SCHEDULER:	tell myself to remember plan (HAND 1 place block D on B).
	tell myself to remember plan (END).
USER:	tell SCHEDULER give final plan.
SCHEDULER:	final plan—

HAND 1	HAND 2
pick up block B.	pick up block D.
wait for HAND 2 to remove block C.	place block D down.
place block B on previous location of C.	pick up block C.
pick up block D.	place block C on previous location of B.
place block D on B.	

23.6.5. Limitations

Since this is mainly a study in robot plan generation, actual manipulator control is considered of secondary importance. The latter involves various mathematical trajectory computations and coordinate transformations, and so on, which should be considered as a separate interface system. This is especially true owing to the machine dependence of hand control. There are also several difficult problems that are not yet solved. One of them, the problem of empty space modeling,[30] has only been dealt with, so far, at a primitive level.

23.7. NEED FOR A HANDS-ON TRAINING FACILITY

At present, training in the programming of robots is conducted in only a very few centers, yet the requirement is forecast to expand in all advanced countries along a rising curve. Neither does any

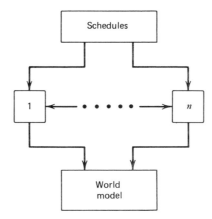

Fig. 23.3. A general structure for a hypothesize-and-test procedure in a distributed environment.

software base exist suitable for the teaching function as contrasted with the use of large expensive robots on the shop floor. Furthermore, even industrial training requires putting its main emphasis not on the techniques of the first generation, which still dominate practice, but rather on those of the second and third generations that will be making their appearance in the factories during the trainee's post-apprentice life.

The remedies already lie at hand, even for the seemingly ambitious task of putting together an effective tutorial work station at which the trainee can be assisted to become proficient in expert systems programming in the hand-eye domain. The following facility would be feasible at a cost per station in the region of $10,000. A local area network of IBM PC micros or equivalent is envisaged. Every component capability listed has, to the author's knowledge, already been checked out in one or another student environment.

23.7.1. Intelligent Tutorial Work Station for Robotics

A number of new software aids and techniques which have been individually validated in academic research laboratories require integration into an inexpensive and versatile programmer's workbench for robotics software. Such a facility in its final form would permit the following modes:

1. Factual instruction of the robot in the objects, relations, and logic of its situation.
2. Availability of procedurally coded shortcuts within the logic system (for example, deduction of movement plans may need to call shortest-path or other numerical optimization routines).
3. Autonomous deduction of rule-structured plans for attaining user-defined goals and subgoals. History trace and diagnostics for display and editing of machine-generated plans.
4. Computer-induction facilities to allow the user to teach the system new plans by examples and counterexamples of relevant situation-action linkages.
5. A heuristic model of plan debugging. Such expert systems cannot be built without a domain specialist as source of know-how. First step: the students must turn themselves into domain specialists through intensive use of prototype systems 1–4 listed above.
6. Generalize the foregoing to the coordinated multirobot case.

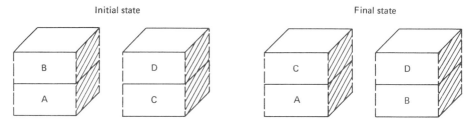

Fig. 23.4. A possible scenario for object-oriented robot planning.

Immediate training applications: hand-eye manipulations and problem solving in a bench-top world, including planning, inductive acquisition and execution of simple placing and packing strategies; elementary image processing and methods for combining and assessing data arriving along different sensory channels.

23.8.   ROBOT LEARNING FROM EXAMPLES

To teach a robot something the robot must be capable of learning it from the tutorial format presented. The simplest and most direct tutorial format is a well-chosen sequence of examples. Hence we want the robot to learn rules from examples. Various vision algorithms and "training by showing" (see Chapter 14, "Vision Systems") are some such cases. Programs for computer induction have recently been developed as aids to generating rule bases for expert systems. The following is a description of one particular induction program available commercially.[35]

ACLS (Analog Concept Learning System) is a program that generates rules, whether classification rules or action rules, by induction from examples. A training set of examples is presented to ACLS as a list of records. All fields of the record except the last are entered with the values of those attributes considered by the user to be relevant to the classification task (*primitive* attributes). The last field is entered with the name of the decision class to which the record is to be assigned (e.g., CIRCLE, ELLIPSE, TRIANGLE, SQUARE, POLYGON, OTHER; or MALIGNANT, BENIGN, IMAGINARY; or ALLOWED, DISALLOWED, SPECIAL-CASE: or just TRUE, FALSE). From these example records ACLS derives a classification rule in the form of a decision tree, branching according to values of the attributes. ACLS also generates a PASCAL conditional expression logically equivalent to the decision tree, and this PASCAL code can be run to classify new test examples. Whenever a new example is found that refutes the current rule, ACLS can be asked to restructure the rule to accommodate the new case and to display, store, or output it as before.

The set of attributes should be chosen by an expert in the given problem domain to be sufficient for classifying the data. In choosing the set of attributes the maxim is: if in doubt, put it IN. Any attribute that is in reality irrelevant to the task will simply be left on one side by ACLS and not included in the tree (and corresponding PASCAL code) that it generates. *It is not necessary that the expert be armed with a clear mental picture of the rule that he himself uses when performing the given task.* His role is to supply the list of *primitives* and to act as an *oracle* by assigning example data to what he considers to be their correct classes. ACLS observes his behavior as a classifier and from this constructs the simplest rule it can, using the primitives supplied, that will assign the same example data to the same classes as he does. As the training set grows with the addition of fresh examples, so the ACLS-synthesized rule grows more sophisticated. As it does so it tends to approximate to a form on which the expert is likely to comment: "That looks like the way that I think I probably do it." The expert is thus enabled to transfer to the machine a judgmental rule that he already had in his head but had not explicitly formulated. The facility thus takes the user a step in the direction of a do-it-yourself kit for building his own expert system.

23.8.1.   Generation of Robot Plans for Simple Configurations of Components

Commercially available robot systems have to date possessed no true ability to learn. It is possible to present them with a previously generated sequence of movements which they will follow faithfully. This is satisfactory for applications such as spot welding on cars, where only one model of car has to be dealt with. But when a robot is presented with components to be assembled that are randomly positioned, for example, on a conveyor belt, this kind of operation is not adequate. It would be advantageous if the robot could learn by example. It would be shown how to assemble the components starting from a number of different configurations. These examples would be used to generate a rule explaining them, and this rule could then be used as a plan to generate robot actions appropriate to the various starting and intermediate configurations of components.

As stated earlier, ACLS is a program that learns by example. Each example is comprised of two parts: a list of the salient features of the situation, the attribute-values, plus the type to which the situation belongs, the class name. Suppose, to take a fanciful case, that a robot were shown sets of described situations in which it should use its umbrella, and others in which the umbrella should not be used. With ACLS the robot could in this way be trained in the use of umbrellas. Examples consist of attribute-value lists (e.g., it is wet, you are not soaked, and you are outdoors) linked with a decision-class name, for example, "use" or "dontuse."

After a variety of examples have been presented, ACLS can be asked to induce a rule to explain them. What does rule-learning mean for a computer? Let us explore this further.

When a person learns a concept, he/she acquires the ability to classify objects into categories. So he/she acquires a classification rule of some kind. He/she can apply this rule even though he/she may not be able to tell you clearly what the rule is. A computer's classification rule takes the form of a program. Here is an example:

```
function decision : dtype;
begin
    case weather of
    blustery : decision := dontuse;
        dry : decision := dontuse;
        wet : case indoors of
            no : decision := use;
            yes : decision := dontuse;
            end;
    end;
end;
```

In human terms the concept enables its possessor to recognize "at a glance" whether circumstances are conducive to using his umbrella. The corresponding program is, in this case, a conditional expression in the PASCAL programming language. An inductive learning algorithm such as ACLS takes as its input specimen sets of circumstances paired with their known class membership and produces as output a hypothesized classification rule, in this case in the form of a PASCAL conditional expression.

ACLS thus derives a classification rule that for purposes of display is cast in the form of a decision tree, branching according to the values of the attributes. In the preceding example the attributes are "weather" and "indoors." The respective sets of values are "wet, dry, blustery" and "yes, no." The set of attributes is chosen by an expert to be representative of the data being classified. ACLS can output the rule either as a decision tree or as a PASCAL program.

23.8.2.   Example of Elementary Plan Generation

Consider first the very simple example given earlier. We are trying to decide whether or not to use an umbrella. We shall go beyond the two attributes ("weather" and "indoors") used previously and judge four factors to be probably relevant to making the decision. These attributes are the weather (which can have one of three values, wet, dry or blustery) and whether we are indoors, soaked, or in the car (each of which is either yes or no).

This specification is given in an attribute file. The general form of this file need not concern us at the moment. The attribute file used for the particular example is as follows:

```
4
weather logical wet dry blustery
inside  logical yes no
soaked  logical yes no
incar   logical yes no
2
use dontuse
```

The numeral 4 warns the system to expect 4 attributes, and similarly the numeral 2 specifies the number of classes. We have now described the problem to ACLS and are in a position to give it some examples. Four such examples follow. The format is that used by ACLS when it displays its current set of examples: note that ACLS only looks at the first seven characters of a string, so that the user's "blustery" appears as "bluster."

no.	weather	inside	soaked	incar	class
primary examples					
1.	dry	yes	no	no	dontuse
2.	wet	no	no	on	use
3.	bluster	no	no	yes	dontuse
4.	wet	yes	no	no	dontuse

The first of these examples says that if the weather is dry and we are inside and neither soaked nor in the car, then do not use the umbrella. The other examples can be read in a similar way.

ACLS can then be asked to induce a rule from these examples. The resulting rule follows:

```
weather
    bluster : dontuse
        dry : dontuse
        wet : indoors
            no : use
            yes : dontuse
```

This rule says that when deciding whether to use an umbrella the first factor to be taken into account is the weather. If it is dry or blustery, then do not use it. If it is wet, then we must take into account whether we are inside or not. If so, then do not use it; otherwise, use it.

ACLS is generally used in an interactive fashion. It can be seen, for instance, that the rule just induced will give "use" as an answer if it is wet and we are outdoors but in the car. To change this, the following example could be given.

```
no.  weather  indoors  soaked  incar  class
primary examples
 5.   wet       no       no     yes   dontuse
```

Taking this additional example into account the rule becomes:

```
weather
bluster : dontuse
    dry : dontuse
    wet : indoors
             no : incar
                     no : use
                    yes : dontuse
            yes : dontuse
```

This process of looking at the current rule, finding an example that contradicts it and then reinducing can continue until the user is satisfied with the rule produced.

In the demonstration to be described later the task of the program was to direct two robots to build an arch from small blocks. The blocks that make up the legs of the arch are identified by a letter—A, B, C, or D. The order in which they are finally assembled is fixed. C must be on top of A and D on top of B. However, the initial configuration is up to the user, and the problem is to unstack the blocks, and reassemble them in the correct order. By using initial block positions as attributes, and the destinations as class values, ACLS can be used to build up a rule from examples. These examples can be either entered at the keyboard or from an example file on disk, or both. It was shown that, by successive application of this rule, the two legs of the arch can be built. A fuller account, in which the inter-robot communication system is described, appears in Section 23.9.

23.8.3.   Inductive Learning for Shape Classification

The ACLS inductive learning program has also been used for classifying television images of "Black Magic" chocolates. The classification task was previously attempted by Unimation Corporation (UK) on behalf of Rowntrees Ltd. at York.

1. The test data consisted of gray-level pictures of the various chocolates. A gray-level picture consisted of 256×256 pixels. There were 12 possible chocolate types.

2. In order to extract identifying shape features from the images it was first necessary to convert the gray-level images to silhouettes.

3. All the silhouettes now passed through a "feature extractor" which computed the values of attributes (features) considered by the user to be possibly useful in distinguishing between the different shapes.

4. The inductive learning algorithm induced classification rules from a training set of examples. The example set was first read into the internal store of ACLS and displayed on the screen before induction of rules.

5. All attributes were expressed as integers. The decision taken at any node in the tree was "less than" or "greater than or equal to" the value specified at that node.

6. Tests on new pictures revealed that the accuracy of the induced rule was at least equal to that of human subjects confronted by the same test material, being approximately 80%. The algorithm had extracted all that there was to be gotten from pure shape-silhouette data. To close the gap, use of gray-level information would be requisite.

7. Inductively generated rules showed large run-time economies relative to conventional statistical pattern-recognition algorithms, performing at least as accurately at the expense of far fewer attribute tests.

These experiments were done at Edinburgh by A. Blake, P. Mowforth, and B. Shepherd. They included comparisons with classical feature-space adaptive algorithms of the type that underlie trainable vision systems such as those implemented for the commercial market by Machine Intelligence Corporation and Automatix. Sufficient complementarity between the two approaches was found to suggest combining them for difficult cases. ACLS-based routines proved uniformly faster at run time. So one possible idea is for the system to be trained on both, but only to invoke run-time evidence from the feature-space classifier as a "second opinion" in those cases where conditions of difficulty are detected by the system.

23.9. INTERCOMMUNICATION OF LEARNED STRATEGIES

This demonstration involved generating a plan to solve a simple problem under conditions that obliged the learning robot to communicate commands to an assistant robot. Refer to Figure 23.5 for the following. Four small blocks, initially stored in any order in an area accessible to the small robot, must be assembled, in the correct order, by the large robot. A beam must be placed across the two towers so constructed to form an arch. The blocks have been arbitrarily labeled A, B, C, and D. The order in which they are to be finally assembled is C on A, D on B, and both A and B on the "ground" in the construction area of the large robot. The storage area for the small robot has four fixed locations, labeled E0, E1, TEMP, and TRANS. E0 and E1 are the locations where the blocks are initially stacked, no more than two high. TRANS is the transfer location from which the large robot picks blocks to be incorporated into the arch. TEMP is a temporary location used when the required block is inaccessible, by virtue of being under another. The beam is stored in a location accessible to the large robot, labeled STORAGE4. The construction area locations, shown in Figure 23.6, are as follows:

C0 for block A
C1 for block B
C2 for block C
C3 for block D
C4 for the beam

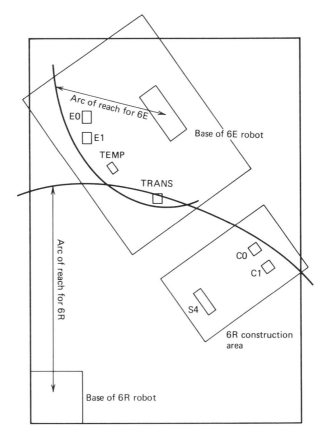

Fig. 23.5. Robot layout.

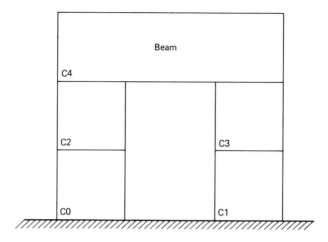

Fig. 23.6. 6R construction area.

23.9.1. Overview of Equipment

The equipment can be divided into two parts: that controlled by an Apple II microcomputer and that controlled by an Acorn Atom microcomputer. The Apple II controlled the small robot, a Systems Control 6E. It communicated with the other computer by a Votrax "Type 'n Talk" speech synthesizer and a heuristic voice recognition card. The Acorn Atom controlled the large robot, a Systems Control 6R. It too "speaks" with a Votrax "Type 'n Talk," and it used an Interstate VRM 1002 for voice recognition. In addition, there was a controller for the 6R. The positions of the two robots were fixed relative to one another and were such that the 6E's storage area was inaccessible to the 6R, and the 6R's construction area was inaccessible to the 6E.

23.9.2. Description of Software

The Apple II software is an extension of ACLS. The user enters the block positions as attributes and the move (i.e., what block to pick up and where to put it) as the class value. Examples can be entered at the keyboard or by an example file, as per usual. However, the USER command has been changed. On Typing U (after the command prompt), the computer displays:

AON   BON   CON   DON   CLASS

The user types what A, B, C, and D are on (e.g., another block, E0, or ARCH, etc.). Then the computer prompts

EXECUTE MOVE?

to which the user can respond YES or NO. If he responds YES, the move defined by the CLASS value is executed by the 6E.

Once a set of examples has been entered, the user can determine by induction a rule to explain these. The rule only plans one move, that is, only one block will be picked up and placed in another location. The system works by successively applying this rule until the task is complete.

The AUTO command allows the user to perform this successive application of the rule. It also provides him with a number of other options, such as training and testing the robot and the learning, testing, and storing of vocabularies for the voice recognition equipment.

23.9.3. Description of the Acorn Atom Software

From the master menu, the user can choose a number of options. He can train the VRM, store vocabularies, and recognize spoken words. He can train the robot. Eleven files are provided in RAM for sequences of robot steps, and they can be downloaded or uploaded from cassette. He can also enter the AUTOMATIC mode, which complements the Apple's AUTO mode. In this mode the Acorn listens for commands from the Apple. It can interpret these commands, concatenate appropriate files of robot steps, and use these to transfer a block from TRANS to the construction area or to place the beam across the two completed towers.

23.9.4. Worked Example

This illustrates how ACLS can be taught to build an arch from a random initial configuration of blocks.

(1)	AON	BON	CON	DON	CLASS
	C	E1	E0	B	ATOARCH
(2)	AON	BON	CON	DON	CLASS
	ARCH	E1	E0	B	CTOA
(3)	AON	BON	CON	DON	CLASS
	ARCH	E1	A	B	DTOTEMP
(4)	AON	BON	CON	DON	CLASS
	ARCH	E1	A	TEMP	BTOARCH
(5)	AON	BON	CON	DON	CLASS
	ARCH	ARCH	A	TEMP	DTOB
(6)	AON	BON	CON	DON	CLASS
	ARCH	ARCH	A	B	STOP

In AUTO mode, a rule induced from these examples is successively applied until the CLASS value is STOP, whereupon control is returned to the user.

It was possible in this way to use an inductive learning program, ACLS, to generate single-move plans for assembling small numbers of components. Communication between two computers, using humanly understandable synthesized speech, was feasible. User intervention by spoken commands was implemented also. An improved implementation of the learning component has recently been described by Dechter and Michie.[39]

Clearly, the style of coordinating computer-controlled devices illustrated by the previously described experiment need not be restricted to the interaction of a robot with another robot. It might be desired, for example, to interface a parts-loading robot to a numerically controlled machine tool. Beyond that, there is a problem of interfacing between one computer with its I/O devices and another, quite generally. The human-factors benefits in terms of the user-transparency of interprocess transactions, and the ability to simplify the interfacing problem almost out of existence, are bound to speak eloquently to technical managers in the world of industrial and military automation, and beyond.

23.10. THE NEXT STAGE

Extrapolation to the future is prompted by the example of other branches of computer-based complex control engineering, for instance, the monitoring and regulation of nuclear power stations, automation of air traffic control, and the computer-controlled networks used by the military for distant early warning of nuclear strike. The lessons of history in these man-machine systems have recently been reviewed by Kopec and Michie.[36] Using this analysis as a guide to the immediate future of factory robotics, one may predict the following steps:

1. Straight-line development of current hardware and software trends, with accompanying escalation of the complexity and user opacity of robotic systems.

2. A mounting series of increasingly disturbing and costly mishaps, foul-ups, and crashes, analogous to the collapse of production in the Hoogovens steel strip-rolling mill reviewed by Kopec and Michie. The collapse was the direct consequence of replacement of an obsolescent by an "improved" automation system. "Improvement" was in the sense of classical software engineering. As knowledge engineering, like all factory-installed systems of today, it would have failed to qualify.

3. Eventual awareness that some doors cannot be opened with a battering ram. It may be better to know the appropriate "Open Sesame." There may then follow a backing off for reflection and replanning at the level of fundamental design.

4. The rise of the user-friendly robot, owing its top-level control to software cast neither in the mold of the classical compact algorithm (too "deep" to sustain explanatory dialogue) nor in that of the classical data base (too "shallow"), but in the expert-systems rule-structured mold.

The evidence and arguments relating the issue of man-machine cognitive compatibility to the choice of the right software structures for machine-executable descriptions and strategies have been reviewed recently.[37]

REFERENCES

1. McDermott, J., XSEL: A computer sales-person's assistant, in Hayes, J. E., Michie, D., and Pao Y-H., Eds., *Machine Intelligence* 10, Horwood, Chichester, and Halsted Press, New York, 1982, pp. 325–337.

2. Clarke, M. R. B., Ed., A quantitative study of King and Pawn against King, *Advances in Computer Chess 1*, Edinburgh University Press, 1977, pp. 108–115.

3. Bratko, I., Proving correctness of strategies in the AL1 assertional language, *Information Processing Letter*, Vol. 7, 1978, pp. 223–230.

4. Kopec, D. and Niblett, T., How hard is the play of the King-Rook-King-Knight ending? Clarke, M. R. B., Ed. *Advances in Computer Chess*, Vol. 3, 1980, pp. 57–81.

5. Shortliffe, E. H., *Computer-Based Medical Consultations: MYCIN*, Elsevier, New York, North Holland, 1976.

6. Pople, H. E., Myers, J. D., and Miller, R. A., DIALOG: a model of diagnostic logic for internal medicine, *Proceedings of the 5th International Joint Conference on Artificial Intelligence* (IJCAI-77), Computer Science Department, Carnegie-Mellon University, Pittsburgh, Pennsylvania, 1977. (This program was subsequently renamed INTERNIST. It is now called CADUCEUS.)

7. Wipke, W. T., Computer-assisted three-dimensional synthetic analysis, in W. T. Wipke, Heller, S. R., and Hyde, E., Eds., *Computer Representation and Manipulation of Chemical Information*, Wiley-Interscience, New York, 1977, pp. 147–174.

8. Michalski, R. S. and Chilausky, R. L., Learning by being told and learning from examples: an experimental comparison of the two methods of knowledge acquisition in the context of developing an expert system for soybean disease diagnosis, in *International Journal of Policy Analysis and Information Systems*, Vol. 4, 1980, pp. 125–161.

9. Buchanan, B. G., Smith, D. H., White, W. C., Gritter, R., Feigenbaum, E. A., Lederberg, J., and Djerassi, C., Applications of Artificial Intelligence for chemical inference, XXII, Automatic rule formation in mass spectrometry by means of the Meta-DENDRAL program, *Journal of the American Chemical Society*, Vol. 98, 1976, pp. 6168–6178.

10. Bratko, I. and Mulec, P., An experiment in automatic learning of diagnostic rules, *Informatica*, Vol. 4, 1980, pp. 18–25.

11. Ambler, A. P., Barrow, H. G., Brown, C. M., Burstall, R. M., and Popplestone, R. J., A versatile system for computer-controlled assembly, *Artificial Intelligence*, Vol. 6, No. 2, 1975, pp. 129–156.

12. *CAP-4 Programmers Manual*, Image Processing Group, University College, London, 1977.

13. Jelinek, J., The informative function and its application in parallel picture processing, *Research Memorandum MIP-R-117*, Machine Intelligence Research Unit, University of Edinburgh, 1977.

14. Armstrong, J. L., Programming a parallel computer for computer vision, *Computer Journal*, Vol. 21, 1978, pp. 215–218.

15. Duff, M. J. B., Review of the CLIP image-processing system, *Proceedings of the National Computer Conference*, 1978, pp. 1055–1060.

16. Jelinek, J., An algebraic theory for parallel processor design, *Computer Journal*, Vol. 22, 1979, pp. 363–375.

17. Zdrahal, Z. and Blake, A., A simple emulator of a parallel processor: user guide, Machine Intelligence Research Unit, University of Edinburgh, 1980.

18. Blake, A. and Ruttledge, H., CAP assembler and driver of CLIP-4 emulator, Machine Intelligence Research Unit, University of Edinburgh, 1980.

19. Blake, A., Edge growing and relaxation in parallel, *Research Memorandum MIP-R-134*, Machine Intelligence Research Unit, University of Edinburgh, 1981.

20. Zdrahal, Z., Bratko, I., and Shapiro, A., Recognition of complex patterns using cellular arrays, *Computer Journal*, Vol. 24, No. 3, 1981, pp. 263–271.

21. Blake, A., A convergent edge relaxation algorithm, *Research Memorandum MIP-R-135*, Machine Intelligence Research Unit, University of Edinburgh, 1982.

22. Blake, A., Fixed point solutions of recursive operations on Boolean arrays, *Computer Journal*, Vol. 25, 1982, pp. 231–234.

23. Kowalski, R. A., *Logic for Problem Solving*, North-Holland, 1979.

24. Michie, D., Ed., *Expert Systems in the Micro-electronic Age*, Edinburgh University Press, 1979.

25. Michie, D., The state of the art in machine learning, in *Introductory Readings in Expert Systems*, Michie, D., Ed., Gordon and Breach, London and New York, 1982, pp. 208–229.

26. Michie, D., Computer chess and the humanisation of technology, *Nature*, Vol. 229, 1982, pp. 391–394.

27. Hewitt, C., Viewing control structures as patterns of passing messages, *Artificial Intelligence*, Vol. 8, 1977.

28. Doran, J. E. and Michie, D. Experiments with the Graph Traverser program, *Proceedings of the Royal Society, A*, Vol. 294, 1966, pp. 235–259.

29. Fikes, R. E. and Nilsson, N. J., STRIPS: a new approach to the application of theorem proving to problem solving, *Artificial Intelligence*, Vol. 2, 1971.

30. Fahlman, S. E., A planning system for robot construction tasks, *Artificial Intelligence*, Vol. 5, 1974, pp. 1–49.

31. Smith, R. G., A framework for distributed problem solving, *Proceedings Sixth International Joint Conference on Artificial Intelligence*, Vol. 2, Tokyo, Japan, 1979, pp. 836–841.

32. Doran, J. E., Planning and robots, in Meltzer, B. and Michie, D., Eds. *Machine Intelligence 5*, American Elsevier, New York, 1970.

33. Corkill, D. D., Hierarchical planning in a distributed environment, *Proceedings Sixth International Joint Conference on Artificial Intelligence*, Vol. 1, Tokyo, Japan, 1979, pp. 168–175.

34. McArthur, D. and Klahr, P., *The ROSS Language Manual*, N-1854-AF, The Rand Corporation, Santa Monica, California, 1982.

35. Paterson, A. and Niblett, T., *ACLS Manual*, Intelligent Terminals Limited, Oxford and Edinburgh, 1982.

36. Kopec, D. and Michie, D., *Mismatch between Machine Representations and Human Concepts*, Commission of the European Economic Community (FAST program), Brussels, 1983.

37. Michie, D., Automating the synthesis of expert knowledge, *ASLIB Proceedings*, Vol. 36, 1984, pp. 337–343.

38. McCabe, F. G., *Micro-PROLOG User's Manual*, Logic Programming Associates, London, 1981.

39. Dechter, R. and Michie, D., Structured induction of plans and programs, *Working Paper*, IBM LA Scientific Center, Los Angeles, and the Turing Institute, Glasgow, 1984.

BIBLIOGRAPHY

Ferguson, R., PROLOG: a step toward the ultimate computer language, *BYTE*, November 1981, pp. 384–399.

Fikes, R. E., Hart, P. E., and Nilsson, N. J., Some new directions in robot problem solving, in Meltzer, B. and Michie, D., Eds., *Machine Intelligence 7*, Edinburgh University Press, 1972.

Hayes, P. J., The naive physics manifesto, Michie, D., Ed., in *Expert Systems in the Micro-electronic Age*, Edinburgh University Press, 1979, pp. 242–270.

Hunt, E. B., Martin, J., and Stone, P., *Experiments in Induction*, Academic Press, New York, 1966.

Juhn, H., *Object recognition with parallel computation of high-level features*, M.S. thesis, Department of Computer Science, University of Illinois, Urbana, 1980.

Mozetik, I., User's manual for the AL1.5 system, *Research Memorandum MIP-R-130*, Machine Intelligence Research Unit, University of Edinburgh, 1980.

Paterson, A., *AL/X Manual*, Intelligent Terminals Limited, Oxford and Edinburgh, 1981.

Quinlan, J. R., Discovering rules by induction from large collections of examples, in Michie, D., Ed., *Expert Systems in the Micro-electronic Age*, Edinburgh University Press, 1979, pp. 168–201.

Quinlan, J. R., Semi-autonomous acquisition of pattern-based knowledge, in Michie, D., Ed., *Introductory Readings in Expert Systems*, 1982, pp. 192–207.

Quinlan, J. R., Learning efficient classification procedures and their application to chess end-games, in Michalski, R. S., Carbonell, G., and Mitchell, T., Eds., *Machine Learning: An Artificial Intelligence Approach*, Tioga, Palo Alto, 1982.

Raggett, D., A survey of computer vision research, *Research Memorandum MIP-R-128*, Machine Intelligence Research Unit, University of Edinburgh, 1980.

Reiter, J. E., AL/X: an expert system using plausible inference, paper presented at British Computer Society Conference on Practical Applications of Knowledge Engineering, June 29, 1980, Intelligent Terminals Limited, Oxford and Edinburgh.

Shapiro, A., Interactive ID3: documentation, Machine Intelligence Research Unit, University of Edinburgh, 1981.

Warren, D. H. D., PROLOG on the DEC system-10, in Michie, D., Ed., *Expert Systems in the Micro-electronic Age*, Edinburgh University Press, 1979, pp. 112–121.

Winston, P. H., The MIT robot, in Meltzer, B., and Michie, D., Eds., *Machine Intelligence, 7*, Edinburgh University Press, 1972.

CHAPTER 24

THE ROLE OF THE COMPUTER IN ROBOT INTELLIGENCE

ULRICH REMBOLD

R. DILLMANN

P. LEVI

University of Karlsruhe
Karlsruhe, West Germany

24.1. INTRODUCTION

Industrial robots have become an important tool in manufacturing. At present they perform successfully a variety of repetitive operations in different industries. Typical assignments are to pick and place parts, to spot and seam weld, to spray paint, and to do simple assembly tasks. In the future the growth potential of the robot will be in assembly. However, to be efficient in this activity, industry first must learn how to design products for robotic assembly. The human assembler has two very versatile effectors and can easily coordinate the movement of his or her arms to do extremely complex operations. In addition, he or she has a well-developed sensory system and an almost unlimited memory capacity. The human is also capable of learning from previously performed operations and planning and executing new tasks with the aid of a large knowledge base. Thus for assembly it will be necessary to provide the robot with human knowledge and a variety of different complex sensors. This can only be done with the help of the computer. This implies that the engineer will be able to develop powerful low-cost computers, sophisticated hardware circuits to perform fast calculations, and many new sensory systems. In addition, the user of robots must have the capability to write his programs in an application-oriented language that can easily be learned and used for a diversity of assembly tasks.

As a summary to Part 4, "Robot Intelligence," this chapter discusses the role of the computer in robot intelligence. The main emphasis is placed on programming systems, expert systems, sensors, laser scanners, and the computer architecture for robot controls.

24.2. PROGRAMMING SYSTEMS

An assembly robot obtains its flexibility through an efficient programming system. Conventional programming such as the teach-in method is very cumbersome to use when points in a three-dimensional space must be described. This method also leads to problems when the effector must move along a complex trajectory or when a moving object must be followed. For this reason numerous programming languages are being developed.

A ranking of different programming languages is shown in Figure 21.1.[1] The majority of these are explicit assembly or compiler languages. Here every movement must be described explicitly by the programmer. The number of statements to program an assembly task will be quite numerous, even for a simple problem. At present, there is only one implicit programming language available, AUTOPASS. It has limited capabilities to describe simple assembly primitives. A typical instruction would be "pick up the bolt, insert it into a hole." Three languages are based on the NC-language concept; they may be used in connection with loading and unloading of NC-machine tools. In this case the programmer only need be familiar with the NC-language concept. The NC languages, however, become very complex when variables must be defined and when sensor data derived from moving objects must be processed.

Figures 24.1 and 24.2 show typical features of programming languages for robots.[1] The arm configuration, number of axes, and the sensor signals they are capable of handling are of interest. In several cases the languages can be used to program several arms or different arm configurations.

PROGRAMMING LANGUAGE	LANGUAGE	NUMBER OF ARMS	OTHER ARM CONFIGURATIONS	SENSORS	VISION
Funky	Point to Point Programming			Touch	
T3	Assembly	1		Limit Switches	
Anorad	NC-Program	1			
Eminly	Assembly	2	x	Touch Proximity	
RCL	Fortran	1			
RPL	Fortran	1		Touch Vision	Location Orientation
Sigla	Assembly	1 - 4	x	Force Torque	
VAL	Assembly	1		Vision	
AL	Algol	2	x	Force Torque	Recognition
HELP	Pascal	1 - 4	x		
Maple	PL/1	1	x	Force Proximity	
MCL	APT	1	x	Touch Vision	Modelling Recognition
PAL	Transformations	1	x		
Autopass	PL/1	1	x	Sensors	Recognition Modelling

Fig. 24.1. Capabilities of different languages.

Figure 24.3 shows desired constructs of programming languages for robots.[1] In addition to the constructs of conventional languages, there should be several new ones specific to robots, for example, instructions describing vector, frame, rotation, and translation. It also should be possible to describe to the robot an effector trajectory and how to handle the synchronization of the work of several arms. The robot must be able to operate the effector and the work tools under program control. In addition, there must be language constructs available that can describe sensor signals to which the robot is capable of reacting.

The many languages currently available suggest that the output of the compiler will be in a standard intermediate code, Figure 24.4. The robot manufacturer in turn will lay out the control system in

ROBOT	MANUFACTURER	ARM CONFIGURATION	NO OF AXES	LANGUAGES
T3	CINNCINNATI MILACRON	R R R R R R	6	T3
PUMA	UNIMATION	R R R R R R	6	RPL VAL
STANFORD	SHEINMANN	R R P R R R	6	AL PAL
IBM ARM	IBM	R P P P R R	7	FUNKY EMILY MAPLE AUTOPASS
PACS ARM	BENDIX	R P P R R R	6	RCL
ALLEGRO	GENERAL ELECTRIC	P P P R R R	6	HELP
ANOMATIC	ANORAD	P P P R	4	ANORAD
SIGMA	OLIVETTI	P P P P	4	SIGLA

R = rotational joint
P = prismatic joint

Fig. 24.2. Different implemented robot languages.

TEACH-IN PROGRAMMING
CONTROL STRUCTURE
SUBROUTINES
NESTED LOOPS
DATA TYPES
COMMENTS
TRAJECTORY CALCULATION
EFFECTOR COMMANDS
TOOL COMMANDS
PARALLEL OPERATION
PROCESS PERIPHERALS
FORCE-TORQUE SENSORS
TOUCH SENSORS
APPROACH SENSORS
VISION SYSTEMS

Fig. 24.3. Features of programming languages for robots.

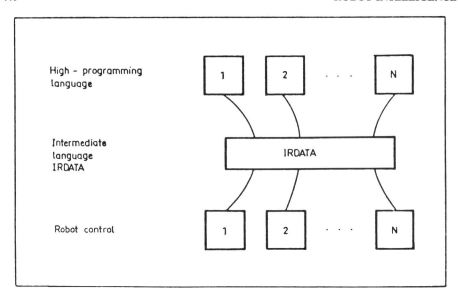

Fig. 24.4.   The use of an intermediate language.

such a manner that the interface of the controller will accept the intermediate code. Thus it would be possible to use for different robots different languages via a standardized interface.

In addition to the language, there must be a powerful programming system available consisting of several software packages and of a low-cost program-development computer. Figure 24.5 shows a comprehensive programming system for assembly robots. The user describes to the robot the object and the workplace with the help of an application-oriented language. This information is processed by a geometry processor and entered into a world model.

In a like manner the movement of the robot is functionally described by implicit instructions, and a syntactical analysis is performed. This program is combined with information from the world model. The result is sent to the AL compiler by way of a generating model. It is also possible to interactively communicate with the AL compiler to enter or edit instructions. The output of the AL

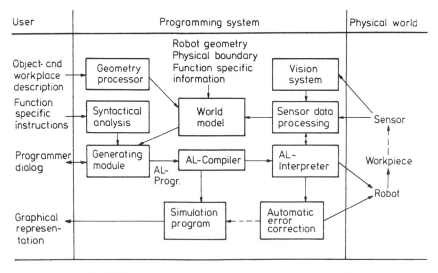

Fig. 24.5.   An advanced programming system based on AL.

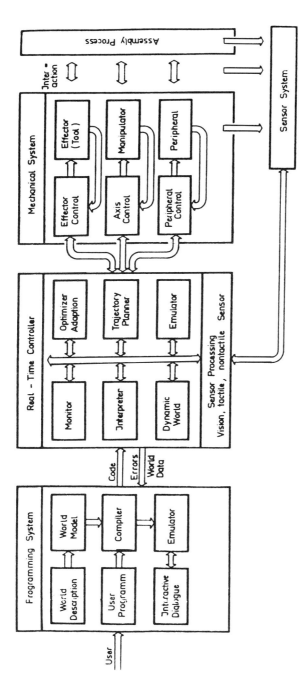

Fig. 24.6. Schematic structure of the assembly robot system.

compiler in form of interpretative code is loaded down to the control computer of the robot. Sensor signals from the robot can be brought back to the sensor data-processing module. In case an object or a workpiece has changed its position, this module will send instructions to the world model to update it. The same information is needed by the AL interpreter to correct the movements of the effector. An available simulation program allows the programmer to display graphically the work environment of the robot and to check its movement and detect possible collision.

The graphical emulation system is part of the programming system or of the real-time controller (see Figure 24.6). In the early stage of robot design and its kinematic attributes (joint, links, end-effectors), the assembly cell as well as its environment can be defined on a graphic display. Trajectory planning (interpolation in Cartesian coordinates) and the corresponding coordinate transformation can be tested and optimized. By adding a program for the simulation of the robot's dynamics, the response of the axis motor drives and their control can be traced to evaluate the dynamics of the robot. For debugging of assembly programs, the simulated robot is interfaced to the programming system that defines multiple moving tasks. With off-line program test facilities, workpieces and robot components can be emulated without the risk of collision. When it is certain that all assembly sequences can be executed without conflict, the program can be transferred to the robot-control computer which executes the program in real time and moves the mechanical manipulator. Verification of the assembly can be performed in this stage.

The simulation of the manipulator equipment (e.g., a two-arm mobile robot) allows software development without the availability of hardware. An additional aid for high-level software-planning tools is the graphical simulation of the assembly process.

A typical graphical simulation system would be implemented on a medium-sized control computer. It interfaces with a graphic display, a vector generator, and a picture refresh memory. With it a disembodied geometric model of the manipulator and its environment consisting of generalized bodies can be generated. The joints of the manipulator are moved under program control along the specified trajectory. A central projection can be done with a three-dimensional geometric model to generate a two-dimensional perspective picture. Different projections of the model can be displayed on the graphic terminal. The system input is either AL code[2] for debugging of assembly programs or a modified software task for experimental research. A teach-in facility to move the model on-line in Cartesian, robot, and tool coordinates by a joystick and a switch is provided. Most of the trajectory control tasks of the emulation system are implemented on the robot-control computer. They run under the direction of a real-time multitask system. The structure of the robot emulation system, the moving tasks, and the interfaces are shown in Figure 24.7.

The development system consists of three basic parts:

The robot-modeling part.
The robot's world-modeling part.
The part to generate motion data.

Further features check for collision and to detect hidden lines. The interactive robot-modeling system allows the generation of a disembodied geometric robot model, including its end effector. A list of generalized bodies is used to aid modeling. The second modeling system is used to describe the robot's world, including all objects to be handled. The geometric-modeling programs generate both a data base of object data and frames that define the position and orientation of each geometric part and the motion axis for each robot link. The motion data, applied as operator to the list of frames, is generated by the trajectory planning tasks. Two interfaces can be used for the specification of the robot workpiece interaction and the desired robot motion. A direct access to the trajectory planning tasks is possible for experimental studies. The second interface for debugging of assembly programs has as input AL code[2]. The motion tasks generate motion data for the selected manipulator configuration and the objects to be moved. Motion errors are indicated by an external module which checks collision. The computing mode is either debug or on-line. The three-dimensional data are displayed on a graphic display system capable of displaying different three-dimensional views (top, front, side) using viewpoint transformation. Zooming and hidden-line removal operations are possible.

24.3. EXPERT SYSTEMS FOR ROBOTS

An expert system is an intelligent computer program that performs decision functions similar to those of a human expert (see Chapter 23, "Expert Systems and Robotics"). Its basic contents are knowledge and production rules that can be used in industrial settings to solve difficult problems. Figure 24.8 shows a simplified block diagram of an expert system.

The knowledge base must be capable of acquiring and processing knowledge "at the right time." It also must be able to explain reasoning and justify conclusions. Typical applications of expert systems in the field of robotics are planning and control of assembly, as well as processing and recovery of assembly errors.

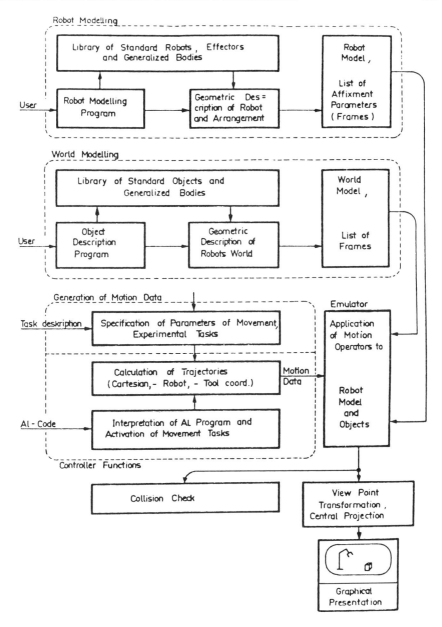

Fig. 24.7. Structure of the robot emulation system.

The error analysis must take into account possible errors in part placements and the tolerances of parts.[3] A checker based on geometric models (e.g., CAD) simulates the effects of the actions and the propagation of errors. The plan may be modified if additional constraints are sensed in order to ensure the success of the robot's actions.

The automatic generation of work plans for a robot is a new research field. The problems that must be solved to generate automatically a complete robot work plan for industrial assembly operations are quite numerous and complex. For this reason most tasks carried out by robots, so far, were planned by people and not by computer programs.

A flexible manufacturing system (FMS), conceived by Westinghouse to forge steam turbine blades

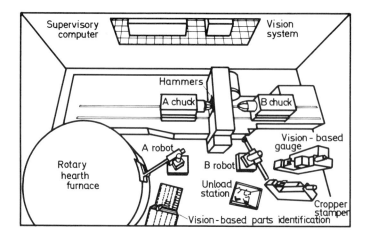

Fig. 24.8. Simplified architecture of an expert system.

for electric power generators, is one of the first advanced robot-equipped cells to use a limited expert system.[4] It will be able to adjust the production automatically to different blade batches and will require no human assistance, either for the operation or for the setup. A rule-based language will facilitate programming of the cell. Two vision-guided robots and several numerically controlled machine tools to produce turbine blades are the principal equipment of the manufacturing cell (Figure 24.9).

The cell will operate in the following manner. Workers deliver pallets of billets to a part identification station. It determines visually the location of the billets on the pallets. A robot then picks up the billets from the pallets and loads them into the rotary hearth furnace for heating. The same robot transfers the hot billets from the furnace to the swagging machine for shaping. The second robot removes the finished preforms and transfers them to a cropper for removing of excess material and for imprinting of an identification number. The robot then delivers the preforms to the optical gaging station. Finally, the robot drops the preforms into a basket for transfer to the presses.

This concept of a manufacturing cell is the first key to a master plan to transform the turbine component plant into a factory of the future. It will be manned by robots and managed by computers. The use of robots allows the cell to handle billets of various sizes under computer control. The vision system does away with the need to use special-purpose fixtures. The rotary furnace accepts billets of various sizes.

The complexity of an expert system for a robot has two reasons.[5] First, there is the complexity of details when the computer drives a robotic cell: the computer is an abstract machine, and the

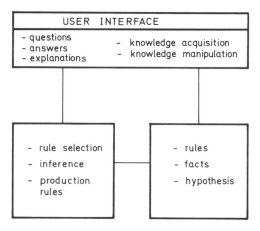

Fig. 24.9. Westinghouse's robotic cell.

robotic cells exist in the real world. Second, there is the complexity of uncertainty. Many manufacturing processes contain unknown stochastic elements (e.g., tolerances). This means that the representation of and the reasoning about assembly encounter significant difficulties. To date they cannot be solved satisfactorily.

Resolving the uncertainty about the real world must be done with the aid of sensors. However, so far, little concrete work in robotics has been performed to understand modeling and the stochastic nature of sensory data. It is very difficult to cope with the large volume of data acquired from a multiple-sensor system (e.g., 58 in the Westinghouse forging cell) and to integrate them into a consistent model to represent the actual state of the world.

24.4. LANGUAGES FOR VISION SYSTEMS

Specific problem areas require programming languages tailored to the particular application. For example, LISP and PROLOG are used in the field of artificial intelligence, ALGOL for scientific problems, and so on. It is a wrong assumption when the languages are separated from the hardware concept since languages are software machines and must be compatible with the hardware.[6] A language for a vision system or a multisensor system must take into consideration the following factors:

1. Image preprocessing (restoration, enhancement).
   (a) Instructions for visual sensors.
   (b) Processing of all pixels.
   (c) Parallel local operation.
   (d) Efficient storing of two- or three-dimensional structured data.
2. Image extraction (feature extraction, object recognition).
   (a) Arithmetic operations.
   (b) Statistical operations.
   (c) Instructions for the teach-in mode.
   (d) Instructions for feature definition.
   (e) Quantitative feature extraction.
3. Image analysis (description).
   (a) Definition of data structures (lists, graphs, etc.).
   (b) Definition of object relations ("above," "below," etc.).
   (c) Composition (decomposition) of relations.
   (d) Knowledge acquirement.
   (e) Model generation (geometric, symbolic, CAD, etc.).
4. Image context (scene analysis).
   (a) Analysis based on stored knowledge.
   (b) Inference rules.
   (c) Reasoning and queries.
   (d) Knowledge management.

The features of points 1 and 2 are partially available in professional image-processing languages. For example, the language RAIL[7] defines 45 vision features such as number of holes, color, diameter, area, and position. The object identification can be done directly by the programmer by defining which of the vision features should be used for the recognition. In addition, special commands for the calibration of the camera and the robot are available.

The features listed under points 3 and 4 have until now no practical implication. Some research laboratories are only beginning to construct symbolic, rule-based languages derived from the results of artificial intelligence research. With the aid of these features it will be possible to develop and to program expert systems for robotics cells.

24.5. COMPUTER ARCHITECTURE

Automatic assembly requires extensive sensor-signal processing to identify the environment of the robot and to control its interaction with its world. Assembly strategies for the robot must be solved in real time. Interpolation and coordinate transformation are considered to be standard primitives. To have a self-contained robot configuration and an interface to the AL-programming systems are objectives for an efficient and flexible robot operation. The higher the level of adaption (intelligence), the faster the data rate.

Many attempts have been made to operate assembly programs in real time. Early systems[8] were configured on a minicomputer (interpreter). They were aided by a more powerful computer for background compilation of programs. The basic robot functions could be controlled by a fast 16-bit processor coupled to a multiprocessor system, thereby allowing modularization of the control task.[9,10]

Several multiprocessor systems have been proposed that are supported by efficient system software.

In 1981 a multiprocessor standard[11] was defined for PC- and NC-programming, including NC-program controllers, multiple-axis controllers, geometric units, and I/O interfaces. To ensure the execution of a problem-oriented assembly program in real time, a symmetric multiprocessor architecture is of advantage. High-speed computation, fault tolerance, parallel task execution, and dynamic system configuration are important features. Parallel task processing on different control levels requires a distributed processor system. A multiple bus system with hardware arbitrators allows parallel data transfers. An arbitrarily segmented global memory, a distributed operating system (each processor contains system operation functions), and the capability to expand up to 20 and more processors makes the system suitable to run complex assembly programs for multirobot configurations. The features of the multiprocessor-based robot controller can be summarized as follows:

Different control levels.

The control hierarchy of the robot is distributed among microprocessors.

Distributed system operation.

Modular symmetric hardware configuration.

Multiple bus system.

Expandability with no need for hard- or software reconfiguration.

Special-purpose modules (sensor I/O, arithmetic, FFT, interpolators).

Parallel task decomposition for problem-oriented programs.

Handling of a world model in global memory.

24.5.1. General Structure of the Robot Real-Time Control System

The general structure of the robot control system is shown in Figure 24.7. The system is subdivided into the following components:

The mechanical manipulator (kinematic chain).

The axis control system.

The end effector control.

The peripheral control.

The sensor system.

The central robot control.

The programming system.

24.5.2. Mechanical Structure of Industrial Robots

The mechanical system consists of a kinematic chain with prismatic and rotary joints. Six joints are required to position the effector (end effector) at any point within the workspace and to allow it to assume any arbitrary orientation. To obtain a high positional accuracy of the effector (positioning repeatability less than 1 mm), each joint must be operated with a high bit resolution.

In the case of path control, the joints must follow defined trajectories without delay. Most of the present general-purpose robots have their joint motions under servo control. If high-speed operation with high path accuracy is desired, the servo control must be extended. Advanced systems have resolved acceleration control with nonlinear decoupling.[12,13]

24.5.3. Axis Control System

Most of the modern general-purpose robots have path control. The end effector of the robot must follow a defined trajectory. The trajectory is planned by the central robot controller under guidance of sensor information in real time. The following parameters will affect the control strategy:

Gravity force.

Centrifugal force.

Coriolis force.

Friction.

Reaction forces.

Numerous control algorithms have been developed to solve these control problems.[14,17] Their implementation is cumbersome because of the high number of arithmetic operations. For the design of control algorithms, therefore, it is necessary to reduce the number of additions and multiplications.

The joint control is performed in joint-space coordinates (robots coordinates). The joint-space

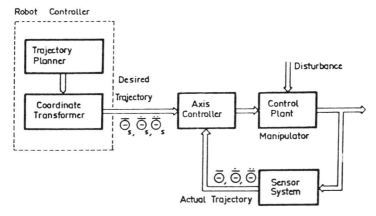

Fig. 24.10. Structure of the robot axis-control system.

trajectories are generated by postprocessors. They transform the robot-independent Cartesian trajectory into robot-dependent joint coordinates (see Figure 24.10). Efforts are made to extend resolved acceleration methods to control the end effector (tool) directly in Cartesian world coordinates.[18]

24.5.4. Control of the End Effector (Tool)

The direct interaction between the robot and the assembly environment is controlled by the end effector. From this point of view the mechanical system of the robot is understood as a positioning device for the end effector. Most of the assembly tasks are compliant operations with active and passive adaption. The end effector is characterized by the following:

The tool.
The piece part to handle.
Position and orientation of the object.
Reaction forces and torques.
The grasping parameters.
The mass of the object.
The geometry of the object.
The friction between finger and object.
The compliant forces.

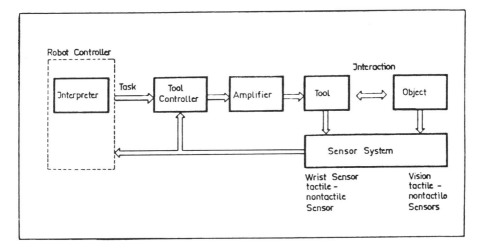

Fig. 24.11. Structure of the end-effector control.

Depending on the assembly operation, special end-effector control modules are in use. Their general structure is shown in Figure 24.11. The control tasks are decomposed by the program interpreter.

24.5.5. Sensor System and Processing of its Data

The sensor environment of the assembly robot is classified as follows:

1. Internal sensor.
   (a) Joint position.
   (b) Velocity.
   (c) Acceleration.
   (d) Force wrist sensor.
   (e) Gripper force.
2. External sensors.
   (a) Approximation sensor.
   (b) Touch sensor.
   (c) Geometric sensor (two-, three-dimensional).
   (d) Vision system.
   (e) Safety sensor.
   (f) Special-purpose sensor.

Internal sensors are part of a closed axis control loop and must sample data with high speed (hot sensors). External sensors work at lower speed and process data to update the robot model. The sensor system consists of the data-acquisition and the data-processing systems. The data-acquisition system samples the process data and generates pattern. The data-processing system identifies the pattern and generates frames for the dynamic world processor.

24.5.6. Central Robot Controller

The central robot controller is usually independent of the robot configuration and performs the following operations:

Interpretation of the program in real time.
Multitask scheduling.
Task decomposition.
Trajectory planning under sensor influence.
Coordinate transformation.
Effector control.
Control of peripherals.
Sensor control and data processing.

Controller function includes the distribution of tasks and primitives to the multiprocessor environment with producer and consumer relations. The processors are of different types, as, for example, the following:

Arithmetic (floating point).
Arithmetic (fixed point).
Sensor.
Logical.
I/O.

Intertask communication is performed between the individual processors. The most important functional tasks are discussed next.

24.5.7. Trajectory Planner

Trajectory planning is the functional task that determines the motion of the manipulator under sensor control. The parameters of the trajectory are defined by the user program. The following motions must be generated with the help of the basic frame equations:

From point to point.
On straight lines.

On defined curves with defined velocity.
To pursue moving objects (tracking).
For assembly.
To avoid collision.
Under sensor control.

Depending on the selected end-effector motion, different interpolation algorithms are used. Typical interpolation methods are (see Figure 24.12) the following:

Linear interpolation for straight lines.
Linear interpolation with transition between the straight-line segments.
Trajectory interpolation by algebraic functions (polynomes).
Interpolation between time-variant points using time functions or gradient strategies.

Cartesian trajectory generation is under development with a data rate of 100 Hz (see Figure 24.13).

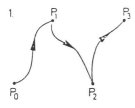

POINT TO POINT PLANNER

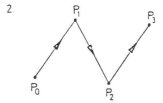

LINEAR INTERPOLATION

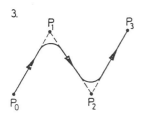

LINEAR INTERPOLATION
WITH SEGMENT TRANSITION
(CIRCULAR, QUADRATIC, CUBIC ETC.)

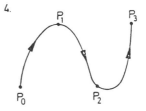

INTERPOLATION WITH POLYNOMES

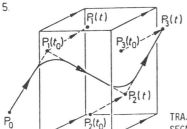

TRAJECTORY PLANNING BETWEEN TIMEVARIANT
SEGMENTS

Fig. 24.12. Different types of robot trajectory generation.

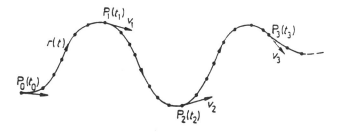

Fig. 24.13. An interpolated robot trajectory with equidistant sample data intervals ($\Delta = 10$ msec).

24.5.8. Coordinate Transformation

For each joint of the robot, the time history of the joint angles must be calculated (transformation of world into robot coordinates) to move effectors along the calculated Cartesian trajectories. The reverse operation is performed simultaneously to know where the robot is actually located in Cartesian coordinates (see Figure 24.14).

For sensor-controlled operations the following dynamic world data are used for trajectory calculation (see Figure 24.15):

Position and orientation of the workpiece.

Distances.

Diameters.

Geometry of the workpiece.

The path of the workpiece (position, velocity).

Contours.

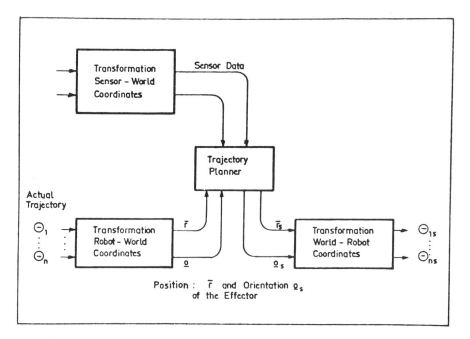

Fig. 24.14. Structure of the coordinate transformation data flow of the robot system.

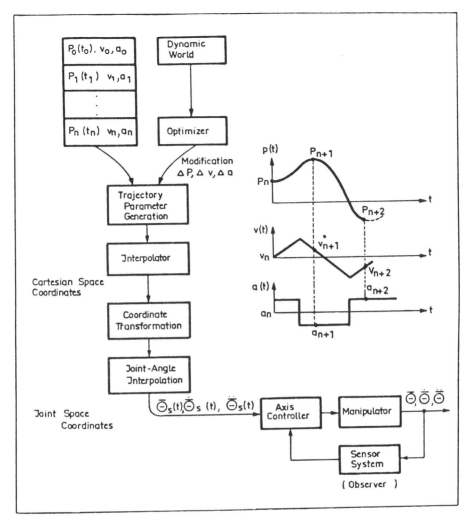

Fig. 24.15. Generation of moving data under sensory influence.

24.5.9. Dynamic World Processor

The central robot controller operates on a world model as reference. This model contains the frames describing the geometric and physical relations between piece parts of the assembly station. The frames of the world model are updated continuously to present the actual state of the robot world as follows:

Variable position and orientation of workpieces.

Moving objects (on a conveyor).

Variable collision space.

Variable objects (contour, weight).

Multirobot arms, conveyors, and manufacturing machines.

The actual world of the robot data is processed by the dynamic world processor. If significant events are detected, an optimizer generates a modification of the program with the aid of a decision task.

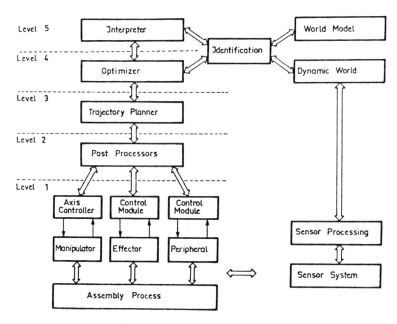

Fig. 24.16. Hierarchical control levels of the real-time robot control system.

24.5.10. Modular Structure of the Robot Controller

The outlined functional tasks are performed by a hierarchy of control levels.[10] From higher control levels data are distributed to the lower-level primitives. Beginning from the lowest level the hierarchy is defined as follows (see Figure 24.16).

Servo Control Level and Effector Control (Level 1)

At this level the drive signals for the actuators are generated to move the joints. Dedicated control algorithms control the individual links. This control level consists of primitives and is highly robot dependent. The input signals are joint trajectories in joint coordinates.

Coordinate Transformation Level (Level 2)

The transformation of a Cartesian trajectory into robot-specific joint coordinates is performed at this level. This trajectory depends on the hardware configuration of the robot.

Trajectory Interpolation Level (Level 3)

At this level the frame equations of the robot are calculated to generate a continuous trajectory in world coordinates. The inputs at this control level are parameters (e.g., trajectory points, speed) that define the conditions that the trajectory must fulfill. Depending on the application different interpolation routines (quadratic, cubic, circular, polynomials, special functions) can be applied. At this level, the entire kinematics of the robot are calculated in Cartesian frame coordinates.

Trajectory Control Level (Level 4)

The central controller is an interpreter that performs task scheduling and task decomposition. Real-time processing, synchronization, and multitasking is controlled at this level. References to the world model, to the dynamic world processor, to a program library with special software tasks (e.g., effector modules), and to the trajectory control level are made. Also, scheduling and dispatching of tasks for the multiprocessor environment are controlled at this level.

24.5.11. Requirements for the Computer Architecture

The development and implementation of the robot-control system is performed under the following conditions:

Clear separation of all control levels.

Modular architecture on the basis of a symmetrically distributed system.

Standardized interfaces between the control modules (hardware and software).

User transparency at all control levels.

Defined format and standardized protocols.

Multibus system to ensure parallel communication.

Decentralized system control.

Dynamic reconfiguration for a polyprocessor environment.

The hardware configuration of the system is shown in Figure 24.17. It contains microprocessors and a multibus system for communication and task synchronization. The communication system is provided with special lines for sensor data, frames, and integer and floating-point data. Typically, 16-bit computers were selected for most of the operation nodes because of their wide addressing space. A fast hardware logic (bus arbitrator) ensures conflict-free communication. The expandability of the system is possible without the need for a hardware and software reconfiguration. The system is driven by a high-level problem-oriented programming system.[19] Multiarm control and integration into a CAD system (expert system) is possible.

24.6. INTEGRATION OF SENSORS INTO THE CONTROL SYSTEM

The flexibility of an assembly robot is based on the application of an efficient problem-oriented robot-programming language, which allows the description of assembly sequences and which can process sensor data. An assembly, which in the majority of applications is a force-compliant operation, can be defined textually and characterized by the interaction between the effector (tool, gripper) and the object. After compilation of the source files, an interpreter activates in real time multiple tasks for the control, adaptation, and handling of sensor data (see Figure 24.7). For defined manipulation of the object, a vision system[20] and nontactile sensors are used. In the case of direct interaction, forces and torques between the effector and the object are of interest. When the object is being handled, the grasp force, the position of fingers, and the orientation of the object between the fingers must be recorded. In the following section we describe a microcomputer-controlled gripper system that allows the end-effector to approach and grasp an object, and allows the measurement of the wrist force-torque vectors with 6 DF. The gripper is developed for small robots, for example, the PUMA 600, and is used for assembly operation. Control tasks are decomposed by an AL interpreter. The integration of the sensor-gripper system into the robot-control hierarchy is discussed.

24.6.1. The Gripper

A mechanical structure of a gripper is shown in Figure 24.18. Two parallel fingers are driven by a DC motor by a gear train. The fingers are exchangeable and can grasp with a defined force, both in the opening and closing direction. The trajectory of the fingertips is curved (radial), which means that the grasping point of the gripper varies depending on the desired grasp opening (grasp angle). Thus it becomes necessary to calculate the position of the gripper as function of the dimension of the object to be grasped.

A potentiometer is used for the measurement of the opening distance of the fingers. The maximum opening distance is 8 cm, and its maximum effective grasp force is 30 N. The gripper is controlled by a microcomputer, which exchanges data with the central robot controller such as position information, finger speed, grasp force, sensor data, and status signals. The primitive operations of the gripper are as follows:

Movement of fingers with defined speed.

Movement of fingers with defined speed to a defined position.

To grasp with defined force and torque limits.

Movement of the fingers with defined speed to a defined position, exertion of a defined force.

To stop the fingers at any closing distance (e.g., to locate the gripper concentric with an object, using touch sensors).

Activation of the sensor system (tactile and nontactile).

Deactivation of the sensor system.

Transfer of sensor data to the robot controller (dynamic world processor).

The first five properties mentioned make a closed-loop control for the control of position, speed, and grasp force necessary. Force and position are controlled by software, whereas the speed control is realized by hardware. For the control of the object to be handled, both tactile and nontactile sensors are used (see Figure 24.19).

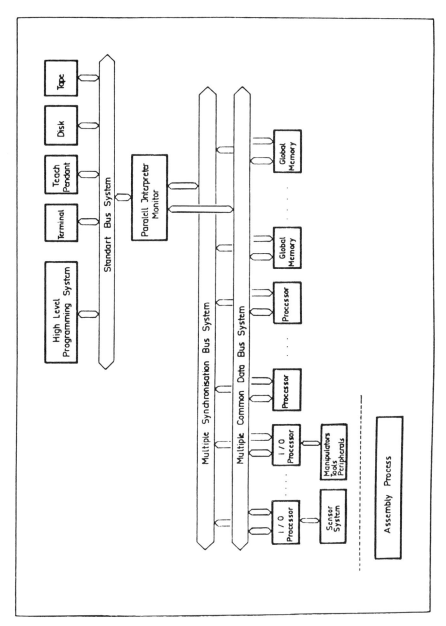

Fig. 24.17. Hardware configuration of the multiprocessor-based robot control.

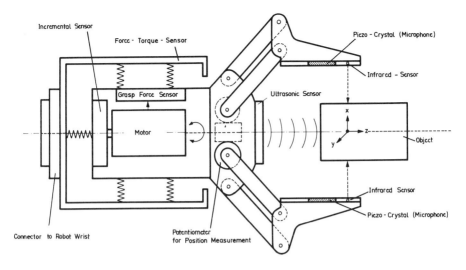

Fig. 24.18. Scheme of the sensor-gripper system.

24.6.2. The Sensor System

The sensor environment of the gripper is a combination of the following:

1. Tactile sensors working as:
   (a) A wrist to sense forces.
   (b) Grasp force sensor.
   (c) Touch sensors.
2. Nontactile sensors applied for:
   (a) Sensing of the approach of an object (ultrasonic).
   (b) Recognition of the object position between the fingers (infrared).

The structure of the sensor and handling of data are discussed in the following section.

Force-Sensing Wrist

In sensing of a complex force-torque vector, the interaction between the effector and the object is determined with the help of elastomechanical devices. Design of a force-sensing wrist requires a compromise between its elasticity to get good signals from strain gages and its structural stiffness to avoid undue deflection which reduces the positioning accuracy of the robot. The problem of the coupling effects originating from the preference directions must be considered. A typical wrist sensor senses all components of the force-torque vectors applied to the gripper (three orthogonal components of force and three components of torque). Forces and torques are measured indirectly by the deformation of eight elastic springs using eight pairs of strain gages. To resolve the measured values into orthogonal components of the local effector-coordinate system (frame), a multiplication with matrices is necessary.

The developed force-sensing wrist allows the registration of forces up to 50 N and torques up to 4.2 Nm with an error less than 5% (caused by coupling effects). The sample data rate is 100 per second (100 Hz). The structure of the hardware configuration is shown in Figure 24.20.

Grasp Force Sensor

Figure 24.21 shows a typical grasp-force-sensing element to define grasping of an object; a predetermined force is transferred by a gear train to the fingers. The motor, which is suspended by a spring mechanism, is moved along the z axis of the gripper by the reaction force. The spring deflection in the z direction is proportional to this grasp force and is measured by a potentiometer. The relation between the deflection z and the motor drive torque M is given by the following motion equation:

$$a z(t) + b z(t) + c \sin z(t) \cdot (d \, |z(t)| + r) = M$$

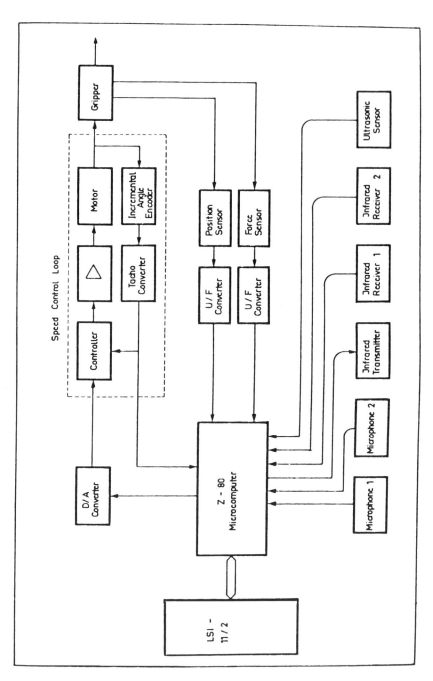

Fig. 24.19. Data flow of the sensor-gripper system.

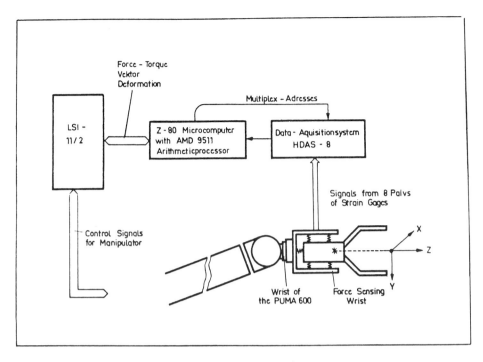

Fig. 24.20. Structure of the wrist torque-force-sensory system.

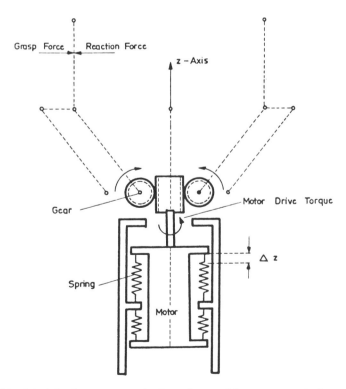

Fig. 24.21. Principle of motor suspension by springs used for measurement of grasp forces.

Here a, b, c, d are constant parameters, and r describes a constant friction. This equation is controlled by a control algorithm implemented on a microcomputer.

Touch Sensor

A piezo crystal working as microphone is used to sense touching of an object by the fingers and to detect possible sliding of the object after grasping. The crystals are isolated by a rubber substrate to filter out noise from the robot's environment and its own motor. The sensor generates suitable signals for many different materials and objects of different size. The sensor aids positioning the gripper centric to an object. Sliding of the object as well as a possible collision with another object can be detected.

Quantitative measurements of the touch vector (location of touch), the contact area, and the slide vector make necessary the use of several piezo crystals arranged in an array. Other useful materials for distributed touch sensing (artificial skin) are not considered here because of their high hysteresis and nonlinearities.

Ultrasonic Sensor

The vertical approach of an object can be controlled with the help of an ultrasonic sensor. This device works as a transmitter and a receiver. It initiates a measurement by transmitting multiple pulse trains at ultrasonic frequencies.

Objects having different reflection characteristics can be detected. A fast counter can be used to measure the time interval between the emitting and receiving of a signal. The interval, in turn, is used to calculate the object's distance. A good sensor has a resolution of 1 mm. Its minimum range depends on its design. This limitation is due to the long transmitting time for a pulse train and the time interval needed for switching the sensor from transmitting mode to receiving mode. By using two transducers (single transmitter, single receiver) and only one object-specific pulse-train frequency, the resolution and the minimum distance can be optimized. The angle of the ultrasonic column is confined by a foam rubber cone ($< 1.5°$). Thus scanning of surfaces (e.g., for batch operation) is possible.

Recognition of Objects between Fingers

To recognize an object between the fingers, two infrared transmitters that generate modulated signals can be used, each of which has two receivers (see Figure 24.22). When a light beam is interrupted, an object is located between the fingers. When both receivers of a transmitter are sensing the reflected light signals, the two surfaces of an object (only objects with clear reflection characteristics) are located parallel to the finger surfaces. If only one receiver gets a signal, the angle can be determined about which the gripper must be rotated to be parallel with the object's surface. The closer the object is to the surface of the finger the higher is the sensitivity of the receiver. In case the object is very close, no signal is detected; this means that the sensor can also be used as nontactile proximity sensor. The sensitivity and the approach distance can be modified by changing the geometry of the sensor and the angle of the transmitted light beam. By using more than two receivers in connection with one transmitter better quantitative data are obtained. The receivers are independent of the intensity of the reflected signal.

Control of the Gripper System

The task to control the gripper system and its sensor environment can be decomposed by the AL interpreter which operates on the central robot controller. Thereby control primitives are activated on a lower operational level on a microprocessor.

The control primitives are stored in PROMS (see Table 24.1). After bootstrapping the task routines the microprocessors are in a polling state waiting for input or output data. If a grasping operation is to be performed, the control primitives will be activated. Viewed from the central robot controller, the peripheral processors are working as I/O modules. They require no additional communication software. The processors are very small (minimal configuration) and contain a local monitor which allows program modification by the user by a terminal interface.

The sensor environment of the presented gripper system consists of simple basic sensors that allow the gripper to approach an object and to detect it between the fingers. The force-sensing wrist aids the manipulation of workpieces (e.g., assembly sequences). For special applications and in particular for quantitative measurements, several tactile and optical sensors may have to be used. They should be arranged in an arrays pattern (artificial skin). By using the peripheral processors, no critical real-time delay problems will rise.

The described modular gripper and sensor system can be integrated into the hierarchical control

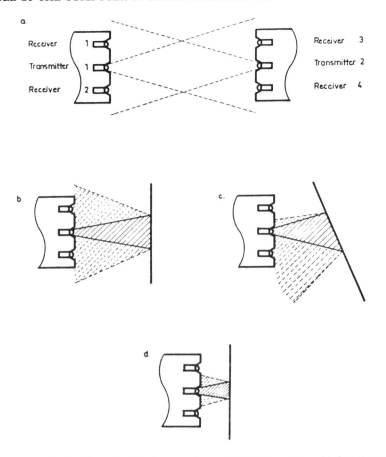

Fig. 24.22. Principle of object detection between fingers. (*a*) Configuration of infrared transmitter and receiver. (*b, c*) Reflection of signals by parallel and nonparallel object surface. (*d*) Sensor configuration for object approach.

system of a robot. For the addition of a sensor or a change of a gripper, the software primitives can be reconfigured.

24.6.3. Laser Scanner for Distance Measurements

The three different measurement techniques used to obtain distances for industrial robots (from 10 cm to 10 m) are based on the phase shift, the time of light, and the triangulation principle. All of these use laser scanners. However, they have severe restrictions: the large dynamic range of the reflected beam (up to 100 dB), the low signal-to-noise ratio (e.g., due to photon noise), and sometimes the slow pixel scanning rate (e.g., 500 msec/pixel).

Laser scanners, which are based on phase-shift measurements (indirect time of light measurement), are described in Reference 21. Here we mention only that the maximum pixel rate that can be obtained by this kind of scanner is about 25 msec/pixel. Modern laser scanners are controlled by computers and are therefore not used as off-line devices, as was done, for example, in Reference 21.

The most ambitious approach to measuring distances is the direct measurement of the time of light. Distance and intensity data can be obtained in about 1 μsec. However, for a distance of 1 m the light time of light is 3.3 nsec. This means that a device must be used whose time resolution is 16.5 psec to get a distance resolution of 5 mm. The only technique that can meet these requirements is coincidence measurement (time spectroscopy), which is used in nuclear science and in high-energy physics. Time-to-amplitude converters (TACs) are used to measure the time relationship between a start and stop signal.[22] The timing resolution of these instruments is defined by the full width half

TABLE 24.1. A LIST OF PRIMITIVES FOR SENSOR-GRIPPER CONTROL

Primitives	Function
POS	Activates position and velocity control algorithme
FORCE	Activates force control algorithme
VEL	Transfer of parameters to velocity control
SENS ON I	Activates microphone and infrared diodes
SENS ON II	Activates force sensing wrist
SENS ON III	Activates ultrasonic sensor
SENS OUT I	Interrupts microphone and infrared diodes
SENS OUT II	Interrupts force sensing wrist
SENS OUT III	Interrupts ultrasonic sensor
FORCEPOS	Activates SENSON I and POS
STOP	Stops finger in shortest time
INF I	Transfers the gripper's state from the sensor data table to the robot controller
INF II	Transfers the force-torque vector to the robot controller
INF III	Transfers data from ultrasonic sensor to the robot controller
POSJU	Adjusts position sensor
FORCEJU	Adjusts force sensor
WRISTJU	Adjusts force sensing wrist

maximum (FWHM) of the timing spectrum they can discriminate. To date, it is possible to obtain with a TAC a timing resolution of about 10 psec. Figure 24.23 indicates a block diagram of a time-of-light laser scanner.

A laser diode (e.g., pulse laser diode, 50 mW) shines light on an object. The laser diode pulser is used to generate the starting signal. A time-pick-off element produces a logic pulse that is to a great extent independent of the shape and amplitude variations of the input pulse. The output signal of this trigger initiates the TAC. The reflected beam impinges on a silicon surface barrier detector (SBD) that emits time jitters as they exist in conventional photomultiplier tubes. The output signal of this detector is preamplified and divided into two signals. A fast filter amplifier further intensifies and forms the shape of the "intensity" signal. This new shape (flat roof) is more suited for an analog-digital conversion (8 bits) than the original shape form (peak).

The "distance" signal is discriminated (constant-fraction discriminator) and delayed, which defines the time signal to stop the TAC. The output of the TAC is precisely (time resolution $\leq$ 10 psec) correlated to the time interval, defined by the start and stop signal. The TAC output is converted to a digital value (12 bits).

The most difficult problem with this kind of laser scanner is the highly experimental dynamic range of about 10^5–1. The timing resolution is very sensitive to the intensity (energy) of the reflected beam. The more energy is deposited in the reflected beam the better is the resolution. Furthermore, in nuclear experiments the dynamic range is limited to about 100–1. This means that the main building

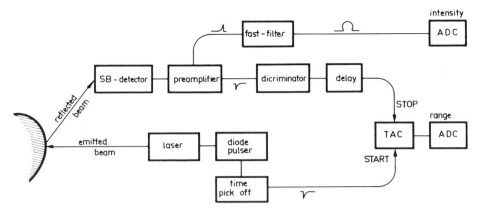

Fig. 24.23. Block diagram of a time-of-light laser scanner.

blocks of a time-of-light measuring system, the preamplifier, the constant-fraction discriminator, and the TAC, must operate efficiently if the reflected intensity is at its maximum and if the dynamic range approaches zero. However, these two assumptions cannot be taken for granted in a harsh industrial environment. The first results obtained in an experimental setup demonstrate that realistic timing resolutions for direct time-of-flight laser scanners are about 5 nsec (distance resolution of 1.5 m). Therefore, only very dedicated applications (e.g., constant specular reflection and constant temperature) are candidates for distance measurements by this approach.

Figure 24.24 depicts the principle of operation of a triangulation laser scanner. One camera of the stereoscopic approach is substitued by a laser source. The control unit of the scanner can also be a multiprocessor system. For example, in Reference 23, 64 parallel operating module processors control the laser and process the data. The range data can be calculated with the help of the horizontal deflection angle δ or the vertical deflection angle γ of the emitted beam, the basis line length d, the focal length f_0 of the camera lens, and the picture plane coordinates x or y, which correspond to the angles α and β of the reflected beam.

By the use of parallel projection (calculation of the distance h between the baseline and the point

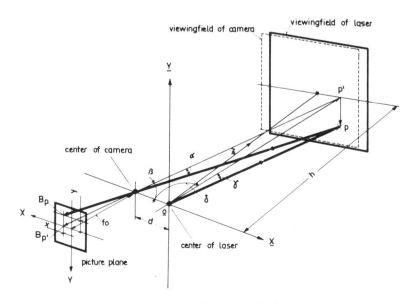

Fig. 24.24. Geometry of a triangulation laser scanner.

p') only two of the four deflection angles are needed (e.g., β and δ). This distance h can be calculated as follows:

$$h = \frac{d}{\operatorname{ctn} \delta + x/f_0}$$

The distance resolution

$$\Delta h[\Delta h(x, \delta) = h(x + x_{\min}, \delta + \delta_{\min}) - h(x, \delta)]$$

can therefore be defined by a linear approximation as

$$h = \left(\frac{h^2}{d}\right) x \left(-\frac{x_{\min}}{f_0} + \frac{f_0 \delta_{\min}}{\sin^2 \delta}\right)$$

Here, $x_{\min}$ gives the minimal detectable x distance in the image plane, and $\delta_{\min}$ defines the minimal horizontal deflection angle that can be measured. The resolution is optimal when the distance is small h^2, the baseline is long d, δ is equal to 90° (normal to object surface), and when $x_{\min}$ and $\delta_{\min}$ are as small as possible.

To cite the results of an experiment, it can be stated that such a system renders a very good resolution of 0.25 mm at a distance of 1 m. The time for the distance evaluation for every pixel can be decreased to about 0.5 μsec (e.g., CCD-linear array, hardware multiplication).

Often this type of laser scanner is connected with the structured-light approach (e.g., line projection, grid projection). However, this approach operates with parts that are presented and preoriented. This renders a unique assignment between every projected object point and the corresponding image point. Thus the distance can be calculated. If there is no such unique assignment, this approach can only be used to classify different object classes. In this case absolute distance measurements are not possible.

REFERENCES

1. Bonner, S. and Kang, G., A Comparative Study of Robot Languages, *Computer,* December 1982.

2. Mujtaba, S. and Goldmann, G., *AL User's Manual,* Stanford University, Stanford, California, 1979.

3. Brooks, R. A., Symbolic Error Analysis and Robot Planning, *International Journal of Robotics Research,* Vol. 1, No. 4, 1982, pp. 29–67.

4. Kinnucan, P., Flexible System Forges Turbine Blades, *High Technology,* June 1983, pp. 20–22.

5. Kempf, K. G., Artificial Intelligence in Robotics, Tutorial on Artificial Intelligence, IJCAI, Karlsruhe, 1983.

6. Duff, M. and Levialdi, S., Eds., *Languages and Architectures for Image Processing,* Academic Press, 1981.

7. *RAIL Reference Manual AUTOVISION,* Automatix Inc., Burlington, Massachusetts, 1981.

8. Binford, T., et. al., Exploratory Study of Computer Integrated Assembly Systems, Progress Report 4, Stanford University, Stanford, California, 1977.

9. Stute, G. and Wörn, H., Mehrprozessorsteuersysteme für Industrieroboter, Proceedings of the 8th ISIR, Stuttgart, 1978.

10. Albus, J., Barbera, A., and Fitzgerald, M., Hierarchical Control for Sensory Interactive Control, Proceedings of the 11th ISIR, Tokyo, 1981.

11. DIN 66264: Mehrprozessor-Steuersystem für Arbeitsmaschinen (MPST), Beuth-Verlag, Berlin, 1981.

12. Luh, J. Y. S., Walker, M. W., and Paul, R. P. C., Resolved-Acceleration Control of Mechanical Manipulators, *IEEE Transactions on Automatic Control,* No. 3, June 1980, pp. 468–474.

13. Freund, E. and Hoyer, H., Das Prinzip nichtlinearer Systemkoppelung mit der Anwendung auf Industrie-roboter, Regelungstechnik, No. 28, Jahrgang, Heft 3, 1980, (in German).

14. Paul, R. C., Modelling Trajectory Calculation and Servoing of a Computer Controlled Arm, Stanford Artificial Intelligence Laboratory, Stanford University, California, A. I. Memo 177, September 1972.

15. Pieper, D. L., The Kinematics of Manipulators under Computer Control, Ph.D. dissertation, Stanford University, 1968.

16. Horn, B. K. P. and Raibert, M. H., Configuration Space Control, Technical Report AI-M-458, MIT, Artificial Intelligence Laboratory, Cambridge, Massachusetts, 1978.

17. Paul, R., et. al., Advanced Industrial Robot Control Systems, Second Report, School of Electrical Engineering, Purdue University, West Lafayette, Indiana, July 1979.

18. Chi-Haur Wu, Paul, R., Resolved Motion Force Control of Robot Manipulator, *IEEE Transactions on Systems, Man, and Cybernetics,* No. 3, June 1982.

19. Blume, C., A Structured Way of Implementing the High Level Programming Language AL on a Mini- and Microcomputer Configuration, Proceedings of the 11th ISIR, Tokyo, 1981, pp. 663–674.

20. Rembold, U., et. al., A Very Fast Vision System for Recognizing Parts and Their Location and Orientation, Proceedings of the 9th ISIR, Washington, 1979, pp. 265–280.

21. Nitzan, D. et. al., The Measurements and Use of Registered Reflectance and Range Data in Scene Analysis, *Proceedings of IEEE,* Vol. 65, No. 2, 1977, pp. 206–220.

22. Knoll, G. F., *Radiation Detection and Measurement,* Wiley, New York, 1979.

23. Levi, P., et. al., ROVIKA: A Multiprocessor-Based Robot Vision System at the University of Karlsruhe, Proceedings of the 3rd Symposium on Microcomputer and Microprocessor Applications, Budapest, 1983, pp. 588–597.

PART 5

SOCIAL AND ECONOMIC ASPECTS

CHAPTER 25
SOCIOECONOMIC IMPACTS OF INDUSTRIAL ROBOTS: AN OVERVIEW

ROBERT U. AYRES

STEVEN M. MILLER

Carnegie-Mellon University
Pittsburgh, Pennsylvania

25.1. RECENT HISTORY AND TRENDS IN THE USE OF ROBOTS

In 1959 the first industrial robot was produced and sold in the United States. A decade later there were only 200 robots in use in the United States, and only a small number used elsewhere in the world. As of January 1983 there were *at least* 32,800 robots in use in more than 19 countries worldwide. Japan leads the world with more than 18,000 units in use, and the United States is the second largest user, with more than 6800 units. Other major users include the USSR, West Germany, Great Britain, Sweden, and France. An estimate of the international robot population, compiled by the Robot Institute of America (RIA), is shown in Table 25.1.

Historical estimates of the growth of the robot populations in the United States and in Japan are shown in Figure 25.1.* In the United States robots came into use very slowly during the 1960s, and their use did not take off until the latter part of the 1970s. The first steps to import an industrial robot into Japan were taken in 1966 when AMF Japan (a subsidiary of AMF, U.S.) showed a film of a Versatran robot to executives of the Tokyo Machine Trading Company.† In December 1967 the first industrial robot to be installed in a Japanese plant for commercial use went into operation at Toyoda Automatic Loom Company, parent company to the Toyota Automobile Group. In 1968 another major step in the use of industrial robots in Japan was taken with the signing of a technology cooperation agreement between Unimation and Kawasaki Aircraft. Fijitsu Fanuc, another important current Japanese manufacturer of industrial robots, got its start in the robot industry by purchasing a Kawasaki-Unimate robot. Since that time, robots have diffused much more quickly in Japan then in the United States.

This chapter reviews the diffusion of robots in the United States between 1976 and 1981 based on a survey and interviews of 54 U.S. firms belonging to the Robot Institute of America which we conducted during the spring of 1981. More than one-third of the U.S. robot population at the end of 1980 was accounted for in the sample. Our data show an expansion in the applicability of robots from large establishments to smaller ones and from mass production to batch and custom production during the five-year period from 1976 to 1981. More than 90% of the users who acquired robots prior to January 1976 were large firms with more than 1000 production workers in the establishment where robots were used, and none had fewer than 500 production workers. Most were engaged in mass production, including the four major U.S. auto makers. Among the firms who acquired robots after 1976, nearly 60% had more than 1000 production workers, and five establishments had fewer than 500. While more mass producers began to use robots, even more of the new entrants were engaged in custom and batch production. Most of the post-1976 entrants engaged in custom production were

---

* Sources for the estimates of the U.S. population between 1970 and 1981 are given in Ayres and Miller.[1] Different sources do not concur on the figures, as indicated. The U.S. estimate for December 1982 comes from Hunt and Hunt.[2] The estimates of the Japanese population come from Paul Aron.[3]

† See Ayres, Miller, and Lynn[4] for a history of the introduction of robots into Japan.

The background research for this paper was supported in part by the Carnegie-Mellon Program on the Social Impacts of Information and Robotics Technologies and by the C-MU Robotics Institute.

TABLE 25.1. INTERNATIONAL ROBOT POPULATION—FEBRUARY 1982

Country	Type A[a]	Type B[b]	Type C[c]	Type D[d,e]	Total
Japan	—	6,899	—	7,347	14,246[f]
US	400	2,000	1,700	600	4,700[g]
USSR	—	—	—	—	3,000
West Germany	290	830	200	100	1,420
Great Britain	356	223	54	80	713
Sweden	250	150	250	50	700
France	120	500			620
Italy	—	—	—	—	353
Czechoslovakia	150	50	100	30	330
Poland	60	115	15	50	240
Norway	20	50	120	20	210
Denmark	11	25	30	0	166
Finland	35	16	43	22	116
Australia			62		62
Netherlands	48	3	5	0	56
Switzerland	10	40	—	—	50
Belgium	22	20	0	0	42
Yugoslavia	2	3	5	0	10
Total	1,774	10,924	2,584	8,299	26,924[h]

Source: Estimates for February 1982: Robot Institute of America (1982).

[a] Type A: Programmable, servo controlled, continuous path;

[b] Type B: Programmable, servo controlled, point-to-point;

[c] Type C: Programmable, nonservo robots for general-purpose use;

[d] Type D: Programmable, nonservo robots for diecasting and molding machines;

[e] Type E: Mechanical Transfer Devices (pick and place)—not shown.

[f] The estimated total for Japan as of December 1982 is 18,000, according to Paul Aron of Daiwa Securities.

[g] The estimated total for the U.S. as of December 1982 is 6800, according to Hunt and Hunt (1982).

[h] World total, as of February 1982, excluding Japan and the United States = 7978. Lower bound on world total as of December 1982 = 18,000 + 6800 + 7978 = 32,778.

aerospace firms who had acquired robots for experimental purposes under a Defense Department program to modernize technology. Of the prospective users considering robot use in January 1981, fewer than 50% had more than 1000 production workers, and only one-fifth were engaged in mass production. The recent increase in available robot models, in conjunction with improvements in capabilities and reductions in cost per unit of capability, will undoubtedly result in a continued expansion of the use of robots by smaller-sized establishments and by more custom and small-batch producers.

The percentage of the robots in our sample accounted for by auto makers declined between 1976 and 1981, indicating growth in other areas of applications. Outside the auto industry, most of the increase in robots was concentrated in two firms, both diversified manufacturers of capital goods. Experienced users accounted for more than 90% of the robots in the sample added between 1976 and 1981, and almost 90% of the increase within all experienced users was accounted for by five firms. In January 1981 six firms accounted for more than 30% of the total number of robots used in the United States. Within the United States, it is clear that the use of robots has spread rapidly within a small number of firms. On the other hand, a growing number of firms have purchased one or a few robots to evaluate them in a new environment. More recent trends suggest that robot use in the United States *will not* remain concentrated in a relatively small number of firms.

A comprehensive overview of patterns of robot use by industry and application in Japan has been published by the Japan Industrial Robot Association.[5] The survey results in the report show that the industries using the most robots are precision machinery, metalworking equipment, electrical equipment, and transportation equipment. The most common applications are the loading and unloading of machine tools, the loading and unloading of assembly machines, and spot welding. In contrast to usage patterns in the United States, robot use is spread across a large number of firms, and a large fraction of the robots are used in small and medium-sized establishments.

25.2. THE METALWORKING SECTOR

Most of the present membership of the RIA and about 90% of current robot users in the United States fall within a group of manufacturing industries referred to as the *metalworking sector*. As the

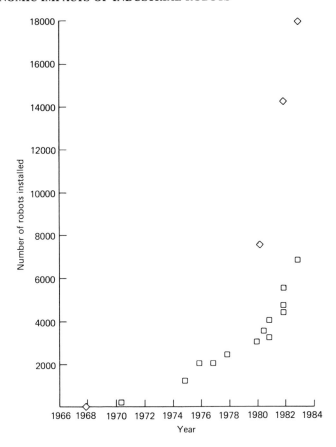

Fig. 25.1. Estimates of robot populations in Japan and the United States. Estimates of U.S. robot population, □; estimates of Japanese robot population, ◇.

name implies, these industries are engaged in the fabrication, finishing, or assembly of products from standard metal shapes, and from mechanical and electronic parts and subassemblies purchased mostly from other metalworking industries. The metalworking sector includes the following groups of industries, designated by standard industrial classification (SIC) codes:*

SIC Code	Major Group Name
34	Fabricated metal products
35	Machinery, except electrical machinery
36	Electrical equipment and machinery
37	Transportation equipment

These groups of industries include more than 85% of the units of metalworking machinery counted in the United States in the Twelfth American Machinist Inventory of Metalworking Equipment. All industries in primary metals (SIC 33) and selected industries in furniture (SIC 25), precision instruments (SIC 38), and miscellaneous manufacturing (SIC 39) are also included in the *American Machinist* magazine definition of the metalworking sector.

Statistics from the Japan Industrial Robot Association confirm that between 80 and 85% of all robots sold during 1978–1980 went to their metalworking sector.[6] The following five companies using the most robots in Japan are all part of the metalworking sector.[7]

---

* This is the system of industrial classification that has been developed over a period of many years under the guidance of the U.S. Department of Commerce. All data collected by the Bureau of Census as part of the economic census, including the Census of Manufacturers, is organized according to the SIC system.

Company	Number of Programmable Manipulators Installed (as of December 1982)
Sanyo Electric Machinery	1063
Sharp Electric Machinery	897
Toyota Motor Company	780
Nissan Motor Company	730
Mitsubishi Electric Machinery	427

The metalworking sector has been described by Vietorisz[8] as "the bell-wether of economic develop-ment" for an industrial society because all of the tools and capital equipment used by all manufacturing industries (including itself), and by all other sectors of the economy are produced within it. It is the place within the industrial system where new knowledge is embodied into a physical form, enabling it to be utilized throughout the entire economic system. Since all new products and processes require the capital goods purchased by these industries, it is not farfetched to claim that much of the knowledge that becomes part of the economic system enters through the metalworking sector. Since capital goods play such a critical role in the creation of new products and processes and in the creation of new wealth, one can argue that the importance of this sector goes beyond the number of people directly employed within it.

There are a limited number of robot applications in manufacturing sectors other than metalworking, though, at present, problems associated with the processing of nonrigid or delicate materials and with very high speed production lines restrict their use. Current and near-term future applications of robotics in the processing of leather, rubber, asbestos, plastics, and food, and in the manufacturing of glass, clothing, and wood products are briefly reviewed in Schraft, Shults, and Nicolaisen.[9] In both Japan and in the United States demand for robots in the nonmanufacturing sectors of the economy currently accounts for a negligible part of the total market. Projections of robot use in nonmanufacturing industries in Japan over the next several decades are given in Reference 5.

25.3. PRODUCTIVITY IMPACTS IN THE METALWORKING SECTOR

We now turn more specifically to the metalworking sector and explain some of its important technological features. We discuss the comparative economics of small-, medium-, and large-volume production and point out why and where there are substantial potentials for productivity improvement within each of these "domains." We estimate how much cost saving might be realized if robotics and computer-aided manufacturing (CAM) technologies could make small- and medium-volume production closer in efficiency to mass production operations. Subsequently, we analyze the magnitude of cost saving that might be realized by substituting robots for workers, and the potential for increasing output by increasing the utilization of machine tools. Our calculations suggest that in low- and mid-volume manufacturing, savings theoretically obtainable from increasing the utilization of capital equipment are roughly an order of magnitude larger than savings that could be realized by eliminating a fraction of production labor cost.

We identify several consequences of achieving dramatic increases in output per unit of capital and labor (or equivalently, of reducing unit cost) in robotic plants. For one thing, fewer plants will be needed to meet current levels of demand. If demand remains at current levels, additional facilities will not be needed, and older, less productive facilities will most likely be shut down. Lower unit costs should result in some increased demand, but we later show that price-induced market growth is unlikely to keep up with increases in capacity. To reap the benefits of robotics fully, managers will have to find new ways of utilizing the added capacity by making greater use of the flexibility of robotic production systems to manufacture new and rapidly evolving products. Thus we suggest that the full benefits of robotics will be realized only if there is a parallel emphasis on product innovation. Finally, we discuss briefly some of the broader economic impacts of improving productivity in the metalworking sector. It appears that the application of robotics to batch manufacturing could result in significantly reduced real cost for capital goods in relation to other factors of production. This could have ripple effects on the prices of manufactured goods throughout the economy, and beneficial long-term effects on the rate of price inflation and on the competitiveness of the manufacturing industry in the United States and other advanced countries.

25.3.1. Low-, Mid-, and High-Volume Production

The appropriate choice of capital and organization within particular establishments is strongly influenced by the average batch size and the length of the average production run in the factory. The more diverse the mix of parts (or products) being produced, the smaller the batch size for a given part. There has always been a conflict between retaining a capability for rapid redesign or reconfiguration of the product and achieving high levels of production and low unit cost. This is sometimes referred to as the *flexibility versus efficiency* trade-off.

The current trade-offs between having the flexibility to produce a range of different products and being efficient enough to produce a large number of products are shown schematically in Figure 25.2. A conventional *job shop* produces a wide range of customized products with low levels of output for each different product. General-purpose machines and highly skilled labor are required. Since each job is unique and very few operations are repeated in quite the same way, it is not possible to automate many of the tasks associated with setting up or operating the machines. For almost 200 years, job shops engaged in custom production have been considered flexible in the sense that manually operated metalworking machines, such as lathes, have been used to make a wide variety of different products on demand. But the unit costs in these labor-paced shops have necessarily been considerably higher than the unit cost of products mass produced with specialized production equipment like that employed in the auto industry.

Batch production covers the wide middle ground between the extremes of custom producing a variety of "one-of-a-kind" products and mass producing a small number of standardized products. The same part may be reproduced in volumes of tens, hundreds, or even thousands. There is still a regular need for general-purpose (i.e., flexible) machines to process batches of different types of products. In any batch-production facility, equipment must be readily adaptable to make a new part within the "family" of parts being manufactured. The ease with which such changes can be made directly governs batch throughput time and costs. At present, it is expensive to change over from one batch to another. This cost is the irreducible penalty of maintaining a desired degree of flexibility with respect to product change.

As the batch size increases and the ratio of processing time to setup time grows larger, it becomes economically feasible to make greater use of automated machines to speed up repetitive processing, to use more specialized tooling to speed up the loading and unloading, and to automate the transfer

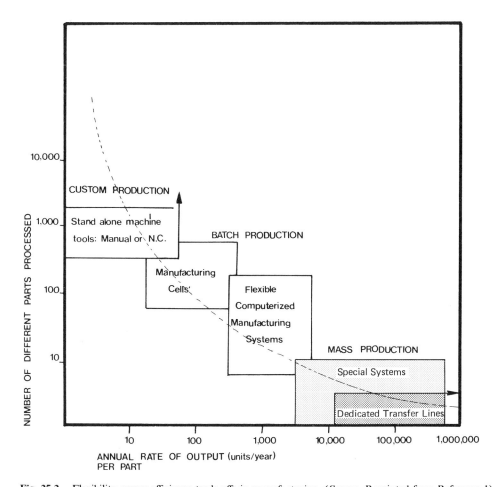

Fig. 25.2. Flexibility versus efficiency trade-offs in manufacturing. (*Source.* Reprinted from Reference 1).

of material between machines. In mass production, highly standardized parts are produced with equipment that is optimized by design to specific operations at fixed rates and dedicated to the particular product. The plant achieves high efficiency by sacrificing flexibility. With conventional types of technologies—dedicated automation for mass production and stand-alone, general-purpose machine tools for batch production—efficiency is achieved at the expense of flexibility, and vice versa, making it impossible to maximize both simultaneously.

25.3.2. Cost Versus Batch Size

Examples of cost curves for technologies typically used for custom, batch, and mass production are shown in Figure 25.3. The curve labeled *piece production* represents a labor-intensive technology typical of custom (or piece) production. Fixed capital requirements are lowest and the unit labor requirement is the highest of the three technologies. The annualized capital charge is assumed to be negligible in proportion to the annual labor cost, so the average cost curve is nearly constant over most of the volume range. The curve labeled *mass production* represents a capital-intensive, highly automated system with relatively small unit labor requirements, typical of that used to mass produce a standard product design. The annualized capital equipment charge is the largest among the three technologies. Average cost is very high at small volumes because of the large fixed investment requirements, but decreases sharply (by a factor of $1/$output) as the volume of output increases. The curve labeled *batch production* represents a semi-automated production technique typical of that used when products are manufactured in mid-sized batches. In comparison to the custom- and mass-production technologies, it represents an intermediate case where average cost is not completely dominated by either variable cost, labor or fixed capital. There is a cost-minimizing choice of technology in the low-volume, mid-volume, and high-volume range, as indicated by the bold line in Figure 25.3. Each of the three technologies is the cost-minimizing choice only within the volume range for which it is intended, and is an "inefficient" choice outside of its appropriate range. For a given product, unit cost would decrease in a regular fashion as output increases over a wide range of volumes, if one considers the lower envelope of the "long-run" cost curve, where the optimal (i.e., cost-minimizing) technology is used for each level of output.

The current relationship between unit cost and the level of output in discrete parts production can be seen roughly by comparing most of the metalworking industries in terms of unit-processing

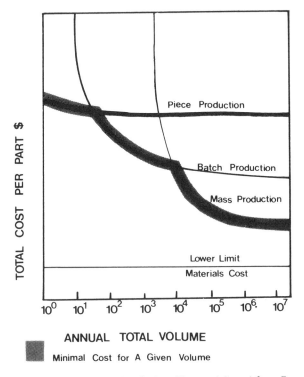

Fig. 25.3. Average cost versus batch size. (*Source.* Adapted from Borzcik.[10])

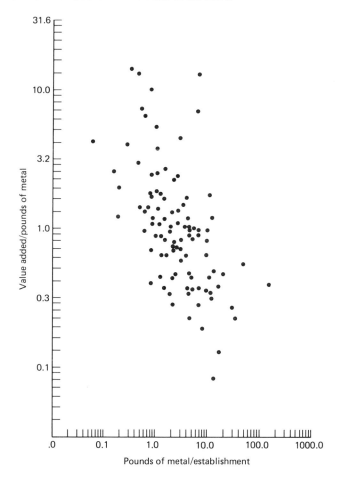

Fig. 25.4. Value added per pound of metal, versus pounds of metal per establishment for metalworking industries, 1977.

cost to the pounds of output produced as shown in Figure 25.4. Unit-processing cost is defined here as value added (total labor cost plus capital cost and profits) per pound of output produced. An estimate of the quantity (weight) of basic metals and of processed metal inputs purchased from other metalworking industries is used as a surrogate measure of the physical quantity of output produced by each industry.* Cost per unit (weight) versus quantity (weight) are computed for 101 industries in SIC 34–37.† The metalworking industries in Figure 25.4 are similar in that the costs of basic metals and of processed inputs purchased from other metalworking industries account for the majority of total material purchases. However, each industry's products vary in complexity—tolerance, geometry, material properties, size, and so on—and differences in the nature of the processing requirements associated with differences in product complexity affect the cost required to process a given amount of output. This is a second major reason for why average cost per unit varies across industries. In

---

* The term *basic metals* refers to inputs of "raw" metal stock—steel, brass, and aluminum in the form of bars, billets, sheets, strips, plates, pipe, tubes, and so on, as well as casting and forgings made of the three basic metals. The term *processed metals* refers to inputs which are themselves the products of the industries in major groups SIC 34–38. In general, these products are basic metals that have been further processed within the metalworking industry. Pounds of metal processed is divided by the number of establishments within the industry to adjust for differences in the number of establishments across industries.

† There are an additional 31 industries in SIC 34–47 which are excluded because of inadequate data on their material inputs.

addition, there are interindustry differences in hourly wage rates and in the mix of activities carried out (fabrication versus assembly). Despite these differences, which account for most of variation in the scatter plot, it is still clear that unit-processing cost (as defined) decreases as the level of output increases.

Unit cost measures for some of these industries are shown in Table 25.2. Within this group, the highest measure of unit-processing cost (14.13) is for SIC 3662, radio and television communication equipment, an extreme case of custom or very small batch production. The lowest measure of unit processing cost is for SIC 3465, auto stampings, an extreme case of mass production. The range between these two extremes is a factor of 60. That is, SIC 3662 spends more than 60 times as much in capital and labor cost for pound of material processed as SIC 3465 does. Much of the unit cost differential between these two industries is due to differences in product complexity. However, if the unit-processing cost of auto stampings is compared to that of a more similar type of product that is custom produced, such as nonferrous forgings, there is still a tenfold difference.

Using the data shown in Figure 25.4, we have estimated the elasticity of unit cost with respect to the level of output by means of regression analysis. Unit-processing cost decreases by 0.436% for each 1% increase in output. Two principal effects drive this decrease. As the level of output increases:

More automated types of machinery with higher throughput rates are used, and machine utilization rates tend to increase.

There tends to be increasing emphasis on using lower-cost materials and on designing for ease of manufacturability.

This implies that if a typical product were produced "one-of-a-kind," unit-processing cost would be 400 times greater than if it were mass produced at a million copies per year. If a typical product

TABLE 25.2. UNIT-PROCESSING COST MEASURES FOR SELECTED METALWORKING INDUSTRIES

Batch Size	SIC Code	Industry Name	Unit-Processing Cost (VA/Pound of Metal)
Custom and Small Batch	3662	Radio and TV communication equipment	14.13
	3761	Guided missiles and space vehicles	13.10
	3573	Electronic computing equipment	10.19
	3721	Aircraft and parts	7.01
	3724	Aircraft engines and parts	4.56
	3565	Industrial patterns	4.29
	3545	Machine tool accessories	4.05
	3541	Machine tools, metalcutting	2.45
	3463	Nonferrous forgings	2.27
Mid-Size Batch	3465	Residential lighting fixtures	1.41
	3561	Pumps and pumping equipment	1.33
	3494	Valves and pipe fittings	1.02
	3568	Power transmission equipment	1.00
	3562	Ball and roller bearings	0.98
	3582	Commercial laundry equipment	0.83
	3564	Blowers and fans	0.72
	3792	Travel trailers	0.71
	3451	Screw machine products	0.70
	3433	Heating equipment	0.64
Large Batch	3531	Construction equipment	0.65
	3523	Farm machinery and equipment	0.63
	3713	Truck and bus bodies	0.47
	3743	Railroad equipment	0.47
Mass	3711	Motor vehicles, car bodies	0.39
	3714	Motor vehicle parts, accessories	0.38
	3644	Noncurrent carrying wiring devices	0.35
	3411	Metal cans	0.27
	3441	Fabricated structural metal products	0.22
	3465	Auto stampings	0.22

Source: Derived from data in Miller.[11]

TABLE 25.3. DISTRIBUTION OF VALUE ADDED AND OUTPUT BY MODE OF PRODUCTION, SIC 34–37

Region and Mode of Production	Major SIC Groups				Total, 34–37
	34	35	36	37	
Percent of Value Added for Industries in Sample					
Custom and small batch	1.1	41.0	58.5	30.6	31.5
Mid-batch	28.3	42.5	30.2	6.9	26.4
Large batch	45.5	16.5	3.1	2.9	16.4
Mass	25.1	0.0	8.2	59.6	25.7
Sample coverage of value added in all industries	94.2	95.8	51.2	97.4	86.0
Percent of Total Output for Industries in Sample					
Custom and small batch	1.0	36.3	55.1	20.1	24.8
Mid-batch	25.3	44.3	31.4	4.9	23.3
Large batch	45.1	19.4	3.8	2.8	19.9
Mass	28.6	0.0	9.7	72.2	35.7
Sample coverage of output in all industries [a]	95.1	96.1	51.0	98.0	88.0

$$\text{Sample coverage} = \frac{\text{combined value added for industries in sample}}{\text{combined value added for all industries}}$$

Source: Miller.[11]

[a] Industries with insufficient data on material inputs are omitted from the sample.

were produced in batches of 100 or 1000, unit-processing cost would still be 55 to 20 times higher than if it were mass produced at a million copies per year. Clearly, producers and consumers are paying a high premium for metal products produced in small batches.

Based on the analysis of unit-processing cost and of the level of output, we have estimated the dominant mode of production within the 101 metalworking industries in our sample. We estimate that mass production industries account for about one-quarter of the value added and for about one-third of the total output in the metalworking sector (Table 25.3). This suggests that much of the value added can be thought of as the cost of flexibility needed for product differentiation and specialization.

25.4. POTENTIAL IMPACTS ON LABOR COST

Existing and likely near-term capabilities of robots make them candidates to replace significant numbers of machine operators and unskilled laborers over the next three decades or so. In general terms, the replacement of several million production workers by robots will decrease direct labor input. Consequently, output per labor hour will most certainly increase, although the impact on total factor productivity depends on the requirements for new capital and other ways that robotic (and CAD/CAM) technology may alter production technology.

To gage roughly how much of an impact the direct substitution of robots for factory workers could have on labor cost, all other factors constant, we compare 1977 production labor costs in the selected industries in Table 25.2 with a reduced total derived from assumptions about the potential for robotic substitution. Within this sample of 29 industries, total wage and benefit payments to production workers comprise less than 18% of the total value of output, on average. The production labor cost proportion of output within each industry understates the total labor cost of the product, however, since purchased materials account for the largest share of costs, and most of the material inputs originate from other industries within the metalworking sector. The "true" labor cost is actually the accumulated sum of direct labor cost within a given industry plus the labor cost "embodied" in the purchased materials. However, only direct labor costs are relevant to technological choice within a given industry here. For now, we consider an industry in isolation, without considering the effects of cost savings that might be passed on to customers.

Potential reductions in production labor payments (see Table 25.4) are based on three scenarios, outlined as follows:

REPLACEMENT SCENARIOS

	Percent Replaced		
Occupation	Low	Medium	High
Fabrication workers	20	50	75
Assembly workers	0	25	75
Inspectors	0	0	75
Supervisors	0	0	75

Fabrication workers include all types of skilled and semiskilled machine operators and setup workers, as well as material handlers, laborers, and miscellaneous types of skilled, semiskilled, and unskilled production workers. It is assumed that maintenance workers will not be replaced. The low scenario represents the current potential for replacing factory workers with insensate robots. The medium scenario represents the near-term potential for replacing factory workers with the emerging generation of sensor-based robots. From a technical standpoint, these two scenarios could be realized within the decade. The high scenario is our own subjective estimate of the long-term potential for eliminating production labor in the millennial "factory of the future." The occupational structure within each industry is considered in the calculation of the reduction in total labor cost. The potential for cost reduction in these scenarios is calculated under the highly restrictive assumptions that the quantity and mix of physical output remain at current levels and that the organization of production remains unchanged, not unreasonable assumptions for the near term, although unrealistic when considering a longer horizon.

In the low- and medium-replacement scenarios, the potential cost reductions appear to be modest, averaging near 2 and 7% respectively. The potential cost reduction is greater in the high-replacement scenario, averaging about 13%. Of course, the upper limit on potential cost savings from eliminating production workers is given by the total production worker portion of output. In the near term,

TABLE 25.4. POTENTIAL IMPACTS OF REDUCING LABOR COST IN SELECTED METAL-WORKING INDUSTRIES

Industry Name	Labor Cost/ Output (%)	Reduction in Total Cost for Replacement Scenario (%)		
		Low	Medium	High
Radio, TV communication equipment	16.5	1.5	5.4	11.8
Guided missiles, space vehicles	14.3	1.5	4.8	9.5
Electronic computing equipment	8.6	1.2	3.3	6.1
Aircraft and parts	15.4	1.7	5.1	10.3
Aircraft engines and parts	19.2	2.1	6.4	12.8
Industrial patterns	42.7	5.9	16.4	30.1
Machine tool accessories	24.5	3.4	9.4	17.3
Machine tools, metalcutting	22.0	3.0	8.5	15.6
Nonferrous forgings	17.2	2.5	6.7	11.7
Residential lighting fixtures	15.9	1.4	5.2	11.4
Pumps and pumping equipment	15.7	2.2	6.0	11.1
Valves and pipe fittings	18.9	2.7	7.3	12.9
Power transmission equipment	23.2	3.2	8.9	16.4
Ball and roller bearings	27.9	3.8	10.7	19.7
Commercial laundry equipment	18.5	2.5	7.1	13.0
Blowers and fans	17.3	2.4	6.6	12.2
Travel trailers	14.2	1.5	4.7	9.5
Screw machine products	25.4	3.7	9.9	17.4
Heating equipment, not electrical	15.3	2.2	5.9	10.5
Construction equipment	16.8	2.3	6.5	11.9
Farm machinery and equipment	16.0	2.2	6.2	11.3
Truck and bus bodies	17.8	1.9	5.9	11.9
Railroad equipment	18.3	2.0	6.1	12.2
Motor vehicles, car bodies	9.8	1.1	3.3	6.5
Motor vehicle parts, accessories	22.2	2.4	7.4	14.9
Noncurrent carrying wiring devices	17.6	1.6	5.7	12.6
Metal cans	13.1	1.9	5.1	8.9
Fabricated structural metal products	17.6	2.5	6.8	12.1
Auto stampings	25.6	3.7	9.9	17.5

considering the low- and medium-replacement scenarios, it does not appear that the direct substitution of robots for factory workers by itself would have a substantial impact on total cost, even in those industries where robot use is most heavily concentrated. Something else must happen if large savings in cost and increases in productivity are to be realized in the near term.

25.5. POTENTIAL IMPACTS ON CAPACITY

Possibly more significant than impacts on labor cost is the impact of robotization and CAM on factory organization and the utilization of machinery and equipment. Throughout the major metalworking industries, there are, on average, more machines than operators. The ratio of machines to operators understates this point, since each machine is available three shifts per day, whereas an operator only works a single shift, or slightly longer, allowing for overtime hours. Comparing the total hours worked by machine operators to the total hours that machines are available, it is clear that, on average, machines are only operated a small portion of the total time they are available. According to our estimates in Table 25.5, the effective utilization of manually operated metal-cutting machines, metal-forming machines, and welding equipment is remarkably low.

The average figures are 13, 14, and 21%, respectively, assuming theoretical "utilization" corresponds to 24 hr per day, 7 days a week with one operator per machine. On average, numerically controlled (NC) metal-cutting machines are utilized more fully than manually controlled cutting and forming machines. However, in 1977, fewer than 3% of the metal-cutting machines were numerically controlled. These estimates measure only the proportion of time an operator is available to run the machine, with allowance for the possibility that the machines is idle part of the time the operator is on duty. According to estimates shown in Figure 25.5, productive cutting time by machine tools as a fraction of theoretical capacity in low- and medium-volume shops is 6 and 8%, respectively, increasing to 22% for high-volume, mass-production operations.

Thus it is clear that most machines, especially in the metalworking industries, are idle most of the time—even before making allowance for setup time, load/unload time, and other adjustments. Even in the most automated industries, such as the production of automobile engines and transmissions, true machine-tool-utilization rates as high as 50% are seldom, if ever, achieved. We believe the major quantifiable economic impact of robotics and CAM will be to expand sharply the effective capacity of production facilities by increasing both the amount of time per year the plant is operating and the throughput per shift. Much of the lost time is due to incomplete use of the second and third shifts and to plant closings for holidays, strikes, and other reasons. Weekends, holidays, and night shifts are less popular than the "normal" 40-hr work week. Consequently, even if labor is available during these periods, it is more expensive.

TABLE 25.5. ESTIMATES OF AVERAGE MACHINE-TOOL UTILIZATION IN THE METALWORKING INDUSTRIES, 1977

Major Group	Metal-Cutting Tools Manual (%)	Metal-Cutting Tools NC[a] (%)	Metal-Forming Tools (%)	Joining (Welding) (%)
34	12.4	18.4	15.8	15.1
35	12.7	20.9	7.4	17.8
36	9.9	29.3	13.0	8.2
37	16.5	14.8	17.5	40.0
Average	12.9	20.0	13.6	21.3

Sources: Machine-tool hours derived from American Machinist (1978). Labor hours derived from Bureau of Labor Statistics (1980).

[a] Numerically controlled (NC). Assumptions: Machine utilization is defined here as follows:

$$\text{utilization} = \frac{\text{total operator hours available}}{\text{total machine hours available}}$$

with

total operator hours available = (number of operators)*(average hours worked per operator per year)

$$\frac{\text{average hours worked}}{\text{per operator per year}} = \frac{\text{total hours worked by production workers}}{\text{total production workers}}$$

total machine hours available = (number of machines) $\times$ (8760 hr/year)

8760 hr/year = (24 hr/day)*(365 days per year)

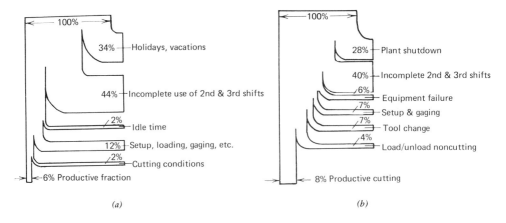

(a)   (b)

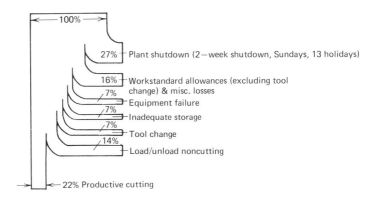

((c)

Fig. 25.5. Breakdown of theoretical capacity in (a) low-volume, (b) mid-volume, and (c) high-volume manufacturing. (Reprinted from Reference 1.)

Average estimates of the days per year that plants are open and of the shifts per day they are actually operating are given in Table 25.6. We deduced these estimates from the breakdowns of theoretical capacity shown in Figure 25.5. These are intended to represent normal operating conditions in a "healthy" economy.* These figures imply that even high-volume plants shut down nearly 80 days per year due to Sundays, holidays, and planned closings for retooling. Mid-volume plants are closed, on average, 102 days (all weekends), and low-volume plants are closed nearly 125 days out of the year (weekends plus three weeks for holidays and shutdown). When open for production, high-volume plants are typically operating more than 22 hr/day, whereas mid-volume and low-volume plants are typically scheduled to operate 10.7 and 8 hr/day, respectively. Clearly, there is considerable potential for increasing output (and thereby decreasing unit cost) by saving long runs for an unmanned third shift (or weekend), using robot operators. During the next two decades, as manufacturers gain experience with unmanned factory operations, the less routine machine setups, repair, maintenance, and inspection tasks could be reserved for the regular day shift. Planned shutdowns required for retooling would be substantially reduced, or possibly eliminated in a robot-integrated factory with flexible production technologies. A summary of the potential for increasing available production time (and hence output)

---

* Actual figures vary with demand. The high-volume manufacturing estimates are based on operations of several machining plants for an automobile producer before the sharp cutbacks in production. According to a survey of plant capacity conducted by the Bureau of Census in 1982, most metalworking industries were only producing 60–80% of their potential output. Lack of sufficient orders was by far the most important reason that operations fell short of practical capacity.

TABLE 25.6.   ESTIMATES OF PLANNED PRODUCTION TIME IN LOW-, MID-, AND HIGH-VOLUME METAL-FABRICATING MANUFACTURING

			Low-Volume	
	High-Volume	Mid-Volume	One-Shift Operation	Two-Shift Operation
Maximum days per year available	365	365	365	365
Days per year open for operation	286	263	241	241
Hours per Day Scheduled For:				
Production	22.2	10.7	8.0	16.0
Preventive maintenance[a]	1.0	1.0	1.0	1.0
Not scheculed (idle)	0.8	12.3	15.0	7.0
Scheduled production time as a fraction of maximum available time	72.6%	32.0%	22.0%	44.0%

Source: Ayres and Miller.[1]

[a] Most factories do not stop production on a daily basis to perform scheduled preventive maintenance. Machines are typically serviced on an "as needed" basis. Assuming 1 hr/day of scheduled maintenance may even be a high estimate. Major machine overhauls and repairs are typically carried out during scheduled plant shutdowns.

in existing facilities is shown in Table 25.7. Output per year could be increased by perhaps 30% in high-volume plants and by almost 200% in mid-volume plants. If low-volume plants are operating only one shift per day, as is suggested by the breakdowns in Table 25.5, recouping lost time could increase output by more than 330%! If we generously assume that low-volume shops are operating on a two-shift basis (which only a portion do), output per year could still be increased by nearly 120%.

In addition to extending the amount of working time per year, robotics, especially when integrated with other CAM technologies, can increase capacity by increasing the number of parts produced per hour. Based on the breakdowns of theoretical capacity in Table 25.5, machines are only productively engaged about 30% of the time that the plant is open and operating. The remaining 70% of scheduled production time is lost for a variety of reasons. Recouping the fraction of scheduled production time that is "lost" to nonproductive uses would further increase the effective capacity of machine tools.

Much of the "lost" time is machine related: equipment limitations, tool changing, and equipment failures. However, a sizable fraction of time lost is due to management and workforce practices, including personal time breaks, late starts, early quits, material handling, excessive machine adjustments, and in-line storage losses due to scheduling inefficiencies. Personal time, late starts, and early quits, and some fraction of the material handling time could be virtually eliminated by replacing workers with

TABLE 25.7.   POTENTIAL PERCENTAGE INCREASES IN OUTPUT FROM UTILIZING LOST TIME

Type of Plant	From Utilizing Days Plant is Closed	From Utilizing Nonscheduled Production Time	Total Percentage Increase in Output
High-volume	28	3	31
Mid-volume	83	115	198
Low-volume (one-shift operation)	148	187	335
Low-volume (two-shift operation)	74	43	117

Source: Derived from estimates of available time in Table 25.6.

TABLE 25.8. POTENTIAL PERCENTAGE OUTPUT INCREASES FROM RECOUPING NONPRODUCTIVE TIME DURING PLANNED OPERATIONS: HIGH-VOLUME MANUFACTURING

Function	Percent of Operating Time	Robots Only		Robots with CAM	
		Potential Percent Reduction	Adjusted Percent	Potential Percent Reduction	Adjusted Percent
Load/unload, noncutting[a]	20	−10	18	−25	15
Workstation allowances	20	−40	12	−80	4
Inadequate storage	10	0	10	−50	5
Tool change[a]	10	0	10	−20	8
Equipment failure	10	0	10	0	10
Productive fraction	30	0	30	0	30
Total[b]	100		90		72
Potential output index	1.00		1.11		1.39

Sources: Ayres and Miller.[1]

[a] Already highly automated in high-volume plants.

[b] Total equals total scheduled production time.

robots. Time losses due to tool changing, equipment failures, excessive machine adjustments, setups, and scheduling inefficiencies will probably not be affected directly by robots, but might be reduced if more aspects of factory work were consolidated and controlled by sensor-based computer systems. For example, sensors monitoring machine performance would eliminate unnecessary adjustments and would speed up diagnosis of machine failures.* If "stand-alone" machines were replaced by a flexible manufacturing system, and parts processing were "rationalized" by adopting group technology, there would be less material handling, and the scheduling of parts and tools would be simplified. Even a substantial fraction of the equipment-related losses could be eliminated in a fully integrated flexible manufacturing system, since the whole system need not be stopped if one station malfunctions. Robots or programmable pallets under the control of a central scheduling computer could reroute parts to other work stations.

It is difficult to discuss the potential improvements in productivity that may be brought about from robotics in isolation of the development of CAM systems and other forms of factory automation. Retrofitting robots into existing production lines will bring about some improvements, such as improving the utilization of a single machine or work station, but we do not expect that it would dramatically improve overall factory performance. Substantial impacts on performance and cost at the factory level require the integration of robots and other forms of factory automation into coordinated manufacturing systems. Also, it becomes more difficult and less meaningful to distinguish between robots and other forms of factory automation as the concept of robotics evolves from programmable manipulators to machines and systems that can "sense, think, and act."†

Our own rough estimates of potential increases in throughput that could be achieved from recouping the "nonproductive" time lost during scheduled operations are shown in Tables 25.8 through 25.10. These estimates are based only on informed judgement but have been reviewed by several industry experts. They are not the result of detailed analysis. We distinguish two levels of improvement: (1) as a result of the use of robots per se; and (2) as a result of integrating robots with CAM systems and other forms of factory automation. We suggest that the installation of robots, without increasing the time normally planned for operations and without extensively adding other forms of automation, would result in a 10% increase in output in high-volume machining operations (not including assembly), and nearly a 15% increase in output in mid- and low-volume production. If robots were used in

---

* In the next few years, time lost to equipment failures could conceivably increase as systems become more automated and more complex. However, we expect improvements in machine reliability and in sensor-based diagnostic systems to improve machine and system reliability and to reduce equipment failures over the next two decades.

† We borrow the broader defintion of robotics as machines that can "sense, think and act" from Professor Raj Reddy, Director of the Carnegie-Mellon Robotics Institute.

TABLE 25.9.   POTENTIAL OUTPUT INCREASES FROM RECOUPING NONPRODUCTIVE TIME DURING PLANNED OPERATIONS: MID-VOLUME MANUFACTURING

Function	Percent of Operating Time	Robots Only		Robots with CAM	
		Potential Percent Reduction	Adjusted Percent	Potential Percent Reduction	Adjusted Percent
Setup and gaging	22	−30	15.4	−65	7.7
Load/unload and noncutting	12	−40	7.2	−60	4.2
Tool change	22	−5	20.9	−15	18.7
Equipment failure	7	0	7	0	7
Idle time	12	0	12	−25	9
Productive fraction	25	0	25	0	25
Total[a]	100		87.5		64.6
Output index	1.0		1.14		1.55

Source: Ayres and Miller.[1]

[a] Total equals total scheduled production time.

conjunction with other forms of factory automation systems, still without increasing the number of days normally planned for operations, output might be increased by nearly 50% in mid- and low-volume production and possibly by 40% in high-volume production.

Our judgments regarding the estimated improvements in Tables 25.8 through 25.10 are mostly applicable to situations of *retrofitting* or incrementally adding technologies within existing plants. Other studies suggest that completely redesigning the factory around new technology could result in substantially greater improvements in throughput during planned production periods. For example, Mayer and Lee,[12] of Ford Motor Company, estimated the combined effects of applying the most advanced concepts to almost all aspects of machine design and control and factory layout in high-volume machining systems. They considered the use of automatic loading/unloading, automatic tool changing, diagnostic sensing, component reliability, unmanned operation, as well as improved line balancing, and faster cutting speeds. Their results suggest that all of these improvements, *without* increasing cutting speeds, could result in a 90% increase in output. If cutting speeds were increased to their upper limits as well, a more than 200% increase in output could be achieved. Their estimates are also based on *current* operating times (no changes in shifts per day, or days per year).

The potential effects of utilizing nonscheduled production time and of recouping time lost to nonproductive uses during scheduled operations are combined in Table 25.11. To utilize all of the time normally not scheduled for production (plant shutdowns, holidays, Sundays) the plant would sometimes have to operate with "skeleton" crews, or even unmanned during some periods, under the control of computer

TABLE 25.10.   POTENTIAL OUTPUT INCREASES FROM RECOUPING NONPRODUCTIVE TIME DURING PLANNED OPERATIONS: LOW-VOLUME MANUFACTURING

Function	Percent of Operating Time	Robots Only		Robots with CAM	
		Potential Percent Reduction	Adjusted Percent	Potential Percent Reduction	Adjusted Percent
Setup, loading, and gaging	55	−25	41.3	−50	27.5
Idle time	9	0	9	−50	4.5
Cutting conditions	9	0	9	−25	6.8
Productive fraction	27	0	27	0	27
Total[a]	100		86.3		65.8
Output index	1.0		1.16		1.52

Source: Ayres and Miller.[1]

[a] Total equals total scheduled production time.

TABLE 25.11. SUMMARY OF POTENTIAL IMPACTS ON CAPACITY

Type of Plant	Base Case	Potential Capacity Increases	
		Robots Only	Robots with CAM
High-Volume			
Available hour index	1.00	1.31	1.31
Throughput index	1.00	1.11	1.39
Output index	1.00	1.45	1.82
Increase in output (%)		45	82
Mid-Volume			
Available hour index	1.00	2.98	2.98
Throughput index	1.00	1.14	1.55
Output index	1.00	3.40	4.62
Increase in output (%)		240	362
Low-Volume: Single Shift			
Available hour index	1.00	4.35	4.35
Throughput index	1.00	1.16	1.52
Output index	1.00	5.05	6.61
Increase in output (%)		405	561
Low-Volume: Double Shift			
Available hour index	1.00	2.17	2.17
Throughput index	1.00	1.16	1.52
Output index	1.00	2.52	3.30
Increase in output (%)		152	230

Sources: Potential increases in output derived from data in Tables 25.7–25.10.

systems. Thus if we assume that hours available for production could be increased to its upper limit, we should consider the case of robots used in conjunction with other CAM technologies. For the "robots with CAM" case, high-volume machining operations show a potential output increase of 80%. Mid-volume manufacturing and low-volume producers already on a double shift show a potential output increase of 360 and 230% respectively. For low-volume producers operating on normal single shifts, the potential increase is 560%. In mid- and low-volume manufacturing, potential increases in

TABLE 25.12. PERCENT DECREASE IN UNIT COST DERIVED FROM ESTIMATE OF OUTPUT ELASTICITY

Percent Increase in Output	Percent Decrease in Unit Cost Assuming Scale Elasticity Equals	
	−0.295[a]	−0.436[b]
50	11.3	16.2
100	18.5	26.1
200	27.7	38.1
300	33.6	45.4
400	37.8	50.4
500	41.0	54.2
1000	50.7	64.8

Source: Miller.[11]

[a] Output elasticity includes the effects of the use of more automated types of machinery with higher throughput rates and increases in machine utilization rates.

[b] Output elasticity includes the effects of both the use of more automated types of machinery with higher throughput rates and increases in machine utilization rates and increasing emphasis on using lower-cost materials and on designing for ease of manufacturability.

output are almost all the result of increasing planned production time. In high-volume machining operations, the contribution of increasing throughput per period is somewhat greater than the contribution of increasing hours planned for production.

Assuming that there is a market for additional goods, an increase in output can be viewed as a reduction in unit cost. Estimates of the percentage decrease in unit cost that would result from severalfold increases in output are given in Table 25.12. In the middle column (smaller price decreases), the effects of substituting material inputs and simplifying manufacturability as the level of output increases are not included in the output elasticity. If output could be increased by 50–100%, as might be the case for typical mass-production plants, we estimate that unit cost would decrease by approximately 10–25%. If output could be increased by a factor of 5, as might be the case for many low-volume plants, we estimate that unit cost would decrease by approximately 40–55%! For comparison purposes, in our discussion of the impacts on labor costs earlier in this chapter, we noted that, on average, total cost would be reduced by roughly 2% if 20% of all fabrication workers were replaced and by roughly 5% if 50% of all fabrication workers and 25% of all assemblers were replaced. Potential cost savings theoretically obtainable from increasing machine utilization appear to be roughly an order of magnitude larger than savings that could be realized by replacing production labor costs under the low- and medium-replacement scenarios. Even if almost all production labor costs were eliminated, the savings that could be realized in low- and mid-volume manufacturing by increasing output would be two to three times greater.

25.6. POTENTIAL IMPACTS ON CAPITAL COSTS

25.6.1. Inventory and Work-In-Process

It has been estimated by Carter[13] that in typical batch-production operations a workpiece is typically machined only 5% of the total time it spends in the factory, and the remaining 95% of the time is spent in transit or in storage. This suggests that a piece requiring 10 hr of machining time would take at least eight days and up to several weeks to pass through the factory. Improved work scheduling, in combination with higher rates of machine utilization could dramatically reduce the time required to move work through the factory.

Inventories of work-in-process and finished goods on hand at the end of the year typically comprise between 10 and 30% of the value of shipments within an industry. Robotics and flexible manufacturing systems will not necessarily decrease the levels of inventory on hand. In fact, they may require increased inventory levels since higher levels of machine utilization will mean higher output rates, and it will be more important to keep a "buffer" supply of work always available to keep the machines busy. However, the financial benefits of reducing the time it takes to move inventory through the plant can be derived from the following example. Suppose a shop with an average of 1 million dollars annual revenues has a 100-day average lead time, which is not atypical. If that lag could be "instantly" cut to 10 days, the firm would be able to put the next 90 days' revenue ($250,000) into the bank, without increasing outlays at all. This amount would effectively be converted from unavailable working capital into cash. The firm would also save significantly on out-of-pocket warehousing and other inventory-related costs.

25.6.2. Capital Cost

In a general-purpose batch production facility, capital equipment and labor are shared among a large number of products since the requirements for a particular product are not large enough to tie up all the available equipment. As discussed earlier, the flexibility of a job shop is achieved at the cost of reduced levels of equipment utilization. This can be thought of as capacity loss due to sharing. Increased machine utilization resulting from the adoption of robotics can be viewed as recouping some of the capacity that was lost as a result of capital sharing. In many instances increases in output would cut fixed cost per unit produced drastically, despite the added expenditures for the robots and for the accompanying manufacturing systems. The other major impact of programmable automation on capital cost would be a reduction in the long lead times incurred in the sharing context, which would reduce some part of the inventory cost, as mentioned earlier.

The capital costs of employing flexible manufacturing systems at present are typically somewhat greater than those of conventional types of automation.* Currently, computer-controlled, flexible automation is expensive. However, this differential might be reduced to some extent or even eliminated if such systems are built out of relatively standardized elements, such as mass-produced robots and mini- and microcomputers, as opposed to being custom built, like most specialized production facilities.

---

* However, the flexible manufacturing systems typically have a greater range of capabilities than the equipment replaced. Although overall capital cost might be higher, cost per unit of capability (if this could be measured) might well be lower for the computer-controlled flexible manufacturing systems.

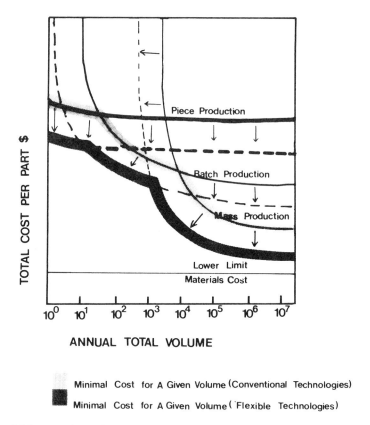

ANNUAL TOTAL VOLUME

■ Minimal Cost for A Given Volume (Conventional Technologies)

■ Minimal Cost for A Given Volume (Flexible Technologies)

Fig. 25.6. Potential impact of robotics and CAM on cost per part and batch processes. (*Source.* Reprinted from Reference 1.)

In general, if capacity could be increased significantly, flexible manufacturing systems could reduce both fixed and variable costs of batch production per unit. Figure 25.6 shows how robotics and CAM promise to shift the existing average unit cost curve envelope of Figure 25.3 closer to the ultimate lower limit (materials cost) over much of the spectrum of the production rates, particularly the mid-to-high-volume range.

25.6.3. The Consequences of Dramatic Improvements in Productive Capabilities

Suppose, as we suggest, unit cost in batch and custom manufacturing could be reduced below current levels, mostly as a result of expanding the capacity (and output) of existing facilities.* There are several consequences for employment, depending on how demand for the product changes as price declines. This relationship is given by a parameter referred to as a product's (own) price elasticity of demand, defined as the percentage increase in demand for a 1% decrease in price.† A distinction is usually drawn between three cases as follows:

---

* Much of the savings could, in principle, be achieved without eliminating labor. However, the higher machine utilization rates can only be achieved by using computers and robots to control the flow of work within the whole factory, eliminating the need for much of the "hands-on labor," which in turn eliminates worker-related slowdowns and bottlenecks. If capacity increases are achieved, it could be profitable to pay some of the current workers just to stay out of the way in order that the machines can be more fully utilized. However, this is unlikely to be the most productive, or socially acceptable, use of human resources. It also depends on being able to sell the additional output.

† The demand for a product depends on its own price, as well as the price of other products that could be used as substitutes. In the following discussion, we assume that only the price of the product in question is varying, and that prices of other products and of other important variables, such as income levels, remain constant.

Inelastic. A 1% reduction in price leads to a less than 1% increase in quantity demanded. As a result, price cuts decrease total revenue and profits.

Unitary. A 1% reduction in price leads to a 1% increase in quantity demanded. Price cuts leave total revenue and profits unchanged.

Elastic. A 1% reduction in price leads to a more than 1% increase in quantity demanded. Price cuts result in an increase in total revenue and profits.

First, take the case of a relatively modest price cut. Suppose, for the sake of argument, that firms within an industry *retrofit* robots into existing facilities. They realize a 10% decrease in production labor cost and a 25% increase in throughput at the cost of a 5% increase in annual capital outlays. Based on these assumptions, labor requirements per unit of output would decrease from 1.0 to 0.72, a 28% decrease. Given this decrease in unit labor requirements, demand within this industry would have to increase by (1.0/0.72 − 1.0) or by 39% to keep the same number of workers employed as previously. To calculate the reductions in unit cost, which we assume are passed on as price cuts, we figure that the cost proportions in this industry are representative of several industries within the fabricated metals sector (SIC 34).* Assuming the improvements and cost mentioned, this industry would realize a 21% reduction in unit cost. The price elasticity necessary to induce enough increase in demand to keep all the displaced people employed is given by

$$\frac{39\% \text{ increase in output}}{21\% \text{ decrease in price per unit}} = -1.86$$

In other words, for each 1% *decrease* in price, there would have to be a corresponding 1.86% *increase* in quantity demanded to generate enough additional employment so that displaced workers could remain within the industry. For comparison purposes, we note that the elasticity for household appliances has been estimated to be roughly −2.0, which makes it one of the most price responsive of all consumer goods. We would not expect capital goods, or most other products of the metalworking industry, to be as price responsive as consumer appliances. Thus even a relatively "modest" cost improvement achieved by installing robots would require that the demand for an industry's products be highly elastic if we were to hope that all displaced workers were to remain within the industry.

Because most metalworking products are either material or capital inputs for other industries (including metalworking itself), cost savings within the metalworking sector have *ripple effects* on the prices of all goods and services throughout the rest of the economy. (For the same reasons, cost increases in metalworking drive up prices throughout the economy.) It follows that demand for most of these goods, with the exception of autos, appliances, and other goods sold directly to consumers, is "derived" indirectly from the demand for final goods and services. For most metal products, it is unlikely that demand would be highly elastic. Unitary elasticity of demand (the percentage increase in quantity demanded equals the percentage decrease in price) is the most reasonable assumption for most intermediate goods, lacking other data. Given a price elasticity of unity, a 21% reduction in price per unit would induce a 21% increase in output, with no change in revenues. With the reduced unit labor requirements (as described), this outcome would still leave 13% of the workers previously employed in the industry displaced.† Demand for some consumer goods including automobiles and appliances is more responsive to price. If the price elasticity of automobiles were −2, a 21% reduction in total cost would induce a 42% increase in sales. Even with the reduced unit labor requirements as given, a 42% increase in sales would yield a 2% increase in employment requirements and produce additional revenue and profits for the manufacturer as well.

Robot use could easily result in more dramatic reductions, both in unit labor requirements and in prices.‡ Suppose all producers within an industry were to refurbish completely their existing facilities, or even build new plants. For the sake of discussion, let us assume the new facilities realized a 20% reduction in production labor cost and achieved a 100% increase in output, at the expense of a 20% increase in annual capital outlays. In this case, unit production cost would drop by 50%. With the new unit labor requirements, output would have to expand by 250% to create enough work to keep all of the displaced workers employed in the same industry, which would require a very high price

---

* We assume the following cost proportions: production labor cost equals 23% of value of shipments, nonproduction labor cost equals 7%, capital cost equals 20%, and materials cost equals 50%.

† 0.72 labor units/unit of output × 1.21 units of output = 0.87 units of labor = 13% decrease from the base case.

‡ As an aside, if demand were to increase by several hundred percent (and in some cases, it might), most established organizations are not prepared to cope with the increased complexity of organizing their business. It would strain the organizational structure, especially information-processing capabilities. As a result, producers sometimes purposefully restrain their technological capabilities to keep their business "manageable."

elasticity (−5). If demand increased by only as much as prices decreased (50%), then employment of production workers would drop 40% from previous levels.

The foregoing examples suggest that if robotic technologies were to be widely used, not all displaced workers could be expected to be reemployed in their current industries as a result of price reductions and increased demand. The effects of price reductions on demand and the net employment effect, balancing job displacement and job creation, will vary considerably among the various metalworking industries, depending on the nature of the product and its market. The logical conclusion is that employment of production workers in most manufacturing industries would decrease, despite substantial improvements in productivity within these industries and possible increases in production. This does not mean, however, that total employment in the economy as a whole would decrease. Substantial reductions in the prices of intermediate and capital goods (for example, 20% in the first example and 50% in the second) should reduce the cost of manufacturing consumer goods and of creating new goods and services, both of which will increase the consumer's real buying power. This will, in turn, stimulate effective demand for other goods and services. This should create new employment opportunities.

There is one important point that has been overlooked in the preceding examples and in economic analysis in general. We implicitly assumed that the only way to utilize the "extra" capacity made available by using robotic systems was to increase the output of the goods that are already produced in that factory (or industry). However, there is an option of making greater use of the expanded capabilities and of the flexibility of robotic production systems to produce a wider range of products and to manufacture new, high-performance products. Thus simply looking at the price elasticity of demand for current products might substantially underestimate the extent to which additional flexible capacity could be utilized. If the benefits of robotics and of other types of programmable manufacturing technologies are to be fully exploited, there needs to be a concurrent emphasis on the development of new products to utilize the expanded capabilities and the greater capacity. A new strategy that places much more emphasis on product performance and less on standardization and cost reduction might require an abrupt shift in many existing corporate strategies.

Will the benefits of robotic technology be fully appreciated and exploited by today's manufacturing management? Almost all of the existing installations of robotics and of flexible manufacturing systems have been motivated by the desire to reduce the cost, principally the labor cost, of produced existing goods. To date, robotic production technologies have had little effect on product design and development and on marketing strategy. If this trend continues, the implications are that the amount of job creation might be small in comparison to displacements. But past trends with respect to the motivations for, and uses of, robotics might be a misleading indicator of future applications. To date, producers have had relatively little experience with programmable manufacturing technologies, and it is to be expected that the initial applications are motivated by some of the more conservative and easily realized goals. However, there are already strong indications that designers and strategists within some of the major manufacturing companies are giving serious attention to integrating developments in robotics and other manufacturing technologies with product development (using computer-aided design, CAD) and overall corporate strategy. If this trend continues, then there would be reason to expect that the widespread use of robotic technologies will directly and indirectly create many more jobs than it displaces.

To summarize the foregoing argument, the primary economic benefit of robotics is likely to be a reduction in the real cost of manufacturing products made in small to medium batches. Capital goods—machine tools and the other types of durable equipment—as well as the parts used within them are largely batch produced.* Thus the price of capital goods in relation to final products can be expected to decline significantly over the next quarter century. This will cause secondary ripple effects on the prices of other manufactured goods and services throughout the economy. This, in turn, will reduce the real price of final output of mass-produced consumer goods, as well as the real price of output of the nonmanufacturing sectors. Final demand would also be stimulated to some extent, depending on the sensitivity of final demand to price. (For consumer goods, high price elasticities tend to be more the rule than the exception.) Lower production costs will also have a beneficial impact on the rate of inflation. Insofar as inflation is caused by "too much money chasing too few goods," an increase in productivity is perhaps the best way to break out of the vicious cycle. Ultimately, such changes will also affect other important macroeconomic variables, including the overall level and composition of employment and the level and distribution of income. These second-order effects, while less immediate, may have greater ultimate importance than the immediate improvements in labor productivity in manufacturing. It is beyond our present scope to attempt to forecast the detailed nature, the magnitude, or the time phasing of these broader economywide economic impacts.

We expect improved robots and substantial reductions in the price of intermediate and capital goods to play an important role in facilitating the development of several capital-intensive growth

---

* Automobiles and consumer appliances, such as refrigerators, air-conditioners, and washing machines, are mass produced, but are not classified as capital goods since they are sold to consumers.

sectors in the economy, including hazardous-waste management, biotechnology, undersea mineral exploration, and space manufacturing. These sectors would also provide employment. It is important to know if the levels of economic growth required to absorb workers displaced by robotics and other forms of technological changes can be achieved in the economy *as it is now structured*. If the required levels of economic growth (and employment) can not be achieved as a result of cost-saving process improvements in manufacturing, resources may have to be reallocated to encourage the creation of new products or services, or the development of new frontiers such as the oceans and space. This would require a reevaluation of traditional policies of stimulating economic growth by encouraging aggregate consumption or aggregate investment.

25.7. EMPLOYMENT IMPACT OF ROBOTIZATION

The limited experience with robots to date is consistent with the point of view that, for the overall economy, industrial robots pose little serious threat to employment in the coming decade. Assuming 1600 robots in use in the United States in 1977, and roughly 6800 by the end of 1982 (Figure 25.1), and that each robot displaces two or three workers on average, robots may have displaced 3000–5000 workers by 1977 and between 14,000 and 20,000 by the end of 1982. The last figure cited (20,000) would represent about one-fifth of 1% of the approximately 9.7 million semiskilled operatives and unskilled laborers employed in manufacturing industries in the United States in 1980. According to Koshiro,[7] there were 14,000 robots in use in Japan toward the end of 1982. He claims that no worker has really been displaced because of robotization because (1) robots have been used to meet labor shortages, particularly in dirty, dangerous, and heavy and/or highly repetitive work, and (2) many robots have been used in industries that have been expanding production.

From this perspective, the effects on employment up to now have been negligible. But, extrapolating the experience of the recent past into the future overlooks the concentration of effects in a relatively narrow sectoral, occupational, and regional setting. In particular, robotization—along with other and frequently related developments—could diminish employment opportunities for semiskilled operatives and unskilled laborers in durable goods industries, especially in the metalworking sector. These considerations imply that the displacement will be of sufficient magnitude, at least in some industries and regions, and for some groups of workers, to be a cause of concern.

25.7.1. Aggregate Trends in Manufacturing Employment

Robots and computer-aided manufacturing (CAM) continue the long-term trend toward mechanization and computerization that have made for relatively static employment in manufacturing industries in the United States, despite increased output. The significance of robotization in the next decade or two is that it will add to the combination of factors that have retarded growth in manufacturing employment. The percentage of the workforce employed in goods-producing industries (agriculture, manufacturing, and mining) has steadily declined from about one-third of the workforce in 1959 to barely one-quarter in 1977. Projections of the Bureau of Labor Statistics indicate that this trend will continue. In addition, the *service content* of each major sector of the economy, as represented by the proportion of the industry's workforce classified as *nonproduction* workers, has steadily increased. In 1980 more than one quarter of the employees in manufacturing and mining were performing managerial, professional, clerical, sales, or supervisory activities.

25.7.2. Occupational Employment in Manufacturing

In 1980 about 14 million people were employed as production workers in manufacturing in the United States. Nearly half of all production workers (and, for that matter, nearly half of all manufacturing workers) are concentrated in the metalworking industries.

Manufacturing production workers are classified into three main categories (Table 25.13): craft (skilled workers), operative (semiskilled workers), and laborers (unskilled workers). Examples of occupational titles within each of these major groupings are shown in Table 25.14. Robots without sensory feedback perform successfully in simple repetitive, well-structured tasks that can be preprogrammed, such as spot welding, spray painting, palletizing, and materials handling. Sensor-based robotic systems can accomplish tasks involving a greater degree of variability and to a limited (but rapidly increasing) degree can select workpieces regardless of the order in which they arrive, assemble small components, and carry out dimensional inspections.

Most of the routine repetitive jobs that currently lend themselves to automation and robotization are performed by the semiskilled (nontransport) operatives who comprise nearly 40% of total manufacturing employment. In several job categories that are most amenable to robotization, almost all of the employment is concentrated in metalworking. For instance, nearly all the 1 million operatives of metalcutting and metalforming machines are employed in SIC 33–38. In addition, almost all assemblers, welders, flame cutters, and production painters are employed in these industries, too. Although the percentage of packers and inspectors, sawyers, and laborers is comparatively small, it is probably

TABLE 25.13. EMPLOYMENT OF PRODUCTION WORKERS, 1980: METALWORKING AND TOTAL MANUFACTURING

Occupation	Employment in Metalworking SIC (33–38)	Total Employment, All Manufacturing	Employment in Metalworking (%)
Total, all occupations	9,964,878	20,361,568	48.9
Production workers, total: (craft workers, operatives, + laborers)	6,688,306	14,190,289	47.1
Craft and related workers, total	2,015,212	3,768,395	53.4
Metalworking craft workers	582,861	668,002	87.2
Other craft workers	1,432,351	3,100,393	46.1
Operatives, total	4,060,916	8,845,318	45.9
Nontransport operatives	3,880,876	8,134,123	47.7
Assemblers	1,311,870	1,661,150	78.9
Metalworking machine operatives	1,030,132	1,069,540	96.3
All other machine operatives	893,701	4,231,988	21.1
Welders and flame cutters	369,558	400,629	92.2
Production painters	79,594	106,178	74.9
Packing and inspection operatives	78,413	587,631	13.3
Sawyers	17,604	76,728	22.9
Transport operatives	180,040	711,195	25.4
Laborers, except farm	612,178	1,576,576	38.8
Nontransport operatives and laborers	4,493,054	9,710,699	46.2

Source: Reprinted from Ayres and Miller.[1]

true that most early robot applications in these jobs will be in metalworking since metal products are most suitable for robot handling.

While the majority of jobs that can be robotized are semiskilled operative jobs, there are already robot applications in heat treating, sheet metal work, and forge and hammer operations, all of which are classed as skilled jobs. As computer-aided design and manufacturing become more integrated, and factories are redesigned to fully exploit robotics and other types of programmable automation, a larger fraction of the so-called skilled metalworking crafts will be within the domain that can be automated.

The *technical potential* for replacing workers by robots has been estimated from an analysis of industry employment by occupation and from survey responses of the potential for substitution within a given occupation. Two levels of robot technology are distinguished: robots similar to those on the market in 1981 (Level 1) and robots with rudimentary sensory capabilities (Level 2). In 1980 there were nearly 6.7 million production workers employed in the metalworking sector in the United States. Of these, nearly 5 million worked within the three broad categories of jobs most amenable to robotization—metalworking craft workers, semiskilled machine operators, and laborers (see Table 25.13). Based on the survey results, we estimate that Level 1 robots theoretically could replace 16% of the workers in these three groups, and that Level 2 robots theoretically could replace 40% of the same population of workers. Thus, if all the potential for job displacement of Level 1 robots were realized in metalworking, more than 800,000 jobs could be eliminated. If Level 2 robots were available and fully exploited, an additional 1.2 million jobs, or a total of nearly 2 million jobs, theoretically could be eliminated. Extrapo-

TABLE 25.14. REPRESENTATIVE OCCUPATION TITLE FOR CRAFT WORKERS, OPERATIVES, AND LABORERS

Metalworking Craft Workers, Except Mechanics

Blacksmiths
Core makers
Forging press operators
Heat treaters, annealers,
    temperers
Layout makers, metal
Machinist
Metalworking machine tool setters
Metal molders
Punch press setters
Sheet metal workers and tinsmiths
Tool and die makers

Metalworking Machine Operatives

Casters
Drill press and boring machine
    operators
Electroplaters
Furnace charges and operators
Grinding, abrading machine oper-
    ators
Lathe machine operators
Machine tool operators, combina-
    tion
Machine tool operators,
    numerically controlled
Machine tool operators,
    tool room
Milling and planing
    machine operators
Metal punch press machine
    operators
Power brake and bending machine
    operators
Punch press operators

Other Craftworkers

Electricians
Plumbers
Pipefitters
Mechanics and installers
Blue-collar supervisors
Heavy equipment operators

Other Machine Operatives

Machine operators for textile products
Machine operators for wood products
Machine operators for
    rubber products and plastics
Machine operators for food products
Machine operators for paper products
Machine operators for
    chemical products

Laborers

Conveyor operators and tenders
Furnace operator helpers
Off-bearers
Loaders, car and truck
Stock handlers

lating the data for metalworking to similar tasks in other manufacturing sectors, it appears that Level 1 robots theoretically could replace about 1.5 million metalworking craft workers, semiskilled machine operators, and laborers, and Level 2 robots theoretically could replace about 4 million out of the current total of 10.4 million of these workers. The time frame for this displacement is *at least* 20 years, however. In the course of the coming decades, the capabilities of robots can be expected to increase further, and with these changes their potential for displacing operatives will increase. On the other hand, not all of the potential displacement will actually be realized, and if the economy grows as anticipated, some of the job loss due to displacement of people by robots could be offset by increases in other manufacturing employment.

By 2010 or so, it is conceivable that more sophisticated robots (Level 3 ?) will replace almost all operative jobs in manufacturing (about 9% of today's workforce), as well as a number of skilled manufacturing jobs and routine nonmanufacturing jobs. Concerted efforts should be made by the private and public sector to redirect the *future* workforce in response to these changes. Even though several million jobs in the current manufacturing workforce are vulnerable to robotization, the transition seems hardly catastrophic on a national scale, provided new job entrants are properly trained, and directed. In our view, the oncoming transition will probably be less dramatic than the impact of office automation over the same period. By 2010, most current operatives would have retired or left their jobs. The jobs would not disappear all at once, and robot manufacturing, programming, and maintenance itself will provide some new jobs, although we think most new jobs will *not* be in manufacturing, despite the rapid growth of the robotics industry itself. New "growth" sectors in the economy, including undersea and space exploration, may also provide many new jobs. The important conclusion is that *young people seeking jobs in the near future will have to learn marketable skills other than welding, machining, and other operative tasks that are now being robotized.* Even though the adjustment problems seem manageable, the potential for social unrest in specific locations cannot be dismissed so lightly. Consider the following points:

1. Nearly half of all the unskilled and semiskilled "operative" workers—the types of jobs which could be replaced by robots—are concentrated in four metalworking industries (SIC 34–37). In the United States almost one-half of all production workers in these four industries are geographically concentrated in the five proximate Great Lakes States (Indiana, Illinois, Michigan, Ohio, and Wisconsin) plus New York and California. In the five Great Lakes States, the metalworking sector also accounts for a large percentage of the total statewide employment in manufacturing. Adjustments in response to the rapid diffusion of robotics may be intensified in these areas. (The adverse impacts of *not* improving the productivity and competitive standing of these industries would also be concentrated in the same few states, of course.)

2. Older established workers will generally be protected by union seniority rules, except in cases where the whole plant closes. Unfortunately, this is happening with increasing frequency. Even in the newest, most efficient plants, some younger workers with less seniority may be "bumped." When either event occurs, the displaced worker starts again at the bottom of the ladder. Thus reemployed "displacees" are also more vulnerable to subsequent layoffs. *A class of perpetually insecure, marginal workers could result.* This would be a potential source of social problems and political dissension.

3. The states where jobs are most likely to be lost to robots are mainly in the North Central region where industry is also most unionized, plants are oldest, and wages are highest. The "Sunbelt" states, where many new jobs are being created, have newer plants and lower wages. Many displacees will have to migrate to other regions. Those unable to upgrade their skills sufficiently might have to accept lower-paying service jobs or join the "underclass" of insecure marginal workers who never became established with a stable employer.

4. There would likely be a disproportionate impact on racial minorities and on women. Nonwhites account for only 11% of the national workforce, but comprise more than 16% of total employment in semiskilled and unskilled manufacturing jobs. Women employed in semiskilled and unskilled manufacturing jobs are less likely to be represented by labor organizations than their male counterparts. *De facto* economic discrimination will accordingly increase.

5. Unions representing the affected categorized workers will probably experience sharp declines in membership and political/economic clout. *A policy of organized resistance to the introduction of labor-saving technologies might seem attractive to fearful workers and their unions, resulting in a severe drag on the productivity of the manufacturing sector.*

Projected employment trends in the metalworking industries provide a point of departure for estimating the potential displacement of employees that could take place in the 1980s. Projections of industry employment for 1990 (see Table 25.15) are made by the Bureau of Labor Statistics (BLS), derived from a macroeconomic model of the economy as a whole. The productivity and employment estimates assume continuing technological change but make no special assumptions about the impact of robots or related technical advances for employment in the 1980s.

Within manufacturing, employment growth rates for the durable goods industries (comprised mostly of the metalworking industries) are higher than growth rates for nondurable goods industries. According to BLS projections, by 1990 more than half of all employment in manufacturing will be in the metalworking industries (SIC 33–38). Machinery, except electrical (SIC 35) and fabricated metal products (SIC 34), are projected to have the highest growth rates in employment among the metalworking industries. At the more detailed industry level, typewriters and office equipment and computers and peripheral equipment, both in SIC 35, are projected to be among the most rapidly growing industries throughout the entire economy. On the basis of BLS projections, we see that the manufacturing industries most likely to be subject to robotization in the 1980s are also those in which above-average increases in employment would otherwise be anticipated.

These figures suggest that if robots are fully implemented over the next decade, increases in manufacturing output will not necessarily be accompanied by increases in employment requirements, especially for operatives and laborers. The extent to which the potential impacts will become translated into actual displacement of people from jobs will also depend on the rate of investment in industry, wage trends, and robot price trends and performance.

Job losses that may have a minor effect on total employment can have a magnified impact if they are concentrated in specific industries. A total loss of a million jobs would have a modest effect for total employment if the displacement were evenly dispersed throughout the country. The loss would amount to less than 1% of the total private employment level, projected to be 104 million in 1990 (low-growth trend). The social problems created by the displacement would obviously be greater if a majority of the job losses occurred in few years and if, as expected, they were concentrated in semiskilled occupations in areas that are already experiencing slow economic growth.

TABLE 25.15. PROJECTED EMPLOYMENT GROWTH VERSUS POTENTIAL ROBOTIC
REPLACEMENT

	Employment, 1980 (1000s)	Incremental Employment Projected for 1990 (1000s)		
		Low	High 2	High 1
Metalworking: SIC 33–38				
Operatives and laborers[a]	4,673.6	+785.5	+901.5	+1373.1
Craft workers[b]	2,015.2	+359.4	+408.0	+608.9
Blue-collar, total	6,688.8	+1144.9	+1309.0	+1982.0
Potential for Robotization				
Level 1	800.0			
Level 2	2,000.0			
Manufacturing, Total: SIC 20–39				
Operatives and laborers	10,421.9	+1296.5	+1447.9	+2234.7
Craft workers	3,768.4	+536.1	+605.6	+888.1
Blue-collar, total	14,190.3	+1832.6	+2053.6	+3122.7
Potential for Robotization				
Level 1	1,500.0			
Level 2	4,000.0			

Sources. 1980 employment and 1990 employment projections: Bureau of Labor Statistics. Reprinted
from Ayres and Miller.[1]

[a] Operatives include transport workers.

[b] Craft workers include mechanics, repairers, and construction workers.

25.7.3. Union Response to Technological Change

The industries that are candidates for extensive robotization are mostly characterized by the presence
of strong unions and well-established collective bargaining procedures. More than one-third of all
wage and salary workers in manufacturing and a significantly higher proportion of production workers—
85% of motor vehicle equipment operatives and 41% of nondurable goods operatives—are represented
by labor organizations. More than 90% of those represented actually belong to unions. Policies for
dealing with the displacement, therefore, are unlikely to be adopted unless organized labor participates
in their formation. There are no reliable statistics that cross-classify union membership by manufacturing
industry, but it appears that almost all of the membership of the United Auto Workers (UAW),
International Association of Machinists (IAM), International Union of Electrical Workers (IUE), United
Electrical Workers (UE), and United Steel Workers (USW) are in the metalworking industries (sectors
33–38). On the other hand, most of the membership of the International Brotherhood of Electrical
Workers (IBEW) work outside of manufacturing.

Prior to the 1980s, labor organizations have shown only a moderate concern with robots. At first
they were used mainly in dirty, monotonous, and unsafe jobs. The general assumption by unions in
the past, and one supported by robot manufacturers in promoting their use, was that the employees
who were displaced could readily be absorbed in job openings created by normal attrition or by growth.
This assumption was reinforced somewhat by the loss of credibility of earlier forecasts of mass displacement of blue-collar workers due to automation. However, accelerated rapid growth in the use of industrial
robots since 1979, in the midst of a general economic slowdown, has rekindled workers' fears about
loss of jobs.

Kuwahara,[14] of the Japan Institute of Labor, reports that Japanese labor and management have
so far been successful in introducing robots without inducing unemployment on an observable scale.
However, he notes that Japan's labor unions seem to be modifying their optimistic attitude toward a
more cautious one in the midst of the continuing economic stagnancy and escalation of international
trade friction. He anticipates that the time will eventually come when the "smooth" adjustment of
employment within Japanese establishments will no longer be possible. Koshiro,[7] also of the Japan
Institute of Labor, reports that in spite of some signs of emerging conflict of interest between labor
and management, it seems the broad consensus among labor, management, scholars, and the government
in Japan is that the new technologies should be applied to humanize life and improve the quality of
work.

As of 1980, between two-thirds and three-fourths of U.S. operatives and laborers were less than
45 years old, which means that barely a third of these workers would be retired in the normal way

by the year 2000. On the average, skilled workers are older, but they are not as likely to be replaced by robots in the near future. This suggests the possibility of retiring some workers earlier than normal. However, provisions for early retirement are less likely to figure as part of the solution to the robot displacement problem in the 1990s than in the earlier collective bargaining agreements dealing with technological change. Demographic changes have been increasing the number of older persons in the unions that would be mainly affected. The United Automobile Workers Union, for example, included 190,000 retirees from the Big Three automobile manufacturers among its members in the late 1970s.

Faced with escalating Social Security taxes and costs, national policy has been shifting from favoring early retirement as an employment-creating system for young persons toward proposals to keep more older persons in the labor force. Recent proposals to raise the age for qualifying for full Social Security benefits from 65 to 68 and the 1978 amendments to the Age Discrimination Act outlawing compulsory retirement for most retirees before age 70 symbolize the shift in public policy. Moreover, dependence on company pensions will become less attractive to employees if inflation continues since they are seldom indexed to changes in the cost of living. Emphasis on retirement gains also raises the possibility of creating intergenerational conflicts within the unions, since the gains for the employed members must be traded off for inflation adjustments or other benefits for older persons who are already retired or who are about to retire.

Unions can be expected to favor two approaches in dealing with the job losses for their members threatened by robotization. The first is to transfer and retrain the displaced employees into other jobs that have been created by attrition or by growth. This type of remedy is likely to be least costly to employers and to constitute a minimum barrier to the introduction of robots. The limitation of this approach is that it assumes a pace of robotization that is consistent with the number of suitable job openings created within the same plant (or in other plants belonging to the firm). Intrafirm transfers may not be possible in small firms or in declining industries, or may be impeded by union rules. As a rule, there are severe restrictions on the transferability of seniority rights for promotion and for protection against job layoffs. Seniority rights in these two critical areas are usually forfeited if the worker transfers out of the bargaining unit.* In some contracts, seniority is specific to particular work areas within the plant. There are even cases where these rights are only retained if the worker remains within a specific occupation within the bargaining unit. Nontransferability of seniority is one of the most effective impediments to labor mobility since it inhibits upgrading of skills, especially among older employees.

The other approach favored by many unions is to attempt to protect threatened jobs by raising the cost of introducing new technology and, in this way, transferring part of the productivity gain from employers to employees. Policies in this category include restrictive work rules, shortening the work week, lengthening paid vacations, or adding paid personal holidays. They also include employment guarantees and employer-financed pensions for older employees who retire early. Another possible measure intended to assure that workers' interests are considered in the decisions affecting job displacement is a requirement for advance notice to be given to unions before the new technology can be introduced. A side effect of these measures, however, is to slow down the introduction of new technology, thus adversely affecting international competitiveness. Typical collective-bargaining contract provisions found in U.S. labor contracts are shown in Table 25.16. According to Koshiro,[7] the Federation of Japanese Automobile Workers Union asked Nissan and all of its related companies in 1981 for an agreement on new technology which requested prior consultation on technological change, protection against layoffs due to the introduction of robots, no demotions or wage reductions due to robots, and education and retraining for workers prior to, and after, the introduction of robots.

25.7.4. Job Openings

In the occupations expected to be primarily affected by robotization, the job openings likely to be created by attrition in the 1980s provide a basis for assessing policies dealing with displacement during the next two decades. Attrition rates for semiskilled workers in metalworking are approximately 3% a year, depending on the sex and age distribution of the persons employed in them. However, these figures substantially underestimate the number of people transferring out of specific occupations, since they include only people who leave the establishment.† A 3% per annum attrition rate suggests an

---

* Seniority rights for other privileges, such as vacation preferences, health care, pensions, or for overtime, are more easily transferred across bargaining units within the same company. If a union such as the UAW has national agreements with a large company, there are exceptional circumstances under which seniority rights can be transferred. In some instances when this has happened in the past, the transplanted worker was greeted with hostility by other workers in the plant to which he transferred.

† Attrition is used to refer to workers who leave the establishment as a result of quits, discharges, permanent disability, death, retirement, or transfers to other companies. The other main source of labor turnover are layoffs (suspensions without pay for more than seven consecutive days initiated by the employer). Together, the attrition rate and the layoff rate comprise the "total separation" rate.

TABLE 25.16. COLLECTIVE BARGAINING PROVISIONS RELEVANT TO TECHNOLOGICAL CHANGE

Type of Provisions	Specific Clauses
Advance notice provisions	Layoffs
	Plant shutdown or relocation
	Technical change
Interplant transfer and relocation allowance provisions	Interplant-transfer provisions
	Preferential hiring
	Relocation allowance
Unemployment compensation provisions	Supplemental unemployment benefit plans
	Severance pay
	Wage-employment guarantee
Seniority and recall related provisions	Retention of older workers
	Merging seniority lists
	Retention of seniority in layoff
Exclusion from job security provisions	Exclusions from job security grievance procedure
	Exclusion from job security arbitration procedure
Work-sharing provisions applicable in slack work periods	Division of work
	Reduction of hours
	Regulation of overtime
Education and training provisions	Leaves of absence for education
	Apprenticeship
	On-the-job training
	Tuition aid for training
Provisions calling for joint labor-management committees	Industrial relations issues
	Productivity issues

Source: Bureau of Labor Statistics.[15]

annual average of about 170,000 job openings among blue-collar workers in the metalworking industries, alone, during the 1980s.* If one-half of these openings could be filled by operatives displaced from other jobs by robots, an average of over 85,000 jobs a year, or nearly 850,000 during the decade, could be filled, while still accommodating some first-time job seekers.

As the number of new entrants into the labor force declines in the 1980s because of the drop in birth rates after the mid-1960s, older and more experienced blue-collar job seekers will face less competition from younger workers than was the case in the 1970s. However, an unresolved question at this point is the extent to which economic growth or continued recession in basic industries such as automobile and steel will increase the numbers of job seekers competing with employees displaced because of robotization. In a declining industry, moreover, openings that would otherwise be created by attrition are often left unfilled. Turnover openings that would arise from occupational mobility can fall off sharply as fewer persons are added to the employment roll, and few quit voluntarily to take other positions.

Many of the blue-collar workers displaced by robots would possess the educational qualifications for more skill training which could lead to a better-paid position in other occupations. The traditional stereotype of a factory operative has been that of a person with limited education, often a "functional illiterate." A generation ago, the typical educational level for operatives was below that of the overall workforce. In the mid-1950s, for instance, the median number of years of schooling completed by operatives was 9.5 years as compared with 11.7 years for all employed civilian workers. However, by 1978, the median for operatives, excluding transport operatives, was 12.1 years. This compared with an overall average of 12.6 years. Operatives, as a group, tend to possess at least a high school education or its equivalent. This can provide a basis for further specific vocational training or for further higher education in a two-year or four-year college.

Changes in national priorities could also expand the range of job openings and outmode projections of the type that the Bureau of Labor Statistics based on the experience of the recent past. A shift in national priorities favoring more adequate home care, income support, and medical services for the

---

* Assume 6.7 million blue-collar workers in metalworking jobs and an average rate of 3% decrease per year over a 10-year period.

elderly, the retarded, and the handicapped would be reflected in new kinds of jobs for persons with the appropriate retraining. Private and public efforts to rehabilitate physical infrastructure (bridges, subways, water/sewer systems, etc.) could create large numbers of job openings that could be filled partially by displaced production workers.

The 40-hour standard work week has remained unchanged in most manufacturing industries for the last 20 years. White-collar workers typically enjoy a shorter work week. For example, two-thirds of all office workers in the finance, insurance, and real estate sectors now work a standard work week of less than 40 hours. A gradual reduction in the standard work week, leaving the hourly wage unchanged, would diminish job losses by spreading the available work over more employees. Clearly, the reduction in annual work hours could be accomplished in various ways. Sabbaticals, now confined to teachers and to some civil servants and steel workers, could be extended generally to production workers. Required sabbaticals, with partial pay, could be used to explore another occupation, to care for babies, or to become a student again. Blue-collar workers returning to, or first entering, a university while on a sabbatical could provide a new market for the services of colleges and universities faced with shrinking enrollments because of low birth rates two decades earlier.

25.8. IMPLICATIONS FOR RETRAINING

In their study of the human resource implications of robotics, Hunt and Hunt[2] conclude:

> *The most remarkable thing about the job displacement and job creation impacts of industrial robots is the skill-twist that emerges so clearly when the jobs eliminated are compared to the jobs created. The jobs eliminated are semi-skilled or unskilled, while the jobs created require significant technical background. We submit this is the true meaning of the robotics revolution.*

Based on the Japanese experience, Kuwahara[14] arrives at the same conclusion:

> *Traditional skills of craftsman are being replaced by machines or diluted into simpler skills. On the other hand, new types of skills have emerged. Major characteristics of these new types of skills are profound knowledge concerning complicated machinery and its functioning, programming ability, and perspectives upon the total machine system. This transformation makes it more and more difficult for aged workers to follow and adapt to the new technology.*

Hunt and Hunt suggest that an efficient human resource management strategy would be to train the former semiskilled operator to operate a machine that will not be robotized and to concentrate the robotics-related training (repair and maintenance) on plant maintenance workers who are already skilled. We add that older, displaced workers without skills would be most effectively used in semiskilled jobs that robots can not perform well, namely, those requiring complex types of sensory-information processing.

Koshiro[7] reports the results of a survey by the Japan Economic Research Institute on the impact of numerical control and microelectronics on the quality of work life in Japan. Most of the companies responded that they want to utilize the skill of older workers in spite of technological progress and that older workers will require more reeducation and retraining to cope with the new technologies. Many firms indicated they would use older workers in the development of advanced software.

25.9. CONCLUSIONS

The increased use of robots forces society to confront the short-term prospects of technological displacement and the longer-term prospects of basic structural shifts in the economy. But robots are only one of several change agents in the work environment. Concurrent advances in product design, metal cutting, metal forming, finishing, assembly, and inspection, under the control of computers, will also modify the mix of skills needed to work in the "factory of the future." Differential growth rates between different industries (e.g., electronics versus steel) may also cause broad shifts in overall levels of employment, skill requirements, and the occupational composition of the work force. Shifts in the industry mix and changes in the composition of the labor force are indirectly influenced by trade and defense policies, too.

As a nation, we are not confronting radical technological changes for the first time. Robots should not be given the credit (or blame) for initiating these changes. This does not make the potential problems associated with the phasing in of robots less important, or less urgent. It does mean that the need to cope with technological change is continuing. Resistance to the use of robots would not affect the likelihood of having a surplus of people whose skills are no longer needed while there is a simultaneous shortage of people with the skills required to develop and support the new technologies. Both mismatches are potentially troublesome.

Experiences from a long history of technological innovation in the U.S. economy suggest that the rate of robot introduction, as well as the social impacts of their use, will depend on factors beyond the control of individual firms. It is important to recognize that the adjustment issues we face as a nation depend on the complex interactions between technological progress and the worldwide economy. But the variable and uncontrollable elements of our economy should not be used as a smokescreen for ignoring the adverse possibilities of labor displacement, or for delaying the implementation of necessary programs for vocational training, retraining, and adjustment.

With a few exceptions, robot users have been reluctant to discuss plans for robot use in the future, even though many manufacturers are testing applications. They argue that such information must be kept confidential for competitive reasons. One result of private industry's uncommunicative attitude about future plans is that very little is being done to warn or prepare those workers whose jobs may be eliminated, or substantially modified, as a direct, or indirect result of introducing robots. In the absence of solid facts, or even informed speculations as to the types of adjustments that might occur, unions, media reporters, and government officials have started to suspect the worst, and ask, How many people will lose their jobs as this new wave of automation sweeps through industry? Private industry undoubtedly has an interest in the public perception of the impacts of robots on the labor force. If the phasing-in of robots on a large scale is handled ineptly and insensitively (or if people even think this is the case), unions and other factions of society might conceivably find enough common interest—based on a fear of technology—to organize a "Neo-Luddite" political attack on robots and other forms of automation. Unions are already advocating various forms of protectionism to blunt the impact of foreign competition. In the most extreme scenario, widespread social dissension could occur, fed by distrust of business and dissatisfaction with the record of a capitalist society in dealing with a cluster of festering social problems.

To develop the necessary human capital at both the institutional and individual level, and to smooth the short-term transitory impacts on the labor force, all the major actors must commit themselves to a cooperative effort to prepare and assist the workers most likely to be affected by the changes to come. In the past, the "band-aid" approach to social welfare problems has not proved especially effective. The only realistic alternative to "band-aids," however, is some sort of preventive medicine. To prevent social trauma due to rapid introduction of robotics effectively, without impeding technological progress itself, requires at least the following:

Identification of vulnerable categories of workers well in advance of actual job elimination.

Long-range planning by industry and government for future employment needs and new job skill requirements.

The provision of *effective* education and training facilities to upgrade workers from skill categories that are, or will be, in surplus supply to skill categories that are scarce.

The provision of *effective* facilities to locate suitable jobs and place workers in them, with relocation assistance if necessary.

In addition, more emphasis must be put on the development and exploitation of new and evolving technologies to replace those mature and declining industries that manufacture standardized "commodity" products. In an advanced industrial economy, standardized products must be replaced by new types of products produced in the "batch" mode because the technology is *not* standardized. This, in turn, implies a continuing rapid rate of technological innovation in both the product and its production techniques. Otherwise, standardization would be inevitable and rapid, and the loss of these industries to overseas platforms with lower material and labor cost would follow. Robotics and CAM are particularly well suited to the batch mode of production, especially when product design is constantly evolving. Thus, robotics and CAM can contribute significantly to the ability of the United States and of other advanced industrial nations to shift their industrial system away from concentration on low-cost mass production of standardized products to batch production of rapidly evolving products.

REFERENCES

1. Ayres, R. U. and Miller, S. M., *Robotics: Applications and Social Implications,* Ballinger Publishing Company, Cambridge, Massachusetts, 1983.

2. Hunt, H. A. and Hunt, T. L., *Human Resource Implications of Robotics,* W. E. Upjohn Institute for Employment Research, Kalamazoo, Michigan, 1983.

3. Aron, P., *Robots Revisited: One Year Later,* Technical Report 25, Daiwa Securities of America, Inc., 28 July 1981.

4. Ayres, R. U., Lynn, L., and Miller, S. M., Technology Transfer in Robotics: US/Japan, in Uyehara, C. H., Ed., *Technological Exchange: The U.S.-Japanese Experience,* Japan-American Society, Washington, D.C., 1982.

5. Japan Industrial Robot Association, *The Robotics Industry of Japan: Today and Tomorrow,* Prentice-Hall, Englewood Cliffs, New Jersey, 1982.

6. Yonemoto, Kanji, The socio-economic impacts of industrial robots in Japan, *Industrial Robot,* Vol. 8, No. 4, December 1981, pp. 238–241.

7. Koshiro, Kazutoshi, The Employment Effect of Microelectronic Technology, in *Highlights in Japanese Industrial Relations: A Selection of Articles for the Japan Labor Institute,* Japan Institute of Labor, Tokyo, 1983, pp. 80–88.

8. Vietorisz, T., *UNIDO Monographs on Industrial Development: Volume 4: Engineering Industry,* United Nations Industrial Development Agency, Vienna, Austria, 1969.

9. Schraft, R. D., Schults, E., and Nicolaisen, P., Possibilities and Limits for the Application of Industrial Robots in New Field, in *10th International Symposium on Industrial Robots, Milan,* IFS Publications, Ltd., Bedford, England, 1980.

10. Borzcik, P. S., Flexible Manufacturing Systems, in Thompson, A. R., Working Group Chairman (Ed.), *Machine Tool Systems Management and Utilization, Machine Tool Task Force Report on the Technology of Machine Tools,* Vol. 2, Lawrence Livermore National Laboratory, October 1980, pp. 62–74.

11. Miller, S. M., *Potential Impacts of Robotics on Manufacturing Cost within Metalworking Industries,* PhD thesis, Carnegie-Mellon University, 1983.

12. Mayer, J. E., and Lee, D., Estimated Requirements For Machine Tools During the 1980–1990 Period, in Thompson, Arthur R. (Ed.), *Machine Tool Systems Management and Utilization, Machine Tool Task Force Report on the Technology of Machine Tools,* Vol. 2, Lawrence Livermore National Laboratory, October 1980, pp. 31–41.

13. Carter, Charles F., Towards Flexible Automation, *Manufacturing Engineering,* Vol. 89, No. 2, August 1982, pp. 75–79.

14. Kuwahara, Yasuo, Living with New Technology—Japan's Experience with Robots, in *Highlights in Japanese Industrial Relations: A Selection of Articles for the Japan Labor Institute,* Japan Institute of Labor, Tokyo, 1983, pp. 75–79.

15. Bureau of Labor Statistics, U.S. Department of Labor, *Characteristics of Major Collective Bargaining Agreements: 1 January 1980,* Government Printing Office, Bulletin 2095, Washington, D.C., 1981.

BIBLIOGRAPHY

AFL-CIO, Industrial Union Department, *Comparative Survey of Major Collective Bargaining Agreements: Manufacturing and Non-Manufacturing,* Technical Report, AFL-CIO, January 1982.

American Machinist, The 12th American Machinist Inventory of Metalworking Equipment, 1976–1978, *American Machinist,* Vol. 122, No. 12, December 1978, pp. 133–148.

Ayres, R. U. and Miller, S. M., Robotics and Conservation of Human Resources, *Technology and Society,* Vol. 45, No. 3, Winter 1982.

Bureau of Labor Statistics, U.S. Dept. of Labor, *Occupational Employment in Manufacturing Industries, 1977,* Government Printing Office, Bulletin 2057, Washington, D.C., 1980.

Hasegawa, Yukio, *How Society Should Accept the Full-Scale Introduction of Industrial Robots,* Technical Report, Japan External Trade Organization, Machinery and Technology Department, 1982.

Hasegawa, Yukio, *Age of Robotization,* Technical Report, Japan External Trade Organization, Machinery and Technology Department, 1982.

Robot Institute of America, *Robot Institute of America Worldwide Robotics Survery and Directory,* Society of Manufacturing Engineers, Dearborn, Michigan, 1982.

CHAPTER 26

THE IMPACT OF ROBOTICS ON EDUCATION AND TRAINING

ANN M. MARTIN

U.S. Department of Education
Washington, D.C.

26.1. THE ROBOTICS REVOLUTION IN PERSPECTIVE

Recent reports and studies indicate that robotics technology and flexible manufacturing systems must be given high priority in the drive to rebuild the U.S. economy with a high-technology base. In brief, they indicate that (1) there is widespread agreement among economists and business and government leaders that American industry must retool rapidly if it is to compete in world markets; (2) industrial robotics must replace some of the aging capital stock in the United States; (3) that these new capital-intensive processes will increase productivity and will enable U.S. firms to become more competitive on world markets. In short, we are entering a new industrial era as the result of a high-technology and robotics "revolution."

The objective of this chapter is to discuss present and anticipated changes in jobs and job requirements in the industrial sector as a result of this revolution, the pace of these changes, and how education and training programs must deal with them. The chapter is organized along two main topics: (1) human resource implications of the robotics revolution, covered in Sections 26.2 and 26.3; (2) education and training implications, in the rest of the chapter.

First, however, it is necessary to put the revolution into some perspective. There is very little hard data on industrial robots and how robotics will change job requirements. The public conception of robots has been shaped largely by hyperbole in the popular press in the last year or so. For example, Peter F. Drucker, writing for the *Wall Street Journal,* reports that as a result of the move toward high technology and the rapid strides in the use of automation in industry, our work force will rely increasingly on specialists, thus alleviating the need for a large segment of the blue-collar work force.[1] Similarly, but in a less serious vein, John Naisbitt in his book *Megatrends* points out that our national talents and human resources predispose us to create an "information society" in which most of us will eventually be engaged in "the creation, processing, and distribution of information." He believes that as our school systems fail us, corporations will become the universities of the future.[2] Henry M. Levin and Russel W. Rumberger, in a *Washington Post* article entitled "High-Tech Requires Few Brains," take a different point of view—that the jobs being created by high technology will demand *less* skill, not more.[3] Meanwhile, Gene Bylinsky writes in a *Fortune* article, "The Race to the Automatic Factory," that we are in a "pre-industrial society," that is, we are at the threshold of a new industrial revolution.[4] He suspects that many companies do not have the skilled personnel required for radical changes in industrial patterns, in that robotics requires new skills that do not necessarily build on established skills.

Although strong growth is expected in robotics, this growth may not be as rapid as many would like to see. There is much evidence to suggest that a revolution in manufacturing processes will not happen overnight and that the word "revolution" may be inappropriate when dealing with *any* manufacturing process technology. Hunt and Hunt in their research report on the human resource implications of robotics point out that traditional production has a long history and will not be easily or immediately replaced.[5] Their study found that the growth rate of numerically controlled machine tools, usually

This article was written by the author in her private capacity. No official support or endorsement by the Department of Education is intended or should be inferred.

regarded as the capital equipment most closely related to robots, averaged only 15% annually for the period from 1965 to 1981 and only 12% annually during the 10-year period from 1972 to 1981 in the United States. Furthermore, after 25 years, only 3–4% of all metal-cutting machine tools are numerically controlled. Hunt and Hunt also report that digital computers, widely heralded as the most significant technological innovation of the 1960s and 1970s, had an annual percentage increase averaging 26% for the period from 1961 to 1979. It remains to be seen whether the change in manufacturing-process technology, as the result of robotics, will be incremental and evolutionary or revolutionary.

In considering the factors that well may place constraints on the rapid application of robotics in industrial technology, two areas out of the several that might be ascribed a significant role seem most important for the purpose of this chapter. These two interrelated areas are social/demographic and human factors.

26.1.1. Social/Demographic Factors

Several social and demographic factors must be considered in conjunction with the speed and degree to which America becomes a high-tech information society. These factors, developing in response to existing social, political, and economic needs, include: (1) rededication of resources to basic industrial sectors of our society; (2) anticipated shifts in the labor force; and (3) fear of massive unemployment because of any type of automation. These countertrends to how we move into a new industrial age will have their own impact on the types of skills required by the labor market in the near and not so near future and how these skills will be developed through education, training, and/or retraining programs.

26.1.2. Rededication of Resources to Basic Industrial Sectors

Amitai Etzioni believes that the changed economics of world energy and the decaying state of the economy's infrastructure have been recognized and that a redirection is being achieved of resources to basic sectors of the society (mining, transportation, construction, steel) that have been neglected and cannot be replaced by high technology and knowledge industries.[6] Our national security requires that we not rely on other nations for shipbuilding, steel manufacturing, and mineral mining, especially coal. The United States must shore up its materials infrastructure and capital goods sectors after a generation of public and private overconsumption and underinvestment. This entails repairing U.S. ports, railroads, highways and waterways, pipelines, building massive plants for synthetic fuels, and substituting for existing machinery energy-efficient models that could possibly be adapted to energy other than oil. These are all primary- and secondary-sector heavy-duty industries. By such rededication of resources, Etzioni has estimated, two out of three workers will work in essential basic industries and related services while only one-third of the work force will engage in high technology.

26.1.3. Pace of Labor Force Shifts

Anticipated changes in the labor force in the various sectors of the economy do occur, but at a very slow pace—less than 1% a year. Far from closing down one sector—and opening another—a marginal and very gradual shift of resources takes place. Etzioni points out that over a period of 20 years (1959–1979), the proportion of workers employed in manufacturing declined at an average annual rate of 0.175%, and is expected to decline another 1.4% by 1990.[7] Between 1959 and 1979, services increased their share of the labor force at an average annual rate of 0.3%; a rise of less than 2% is expected by 1990. Although each percentage point represents man thousands of people, in actuality there is only a gradual shift in the total employment mix.

26.1.4. Fear of Massive Unemployment

Fear of massive unemployment with the introduction of new production processes is as old as the industrial era. Both labor and unions have often obstructed or delayed the introduction of automated processes out of concern for the loss of jobs. These fears are particularly acute during major recessions. For example, the automation problem caused national concern in the early 1960s after a slow recovery from the sharp recession of 1958–1959. There were grim predictions that automation was causing permanent unemployment in the auto and other industries. A national commission was appointed to study the problem, and in 1966, with the labor force almost fully employed, the commission rendered its final reports. It concluded, to no one's surprise, that a sluggish economy was the major cause of unemployment rather than automation. Nevertheless, pressures from labor and unions to preserve and protect jobs had an impact on the introduction of automated processes.

26.1.5. Human Factors

Three human factors stand out as important in the application of industrial robotics in the United States: (1) the lack of skilled and trained professional and paraprofessional personnel; (2) the lack of management expertise to cope with a technological age and environment; and (3) the lack of management commitment to robotics.

26.1.6. Lack of Skilled/Trained Professional and Paraprofessional Personnel

American industry today and for the immediate future lacks trained personnel to implement robotics technology and to maintain and support that technology once installed. Graduate engineers in robotics are scarce and their experience is generally minimal.

American universities produce engineers who are overspecialized rather than generalists who understand manufacturing technology and how to apply it. Although educational programs for skilled robotics technicians (a two-year degree) are expanding rapidly, the training of graduate engineers is much more uncertain. Universities must upgrade their personnel and facilities to meet the challenge of robotics education. Few current faculty have training experience in robotics, and only a handful of PhDs are produced each year with a background in robotics. Textbooks are scarce, and the sources of robotics information are scattered among many conferences and journals. Professors must retrain themselves before they can teach robotics to their students. Industry must play a vital role in robotics education by lending personnel and experience to local universities.

26.1.7. Lack of Management Expertise in Coping with Robotics

Flexible manufacturing systems (FMS) are spreading throughout Japanese industry, and Panasonic, Mitsubishi, and other consumer and industrial goods producers are installing the new systems. So far, nothing comparable has happened in the United States. In 1983 American industry can boast about 30 flexible manufacturing systems in place; in Japan one large industrial company, Toyota Machine Tool Company, reportedly has more than 30.

The Japanese lead underscores the growing concern in the United States that American managers are too remote from technical disciplines to appreciate the potential of such new technologies and more engrossed with short-run financial results than long-term investment. More often than not, American executives regard flexible machining systems only in relation to the narrowly defined functions of the conventional tools it might replace—not for its potential to provide a different and far more efficient organization of the manufacturing process. More than likely, they tend to cling to obsolete machinery.

Managers' expertise in capital-intensive production planning is no less antiquated. Dedicated to short-run objectives and simultaneously holding down costs, they often fail to effect maximum plant efficiency. From all evidence, they are more obsessed with cutting back direct labor costs rather than interested in reorganizing their work force or investigating the extent to which new technologies could lower labor costs.

26.1.8. Lack of Management Commitment to Robotics

Management commitment is needed for the successful implementation of industrial robots. This commitment is rare and often limited. For example, pilot installations of robots have almost invariably identified a particular department or area of the factory that can operate *in isolation,* thus assuring a relatively easy introduction of robots which may, in turn, also assure their success. Those types of pilot installations, unfortunately, are limited in both scope and efficiency. It is becoming increasingly evident that American user firms must rethink and restructure the factory to accommodate robots. The traditional emphasis of American industry on short-run payback does not facilitate such rethinking, however, and there are few indications that this emphasis will change.

A.   HUMAN RESOURCE IMPLICATIONS

26.2.   ROBOTS' POTENTIAL TO DISPLACE WORKERS

A recent Carnegie-Mellon University study identified prime occupations vulnerable to robotics.[8] It analyzed the types and percentages of tasks robots could perform at two levels of sophistication. "Level 1" robots, those able to perform relatively simple operations, will be able to perform such tasks as production painting, pick-and-place operations, and spot welding. "Level 2" robots, which will have sensory capabilities such as vision and touch, will be able to perform precision tasks such as assembly and machining. Production facilities may be computer controlled with robots moving the workpiece from machine to machine. Applying estimated percentages of tasks robots could perform to the current number of workers involved in the vulnerable occupations, the study concluded that Level 2 robots theoretically could replace 2.9 million production workers. Table 26.1 summarizes a major finding.

TABLE 26.1. ESTIMATES OF THE NUMBER OF JOBS ROBOTS COULD DO IN ALL MANUFACTURING

Occupation	Total Employed, 1979 (All Manufacturing)	Percentage of Jobs That Could Be Done		Number of Jobs That Could Be Done	
		Level 1[a]	Level 2[b]	Level 1[a]	Level 2[b]
Assemblers	1,289,000	10	30	128,900	386,700
Checkers, examiners and inspectors	746,000	15	35	111,900	261,100
Packers and wrappers (except meat and produce)	626,000	15	40	93,900	250,400
Production painters	185,000	45	65	83,250	120,250
Welders and flamecutters	713,000	25	50	178,250	356,600
Machine operators	3,027,000	20	50	605,400	1,513,500
	6,586,000			1,201,600	2,888,450

Source: The Impacts of Robotics on the Workforce and Workplace, Department of Engineering and Public Policy, School of Urban and Public Affairs, Carnegie-Mellon University, 1981.

[a] Level 1 robots have no sensory input.

[b] Level 2 robots have sensory capabilities such as rudimentary vision and touch.

Converting the projected number of robots into numbers of displaced workers is misleading, however, because automation affects different tasks in different ways. Analyses are needed that relate effect of the technology on specific segments of the work force to establish a time frame for possible displacement. In addition, detailed analyses must develop strategies for both the development and the diffusion of the technology.

For example, if an entire production cell is computer controlled, then human workers will not be needed except for maintenance, provision of the necessary material inputs, and for movement to output. If off-line reprogramming capability becomes available, then human operators will not even be needed to switch to the next batch. Such flexible automated systems will ultimately be linked together and will lead to the automated factory of the future.

However, both flexible cells or systems and the automated factory are still in the future. Bela Gold, an economist at Case Western Reserve who has studied technological change for more than 20 years, expects this "factory of the future" to lie beyond the year 2000.[9]

Hunt and Hunt do not concur with the Ayres and Miller theory that suggests that the gradual job displacement by occupation is proof that our public institutions are currently training their clientele in obsolete skills.[10] They note that Ayres and Miller offer no evidence about the nature of emerging occupations. If policy responses to the challenges of the future are to be formulated, including the possible effects of robotics technology on the work force, then the assessment must be based upon the most probable events expected to occur within a definite time frame. For these reasons, new methods of occupational forecasting need to be devised and employed to assess the impact of robotics technology and other technologies upon the work forces of present industries and employers, and to provide reliable forecast information for guiding the development of education, training, and retraining programs for transition to the year 2000.

26.2.1. Job Displacement

Before discussing the displacement effects of robots, it is important to ensure that the meaning of the term "displacement" is clear. Displacement refers to the elimination of particular jobs, not to the layoff of individual workers. Clearly it is possible that using a robot for a particular job could lead to the layoff of the occupant of that job, but this is not necessarily the case. Layoff refers to the involuntary separation of a worker from his job; displacement refers to the elimination of the job itself.

There is general agreement that although job displacement due to robots will not be a serious problem before 1990, there are specific areas that will be significantly affected and that can already be identified. Chief among these will be painting and welding jobs, for which today's robots are already well suited. Metalworking machine operations and assemblers will also be affected, but to a lesser extent. It is not expected that job displacement will lead to significant job loss among the currently employed. Turnover rates historically have been sufficient to offset the reduction in force that might occur. It will be the new entrants to the labor market who will find more and more factory gates closed. Thus if robotics technology reduces the work force, the less experienced, less educated segment of the labor force will be the most affected.

26.2.2. Job Creation

There is general agreement among forecasters that by 1990 there will be new jobs in four broad areas: robot manufacturing, suppliers to robot manufacturers, robot systems engineering, and corporate robot users (autos and all other manufacturing). The jobs in the corporate robot users category are defined by the maintenance requirements for robots, whereas the jobs in the robot systems engineering category are defined by the applications engineering requirements for robot systems, without regard to industry of employment.

The largest single group of jobs created by robotics will be robotics technicians, a term coming into general usage that refers to an individual with the training or experience to test, program, install, troubleshoot, or maintain industrial robots. It is anticipated that most robot technicians will be trained in two-year community college programs. It is doubtful that these technicians will be in short supply since the nation's community college system gives every indication that they are prepared to train whatever numbers are needed. In fact, the opposite may be the case.

Hunt and Hunt developed a current occupational profile of U.S. robot manufacturers, the motor vehicles and equipment industry, manufacturing plants, and all industry (see Table 26.2).[11] They found that slightly more than two-thirds of the workers in robot manufacturing are in the traditional "white collar" areas of professional, technical, administrative, sales and clerical workers, whereas only one-third are in the traditional "blue collar" areas of skilled workers, production operatives, and laborers. They reported that (1) this is indicative of a young high-technology industry with low sales, which tends to assemble rather than manufacture parts, and (2) this is also indicative of a specialized product with specific requirements for engineering design, programming, and installation for each application.

It is clear from the preceding discussion that a relatively large number of graduate engineers will be needed if robotics technology is to be developed. These will be mostly electrical, mechanical, and industrial engineers. A shortage of electrical engineers already exists, and it is possible that industrial engineers will also be in short supply. Thus the newly emerging robotics industry starts from a deficit position and will be forced to compete for a limited number of qualified professionals. Thus a shortage of engineers could compromise the expansion of robotics technology.

The second most prevalent occupation, engineering technicians, represents 15.7% of the work force. The majority of these jobs could be called "robotics technicians," although there are also mechanical and engineering technicians and electrical and electronic technicians in this category.

TABLE 26.2. CURRENT U.S. OCCUPATIONAL PROFILES—ROBOT MANUFACTURING, MOTOR VEHICLES AND EQUIPMENT, ALL MANUFACTURING, AND ALL INDUSTRIES[a]

Occupation	Employment Distribution (%)			
	Robot Manufacturing	Motor Vehicles and Equipment	All Manufacturing	All Industries
Engineers	23.7	2.3	2.8	1.2
Engineering technicians	15.7	1.2	2.2	1.4
All other professional and technical workers	4.2	2.4	4.0	13.5
Managers, officials, proprietors	6.8	3.3	5.9	8.1
Sales workers	3.4	0.5	2.2	6.3
Clerical workers	13.9	6.2	11.3	19.9
Skilled craft and related workers	8.4	20.8	18.5	11.8
Semiskilled metalworking operatives	4.2	15.8	7.2	1.7
Assemblers and all other operatives	19.0	38.6	36.2	13.1
Service workers	—	2.8	2.0	15.8
Laborers	0.7	6.1	7.7	6.0
Farmers and farm workers	—	—	—	1.0
Total	100.0	100.0	100.0	100.0

Source: Hunt, H. A. and Hunt, T. L., *Robotics: Human Resource Implications for Michigan,* W. E. Upjohn Institute for Employment Research, Michigan, November 1982, p. 77. (Final Report under Contract No. 99–1–1818–17–23–0, Michigan Occupational Information Coordinating Committee.)

[a] Columns may not add to total due to rounding.

The concentration of jobs in the technical areas is offset by a relative lack of jobs in the production occupations typical of more conventional industries. Table 26.2 shows a marked lack of craft workers, semiskilled metalworking operatives, assemblers, and laborers when compared to other manufacturing. Clearly, this reflects the low level of robot production, but it also reveals the high-technology component of robotics.

26.3. SKILL SHIFTS ARE INEVITABLE

A National Alliance of Business study found that the skills of workers likely to be displaced by automated manufacturing technology differ greatly from those required for high-growth jobs.[12] What was evident, according to the study, is that the displaced workers often must undergo significant retraining and/or reeducation before being able to qualify even for entry-level positions in some of the occupations where the largest number of jobs will be.

Some disparity in skills would exist even if the shift were mainly within the manufacturing sector. Jobs created by automation will require technical skills that production workers may not possess. For example, the results of Hunt and Hunt's study on robotics effects in automobile plants in Michigan show significant movement of factory jobs from production line to professional categories. Thus significant retraining would be required in this area of displaced workers on the production line. Furthermore, most new jobs created by robotics will require a variety of technical skills not needed for today's jobs. Such a skill shift could present problems for people training for manufacturing occupations since their skills could be obsolete in a few years. Failure to recognize a potential mismatch between skills and jobs could result in both unemployment and a shortage of workers trained to program, operate, and maintain automated machinery.

26.3.1. Skill Twist

The most remarkable aspect of the job displacement and job creation impacts of industrial robots is the skill twist that emerges so clearly when the jobs eliminated are compared to the jobs created. The jobs eliminated are semi- or unskilled, whereas the jobs created require significant technical background. This skill twist is the most salient factor in the impact of robotics on education and training.

26.3.2. Transfer of Skills

The basic research efforts of Stump into uncovering the human performance requirements (knowledges, skills, and attitudes) that are consistent across a variety of occupations should be reexamined and expanded in view of the skill shifts created by the implementation of robotics technology.[13] Stump raises such questions as, "What are transferable skills?" "What characteristics of jobs should be considered common?" His research has explored the development of classification and analysis systems that permit the identification of potential transfer of skill from one occupation to another—also, the determination of the similarities of performance requirements of jobs and the nature of the process of transferring performance from one program to another or from one job to another.

It may not be necessary to develop different training programs for each emerging occupation. According to Dr. Leonard Greenhalgh, Assistant Professor of Business Administration at Dartmouth College, these new jobs will have certain basic similarities. It will thus be the duty of the public schools to teach these basic core requirements and the company that hires the graduate will train the person for its specific needs.

B. EDUCATION AND TRAINING IMPLICATIONS

26.4. SHORTAGE OF SKILLED PERSONNEL AT ALL LEVELS

One of the most consistently identified barriers to efficient applications of robotics technology is the shortage of engineers, technicians, and other skilled personnel trained in automated manufacturing systems.

Industry executives recognize the problem. Management in automation users such as Ford, International Harvester, and Deere and suppliers such as Robogate, Prab, and Autoplace specifically cite a lack of manufacturing and systems engineers and technicians trained to implement and maintain the new manufacturing technology. This problem is being linked to our educational system. The Machine Tool Task Force study on machine tool state of the art, for example, found more programs are needed that offer courses on programming languages for numerical controls, machine shop planning, and computer graphics for machine tools.[14] Furthermore, the Task Force reported that education and training in manufacturing of machine tools has been neglected during the last three or four decades.

Fortunately, the seriousness of the problem is starting to attract the kind of attention that leads to effective action. Many universities, community colleges, technical institutes, and vocational schools are establishing or upgrading technical programs designed to produce individuals with the new skills now being called for by industrial robotics. Funding by private institutions and state and federal government agencies is also being applied in support of much-needed studies aimed at determining qualifications criteria, specific training goals, and establishing appropriate curricula. Although, as we have seen, information on America's current and future manpower needs in robotics is not very available or precise, nonetheless, certain needs have become obvious since the beginning of the Robot Revolution. These needs include:

Personnel to maintain robots and robot systems
Applications engineers
Researchers
Development teams
College teachers
Large-scale retraining of current personnel
General public education

To fulfill these needs, there exist a number of providers of manpower, education, training, and retraining:

Universities
Four-year colleges
Two-year colleges
Technical schools
Vocational schools
Private sector training centers
Professional organizations [such as Robotics International of the Society of Manufacturing Engineers (RI/SME)]
Short courses
Self-education

The following material covers a sampling of some programs now being implemented in a range of types and levels of institutions and organizations committed to bringing education and training on line with the increased utilization of advanced manufacturing technologies. The material includes a job analysis for a robotics technician provided by the Occupational Analysis Division of the U.S. Department of Labor, and a matrix of occupational competencies for a robotics service technician supplied by Piedmont Technical College, South Carolina.

26.5.   ROBOTICS AND THE UNIVERSITY

Historically, there has been considerable interaction at the research level between universities and the robot industry. Stanford and the Massachusetts Institute of Technology (and their associated local research arms) have been responsible for major breakthroughs in the following areas:

Robot design and kinematics
Force sensing and control
Compliance
Simulation of robot systems
Computer vision
Robot programming languages

These research achievements have been translated into the industrial robots sold today, and the graduates of university robotics programs work for many robot companies.

Universities play an important role in producing research and development staff and in teaching robotics to both new students and old.

26.6.   PREPARATION OF ENGINEERS

26.6.1.   University of Southern California

At the University of Southern California (USC),[15] the Electrical Engineering (EE) Department took the lead in establishing a robotics curriculum. The department faculty recognized that robotics is inherently interdisciplinary, cutting across the boundaries of traditional engineering departments and across the traditional disciplines of computer science, electrical, industrial, and mechanical engineering. They also recognized that robotics could belong to many departments, that is, mechanical engineering, industrial engineering, computer science, and others. (A similar situation existed when computer science was claimed by both electrical engineering and mathematics.) For these reasons, the EE Department at USC encouraged other departments to join in a cooperative effort, now guided by a Steering Committee on Robotics Education (SCORE). SCORE includes members of all the academic departments of the School of Engineering, meets monthly, and reports directly to the Dean. Among the duties of the Steering Committee are the following:

Coordinating robotics courses across departments.

Acquiring and maintaining laboratory facilities for robotics education.

Informing the faculty of seminars and conferences.

Maintaining liaison with local industries regarding their educational needs.

Curriculum

USC now offers four specific courses in robotics:

EE 545   Introduction to Robotics

EE 546L   Basic Robotics Laboratory

EE/CS 547   Software Methods in Robotics

MM/EE 548   Analytical Methods in Robotics

These courses are supplemented by numerous related courses throughout the School of Engineering in fields such as:

Mechanics
Control theory
CAD/CAM
Engineering economics
Computer vision
Artificial intelligence

After completing formal courses, promising students are encouraged to continue work in robotics by means of seminars and individual research. The implementation of a degree program in robotics is still under consideration. Students now take their degrees in a traditional department with a specialization in robotics.

26.6.2.   Purdue University

A somewhat different approach is followed at Purdue University,[16] although, here too, no degree program in robotics is available yet. Three main courses in robotics are offered separately by each of the Electrical, Industrial, and Mechanical Engineering schools. Each course emphasizes the unique professional aspects of the school. Thus in Electrical Engineering the course is Robot Manipulator Programming and Control; in Industrial Engineering, Industrial Robotics; and in Mechanical Engineering, Analysis and Design of Manipulators (Tables 26.3 to 26.5). Students specializing in robotics are encouraged to take at least two of the courses. Thus students are provided with a multifaceted view of the subject of robotics and the interdisciplinary objective can be better served. The courses are supported by extensive laboratory experiments, research projects, and related courses in computer science, artificial intelligence, computerized manufacturing, computer-aided design, and image processing. A strong influence on robotics education at Purdue comes from the Computer Integrated Design, Manufacture and Automation Center (CIDMAC), a university-industry collaborative research center. The center supports interdisciplinary research by the various engineering schools and the School of Management, attempting to integrate CAD, CAM, and robotics.

26.6.3.   GMI Engineering and Management Institute

The robotic facilities at GMI Engineering and Management Institute have been developed with several objectives in mind.[17] The primary objective is to educate both undergraduate engineering students and practicing engineers through continuing education (CE) programs. This philosophy not only covers

TABLE 26.3.   COURSE OUTLINE FOR PURDUE'S EE 569: ROBOT MANIPULATOR PROGRAMMING AND CONTROL

Topic

Robotics and automation	Region growing
Introduction to manipulation	Range
Homogeneous transformations	Binary vision systems
Coordinate frames	Focus feature method
General rotation	Robot task description
Kinematic equations	Joint coordinate motion
Manipulator kinematics	Cartesian coordinate motion
The Stanford arm	Lagrangian mechanics
Differential relationships	Simplified dynamics
The manipulator Jacobian	Position control
Manipulator degeneracies	Torque control
Solution to kinematic equations	Static forces
The Stanford arm solution	Compliance
The SCAR robot	Programming
Introduction to vision	Functionally defined motion
Perception of simple solids	Touch feedback
Segmentation of scenes	Productivity
Representation of complex shapes	Social aspects
Edge detection	Factory of the future

Texts

Robot Manipulators: Mathematics, Programming and Control, R. P. Paul, MIT Press.
Machine Perception, R. Nevatia, Prentice-Hall.

Laboratory Experiments

1. General operating procedures, joint and Cartesian motion, simple program teach.
2. Symbolic robot manipulator programming.
3. Program loops, palletizing.
4. Tests and program branching.
5. Compound transformations, conveyor tracking.
6. Functionally defined trajectories.
7. Manipulator kinematics joint to Cartesian coordinates program.
8. Manipulator kinematics Cartesian to joint coordinates program.
9. Resolved motion rate control and the manipulator Jacobian.
10. Manipulator force control and model updating.
11. Manipulator touch sensor control and model updating.
12. Manipulator vision control and model updating.
13. Robot assembly.
14. Two-manipulator task control.
15. Robot communications.

Source: Nof, S. Y.[16]

the fundamentals in classroom experience, but also provides considerable hands-on experience through the use of the eight industrial robots in the lab.

An Introduction to Robotics course has been designed to be of direct benefit to practicing engineers. The robotics course, ME 475, is a three-credit course consisting of two hours of lecture and two hours of lab per week. Lecture includes such topics as robot classifications, components, tooling, matching the robot to the workplace, planning and implementation, reliability and maintenance, economic considerations, safety, and future developments. (See Table 26.6 for the course outline.)

At present, practical experience is obtained during the lab time as the students learn to program eight types of industrial robots in the robotics lab. The "Introduction to Robotics" course is at present offered to Junior-level mechanical engineering students and to Senior-level electrical and industrial engineering students.

TABLE 26.4. COURSE OUTLINE FOR PURDUE'S IE 575: INDUSTRIAL ROBOTICS

Objective

To study the design, control and operation of programmable robot systems in industry. Specific objectives are:

1. Review control theory as a basis for robotic manufacturing.
2. Analyze problems of design and operation of industrial robot systems.
3. Learn, through laboratory experiments, the programming and use of industrial robots.
4. Examine new technologies of computer control, robotics and programmable assembly.

   Recommended Texts

1. Boothroyd, G., Poli, C., Murch, L. E., *Automatic Assembly,* M. Dekker, 1982.
2. Engelberger, J. F., *Robotics in Practice,* AMACOM, 1980.
3. Warnecke, H. J. and Schraft, R. D., *Industrial Robots,* I.F.S. Publications, 1982.

Laboratory Experiments

1. Lab introduction to robots; safety
2. T^3 teaching
3. Load-unload tasks
4. Palletizing
5. Part insertion
6. RTM exercises
7. IBM AML basic program
8. AML advanced programs
9. Compliance
10. Assembly
11. Recovery
12. PTP and CP in path control
13. Multiple robot control

Topics

1. Control fundamentals.
   a. Review of control theory and models.
   b. Control of motion, position and operation.
2. Industrial robot hardware.
   a. Basic structure, control and performance specifications.
   b. Control hardware and sensors.
   c. Design considerations of control systems and mechanical components.
3. Robot programming and application planning.
   a. Teaching (T^3).
   b. Programming (AML).
   c. Robot programming languages.
   d. Task level programming.
   e. Task evaluation (RTM).
4. Control of robotic systems.
   a. Adaptive and optimization control.
   b. Distribution and hierarchical control.
   c. Operational control.
   d. Algorithms for industrial task control (conveyor tracking, welding, bin picking, inspection).
5. Programmable assembly.
   a. Components and control.
   b. Planning assembly cells.
   c. Economic and other design considerations.
   d. Product design for assembly.

Source: Nof, S. Y.[16]

TABLE 26.5.   COURSE OUTLINE FOR PURDUE'S ME 597T: ANALYSIS AND DESIGN OF MANIPULATORS

Topics

1. Kinematics and dynamics   7 weeks
   Kinematic configurations, position solution strategies, and simulation techniques. Force and torque solutions and rigid body dynamics.
2. Sensors and actuators   3 weeks
   Selection and placement of various sensors and motors including position, velocity, acceleration, force and torque sensors, and electric, hydraulic, and pneumatic motors.
3. Control methods   3 weeks
   Discussion of various control strategies including servo, digital, hierarchical, and adaptive control.
4. Special topics   2 weeks
   Will include design considerations, parts assembly, flexible linkages, and vision.
   ** Robot simulation programs, graphic displays, and hardware will be used throughout the course to demonstrate the concepts and ideas presented.

Source: Nof, S. Y.[16]

Laboratory/Research Projects

Upon completion of the introductory course, many of the students are given work assignments during the co-op phase of their education that are directly related to robotic activities.

New Courses in the Offering

Besides the introductory course, there are two more courses offered in robotics: (1) "Robotic Applications," which deals with applications, sensors, and interfacing, and (2) "Robotic Interfacing and Commu-

TABLE 26.6.   ME 475 INTRODUCTION TO ROBOTICS—OUTLINE

WEEK 1	Session 1	Course introduction, Robot definition
	Session 2	Robot classification: (1) program control and (2) path generated, (3) design variations, (4) functional type, (5) technology type, (6) method of programming
WEEK 2	Session 3	Robot classification (cont'd)
	Session 4	Generic programming
WEEK 3	Session 5	Robot components (mechanical unit, power source, control system, and programming and tooling)
	Session 6	Robot components (cont'd)
WEEK 4	Session 7	Hydraulic, pneumatic systems, and servo valves
	Session 8	Positional and velocity sensors (encoders, resolvers, tachometers)
WEEK 5	Session 9	Description of analog and digital control systems
	Session 10	Robot operating characteristics (operating speeds, repeatability, load capacity, work envelope, memory, programming, limitations, mounting restrictions)
WEEK 6	Session 11	Robot operating characteristics (cont'd)
	Session 12	End effectors (tools and grippers)
WEEK 7	Session 13	EXAM
	Session 14	Matching robots to workplace (workplace configurations, floor layout, material-handling requirements, backup, accessibility for maintenance and operators, and facilities)
WEEK 8	Session 15	Matching robots to workplace (cont'd)
	Session 16	Reliability and maintenance
WEEK 9	Session 17	Robot planning and implementation
	Session 18	Advantages of robots
WEEK 10	Session 19	Economic considerations
	Session 20	Safety
WEEK 11	Session 21	Future developments
	Session 22	EXAM
WEEK 12	Session 23	Review
		FINAL EXAMINATION

Source: GMI Engineering and Management Institute, Flint, Michigan, 1983.

nication," offered by the Electrical Engineering Department, which includes such topics as analog and digital electronics, controls, vision, speech, and networks.

Continuing Education (CE) Robotic Programs

At present there are two continuing engineering (CE) education courses being taught at GMI's facilities. The first one is a three-day robotics course offered by GMI's CE department and is entitled "Robotics in Manufacturing." This program consists of 14 hours of lecture on basics such as classifications, systems, components, advantages, justification, planning and implementation, operating characteristics matching to the workplace, end effectors, programming, and future developments. Ten hours are devoted to lab time, with four two-hour blocks given to general instruction on operating and programming procedures for the various robots.

Future CE Program

New robotic courses are being developed for regular engineering students, and corresponding continuing engineering education courses are planned to be offered to the public.

26.7. ROBOTICS AND THE TWO-YEAR COLLEGES

Technical education is defined by the U.S. Department of Labor as "programs that prepare individuals at the technical or specialist level in a specialized field of technology in the physical, related engineering, biological, and/or social sciences."

The education of technicians has generally required that they learn and be able to apply the underlying scientific principles of their special field and the mathematics that support the principles. Technicians also must have enough knowledge and judgment to perform specific portions of the work of the professionals they support. This often involves the supervision of skilled or semiskilled workers.

Preparation of technicians is different from that of skilled workers or craftsmen, who generally need to master much less theoretical and more practical science and mathematics related to the special skills needed to perform their jobs. The current developments in industrial robotics will increase the need for technicians while decreasing the need for semiskilled workers and craftsmen. The largest number of robotics technicians will be taught in formal two-year programs.

26.7.1. Robot Technician Job Description

Robot technician is a generic term that describes individuals who have sufficient familiarity with robotics technology to be capable of testing, programming, installing, troubleshooting, and maintaining industrial robots. The Occupational Analysis and Testing Section of the U.S. Department of Labor has recently (1983) developed a job description for *Robot Technician* and assigned a permanent Dictionary of Occupational Titles Code, 636.261–580, to this job title (see Table 26.7). This is a new code under the category of "Miscellaneous Occupations in Machine Installation and Repair." New jobs in robotics will be assigned in this area. The fourth digit of the occupational code (2) and the sixth digit (1) refer to the skill levels of Analyzing and Precision Working. This indicates that a relatively high skill level has been assigned to this job, which can be taken into consideration when establishing training requirements.

The Department of Labor has also established a code of 05.05.09 for this job in the Guide for Occupational Exploration. This code, which comes under the category of "Craft Technology," can be used for guidance in establishing job relationships or in job counseling.

26.7.2. Preparation of Technicians

South Carolina's Technical Education System (TEC)

A system of 16 technical colleges provides courses specifically geared to the South Carolina job market. Currently, TEC is emphasizing its "Design for the Eighties" program which focuses on training for emerging high-technology industries.

Piedmont Technical College

Designated the Robotics Resource Center for the South Carolina Technical Education System and created under the "Design for the Eighties" program, this center has attracted national attention from industry, education, and the media. One Cincinnati-Milacron T-3 robot is located in the center on consignment, and a Seiko watchmaker has been donated. The center conducts comprehensive sessions for instructors in the State TEC System as well as robotic workshops for business and industry in South Carolina. Programs have been developed that utilize nationally known experts in the robotics field (see Figure 26.1).

TABLE 26.7. JOB ANALYSIS AND CLASSIFICATION—ROBOTICS TECHNICIAN

638, 261–580 ROBOT TECHNICIAN (and ind.)
Robotic Repairperson, Robotic Maintenance Person

Repairs and maintains automated robotic equipment consisting of mechanical, pneumatic, hydraulic, computer hardware, and computer software components. Confers with robot manufacturers' representatives and service people to assist in diagnosing problems and maintaining robotic equipment and systems of which they are a part. Repairs and/or replaces faulty or defective parts or components of the robot or system of which the robot is part, including mechanical, pneumatic, or hydraulic parts, and computer integrated circuit boards or other electronic components. Capable of executing diagnostic robot routines per manufacturers' instructions including returning robot program to null or initial position. Capable of programming robot to teach mode for the purpose of verifying repairs or maintenance. May be capable of complete "teach box" operation to set up robot for production operations as instructed. Capable of understanding mechanical drawing, electrical drawing, hydraulic or pneumatic network drawings, and computer software flow diagrams.

Skills

1. Manual dexterity in a precision environment.
2. Read and interpret mechanical, electrical, pneumatic, hydraulic schematics, and computer flow diagram.

   Knowledge

1. Knowledge of electrical, mechanical, and hydraulic mechanics, and computer hardware and software.

   Abilities

1. Reading at a twelfth-grade level
2. Distinguish total spectrum of colors
3. Visualize objects in space
4. Eye/hand coordination
5. Analytical
6. Integrate

Prepared by: Occupational Analysis Field Center, Raleigh, North Carolina, Bureau of Employment Security, U.S. Department of Labor.

TEC has developed a Dacum matrix of occupational competencies for a robotics service technician and a robotics application technician. Table 26.8 presents the competencies for a robotics service technician.

Macomb County Community College

Macomb, in Warren, Michigan, is generally acknowledged as the originator of the robotics technician curriculum in the United States.[18] The beginning of Macomb's program dates to 1978 when it added a specialty in robotics to its fluid power technology associate degree program.

The current robotics technician program is designed to provide students the opportunity to develop basic skills in hydraulics, pneumatics, electronics, and mechanical devices. Students then learn to analyze, test, and troubleshoot hydraulics, pneumatics, and electrical circuits. The operation and maintenance of servo valves, function generators, and logic control devices is also covered. Throughout the training program, students are involved in laboratory projects that have them working with the actual industrial equipment.

Upon completion of this program, students will have earned 50 credits toward an Associate Degree and be prepared for jobs as robot installers, robot programmers, robot application technicians, robot design technicians, and robot mechanics.

26.8. UNIVERSITY-INDUSTRY COLLABORATIVE EFFORTS

26.8.1. Industrial Assistance to Universities

Solving some of the ongoing engineering-education problems may be best achieved through cooperative industry-university efforts. Many of these already exist; for example, General Electric, General Motors, and Boeing recently launched a new productivity center at Rennsselaer Polytechnical Institute; Control Data Corporation is committed to developing a Purdue University Center for research in computer-aided design and manufacturing; at least six other universities have industry-sponsored manufacturing

Fig. 26.1. Students at the South Carolina robotics center become acquainted with their new partner on the production line.

research centers. Engineering education is receiving further assistance from industry to address faculty shortages in engineering schools. For example, Exxon Educational Foundation is providing $15 million in grants to 66 institutions to create 100 teaching fellowships and 100 salary-support grants; eight high-tech conglomerates are financing the American Society of Engineering Education to study faculty shortages; and the Ford Motor Company is sponsoring a program to develop a new curriculum with the Society of Manufacturing Engineering. In addition, the Massachusetts High Technology Council has proposed that its 125 members contribute up to 2% of their yearly research budgets toward funding more engineering faculty and buying computer equipment. This funding could reach $14 million in 1983.

Michigan's Industrial Technology Institute

An Industrial Technology Institute has been established in Michigan with the shared commitment of the Michigan state government, the business community, labor, and Michigan's education network. This nonprofit institute is designed to use the resources of Michigan's higher education system, while working with private industry, to develop leading-edge training and retraining programs needed to integrate advanced robotics technology into the workplace. The Institute is supported by private corporations, foundations, individuals, and governmental sources.

26.9.   ROBOTICS AND PRIVATE-SECTOR TRAINING

26.9.1.   Engineering Education and Training in Industry

In addition to university-industry cooperation efforts, industry has taken the initiative to do a great deal of teaching itself. As the pace of technology change accelerates, more training will occur on the job where new knowledge is generated and skills have been developed. For example, Boeing is investing in training engineering graduates to use their computer-automated systems, General Motors has its own accredited engineering institute, and IBM has grant and fellowship programs as well as a post-doctoral program that brings people into their research labs for one or two years.

26.9.2.   Training for Users of Robots

David M. Osborne[19] has asserted that proper training, before and after a robot installation, is an important factor in the economic use of robots. The main areas in which he sees training affecting

TABLE 26.8. A DACUM MATRIX OF OCCUPATIONAL COMPETENCIES—ROBOTICS SERVICE TECHNICIAN: A ROBOTICS SERVICE TECHNICIAN WILL IDENTIFY PROBLEMS AND TAKE APPROPRIATE ACTION TO REMEDY THE PROBLEM QUICKLY

	1	2	3	4	5	6	7	8
A. Communicate	A-1 Communicate orally	A-2 Listen	A-3 Read effectively	A-4 Write technical reports	A-5 Train	A-6 Follow up		
B. Implement and maintain robot	B-1 Develop thorough operating ability of robot	B-2 Practice safety	B-3 Evaluate robot performance	B-4 Apply programming techniques	B-5 Implement and maintain end effectors	B-6 Implement and maintain sensors	B-7 Implement a preventive maintenance program	
C. Troubleshoot electronics systems	C-1 Demonstrate electrical fundamentals	C-2 Demonstrate electronic fundamentals	C-3 Demonstrate computer fundamentals	C-4 Demonstrate programmable controller fundamentals	C-5 Demonstrate servo system fundamentals	C-6 Read electrical schematics	C-7 Demonstrate electronic test equipment	C-8 Utilize diagnostic aids
D. Troubleshoot hydraulic and pneumatic systems	D-1 Demonstrate hydraulic fundamentals	D-2 Demonstrate pneumatic fundamentals	D-3 Demonstrate servo system fundamentals	D-4 Read schematics	D-5 Operate test equipment			
E. Implement and maintain interface elements	E-1 Practice safety	E-2 Evaluate system performance	E-3 Implement and maintain peripheral equipment	E-4 Implement and maintain programmable controller	E-5 Implement and maintain communications links	E-6 Implement and maintain interlocks	E-7 Implement a preventive maintenance program	
F. Utilize problem-solving skills	F-1 Apply problem-solving logic	F-2 Apply statistical methods	F-3 Maintain records	F-4 Determine cause of failure				

DACUM CONSULTANT: Jack Harris, Stark Technical College

Panel members: Wayne Mechlin, Cincinnati Milacron; L. E. Ording, Ford Motor Company; Dick Wetherill, Caterpillar Tractor Co. Sponsored by the S.C. State Board for Technical and Comprehensive Education and Piedmont Technical College.

the use of robots are (1) removing the robot's false mystique, (2) properly utilizing the robot's potential, and (3) maintaining the robot for greatest productive life.

Osborne believes that for the best robot installation people must be trained from areas of the company that may seem far from the actual use of the robot. Although each case is different, employees a company may expect to train include:

Process engineers

Tooling engineers

Material-handling engineers

Plant engineers

Engineering staff designing the product

Maintenance staff

Production operators

Production supervisors

Personnel staff

Purchasing staff

Plant personnel working near the robot that may become uneasy at its presence

While it is obvious that the people who will actually implement the robot and will use it daily as a part of their work must receive training, it is a relatively new concept that those who will design the product on which the robot will work must become familiar with its capabilities.

Osborne proposes the following five levels of training to be considered by users who plan for the installation of a robot:

1. Extensive technical training.
2. Simple technical training.
3. Operation training.
4. Features/economies training.
5. Awareness training.

Extensive technical training would include the most intricate details of the working of the robot, e.g., the details for the customer to maintain the robot properly without the help of the vendor. This training might not be as necessary if the user is to have a service contract with the vendor, but must still be considered for the manufacturing engineering staff that will rely on the robot's ability to accomplish a task. Simple technical training might include the development of skills such as the repair of the robot and its ability to interact with other equipment.

Operational training would include such features as the programming and the day-to-day operation of the robot. Education on the features and the economies of robot operation could include the ability of various robots to accomplish the different tasks of the user and the relative merits the robot has for the economic operation of the plant.

Awareness training is a simple explanation of the facts about robots that will allow everyone to judge a robot properly on its real features.

Much of the training that is needed by the customer can be provided by the robot vendor. A suggested schedule of training that could be expected to fill all of the user needs is listed as follows:

Electrical maintenance

Mechanical maintenance

Scheduled maintenance

Robot use engineering

Nontechnical features

Demonstrations and literature for awareness training

Some of the individual classes might be combined to allow several aspects of robotic technology to be taught at once.

26.10.   ROBOTICS AND RETRAINING PROGRAMS FOR DISPLACED WORKERS

The need for large-scale retraining programs depends on how rapidly displacement occurs, what types of jobs are created, and the location of these jobs. These factors may also determine who will bear the retraining responsibility. It is clear that employers would be more willing to retrain workers for new jobs or for other jobs within the plant. Similarly, unions would be more likely to participate if new jobs stayed within the bargaining unit. But the responsibility for retraining workers when the employer has no new jobs to offer is less clear.

CETA Retraining Program (Michigan)

A Comprehensive Employment and Training Act (CETA) demonstration project sponsored by the Downriver Community Conference in Michigan is trying to solve some of the problems of workers displaced by the new automation. The project offers training and placement assistance to workers laid off because of increasing import sales and the recession. During 1983 the project included a robotics training program which demonstrated one role for vocational education institutions in retraining displaced workers. A two-year community college course was condensed to 36 weeks as workers were retrained to build and service robots. At the completion of the training program, CETA placement services and industry contacts were used to find the workers new jobs.

Cross-Training (Pennsylvania)

The Community College of Allegheny County, western Pennsylvania, has established a program to train unemployed steelworkers in a millwright-to-stationary-engineer cross-training program. The cross-training concept is to provide a training program that will transfer experience and skill to new and growing job areas for displaced workers with experience and skills in an occupational area that is becoming obsolete or where jobs have been displaced. Assistance in developing cross-training programs in western Pennsylvania has been provided through the cooperative efforts of Westinghouse Electric Corporation, D'Appalonia Engineering, Carnegie-Mellon University, and the United Steelworkers Union.

Meanwhile a robotics center established at the community college is preparing to upgrade factory and unemployed steelworkers to qualify for manufacturing technology maintenance and repair jobs (see Fig. 26.2). Here again, the cross-training course has been developed for experienced workers who have transferable skills. The Robotics Center has identified 23 western Pennsylvania firms that have a need for automation maintenance. The technicians who complete the program will be prepared to work in such fields as hydraulics, pneumatics, electronics, mechanics, and microcomputer programming. Many factory and steelworkers already have some of the skills needed for the new automation maintenance and repair jobs which they acquired when repairing factory and mill equipment and machinery.

26.11. UNION-EMPLOYER RETRAINING PROGRAMS

Trade unions with members in industries vulnerable to CAD/CAM are generally approaching the question of automation with caution. Current economic conditions, combined with growing levels of automation, increase the importance of job security. Although relatively few workers have specific contract protection, programmable automation has already become a larger factor in contract talks and contract arguments. The ability of workers to obtain reasonable protection from displacement will affect the level of opposition and thus the rate of adoption.

26.11.1. Two Innovative Union-Employer Ventures

Ford and General Motors are currently involved in innovative programs aimed at retraining workers with outmoded skills.

Ford Motor Company and the UAW

The Ford Motor Company in conjunction with the United Auto Workers (UAW) has initiated a National Development and Training Center program. This program is available to any unemployed UAW-Ford worker or any UAW-Ford currently employed hourly worker. The Center and program will be based at the Henry Ford Community College in Dearborn, Michigan.

The Center's programs will be directed to education and training requested by the worker as opposed to job training offered by the company, and will include training in new jobs areas such as robotics. Funding for the program comes from five cents per hour provided in the new UAW-Ford contract.

General Motors and the UAW

With a new provision of $4 million in the UAW-General Motors contract and $6 million from the State of California Work Site Education and Training Act program and from the federal government, a program has been launched to find new jobs for unemployed GM workers from the Fremont and South Gate company assembly plants.

Fig. 26.2. Westinghouse and the Community College of Allegheny County in Pittsburgh, Pennsylvania, are collaborating on a program of *cross-training* to upgrade experienced workers for maintenance and repair of advanced automated assembly equipment, such as Westinghouse's Adaptable Programmable Assembly System (APAS).

The California program, unlike that in Michigan, will provide training, both private and public, in demand occupations. These demand occupations will be given priority and will include smog-control, auto mechanics, aerospace equipment assemblies, data processing equipment repairers, and heavy-equipment mechanics.

26.12. ROBOTICS AND VOCATIONAL SCHOOLS

Patricia A. Cole in a report for the *GAO Review* called "Vocational Education and the Robotics Revolution" has stressed that as the parameters of the impact of this new technology become apparent, so do the challenges facing those who must prepare the work force.[20] The skills associated with programmable automation require new highly technical training programs. However, these new programs may not need to train as many people as have the traditional vocational programs, such as welding or drafting. Thus the vocational education system will have to provide very expensive training for a relatively small number of students. At the same time, the vocational education system will be confronted with the task of providing trainees for the nonmanufacturing sectors, many of which will also feel the effects of microelectronics.

Cole points out that educators will have to reexamine the vocational education system, particularly at the secondary school level, not only to decide where to place future emphasis, but also to prevent students from training for occupations that could soon be obsolete. And reexamination must extend beyond vocational training. Secondary education also must evaluate the adequacy of existing mathematics and science courses as these disciplines become prerequisites for obtaining highly skilled technical jobs in automated technology industries.

26.12.1. Remedial Education Needs

Thus it is well to remember that vocational education does not exist in a vacuum. Its best efforts to produce highly skilled workers and technicians, in the numbers required, cannot succeed if our public schools continue to turn out substantial numbers of graduates who are functionally and technologically illiterate. The amount of remedial education that vocational and technical education teachers must do is staggering. Perhaps the best "vocational education" a high school graduate can have is a solid grasp of the academic fundamentals, with a solid base in mathematics and the physical sciences.

Thus the tasks of vocational-technical educators are becoming increasingly more difficult and complex. They must respond to the enormous and growing need to train and retrain workers affected by industrial dislocation, while at the same time preparing a new generation of workers for employment in a labor market being changed by rapid advances in technology.

26.13. EQUITY AND EFFICIENCY

Currently women—especially minority women—play a small role in high-growth occupations. Among technicians, for example, there were 169,000 males in 1981, but only 48,000 females, according to the Bureau of Labor Statistics. This is less than 25% of the total. Less than 5% were minority women. Similar statistics exist for computer specialists and engineers. Furthermore, the evidence from many studies indicates that women (and to a great extent minority males) have had a quantitatively different preparation in math and science than have white males.

A small number of girls, women, and minorities are currently enrolled in vocational or technical education programs that prepare students for technological careers or other high-growth industries. For girls and women students, the problem of entrance into these new fields is complex: sex-stereotyped elementary education leads very few girls to enroll in math and science courses that would prepare them to enter technological instruction; inadequate career counseling linking school curriculum to technical jobs, and the lack of support services (i.e., child care and transportation) in vocational-technical schools, prohibits access for adult women to appropriate training. Issues of occupational segregation, sex bias, female poverty, and the automation of many traditional female jobs make the access of women to solid science and technology training an important and urgent educational goal.

Pat Choate, in a report for the Northeast-Midwest Institute called *Retooling the American Work Force: Toward a National Training Strategy,* points out that the growth of the American workforce is slowing dramatically as the "baby boom" generation matures.[21] He notes that women are *the* major source of new workers in the 1980s—two of every three new workers—but that no institution has integrated women fully into its professional, managerial, technical, or production jobs.

In a research report to the American Society for Training and Development, "Improving Access and Mobility of People in Organizations," Ann Martin draws parallel conclusions and affirms that completing this process must be a major item on the nation's economic and social agenda during the remainder of the 1980s.[22] The report stresses that specific programs are needed to recruit, train, and place adult women and minorities in nontraditional occupations.

26.14. CONCLUSIONS

The rapid advance of manufacturing technology has increased the need for specialists in areas such as robotics and CAD/CAM. Both users and suppliers complain that the shortage of trained engineers to plan for, implement, and maintain this technology is impeding its adoption.

Engineers who can develop the software needed to make industrial robotics work are scarce and much in demand. Also in demand are production and manufacturing engineers who can design a plant to accommodate automated equipment in the most productive manner possible.

Programmable automation and its accompanying skill shifts are also significantly altering the training needs of the nation's blue-collar workforce. Displaced workers require extensive retraining, curricula for new trainees is being restructured, and students are being encouraged into new robotics technician occupations.

A 1983 report of the National Commission on Excellence in Education entitled *A Nation At Risk* describes a "rising tide" of mediocrity in the United States and includes a litany of revealing statistics about the decline of educational quality.[23] The report, more than anything else, suggests that what is needed is not only a revitalization of the decaying state of capital goods and works but also a renaissance of the American mind in the classroom, training program, and workplace and a renewal of personal, moral, and work values.

Historically, the entire course of American development was based on what used to be called the economy of high-wages principle, for example, that high-wage labor undersells low-wage labor because it is high-productivity labor. The American worker was better paid, better trained, and more highly educated ("human capital") with values that led him or her to work hard ("social capital"). The most important thing in the development of the wealth and productivity of the United States was that technological innovation was being made and direct investment was occurring. This innovation was in the area of industrial technology—steel-making, automotive production, the assembly line, and most important of all, in expensive machinery embodying better technology at lower cost. Industrial technology was the capital that American labor was working with then, and will, in the form of high technology, be working with today and tomorrow. Increase in American productivity will depend not only on the availability of capital to develop and expand direct investment in high technology but also on the education, training, and retraining needed to support, maintain, and expand that technology.

Educational and training institutions in the United States are beginning to ask the vital questions necessary to improve labor productivity—questions such as "What kinds of professional and technical education must we produce to enable American labor to work with the new high-technology capital? What kinds of education—professional and technical education in particular—are being offered in Europe?" "Should we be developing a more advanced professional and technical education system for this country?" "What type of planning needs to be done and what educational inputs are necessary to train or retrain the human resource needed for an industrial technology boom?" "How will we alter our programs to include more women, minorities, and handicapped who will comprise a larger proportion of the workforce of the next decade?"

Federal support for education and vocational training has declined, however, leaving questions unanswered about which sectors of the economy should be responsible for the training and retraining of new technical skills and for education programs in engineering, computer science, and other disciplines, for which a bottleneck to industrial technology exists.

This chapter has attempted to emphasize some promising beginnings in education, training, and retraining programs that begin to answer some of these questions and to outline strategies and a range of education and training needs that must still be met.

REFERENCES AND NOTES

1. Drucker, P. F., The re-industrialization of America, *The Wall Street Journal,* June 13, 1980, p. 10.

2. Naisbitt, J., *Megatrends: Ten New Directions Transforming Our Lives,* Warner, 1982.

3. Levin, H. M. and Rumberger, R. W., High-Tech Requires Few Brains, *Washington Post,* January 30, 1983, p. 30.

4. Bylinski, G., The race to the automatic factory, *Fortune,* February 21, 1983, pp. 50–64.

5. Hunt, H. A. and Hunt, T. L., *Robotics: Human Resource Implications for Michigan,* W. E. Upjohn Institute for Employment Research, Michigan, November 1982. (Final Report under Contract No. 99-1-1818-17-23-0, Michigan Occupational Information Coordinating Committee.)

6. Etzioni, Amitai, *An Immodest Agenda: Rebuilding America Before the 21st Century,* McGraw-Hill, New York, 1983.

7. Etzioni, Amitai, Prematurely Burying our Industrial Society, *The New York Times,* June 28, 1982.

8. Ayres R. and Miller, S., The impacts of robotics on the workforce and workplace, Carnegie-Mellon University, Department of Engineering and Public Policy, 1981. The study reports the results of the original student project in which Ayres and Miller were the principal investigators. Its estimates of potential job displacement caused by robots were based on a survey of corporate users. It concludes that today's robots could theoretically replace 1 million operatives in manufacturing and that the next generation of robots could potentially replace an additional 3 million operatives in manufacturing.

9. Gold, Bela, Robotics, Programmable Automation and Increasing Competitiveness, in *Exploratory Workshop on the Social Impacts of Robotics: Summary and Issues,* Congress of the United States, Office of Technology Assessment, U.S. Government Printing Office, Washington, D.C., July 1981, pp. 91–117. The study concludes that the actual economic impact of major technological changes have usually been less than expected owing to an overconcentration on the change itself which neglects the total production framework and its many interactions.

10. Hunt and Hunt, *ibid.*

11. *Ibid.,* p. 77.

12. National Alliance of Business, *Worker Adjustment to Plant Shut Downs and Mass Layoffs: An Analysis of Program Experience and Policy Options,* Berth M. C. and Reisner F., prepared under contract to NAB, 1981.

13. Stump, R. W. and Ashley, W. L., Occupational adaptability and transferable skills: preparation for tomorrows careers, in Springer, J., Ed., *Career and Human Resource Development,* ASTD Research Series, Madison, Wisconsin, 1980.

14. Machine Tool Task Force, *Machine Tool Systems Management and Utilization,* Vol. 2, University of California, Livermore, California, 1980. The Machine Tool Task Force was an international, multidisciplined, two-year effort that involved the participation of 122 experts in the specialized technologies of machine tools and in the management of machine tool operations. The major components of its assignment were to investigate the state of the art of machine technology, to identify promising future directions of that technology for both the U.S. government and private industry, and to disseminate the findings of its research.

15. Soroka, B. I., How should the university teach robotics? *Proceedings of the 13th International Symposium on Industrial Robotics,* Chicago, Illinois, April 1983.

16. Nof, S. Y., Robotics education at Purdue University, *Proceedings of the IBM University Study Conference,* Raleigh, North Carolina, October 1983.

17. Hammond, G. C., Robot education for engineers in manufacturing, *Proceedings of the 13th International Symposium on Industrial Robotics,* Chicago, Illinois, April 1983.

18. Schreiber, R. R., Meeting the demand for robotics technicians, in *Robotics Today,* Summer 1981, reprinted in *Robotics Today 1982 Annual Edition,* Society of Manufacturing Engineers, Dearborn, Michigan, 1982, pp. 78–79. This describes the development of the first U.S. robotics technician curriculum in the Macomb Community College, Warren, Michigan, in 1978.

19. Osborne, D. M., Training, the key to success in the use of robots, *Proceedings of the 13th International Symposium on Industrial Robotics,* Chicago, Illinois, April 1983.

20. Cole, P. A., Vocational education and the robotics revolution, *The GAO Review,* Vol. 18, No. 2, 1983, pp. 21–25.

21. Choate, P., *Retooling the American Work Force: Toward a National Training Strategy,* Northeast-Midwest Institute, Washington, D.C., July 1982.

22. Martin, A. M., Improving access and mobility of people in organizations, in *Critical Research Issues in Human Resource Development,* American Society for Training and Development, Washington, D.C., 1982, pp. 27–70.

23. The National Commission on Excellence in Education, *A Nation at Risk: The Imperative for Educational Reform,* Washington, D.C., 1983.

CHAPTER 27

INDUSTRIAL ROBOT STANDARDIZATION

YUKIO HASEGAWA

Waseda University
Tokyo, Japan

27.1. SIGNIFICANCE OF STANDARDIZATION

Industrial robots are new machines used mainly to substitute for human workers in many working places. The machines are backed up by such new technologies as computer engineering, mechatronics, and sensory devices and are making progress very rapidly like young creatures.

Although there are some opinions expressed that it is too early to start standardization activities, when we think of the age of robotization and the industry for manufacturing them, the author believes that the necessity of industrial robot standardization is very crucial for the following reasons:

1. Establishment of international measure of industrial robot statistics. Several years ago, the Robot Institute of America started to make robot population statistics in cooperation with industrial robot associations in major countries. However, precise comparison of the statistics is very difficult because no internationally authorized definition and classification of industrial robots has been established yet.

2. International exchange of academic and technical information. Since the first International Symposium on Industrial Robots (I.S.I.R), held in 1967 in the United States in Chicago, more than a dozen meetings of I.S.I.R. have been held in the United States, Europe, and Japan. In addition, other robot events and the distribution of materials have been increasingly active. However, because there is no international standard of terminology, symbols, or rules to measure performances of the robots, exchange of information has been confusing.

3. International trade of industrial robots. When we export and import industrial robots to and from other countries, confusion is possibile in understanding of catalog specifications, connecting power cables and pipes, installing machines, and so forth.

4. Ensuring safety. Today, the driving mechanism of passenger cars is very widely internationally standardized. We can easily drive foreign cars. But in the world of industrial robots, the design of control panels has not yet been standardized. Imagine a scene where one operator controls five robots made in different countries. In case of emergency he may easily confuse the different controls. This situation presents the possibility of serious accidents through mistakes with the robots' control.

5. Education and training. In the recent results of surveys by the U.S. and Japanese governments about the impact of robot introduction on workers, it was commonly reported that reeducation and retraining are the most serious problems for smooth reassignment of the displaced operators. The development of easy methods of robot control appears to be especially critical for older workers and women. That is also promoted by our standardization effort.

27.2. INTERNATIONAL STANDARDIZATION ACTIVITIES

27.2.1. Present Status of Industrial Robot Standardization

Figure 27.1 shows the results of a survey by the Robot Institute of America regarding industrial robot standardization activities in each country in the world. The survey reports that standardization has been started in eight and not yet in 11 robot-using countries. The survey was done in 1981, and the author hopes that more countries will start the standardization process.

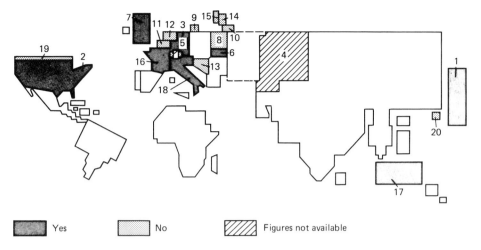

Fig. 27.1. Robot standardization activities around the world. 1. Japan; 2. United States; 3. West Germany; 4. USSR; 5. Switzerland; 6. Czechoslavakia (Research Institute project); 7. Great Britain; 8. Poland; 9. Denmark; 10. Finland; 11. Belgium; 12. Netherlands; 13. Yugoslavia; 14. Sweden; 15. Norway; 16. France; 17. Australia; 18. Italy; 19. Canada; 20. Taiwan. (Survey by Robot Institute of America, 1981.)

TABLE 27.1. INDUSTRIAL ROBOT STANDARDS ISSUED IN JAPAN

Items	Issued Standard and Number		Year
Terms and Symbols			
1. Basic terms	Glossary of terms for industrial robots	JISB 0134	1979
2. Related terms			
3. Symbols	Symbols for industrial robots	JISB 0138	1980
4. Signs			
Function			
5. Motion function indication	Methods for indicating characteristics of industrial robots	JISB 8431	1981
6. Motion function measurement	Methods for measuring industrial robot function and characteristics	JISB 8432	1981
7. Elements	Standard of wiring and piping of industrial robots	JIRAS 1006	1980
8. Signal system	Standard of electric signals of industrial robots	JIRAS 1004	1981
9. Environmental test			
Application Terminology			
10. Modules	Standard of industrial robot modules	JIRAS 1005	1982
11. Control	Standard of industrial robot control panel	JIRAS 1003	1981
12. Safety	Safety standard on industrial robots	JIS Draft	1982
13. Robot language	(Under study)		

TABLE 27.2. GLOSSARY OF TERMS

Overshoot

The degree to which a system response to a step change in reference input goes beyond the desired value.

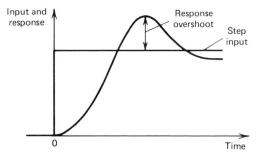

Passive Accommodation

Compliant behavior of a robot's endpoint in response to forces exerted on it. No sensors, controls, or actuators are involved. The remote center compliance provides this in a coordinate system acting at the tip of a gripped part.

Point-to-Point Control

A control scheme whereby the inputs or commands specify only a limited number of points along a desired path of motion. The control system determines the intervening path segments.

Position Control

Control by a system in which the input command is the desired position of a body.

Process Control

Control of the product and associated variables of processes (such as oil refining, chemical manufacture, water supply, and electrical power generation) which are continuous in time.

Source. International Standard Organization.

27.2.2. International Standard Organization (ISO) Activities

ISO began international standardization of industrial robot activities in 1978 under Working Group 2, which oversees standards of machine tools, in a technical committee (TC 97) that handles standardization of electronic computer and information-processing standardization.

The working group, of which the author is the Japanese representative, is composed of specialists from several countries. The following items have been the subject of numerous meetings held in European countries:

1. Definition.
2. Classification.
3. Graphic representation.
4. Glossary of terms.
5. Performance to be tested.
6. Safety.

France, Germany, Japan, Sweden, and the United States have submitted materials to the working group, which is scheduled to be reorganized and reinforced because of the importance of the project.

27.2.3. Standardization Activities in Each Country

The author has collected information that the following activities for industrial robot standardization have been executed in each country:

1. France: A host to the ISO working group for industrial robot standardization, France has submitted its proposal for industrial robot definition, classification, graphic representation, and glossary of terms. The activities are organized by the Association Francaise de Normalisation.

TABLE 27.3. SYMBOLS USED TO REPRESENT THE MECHANICAL STRUCTURE OF AN INDUSTRIAL ROBOT

	Front View	Side View	Top View	Isometric View
Prismatic joint telescopic				
Prismatic joint transverse				
Rotary joint pivot				
Rotary joint hinge (a)				
Rotary joint hinge (b)				
Example of a distributed joint				
Coupling device				
Base				
Ground				
Base with ground				
Example of a concurrent axes subassembly				

Source: International Standard Organization.

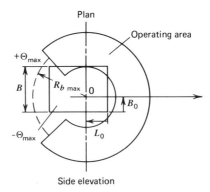

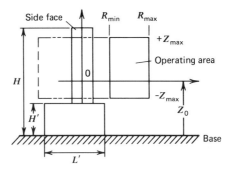

Fig. 27.2. Graphic symbols for a cylindrical coordinate robot. R_{max} and R_{min}: length between origin and limit of "out" ("in"). $+\theta_{max}$ and $-\theta_{max}$: maximum turning angle of arm in positive (negative) direction. $+Z_{max}$ and $-Z_{max}$: length between origin and upper (lower) limit. $R_{b\,max}$: length between origin and outer limit of tail end of arm. Z_0: length between position of arm corresponding to origin and ground basis. B: width of body. L: length of body. H: height of body. H': height of base. (*Source.* International Standards Organization.)

2. West Germany: The major standardization activities of industrial robots has been done by VDI (Society of German Engineers) and some of their work is as follows: definition of industrial robots, characteristics of handling devices, definition of safety statements, definition of assembly and handling functions, and a glossary of terms.

3. Japan: Industrial robot standardization efforts are for the most part undertaken through the close cooperation of the Industrial Science and Technology Agency of the Ministry of International Trade and Industry and the Japan Industrial Robot Association. Research and development for standards was begun in 1974, and the assignments were done by committees set up by the Japan Industrial Robot Association. The association also started a committee to support ISO industrial robot standardization activities in 1982, which is becoming more important internationally. The major Japanese results are shown in Table 27.1.

4. Sweden: SIS (Standardiseringskommissionen i Sverige) submitted their proposal of a glossary of terms to the ISO working group in 1983.

5. United States: Industrial robot standardization activities are managed by the National Bureau of Standards (NBS), and the Robot Institute of America and Robotic International of SME are eagerly cooperating with NBS. They have recently completed standards for safety procedures for industrial robots in the workplace, the interface between the wrist and end effector of industrial robots, and for the electrical interlock between robots and simple sensors.

27.3. STANDARDIZATION EXAMPLES

Although still in the process of international study, the following examples introduce a part of a scenario of industrial robot standardization in the future. The author believes that a set of international authorization standards will greatly contribute to the smooth incorporation of industrial robots into society.

TABLE 27.4. FUNCTIONAL SYMBOL STANDARD OF MATERIAL-HANDLING MOTIONS

Subfunctions of Material-Handling
Motion Functions

Function Symbol	Definition
  Rotation	Moving an object from one defined orientation into another defined orientation about an axis passing through a reference point on the object coordinate system. The position of the object reference point remains unaltered.
  Swiveling	Moving an object from one predetermined into another predetermined orientation and position through rotation about an axis outside the object.
  Translation	Moving an object from one predetermined into another predetermined position by translational motion along a straight line. Orientation of the object remains unaltered.
  Orienting	Moving an object from an undefined into a predetermined orientation. The position of the object is not taken into account.
  Positioning	Moving an object from an undefined into a predetermined position. The orientation of the object is not taken into account.
  Ordering	Moving objects from an undefined into a predetermined orientation and position or direction of motion.
  Guiding	Moving objects from one predetermined into another predetermined position along a defined path. Orientation of the objects is defined at each point.
  Transferring	Moving objects from one predetermined into another predetermined position along an undefined path. The degree of orientation remains unaltered.
  Conveying	Moving objects (or bulk goods) from one random into another random position. The path of motion and orientation of the objects during motion need not necessarily be defined. (See VDI 2411.)

Source: International Standard Organization.

1. Glossary of terms. Table 27.2 shows a partial example of a glossary of terms. The terms have been proposed by the United States to the ISO working group.

2. Symbols representing the mechanical structure of an industrial robot. Table 27.3 shows symbolized elements of an industrial robot mechanical structure. By using these symbols we can express and identify the structure of a particular industrial robot. The symbols have been proposed by France to the ISO working group.

3. Graphic symbols standard. Figure 27.2 shows graphic symbols for a cylindrical coordinate robot. By using this type of standardized expression we can identify specifications of the robot without overlooking any important item. The symbols have been proposed by Japan to the ISO working group.

4. Functional symbol standard of material-handling motions. Table 27.4 shows the functional symbol standard of material-handling motions, which allows a more detailed expression of the material-handling motions from a functional standpoint.

27.4. FUTURE DIRECTION OF STANDARDIZATION

At the present stage, standardization of industrial robots has begun with fundamental items such as glossary of terms, symbols, and hardware elements.

The author anticipates that the themes of standardization will proceed to softwares such as language and programs and then will be expanded to the upper level of objects, process patterns, and structure of robotized production systems, and so forth. Wider diffusion of industrial robots will require greater standardization in the future.

REFERENCES

1. Research and Survey Report on Industrial Robot Standardization (I)–(IV), Japan Industrial Robot Association, 1975–1980.

2. *Worldwide Robotics Survey and Directory,* Robot Institute of America, 1982.

3. ISO/TC, 97/SC8/WG2, Documents (N1)–(N29), International Standard Organization, 1982–83.

BIBLIOGRAPHY

1. *Standards for Computer-Aided Manufacturing,* Final Report, AFML-TR-77-145, National Bureau of Standards, January 1978.

2. Evans, J. M., Barbera, A. J., and Albus, J. S., Standards and Control Technology for Industrial Robots, *Proceedings of the 7th International Symposium on Industrial Robots,* Tokyo, October 1977, pp. 479–486.

3. Evans, J., CAM Standards Directions, SME Paper No. MS78-483, 1978.

4. Hasegawa, K. and Kaneko, T., Study on the Standardization of Terms and Symbols Relating to Industrial Robots in Japan, *Proceedings of the 3rd CISM-IFTOMM Symposium on Theory and Practice of Robots and Manipulators,* Udine, Italy, September 1978, pp. 471–478.

5. Kozyrev, Y. G., Constructing a Standard Series of Industrial Robots, *Machines and Tooling,* Vol. 49, No. 7, 1978, pp. 3–10.

6. Ozaki, S., Ito, K., and Inagaki, S., Standardization of Characterization Relating to Industrial Robots, *Proceedings of the 7th International Symposium on Industrial Robots,* Tokyo, Japan, October 1977, pp. 453–459.

7. Inagaki, S., What is the Standardization for Industrial Robots? *Industrial Robot,* Vol. 7, No. 1, March 1980, pp. 46–49.

8. Sugimoto, N., Safety Engineering on Industrial Robots and Their Draft Standard Safety Requirements, *Proceedings of the 7th International Symposium on Industrial Robots,* Tokyo, Japan, October 1977, pp. 461–470.

CHAPTER **28**

ROBOTICS IN EASTERN EUROPE

TIBOR VÁMOS
JÓZSEF MARTON

Hungarian Academy of Sciences
Budapest, Hungary

28.1. SOCIOECONOMICS

At present each East European member of the COMECON* has a program for robotics that is approved and financed or subsidized by the government; most are very ambitious and of extremely wide range. This development is influenced of course by world fashion, too, with one eye on Japan and the other on Europe. Robotization, nevertheless, is a real and emerging need in these countries, perhaps more so than in any other countries, and it can be supported by many traditional industrial and scientific bases.

Eastern European countries have a chronic labor shortage, especially where labor is troublesome or its social status is low. The problem in this situation—contrary to the Western economy where unemployment is one of the major troubles—is political: the system does not permit any kind of unemployment. The effects are on one hand a heavy taxation of the better working part of the population, a lower general productivity, and many people whose accomplishment is much lower than the average. On the other hand, several social arrangements are much more generous than those in countries with a gross national product (GNP) two to three times higher per capita. For example, retirement age is 60 for men and 55 for women; a three-year reduced-pay maternity leave is warranted for each child, utilized by one of the parents—after a half-year fully paid birth leave!

One major consequence of this policy is a "spoiling" of the working population rather similar to that in the United States or anywhere in Western Europe: low-esteemed working places are empty, and the wages are relatively exorbitant for them. This deforms the social labor structure: there is less interest in higher education and a degradation of professional ethics if an intellectual earns two or three times less than a nearly illiterate garbage collector.

There is another difference in the policy concerning these workplaces with inferior classification: the Eastern European governments, with some exceptions, are not stimulating Gästarbeiter† immigration and have no significant minorities to serve as reserves for these jobs. The rising social strains connected with this situation are considered to overwhelm the advantages.

The population in the majority of the COMECON countries has generally shown a zero-level growth for many years—sometimes a small decrease. This is especially significant in that it holds not only for families with higher education but also for those layers of the society that are more prepared for doing any kind of disciplined work. The disadvantages of such an absolute job security system are well known, but in these countries unemployment is considered more expensive for the society, not primarily because of the unemployment dole but because of demoralization and infection of the whole population.

---

* COMECON—*Co*uncil of *M*utual *Econ*omic Assistance. Regular members: Bulgaria, Czechoslovakia, German Democratic Republic, Hungary, Cuba, Mongolia, Poland, Rumania, and USSR. Observers and participants in some agreements: Angola, Ethiopia, Yemen, Yugoslavia, Korean People's Republic, Laos, Iraq, Finland, and Mexico.

† *Gästarbeiter* is the German name for the foreign workers who have come into the Western European countries from the South European countries (Turkey, Greece, Yugoslavia, Italy, Spain) during the years of prosperity, mainly in the last two decades.

Robotization should be a remedy for lack of or low discipline in the labor force at the lower end of work distribution; incentives of more income differentation are supposed to further help problems of the ensured workplace policy.

Economic justification for robots is also different in Eastern Europe. Nominal wages—if they are calculated on the official exchange rates—are rather unrealistic. Some countries, as for example, the USSR or Czechoslovakia, have a much lower exchange rate than the real value of the Western currency; others, as for example, Hungary, a higher rate to promote exports and limit the imports. The social benefits, practically no taxes, are also components of a hardly comparable price and wage system— although this is also in permanent transformation, approaching more the practice of the more industrialized countries. The consequence of these circumstances in that the justification for robots takes into consideration two major factors: the loss in production and depreciation of capital investment not used efficiently (more shifts, etc.) if they are not served by robots; and the savings in social benefits, especially housing, infrastructure. This latter is important because the fast urbanization of these mostly developing countries has made the settlement restructuring an additional burden on the economy.

28.2. SCIENTIFIC AND TECHNICAL BACKGROUND

Mathematical-mechanical scientific heritage has determined the course of robotics and many other technological innovations in Eastern Europe. The modern science of the dynamics of motion stems mostly from a Franco-Russian school of the turn of this century and thereafter. We refer here to Euler, Poincaré, Ljapunov, Krylow, Bogoljubow, Andronow, Chaikin, Witt, and Lurje. This outstanding school survived several generations, and their present representatives have found the field of robot-manipulation techniques a bonanza for application of their skills oriented toward mathematical mechanics. The results of their activity are summarized in numerous monographs (see References 1–30).

Most of the publications and investigations are related to a hypothetical or a laboratory environment, the models are mostly abstract, and the calculations—although demonstrated on some general-purpose computers—are not in real-time. The emphasis has been more on the mathematical-algorithmic phase— which is also very important—but until recently much less on the real-time aspects. Cross references are rather rare in the literature, although by having faster and longer word-length microprocessors and pursuing faster robot movement the practical need and possibility increase for these more accurate force, torque, and motion calculations.

Relevant research has been done in the field of problems of motion description and analysis. One of the most outstanding projects is the work of the Ochotsimsky groups: the analysis of the requirements of a highly stable walking robot.[31] The enormous variation possibilities of a multileg-multilimb walking procedure was reduced to a six- or eight-leg/two-limb solution and proved that nature is a good

TABLE 28.1. METAL AND ENGINEERING PRODUCTION AS PERCENTAGE OF TOTAL INDUSTRIAL PRODUCTION (1978)

Czechoslovakia	34.0
Hungary	33.6
Poland	31.6
Belgium	30.0
France	37.8
Germany, Federal Republic	37.9[a]
Italy	34.4
Netherlands	29.7[a]
United Kingdom	24.4[a]
Austria	27.8
Finland	26.3
Norway	27.8
Sweden	41.5
Spain	24.8[a]
Yugoslavia	29.4
Canada	36.3

Source.   United Nations Yearbook of Industrial Statistics, 1978 Edition, Vol. 1.

[a] Figure for 1976 is available only.

TABLE 28.2. SHARE OF EMPLOYMENT IN METAL AND ENGINEERING PRODUCTS IN TOTAL INDUSTRIAL EMPLOYMENT (1978)

Bulgaria	27.0
Czechoslovakia	39.4
German Democratic Republic	41.0
Hungary	30.7
Poland	32.4
Rumania	34.4
Belgium	22.7
Denmark	19.5
France	30.7
Germany, Federal Republic	36.0
United Kingdom	33.8
Austria	32.0
Finland	21.2
Norway	23.6
Sweden	34.0
Canada	17.4
United States	31.5
Japan	16.6

Sources. OECD, *Labour Force Statistics, Basic Statistic, 1976–1978.* CMEA, *Statistical Yearbook of Member Countries of the Council for Mutual Economic Assistance,* 1979.

engineer as usual: the sequence of movements that warrants a stable and safe movement on uneven surfaces, even broken by clefts, is very similar to that used by most animals. In this topic a very fruitful Soviet-American cooperation has existed for many years.

The first relevant studies on manipulator techniques were published in the 1930s, half a century ago, and here we must mention the name of the late Artobolevsky, who first analyzed the open-chain kinematics of a multijoint manipulator.[1,32] He can be considered as the father of later Soviet robotics. Based on his research, several types of sophisticated manipulators were used in nuclear and other dangerous environments in the 1940s and early 1950s. Special attention was devoted to underwater robotics in these early times, too.[28]

Another rather early success was reached in prosthetics control: the first IFAC Congress in Moscow held in 1960 was not only historical because of the publications of R. Kalman and Pontriagin on their basic control theory contributions but also because of the demonstration of a biovoltage-controlled prosthetic arm by Kobrinsky,[33] which concentrated on the upper extremities, and, not much later, by a Yugoslavian team, led by R. Tomovic and M. Vukobratovic, dealing with the upper and lower extremities.[34]

28.2.1. The Effect of the Machine-Tool Industry

The other basis for the robot industry is the traditional, highly developed machine-tool industry. A typical feature of the industrial structure of the Eastern European countries is a relatively high ratio of machine-tool industry compared to the general developmental stage (Tables 28.1, 28.2, 28.3).* This basis indicates a somewhat different start compared to those countries where electronics and precision mechanics were the major resources. Just this difference explains a relevant drawback: The industrial robot at first glance looks like a new version and application of numerically controlled machine tools, being even more a machine than it is a real tool. However, its dynamic requirements are far more delicate than those of the several most demanding products of conventional machine tools: precision,

---

* More recent statistical sources (Annual Review of Engineering, Industry, and Automation 1980) reflect insignificant changes in percentages.

TABLE 28.3. MANUFACTURING OF
METALCUTTING MACHINE TOOLS (1978)

	Number Units
Bulgaria	15,315
Czechoslovakia	34,247
German Democratic Republic	19,328
Hungary	12,689
Poland	27,545
Rumania	27,197
USSR	237,885

Sources. United Nations, *Yearbook of Industrial Statistics,* 1978 Edition, Vol. 2. CMEA, *Statistical Yearbook of the Member Countries of the Council for Mutual Economic Assistance, 1978.*

speed, acceleration and deceleration, range of action, and so on. These requirements are not to be solved by high inertia masses and relatively free variable dimensions but, first of all, by intrinsically dynamic, active methods, for example, highly sophisticated controls and those means more typical of precision mechanics: low inertia, high momentum, small-volume drives, combined with high-resolution positioning, very high-quality magnetic materials, gears, and so on. The technological bottlenecks of robot construction are rather different from conventional machine tools. As the integration of manufacturing progresses and more flexible solutions emerge, these differences become narrow, but the starting differences were significant.

Analyzing the origins, one notes that Cincinnati Milacron is one of the major vendors for robots in the United States. Although this is true, the background structure, the inclination, of Cincinnati toward the electronic technologies is different from the machine-tool industry and even more from what the application of heavy-duty robots requires; in this respect the technical gap is much smaller than in other fields, for example, assembly.

The reliance on more conventional industries, such as machine tools and electric motors, traditional drives, pneumatics and hydraulics, and long lacking the impetus of revolutionary technologies, was one of the major causes of the slow and inadequate start of industrial robot production in Eastern Europe. The experience gained in this respect led industrial managers in the COMECON countries to investigate Western solutions and to an attempt not to copy the types already successful on the market, but to try buying licences.

28.2.2. Way of Technical Approach

A result typical of the situation in Soviet technology—which differs entirely from that of any other COMECON country—is a dual face of gap and connection between high-level science and relatively low-level technology which can be very well demonstrated by the example of the Lunokhod. The Lunokhod is a very typical Soviet engineering design: a rather simple and extremely robust device avoiding any sophistication that could decrease reliability. The two Lunokhods each worked four months long, rolling through an uneven surface of about 3 by 37 kms. No high-level control was applied to them: a simple command system broadcast from the earth base was received by the movement control for the two forward and one backward speeds, turns right, left, and on the spot, and relayed back control signals of the movement, of the wheel temperatures, wheel-drive currents, and the solidity of the soil. The entire mechanics of the movement was based on the eight driven wheels, nothing more; the movement control, as described, was completely detached from any other operations that were the duties of the robot. These design principles may remind one of the famous and ingenious Russian devices of the Second World War when the Kalashnikov machine guns, the Katyusha rocket organ and the T-34 tanks showed performance superior to the highly sophisticated German war machines.

To balance this statement it must be realized that following the advent of highly reliable electronics the validity of this philosophy was transformed. Another remark that should be repeated several times is that the technical philosophy and educational tradition of the region discussed here is multifarious; it contains a real and tradition-keeping German part, a Central European part that was detached for a long time from the East, and has had and maintains now, too, a tradition of mixing and melting several parallel influences in a rather pragmatic practice.

As emphasized earlier, in computer control this region is a latecomer and has not produced any relevant new results in the robot software field. This very sincere remark does not indicate that there are no institutions and people who could not develop any kind of new robot-control language and implement it for any system now, but, realizing the situation, most of the pragmatic research and development managers do not want to invest in reinventing something that would be an alternate to software already distributed and practically standardized worldwide—no "socialist" Fortran or even ADA is envisaged. In spite of this general philosophy, one can find some isolated intellectual islands.

28.2.3. Intelligent Robots

There is no European socialist country where some efforts have not been made. We give an overview based on the Hungarian results,[35-54] as the history is mostly uniform: algorithmic-theoretical results in pattern recognition, the application or the invention of some new efficient local operators (as the Mérö operator[37]), linguistic constructions, high-level, robot-operation description, assembly-oriented languages. Another fine result is a fast and user-friendly three-dimensional representation on graphic display, a tool for modeling and robot-motion planning which can be used by a low-skilled operator. Much attention was devoted to the avoidance problem: how can forbidden regions (obstacles) be avoided in an automatic, inspection-helped way in real time, considering the robot's articulation, the object that is held by the robot, and the unplanned changes within the robot's reach. Two, sometimes combined, approaches are used. The first approach cuts the entire movement space of the robot into regions. The shape of the regions depends on the type of robot motion (rectangular or polar or combined). Then by calculations and inspection the regions are classified into free and forbidden, and this puts constraints to the motion control. The other approach approximates the covered trajectories by separating the degrees of freedom into two joint groups. This is a limitation on parallel motion of all joints (possibly only in some regions), and by that the more accurate free path can easily be calculated.[53]

Other efforts are demonstrated in an experimental vision-controlled pick-and-place system, situation modeling (model and real situation), fast skeleton movement-animation.[47]

Multidimensional numerical control has a long tradition in these countries, a parallel computation of six axes and more are easily realizable, for example, by the Hungarian NC controllers, all patented original design. Several relevant achievements mark the NC programming background, too—their development is following the recent international standards. That means that trajectory calculations and controls are well combined and the application of some higher-level control languages such as, for example, VAL or others, is well established.[54] Research is concentrated on AUTOPASS, AL, and AML-like assembly languages. Some work has started in the application of AI principles, not only in the old paradigm of motion planning but also in development of expert systems on robot applications.

28.3. TENDENCIES IN ROBOT DEVELOPMENT AND APPLICATION

28.3.1. Some Statistics

The last few years have introduced a dramatic change in the picture. First, it is demonstrated by a few figures for the current five-year plans: the number of industrial robots working in the Soviet industry amounts to an estimated 8500. By the end of this plan in 1985 this should be increased to 40,000. A long-range goal for the end of the 1980s is about 120,000. The plans of the GDR contain similar figures for a country with a population of less than 7% of that of the USSR: 40,000–45,000 robots working in the industry by 1985!

Looking at statistics, an important question is always the nomenclature. It is well known that the Japanese statistics contain those limited-sequence automatic manipulators that are not considered industrial robots in the American surveys. These statistics are closer to the Japanese notion than to the American. However, a warning should be added: since these stocks will consist of specimens of a newer generation, this involves, in most cases, the application of more sophisticated types.

As indicated in the discussion on socioeconomic aspects in the recent past, most European socialist governments started ambitious plans in robotics and have put a special emphasis on robotization. The structure of the government intervention is very different, depending on the control system of national economy in the country. In some countries this is highly centralized and directed; in others, as, for example, in Hungary, economic control is based on independence of the companies, not interfering in internal decisions, yet using leverage by economic means: taxes, credits, customs, and so on.

At the Budapest meeting of COMECON governments in June 1982 special agendas and resolutions were reserved just for robotization, recommendations on accelerated research and development, standardization, and distribution of profiles for production and application.

A catalog of 1981[55] contains 102 types of industrial robots produced in the COMECON countries. Some statistics from this survey, presented in Tables 28.4 to 28.8, together with the following discussion, illustrate the situation about two years ago. The catalog is not complete; moreover, some items of the list are either obsolete or only prototypes. Table 28.4 demonstrates that this catalog is far from being complete, since the estimated number of robot types* in GDR is about 20.

---

* Types are considered to be different according to manufacturer's designation. For example, building-block systems of modular robots are considered as one type.

TABLE 28.4. DISTRIBUTION
OF INDUSTRIAL ROBOT
TYPES BY COUNTRIES

Country	Robot Types
Bulgaria	11
Hungary	6
GDR	6
Poland	15
Czechoslovakia	8
Soviet Union	56
	102

TABLE   28.5.
DISTRIBUTION OF
INDUSTRIAL ROBOT
TYPES BY LOAD

Load Range (kg)	Types
0–0.1	5
0.1–0.5	6
0.5–1.0	5
1.0–5.0	26
5–10	14
10–50	32
50–100	9
>100	5
	102

Most earlier applications were concentrated in heavy-load tasks, and small special-purpose pneumatic pick-and-place units were used mostly in precision mechanics assembly, for mass production. The low number of electric types reflects many aspects of our earlier discussion. The picture is now changing dramatically.

The application fields are not different from those of any other country: mostly pick-and-place, especially in heavy industry (forging, molding, etc.); feeding machine tools; painting; and welding. As described here several times, since the machine-tool industry has been given prominence in the manufacturing structure of these countries, flexible automation systems are rapidly evolving from earlier CAD-CAM experiments. Most COMECON countries have several flexible manufacturing cell prototypes and even production systems, groups of machine tools served by robots, the whole cell operated by distributed CNC multimicroprocessor controllers for each machine component and a more powerful control mini (mostly of PDP-11 class) for the programming and supervision of the entire cell. The micros and minis are produced for the most part by the country or by another COMECON member.

TABLE   28.6.
DISTRIBUTION OF
INDUSTRIAL
ROBOT TYPES BY
THE ACTUATING
POWER

Power	Types
Pneumatic	49
Hydraulic	38
Electric	15
	102

**TABLE 28.7. DISTRIBUTION OF
INDUSTRIAL ROBOT TYPES BY THEIR
KINEMATIC ARRANGEMENT**

Kinematic Arrangement	Types
Cylindric coordinate (like Versatran)	63
Spherical coordinate (like UNIMATE)	8
Joint coordinates (anthropomorphic like ASEA or PUMA)	11
Overhead	15
Others	5
	102

The standards are international, mostly minis PDP-11 compatible; the systems use LSI-11, Intel 8080, 8086, Z80, M 68000 that are compatible micros, the assortment of 16-bit micros just about to broaden. A few characteristic examples of robots and their applications in Eastern Europe are illustrated in Chapter 47, Robot Applications in Eastern Europe.

28.3.2. The Robotization Environment

In accelerating the progress of robot development several countries pursue a combined policy of new development and purchase of licenses. Poland produces ASEA-robots under a license agreement. The most extensive and developed license policy can be observed in Bulgaria: at first they had a Versatran license, but a long-range cooperation arrangement with FUJITSU-FANUC led to the mass manufacture of several FANUC machine tools and robots, and then to a new big factory devoted to this cooperation. In addition, Hungary has purchased a Soviet license.

The fast change in the attitudes and programs are due largely to general modernization efforts, to a conviction that robotization can be an efficient vehicle of higher manufacturing culture, and, last but not least, to demographic, social work-distribution trends foreseen for the next decades. All this has been reflected not only in robot manufacturing, but also in the realization that manufacturing itself needs a much better and more specialized industrial infrastructure, especially in the field of components. The previously cited catalog reflects this, too; it comprises not only robot types but drives, special actuators, and the like as well. The need for high-performance harmonic drives and motor-type controls as single-card realizations is not only expressed but carefully specified. A major emphasis is laid on the environmental conditions.

Education on every level (skilled workers, technicians, engineers), demonstration centers, and training for different applications have been founded and will be provided in increasing numbers. Facilities, brochures, publications, and industrial managers' training courses giving information on the possibilities and future trends are an important part of the programs. It is well understood that robotization is not a single product for introduction and distribution but a change in organization, attitudes, and working conditions. Without an appropriate environment it remains only a curiosity or a practical addition to existing production, but not a real revolutionary change.

Robot exhibitions are regular now in Plovdiv (Bulgaria), Brno (Czechoslovakia), and in Moscow. Regular meetings are devoted to the scientific problems of robotics, some of them nationally organized

**TABLE 28.8. DISTRIBUTION OF
INDUSTRIAL ROBOT TYPES BY THEIR
INTENDED APPLICATION**

Application Fields	Types
Small, special-purpose assembly	15
Machine-tool handling, specialized	24
Machine-tool and press handling	29
Hot forging	5
Plastics or hot metal forming	8
Painting and/or welding	7
General purpose	6
Others, special	8
	102

but with international audience, some in the framework of the COMECON or in cooperation with the Academies of Sciences. In Bulgaria a regular summer school is devoted every year to this topic for young people interested in robotics. In the German Democratic Republic a four-day obligatory seminar was recently organized for high-level industry executives on the possible applications of robots, future plans, human aspects, and so on.

Between 1979 and 1985 150 robot standards should be accepted in the Soviet Union, 32 of them on the federal level. The Soviet robot program comprises 265 organizations, 45 ministries (which are more than big industrial trusts), and other authorities controlling 20 academic research institutes, 26 academic university centers, and 219 industry institutes. The programs encompass a broad spectrum: research and development, design, technology in every aspect, application know-how, support, service, training, and social implications.

28.4. CONCLUSION

The historical evolution of the central and eastern part of Europe has been somewhat different from that of the Western Hemisphere, and, depending on the country, the general economic and industrial progress lagged behind by 50–200 years up to the eighteenth and nineteenth centuries. On the other hand, the intellectual gap was always much less, and in some periods and in some respects this region added much to the general human and technological development. This historical inertia is also reflected in the present stage of robot development and application.

There may be a few critical technological components of high demand that are more easily available on the Western market than manufactured in small volumes, at high prices, with lower yield, in some special laboratory or otherwise dedicated environment in Eastern Europe, but the crucial issue is really the historical inertia that cannot be offset by commercial imports, never can change real balances or broad gaps of technology. This issue encompasses the general culture of manufacturing, reliability, economy, and the systematic environment of application—everything that is really needed for a more cooperative, less dangerous world, which has always been brought together by peaceful connections of commerce and exchange and always been separated by prejudices and impatience in understanding other peoples' beliefs and cultures. A long-range policy of equalization and cooperation is therefore in the interest of the entire human race.

REFERENCES

These references include books on robotics and closely related subjects in Russian. For better identification, titles are first given in transliteration, then in translation (the second in parantheses).

1. Artobolevskij, I. I., *Teoriya mekhanizmov i mashin* (Theory of Mechanisms and Machines), Nauka, Moskva, 1952.

2. Artobolevskij, I. I., *Teoriya i ustrojstvo manipulyatorov* (Theory and Equipment of Manipulators), Nauka, Moskva, 1973.

3. Aksenov, E. P., *Teoriya dvizheniya iskustvennyh sputnikov Zemli* (Theory of Satellite Motion), Nauka, Moskva, 1977.

4. Andreenko, S. N., Vorosilov, M. S., and Petrov, B. A., *Proektirovanie privodov manipulyatorov* (Design of Manipulator Drives), Masinostroenie, Leningrad, 1975.

5. Belyanin, P. N., *Promyslennye roboty* (Industrial Robots), Masinostroenie, Moskva, 1975.

6. Ignat'ev, M. B., Kulakov, F. M., and Pokroskij, A. M., *Algoritmy upravleniya robotami-manipulyatorami* (Control Algorithms for Robot-Manipulators), Masinostroenie, Leningrad, 1972.

7. Katys, G. P., *Informatsionnye roboty u manipulatory* (Information Robots and Manipulators), Mir, Moskva, 1976.

8. Korenev, G. V., *Tsel' i prisposoblyaemost' dvizheniya* (Motion Aim and Adaptability), Nauka, Moskva, 1974.

9. Korenev, G. V., *Vvedenie v mekhanizm tseloveka* (Introduction to Human Mechanisms), Nauka, Moskva, 1977.

10. Korenev, G. V., *Otserki mekhaniki tselenapravlennogo divzheniya* (Essay on the Mechanics of Purposeful Motion), Nauka, Moskva, 1980.

11. Kulakov, F. M., *Supervizornoe upravlenie manipulyacionnymi robotami* (Supervisory Control of Manipulating Robots), Nauka, Moskva, 1980.

12. Kulesov, V. S., and Lakota, N. A., *Dinamika sistem upravleniya manipulyatorami* (Dynamics of Manipulator Control Systems), Energiya, Moskva, 1971.

13. Kulesov, V. S., *Proektirovanie sledyashchikh sistem dvustoronnego dejstviya* (Design of Force-Reflecting Servo Systems), Masinostroenie, Moskva, 1980.

14. Medvedev, V. S., Leskov, A. G., and Jushchenko, A. S., *Sistemȳ upravleniya manipulyatsionnȳkh robotov* (Control Systems of Manipulating Robots), Nauka, Moszkva, 1978.

15. Morecki, A., *Manipulatory bioniczne* (in Polish) (Bionic Manipulators), Panstwowe Wydawnistwo Naukowe, Warsawa, 1976.

16. Popov, E. P., *Robotȳ-manipulyatorȳ* (Robots and Manipulators), Znanie, Moskva, 1974.

17. Popov, E. V. and Firdman, G. R., *Algoritmicheskie osnovȳ intellektual'nȳkh robotov i iskusstvennogo intellekta* (Algorithmical Bases of Intelligent Robots and Artificial Intelligence), Nauka, Moskva, 1976.

18. Popov, E. P. and Ignat'ev, M. B., *Distantsionno upravlyaemȳe robotȳ-manipulyatorȳ* (Remote Control of Robot-Manipulators), Mir, Moskva, 1976.

19. Popov, E. P., Vereshchagin, A. F., and Zenkevits, S. L., *Manipulyatsionnȳe robotȳ. Dinamika i algoritmȳ* (Manipulating Robots: Dynamics and Algorithms), Nauka, Moskva, 1978.

20. Timofeev, A. V., *Robotȳ i iskustvennȳj intellekt* (Robots and Artificial Intelligence), Nauka, Moskva, 1978.

21. Timofeev, A. V., *Postroenie adaptivnȳh sistem upravleniya programmnȳm dvizheniem* (Adaptive Control Systems for Programmed Motion), Energiya, Leningrad, 1980.

22. Vukobratovics, M., *Sagayushchie robotȳ i antropomorfnȳe mekhanizmȳ* (Walking Robots and Anthropomorph Mechanisms), Mir, Moskva, 1976.

23. Yurevich, E. I., Ed., *Promȳshlennȳe robotȳ, Nautsno-tekhnitseskij sbornik No. 1* (Industrial Robots, Collected Papers), Masinostroenie, Leningrad, 1977.

24. *Ibid*, No. 2, 1979.

25. *Ibid*, No. 3, 1982.

26. Yurevich, E. I., Avetikov, B. G., Korȳtko, O. B., Eds., *Ustrojstvo promȳslennȳkh robotov* (Industrial Robot Equipment), Masinostroenie, Leningrad, 1980.

27. Yurevich, E. I., *Upravlenie robotami ot EVM* (Computer Control of Robots), Energiya, Leningrad, 1980.

28. Yastrebov, V. C., *Teleupravlyaemȳe podvoquȳe apparatȳ* (Remote-Controlled Underwater Equipment), Sudostroenie, Moskva, 1973.

29. Yastrebov, V. S., *Podvodnȳe robotȳ* (Underwater Robots), Sudostroenie, Leningrad, 1977.

30. Yastrebov, V. S. and Filatov, A. M., *Sistemȳ upravleniya dvizheniem robota* (Robot Motion Control Systems) Masinostroenie, Moskva, 1979.

31. Ochotsimky, D. E. et al., *Integrated Walking Robot Simulation and Modelling,* Proceedings of the 7th Congress of the IFAC, Helsinki, 1978.

32. Artobolevskij, I. I., *A robottechnika tudományos problémái* (The Scientific Problems of Robotics), Müszaki Tudomány, Vol. 54, No. 3–4, 1977, pp. 305–316.

33. Kobrinski, A. E., Bolkhovition, S. V., Voskoboinikova, L. M., Ioffe, D. M., Polyan, E. P., Popov, B. P., Slavetski, Ya. L., Sysin, A. Ya., and Yakobsen, Ya. S., *Problems of Bioelectronic Control,* Proceedings of the 1st International Congress of IFAC, Moscow, 1960.

34. Tomovic, R., *Human Hand as a Feedback System,* Proceedings of the 1st International Congress of IFAC, Moscow, 1960.

35. Vámos, T. and Vassy, Z., *Industrial Pattern Recognition Experiment—A Syntax Aided Approach,* Proceedings of the 1st International Joint Conference on Pattern Recognition, Washington, D.C., 1973.

36. Vámos, T. and Vassy, Z., *The Budapest Robot—Pragmatic Intelligence,* Proceedings of the 6th World Congress of IFAC, Boston, Massachusetts, 1975.

37. Vámos, T. and Mérö, L., *Real-Time Edge-Detection Using Local Operators,* Proceedings of the 3rd IJCPR, Coronado, California, 1976.

38. Vámos, T., Industrial Objects and Machine Parts Recognition, in Fu, K. S., Ed., *Applications of Syntactic Pattern Recognition,* Springer-Verlag, Heidelberg, 1977.

39. Vámos, T., Visual Recognition of Artifical Objects, *Problems of Control and Information Theory,* Vol. 6, 1977.

40. Vámos, T. and Galló, V., *A Program for Grammatical Analysis of Arbitrary Geometrical Patterns,* Proceedings of the American-Hungarian Joint Seminar on Pattern Recognition, Budapest, Hungary, 1977.

41. Vámos, T., Báthor, M., and Mérö, L., *A Knowledge-Based Interactive Robot Vision System,* Proceedings of the 6th IJCAI, Vol. 2, Tokyo, 1979.

42. Vámos, T. and Báthor, M., *3D Complex Object Recognition Using Programmed Illumination,* Proceedings of the 5th ICPR, Florida, 1980.

43. Vámos, T., *Research Works in the Field of Intelligent Robots and Possible Application,* MANUFA-CONT, Budapest, 1980.

44. Vámos, T., Báthor, M., and Siegler, A., A Knowledge-Based Interactive Robot-Vision System, in Bolc, L., Kulpa, Z., Goos, G., and Hartmanis, J., Eds., *Lecture Notes in Computer Science: Digital Image Processing Systems,* Vol. 109, Springer-Verlag, 1981.

45. Vámos, T. and Siegler, A., *Intelligent Robot Action Planning,* Preprint of the 8th World Congress of IFAC, Kyoto, Japan, 1981.

46. Mérö, L. and Vámos, T., Medium Level Vision, in Rosenfield, A. and Kanal, L., Eds., *Progress in Pattern Recognition,* Vol. 1, North Holland, New York, 1981.

47. Vámos, T. and Báthor, M., *A robot vision lab concept,* in Michie, D., Ed., *Machine Intelligence,* Vol. 10, Horwood Ltd., Chichester, Wiley, New York, 1982.

48. Siegler, A. and Báthor, M., *Graphic Simulation for Robot Programming,* Proceedings of the AUTO-MAN, Brighton, U.K., 1981.

49. Siegler, A., Kinematics and microcomputer control of a 6 degree-of-freedom manipulator, Research Report, Cambridge University Engineering Department, CUED-CMS 185/1979.

50. Chetverikov, D., *Experiments in Rotation-Invariant Texture Discrimination Using Anisotropy Features,* Proceedings of the 6th ICPR, Munich, 1982.

51. Mérö, L., Chetverikov, D., and Báthor, M., *Bus-Body SHEET Identification-Oriented Two-Dimensional Recognition System,* MANUFACONT, Budaptest, 1980.

52. Chetverikov, D., *Textural Anisotropy Features for Texture Analysis,* IEEE Conference on Pattern Recognition and Image Processing, Dallas, Texas, 1981.

53. Jakubik, P., *Supervisory Control of Multijointed Robot Arms,* Dr. Techn. thesis, Technical University of Budapest, 1982.

54. Marton, J. and Jakubik, P., Laboratory and Practical Experiences with VAL Robot Control Language, *Mérés és Automatika,* Vol. 31, No. 1, 1983.

55. Promȳshlennȳe Robotȳ, Katalog (Industrial Robots Catalog), Vasilev, V. C., Ed., Moskva, 1981.

PART 6

APPLICATION PLANNING: TECHNIQUES

CHAPTER 29
PRODUCT DESIGN AND PRODUCTION PLANNING

WILLIAM R. TANNER

Productivity Systems, Inc.
Farmington, Michigan

29.1. INTRODUCTION

The application of industrial robots to manufacturing operations is generally done under one of two sets of circumstances. The first is the situation involving a new facility, process, or product; here, robots are incorporated into the initial plans and are implemented routinely along with other equipment and facilities. The second, more common, situation involves the application of robots to existing processes and operations, often in response to management direction or upon a suggestion from a supplier of robotic equipment. Here, the robot must be integrated into ongoing operations, and changes to product, process, equipment, or facility which may be necessary are often difficult to accomplish.

To assure success in either case, the application of industrial robots must be approached in a systematic manner. Launching a robotic production system is best done in a multistep process that involves not only the robot, but also the product, production equipment, layout, scheduling, material flow, and a number of other related factors. Where robots are being integrated into existing operations, there are five discrete steps in this process:

Initial survey
Qualification
Selection
Engineering
Implementation

Where robots are incorporated into a new operation, the first three steps are not specifically followed. This chapter addresses the first four of these steps. Emphasis is on practical, simple-to-use rules and principles. Readers are referred to other chapters in this section for more detailed techniques.

29.2. INITIAL SURVEY

The selection of the manufacturing process or processes to which a robot is to be applied should not be done arbitrarily. It requires the careful identification and consideration of all potential operations and begins with an initial survey of the entire manufacturing facility involved. The objective of the initial survey is to generate a "shopping list" of opportunities, that is, operations to which robots might be applied.

During the initial survey, one should look for tasks that meet the following criteria:

1. An operation under consideration must be physically possible for a robot.
2. An operation under consideration must not require judgment by a robot.
3. An operation under consideration must justify the use of a robot.

At this point in the process, a detailed analysis of each operation relative to these three criteria should not be undertaken; rather, a few simple "rules of thumb" should be applied.

With regard to physical constraints, *generally avoid* operations where:

Cycle time is less than 5 sec.

Working volume exceeds 30 m³.

Load to be handled exceeds 500 kg.

Positioning precision must be better than ±0.1 mm.

"Randomness" in workpiece position and orientation cannot be eliminated.

"Randomness" in process cannot be eliminated.

Work lot size is typically less than 25 pieces.

Number of different workpieces per process is typically greater than 10.

With regard to judgment requirements, *generally avoid* operations where:

The robot's alternative actions cannot be readily identified.

The robot may be required to execute, at random, any one of more than five alternative actions.

Specific, quantified workpiece and process standards do not exist.

The critical properties of the workpiece and the process cannot be measured.

Workpiece identification, condition, and orientation may be ambiguous.

With regard to justification, the economic attractiveness of a potential robot application, as measured by return on capital or by payback period, is usually of primary importance. (See Chapter 33, Evaluation and Economic Justification.)

In the incorporation of robots into existing operations, the major source of savings is the reduction of labor cost resulting from displacement of workers. A rough estimate of these savings can be made based upon the potential labor that could be displaced; along with an estimate of the cost of the robot installation, a rate of return or a payback period could be calculated. For purposes of the initial survey, however, a simple "rule of thumb" may suffice:

A single robot installation can be justified by the displacement of two workers (assuming a 40-hour work week for each worker).

During the initial survey, value judgments regarding the relative merits of potential applications should be avoided. The purpose of this step is to develop objectively a list of opportunities that are technically and economically feasible and which will next be screened and prioritized.

29.3. QUALIFICATION

The second step in the launching of a robotic production system is the qualification of the operations identified in the first step as potential robot applications. Although some screening was done in that step, there are likely to be operations on the list that are not, upon further scrutiny, technically or economically feasible for robotics. Also, all operations on the list will not be of equal importance or complexity, nor can robot production systems be implemented on all of them simultaneously; thus the qualification step will also involve prioritizing the qualified applications.

Qualification and prioritization will be an iterative process. The first element of the process involves the review of each listed operation to answer the basic question, "Can I use a robot?" There are seven factors that should be considered at this time in deciding whether or not a potential exists to apply a robot on that particular operation:

Complexity of the operation

Degree of disorder

Production rate

Production volume

Justification

Long-term potential

Acceptance

For each of these factors, a simple rule has been presented. Reviewing each operation on the list, one finds that the process will eliminate those where a robot should not be used unless the rules given can all be clearly applied to the operation in question.

Regarding complexity, although simple robots exist and are well suited to simple tasks, there are operations where a cylinder, a valve, and a couple of limit switches are sufficient. In other cases, a gravity chute may suffice to transfer and even reorient a part from one location and attitude to another.

At the other end of the scale, operations that require judgment or qualitative evaluation should be avoided. Checking and accepting or rejecting parts on the basis of a measurable standard can be done with a robot. If, however, the only feasible measuring system is human sight or touch, then a robot is out of the question.

In the same vein, operations that involve a combination of sensory perception and manipulation should also be avoided. An example would be a machine tool loading operation that requires that the part being loaded into the chuck be rotated until engagement of a notch with a key is felt, after which the part is fully inserted. While the development of a "hand" for the robot capable of doing this is technically feasible, the complexity of the hand and its potential unreliability will certainly reduce the overall probability of success. Of even greater complexity are operations requiring visual determination of random spindle orientation and orienting the part to match.

The rule to apply here is:

Avoid both extremes of complexity.

Robots cannot operate in a disorderly environment. Parts to be handled or worked on must be in a known place and have a known orientation. For a simple robot, this must be always the same position and attitude. For a more complex robot, parts might be presented in an array; however, the overall position and orientation of the array must always be the same. On a conveyor, part position and orientation must be the same, and conveyor speed must be known.

Sensor-equipped robots (vision, touch) can tolerate some degree of disorder; however, there are definite limitations to the adaptability of such robots today. A vision system, for example, enables a robot to locate a part on a conveyor belt and to position its arm and orient its hand to grasp the part properly. It will not, however, enable a robot to quickly remove a part, correctly oriented, from a bin of parts or from a group of overlapping parts on a conveyor belt.

A touch sensor enables a robot to find the top part on a stack. It does not, however, direct the robot to the same place on each part if the stack is not uniform or is not always in the same position relative to the robot.

The rule to apply here is:

Repeatability is necessary, disorder must be eliminated or made unambiguous.

When we consider production rate and cycle time, we find that small nonservo (pick-and-place) robots can operate at relatively high speeds. There is a limit, though, on the capability of even these devices. A typical pick-and-place cycle takes several seconds. A rate requiring pickup, transfer, and placement of a part in less than about 5 sec cannot be consistently supported by a robot.

Operations that require more complex manipulation or involve parts weighing pounds rather than ounces require even more time with a robot. The larger servo-controlled robots are able to move at speeds up to 1300 mm/sec; however, as speeds are increased, positioning repeatability tends to decrease. For a rough estimate of cycle time, 1 sec per move or major change of part orientation should be allowed. In addition, at least a half-second should be allowed at each end of the path to assure repeatable positioning. In handling a part, allow another half-second each time the gripper or handling device is actuated. (More accurate techniques are covered in Chapter 30, Robot Ergonomics: Optimizing Robot Work.)

The rule to apply here is:

Robots are generally no faster than people, as measured by cycle time; however, robots maintain the same pace whereas people do not.

There are two factors related to production volumes to consider. In batch manufacturing, the typical batch size must be considered. In single-part volume manufacturing, the overall length of the production run is important.

In small-batch manufacturing, changeover time is significant. Recalling that a robot needs an orderly environment, we find that part orienting and locating devices may need to be changed or adjusted before each new batch is run. A robot's end-of-arm tooling and program may also have to be changed for each new batch of parts. Generally, people do not require precise part location—their hands are instantly adaptable and their "reprogramming" is intuitive. A robot becomes impractical when its changeover time from batch to batch approaches 10% of the total time required to manufacture the batch of parts.

If a single part is to be manufactured at high annual volumes for a number of years, special-purpose automation should be considered as an alternative to robots. Per operation, a special-purpose device is probably less costly than the more flexible, programmable automation device, or robot. Single-function, special-purpose devices may also be faster and more accurate than robots. Where flexibility is required or obsolescence is likely, robots should be considered; where these are not factors, special-purpose automation may be more efficient and cost effective.

The rule to apply here is:

For very short runs (about 25 pieces or less), use people; for very long runs (several million per year of a single part), use special-purpose automation; use robots in between.

The application of an industrial robot can represent a significant investment in capital and in effort. Economic justification must therefore be carefully considered: on the balance sheet, increased productivity; reduced scrap losses or rework costs; labor cost reduction; improved quality; improvement in working conditions; avoiding human exposure to hazardous, unhealthy, or unpleasant environments; and reduction of indirect costs are among the plus factors. Offsetting factors include capital investment; facility, tooling, and rearrangement costs; operating expenses and maintenance cost; special tools, test equipment, and spare parts; and cost of downtime or backup expense.

Ballpark estimates of the potential costs and savings should be made whether or not a reasonable return on the investment can be expected. The savings can be roughly estimated by multiplying the number of direct labor heads displaced per shift, times the number of production shifts per day, times the fully burdened annual wage rate. The costs can be roughly estimated by multiplying the basic cost of the robot planned for the operation times 2.5.

"Management direction," "following the crowd," and emotion are no substitutes for economic justification and, in the long run, will not support the application of robots. In some cases, safety or working conditions may override economics; however, these are usually exceptional circumstances.

The rule to apply here is:

If ballpark costs do not exceed ballpark savings by more than a factor of 2, the application can probably be economically justified.

Another consideration is the long-term potential for industrial robots in the particular manufacturing facility. Both the number of potential applications and their expected duration must be taken into account.

Because of its flexibility, a robot can usually be used on a new application if the original operation is discontinued. Since the useful life of a robot may be as long as 10 years, several such reassignments may be made. Unless the first application of the robot is to be of relatively long duration, it's possible that reapplication must be considered. In the process of justifying the initial investment, the cost of reapplying the robot should also be included. If the initial application is of significantly shorter duration than the robot's useful life and no follow-on applications can be foreseen, it can seldom be justified.

As with any electromechanical device, an industrial robot requires some special knowledge and skills to program, operate, and maintain. An inventory of spare parts should be kept on hand. Auxiliary equipment for programming and maintenance or repair may also be required. Training of personnel, spare parts inventory, special tools, test equipment, and the like may represent a sizeable investment. The difference between the amount invested in these items to support a single robot or to support half a dozen or more robots is insignificant.

Maintenance and programming skills and reaction time in case of problems tend to deteriorate without use. Few opportunities will normally arise to exercise these skills in support of a single robot. Under these conditions, the abilities may eventually be lost and any serious difficulty with the robot may then result in its removal.

The rule to apply here is:

If there are not feasible opportunities for more than one robot installation, the single installation is seldom warranted; don't put just one robot into a plant.

Not everyone welcomes robots with open arms. Production workers are concerned with the possible loss of jobs. Factory management is concerned with the possible loss of production. Maintenance personnel are concerned with the new technology. Company management is concerned with effects on costs and profit. Collectively, all of these concerns may be reflected in a general attitude that "Robots are OK, but not here."

It is essential to know whether a robot will be given a fair chance. Reassignment of workers displaced by a robot can be disruptive. Training of personnel to program and maintain the robot can upset maintenance schedules and personnel assignments, and new skills may even have to be developed. The installation and startup can interrupt production schedules, as can occasional breakdowns of the robot or related equipment. Unless everyone involved is aware of these factors and is willing to accept them, the probability of success is poor.

The rule to apply here is:

A robot must be accepted by people, not only on general principles, but on the specific operation under consideration.

The foregoing screening process will, no doubt, eliminate a number of operations from the "shopping list." Those remaining should be operations that qualify as technically and economically feasible for the application of robots. These operations should now be prioritized, in preparation for the selection step. The prioritizing of operations and subsequent selection of an initial robot application can be facilitated by the use of an operation scoring system. The elements of the scoring system might include:

Complexity of the task.

Complexity of end-of-arm tooling, part orienters, feeders, fixtures, and so on.

Changes required to facilities and related equipment.

Changes required to product and/or process.

Frequency of changeovers, if any.

Impact on related operations.

Impact on work force.

Cost and savings potential.

Anticipated duration of the operation.

For each of the elements involved in the prioritization, a set of measures and a score range is established, with the more important elements having a higher range of points than the less important elements. A typical set of elements, measures, and score ranges is shown in Table 29.1.

Using this scoring system, each operation on the "shopping list" can be rated and prioritized; the operation with the highest score will be the prime candidate for the first application. Other factors such as timing, management direction, experience, and human relations might also be considered; however, subjectivity in establishing priorities should be minimized.

29.4. SELECTION

The third step in the launching of a robotic production system is the selection of the operation for which the robot will be implemented. If the initial survey was made with care, and if the qualification and prioritization have been objectively done, this step will be virtually automatic. If a scoring system is used in establishing priorities, then the selection should be made from among the two or three operations with the highest scores. Some consideration might be given to conditions and circumstances that were not measured in the second step; however, it is again important that subjectivity and arbitrariness be minimized.

It is imperative to review, on-site, the top few candidate operations before making the final selection to ascertain that they are, indeed, technically feasible and justifiable. For the first robot application, *the cardinal rule is: Keep it simple.* It is wise to forego some degree of economic or other benefit for the sake of simplicity; even the most elementary of potential applications is likely to be more complex than anticipated. Avoid the temptation to solve a difficult technological problem; those opportunities will come with later installations. Resist the pressure of uninformed suggestions from managers or others trying to be helpful. Do not try to solve some production or manufacturing problem with the robot; if conventional approaches did not succeed, the robot is also likely to fail. Be aware of the robot's limitations; do not choose an application that requires 100% of some capability, such as reach, load capacity, speed, or memory. Once the operation to which the robot will first be applied has been selected, the engineering of the application can begin. The prioritized, qualified "shopping list" should be retained as a source of further robot applications.

29.5. ENGINEERING

The fourth step in the launching of a robotic production system is the system engineering. There are a number of engineering activities involved, some of which must be done sequentially and some of which can be performed concurrently. The first activity is to return to the chosen workplace and thoroughly study the job to make sure that everything that must be done is identified and planned. During this study phase there are a number of considerations that must be addressed:

Alternatives to the robot—can the desired result be better accomplished by a special-purpose device, by basic facility changes, or by restructuring the operation?

Alternative robot attitudes—are there any advantages to mounting the robot in other than the usual feet-on-the-floor attitude, such as overhead?

Alternatives to existing process—are there advantages in reversing the usual "bring the tool to the work" approach and having the robot carry the work to the tool?

Backup—what arrangements, equipment, or actions will be taken to back up the robot during downtimes?

TABLE 29.1. SCORES FOR RANKING ROBOT POTENTIAL APPLICATIONS

Element	Measured by	Score Range
1. Complexity of task	Number of parts	1–10 to 5–5
	Number of operations	1–10 to 5–5
	Number of batches	1–5 to 5–1
2. Complexity of tooling and peripherals	Number of parts	1–10 to 5–5
	Part orientation at delivery	Single, oriented–10 Matrix–5 Bulk, random–0
	Ease of orienting parts	Easy–10 to difficult–5
3. Facility and equipment	Relocation required	No–2 Yes–0
	Utilities availability	At site–5 Nearby–3 Not available–0
	Floor loading	Adequate–3 Need new–0
4. Product and/or process changes	Product changes required	No–10 to minor–5 to major–0
	Process changes required	No–10 to minor–5 to major–0
5. Impact on related operations	Synchronized with previous operation	No–3 Yes–0
	Synchronized with following operation	No–3 Yes–0
	"Bottleneck"	No–3 Yes–0
	Backup/buffer	Easy–3 Hard–1 No way–0
6. Impact on work force	Monotonous, repetitious	Yes–3 No–0
	Bad environment	Yes–3 No–0
	Safety hazard	Yes–5 No–0
	Fast pace or heavy load	Yes–3 No–0
	Labor turnover	High–3 Low–1
7. Risk of unforeseen or random problems	Number of potential different occurrences	1–10 to 10–1
	Attitude/expectations of management	Understanding, reasonable–10 to tough, unrealistic–0
8. Potential benefits	Labor savings, per shift	One point per 0.1 head
	Production shifts per day	Five points per shift
	Quality improvement	Yes–5 No–0
	Productivity improvement	One point per percent increase
	Reduced repair and rework	Yes–5 No–0

Environment—does the robot need special protection from excessive heat or cold, abrasive particulates, shock and vibration, fire, or explosion hazards, and so on?

Space—will the robot occupy significantly more space than the alternatives and, if so, what difficulties might this cause?

Layout—will the robot create problems with accessibility to it, other equipment, and the workplace for material handling, maintenance, inspection, and so on?

Safety—how will the installation be done so as to protect people from the robot and vice versa?

Unusual, intermittent, random occurrences—have all the things that could possibly go wrong with the operation been anticipated, and have contingency plans been made for each?

The detailed study of the operation chosen has as its objective a thorough familiarity with the operational requirements, sequence, and pace, as well as with the occasional random disruptions that seem to occur in any process. Because the robot is not adept at handling disruptions or disturbances in its normal routine, approaches must be developed to minimize their occurrence and impact, to prevent damage when they happen, and to recover rapidly afterward. Significant data to be gathered about the operation include the following:

Number and description of elemental steps in the operation.

Size, shape, weight, and the like of parts or tools handled.

Part orientation at delivery, acquisition, in-process, and at disposal.

Method and frequency of delivery and removal of parts.

If batch production, lot sizes, characteristics of all parts in family, frequency of changes, and changeover time for related equipment.

Production requirements per hour, day, and so on.

Cycle times—floor to floor and elemental.

Inspection requirements, defect disposition.

At this point, a layout drawing of the installation is made. Typically this starts with a scale layout of the existing area onto which the robot and its work envelope are superimposed. Locations for incoming and outgoing material, buffers, and intermediate positions of parts, if necessary, are determined. From this layout, potential interference points can be located and equipment relocations, if any, can be developed. Sources and routing for utilities, such as electrical power, compressed air, and cooling water, are also shown on the layout.

Simultaneously with preparation of the layout, a detailed description of the robot's task is written. This task description must be broken down to a level comparable to the individual steps of the robot's program; it will, in fact, become the basic documentation of that program. Elemental times are estimated for each step so that an approximate cycle time can be established for the entire task.

Working with the layout and the task description, the robot's program is optimized. The objectives are to minimize the number of program steps and robot moves to attain the shortest cycle time. Often, rearranging incoming and outgoing material locations or even the positions of the equipment in the work station can significantly affect the cycle time. Thus both the layout and the task description are necessary elements of this process. Product and/or process changes may also be necessary or desirable. If so, these should be identified and described at this time, with actual engineering to follow.

Up to this point the final selection of a specific model of robot should not be made. Ideally several robot models will be capable of performing this operation, and alternative layouts should be made for each. The task description is basic to the process and should be common for all robot models considered. The selection of the desired robot is now made, based upon best fit to the layout; performance advantages, if any; price, delivery, support, and other similar considerations.

Once the robot model has been chosen and the layout has been optimized, personnel and equipment access points are determined and hazard-guarding (safety barrier) locations are established. It is necessary that an area encompassing the robot's entire working volume be guarded against accidental intrusion by people. Although some installations use active intrusion devices such as light curtains or safety mats interlocked with the robot's control to stop the robot when a person enters the area, passive systems such as fences, walls, or guard rails are more dependable.

Working with the task description, the interlocks between the robot and related equipment are determined. The robot will not directly control the other equipment, that is, the robot's controller will not directly operate other machines in the work cell. The robot will, however, initiate other machine cycles, and its operation will, in turn, be initiated by other machines or devices. For each of the different inputs and outputs, an i/o port on the robot control must be hard-wired to or from some other device. In more complex operations, the robot control may not have sufficient input/output capacity, and an external i/o device such as a programmable controller may be required. When determining inputs and outputs, it is also important to consider the backup method to be used. If manual backup is to be employed, then a manual control station may also have to be provided. This control station must have a "manual/automatic" mode selector, which should be a lockable selection switch.

Following the task description, robot selection, and layout finalization, there are several engineering tasks to be performed. These should all be undertaken simultaneously because they are very much interdependent. These engineering tasks include the following:

End-of-arm tooling design.

Parts feeders, orienters, and positioners design.

Equipment modifications.

Part (product) redesign.

Process revisions.

29.5.1. End-of-Arm Tooling

Typically, a robot is purchased without the end-of-arm tooling unique to its intended task. The robot supplier may furnish a "standard gripper" actuating mechanism, or a suitable device may be obtained from another source; however, adaptation of a standard mechanism may still require some design effort. Likewise, a standard power tool such as a screwdriver or grinder or a spray gun or welding torch which is to be mounted on the end of the robot arm will require the design of mounting hardware, such as brackets, adaptors, and so on. The lack of standard robot/tooling mechanical interfaces means that little "off-the-shelf" hardware is available.

End-of-arm devices lack the dexterity of a human hand; thus, in the case of batch manufacturing, several interchangeable tools may be required. A multifunctional tool for such tasks must represent a practical compromise between simplicity, for reliability, and flexibility, to perform a number of functions or handle a number of different parts. Interchangeable tools should be designed for ease of removal and installation and for repeatable, precise location on the robot arm to avoid the necessity to reprogram the robot with each tool change. In some cases, automatic exchange of tools by the robot may be possible through the use of quick-disconnects, collet/drawbar arrangements similar to preset machine tool holders, tool racks, and the like. (More discussion of end-of-arm tooling is in Chapter 28.)

29.5.2. Part Feeders

Another engineering requirement may be for parts feeders, orienters, and positioners, or other parts-acquisition systems. As noted earlier, today's robots require an ordered, repeatable environment and cannot easily acquire randomly oriented parts delivered in bulk. There are several solutions to this problem, including trays or dunnage that contain parts in positive locations and orientations; mechanical feeder/orienter devices; manual transfer of parts from bulk containers into feeder systems; and sensor-based acquisition systems, such as vision or tactile sensing. Table 29.2 summarizes typical mechanisms for part feeding and their functions.

The mechanically simplest approach is to use parts' containers that retain individual parts in specific locations. A robot with microprocessor or computer control and sufficient memory can be programmed to move to each location in the container, in sequence, to acquire a part. Multiple layers of parts may be packed in this manner, with the robot also programmed to remove empty trays or separators between layers. The only requirement in the workplace is to provide locators for repeatably positioning the containers. The multiple pickup points, in addition to requiring a computer control or large memory capacity, may increase the average cycle time for the operation. And, if the robot lacks the capability to acquire parts from a matrix array, some other approach must be taken.

Another parts-presentation approach is to use mechanical feeder/orienters. These, for small parts, may be centrifugal or vibratory feeders which automatically orient parts in a feeder track. Larger parts may be handled with hoppers and gravity chutes or elevating conveyors and chutes, which also present parts in proper orientation at a specific pickup point. Usually, these devices are adaptations of standard, commercially available equipment. Advantages of this approach are that the single acquisition point for each part minimizes nonproductive motions, and the orienters can often present the parts in attitudes that require little manipulation by the robot after pickup. Disadvantages are difficulties in orienting and feeding some parts, relatively high cost of mechanical feeders, lack of flexibility to handle a variety of parts (as in batch manufacturing), potential damage to delicate, fragile, highly finished, or high-accuracy parts, and inability to handle large, heavy, or awkward-shaped objects.

A third approach is the manual transfer of parts from bulk containers to mechanical feeders such as gravity chutes or indexing conveyors. An obvious disadvantage of this approach is the use of manual labor, especially to perform the very sort of routine, nonrewarding tasks to which robots should be applied. Advantages are relatively low capital investment requirements, ability to handle difficult or critical parts, and flexibility to accommodate a variety of similar parts, as in batch manufacturing.

TABLE 29.2. TYPICAL PART FEEDING DEVICES AND THEIR FUNCTIONS

Feeding Device	Feeding Functions[a]						
	Transfer	Order	Orient	Position	Metering	Binning	Magazining
Bins or hoppers							
Feed hopper	•				•	•	
Tote bin						•	
Storage conveyor	•					•	
Magazines							
Pallet magazine			•				•
Drum magazine	•		•				•
Spiral magazine	•		•				•
Channel magazine	•		•				•
Transfer devices							
Belt/roller conveyor	•						•
Vibratory conveyor	•						
Walking beam conveyor	•						
Rotary index table	•			•	•		•
Pick-and-place			•	•	•		
Ordering devices							
Vibratory bowl feeder	•	•				•	
Rotary feeder	•	•				•	
Elevating feeder conveyor	•	•				•	
Centerboard hopper feeder	•	•				•	
Metering devices							
Pop-up pusher	•				•		
Screw feeder	•				•		
Escapement	•	•			•	•	

Source: Reference 1.

[a] Definition of functions: (1) Transfer: movement of parts to feed point. (2) Order: bringing random parts to a predetermined position *and* orientation. (3) Orient: bringing parts in a known orientation or direction. (4) Position: placing a part in an exact position. (5) Metering: physical separation of parts at feed point. (6) Binning: random storage of parts for feeding. (7) Magazining: storage of parts in a definite orientation.

A fourth approach is the use of sensors such as vision or tactile feedback devices to modify the robot's programmed motions, enabling it to acquire somewhat randomly oriented parts. Advantages are a reduction in the extent of mechanical orientation required and a potential to work with a variety of randomly mixed parts or to accommodate batch manufacturing lot changes with little or no physical changes required. Disadvantages are relatively high cost, compared to simple mechanical feeder/orienters; the possible need for special lighting; relatively slow processing time; and difficulty with touching or overlapping parts or with three-dimensional space (such as bins). The solutions to parts presentation for the robot often combine all of the approaches described, as well as others, such as automatic, single-part-at-a-time delivery from a previous operation by means of a conveyor or shuttle device.

29.5.3. Equipment Modification

A third engineering requirement may be the modification of existing equipment with which the robot will operate. Typical modifications may include adding a cylinder and solenoid valve to a machine-tool splash guard for automatic, rather than manual, opening and closing. Other changes may be made to guards and housings for improved access by the robot. Machine-tool chucks and collets may be modified to increase clearances or to provide leads or chamfers for easier insertion of parts.

Powered clamping and shuttle devices may be substituted for manually actuated mechanisms. In machine-tool operations, coolant/cutting fluid systems may be changed or chip blow-off systems added to automatically remove cuttings (chips) from the work and work holders. Assembly operations may require the development of simple jigs and fixtures in which to place parts during the process (remember that robots are generally single-handed devices and cannot hold something in one hand while adding components to it with the other). Likewise, manual tools such as screwdrivers and wrenches will have to be replaced with automatic power tools.

29.5.4. Part/Product Redesign

Part orienting and feeding and/or part handling by the robot's end-of-arm tooling may require some redesign of a product. Ideally, the product should be designed so that it has only one steady-state orientation, that is, it should be self-orienting. As an alternative, the product should be designed so that its orientation for acquisition is not critical (for example, a flat disk or washer shape). A family of parts which are all to be handled by the robot should have some common feature by which they are grasped; this feature should be of the same size and in the same location on all products in the family.

Vacuum pickups are simple, fast, and inexpensive. Product designs that incorporate surfaces or features to which a vacuum pickup can be applied facilitate easier handling by the robot. Products should be designed so that the robot's task (such as load, unload, insert, and assemble) requires a minimum of discrete motions; complex motions, especially those that require the coordinated movements of two or more robot axes (such as a helical movement of the part), should be avoided. Tolerances should be "opened up" as much as possible. Chamfers should be provided on inserted parts to aid in alignment. Parts should be self-aligning or self-locating, if possible. Parts that are to be mechanically or gravity oriented and fed to the robot should be designed so that they do not jam, tangle, or overlap.

Because product redesign is costly and time-consuming, it should not be undertaken lightly, but should be considered only when its potential benefits significantly outweigh its cost. In the design of new products, however, incorporation of features that facilitate the use of robots should add little or nothing to the cost and should, thus, be encouraged. In Table 29.3 and Figure 29.1 rules and principles are provided for design of parts and products for automatic assembly. Such rules can guide designers in the design of parts for robotic handling.

29.5.5. Process Revisions

Another engineering requirement may be the modification of the process with which the robot is involved. Process revision may include changing an operational sequence so that critical part orientation is not required. Process revision may involve moving several machines into an area and setting up a

---

TABLE 29.3. PART/PRODUCT REDESIGN FOR AUTOMATIC ASSEMBLY

A. *Rules for Product Design*

1. Minimize the number of parts.
2. Product must have suitable base part on which to build.
3. Base part should ensure precise, stable positioning in horizontal plane.
4. If possible, assembly should be done in layers, from above.
5. Provide chamfers or tapers where possible to aid correct guidance and positioning.
6. Avoid expensive and time-consuming fastening operations, e.g., screwing, soldering.

B. *Rules for Part Design*

1. Avoid projections, holes, or slots that may cause entanglement in feeders.
2. Symmetrical parts are preferred because less orientation is needed and feeding is more efficient.
3. If symmetry cannot be achieved, design appropriate asymetrical features to aid part orienting.

C. *Design Features that Determine Cost-Effectiveness of Automatic Assembly*

1. Frequency of simultaneous operations.
2. Orienting efficiency.
3. Feeder required.
4. Maximum feed rate possible.
5. Difficulty rating for automatic handling.
6. Difficulty rating for insertions required.
7. Assembly operations required (number and type).
8. Total number of component parts per assembly.

Source: References 2 and 3.

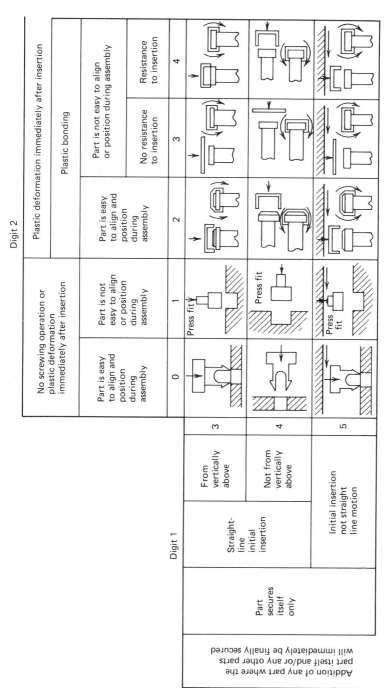

Fig. 29.1. Coding system for automatic assembly processes. (*Source:* Reference 2.)

machining cell to take advantage of initial part orienting and to increase the robot's utilization. It may involve linking of operations with conveyors so as to retain part orientation for the robot, or the incorporation of compartmentalized pallets or dunnage to retain orientation between operations. Process revision may involve the rescheduling of batch operations to increase lot size or to minimize changeover between batches.

Process revisions can often be accomplished at minimal cost, particularly those that involve only scheduling, and can sometimes significantly increase the efficiency of the robotic production system. Like product changes, process revisions should not be undertaken lightly, however, but should be carefully examined for cost-effectiveness.

The first four steps of launching a robotic production system, *initial survey, qualification, selection,* and *engineering,* which have been described in this chapter, should, if followed carefully and thoroughly, make the fifth and last step, *implementation,* relatively easy and trouble free.

REFERENCES

1. Warnecke, H. J. and Schraft, R. D., *Industrial Robots Application Experience,* IFS Publications, 1982.

2. Boothroyd, G., Poli, C., and Murch, L. E., *Automatic Assembly,* Marcel Dekker, New York, 1982.

3. Boothroyd, G. and Dewhurst, P., *Design for Assembly,* Department of Mechanical Engineering, University of Massachusetts, 1983.

4. Bailey, J. R., Product design for robotic assembly, *Proceedings of the 13th International Symposium on Industrial Robots,* Chicago, Illinois, April 1983, pp. 1144–1157.

5. Pham, D. T., On designing components for automatic assembly, *Proceedings of the 3rd International Conference on Assembly Automation,* Stuttgart, West Germany, May 1982, pp. 205–214.

6. Tanner, W. R. and Spiotta, R. H., Industrial robots today, *Machine Tool Blue Book,* Vol. 75, No. 3, March 1980, pp. 58–75. (Analyzes applications in which robots can be efficient and economical.)

7. Engelberger, J. F., Production problems solved by robots, *SME Paper* No. MS74–167, 1974. (Describes production applications problems that robots can and cannot solve.)

8. Estes, V. E., An organized approach to implementing robots, *Proceedings of the 16th Annual Meeting of the Numerical Control Society,* Los Angeles, California, March 1979, pp. 287–307. (Describes the approach taken by General Electric Consulting Services to implement robotics successfully in the company.)

CHAPTER 30

ROBOT ERGONOMICS: OPTIMIZING ROBOT WORK

SHIMON Y. NOF

Purdue University
West Lafayette, Indiana

30.1. THE ROLE OF ROBOT ERGONOMICS

The word *ergonomics,* in Greek, means "the natural laws of work." Traditionally, it has meant the study of the anatomical, physiological, and psychological aspects of humans in working environments for the purpose of optimizing efficiency, health, safety, and comfort associated with work systems. Correct and effective introduction of robots to industrial work requires use of ergonomics. However, planning the work of robots themselves in industry brings about a completely new dimension.

For the first time, as far as we can tell, we have the ability to design not only the work system, its components and environment, but also the structure and capabilities of the *operator*—the robot.

The purpose of this chapter is to explain a number of techniques that have been developed and applied in recent years for planning various applications of robots in industry. Some of the techniques bear strong similarity to the original, human-oriented ergonomics techniques. But all these techniques have been developed or adapted for the unique requirements of industrial robots. Their common objective: to provide tools for the study of relevant aspects of robots in working environments for the purpose of optimizing overall performance of the work system. Specifically, robot work should be optimized to: (1) minimize the time per unit of work produced; (2) minimize the amount of effort and energy expanded by operators; (3) minimize the amount of waste, scrap, and rework; (4) maximize quality of work produced; (5) maximize safety.

A general ergonomics procedure for optimizing industrial robot work is depicted in Figure 30.1. For given job requirements, it entails the analysis and evaluation of whether a human or a robot should be employed for the job. If a robot—the best combination of robot models and work method, implying also the best workplace, should be selected. In integrated human and robot systems, the best combination must be designed. The subsequent sections cover the ergonomics techniques that are useful to follow the foregoing procedure in practice. The following topics are covered:

Analysis of work characteristics.

Work methods analysis.

Workplace design.

Performance measurement.

Integrated human and robot ergonomics.

30.2. ANALYSIS OF WORK CHARACTERISTICS

To implement the ergonomic procedure shown in Figure 30.1 effectively, a general list of considerations such as those proposed by Ottinger[2] and elsewhere can be prepared, as shown in Table 30.1. In addition to such a general analysis, it is also necessary to know the detailed characteristics and skills of today's industrial robots, as well as those of humans. A series of the Robot-Man Charts can serve this purpose well.

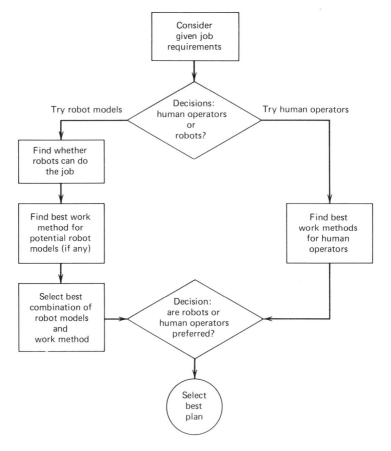

Fig. 30.1. Ergonomics procedure for optimizing industrial robot work (*Source:* Reference 1.).

30.2.1. The Robot-Man Charts

These charts, originally prepared by Nof, Knight and Salvendy,[3] are developed with two functions in mind, namely, (1) to aid engineers in determining whether a robot can perform a job, and (2) to serve as a guideline and reference for robot specifications. (See Table 30.2.) The Robot-Man Charts can also be useful in job design of combined systems that integrate both robots and human operators.

The Robot-Man Charts contain three main types of work characteristics.

1. Physical skills and characteristics, including manipulation, body dimensions, strength and power, consistency, overload/underload performance, and environmental constraints. Table 30.2a provides details of this category. Typical ranges of maximum motion capabilities (TRMM) are given for several categories of body movement and speed, and arm and wrist motions.

To clarify the meaning of "typical ranges of maximum motion capabilities" (TRMM), consider the following example. For robot arm right-left traverse, the table lists: "Maximum movement range of 100 to 6000 mm at a maximum velocity range of 100–1500 mm/sec." This means that for the surveyed population of robot models, it was found that a maximum arm right-left linear motion is typically 100 mm (for some models), and up to 6000 mm (for some other models). The maximum velocity values for right-left traverse were found to be from 100 mm/sec up to 1500 mm/sec.

2. Mental and communicative characteristics: The Robot-Man Charts contain mental and communicative system attributes for robots and humans, listed in Table 30.2b.

3. Energy considerations: A comparison of representative values of energy-related characteristics, such as power requirements and energy efficiency, for robots and humans, is given in Table 30.2c.

TABLE 30.1.   CONSIDERATIONS IN PLANNING ROBOT WORK

Consideration Item	Consideration Variables
1. Payload	At full and reduced speed At full and partial extension
2. Reach	Horizontal; vertical With tool or part at various orientations
3. Stroke	Horizontal; vertical With tool or part
4. Memory capacity	Standard; expansions; different controllers Single or multiple programs
5. Programming skills requirement	Simple—air logic or stepping drum Advanced—microprocessor, computer
6. Controls flexibility	Branching; palletizing; tracking; sensor feedback, editing; subroutines; communication
7. Cycle time	Current concept Revised concept: new method, equipment, accessories, layout; more parts, less errors, etc.
8. Maintenance skills requirement	Simple: pneumatic or hydraulic Advanced: hydraulic or electric
9. Vendor considerations	Corporate strength, quality, service, training
10. System cost	Robot manufacturer System house In-house
11. Human factors	Management Engineering/production Maintenance
12. Down time contingency plan	Automatic takeover Manual intervention Shutdown until fixed

Certain significant features that distinguish robots and human operators in the context of ergonomic studies can be observed, and can be effectively utilized to select jobs that robots can do well:[3]

1. The more similar are two jobs, the easier it is to transfer either robot or human from one job to the other. For humans, such transfer is almost entirely a question of learning or retraining. For robots, however, as job similarity decreases, robot reconfiguration, as well as reprogramming (retraining), becomes necessary for economical interjob transfer.

2. Humans possess a set of basic skills and experiences accumulated over the years, and therefore may require less detail in their job description. Today's robots, on the other hand, perform each new task essentially from scratch and require a high degree of detail for every micromotion and microactivity.

3. Robots do not have any significant individual differences within a given model. Thus an optimized job method may have more generality with robot operators than with human operators.

4. Robot sensing, manipulative, and decision-making abilities can be designed for a given specialized task to a much greater degree than can a human's abilities. Of course, this specialization may entail the cost of decreased transferability from one task to another.

5. Robots are unaffected by social and psychological effects (such as boredom) that often impose constraints upon the engineer attempting to design a job for a human operator.

Job Selection for Robots

In planning work in industry, two decisions must be made:

1. Selection: Who should perform a given task or set of tasks—a human operator or a robot?
2. Specification: What are the specifications of the job and the skills? If a robot was indicated in the previous decision (1), complete robot specifications are also sought.

TABLE 30.2. ROBOT-MAN CHARTS

Characteristics	Robot	Human
	a. Comparison of Robot and Human Physical Skills and Characteristics	

1. Manipulation
   A. Body

   a. One of four types: Uni- or multiprismatic Uni- or multirevolute Combined revolute/prismatic Mobile

   a. A mobile carrier (feet) combined with 3 DF wristlike (roll, pitch, yaw) capability at waist.

   b. Typical maximum movement and velocity capabilities:
   Right-left traverse
   5–18 m at
   500–1200 mm/sec
   Out-in traverse
   3–15 m
   500–1200 mm/sec

   b. Examples[a] of waist movement:
   Role: ≃ 180°
   Pitch: ≃ 150°
   Yaw: ≃ 90°

   B. Arm

   a. One of four primary types: Rectangular Cylindrical Spherical Articulated

   a. Articulated arm comprised of shoulder and elbow revolute joints.

   b. One or more arms, with incremental usefulness per each additional arm.

   b. Two arms, cannot operate independently (at least not totally).

   c. Typical maximum movement and velocity capabilities:
   Out-in traverse
   300–3000 mm
   100–4500 mm/sec
   Right-left traverse
   100–6000 mm
   100–1500 mm/sec
   Up-down traverse
   50–4800 mm
   50–5000 mm/sec
   Right-left rotation
   50–380°[b]
   5–240°/sec
   Up-down rotation
   25–330°
   10–170°/sec

   c. Examples of typical movement and velocity parameters:
   Maximum velocity: 1500 mm/sec in linear movement.
   Average standing lateral reach: 625 mm
   Right-left traverse range: 432–876 mm
   Up-down traverse range: 1016–1828 mm
   Right-left rotation (horizontal arm) range: 165–225°
   Average up-down rotation: 249°

   C. Wrist

   a. One of three types: Prismatic Revolute Combined prismatic/revolute
   Commonly, wrists have 1–3 rotational DF: roll, pitch, yaw; however, an example of right-left and up-down traverse was observed.

   a. Consists of three rotational degrees of freedom: roll, pitch, yaw.

   b. Typical maximum movement and velocity capabilities:
   Roll
   100–575°[c]
   35–600°/sec
   Pitch
   40–360°
   30–320°/sec

   b. Examples of movement capabilities:
   Roll: ≃ 180°
   Pitch: ≃ 180°
   Yaw: ≃ 90°

TABLE 30.2   ROBOT-MAN CHARTS (Cont.)

Characteristics	Robot	Human
	Yaw 100–530° 30–300°/sec Right-left traverse (uncommon) 1000 mm 4800 mm/sec Up-down traverse (uncommon) 150 mm 400 mm/sec	
D. End effector	a. The robot is affixed with either a hand or a tool at the end of the wrist. The end effector can be complex enough to be considered a small manipulator in itself. b. Can be designed to various dimensions.	a. Consists of essentially 4 DF in an articulated configuration. Five fingers per arm each have three pitch revolute and one yaw revolute joints. b. Typical hand dimensions: Length: 163–208 mm Breadth: 68–97 mm (at thumb) Depth: 20–33 mm (at metacarpal)
2. Body dimensions	a. Main body: Height: 0.10–2.0 m Length (arm): 0.2–2.0 m Width: 0.1–1.5 m Weight: 5–8000 kg b. Floor area required: from none for ceiling-mounted models to several square meters for large models.	a. Main body (typical adult): Height: 1.5–1.9 m Length (arm): 754–947 mm Width: 478–579 mm Weight: 45–100 kg b. Typically about 1 m² working radius:
3. Strength and power	a. 0.1–1000 kg of useful load during operation at normal speed; reduced at above normal speeds. b. Power relative to useful load.	a. Maximum arm load: < 30 kg; varies drastically with type of movement, direction of load, etc. b. Power: 2 hp ≃ 10 sec 0.5 hp ≃ 120 sec 0.2 hp ≃ continuous 5 kc/min subject to fatigue; may differ between static and dynamic conditions.
4. Consistency	Absolute consistency if no malfunctions.	a. Low b. May improve with practice and redundant knowledge of results. c. Subject to fatigue: physiological and psychological. d. May require external monitoring of performance.
5. Overload/underload performance	a. Constant performance up to a designed limit, and then a drastic failure. b. No underload effects on performance.	a. Performance declines smoothly under a failure. b. Boredom under local effects is significant.
6. Environmental constraints	a. Ambient temperature from −10°C to 60°C. b. Relative humidity up to 90%.	a. Ambient temperature range 15–30°C. b. Humidity effects are weak.

TABLE 30.2   ROBOT-MAN CHARTS (Cont.)

Characteristics	Robot	Human
	c. Can be fitted to hostile environments.	c. Sensitive to various noxious stimuli and toxins, altitude, and airflow.

b.   Comparison of Robot and Human Mental and Communicative Skills

Characteristics	Robot	Human
1. Computational capability	a. Fast, e.g., up to 10 Kbits/ sec for a small minicomputer control.	a. Slow—5 bits/sec.
	b. Not affected by meaning and connotation of signals.	b. Affected by meaning and connotation of signals.
	c. No evaluation of quality of information unless provided by program.	c. Evaluates reliability of information.
	d. Error detection depends on program.	d. Good error detection correction at cost of redundancy.
	e. Very good computational and algorithmic capability by computer.	e. Heuristic rather than algorithmic.
	f. Negligible time lag.	f. Time lags increased, 1–3 sec.
	g. Ability to accept information is very high, limited only by the channel rate.	g. Limited ability to accept information (10–20 bits/sec).
	h. Good ability to select and execute responses.	h. Very limited response selection/execution (1/sec); responses may be "grouped" with practice.
	i. No compatability limitations.	i. Subject to various compatibility effects.
	j. If programmable—not difficult to reprogram.	j. Difficult to program.
	k. Random program selection can be provided.	k. Various sequence/transfer effects.
	l. Command repertoire limited by computer compiler or control scheme.	l. Command repertoire limited to experience and training.
2. Memory	a. Memory capability from 20 commands to 2000 commands, and can be extended by secondary memory such as cassettes.	a. No indication of capacity limitations.
	b. Memory partitioning can be used to improve efficiency.	b. Not applicable.
	c. Can forget completely but only on command.	c. Directed forgetting very limited.
	d. "Skills" must be specified in programs.	d. Memory contains basic skills accumulated by experience.
		e. Slow storage access/retrieval.
		f. Very limited working register: ≃ 5 items.
3. Intelligence	a. No judgment ability of unanticipated events.	a. Can use judgment to deal with unpredicted problems.
	b. Decision making limited by computer program.	b. Can anticipate problems.
4. Reasoning	a. Good deductive capability, poor inductive capability.	a. Inductive.
	b. Limited to the programming ability of the human programmer.	b. Not applicable.

TABLE 30.2 ROBOT-MAN CHARTS (Cont.)

Characteristics	Robot	Human
5. Signal processing	a. Up to 24 input/output channels, and can be increased, multitasking can be provided. b. Limited by refractory period (recovery from signal interrupt).	a. Single channel, can switch between tasks. b. Refractory period up to 0.3 sec.
6. Brain-muscle combination	a. Combinations of large, medium, and small "muscles" with various size memory, velocity and path control, and computer control can be designed.	a. Fixed arrangement.
7. Training	a. Requires training through teaching and programming by an experienced human. b. Training doesn't have to be individualized. c. No need to retrain once the program taught is correct. d. Immediate transfer of skills ("zeroing") can be provided.	a. Requires human teacher or materials developed by humans. b. Usually individualized is best. c. Retraining often needed owing to forgetting. d. Zeroing usually not possible.
8. Social and psychological needs	a. None.	a. Emotional sensitivity to task structure—simplified/enriched; whole/part. b. Social value effects.
9. Sensing	a. Limited range can be optimized over the relevant needs. b. Can be designed to be relatively constant over the designed range.	a. Very wide range of operation (10^{12} units). b. Logarithmic: vision: 1. visual angle threshold—0.7 min 2. brightness threshold—4.1 $\mu\mu\ell$ 3. response rate for successive stimuli $\simeq 0.1$ sec audition: 1. threshold—0.002 dynes/m^2 tactile: 1. Threshold—3 g/mm^2
	c. The set of sensed characteristics can be selected. Main senses are vision and tactile (touch). d. Signal interference ("noise") may create a problem. e. Very good absolute judgment can be applied. f. Comparative judgment limited by program.	c. Limited set of characteristics can be sensed. d. Good noise immunity (built-in filters). e. Very poor absolute judgment (5–10 items). f. Very good comparative judgment.
10. Interoperator communication	Very efficient and fast intermachine communication can be provided.	Sensitive to many problems, e.g., misunderstanding.
11. Reaction speed	Ranges from long to negligible delay from receipt of signal to start of movement.	Reaction speed ¼–⅓ sec.

TABLE 30.2   ROBOT-MAN CHARTS (Cont.)

Characteristics	Robot	Human
12. Self-diagnosis	Self-diagnosis for adjustment and maintenance can be provided.	Self-diagnosis may know when efficiency is low.
13. Individual differences	Only if designed to be different.	100–150% variation may be expected.

c.   Comparison of Robot and Human Energy Considerations

1. Power requirements	Power source 220/440 V, 3 phase, 50/60 Hz, 0.5–30 KVA. Limited portability.	Power (energy) source is food.
2. Utilities	Hydraulic pressure: 30–200 kg/cm² Compressed air: 4–6 kg/cm²	Air: Oxygen consumption 2–9 liters/min.
3. Fatigue, downtime, and life expectancy	a. No fatigue during periods between maintenance. b. Preventive maintenance required periodically. c. Expected usefulness of 40,000 hr (about 20 one-shift years). d. No personal requirements.	a. Within power ratings, primarily cognitive fatigue (20% in first 2 hr; logarithmic decline). b. Needs daily rest, vacation. c. Requires work breaks. d. Various personal problems (absenteeism, injuries, health).
4. Energy efficiency	a. Relatively high, e.g. (120–135 kg)/(2.5–30 KVA). b. Relatively constant regardless of workload.	a. Relatively low, 10–25%. b. Improves if work is distributed rather than massed.

[a] Where possible, fifth and ninety-fifth percentile figures from Woodson[7] are used to present min. and max. values. Otherwise, a general average value is given.
[b] A continuous right-left rotation is available.
[c] A continuous roll movement is available.

For example, Figure 30.2 depicts an analysis relative to reachability. In (*a*), a comparison is made between human and robot reachability. In (*b*), alternative robot models are compared. Usually three cases can be identified in the process of job selection for robots:

1. *A human operator must perform the job* because the task is too complex to be performed economically by any available robot.

2. *A robot must perform the job* because of safety reasons, space limitation, or special accuracy requirements.

3. *A robot can replace a human operator* on an existing job, and the shift to robot operation could result in improvements such as higher consistency and better quality. Labor shortages in certain types of jobs may also result in robot assignments.

In the first two cases, the selection is clear. In the third case the main concern is whether a robot can at all perform a given task. The Robot-Man Charts provide a means of identifying job dimensions that can or cannot be done by robots or humans.

Another approach for assessing different dimensions in the problem is a systematic comparison between robot time and motion (RTM) task method elements for a robot, and methods time measurement (MTM) elements for a human operator. (See Section 30.5 and Table 30.10.) Additional information for this decision can be obtained from a data base of robot work abilities.

30.2.2.   Robot Anatomy and Distribution of Work Abilities

A thorough examination of industrial robots and their controls[4] provides anatomy of the basic structure and controls of robots and reveals their resulting limitations, particularly in the area of sensor ability

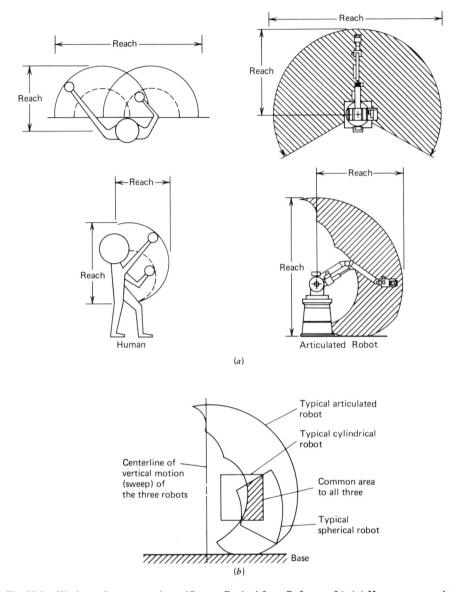

Fig. 30.2. Work envelope comparisons (*Source:* Revised from Reference 2.). (*a*) Human versus robot. (*b*) Different robot models.

and task interactions. Other sources, such as Jablonowski,[5] provide definitions of the important robot specifications. To determine the current abilities of industrial robots, literature describing 282 models was summarized and analyzed in two forms:[6] characteristic frequency distributions and motion/velocity graphs. (Sources for this 1982 survey included numerous 1982 industrial robot specification guides.)

Distributions of Work Characteristics

A series of frequency distributions were developed for the following robot parameters:

Arm structure and size type.
Lifting capacity at end effector.

Actuator type.

Degrees of freedom.

Control mode.

Sensory ability.

Repeatability by actuator type.

These distributions were categorized for three cases: (1) Japanese models; (2) models marketed in the United States, including many European robots but excluding the Japanese models covered by the first category (hereafter, "U.S. market"); and (3) the combined total of all models from the two previous categories. Of the 282 models surveyed, 183 were Japanese models and 99 were U.S. market models. Figure 30.3*a–j* contains the distributions for the surveyed robot parameters. (Figure 30.3 *k–p* shows the result of an update survey that was conducted in 1984, which we discuss later in this chapter.)

1. In Figure 30.3*a* it can be seen that the Japanese have concentrated much more on the rectangular models (52% of all their models) whereas the U.S. market has more articulated models (48% of U.S. market models). By far, the most common model size is "medium" (maximum dimension of robot envelope from 2 to 5 m) with the next being "small" (maximum dimension of robot envelope from 1 to 2 m). *Note:* This analysis is for models and not actual numbers of each, which are certainly different.

2. More U.S. market models have relatively greater lifting capacity compared to the Japanese models, although both concentrate their capacity in the under 40-kg range.

3. Whereas most U.S. models employ hydraulic systems, it is observed that the Japanese have built more electric-actuated models. From this survey it is apparent that, recently, U.S. and European manufacturers are also producing more electric models owing to the need for improved accuracy capability for assembly and other precision tasks. Overall, the pneumatic-actuated model is the least available.

4. In the United States a 6 DF (degrees of freedom) robot is most frequent, with 5 DF models placing second. Japanese models tend toward 5 DF by a margin of double the nearest categories of 3, 4, and 6 DF. This leads to an observation of generally less articulated models in Japan relative to the United States and supports the assumption that many robot applications can employ simpler models.

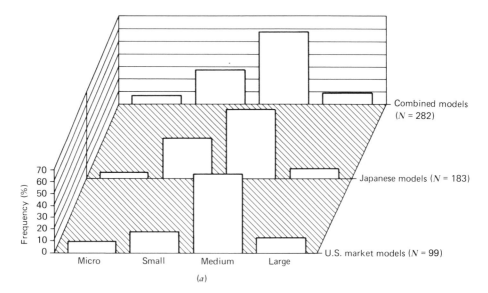

Fig. 30.3. Frequency distributions of 1982 industrial robots work characteristics. (*a*) Robot size: micro, $x \leq 1$; small, $1 < x \leq 2$; medium, $2 < x \leq 5$; large $x > 5$; x = maximum dimension of robot envelope (m).

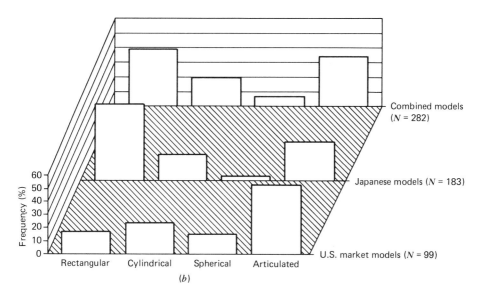

Fig. 30.3. (*b*) Arm structure.

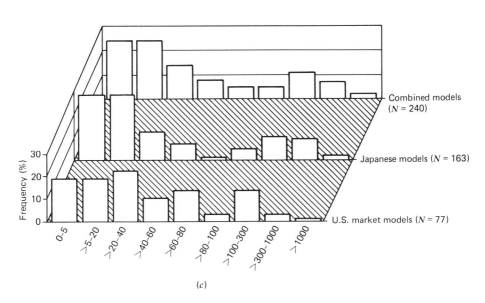

Fig. 30.3. (*c*) Lift capacity at end effector (kg).

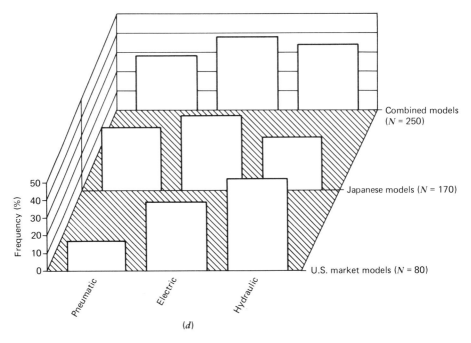

Fig. 30.3. (*d*) Actuator type.

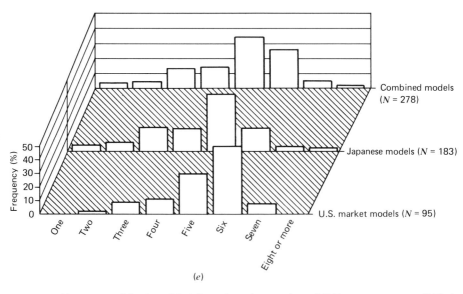

Fig. 30.3. (*e*) Degrees of freedom (Modular robots that can have 1 DF, or more, are available in the United States but are not included in the survey).

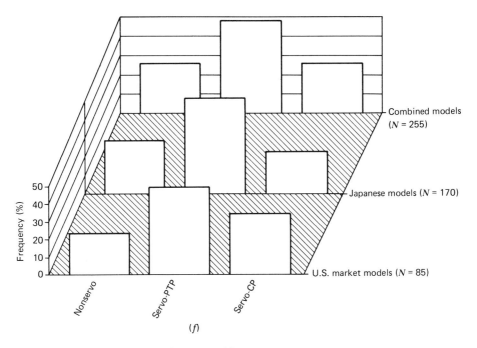

Fig. 30.3. (*f*) Control mode.

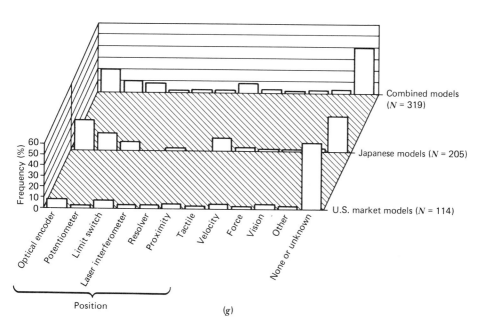

Fig. 30.3. (*g*) Sensor ability (Note that the "None or unknown" category is artificially high, since many models not specifying sensor use were known to employ them in actuality).

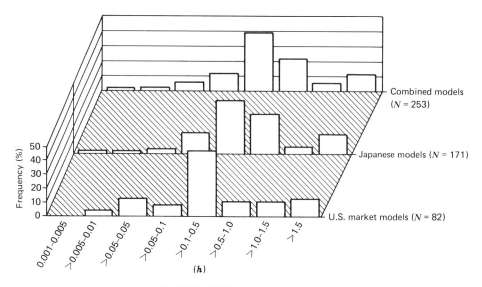

Fig. 30.3. (*h*) Repeatability (±mm).

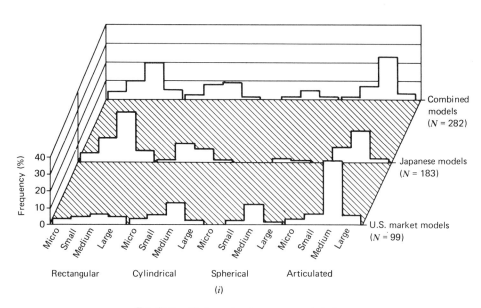

Fig. 30.3. (*i*) Arm structure versus robot size.

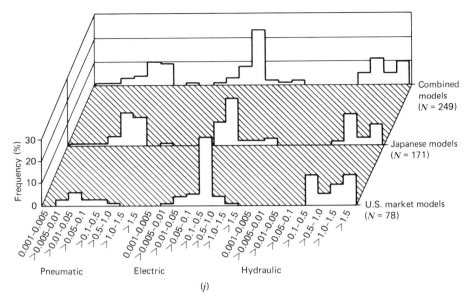

Fig. 30.3. (*j*) Actuator type versus repeatability (±mm).

Frequency distributions of 1984 industrial robot work characteristics are shown in parts (*k*)–(*p*).

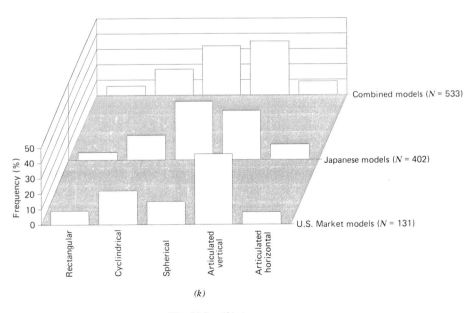

Fig. 30.3. (*k*) Arm structure.

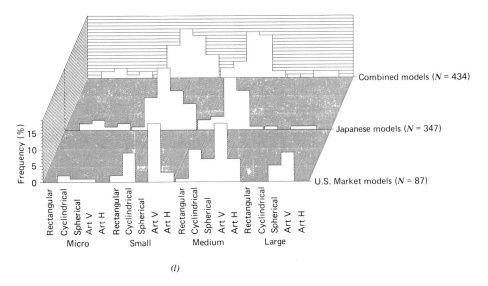

Fig. 30.3. (*l*) Robot size versus arm structure.

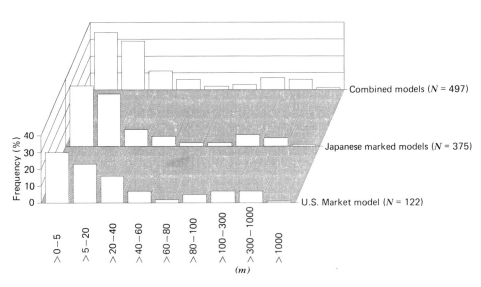

Fig. 30.3. (*m*) Lift capacity (kg).

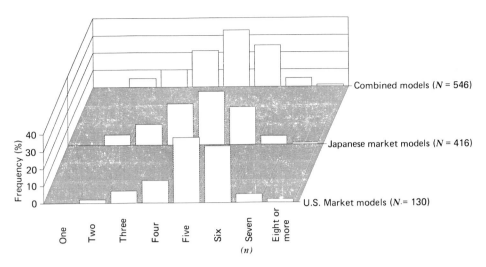

Fig. 30.3. (*n*) Degrees of freedom.

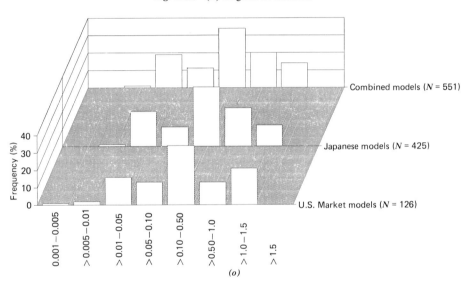

Fig. 30.3. (*o*) Repeatability (±mm).

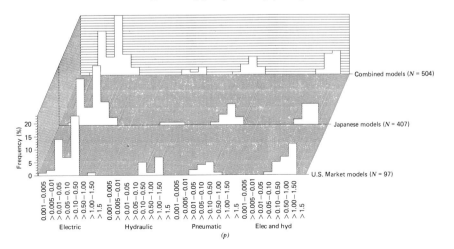

Fig. 30.3. (*p*) Actuator type versus repeatability.

5. Servo-PTP (point-to-point) models are definitely in the majority in both markets. However, when the other two categories are examined, it can again be seen that the Japanese have opted for the lower-technology nonservo category, at least in number of models, as opposed to the U.S. market, which has, by percentage, more servo-CP (continuous path) models, requiring higher technology.

6. In both markets the most common repeatability value was between 0.1 and 0.5 mm. This category is by far the most common in the United States. However, in Japan a large percentage is also designed for the 0.5–1.0-mm category, second most common. Since it is known that the actuator type has a great deal to do with repeatability, that is, the ability of the end effector to repeat a position, repeatability was analyzed relative to the actuator type (Figure 30.3*j*). It is noted that repeatability decreases, that is, the positioning tolerance becomes larger, as the actuator type tends towards hydraulic, whereas in the opposite direction of achieving greater repeatability, electric actuators are more prevalent. The best repeatability surveyed was 0.002 mm, present in a pneumatically actuated model. However, this is misleading since that repeatability is gained through an external mechanical stop.

Motion-Velocity Graphs

In analyzing a prospective robot task, it is useful to have knowledge of translational and rotational capabilities and the velocities at which they can be performed. Figure 30.4*a–h* shows regions of maximum movement and velocity combinations for common arm and wrist motions. As in the preceding section, U.S. market (solid-line regions) and Japanese (broken-line regions) models were examined. However, all information was condensed to a single graph for each motion category to aid the comparison.

Individual points inside the region shown in the graphs were generated for individual robot models (whenever information was available). For example a Cybotech V-80 arm can rotate up to 270° right to left at up to 1 rad/sec (57°/sec). This case is included as one point inside the region in Figure 30.4. The regions shown were developed by connecting extreme perimeter points. Thus a region illustrates overall ranges. Since motion/velocity combinations are not equally distributed inside a region, centroids were computed for each graph. The centroids are indicated in the figures and listed in Table 30.3.

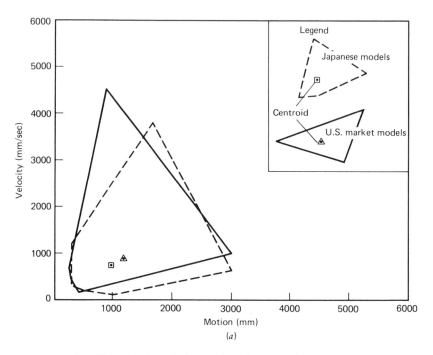

Fig. 30.4. Motion-velocity graphs. (*a*) Arm out-in traverse.

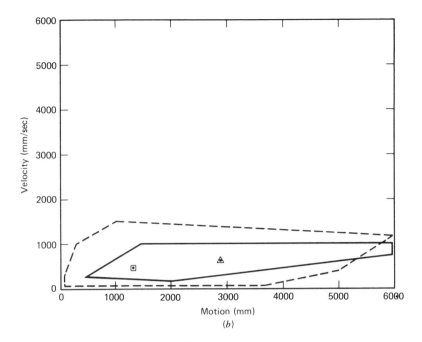

Fig. 30.4. (*b*) Arm right-left traverse.

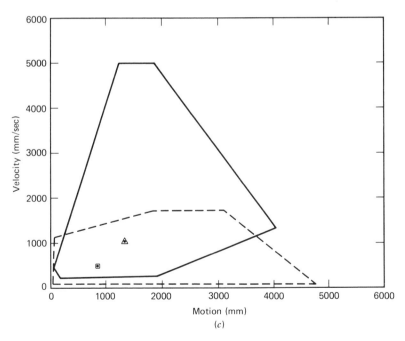

Fig. 30.4. (*c*) Arm up-down traverse.

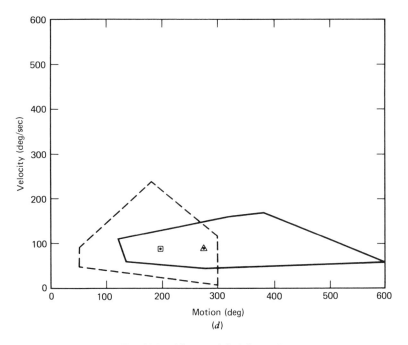

Fig. 30.4. (d) Arm right-left rotation.

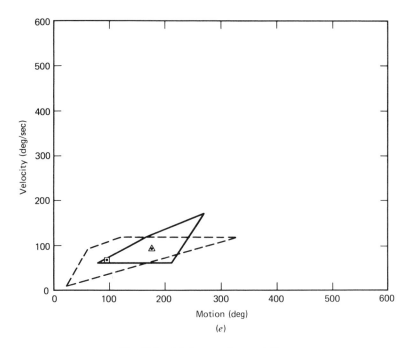

Fig. 30.4. (e) Arm up-down rotation.

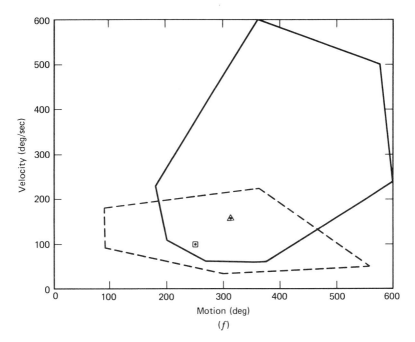

Fig. 30.4. (*f*) Wrist roll.

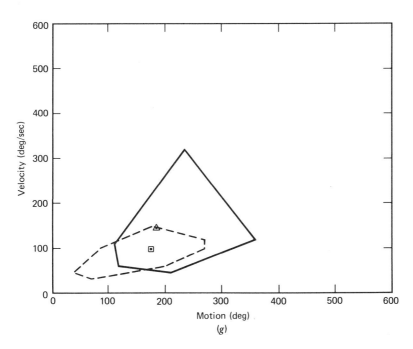

Fig. 30.4. (*g*) Wrist pitch.

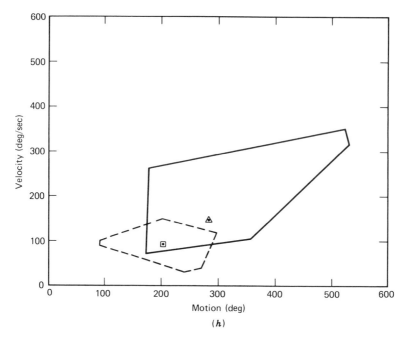

Fig. 30.4. (*h*) Wrist yaw.

Based on the graphs the following can be observed:

1. Motion/velocity capability ranges differ considerably between the two markets. Only in the case of arm out-in traverse are both the centroids and regions similar. In all other cases considerable differences exist.

2. Generally, the centroids calculated for wrist motions in U.S. market models indicate that, among those robots, models with larger maximum velocity are available. The same is true with regard to the maximum movement range. For instance, in the wrist-yaw combinations, up to over 500° of yaw at a velocity of about 300°/sec are available in the U.S. market. This is compared with maximum combinations of 300° at 120°/sec, and 200° at 150°/sec in the Japanese market.

3. On the other hand, Japanese robot models have the advantage of a relatively larger variety of model motion/velocity combinations in terms of arm right-left traverse, arm up-down rotation. A clear advantage to Japanese robot models is in the availability of larger capabilities in terms of arm right-left rotation velocity. As Japanese models have relatively smaller arms on the average, consequently lower inertias, higher rotation velocities could be designed.

1984 Update Survey of Robot Work Characteristics

During 1984 an update survey was initiated, mainly to establish trends in robot work ability developments. Specification guides for over 700 models (1983 and 1984) were analyzed; at the time of printing, partial results for 551 Japanese and U.S. market models were available, and are shown in Figure 30.3*k–p*. (It can be seen that complete information on all the characteristics was not available for all robots in the survey.) Based on the update survey, here are several observations:

1. While in the U.S. market robot models of medium size continue to be most common (42%), they are now followed closely by small models (38%). In contrast, the most common size in Japanese models is now small (52%), with medium at only 36%. A change is indicated also in arm structure: In both markets rectangular is now the least common, and articulated (vertical plus horizontal) the most popular. In the current Japanese models, however, the spherical exceeds the articulated vertical model. Figure 30.3*l* provides the robot structure distribution within each arm size class.

2. The distribution in terms of lift capacity remains relatively unchanged, as is shown in Figure 30.3*m*.

3. In terms of actuation, both markets now overwhelmingly employ electric actuators: about 60% in Japanese models, 48% in the U.S. market. In both markets models that combine electric and hydraulic actuators, and AC (rather than DC) electric motors, have emerged.

TABLE 30.3. MOTION/VELOCITY CHARACTERISTICS OF ROBOT ARMS AND WRISTS[a]

	U.S. Market Models	Japanese Models	Combined Models
Arm Out-In Traverse			
N	32	49	81
$\bar{x}$	1196.94	977.04	1063.91
$\bar{y}$	942.22	741.43	820.75
TMMR			300–3000
TMVR			100–4500
Arm Right-Left Traverse			
N	8	85	93
$\bar{x}$	2877.88	1320.00	1454.01
$\bar{y}$	635.75	466.71	481.25
TMMR			100–6000
TMVR			100–1500
Arm Up-Down Traverse			
N	33	118	151
$\bar{x}$	1336.39	853.05	958.68
$\bar{y}$	1036.85	489.36	609.01
TMMR			50–4800
TMVR			50–5000
Arm Right-Left Rotation			
N	28	44	72
$\bar{x}$	275.25	196.91	227.38
$\bar{y}$	90.32	89.23	89.65
TMMR			50–380[b]
TMVR			50–240
Arm Up-Down Rotation			
N	7	11	18
$\bar{x}$	176.57	95.45	127.00
$\bar{y}$	93.57	63.64	75.28
TMMR			25–330
TMVR			10–170
Wrist Roll			
N	33	79	112
$\bar{x}$	312.58	250.13	268.53
$\bar{y}$	158.18	99.05	116.47
TMMR			100–575[c]
TMVR			35–600
Wrist Pitch			
N	29	37	66
$\bar{x}$	183.97	174.86	178.86
$\bar{y}$	144.93	97.84	118.53
TMMR			40–360
TMVR			30–320
Wrist Yaw			
N	18	30	48
$\bar{x}$	281.94	203.07	232.65
$\bar{y}$	140.17	92.90	110.63
TMMR			100–530
TMVR			30–300

[a] N = number of observations.

$\bar{x}$ = mean value of maximum motion in degrees for rotation and mm for traverse.

$\bar{y}$ = mean value of maximum speed in degrees/sec for rotation and mm/sec for traverse.

TMMR = Typical maximum motion range ⎫ components of TRMM in Table 30.2.
TMVR = Typical maximum velocity range ⎭

[b] A continuous right-left rotation is available.

[c] A continuous roll movement is available.

4. In Japan, a 5 DF robot continues to be the most frequent; in the U.S. market, an adjustment has taken place and now the 5 DF robot is also most frequent. On the other hand, it can be noted from Figure 30.3n that in the Japanese models the 4 and 6 DF models are now close in popularity to the 5 DF class.

5. In both markets repeatability of 0.1–0.5 mm is still most common (see Fig. 30.3o). The most significant trend is the increase, relative to the previous survey, in the number of Japanese models with repeatability of 0.01–0.05 mm. The general trend of improved repeatability can be explained by better design of robot mechanism and control. It is also strongly associated with the frequency increase in electric actuation. The distribution of repeatability within each actuator class (see Fig. 30.3p) remains relatively unchanged compared to the previous study. However, more electric robot models can now provide relatively better repeatability, in the range 0.01–0.05 mm, and the repeatability of hydraulic robots has improved: none are specified with repeatability of above 1.5 mm in both markets.

6. The current distribution of maximum nominal velocity is quite similar in both markets, with 10–50 ips (250–1250 mm/s) the most common range (65% in Japanese models, 53% in the U.S. market). However, while in Japanese robot models very few, only 5%, can exceed 90 ips (250 mm/s), in the U.S. market 15% can now move at up to 90–130 ips (2250–3250 mm/s), and 9% can exceed even this range.

Through careful robot task planning, limits can be derived for various necessary performance parameters. By examining the Robot-Man Charts and motion/velocity graphs for parameters of importance, the applicability of a robot for a particular task can be better ascertained. In many cases it may be seen that a human is much better suited to the performance of a given task. If a robot is appropriate, the charts can help in visualizing those systems that may best suit the application.

Although in this section primary emphasis has been placed on choosing robots to perform a task, engineers can redesign a workstation and task to better accommodate available automation. Work method and workplace analyses techniques are helpful in these situations.

30.3. PLANNING WORK METHODS AND PROCESSES

A good work method determines how well limited resources such as time, energy, and materials are being utilized and has major influence on the quality of the product or output. In planning robot applications, there are two levels of consideration: the macro and the micro levels. The macro level includes the production process planning and system planning, as described in Chapters 29 and 34, and ergonomic studies as described in Section 30.2. In the micro level ergonomics techniques are applied to plan the detailed work method. The latter is supposed to include the following items:

Type and quantity of required parts and materials.

Type and quantity of required equipment.

Type and quantity of operators (human and robot).

Specification of tasks and operations to be performed.

Sequence and precedence of operations.

Layout of equipment and work stations.

Work flow in the layout.

A strategy for designing new methods or for improving existing methods is comprised of seven steps:[8]

1. Determining the purpose of the method.
2. Conceptualizing ideal methods.
3. Identifying constraints and regularity.
4. Outlining several practical methods, using principles such as those in Table 30.4.
5. Selecting the best method outline by evaluating the alternatives using criteria such as hazard, economics, and control.
6. Formulating the details of the selected method outline.
7. Analyzing the proposed method for further improvement.

Work methods must be documented for records, ongoing improvement, time study, and training. There are several tools for methods documentation, as well as for gathering and analyzing information about work methods (see, for instance, Clark and Close[9]). Examples are: process chart, workplace chart, multiple activity chart, and product flow sequence chart.

TABLE 30.4 PRINCIPLES FOR METHODS DESIGN

1. Design only to accomplish necessary purposes in the most ideal way.
2. Consider all system elements and dimensions.
3. Design for regularity before considering exceptions.
4. Focus on what should be, rather than on what is.
5. Consider layout and equipment design.
6. Eliminate or minimize all motions.
7. Consider best operator's position:
    Keep people's back straight and their hands close to their body.
    Keep robots close to points of operation to minimize motion distance.
8. Handle objects and record information only if needed, and then only once.
9. Minimize use of all resources: time, energy, materials, money.
10. Follow motion economy principles.

Source: Revised from Nadler, Reference 8.

30.3.1. Motion Study

Work performance is usually accomplished by a combination of motions, and this is certainly true for robot work. The effectiveness of the movement can be measured in terms of time and accuracy. Motion study applied various techniques to examine thoroughly all the motions that comprise a work method. Based on this examination, alternative methods can be evaluated, compared, and improved. One of the most useful practices in motion study is utilization of video cameras. Although the various charts identified before are commonly used in motion study, the use of movies has several advantages: it provides data acquisition and records facts that are unobtainable by other means; can provide quick study results; allows individual analysis of simultaneous activities; allows detailed analysis away from actual work area; provides a permanent record for future reference and training. One disadvantage, however, is that this technique can be applied only if there is access to an existing system.

Motion Study of Free Forging[10]

Two memo-motion cameras were used to record the details of free-forging group operations performed by operators. The objective was to plan robotization alternatives in response to poor working conditions in this process, high frequency of injuries, and as a result, shortage of labor. Operation charts, work load charts, and process flow charts were prepared. In addition, it was found that for smooth free-forging group operations a large amount of information handling among operators is necessary. A detailed analysis of motions and motion patterns was prepared (Fig. 30.5). Based on the operation analysis data and following the strategy for method design described, the project team prepared several robotization alternatives which guided subsequent implementations. (See one proposed alternative in Figure 30.6.)

30.3.2. Job and Skills Analysis

Job and skills analysis methods have been utilized for cost reduction programs in man-man and conventional man-machine work environments and for the effective selection and training of personnel. The job analysis focuses on *what* to do, while the skills analysis focuses on the *how*. The method, as it has been traditionally applied to human work, includes three major stages:

1. Examine the task to be analyzed to understand *what* and *how* the operator has to do the job. If it is not possible to observe an existing task, assess how the task would be performed.
2. Utilizing Table 30.5, document the *what* and the *how* of task performance.
3. From the documentation in Table 30.5, examine in a systematic way the possibilities of performing the task in different ways, including the following guidelines that are relevant for human operators.
    (a) Use of kinesthetic (sense of direction) instead of visual senses.
    (b) Reduction of factors complicating the task performance ("noise").
    (c) Reduction of, or utilization of, perceptual load associated with task performance.
    (d) Utilization of principles of motion economy and workplace design to simplify the task.
    (e) Resequence, eliminate, or combine elements.

Alternative ways are compared by the time it takes to perform them, complexity, quality, error probability, cost, safety, training, skill requirements, and so on. Based on such analysis the best method to perform a task can be selected and implemented.

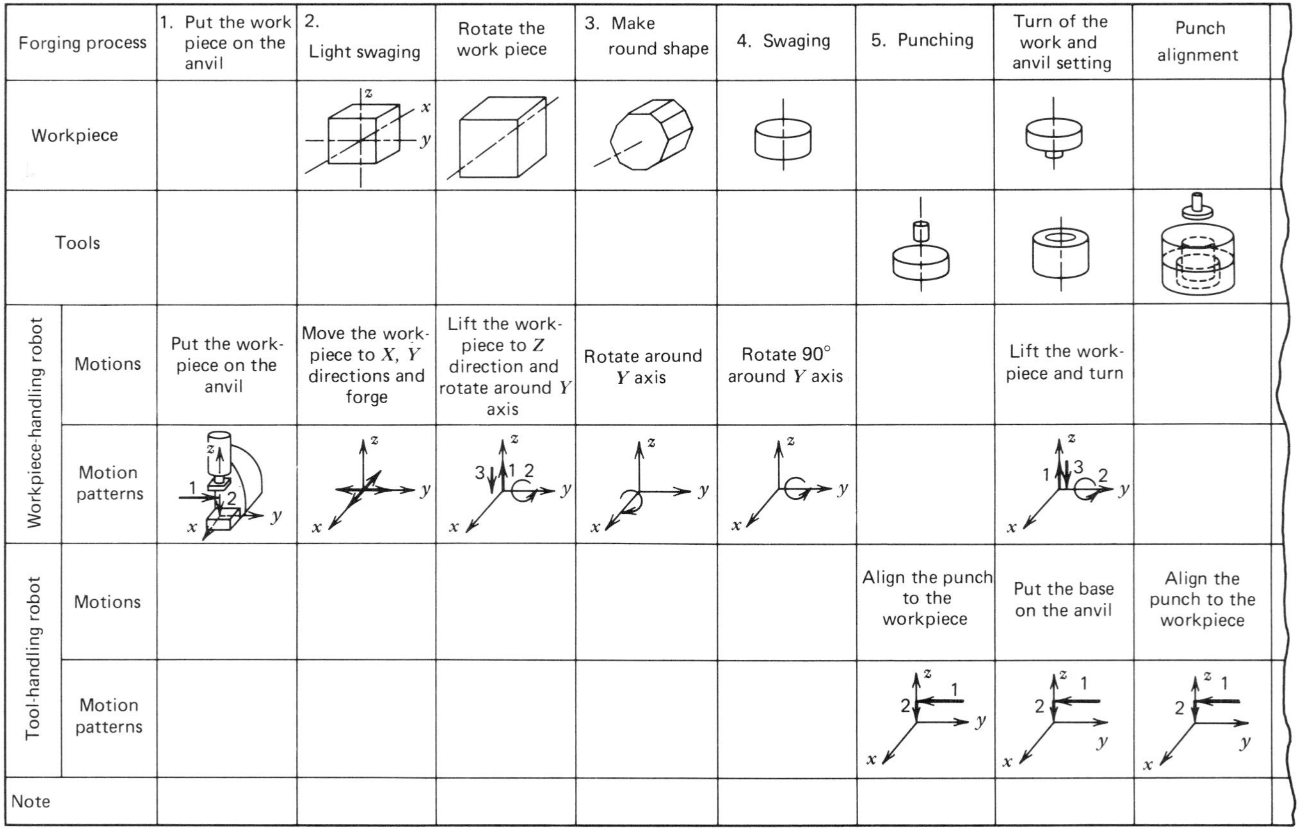

Forging process	1. Put the work piece on the anvil	2. Light swaging	Rotate the work piece	3. Make round shape	4. Swaging	5. Punching	Turn of the work and anvil setting	Punch alignment
Workpiece								
Tools								
Workpiece-handling robot — Motions	Put the work-piece on the anvil	Move the work-piece to X, Y directions and forge	Lift the work-piece to Z direction and rotate around Y axis	Rotate around Y axis	Rotate 90° around Y axis		Lift the work-piece and turn	
Workpiece-handling robot — Motion patterns								
Tool-handling robot — Motions						Align the punch to the workpiece	Put the base on the anvil	Align the punch to the workpiece
Tool-handling robot — Motion patterns								
Note								

Fig. 30.5. Details of motions and motion patterns in robotic free forging (*Source:* Reference 10.).

574

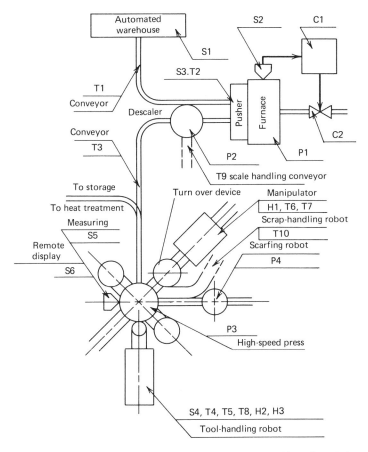

Fig. 30.6. One robotization alternative for free-forging operations. T: tool handling; H: hammer operation; S: instruction; P: process.[10]

The features that distinguish robot and human operators, as discussed, necessitate modification of the original human job and skills analysis method, as originally suggested by Nof et al.[3] As before, a task is broken down to elements that are specified with their times and requirements. However, the columns are modified as shown in Table 30.6.

Limbs Column

The human left/right hand column is replaced by a general limb column, since the robot may be designed to have any number and variety of limbs, for example, arms, grippers, and special tool hands.

Memory and Program Column

A column for memory and decision details is added. These details will determine what type of computer memory and processing capability are needed, if at all. Humans have their own memory, but decisions that are self-explanatory and often trivial for humans must be completely specified for robots. Robots may be designed, as explained earlier, with no programmability, that is, with a fixed sequence; or a variable sequence that is fixed differently from time to time; or with computer and feedback mechanisms that control the robot operations. The cost of a robot increases, of course, with the degree of programmability and sophistication of computer control. Higher levels of robot control require additional investment in hardware. Furthermore, limited controllers may be based on pneumatic control which requires relatively simple human skills to resequence. More complex operations may require computer control with associated system software and control programmers. Therefore, a work method that permits a fixed sequence will need a simple, less expensive robot, and be preferred to one that needs periodic resequencing, or complex programming.

TABLE 30.5. JOB AND SKILLS ANALYSIS UTILIZED FOR HUMAN PERFORMANCE

Element	Time	Left Hand, Right Hand	Vision	Other Senses	Comments
An operationally definable task component that has a clear beginning and end points; e.g., moving an item from point A to point B.	Time to perform the given element.	Describes which part of the element is performed by which hand. (Separate columns are used for left, right hand.)	Indicates the extent to which vision is used for the performance of the element; e.g., confirm by vision the current position of an object.	Indicates the extent to which other senses such as touch, hearing, etc., kinesthesis, hearing, etc., are used for the performance of the element.	Components of the element where some special precaution must be exercised; e.g., in joining two parts, the holes in component A must match the holes in component B.

Source: Reference 3.

TABLE 30.6. JOB AND SKILLS ANALYSIS FOR ROBOT PERFORMANCE

Element	Time	Senses	Limbs	Memory and Program	Comments
As in Table 30.5, except each element is a detailed micromotion.	Time to perform the given element.	Each of the major robot senses, namely, vision, touch, force compliance, is treated in the same manner that vision is treated in Table 30.5. (A separate column can be used for each.)	Each of the robot limbs is treated individually (in separate columns) as left/right human arms are treated in Table 30.5.	Specific memory requirements needed for the element such as word length and storage capacity. Program required includes reference information and logic for decision making in the element.	Comments about special requirements such as engineering tolerances, electricity, air pressure, and other utilities.

Source: Reference 3.

577

Comment Column

This column is for details about additional requirements such as position tolerances and utilities. Special precautions which are typical in the human-oriented analysis are probably not necessary here because they should appear in written decision logic information.

As indicated before, robots possess no basic knowledge or experience and therefore necessitate much detail in the task specifications. Therefore, elements will most commonly specify micromotions with their time measured in seconds or (minute/100).

Once a task is specified, its analysis basically follows the three stages described in the human-oriented method, that is: (1) examine task elements; (2) document the *what* and *how* of all elements; (3) systematically examine and evaluate alternative ways. However, since it is possible to select a robot and design its capabilities to best suit the task requirements, the performance evaluation in the last stage should be expanded as follows: From the documentation in Table 30.6 examine systematically, using robot motion-economy principles, the possibilities of performing the task in different ways and of using different robots.

Job and Skills Analysis of Robotic Assembly Task

We use an automobile water pump assembly task to illustrate the job and skills analysis approach. Each water pump consists of a pump base, a pump top, a gasket to be inserted between the top and base parts, and six identical screws used to fasten the foregoing three parts together. First, the analysis approach can be applied to a human operator and then to a robot. Assuming a robot is preferred, the task can be performed in a workplace as shown in Figure 30.7 by a Stanford Arm robot, as described by Bolles and Paul:[11]

1. Visually locate the pump base.
2. Determine the base-grasping position by touch sensing.
3. Bring and place the base on its plate.
4. Insert the two aligning pins in the base.
5. Put a gasket on the base, guided by pins.
6. Visually check the position of the gasket.
7. Place the top on the base and gasket.
8. Screw in the first two screws.
9. Take out the aligning pins and place in their standard position.
10. Screw in the last four screws.
11. Check the force required to turn the pump rotor.

Subtask 7, placing a top part on a base, is divided into 12 elements, as shown in Table 30.7. Each of the elements is specified according to the method with its required senses, limbs, memory and program, and special comments. The time estimates per element given in the table are based on the actual, original operation. The total time is 12.5 sec.

The original method can be improved by two of the principles described previously: replacement of senses and elimination of elements. Since vision is required by earlier subtasks, some touch and force sensor requirements may be removed. Reviewing the 12 task elements in Table 30.7 it is evident that in elements 3 through 7 the robot arm reaches the bearing twice. The first time the robot only identifies the precise location of the top part in order to align it later. If it is feasible with the vision system used, both the location and orientation of the top part could be found by the camera at the same time. Then it would be possible to eliminate all elements from the end of element 3, "save precise bearing location," to and including element 9. The result, shown in Table 30.8, could yield a savings of 4 sec, almost one-third of the total original time.

On the other hand, if vision were not needed in other subtasks, the method could apply a relatively simple fixture to locate the top part at a fixed and preprogrammed position. With the top part at a fixed and preprogrammed position and orientation, the arm could reach directly to grip the top by the bearing (first part of element 3). Both the camera and the Cartesian coordinate offsetting capabilities could be eliminated, and a simpler, less expensive computer could be used to control the robot. Again, elimination of the end of element 3 through element 9 in the task method could be achieved. Further improvement can be obtained by another fixture for positioning the pump base.

The time required for the improved work method can now be considered with the cost of robot and workstation to compare with other robots and with the manual work method for water pump assembly. If replacement by robots can be justified, then the selected robot and work method combination will yield optimized work results.

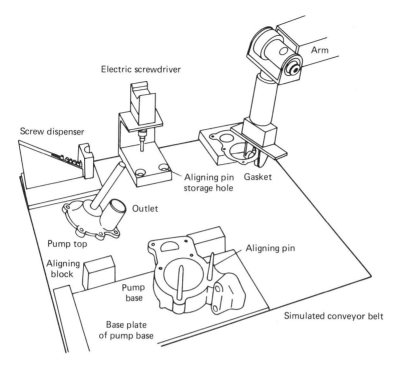

Fig. 30.7. Workplace for robotic assembly of water pump.

30.3.3. Motion Economy Principles

Principles of motion economy were proposed first by Frank Gilbreth in 1923 and later formalized and expanded by Barnes[12] and others. The purpose of these principles is to guide the development, troubleshooting, and improvement of work methods and workplaces, with special attention to human operators. However, some of the principles can be adopted or adapted for robot work, although others are not usfeul. The following sections comprise a list of important principles for human operators and their relevance to robot work.

A. Principles Related to Operator

A1. *Hands and arms should follow smooth, continuous motion patterns.*
Reason: sudden stop and change of direction break the human work rhythm and require higher-level attention and control.
Robot work: this principle may be useful for robot work when process quality considerations require it, for example, in painting, to eliminate jerky strokes (see principle D2).

A2. *Human hands should move simultaneously and symmetrically, beginning and ending their motions together.*
Robot work: this principle is irrelevant for robots, since coordinated robot arms can operate well under distributed computer controls.

A3. *Minimize motion distance, within practical limits.*
Reason: to reduce motion time and improve accuracy by not operating overextended limbs.
Relevant for both humans and robots.

A4. *Both hands should be used for productive work.*
Robot work: this principle is irrelevant for robots, since the number of arms can be chosen to maximize utilization.

A5. *Work requiring the use of eyes should be limited to the field of normal vision.*
Reason: in humans, vision guides the hands.
Robot work: directly relevant to robots whenever vision is required.

TABLE 30.7. ORIGINAL WORK METHOD FOR PLACING WATER PUMP TOP ON BASE

Number	Element	Time (sec)	Senses	Limbs	Memory and Program	Comments
1	Reach approach position, 10 cm			Arm		(from previous subtask)
2	Open fingers 8 cm; reach over main bearing, 20 cm	2.0		Two fingers	(General bearings' position found previously by camera)	Coordinated motion required
3	Center on bearing by closing fingers;	1.0	Touch (≤ 0.5 kg)	"	Until one finger touches	
	Continue closing grip.		Touch (≤ 0.5 kg)	"	a. Until both fingers touch b. If fingers close to 2 cm print error	
	Save precise bearing location					Cartesian coordinate off-setting required
4	Open fingers to 8 cm	0.5		"		
5	Reach to top outlet 12 cm	1.0		Arm	Outlet general position relative to bearing is stored in advance	
6	Close fingers to align the top part	0.5		Two fingers	Bearing-outlet alignment line stored in advance	
7	Open fingers to 8 cm	0.5		"		
8	Reach to bearing, 12 cm	1.0		"	Bearing location was saved in (3)	
9	Close fingers symmetrically to grip top by bearing	0.5	Touch (≤ 0.5 kg)	"	"	Support by leaning rotor on hand
10	Move top above and down over locating pins, 30 cm	3.0		Arm		
11	Move down 5 cm to insert top on pins	2.0	Compliance	"	If resistance force > 0, stop	
12	Move down 2 cm to seat top on base	0.5		"		Positioning accuracy required: ± 0.3 mm
		12.5				

TABLE 30.8. IMPROVED WORK METHOD FOR PLACING WATER PUMP TOP ON BASE

Number	Element	Time (sec)	Senses	Limbs	Memory and Program	Comments
1	Reach approach position, 10 cm			Arm		(from previous subtask)
2	Open fingers 8 cm; reach over main bearing, 20 cm	2.0		Two fingers	(General bearing's position found previously by camera)	Coordinated motion required
3	Center on bearing by closing fingers; Continue closing grip	1.0	Touch (≤ 0.5 kg) Touch (≤ 0.5 kg)	"	Until one finger touches a. Until both fingers touch b. If fingers close to 2 cm print error	
4	Move to fixture, 30 cm	3.0		Arm	Fixture position is known, fixed	
5	Move in 5 cm to comply with fixture	0.5	Compliance	"	If no resistance after 5 cm, error	
6	Move down 2 cm to seat top on base	0.5		"		
		$\overline{7.0}$				

581

A6.   *Actions should be distributed and assigned to body muscles according to their abilities.*
Reason: upper limbs are fast, accurate, and more flexible; lower limbs have strength and stability. Distribution among different muscle groups provides rest periods.
Robot work: this principle is only partially useful for robots—the robot work abilities should be specified according to the precise task requirements.

A7.   *Muscular force required for motions should be minimized.*
Body momentum should be utilized to advantage; on the other hand, momentum should be minimized to reduce the force required to overcome it, for example, it is better to slide parts than to carry them. Usually, a human arm weighing about 5 kg handles parts or tools weighing much less.
Robot work: this principle applies to robot for similar reason.

B.   Principles Related to the Work Environment

B1.   *Workplace should be designed for motion economy.*
That is, tools, parts, and materials should be placed in fixed, reachable positions and in the sequence they are used by a human operator. Additionally, work surface height should be designed to allow for human operator's sitting or standing.
Robot work: the first part of the principle applies directly to robots. The height of a work surface should be designed within the robot work envelope.

B2.   *Tools and equipment should be designed and selected with ergonomics guidelines.*
This principle is concerned with the dimension and shape of items handled by human operators, with safety, effectiveness, and comfort in mind.
Robot work: this principle is directly useful for robot end-of-arm tooling.

B3.   *Materials-handling equipment should be selected to minimize handling time and weight.*
For instance, parts should be brought as close to possible to the point of use; fixed items position and orientation simplify pickup and delivery.
Robot work: this principle is directly applicable to robots.

C.   Principles Concerning Time Conservation

C1.   *All hesitations, temporary delays, or stops should be questioned.*
Regular, unavoidable delays should be used to schedule additional work.
Robot work: this principle holds for both humans and robots. For instance, in a foundry delays caused by metal cooling-off period can be utilized by human or robot for gate removal; in machine tending, when one machine is performing a process the operator can load/unload another machine.

C2.   *The number of motions should be minimized.*
Elimination and combination of work elements are the two most common methods to achieve this goal.
Robot work: this principle is applicable for both humans and robots.

C3.   *Working on more than one part at a time should be attempted.*
This principle refers to two-armed human operators.
Robot work: it can be generalized to multiarm robots. Again, the number of robot arms or hands can be chosen for the most effective work method.

D.   Principles Concerning Robots

These new principles are based mainly on the fact that robot work abilities can be designed and optimized to best fit the task objectives.

D1.   *Reduce the robot's structural complexity.*
That is, minimize the number of arms and arm joints, which are determined by the number of hand orientations; reduce the robot dimensions, which are determined by the distance the robot must reach; reduce the load that the robot must carry. These reductions will result in a requirement for a less expensive robot, lower energy consumption, simpler maintenance, smaller work space.

D2.   *Simplify the necessary motion path.*
Point-to-point motion requires simpler control of positioning and velocity compared to continuous-path motion.

D3.   *Minimize the number of sensors.*
Each sensor adds to installation and operating costs by additional hardware, information processing, and repairs. The use of a robot with no senses is preferred to one sense, one sense is preferred to two, and so on.

D4. *Use local feedback, if at all necessary, instead of nonlocal feedback.*
For example, use of aligning pins for compliance provides quick local feedback. Use of touch or force sensors requires wiring and processing by a control system. Local feedback may add operations and may increase the overall process time, but usually needs no wiring and no information processing.

D5. *Use touch or force sensors, if at all necessary, instead of vision systems.*
The cost of the latter is significantly higher and requires more complex computer support. However, when vision is necessary as part of an operation, attempt to minimize the number of lenses required, since this will simplify information-processing requirements and shorten the visual sensing time.

D6. *Take advantage of robot abilities that have already been determined to be required, to reduce the cost and time of the job method.*
In other words, if a sensor or a computer must be provided for a certain task function, utilize it to improve the performance of other functions.

30.4. ROBOT WORKPLACE DESIGN

The best workplace is the workplace that supports and enables the most effective accomplishment of the chosen work method. The layout of a workplace determines, in general, four main work characteristics:

1. The distance of movements and the motions that must be carried out to fulfill the task.
2. The amount of storage space.
3. The delays caused by interference with various components operating in the workplace.
4. The feelings and attitudes of operators toward their work. The latter, obviously, is not a concern for robot operators.

Typical configurations of robotic workplaces are shown in Figure 30.8. General information about the design of a workplace layout can be found in work by Apple[13] and Francis and White.[14] In traditional ergonomics, a workplace is designed for human operators, as developed, for instance, by Bonney and Case.[15] Hence, there are anthropometric, biomechanical, and other human factors considered. In analogy, with regard to robotic workplaces, robot dimension and other physical properties such as reachability, accuracy, gripper size, and orientation will determine the design of the workplace. Additionally, requirements as those shown in Table 30.9 must be considered: whether tasks are variable; handled by one or more robots; whether workplace resources are shared by the robots; the nature and size of components stationed in the workplace; the characteristics of parts flow into, within, and outside of the workplace. Because all robot operations are controlled by a computer, several robots can interact while concurrently performing a common task and even share resources[16]—a difficult requirement when people must interact under a tight-sequencing control. Thus a robotic workplace can be optimized with regard to the layout, and also planned for effective control of operations.

30.4.1. Workplace Design Techniques

Several studies describe tools and techniques to aid the workplace designer in emphasizing layout considerations and equipment selection. The requirements for robot workplace layout design are analyzed in Kjellberg[18] and Hanify.[31] Systematic analysis for planning the robotic workplace, including considerations of control strategies, is explained by Chang and Goldman,[19] Fisher et al.,[20] Severiano and Gruver,[21] Young and Yuan,[22] and Nof.[23,24] In Sarin and Wilhelm[25] mathematical programming models are formulated for the optimization of some of the workplace configurations described in Table 30.9. Other useful, quantitative models are discussed in Chapter 31.

Several applications of computer graphics systems have also been described, such as in Schraft and Schmidt,[26] Heginbotham et al,[27] Findler and Shaw,[28] Kretch,[29] and Fors and Donaghey.[30] For instance, in Kretch[29] the graphics package PLACE is described. Its purpose is to create, analyze, and edit robot cell descriptions, including tools, parts, end effectors, conveyors, and NC equipment. Primitive cell descriptors are generated by a CAD system and then combined to create complex configurations. Kinematic relationships of devices are also part of the input. The advantage of a graphic simulator is that various software functions allow the user to edit and manipulate the cell components, and obtain some evaluation of the cell spatial performance. In PLACE, for instance, evaluation is in terms of the geometric motion capability. As a result, users have been able to create and test useful robot workplace designs that have been implemented. In Chapter 20 another graphics CAD system, GRASP, is described for similar purposes (see Figure 20.1).

Another simulation approach is described in Nof, Meier, and Deisenroth,[32] which uses a computer-controlled scale model of a robot arm to study a load/unload operation on machining centers (see

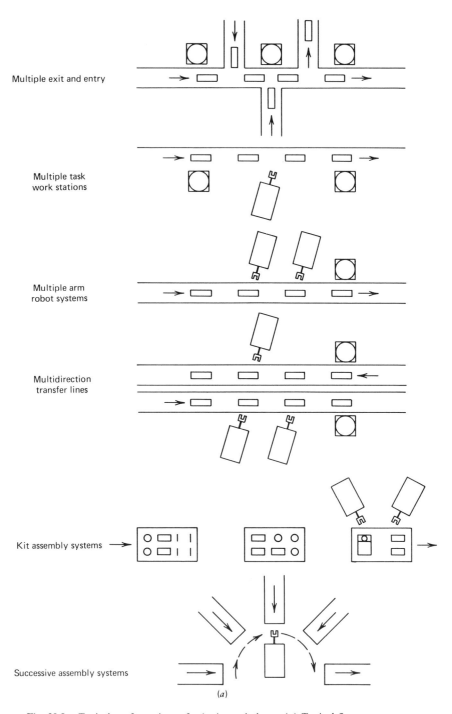

Multiple exit and entry

Multiple task
work stations

Multiple arm
robot systems

Multidirection
transfer lines

Kit assembly systems

Successive assembly systems

(a)

Fig. 30.8. Typical configurations of robotic workplaces. (a) Typical flow arrangements.

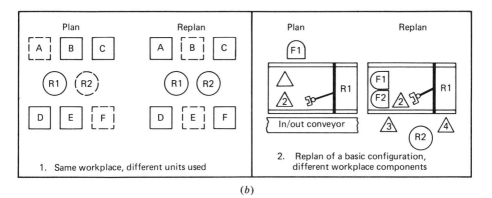

(b)

Fig. 30.8. (b) Two examples of dynamic workplace reconfiguration. $\square$, machine; $\triangle$, fixture; $\bigcirc$, feeder; $\bigcirc$, robot.

LAYOUT EXAMPLE #1

CONVEYOR CONFIGURATIONS

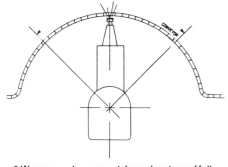

- Straight line conveyor limits access by robot, by in & out movement of robot.

LAYOUT EXAMPLE #2

CONVEYOR CONFIGURATIONS

- Wrap around conveyor takes advantage of full radial motion of robot, (90 to 300 degrees.)

LAYOUT EXAMPLE #3

PUNCH PRESS SYSTEM

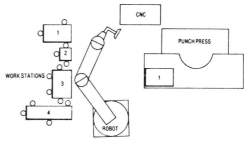

- Various sheet metal blanks stored in designated work stations.
- Robot picks part from station sequentially as long as all stations are full.
- When part is removed from station "cnc" control is notified which program to use to operate the punch press.
- All parts look same to the robot. Parts all bank off two edges when located in machine.
- Vacuum cup hand used for versatility.
- Random access programming required in robot.

(c)

Fig. 30.8. (c) Nine examples of application considerations in planning robot workplace. (Courtesy: General Electric.)

LAYOUT EXAMPLE #4

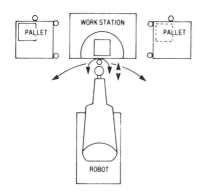

- At least 3 axes of motion required.
- Extensive axes motions complicate programming task.

LAYOUT EXAMPLE #5

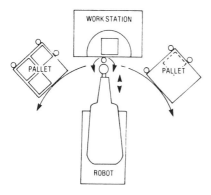

- At least 3 axes of motion required.
- Axes motions reduced somewhat, but programming still not simple.

LAYOUT EXAMPLE #6

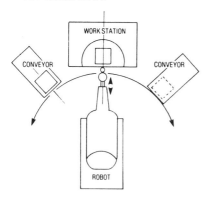

- At least two axes of motion required.
- Simplest of last three work stations to program.

LAYOUT EXAMPLE #7

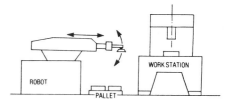

- Variation in heighth between work station and pallet will require additional robot axes as shown.
- Increases programming effort.

LAYOUT EXAMPLE #8

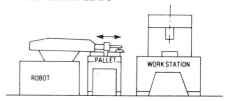

- Axes of robot motion reduced by making work station heighth and part pickup, *same!*
- Simpler programming task.

LAYOUT EXAMPLE #9

"SAFETY"

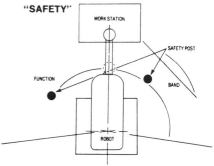

- You never trust a robot within its mechanical function parameters.
- Overpowering restriction posts or other barriers are the only acceptable protection for humans in the area.
- Interlock gates should be used to keep personnel out of robot area.
- Interlock footpads can also be used.

Fig. 30.8. (*c*) (*Continued*)

1.

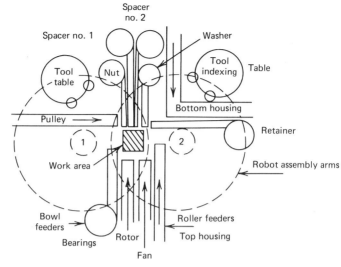

2.

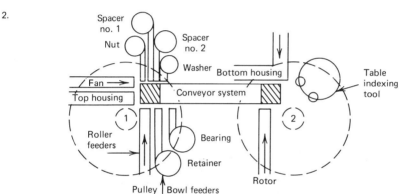

(d)

Fig. 30.8. (d) Two alternatives for a two-robot workplace; (1) A single unit worked on by two synchronized arms; (2) two arms working on different parts of the product. (*Source:* Reference 17.).

TABLE 30.9. CLASSIFICATION OF ROBOTIC WORKPLACES

Robots	Tasks	Stations Inside Workplace: Machines, Feeders, Other Equipment	Input/Output Flow
Simple robot	Single or multiple tasks	Uniform or different station area	Single or multiple entry, exit
	Fixed or variable motion sequence	Fixed or variable equipment	One or multi-direction of flow
Multiple robots	Single or multiple tasks per robot per team	With or without equipment sharing (for multiple robots)	With or without part grouping (binning, kitting, magazining, palletizing)
	With or without task interaction		
Multiple robot cells	With or without cell overlap		

Figure 30.9). Like graphics simulators, a physical workplace simulator is very useful for visually demon-strating complex operations. It can also be used to analyze, experiment, and evaluate the workplace. Additionally, it can be used for some training of operators in a simulated approach and for debugging the system's control software.

30.5. PERFORMANCE MEASUREMENT

Performance measurement, including work measurement, performance prediction, and performance evaluation, is the ergonomics function that accompanies a work system throughout its life cycle.

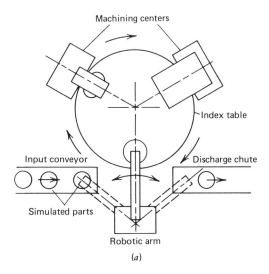

(a)

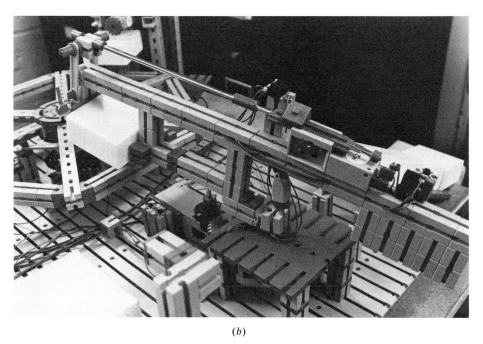

(b)

Fig. 30.9. Robot simulation for workplace design. (a) Workplace layout diagram for robot load/unload of a machining cell. (b) Detailed view of physical simulator. (c) Graphic simulation for workplace design.

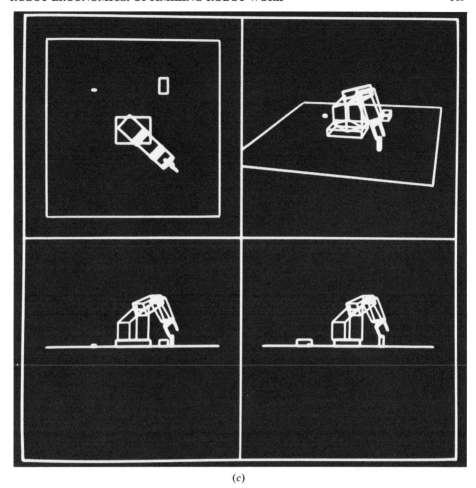

(c)

Fig. 30.9. (*Continued*)

1. In the planning and design stages, performance prediction is required to evaluate the technical and economic feasibility of a proposed plan, and to compare and select the best out of a set of feasible alternatives.

2. During the development and installation stages performance measurement provides a yardstick for progress toward effective implementation.

3. During the regular, ongoing operations, performance evaluation serves to set and revise work standards, troubleshoot bottlenecks and conflicts, train workers, and estimate cost and duration of new work orders. Another vital function of performance measurement at this stage is the examination of new work methods, technologies, and equipment that can be used to upgrade, expand and modernize the existing operations.

30.5.1. Work Measurement, Time And Motion Studies

Robot designers have long been concerned with planning robot motions in an optimal way, that is, to follow accurately specified trajectories while avoiding collisions and moving at the minimum amount of time (for example, see Brady[33]). For instance, early studies by Pieper and Roth[34] investigated the minimum cycle time problem of a robot moving among obstacles by analyzing alternative combinations of simultaneous motions of different joints. Others, Kahn and Roth,[35] Luh and Walker,[36] Lynch,[37] Luh and Lin,[38] examined additional problems of motion control optimization. From a work method point of view, however, cycle times can be reduced by carefully following the motion principles described before. For instance, in Birk and Franklin[39] proper placement of workpieces led to improved performance. Verplank,[40] Vertut,[41] Rogers,[42] and Kondoleon,[17] among others, performed time and motion studies to evaluate, select, and compare alternative robot system designs.

30.5.2. The RTM Method

The RTM, robot time and motion method (Paul and Nof,[43] Nof and Paul,[44] Nof and Lechtman[45]) for predetermined robot cycle times, is based on standard elements of fundamental robot work motions. RTM is analogous to the MTM (methods time measurement) technique (Maynard et al.[46] and Antis et al.[47]), which has long been in use for human work analysis. Both methods enable users to estimate the cycle time for given work methods without having to first implement the work method and measure its performance. Therefore these methods can be highly useful for selection of equipment as well as work methods, without even having initially to purchase and commit to any equipment. MTM users, however, must consider human individual variability and allow for pacing effects. RTM, on the other hand, can rely on the consistency of robots and apply computational models based on physical parameters of each particular robot model.

The RTM methodology provides a high-level, user-friendly technique with the following capabilities:

Systematically specifying a work method for a given robot in a simple, straightforward manner.

Applying computer aids to evaluate a specified method by time to perform, number of steps, positioning tolerances, and other requirements so that alternative work methods can be compared.

Repeating methods evaluation for alternative robot models until the best combination is established.

The RTM system is comprised of three major components: RTM elements, robot performance models, and an RTM analyzer. The system has been implemented, experimented with, and applied with several robot models, including the Stanford Arm equipped with touch and force sensing, Cincinnati Milacron's T3, PUMA, Unimates, and others. Several companies have adopted and applied the RTM methodology, originally developed at Purdue University, and there has been one development of a commercial product (ROFAC) based on it (Hershey et al.[48]).

The RTM user can apply 10 general work elements to specify any robot work, by breaking the method down to its basic steps. The RTM elements, shown in Table 30.10A, are divided into four major groups:

RTM Group 1:   Movement elements—REACH, MOVE, and ORIENT
RTM Group 2:   Sensing elements—STOP-ON-ERROR, STOP-ON-FORCE/TOUCH
               and VISION
RTM Group 3:   Gripper or tool elements—GRASP and RELEASE
RTM Group 4:   Delay elements—PROCESS-TIME-DELAY and TIME-DELAY.

By applying these elements with their parameters as shown in Table 30.10A, the alternative robot work methods can be analyzed, evaluated, compared, and improved. Table 30.10B provides a comparison between RTM and MTM work elements. It can be used to evaluate systematically possible conversion from manual to robotic work methods.

RTM Performance Models

There are two possible ways to approach the modeling of robot work: by approximating empirical data gained from laboratory experiments, and by the engineering design of the robot operation. The first way is exemplified by the use of element tables and regression equations; the other by the use of motion control models and motion path analysis.

Element Tables

The simplest modeling approach, which follows the original MTM approach for human work methods, applies a set of tables with estimates for each element according to particular parameters. Tables are developed based on laboratory experiments with the robot type for which data are prepared. For instance, Table 30.11 shows RTM element tables for the Stanford Arm. Table 30.12 contains time data for elements REACH (R1) or MOVE (M1), and for ORIENT (OR1) by the T3. Note that for the T3 REACH and MOVE elements are identical, since carried weight does not affect performance time.

The table approach is relatively simple, although it requires extensive laboratory experimentation to develop the table values for each robot model family. However, once tables are established for a particular robot type, they can be applied by everybody. Times for motions at distances not in the table can easily be interpolated. It is important to note that despite the relative simplicity of this approach, it has been found to be quite satisfactory for its prediction purposes. In laboratory experiments, variations between predictions based on the table approach and actual, measured time values of complete tasks have been within the range of about ±5 to 10%.

TABLE 30.10A.   RTM SYMBOLS AND ELEMENTS

Element Number	Symbol		Definition of Element	Element Parameters
1	Rn		*n-segment reach:* Move unloaded manipulator along a path comprised of *n* segments	
				Displacement (linear or angular) and velocity or
2	Mn		*n-segment move:* Move object along path comprised of *n* segments	Path geometry and velocity
3	ORn		*n-segment orientation:* Move manipulator mainly to reorient	
4	SEi		*stop on position error*	Error bound
4.1		SE1	Bring the manipulator to rest immediately without waiting to null out joints errors	
4.2		SE2	Bring the manipulator to rest within a specified position error tolerance	
5	SFi		*Stop on force or moment*	Force, torque, and touch values
5.1		SF1	Stop the manipulator when force conditions are met	
5.2		SF2	Stop the manipulator when torque conditions are met	
5.3		SF3	Stop the manipulator when either torque or force conditions are met	
5.4		SF4	Stop the manipulator when touch conditions are met	
6	VI		Vision operation	Time function
7	GRi		*Grasp an object*	
7.1		GR1	Simple grasp of object by closing fingers	
7.2		GR2	Grasp object while centering hand over it	Distance to close/open fingers
7.3		GR3	Grasp object by closing one finger at a time	
8	RE		Release object by opening fingers	
9	T		Process time delay when the robot is part of the process	Time function
10	D		Time delay when the robot is waiting for a process completion	Time function

Regression Equations

A more advanced approach than element tables utilizes regression equations, which are also generated based on experimental laboratory data. This approach is considered useful when observations are random, or when a complicated functional relationship must be approximated. Both conditions can be true in the case of robot work. Experiments have shown that even though robots, as machines, are highly consistent in their operation, dynamic variations in electric power or hydraulic pressure may result in some random deviations. Another potential source of random deviations is, of course, measurement error.

TABLE 30.10B. RTM EQUIVALENT WORK ELEMENTS FOR MTM

MTM Work Element	Equivalent RTM Work Elements
RA	Rn, SE
RB	Rn, SF, SF
RC	†
RD	R, SE, V, R, SF
RE	R, SE
MA	†
MB	Mn
MC	Mn, SF, SF
G1A	GR
G1B	SF, ST
G1C	†
G2	†
G3	SE, SF, RE
G4	†
G5	GR
P1	SF
P1 and P3S	SF, M1, SF
P3SS and P3NS	SF, M1, M1, SF

† Impossible to perform by a robot.

In experimental work with Unimate 4000B, for example, the equation for RTM element M1 was found (with 95% confidence level, as shown in Figure 30.10) to be

$$T(M1) = 0.423 + 0.009 \times L$$

where L is in centimeters, $T(M1)$ is in seconds.

Regression equations developed for the RTM system have been found to yield a predictive accuracy similar to that of the table approach.

Motion Control Models

A different modeling approach predicts motion time according to the pattern by which the robot velocity is controlled. For the T³, like several other robot models, the velocity is controlled as shown in Figure 30.11. From kinematic relations, and experimental work for very short motions, the following equations were developed for the T³:

$$
\begin{aligned}
T(\text{R1 or M1, in seconds}) &= \frac{S}{V} + \frac{1}{4} \quad \text{for } S > \frac{V}{4} \\
&= \frac{S}{V} + 0.365 \quad \text{for } S > \frac{V}{2.857} \\
&= 0.82 \quad \text{for } 17.50 \leq S \leq 43.75 \text{ cm} \\
&= 0.610 \quad \text{for } 6.25 \leq S < 17.5 \text{ cm} \\
&= 0.413 \quad \text{for } S < 6.25 \text{ cm}
\end{aligned}
\tag{30.1}
$$

where S is the total distance moved in centimeters and V is the user-specified velocity in centimeters per second.

For multisegment motions (see Figure 30.11c), the analysis yielded the following general relationship:

$$T(\text{Rn or Mn, in seconds}) = 0.24 + \frac{n}{8} - \frac{1}{8} \sum_{i=2}^{n} \frac{V_{(i-1)}}{V_i} + \sum_{i=1}^{n} \frac{S_i}{V_i} \tag{30.2}$$

where n is the number of segments.

TABLE 30.11. RTM TABLES FOR THE STANFORD ARM

1. REACH (R_n)

Distance to Move (cm)	Time[a] (sec) Number of Path Segments				
	R1	R2	R3	R4	R5
1.0	0.38	0.76	1.13	1.51	1.89
30.0	0.45	0.83	1.19	1.58	1.91
100.0	0.72	0.94	1.35	1.73	1.98

2. MOVE (M_n)

Distance to Move (cm)	Time[a] (sec) Number of Path Segments					Weight Allowance	
	M1	M2	M3	M4	M5	Weight, up to (Kg)	Factor
1.0	0.38	0.76	1.13	1.51	1.89	0.5	1.0
30.0	0.45	0.83	1.19	1.58	1.91	1.0	1.5
100.0	0.72	0.94	1.35	1.73	1.98	2.0	2.0
						3.0	3.0

3. STOP-ON-ERROR (SE2)

Error Bound (cm)	Time (sec)
1.0	0.13
0.2	0.29
0.05	0.83

4. STOP-ON-FORCE/TOUCH (SF1/SF4)

Distance to Move (cm)	Time (sec)
0.5	0.11
3.0	0.90
15.0	4.50

5. GRASP (GR1)

Distance to Close (cm)	Time (sec)
0.5	0.14
3.0	0.36

6. RELEASE (RE)

Distance to Open (cm)	Time (sec)
0.5	0.14
3.0	0.36

[a] All motions are at the normal arm operating velocity.

Other factors that are considered during the RTM computation are limitations on motions that entail excessive rotation and combined rotation and linear motions. The predictive accuracy of this modeling approach, as found in laboratory experiments, has been in the range of −2 to +3%.

Path Geometry

Robot motions can also be modeled by specifying the complete motion path geometry. Based on robot joint and link velocities, the motion time can be computed. In essence, the distance, linear or rotational, that each joint or link must move relative to its preceding link to reach a specified position for the end effector can be figured out. The time it takes to complete the motion will depend on the longest individual link or joint motion time. The predictive accuracy of path geometry programs, as found in laboratory experiments, is between −2 and +12%. It should be noted that the detailed input data required for the specification of each motion path are usually not available during advanced planning stages.

Sensors Performance

For work elements of RTM group 2, sensing elements, there is a high dependency on the hardware used. In addition to the sensory elements shown earlier for the Stanford Arm, Table 30.13 illustrates the performance of an example vision system.

TABLE 30.12. RTM TABLES FOR T³ REACH/MOVE AND ORIENT

1. REACH (R1) or MOVE (M1)

Distance to Move (*cm*)	Time (*sec*) at Velocity (*cm/sec*)				
	5.0	12.5	25.0	50.0	100.0
1	0.4	0.4	0.4	0.4	0.4
30	6.4	2.8	1.6	1.0	0.8
100	21.3	8.7	4.5	2.4	1.4

2. ORIENT (OR1)

Angle to Move (*deg*)	Time (*sec*) at Velocity (*cm/sec*)				
	5.0	12.5	25.0	50.0	100.0
15	3.0	1.4	0.8	0.6	0.6
60	10.8	4.6	2.5	1.4	0.9
120	21.3	8.7	4.6	2.5	1.4

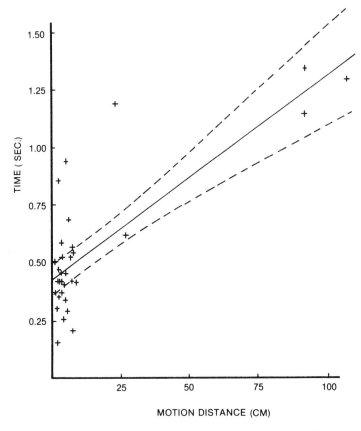

Fig. 30.10. RTM element M1 regression line for Unimate 4000B.

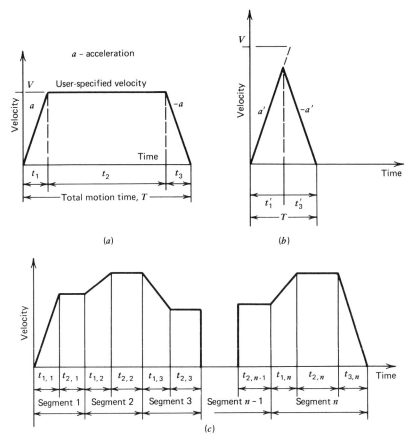

Fig. 30.11. Velocity patterns. (*a*) Velocity pattern of regular, one-segment T³ motions. (*b*) Velocity pattern of short, one-segment T³ motions. (*c*) Velocity pattern of general, multisegment T³ motions.

TABLE 30.13.   BINARY VISION SYSTEM PERFORMANCE FOR MIC VS-100[53]

General Operation Processing Time	Measured Processing Time for Typical Task (*sec*)
1. Read in the image and process it	
a. Using software run-length encoding:	
484 msec + 0.97 msec/segment	0.583
b. Using hardware run-length encoding:	
267 msec + 0.53 msec/segment	0.321
2. Connectivity analysis	
66 msec + 1.8 msec/segment + 3.3 msec/blob	0.270
3. Perimeter accumulation	
0.86 msec/segment + 2.3 msec/blob	0.102
4. Accumulation of second moments	
2.3 msec/segment	0.235
5. Perimeter and radius calculation	
5.2 msec/segment	0.500

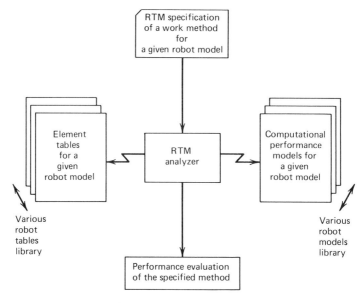

Fig. 30.12. RTM analyzer.

The RTM Analyzer

The RTM analyzer has been developed to provide a means of systematically specifying robot work methods with direct computation of performance measures. The general structure of the analyzer is shown in Figure 30.12.

The input to the RTM analyzer includes control data (e.g., task and subtask titles, type of robot,

TABLE 30.14. SUMMARY OF RTM SYSTEM'S STATEMENTS

Statement Type	Statement Structure
1. Subtask title	SUBT, (no.), (title), (comment)
2. REPEAT control card	REP, no. of first, TO, no. of last, no. times, (comment) operation operation to repeat [a]
3. PARALLEL control card	PAR, no. of first, TO, no. of last, (comment) operation operation
4. Conditional branching	IF, (condition name. condition. value), GOTO, operation no. or subtask number
5. Control transfer	GOTO, operation no., (comment) or subtask no.

6. Movement elements
 (Rn, Mn, ORn)
    a. By detailed commands:
        Position
        initialization
        Followed by: End point
        of segments
    b. By displacement
	
	 (Joints parameters)
	 (operation no.) (RTM symbol), (comment)
	 (velocity), (joints parameters)
	
	 (operation no.), (RTM symbol), A-Angular, (velocity), (displacement [a])
	                            or D-Linear

7. All other RTM elements	(operation no.), (RTM symbol), (operation parameter), (comment)
8. END Card	END
9. CONDITION initialization	COND (condition name), (set of initial values [a]) END

[a] These can be generated randomly.

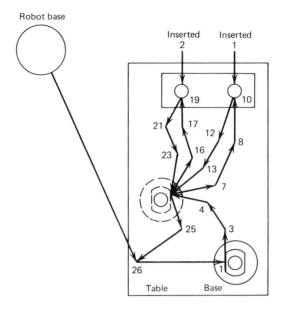

Fig. 30.13. Layout of insertion task (numbers correspond to operation numbers in Table 30.15).

and type of RTM model to apply) and operation statements. The statements specify (1) robot operations, each represented by RTM element and its parameters; (2) control logic, which provides capabilities of REPEAT blocks, PARALLEL blocks, conditional branching, and probabilistic generation of conditions. A summary of RTM statements is given in Table 30.14. An illustration of an RTM application for a simple insertion task is shown in Figure 30.13 and in Tables 30.15, 30.16.

A simplified RTM, a subset of RTM, was also developed for point operations such as spot welding, drilling, and riveting by Lechtman and Nof.[49] In point operations the robot carries a tool or a part during the whole operation; the same operation type is performed repeatedly along one or more paths, and the execution time of the tool at each point is well known and can be calculated rather than estimated. Taking advantage of these characteristics, the specification and computation can be simplified with little loss of accuracy.

30.5.3. Performance Simulators

A number of performance simulators have been developed for the analysis and evaluation of robotic work (Kuno, et al.,[50] Medeiros, et al.,[51] and Robinson and Nof[52]). These kinds of simulators are less concerned with the geometric arrangement of the robot workplace than are the graphics simulators described in Section 30.4. Here the focus is on operational control, cycle times, congestion, utilization, and productivity. Additional discussion about such simulators can be found in Chapter 31.

A good example of the use of simulation for robot work analysis is described by Kondoleon.[17] He studied alternative configurations for robotic assembly of automobile alternators, including a single-arm programmable station with several alternative work methods; two synchronized arms co-working on the same unit; two arms working on different parts of the product. The objective was to increase cycle time efficiency, mainly by reducing the tool change time. An interesting alternative that was examined entailed the building of several assemblies together by specialized tooling, thus spreading the tool change time over many units. For instance, it was found by the simulation that at a higher cost of fixturing, building six units together reduced the cycle time per unit by about 40%. On the other hand, building twice as many units (12) together yielded a reduction of only about 44%. These reductions held approximately the same for five alternative methods. However, the preferred method of complete assembly of units at one station (for 1, 6, or 12 at a time) required a cycle time about 20% shorter than the worst alternative considered, and about 13% shorter than the average cycle time of all five alternatives. Thus, a performance simulation combining the details of work method, workplace, and control strategy can provide highly useful information to designers.

30.5.4. Robot Learning

Learning is the process by which the time or the cost per cycle decreases as the number of performed, repetitive cycles increases (Hancock[54] and Nanda and Adler[55]). The learning process has been observed

TABLE 30.15. RTM INPUT DATA FOR INSERTION TASK BY THE T³

Statement	Comments
T³	Robot type
Insertion Task	Task title
*Conditions	
. . . .	Input of condition signals, if
*	any were used
[REPEAT 1 To 26 SEVEN TIMES]	Repetition command could be stated here
1 R1 5, 7.5	Reach 7.5 cm at 5 cm/sec to start position above base
2 GR1	Grasp base
3 M1 5, 5.0	Raise base
4 M1 25, 48.0	Move to above assembly position
5 M2 25, 12.5, 5, 5.0	Bring base down in two segments movement
6 RE	Release it
7 R1 25, 23.0	Rise
8 R1 25, 55.0	Reach fixture
9 D 2	Wait 2 sec (for continuation)
[IF (SIGNAL. NE. VALUE) GO TO END]	Conditional branching could be used
10 R1 5, 15.0	Bring fingers above peg 1
11 GR1	Grasp peg 1
12 M1 5, 15.0	Raise peg 1
13 M1 25, 55.0	Move it to above base
14 M2 25, 11.0, 5, 9.0	Insert peg in two steps
15 RE	Release peg 1
16 R1 25, 22.0	Rise
17 R1 25, 50.0	Reach above peg 2
18 D 2	Wait 2 sec
19 R1 5, 12.5	Bring fingers above peg 2
20 GR1	Grasp peg 2
21 M1 5, 12.5	Raise it
22 M1 25, 50.0	Move it above base
23 M2 25, 10.0, 5, 5.0	Insert it in two steps into peg 1
24 RE	Release peg 2
25 R1 25, 17.5	Rise
26 R1 25, 60.0	Reach start point

in both individuals and in organizations, and in both cases follows the typical learning curve (see Figure 30.14). The general form of a learning curve can be:

$$T(n) = T(1) \times n^{-A}$$

where T = time (or cost) per cycle
  $T(1)$ = time (or cost) of the first cycle
  n = the number of cycles performed
  A = an improvement constant, determined by the learning rate.

In human learning, the main factors are (1) person's age; (2) the amount of previous experience in learning; (3) personal physical and psychological capabilities; (4) the job complexity in terms of cycle length, amount of uncertainty, and degree of similarity to previous jobs. Learning by an organiza-

TABLE 30.16. RTM ANALYSIS RESULTS FOR INSERTION TASK BY T³

Operation Number	RTM Element	Distance (cm)	Angle (degrees)	Velocity $V\left[\dfrac{cm}{sec}\right]$	Calculated Time (msec)	Time from Tables (msec)	Measured Time (msec)
1	R1	8.25	—	5	2000	1854	2003
2	GR1				10	10	36
3	M1	5.10	—	5	1385	1365	1390
4	M1	48.18	—	25	2292	2169	2299
5	M2	13.18 ⎫ 5.75 ⎬	—	25 ⎫ 5 ⎬	1543	2173	1526
6	RE				10	10	36
7	R1	23.5	—	25	1305	1268	1293
8	R1	55.75	—	25	2594	2564	2590
9	D				2000	2000	2000
10	R1	15.00	—	5	3360	3372	3368
11	GR1				10	10	37
12	M1	15.00	—	5	3360	3372	3367
13	M1	55.75	—	25	2595	2564	2590
14	M2	10.75 ⎫ 9.25 ⎬	—	25 ⎫ 5 ⎬	2120	3041	2099
15	RE				10	10	37
16	R1	21.75	—	25	1233	1268	1243
17	R1	50.25	—	25	2373	2169	2371
18	D				2000	2000	2000
19	R1	13.50	—	5	3075	3372	3076
20	GR1				10	10	37
21	M1	13.50	—	5	3075	3372	3075
22	M1	50.25	—	25	2372	2366	2370
23	M2	10.75 ⎫ 5.00 ⎬	—	25 ⎫ 5 ⎬	1290	2173	1271
24	RE				10	10	37
25	R1	17.50	—	25	1067	1122	1073
26	R1	59.25	—	25	2731	2761	2724
Total (msec)					43830	46405	43948
Difference from measured					0%	5.6%	

599

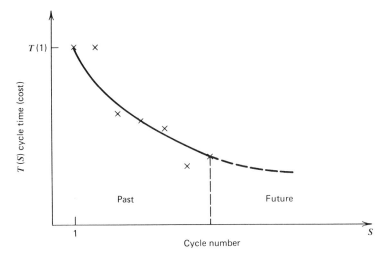

Fig. 30.14. The learning curve.

tion is characterized by production progress functions that indicate the rate at which the organization learns to produce its products. The main factors that have been found to cause such improvements are (1) organizational improvement; (2) improvements introduced in work methods; (3) training programs to improve employees' skills; (4) improved production technology and new equipment. Although operators differ in their learning rate by the factors of human learning, organizations differ both by the foregoing factors and by the nature of their major activity.

In both individual and organization learning the objective of learning curves as performance measurement tools is to provide a model relating past experience with future performance. Such information is valuable in planning the system's operations effectively, for example, time and cost estimates of new orders, estimation of training costs, and performance monitoring and evaluation.

Robot learning can be found in three main areas in industry:

1. Human operators learning to accept and work with robots.
2. Organizations learning to introduce and utilize robots.
3. Robots learn about their operation.

The first two areas follow, in principle, the previous discussion about human learning and are further discussed in Section 30.6, "Integrated Human and Robot Ergonomics." Robots themselves learn, or acquire knowledge, according to Seltzer,[56] by three different methods:

1. Being taught by an operator by means of a teach box.
2. Being taught by means of an off-line geometric data base and programs.
3. Learning from on-line experience.

Robot teaching and off-line programming are covered in detail in Chapters 19 and 20. Other robotic knowledge acquisition approaches are covered in Chapters 22 and 23. In the area of vision, Gleason and Agin[57] and others have shown that pattern recognition techniques can be combined with simple learning algorithms into a machine vision system that can, in certain cases, program itself to recognize objects. Contact sensing can also be used to guide a variable process by compliance (see details in Chapter 64). A new learning technique based on stochastic control theory is described by Whitney and Junkel[58] for long-term control. According to this technique, which is applicable for both contact and noncontact sensing, ongoing measurement of force or displacement is performed in the robot's environment for feedback. Random errors occur in the robot environment because of factors such as inaccuracies or wear in jigs, tolerances in parts, errors in sensors, and thermal drift. Applying the stochastic control technique, the robot can utilize the feedback to learn, accumulate knowledge, about world parameters such as weight of grasped objects and force required to join parts. Thus it can automatically learn to identify random events when they occur, and how to react effectively to them by bridging the gap between originally taught points and actual destinations. Figure 30.15 depicts how the technique can be applied by the robot controller to improve its performance by learning.

The latter type of intelligent learning by a robot can significantly improve the effectiveness of robot work, and may in the future completely revise the need for accurate robot teaching and program-

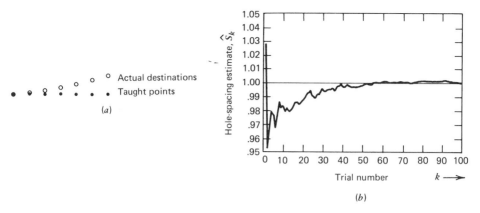

(b)

Fig. 30.15. Robot learning: performance is improved by the stochastic control technique of Whitney and Junkel.[59] (a) Deviation over time between preprogrammed and actually needed destinations. (b) Hole-spacing estimates versus time.

ming. From the point of view of ergonomics it is somewhat analogous to human learning. The objectives of learning curves as a performance measurement and evaluation tool are certainly applicable in this case.

30.6. INTEGRATED HUMAN AND ROBOT ERGONOMICS

A vital area for ergonomic study is the integration of humans and robots in work systems. Although industry has tended to separate employees from robot activities as much as possible, mainly for safety reasons, there are several important issues to consider.

30.6.1. The Roles of Human Operators in Robotic Systems

Except for unmanned production facilities, people will always work together with robots, with varying degrees of participation. Parsons and Kearsley[59] offer the acronym SIMBIOSIS to the roles of humans in robotic systems:

Surveillance—monitoring of robots.
Intervention—setup, startup, shutdown, programming, and correcting.
Maintenance.
Backup—substituting manual work for robotic at breakdown or changeover.
Input and output—manual handling before and after the robotic operation.
Supervision—management, planning, and exception handling.
Inspection—quality control and assurance beyond automatic inspection.
Synergy—combination of humans and robots in operation, for example, assembly or supervisory control of robots by humans.

In all the foregoing roles, the objective is to optimize the overall work performance. The idea is to plan a robotic system with the required degree of integration to best utilize the respective advantages of humans and robots working in concert. Human factors considerations in planning robot systems for these roles include job and workplace design, work environment, training, safety, pacing, and planning of supervisory control. These considerations are further studied in Chapter 32 (also published in reference 65), and in Parsons and Kearsley,[59] Howard,[60] Amram,[61] and Luria.[62]

30.6.2. Learning to Work with Robots

Both individuals and organizations using robots must learn to work effectively with the robots. Aoki[63] describes the progress of his company in utilizing robots. Particularly, as shown in Figure 30.16, robot performance in die casting was improved over time, following the learning curve model. The workers planning and operating the robot first improved its work method and program; then they improved the program (and method) further and introduced improvements in hardware. This is basically a process of an organizational learning based on learning by a group of individuals to better use robots. Many other companies report similar progress in adopting robot operators, similar to adopting any new technology.

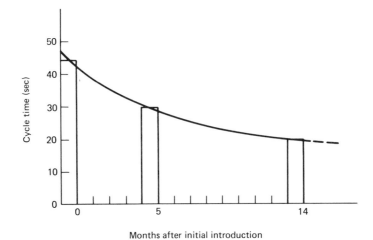

Fig. 30.16.   Reduction of robot cycle time at die-casting operation. (*Source.*   Data taken from Reference 64.)

In a prototype study (Argote, et al.[64]) the objective was to investigate how employees, as individuals, perceive and accept a new robot, the first in their company. Workers were interviewed 2.5 months before and 2.5 months after the robot introduction. As can be expected, with time and experience workers increased their understanding about what a robot really is. However, with time, workers' beliefs about robots, for instance, the potential hazards associated with robots, became more complex and pessimistic. Additionally, an increase in stress was indicated among workers interacting directly with the robot. Further research is needed on this problem; however, the researchers saw their findings as another indication that effective strategies for correct introduction of robots to the factory are vital to the success of robot implementations. More about human factors and robots is covered in Chapter 32 and about humans in supervisory control of remote robot applications in Chapter 17.

ACKNOWLEDGMENT

The education I received from Professor H. Gershoni, who taught me systems integration through methods engineering at the Technion, stimulated my interest in the general area that I have termed "Robot Ergonomics." For this I am grateful to him. Thanks are also due to the students who worked with me enthusiastically in this area over the years: Hannan Lechtman and Andy Robinson, who developed the RTM software and robot simulation; Ed Fisher and Richard Penington, who carried out the extensive statistical robot surveys; and Oded Maimon and Bob Wilhelm, who worked on robot cell design and control. Their contributions were valuable to the development of this chapter.

REFERENCES

1. Nof, S. Y., Decision aids for planning industrial robot operations, *Proceedings of the IIE Annual Conference,* New Orleans, Louisiana, May 1982, pp. 46–55.

2. Ottinger, L. V., Evaluating potential robot applications in a system context, *Industrial Engineering,* January 1982.

3. Nof, S. Y., Knight, J. L., and Salvendy, G., Effective Utilization of Industrial Robots—A Job and Skill Analysis Approach, *AIIE Transactions,* Vol. 12, No. 3, September 1980, pp. 216–225.

4. Luh, J. Y. S., An anatomy of industrial robots and their controls, *IEEE Transactions on Automatic Control,* Vol. 28, No. 2, February 1983, pp. 133–153.

5. Jablonowski, J., Robots: Looking over the specifications, *American Machinist, Special Report 745,* May 1982.

6. Nof, S. Y. and Fisher, E. L., Analysis of robot work characteristics, *The Industrial Robot,* September 1982, pp. 166–171.

7. Woodson, W. E., *Human Factors Design Handbook,* McGraw-Hill, New York, 1981.

8. Nadler, G., *The Planning and Design Professions: An Operational Theory,* Wiley, New York, 1981.

9. Clark, D. O. and Close, G. C., *Motion Study,* in Salvendy, G., Ed., *Handbook of Industrial Engineering,* Wiley, New York, 1982.

10. Hasegawa, Y., Analysis of complicated operations for robotization, SME Paper No. MS79–287, 1979.

11. Bolles, R. and Paul, R., The use of sensory feedback in a programmable assembly system, Stanford Artificial Intelligence Laboratory, Memo AIM-220, October 1973.

12. Barnes, R. M., *Motion and Time Study*, 6th ed., Wiley, New York, 1968.

13. Apple, J. M., *Plant Layout and Material Handling*, 3rd ed., Wiley, New York, 1977.

14. Francis, R. L. and White, J. A., *Facility Layout and Location*, Prentice-Hall, Englewood Cliffs, New Jersey, 1974.

15. Bonney, M. C. and Case, K., SAMMIE computer aided work place and work task design system, *CAD/CAM*, February/March 1978, pp. 3–4.

16. Maimon, O. Z. and Nof, S. Y., Activity controller for a multiple robot assembly cell, *Proceedings of ASME Winter Annual Meeting*, Boston, Massachusetts, 1983.

17. Kondoleon, A. S., Cycle time analysis of robot assembly systems, SME Paper No. MS79-286, 1979.

18. Kjellberg, T., A method to analyze handling problems in robot applications, *Proceedings of the 2nd Conference on Industrial Robot Technology*, Birmingham, U.K., March 1974, pp. E5:59–76.

19. Chang, C. A. and Goldman, J., An off-line design procedure for the work station of assembly robots, *Proceedings of the Fall IIE Conference*, 1982.

20. Fisher, E. L., Nof, S. Y., and Seidmann, A., Robot system analysis: basic concepts and survey of methods, *Proceedings of the Fall Industrial Engineering Conference*, November 1982, pp. 385–395.

21. Severiano, J. W. and Gruver, W. A., Optimization of material and information flow within robotic workcells and systems, SME Paper No. MS82–131, 1982.

22. Young, R. E. and Yuan, S. C., Development of robot-centered work station, *Proceedings of the Fall IIE Conference*, 1982.

23. Nof, S. Y., Computer aided planning of robotic assembly, *Proceedings of the AUTOFACT Europe*, Geneva, Switzerland, September 1983.

24. Nof, S. Y., An expert system for planning/replanning programmable production facilities, *International Journal of Production Research*, Vol. 22, No. 5, 1984 pp. 895–903.

25. Sarin, S. C. and Wilhelm, W. E., Prototype models for two-dimensional layout design of robot systems, *IIE Transactions*, 1984.

26. Schraft, R. D. and Schmidt, U., A computer aided method for the selection of an industrial robot for the automation of a working place, *Proceedings of the 3rd Annual Conference on Industrial Robot Technology*, Nottingham, U.K., March 1976.

27. Heginbotham, W. B., Donner, M., and Case, K., Robot application simulation, *The Industrial Robot*, June 1979, pp. 76–80.

28. Findler, N. V. and Shaw, J. N., Multi-Pierre—A learning robot system, *Computers and Graphics*, Vol. 3, 1978, pp. 107–111.

29. Kretch, S. J., Robotic Animation, *Mechanical Engineering*, August 1982.

30. Fors, N. and Donaghey, C. E., Improvements to Multiple Robot Simulation Language, *5th National Conference on Computers in Industrial Engineering*, Orlando, Florida, March 1983.

31. Hanify, D. W., Industrial robot analysis—working place studies, SME Paper No. MS75–241, 1975.

32. Nof, S. Y., Meier, W. L., and Deisenroth, M. P., Computerized physical simulators are developed to solve IE problems, *Industrial Engineering*, October 1980, pp. 70–75.

33. Brady, M., et al., Eds., *Robot Motion Planning and Control*, MIT Press, Cambridge, Massachusetts, 1982.

34. Pieper, D. L. and Roth, B., The kinematics of manipulators under computer control, *Proceedings of the 2nd International Conference on the Theory of Machines and Mechanisms*, Warsaw, Poland, September 1969.

35. Kahn, M. E. and Roth, B., The near-minimum-time control of open-loop articulated kinematic chain, *ASME Transactions, Journal of Dynamic Systems, Measurement and Control*, Vol. 93, 1971, pp. 164–172.

36. Luh, J. Y. S. and Walker, M. W., Minimum-time along the path for a mechanical arm, *Proceedings of the 1977 IEEE Conference on Decision and Control*, New Orleans, Louisiana, 1977, pp. 755–759.

37. Lynch, P. M., Minimum time, sequential axis operation of a cylindrical two-axis manipulator, *Proceedings of the Joint Automatic Control Conference*, Charlottesville, Virginia, 1981, Paper WP-2A.

38. Luh, J. Y. S. and Lin, C. S., Optimum path planning for mechanical manipulators, *ASME Transactions, Journal of Dynamic Systems, Measurement and Control,* Vol. 103, June 1981, pp. 142–151.

39. Birk, J. R. and Franklin, D. E., Pitching workpieces to minimize the cycle time of industrial robots, *Industrial Robot,* Vol. 1, No. 5, September 1974, pp. 217–222.

40. Verplank, W. L., Research on remote manipulation of NASA/Ames research center, NBS Special Publication 459, U.S. Department of Commerce, 1976, pp. 91–96.

41. Vertut, J., Experience and remarks on manipulator evaluation, NBS Special Publication 459, U.S. Department of Commerce, 1976, pp. 97–112.

42. Rogers, P. F., Time and motion method for industrial robots, *Industrial Robot,* Vol. 5, December 1978, pp. 187–192.

43. Paul, R. L. and Nof, S. Y., Work Methods Measurement—A Comparison between Robot and Human Task Performance, *International Journal of Production Research,* Vol. 17, No. 3, 1979, pp. 277–303.

44. Nof, S. Y. and Paul, R. L., A Method for Advanced Planning of Assembly by Robots, *Proceedings of SME Autofact-West,* Anaheim, California, November, 1980, pp. 425–435.

45. Nof, S. Y. and Lechtman, H., Analysis of Industrial Robot Work by the RTM Method, *Industrial Engineering,* April 1982.

46. Maynard, H. B., Stegemerten, G. J., and Schwab, J. L., *Methods-Time Measurement,* McGraw-Hill, New York, 1948.

47. Antis, W., Honeycutt, J. M., and Kock, E. N., *The Basic Motions of MTM,* 4th ed. (The Maynard Foundations), 1973.

48. Hershey, R. L., Letzt, A. M., and Nof, S. Y., Computerized methods for predicting robot performance, *Autofact 5,* Detroit, Michigan, November 1983, pp. 3.9–13.16.

49. Lechtman, H. and Nof, S. Y., Performance time models for robot point operations, *International Journal of Production Research,* Vol. 21, No. 5, 1983, pp. 659–673.

50. Kuno, T., Matsunari, F., Moribe, H., and Ikeda, T., Robot performance simulator, *Proceedings of the 9th ISIR,* Washington, D.C., March 1979, pp. 323–330.

51. Medeiros, D. J., Sadowski, R. P., Starks, D. W., and Smith, B. S., A modular approach to simulation of robotic systems, *Proceedings of the 1980 Winter Simulation Conference,* pp. 207–214.

52. Robinson, A. P. and Nof, S. Y., SINDECS-R: A robotic work cell simulator, *Proceedings of the 1983 Winter Simulation Conference,* pp. 350–355.

53. VS-100 Reference Manual, Machine Intelligence Corp., 1980.

54. Hancock, W. M., The learning curve, in Maynard, H. B., Ed., *Industrial Engineering Handbook,* McGraw-Hill, New York, 1971.

55. Nanda, R. and Adler, G. L., *Learning Curves, Theory and Application,* American Institute of Industrial Engineers, 1977.

56. Seltzer, D. S., Use of sensory information for improved robot learning, SME Paper No. MS79–799, Autofact, Detroit, November 1979.

57. Gleason, G. J. and Agin, G. J., A modular vision system for sensor-controlled manipulation and inspection, *Proceedings of the 9th ISIR,* March 1979.

58. Whitney, D. E. and Junkel, E. F., Applying stochastic control theory to robot sensing, teaching, and long term control, *Proceedings of the Joint Automatic Control Conference,* Alexandria, Virginia, June 1982, pp. 1175–1183.

59. Parsons, H. M. and Kearsley, G. P., Robotics and human factors: Current status and future prospects, *Human Factors,* Vol. 24, No. 5, 1982, pp. 535–552.

60. Howard, J. M., Focus on the human factors in applying robotic systems, *Robotics Today,* December 1982, pp. 32–34.

61. Amram, F. M., Robotics: The human touch, *Robotics Today,* April 1982, p. 28.

62. Luria, D. D., Technology, employment, and the factory of the future, *Proceedings of Autofact,* Detroit, Michigan, 1982, pp. 18–181.

63. Aoki, K., High speed and flexible automated assembly line—Why has automation successfully advanced in Japan?, *Proceedings of the 4th International Conference on Production Research,* Tokyo, 1980.

64. Argote, L., Goodman, P. S., and Schkade, D., The human side of robotics: Results from a prototype study on how workers react to a robot, Technical Report CMU-RI-TR-83-11, Carnegie-Mellon University, Pittsburgh, Pennsylvania, May 1983.

65. Salvendy, G., Review and reappraisal of human aspects in planning robotic systems, *Behaviour and Information Technology,* Vol. 2, No. 3, 1983, pp. 263–287.

CHAPTER **31**

QUANTITATIVE TECHNIQUES FOR ROBOTIC SYSTEMS ANALYSIS

RAJAN SURI

Harvard University
Cambridge, Massachusetts

TERMINOLOGY

Symbol Section Reference and Brief Explanation

$\bar{a}$ = 31.7.2: mean time between arrivals

B = 31.4.1: budget limit

c_j = 31.4.1: cost of WS_j

C_s = 31.7.2: coefficient of variation of service time (standard deviation divided by mean)

D = 31.7.2: deterministic (i.e., constant time) process

e_i = 31.4.1: existing annual cost of performing OP_i

$\bar{f}$ = 31.7.6: mean time to failure (MTTF)

$F(t)$ = 31.7.6: cumulative distribution function (CDF) for failure

G = 31.7.2: general (arbitrarily specified) distribution

h_j = 31.4.1: annual number of hours WS_j is available

M = 31.7.2: Markovian (or exponential) distribution

N_j = 31.4.1: number of WS_j that will be purchased

n_{TOT} = 31.7.2: average total number of jobs in the facility

n_w = 31.7.2: average number of jobs waiting for service

$O(j)$ = 31.4.1: indexes of operations that can be performed by WS_j

OP_i = 31.4.1: operation i

$\bar{r}$ = 31.7.6: mean time to repair (MTTR)

r_i = 31.4.1: annual rate for OP_i

R_i = 31.7.6: reliability of component i

$R_{m\,|\,n}$ = 31.7.6: reliability of "m out of n" structured system

R_p = 31.7.6: reliability of parallel structured system

R_s = 31.7.6: reliability of series structured system

R_{TMR} = 31.7.6: reliability of triple modular redundant system

$R(t)$ = 31.7.6: cumulative distribution function (CDF) for repair

$\bar{s}$ = 31.7.2: mean service time of a job

s_a = 31.7.6: standard deviation of times system is available

s_f = 31.7.6: standard deviation of time to failure

s_r = 31.7.6: standard deviation of time to repair

$\overline{T}_a$ = 31.7.6: mean of times system is available

t_{ij} = 31.4.1: time required to perform a single OP_i on WS_j

t_{TOT} = 31.7.2: average total time spent by a job from arrival to departure

t_w = 31.7.2: average time spent by a job while waiting for a processor

u = 31.7.2: utilization of processor

v_{ij} = 31.4.1: annual cost of performing OP_i on WS_j

$W(i)$ = 31.4.1: indices of workstations that can perform OP_i

WS_j = 31.4.1: workstation j

X_i = 31.4.1: decision whether OP_i is chosen for automation

x_{ij} = 31.4.1: proportion of OP_i requirements to be performed at WS_j

31.1. INTRODUCTION

31.1.1. Scope of the Chapter

There are many quantitative techniques that can be used as effective aids during the planning, design, and operation of a robotic system. This chapter reviews a number of such techniques with the aim of enabling readers to identify quickly which may be suitably used for their situation. An overview is given of the main methods, and for readers interested in further details, references to more comprehensive works are included.

At the outset, the reader should be aware of the types of issues that are addressed in this chapter. First, and foremost, the chapter is concerned with *macro* system design, rather than *micro* system design. At the micro level, the designer would be concerned with details of individual robot workstations: such issues are addressed in Chapter 30. At the macro level, the designer addresses a system having (possibly) multiple robot workstations, and issues of concern include choice of alternative workstations and combinations of workstations, configuration of the workstations and material-handling system, evaluating the performance of the system of interacting workstations, and integration of this robotic facility into the rest of the manufacturing plant. These are the kinds of issues addressed here.

Second, it is useful to make the distinction between *analysis* techniques and *synthesis* techniques. In the former, the designer has a particular system configuration planned and wishes to analyze its performance, whereas in the latter the technique presents the designer with a candidate plan to be evaluated. Although both analysis and synthesis techniques are covered here, it should be mentioned that more emphasis is placed on the former. The reason is simply that the state of the art is less advanced in synthesis techniques for robotic system design.

In summary, this section of the handbook deals with the *systems* aspects of multistation robotic systems. The need for an entire chapter devoted to this aspect arises from the fact that even a small-sized robotic facility is a complex system, consisting of many interconnected components of hardware and software, as well as many limited resources such as buffer storage space, part feeders or orienters, end effectors, and material-handling equipment. Designing such a system and operating it efficiently can be difficult tasks because of the interaction between the components, which makes it hard to predict the overall system performance. It is therefore important to use sophisticated techniques to analyze the system design and operation. This chapter gives the designer an overview of available techniques for this task.

Space considerations limit the extent to which each technique can be discussed. As a general rule, techniques that have been widely studied in the context of production and operations management are treated briefly, and references to well-known publications are given for them. Techniques that consider problems particular to robotic systems, or new problems introduced by these systems, are given more coverage.

31.1.2. Phases of the Design Process

In the planning, design, and acquisition of a robotic system an organization goes through several of the following distinct phases of activity:

1. **System Planning Phase.** This involves a feasibility study and/or a preliminary system design with the primary aim to establish the feasibility of the project, with coarse estimates of the strategic, financial, logistical, and operational aspects of the project. As a result, a few alternatives are earmarked for more detailed study.

2. **Process Planning and Configuration Design Phase.** During this phase, for each of the foregoing alternatives selected, the operations to be performed by the system are identified, and detailed system design is carried out in terms of equipment selection and placement and workstation design.

3. **Installation Phase.** This includes the system procurement, installation, and debugging.

4. **System Operation Phase.** After the system is installed, it must be operated in a manner that meets production requirements and other management criteria (such as limits on work

in process). Issues of reconfiguration due to failure, or due to unanticipated production require-ments, constantly need to be addressed.

5. **Ongoing Tasks.** Even after a system has been made operational, there are long-term tasks that must be addressed in addition to the day-to-day operating tasks. Examples are gradual performance tuning of the system, identifying and evaluating potential modifications to the system, and planning/designing potential expansions of the system.

In each of the activity phases listed, the task of the designer can be aided by quantitative techniques. Details of the design requirements for each phase, as well as appropriate quantitative techniques for addressing those requirements, are given in Sections 31.2 through 31.6 of this chapter.

31.1.3. Overview of Quantitative Techniques

The techniques covered here range over a wide spectrum of methods, although, broadly speaking, all of these methods are related to the area of operations research/management science (OR/MS). Nevertheless, the unfamiliar reader may be overwhelmed by the diversity of methods discussed. It is useful therefore to place the methods in a simple classification scheme which aids the reader in identifying the critical elements addressed by each technique and in comparing and contrasting the techniques. This scheme is introduced here and is used throughout the chapter.

The scheme is based on two attributes. The *first* attribute states whether the technique is based on a *deterministic* or *probabilistic* model. In the latter case, uncertainties prevalent in the situation being studied are explicitly incorporated into a model of the situation. In the former case, such uncertainties are either neglected altogether, or modeled indirectly (e.g., by using "safety factors"). The *second* attribute states whether the technique uses a *static* or *dynamic* model. In the latter case, the technique explicitly models the evolution and interaction of the system components over time, thus enabling study of phenomena that are prevalent at some times and not at others, and of effects due to particular sequences of behavior. In the former case these details are replaced by aggregate models of behavior that (attempt to) summarize the system performance over the entire time period. For consistency with accepted terminology, in the case of a probabilistic model the term *steady state* is used rather than *static*.

With these attributes it is seen that all the quantitative techniques described here fall into just four categories (Table 31.1). In order of increasing complexity these categories are as follows:

TABLE 31.1. CATEGORIZATION OF QUANTITATIVE TECHNIQUES: AN EXAMPLE OF EACH CATEGORY

	Attribute 2	
Attribute 1	Static (or Steady-State)	Dynamic
Deterministic	Linear programming	Scheduling
Probabilistic	Queueing-network algorithms	Simulation

Deterministic/static
Probabilistic/steady-state
Deterministic/dynamic
Probabilistic/dynamic

An example of a technique in each category can be found in the table.

In the rest of the chapter, any quantitative technique introduced is classified in this scheme. This enables the reader to decide whether, at first glance, the technique addresses these two important aspects in the manner desired. It also enables a quick comparison between other techniques in the same category.

31.1.4. Matching the Techniques to the Project Implementation Phase

The four categories of methods are also listed in order of increasing accuracy. The conservative designer or manager may therefore be tempted always to use the most accurate technique. However, this is not necessarily the best policy. The accuracy arises from a greater level of detail in the model and greater thoroughness in its analysis. An accurate technique invariably requires more detailed knowledge about the system to set up the model correctly, thus requiring a lot of time to be spent on determining the precise model structure and model parameters, as well as values for these parameters. It is more than likely that exercising this technique will also be expensive in terms of computer resources. Therefore, at the preliminary stage of the design process, when many widely differing system alternatives are to be explored, it may be impractical or even infeasible to use such a technique. In contrast, a simpler model may be very easy to set up, and may allow a wide range of alternatives to be explored at relatively little effort and cost. Since this technique may be more coarse, it should be used primarily to eliminate alternatives and to identify potential designs for more detailed investigation. Nevertheless, at the feasibility phase or preliminary design phase, such a technique may be sufficient to identify "ball park" values of the significant variables, such as total investment, rate of return, and production capacity.

To make this point concrete, consider the use of a queueing model (probabilistic/steady-state) instead of a detailed simulation (probabilistic/dynamic) for a multi-workstation system. (These models are described later.) The former model might require only 20 input data items, and execute in under a second on a minicomputer. The latter might require 500 input data items, and take many minutes to execute. Clearly the former is far more suited to rapid interactive design for exploring a wide range of alternatives. Furthermore, as discussed in later sections, the queueing model may supply enough information for an initial decision.

31.1.5. Structure of the Chapter

The remainder of the chapter is devoted to describing specific quantitative techniques. It is envisioned that the reader is a system designer or manager engaged in one or more phases of implementation of the robotic system. At each phase it would be useful to know the issues to be addressed and the relevant techniques that can aid in addressing these issues. The first part of the chapter is therefore divided into sections, one for each phase of the implementation process, which enables the reader to focus on the phase that is currently of importance. Within each section, there are three main subsections. The first summarizes major requirements to be met, and issues to be addressed, by the particular phase. This subsection should serve as a useful checklist for the designer. The second subsection indicates typical performance measures used in studying decision alternatives for this phase and explains their significance. The final subsection then describes quantitative techniques relevant to studying the requirements, issues, and performance measures detailed in the previous subsections.

The last part of the chapter (Section D) discusses issues relevant to the entire design/operation process.

A. SYSTEM PLANNING PHASE

This section addresses the preliminary system design phase during which the concern is to establish feasibility of the project with initial estimates for system size, cost, return on investment, and other variables of interest to upper management.

31.2. DESIGN REQUIREMENTS AND ISSUES

As the first step in fulfilling the typical requirements of this phase of the design process, as well as issues that must be addressed, the major parameters of concern should be identified and qualified to provide guidelines for the scope of the planning study. Such parameters would include the following:

1. **Management Objectives.**   What is the reason (or the set of possible reasons) for the organization to consider a robotic system? Knowledge of this is important so that the planning phase can result in performance indicators that relate to the reason(s). Typical reasons could be one or more of the following: reduced labor costs, improved quality, worker safety, increased product flexibility. (Some reasons may not be known in advance and may become apparent only as the analysis proceeds. These will then have to be incorporated into the models and methods being used by the study.)

2. **Location.**   If a new plant is being constructed to house the automated facility, what is its best location with respect to material supply and demand and with respect to overall costs? If the automated facility is being placed in an existing plant, what is its best location within the plant, again with respect to material movement?

3. **Operations to be Considered.**   What are the operations that will be considered for potential robot automation? For example, in a manufacturing facility, is it only a given operation for a

given product that is being studied, or all possible operations for all the products (or any limited combination in between)? At this preliminary stage, it is not necessary to study each operation in detail to decide whether to include it. Rather, various gross characteristics of the operations/products should be used to select or eliminate candidates. Group technology codes can assist in this process.[12-14]

4. **System Size.** What are rough limits on system size dictated by floor space, number of elements, overall system complexity, or other criteria?

5. **Capital Budget.** What is the range of dollar amounts that is being considered as the capital outlay for this system?

The main aim of identifying these parameters is to narrow the scope of the preliminary design study. In some situations management may not wish to limit its alternatives and might require many of the parameters to be determined during the study. However, the more that can be done to elicit estimates such as the preceding using *qualitative* arguments, the more focused and productive will be the quantitative study described.

Given these general guidelines, specific requirements of this preliminary phase are, typically, to identify the following:

1. One or more alternative systems that could satisfy management objectives.

2. For each alternative, estimates of performance indicators relevant to upper management.

3. For each alternative, the operations to be automated and the typical equipment that might be used.

31.3. PERFORMANCE MEASURES

In comparing the alternatives with each other, and with the technology currently in use, several indicators of performance are used. Prime among these are the following:

1. **Amount of Investment.** What will be the total cost of the system, and how will the investment be phased in time?

2. **Return.** What will be the return on investment (ROI) for the project? Some organizations use measures such as internal rate of return (IRR), net present value (NPV), and payback period as alternatives, or as additional measures. Organizations will usually have a standard procedure for evaluating these financial measures, but if there is some doubt, an extensive discussion can be found in the work of Canada.[6]

3. **Operational Flexibility.** How easily will the system adapt to day-to-day perturbations in workload and variety of tasks and to subsystem failures?

4. **Capacity Flexibility.** What will be the system's capacity to accommodate lasting changes in production requirements?

5. **Strategic Flexibility.** To what extent can the system be adapted to changes in operation, for example, to assemble a different product mix?

31.4. QUANTITATIVE TECHNIQUES FOR SYSTEM PLANNING

In addition to the description of techniques that are useful in the planning phase given in this section, reviews and literature surveys can be found in Fisher, Nof, and Seidmann,[1] Fleischer,[2] and Nof.[3]

There are, clearly, many elements of the preliminary design phase that are not unique to robotic systems, and these elements might already be familiar to system planners/designers. One such element is *strategic planning*.[15-18] Depending on the size and scope of the robotic system being considered, management may wish to use methods of strategic planning along with the other tools discussed in this section. Another important element is *forecasting*.[19-22] The rate of operation required of the system is usually dependent on certain exogenous variables such as demand for a product. It is therefore essential to forecast the relevant exogenous variables for a given time period into the future. Uncertainty in forecasts can be dealt with rigorously, using *decision analysis*.[23,33] A fourth element is facility location analyzed by the techniques of *location analysis*.[44,45] Since these aspects are not special to robotic systems, the reader desiring more information is referred to the well-known works cited previously. (There is an aspect of location analysis that is particularly relevant to automated facilities. Further discussion and references are in Section 31.7.1.)

The task of selecting which operations should be automated and what equipment should be used to automate these operations can be a very complex one and, of course, is particular to the concept of automation. It is for this task that we introduce the first set of quantitative techniques. These fall in the deterministic/static category and are linear programming (LP), integer programming (IP), and mixed integer programming (MIP). The techniques are introduced in the context of a simple example.

31.4.1. Planning Example

Suppose the main objective of automation is to maximize the annual cost saving, given that a limit exists on the capital to be invested in the robotic system. Let

$\{OP_1, OP_2, \ldots, OP_I\}$ be the set of candidate operations for possible automation. The guidelines in Section 31.2 should have helped to define and limit this set.

$\{WS_1, WS_2, \ldots, WS_J\}$ be the set of possible workstations that could be used for automation of one or more of the foregoing operations. Again, by limiting the set of candidate operations, it is easier to review capabilities of available equipment and identify alternative equipment that is suitable for these operations.

r_i be the annual rate at which OP_i must be performed (e.g., number of times per year).

c_j be the cost of WS_j.

h_j be the annual number of hours that WS_j is expected to be available, taking into account the number of shifts, reliability, and other stoppages.*

$O(j)$ and $W(i)$ be sets of indexes: $O(j)$ contains indexes of operations that can be performed by WS_j, and $W(i)$ contains indexes of workstations that can perform OP_i.

t_{ij} be, for j in $W(i)$, the time required to perform a single OP_i on WS_j.

e_i be the existing annual cost of performing OP_i using current methods: this includes variable costs (e.g., labor and materials) as well as related overheads.

B be the capital budget limit.

v_{ij} be, for j in $W(i)$, the annual cost of performing OP_i on WS_j: this includes variable costs and related overheads, but *not* recovery of capital investment.

Several parameters require an economic model of the system. Relevant approaches for robotic systems are described in Benedetti,[4] Boothroyd,[5] Ciborra and Romano,[7] Fleischer,[8] Heginbotham,[9] Owen,[10] and Whitney et. al.[11]

Next, the *decisions* to be made for the preliminary design are represented by variables:

$X_i = 1$ if OP_i is chosen for automation ($= 0$ if not).

x_{ij} = proportion of annual OP_i requirements that will be performed at WS_j.

N_j = number of WS_j that will be purchased.

31.4.2. Linear Programming[25,28,29,32]

With decision variables and data items as given, the preliminary design problem can be formulated as the following *linear programming* (LP) problem

$$\max \sum_{i=1}^{I} e_i X_i - \sum_{i=1}^{I} \sum_{j \in W(i)} v_{ij} x_{ij} \tag{31.1}$$

subject to

$$\sum_{j=1}^{J} c_j N_j \leq B \tag{31.2}$$

$$\sum_{j \in W(i)} x_{ij} = X_i \quad \text{for all } i \tag{31.3}$$

$$\sum_{i \in O(j)} r_i t_{ij} x_{ij} \leq h_j N_j \quad \text{for all } j \tag{31.4}$$

$$X_i \leq 1 \quad \text{for all } i \tag{31.5}$$

$$x_{ij}, X_i, N_j \geq 0 \quad \text{for all } i, j \tag{31.6}$$

Equation (31.1) states the objective, which is to maximize the annual savings in cost, while (31.2) is the budget constraint. Equation (31.3) ensures that if an operation is selected (i.e., $X_i = 1$), then the proportions of that operation's requirement assigned to various workstations add up to unity. Equation (31.4) is the available capacity on workstations of type j. Equations (31.5) and (31.6) ensure that the decision variables are nonnegative and that X_i does not exceed unity. Observe, however, that in this formulation the decision variables are not restricted to integer values, and thus a solution with $X_i = 0.6$ and $N_j = 2.3$ could be obtained. Such a solution would be meaningful only if partial automation of an operation were acceptable ($X_i = 0.6$ says that 60% of the annual requirement of OP_i should be performed on the new system, and 40% should continue to be performed in the existing way). Also, if it were possible to buy part of a workstation, for example, if three WS_j were purchased and another department used 70% of one of them, then $N_j = 2.3$ would be meaningful. In practice, however, such fractions are usually not reasonable, and this is the shortcoming of the LP formulation.

Still, the advantages of the LP formulation are numerous. Theory and algorithms for LP are well known and well understood,[25,28,29,32] and efficient software programs exist even for very large problems (e.g., $I = 1000$ and $J = 50$). The LP solution serves as an upper bound (the best value that might be attained) and thus can suffice in eliminating alternatives, for example, alternatives for which even the LP value is not satisfactory. Often, the LP solution is close to the optimal integer solution, and so it can be used as an approximation to the true solution. In a preliminary analysis this may be enough to identify feasibility of the project and typical values of financial and operational variables.

31.4.3. Integer Programming[27,34]

Refinement of the LP formulation involves adding the following constraints which prevent splitting of operations or workstations:

$$X_i = 0 \text{ for } 1 \quad \text{for all } i \tag{31.7}$$

$$N_j = 0, 1, 2, \cdot \ldots \text{ (nonnegative integer) for all } j \tag{31.8}$$

$$x_{ij} = 0 \text{ or } 1 \quad \text{for all } i, j \tag{31.9}$$

Equations (31.1) through (31.9) now constitute an *integer prorgamming* (IP) problem. Note that Equation (31.9) has added the additional constraint that if an operation is selected for automation, it should be done only on one type of workstation. This restriction is discussed further later, but in many situations it is the alternative preferred by system managers anyway. The preceding IP problem is considerably more complex than the LP. Typical computer time can easily be that for 1000 times solving the LP, depending on the size of the problem. Still, this IP constitutes a well-known problem, and several algorithms and software packages exist.[34]

31.4.4. Mixed Integer Programming[34]

A third formulation is obtained by removing the constraint of Equation (31.9), in other words, requiring only whole operations or whole workstations to be selected, but if an operation is selected, it may be performed at more than one type of workstation. The reason that this may be discouraged, as in Equation (31.9), is that it leads to additional costs associated with designing and procuring different end effectors, feeders, orienters, and so on for this operation on each type of workstation. However, in some contexts it may be necessary to split an operation, since this is the only way to fill spare capacity on two types of workstations. The resulting formulation, Equations (31.1) through (31.8), constitutes a *mixed integer programming* (MIP) problem. The level of complexity is somewhat less than that of the foregoing IP, but still it remains closer in magnitude to that of the IP than the LP. Again, this formulation can be solved using well-known algorithms and software.[24,34]

31.4.5. Enhancement of the Model

All the preceding formulations neglected several factors, primarily for clarity of the examples. Depending on the level of detail required of the preliminary analysis, several more factors can be incorporated into the model. Examples of such factors follow.

Alternative objective functions may be desired, such as others stated in Section 31.3. Differences in operation times due to sequencing can be incorporated: for instance, load/unload times and end-effector change times may depend on which operation follows the current one.[26] Additional constraints, such as a limit on the number of workstations or available floor space, can be included. Cost savings due to sharing of resources, such as end effectors, by more than one operation on the same workstation can also be studied.[36] If the operations being considered constitute processing for various product lines, then it may be desirable to automate either all the operations for a product, or none of them. Then the decisions become whether or not to automate a group of operations, and Equations (31.1)–(31.9) can be suitably modified.[36]

31.4.6. Suboptimal and Heuristic Techniques

When the model is enhanced by adding in detail factors, its solution becomes considerably more difficult to obtain. (In principle the solution can always be found, but in practice the computer time required for each solution may be unacceptably high.) So, for these more complex problems suboptimal or heuristic approaches are used. A *suboptimal* solution refers to one obtained by prematurely stopping an optimizing technique (a technique that would obtain the optimum given enough computer time) and then picking the best feasible solution obtained so far. A *heuristic* technique is one that uses a set of rules to generate a solution.[37] In this case, there is no guarantee the solution will be optimal or even good; however, such techniques are supported by much empirical evidence showing that they perform well. The advantage of using a suboptimizing method is that it is systematic and based on sound theory, and usually also provides bounds on the difference between the chosen solution and the optimum. The advantage of a heuristic method is that it can be extremely efficient and can take into account constraints that do not fit easily into the framework of mathematical programming methods. Good suboptimal solution methods for robotic system planning have been developed by Whitney et. al.,[36] and Graves and Lamar.[26] For large problems with many candidate operations, Whitney and Suri[36] present a very fast suboptimizing algorithm. Heuristic methods have also been developed for this task. An algorithm by Whitney and Suri[36] has been empirically shown to perform well, and in addition it can be used when there is resource sharing (as above) or many other unusual constraints, which the other procedures are not able to model effectively. For extremely large problems, involving automation of entire plants, where the number of operations could be in the tens of thousands, it may be worthwhile considering a different formulation of the objectives, in terms of constraints alone, since in this case considerable theory can again be brought to bear on solving the problem efficiently.[35]

31.4.7. Parametric Studies

Much more insight is gained during the planning stage by solving a model for a range of values of an input parameter. For instance, in the planning example of Section 31.4.1, the model can be solved for a range of values of B (the budget constraint). Suppose the IP model (Section 31.4.3) is used, and the optimal solution (i.e., the annual cost savings) is plotted as a function of B. A typical graph might be as shown in Figure 31.1, which serves to illustrate some important points. Note first that the function is quite irregular in its behavior—this is due to the integer nature of the problem. Next, for any solution, the ROI is given by the slope of the line joining the origin to the point x representing the solution. This shows that the highest ROI is obtained for an investment of 40,000 dollars (line A), but this option provides an annual savings of just 15,000 dollars. On the other hand, the highest annual savings is obtained when 240,000 dollars is invested, but this decision gives the lowest ROI (line C)! One approach to resolving this trade-off could be to specify the minimum acceptable ROI, say, 20% (line B), and to maximize the annual savings subject to this ROI constraint. In that case the best decision is the one where 120,000 dollars are invested. Thus the parametric study in Figure 31.1 provides useful insight for management.

Another type of parametric analysis that is helpful is a study of the *surge capacity* of a given system. If the models of Sections 31.4.2–31.4.4 give rise to several candidate designs, all with comparable cost and performance measures, this additional study may help to pick out a superior candidate. For a given system (i.e., choice of workstation), the model is solved for successively increasing production requirements of a given operation or product line. The production requirement is increased until the solution becomes infeasible (the system capacity is just exceeded), giving rise to a measure of surge capacity. A typical result, for a system with three products A, B, C, might be that the system has a surge capacity of 35% with respect to product A alone, and 10% with respect to all three products simultaneously. This provides management with another dimension on which to judge alternative candidates.

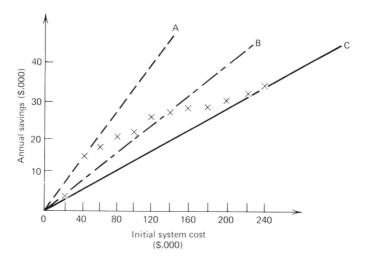

Fig. 31.1. Parametric study of annual savings as a function of system cost.

31.4.8. Limitations and Benefits of Planning Models Described

All the approaches so far still neglect certain other factors, to keep the problems simple and the computer solution times reasonable. Layout of the workstations is not considered—it is assumed that a feasible layout will be found for the selected design. Intermediate buffer spaces are not modeled. There is no limit on the number of operations assigned to a station—it may be that some robots have a limited repertoire of tasks they can perform. Intricate details of sequencing and/or grouping of operations are not considered—it is possible that time and cost savings could be achieved by these methods.

Despite these limitations of the models, these techniques are useful in the preliminary design phase, as they give approximate values of the main performance indicators, and enable the designer to select a few alternatives for more detailed study as described in the next section. Above all, these quantitative tools, if implemented effectively as software decision aids, not only help the organization in exploring wide ranges of parameters for the system, but also in so doing enable the designer to learn at first hand the elements that are significant for the more detailed project that lies ahead.

B. CONFIGURATION DESIGN PHASE

This phase begins when a preliminary set of operations has been selected for automation, along with a preliminary set of workstations to comprise the robotic system. It may be that two or three such operation/system combinations have been selected for further study. The aim of this phase is to provide a detailed system design for each of the candidate operation/system choices. While the planning phase used gross characteristics of the operations and workstations, this phase requires further refinement of the data on these items. Operations must be studied in detail, and if necessary modified to be compatible with available automation technology. The capabilities of selected workstations must be understood, and the types of accessory equipment required (such as end effectors, fixtures, partfeeders, etc.) must be identified. These points are described in Chapter 29, so in the remainder of this section it is assumed that the operations and workstations have been studied, and necessary information on them has been obtained.

31.5. DESIGN REQUIREMENTS AND ISSUES

The main requirement of this stage is to design the detailed system configuration, which involves deciding the following:

1. **Equipment Location.** Includes location of workstations and other facilities such as tool cribs, supervisors, and maintenance personnel.
2. **Material-Handling Equipment.** Includes types of equipment to be used for moving workpieces between workstations (e.g., conveyors, robots, wire-guided carts) and the layout and capacity of this subsystem (transport speed, number of transporters).
3. **Buffer Storages.** Location and sizes of intermediate storages in the system.
4. **Accessories.** Numbers and types of end effectors, part feeders, part orienters, part magazines, fixtures, tool holders, and so on.

31.6. PERFORMANCE MEASURES

The performance indicators used for judging the detailed configuration design are many. First of all, the organization would be interested in *refined estimates for all the measures* stated in Section 31.3 since these are the leading financial and operational indicators. The reader is advised to review those before reading on. In addition, the configuration design would be evaluated using the following more detailed measures:

1. **Production Rates.** For performing each operation, and/or net values for a product line.
2. **Turnaround Time** (for a manufacturing operation). This is the average time it takes from the entry of a workpiece into the system to the exit of the same piece from the system after all operations on it have been performed.
3. **Work-in-Process (WIP) Inventory.** The average level of inventory accumulated at various parts of the sytem.
4. **Equipment Utilization.** The time that each piece of equipment is being productively used, as a proportion of available time.

5. **Queue Sizes.** The number of workpieces or tasks that are waiting for a particular workstation or other equipment (e.g., transporter).

6. **Blocked Times.** The amount of time that a given item of equipment is ready to perform a task but is "blocked" for reasons such as its output buffer is full, or the end effector it needs is being used by another robot, or a transporter is required to remove the current workpiece.

7. **System Reliability.** An evaluation of how the reliability of the components will interact in the given configuration, leading to a measure of overall system reliability.

These indicators are used, not just as means for evaluating a design, but as aids to diagnosing problems in a design and perhaps indicating directions for improvement. The configuration design process is inevitably an iterative one where each candidate design is successively refined by modification and reevaluation.

31.7. QUANTITATIVE TECHNIQUES FOR CONFIGURATION DESIGN

This section discusses techniques that can assist the system designer during the configuration design phase.

31.7.1. Location Analysis

In the preliminary design phase the location of various system components was not considered. The first step of refining a candidate design is to work out the details of the placement of workstations, buffers, material-handling equipment, accessories, and so on. Study of such decisions belongs in the domain of *location analysis,* [44,45] also studied under the titles of *facility location* and *layout planning.* The quantitative techniques arising from this field belong mainly in the deterministic/static category. Locational problems encountered in designing an automated facility can be broadly placed into three classes:

1. Location of the automated facility with respect to existing facilities.
2. Location of equipment within the facility.
3. Location of workpieces, accessories, and tools within a workstation.

The first problem is part of the system planning effort, and, for reasons stated in Section 31.4, not dealt with in detail in this chapter. However, some of the techniques to be described would be relevant for the first problem too. The third problem belongs in the realm of workstation design, discussed in Chapter 30. Thus this section focuses on the second problem, with occasional remarks on the first problem.

In deciding the location of equipment (or of the entire facility), the following are typical factors that must be considered:

1. **Transport Time** (*or costs*). In locating a workstation or designing a material-handling system the time to transport workpieces may be a consideration. (In locating an entire facility, where distance to other plants is significant, transport costs could be a factor).

2. **Workstation Dimensions.** This can be important if the workstation size is significant compared to the facility size.

3. **Transporter Reach.** Some material transport systems (such as a radial arm) have limits on their reach. Others (such as under-floor tow-line systems) might have an existing layout, or limits/costs associated with length of a new track.

4. **Fixed Costs.** These may depend on the site chosen for the equipment (or facility).

5. **Capacity Limits.** Material movement between workstations may be limited by transporter capacity.

The simplest approach to location analysis involves *planar location models.* These involve simplifying assumptions such as any point in the plane is a valid location, fixed costs are negligible, and distribution problems can be ignored. A final assumption concerns the way in which "distance" is measured between two locations. A wide variety of measures are permitted, but the measure always depends only on the relative coordinates of one point with respect to the other. (In particular, it cannot depend on a given network of paths.) In spite of these assumptions, planar models are often used because they are easy to solve and can provide useful insight.[46] An introduction to such models can be found in Francis and White,[44] and recent surveys with detailed references are in Hearn and Vijay,[48] and in Francis, McGinnis, and White.[45]

An important concern in the design of an automated facility is the layout of the automated material-handling system. In the study of this problem, related work on *conveyor theory*[53] and on *automated storage and retrieval systems (AS/RS)*[39,40,44,56] can prove useful. A recent survey of quantitative techniques for material handling is in Matson and White.[51]

If the transport between locations must take place by way of a given network, such as an existing track layout for a transport system or a network of aisles, then planar models may not be accurate enough and a *network location model* should be used. Introductions to such models can be found in Handler and Mirchandani[47] and Minieka,[52] and an up-to-date survey of solution techniques is in Tansel, Francis, and Lowe.[57]

The most realistic location problems are solved using *discrete location models*. These limit the decisions to a given set of locations and allow the inclusion of fixed costs, which can depend on the site. They also admit many other features, such as capacity limits, limits on the number of sites chosen, and distribution decisions. Surveys of work on these models can be found in Francis, McGinnis, and White,[45] and Krarup and Pruzan.[50] Some basic formulations and extensions are given in Akine and Khumawala,[38] Elshafei,[42] Khumawala,[49] and Ross and Soland.[55] The disadvantage of including these details in the model is, of course, that the model rapidly becomes very difficult to solve as the number of candidate decisions increases. Model formulation requires much more input data to be gathered, and solution requires very sophisticated computer software.[38,41,43,49,54] If, however, the scope of the problem is such that potential savings could be high, it merits investigation by these more sophisticated models.

31.7.2. Queueing Models

The next level of detail for evaluating a design can be provided by the techniques of *queueing theory*.[61-66,70] These are quantitative methods that explicitly model some of the variability encountered in day-to-day system operation, as well as modeling the system dynamics. As we show, however, usually only aggregate indicators relating to dynamic performance are obtained, so this method falls into the category of probabilistic/steady-state models.

The main improvement of a queueing model over all the capacity models described in preceding sections is that it models not only the processing time required for an operation, but also the time that jobs wait for a processor* as they move through any facility. In practice, the arrivals of jobs to any processor are not evenly spaced in time, and also the processing time required may vary for different jobs. This *dynamic imbalance* of flow rates can produce effects not predicted by the deterministic/static planning model: that model simply ensured that the *average* processing rate exceeded the *average* demand for processing. Examples are given later where the queueing model reveals deficiencies in a design that appears feasible from a deterministic/static viewpoint.

The performance indicators available from a queueing model include the following (these symbols are used throughout this section):

u = utilization of the processor(s).

n_W = average number of jobs waiting for service (not including the job being processed): this is useful for designing buffer space.

n_{TOT} = average total number of jobs in the facility (waiting and being served): this measures the WIP inventory.

t_W = average time spent by a job while waiting for a processor: this measures the "wasted" time for each workpiece.

t_{TOT} = average total time spent by a job from arrival to departure: this measures the turnaround time of the facility.

Other measures are available too, but those listed are the most useful for a system designer.

Single-Queue Systems

The simplest branch of queueing theory is concerned with study of a single-queue facility (Figure 31.2). This consists of a number of identical processors, each of which can process any of the arriving jobs (one at a time). Jobs arrive at the facility from some external source, and if all processors are busy with a job, the arriving job must wait. When some processor finishes its current job, it picks another job from those that are waiting, according to a prescribed priority scheme. Such a queue is characterized by a standard shorthand form: the A/S/n notation. Here A is a symbol that identifies the *arrival process*, that is, statistics of the time between arrivals of jobs at the facility. Similarly, S identifies the *service process*, that is, the statistics of the time that it takes to complete a job. Finally,

---

* The term *processor* denotes an entity that processes tasks. Examples are workstations, transporters, repair personnel.

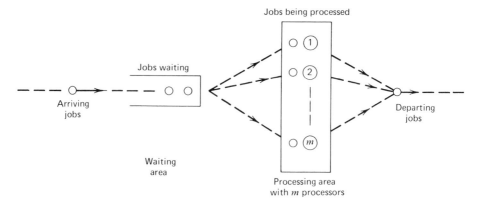

Fig. 31.2. A single-queue facility.

n is the number of processors. (Extensions of this notation exist[65] but are not considered here.) Standard "values" for the symbols A and S are:

M = Markovian (or exponential distribution*).
D = deterministic (i.e., constant time).
G = general (the analysis is done for an arbitrarily specified distribution).

Thus M/D/2 denotes a facility where the time between arrivals has an exponential distribution, all jobs take the same amount of time to be processed, and there are two processors.
    The following symbols are also used in the discussion that follows:

$\bar{a}$ = mean time between arrivals to the queue.
$\bar{s}$ = mean service time of a job at a processor.
C_s = coefficient of variation of the service time (standard deviation divided by mean).

Little's Law

This is the most basic and widely used result in queueing theory, and it relates arrival times, total system times, and number of jobs in a system. It was first derived under restricted conditions;[67] later Stidham[69] showed that the result holds under fairly general conditions, which can be assumed to hold in a practical facility. The result states simply that for a queueing system

$$t_{TOT} = n_{TOT}\, \bar{a} \qquad\qquad (31.10)$$

Thus if any two of these quantities are known, the third is automatically determined. Letting the "system" now be only the waiting line (not the jobs in service) and applying Little's law to this "system," another useful result is obtained:

$$t_W = n_W\, \bar{a} \qquad\qquad (31.11)$$

For a facility with a single processor, it is also useful to know that the following relations hold for general arrival- and service-time distributions, that is, for the G/G/1 queue:[65]

$$u = \frac{\bar{s}}{\bar{a}} \qquad\qquad G/G/1 \qquad\qquad (31.12)$$

$$n_{TOT} = u + n_W \qquad G/G/1 \qquad\qquad (31.13)$$

$$t_{TOT} = \bar{s} + t_W \qquad G/G/1 \qquad\qquad (31.14)$$

The reader will find it a useful exercise to find intuitive explanations for these three results.

---

* The exponential probability distribution is widely used in queueing and reliability theory, hence referred to frequently in this section. Descriptions of this and other common distributions can be found in standard texts on probability.[58-60]

M/G/1 Queues

One type of single-queue model that can be adapted to many situations is the M/G/1 queue. Here jobs arrive at a workstation from other facilities, and the randomness in the time between arrivals is found to be suitably modeled by an exponential distribution with mean $\bar{a}$. The distribution of service times required by jobs can be arbitrary (but known), and there is one processor. Formulas for the performance of such a facility are given here, and their use is illustrated next. The first performance indicator of interest is the utilization u whose value is given by the general Equation (31.12). The next measure is given in terms of u and the known value of C_s:[65]

$$n_W = u^2 \frac{(1 + C_s^2)}{2(1 - u)} \qquad \text{M/G/1} \qquad (31.15)$$

The value of t_W then follows from Little's law, Equation (31.11), while n_{TOT} and t_{TOT} follow by application of Equations (31.13)–(31.14). Notice that the only derivation required that is particular to the M/G/1 case is that of n_W; these other measures follow by application of general results.

Example for Robotic System

We use the M/G/1 model to illustrate a situation commonly found in robotic systems, the M/D/1 queue. The arrival process is as explained before. All jobs require precisely the same operation to be done at this station, which take $\bar{s}$ time units (no variation, so $C_s = 0$). There is only one robot at the workstation. The utilization u is given by Equation 31.12, and the value of n_W follows from Equation (31.15):

$$n_W = \frac{u^2}{2(1 - u)} \qquad \text{M/D/1} \qquad (31.16)$$

and t_W, n_{TOT}, and t_{TOT} follow as before.

The value of u is the same as that predicted by a deterministic/static capacity model. A measure not provided by that model, though, is n_W, which is graphed as a function of u in Figure 31.3. Note how steeply the function rises as u approaches unity. This brings out an important distinction between the deterministic model and the queueing model. Suppose initial estimates are $\bar{a} = 10$ sec and $\bar{s} = 7$ sec, giving $u = 0.7$ and $n_W = 0.8$. Now suppose it is acknowledged that the estimates of $\bar{a}$ and $\bar{s}$ may each have up to 10% error, so that, in the worst case, the actual system will have $\bar{a} = 9$ and $\bar{s} = 7.7$. In the deterministic model this gives $u = 0.855$, which seems feasible. But the queueing model, while giving the same value for u, also shows that $n_W = 2.5$. This is a threefold effect on the estimated

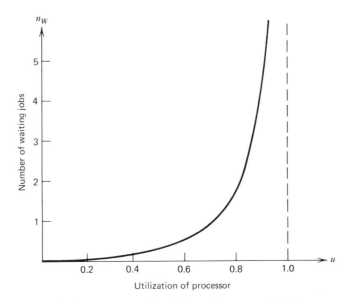

Fig. 31.3. Average number of jobs waiting at an M/D/1 queue.

number of jobs waiting! The WIP is also more than doubled (n_{TOT} changes from 1.5 to 3.4). Thus, even in this simple example the queueing model has pointed out major effects of the (apparently) minor parameter variations.

Multi-Processor Queues

Another useful model for multi-robot systems is the M/M/m queue. Here the arrival of jobs is the same as in the previous example; however, the jobs can be of many different types, each type requiring different processing time. The time spent by a job at a processor is therefore variable, and in some situations is modeled well by assuming an exponential distribution. There are m identical processors, and any job can be done at any processor. For an m-processor facility, the general analogs of (3.2.3–4) are:[66]

$$u = \frac{\bar{s}}{m\bar{a}} \qquad \text{M/M/m} \qquad (31.17)$$

$$n_{TOT} = mu + n_W \qquad \text{M/M/m} \qquad (31.18)$$

where u is the utilization of any individual processor. For the M/M/m case it turns out that[70]

$$n_W = \frac{u}{(1-u)^2} \frac{(mu)^m}{m!} p_0 \qquad \text{M/M/m} \qquad (31.19)$$

where

$$p_0 = 1 / \left[\sum_{k=0}^{m-1} \frac{(mu)^k}{k!} + \frac{1}{1-u} \frac{(mu)^m}{m!} \right] \qquad \text{M/M/m} \qquad (31.20)$$

(p_0 is the proportion of time there are no jobs in the facility). These expressions can be simplified for the M/M/2 case to[70]

$$p_0 = \frac{1-u}{1+u} \qquad \text{M/M/2} \qquad (31.21)$$

$$n_W = \frac{2u^3}{1-u^2} \qquad \text{M/M/2} \qquad (31.22)$$

Example of Pooling Facilities

The preceding results form a useful basis for comparing alternative designs with and without *pooling* of processors. Consider the problem of designing a system to process two types of jobs, using two robots. One possible design, which is likely to be the less expensive one, is to dedicate each robot to processing one type of job. The second design, which could be more expensive owing to duplicated end effectors, part orienters, robot capabilities, and so on, would be to let each robot be capable of processing either type of job. Suppose the time between arrivals, considering both types of jobs, has mean a^*, and that for each type of job alone has mean $2a^*$. Both types of jobs have the same mean service time $\bar{s}$. Then the first design has two identical M/M/1 queues, and letting $\bar{a} = 2a^*$ in Equation (31.3), it is seen that each queue has

$$u = \frac{\bar{s}}{2a^*} \qquad (31.23)$$

The second design has one M/M/2 queue, and setting $\bar{a} = a^*$ and $m = 2$ in Equation (31.15) gives exactly the same value for u as above. This value would also be obtained from a deterministic model for each design, and so that type of model does not distinguish between the designs. An experienced designer would recognize that the second design is preferable from a reliability point of view, since both types of jobs can still be processed if one robot fails. However, there is another important difference in the designs. In the first design each queue has*

$$n_W = \frac{u^2}{1-u} \qquad \text{M/M/1} \qquad (31.24)$$

---

* M/M/1 is a special case of M/G/1 where the general service time is chosen to be the exponential distribution, for which the mean equals the standard deviation, so putting $C_s = 1$ in Equation (31.6) gives this equation.

So the total number of jobs waiting at the two stations in design 1 is

$$n_W(1) = \frac{2u^2}{1-u} \tag{31.25}$$

In the second design Equation (31.13) shows that this value is

$$n_W(2) = \frac{2u^3}{1-u^2} \tag{31.26}$$

So the first design has

$$n_W(1) - n_W(2) = \frac{2u^2}{1-u^2} \tag{31.27}$$

more jobs waiting on average. For a utilization of 75%, these numbers are $n_W(1) = 4.5$, $n_W(2) = 1.9$, or an excess of 2.6 waiting jobs in the first design. Similarly, the total WIP (i.e., n_{TOT}) is 6.0 versus 3.4, and the turnaround time is $6.0a^*$ versus $3.4a^*$ time units. Clearly the queueing model points out a number of advantages of the more expensive design, in which the two robots have been *pooled* to form a facility that can process all jobs.

Many extensions of the foregoing models can be used to give insight into operation procedures used at a facility. Details can be found in Shanthikumar.[68]

Although the distributions assumed (such as M) for the arrival and service processes may not be exact models of the real system, performance predictions from these queueing models tend to be quite robust. More is said about this in the next section.

31.7.3. Queueing Network Models

In a multirobot system, there may be many types of workstations, and several of each type of workstation. The system may also be capable of handling a number of different types of jobs simultaneously. Each job enters the system, and visits various workstations, depending on its requirements. In this situation, a *queueing network model* may be useful. (To each job, the system appears as a network of queues through which it must traverse.) Such models are of three main types: open, closed, and mixed. In an *open* model each type of job arrives from an external source, independently, according to a prescribed arrival process. Jobs circulate around the system, visiting stations according to rules specific to each job type. The total number of jobs in the system varies with the arrival and service events. In a *closed* model there is a fixed number of each type of job in the system. This can be the case if there is a limited number of fixtures available for each job type. Only when a job of that type is completed and dismounted from its fixture can another job of that type be mounted on this fixture and put in the system. In a *mixed* model some job types have open arrivals while others have closed arrivals.

The study of queueing network models began in the operations management context, but these models were found extremely useful in the fields of computer systems and communication networks, so a number of significant developments were made in those contexts: reviews of these developments are in Koenigsberg[75] and Trivedi.[83] More recently, these models have been used for studying automated systems, particularly of the flexible manufacturing type,[71,74,76-79] and it is clear that they can be used for multirobot systems such as the one described. A tutorial introduction of these models from a manufacturing system designer's point of view can be found in Suri and Hildebrant.[81]

Two examples are given to illustrate the insight obtained by a queueing network model. In both it is assumed that jobs require expensive fixtures, and so the number of fixtures is a significant design parameter in terms of system cost.

Example of a Balanced System

In the first example all jobs use the same type of fixture, but there are variations in the processing requirements of each job. The system is designed and operated in a *balanced* manner, in which each processor (e.g., workstation or transporter) is assigned the same workload on average, resulting in equal utilizations of all processors. Now suppose there are 10 processors (including the stations where jobs are fixtured and defixtured). How many fixtures should be purchased? A superficial analysis might be to say there should be one fixture per processor (to keep each processor busy) plus 30% more "for good measure." However, it can be shown[80] that in a balanced network the utilization of each processor will be

$$u = \frac{N}{M+N+1} \tag{31.28}$$

where M is the number of processors and N is the number of fixtures. So $M = 10$ and $N = 13$ (chosen by the foregoing rule) give a utilization of only 59%. If a design specification of 75% utilization is to be met, then *27* fixtures are needed!

The preceding example highlights two points. The first is that the deterministic capacity analysis gives little assistance for the decision regarding fixtures. The second is that doubling the number of fixtures multiplies the utilization only by 1.25, illustrating the highly nonlinear nature of the queueing phenomena being modeled. It is instructive to review why such a large number of fixtures are required here. The technique is modeling a high variability in the processing demanded by different jobs. Although, on average, all processors are equally used, at a given time several jobs may be waiting for one processor while several other processors are lying idle. It is this *dynamic imbalance* in the usage of resources that is captured in the aggregate measures provided by the queueing model.

Example of Splitting Operations

The second example involves a system with three workstations and four types of jobs. Each job type uses a different fixture type. Two different designs are possible for the system (see Tables 31.2 and 31.3). In the first, job type A is processed entirely at one workstation, whereas in the second, its processing is split up. The totals in the last row of each table show that, from a deterministic point of view, both designs are the same. Calculations based on queueing network algorithms[81] show, however, that to achieve the desired production rates, the number of fixtures of each type differs for the two designs, see Table 31.4. In the second case the total number of fixtures is 20% smaller, which might be a significant saving. Again, this insight was not available without the queueing model.

31.7.4. Accuracy of Queueing Models

The theory behind some queueing models (Section 31.7.2) and queueing network models (Section 31.7.3) is based on several assumptions, which are often not satisfied by a practical robotic system. An example is the assumption of exponentially distributed processing times. Still, quite reasonable predictions of system performance can be obtained using these models. This has been explained to some extent by a new approach to queueing called *operational analysis*.[72,73] A rigorous study of the alternative assumptions of operational analysis has recently been done[80] which justifies the use of these models in practical systems. Another analysis shows that the queueing network models give reasonable predictions even when there are substantial errors in the original data.[82]

All the foregoing queueing models do not work well when there is significant *blocking* (defined in Section 31.6). This is illustrated in Section 31.10.4. Extensions and approximations of queueing

TABLE 31.2. PROCESSING TIMES FOR DESIGN 1

| Workstation Type | Processing Times in Minutes | | | | Total Time Used at Workstation[a] |
	Job Type A	Job Type B	Job Type C	Job Type D	
1	60	7	7	7	81
2	0	27	27	27	81
3	0	27	27	27	81
Total[b]	60	61	61	61	

[a] This is the total time used at a workstation for unit production of all job types.

[b] This is the total processing time for one workpiece of that job type.

TABLE 31.3. PROCESSING TIMES FOR DESIGN 2

| Workstation Type | Processing Times in Minutes | | | | Total Time Used at Workstation[a] |
	Job Type A	Job Type B	Job Type C	Job Type D	
1	20	7	27	27	81
2	20	27	7	27	81
3	20	27	27	7	81
Total[b]	60	61	61	61	

[a] This is the total time used at a workstation for unit production of all job types.

[b] This is the total processing time for one workpiece of that job type.

TABLE 31.4. COMPARISON OF NUMBER OF FIXTURES REQUIRED FOR TWO DESIGNS

Job Type	Number of Fixtures Required[a]	
	Design 1	Design 2
A	1	2
B	3	2
C	3	2
D	3	2
Total	10	8

[a] This is to ensure a production of 10 pieces per day of each job type. (One day is assumed to have two 8-hr shifts.)

models have been proposed to deal with blocking due to shortage of buffer space,[84-89] or sharing of end effectors.[87] Alternatively, a detailed simulation may be required to study such problems, as described next.

31.7.5. Simulation

Before finalizing a design it is highly advisable to check it and fine-tune it using a detailed simulation, particularly if several workstations are involved. Since techniques used are similar to those used during system operation, the reader should see Sections 31.10.2 and 31.10.4.

The question of when to use a queueing model and when to use simulation is an important one. Simulation should be employed whenever detailed operating rules must be studied, whereas queueing models are useful for studying a wider range of designs. (See the discussion in Section 31.1 on the trade-offs between these models.) Further discussion of this issue can also be found in Suri and Hildebrant.[81]

31.7.6. Reliability Theory

An issue, critical from a system manager's viewpoint, is how often (and for how long) an automated system will be unavailable due to component failures. The quantitative study of this issue belongs in the realm of reliability theory.[90,91,95,97]

Single-Component Systems

Consider first the behavior of a single-component system which may fail and, if failed, may be repaired, with

$\bar{f}$ = mean time to failure (MTTF).
$\bar{r}$ = mean time to repair (MTTR).
s_f = standard deviation of time to failure.
s_r = standard deviation of time to repair.

The statistics of the failure and repair processes are given by their cumulative distribution functions (CDFs) $F(t)$ and $R(t)$ as follows:

$F(t)$ = probability that a failure occurs by time t since the last repair.
$R(t)$ = probability that a repair is completed by time t since the last failure.

It is assumed that *each* failure [or repair] is an independent random process, with the same CDF $F(t)$ [or $R(t)$].

A basic result for such a system concerns the amount of time T_a that it is available (i.e., not failed). For a total observation time T, where T is very large compared with $\bar{f}$ or $\bar{r}$, T_a can be considered to be normally distributed with mean $\bar{T}_a$ and standard deviation s_a where[94]

$$\bar{T}_a = \frac{\bar{f}}{\bar{f} + \bar{r}} T \tag{31.29}$$

$$s_a^2 = \frac{(\bar{f} s_r)^2 + (\bar{r} s_f)^2}{(\bar{f} + \bar{r})^3} T \tag{31.30}$$

Note that this result holds for general distributions $F(t)$ and $R(t)$. The long-term proportion of time the system is available is thus seen to be $\bar{T}_a/T$ or $\bar{f}/(\bar{f} + \bar{r})$.

Multicomponent Systems

The next step in reliability analysis is to consider a system consisting of n components, with R_i being the reliability of the ith component. (Here *reliability* is defined as the proportion of time a component/system is functioning.) It is assumed that the functioning of components is independent of each other. A *series*-structured system is one in which the entire system fails if any component fails. Its reliability R_s is then[97]

$$R_s = \prod_{i=1}^{n} R_i \tag{31.31}$$

A *parallel*-structured system is one that fails only when all components fail, and its reliability R_p is[97]

$$R_p = 1 - \prod_{i=1}^{n} (1 - R_i) \tag{31.32}$$

The effect of structure on reliability can be quite dramatic, as illustrated by a system consisting of five components, each with reliability of 90%. If all five are required to perform a task (high interdependence), system reliability comes down to 59% (see R_s above), whereas if only one of the five is required for task performance (high redundancy), reliability goes up to 99.999%!

The preceding two basic formulas can easily be used to compute the reliability of a more complex system, as long as it is composed of a hierarchy of only series and parallel substructures, by beginning at the component level and using the appropriate formula to combine the reliability of each level of the hierarchy.[97]

Another commonly used system structure requires any m (or more) out of n components to function for correct operation of the system. In this case if all components have reliability R, the system reliability $R_{m\,|\,n}$ is given by[97]

$$R_{m\,|\,n} = \sum_{k=m}^{n} \frac{n!}{k!(n-k)!} R^k (1-R)^{n-k} \tag{31.33}$$

As an application of this equation, consider a safety device consisting of three components each of reliability R, which functions correctly as long as two or more components are working. (This is called triple modular redundancy[97] or TMR.) The reliability R_{TMR} of the safety device is then

$$R_{TMR} = \frac{3!}{2!1!} R^2 (1-R) + \frac{3!}{3!0!} R^3$$
$$= R^2 (3 - 2R) \tag{31.34}$$

With component reliability of 90%, this device achieves a reliability of 97.2%. Instead, if it consisted of just two components, both of which were required to function correctly, its reliability would be only 81%. The value of the third component in a TMR system is clear.

Machine Repairman Model

This model is useful for studying the amount of resources to be devoted to maintenance personnel. Let there be M machines in a system, and for each machine let $\bar{f}$ and $\bar{r}$ be as before. (The value for $\bar{r}$ assumes all the maintenance resources are concentrated on repairing one machine at a time.) It is usual to assume exponentially distributed failure and repair times. Then the proportion of time k machines are failed p_k is given by[97]

$$p_k = \left(\frac{\bar{r}}{\bar{f}}\right)^k \frac{M!}{(M-k)!} p_0 \tag{31.35}$$

with

$$p_0 = \left[\sum_{k=0}^{M} \left(\frac{\bar{r}}{\bar{f}}\right)^k \frac{M!}{(M-k)!}\right]^{-1} \tag{31.36}$$

For a system with four machines, each with $\bar{f} = 10$ hr, and a single repairman, with $\bar{r} = 1$ hr, the proportion of time that all machines are working p_0 is calculated to be 64.7%.

These examples cover only the simpler types of reliability analyses. Generalizations involve systems that cannot be decomposed into series or parallel subsystems, repair models that involve fault detection followed by fault correction, inclusion of irrecoverable failures, and more general probability distributions. Details can be found in Stiffler et. al.[96] and Trivedi.[97] In most of these cases the analysis becomes quite tedious, and computer-based reliability analysis packages such as CARE[92,96] and ARIES[93] have been developed to assist in this task.

C. SYSTEM OPERATION PHASE

Following the approval of a detailed configuration design, a robotic system will be purchased, perhaps by putting out a request for proposal (RFP) and receiving competitive bids, or by "shopping around" for components. The system will then be installed, typically going through a testing and debugging period during the installation. It is not the intention of this chapter to cover either the RFP/purchase activities or the installation/testing activities. This section considers the next phase of activity, when the system is declared operational. The aim now is to ensure that the system performance lives up to the specifications that were laid down. Many points need attention during day-to-day operation of the system: details that were neglected at the design phase, such as how to cope with short-term changes in material supply, production targets, or workstation availability. These are the subject of this section.

31.8. REQUIREMENTS AND ISSUES

The main requirements at this stage can be broadly stated as, first, to meet the production targets, and, second, to operate the system efficiently and effectively. There are many issues to be addressed to meet these requirements:

1. **Batch Sizes.** Need for operation in batches or lots may be dictated by system capacity, setup times, material availability, part magazines, or end-effector availability. If batch operation is necessary, then the size, constitution, and timing of each batch must be determined.

2. **Scheduling and Sequencing.** Within each batch, tasks must be scheduled for different workstations. At each workstation the detailed sequence of operations must be decided for each task.

3. **Dynamic Work Allocation.** During system operation minor disturbances (or errors in operation time estimates) may cause imbalances in the work allocated to each station. There must be ways to correct this dynamically, that is, while the system is operating.

4. **Reacting to Disruptions.** In contrast to the minor disturbances just mentioned, significant disruptions in system operation can be caused by equipment failures, nonavailability of material, or sudden changes in production requirements. Strategies for dealing with these situations must be available.

5. **Maintenance.** Scheduling of preventive maintenance, as well as scheduling of emergency maintenance, must be considered.

To decide between alternative strategies for addressing these issues, the impact of each strategy on several performance indicators should be considered. These indicators are discussed next.

31.9. PERFORMANCE MEASURES

The measures used to evaluate system performance at this stage are essentially the same as those used in the detailed configuration design phase. This is because that phase should have considered most of these indicators before settling on a final design. The difference is that while the final design phase would have used a simulation, or some elaborate model, to approximate the system behavior, in this phase the actual system is being used. This necessarily forces a greater level of detail on the issues to be considered and the models to be used. Nevertheless, the performance indicators remain the same as in Section 31.6. The reader therefore may wish to review Section 31.6 before proceeding.

31.10. QUANTITATIVE TECHNIQUES FOR SYSTEM OPERATION

In the planning and control of day-to-day system operation, many quantitative methods can be used as aids to decision making. Running a production or assembly facility smoothly is not a new problem introduced by robotic technology—most organizations would be familiar with typical problems encountered. Robotic systems have certain features, however, that distinguish them from conventional facilities.

This section concentrates on methods that are particular to operation of robotic systems: techniques that are already widely used in production and operations management are mentioned only briefly and references to well-known publications are given for them.

31.10.1. Batching, Scheduling, and Sequencing

Batching

The need for batches (or lots) arises when a system cannot easily perform a variety of tasks in random order. Typically, a particular system configuration is set up, and tasks belonging to a certain family can be performed. A new setup is necessary before tasks of a different family can be performed. Changing the setup may be relatively expensive and time-consuming, so it is preferable to operate in a mode where a batch of tasks of one family are performed, followed by another batch from another family, and so on.

Determining the mix of tasks in a batch and their number (also known as lot size) is an age-old problem in manufacturing, and many good references are available;[98-109] these approaches lie mainly in the deterministic/dynamic category. In the context of automated systems, this problem has some special characteristics. First, there may be some unusual constraints arising from capacities of part magazines, or from different tasks sharing a limited set of end effectors or tools. Second, the flexibility of the system makes the setup times relatively small, and so batch sizes (and WIP) can be quite small. Often this means that the batch mix and size will be decided in real time, from the tasks currently waiting to be done. Both these characteristics make it difficult for the mathematical programming methods conventionally used to be applied for the robotic system problem. Recently some new algorithms have been proposed which appear to be promising and efficient.[110] The algorithms use a heuristic approach which enables them to incorporate the difficult constraints. They also use a technique known as a *rolling batch* which allows future batch composition to be modified to account for information on system performance, material availability, due dates of jobs, and so forth. These methods therefore extend into the probabilistic/dynamic category.

Scheduling Algorithms and Rules

The problem of scheduling tasks to machines within a batch, or of scheduling batches of tasks onto the system, can be a difficult one, especially for a complex system with many workstations of different types. Again, scheduling is a well-known problem in manufacturing for which many theoretical results exist.[111,114,115,118,121,124] Most of these results use a deterministic/dynamic model.[120] Useful reviews can be found in Gelders and Van Wassenhove,[119] Graves,[120] and Schrage.[123]

In an automated shop floor environment it is necessary to react quickly to disturbances and changes as mentioned before, so probabilistic/dynamic models would give a better analysis of the situation. For simple systems, theoretical analysis of such models is possible using Markov decision processes,[127,129] and scheduling problems have been analyzed for some manufacturing systems.[128,130] However, practical automated systems often have several workstations and part types and do not obey the Markovian assumptions. For such models rigorous solutions are hard to obtain, so instead a large body of heuristic *scheduling rules* have been developed. An example of a simple scheduling rule is the shortest processing time (SPT) rule. This says that if a number of jobs are waiting for a workstation, then the job with the smallest processing time should be scheduled first.

The performance of a particular scheduling rule, compared to that of an alternative rule, is dependent on the specific system characteristics and the measures of performance being used. Ideally then, alternative scheduling rules should be tested on a detailed simulation of the system (see Section 31.10.2). If this is not feasible, the system operator can refer to general guidelines obtained by researchers who have performed many experiments with these rules.[112,116,117,122,125,126] These references also provide a useful source for the different rules and performance indicators used to study them.

Operation Sequences

At each workstation, the detailed sequence of operations to be performed for a particular task can also be optimized to minimize the time for the task. In contrast to the preceding scheduling problem, in this case it makes sense to study the sequencing in a deterministic/dynamic framework. This is because once the sequence has been decided, that same sequence of operations will always be used to perform that task. The element of uncertainty can thus be removed from the model. Although well-known quantitative methods exist for deterministic sequencing problems, this aspect belongs in the realm of individual workstation design, which is covered in Chapter 30.

31.10.2. Computer Simulation

The most accurate model of robotic system performance is obtained by using a computer-based *discrete event simulation*,[136,137,139,146] Such a simulation views the system operation as a succession of events, and, in principle, it can mimic system behavior in as much detail as is desired. Characteristics of all the hardware components (workstations, buffers, transporters) and software components (control programs, scheduling algorithms) can be incorporated in the model. System dynamics and interactions, as well as failures and other uncertainties, can all be modeled, so this technique belongs in the probabilistic/dynamic category.

There are three main approaches to simulating a given robotic system:

1. **"Canned" Package Approach.** This involves adapting a manufacturing system simulation package[127,140,142,144,147] to fit the system—often the easiest alternative and quickest to implement. However, it may not be able to model specific characteristics of the given system or allow much freedom to explore alternative control strategies.

2. **Simulation Language Approach.** Here a discrete event simulation language[132,133,138,141,145] is used to write a simulation of the given system. The language makes it simple to model such systems, and characteristics can be modeled in as much detail as desired. However, a simulation language must first be purchased and learned. If portability between different computer systems is important for the organization, this must be ascertained for the language. Finally if other language subroutines are to be used (e.g., a FORTRAN linear programming package for decision making in the simulated system), then the simulation language must have interface capabilities for other languages. Still, modern simulation packages are implemented on most computer systems and have good interface capabilities, so this is often the best overall approach.

3. **Programming Language Approach.** In this alternative a common programming language, already in use at the organization (such as FORTRAN or PASCAL) is used to write a simulation of the system. The main advantages are no initial cost, wide portability, and ease of interfacing. The disadvantage is that all the basic components of a simulation (e.g., scheduling of events, statistics collection) must be written from scratch.

Further discussion of these trade-offs, along with descriptions of available packages, can be found in Bevans.[131]

Parameter Optimization Using Simulation

Since a simulation is capable of modeling as much detail of the real system as desired, it can be used (in principle) to study any or all of the issues in Section 31.8, and any or all of the performance indicators in Section 31.9 (for example, see Nof, Halevi, and Bobasch).[143] To optimize some performance measure with respect to a number of decision parameters, a *baseline* simulation is performed at a chosen set of parameter values, then each parameter is changed a small amount, one at a time, and a new simulation is performed after each change. This gives an estimate of the gradient vector (sensitivity) of the measure with respect to each parameter. This gradient can then be used along with any mathematical programming procedure[27-30,32,34,35] to get a new set of parameters, and in this manner to iteratively optimize that measure. The disadvantage is that, for N parameters, each iteration step requires N new simulations. A simulation of an automated facility can take (typically) 30 min on a PDP 11/34 computer.[135] This method can therefore be computationally very demanding.

Recently, an efficient alternative to this "brute-force" simulation approach has been developed. The alternative retains the precision of a detailed simulation model, while incorporating some efficiency by using analysis. The efficiency obtained is $N : 1$ where N is the number of parameters as before. Even for 20 parameters, then, the savings can be considerable. The approach is described under "fine-tuning" (Section 31.10.4).

The same system that required 30 min to simulate (before), can be analyzed in under 10 sec on a PDP-11 using a simpler queueing model.[81] The choice of when to use a detailed simulation and when to use an aggregated queueing model is therefore important, see Sections 31.1 and 31.7.5.

31.10.3. Hierarchical Approaches

Since a robotic system is usually part of a larger manufacturing environment, the inputs and outputs of material to the system must match the overall plant material requirement plan and master production plan.[148,153,154] These plans specify various availability dates for raw material and due dates for completed pieces, as well as quantities to be produced. At the same time, while trying to meet the overall plans, the system manager must satisfy many other constraints, such as limited numbers of part magazines and end effectors, workstation time, and amount of work in process.

The task of meeting all the production requirements, while using the robotic system resources efficiently, is clearly complex. To make its solution tractable, it is usually divided into a number of stages or levels[157] that correspond to a hierarchy of decisions. Several alternative approaches have been proposed for partitioning this problem in the context of automated manufacturing. For example, in Hildebrant and Suri[151] the top level makes trade-offs between alternative resource allocations, in the face of long-term uncertainty, by incorporating the uncertainty in an aggregate way, and then solving a deterministic nonlinear optimization problem. The middle level explicitly models medium-term uncertainties using a queueing model, and the lowest level uses scheduling rules for the actual dispatching of jobs. In an alternative approach, Kimemia[152] explicitly models the long-term uncertainty in the top level, but then must resort to an approximate-solution procedure since the model becomes intractable for exact solution. The lower levels use flow models and scheduling rules. Other approaches can be found in Graves,[149] Hax and Meal,[150] Stecke,[155] and Suri.[156]

It can be seen that the hierarchical approaches combine static and dynamic models, as well as deterministic and probabilistic models, using different models at different levels of the formulation. Further discussion on the design and implementation of decision aids, in an appropriate hierarchical structure, is given in Section D of this chapter.

31.10.4. System Monitoring and Fine-Tuning

It is important to have a means for monitoring the performance of the robotic system relative to management goals and for ascertaining the economic (and other) returns from the automated operations. This is best done by a mechanism that collects real-time data during system operation. Such monitoring systems are relatively inexpensive given current microcomputer technology.[158] Analyzing and summarizing the detailed data can then be done according to management needs, using a standard management information system (MIS).

Although conventional MIS packages produce useful statistics of system performance (such as equipment utilization, downtimes, production rates), if a problem exists (e.g., low utilization of a workstation) these statistics do not necessarily give insight as to how the problem can be rectified. Recently a new analytic technique has been developed that derives *much more information and insight* from the real-time data. The technique, called perturbation analysis (P/A), is equally applicable to simulation output, and has been successfully applied for design and real-time control to automated systems.[159]

P/A operates by using recently developed analytical methods[159,161] to implement calculations on data as it is obtained from a monitoring system (or from a simulation)—the amount of additional computer time/memory used for this is negligible compared with the needs of the monitoring system or simulation itself. As a result of these calculations, P/A applies the sensitivity of system performance measures to all the decision parameters. That is, it answers questions such as, What would have been the production of all part types today if there were one more fixture for part type A?, and it does so for all decision parameters of interest. Thus effectively, it computes gradient vectors of performance measures with respect to decision variables during regular monitoring or simulation. For simulation studies of N parameters, this provides an *N-fold increase in efficiency* over the "brute-force" method mentioned in Section 31.10.2, while for a monitoring system, since it operates directly on actual data, it is free from the restrictions and assumptions of most analytical models. It should therefore be of interest to designers and managers of automated systems and greatly enhance the information obtained from monitoring systems (see Figure 31.4).

The P/A approach has been successfully applied to optimization of serial production lines[160] and flexible manufacturing systems.[162] As an illustration of its use, consider the design of a three-station system where blocking can occur since each station has an input buffer with a limit of five workpieces. The aim is to find the number of fixtures that maximizes the production rate of the system. Figure 31.5 displays three graphs: the first, obtained using the simple queueing-network model described in Section 31.7.4; the second, using detailed simulation; and the third using the marked customer method,[163]

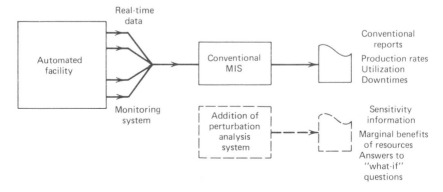

Fig. 31.4. Enhancement of conventional MIS with perturbation analysis system.

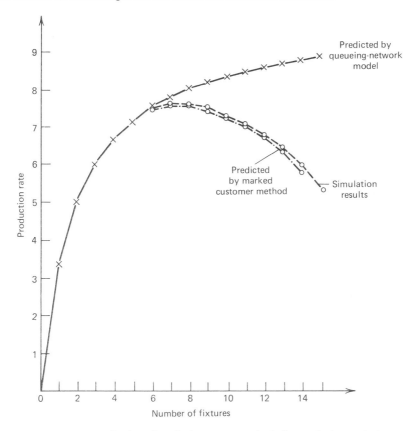

Fig. 31.5. Application of marked customer method of perturbation analysis.

which is a technique based on P/A. The queueing model fails to predict the blocking phenomenon (as explained in Section 31.7.4), but the marked customer method predictions are in close agreement with the simulation. This method becomes computationally very attractive for a facility that is processing many different parts.

31.10.5. Maintenance Strategies

Broadly speaking, in automated systems the issue of maintenance strategies can be considered for two types of situations. The first concerns preventive maintenance and addresses a single piece of equipment. It involves trading off the cost of a scheduled interruption for preventive maintenance with the cost of disruption due to an unexpected breakdown and its repair. The second situation concerns allocation of maintenance persons and resources when several maintenance tasks need to be performed. Analyses dealing with each type of situation can be found in the references.[164-170]

31.10.6. System Modification or Expansion

Even after a robotic system is operational, management will continue to make decisions that have far-reaching consequences for the system. Examples of such decisions are the following:

Parts-mix changes, for example, allocating a new part type (or part types) for production on the system.
System modification/expansion, for example, adding a new workstation or changing the layout.

These decisions involve complex trade-offs between economic investments and resulting changes in system performance. The trade-offs are of the same nature as those studied during system planning or configuration design (Sections A and B). The main difference is that most of the system parameters have been decided, and only a limited set of modifications is under study. Keeping this in mind, all the techniques in Sections A and B could be used for studying these decisions.

D. HARDWARE AND SOFTWARE STRUCTURES FOR DECISION SUPPORT

It has been seen how quantitative techniques can aid in the design and operation of robotic systems. Owing to the complexity of the task of operating such systems efficiently, the capacity of these advanced systems is often underutilized after their installation. Indeed, the complexity of operating a multistation automated facility should not be underestimated; even experienced shop floor supervisors find that running such a system can be very difficult.[175]

This section underscores the point that a decision support system (DSS) can be designed to enable an organization to achieve maximum benefit from a robotic system.[171-174] The structure of this DSS is presented in terms of the organizational activities involved in running the robotic facility, and it is shown how this structure should be implemented using appropriate hardware and software components. The concepts here are summarized from Suri and Whitney,[175] and for details the reader should see the original reference.

To structure the DSS, it is useful to place the robotic facility in the overall context of the organization, since successful functioning of the facility will require ongoing activities at all levels of the organization. The various activities required are best understood in terms of the classical three-level view of organizational operation.

The first level consists of long-term decision making, typically done by higher management. This involves establishing policies, production goals, economic goals, and making decisions that have long-term effects. The second level involves medium-term decisions, such as getting the production targets for each part for the next month. These decisions are typically made by the robotic system line manager, aided by decision-support software. The third level involves short-term decisions, such as which workpiece should be introduced next into the system. Under normal circumstances, these decisions are made by the robotic system control computer(s). However, when an exception occurs, such as a workstation failure, the line supervisor may decide to take over some of this decision making, again aided by the decision-support software.

A summary of the three levels of decision making and associated software, hardware, and management tasks is given in Table 31.5 and they are described next. The aim here is to give an understanding of the issues involved in operating a robotic facility, and the typical software decision aids that should be available to the facility managers/supervisors. The detailed architecture of the software and hardware components are not discussed here; examples of suitable hardware/software components can be found in references cited throughout this chapter.

It is assumed that the robotic facility is part of a larger manufacturing environment, so that the following functions are already being performed at the corporate (or plantwide) level.

Plantwide material requirement planning (MRP).
Plantwide production plan.
Plantwide data base management and information system.

These plantwide functions typically will set overall targets and production goals for a long time horizon. This information usually will reside in a mainframe corporate/plant computer and will serve as inputs of the three levels of operations described in these sections.

31.11. FIRST-LEVEL OPERATIONS

These should encompass the following operational areas:

1. Strategic decision making for the robotic facility.
2. Evaluating performance of the facility.
3. Ancillary support for facility operation.

The execution of activities at this level typically will be supported by software on a mainframe computer. In some organizations, a feasible alternative is to have a separate medium-sized computer for these activities, which can be considered a *DSS computer.*

Software components to assist in strategic decision making and performance evaluation can be

TABLE 31.5. LEVELS OF DECISION MAKING FOR ROBOTIC FACILITY OPERATION

Time Horizon	Management Level	Typical Tasks	Typical Decision Support Software Used	Hardware Used
Long term (months/years)	Upper management	Part-mix changes System modification/ expansion	Part selection program Queueing models Simulation	Mainframe computer or DSS computer
Medium term (days/weeks)	Robotic facility manager	Divide production into batches Maximize machine utilization Respond to disturbances in production plan/ material availability	Batching and balancing programs Simulation	DSS computer or facility's computer
Short term (minutes/hours)	Robotic facility line supervisor (exceptions only)	Work order scheduling and dispatching Tool management React to system failures	Work order dispatching program Operation and tool reallocation program Simulation	Facility's computer

based on the quantitative techniques described in this chapter. Ancillary support includes such items as extended part-programming and program-verification tools.

31.12. SECOND-LEVEL OPERATIONS

This level encompasses decisions typically made by the robotic facility manager over a time horizon of several days or weeks. The main tasks to be performed at this level are:

1. Dividing overall production targets into batches of parts.
2. Within each batch, assigning system resources in a manner that maximizes resource utilization.
3. Responding to changes in upper-level production plans or material availability.

The issues involved in each of these tasks have been described in previous sections. Also mentioned were quantitative techniques on which software tools could be based, which would aid the manager or line supervisor in decision making. These software decision aids typically reside on the facility's control computer, or if this is not feasible, then on a DSS computer (as defined in the previous section.)

31.13. THIRD-LEVEL OPERATIONS

This level is concerned with the detailed decision making required for real-time operation of the facility including the material-handling system (MHS). The time horizon here is typically a few minutes or hours, and the decisions involved are as follows:

1. Word order scheduling and dispatching: which workpiece to introduce next into the facility, and when.
2. Movement of workpieces and MHS: which workstations to send this workpiece to next, which transporter to send to pick up this workpiece, and so on.
3. Tool management.
4. System monitoring and diagnostics.
5. Reacting to disruptions such as failure of one or more system components, or a sudden change in production requirements.

During normal system operation, most of these decisions are made by software in the facility's control computer. However, when an exception occurs, such as failure of a workstation, the line supervisor will usually take charge of the decision making. If it is going to take a long time to repair, he may, for example, decide to reallocate its production to other workstations. This involves a complex sequence of trade-offs between production rates and workstation capabilities, magazine capacities, end-effector availability, and so on. Again, the supervisor's task can be simplified considerably by employing various software decision aids. These would be based on the techniques described in Section C of this chapter. The aids should typically reside on the facility's control computer to enable rapid implementation of the changed decisions, but in some systems the architecture could involve use of a separate DSS computer, as described.

31.14. INTEGRATION OF OPERATIONAL LEVELS

The preceding sections described the various levels of decision making relevant to successful and efficient facility operation. Figure 31.6 summarizes the decisions involved at each level (only the major decisions are shown, for clarity).

Of importance equal to the decision making *within* each level, is the question of communication *between* the levels. The organization should be sure of the answers to these questions before becoming "locked in" a particular system architecture.

1. How will data (such as control programs) be moved from the mainframe computer to the facility's computer?
2. How will information (such as system performance) be communicated from the facility's computer to the mainframe computer?
3. Will a separate DSS computer be used, and if so, how will it communicate with the preceding two computers?

Integration of the operational levels is also an important ability to be incorporated within the DSS software to be used with the facility. For example, it should be possible to test any decision made at a higher level (e.g., workstation selection) by trying out all the lower levels (such as batching and balancing, and detailed simulation) and thus evaluating that decision in detail. In this respect, it

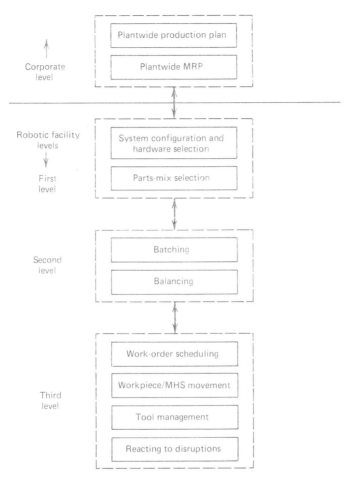

Fig. 31.6. Integration of operational levels.

should be noted that all the decision aids for lower-level decisions are part of the decision aids for a higher level. Thus, for example, simulation should also be thought of as a decision aid for the batching and balancing problem, even though this was not mentioned explicitly in the section on batching and balancing.

In summary, appropriate software and hardware components can be integrated into the organizational hierarchy to help achieve the objective of maximizing the benefit from a robotic facility. The resulting decision support system should be thought of as an integral part of the facility, and the design, purchase, and operation of a robotic facility should explicitly include provision for all the components of such a DSS. If a robotic facility is not supplied with an adequate DSS, the creation of one should receive top priority. Without this support system, the organization may find itself with a highly sophisticated manufacturing system, but without the ability to use the sophistication effectively.[175]

REFERENCES

General Reviews and Surveys

1. Fisher, E. L., Nof, S. Y., and Seidmann, A., Robot System Analysis: Basic Concepts and Survey of Methods, *Proceedings of the IIE Conferences,* Cincinnati, Ohio, November 1982.

2. Fleischer, G. A., A Generalized Methodology for Assessing the Economic Consequences of Acquiring Robots for Repetitive Operations, *Proceedings of the IIE Spring Annual Conference,* New Orleans, Louisiana, May 1982, pp. 130–139.

3. Nof, S. Y., Decision Aids for Planning Industrial Robot Operations, *Proceedings of the IIE Conference,* New Orleans, Louisiana, May 1982.

Economic Aspects

4. Benedetti, M., The Economics of Robots in Industrial Applications, *The Industrial Robot,* September 1977, pp. 109–118.
5. Boothroyd, G., Economics of Assembly Systems, *Journal of Manufacturing Systems,* Vol. 1, No. 1, 1982, pp. 111–127.
6. Canada, J. R., *Intermediate Economic Analysis for Management and Engineering,* Prentice-Hall, Englewood Cliffs, New Jersey, 1971.
7. Ciborra, C. and Romano, P., Economic Evaluation of Industrial Robots, *Proceedings of the 8th ISIR,* Stuttgart, June 1978, pp. 15–23.
8. Fleischer, G. A., A Generalized Methodology for Assessing the Economic Consequences of Acquiring Robots for Repetitive Operations, *Proceedings of the IIE Spring Annual Conference,* New Orleans, Louisiana, May 1982, pp. 130–139.
9. Heginbotham, W. B., Can Robots Beat Inflation? *SME Paper,* No. MS77-756, 1977.
10. Owen, A. E., Economic Criteria for Robot Justification, *The Industrial Robot,* September 1980, pp. 176–177.
11. Whitney, D. E. et al., Design and Control of Adaptable-Programmable Assembly Systems, Final Report, C. S. Draper Laboratories, Cambridge, Massachusetts, December 1980.

Group Technology

12. Burbidge, J. L., *The Introduction of Group Technology,* Wiley, New York, 1975.
13. Burbidge, J. L., *Group Technology in the Engineering Industry,* Mechanical Engineering Publications Ltd., London, 1979.
14. Gallagher, C. D. and Knight, W. A., *Group Technology,* Butterworth and Co., London, 1973.

Strategic Planning

15. Freidenfelds, J., *Capacity Expansion,* North Holland, 1982.
16. Fundenberg, D. and Tirole, J., *Capital as a Commitment: Strategy Investment in Continuous Time,* MIT, Cambridge, Massachusetts, May 1981.
17. Porter, M. E., *Competitive Strategy: Techniques for Analyzing Industries and Competitors,* Free Press, 1980.
18. Spence, A. M., Entry, Capacity, Investment and Oligopoly Pricing, *The Bell Journal of Economics,* 8, Autumn 1977, pp. 534–544.

Forecasting

19. Bunn, D. W., A Comparison of Several Adaptive Forecasting Procedures, *Omega,* Vol. 8, 1980, pp. 485–491.
20. Hollier, R. H., Khir, M., and Storey, R. R., A Comparison of Short-Term Adaptive Forecasting Methods, *Omega,* Vol. 9, 1981, pp. 96–98.
21. Makridakis, S. and Wheelwright, S. C., Eds., *Forecasting,* North Holland, 1979.
22. Martino, J. P., *Technological Forecasting for Decision Making,* North Holland, 1982.

Mathematical Programming and Decision Analysis

23. Behn, R. D. and Vaupel, J. W., *Quick Analysis for Busy Decision Makers,* Basic Books, New York, 1982.
24. Brown, R. W., Northup, W. D., and Shapiro, J. F., LOGS: An Optimization System for Logistics Planning, MIT Operations Research Center Working Paper No. OR107-81, May 1982.
25. Dantzig, G. B., *Linear Programming and Extensions,* Princeton University Press, 1963.
26. Graves, S. C. and Lamar, B. W., A Mathematical Programming Procedure for Manufacturing System Design and Evaluation, *Proceedings of the IEEE International Conference on Circuits and Computers,* 1980.
27. Hillier, F. H. and Lieberman, G. J., *Operations Research,* Holden-Day, San Francisco, California, 1974.

28. Lasdon, L. S., *Optimization Theory for Large Systems*, Macmillan, New York, 1970.

29. Luenberger, D. G., *Introduction to Linear and Nonlinear Programming*, Addison-Wesley, Reading, Massachusetts, 1965.

30. Moder, J. J. and Elmaghraby, S. E., Eds., *Handbook of Operations Research*, Van Nostrand, New York, 1978.

31. Muller-Merbach, H., Heuristics and Their Design: A Survey, *European Journal of Operational Research*, Vol. 8, 1981, pp. 1–23.

32. Pfaffenberger, R. C. and Walker, D. A., *Mathematical Programming for Economics and Business*, Iowa State University Press, Ames, Iowa, 1976.

33. Raiffa, H., *Decision Analysis*, Addison-Wesley, Reading, Massachusetts, 1970.

34. Shapiro, J. F., *Mathematical Programming*, Wiley, New York, 1979.

35. Suri, R., *Resource Management Concepts for Large Systems*, Pergamon Press, New York, 1981.

36. Whitney, C. K. and Suri, R., FMS Part Selection, TIMS/ORSA, Chicago, Illinois, April 1983.

37. Whitney, D. E. et al., Design and Control of Adaptable-Programmable Assembly Systems, Final Report, C. S. Draper Laboratories, Cambridge, Massachusetts, December 1980.

Location Analysis

38. Akine, U. and Khumawala, B. M., An Efficient Branch and Bound Algorithm for the Capacitated Warehouse Location Problem, *Management Science*, Vol. 23, 1977, pp. 585–594.

39. *Case Studies Featuring Practical Applications of Automated Storage/Retrieval Systems*, The Material Handling Institute, Pittsburgh, Pennsylvania, 1980.

40. *Considerations for Planning and Implementing an Automated Storage/Retrieval System*, The Material Handling Institute, Pittsburgh, Pennsylvania, 1977.

41. Cornuejols, G., Fisher, M., and Nemhauser, G., Location of Bank Accounts to Optimize Float: An Analytic Study of Exact and Approximate Algorithms, *Management Science*, Vol. 23, 1977, pp. 789–810.

42. Elshafei, A. W., Hospital Layout as a Quadratic Assignment Problem, *Operational Research Quarterly*, Vol. 28, 1977, pp. 167–179.

43. Erlenkotter, D., A Dual Based Procedure for Uncapacitated Facility Location, *Operations Research*, Vol. 26, No. 6, November–December 1978, pp. 992–1009.

44. Francis, R. L. and White, J. A., *Facility Layout and Location: An Analytical Approach*, Prentice-Hall, Englewood Cliffs, New Jersey, 1974.

45. Francis, R. L., McGinnis, L. F., and White, J. A., Location Analysis, *European Journal of Operational Research*, Vol. 12, 1983, pp. 220–252.

46. Geoffrion, A. M., The Purpose of Mathematical Programming is Insight, Not Numbers, *Interfaces*, Vol. 7, 1976, pp. 81–92.

47. Handler, G. Y. and Mirchandani, P. B., *Location on Networks*, M.I.T. Press, Cambridge, Massachusetts, 1979.

48. Hearn, D. W. and Vijay, J., Efficient Algorithms for the (Weighted) Minimum Circle Problem, *Operations Research*, Vol. 30, 1982, pp. 777–795.

49. Khumawala, B. M., An Efficient Branch and Bound Algorithm for the Warehouse Location Problem, *Management Science*, Vol. 18, No. 12, August 1972, pp. B718–B731.

50. Krarup, J. and Pruzan, P., Selected Families of Location Problems, *Annal of Discrete Mathematics*, Vol. 5, 1979, pp. 327–387.

51. Matson, J. O. and White, J. A., Operational Research and Material Handling, *European Journal of Operational Research*, Vol. 11, 1982, pp. 309–318.

52. Minieka, E., *Optimization Algorithms for Networks and Graphs*, Marcel Dekker, New York, 1978.

53. Muth, E. J. and White, J. A., Conveyor Theory: A Survey, *AIIE Transactions*, Vol. 11, No. 4, 1979, pp. 270–277.

54. Nauss, R. M., An Improved Algorithm for the Capacitated Facility Location Problem, *Journal of the Operational Research Society*, Vol. 29, 1978, pp. 1145–1201.

55. Ross, G. T. and Soland, R. M., Modeling Facility Location Problems as Generalized Assignment Problems, *Management Science*, Vol. 24, 1978, pp. 345–357.

56. Schwarz, L. B., Graves, S. C., and Hausman, W. H., Scheduling Policies for Automatic Warehousing Systems: Simulation Results, *AIIE Transactions*, Vol. 10, 1978, pp. 260–270.

57. Tansel, B. C., Francis, R. L., and Lowe, T. J., Location on Networks: A Survey (Parts I and II), *Management Science*, Vol. 29, No. 4, April 1983, pp. 482–511.

Basic Probability

58. Clarke, A. B. and Disney, R. L., *Probability and Random Processes for Engineers and Scientists,* Wiley, New York, 1970.

59. Meyer, P. L., *Introductory Probability and Statistical Applications,* Addison-Wesley, Reading, Massachusetts, 1972.

60. Papoulis, A., *Probability, Random Variables, and Stochastic Processes,* McGraw-Hill, New York, 1965.

Queueing Theory

61. Bhat, U. N., *Elements of Applied Stochastic Processes,* Wiley, New York, 1972.

62. Cooper, R. B., *Introduction to Queueing Theory,* Macmillan, New York, 1972.

63. Cox, D. R. and Smith, W. L., *Queues,* Methuen, 1961.

64. Gross, D. and Harris, C. M., *Fundamentals of Queueing Theory,* Wiley, New York, 1974.

65. Kleinrock, L., *Queueing Systems,* Vol. 1, Wiley, New York, 1975.

66. Kleinrock, L., *Queueing Systems,* Vol. 2, Wiley, New York, 1976.

67. Little, J. D. C., A Proof of the Queueing Formula $L = \lambda W$, *Operations Research,* Vol. 9, 1961, pp. 383–387.

68. Shanthikumar, J. G., Approximate Queueing Models of Dynamic Job Shops, Ph.D. Thesis, Department of Industrial Engineering, University of Toronto, 1979.

69. Stidham, S., A Last Word on $L = \lambda W$, *Operations Research,* Vol. 22, 1974, pp. 417–421.

70. Trivedi, K. S., *Probability and Statistics with Reliability, Queueing, and Computer Science Applications,* Prentice-Hall, Englewood Cliffs, New Jersey, 1982.

Queueing Networks

71. Buzacott, J. A. and Yao, D. D. W., Flexible Manufacturing Systems: A Review of Models, Working Paper No. 7, Department of Industrial Engineering, University of Toronto, March 1982.

72. Buzen, J. P. and Denning, P. J., Operational Treatment of Queue Distributions and Mean Value Analysis, *Computer Performance,* Vol. 1, No. 1, June 1980, pp. 6–15.

73. Denning, P. J. and Buzen, J. P., The Operational Analysis of Queueing Network Models, *Computer Surveys,* Vol. 10, No. 3, September 1978, pp. 225–262.

74. Hildebrant, R. R., Scheduling Flexible Machining Systems Using Mean Value Analysis, *Proceedings of the IEEE Conference on Decision and Control,* Albuquerque, New Mexico, 1980.

75. Koenigsberg, E., Twenty Five Years of Cyclic Queues and Closed Queue Networks: A Review, *Journal of the Operational Research Society,* Vol. 33, 1982, pp. 605–619.

76. Solberg, J. J., A Mathematical Model of Computerized Manufacturing Systems, *Proceedings of the 4th International Conference on Production Research,* Tokyo, Japan, 1977.

77. Solberg, J. J., *CAN-Q User's Guide,* Report No. 9 (Revised), School of Industrial Engineering, Purdue University, West Lafayette, Indiana, 1980.

78. Stecke, K. E., Production Planning Problems for Flexible Manufacturing Systems, Ph.D. Dissertation, School of Industrial Engineering, Purdue University, West Lafayette, Indiana, 1981.

79. Suri, R., New Techniques for Modelling and Control of Flexible Automated Manufacturing Systems, *Proceedings of the IFAC 8th Triennial World Congress,* Vol. 14, Kyoto, Japan, 1981, pp. 175–181.

80. Suri, R., Robustness of Queueing Network Formulae, *Journal of ACM,* July 1983.

81. Suri, R. and Hildebrant, R. R., Using Mean Value Analysis, *Journal of Manufacturing Systems,* Vol. 3, No. 1, 1984.

82. Tay, Y. C. and Suri, R., *Sample Statistics and Error Bounds for Performance Predictions in Queueing Networks,* Harvard University, Aiken Computation Laboratory Report TR-07-83, Cambridge, Massachusetts, February 1983.

83. Trivedi, K. S., *Probability and Statistics with Reliability, Queueing, and Computer Science Applications,* Prentice-Hall, Englewood Cliffs, New Jersey, 1982.

Analytical Models of Blocking

84. Buzacott, J. A., The Production Capacity of Job Shops with Limited Storage Space, *International Journal of Production Research,* Vol. 14, No. 5, 1976, pp. 597–605.

85. Buzacott, J. A. and Hanifin, L. E., Models of Automatic Transfer Lines with Inventory Banks— A Review and Comparison, *AIIE Transactions,* Vol. 10, No. 2, 1978, pp. 197–207.

86. Buzacott, J. A. and Shanthikumar, J. G., Models for Understanding Flexible Manufacturing Systems, *AIIE Transactions,* December 1980, pp. 339–350.

87. Diehl, G. and Suri, R., Queueing Models for Efficient Analysis of Tool Sharing and Blocking in Automated Manufacturing Systems, TIMS/ORSA, Chicago, Illinois, April 1983.

88. Gershwin, S. B., An Efficient Decomposition Method for the Approximate Evaluation of Tandem Queues with Finite Storage Space and Blocking, Technical Report LIDS-P-1309, MIT, Cambridge, Massachusetts, October 1983.

89. Suri, R. and Diehl, G. W., A Variable Buffer-Size Model and Its Use in Analyzing Closed Queueing Networks with Blocking, *Management Science,* submitted, 1983.

Reliability

90. Amstadter, B. L., *Reliability Mathematics,* McGraw-Hill, New York, 1971.

91. Barlow, R. E. and Proschan, F., *Statistical Theory of Reliability and Life Testing,* Holt, Rinehart and Winston, 1975.

92. Mathur, F. P., Automation of Reliability Evaluation Procedures Through CARE—The Computer-Aided Reliability Estimation Program, *AFIPS Conference Proceedings,* Vol. 41, Fall Joint Computer Conference, 1972, pp. 65–82.

93. Ng, Y-W., Reliability Modeling and Analysis for Fault-Tolerant Computers, Ph.D. Dissertation, Computer Science Department, University of California at Los Angeles, 1976.

94. Parzen, E., *Stochastic Processes,* Holden-Day, San Francisco, California, 1962.

95. Shooman, M. L., *Probabilistic Reliability: An Engineering Approach,* McGraw-Hill, New York, 1968.

96. Stiffler, J. J. et. al., CARE III Final Report, Phase I, NASA Contractor Report 159122, November 1979.

97. Trivedi, K. S., *Probability and Statistics with Reliability, Queueing, and Computer Science Applications,* Prentice-Hall, Englewood Cliffs, New Jersey, 1982.

Batching

98. Afentakis, P., Gavish, B., and Karmarkar, U., Exact Solutions to the Lot-Sizing Problem in Multistage Assembly Systems, *Management Science,* to appear.

99. Berry, W. L., Lot Sizing Procedure for Requirement Planning Systems: A Framework for Analysis, *Production Inventory Management,* Vol. 13, 1972, pp. 19–34.

100. Crowston, W. B. and Wagner, M. H., Dynamic Lot Size Models for Multi-Stage Assembly Systems, *Management Science,* Vol. 20, 1973, pp. 14–21.

101. Doll, D. C. and Whybark, D. C., An Iterative Procedure for the Single-Machine Multi-Product Lot Scheduling Problem, *Management Science,* Vol. 20, 1973, pp. 50–55.

102. Elmaghraby, S. E., The Economic Lot Scheduling Problem (ELSP): Review and Extensions, *Management Science,* Vol. 24, 1978, pp. 587–598.

103. Goyal, S. K., Determination of Optimum Packaging Frequency of Items Jointly Replenished, *Management Science,* Vol. 21, 1973, pp. 436–443.

104. Graves, S. C., Multistage Lot-Sizing: An Iterative Procedure, Technical Report No. 164, Operations Research Center, MIT, Cambridge, Massachusetts, 1979.

105. Peterson, R. and Silver, E. A., *Decision Systems for Inventory Management and Production Planning,* Wiley, New York, 1979.

106. Silver, E. A., A Simple Method of Determining Order Quantities in Joint Replenishments Under Deterministic Demand, *Management Science,* Vol. 22, 1976, pp. 1351–1361.

107. Silver, E. A. and Meal, H. C., A Heuristic for Selecting Lot Size Quantities for the Case of a Deterministic Time-Varying Demand Rate and Discrete Opportunities for Replenishment, *Production Inventory Management,* Vol. 14, 1973, pp. 64–74.

108. Van Nunen, J. A. E. E. and Wessles, J., Multi-Item Lot Size Determination and Scheduling Under Capacity Constraints, *European Journal of Operational Research,* Vol. 2, 1978, pp. 36–41.

109. Wagner, H. M. and Whitin, T., Dynamic Version of the Economic Lot Size Model, *Management Science,* Vol. 5, 1958, pp. 89–96.

110. Whitney, C. K., FMS Batching and Balancing, TIMS/ORSA, Chicago, April 1983.

Scheduling

111. Baker, K. R., *Introduction to Sequencing and Scheduling,* Wiley, New York, 1974.
112. Baker, K. R., A Comparative Study of Flow Shop Algorithms, *Operations Research,* Vol. 23, 1975, pp. 62–73.
113. Berry, W. L. and Rao, V., Critical Ratio Scheduling: An Experimental Analysis, *Management Science,* Vol. 22, 1975, pp. 192–201.
114. Coffman, E. G., Jr., *Computer and Job Shop Scheduling Theory,* Wiley, New York, 1976.
115. Conway, R., Maxwell, W. L., and Miller, L. W., *Theory of Scheduling,* Addison-Wesley, Reading, Massachusetts, 1967.
116. Dannenbring, D. G., An Evaluation of Flow Shop Sequencing Heuristics, *Management Science,* Vol. 23, 1977, pp. 1174–1182.
117. Dar-El, E. M. and Wysk, R. A., Job Shop Scheduling—A Systematic Approach, *Journal of Manufacturing Systems,* Vol. 1, No. 1, pp. 77–88.
118. Fisher, M. L., Optimal Solution of Scheduling Problems Using Lagrange Multipliers, Part I, *Operations Research,* Vol. 21, 1973, pp. 1114–1127.
119. Gelders, L. F. and Van Wassenhove, L. N., Production Planning: A Reviw, *European Journal of Operational Research,* Vol. 7, 1981, pp. 101–110.
120. Graves, S. C., A Review of Production Scheduling, *Operations Research,* Vol. 29, No. 4, 1981, pp. 646–675.
121. Muth, J. F. and Thompson, G. L., Eds., *Industrial Scheduling,* Prentice-Hall, Englewood Cliffs, New Jersey, 1963.
122. Putman, A. O., Everdell, R., Dorman, G. H., Cronan, R. R., and Lindgren, J., Updating Critical Ratio and Slack-Time Priority Scheduling Rules, *Production Inventory Management,* Vol. 12, pp. 51–72.
123. Schrage, L., Scheduling, in Belzer, J., Holzman, A. G., and Kent, A., Eds., *Encyclopedia of Computer Science and Technology,* Marcel Dekker, New York, 1979.
124. Schrage, L. and Baker, K. R., Dynamic Programming Solution of Sequencing Problems with Precedence Constraints, *Operations Research,* Vol. 26, 1978, pp. 444–449.
125. Stecke, K. E. and Solberg, J. J., Loading and Control Policies for a Flexible Manufacturing System, *International Journal of Production Research,* Vol. 19, No. 5, 1981, pp. 481–490.
126. Vaithianathan, R. and Roberts, K. L., On Scheduling in a GT Environment, *Journal of Manufacturing Systems,* Vol. 1, No. 2, 1982, pp. 149–155.

Markov Decision Processes

127. Jewell, W. J., Markov-Renewal Programming, *Operations Research,* Vol. 11, 1963, pp. 938–971.
128. Kimemia, J. and Gershwin, S. B., Computation of Production Control Policies by a Dynamic Programming Technique, *Proceedings of the IEEE Large Scale Systems Symposium,* Virginia Beach, Virginia, October 1982, pp. 393–397.
129. Ross, S. M., *Applied Probability Models with Optimization Applications,* Holden-Day, San Francisco, California, 1970.
130. Seidmann, A. and Schweitzer, P. J., *Real-Time On-Line Control of a FMS Cell,* Working Paper QM8217, Graduate School of Management, University of Rochester, New York, 1982.

Simulation

131. Bevans, J. P., First Choose an FMS Simulator, *American Machinist,* May 1982, pp. 143–145.
132. Clementson, A. T., Extended Control and Simulation Language II, *User's Manual,* University of Birmingham, England, 1972.
133. Dahl, O. J. and Nygaard, K., SIMULA—An ALGOL-Based Simulation Language, *C.ACM 9,* No. 9, 1966, pp. 671–678.
134. ElMaraghy, H. A., Simulation and Graphical Animation of Advanced Manufacturing Systems, *Journal of Manufacturing Systems,* Vol. 1, No. 1, 1982, pp. 53–64.
135. ElMaraghy, H. A. and Ho, N. C., A Flexible System for Computer Control of Manufacturing, *Computers in Mechanical Engineering,* August 1982, pp. 16–23.
136. Fishman, G. S., *Principles of Discrete Event Simulation,* Wiley, New York, 1978.
137. Gaver, D. P., Simulation Theory, in Moder, J. J. and Elmaghraby, S. E., Eds., *Handbook of Operations Research,* Van Nostrand, New York, 1978, pp. 545–565.

138. Gordon, G., *The Application of GPSS V to Discrete System Simulation,* Prentice-Hall, Englewood Cliffs, New Jersey, 1975.

139. Gordon, G., Simulation-Computation, in Moder, J. J. and Elmaghraby, S. E., Eds., *Handbook of Operations Research,* Van Nostrand, New York, 1978, pp. 566–585.

140. Hutchinson, G. K. and Hughes, J. J., A Generalized Model of Flexible Manufacturing Systems, *Proceedings of the Multistation Digitally Controlled Manufacturing Systems Workshop,* University of Wisconsin—Milwaukee, January 1977.

141. Kiviat, P. J., Villanueva, R., and Markowitz, H. M., *The SIMSCRIPT II Programming Language,* Prentice-Hall, Englewood Cliffs, New Jersey, 1969.

142. Lenz, J. E. and Talavage, J. J., General Computerized Manufacturing Systems Simulator (GCMS), Report No. 7, School of Industrial Engineering, Purdue University, West Lafayette, Indiana, August 1977.

143. Nof, S. Y., Halevi, G., and Bobasch, M., Simulation Helps Minimize WIP, *American Machinist,* February 1982, pp. 126–130.

144. Phillips, D. T. and Handwerker, M., GEMS: A Generalized Manufacturing Simulator, *Proceedings of the 12th International Conference on Systems Science,* Honolulu, Hawaii, January 1979.

145. Pritsker, A. A. B., and Pegden, C. D., *Introduction to Simulation and SLAM,* Wiley, New York, 1979.

146. Rubinstein, R. Y., *Simulation and the Monte Carlo Method,* Wiley, New York, 1981.

147. Runner, J. A. and Leimkuhler, F. F., *CAMSAM: A Simulation Analysis Model for Computer-Aided Manufacturing Systems,* Report No. 13, School of Industrial Engineering, Purdue University, West Lafayette, Indiana, 1978.

Hierarchical Approaches

148. Gelders, L. F. and Van Wassenhove, L. N., Production Planning: A Review, *European Journal of Operational Research,* Vol. 7, 1981, pp. 101–110.

149. Graves, S. C., Using Lagrangean Techniques to Solve Hierarchical Production Planning Problems, *Management Science,* Vol. 28, No. 3, March 1982, pp. 260–275.

150. Hax, A. C. and Meal, H. C., Hierarchical Integration of Production Planning and Scheduling, in Geisler, M. A., Ed., *Logistics,* North Holland, 1975.

151. Hildebrant, R. R. and Suri, R., Methodology and Multi-Level Algorithm Structure for Scheduling and Real-Time Control of Flexible Manufacturing Systems, *Proceedings of the 3rd International Symposium on Large Engineering Systems,* Memorial University of Newfoundland, July 1980, pp. 239–244.

152. Kimemia, J., *Hierarchical Control of Production in Flexible Manufacturing Systems,* Ph.D. Thesis, Department of Electrical Engineering and Computer Science, MIT, Cambridge, Massachusetts, April 1982.

153. Orlicky, J. A., *Material Requirements Planning,* McGraw-Hill, New York, 1975.

154. Peterson, R. and Silver, E. A., *Decision Systems for Inventory Management and Production Planning,* Wiley, New York, 1979.

155. Stecke, K. E., A Hierarchical Approach to Production Planning in Flexible Manufacturing Systems, *Proceedings of the 20th Allerton Conference on Communication, Control and Computing,* Monticello, Illinois, October 1982, pp. 426–433.

156. Suri, R., New Techniques for Modelling and Control of Flexible Automated Manufacturing Systems, *Proceedings of the IFAC 8th Triennial World Congress,* Vol. 14, Kyoto, Japan, 1981, pp. 175–181.

157. Suri, R. and Whitney, C. K., Decision Support Requirements for Flexible Manufacturing, *Journal of Manufacturing Systems,* Vol. 3, No. 1, 1984.

Monitoring and Fine-Tuning

158. Behee, R. D., Measuring Multiple Channels of Transient Data with a Microcomputer-Based Acquisition System, *Computers in Mechanical Engineering,* January 1983, pp. 18–23.

159. Ho, Y. C., Ed., *SPEEDS: A New Technique for the Analysis and Optimization of Queueing Networks,* Technical Report No. 675, Division of Applied Sciences, Harvard University, Cambridge, Massachusetts, February 1983.

160. Ho, Y. C., Eyler, M. A., and Chien, T. T., A New Approach to Determine Parameter Sensitivities on Transfer Lines, *Management Science,* Vol. 29, No. 6, June 1983, pp. 700–714.

161. Suri, R., Infinitesimal Perturbation Analysis of Discrete Event Dynamic Systems: A General

Theory, *Proceedings of the 22nd IEEE Conference on Decision and Control,* San Antonio, Texas, December 1983.

162. Suri, R. and Dille, J. W., Optimization of Flexible Manufacturing Systems Using New Perturbation Analysis, *Proceedings of the ORSA/TIMS FMS Conference,* Ann Arbor, Michigan, August 1984.

163. Suri, R. and Cao, X., The Phantom Customer and Marked Customer Methods for Optimization of Closed Queueing Networks with Blocking and General Service Times, *ACM Performance Evaluation Review,* August 1983, pp. 243–256.

Maintenance

164. Assaf, D. and Levikson, B., On Optimal Replacement Policies, *Management Science,* Vol. 28, No. 11, November 1982, pp. 1304–1312.

165. Barlow, R. E. and Hunter, L. C., Optimum Preventive Maintenance Policies, *Operations Research,* Vol. 8, 1960, pp. 90–100.

166. Derman, C., Lieberman, G. J., and Ross, S. M., A Renewal Decision Problem, *Management Science,* Vol. 24, 1978, pp. 554–561.

167. Gertsbakh, I. B., *Models of Preventive Maintenance,* North-Holland, 1977.

168. Haurie, A. and L'Ecuyer, P., A Stochastic Control Approach to Group Preventive Replacement in a Multicomponent System, *IEEE Transactions on Automatic Control,* Vol. AC-27, No. 2, April 1982, pp. 387–393.

169. Nahmias, A., Managing Repairable Item Inventory Systems: A Review, in Schwarz, L. B., Ed., *Multi-Level Production/Inventory Control Systems: Theory and Practice,* North Holland, 1981, pp. 253–277.

170. Pierskala, W. P. and Volker, J. A., A Survey of Maintenance Models: The Control and Surveillance of Deteriorating Systems, *Naval Research Logistics Quarterly,* Vol. 18, No. 3, 1976.

Decision Support

171. Baxter, J. D., Line Managers Move from MIS to DSS for Decision-Making Help, *Iron Age,* September 28, 1981, pp. 71–73.

172. Chen, P. H. and Talavage, J., Production Decision Support System for Computerized Manufacturing Systems, *Journal of Manufacturing Systems,* Vol. 1, No. 2, 1982, pp. 157–168.

173. Nof, S. Y., Theory and Practice in Decision Support for Manufacturing, in Holsapple, C. W. and Whinston, A. B., Eds., *Data Base Management: Theory and Applications,* D. Reidel, 1983, pp. 325–348.

174. Nof, S. Y. and Gurecki, R., MDSS: Manufacturing Decision Support System, *Proceedings of the AIIE Spring Annual Conference,* May 1980, pp. 274–283.

175. Suri, R. and Whitney, C. K., Decision Support Requirements for Flexible Manufacturing, *Journal of Manufacturing Systems,* Vol. 3, No. 1, 1984.

176. Wagner, G. R., Mind Support Systems, *ICP Interface Manufacturing and Engineering,* Spring 1982, pp. 19–22.

CHAPTER 32

HUMAN FACTORS IN PLANNING ROBOTIC SYSTEMS

GAVRIEL SALVENDY

Purdue University
West Lafayette, Indiana

32.1. INTRODUCTION

The purpose of this chapter is to acquaint practitioners with the nature and characteristics of the human factors that impact effective planning, design, control, and operation of industrial robotic systems. A good understanding of the human element in robotics systems contributes to wider adoption and more effective utilization of industrial robots than currently is feasible. The current robotic technology (Part 3) is developed to a level that would currently enable American industries to utilize more than one million robots. Present utilization of only 1% of that number is largely attributable to insufficient consideration given to human factors in the analysis, design, operation, control, and implementation of industrial robotic systems. Human factors issues—which impact effective implementation and utilization of industrial robotic systems that are economically viable, humanly acceptable, and result in increased productivity and quality of life—must give due consideration to the social, safety, human performance, and motivational issues in the analysis, design, implementation, control, and operation of industrial robotics systems. Although these issues are conceptually integrated, for operational purposes each is discussed separately in the chapter.

32.2. SOCIAL ISSUES

There are at least two major social factors that impact the effective and widespread utilization of industrial robots: worker displacement and worker retraining.

32.2.1. Worker Displacement

The extent of worker displacement due to automation is difficult to ascertain. Historical data in relation to the application of automation in manufacturing are not particularly reliable indicants of future trends. Technological change does not necessarily create jobs or avoid job displacements.[1] Senker[1] states that there are two phases in major technological revolutions. In the initial phase new technology primarily generates employment. The latter phase, or mature phase, tends to displace labor. Senker asserts that the mature phase has been reached in the "electronics technological revolution." The extension of this, as it applies to industrial robots, is that the low cost and high reliability of microprocessors aid in decreasing robotics costs and concurrently enlarge their range of applicability. The result is the expansion of production without a proportionate increase in employment.[1] In the past, increased product demand has caused an increase in manual and expanded work-force demand. The impact of automation has therefore been masked.

To demonstrate a possible net decrease in work as a result of robotization of manual production operations, a simple material-handling operation is presented. If these manual material-handling systems are required to feed three numerical control (NC) machines, a great deal of cost is incurred for manual labor and indirect costs due to manual labor. A worker may incur a total first-shift cost of nearly $30.00 per hour. Tote bins typically cost $125 to $150.[2] Other expenses may include forklift operators to move pallets of tote bins, and so on.

Suppose these NC machines were arranged in a manufacturing cell; owing to the electronics and software capabilities currently available, one robot is capable of tending each machine even though each may perform a different operation. It must be realized that some type of materials-feeding system must be in effect or designed for the robot. It may be possible to eliminate the tote bins which may assume production space. Forklifts and their operators may be modified or eliminated, and the material-handling personnel may also be displaced. Over an extended time period, humans can work only one shift per day; since a robot is capable of working more than one shift, it may replace more than one worker.[2] In this particular situation, it can be seen that direct and indirect cost savings may be quite significant, and robots can contribute to disproportionate displacement per job.

The impact of similar situations for unskilled and semiskilled workers is fewer hours of work, which translates to fewer jobs and less job security. This has spurred organizations such as the United Auto Workers (UAW) to develop positions on integrated automation, chiefly concerning industrial robots. Precarious as it is to permit one specific organization to speak for all production employees, the UAW does encompass a larger proportion of employees in an industry that utilizes the greatest number of industrial robots. The UAW does not place a specific emphasis on robots but instead considers them another technological advancement it must consider.[3] The union also recognizes that enhanced productivity is necessary for long-term economic viability.[2,3] It is, however, aware of possible detrimental impacts upon its membership primarily due to job insecurity. The union believes that technological advancement is acceptable and is encouraged as long as the current work force retains job security.[2] The UAW is well aware of the Japanese workers' "lifetime employment" status;[2] the UAW's response to job security and robotics follows along these lines:

Management provides advance notice of new technology to enable discussion.

Introduction of new technology should displace as few workers as possible by using normal work-force attrition.

When increases in productivity outpace attrition rate, the protection of workers against displacement is an appropriate first charge against productivity.

Bargaining-unit integrity must be maintained; bargaining-unit work must not be transferred to out-of-unit employees.

In-unit employees must be given adequate training to perform jobs introduced by new technology.

Work time must be reduced to afford adequate job opportunities to all who want to work.[2,3]

Douglass Frasier asserts that to achieve these objectives, hours of employment depend on paid work needed to produce a desired output; the number of job slots depends upon the need to equate that output. To equate the two, we need to reduce the number of work hours per job.[2] The UAW asserts that a work-time decrease is an alternative to bargaining for higher wages.[2,3,4] Further, the union is adamant that there must explicitly be pay for lost work hours.[2] The extent to which these issues affect the entire work force is impossible to determine.

Quality of work life (QWL) is also affected by the introduction of robotics. In reference to job security, it is possible that robots may ultimately improve job security. Robot adaptability enables the robot to be assigned and tooled to many production tasks; the degree to which those tasks are similar to tasks performed by humans increases the likelihood that a human and robot are interchangeable in task performance.[5] In the case of consumer items where market fluctuations may be quite drastic, robots can offer a distinct advantage in production assemblies.[4] In QWL terms, a company may initiate "robot layoffs" due to downward fluctuation in demand and temporarily assign humans to the assembly line, thus minimizing the displacement effects of a market downturn.

It is often argued that technological change (i.e., industrial robots) will create jobs. There is a wide range of estimates on the extent to which displaced jobs will be compensated by newly developed jobs due to increases and complex robotic utilization.[6] This may well be the case, but it is naive and myopic to compare only quantities. To discuss adequately job displacement and job creation as one being a counteractant for the other, the type of job created must be compared to the type of job eliminated. The literature evidences a discrepancy in skill level between those jobs that will become available and those that will be eliminated. There is and probably will be a significant demand for highly trained personnel in computer programming, mechanical engineering, electronic design, and so forth, all highly skilled positions to implement, utilize, and/or maintain the industrial robot.[6] In all probability, those to be displaced will be workers in unskilled and semiskilled jobs. A large void in skill level exists between those jobs eliminated and those jobs created. Companies in the United States who currently utilize robots appear not to attempt any major effort at retraining displaced employees.[7] The United Auto Workers observes that benefits derived from automation (robotics) are applied to a smaller hourly work force.[3]

32.2.2. Worker Retraining

One of the most acute problems associated with introducing and utilizing flexible manufacturing and industrial robots is that the skill requirements in these new technologies do not capitalize and build on the skills, perception, and knowledge accumulated by the industrial worker.[8] This implies that acquired industrial skills, which were widely utilized in the premicroelectronics-automation era,[9] are completely lost and have become redundant for the industrial robot revolution era.

This has two major implications. First, it must be assessed who can be retrained for the new skills. This can be achieved by analyzing skills and knowledge requirements for robotics jobs. From this analysis, either work samples or tests that simulate the job can be developed. After assessing the reliability and validity of these tests, the samples can be administered to displaced workers to assess the likelihood of their success in mastering new skills.[10] Based on this evaluation and on the nature of human abilities, it may be estimated that more than one-half of these displaced workers will not possess employable abilities for the new robot-oriented and computer-based manufacturing work environment. If we do not provide careful manpower planning, we may end up with more than 20 million unemployed Americans by the year 2000, an intolerable social and economic situation. To eliminate or reduce this situation, the industrial robotics systems must be so designed, developed, and operated to capitalize on (as far as possible) acquired and used human skills.

32.3. ORGANIZATION DESIGN AND JOB DESIGN ISSUES

The introduction of industrial robots to the workplace changes the requirements for the design of new organizational structures that link computational hierarchy through behavioral hierarchy to organizational hierarchy (Figure 32.1).

The command and control structure for successful organizations of great complexity is invariably hierarchical, wherein goals, or tasks, selected at the highest level are decomposed into sequences of subtasks that are passed to one or more operational units at the next lower level in the hierarchy. Each of these lower level units decomposes its input command in the context of feedback information obtained from other units at the same or lower levels, or from the external environment, and issues sequences of sub-subtasks to a set of subordinates at the next lower level. This same procedure is repeated at each successive hierarchical level until at the bottom of the hierarchy there is generated a set of sequences of primitive actions that drive individual actuators such as motors, servo valves, hydraulic pistons, or individual muscles. This basic scheme can be seen in the organizational hierarchy on the left of Figure 32.1.

A single chain of command through the organizational hierarchy on the left is shown as the computational hierarchy in the center of Figure 32.1. This computational hierarchy consists of three parallel hierarchies: a task decomposition hierarchy, a sensory processing hierarchy, and a world model hierarchy. The sensory processing hierarchy consists of a series of computational units, each of which extracts the particular features and information patterns needed by the task decomposition unit at that level. Feedback from the sensory processing hierarchy enter each level of the task decomposition hierarchy. This feedback information comes from the same or lower levels of the hierarchy or from the external environment. It is used by the modules in task decomposition hierarchy to sequence their outputs and to modify their decomposition function to accomplish the higher-level goal in spite of perturbations and unexpected events in the environment.

The world model hierarchy consists of a set of knowledge bases that generate expectations against which the sensory-processing modules can compare the observed sensory data stream. Expectations are based on stored information which is accessed by the task being executed at any particular time. The sensory-processing units can use this information to select the particular processing algorithms that are appropriate to the expected sensory data and can inform the task decomposition units of whatever differences, or errors, exist between the observed and expected data. The task decomposition unit can then respond, either by altering the action to bring the observed sensory data into correspondence with the expectation, or by altering the input to the world model to bring the expectation into correspondence with the observation.

Each computational unit in the task decomposition, sensory processing, and world-modeling hierarchies can be represented as a finite-state machine. At each time increment each unit reads its input and, based on its present internal state, computes an output with a very short time delay.

If the output of each unit in the task decomposition hierarchy is described as a vector, and plotted versus time in a vector space, a behavioral hierarchy such as is shown on the right side of Figure 32.2 results. In this illustration a high-level goal, or task (BUILD SUBASSEMBLY ABCD), is input to the highest level in a robot control hierarchy. The H5 task decomposition unit breaks this task down into a series of subtasks, of which (ASSEMBLE AB) is the first. This "complex" subtask command is then sent to the H4 task decomposition unit. H4 decomposes this "complex" subtask into a sequence of "simple" subtasks (FETCH A), (FETCH B), (MATE B to A), (FASTEN B to A). The H3 unit, subsequently decomposes each of the "simple" subtasks into a string of "elemental moves" of the form (REACH TO A), (GRASP), (MOVE to X), (RELEASE), and so on. The H2 decomposition unit then computes a string of trajectory segments in a coordinate system fixed in the work space, or in the robot gripper, or in the workpiece itself. These trajectory segments may include acceleration, velocity, and deceleration profiles for the robot motion. In H1 each of these trajectory segments is transformed into joint angle movements, and the joint actuators are served to execute the commanded motions.

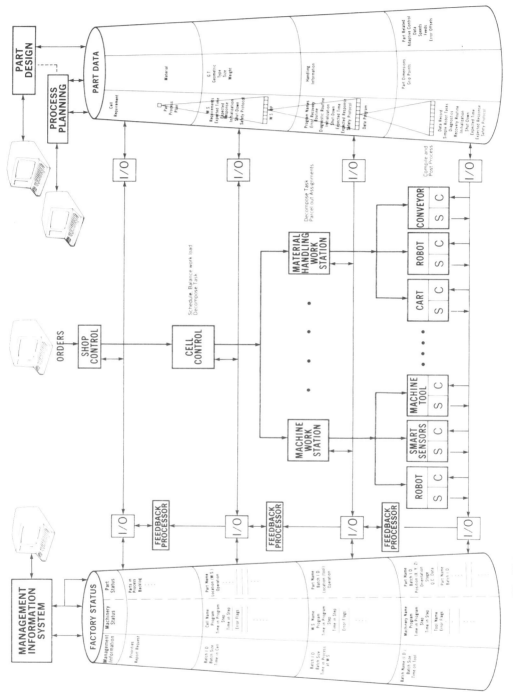

Fig. 32.1. A generic system that can be applied to a wide variety of automatic manufacturing facilities and can be extended to much larger applications. (*Source.* National Bureau of Standards.)

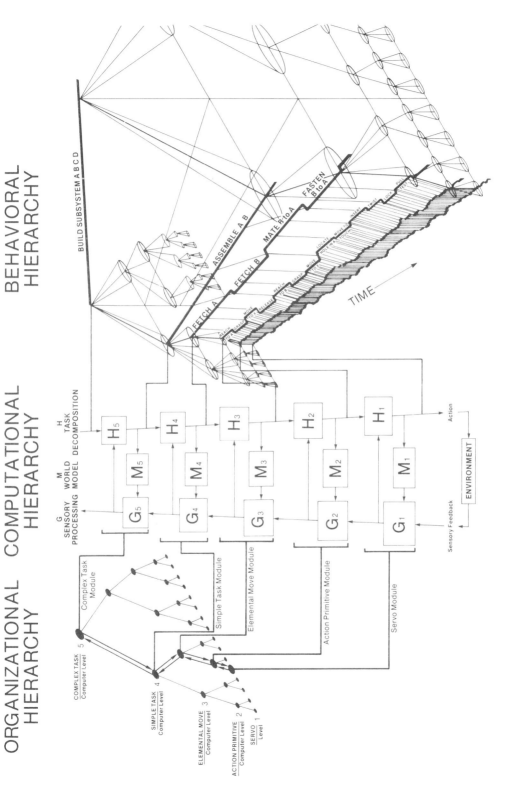

ORGANIZATIONAL COMPUTATIONAL BEHAVIORAL
HIERARCHY HIERARCHY HIERARCHY

Fig. 32.2. The nature and interrelationship of organizational, computational and behavioral hierarchies. (*Source.* National Bureau of Standards.)

643

At each level, the G units select the appropriate feedback information needed by the H modules in the task-decomposition hierarchy. The M units generate predictions, or expected values, of the sensory data based on the stored knowledge about the environment in the context of the task being executed.

The operational effects of this hierarchical control are illustrated in Figure 32.2. This shows the information flow from the robots in a computerized flexible manufacturing system. Such organizational structures create a supervisory control in which the production processes and productivity are controlled by the operator by computer terminals.

The computing architecture shown in Figure 32.1 is intended as a generic system that can be applied to a wide variety of automatic manufacturing facilities and can be extended to much larger applications. The basic structure is hierarchical, with the computational load distributed evenly over various computational units at various levels of the hierarchy. At the lowest level in this hierarchy are individual robots, N/C machining centers, smart sensors, robot carts, conveyors, and automatic storage systems, each of which may have its own internal hierarchical control system. These machines are organized into workstations under the control of a workstation control unit. Several cell control units may be organized under and receive input commands from a shop control unit, and so on. This hierarchical structure can be extended to as many levels with as many modules per level as are necessary, depending on the complexity of the factory.

On the right side of Figure 32.2 is shown a data base which contains the part programs for the machine tools, the part-handling programs for the robots, the materials requirements, dimensions, and tolerances derived from the part design data base, and the algorithms and process plans required for routing, scheduling, tooling, and fixturing. These data are generated by a Computer-Aided-Design (CAD) system and a Computer-Aided-Process-Planning (CAPP) system. This data base is hierarchically structured so that the information required at the different hierarchical levels is readily available when needed.

On the left is a second data base which contains the current status of the factory. Each part in process in the factory has a file in this data base which contains information as to the part's position and orientation, its stage of completion, the batch of parts that it is with, and quality control information. This data base is also hierarchically structured. At the lowest level, the position of each part is referenced to a particular tray or table top. At the next higher level, the workstation, the position of each part refers to which tray the part is in. At the cell level, position refers to which workstation holds the part. The feedback processors on the left scan each level of the data base and extract the information of interest to the next higher level. A management information system makes it possible to query this data base at any level and determine the status of any part or job in the shop. It can also set or alter priorities on various jobs.

This resulting organizational design raises a number of critical questions such as:

What is the optimal allocation of functions between human supervisory control and the computer?

What is the relationship between the number of machines controlled by one supervisor and the productivity of the overall system? What is the optimal number of machines that a supervisor should control?

What is the impact of work isolation of the supervisor in a computer-controlled work environment on the quality of working life and mental health of the operator?

In allocating functions between computer and humans, emphasis must be placed on optimizing human arousal, job satisfaction, and productivity.

Evidence pertaining to job design, Table 32.1, indicates that the numbers of people who prefer to work at and are more satisfied and productive in performing the task in simplified mode are equal to those who prefer enriched jobs; but 10% of the labor force does not like work of any type. When 270 shop floor workers performed their work in both enriched and simplified modes, it was evident that the numbers of people who preferred simplified jobs and those who preferred enriched jobs were equal. It is typically the older worker who prefers simplified jobs, whereas the younger workers prefer enriched jobs. In this study, 9% of the labor force did not like work of any type.

In the simplified job design, the operator performs only very small components of the total job without having decision latitude about task performance. These simplified jobs can be enlarged either vertically or horizontally. Thus the operator may either do more of the same thing, thus enlarging the job vertically, or additional tasks may be added for the task performance, thus enlarging the task horizontally, which results in job enrichment. It is typically the older worker (past 45 years of age) who prefers to work at simplified jobs whereas the younger worker prefers to work at, and is both more satisfied and productive in, enriched jobs. The overwhelming majority of computer-based supervisory control tasks are manned by younger operators (below the age of 45 years). Hence in allocating the function between human and computer, the division should be made such that the task content of the human is sufficiently enriched to provide for psychological growth of the individual.

TABLE 32.1.   WORKER JOB SATISFACTION IN RELATION TO JOB SIMPLICITY

Variable Measured	Satisfied with Enriched Jobs (Dissatisfied with Simplified Jobs)	Dissatisfied with Enriched Jobs (Satisfied with Simplified Jobs)	Do Not Like Work of Any Type
1. Percent of labor force	47	44	9
2. Percent of labor force in category 1 who are dissatisfied	4	87	100
3. Productivity of the labor force in category 1	91	92	84
4. Percent of labor force over 45 years of age	11	82	50

Source.   Salvendy.[26]

In making the allocation of functions, it should be noted (Figure 32.3) that an optimal arousal level exists for maximizing productivity and job satisfaction: when the arousal level is too low, boredom sets in; when the arousal is too high, mental overload occurs.

32.4.   SAFETY ISSUES

Humans can interact with industrial robots in the following ways: as supervisors, as co-workers, and preparing and setting up as maintenance robots. The only safe way to design industrial robot systems is to keep operators physically away from the robot.

For example, in West Germany, robot manufacturers have spent one-third of the total robot programming time on progamming for safety. But, even in these carefully designed situations, accidents and injury to the operator do occur. Hence, it is safest and most effective for the human to exercise supervisory control through computer-based information networks—a subject which is discussed later in this chapter. Potential injury to the setup and service personnel is apparent for maintenance and repair work. Potential accidents can be reduced (but not entirely eliminated) when careful consideration is given to safe job design for the maintenance personnel.

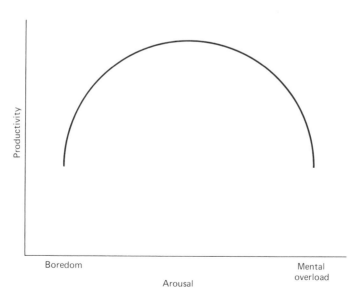

Fig. 32.3.   When task performance requires low arousal level it then results in low human attention and increased job productivity. High arousal level results in mental overload and decreased productivity. For each job and each individual an optimal level of arousal exists that results in maximum productivity.

Based on safety studies of industrial robots in Japan[37] the percentage distribution of near-accidents caused by industrial robots is illustrated as follows:

NEAR-ACCIDENTS CAUSED BY INDUSTRIAL ROBOTS

Cause	Percent
Erroneous action of robot in normal operation	5.6
Erroneous action of peripheral equipment in normal operation	5.6
Careless approach to robot by human	11.2
Erroneous action of robot in teaching and test operation	16.6
Erroneous action of peripheral equipment during teaching and test operation	16.6
Erroneous action during manual operation	16.6
Erroneous action during checking, regulation and repair	16.6
Other	11.2

From this summary it can be seen that about 28% of near-accidents were human related, 61% were equipment related, and 11% were not classified. The 61% that were equipment related owing to the low reliability of robot systems are illustrated as follows:

RELIABILITY OF ROBOTS

(a) Trouble in Robots (%)		(b) Mean Time between Failure of Robots	
Faults of control system	66.9	Under 100 hr	28.7%
Faults of robot body	23.5	100–250 hr	12.2
Faults of welding gun and tooling parts	18.5	250–500 hr	19.5
Runaway	11.1	500–1000 hr	14.7
Programming and other operational errors	19.9	1000–1500 hr	10.4
Precision deficiency, deterioration	16.1	1500–2000 hr	4.9
Incompatibility of jigs and other tools	45.5	2000–2500 hr	1.2
Other	2.5	Over 2500 hr	8.5

The methodology of fault tree analysis illustrated in Figure 32.4 shows the sequence and link of events in robot-related accidents.

32.5. HUMAN INDUSTRIAL WORK PERFORMANCE

A number of human performance capabilities impact the effective design and operation of robotic systems: design of controls, human variability in work performance, information-processing capabilities, task pacing, and job satisfaction. Although these variables are conceptually integrated, for purposes of presentation each of them is discussed separately.

32.5.1. Design of Controls

In designing monitoring systems for robots that are compatible with human performance capabilities, one must be concerned with two main questions, namely, which control is best to use for which purpose, and, given the selection of a certain control, the determination of the appropriate and applicable range for size, displacement, and resistance for each control. The values for these parameters are presented in Tables 32.2–32.3.

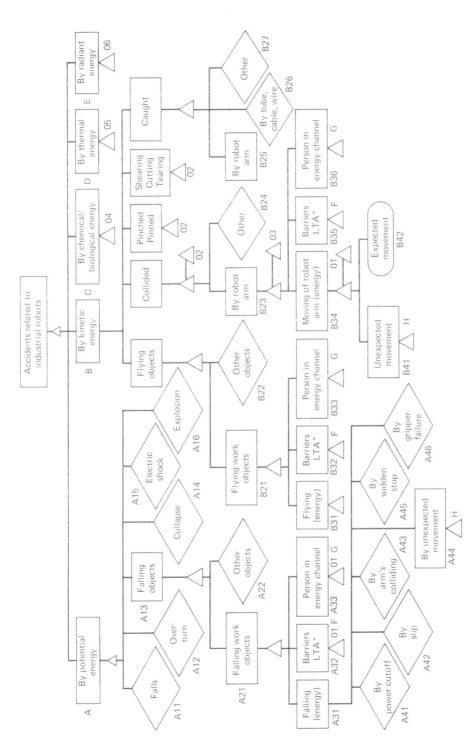

Fig. 32.4. Fault tree for robot-related accidents.* LTA, less than adequate. (*Source.* Reference 37.)

TABLE 32.2. COMPARISON OF THE CHARACTERISTICS OF COMMON CONTROLS

Characteristic	Hand Push Button	Foot Push Button	Toggle Switch	Rotary Switch	Knob	Crank	Lever	Hand-wheel	Pedal
Space required	Small	Large	Small	Medium	Small-medium	Medium-large	Medium-large	Large	Large
Effectiveness of coding	Fair-good	Poor	Fair	Good	Good	Fair	Good	Fair	Poor
Ease of visual identification of control position	Poor[a]	Poor	Fair-good	Fair-good	Fair-good[b]	Poor[c]	Fair-good	Poor-fair	Poor
Ease of nonvisual identification of control position	Fair	Poor	Good	Fair-good	Poor-good	Poor[c]	Poor-fair	Poor-fair	Poor-fair
Ease of check reading in array of like controls	Poor[a]	Poor	Good	Good	Good[b]	Poor[c]	Good	Poor	Poor
Ease of operation in array of like controls	Good	Poor	Good	Poor	Poor	Poor	Good	Poor	Poor
Effectiveness in combined control	Good	Poor	Good	Fair	Good[d]	Poor	Good	Good	Poor

Source. Reproduced with permission from McCormick, E. J. and Sanders, M. S., *Human Factors in Engineering and Design*, 5th ed., McGraw-Hill, New York, 1982.

[a] Except when control is backlighted and light comes on when control is activated.

[b] Applicable only when control makes less than one rotation and when round knobs have pointer attached.

[c] Assumes control makes more than one rotation.

[d] Effective primarily when mounted concentrically on one axis with other knobs.

32.5.2. Range of Human Performance Abilities

Variation in human performance levels occurs both among different operators and within a single operator over a period of time. This variation arises from the following three general classes of operator characteristics:

Experience and training.
Enduring mental and physical characteristics.
Transitory mental and physical characteristics prevailing at the time of task performance.

Transitory characteristics are influenced by many specific factors, including the following:

Motivation
Temporary illness
Fatigue
Stress
Alcohol and other drugs
Hours of work (e.g., overtime, shift worked)
Physical, social, and psychological work environments
Food intake

Human performance also is influenced by task characteristics such as equipment variability, defects and malfunctions, and, especially among different operators, the methods employed by operators to perform their tasks.

The combined impact of these various factors on the performance variability of an individual operator (i.e., within-operator variability) has been documented among blue-collar workers in manufacturing industries. These studies[9] indicate that reliability* of production output varies from .7 to .9, with a mean of .8. This implies that about 64% (i.e., $8^2 \times 100$) of an operator's performance in one week can be predicted by his or her performance observed during a prior week. Conversely, 36% of the operator's performance cannot be explained in this manner, but is apparently explained by such factors as those previously listed.

It should be noted that individual variability within a working day is markedly smaller than between working days. Furthermore, performance variability within a workday is smallest from mid-morning to early afternoon (Figure 32.5). During this period, performance fluctuation around a mean level is only about 5% (of the mean), but this variability increases markedly on either side of the mid-morning to early afternoon period. These patterns of within-operator variability, as well as warm-up and slowdown at the beginning and end of the workday, must be accounted in the design, control, and operation of robotics systems in which the human is a part.

Based on many studies, it is well known that human performance variability among operators is much larger than that observed within the same operator over successive observations. Generally a performance range of 2 to 1 encompasses 95% of the working population.[12] However, in practical work situations the range encountered is likely to be much smaller than this because of preemployment selection, attrition of some low-performance operators, and peer pressures that may limit the output of high-ability operators. Thus, when these limiting factors are not operating, in a group of 200 workers, if the highest-performing 5 and the lowest-performing 5 are not considered, then in the remaining 190 operators, the highest-performing will not perform more than twice as well as the lowest-performing (and, conversely, the lowest-performing operator will do at least half as well as the highest-performing operator). The recognition of this range of performance levels is critical to the design of robotics systems and to the development of effective production planning and control techniques.

32.5.3. Human Information-Processing Memory and Decision-Making Capabilities

The operator's ability to perform these crucial mental activities, and therefore the ability to perform tasks effectively, rests upon fundamental cognitive processes and functions. These basic mental functions and processes (or stages) appear in Figure 32.6, which represents an information-processing model of the human operator. In this model, the operator is continuously presented with information to accomplish his or her work objectives. The operator is viewed as a channel through which information flows. In the model of Figure 32.6 three major information-processing stages are shown: perception, decision

---

* The reliability coefficient is a measure of consistency determined by the extent to which two successive samples of same-task performance provide similar results. Thus, for example, reliability of performance may be obtained by correlating one week's performance with another's.

TABLE 32.3. SUMMARY OF SELECTED DATA REGARDING DESIGN RECOMMENDATIONS FOR CONTROL DEVICES

Device	Size (in)		Displacement		Resistance	
	Minimum	Maximum	Minimum	Maximum	Minimum	Maximum
Hand push button						
Fingertip operation	½	None	⅛ in	15 in	10 oz	40 oz
Foot push button						
Normal operation	½	None	½ in			
Wearing boots			1 in			
Ankle flexion only				2¼ in		
Leg movement				4 in		
Will *not* rest on control					4 lb	20 lb
May rest on control					10 lb	20 lb
Toggle switch			30°	120°	10 oz	40 oz
Control tip diameter	⅛	1				
Lever arm length	½	2				
Rotary selector switch					10 oz	40 oz
Length	1	3				
Width	½	1				
Depth	½					
Visual positioning			15°	40° [a]		
Nonvisual positioning			30°	40° [a]		
Knob, continuous adjustment						4½–6 in/oz
Finger-thumb						
Depth	½	1				
Diameter	⅜	4				
Hand/palm diameter	1½	3				
Crank [b]						
For light loads, radius	½	4½				
For heavy loads, radius	½	20				
Rapid, steady turning						
<3–5 in radius					2 lb	5 lb
5–8 in radius					5 lb	10 lb
8-in radius					?	?
For precise settings					2½ lb	8 lb

	Minimum	Maximum	Displacement	Resistance, minimum	Resistance, maximum
Levers[d]					
Fore-aft (one hand)			14 in	12 oz	32 oz
Lateral (one hand)			38 in	2 lb	20–100 lb
Finger grasp, diam.	½	3			
Hand grasp, diam.	1½	3			
Handwheel[b]					
Diameter	7	21	90°–120°	5 lb	30 lb[c]
Rim thickness	¾	2			
Pedal					
Length	3½				
Width	1				
Normal use			½ in		
Heavy boots			1 in		
Ankle flexion			2½ in		
Leg movement			7 in		
Will *not* rest on control				4 lb	10 lb
May rest on control				10 lb	180 lb

Reproduced with permission from McCormick, E. J. and Sanders, M. S., *Human Factors in Engineering and Design*, 5th ed., McGraw-Hill, New York, 1982.

[a] When special requirements demand large separations, maximum should be 90°.

[b] Displacement of knobs, cranks, and handwheels should be determined by desired control-display ratio.

[c] For two-handed operation, maximum resistance of handwheel can be up to 50 lb.

[d] Length depends on situation, including mechanical advantage required. For long movements, longer levers are desirable (so movement is more linear).

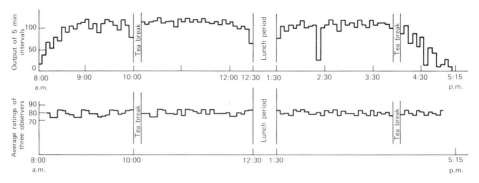

Fig. 32.5. Comparison of output curve and ratings made on one operator during a repetitive manual operation (thread-roll bulb holder) during a working day, utilizing continuous time studies. Similar results were obtained for other manual repetitive tasks and for other operators. (*Source.* Dudley, N. A., Work Measurement: Some research studies, London: Macmillan, 1968.)

making, and response control. Also shown are three memory systems (sensory, short-term, and long-term and response control) which depend upon, and are limited by, the information-processing capacities of these three major stages and the storage characteristics of the three memory systems.

Limits of human performance arise from two characteristics of the major information-processing stages: (1) they require a minimum time in which to perform their functions, and (2) they have limits as to the amount of information they can process per unit time. If information arrives too rapidly, a stage may become overloaded and unable to operate effectively. This limit to the rate at which a stage can handle (i.e., transmit) information is its channel capacity.

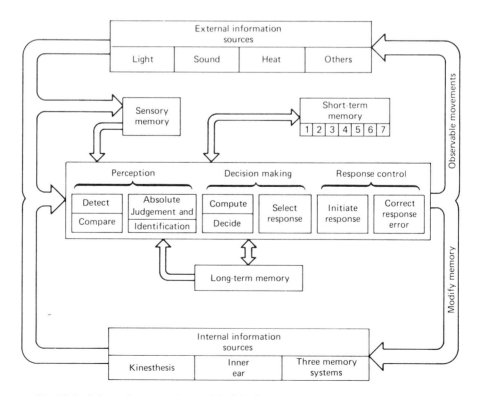

Fig. 32.6. Information processing model of the human operator. (*Source.* Reference 36.)

This limit can be reached in three ways. First, a task may be inherently difficult and present information to a particular stage at an excessive rate. Psychomotor performance can improve if the processing capacity of the affected stage(s) increases. Kalsbeck and Sykes[13] studied the task of handwriting and found evidence for increasing capacity limits of the response-control stage.

Inexperienced operators are prone to a second source of stage overload. Typically, much of the information available to an operator is either irrelevant or redundant. A novice operator will fail to recognize this and attempt to process more information than necessary. This results in overload and consequently low performance. For example, an operator may attend to (i.e., process) many small and irrelevant details in the appearance of a workpiece and thus fail to detect a critical flaw.

The process whereby an operator comes to attend to only essential information in a task is a critical mechanism underlying the development of skill. To take advantage of redundancy, an internal model of the task being worked on must be developed by the operator. This internal cognitive model uses available information to make predictions about future task requirements. Thus, in the preceding example, the operator, having acquired knowledge of path regularities in the movement of the workpieces, was able to predict future workpiece locations and regularities in the movement of the workpieces, thus avoiding processing unnecessary information. The availability of an accurate internal mode of the task is the most significant advantage enjoyed by a skilled operator over a novice counterpart.

The third way in which overload may occur is when two tasks compete for an operator's attention and simultaneously present information to the same limited-capacity stage. In this case, the operator may choose to process the information from only one task, thus drastically degrading performance in the other task. Or, the operator may choose to process some information from each task, thus producing milder degradation in both cases.

The foregoing model of the industrial operator as a series of stages sensitive to information flow rate (i.e., transmission rate) emphasizes the need for information-selection mechanisms to protect the operator from overload. An internal model enhances the ability to select properly only the essential information. The skilled operator is one who efficiently selects only needed information for processing during psychomotor activity.

The information-processing model contains three memory systems. These systems contribute several essential functions in psychomotor performance. They act as buffers to store temporarily (from 1 to 2 sec) rapidly arriving sensory information (sensory memory). They temporarily store up to seven "chunks" (words, names, digits, etc.) of information (short-term memory). Finally, they provide long-term storage that underlies learning and improvement in psychomotor performance (long-term memory).

Industrial work often requires the operator to time-share, or simultaneously perform, several separate subtasks. This time-sharing demand occurs on three levels. First, in even simple tasks, the operator must receive information, make decisions, and control response movements. Efficient performance may require that these activities occur parallel to one another. Second, more complex tasks often require the operator to make several separate responses simultaneously, for example, concurrent, but separate, hand motions. Third, the operator may be asked to perform two quite separate tasks at once. How efficiently can activities at each of these levels overlap?

At the first level, evidence suggests that information reception can efficiently overlap both decision making and response control. However, these latter two functions interfere with each other. More specifically, the initiation and correction of movements interferes with decision making. These response functions occur primarily in the second phase of movement control. Hence performance can be enhanced by eliminating or minimizing second-phase control. This can be done by terminating movements with mechanical stops rather than with closed-loop, operator guidance.

Time-sharing at the second level can be enhanced if the same mental function, information reception, decision making, or response control is not needed simultaneously by both activities. The refractory period* of the central decision-making stage requires that successive inputs to this process be separated by at least 300 msec. For example, if the operator is required to identify and respond to two successive signals, those signals should not occur within 300 msec of each other.

The processes involved in closed-loop movement control, including monitoring, selecting an appropriate corrective response, and initiating the correction, impose particularly high information-processing demands. Hence, when these processes are required by two simultaneous subtasks (e.g., independently moving each gripper), information overload and consequent interference between the subtasks can be expected. For example, elements "position" and "grasp" both impose high information-processing loads because they require significant second-phase, closed-loop control. Therefore they cannot be effectively time-shared. At the other extreme, elements "reach" and "move" (which do not generally involve precise, closed-loop movement control) generally can be carried out parallel with other elements.

Another critical factor in time-sharing efficiency is response-response compatibility. Some combinations of responses can be performed more easily than others. In executing simultaneous movements, performance is best when the hands (or feet) move in the same direction (e.g., both forward). Next best is complementary movements (e.g., one forward, one backward). Performance is worse for perpen-

---

* The period during which the operator is unable to process any new information.

dicular arrangements (one forward, one sideways). Similarly, responses that start at the same time are easier to time-share than those that do not. Selecting, initiating, or monitoring parallel (or successive) responses that have similar characteristics apparently requires less information processing than occurs in the case of unrelated movements. Symmetrical relationships between movements enhance this similarity effect even when the movements are made in opposite directions.

Time-sharing efficiency will be greatly enhanced when highly compatible stimulus-response (S-R) relationships are used. High S-R compatibility reduces the load on the decision-making stage responsible for selecting responses. Responses may almost become "self-selecting" with the most compatible mappings. This most readily occurs with tactile signals. For example, a vibrating machine control provides a highly compatible signal for the response of grasping the control more firmly. The operator may do this almost immediately with no disruption of other movement activities.

Finally, at the most complicated level, performing two separate tasks at once, performance depends on a wide variety of factors including the priorities that the operator attaches to the competing tasks. Typically, when an easy task was combined with a more difficult one, a greater percentage decline in performance was found for the easier task.[14]

Time-sharing efficiency improves with task experience for a variety of reasons. First, there is evidence that time-sharing is a general ability that can be enhanced by training. Operators who efficiently time-share one pair of tasks often are superior at time-sharing other task pairs. Second, as operators become well trained, tasks impose lower information-processing loads and even appear to become "automatic." Several reasons for this have been considered, including the following:

An internal task model frees the operator from processing redundant information.

Kinesthetic information, which may be processed faster than visual information (Table 32.4.) and which often is highly S-R compatible, is gradually substituted for visual information.

Certain information-processing steps (e.g., "check" operation) may be minimized or deleted entirely.

More efficient movement sequences involving less second-phase, close-loop control are developed.

Because the "automatic" time-shared tasks each impose lower information-processing demands upon the operator, there is less likelihood of overload, and efficient time-sharing is possible.

The preceding discussion has focused on time-sharing difficulties resulting from "central" (i.e., cognitive) interference between two activities. Obviously, tasks also may interfere with each other because of "structural" interaction: if one task requires the operator to look to the right while the other requires the operator to look left, the tasks will be mutually interfering. Such structural interference often is quite difficult to distinguish from central interference. This represents a primary difficulty in attempting to use "secondary-task" methods to assess mental workload.

Decision Making

Decision making refers to the processes whereby operators evaluate information made available by the initial perceptual processing. Decision making results in the selection of an intended course of action. Two decision-making characteristics are especially important: how much time decision making requires and how accurate decisions are.

Decision delays stem from two sources, capacity limitations and refractory limitations. Capacity limitations arise because decision-making stages can process information at only a limited rate. The amount of information transmission involved in a decision increases logarithmically with the number of possible stimuli that might be presented and the number of alternative responses from which the operator might select. In general, doubling the number of possible stimuli and responses increases the information transmitted in the decision by one bit.

Hick[15] showed that the rate of information flow per unit time remains constant at about 1 bit/ 220 msec. However, if the operator exceeds these margins by trying to go too fast, accuracy drops

TABLE 32.4. MINIMUM
REACTION TIMES (Kp) FOR
VARIOUS SIMULATION
MODALITIES

Simulation Modality	Reaction Time (msec)
Visual	150–225
Auditory	120–185
Tactual	115–190

very rapidly, and the rate of information transmission will fall. This occurs when the operator tries to increase the speed more than about 20%.

The other source of decision-making delay is a fixed delay of about 300 msec that must separate successive decisions. This is the so-called psychological refractory period. If information is presented to the decision-making stage within 300 msec of a previous decision, decision making will be delayed until the psychological refractory period has elapsed. This refractory delay does not decline with practice.

Decision accuracy depends not only on the operator's speed and accuracy strategy, but also on built-in biases. Some of these biases are listed in Table 32.5.

32.5.4. Paced Work

Various subjective estimates suggest that more than 50 million people worldwide are working on machine-paced (M/P) tasks. Hence, improving the working conditions on M/P tasks has spurred the interest of many researchers. In effect, more than 100 scientific papers have been published on this subject.[16] However, there is a high degree of diversity in the results obtained from these experiments. It is quite common to find a situation where results obtained from one study are completely contradicted in another study. This makes it very difficult to draw inferences that can be used to improve the working conditions on M/P tasks.

If there had not been a distinct economic advantage in utilizing M/P work, there might not have been more than 50 million people working in this area. It also would be true to say that if there had not been some disadvantages for the human working on M/P tasks, there might not have been over 100 publications in this area.

The economic advantages and disadvantages of using M/P tasks are reviewed in Table 32.6. The frequently referred to psychological disadvantages of M/P work include the following: (1) M/P work does not provide psychological growth for the workers; (2) M/P work causes boredom and job dissatisfaction.

We should maintain M/P tasks in the work environment only if we can simultaneously maintain the economic benefits and alleviate the human disadvantages of working on an M/P task.

From the research findings relating to the comparative merits of M/P and self-paced (S/P) work, the following emerges.

1. It becomes evident when utilizing young subjects on a pump ergometer, a bicycle ergometer, and a Harvard step test that the human body's efficiency is higher in S/P than in M/P work[17]; however, for older subjects, using an arm ergometer, the highest efficiency occurred in M/P work. On the same task for younger subjects, the highest efficiency occurred in S/P work.[18]

2. Perceptual load associated with task performance plays a significant role in evaluating M/P and S/P tasks. When the perceptual load of the task is low, there are no significant differences in stress levels between M/P and S/P tasks; however, for tasks with high perceptual load the stress is significantly lower during M/P than S/P task performance, and the error rates are higher.[19] These results are attributed to two facts, namely, (1) high stress during task performance is associated with high achievement motivation, high production performance, and low error rate[20]; and (2) during the

TABLE 32.5. SOME HUMAN BIASES THAT MAY AFFECT PERFORMANCE

Quantity Estimation	Bias
Horizontal distance	Underestimate
Height	Overestimate when looking down
	Underestimate when looking up
Speed	Overestimate if object accelerating
	Underestimate if object decelerating
Angle	Underestimate acute angles
	Overestimate obtuse angles
Temperature	Overestimate heat
	Underestimate cold
Weight	Overestimate if bulky
	Underestimate if compact
Numerosity	Consistently underestimate
Probability	Overestimate pleasant event likelihood
	Underestimate unpleasant event likelihood

TABLE 32.6. ECONOMIC ADVANTAGES AND DISADVANTAGES OF MACHINE-PACED WORK

Advantages
1. Reduces overhead cost through economic use of high technology, reduction of stock in progress, reduction in factory floor space, reduction in supervision cost.
2. Reduces direct cost through decreased training time, lower hourly wages, high production return per unit of wages.
3. Contributes to national productivity through provision of employment for less capable workers, reduction in the production costs of goods and services.

Disadvantages
1. Does not provide for each worker's maximal work capacity.
2. Economically viable only for high-volume production.
3. Does not provide for the psychological growth of workers.

S/P task the operator had to keep track of the quantity of work output, whereas in M/P tasks the work output was controlled by the machine. It was hypothesized that this additional task of keeping track of work output imposes additional mental load and increases the stress associated with high mental load task performance. In a study by Knight and Salvendy,[21] subjects performed a task with high perceptual load both in the M/P and S/P modes. In the S/P mode the subjects performed the task with a variety of different performance feedbacks. Table 32.7 shows that the stress associated with task performance is the function of performance feedback. The more precise the performance feedback, the lower the stress associated with task performance.

3. In man-computer interactive work, the attentional work environment has a much greater impact on the stress associated with task performance than do the stresses associated with M/P and S/P task performance. This is illustrated in Table 32.8. In this experiment,[22] subjects were asked to perform the same task, both M/P and S/P, in both financial and nonfinancial work environments. A waiting, or anticipation, period of 4–6 sec was introduced between each work cycle. The physiological measures presented in Table 32.8 reflect on the deceleration and acceleration of the heartbeats during the waiting period. The external attention task required visual input from a VDT terminal, whereas the internal attentional task required arithmetic calculations.

4. There is some experimental evidence to suggest[23] that two-thirds of the blue-collar labor force prefer to work in an S/P work environment whereas one-third of the labor force prefers to work in an M/P setting. These job preferences correlate very closely with job satisfaction and productivity. The psychological profiles of those who prefer M/P work versus those who prefer S/P work are illustrated in Table 32.9.

32.6. SUPERVISORY CONTROL OF ROBOTICS SYSTEMS

Humans can supervise industrial robots in one of the following two ways:

1. When humans are working adjacent to industrial robots. This type of supervision is strongly discouraged on grounds of lack of safety and the potential psychological and mental health implications of working adjacent to robots. The only partial justification for this work arrangement may possibly exist for low-reliability robots that require a high degree of operator attention.

TABLE 32.7. EFFECTS OF PERFORMANCE FEEDBACK ON THE REDUCTION OF STRESS IN SELF-PACED WORK

Work Condition	Stress Index[a]
Self-paced	
No feedback	100
Cycle feedback	90
Time feedback	86
Combined time and cycle feedback	57
Machine-paced	62

[a] A difference in the stress index of six or more units is statistically significant at 5% level.

TABLE 32.8. ACCELERATION AND
DECELERATION IN HEARTBEATS
AS A FUNCTION OF JOB CONTENT,
JOB DESIGN, AND THE NATURE OF
INCENTIVES

	External Attention		Internal Attention	
	M/P	S/P	M/P	S/P
No incentive	66	63	71	69
Financial incentive	68	62	75	72

2. When robots form a part of a flexible manufacturing system, such as that illustrated in Figure 32.7. In these cases, the entire system is, typically, jointly supervised by computer and human. The human supervisor sits in front of a computer terminal and has an impact on certain parts of the total systems functioning.

This second mode of supervisory control is the way current and future industrial robots should be supervised, and, as such, is the subject of discussion in this section.

32.6.1. Concept of Supervisory Control

An important point to consider in advocating the role of the human in supervising FMS comes from reliability data. Figure 32.8 compares the reliability of completely automatic systems at different levels of redundancy, against a single-redundant system in which one of the components is a well-trained human.

The human should feel in control of the plant and thus the computer software should be at his disposal. He should have the option to override the computer if he feels that it is necessary. This is because the human is much more flexible to novel situations than the computer. An important point is that the human's role should be coherent. This coherence of the human's role must be assured in the initial stages of the design process when system tasks are allocated between man and computer (Figure 32.3). As Figure 32.3 indicates an inverted "U" relationship exists between arousal level, productivity, and job satisfaction.

If the operator supervising an FMS through the computer is given too little to do, boredom results, which leads to degraded performance and less productivity. On the other hand, if the operator is given too much to do, mental overload occurs, which also leads to decreased performance and less productivity.

Thus some level between the two extremes of Figure 32.3 will result in maximum performance. Salvendy[26] has hypothesized that enriched jobs result in lower fatigue and psychological stress than simplified jobs. The rationale is that simplified jobs have less decision making than enriched jobs.

In assigning functions to humans in a FMS, many decision-making responsibilities correspond to an enriched job, whereas low arousal levels and minimum decision making for the operator correspond to a simplified job. Thus, in allocating responsibilities between human and computer in FMS, one must be cognizant of levels of arousal for the human and the degree of decision making allocated between human and computer.

Figure 32.9 shows that allocation of functions takes place after separation of functions. At this

TABLE 32.9. PSYCHOLOGICAL
PROFILES OF OPERATORS WHO
PREFER S/P AND THOSE WHO
PREFER M/P WORK

M/P Work	S/P Work
Less intelligent	More intelligent
Humble	Assertive
Practical	Imaginative
Forthright	Shrewd
Group-dependent	Self-sufficient

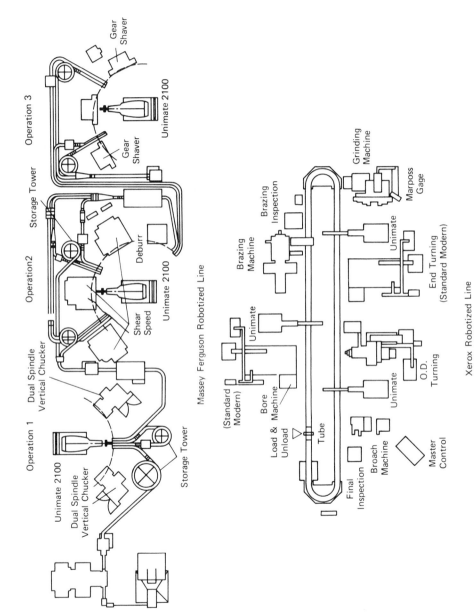

Fig. 32.7. Examples of robotized flexible conveyor lines where parts must go through each of the machines and where the entire system is controlled by a combined decision support of a computer and a supervisor, who interacts with the computer terminal. (*Source.* Reference 24.)

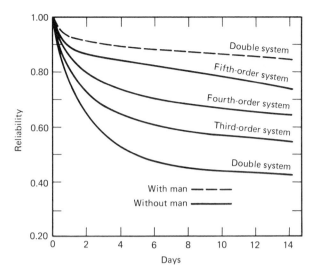

Fig. 32.8. The reliability of a double-redundant navigation system in which one redundant component is a human (dashed lines), as compared with the reliability of systems with various orders of redundancy in which all components are machines (solid lines). (*Source.* Reference 25.)

point, tasks that the human performs and tasks that the computer performs are separated. Also, there can be some overlap between human and computer to increase reliability and efficiency in the FMS. Tables that show primarily what the human does and what the computer does are useful in assigning tasks to computer or to man. Thus we have developed human/computer comparison tables to help decide which tasks go primarily to the computer or primarily to the human in the FMS.[38] At the initial stages of allocation, one must design function tasks that are meaningful to the operator.

32.6.2. Models of Supervisory Control

The operator in FMS may shift his attention among many machines, rendering to each in turn as much attention as is necessary to service it properly or keep it under control. The human tends to have more responsibility for multiple and diverse tasks. It is appropriate to view the human as a time-shared computer with various distributions of processing times and a priority structure that allows preemption of tasks. This can be done by queueing theory formulation.[27,29,30]

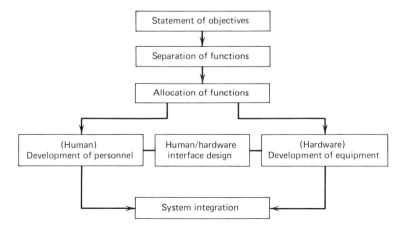

Fig. 32.9. Levels of allocation of functions in a system. (*Source.* Reference 28.)

Some investigators have developed models of human decision making in multitask situations. Senders[28] has modeled instrument-monitoring behavior of humans. He assumed that the human used his limited input capacity to sequentially observe a number of instruments in a random order. The fraction of time spent observing a particular instrument was a measure of work load.

Tulga and Sheridan[31] developed a multitask dynamic decision paradigm with such parameters as interarrival rate of task, time before tasks hit the deadline, task duration, the productivity of the human for performing tasks, and task value densities. In this experiment, a number of task-sharing finite completion times and different payoffs appear on a screen. The subject must decide which task to perform at various times to maximize the payoff. When the human performs a variety of tasks by the aid of computer, allocation of responsibility for different tasks is important for optimum performance.

Several investigators[27,29] have studied multitask decision making where the human is required to allocate his attention between control tasks and discrete tasks. In a queueing theory formulation, the human "server" serviced various tasks, arriving at exponentially distributed interarrival times. The growing model predicted the mean waiting time for each task as well as the mean fraction of attention devoted to each task.

For optimal performance of different tasks, Govindaraj and Rouse[32] developed a model with a number of parameters: (1) ratio of weights on control to weights on later error; (2) ratio of nominal weights on control to weights on control over discrete task intervals; and (3) threshold on changes in control. The model could be useful in evaluating displays where the future reference is known for a certain distance. If the discrete task characteristics are known, the amount of time required for the discrete tasks and when they should be performed can be determined. An experiment was conducted where subjects controlled an airplane symbol over a map, shown a fixed distance into the future. Results revealed that the model compared favorably with experimental data.

32.7. MAJOR ISSUES TO BE CONSIDERED FOR THE EFFECTIVE INTEGRATION OF HUMAN FACTORS PRINCIPLES IN PLANNING ROBOTIC SYSTEMS

According to the U.S. Congress Office of Technology Assessment,[33] there are a number of institutional and organizational barriers to the use of informational technology, which also has bearing on the use of industrial robots. These include high initial cost, the lack of high-quality programming, and the dearth of local personnel with adequate training. In this connection Rosenbrock[8] addresses two vital behavioral and social issues, namely, (1) the skills that robots call for will usually be new, yet there is no reason why they should not be based on older skills and developed from them; and (2) industrial robots will aid us to carry on rapidly the process of breaking jobs into their fragments and performing some of these fragments by machines, leaving other fragments to be done by humans. This has broad implications for the training and retraining of personnel and for the design of the psychological contents of jobs. Many of these implications associated with the introduction of industrial robots have effectively been managed in Japan.[34] In this regard, seven positive and seven negative aspects of the social impact of robots are summarized in Table 32.10.

The most significant study yet conducted on the impact of robotics on supervision, management, and organization was published in 1983[35] in both Japanese and English. This 300-page report summarizes a large-scale study undertaken by the Japan Management Association. In this study, more than 20 "robotized" Japanese companies were selected as case studies to determine just what kind of impact on management the introduction of robots to the workplace has had to date. For instance, what

TABLE 32.10. SOCIAL IMPACTS OF ROBOT DIFFUSION

Positive Impacts	Negative Impacts
1. Promotion of worker's welfare	1. Unemployment problems
2. Improvement of productivity	2. Elimination of pride in old skills
3. Increase in safety of workers	3. Shortage of engineers and newly trained skilled workers
4. Release of workers from time restrictions	4. Production capacity nonproportional to the size of the labor force
5. Ease in maintaining quality standard	5. Decrease in flow of labor force from underdeveloped to developed countries
6. Ease of production scheduling	6. Safety and psychological problems of robot interaction with human
7. Creation of new high-level jobs	7. Great movement of labor population from the second to the third sector of industry

kind of problems did these companies encounter during the introduction process and how did they go about overcoming them? What have they experienced following the robotization of their respective manufacturing processes? How have they coped with day-to-day worker-related problems? How has robotization of their various operations affected them overall? What type of utilization is envisioned for the future? These and other key points are taken up and discussed in detail in this new study on robotics in Japan. The various case studies and other data contained in this report should prove an effective tool for any manufacturer considering the robotizing of its operations.

32.7.1. Study Objectives

The 1980s are expected to witness more widespread, advanced use of industrial robots, and the impact this technology and its extensive applications will have is certain to be far-reaching. Accelerating this trend toward industrial robotization are the following several factors:

>   Improvement of productivity
>   Prevention of labor accidents and occupational hazards
>   Conservation of materials and energy
>   Improvement of production control
>   Improvement of working environment
>   Coping with the shortage of skilled labor

These are some of the advantages of introducing robots into the manufacturing process. At the same time, however, a company that intends to robotize its manufacturing process cannot avoid coming to grips with some very serious problems in terms of management, such as labor-management relationships, work displacement, surplus manpower, retraining, optimum investment levels, and assessment of the effects of robotization.

This study is an in-depth analysis and assessment of the technologies, applications, labor-management issues, and demand trends of robotics. As such, it should provide some significant insights into the issues surrounding technological innovations and management, thus proving an effective tool for corporate managers, labor union leaders, government-related agencies, and research institutes worldwide who are seriously considering introducing robot technology.

32.7.2. Study Method

Interviews were conducted with top-level managers from more than 20 companies at varying stages of robotization from a variety of industries. This interview process has been supported by extensive and in-depth independent research to generate a complete and comprehensive picture of the present and future aspects of robot utilization in Japan, with special emphasis on management. The companies interviewed were chosen from among the general machinery, transportation machinery, electric equipment, and precision instrument industries, with equal numbers selected from among different corporate sizes (large, medium, and small).

32.7.3. Study Results

The study summarizes the social and economic factors contributing to the spread of robots and the results derived from robot utilization, as follows:[35]

> *A close look at the situation in Japanese industry shows that industrial robots are being utilized primarily because*
>
> *(a) there is a shortage of skilled labor at those worksites called "hazardous or otherwise undesirable working environments"—what skilled workers there are, are mostly older workers;*
>
> *(b) competition between companies is intensifying as productivity rises and quality improves— market needs have become more advanced;*
>
> *(c) regulations for the prevention of labor accidents are being strengthened—Labor Ministry requirements.*
>
> *The development of industrial robots in Japan is proceeding apace with the needs of the manufacturing industry. Robot utilization is fulfilling the above requirements and can be said to be achieving (a) increased productivity and improved adaptability vis-a-vis the product changes that occur in multi-product, small-batch production processes; (b) improved and consistent quality; (c) more efficient equipment investments (reduced time until returns on investments are realized); (d) worker protection and prevention of labor accidents; (e) labor savings; (f) more efficient production*

planning—more stable quality means that the amount of stock on hand can be reduced, thus lowering stock costs and making production planning easier; and (g) elimination of problems related to skilled labor shortages.

There are also quite a few companies that have attained results not anticipated prior to the introduction of industrial robots. Some examples of these are:

(a) Improved worker attitudes—workers, stimulated by the new technology, have been motivated to take a more active part in the improvement of work processes, submitting suggestions concerning how best to utilize that technology;

(b) Improvements in those work processes before and after the ones making use of industrial robots—improved peripheral equipment and overall production technology to keep pace with the robots;

(c) The realization of total production systems—the installation of robots makes it possible to encode the job knowhow of skilled laborers, thus approaching a total production system for the entire assembly lines;

(d) The stabilization of production output—output can be stabilized regardless of the number of workers who show up for work each day, or their degree of skill;

(e) More reliable technological capabilities have led to increased product orders.

The strategic implications of industrial robots for those enterprises utilizing them in their manufacturing processes can, in broad terms, be summarized as follows:

(a) By enhancing productivity and improving product quality, companies are capable of increasing their shares of markets and thus improving their market positions;

(b) Companies utilizing industrial robots find it possible to open new lines of business (including the development of new products) and enter into new markets;

(c) Industrial robots assure companies of stable labor power and provide labor itself with improved benefits.

Based on results such as these, it seems safe to conclude that robot utilization not only provides the user company with a competitive edge in the market, but also plays a major role in stabilizing its labor situation.

Nevertheless, in order to realize this kind of performance on the part of industrial robots, the managers of user companies have a number of areas which they must give proper consideration to. Principal among these are:

(a) The re-education and re-training of employees in line with the introduction of robots;

(b) The carrying out of robot engineering for the purpose of installing and operating industrial robots; and

(c) The implementation of measures aimed at improving the worksite and ensuring worker safety.

There are still any number of jobs currently being done by humans that can be classified as dangerous or heavy labor both within and outside of the manufacturing industry. However, the installation and utilization of industrial robots to perform these kinds of jobs is seen as continuing. It is also expected that robots will begin to serve in the realm of social services as well before too long.

ACKNOWLEDGMENT

The author is grateful to Professor Shimon Y. Nof, who has inspired the author to write on this subject matter; to Peter J. Faber for his contributions to the literature search on the social impacts of industrial robots; and to Shue-Ling Hwang for her contribution to the literature search on supervisory control. Parts of the section on industrial work performance is taken from Salvendy, G. and Knight, J. K., "Psychomotor Work Performance," in *Handbook of Industrial Engineering* (Editor: G. Salvendy), New York: John Wiley, 1982. A version of this chapter is published in *Behavioral and Information Technology,* Vol. 2, No. 3, 1983. The author is grateful to the publishers of *Behaviour and Information Technology* for the permission granted to reproduce significant parts of the article in this chapter.

REFERENCES

1. Senker, P., Social Implications of Automation, *The Industrial Robot,* Lolswold Press, Oxford, England, Vol. 6, No. 2, June 3, 1979, pp. 59–61.

2. Mangold, V., The Industrial Robot as Transfer Device, *Robotics Age,* Robotics Publishing Corporation, Houston, Texas, July/August 1981, pp. 20–26.

3. Weekley, T. L., A View of the United Automobile Aerospace and Agricultural Implement Workers of America (UAW) Stand on Industrial Robots, SME Technical Paper MS79–776, Dearborn, Michigan, 1979.

4. Sugarman, R., The Blue Collar Robot, *IEEE Spectrum,* Institute of Electrical and Electronics Engineers, New York, Vol. 17, No. 9, September 1980, pp. 52–57.

5. Nof, S. Y., Knight, J. L., Salvendy, G., Effective Utilization of Industrial Robots—A Job and Skills Analysis Approach, *AIIE Transactions,* Vol. 12, No. 3, September 1980, pp. 216–225.

6. Albus, J., Industrial Robot Technology and Productivity Improvement, in OTA, *Exploration Workshop on the Social Impact of Robotics,* U.S. Congress Number 90–240 0–82–2, 1982.

7. Aron, P., Robots Revisited: One Year Later, Office of Technology Assessment, *Exploratory Workshop on the Social Impact of Robotics,* U.S. Congress Number 90–240 0–82–2, 1982.

8. Rosenbrock, H. H., Robots and People, *Measurement and Control,* Vol. 15, March, 1982, pp. 105–112.

9. Salvendy, G. and Seymour, W. D., *Prediction and Development of Industrial Work Performance,* Wiley, New York, 1973, pp. 105–125.

10. Borman, W. C. and Peterson, N. G., Selection and Training of Personnel, Chapter 5.2 in Salvendy, G., Ed., *Handbook of Industrial Engineering,* Wiley, New York, 1982.

11. Warnecke, H. J. and Schraft, R. D., *Industrial Robots: Application Experience,* Bedford, England: I.F.S. Publication Ltd., 1982.

12. Wechsler, D., *The Range of Human Capabilities,* 2nd ed., Williams and Wilkins, Baltimore, 1952.

13. Kalsbeck, J. W. H. and Sykes, R. N., Objective Measurement of Mental Load, in Sanders, A. F., Ed., *Attention and Performance,* North-Holland, Amsterdam, 1970.

14. Kantowitz, B. H. and Knight, J. L., Testing Tapping Time-Sharing: II. Auditory Secondary Task, *Acta Psychologica,* Vol. 40, 1976, pp. 343–362.

15. Hick, W. E., On the Rate of Gain of Information, *Quarterly Journal of Experimental Psychology,* Vol. 4, 1952, pp. 11–26.

16. Salvendy, G. and Smith, M. J., Eds., *Machine Pacing and Occupational Stress,* Taylor and Francis, London, 1981.

17. Salvendy, G., Physiological and Psychological Aspects of Paced Performance, *Acta Physiologica,* Vol. 42, No. 3, 1973, pp. 267–275.

18. Salvendy, G. and Pilitsis, J., Psychophysiological Aspects of Paced and Unpaced Performance as Influenced by Age, *Ergonomics,* Vol. 14, 1971, pp. 703–711.

19. Salvendy, G. and Humphreys, A. P., Effects of Personality, Perceptual Difficulty, and Pacing of a Task on Productivity, Job Satisfaction, and Physiological Stress, *Perceptual and Motor Skills,* Vol. 49, 1979, pp. 219–222.

20. Salvendy, G. and Stewart, G. K., The Prediction of Operator Performance on the Basis of Performance Test and Biological Measures, *Proceedings of the Third International Conference on Production Research,* Amherst, Massachusetts, August 4–8, 1975.

21. Knight, J. L. and Salvendy, G., Effects of Task Feedback and Stringency of External Pacing on Mental Load and Work Performance, *Ergonomics,* Vol. 24, No. 10, 1981, pp. 757–764.

22. Sharit, J. and Salvendy, G. External and Internal Attentional Environments: II. Reconsideration of the Relationship between Sinus Arrhythmia and Informational Load, *Ergonomics,* Vol. 25, No. 2, 1982, pp. 121–132.

23. Salvendy, G., McCabe, G. P., Sanders, S. G., Knight, J., and McCormick, E. J., Impact of Personality and Intelligence on Job Satisfaction of Assembly Line and Bench Work—An Industrial Study, *Applied Ergonomics,* Vol. 13, No. 4, December 1982, pp. 293–299.

24. DuPont-Gatelmond, C., A Survey of Flexible Manufacturing Systems, *Journal of Manufacturing Systems,* Vol. 1, No. 1, 1982, pp. 1–15.

25. Grodsky, M. A., Risk and Reliability, *Aerospace Engineering,* January 1962, pp. 28–33.

26. Salvendy, G. An Industrial Dilemma: Simplified versus Enlarged Jobs, in Murumatsu, R. and Dudley, N. A., Eds., *Production and Industrial Systems,* Taylor and Francis Ltd., London, 1978.

27. Rouse, W. B., Human-Computer Interaction in Multitask Situations, *IEEE Transactions on Systems, Man, and Cybernetics,* SMC-7, No. 5, May 1977, pp. 384–392.

28. McCormick, E. J. and Sanders, M. S., *Human Factors in Engineering and Design,* McGraw-Hill, New York, 1982.

29. Chu, Y. Y. and Rouse, W. B., Adaptive Allocation of Decision Making Responsibility between

Human and Computer in Multitask Situations, *IEEE Transactions on Systems, Man, and Cybernetics,* SMC-9, No. 12, December 1979, pp. 769–778.

30. Walden, R. S. and Rouse, W. B., A Queueing Model of Pilot Decision Making in A Multitask Flight Management Task, *IEEE Transactions on Systems, Man, and Cybernetics,* SMC-8, No. 12, December 1978, pp. 867–875.

31. Tulga, M. K. and Sheridan, T. B., Dynamic Decisions and Work Load in Multitask Supervisory Control, *IEEE Transactions on Systems, Man, and Cybernetics,* SMC-10, No. 5, May 1980, pp. 217–232.

32. Govindaraj, T. and Rouse, W. B., Modeling the Human Controller in Environments that Include Continuous and Discrete Tasks, *IEEE Transactions on Systems, Man, and Cybernetics,* SML-11, No. 6, 1981, pp. 410–417.

33. Office of Technology Report, Information Technology and Its Impact on American Education, U.S. Government Printing Office, GPO Stock No. 052–003–00888–2, 1982.

34. Hasegawa, Y., How Robots Have Been Introduced into the Japanese Society, presented at the Micro-electronics International Symposium, Osaka, Japan, August 17–19, 1982.

35. FUJI Corporation, Robotics and the Manager, Business Building 5–29–7 Jingu-mae, Shibuya-ku, Tokyo 150, Japan, 1983.

36. Salvendy, G. and Knight, J. R., Psychomotor Work Capabilities, Chapter 6.1 in Salvendy, G. Ed., *Handbook of Industrial Engineering,* Wiley, New York, 1982.

37. Sugimoto, N. and Kawaguchi, K., *Proceedings of the 13th International Symposium on Industrial Robots,* Chicago, Illinois, April 17–21, 1983.

38. Hwang, S. L., Barfield, W., Chang, T-C., and Salvendy, G., "Integration of Human and Computers in Flexible Manufacturing Systems," *International Journal of Production Research,* Vol. 22, 1984, pp. 841–856.

BIBLIOGRAPHY

Ayres, R. V. and Miller, S. M. *Robotics Applications and Social Implications,* Cambridge, Massachusetts, 1983.

McCormick, E. J. and Sanders, M. S., *Human Factors in Engineering and Design,* 5th ed., McGraw-Hill, New York, 1982.

Parsons, H. M. and Kearsley, G. P., Robotics and Human Factors: Current Status and Future Prospects, *Human Factors,* Vol. 24, 1982, pp. 535–552.

Office of Technology Assessment, Exploratory Workshop on the Social Impact of Robotics, U.S. Superintendent of Document Catalogues, No. 90–240–0–82–2, Washington, D.C., 1982.

Office of Technology Assessment, Blumenthal, M. S., et al., *Computerized Manufacturing Automation: Employment, Education, and the Workplace,* Government Printing Office, Washington, D.C., 1984.

Salvendy, G., Ed., *Handbook of Industrial Engineering,* Wiley, New York, 1982.

Salvendy, G., Ed., *Handbook of Human Factors,* New York: Wiley. In print.

Warnecke, H. J. and Schraft, R. D., *Industrial Robots: Application Experience,* I.F.S. Publication Ltd., Bedford, England, 1982.

CHAPTER 33
EVALUATION AND ECONOMIC JUSTIFICATION

YUKIO HASEGAWA

Waseda University
Tokyo, Japan

33.1. INTRODUCTION

Investment in new technology requires careful planning and uses cost-evaluation techniques similar to those for traditional capitalized equipment. However, through robotics and systems integration, the potential for cost savings, flexibility, and improved throughput is much higher for certain applications.

It should be remembered that the objective of any robotization project is not to emulate existing methods and systems, simply replacing humans by robots, but to develop a new, integrated system providing the following:

1. Flexibility.
2. Increased productivity.
3. Reduced operating costs.
4. Increased product quality.
5. Elimination of health and safety hazards.

Economic evaluation provides the decision framework to compare the benefits of automation through robotics with the present system and with other alternatives. The economic justification is based on the comparison between the capital cost and operating expenses of the robot installation that is being considered and the cash flow benefits projected.

The purpose of this chapter is to describe the complete procedure for economic evaluation and justification of proposed robotization projects. Although investment in a robotization project is similar to other capitalized equipment projects, four major differences stand out.

1. Robots can replace human labor and yield not only labor and benefit savings but also significant reduction in requirements for employee services and facilities.
2. Robots can provide such flexibility in production capability that the capacity of a company to respond effectively to future market changes has a clear economic value, but this capacity is usually difficult to measure.
3. As components of computerized production systems, robots force their users to rethink and systematically define and integrate the functions of their operation. This in itself carries major economic benefits and frequently lets a company "clean up its act."
4. A robot is by definition reprogrammable and reusable and has a useful life that can often be longer than the life of a planned production facility.

33.1.2. Justifying Industrial Robot Applications

A primary decision issue is whether a robot is indeed the best solution for a particular application. The justification used by companies that have applied robots generally follows the five benefit areas listed earlier. The results of recent surveys on such justification factors are found in Table 33.1.

TABLE 33.1.   JUSTIFICATION FOR USING ROBOTS AS FOUND IN COMPANIES THAT ARE USING ROBOTS

Rank	U.S. Companies	Japanese Companies[a]
1	Reduced labor cost	Economic advantage
2	Improved product quality	Increased worker safety
3	Elimination of dangerous jobs	Universalization of production system
4	Increased output rate	Stable product quality
5	Increased product flexibility	Labor shortage
6	Reduced material waste	
7	Compliance with OSHA regulations	
8	Reduced labor turnover	
9	Reduced capital cost	

Source: Reference 21.

[a] Survey in Japan consisted of only five categories.

33.2.   GENERAL PROCEDURE OF ROBOTIZATION PROJECT EVALUATION AND JUSTIFICATION

Figure 33.1 summarizes the series of steps used to evaluate fully the economics of a robotization project. The step numbers presented in the figure correspond to the detailed discussion that follows. The evaluation and analysis of a robotization project can be divided to a precost and a cost-analysis phase.

33.2.1.   Precost Phase

The precost phase focuses basically on the *feasibility* of a proposed robotization project: feasibility in terms of technical capability to perform the necessary job and in terms of production capacity and utilization relative to predicted production schedules. This phase follows six steps.

STEP 1:   Determination of alternative manufacturing methods.
STEP 2:   Feasibility study.
STEP 3:   Select which job to robotize.
STEP 4:   Noneconomic considerations.
STEP 5:   Data requisition and operational analysis.
    (a)   Projected production volumes.
    (b)   Parts throughput requirements.
    (c)   Projected daily production hours.
    (d)   Robot utilization.
    (e)   Capacity-volume requirements.
STEP 6:   Decisions concerning future applications.

33.2.2.   Cost-Analysis Phase

This phase focuses on detailed cost analysis for investment justification and includes five general steps.

STEP 7:   Period evaluation, depreciation, and tax data requirements.
STEP 8:   Project cost analysis.
    (a)   Labor considerations.
    (b)   Acquisition and start-up costs.
    (c)   Operating expenses.
STEP 9:   Evaluation techniques.
    (a)   Minimum cost rule method.
    (b)   Capital recovery method.
    (c)   Rate of return on investment method.
    (d)   Permissible investment amount method.

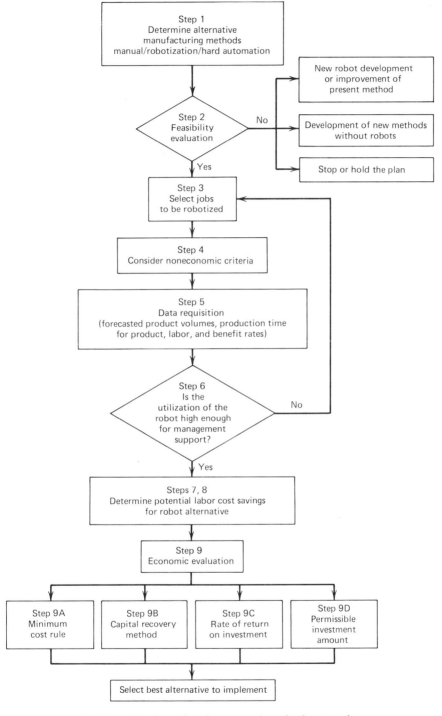

Fig. 33.1. Flowchart of project economic evaluation procedure.

Step 10:   Additional economic considerations.

   (a) Effects of interest rates.

   (b) Performance gaps.

   (c) Use economic evaluations for reference only.

   (d) Two types of replacement mistakes.

   (e) Effect of recession and inflation.

STEP 11:   Recommendations based on analysis results.

33.3.   PRECOST PLANNING ANALYSIS

Prior to any thorough economic evaluation of capitalized equipment, several initial considerations and planning studies must be carried out. The first six steps of the project cost analysis are related to determining the best manufacturing method, selecting the best jobs to robotize, and the feasibility of these options. Noneconomic considerations must be studied as well as pertinent data collected concerning product volumes and operation times.

33.3.1.   STEP 1:   Determination of Alternative Manufacturing Methods

The three main alternative manufacturing methods, namely, manual labor, flexible automation and robots, and hard automation, are compared in Figure 33.2. These alternatives are economically compared by their production unit cost at varying production volumes. Flexible, programmable automation and robots are most effective for medium production volumes. These medium volumes can range, depending on the particular products, from a few tens or hundreds of products per year per part type to thousands of products per year. For annual volumes of hundreds of thousands, hard automation is usually preferred.

In the area of assembly, Boothroyd[6] compared alternative assembly systems, including manual assembly with transfer devices or with special-purpose parts feeders; special-purpose automatic assembly with indexing transfer or with free transfer; programmable robotic assembly with fixtures or with moving conveyors. Effects of the number of parts per assembly, annual volume, and style/design variations have been studied. Results are illustrated in Figure 33.3.

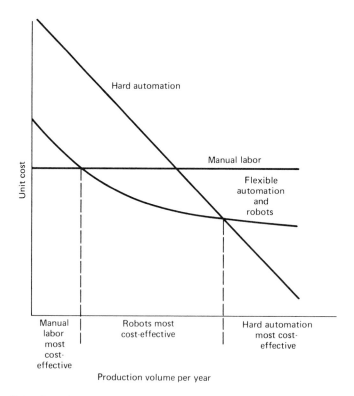

Fig. 33.2.   Comparison of manufacturing methods for different production volumes.

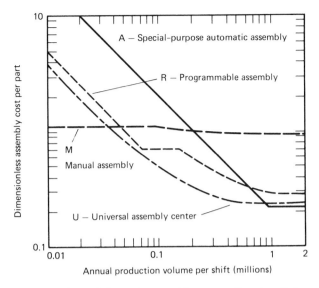

Fig. 33.3. Comparison of alternative assembly systems (based on Reference 6.)

33.3.2. STEP 2: Feasibility Study

At first it is very important to check the feasibility of the robotization plan carefully. Generally speaking, there have been some cases that have passed (by mistake) the economic justification, but then still had a problem of feasibility. The author imagines that the reason there have been many failures in robotization plans is that the robotization system design includes many complicated conditions that are more difficult than the conventional production systems design and are not yet fully understood.

In the feasibility study of the alternatives we must consider the following items:

1. Is it possible to do the job with the planned procedure?
2. Is it possible to do the job within the given cycle time?
3. Is it possible to ensure reliability as a component of the total system?
4. Is the system sufficiently staffed and operated by assigned engineers and operators?
5. Is it possible to maintain safety?
6. Is it possible to keep the designed quality level?
7. Can inventory be reduced?
8. Can material handling be reduced in the plant?
9. Are the material-handling systems adequate?
10. Is the product designed for robot-handling and automated assembly?
11. Can the product be routed in a smooth batch-lot flow operation?

The alternatives that have passed the feasibility screening are moved to the next level of evaluation. But if the plan is not passed, as shown in Figure 33.1, we must search for other solutions, such as developing a new robot, improving the proposed robot, or developing other alternatives without robots.

33.3.3. STEP 3: Select Which Job to Robotize

Job selection for a single robot or a group of robots is a difficult task. In general, the following five job grouping strategies can be used to determine feasible job assignments.

1. Component products belonging to the same product family.
2. Products at present being manufactured near each other.
3. Products that consist of similar components and could share part-feeding devices.
4. Products that are of similar size, dimensions, weight, and number of components.
5. Products with a rather simple design that can be manufactured within a short cycle time.

33.3.4. STEP 4:   Noneconomic and Intangible Considerations.

Several noneconomic issues should be addressed in regard to specific company characteristics, policy, social responsibility, and management's direction, such as:

1. Will the robotization meet the general direction of the company's automatization?
2. Will the robotization meet the fundamental policy of the standardization of equipment and facilities?
3. Will the plan be able to meet future product model change or production plan change?
4. Will the plan promote improved quality of working life for workers?
5. Will the plan influence good company reputation?
6. Will the plan raise the morale of employees?
7. Will the plan promote technical progress of the company?

Table 33.2 lists several other intangible benefits associated with robotization.

Several special differences between robots and other capitalization equipment also provide numerous intangible benefits.

1. Robots are reusable.
2. Unlike hard automation, robots are multipurpose and can be reprogrammed for many different tasks.
3. Because of reprogrammability, the useful life of the robotic system can often be three or more times longer than that of fixed (hard) automation devices.
4. Tooling costs for robotic systems also tend to be lower owing to the programming capability around certain physical constraints.

TABLE 33.2. INTANGIBLE BENEFITS OF ROBOTIZATION[a]

Robotization Can Improve	Robotization Can Reduce or Eliminate
Flexibility	Hazardous, tedious jobs
Plant modernization	Safety violations and accidents
Labor skills of employees	Personnel costs for training
Job satisfaction	Clerical costs
Methods and operations	Cafeteria costs
Manufacturing productivity capacity	Need for restrooms
	Need for parking spaces
Reaction to market fluctuations	Burden, direct, and other overhead costs
Product quality	
Business opportunities	Manual material handling
Share of market	Inventory levels
Profitability	Scrap and errors
Competitive position	New-product launch time
Growth opportunities	
Handling of short product-life cycles	
Handling of potential labor shortages	
Space utility of plant	
Level of management	

[a] Analyzing the amount of change in each of these categories in response to robotization and assigning quantitative values to the intangible factors is necessary if they are to be included in the financial analysis. Otherwise, they can only be used as weighting factors when determining the best alternative.

5. Production operations can often be started up much sooner because of the lesser construction and tooling constraints.

6. Modernization in the plant can be implemented by eliminating discontinued automation systems.

Questions often arise concerning long-range unmeasurable effects of robotization on economic issues. A few such issues include:

1. Will the robotization raise product value and price?
2. Will the robotization expand the sales volume?
3. Will the plan decrease the production cost?
4. Will the robotization decrease the initial investment amount?
5. Will the robotization reduce lead time for products?
6. Can manufacturing costs be reduced?
7. Can inventory costs be reduced?
8. Will robotization reduce direct and indirect labor costs or just shift workers' skills?
9. Can the burden (overhead) rate be reduced?
10. Will the robot be fully utilized?
11. Will setup time and costs be reduced?
12. Can material-handling costs be reduced?
13. Will damage and scrap costs be reduced?

33.4. CASE STUDY ILLUSTRATION: BRAKE DRUM ASSEMBLY OPERATION

An assembly operation is analyzed as a case study for illustration purposes throughout the remainder of this chapter. This is a hypothetical example, and although it involves assembly, the analysis procedure is general in nature and applicable to other application areas as well. Suppose the operation is responsible for all parts and assemblies used in both front and rear brake drums manufactured by a given company. The brake components are shown in Figure 33.4.

Presently only one type of brake drum is being produced; however next year a new Version II model will be introduced. In five years, the original version will be phased out of operation at this particular plant. Other product types may be announced in the future, but this case considers only the assembly process and cost structure of Versions I and II. Version II will contain a longer bleeder valve in both the front and rear wheel cylinder.

Parts are produced and inventoried in the general parts warehouse until needed on the final assembly line. Management will alter the production schedule beginning next year to a just-in-time scheduling approach. This control technique assigns a master production schedule for the month and a constant production of parts per day to meet the next day's final assembly schedule. Therefore the robot, if implemented, would produce only one day's supply of parts, and in-process storage would be essentially eliminated.

33.4.1. STEP 5: Data Requisition and Operational Analysis

To analyze the economic effects of the robot operation fully, much data and subsequent analysis concerning the plant's operating and long-range manufacturing plans are required, including:

1. Projected production volumes.
2. Parts throughput requirements.
3. Projected daily production hours.
4. Robot utilization.
5. Capacity volume requirements.

Projected Production Volumes

Projected volumes for the two types of brake drums were obtained from the cost engineering department. They indicate vast fluctuations in product and assembly production schedules as shown in Table 33.3.

Parts Throughput Requirements

The volumes of each of the three subassemblies used on both brake drum versions are listed in Table 33.4. These volumes are based on quantities required for assembly of new brake drums plus additional quantities specified by the sales department. The total quantities are used later to determine parts

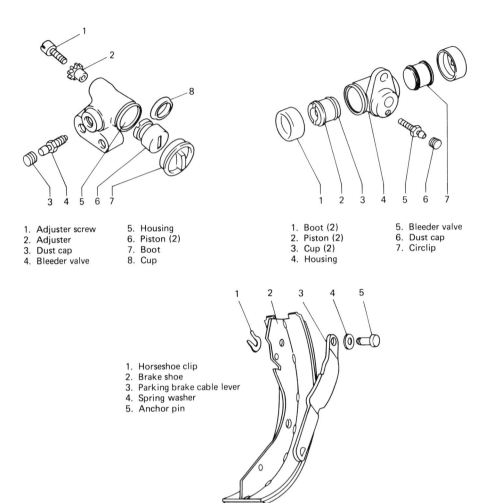

1. Adjuster screw
2. Adjuster
3. Dust cap
4. Bleeder valve
5. Housing
6. Piston (2)
7. Boot
8. Cup

1. Boot (2)
2. Piston (2)
3. Cup (2)
4. Housing
5. Bleeder valve
6. Dust cap
7. Circlip

1. Horseshoe clip
2. Brake shoe
3. Parking brake cable lever
4. Spring washer
5. Anchor pin

Fig. 33.4. Subassemblies and components of the brake drum example. (*a*) Front wheel cylinder; (*b*) rear wheel cylinder; (*c*) brake shoe.

TABLE 33.3. BRAKE DRUM EXAMPLE PRODUCTION VOLUMES

	Year						
Brake Drums	0[a]	1	2	3	4	5	6
Version I							
Annual	20110	15590	8430	7120	5700	0	0
Daily[b]	80	62	34	29	23	0	0
Version II							
Annual	0	4535	12350	16300	7660	6700	6220
Daily	0	18	49	65	31	27	25
Total							
Annual	20110	20125	20780	23420	13360	6700	6220
Daily	80	80	83	94	54	27	25

[a] Year 0 data (before robot) is shown for later use in cost analysis.

[b] Based on approximately 250 days per year; two shifts per day.

TABLE 33.4. ASSEMBLIES PRODUCTION VOLUMES FOR BRAKE DRUM EXAMPLE[a]

	Year						
	0[b]	1	2	3	4	5	6
Front wheel cylinder							
Annual	60340	47670	62350	70270	40090	20100	18660
Daily[c]	241	191	249	281	160	80	75
Rear wheel cylinder							
Annual	40250	40250	32000	41750	46750	26750	12500
Daily	161	161	128	167	187	107	50
Brake shoe							
Annual	120690	95330	124700	140540	80180	40200	32320
Daily	483	381	499	562	321	161	149

[a] The table represents the total number of assemblies considered to be assembled by the robot.

[b] Year 0 data is shown for later use in cost analysis.

[c] Based on approximately 250 days per year.

throughput per day, number of production hours required to produce a single day's supply of assemblies, and percent utilization per part for the robot.

Projected Daily Production Hours

The RTM analysis technique was used to estimate robotic assembly times for the three assemblies. These values are compared with the manual assembly times estimated by the MTM method as shown in Table 33.5.

Although the robot is slower in assembling both the rear wheel cylinder and brake shoe, it is usually more efficient in the long run because service is not interrupted for lunch, coffee breaks, and vacations as for its human counterparts.

Using the daily production volumes and assembly time for each of the subassemblies, the necessary hours per day for production can be determined as shown in Table 33.6.

Robot Utilization

To determine the hypothetical utilization of the equipment, the hours per day in production of each product, based on the average daily production runs obtained previously, are divided by the assumed

TABLE 33.5. ASSEMBLY TIME REQUIREMENT FOR BRAKE DRUM EXAMPLE

Assembly	By Manual (sec)	By Robot (sec)
Front wheel cylinder	39.628	26.000
Rear wheel cylinder	19.811	31.312
Brake shoe	17.826	20.744

TABLE 33.6. PROJECTED ASSEMBLY HOURS PER DAY FOR BRAKE DRUM EXAMPLE

	Year					
	1	2	3	4	5	6
Assembly						
Front wheel cylinder	1.38	1.80	2.03	1.16	0.58	0.54
Rear wheel cylinder	1.40	1.11	1.45	1.63	0.93	0.53
Brake shoe	2.19	2.87	3.24	1.85	0.93	0.86
Total assembly (hours/day)	4.97	5.78	6.72	4.64	2.44	1.93

available production time for the robot. In the brake drum example, assuming 18 hours of available time daily, the utilization results are shown in Table 33.7. From these values, one can find the remaining available time for additional products for increased production. On the other hand, if it is found that the robot is overutilized (about 90% or more of the available time), it would indicate that a more efficient alternative must be found.

Capacity-Volume Requirements

As a reference consideration, it is useful to check the number of assemblies that *could* be made in a two-shift day by either the manual or robot operation. This information provides another comparison between the two approaches and indicates how much additional production volume can be assumed by the facility. Table 33.8 shows the capacity-volume potentials for the brake drum example. The manual values are based on 13.5 productive hours per day during a two-shift operation. This assumes 87.5% efficiency during the first shift, 81.25% efficiency during the second shift, per time dedicated to lunch and coffee breaks, departmental meetings, and vacations.

33.4.2. STEP 6: Decisions Concerning Future Applications

Underutilized robots cannot be cost justified owing to the high initial startup expenses and low labor savings. Additional applications or planned future growth are required to drive the potential cost-effectiveness up; however, there is also an increase in tooling and feeder costs associated with each new application. As determined in the robot utilization study of the brake drum example, for instance, the assembly robot is only utilized approximately one quarter of its potential capacity.

33.5. COST ANALYSIS PHASE

33.5.1. STEP 7: Period Evaluation, Depreciation, and Tax Data Requirements

If, based on the previous six steps, it is found that the proposed robotization is technically feasible and preferred to other alternatives, a detailed economic evaluation can begin. Determination of the evaluation period, property tax rates, income tax rates, and depreciation method is essential before further analysis can be attempted. The evaluation period defines the project life for the analysis. The tax values depend on the plant's location and the tax rates for the particular state or county. Several depreciation methods are available for analysis and are shown in Table 33.9.

For the brake drum example, the following values are assumed:

1. Evaluation period = 6 years.
2. Property tax rate = 6% per year.
3. Income tax rate = 50% per year.
4. Salvage value excluding tooling = $1000.

TABLE 33.7. DAILY ROBOT UTILIZATION (PERCENT) FOR BRAKE DRUM EXAMPLE

	Year					
	1	2	3	4	5	6
Total utilization	27.6	32.1	37.3	25.8	13.6	10.7

TABLE 33.8. CAPACITY-VOLUME POTENTIALS FOR BRAKE DRUM EXAMPLE (IN UNITS PER TWO-SHIFT DAY)

Assembly	Manual[a]	Robot[b]
Front wheel cylinder	1226	2492
Rear wheel cylinder	2453	2069
Brake shoe	2726	3123

[a] 13.5 productive hours assumed per day

[b] 18.0 productive hours assumed per day

TABLE 33.9. DEPRECIATION METHODS

Depreciation Method	Calculation
Straight line	$D_k = \left(\dfrac{P - F}{n}\right)$
Declining balance	$D_k = R(1 - R)^{k-1} \cdot P$
Sum of years digits	$D_k = \left(\dfrac{n - k + 1}{n(n + 1)/2}\right) \cdot (P - F)$

Variables: D = depreciation per year k
P = initial cost ($\$$)
F = salvage value ($\$$)
n = project life (years)
R = depreciation rate (%)
k = year (1,2, . . . , n)

5. Initial cost = $\$28,000$.
6. Depreciation method = sum of years digits.
   For example, $D_0 = 0$.

$$D_1 = \left(\frac{6 - 1 + 1}{6(6 + 1)/2}\right) \cdot (28,000 - 1000) = \$7715$$

Industry has tended to justify robotization mainly on the basis of labor displacement. In addition, project cost analysis must involve the comparison of acquisition and operating costs, over the projected life of the equipment, to the projected revenues.

33.5.2. STEP 8: Project Cost Analysis

Project cost analysis includes:

Labor considerations.
Acquisition and start-up costs.
Operating expenses.

Labor Considerations

As shown in Figure 33.5, direct labor hourly costs have risen exponentially over the last two decades and are expected to continue to increase even more sharply. In contrast, robot hourly cost has remained and will continue to be relatively constant. An important factor in this phenomenon is the fact that production and operation costs of robots decrease with the increase in robot population.

The differences in labor costs among different nations may greatly affect economic decisions. An inexpensive labor force will result in fewer robot installations owing to the relatively high acquisition cost for the robot and relatively small, if any, direct labor cost savings.

In the brake drum example, three assumptions are to be made concerning manual versus robotic assembly of brake drum components:

1. Each subassembly is currently being assembled manually by different workers.
2. If a robot is used to assemble the parts, then there will be a reduction in workload for each of the different workers.
3. This study is reviewing the potential reduced labor hours per year for the assembly operation rather than total reduced headcount associated with installing the robotic system.

Table 33.10 shows the manual workload reduction justification and associated labor and benefit cost savings for years 1–6.

If robotization substitutes for manual labor to the point that the work force can be reduced, then headcount considerations may lead to potential savings in reduced size of employee facilities, parking lots, and the like.

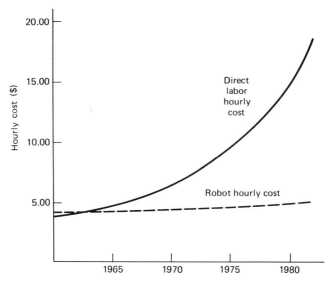

Fig. 33.5. Hourly cost of robot and direct labor in the United States. Direct labor cost includes fringe benefits; robot cost includes support (maintenance, etc.)

Acquisition and Start-Up Costs

The costs associated with robotic installation and start-up are:

Acquisition: robot, controller.
Accessories: part feeders, testers, conveyors, tables, bins, other equipment—by acquisition or development.
Engineering cost.
Programming cost.
Installation costs: utility hookup, foundations, setup labor.
Tooling costs: end effectors, grippers, fixtures.
Training costs.
Related expenses: insurance and others.

Typical robot system cost breakdowns are shown in Figure 33.6.[27] Accessory cost is highly dependent on the specific application, work method, and part design. For illustration, the equipment expenses associated with the brake drum example are listed in Table 33.11.

TABLE 33.10. SAVINGS IN LABOR AND BENEFIT (L&B) COSTS IN BRAKE DRUM EXAMPLE

	Year					
	1	2	3	4	5	6
Manual production						
Total hours/day	4.7	6.2	6.9	4.0	2.0	1.8
Total hours/yr	1,175	1,550	1,725	1,000	500	450
Labor rates ($/hr)	8.50	9.40	10.30	11.40	12.50	13.7
Benefit rates ($/yr)	10,000.00	12,000.00	14,000.00	16,000.00	18,000.00	20,000.00
Applied benefit rates ($/hr)	2.50	3.00	3.50	4.00	4.50	5.0
Total L&B savings ($/yr)	12,925	19,220	23,805	15,400	8,500	8,415

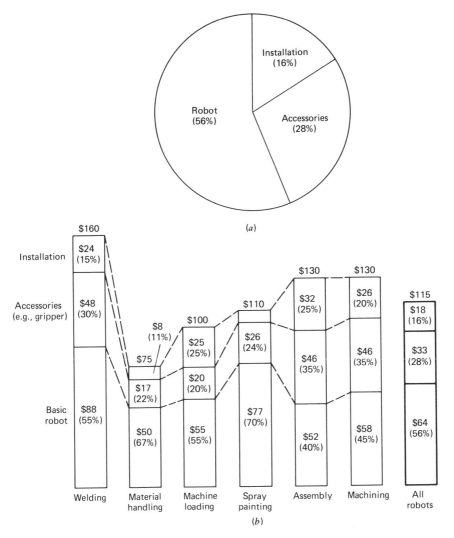

Fig. 33.6. Typical robot system cost breakdown. (*a*) For all robots. (*b*) By application (in thousands of dollars). (*Source.* Reference 27.)

Operating Expenses

The operating expenses associated with a robot typically include:

    Direct labor to tend robot.
    Supply parts to feeders.
    Adjust tools for changeovers.
    Maintenance.

Consider the brake drum example. To determine the required time for a worker to spend tending the robot, two assumptions are necessary:

1. Sixteen hours per day labor time is assumed for a two-shift operation.
2. Five percent of one worker's shift will be used to tend the robot. This 5% value is for the entire shift and does not take into consideration worker inefficiency.

Therefore 0.80 hr per day (5% of 16 hr per day) will be spent by one worker to tend the robot. Table 33.12 shows the costs associated with tending the robot.

TABLE 33.11. CAPITAL EXPENSES FOR BRAKE DRUM EXAMPLE

Equipment	Estimated Cost
Robot and controller	$ 28,000
Bowl feeder for housing	20,000
Bowl feeder for piston, piston and cup	18,000
Bowl feeder for boot	18,000
Bowl feeder for cup	12,000
Bowl feeder for bleeder valve	7,000
Bowl feeder for dust cap	6,000
Bowl feeder for adjuster; circlip	24,000
Bowl feeder for adjuster screw; anchor pin	7,000
"Coin changer" for horseshoe clip	2,000
"Coin changer" for spring washer	2,000
Chute rail feeder for brake cable lever (in-house)	12,000
Chute rail feeder for brake shoe (in-house)	18,000
Total equipment cost for three assemblies	$174,000

TABLE 33.12. ROBOT TENDING COSTS FOR BRAKE DRUM EXAMPLE

	Year					
	1	2	3	4	5	6
Labor rates ($/hr)	8.50	9.40	10.30	11.40	12.50	13.70
Applied benefit rates ($/hr)	2.50	3.00	3.50	4.00	4.50	5.00
Total L&B rates ($/hr)	11.00	12.40	13.80	15.40	17.00	18.70
Tending (hrs/yr)[a]	200	200	200	200	200	200
Total tending L&B costs	2200	2480	2760	3080	3400	3740

[a] (0.8 hr/day × 250 days/year)

Direct Maintenance

For maintenance cost calculations, suppose the robot is 98% reliable. Therefore to determine planned maintenance downtime, it is assumed that 2% of the operation time per year will be used for preventive maintenance and repairs. Table 33.13 projects the planned maintenance costs associated with the robot for years 1–6.

33.5.3. STEP 9: Evaluation Techniques

Overview of Economic Evaluation Methods

Several evaluation methods are available for the economic analysis of competing alternatives. Although all of these methods are usually equivalent in determining which alternative investment is preferred,

TABLE 33.13. MAINTENANCE COSTS FOR BRAKE DRUM EXAMPLE

	Year					
	1	2	3	4	5	6
Hours/day in operation	4.97	5.78	6.72	4.64	2.44	1.83
Hours/year in operation	1242.50	1445.00	1680.00	1160.00	610.00	457.50
Hours/year estimated downtime (2%)	25	29	34	23	12	9
Maintenance rates ($/hr)	15.80	16.90	17.90	19.60	21.20	22.20
Planned annual maintenance costs ($/yr)	395.00	490.00	610.00	450.00	253.00	200.00

different companies may prefer different methods under different conditions. Table 33.14 shows four general types of such cost methods:

1. **Minimum Cost Rule.** Calculate the equipment cost such as operating cost, capital cost, and so forth for each alternative, and decide desirability of investment by comparing the totals.

2. **Capital Recovery Method.** Calculate the possible term to pay back for the invested capital, and decide the desirability of the alternatives by length of the payback term.

3. **Rate of Return on Investment.** Decide desirability of alternatives by comparing profit ratio with the investment. The method is suitable for determining priority of alternatives.

4. **Permissible Investment Amount Method.** Calculate the permissible maximum amount of investment for saving the labor equivalent to one person.

These methods, in turn, pertain to several different calculation techniques which are addressed in the following sections.

TABLE 33.14. SUMMARY OF ECONOMIC EVALUATION METHODS

Calculations Method	Outline
1. Minimum Cost Rule Method	
Simple cost comparison method	Make simple comparison of total cost in a term
Present value comparison method	Make comparison of converted present value of investment and cost
Return on investment method	Make comparison of mean yearly cost of investment and total cost
Adverse minimum cost method	In the case when operating costs differ yearly, calculate adverse minimum of operating cost and capital cost, and compare alternatives by number of years of economic equipment utilization
Old MAPI method	Calculate adverse minimum cost of challenge and defend equipment, and decide investment by comparing both costs
2. Capital Recovery Method	Calculate term of the capital recovery and decide by length of return period
3. Rate of Return on Investment	
Simple rate of return method (the first year)	Compare profit rate of return with the investment of the first fiscal year with the target rate and make decision
Discounted cash flow method	Calculate the profit rate that makes equal the cash flow in (income minus expense and tax) and investment, and decide
Earning ratio method	Calculate the profit rate by using the table which makes equal present value of the investment and the profit, and decide
Mean rate of return method	Make comparison of mean profit and mean book value of equipment in each year and compare the alternatives
New MAPI method	Improve the defects of Old MAPI, and decide the investment by the measure
4. Permissible Investment Amount Method	Calculate permissible amount of investment for one manpower saving on labor cost and decide the investment from comparison with amount to be invested

Minimum Cost Rule Method

A typical minimum cost rule method is the *present value* method. The net present value of a proposed project can be found in cash flow tables, or from the formula

$$PV = \sum_{k=0}^{n} \frac{(-AC_k + S_k - RC_k)}{(1 + i)^k} + \frac{L}{(1 + i)^n}$$

where PV = present value
   AC_k = the acquisition cost in year k
   S_k = the potential savings in year k
   RC_k = the running cost in year k
   L = forecast of remaining (salvage) value
   i = interest rate
   n = project life (years)

Computation of the yearly running cost is explained in Table 33.15. Tables 33.16a and b illustrate one way of calculation with respect to the brake drum example. In Table 33.16a the cash flow for the current, manual system is summarized, and it is assumed that no capital expenditures, maintenance costs, or cost benefits are expended relative to the robotic alternative. The cash flow for the latter is shown in Table 33.16b. The tax rate is assumed at 50% in both tables. The total present worth for each alternative is shown, and should next be compared to the present value of the projected revenues. In comparing several alternatives, that with the highest total present value should be preferred.

If the evaluation period, or project life, n, is not known, the following formula can be used:

$$\sum_{k=0}^{n} \frac{AC_k}{(1 + i)^k} \leq \sum_{k=0}^{n} \frac{(S_k - RC_k)}{(1 + i)^k}$$

Minimum n is calculated, and if the inequality is satisfied, then the project with the shortest payback period will be preferred.

If the yearly savings S_k are not clear, one can use the following equation to compute the life cycle cost (LCC) of the proposed alternatives:

$$LCC = \sum_{k=0}^{n} \frac{(AC_k + RC_k)}{(1 + i)^k} - \frac{L}{(1 + i)^n}$$

The alternative for which this cost is the smallest should be preferred.

TABLE 33.15.   COMPUTATION OF RUNNING COSTS

Operating Cost	Power cost	(power cost per hour) × (yearly net working hours)[a]
	Supplies cost	(supplies cost per hour) × (yearly net working hours)
Maintenance	Preventive maintenance cost	(average preventive maintenance cost) × (frequency of preventive maintenance in a year)
	Repair cost	(average repair cost) × (frequency of out-of-order in a year)
Setup cost		(average setup cost) × (frequency of setups in a year)
Downtime	Out-of-order time loss[c]	(average time loss due to out-of-order) × (frequency of out-of-order in a year)
loss[b]	Setup time	(average time for setup) × (frequency of setups in a year)
Scrap and rework cost		(average scrap and rework cost) × (production in a year) × (defect rate)

[a] Note: "Net working hours" is not the same as "net operating hours."

[b] Do not count as downtime lost time that it is possible to make up for during working hours.

[c] In calculating the cost due to downtime loss, consider actual downtime loss, plus cost of substitute operations to make up for lost time (if any).

TABLE 33.16a. MANUAL ASSEMBLY CASH FLOW

Year (1)	Capital Expenditure (2)	Labor Expense (3)	Maintenance Expense (4)	Operating Expense (5)	Cost Benefits (6)	Total Cash (7)	After Tax Cash (8)	Discount Cash After Tax (at 20%) (9)
0	0	−12,000	0	−12,000	0	−12,000	− 6,000	− 6,000
1		−12,925		−12,925		−12,925	− 6,463	− 5,380
2		−19,220		−19,220		−19,220	− 9,610	− 6,670
3		−23,805		−23,805		−23,805	−11,903	− 6,890
4		−15,400		−15,400		−15,400	− 7,700	− 3,710
5		− 8,500		− 8,500		− 8,500	− 4,250	− 1,710
6		− 8,415		− 8,415		− 8,415	− 4,208	− 1,410
Total								−31,770

TABLE 33.16b. ROBOT ASSEMBLY CASH FLOW

Year (1)	Capital Expenditure (2)	Depreciation (3)	Labor Expense (4)	Maintenance Expense (5)	Operating Expense (6)	Cost Benefits[a] (7)	Total Cash (8)	After Tax Cash (9)	Discount Cash After Tax (at 20%) (10)
0	−174,000	0	0	0	0	0	−174,000	−87,000	−87,000
1		−7,715	−2,200	−395	−2,595	+12,925	+ 10,330	5,165	4,300
2		−6,430	−2,480	−490	−2,970	+19,220	+ 16,250	8,125	5,640
3		−5,143	−2,760	−610	−3,370	+23,805	+ 20,435	10,218	5,920
4		−3,857	−3,080	−450	−3,530	+15,400	+ 11,870	5,935	2,860
5		−2,570	−3,400	−253	−3,653	+ 8,500	+ 4,847	2,424	970
6		−1,283	−3,740	−200	−3,940	+ 8,415	+ 4,475	2,238	750
Total									66,560

[a] Values for column 7 show savings in total labor and benefits, from Table 33.10

Capital Recovery Method

Management is usually concerned with the time necessary for the project to recover its capital investment. The years-payback analysis is a useful tool in determining when this will occur. The simple formula used to determine payback is

$$P = \frac{I}{S - E}$$

where P is the number of years for payback, I is the total investment of the robot and tooling, S is the total annual labor savings, and E is the total annual expense for the robot, including maintenance and tending labor costs.

For the brake drum example, the analysis is as follows, assuming $I = \$174,000$ investment for robot and tooling, $S = \$14,700$, and $E = \$3330$ (S and E are taken as simple annual averages from the foregoing figures for labor savings and direct labor expenses, respectively):

$$P = \frac{174000}{14700 - 3330} = 15.3 \text{ years}$$

The payback for assembling the three subassemblies using the robot is very high, as could be expected from the low utilization that was indicated before. However, if additional applications were determined for the robot, then increased savings would result, and a shorter payback period could be expected.

Rate of Return Analysis

In this approach the rate of return of each alternative is calculated. The one with the highest incremental rate of return should be preferred.

When the interest rate, specified before as i, is not clear, one can use this equation:

$$\sum_{k=0}^{n} \frac{AC_k}{(1 + r)^k} = \sum_{k=0}^{n} \frac{(S_k - RC_k)}{(1 + r)^k}$$

The rate of return r is calculated. The alternative with the highest incremental r will be chosen out of those alternatives that have r higher than the minimum acceptable rate of return.

Permissible Investment Amount

This technique determines the investment limits based on a cost savings of one headcount reduction. Consider the brake drum example. Suppose the proposed robot could be fully utilized for a two-shift operation, thus replacing completely two assembly operators. Then

$$\frac{16 \text{ hours}}{\text{day}} \times \frac{250 \text{ days}}{\text{year}} \times 6 \text{ years} = 24,000 \text{ hours saved}$$

At the applied average labor and benefit rate of $\$14.60/\text{hr}$ the total cost savings is $\$350,400.00$. Therefore the potential investment limit for the robot, tooling, accessories, and operating expenses over the life of the project is $\$350,400.00$ based on reducing total headcount in assembly by two persons.

33.5.4. STEP 10: Additional Economic Considerations

Even after a successful feasibility study and economic evaluation, great care should be taken to avoid common pitfalls. Many companies have experienced the mistake of the following:

1. Investing with expensive money.
2. Basing a decision on the inferiority of workers or old equipment relative to the proposed robot.
3. Determining the investment strictly from the economic figures.
4. Establishing incorrect replacement attitudes concerning new automation equipment and techniques.

Be Careful with Expensive Money

There are several sources of money for investment, each with a different type of interest rate:

Bank-loaned money.
Money by depreciation.
Money from reserved profit.
Money from issuing bonds.
Other sources.

Since interest rates are sensitive and fluctuate, a company may find itself with a bad investment. Also, by borrowing money from outside the company, the financial stability of the company is lowered. There is no such problem if the company can afford to use its own money.

Performance Gaps Bias

Economic justification of a robotized system is frequently achieved by proving a significant performance gap between the new robot(s) (challenger) and the previous worker(s) or equipment(s) (defender). Often, however, such a comparison can be highly biased. To reach an attractive result, a company can keep a passive attitude for a long period of no replacement, then get a big gap of performance between the new challenger and the old defender. In Figure 33.7, by lengthening the duration of replacement cycle, keeping a passive attitude, the sawteeth of economical calculation become bigger. If we are not diligent enough to modernize the facilities, we can obtain an attractive result of economic calculation superficially. But, on the other hand, the company that continuously makes efforts to keep the equipment up to date cannot achieve good results in the calculation, and they need to modify the operating conditions (e.g., increase the number of shifts per day), or use inexpensive money, and so forth.

Use Economic Evaluation Figures for Reference Only

The result of economic calculation gives only referential information for deciding the priority of the investment. However, we must not decide on the investment strictly based on the calculation results. The reason: In most cases, the final decision on any investment is made by comparison with many other investment alternatives by higher management judgment, and not mechanically by the calculation. It is recommended that deciding the priority of investments be based on the result of careful economic calculation. However, final decision on investment must be made after referring to available funds, quality, and other management considerations.

Two Types of Replacement Mistakes

Similar to the two types of mistakes in quality control in industry and in replacement of capital equipment, attitude of replacement may also display two types of mistakes:

1. Too loose an investment-attitude mistake. It is a mistake to decide to invest in spite of uneconomic conditions.
2. Too tight an investment-attitude mistake. It is a mistake to decide not to invest in spite of good economic conditions.

In capital investment, as a general tendency, the person who has made the first mistake had probably carried out a very strict investigation. In the case of the second mistake, probably no particular survey had been performed yet. Therefore, it is natural that people generally take a rather conservative attitude toward the facilities-replacement economic decisions.

Effects of Inflation and Recession

Inflation and recession each have a significant influence on the possibility to justify robotization and should be considered as part of the economic evaluation. To account for inflation, the present worth of the investment should be converted to its respective future worth in year k at a given inflation rate. The future worth FW amount represents the amount the investor would have to pay for goods in the future. Similarly, a future worth must be calculated for revenues. The future worth of a capital purchase in the future FW, whose present cost is known PW, can be figured from standard tables for forecasted interest rate i at year k,

$$FW_k = PW \left(\frac{F}{P, \, k, \, i\%} \right)$$

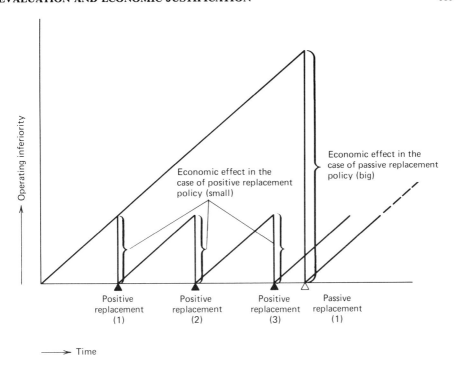

Fig. 33.7. Economic effect of replacement over time.

These future worth amounts A should then be incorporated into an equivalent cash flow to represent the dollars required to make the desired purchases over time until year n. This can be done using the equation

$$A\left(\frac{F}{A, n, i\%}\right) = \sum_{k=0}^{n} PW_i \left(\frac{F}{P, k, i\%}\right)$$

The effects of inflation on robot projects are studied in detail by Heginbotham.[16] Figure 33.8 illustrates the unfavorable affects of increasing inflation rates.

Recession typically means that there is less demand for production, and as a result the capacity of a production facility will be underutilized. The direct result of underutilized equipment is an unfavorable rate of return. Figure 33.9 illustrates the effect of reduced working capacity due to recession on the project's rate of return.

35.5.5. STEP 11: Recommendations Based on Analysis Results

The capital recovery period and rate of return on investment are two primary techniques used in industry to base decisions on alternative manufacturing methods. Often the values are satisfactory and decisions concerning investment are simple. However, sometimes the payback and investment rate do not meet management's expectations because of underutilization of equipment and high start-up costs.

In the brake drum example, the payback period was 15.3 years, and a rate of return investment could not be calculated because the equipment was underutilized (operating an average of 4.5 hr per day).

When this occurs, two recommendations are possible to make the robot alternative more financially attractive:

1. Increase utilization by producing more subassemblies by the robot.
2. Decrease the tooling expense associated with start-up.

The first recommendation offers a good solution for increasing the potential savings over the life of the project but is restricted by the need to keep capital and tooling expenses low. To the contrary, the second recommendation can offer assistance in getting the project implemented at a more favorable expense by utilizing in-house craftsmen to design and fabricate the necessary feeding and other peripheral

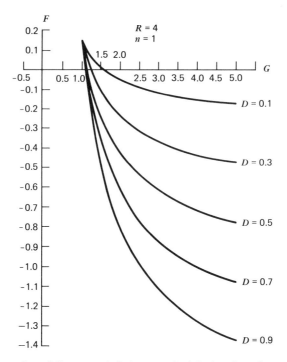

Fig. 33.8. Investment favorability versus inflation rate. As inflation rises, the project investment becomes less favorable. G = inflation magnitude; D = factor to convert G to percentage interest rate; F = investment favorability factor; R = number of product changeovers per year; n = number of shifts per day. (*Source.* Reference 16.)

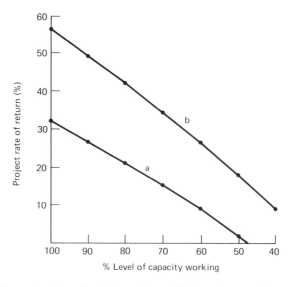

Fig. 33.9. Effect of recession (lower level of capacity working) on the rate of return of a robotization project. a = no government support; b = $33\frac{1}{3}$% government support. (*Source.* Reference 13.)

devices. Designing in-house alternative feeding methods or simple devices can produce good results in terms of efficiency at a much lower investment.

ACKNOWLEDGMENT

The author acknowledges significant and valuable contributions to this chapter by Cristy Sellers, a graduate student at Purdue University, West Lafayette, Indiana, who previously has worked in industry in robots cost justification.

REFERENCES

1. Abraham, R. G., Requirements Analysis and Justification of Intelligent Robots, *SME Paper* No. MS75–247, 1975.

2. Behuniak, J. A., Economic Analysis of Robot Applications, *SME Paper* No. MS79–777, 1979.

3. Benedetti, M., The Economics of Robots in Industrial Applications, *Industrial Robot,* Vol. 4, No. 3, September 1977, pp. 109–118.

4. Blanchard, B. S. and Fabrycky, W. J., *System Engineering and Analysis,* Prentice-Hall, Inc., Englewood Cliffs, New Jersey, 1981.

5. Blank, L. T. and Tarquin, A. J., *Engineering Economy,* McGraw-Hill, New York, 1983.

6. Boothroyd, G., Economics of Assembly Systems, *Journal of Manufacturing Systems,* Vol. 1, No. 1, 1982, pp. 111–126.

7. Bublick, T., The Justification of An Industrial Robot, *SME Paper* No. FC77–630, 1977.

8. Compton, P. A., The Economic Justification of Automatic Assembly, *Proceedings of the 1st International Conference on Assembly Automation,* Brighton, U.K., March 1980, pp. 125–136.

9. Engelberger, J. F., Robots and Automobiles: Applications, Economics and The Future, *Society of Automotive Engineers Paper* No. 800377, March 1980.

10. Engelberger, J. F., Robots Make Economic and Social Sense, *Atlanta Economic Review,* July–August 1977, pp. 4–8.

11. Engelberger, J. F., Economic and Sociological Impact of Industrial Robots, *Proceedings of the 1st International Symposium on Industrial Robots,* April 1970, pp. 7–12.

12. Govsievich, R. E., Determining the Economic Effectiveness of Industrial Robots, *Machines and Tooling,* Vol. 49, No. 8, 1978, pp. 11–14.

13. Grieve, R. J., Lowe, P. H., and Kelly, M. P., Robots: The Economic Justification, *Proceedings of AUTOFACT Europe,* Geneva, Switzerland, September 1983, pp. 2/1–12.

14. Hanify, D. W., The Application of Economic Risk Analysis to Industrial Robots, *Proceedings of the 3rd International Symposium on Industrial Robots,* Zurich, Switzerland, May 1973, pp. 41–50.

15. Hasegawa, Y., Economic Evaluation of Industrial Robot Operation Systems, Engineering Project Research Report, Japan Industrial Robot Association, 1974.

16. Heginbotham, W. B., Can Robots Beat Inflation?, *SME Paper* No. MS77–756, November 1977.

17. Heginbotham, W. B., The Basic Economics of Industrial Mechanization and Automation, *International Journal of Production Research,* Vol. 11, 1973, pp. 147–154.

18. Hughes, T., Capital Equipment Strategy Up To Date?, *Production Engineering,* May 1983.

19. Lassi, K. G., Technical and Economical Considerations Concerning Industrial Robots, *Industrial Robot,* Vol. 2, No. 2, June 1975, pp. 56–61.

20. Mosher, R. R., The Place of Industrial Robots in Manufacturing, *SME Paper* No. MS74–149, 1974.

21. Nof, S. Y., Decision Aids for Planning Industrial Robot Operations, *Proceedings of the 1982 Annual Industrial Engineering Conference,* Institute of Industrial Engineers, 1982.

22. Owen, A. E., Economic Criterion for Robot Justification, *Industrial Robot,* Vol. 7, No. 3, September 1980, pp. 176–177.

23. Technical Insights, Inc., New Jersey, Industrial Robots . . . Key to Higher Productivity, Lower Cost, Report on robot applications design and economic analysis, 1980.

24. Thuesen, H. G., Fabrycky, W. J., Thuesen, G. J., *Engineering Economy,* 4th ed., Prentice-Hall, Englewood Cliffs, New Jersey, 1977.

25. Van Blois, J. P. and Philip, P. A., Robotic Justification: The Domino Effect, *Production Engineering,* April 1983.

26. Van Blois, J. P., Strategic Robot Justification: A Fresh Approach, *Robotics Today,* April 1983, pp. 44–48.

27. *Industrial Robots: A Summary and Forecast,* Tech Tran Co., Naperville, Illinois, 1983.

PART 7
APPLICATION PLANNING: INTEGRATION

CHAPTER 34

PLANNING ROBOTIC PRODUCTION SYSTEMS

LANE A. HAUTAU

GMF Robotics Corporation
Troy, Michigan

FRANK A. DiPIETRO

General Motors Corporation

PREFACE

It is a pleasure to put my stamp of approval on Lane Hautau's contribution to this Handbook of Industrial Robotics and on the idea behind the handbook itself. When we consider the present condition of our national economy, the robotics industry may very well represent our best foot forward into the future.

Many industries have been retrenching because of international competition and poor productivity. For example, the United States is pretty much out of the camera-making business, has lost much of the electronics business, is suffering greatly in the steel area, and almost one of three cars sold in the United States is made in Japan. While some industries have not yet found a way to recover, General Motors and the other auto manufacturers have decided to take action and respond to the market needs vigorously.

A formidable part of the strategy is the deployment of robot battalions on the manufacturing front to increase productivity, assure consistent quality, and help keep costs down.

Nobody imagines that robots are the whole answer to these challenges. But they are an important part of the answer. One of the reasons robots are accepted now, not only by managements but by the general public (including the work force) as well, is that people have come to realize that robots are essential to our planning for a better future; they are productive, they are flexible, they can be reprogrammed for maximum versatility, they are economical. They are, in short, indispensable to the counterattack against competition.

That is why I say that the robotics industry may well be our best foot forward into the future. It is one industry in which we can maintain a technological advantage. Maintaining that advantage is critically important to us all and can only be done by making sure that the best, most useful robotics information gets into the right hands and heads as soon as it becomes available.

Informational handbooks like this one are a practical method of getting robotics information to the right people. Lane Hautau's chapter on planning robotic production is a good example of the kind of useful, state-of-the-art information needed to meet our robotic challenge.

34.1. INTRODUCTION AND PRODUCT CONSIDERATIONS

Production engineering's most effective approach to robotics process planning is a systems approach. In other words, we must look at all the givens collectively: the *product,* the *tooling,* the *process,* the *volume,* and then set up an efficient overall strategy.

Planning a highly automated and robotic system cannot take place in a vacuum or piecemeal. The product and process must constantly be reassessed in terms of the most effective *resultant.* The final result of this systems approach may be an elaborate automated processing plan, as in Figure 34.1, but each system is a combination of subsystems that make up a truly large, totally automated

---

This chapter was written when the author was Senior Manufacturing Project Engineer and Coordinating Engineering Group Manager with the Fisher Body Division of General Motors Corporation.

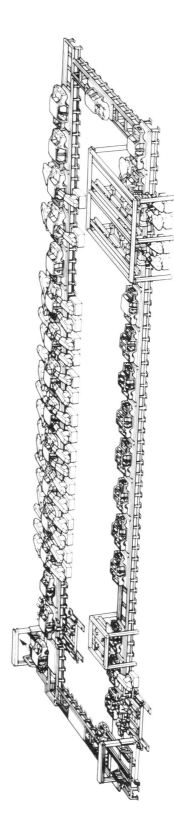

Fig. 34.1. Automated robotic body shop.

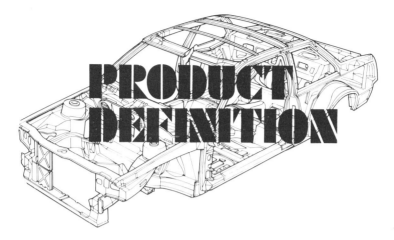

Fig. 34.2. Product definition relative to robotic welding.

arrangement, with a variety of automated islands supported by various shuttles and feeder lines. (For a provocative discussion of this subject see Reference 1.)

Among other things, what we would like to accomplish is to "de-myth" the feeling that this is a tough, difficult, or excessively challenging operation. It all must start with a product definition, and the product considered in this chapter is an automobile body (Figure 34.2). The entire body must be carefully reviewed during the design phase if it is to be fabricated in a highly automated robotic body shop.

For instance, body joint design requirements should be carefully reviewed. They hold the keys to establishing an efficient assembly flow. An example of this might be the treatment of the traditional quarter-inner-panel-to-quarter-outer-panel-joining flange in the lock pillar area. Traditionally, a flange like this is designed primarily to suit fabrication and shipping conditions, as automated joining techniques in assembly were not involved in this equation. However, now that automated assembly is possible on a grand scale, the means to facilitate this end must be plowed back into the product.

A good example of this philosophy is the development of the double-weld gun (Figure 34.3), which can only be carried and maneuvered successfully by a robot. This obvious productivity gain cannot be fully realized unless traditional flange irregularities can be ironed out, so the bulky dual guns can move swiftly over the surface. This is important because precious seconds are lost in maneuvering around these cranks, and this translates into a work-loss content in station.

It must be stressed that it is sometimes advisable to rearrange the sequence of a particular robotic operation. In Figure 34.4, for example, note the welding required deep inside the body. This suggests

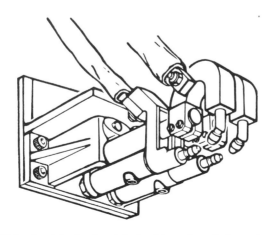

Fig. 34.3. Specially developed dual-weld gun for robotic spot welding.

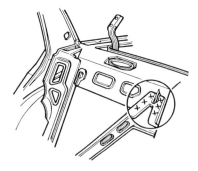

NOT RECOMMENDED

• SPOT WELDS NEAR CENTER LINE OF BODY
  ARE VERY DIFFICULT TO ACCESS WITH
  AUTOMATED SYSTEMS.

Fig. 34.4. Normal welding requirements: rear seat back brace to back window panel.

that in the event of trouble, prohibitive downtime penalties could develop, since much of the robot arm is not easily accessible.

This situation is best dealt with by eliminating the entire product design concept and replacing it with a robotic subassembly reinforcement operation that will accomplish the same thing, but without the original risk on line! A separate reinforcement should be used to provide rear seat stiffness (Figure 34.5). A separate reinforcement eliminates the need for spot-welding a cargo barrier brace near the centerline of the body.

Additionally, attempts should be made to design features into major body panels, rather than providing added small parts—as in Figure 34.6—with their accompanying loading costs. Figure 34.7 shows how this small part might be lanced out of the major panel. In the last figure in this sequence (Figure 34.8) we see an example of automatic pickup and load by a small fixture attached to the robot's weld gun. In many instances, moreover, small parts are not accessible for robotic welding.

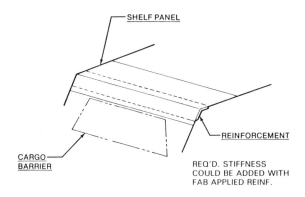

PREFERRED

• UTILIZE REINFORCEMENT TO GET REQUIRED
  SHELF STIFFNESS.

• THIS REINFORCEMENT COULD BE ADDED
  IN FAB OPERATION OR SUB ASSEMBLY
  OPERATION.

• CARGO BARRIER COULD BE PART OF
  SEAT BACK ASSEMBLY.

Fig. 34.5. Preferred design and processing for rear seat back brace with added reinforcement.

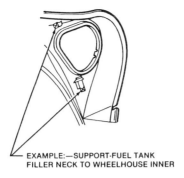

EXAMPLE:—SUPPORT-FUEL TANK
FILLER NECK TO WHEELHOUSE INNER

NOT RECOMMENDED

- SMALL PART LOADS CAUSE
  DIFFICULTIES IN ASSEMBLY
  OPERATIONS.

Fig. 34.6. Small parts loading: fuel tank support clips.

34.2. PROCESS CONSIDERATION AND SYSTEM PLANNING

Now let us explore processing alternatives that can make a robotic operation more attractive from a maintenance standpoint. The door opening, quarter window, front and rear header flanges, and so on are typical places on the body to be considered for automated welding (Figure 34.9). Normally, the rear quarter window operation would be performed by a typical operator with a single manual weld gun (Figure 34.10). A review of this operation indicates that the engineer could perform the welding with one robot, and still have some cycle time remaining.

34.2.1. Reshuffling the Process

In continuing his review, the applications engineer must be in a position to reprocess the line and/or restructure the design to cater to *high-volume* automated robotics. When processing an operation that requires a roll of this magnitude (270°+), we generally find not only that the last axis of the robot is the weakest link (for obvious design reasons), but also that the weld gun cables are twisted beyond their capabilities, causing a serious maintenance problem. In summary, there are three problems:

1. Cycle time imbalance (time left over that would allow the robot to take some work off another robot—for instance, the door line robot).

2. Challenging torque maneuvers for certain axes.

3. Excessive weld cable twisting.

By reshuffling the operations the engineer can overcome these problems directly.

To begin with, a part of the quarter window and part of the front body hinge pillar can be welded by the same robot. The first robot can weld the body hinge pillar (as shown by the dashed line in Figure 34.11) and move to the quarter window and weld the front and lower surfaces. A second robot does the top of the rocker and front lock pillar as a complete operation. A third robot does the quarter window rear flange, then turns around to get the side of the back window opening.

To keep from twisting the weld gun cables, it is just as easy to make the move down the hinge pillar, and then to the front and bottom of the quarter window. Although a smaller, easier move could have been made by keeping the robot inside the window, this—as pointed out—would have resulted in tangling the weld cables. In addition, too much welding time was required to process the whole door opening—shown as the dotted line in Figure 34.11—with one robot. By splitting the operation, the need to purchase an additional robot was eliminated.

On the other hand, the across-car roof header type operation (Figure 34.12) should be kept as a separate operation, regardless of any remaining cycle time. This gives us a convenient place for the back light side by combining its welding with the quarter window rear welding (operation 3, Figure 34.12). The robot will have trouble reaching across the car to the opposite side of the body while maintaining a nice, smooth flange. Remember, the body is only tack-welded at this point, and dimensional

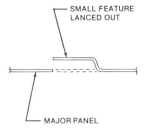

PREFERRED

• DESIGN FEATURES INTO THE
  MAJOR PANELS THAT DO THE
  FUNCTION OF THE SMALLER
  PART.

Fig. 34.7. Small part load elimination: feature developed out of main product.

integrity must be preserved. Consequently, the process engineer should dedicate the weld gun, bracket, weld time, and cabling to this single operation (operation 4, Figure 34.12) and nothing else.

34.2.2. Types of Robots—Brief Overview

When planning robotic systems, the process engineer should have a working knowledge of the kinematics, or what can be referred to as the basic mechanical geometry of all the different specific types of robots. Following are four general categories that include the different types of basic geometry. These

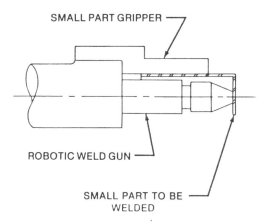

ALTERNATE DESIGN

• DESIGN SEVERAL PARTS TO
  ACCOMMODATE A SINGLE
  GRIPPER DESIGN WHICH WILL
  ALLOW AUTOMATIC PICK-UP
  OR HOPPER FEEDS OF THE
  SEVERAL SMALL PARTS.

Fig. 34.8. Small parts gripper/loader: simple fixture attachment eliminating separate load sequence.

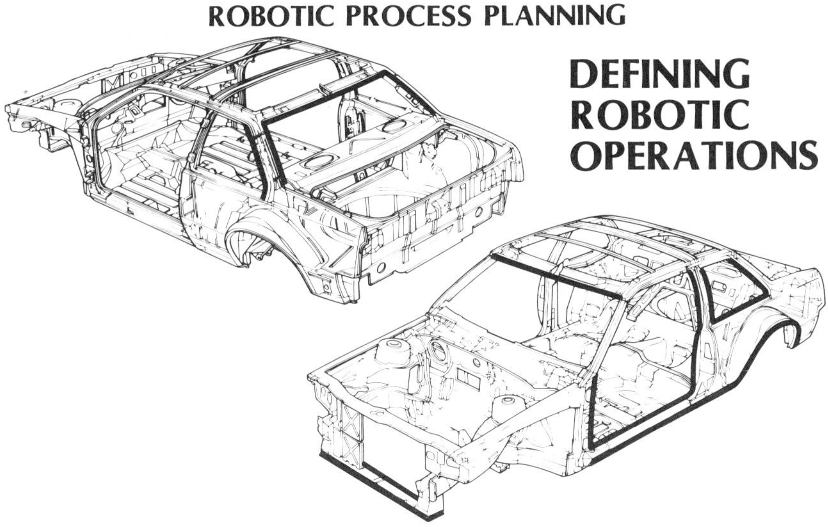

ROBOTIC PROCESS PLANNING

DEFINING ROBOTIC OPERATIONS

Fig. 34.9. Typical flanges available for robotic welding.

categories are not all-inclusive, but are listed to give the reader a general knowledge of the kinematics involved (also see Reference 2, p. 30).

A typical *rectilinear coordinate* robot (Figure 34.13) moves in straight lines, up and down plus in and out. Since these tend to be simple robots, they may lack control logic for coordinated axis drives. That is, one axis drives by itself until it stops, then another takes over until *it* stops.

A second type of robot, built about a column, operates in a cylindrical work envelope (Figure 34.14) and is sometimes called *columnar*.

A third type of robot geometry is the *spherical* (Figure 34.15), which is typical of some Unimate models and is sometimes called the *world coordinate* system, *polar* or *prismatic*.

The fourth category is the articulated arm type (Figure 34.16), typical of Cincinnati Milacron, ASEA, and some GMF robots. Articulated arm robots share many attributes of the other three types of robots, in that they can be made to configure themselves differently, simply by asking the robot's computer to work in the different coordinate systems. If *rectilinear coordinates* are selected, the computer will assist in making square moves. If *cylindrical coordinates* are selected, the robot moves accordingly. Another coordinate system intrinsic to this design is the *hand coordinate* system. In this system the robot's last axis is gun-sighted in a certain direction, and when so directed, the robot will drive the face plate along that line.

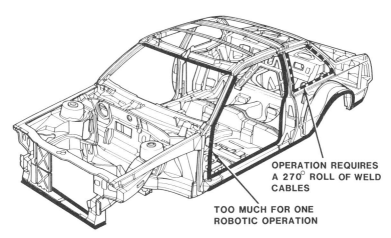

OPERATION REQUIRES
A 270° ROLL OF WELD
CABLES

TOO MUCH FOR ONE
ROBOTIC OPERATION

Fig. 34.10. Typical processing of door opening and rear quarter window.

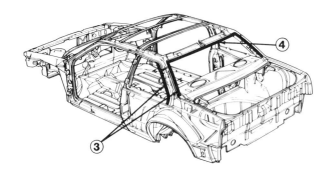

Fig. 34.11. Robotic reprocessing.

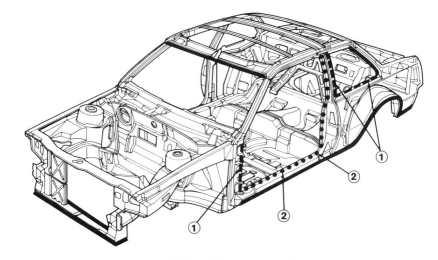

DEFINING ROBOTIC OPERATIONS

Fig. 34.12. Robotic reprocessing.

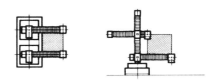

RECTANGULAR COORDINATE ROBOT

Fig. 34.13

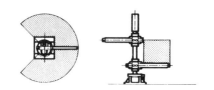

CYLINDRICAL COORDINATE ROBOT

Fig. 34.14

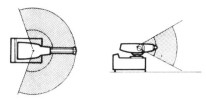

SPHERICAL COORDINATE ROBOT

Fig. 34.15

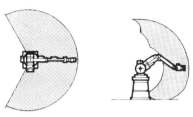

JOINTED ARM ROBOT

Fig. 34.16

All these features are handy in programming. It is possible on some robots to switch back and forth among the various coordinate systems during programming, thereby allowing the engineer to take advantage of the obvious efficiencies within these systems. Conversely, it is just about impossible to program sophisticated robots in manual, which is the system where each axis is individually driven. Computer assistance is required to move this robot efficiently. Consequently, when initially planning the system, the processing/applications engineer must decide which type of robotic kinematics *best* suits the particular conditions, as the kinematic features *of these four categories* are, in fact, *shared* by many robotic designs.

34.2.3. Robot Manufacturers' Specifications

The robot vendors supply an abundance of information (see Figure 34.17), such as motion and travel diagrams, repeatability tolerances, maximum operating loads, maximum/loads axis and overall speeds, and suggested applications.

34.2.4. Laying Out the System

When developing a plan of attack, some sort of rough system layout should take form (Figure 34.18). This advanced planning layout should contain key givens, such as style selection requirements, total number of spots, robot positions sequence of operations, total station time, shuttle time, floor space requirements, buffer margins, and any unused available operation time.

It is important when laying out a system that the process planner spell out all this information so everyone understands what piece of the timing picture he owns. There may be several suppliers working on individual parts of this equation. Trying to put this symphony together is the job of the process planner, and the information must flow to everyone if it is to play reasonably well.

It also is important to rough out, for the plant, each station within the planning system layout (Figure 34.19). Work envelopes for the robot, what the plan is for picking up parts, and where all the various support equipment should be must be known up front. In this way it is possible to make sure there is room for all the equipment that will be required in the station. It is also necessary to make sure there is room to get the robot in and out so work can be performed on the equipment, since consideration *must* be given to downtime. If the robot is going to be down for an hour, the manufacturing process must be able to work around it.

Interference zones (Figure 34.20) should be avoided, even if it is possible to provide a software or hardware limit to accommodate them. The truth of the matter is these zones are generally not programmed out of the system. If robots can clobber each other they probably will, even if there is a software limit. For instance, there may be a situation where a man pulls the robot out *in manual* and it accidentally crosses the interference zone of another robot running in *automatic*. The robot

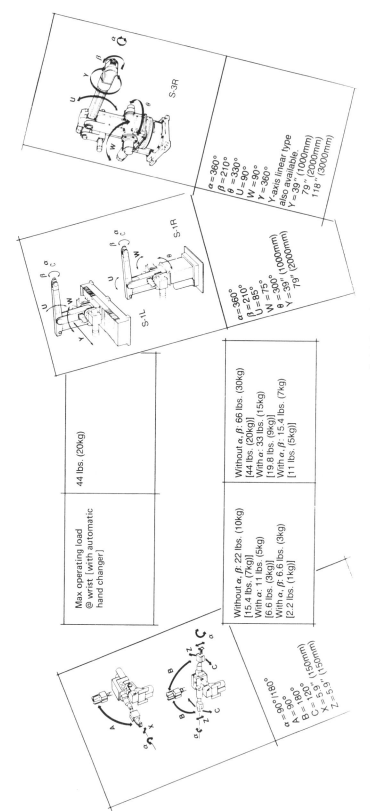

α = 360°
β = 210°
θ = 330°
U = 90°
W = 90°
γ = 360°
Y-axis linear type
also available.
Y = 39″ (1000mm)
     79″ (2000mm)
     118″ (3000mm)

S-3R

α = 360°
β = 210°
U = 85°
W = 75°
θ = 300° (1000mm)
Y = 39″ (1000mm)
     79″ (2000mm)

S-1R

S-1L

Max operating load
@ wrist [with automatic
hand changer]

44 lbs. (20kg)

Without α, β: 22 lbs. (10kg)
With α: 11 lbs. (5kg)
[15.4 lbs. (7kg)]
[6.6 lbs. (3kg)]
With α, β: 6.6 lbs. (3kg)
[2.2 lbs. (1kg)]

Without α, β: 66 lbs. (30kg)
[44 lbs. (20kg)]
With α: 33 lbs. (15kg)
[19.8 lbs. (9kg)]
With α, β: 15.4 lbs. (7kg)
[11 lbs. (5kg)]

α = 90°/180°
A = 90°
B = 180°
C = 120° (150mm)
X = 5.9″ (150mm)
Z = 5.9″ (150mm)

Fig. 34.17. Typical robotic vendor-supplied information.

TYPICAL STOP-GO, SQUARE OUT, AUTOMATION LOOP, SERVICING
A CONVENTIONAL CONTINUOUSLY MOVING MAIN LINE

MAINTENANCE STATION

STOP-GO LOOP

ACCUMULATION

MAIN LINE (CONTINUOUS)

CARLINE	STYLE	TOTAL SPOTS	AVERAGE SPOTS
CHEV-PONT	27	418	
"	35	482	438.5
"	69	476	
"	77	442	
○ = INDICATES SHOT PIN			
• = INDICATES QUEUE			
* = INDICATES DUAL GUNS			
I = INDICATES PILLARS			
P.M. = INDICATES PREVENTIVE MAINTENANCE			

MAIN LINE RUNS AT 62 J.P.H.
58.06 SECOND   STATION
8.00 SECOND   SHUTTLE TIME
50.06 SECOND   ACTUAL STATION TIME
5.00 SECOND   MARGIN
45.06 SECOND AVAILABLE FOR WELDING

Fig. 34.18. System layout: overall advance planning information.

running in automatic may not be communicating with a manual or shut-down mode, and this situation could cause a serious collision. The point is, it is difficult, if not impossible, to cover all contingencies on interference zones. There is always a tendency to pack robots in. Ideally, they should be spread out, without any overlapping of work envelopes.

34.2.5. System Layout

The proposed system can take various forms. For instance, Figure 34.21 is a perspective view of a proposed glass installation operation. It is an illustration of both an on-line and off-line feeder operation working together in one station. The robot first picks a rear quarter window glass out of a rack—an operation called depalletizing. (A highly underutilized feature of robotics in this country. For further explanations see Reference 2, pp. 260–261.) Then the robot tracks linearly, places the glass on a fixture, where an adhesive is applied to it. The robot gripper, disengaging from that particular piece of glass, then rotates, slides linearly again, and installs the glass that had been previously worked on.

SYSTEM PLANNING LAYOUT

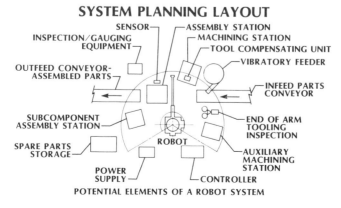

SENSOR
INSPECTION/GAUGING EQUIPMENT
ASSEMBLY STATION
MACHINING STATION
TOOL COMPENSATING UNIT
VIBRATORY FEEDER
OUTFEED CONVEYOR-ASSEMBLED PARTS
INFEED PARTS CONVEYOR
SUBCOMPONENT ASSEMBLY STATION
END OF ARM TOOLING INSPECTION
ROBOT
SPARE PARTS STORAGE
AUXILIARY MACHINING STATION
POWER SUPPLY
CONTROLLER
POTENTIAL ELEMENTS OF A ROBOT SYSTEM

Fig. 34.19. System planning layout: detail station information.

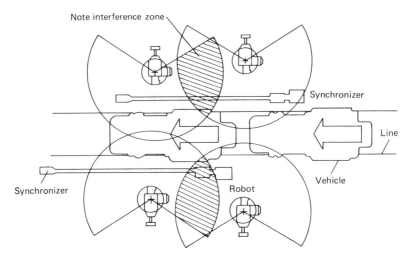

Fig. 34.20. Interference zones: on-line robot stations.

34.2.6. System Planning

Stop-Go Versus Moving Line

Figure 34.17 is an example of a typical stop-go, square-out, automation loop that services a conventional moving line. One might ask, why *stop-go* over a continuously *moving line?*[7] True, the robot can track a moving target and perform an operation it has been taught while in a stationary position. However, this method of processing the system and programming the robots is truly a three-dimensional chess game that requires many hours of planning time to accomplish.[4] Fundamentally, it is much easier to plunk the workpiece in front of the robot and design the workstation accordingly. Also, once the transfer systems have been redesigned in this way, the system will continue to serve what robots do best, whereas the line-tracking systems require painstaking process planning every time.

34.2.7. Location of the Robot to the Workpiece

The Grid Approach

When positioning the robots in a multiple system, it is advisable to place them on a grid (Figure 34.22). When reviewing a 30-robot system that has robots placed on some kind of a prefixed grid, it

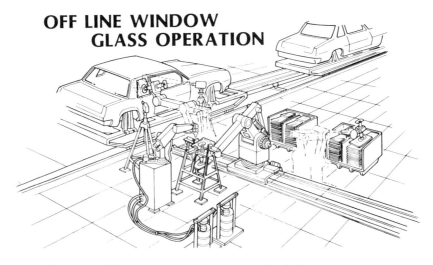

Fig. 34.21. Off-line quarter window glass operation feeding an on-line station.

RECOMMENDED FISHER STANDARD ROBOT POSITIONS

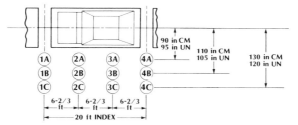

OPERATION DISCRIPTIONS:

1A	TIE BAR	3A	WHEELHOUSE INNER TO UNDERBODY
1B	MOTOR COMPARTMENT	3B	WHEELHOUSE TO QUARTER
1C	MOTOR COMPARTMENT SIDE	3C	SAIL AND ROOF DRIP
2A	FRONT HEADERS	4A	REAR END
2B	ROCKER PANELS	4B	FILLER, REAR END AND BACK BODY OPENING
2C	HINGE PILLARS	4C	QUARTER FILLER

Fig. 34.22. Grid approach to robotic positioning.

helps immensely when one needs to hopscotch operations, robots, and/or periphery equipment around.

For instance, if the plant reports that there is a support column running through that particular robot position on the system layout, and the engineer is about three-quarters of the way through his timing and program planning, robots on a grid can be moved around easily on paper. This information can be communicated easily back and forth by telephone with an engineering job shop, the plant and the robot manufacturer.

Then, too, still more problems arise when one takes a big system and attempts to install it in the plant. Other equipment may be found to be interfering with the planned operation, or one robot will not want to work with another robot in the same station. For example: The rocker welder robot does not work with the quarter window welder because the weld gun hoses drag over one another. If they have been laid out on a grid, the guns, the entire operation, and the operation tapes can be moved to another robot, say, in position 2B, and still maintain the original intent because the system was not developed with a number of odd, dedicated positions. Normally, it does not make much difference, plus or minus a few inches, where the robots are placed. Generally, the end-effector bracket design can take up this differential. On multiple installations the engineer should not waste time splitting hairs over exact position because he must realize that he cannot depend on all equipment being installed in precise locations.

34.2.8. Process Capability

In most robotic station operations there is a need for accuracy. The reason for this is that the robot itself is only repeatable to within a certain tolerance—say, plus or minus a millimeter. The product is no different: it has its own tolerance problems, and there can be as much as a 2-mm variation at any one point from a fixed gage point on the cart. The cart repeatability to the robot's position provides another half-millimeter variation. Therefore, in Figure 34.23, from tool center point to workpiece, there is a possible 2–3 mill tolerance stack-up. It may be necessary to overcome this through some means (i.e., precision guides, vision, tactile sensing, etc.). The point is, it is not easy on a moving line or a stop-go line to get the robot to do what it is programmed to do *accurately!*

This has nothing to do with the quality of the product. It has to do with the inaccuracies resulting from the individual stack-up of all these tolerances.

34.3. PRODUCTION OPERATIONS ANALYSIS

34.3.1. Productivity Assessment

The productivity figure is the figure that must be produced consistently per production hour. Another way of saying this is, this is the "pay" rate. This figure may be, and usually is, higher than the production rate of the plant where it is to be used for the following reasons:

1. It must be able to produce the "peak" rate of a particular style or combination of styles.
2. It may have to "make up" assemblies, due to breakdown or equipment lag.
3. It may have to feed a "bank."

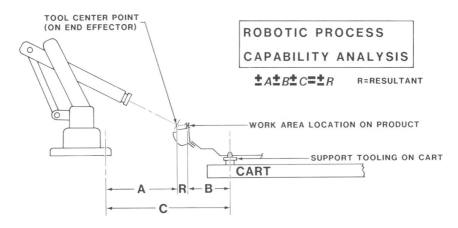

Fig. 34.23. Robotic process capability.

This number must be determined by the process engineer and/or be negotiated with plant engineers for their particular requirements.

Swapping robotic for manual operations on a one-for-one basis is generally only marginally cost-effective. It is necessary to start thinking about robots as devices in their own right. Put another way, *what can this device do to increase productivity that a manual operation cannot do?*

34.3.2. Operation Times

The next step in planning is to determine the time that is available for an operation. When setting up a basic station time the engineer must determine what the maximum line speed or piece-per-hour rate might be in the example shown (Figure 34.24). If the line moves at 75 jobs per hour (JPH) there is a 48-sec total station time. Generally, the stop-go shuttle movement, over 12–24 ft (4–8 m), would require about 8 sec, leaving a total available time for the operation of 40 sec in this case. This leaves a total of 36 sec available for processing the operation.

The reason for taking 10% off is that it is difficult to plan so close to the correct station time. There are too many contingencies to consider. As the process engineer gets sharper at it, obviously he is going to be able to cut into that time a little. But a 4-sec station margin is not very much and can be justifiably carried in the longer station times. On the other hand, there are 36 sec that can be filled with the operations necessary to process the product.

In the same manner, the system's efficiency rating should be developed as previously discussed. Generally, off-line automatic operations are set to run at least 10–15% faster than the main line they

SETTING UP YOUR BASIC STATION TIME

DETERMINING AVAILABLE OPERATION TIME

IF LINE MOVES AT 75 JPH, THERE ARE:

48	sec	**Total Station Time**
- 8	sec	**Shuttle Movement**
40	sec	**Total Time Available For Operation**
- 4	sec	**Margin**
36	sec	**Time Available For Processing Operation**

Fig. 34.24. Operation time.

AUTOMATED BODY SIDE INNER ASSEMBLY

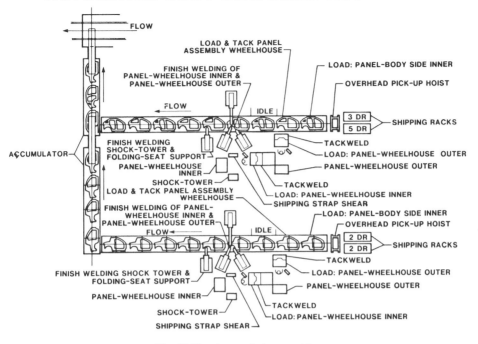

Fig. 34.25. Accumulation provisions.

feed. This extra capacity is sometimes fed into a bank or accumulator (Figure 34.25) to cover downtime possibilities.

34.3.3. The Effect of Acceleration and Deceleration

Why worry about acceleration (Acc) and deceleration (Dec) of the robot? First, it takes approximately $\frac{1}{4}$ sec for most robots to reach terminal velocity [5 in./sec to 30 in./sec (13 cm/sec to 76 cm/sec)] and another $\frac{1}{4}$ sec to stop. This is because the laws of physics are such that a given mass, such as the robot arm, can only be accelerated within a certain time, given a reasonable power system (Reference 2, p. 36). The astounding part is that most busy robotic moves, when tightly processed, never reach terminal velocity—not even a terminal velocity of 5 in. per second (13 cm/sec). The robot may have the capability of moving 60–120 in./sec (152–350 cm/sec), but if it is being used for anything other than certain "pick" and "put" operations, this velocity is never reached totally. When working within a 40–100+/sec operation time, it is certainly not necessary to be roaring around at 60 in. per second (152 cm/sec) on individual moves. Generally, the chart of Figure 34.26 reflects what a move curve winds up looking like for Acc and Dec times for what is estimated to be 90–95% of all robotic moves. For general purposes then, moves under 4 in. (10 cm) take roughly $\frac{1}{2}$ sec. Armed with that, it is possible to start to visualize individual move times.*

34.3.4. General Move Times

If the robot is going to move a longer distance, like 10 in. (25 cm), it can be programmed at 10 in./sec (25 cm/sec), and Acc and Dec time can be discarded. The robot arm is moving far enough so that it is no longer necessary to consider the detail implications of Acc and Dec. Moves of longer distances (10 in. and up, above 25 cm) can be considered quite simply as distance divided by the program speed to get the appropriate time. Ten inches at 10 in./sec (25 cm at 25 cm/sec) equals appoximately 1 sec. Now, if anybody wants to challenge this, we will admit that it is really insignificantly

---

* Readers should be aware that time estimations considered here are rough and may be suitable only for rough estimates. For more accurate time-analysis, methods, such as described in Chap. 30, Robot Ergonomics: Optimizing Robot Work, should be applied.

CALCULATING OPERATIONS TIMES

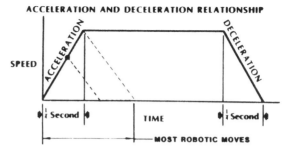

Fig. 34.26. Acceleration and deceleration relationships and their respective control over robotic moves.

longer because of the addition of Acc and Dec. However, it is possible to change the program speed slightly when programming. Consequently, the program speed can be increased up to 12 in./sec (30 cm/sec) for a 10-in. (25-cm) move and still come pretty close to covering the additional Acc and Dec time—and get back to a 1-sec move.

The object here is to avoid getting into a lot of detailed mathematics and decimal places on basic moves. The important point to consider for longer moves is to make the program speed synonymous with distance. That is, if the robot is moving 20 in. (51 cm), consider 20 in./sec (51 cm/sec). If it is moving 30 in. (76 cm) consider 30 in./sec (76 cm/sec), and then it is back to 1-sec packages. This approach is not touted because it is convenient. Long moves are easy to speed up whereas short moves never reach high terminal velocity because they are governed by the threshold of Acc and Dec.

34.3.5. Compartmentalize Times

Next, isolate end-effector weld times, grip time, or flow rates so that they can also be plugged in as a single package. If we are going to move and weld as part of an operation, then the time required to place a spot weld is a combination of the time the robot takes to get to the spot-weld location, plus the time it takes to make a weld. Let us assume we are moving under 3–4 in. (1.9 cm) and it takes approximately 0.5 sec, and the time to make a particular weld also takes approximately 0.5 sec. If the average time to move and weld is known, it is possible to generalize about how long it takes to perform several welds. Then the moves in and out can be added to this figure to establish total operation time.

34.3.6. Straight-Line and Attitude Moves

Figure 34.27 shows an example of typical straight-line (on the left) and attitude or roll (on the right) moves. First let us discuss straight-line moves. If the distance in Figure 34.28 from point 1 to 2 is

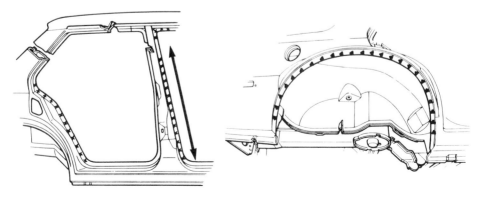

Fig. 34.27. Basic straight-line and attitude moves.

GENERAL TIMING GUIDE FOR ALL ROBOTIC OPERATIONS FROM A MOVE FROM START CYCLE THROUGH THE OPERATION AND BACK TO START CYCLE

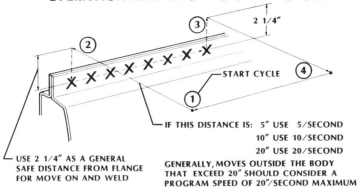

IF THIS DISTANCE IS: 5" USE 5/SECOND
10" USE 10/SECOND
20" USE 20/SECOND

USE 2 1/4" AS A GENERAL SAFE DISTANCE FROM FLANGE FOR MOVE ON AND WELD

GENERALLY, MOVES OUTSIDE THE BODY THAT EXCEED 20" SHOULD CONSIDER A PROGRAM SPEED OF 20"/SECOND MAXIMUM

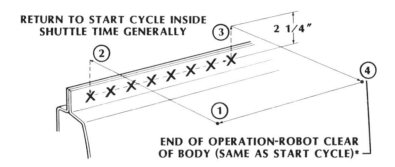

RETURN TO START CYCLE INSIDE SHUTTLE TIME GENERALLY

END OF OPERATION-ROBOT CLEAR OF BODY (SAME AS START CYCLE)*

NOTE: DOUBLE THE TIME FOR PROGRAMS THAT NEED A DELICATE TOUCH

*ONLY IF THIS TIME TO RETURN TO START CYCLE IS LESS THAN THE MACHINE INDEX OR SHUTTLE TIME

Fig. 34.28. Timing guide for straight-line moves.

only 5 in. (13 cm), then one would select a program speed of 5 in./sec (13 cm/sec) and consider it a 1-sec move. A 10-in. (25 cm) move would be programmed at 10 in./sec (25 cm/sec) and so on. In the event it became necessary to program 20 in./sec (51 cm/sec), it would be because the end of the robot was 20 in. (51 cm) away from the workpiece and could not get any closer. Perhaps when the shuttle indexed the body or part, the tool end effector was kept at that distance for clearance purposes.

The object should be to stay as close as possible to the work area at the start cycle. In fact, it may be preferable to drop the weld gun or robot end-effector into the part and reverse-program it to get the most efficient program. Then, reverse the process again and have the robot go forward, programming the same points over the top of each other, and then erase the first points. Utilizing this technique, a very efficient move into the work area has been effected.

Moreover, Figure 34.28 indicates that a move of less than 20 in. (51 cm) from 1 to 2 can be accomplished in 1 sec. A modern robot can move under 4 in. (10 cm) in 0.5 sec, and a typical spot weld can be accomplished in approximately another 0.5 sec. So, it is possible to move in (1 sec), drop down on flange and make a weld (1 sec), and then keep on moving and welding. Each one of these moves and welds takes 1 sec. The rough edges have been knocked off, but we really did not lose anything, and we came up with an operation that is sound.

Since 10% has been taken off the top for a safe margin, there should be enough margin to cover extraneous moves. Extraneous moves might be wrinkles in the flange, jumping around gage slots, and the like, and slight attitude shifts that were not entirely visualized up front.

MOVEMENT ALONG AN ARC

WHEN WELDING ALONG AN ARC THE VELOCITY WILL BE IN DEGREES/SECONDS

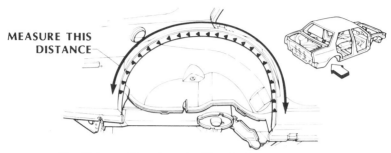

The Distance Moved can be Directly Compared to the Linear Distance for Timing Purposes

Fig. 34.29.   Timing attitude moves.

34.3.7.   Attitude Moves

Now let us take a look at timing some attitude moves that are sometimes difficult to understand and visualize. First, *attitude moves,* for the purposes of this discussion, are defined as all maneuvers that are nonlinear. The simplest of these is the movement along an arc.

According to the "book," "when the tool center point is moving along an arc the velocity will be in degrees per second." True enough, but perhaps a more understandable approach is to simply convert radial movement to linear terms. Then the distance moved *can* be directly compared to linear distance for timing purposes.

Figure 34.29 shows a typical rear quarter wheelhouse opening that would require radical (attitude) movement. What is developing in this type of operation is that the robot's roll axis is pirouetting with almost no linear movement at all. The engineer must determine how much torque this axis can tolerate, how far away the tool center point is from the center-line of roll, and radially how much

MASS CONSIDERATIONS

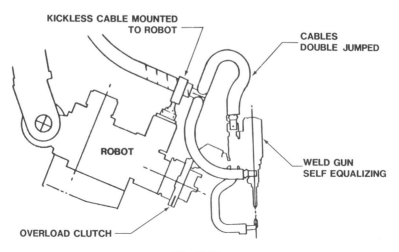

Fig. 34.30

mass is being moved. This can be calculated and compared to what that particular robot axis can handle in degrees per second.

However, few ratings of this nature exist, and even when they do, they are not easily understood. What the robot really wants to do is to not move the mass any faster than it would if it were making a straight-line move. Therefore, take a tape and measure the wheelhouse periphery as in Figure 34.29. Stretch the tape out and figure this as a linear distance. In this situation it is possible to develop the move in the same way that the robot makes a straight-line move, and time it accordingly. This may sound simplistic, but it in fact gets us closer to the truth than following other, more complicated calculations.

34.3.8. Torque/Mass Considerations

Torque/mass considerations can be roughed out just as easily. Figure 34.30 shows a typical weld gun dressing and all its accompanying hardware, the weight of which must be factored out and compared to the weight-carrying capabilities of the particular robot. Next, we need to establish the overall center of gravity, as in Figure 34.31 plus or minus a couple of inches. Finally, the "effective load" can be developed as in Figure 34.32. Normally, this calculation would be determined in hypotenuselike resultants. However, this shortcuts the challenging demands placed on the robot's actual kinematics. Consequently, it is highly advisable to add in any offset mass directly times the weight.

34.3.9. Establishing an Accumulative Total Station Time

Once the basic timing packages are developed, they can be plugged into the computer to get an accurate total time, an example of which is shown in Figure 34.33. This is particularly helpful in marginal operations. For example, if there is a wheelhouse operation that the process engineer would like to do in one station with a single robot, and it looks like he does not have the time to make two of the weld spots, he must look for some other way of combining the operations—since he obviously cannot introduce another robot for just two spots. One suggestion, if it cannot be done with a single weld gun, is to cautiously move into a double gun. Although the double gun is slightly slower, it can make two spots at a time, resulting in about a 50% improvement in overall operation efficiency; and it solves the timing problem.

The next step is to total the times to perform each of the operations that have been discussed. Once we compartmentalize all these times, a computer program will allow an engineer to plug in the move sequence, metal combination, metal thickness, weld gun stroke, gripper close function, sealant flow rate, and other pertinent information to develop a printout of the detailed station-by-station precedence.

TORQUE & WEIGHT CALCULATIONS

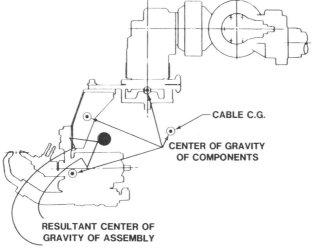

CABLE C.G.

CENTER OF GRAVITY
OF COMPONENTS

RESULTANT CENTER OF
GRAVITY OF ASSEMBLY

Fig. 34.31

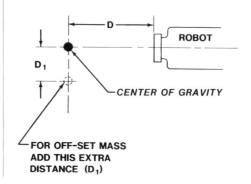

Fig. 34.32

34.3.10. Downtime Considerations

It is necessary to look for some place to accumulate *downtime*. It is, therefore, recommended that in most robotic operations allowance be made for some small accumulation. If there is a pair of robots it is one thing, but if there are 30 robots in a row, like the massive installations in some auto plants, the downtime factorial must be considered. For instance, even if all robots ran at 98–99% uptime but are all interdependent (in a factorial nature), we probably would never run as a dedicated line system or very seldom have *all* robots running. So consideration must be given to the implications of slight downtime for each robot.

34.3.11. Banking Provisions

Studies indicate that 97–98% of most downtime is less than 5 min, so a small bank of only three to four jobs is all that is required to take care of most of these downtime problems (refer to Figure 34.18). How this is accomplished depends on the size and type of the system. A small system of only 10 to 15 pairs of robot can accumulate at the end of the line. Large systems must utilize the "rubber band" effect in that small accumulation areas should be planned in between large automated islands.

34.3.12. Operation Verification

Along with establishing operation times, end-effector weights, and so on, it is necessary to verify the program's physical nature. Methods for verifying robot programs are *graphic*, the *three-dimensional drafting layout*, and *scale modeling*. The robot itself can be used at times, but more often than not this proves to be time-consuming and cumbersome.

Computer Graphics and Robot Process Planning

Generally a *computer graphic* system offers the speediest and most flexible system of all. Computer graphics is a very promising system, and it looks to be the most important advancement upcoming in robotic process planning for the future. Unfortunately, system compatibility and proprietary rights to manufacturers' algorithm formulas are hampering development of a truly flexible system.

To use this system, start by positioning a robot near the job area, as in Figure 34.34, on an area grid. This quickly establishes in and out, fore and aft, and up and down positions. On something as tricky as a wheelhouse opening weld, the engineer must also determine the beginning and ending of the operation limits. The half-moon pictures on the computer graphics illustration in Figure 34.35 represent the limits of each axis. It can easily be determined how close the robot is to the individual limit of an axis.

Computer graphics offers the light-pen capability of independently moving the robot around, moving the part or body around, and simultaneously moving the end effector. If the engineer can combine all three of these capabilities, he truly has a medium that can be used to plan effectively a 30-robot system in an afternoon—an accomplishment that now takes weeks to complete.

FISHER PROGRAM TO TIME ROBOT MOVES AND WELDING

PROGRAM NAME: ROBOT

ENTER GIVEN:

WELD SEQUENCE: 1
METAL CONDITION = 3
METAL THICKNESS = 0.880990 (MM)
GUN STROKE = 1.000 (INCHES)
WELD TIME = 0.448 SEC. S
WELD AMPERES = 9500

SAMPLE PROGRAM

			(INCHES)
SQUEEZE TIME IS		0.116	
HOLD TIME IS		0.016	
GUN STROKE IS		1.000	
SQUEEZE DELAY TIME IS		0.183	
WELD TIME IS		0.133	
TOTAL WELD TIME IS		0.448	
WELD AMPERES IS		9500	
ACCUMULATIVE TIME =		35.950	
**MOVE NUMBER =		47	**
FROM	207.24,	22.04,	-28.97
TO POINT	207.24,	22.04,	-28.97
LINEAR MOVE DISTANCE =		0.000	
MAX. LINEAR VELOCITY =		0.0	
BACKUP TIME =		0.59	
MOVE & BACKUP TIME		0.590	
TOTAL TIME =		0.590	
ACCUMULATIVE TIME =		36.540	
**MOVE NUMBER =		48	**
FROM	207.24,	22.04,	-28.97
TO POINT	207.24,	31.49,	-28.97
LINEAR MOVE DISTANCE =		9.450	
MAX. LINEAR VELOCITY =		30.0	
NO BACKUP.			
MOVE & BACKUP TIME		0.565	
TOTAL TIME =		0.565	
ACCUMULATIVE TIME =		37.105	
**MOVE NUMBER =		49	**
FROM	207.24,	31.49,	-28.97
TO POINT	118.11,	31.49,	39.37
LINEAR MOVE DISTANCE =		112.314	
MAX. LINEAR VELOCITY =		30.0	
NO BACKUP.			

Fig. 34.33. Accumulative total station timing program.

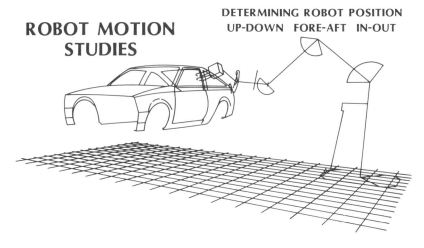

Fig. 34.34.   Use of computer graphics in robotic process.

34.4.   END-EFFECTOR AND AUXILIARY EQUIPMENT

34.4.1.   End-Effector Attitude

There is still another challenge. The robot has been positioned, and the limits of geometry have been tested. Now the planning engineer must set both the attitude and position of the end-effector bracket (Figure 34.36) and that distance to the end-effector tool center point (T.C.P.). The tool center point dimension (Figure 34.37) is the distance from the end of the robot face plate to a point in front of the end-effector where the work will be performed. It should be located, whenever possible, along the roll axis for ease of programming. On high-tech robots that distance can be plugged into the robot's computer, and the computer will help in touching up attitude moves. Setting the attitude of the bracket on something like the backheader of a car body is sometimes very tricky. The engineer must utilize every means at his disposal.

When laying out a system, it may not be desirable to stop with computer graphics. Difficult operations may require scale modeling. At Fisher Body, where engineers have access to scale-model bodies, usually $\frac{3}{8}$ or $\frac{1}{2}$ in. 0.9 to 1.3 cm scale or scale-model robots made by robot manufacturers are utilized. With these tools it is very easy to take weld guns, sealing guns, and brackets, reduce them accordingly, and place them on card stock to test and adjust the end-effector support bracket designs.

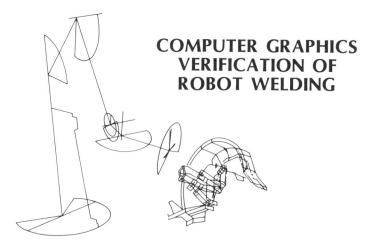

Fig. 34.35.   Computer graphic limit determinates.

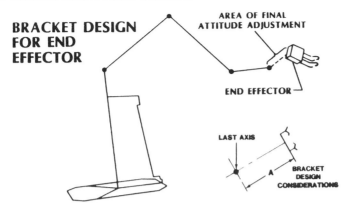

Fig. 34.36. Setting end-effect or attitude and distance from last axis-mounting plate.

This is especially helpful when processing operations utilizing a moving target or line tracking. When planning line tracking or moving-target operations, time and distance must be correlated. The robot stands in a fixed position, working on the product as it enters the station and keeps on working until it leaves the station. Accurate planning and timing of a proposed station is a three-dimensional chess game at best, and modeling has proven to be a great asset for review of detailed work in window correlation, as in Figure 34.38.

Another method for verifying a robot program is the traditional three-view layout (Figure 34.39) that allows the engineer to see most of the processing parameters in plan view. Much of the time this method is adequate for dispatching simple operations. However, difficult brackets that are attached to the end-effector may require delicate attitude positioning and force the engineer back to actual modeling of the operation.

34.4.2. End-Effectors and Associated Equipment

We turn to end-effectors and some associated equipment utilized with robots. An *end-effector* is the device attached to the end of the robot arm. Obviously, this includes many kinds of grippers and devices other than weld guns as in Figure 34.40.

Antishock Clutch

When a weld gun is attached on the end of the robot arm, there is more than just the weld gun. There may also be a clutch (Figure 34.41) and mounting bracket. A clutch is utilized because it is possible to get the wrong body style in front of the robot. The robot charges out and crashes into the body. Robots deal with the force of tons in certain attitudes, in spite of the fact that some of the biggest ones in use can only lift 250 lb (113 kg). They can put a 1-ton blow on the part or body. If

END EFFECTOR CENTER POINT

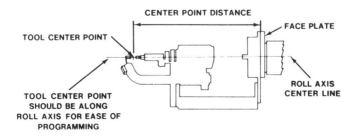

Fig. 34.37. Tool center point.

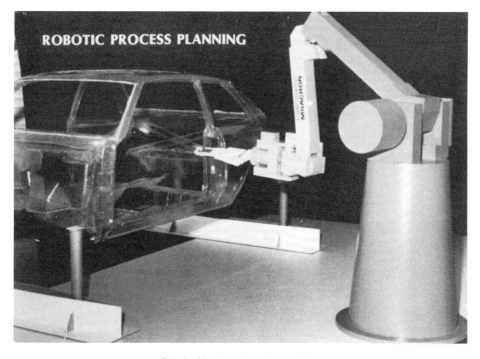

Fig. 34.38. Actual scale modeling.

this situation can be stopped by tripping a clutch, it is possible to keep from scrapping many finished bodies, parts, and/or the end-effector tooling.

Specialized Robotic Tooling

One answer in a body shop was the double-weld gun or even the triple-weld gun (Figure 34.42). The General Motors Lordstown plant was running 110 jobs per hour, and this allowed only 22 sec to weld 33 rocker spots. It was either spread that over six robots or put in one triple-weld gun.

SYSTEM LAYOUT VERIFICATION

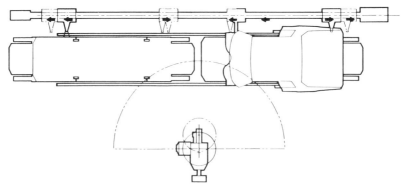

TRADITIONAL DRAFTING LAYOUT

Fig. 34.39

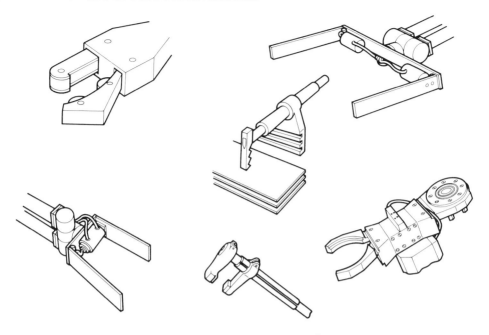

Fig. 34.40. Typical end-effector designs.

The gun welds only lower rocker flanges, and that type of weld is pretty straightforward and does not require challenging attitude moves. Since the moves were not difficult and it was possible to cable each weld gun separately, the gun worked well and negated the need for four more robots.

Auxiliary Equipment

Many engineers are not familiar with the other pieces of auxiliary equipment that make up a station, so there is a tendency to overlook them when station planning. In spot welding, for instance (Figure 34.43), there are items like kickless cables, counter balancers, transformers, jumpers, densification packages, and so on.

The engineer must be familiar with these pieces of equipment because it is necessary to help the

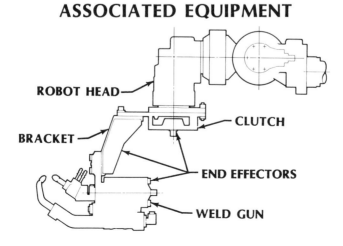

Fig. 34.41. Antishock clutch.

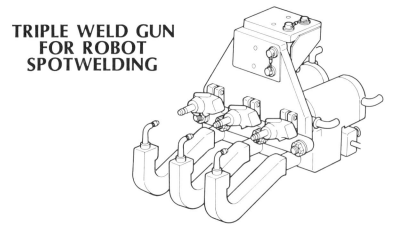

Fig. 34.42. Specialized triple-weld gun.

plant position and order the right equipment for a system. He may want to mount the transformer on the robot arm or, in the case of pick-and-put operations, give consideration to unique types of grippers and prepositioning devices. Also, he must consider input and output signal requirements that will make this equipment function properly. The weight of each device must be known and carefully considered. We call attention to other factors and other pieces of equipment because they must be considered when planning a station.

Associated Equipment

Figure 34.44 shows an example of a piece of associated equipment that is mounted off-line but in-station. This device is a fixture that cleans and dresses welding tips, but it is a necessary part of the station and provisions for its placement must be laid out.

It must be remembered that when dealing with flexible automated robotic systems of this kind, the engineer must reexamine all the peripheral support equipment.

34.5. SYSTEM INSTALLATION

In the installation phase it is recommended that the robot location be carefully dimensioned as in Figure 34.45 and accurately positioned. The station must be laid out in detail. For multiple robotic

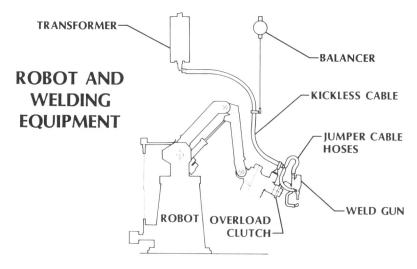

Fig. 34.43. Typical auxiliary equipment considerations.

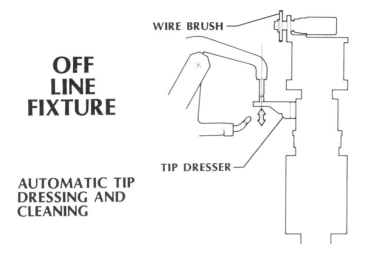

OFF LINE FIXTURE

AUTOMATIC TIP DRESSING AND CLEANING

Fig. 34.44. Typical associated equipment.

operation, the plant generally will be more than happy to let the application engineering department do this because the layout must show the complete system.

34.5.1. Project Responsibilities

After a system is laid out, it is important to establish project responsibilities as in Figure 34.46. This includes the work to be completed by the vendors as well as the local plant.

After the planning stages are roughed out and project status has been granted, the project leader should meet with all those concerned at least once a month and go through the entire system (Figure 34.47).

Schedules are important (Figure 34.48). It is strongly recommended that the system planning engineer get involved early with the project and, most important, *project* timing. When to order the equipment, when application engineering is to write the assignments, and when the manufacturer is supposed to come in and do their programming are all items that must be scheduled. Frequent checks must be made to be sure every bit of that planning happens on time. Setting up a systematic schedule obviously

INSTALLATION LAYOUT FOR INDIVIDUAL STATION

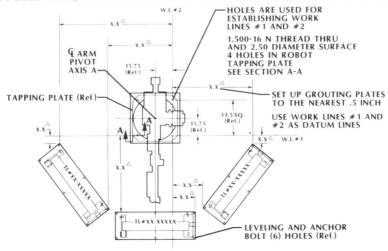

Fig. 34.45. Installation layout with dimension.

ESTABLISH PROJECT RESPONSIBILITIES

"Turnkey" Vendor
   Design and install cart shuttle system.
   Design and install body transfer (on and off cart) and controls.
   Install robots and related equipment to specifications.
   Design and construct pedestals for cart pallet from engineering locations. Program all robots to
       manufacturing processing.

Engineering Responsibilities
   Provide all weld studies and operation description sheets.
   Process welding operations—vendor to verify.
   Design and construct weld guns, brackets, and antishock clutches.

Joint Vendor and Engineering Responsibilities
   Establish location of robots and height of body off pallet relative to 100 mm line.
   Engineering to establish—vendor to verify.

Assembly Plant Responsibilities
   Install transformers and welder controls.
   Install weld guns per engineering assignments.

Fig. 34.46

is the project manager's business, and he should keep that responsibility even on major outside vendor "turnkey" projects.

34.5.2.   Project Tracking Sheets

During the installation phase, we suggest the creation of "project tracking sheets," where it is possible to state problems, indicate the man's name that must respond, put down the date when a speedy resolution is required, and, also, to create a place to list comments beside the problem. Figure 34.49 shows a design of a suggested format that allows the project manager to move around efficiently by the mails—without lengthy letters among the different vendors, suppliers, divisions, and activities—cooperating together on a major project and still getting answers quickly. For example, it is possible to be in the plant and yet deal with the general office, the robot supplier, and perhaps the conveyor people, all at the same time, on a given problem.

   Whether this sample project tracking sheet is used or not, a method of tracking the program must be developed and utilized to bring together the various combinations of input expertise from the different sources.

PROJECT STATUS REVIEW

Fig. 34.47

MASTER TIMING SCHEDULE
1982 "J" BODY   TIMING SCHEDULE RESPOT LINE SOUTHGATE

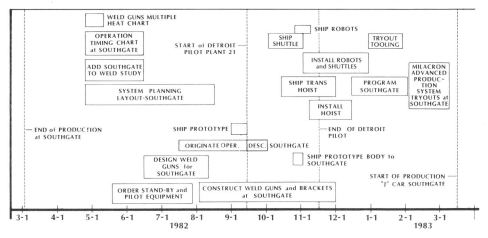

Fig. 34.48.   Typical timing schedule.

34.5.3.   Proper Training

After installation, it is always necessary to go over the system and make some adjustments. It is essential to make sure that proper training is provided for those who will operate and maintain the station. Therefore *proper training is an absolute must!*

Some serious mistakes have been made on this issue, from an engineering, programming, and maintenance standpoint. Often planners overlook the fact that when a plant stays idle during an extended

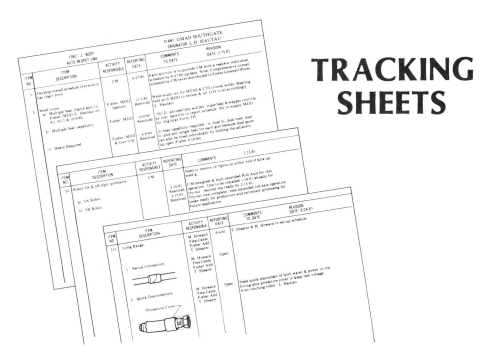

Fig. 34.49

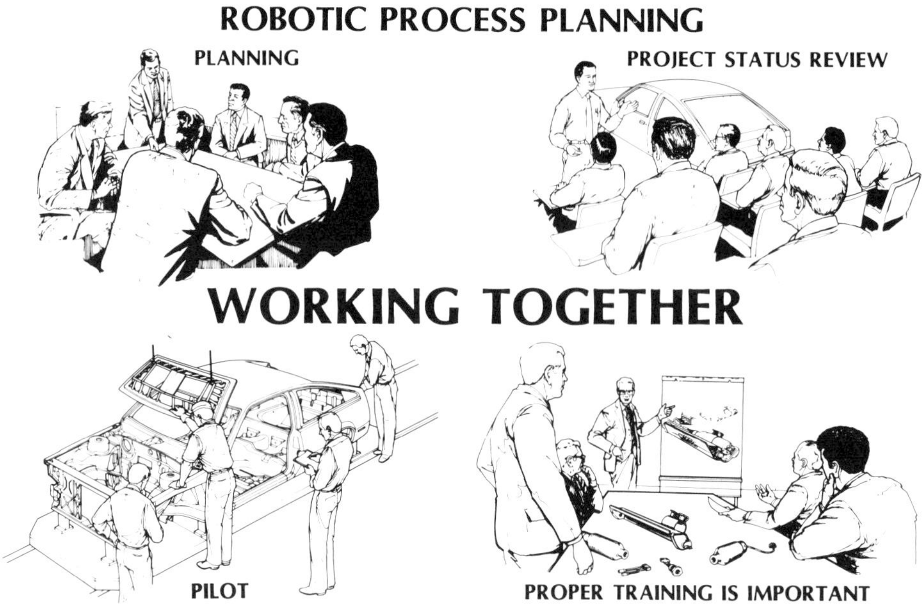

ROBOTIC PROCESS PLANNING

PLANNING PROJECT STATUS REVIEW

WORKING TOGETHER

PILOT PROPER TRAINING IS IMPORTANT

Fig. 34.50. Quality circles.

conversion period it loses some of its people and expertise. In addition, not only is a new product being introduced, but also an entirely different means of production is being implemented. The result is that the expertise planned for is not always there at the time.

In the past there have been gross underestimates of what it takes to run these systems. A planning engineer should not hestitate to tell a plant that, if they have 20 or more new robots in a system, it will require the assistance of an engineering person to manage that system for at least the first six months. After that transitional period, the maintenance people will become familiar enough with the new process and equipment to be successful at running the system on their own.

At first, however, the plant needs somebody who knows what cycles-per-second means, what circular mills on a kickless cable are all about, and what a TR66 transformer can and cannot do. It needs someone who understands the proper use of cupal weld caps and is willing to set weld schedules up properly—knowing that extra cycles to each new schedule within a station can cause the operation to run beyond the station time allotted.

34.5.4. Installation and Start-Up

In piloting a new robotic system, an engineer should not attempt to put that robot system on line and bring it up to start of production readiness without some appropriate tryout and debugging time allowed. A couple of weeks may be all that is necessary on a small system. Multiple systems, however, may require *three months* to make sure the system is at least functionally operational and properly programmed ahead of actual start-up. The last thing a plant needs is to program robots or try to sweeten the timing as progress begins through start-up of an initial acceleration.

The average robotic operation is generally programmed in less than 4 hr, but there may be as much as 20 hr worth of touch-up work still required on marginal or difficult operations. There are a lot of things that cannot be seen until the plant reaches for that last five jobs per hour.

During installation and start-up it is necessary to spend some time on quality circles, or value management teams, or whatever your organization wants to call them. It pays to sit down for an hour a day after work with the production workers and the plant management and go over trouble spots. It really will make things go a lot easier (Figure 34.50).

ACKNOWLEDGMENTS

I gratefully acknowledge the help of Fisher Body's Tom Murphy, who helped pull much of the material together. Fisher Body's fantastic Illustration Department deserves credit for the excellent graphics provided through the entire chapter. Also, I would like to acknowledge Jack Saunders of GMF Robotics, who graciously agreed to edit the rough draft of this material and put it into meaningful form. Finally, I would like to thank Shelly Moss for patience in typing and processing the final drafts.

REFERENCES

1. Friedrich, O., Economy and Business, *Time,* December 8, 1980.
2. Engelberger, J. F., *Robotics in Practice,* American Management Association, 1980.
3. Tanner, W. R., Ed., *Industrial Robots,* Vol. 1, *Fundamentals,* 2nd ed., Society of Manufacturing Engineers, 1981.
4. DiPietro, F. A., Line Tracking Robots for Body Spot Welding, *Proceedings of the 9th International Symposium on Industrial Robots,* March 1979.
5. Tanner, W. R., Ed., *Industrial Robots,* Vol. 2, *Applications,* 2nd ed., Society of Manufacturing Engineers, 1981.
6. *Proceedings of Robots VI,* Robotics International of SME, 1982.
7. Dawson, B. L., *Moving Line Applications with a Computer Controlled Robot,* in Tanner, W. R., Ed., *Industrial Robots,* Vol. 1, 2nd ed., Society of Manufacturing Engineers, 1981.

CHAPTER 35

INDUSTRIAL ROBOTS: RELIABILITY, MAINTENANCE, AND SAFETY

GEORGE E. MUNSON

Robot Systems, Inc.
Norcross, Georgia

35.1. INTRODUCTION

Reliability, maintainability, and safety in operation are essential to achieving high productivity and utilization factors from industrial equipment. And, as we demonstrate, these issues are inextricably bound together.

Obviously, each piece of equipment on the factory floor represents a means to an end—the end being the manufacture of goods of high quality, in sufficient quantities, in a timely fashion, and at competitive costs. Whether the equipment is a machine tool, a press, material-handling devices, inspection or test gages—or an industrial robot—it must have a high availability factor, it must be easily maintained, and it must operate safely in its environment. And the integration of all these elements must be executed in a manner that takes into consideration each of these factors for each piece of equipment. To do otherwise carries the risk of installing a handsomely engineered system that fails to meet the stated objectives.

These ingredients particularly apply to the industrial robot.

35.1.1. Reliability

In the vast majority of instances the robot represents a link in the manufacturing process wherein it services high-cost capital equipment. In this role it governs the degree to which this equipment is utilized. Of necessity, then, the robot must be inherently reliable and must be designed to yield a very high availability factor, that is, be easily and quickly repaired. Fortunately, as later shown, an availability factor for the robot in excess of 97% is typical.

35.1.2. Maintenance

Maintenance and maintainability take on new meaning in the age of automation. The complexities of state-of-the-art computer-controlled machinery, and the removal from the workplace of human operators with all of their sensory and judgmental capabilities, require rigidly enforced programs for proper and regular equipment maintenance. The "fix-it-when-it-breaks" philosophy is no longer acceptable and was probably bad economics anyway. Certainly it is today.

Just as unacceptable is equipment that is not easily maintained, trouble shot, and repaired. Availability, or *uptime,* is a function not only of the inherent equipment reliability but also of the ease and speed of repair when a downtime incident occurs. Even if a machine has a mean time between failure of 2000 hr, its design is questionable if it takes a week to diagnose and repair the problem, and the lost production at that time would be intolerable.

35.1.3. Safety

Safe operation of equipment must be viewed from several angles. First and foremost, human operators must be adequately protected from hazardous conditions in *all* modes of operation of the equipment

---

This chapter was written when the author was Vice President of Marketing, Unimation, Inc.

during setup, when maintenance is being performed, and during normal running. Second, the equipment must be protected from itself. Precautions should be taken that a malfunction will not, in turn, result in damage or breakage. And, third, the equipment should be so integrated that damage to other machinery and devices does not result from a malfunction.

The prolific author Isaac Asimov[1] whimsically formulated "the three laws of robotics" in the early 1940s. Whimsically, because as he stated in his foreword to J. F. Engelberger's book, *Robotics in Practice,*[2] "I did not at that time seriously believe that I would live to see robots in action and robotics becoming a booming industry. . . ." Interestingly, the three laws are quite validly applied to this booming industry, and the robot designer and user will do well to take heed:

1. A robot must never harm a human being or through inaction allow a human being to come to harm.

2. A robot must always obey a human being unless this is in conflict with the first law.

3. A robot must not allow itself to come to harm unless this is in conflict with either the first or second laws of robotics.

A final thought on the RMS (reliability, maintenance, and safety) of robotics: The technology is advancing at a fast pace. With it are pressures from the marketplace for new products and ever-increasing sophistication. We must be cautious that the RMS not be overlooked in the exuberance of meeting these pressures.

35.1.4. The Environment

Equipment reliability starts with design. However, the environment in which the machine is to operate must be thoroughly known and understood. This is particularly challenging in the field of robotics because the robot is expected to be comfortable in an ever-broadening range of environments from metalcutting to meat processing. These environments include extremes of temperature, humidity, atmospheric contaminants and precipitants, radiant heat, shock and vibration, electrical noise, and so on. In addition, liquid sprays, often corrosive, are encountered along with abrasive particles, explosive atmospheres, and a variety of chemicals. Figure 35.1 lists some of these environmental factors.[2]

Foundries of all types pose rather adverse conditions. Figure 35.2 shows a robot servicing a die casting machine. Of necessity it is positioned right at the parting lines of the dies where molten metal (zinc, aluminum, magnesium) often spurts directly at the robot. Heat is radiated from the melt pot. And, in most instances, the robot must dip the casting into a corrosive liquid to cool it. In investment casting, the robot and all of its parts are exposed to slurries and sand mixes in an atmosphere that is ladened with highly abrasive silica dust. Figure 35.3 shows such an application. It is common in such instances to slightly pressurize various parts of the robot to avoid ingress of the contaminants and to pipe clean air to the heat exchanger inlet. Where necessary the electronic control cabinet is sealed and air-conditioned or, as shown in Figure 35.4, the cabinet is located remotely and outside of the processing room.

Forging and heat-treating operations pose other threats. Extremely high shock is encountered in forge shops. The robot must not only be anchored securely, but also often with isolating pads beneath it. In addition, it is not unusual for the robot to hold onto the billet while it is being hammered. A rugged gripper (end-of-arm tooling) is essential as a built-in compliance. Radiant heat abounds in handling red hot billets in heat-treating applications. Radiant shields, strategically located on the robot's arm or between the robot and the source, are often employed. Sometimes the robot is taught to periodically dip its gripper into cooling water. Figure 35.5 shows the kind of heat exposure that can be encountered. Obviously, in such instances, it is undesirable, if not totally impractical, to have electrical/electronic devices or hydraulic fluids located at the robot's arm extremity or wrist.

1. Ambient temperature: up to 120°F without cooling air.
2. Radiant heating: source temperature up to 2000°F.
3. Shock: excursions up to ½ in., repetitions to 2 per second.
4. Electrical noise: line drop-outs, motor starting transients; RF heating.
5. Liquid sprays: water and other coolants, often corrosive.
6. Fumes and vapors: process chemicals, steam cleaning.
7. Particulate matter: sand, metallic dust, hot slag.
8. Fire and explosion risk: open flame, explosive gas and vapor mixtures.

Fig. 35.1. Hazards in the INDUSTRIAL ENVIRONMENT.

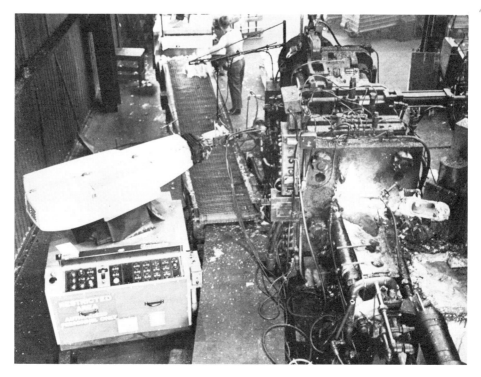

Fig. 35.2. Robot exposed to heat and molten metal in die casting. (Photo courtesy of Unimation, Inc.)

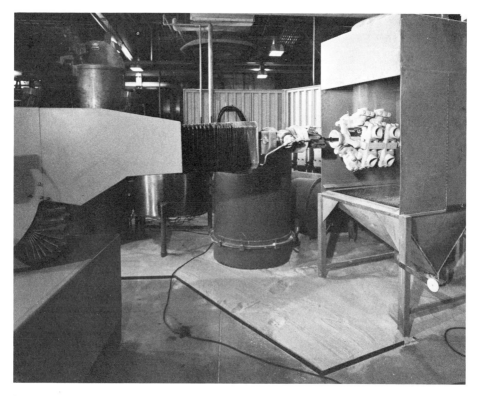

Fig. 35.3. Robot holds investment casting mold in raw of sand. (Photo courtesy of Unimation, Inc.)

Fig. 35.4. To avoid atmospheric contamination, Robot control cabinets are located remotely. (Photo courtesy of Unimation, Inc.)

Fig. 35.5. Billets of 2000°F (1100°C) are handled by a robot. Note the end-of-arm tooling quench tank below the robot arm. (Photo courtesy of Unimation, Inc.)

The hazards in metalcutting are commonly an abundance of chips and of coolants. Figure 35.6 is quite illustrative. Note the protective covers on the robot's arm.

As new applications arise, so do new challenges. For example, robots are now being used in meat processing, a job where daily drenching with water and steam is part of the routine. They are also being asked to handle frozen food packages in giant freezers at −10°F (−23°C). There seems to be no end to the roboticist's challenges in designing for high reliability. Spot welding, arc welding, glass forming—all present their own set of environmental hazards to be dealt with.

Not only must the robot live and work in adverse environments, it must also be designed so that it does not contaminate the environment. When we talk about the reliability of the robot in doing its job, we are addressing getting the job done right. Therefore it must not generate contaminants or other environmental hazards to the manufacturing process. For example, many delicate and precision devices must be assembled in clean rooms where it would be disastrous if bearing-wear-generated minute particles of lubricants were to leak or drip into the environment. Similar restraints are imposed in food industry applications or in handling fabrics and the like.

A design for reliability, then, must take into consideration not only the integrity of each and every part but also how, individually and collectively, the parts will perform in the variety of environments imposed upon them.

35.1.5. The System

In addressing the issue of reliability, it is important to recognize that the robot is not a stand-alone machine. It is a tool and, as such, is always integrated into some kind of system that employs a few or many other machines and devices. At the very least it is intimately associated with some kind of end-of-arm tooling, probably some conveyors, parts feeders, and orienting devices, and, of course, the workpiece itself. In addition, there may be machine tools, trim presses, cooling or processing tanks, other robots or automation, inspection devices, and a central or master control or computer. How and how well the robot interfaces and interacts with all of this equipment, both in terms of manipulative actions and control interlocks, will determine the overall reliability of the installation. Although it is not our intent to discuss the ramifications or reliability of all of this equipment individually or collectively, it is important to recognize that all of these elements impinge on the design of a reliable, serviceable, safe operating system.

A vivid example is the spot welding of automobile bodies by a battery of robots as shown in Figure 35.7. The drama and excitement of this system in action results from the visual impact of all

Fig. 35.6. Robot arm booted to protect against chips in metalcutting. (Photo courtesy of Unimation, Inc.)

of the robots feverishly placing welds all over the car body before the next in line comes into position.

Close inspection reveals a bit more than just robots at work. There is the conveyor that transports the bodies from station to station past the robots. The precision with which it positions the bodies is essential to putting each of, perhaps, 3000 welds in the right place (or in any place at all). Every car body must be identified automatically as to style and this information passed from station to station so that the central control, the master choreographer, can tell each robot which of its programs is presently required (Figure 35.8). Interlocks must provide every workstation with knowledge of whether the conveyor is moving or in location for welding. Each robot must tell the central control whether it is done with its work and in a safe position before the conveyor can be indexed. The weld gun on each robot must have a proper flow of cooling water; the tips must not stick to the metal (periodic maintenance, manual or automatic); the weld gun controller must generate the proper current and dwell times; and the robot must signal the gun controller when to close the tips, and even with what pressure. And, yes, the robot must position the gun tips accurately. There is more. But the point is made. Acceptable quality car bodies will not come off the end of the line unless *everything* works— reliably. Downtime costs are measured in thousands of dollars per minute.

In many instances, means to assure continuous throughput are provided in anticipation of malfunctions or failures. In some cases, manual (human) backup is employed. In the case cited, backup robots are provided down the line to pick up the work of down stations. This is detected and the necessary information dispatched by the central control, including alarms, to draw the attention of the human overseers. A well-founded system will include diagnostics and telltale indicators to pinpoint the problem area for quick reaction time and repairs.

In other cases buffer techniques are applied to maintain throughput. Figure 35.9 shows a four-work-cell machining system where buffer storage is employed between cells. Should one work cell require maintenance or go down and require repair, parts coming from upstream cells can continue to function by taking parts out of storage. Figure 35.10 shows one of the work cells with pallet type buffer storage. Through the central control, seen in the background, robot subroutines are called up to put parts into storage or take them out as needed.

These examples are given to emphasize the need to assess achievable reliability and availability in terms of the work cell (island of automation) or a multiplicity of linked work cells. We show later how this assessment must also include maintenance and safety considerations.

Fig. 35.7. A robotic system in automobile body assembly: spot welding. (Photo courtesy of Unimation, Inc.)

35.2. RELIABILITY

Little has been published on the theoretical or actual level of reliability achieved by robots. However, after the accumulation of millions of hours of on-line operation and 10 man-years or more of work by individual machines, their availability (uptime) factor is known, as well as attendant maintenance costs. The results are impressive and, in fact, overshadow historical experience with most industrial equipment. In almost all documented cases the robot has been available for production work at least 97% of the time. Instances have been reported where 99.5% has been realized.

These results are a tribute not only to the quality level of design and manufacture of the robot but also to an understanding by the manufacturer that the robot *must* be capable of such performance. Otherwise its viability, acceptability, and economic advantages are nonexistent.

35.2.1. Definition of Terms

So far in this discourse several terms have been used that are closely related to one another—*reliability, availability, maintainability,* and *uptime* (or *downtime* for those of a negative bent of mind). Lest the reader become confused, we would do well to give these terms definition.

The *reliability* of a product is generally defined as the probability that the product will give satisfactory performance for a specified period of time when used under specified conditions.[3] As a general rule, a simple relationship exists between the reliability of an equipment and its mean time between failure (MTBF). This relationship is the *exponential* case, which holds when the failure of the equipment is constant during its service life, shown by the equation:

$$R \text{ (for } t \text{ hours)} = e^{-t/MTBF}$$
$$R = \text{reliability factor}$$

Because of this relationship, reliability may be expressed in terms of an allowable MTBF. Figure 35.11 illustrates this function.

There are other refinements. For example, there is the concept of *operational reliability* which

Fig. 35.8. Programmable controller coordinates all elements of a robotic spot welding line. (Photo courtesy of Unimation, Inc.)

takes into consideration not only the inherent reliability of the product as determined by the design and development programs, manufacturing process, and test procedures, but also a reliability degradation factor attributable to shipping, handling, storage, installation, operation, maintenance, and the like. However, it is not within the scope of the discussion to delve into such refinements.

Availability is the probability that, at any point in time, the equipment will be ready to operate at a specified level of performance.[3] It is a measure of how often the equipment is ready when needed. Refinement of this factor takes into consideration both *intrinsic* availability achieved in design (including maintainability) and a degradation factor relating to such things as sufficiency of spare parts provisioning, qualification of maintenance personnel, and adequacy of test and repair facilities. For our purposes *availability* and *uptime* (percent) can be considered synonymous.

By definition, availability depends on both MTBF and the mean time to restore (repair). This is commonly expressed as MTTR. The relationship is expressed by the following equation:

$$A = \frac{MTBF}{MTBF + MTTR} = \frac{1}{1 + MTTR/MTBF}$$

A = availability factor

If the ratio MTTR/MTBF is known, equipment availability can be derived from Figure 35.12. Accordingly, if equipment is required to have a certain availability factor and its MTBF is known, then the required restoration or repair time can be determined. This can be a valuable tool in evaluating the equipment's viability. In production circumstances, equipment having an MTBF of 500 hr and a restoration time of 4 hr may be much more acceptable than equipment having a 5000-hr MTBF but a 40-hr MTTR.

Mean time to restore is, then, a measure of overall *maintainability*. The restoration time will depend on how quickly, easily, and accurately a malfunction can be diagnosed and corrected.

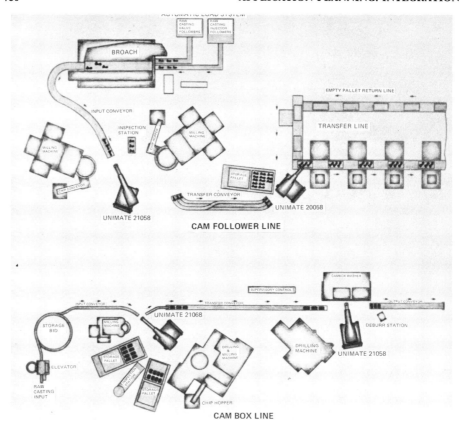

Fig. 35.9. Schematic. Four robots integrate machinery operations in a flexible system. (Photo courtesy of Unimation, Inc.)

35.2.2. Reliability Targets

In one known case, a robot manufacturer embarked on a comprehensive reliability program to achieve a target availability factor of 98% or better.[2] This study was exhaustive, and the theoretical contribution of every component toward a total system failure rate was taken into account. The methods used were those developed by U.S. government agencies to aid in the prediction of reliability of space vehicle systems.

The first step was to conduct a rigorous reliability feasibility study of all electronic and mechanical parts and "nonparts." The nonpart failure rate relates to cost constraints, tolerance buildup, user abuse, environmental problems, and so on. The study results are shown in Figure 35.13 and indicate a theoretical MTBF of 508 hr.

This, then, set the objective and carried with it the requirement that a mean time to restore of not more than 10.2 hr be achievable. The manufacturer set up a management system designed to bring individual components up to standard and assure statistically that the system, as shipped, would meet the goal. Figure 35.14 shows the reliability control points in the equipment's life cycle and includes field experience feedback.

The MTBF was eventually brought to 415 hr. Average MTTR turned out to be in the range of 4.8 hr. This yielded an availability factor of 98.8.

35.2.3. Future Outlook

Robot sophistication and complexity is increasing, suggesting reduced reliability in future generations. However, increasing experience in their manufacture along with ever-increasing reliability of their (solid-state) electronics will, quite likely, result in an improvement in robot reliability.

Fig. 35.10. One of the work cells. Note buffer storage pallets and central control. (Photo courtesy of Unimation, Inc.)

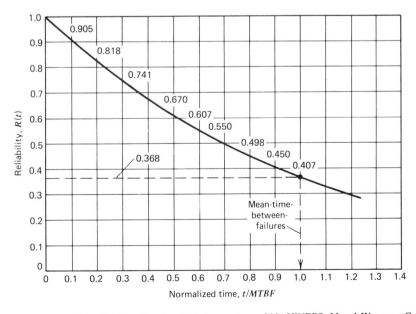

Fig. 35.11. Exponential reliability function. (Photo courtesy of NAVWEPS, Naval Weapons Center 00–65–502.)

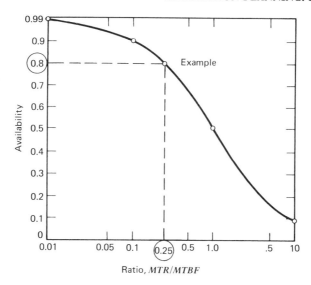

Fig. 35.12. Availability as a function of mean time between failures. (Photo courtesy of NAVWEPS, Naval Weapons Center 00–65–502.)

35.3. MAINTENANCE

Perhaps no other single cost factor in a manufacturing operation has been as neglected, misunderstood, and mismanaged as maintenance. It has been estimated that more than $200 billion is spent annually in American industry and at least $60 billion has been wasted.[4] More than that, poorly maintained equipment results in poor quality products, disrupted production schedules, delayed deliveries, and lost customers. In almost all cases these results can be attributed to the "fix-it-when-it breaks" philosophy.

As already noted, the operational reliability of equipment is a function of its intrinsic reliability and a degradation factor related to, among other things, usage. Degradation can only be minimized by proper usage and *preventive maintenance*. A sound, well-planned and well-managed maintenance program yields cost benefits in a number of ways. The mere fact that a program exists and is properly administered enforces disciplines that can have profound cost impacts. For example, most maintenance operations do not have any spare parts inventory control. In the case of one large plant, it was reported that of $70 million worth of spare parts only $35 million could be accounted for.[4] The rest was apparently just lying around—somewhere.

The case for preventive maintenance is pervasive.

Failure classification	Failure rate (x 10⁻⁶)	MTBF (hours)
Part failures only:		
Electronic/Electrical	555	1800
Mechanical/Hydraulic	673	1485
Non-part failures:		
Electronic/Electrical	267	3745
Mechanical/Hydraulic	475	2100
System failures:		
Parts only	1228	815
Non-tolerance	742	1350
Combined	1970	508

Estimated reliability feasibility,
Unimate 2000: MTBF = 500 Hours

Fig. 35.13. Unimate system reliability estimate. (Photo courtesy of Unimation, Inc.)

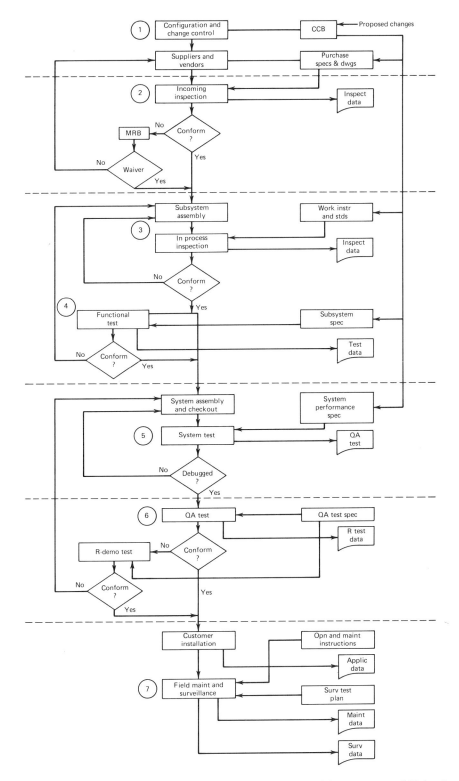

Fig. 35.14. Reliability control points in the UNIMATE life cycle. (Photo courtesy of Unimation, Inc.)

System Complexity. The trend toward increased complexity of automation and automatic in-process functions mandates preventive maintenance. Dependence on human operators to intervene and keep things going is no longer an option. Automatic systems are unforgiving.

Availability Enhancement. Performance degradation is avoided or forestalled and wear-out or burn-out avoided. The domino effect is prevented.

Enforced Discipline. Execution of a planned maintenance program requires scheduling coordination between the maintenance department, production, and engineering. Spare parts must be inventoried and readily available as well as diagnostic equipment and documentation. Trained personnel must be available. The emphasis must be on keeping the equipment going, not fixing it.

Adequate Skills. Maintenance personnel must be properly trained. Traditional lines of responsibility and crafts must be evaluated so that inefficiencies are not built in (e.g., electrician versus millwrights, etc.). A well-managed program will insure that personnel skills are maintained through refresher courses and the like.

Investment Protection. Capital equipment carries a high price tag, and the planned-for return on investment is predicted on continuous high performance. Lack of proper maintenance can destroy profits.

Quality Attitudes. Worker attitudes will determine the quality of performance in any department in the plant. The perceived attitude of management toward good maintenance programs will significantly influence the results and permeate the workplace. Esprit de corps can be a real force in achieving high quality of work and high productivity.

High Throughput. Unscheduled downtime wreaks havoc on the production floor and disrupts timely deliveries. The best—the only—way to minimize these occurrences is through a disciplined preventive maintenance program.

The cost of maintenance can be high. The cost of no maintenance will be higher. It has been estimated that a well-run program can cost as much as 10% of sales.[4] This strongly suggests that the manager must be a businessperson (surrounded by technically competent people) who understands how to run a business. For, indeed, maintenance is a business.

35.3.1. Robot Maintenance

Most robots require relatively little maintenance, but what is specified is essential to their continued performance and longevity. An installed robot can be looked upon as being made up of three elements.

The manipulator is, of course, the functional structure that physically performs the task. It is a jointed mechanical device most often powered by hydraulic motors or actuators (sometimes pneumatic) or electric motors. More complete descriptions of the various robot designs are contained in Chapter 5, Mechanical Design of the Robot System. In one form or another the arm joints and power transmission mechanisms are made up of linkages and gears, drive belts or chains, sliding or rotary bearings, and various types of seals, and so forth.

Although most of the component parts and subassemblies are quite familiar to maintenance personnel, their design and application pose some interesting, and sometimes challenging, aspects. For example, gear trains must be designed for absolute minimum backlash since the repeatability of positioning at the wrist extremity is dependent on the amount of backlash and friction (or binding) in the system. Hydraulic and pneumatic actuators must have very low breakaway and sliding friction for similar reasons. In addition, they must have virtually zero internal and external leakage. The dynamic performance of the machine and its ability to hold a position without drifting from it is adversely affected by leakage. Hence seal design and material, together with accurate mounting alignment of the actuator, are critical.

In most cases, preventive maintenance procedures for these components and subassemblies involve periodic visual and/or operational checks to see if any adjustments are needed or signs of undue wear apparent. Most manufacturers will provide a checklist and corrective instructions based on observed performance. In the long term, part replacement recommendations will be indicated. Figure 35.15 is one page of an eight-page preventive maintenance checklist intended to be performed at 1000-hr intervals. Figure 35.16 shows a recommended parts replacement list, extending to 20,000 hr of operation.

Servoed hydraulic robots employ servo valves which are quite sophisticated. Their reliability is generally very high, with their MTBF measured in thousands of hours. However, they are very susceptible to contaminants in the fluid media. Therefore extreme care must be taken to avoid contamination. If spare fluid is purchased from the robot manufacturer, he/she usually recommends that transference of the fluid to the robot's hydraulic reservoir be directly from the container in which the fluid is supplied. If the user procures his/her own fluid, it must conform to the robot manufacturer's specifications and will probably have to be specially filtered before use. This is a *must* since the fluid is usually contaminated by the container it was shipped in. And this is especially true if fluid is delivered in bulk.

Step Description	Checked	Corrective Action

Power Distribution and Interlock:

6. Check remote STOP function if used, by opening NC switch between A9 and B9 on customer access panel.

7. Check remote HOLD if used, by opening NC switch between A8 and B5 on customer access panel.

8. Check door interlock switch(es) if installed.

9. Check servo power relay by removing 4CR from relay bank. Unimate will be in HOLD.

ELECTRONIC (Power ON)

1. 918 W Board Lights.
   (a) Check that the encoder lamp monitor is functioning with Cycle Start out and Mode switch in REPEAT by pulling an encoder bulb.
   (b) Parity Error light should remain OFF in REPEAT with CYCLE START ON.
   (c) Check IT_2 light is ON at all times *except* when in REPEAT, AUTO with CYCLE START ON, and *not* in HOLD.
   (d) Check DC Indicator light is ON as long as AC power is being supplied to the power supply.

2. 918 T Board Light
   (a) Check Total Position Coincidence light will light when all motions reach TOTAL (position) Coincidence.

3. 918 D/C Board Light
   (a) Check True Total Coincidence light. (And so on.)

Fig. 35.15. Sample page. Preventive maintenance check list. (Photo courtesy of Unimation, Inc.)

Typically the filtration system in the robot includes a main filter, often a return-line filter, and individual filters in each of the servo valves. The overall filtration level is in the order of 35 μ absolute. One manufacturer offers a preventive maintenance service of analyzing fluid samples drawn from the machine to be sure contamination is not incipient. He even provides a convenient tap to draw the fluid from and clean sample bottles for that purpose. Obviously, great care must be taken when hydraulic lines are opened for any reason to avoid introducing contaminants into the system.

Pneumatically powered robots have similar requirements for media cleanliness. In addition, adequate means for water separation is quite important. Plant compressed air supplies are notorious for the amount of condensate (water) in them. Water separators should be of a self-purging type but should be checked at frequent intervals. They should be located directly at the inlet to the machine. Undoubtedly, there would be additional filters internal to the machine that would require servicing.

Usually there are a few lubrication points in the manipulator that require infrequent but regular attention. In some cases a lubricant must be applied to bearing surfaces, but generally there are reservoirs to be filled such as oil bowls or grease to be injected through appropriate fittings.

Most robot systems are air cooled and have air filters at the inlet. These require fairly frequent servicing to avoid overheating of the machine and the damaging effects that can result. Although the manufacturer will provide recommended servicing intervals, the requirements will vary considerably depending on the environment. Very dusty or oil-ladened atmospheres can clog filters in short order. Care must be taken that the inlet air temperature does not exceed the manufacturer's ratings of the machine and, where necessary, cool (but not necessarily air-conditioned) clean air may have to be piped into the robot.

The second element of the robot is its control section, which may be integrated into the manipulator or a separate, free-standing console, suitably connected to the manipulator. The controls are either

Parts to be Replaced		INTERVALS AT WHICH PARTS ARE TO BE REPLACED																			
Part No.	Description	1000	2000	3000	4000	5000	6000	7000	8000	9000	10,000	11,000	12,000	13,000	14,000	15,000	16,000	17,000	18,000	19,000	20,000
403BD1	Prev-main. kit	X	X	X	X	X	X	X	X	X	X	X	X	X	X	X	X	X	X	X	X
377H3B	Encoder lamp										X										X
922N1	Hand gear train										X										X
121AH2 121AH3	U cup feedthru							X							X						X
922W1	Rear drive bend										X										X
922V1	Rear drive yaw										X										X
127A1	Servo valves										X										X
313BH	Relief valve										X										X
825H1	Accumulator										X										X
1912AG5	Encoder cable										X										X
825F1 *	Scavenge pump										X										X
403BA1	Hyd. pump overhaul kit										X										X
182AU1	Press switch															X					
318L2	Air regulator															X					
318L3	Air splenoid															X					
313BD1	Unload valve																				X

Fig. 35.16. Sample of a long term parts replacement schedule. (Courtesy of Unimation, Inc.)

electrical, electronic, or pneumatic. They are either air-cooled or air-conditioned and in either case will have air filters requiring preventive maintenance care. In the case of pneumatic logic controls, filters for the pressurized air will require regular attention.

In general, maintenance of the control will involve functional checks and electrical/electronic adjustments, also observations of the mounting security of components, potential fraying of wires and cables, and the condition of connections and connectors.

Functional checks involve operating all manual controls and switches in a prescribed manner as well as checking power supply voltages and the like. Recognizing that many of the manual controls are left untouched during production running, it is important to check their function periodically so that they are *available* for use when needed. Proper voltage levels and control settings are essential to maintaining peak performance and productivity.

Especially critical are the servo control settings that control acceleration, deceleration, and velocity parameters. Improper settings can adversely affect machine cycle times or induce wear and breakage, or both. In association with these settings is proper nulling of servo valves (in the case of hydraulically powered robots). If valves are not properly nulled, proper control settings cannot be made and "ragged" motions will result.

Obviously, the required maintenance procedures for the robot's control and memory will vary widely depending on the particular design and complexity. It cannot be stressed enough that such procedures should be preventive in nature and should determine the *availability* of all machine functions. The degree to which the manufacturer gives this attention in his manuals, training courses, and maintenance checklists will be indicative of the quality of the product in terms of performance and maintainability.

The third element of the robot is its end-of-arm tooling. Because it is just that—tooling—and varies from application to application, it is usually not included in discussions of robot reliability and maintainability. Yet the perceived reliability and performance of the robot hinges on its dependability. Robot tooling can also be one of the most challenging aspects of robotic application to the work place. Chapter 37, End-of-Arm Tooling, describes the analysis and design of end effectors. Additional discussion is included here from the aspects of reliability and maintenance.

End-of-arm tooling takes the form of gripping devices of all sorts or process tools such as paint spray heads, welding torches or guns, riveters, drills, and grinding wheels.

Grippers usually provide a clamping action to grip the work through a pneumatically actuated mechanism. The simplest form involves handling round or cylindrical parts, one at a time. Dual grippers capable of handling two parts at a time are common, especially in machine tool load/unload operations. More complex configurations are required for irregular-shaped parts or for applications where the workpiece changes shape through progressive operations. Sometimes two or even three independently actuated actions must be built into the gripper. Figure 35.17 shows a variety of designs.

Vacuum grippers are quite common. Magnetic methods are also employed but to a lesser degree. Also, highly sophisticated or "intelligent" grippers are being developed that provide tactile capabilities such as force feedback and/or programmable characteristics to accommodate a variety of workpiece sizes.

The design criteria for robot grippers is rather demanding.

They must be strong and durable. They are susceptible to damage or distortion due to robot programming errors, stuck parts, "crashes."

They must be as light as possible (not very compatible with the first requirement). Every pound of gripper weight is a pound less of payload (workpiece) that can be handled by the robot.

They must have dimensional stability and be able to hold the workpiece orientation under high *g* forces (acceleration and deceleration of the robot). The inherent repeatability of the robot is meaningless if positional accuracy is lost in the gripper.

They must often have some built-in compliance or automatic alignment capability to accommodate positioning tolerances.

They must be fast acting. Clamping and unclamping motions are almost always additive to the work cycle and directly affect production rate.

And, finally, they must be maintainable. Gripping surfaces wear. Sliding bearing surfaces are subject to foreign material buildup and damage. Linkages loosen up. In the case of vacuum cups their edges wear and begin to leak.

A well-designed gripper will meet these criteria, but normal wear and tear can be expected. Therefore preventive maintenance is essential but usually quite simple. Regular lubrication may be required. Worn parts should be replaced. This is particularly true of the gripping surfaces, whether they be composition pads, hardened inserts, or whatever. (They should be designed for easy replacement.) Vacuum cups should be inspected for damage and replaced. If venturis are used, air supply pressure must be checked and adjusted for peak performance. In all cases the design should provide for quick

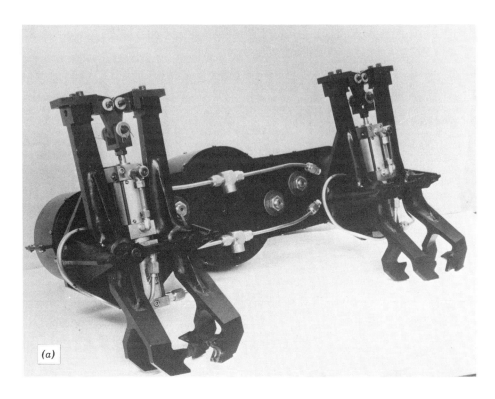

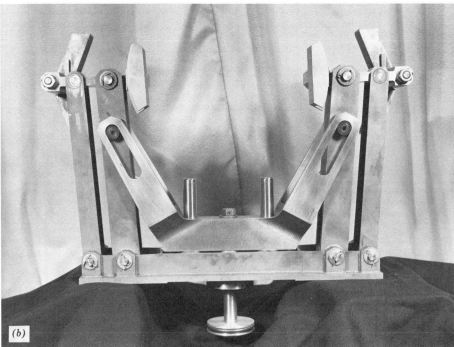

Fig. 35.17. End-of-arm tooling. Design complexity is application dependent. (*a*) Dual gripper; (*b*) Outer diameter/Inner diameter gripper; (*c*) Simple vacuum gripper; (*d*) Inflatable bladders gripper for outer diameter. (Photo courtesy of Unimation, Inc.)

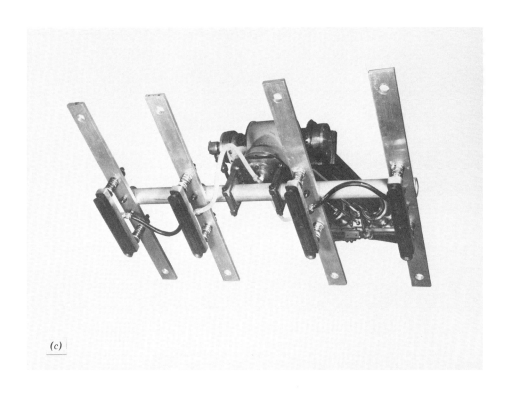

(c)

(d)

Fig. 35.17. (*Continued*)

and easy replacement (preferably) without requiring reprogramming of the robot as a result of maintenance measures.

Most of the same demands are placed on process tools, which can often be more prone to causing downtime or producing poor quality work than are grippers. Hence, regular maintenance is critical, the nature of which will depend on the particular tool. In fact, it may be necessary or desirable to build into the robot's program maintenance measures. For example, spot weld gun tips must be periodically dressed to produce high-strength welds and to avoid sticking to the workpiece. Special dressing wheels can be so located that the robot can pass the tips over the wheels. A similar situation exists with arc-welding torches, which tend to clog. Means for cleaning them automatically are easily devised. And so it is also for paint spray heads.

From this it should be clear that the end-of-arm tooling is a critical element in the reliability chain and must be included in a comprehensive maintenance program.

35.3.2. Planning for Maintenance and Repair

Prerequisite to any well-planned and executed preventive maintenance program is the availability of skilled personnel, and this implies *trained* personnel: personnel trained not only to maintain, troubleshoot and repair, but also trained in how the machine operates and how to operate it. The more familiar the maintenance personnel is with what the robot can do and what is expected of it, the more proficient they will be in tending to its needs.

Also, as one can readily understand from our discussion of *availability* and MTTR, it is essential that these skills reside in-house. Any reliable robot manufacturer will have a field service organization of highly trained technicians capable of supporting the customer in every way. But the user will be well counseled to become virtually self-supporting and defer to outside resources only when confronted with unusual problems. Otherwise, the time to restore equipment to a running condition following a failure will, in most cases, be intolerably long. The most timely use of the manufacturer's technician is during the installation and start-up phases of the system. Even then, it should be a team effort during which the customer personnel receive on-the-job training to reinforce their "basic" training.

Selection of the personnel to be trained is an important part of planning the whole job. In unionized organizations it is even more critical because it may be necessary to negotiate which craftspersons and tradespersons will be assigned to what tasks. This is sometimes a controversial issue and deserves some discussion.

As we have already seen, the robot is made up of a variety of mechanical components and also a variety of electrical/electronic control devices, all of which are inextricably entwined to produce a working machine, not in the least unlike NC and CNC machine tools and other high-technology production equipment. It is argued that to service such equipment with several different classes of skilled people—electricians, millwrights, electronic technicians, hydraulic tradespersons—having separate and divided responsibilities—is like the proverbial camel—built by a committee. Others will argue the merits of maintaining these different disciplines as the only way to insure a team of experts and not a group of jacks-of-all-trades. There are merits in both arguments, and the solution most often will be determined on the basis of what works best for the individual, plant, or company. It has worked both ways with varying degrees of success. However, for the benefit of efficiency, proficiency, and fast response time, the trend is toward integrated skills and responsibilities.

35.3.3. Training for Maintenance and Repair

Training starts with management. More precisely, it starts with an understanding by management of the need and a commitment to meeting that need. Furthermore, it should be fostered in an atmosphere of defined objectives and mutual goals. Thus good communications between all departments and at all levels is essential. This may sound trite, but time and again projects have failed, or have been less than the success they might have been, because of poor communication and lack of a well-conceived plan. This is especially true when introducing new technologies into the workplace.

Training, then, starts (or should) with an understanding of the project—what it is and why it is—by all concerned. Lack of understanding creates a threatening atmosphere, and this is counterproductive (Figure 35.18). The threats are in the eyes of the beholder: For upper or top management, it is the threat of a bad investment; for manufacturing engineering it is the threat of technological complexity; for production managers it is the threat of interrupted schedules and delayed shipments; for line supervisors it is the threat of change; for production workers it is the threat of lost security; and for the maintenance department it is the threat of unmerciful demands in an atmosphere of confusion.

These threats can be substantially eliminated by good communication—starting off with a meeting of those who will be directly involved to describe and discuss what is being planned and why it is needed. Such a meeting can easily be supported by audio/visual presentations (often with the help of the selected suppliers) that graphically illustrate the equipment and how it will be used. This is the kick-off, to be followed by a series of "working" meetings through which detailed plans are developed, concerns expressed, potential problems aired, and plans of action laid. By involving personnel, at all

- ° MANAGEMENT: FEAR OF A BAD INVESTMENT
- ° MANUFACTURING ENGINEERING: FEAR OF COMPLEXITY
- ° PRODUCTION: FEAR OF LOST PRODUCTION
- ° LINE SUPERVISION: FEAR OF CHANGE
- ° WORKERS: FEAR OF JOB INSECURITY
- ° MAINTENANCE: FEAR OF ADDED PRESSURE

Fig. 35.18. Fear of the unknown. Good communications lead to success.

levels, surprises can be minimized and dedication to success is generated. In such a cooperative working environment, the often-neglected maintenance department will become an integral part of the team, ready and able to fulfill its function when and as needed.

Once the maintenance manager has his or her charter he or she can begin to plan for the necessary training of his or her personnel. Early on he or she will have to determine what skills are needed, what skills are lacking, who will be assigned what tasks, and how they will be trained. A reliable equipment manufacturer will offer appropriate training courses and will work closely with his or her customer in meeting training needs. One such course is outlined in Figure 35.19. Note that it includes recommendations for the type of people to be trained. In some cases more specific requirements must be met, in which case tailor-made recommendations are available. Figure 35.20 is one such example for automobile spot welding applications.

Sending the selected personnel to the vendor's school (or arranging for him or her to conduct training in the user's facility) is just the beginning in the training program. The wise manager will look beyond this initial step to how he or she will train and upgrade personnel on a continuing basis to sharpen skills and to provide for personnel turnover. Thus "continuing education" programs within the plant are most important. Here again, the vendor can help through his or her regular schools and in many cases through "learner paced" courses of instruction that he or she may offer, much like the familiar home study courses (see Figure 35.21). These refresher courses are an invaluable means to maintain skills. Repair proficiency is inversely proportional to equipment reliability simply because the required skills are used infrequently.

In addition to attending training schools and in-plant training programs, the selected personnel should become involved with the installation of the equipment. In this way they will not only gain additional knowledge but will become thoroughly familiar with its operation in the specific application. Inevitably there are nuances to be learned, and these should be documented for the benefit of all on a continuing basis.

All of this assumes that adequate instruction manuals and related documentation are available from the equipment manufacturer and *that they are readily available* to those who will need them. It is amazing, indeed, ludicrous, how often these all-important references are nowhere to be found or in a location distant from where they are needed most.

35.3.4. Maintenance Program

Well in advance of the installation and start-up of the equipment, the maintenance manager will want to establish a program of preventive maintenance, coordinate it with the production manager, and provision for its support. This should be looked upon as a *dynamic* document that is adjusted to experience.

Coordination with production is essential, as is a mutual understanding of the needs and purposes. The best maintenance program in the world will be meaningless if production refuses to allot time for its execution. If a one-shift operation is involved, scheduling should be no problem. If it is a three-shift operation, then either weekend maintenance must be planned for or downtime scheduled on one of the shifts. This *must* be negotiated, and the plan must be adhered to.

Initially, what work is to be performed and at what intervals will be best suggested by the vendor, especially where the equipment is the first of its kind in the plant (see Figure 35.15). The maintenance manager should consult with the supplier and take full advantage of his or her experience. Thereafter, the list of items to be done and the schedule can be adjusted to the perceived need. As noted earlier, replacement or cleaning of air filters may have to be done more often in some environments than suggested by the vendor. Time may show that other items may be looked after at less frequent intervals than recommended, but only after sufficient running time and performance results have been accumulated.

This suggests another absolutely essential ingredient in an effective preventive maintenance and repair program—*document all work.* In the short term this is invaluable information for people working on different shifts. In the long term it provides a data base for establishing trends and fine-tuning the program (a management tool). Just as important is that it provides the equipment manufacturer with

```
                         UNIMATE® Industrial Robots
                      UNIMATE Series 1030/2030/4030
                            TRAINING COURSE UA-I
                     OPERATION, PROGRAMMING, and MAINTENANCE


A 4-1/2 day course designed for customers needing instruction in the areas of
operation, programming, maintenance, service, and adjustment of the 1030, 2030,
or 4030 Series UNIMATE Industrial Robots.

We recommend that at least one or more of the following individuals, on each
shift, complete this course:

     1.   Operator/Programmer/Set-up Person
     2.   Electrical/Electronic Maintenance Technician
     3.   Hydraulic/Mechanical Maintenance Technician
     4.   Equipment Maintenance Foreman/Supervisor
     5.   Production Foreman/Supervisor

Additional classifications that can benefit from the course are Systems, Pro-
ject, Application, Process Engineers, and other individuals involved in the
implementation of the robots.  These job titles and their responsibilities will
vary from firm to firm.  Therefore, please contact the Technical Training/
Publications Department if there is a question regarding who should attend.


Program Includes:

     1.   Installation and safety procedures
     2.   Familiarization of standard control panel, teach control, auxiliary
          functions, and options as applicable
     3.   Familiarization of the hydromechanical and electrical systems and their
          interaction
     4.   Basic troubleshooting of typical problems using block diagrams, status
          indicators, and a troubleshooting chart
     5.   Programming procedures, including exercises to understand:
          a.   advantages of a written program
          b.   use of accuracies
          c.   use of auxiliary functions as applicable
     6.   Electrical and mechanical adjustments to enable the student to:
          a.   zero encoders
          b.   null servo valves
          c.   adjust servo power amplifier board
          d.   adjust geartrain backlash
     7.   Preventive maintenance procedures using the 1000-hour checklist
     8.   Use of Tester, Cassette Recorder, and Editor in:
          a.   programming
          b.   troubleshooting
          c.   maintenance

                                   NOTE

          Maximum benefit will result when completion of the course is
          just prior to receipt of the UNIMATE Industrial Robot.

21 Oct 1982
```

Fig. 35.19.   Outline of typical robot training course. (Courtesy of Unimation, Inc.)

a history, should there be a chronic problem requiring his or her assistance. In large operations the computer is employed to collate and correlate the data. Even modest-sized operations can benefit from its use.

The value of this information, computerized or not, simply cannot be underestimated. Unexpected interruptions in production cause chaos, create losses, and can be tumultuous. What is worse is being unable to pinpoint and track the problem and ultimately eliminate it. The historical record is an invaluable tool to solving problems or, better yet, avoiding them.

Next in developing the maintenance plan is to provision for special tools, diagnostic equipment, service kits, and spare parts. Once again, the maintenance manager should enlist help and guidance from the vendor. In fact, the equipment manufacturer should have delivered this information as part of the documentation package.

Ordinarily, the investment in special tools will be minimal, but the cost without them can be measured in tens of thousands of dollars. As the old adage goes—time is money. The most common "special" tools are circuit card pullers, torque wrenches, seal compressors, accumulator charging adap-

Considerations when organizing the
Operation-Maintenance Department of
an "Automated" Factory

People assigned to the above department will not be responsible for plant
maintenance. The Plant Maintenance Department would be responsible for the
physical building and its systems, such as electrical, water, heat, air-conditioning, pneumatic, gas, sewers, etc. The responsibility for these systems
as applicable would end at the point where it is attached to a disconnect/shut
off device at the automated unit.

From this disconnect/shut off device, thru and to the working end of the automated unit would be the responsibility of the "Automated Machine, Operation/
Maintenance Department." Personnel in this department would have one or more
of the following "categories" of responsibilities:

POSITION CATEGORY OF RESPONSIBILITY (Relative to company organization, Union
Contracts, etc.):
Category A - Overall Responsibility for:

1. "Welding machine" (an industrial robot, its attached weld gun, and
weld controller).
2. The ancillary equipment.
Category B - Specific responsibility for one system (e.g. hydraulic, pneumatic, electrical, electronic, etc., for the welding machine and ancillary
equipment).
Category C - Specific responsibility for completing scheduled preventive
maintenance checks. The need for this type of position is normally relative to
the installation size (the larger, the greater the need) and can include one or
both of the following:
1. Servicing of "fluids, filters, and lube" only, or include
2. Minor adjustments and repairs.
If "1" is selected then "2" would be the responsibility of Category A or Category B above.

POSITION LEVELS OF RESPONSIBILITY (Relative to education, training, and
experience):
Level 1 is for persons with limited responsibility for a single system or
function of the "unit." They should be able to operate and/or program the unit
and complete the "Daily Start Up Checkout" and "Operator Maintenance" as required. Additionally, recognition of typical fault symptoms for description
to Level 2 and/or 3 people is also required except for Category "C,1" people.
Level 2 is for persons who have nearly full responsibility for all systems
and functions of the "unit." They have the same responsibilities as Level 1
people, and additionally perform maintenance, adjustment, and removal/replace-
ment of certain specified components that do not require the expertise of
Level 3. They will also complete the "300 hour/monthly Preventive Maintenance
Check List."
Level 3 is for people who have full responsibility for all systems and
functions relative to the "unit" including Levels 1 and 2 above. Because of
their broader and greater experience and/or training, they will be certified to
perform the more complex or technical maintenance adjustments and removal/
replacement procedures.

Fig. 35.20. Example of skill selection recommendations for training. (Courtesy of Unimation, Inc.)

ters, alignment fixtures, and the like. The list, perforce, will be peculiar to the specific equipment involved. Figure 35.22 is an illustration of such lists.

In addition to special tools, the department must be equipped with proper diagnostic equipment. In some cases this will only involve multimeters, gages, and similar devices probably already available in the department. In other cases special diagnostic tools (see Figure 35.23) must be purchased.

Another item that should be stocked by the maintenance department is preventive maintenance kits, usually available from the equipment manufacturer. When supplied and stocked as kits, a long list of items need not be pulled from inventory before going to the job site. The kit can be obtained with the security that everything needed will be supplied without repeated trips to the stock room. Figure 35.24 lists such a kit of parts. Note that it even includes paper towels and a copy of the manufacturer's suggested **PM** (preventive maintenance) checklist.

Finally, an adequate selection of spare parts must be readily available. As was stressed earlier, the secret to *high availability* is *short mean time to restore*. And that can only be achieved by having replacement parts on hand. The most highly skilled technician will be "dead in the water" without this inventory. Most manufacturers will recommend what this inventory should be for the model of machine purchased and what quantities, based on the number of machines in the installation or plant and the statistical (MTBF) likelihood of the need. The total acquisition cost of this inventory may

Fig. 35.21. Maintenance of skills through "learner paced" audio/visual courses of instruction. (Photo courtesy of Unimation, Inc.)

be 12% of the robot cost when only one or two are purchased. This can drop to 5% or even less for a large (10 or more) number of robots. Hence control of this inventory is an important part of maintenance operations.

Earlier in this chapter, examples were cited of large dollar values of spare parts inventory being scattered about and virtually inaccessible. This is an expensive waste and is unconscionable. Spare parts must be in a secure area and stocked in an organized manner under some inventory control system that not only makes them readily available but will also *flag a reorder point.* This need is often overlooked with the resultant unavailability and excessive downtime—in spite of having thousands of dollars of inventory on the shelves.

Although the initial cost of spare parts is an unavoidable investment, there are ways to minimize operating costs and avoid ballooning of this expense. Control of the inventory is most important, as already cited. A second way is to expedite return of the replaced part to the vendor for repair or replacement and rapid return. This will avoid or minimize the tendency for the total inventory to escalate.

Item	Description
Torque Wrench	0 to 500 Inch-pounds
Torque Wrench	30 to 250 Foot-pounds
Pin Extraction Tool	Unimation, 106AH1
Multimeter	Simpson, Model 260 or equivalent
Feeler Gage	—
1-Inch Micrometer	—
Tester	Unimation, Model 502A2/A3
Standard Hand Tools	—
Grade C Loctite	Unimation, 101D2
Down Support Tool	Unimation, 106BG2
Bladder Insertion Tool	Unimation, 106V2
Rotary Seal Compressor	Unimation, 106AS1
Rotary Piston Spanner	Unimation, 106BY1
Out-In Servo Valve Nulling Tool	Unimation, 106BA1

Fig. 35.22. A typical list of special tools for maintenance and repair. (Courtesy of Unimation, Inc.)

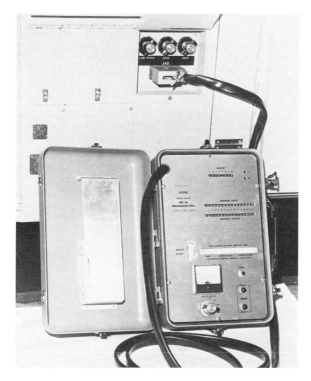

Fig. 35.23. Robot diagnostic tool for maintenance and repair. (Photo courtesy of Unimation, Inc.)

Third, take full advantage of the vendor's repair-or-replace policy. In many cases the manufacturer will repair a failed part to a like-new condition at a fraction (less than one-third) of the new part price. In some cases he or she will, in an emergency downtime situation, immediately dispatch a replacement part at the same (repair) cost, providing he/she receives the defective part in return. What often defeats this cost-saving opportunity is failure of the maintenance department to see that the bad part is returned to the vendor. Once again—discipline and control.

One final point on the subject of spare parts provisioning: Do not forget the end-of-arm tooling.

Item Number	Description	Unimation Part Number
1	Preventive Maintenance Kit, 4000B	403BD1
a.	Air Filter, Oil Cooler	318E6
b.	Air Filter, Electronic Cabinet	318AF1
c.	Hydraulic Oil Filter Kit, 4000B	403CY1
d.	3-Ounce Jar Lubriplate, 130-AA	99H1
e.	1-Pound Can "Gearshield X"	99J1
f.	Lintless Paper Towel Package	—
g.	28-Ounce Spray Bottle "Fantastic" Cleaner	—
h.	Oil Sample Bottle	720Z1
i.	Oil Sampling Procedure	402AL1
j.	Preventive Maintenance Check List	402H1
k.	1-Quart Can Extension Rod Lubricant	99AA1
2	UNIMATE Hydraulic Fluid, 1-Gallon Container (4 each)	99S2
3	Transfer Block Seal, Air	121H3

Fig. 35.24. Typical preventive maintenance parts kit. (Courtesy of Unimation, Inc.)

A perfectly running robot is rendered useless without a working "hand." Not only must the gripper or tool be properly maintained, but also replacement parts must be in inventory. In particularly critical or hazardous operations an entire spare assembly may be most judicious.

Figure 35.25 is a *plan for maintenance* checklist that summarizes this discussion. It is intended to emphasize the need for a well-founded plan and means for execution. Maintenance is a business and should be run like a business. It's been estimated that automated systems carry an annual maintenance cost of 10% of their acquisition cost. Available data indicate that this figure is about 11% for industrial robots.

Furthermore, preventive maintenance is becoming a major contributor to profit and productivity-increasing strategies. For example, the just-in-time methodology for reducing manufacturing inventories to their lowest possible level mandates that every single machine in the manufacturing process function perfectly all the time. Under these conditions, poor maintenance programs will surface rapidly.

35.4. SAFETY

The industrial robot is a very effective safety problem solver. At the same time it poses some safety issues that must be given close attention when designing a robotic work cell.

35.4.1. A Safety Solution

By its very (anthropomorphic) nature, the robot is able to assume dangerous tasks heretofore assigned to human operators and to live comfortably in atmospheres that are debilitating to humans. Indeed, a large number of jobs it was given in its early stages of development were selected solely to reduce human exposure to hazardous and life-threatening conditions, for example, power press loading, die casting, and injection molding machine tending. Added impetuous was provided by the OSHAct of 1971 which, in seeking ways to minimize risk to the worker, imposed costly safety standards for protection (and rightly so). The robot solution to safety not only alleviates risk to life and limb, but also, unlike most safety measures, yields useful work.[5] Thus the robot can be a more cost-effective means for making the workplace safe. This aspect should not be overlooked when evaluating the economics of robotic applications.

35.4.2. A Safety Problem

Throughout this chapter we have placed emphasis on viewing a robotic installation as a system made up of several component parts rather than taking a narrow view and considering the robot only. This is particularly important when addressing the issue of safety and accident prevention.

Typically, the robot and its end-of-arm tooling interfaces with other machinery, conveyors, fixtures, load/unload stations, and often with a central control or computer. In some instances it also interfaces with human operators. For example, the operator may periodically load parts into a magazine feeding the robot, or he/she may load and unload an arc welding fixture, Figure 35.26.

Obviously, then, a safety evaluation must consider all of these aspects, not only as it relates to life and limb but also to the protection of expensive equipment. Further, risk assessment must take into consideration the various modes of operation of the system—normal working, programming, maintenance, and so on—since varying safety related conditions will prevail.

Safety conscious managements, incited by the OSHAct, have developed accident prevention guide-

GET INVOLVED WITH THE PROJECT AT THE START

PLAN FOR TRAINING

IDENTIFY REQUIRED SKILLS

EXECUTE TRAINING PLAN

INVOLVE PERSONNEL AT INSTALLATION

DEVELOP AND COORDINATE MAINTENANCE PLAN

PROVISION FOR SPARE PARTS, TOOLS, DIAGNOSTICS

DEVELOP WORK DOCUMENTATION PLAN

MAINTENANCE IS A BUSINESS -- RUN IT LIKE ONE

Fig. 35.25. A plan for maintenance check list.

Fig. 35.26. Robot/operator interface. Operator loads parts fixture while robot welds assembly. (Photo courtesy of Unimation, Inc.)

lines and techniques for most equipment found on the factory floor, giving due consideration to the potential for human entanglement, shearing actions, trapping and pinch points, molten metal ejection, electrical shock, high heat, and the like. All of these precautions are applicable to robotic work cells and to the robot itself. However, there are unique characteristics of the robot and robotic interfaces that demand special attention.

35.4.3. The Robot

The Robot Industries Association (and other associations around the world) defines a robot as a position-controlled *"reprogrammable multifunctional* manipulator designed to move material, parts, tools or specialized devices through *variable* programmed motions for the performance of a variety of tasks."

From a safety standpoint, the three words in italics are key—*reprogrammable, multifunctional, variable*—because they define a variable mission machine with the implication of "planned unpredictability." For example, one might watch an operating robot for half an hour doing the same thing over and over. Suddenly it takes a different path because it has received a part-reject signal and has been instructed to remove the piece from the production flow.

Also, in many applications, the task does not require the robot to reach the extremities of which it is capable. Its repetitive operation can lull one into erroneously believing what he sees are the extremities.

At the onset, then, these basic characteristics must be guarded against:

Space intrusion by the robot into unsecured areas.

The "hypnotic" syndrome of repetitive operation.

The unseen communication links (interlocks/interfaces) calling for alternate actions and subroutines.

35.4.4. Robot Safety Features

A well-engineered robot will include, to the greatest extent possible, practical safety features that take into account all modes of operation—normal working, programming, and maintenance.[6] Some features are common to all robots; others are peculiar to the type of robot, particularly with regard to its motive power.

In the normal working mode (assuming proper safeguards against human intrusion into the work area) most safety features are for the protection of the equipment. Typically, these will include electrical interlocks between the robot and the machinery with which it works, signifying safe or "ready" conditions and the like. Interrogation of these signals will be part of the robot's program and placed strategically in the proper sequence. Thus the robot will not reach into a press unless a signal indicates the press is open. Similarly, the press is actuated by the robot only when its arm is clear. Or, the robot will pick up a workpiece from an orienting fixture only if a sensor signifies part presence and in proper position.

In some cases optical or infrared sensors are used, as in sensing that a part has been removed and, indeed, in the robot's gripper before cycling a die casting machine, lest damage be done to costly dies. In other cases redundancy is used independent of the robot's program. For example, before an automobile spot welding line conveyor can shuttle, all robots on the line must signal (by limit switch sensing) that they are clear and the arm is retracted. Without this safety provision car bodies have been destroyed along with robot arms.

All of this may sound mundane, but one does well to play the "what if" game when designing an interlock system to avoid costly equipment damage and downtime. More than one robot arm has been crushed by a press that double-stroked and tools have been broken by improperly seated parts. Obviously, too, the sensing units used must be highly reliable (and maintained) and selected to fail to a safe condition.

Also, the entire system should be analyzed in process interrupt, emergency shutdown, and power failure modes in an attempt to avoid damaging results. For example, most robots can be put (by an operator or automatically) into "HOLD" condition, which essentially stops all motion instantaneously; or into a "STEP" mode in which it will complete the command currently being executed, but not proceed further. These are desirable operational features. But, once again, care must be taken in the interlock design that the consequences of such action will not cause "crashes" between equipment because of cycling that has already been initiated.

Similar analyses should be made in the event of an emergency shutdown or power failure. This is not easy because of the randomness of such occurrences and because of transient actions (e.g., transient relay cycling) which may take place. One manifestation of this is power interruption due to a safety system response to human intrusion into the work area. While personnel safety is of prime importance, methods (even procedural in nature) should be devised to minimize consequential equipment damage.

A well-designed robot will include other safety features, related to the equipment complexity and sophistication, to minimize the effects of malfunctions such as, in the case of computer-controlled machines, parity checks, checksums, cyclic redundancy, error detecting, and the like. Also, they will include "software" stops, electrical stops, and "hard" stops, adjustable to the work cell layout.

The robot setup and programming mode presents an additional set of conditions that is directly related to human safety. In most instances, personnel will have to work within the robot's sphere of influence to teach it its task. In the case of computer-controlled robots, the program sequence can be developed off-line through a keyboard terminal. However, spatial locations requiring precise positioning are done with some form of teach pendant and usually require the operator to be close to the manipulator arm extremity. Since most robots in this class are either electrically or hydraulically driven, the following discussion concentrates on large machines of these types.

Since these machines are capable of carrying payloads of 300 lb or more at velocities on the order of 60 in./sec, it is obvious that severe injury could be inflicted upon anyone in its path. Therefore, safety features in the programming or teach mode are essential to good robot design.

First and foremost, the speed and power that the control system is able to deliver in the teach mode must be limited. It should be reduced to a point where an operator can get out of the arm's way fast enough if it should move unexpectedly, and if he is entrapped, that its force not cause injury. In normal (teach) operation, speed is usually limited to less than one-tenth of full speed, but this is usually accomplished through electrical means. A failure can still result in full power being applied. One technique used to eliminate this potential in hydraulic machines is to restrict the fluid flow in a fail-safe manner and, as an added precaution, to incorporate a hydraulic "fuse" which will rapidly sense excess flow and shut the machine down. Such systems have been demonstrated to limit arm travel in a catastrophic failure mode to 0.5 in. In the case of electrically powered machines, this has been accomplished by switching to a low power source in the teach condition. (Regarding our discussions on maintenance, testing of these features should be part of the PM checklist if at all possible.)

Such design approaches attempt to reduce to the absolute minimum the number of components whose failure could cause injury. To illustrate, take the case of a servo-controlled hydraulically powered robot. It is likely that even in TEACH it is still under servo control. Therefore component failure could yield unexpected motion. If at the same time, electrical/electronic circuits failed that normally limit speed (and power), injury could result. Or if a servo valve failed internally to a "hard-over" condition, maximum fluid flow would occur. Obviously, there are a number of events in the safety chain, any one of which could cause an accident. By using an absolute flow restrictor and sensor to turn the machine off, much of this chain is bypassed.

This is not to say that redundancy should not be employed. To the contrary, this is good design practice—*as long as redundancy in design does not reduce the overall reliability of the safety net.*

Other robot safety features include "dead-man" switches on the teach pendant, remote emergency stop buttons, and so forth, in addition to those already mentioned. Figure 35.27 summarizes the most common robot safety provisions. Figure 35.28 shows a typical teach pendant.

In addition to safety design features of the robot, there are some other issues related to setup and teaching that should be briefly mentioned here, in keeping with our holistic approach to the robotic work cell.

First, recognize that the very nature of end-of-arm tooling requires great operator care because of pinch points, and so forth. Also, through operator error or equipment failure there is risk of dropping a workpiece. Unless the operator can stand completely clear of these risks, safety precautions are almost entirely procedural and rest with the knowledge and training of the operator.

Second, other machinery in the work cell should be disarmed when programming the robot, and interlocking, whether directly through the robot's control or a central control, must be appropriately designed so that inadvertent or unexpected cycling of this machinery cannot occur.

Third, program tryout should only be done under the same safety precautions as for "NORMAL RUNNING." Even though checkout may be done at reduced speeds and in step-at-a-time modes, and so on, undoubtedly the most important safety features prevailing in TEACH will be disarmed in playback.

Fourth, during programming it is simply good practice to have two people involved—one doing the programming and the other observing and standing by an emergency stop button.

We have discussed robot safety features in the normal working and programming modes. A third mode remains—maintenance. Maintenance work, and particularly troubleshooting and repair, must be performed under rigid safety precautions. Once again, these will be highly procedural, and strict adherence to the robot manufacturer's instructions in his equipment manuals, as well as to those set

FEATURE	
POWER DISCONNECT SWITCH	Removes all power at machine junction box
LINE INDICATOR	Indicates incoming power is connected at junction box
POWER ON BUTTON	Energizes all machine power
CONTROL POWER ONLY BUTTON	Applies power to control section only
ARM POWER ONLY BUTTON	Applies power to manipulator only
STOP BUTTON, LATCHING	Removes manipulator and control power
HOLD/RUN BUTTON	Stops arm motion, leaves power on
TEACH PENDANT TRIGGER	Must be held in by operator for arm power in TEACH
STEP BUTTON	Permits program execution one step-at-a-time
SLOW SPEED CONTROL	Permits program execution at reduced speeds
TEACH/PLAYBACK MODE SELECTOR	Provides operator with control over operating mode
PROGRAM RESET	Drops system out of playback mode
CONDITION INDICATORS AND MESSAGES	Provides visual indication by lights or display screens of system condition
PARITY CHECKS, ERROR DETECTING ETC.	Computer techniques for self checking a variety of functions
SERVO MOTOR BRAKE	Maintains arm position at stand-still
HYDRAULIC "FUSE"	Protects against high speed motion/force in TEACH
SOFTWARE STOPS	Computer controlled travel limit
HARDWARE STOPS	Absolute travel limit control
MANUAL/AUTOMATIC DUMP	Provides means to relieve hydraulic/pneumatic pressure
REMOTE CONNECTIONS	Permits remoting of essential machine/safety functions

Fig. 35.27. Some robot safety features.

Fig. 35.28. A robot TEACH pendant. (Photo courtesy of Unimation, Inc.)

up by the user maintenance department, must be observed. Figure 35.29 sets the tone in one manufacturer's manual. Following these guidelines are several pages of general "Do's and Do Not's," and throughout the manual WARNINGS and CAUTIONS are highlighted. They are particularly prevalent in the troubleshooting and parts replacement sections of the manual, in the interest of human safety.

Maintenance safety features built into the robot will depend on the type of motive power. In all cases, however, it is assumed that the equipment has been designed to applicable electrical, hydraulic, and/or pneumatic standards.

Some other features designed into the robot specifically to facilitate maintenance are manual or automatic hydraulic or pneumatic pressure dumping, locking brakes on electric drive motors, and control system POWER-ON only (i.e., no manipulator power). In at least one case of a servoed robot, the manufacturer provides a manual control pendant (Figure 35.30) which permits moving of the manipulator articulations by an operator without power to the robot's control system. In addition to these built-in features, most robot manufacturers offer maintenance aids (tools and fixture) to facilitate maintenance safety.

35.4.5. Work Cell Safeguards

Accident prevention requires an appraisal of the potential or actual hazards in the workplace, and therefore depends on the equipment and processes involved. Therefore an in-depth assessment of the risks must be made with at least three objectives:

    To identify the hazards.
    To protect against the hazards.
    To avoid costly, unnecessary complexity.

For these reasons risk assessment has become rather sophisticated, and there are a number of publications that treat it in depth.

In the case of the robotic work cell, the first thing to evaluate is whether or not the robot is capable of causing injury. Most of our discussion has been with large robots in mind, but there are many small machines that pose little or no danger to life and limb. Obviously, they would not require elaborate safety measures. Likewise, some operations are not very complex, and risk assessment of interlocks and interfaces and the consequences of their failure is a simple matter.

In complex systems involving large robots and heavy machinery, risk assessment becomes challenging and requires a methodical approach—and a large dose of common sense. The Machine Tool Trades

SAFETY SUMMARY

I. The following is mandatory reading for all personnel who find themselves in or around the area that the UNIMATE can reach. (See Figure 2-2.)

Personnel cannot be expected to knowledgeably and safely apply, operate, or maintain a UNIMATE by just making the appropriate equipment manual available; no more than a man can knowledgeably and safely fly an airplane, operate a lathe, or repair a complex transfer device by reading a "How To......" manual.

This equipment manual is written under the assumption the user has attended the UNIMATE Training Course conducted by the Unimation Inc. Department of Training and has a basic working knowledge of the UNIMATE.

This equipment manual is NOT to be considered a self-teaching vehicle.

Failure to comply with the above and subsequent warnings can result in <u>serious injury</u> to personnel and/or <u>major damage</u> to the UNIMATE.

II. There are three levels of special notation used in this equipment manual. In descending order of importance, they are:

WARNING

Used to signify that when the statement is not complied with, <u>serious injuries</u> will occur to personnel and/or <u>major damage</u> will be inflicted on the UNIMATE.

CAUTION

Used to signify that when the statement is not complied with, the UNIMATE may be inflicted with minor to near major damage.

Note

Used to give supplementary information or to emphasize a point, procedure, functional check, etc.

Fig. 35.29. Safety is highlighted in the equipment manual. (Courtesy of Unimation, Inc.)

Association in the United Kingdom has published an excellent guide entitled *Safeguarding Industrial Robots*[7] and in it is outlined a framework for risk assessment as follows:

1. Determine the mode of operation, that is, normal working, programming, maintenance.
2. Carry out a hazard analysis to determine potential areas of doing harm.
3. Determine whether "designed" or "aberrant" behavior is to be considered.
4. Determine if hazards are liable to lead to injury.
5. If so, then consider whether there are any recognized methods of guarding the particular machine concerned. At present, such standards may well be available for the associated machinery but probably not for the robot.
6. Consider whether such standards are appropriate, particularly in the context of machines being used in conjunction with robots. For example, the risk assessment could be different for a machine with a human operator than for one that is associated with a robot. One factor that will affect this risk is whether or not the human operator will take over from the robot during, for example, robot failure.
7. If no standard is available, consider what the logical steps should be to establish a reasonable standard for the particular application.

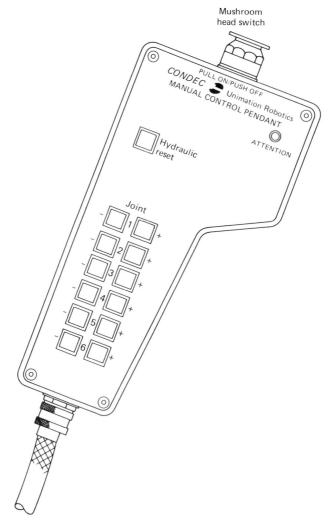

Fig. 35.30. A manual control pendant to drive manipulator joints without servo control power. (Courtesy of Unimation, Inc.)

8. Determine whether a fixed guard can be used.

9. If fixed guards cannot be used, then consider the use of interlocked guards. Determine the type of interlocking system appropriate to the circumstances. The interlocking system should give a reasonable level of integrity appropriate to the risk in question and should enable regular effective maintenance checks to be made. Any "monitoring" system should be carefully examined to assess its effectiveness.

10. When analyzing the system under "aberrant" behavior, a similar process of examination of the hazards and then a risk assessment is carried out. The hazards may be more difficult to determine because they may only exist on failure of part of the machine system such as a control system malfunction on the robot. An alternative to guarding in such circumstances might be to improve the safety integrity of the control system in question, retaining the interlocking safeguards proposed for "designed" behavior.

11. After particular measures have been taken, a reassessment of the system integrity will be necessary. If the hazards are minimized/prevented, reassessment of risks will also be necessary. In most cases and particularly where the risks are high, it is preferable to assume "worst case" when designing the safeguarding system and work on the premise that specific malfunc-

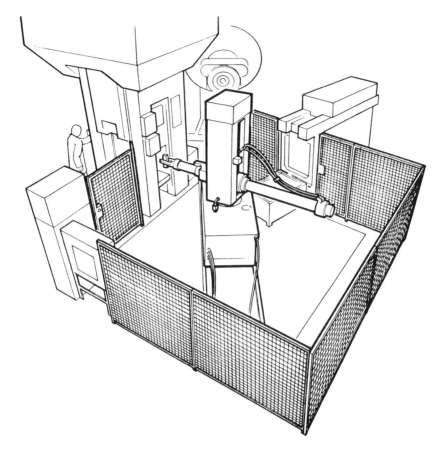

Fig. 35.31. Typical safety fence work cell enclosure. (Photo courtesy of Machine Tool Trades Association.)

tions will occur. It is not prudent to rely solely on the digital programmable electronic system of, say, the robot for all safeguarding features unless a very detailed assessment has been carried out, which may be beyond the competence of the average user.

12. After the analysis has been carried out for normal working, programming, and maintenance—any safeguarding interlocks considered necessary for any one of these modes must be compatible with the requirements of the other from both a functional and a safety integrity point of view—consideration should also be given to emergency stop controls and whether adequate integrity is achieved.

13. The need cannot be overemphasized for documentation concerned with the analysis, decisions, and systems of work, and so on relating to hazards analysis, risk assessment, safety integrity assessments, maintenance requirements, and so on.

35.4.6. Design for Safety

Safeguards should be considered an integral part of the work cell(s) design and provided for at the planning stages. To do otherwise could incur extra expense later on and might compromise safety effectiveness. A generalized approach is outlined:

1. Develop an installation layout showing all equipment. Plan *and* elevation views will usually be required.

2. Lay out a safety enclosure (fencing) around the work cell to preclude human and machinery (e.g., forklift trucks) intrusion. Utilize nonmoving work cell machinery as part of the barrier where possible. Reappraise the layout to see that electrical and control panels and consoles

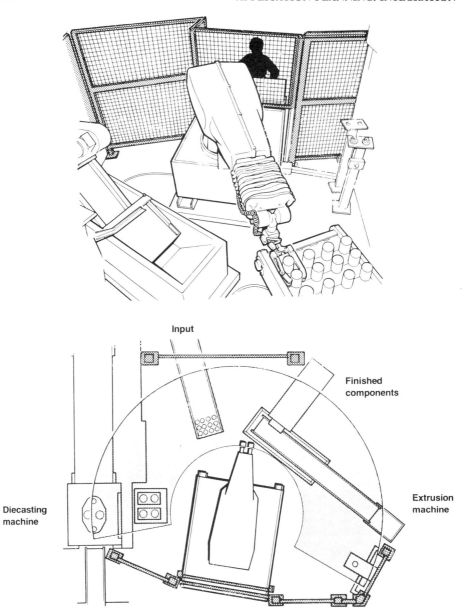

Fig. 35.32. Safety enclosure formed, in part, by work cell machines. (Photo courtesy Machine Tool Trades Association.)

are strategically and conveniently located—especially start-stop controls. Provide for access into the area as required for tool/setup changes, arrival and removal of pallets of workpieces, manual load/unload stations, and so on, and maintenance activities.

3. Evaluate whether restricted motion (hard stops) of the robot will facilitate or improve safety precautions or will conserve floor space. Adjust the layout accordingly.

4. Design the interlock/interface system between all elements of the work cell as required, *including* safety interlocks between the equipment and between the equipment and barrier-access gates. Evaluate the need for primary and secondary or redundant safety interlocks (e.g., in addition to interlocked safety gates designed to stop operations, trip devices such as optical sensors or

Fig. 35.33. Safety fencing in an A. O. Smith Press line. (Top: without fencing.)

electromagnetic shields). Also consider key locks on gates to limit access to authorized personnel only.

5. Review the design to be sure safety provisions are fail-safe.

6. Consider the use of visuals (signs, flashing lights) and audio devices to indicate condition of the operation and to sound alarms.

7. Review all aspects of the final design with sign-offs by manufacturing engineering, production, maintenance, safety officers, and any others deemed appropriate.

This is a suggested approach. Others can be developed with equal or better effectiveness. The important point is that a systematic process be followed, working with a checklist of items to be given attention and a constructive "what if" attitude from all departments involved.

There are other precautions that good management will give attention to, such as adequate lighting where and when needed, safety stripes, warning signs, fire protection, most of which enhance accident prevention at nominal cost. One user is known to have painted similar equipment different colors, including their associated control cabinets, simply to reduce potential confusion and operator mistakes.

You will note that the user of safety barriers or fencing is a foregone conclusion. Regardless of any other accident prevention measures, fencing is almost always a must. There are several reasons for this:

Their very existence (painted appropriately) is a warning even to someone unfamiliar with what they enclose and to the casual observer. Remember the "hypnotic" syndrome.

Safety barriers preclude material-handling equipment and other vehicles in the plant from being inadvertently moved into a danger zone.

Properly interlocked barrier gates tend to enforce procedural discipline when authorized personnel need to gain access to the work area. Tampering with the equipment when not operating is minimized.

Properly designed, barriers will eliminate or at least reduce the potential hazard of objects (work-pieces) rolling or even flying out of the work area.

Sometimes there are other types of barriers or safety screens required that are unique to the application. Most common is the need for eye protection for workers and passersby in robotic arc welding. This will usually take the form of curtains of an approved composition that can easily be drawn around the welding area. Painting applications require similar precautions. Since some kind of spray booth and exhaust system is prerequisite to the job, at least part of the safety issue will automatically be satisfied.

New situations will constantly arise as robotic applications expand. For example, robots are beginning to be applied in laser beam welding and cutting. In each case risk assessment is necessary and accident preventing measures devised. As before, good engineering and common sense will yield the solutions.

Figures 35.31 and 35.32 show typical safety fence guard arrangements.[7] Note in Figure 35.32 that the die casting and extrusion machines form part of the barrier. Obviously, it is essential that the fencing abut the machinery so that a person cannot squeeze through. Similar precautions must

Fig. 35.34. A fenced robotic work cell designed by Massey Ferguson.

be observed in determining the height of the fencing and the gap at floor level. Figure 35.33 shows the safety fencing in place in part of a robotic press transfer line. Figure 35.34 illustrates a typical metalcutting work cell. Note the control panel just outside the fenced area. These controls allow an operator to systematically shut down the machines before entering the work area, thereby protecting the equipment from itself as well as providing human safety.

In the final analysis, a safety engineered system will only be as safe as *people* permit it to be. Part of the commissioning of the installation should include a safety check of all of the built-in features— of the robot, of the related machinery, of the control and safety interlock system, and of the barrier access gates and alarms. This should be a supervised evaluation that follows procedural documentation that is to be posted and always followed. Thereafter, the safety system should be periodically tested for functionality and to be sure that no aspect has been aborted, intentionally or unintentionally, with the passage of time.

35.5. SUMMARY

At the outset we stated that the "RMS" of robotics—*reliability, maintenance,* and *safety*—are, collectively, essential to producing high-quality goods in a timely, cost-effective manner. If this were not obvious then, hopefully it is now.

Reliability must be inherent in the robot design. Reliability is measured in terms of mean time between failure *and* serviceability and maintainability.

A preventive *maintenance* program, backed up by management commitment, trained personnel, and spare parts, is the only way to assure a high availability factor of reliable equipment.

Safety must be designed into the equipment and into the system to the greatest extent humanly possible. A safe system is a productive system.

REFERENCES

1. Asimov, I., *I, Robot,* Doubleday, New York, 1950.
2. Engelberger, J. F., *Robotics In Practice,* AMACOM, New York, 1980.
3. Kapur, K. C., and Lamberson, L. R., *Reliability in Engineering Design,* Wiley, 1977.
4. The High Cost of Bad Maintenance, *Dun's Review,* August 1979, pp. 51–52.
5. Willson, R. D., How Robots Save Lives, *Society of Manufacturing Engineers,* MS82–130, 1982.
6. Barrett, R. J., Bell, R., Hudson, P. H., *Planning for Robot Installation and Maintenance: A Safety Framework, Proceedings of the 4th British Robot Association Annual Conference,* Brighton, U.K., May 18–21, 1981.
7. *Safeguarding Industrial Robots, Part 1, Basic Principles,* The Machine Tool Trades Association, 63 Bayswater Road, London W2 3PH, 1982.

BIBLIOGRAPHY

Reliability

Pollard, B. W., "RAM" for Robots (Reliability, Availability, Maintainability), Society of Manufacturing Engineers, MS80–692, 1980.

Engelberger, J. F., Industrial Robots: Reliability and Serviceability, presented at a conference on robots in Munich, Germany, November 1972.

Engelberger, J. F., Designing Robots for Industrial Environment, Society of Manufacturing Engineers, MR76–600, 1976.

Kapur, K. C., Reliability and Maintainability, in Salvendy, G., Ed., *Handbook of Industrial Engineering,* Wiley, 1982, Ch. 8.5.

Maintenance

The Race to the Automatic Factory, *Fortune,* February 21, 1983, pp. 52–64.

Macri, F. C., Analysis of First UTD (Universal Transfer Device) Installation Failures, Society of Manufacturing Engineers, MS77–735, 1977.

Howard, John M., Human Factors Issues in the Factory Integration of Robotics, Society of Manufacturing Engineers, MS528–127, 1982.

Industry's Man in the Middle, *Iron Age,* January, 21, 1983, pp. 36–38.

Preventive Maintenance: An Essential Tool for Profit, *Production,* July 1979, pp. 83–87.

Safety

ICAM Robotics Application Guide, Technical Report AFWAL-TR-80-4042, Vol. 2, Wright Patterson Air Force Base, Ohio, April 1980.

Worn, H., Safety Equipment for Industrial Robots, *Robots IV Conference,* 1979.

Robot Safety: In a State of Flux and a Jungle, *Robot New International,* December 1982.

Hasegawa, Y., Industrial robot application model design for labor saving and safety promotion in press operations, *Proceedings of the 4th ISIR,* 1974.

Hasegawa, Y., A summary report on FOLS (Foundry Labor Saving and Safety Promotion) Research Project (in Japanese), *Robot,* No. 13, 1976.

Hasegawa, Y. and Sugimoto, N., Industrial safety and robots, *Proceedings of the 12th ISIR,* 1982, pp. 9–20.

McKinnon, R., Robots—are they automatically safe? *Protection,* Vol. 17, No. 5, 1980.

Worn, H., Sicherheitssysteme bei industrierobotern, *Sicherheitstechnik* (in German), No. 5, 1981.

CHAPTER 36

MODULAR ROBOTS IMPLEMENTATION

MARIO SCIAKY

Sciaky S. A.,
Vitry-Sur-Seine, France

36.1. THE CONCEPT OF ROBOT MODULARITY

Modular robots are built of standard independent building blocks and are controlled by one general control system. Each modular mechanism has its own drive unit, power, and communication links. Different modules can be combined by standard interface to provide a variety of kinematic structures designed to best solve a given application requirement. A simple example: When only two degrees of freedom in movement are needed, there is no need to purchase a universal robot with 5 or 6 DF.

There are strong motivations to use modular robots by both users and manufacturers of robots. For robot users the reasons are as follows:

1. Efficient application—robot is built to do what is required and is not universal.
2. High utilization of all robot capabilities—no redundancy in the robot because it is not overdesigned.
3. Capital investment is in what is required—again, no overdesign.
4. Standardization in training—standard modules become familiar.
5. Standardization in maintenance—standard modules become familiar, and standard replacement parts can be used.
6. Expandability or revision of robot structure is possible with additional modules when new requirements warrant it.

The robot manufacturer is motivated by the following:

1. The ability to satisfy customer needs per preceding list.
2. Advantages in production and inventory control by using standard modules and by using off-the-shelf modules.
3. Advantages in training and maintenance.
4. Advantages in marketing.

36.2. DESIGN OF MODULAR ROBOTS

Each module of a modular robot system is a self-contained autonomous mechanism that is designed to combine quickly and with minimal adjustment to several other modules to provide a specific robot requirement. The modularity should mean that individual modules are designed to be simpler than a full, universal robot and, therefore, are more reliable. Figure 36.1 illustrates how seven basic modules can be combined in different possible arrangements into 39 different kinematic structures.

On the critical side, the major difficulties in modular robot design are found in the (1) efficient transmission of power and communication, and (2) effective design of a generalized control system.

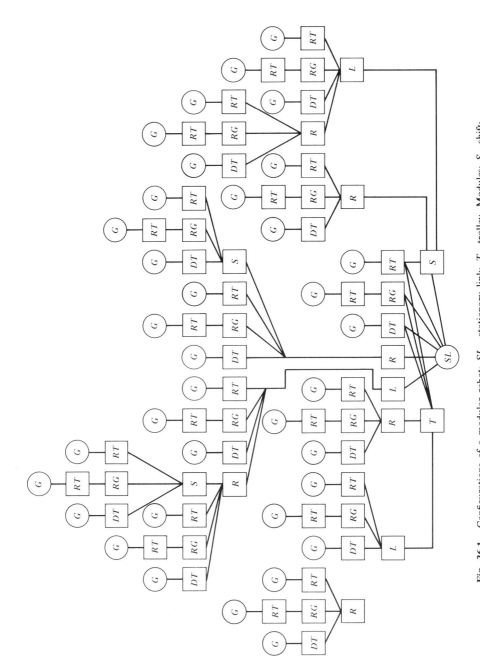

Fig. 36.1. Configurations of a modular robot: *SL*—stationary link; *T*—trolley. Modules: *S*—shift; *L*—lift; *R*—rotation; *DT*—double-turning motion; *RG*—rolling; *RT*—radial travel (swing); *G*—gripper. (*Source:* Reference 1.)

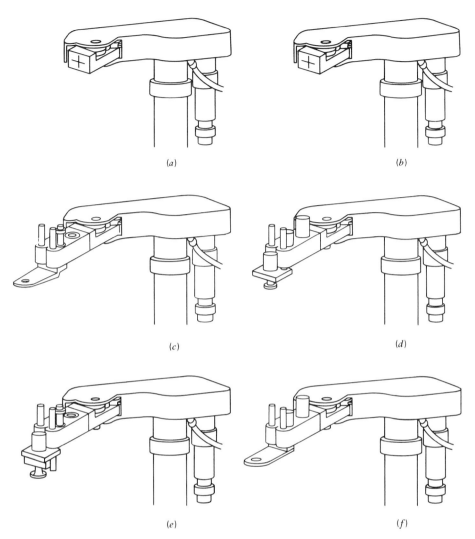

Fig. 36.2. Five variations of a modularized, electric robot. (*a*) Basic structure. (*b*) Two axes: to automate screw drivers, soldering machines, spot welders, sealing material feeders, and the like. (*c*) Two axes + Z motion by pneumatic cylinder: simple pick-and-place or insertion of round parts— attach tools such as nut runners. (*d*) Four axes: for complicated assembly. (*e*) Two axes + Z motion by pneumatic cylinder + wrist axis: applications where the direction of the workpieces varies. (*f*) Three axes with programmable Z axis for applications where variations in height must be made. (Courtesy: Hirata.)

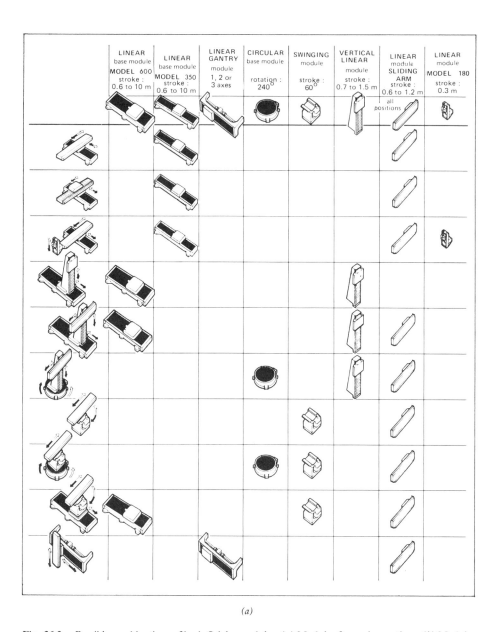

(a)

Fig. 36.3. Possible combinations of basic Sciaky modules. (*a*) Modules for major motions. (*b*) Modules for wrist motions (adaptable to all modules).

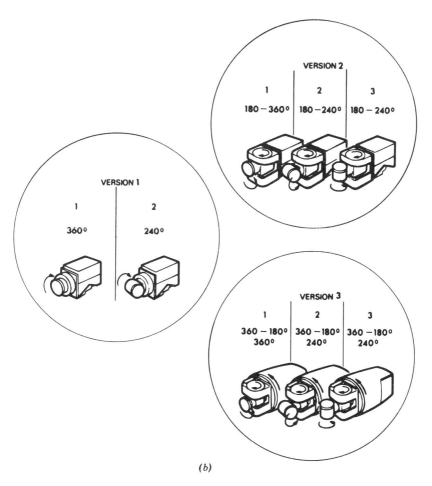

(b)

Fig. 36.3. (*Continued*)

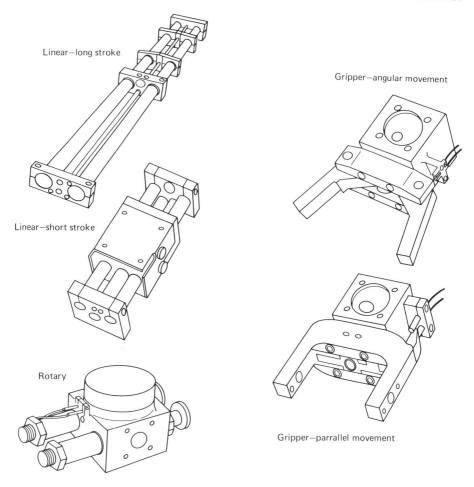

Linear—long stroke

Gripper—angular movement

Linear—short stroke

Rotary

Gripper—parrallel movement

Fig. 36.4. Pneumatic or hydraulic modules for operation under programmable controller. (Courtesy: Fibro Manta.)

At present, there are in the market several families of modular robots with pneumatic, hydraulic, and electric actuation.

36.2.1. Electric Actuation

Electrically actuated modules are preferred for smooth motions with relatively low payload and torque, as typically found in assembly. Figure 36.2 shows a family of five robot versions built of modularized components that are driven basically by DC motors. The arm structure is such that a high degree of flexibility is maintained in horizontal motions, and at the same time vertical rigidity is maintained. The robots in Figure 36.2 are specified to reach speeds of up to 1.4 m/sec. The basic structure with 2 DF can handle parts weighing up to 7 kg. But this payload is reduced significantly when more modules are combined; with 4 DF the payload is reduced about one-third. The controller is a microcomputer, and repeatability is specified as ±0.05 mm.

36.2.2. Hydraulic Actuation

Figure 36.3 shows hydraulic, modular components and how they can be combined. This type of module is suitable for high torque, heavy payload operations. In the next section implementation examples of these modular components are discussed. The speed of robots built of such modules ranges typically between 0.5 and 0.8 m/sec, with a payload of about 50 kg and repeatability of ±1 mm.

36.2.3. Pneumatic Actuation

Modular components for pneumatic actuation exist too, as exemplified in Figure 36.4. Typical combination are 1–4 DF, with load capacities reaching about 10 kg (with pneumatic actuation). Repeatability is usually very high, ±0.005 mm by external mechanical stops.

36.3. IMPLEMENTATIONS OF MODULAR ROBOTS

Modular robots have been applied in a large variety of applications, mainly point-to-point. In Figure 36.5 three examples of pneumatic/hydraulic implementations are described. In each case components are chosen for the specific task at hand, and without any excess capability.

36.3.1. Modular Robots in Assembly

Great interest has been shown in the potential role of robots in programmable assembly. This is clearly indicated by the chapters on assembly (Part 11). Applications of modular robots in assembly are described in Drexel[2] and Romeo and Camera[4] and illustrated in Figure 36.6. On the other hand, progress has also been made in the field called modular assembly (e.g., Riley[3]). The approach here is to develop standard, modular assembly machines capable of performing a broad spectrum of assembly tasks. Such systems could be cost justified for volumes of 500,000 to 1 million units per year. According to Riley,[3] a modular system should be evaluated by considering four key areas:

1. Has modularity been achieved at the price of durability?
2. Does the modularity in the proposed machine allow adaptation to ongoing product changes without major disruptions to regular production schedules?
3. When major product changes do occur, to what degree will the system be reusable?
4. Can the modular machine be successfully maintained by less skilled labor?

These operating considerations will assist users in evaluating the modular approach *before* implementation.

36.3.2. Modular Robots in Spot Welding

Typically, a universal robot has 5 or 6 DF, or six programmable axes, to be able to reach virtually any assigned point. In practice, however, there are applications where six axes are not necessary and others where six axes are not sufficient. In fact, it may be necessary to use—and thus to program—only one or two axes if the geometry of the part and the required target point distribution permit; the unused axes are superfluous. It is often advantageous to dissociate the various robot movements from one another and to keep only those that are indispensable for the job. This can be seen, for example, in all phases of sheet metal spot welding: subassembly preparation shops, geometry conformation and tack welding stations, and final assembly lines. A modular approach to welding gun movement offers numerous advantages by reducing the bulk and complexity of the equipment, maintenance burden, floor space, and investment while increasing adaptability and flexibility with reuseable modular elements. Part program creation and introduction is simplified.

A modular axis system may consist of linear and rotary base modules, a combination of intermediate arm modules, and wrist modules each having one to three rotations. Figure 36.7 shows how universal six-axis robots can be assembled from the basic modules shown before in Figure 36.3. Figure 36.8 illustrates some of the possible combinations having fewer than six axes, useful in spot welding stations.

Frequently, only one linear axis is needed to position the welding gun when making a line of spot welds. A programmable carriage (Figure 36.9) can be used to carry either the welding gun or the part along a horizontal, vertical, or inclined trajectory. A wrist module with the required number of axes for the part geometry is fitted to the sliding table. The carriage must, however, have the same load-handling capacities as a multiaxis robot.

36.3.3. Selection of Modular Robot Configuration

In addition to the previous considerations, geometrical requirements and performance conditions dictate the selection of a kinematic configuration. If the modular approach is taken, the best combination of modules can be selected and configured. The following are examples of such requirements:

Tack Welding Station. Spot welds are required throughout the car body, that is, on both sides, front, and rear (trunk). Normally, six robots are necessary, two working on each side and one on

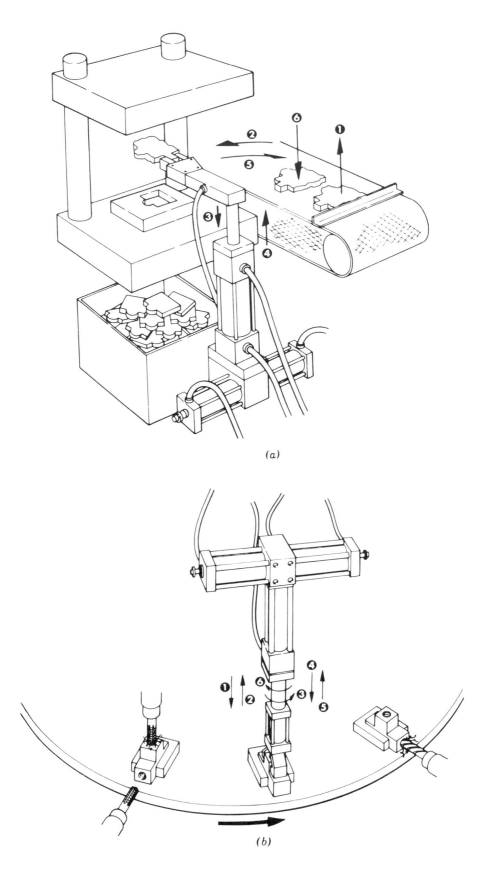

(a)

(b)

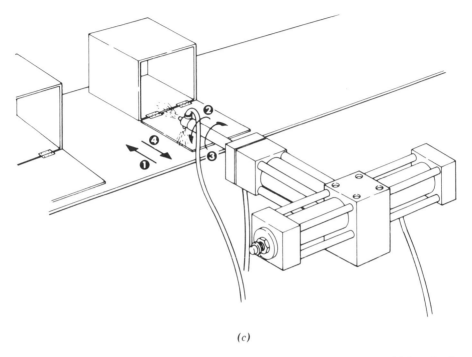

(c)

Fig. 36.5. Implementation examples of pneumatic/hydraulic modular components. (a) Press loading: A multimotion actuator equipped with a cylinder-operated gripper loading an automatic trim press. After the gripper closes on the sprue, the actuator rod extends upward and then rotates to locate part over the trim die. The rod then retracts, lowering part into the die. The gripper is opened, and the actuator extends, rotates, and retracts to be in position to pick up the next part. (Illustrations omit external piston rod supports as normally used in conjunction with air and hydraulic cylinders. External stops for rotary motion must also be considered, depending on internal forces created by working loads.) (b) Parts Turn Around: Eliminates the need for operator to remove and relocate parts before machining a surface that cannot be reached in the initial part's position. Top and one end of pipe tees are tapped. Tees must then be turned 180° before drilling and tapping other end. A multimotion actuator extends to part. Cylinder-operated fingers grab the part which has been unclamped from the fixture. The actuator then retracts and rotates 180°—extends and locates part in fixture again. After a part has been clamped in fixture, fingers release grip, and the actuator retracts and rotates to be ready for the next part. (c) Spraying: An actuator spray painting a container internally with its lid in open position. Operation sequence: Linear section extends while applying paint to lid and one side. At end of linear motion actuator rotates applying paint on three remaining sides and bottom. Paint is shut off, actuator retracts and rotates simultaneously to be ready for next container. (Courtesy: PHD, Inc.)

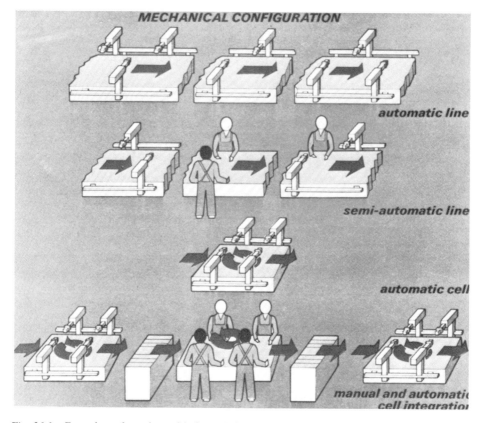

Fig. 36.6. Example configurations with the modular assembly robot Allegro. (Courtesy: General Electric.)

each end (see Figure 36.10). An ideal choice for the lateral robots (see details in Chapter 48, Robots in Spot Welding) would be nongantry Cartesian coordinate robots. The two end robots, since they must be suspended from above, should preferably be hydraulic, articulated robots; these will be able to reach even the bottom of the trunk.

Body Side or Floor Frame. Spot welding can be done by lateral or center access through large openings. Here Cartesian coordinate robots are the preferred choice, as shown in the section of linear modules in Figure 36.11.

Underbody Operations. The main requirement is to perform all welding from the outside, using heavy welding guns with a deep throat. Cartesian coordinate modular robots are selected, as shown in Figure 36.12, since they can be installed using a short kinematic chain that increases load capacity and accuracy. A gantry structure, as shown in Figure 36.13, has the advantage of welding spots on both sides of the line symmetrically. Robots can also be mounted on a gantry with pivoting arms, for increased flexibility, as depicted in Figure 36.14.

Welding along Vertical-Plane Trajectories. In this case, illustrated in Figure 36.15, gantry robots are difficult to use because of the need to access around and under the car body. Side-mounted structures are preferred. In the implementation depicted in Figure 36.15, a linear, vertical module positioned on a horizontal track is combined with a horizontal sweep module, for a total of 3 DF, to accomplish the necessary task.

36.4. SUMMARY

Currently, only about 5% of all robot models are modular. However, as explained in Section 36.1, there is strong motivation to design and implement robots using the modular approach. With increased experience and confidence in robot applications, and with improved performance of modular mechanisms, it is expected that many more modular robot families will be developed and successfully applied.

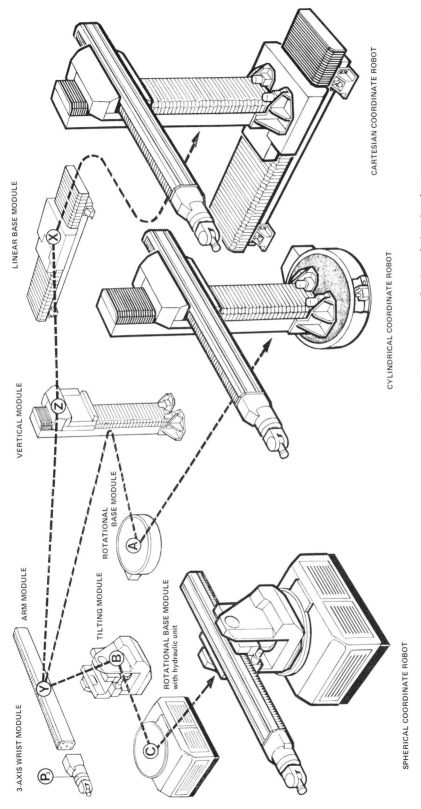

3-AXIS WRIST MODULE

ARM MODULE

VERTICAL MODULE

LINEAR BASE MODULE

TILTING MODULE

ROTATIONAL BASE MODULE

ROTATIONAL BASE MODULE with hydraulic unit

SPHERICAL COORDINATE ROBOT

CYLINDRICAL COORDINATE ROBOT

CARTESIAN COORDINATE ROBOT

Fig. 36.7. Use of modular elements to construct three different types of universal six-axis robots. (Corresponding photos on the next page.)

769

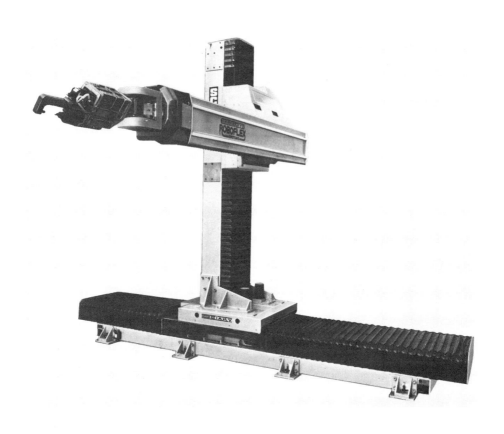

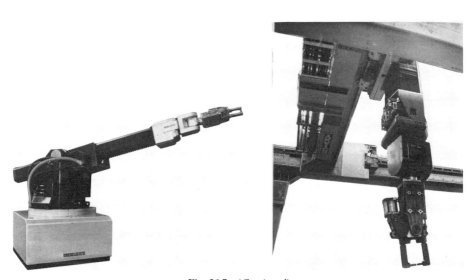

Fig. 36.7 (*Continued*)

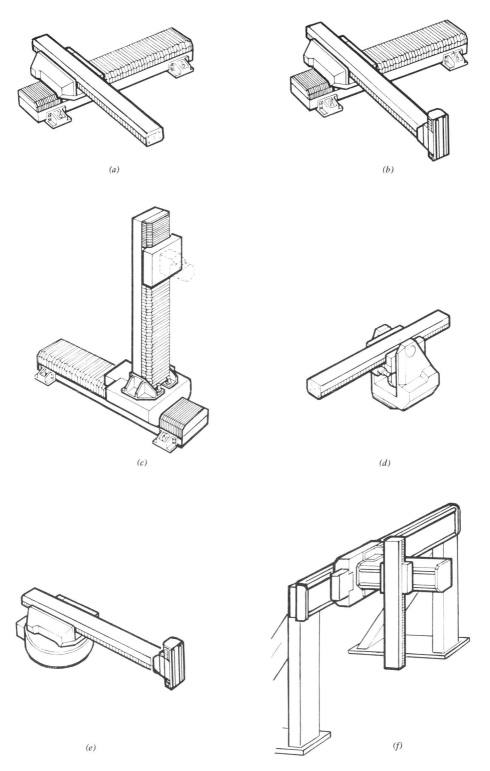

Fig. 36.8. Examples of modular robots having fewer than six axes (shown without wrist modules): (a) two-axis Cartesian in horizontal plane; (b) three-axis Cartesian; (c) two-axis Cartesian in vertical plane; (d) two-axis polar in vertical plane; (e) three-axis cylindrical; (f) three-axis overhead Cartesian.

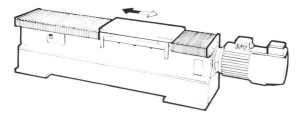

Fig. 36.9. Programmable carriage for one-dimensional displacement of welding head or tooling.

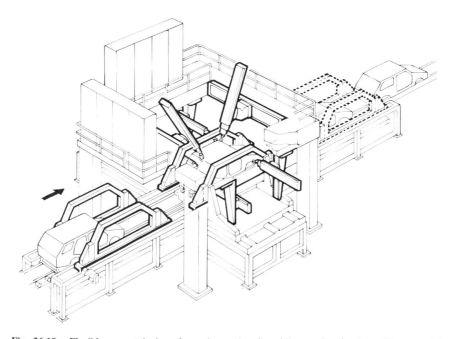

Fig. 36.10. Flexible geometrical conformation and tack welding station for five different models.

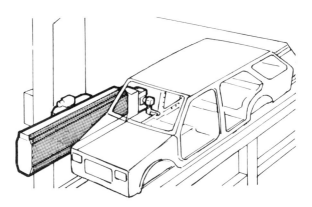

Fig. 36.11. Floor frame welding in a plane parallel to assembly line flow.

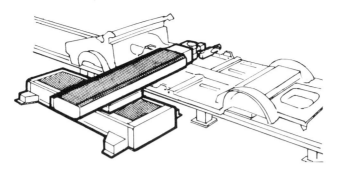

Fig. 36.12. Two-axis Cartesian coordinate robot with two-axis wrist module for underbody assembly.

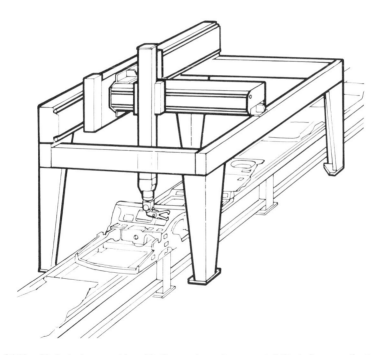

Fig. 36.13. Underbody assembly with three-axis gantry-mounted Cartesian coordinate robot.

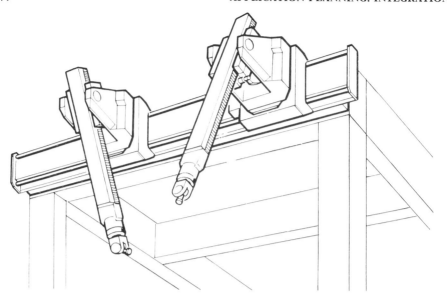

Fig. 36.14. Pivoting gantry-mounted robots for underbody work.

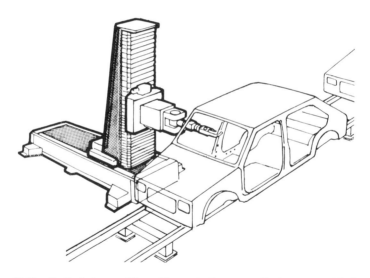

Fig. 36.15. Vertical plane welding with penetration perpendicular to assembly line flow.

REFERENCES

1. Surnin, B. N., Design features of modular type robots, *Machines and Tooling,* Vol. 49, No. 7, 1978, pp. 13–16.
2. Drexel, P., Modular, flexible assembly system "FMS" from Rosch, *Proceedings of the 3rd Assembly Automation Conference,* Stuttgart, West Germany, May 1982, pp. 104–154.
3. Riley, F. T., The use of modular, flexible assembly systems as a half-way path between special design and robots, *Proceedings of the 3rd Assembly Automation Conference,* Stuttgart, West Germany, May 1982, pp. 445–452.
4. Romeo, G., and Camera, A., Robots for flexible assembly systems, *Robotics Today,* Fall 1980.
5. Stauffer, R. N., The Fibro/Manca part handling system, *Robotics Today,* Fall 1979.

CHAPTER 37

END-OF-ARM TOOLING

RONALD D. POTTER

Robot Systems, Inc.
Norcross, Georgia

37.1. INTRODUCTION

End-of-arm tools, also called "end-effectors," give robots the ability to pick up and transfer parts and/or handle a multitude of differing tools to perform work on parts. Robots have been fitted with grippers to load and unload parts from a variety of machines and processes, such as forging presses, injection molding machines, and die casting machines. They have also been fitted with tools to perform work on parts, such as spot welding guns, drills, routers, grinding and cutting tools, and other types of tools to help fabricate and form parts, such as arc welding torches and ladles for pouring molten metal. Tools for assembling parts, such as automatic screwdrivers and nutrunners, have also been attached to robots, as well as tools for performing finishing operations such as paint spray guns and special inspection devices, such as linear variable, differential transformers (LVDTs) and laser gages, to perform quality control functions. There are very few limitations on the type of hand or tool that can be attached to the end of a robot's arm. Unlike the human hand, which is fairly standard with five fingers and a relatively uniform size, a robot's hand can be anything and is normally a unique, one-of-a-kind device designed for a specific application.

End-of-arm tooling is a critical part of an industrial robot system, as it is the part of the system that actually links the robot to the workpiece. The success or failure of an application is very dependent on how well the end-of-arm tooling is conceived, designed, and implemented. In most industrial robot applications, the end-of-arm tooling must be custom designed to match the process requirements. Defined by the part and process, end-of-arm tooling cannot be viewed separately from the other system elements. Throughout the design stages, the interrelationship of all system components with the part and process must be considered. Since the possibilities for end-of-arm tooling are even more diverse than the number of different types of manufacturing processes and machines that exist, it is very difficult to generalize or attempt to restrict the imagination in developing end-of-arm tooling, as infinite as the possibilities are. However, this development should follow a systematic approach to ensure that none of the critical factors in designing tooling is overlooked and that the optimum robot system results.

37.2. SYSTEMATIC APPROACH TO DEVELOPING END-OF-ARM TOOLING

Development of end-of-arm tooling for a particular application should occur at a specific time as one of the sequential steps in the development of the entire robot system. These sequential steps ensure that optimum productivity and efficiency are achieved in the system. Two general rules should be kept in mind when developing a robot system and end-of-arm tooling:

1. Do not attempt to mimic human operations. A human operator, when performing an industrial task, cannot be realistically compared to a robot. Although a human has much more sophisticated sensory capabilities than a robot (i.e., sight, hearing, tactile senses, etc.), a robot does not possess the inherent physical limitations of a human in other areas. A robot, not equipped with a relatively small five-fingered hand, has much greater capabilities in handling heavier weights for longer periods of time in harsher environments than a human. Thus, do not limit the capabilities of the robot system by simply trying to duplicate human capabilities.

2. Do not select the robot first and then try to fit it with an end-of-arm tool and put it to work; select the most appropriate robot for the application as another of the sequential steps in overall system development.

The following systematic approach details the timing of end-of-arm tooling development and robot selection in the system development, as shown in Figure 37.1.

Sequential Steps in Developing a Robot System, Including Timing of End-of-Arm Tooling Analysis

STEP 1:   Understand the *process* thoroughly. Consider what modifications must be made to the process to automate it with any generic robot. Look for ways of improving the efficiency and productivity of the process by altering the present method of manual operation. A more detailed discussion of productivity considerations is presented later in this chapter.

STEP 2:   Analyze the *production equipment* used in the process. Consider what modifications must be made to the equipment to automate it with any generic robot. For example, provisions may have to be made for automatic clamping, sensors for malfunctions, removal of protective guards, interfaces to controls for automatic start/stop, changeover, relocation, and clearances.

STEP 3:   Analyze the *sensors* and *peripheral equipment* that are required to produce an automatic system. Sensors in their various forms provide the paths of communication between all elements of the system, including the end-of-arm tooling. Define all the various conditions that must be sensed in the system and make provisions for them. Peripheral equipment such as parts presentation devices, holding fixtures, conveyors, and inspection stations can be provided to assist the robot in performing the task. Divide up the tasks and do not make the robot do everything. Use peripheral equipment to simplify the design of the end-of-arm tooling and overlap actions to optimize cycle time considerations. As the tasks are divided between robot and peripheral equipment and sensor requirements are determined, the *performance requirements* of the end-of-arm tooling can be defined.

STEP 4:   Conceptualize the *end-of-arm tooling*. At this point, the robot make or model has not been selected. The considerations involved in the preliminary concept of end-of-arm tooling are presented later in this chapter.

STEP 5:   Analyze the *memory type* and *capacity* required for the system. In addition to the robot controller, a programmable controller is normally used to control peripheral equipment and sequence external events. Considerations should be made concerning batch run sizes, number of steps per program, number of programs, changeovers, and so on, and an appropriate system controller should be selected.

STEP 6:   Analyze the *robot type* and *options* best suited for the application. The selection of a particular robot make or model should be made based on the following technical criteria:

   1.   Type of robot (nonservo point-to-point, servo-controlled point-to-point, servo-controlled continuous path).

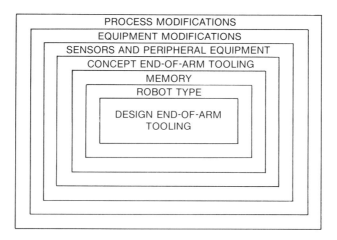

Fig. 37.1.   Systematic approach to developing end-of-arm tooling.

2. Work envelope.
3. Load capacity (including end-of-arm tooling weight).
4. Cycle time.
5. Repeatability.
6. Drive system (pneumatic, hydraulic, or electric).
7. Unique hardware or software capabilities. In addition, other nontechnical considerations such as cost, reliability, and service should be considered in selecting a robot for a particular application.

STEP 7: Final concept and preliminary design of *end-of-arm tooling* to match the selected robot, peripheral equipment, and other system elements. At this point the preliminary concept of the end-of-arm tooling should be analyzed in relation to the robot tool mounting plate and work envelope and modified accordingly.

This preceding sequence of events allows the end-of-arm tooling to be conceptualized and designed only after proper consideration has been given to required modifications in the process and manufacturing equipment, analysis of sensor and peripheral equipment needs, and robot selection.

37.3. DEFINITION OF END-OF-ARM TOOLING

End-of-arm tooling is defined as the subsystem of an industrial robot system that links the mechanical portion of the robot (manipulator) to the part being handled or worked on. An industrial robot is essentially a mechanical arm with a flat tool mounting plate at its end that can be moved to any spatial point within its reach. End-of-arm tooling in the form of specialized devices to pick up parts or hold tools to work on parts is physically attached to the robot's tool mounting plate to link the robot to the workpiece.

37.4. ELEMENTS OF END-OF-ARM TOOLING

End-of-arm tooling is commonly made up of four distinct elements, as shown in Figure 37.2, which provide for (1) attachment of the hand or tool to the robot tool mounting plate, (2) power for actuation of tooling motions, (3) mechanical linkages, and (4) sensors integrated into the tooling.

Mounting Plate

The means of attaching the end-of-arm tooling to an industrial robot is provided by a tool mounting plate located at the end of the last axis of motion on the robot. This tool mounting plate contains either threaded or clearance holes arranged in a pattern for attaching tooling. For a fixed mounting of a gripper or tool, an adapter plate with a hole pattern matching the robot tool mounting plate can be provided. The remainder of the adapter plate provides a mounting surface for the gripper or tool at the proper distance and orientation from the robot tool mounting plate. If the task of the robot requires it to automatically interchange hands or tools, a coupling device can be provided. An adapter plate is thus attached to each of the grippers or tools to be used, with a common lock-in position for pickup by the coupling device. The coupling device may also contain the power source

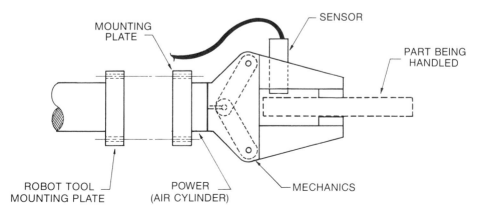

Fig. 37.2. Elements of end-of-arm tooling.

for the grippers or tools and automatically connect the power when it picks up the tooling. Figures 37.3, 37.4, and 37.5 illustrate this power connection tool change application. An alternative to this approach is for each tool to have its own power line permanently connected, and the robot simply pick up the various tools mounted to adapter plates with common lock-in points.

Power

Power for actuation of tooling motions can be either pneumatic, hydraulic, or electrical, or the tooling may not require power, as in the case of hooks or scoops. Generally, pneumatic power is used where possible because of its ease of installation and maintenance, low cost, and light weight. Higher-pressure hydraulic power is used where greater forces are required in the tooling motions. However, contamination of parts due to leakage of hydraulic fluid often restricts its application as a power source for tooling. Although it is quieter, electrical power is used less frequently for tooling power, especially in part-handling applications, because of its lower applied force. Several light payload assembly robots utilize electrical tooling power because of its control capability. In matching a robot to end-of-arm tooling,

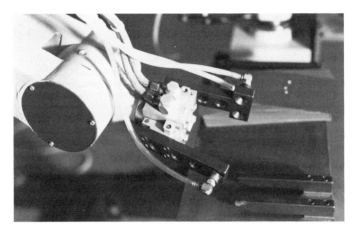

Fig. 37.3. Pickup hand for tool change. Power for tool actuation is ported through the fingers for connection to the various tools to be picked up.

Fig. 37.4. Tool in rack ready to be picked up by robot. Note cone-shaped power connection ports in tool mounting block.

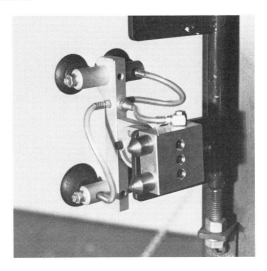

Fig. 37.5. Another tool with power connection block ready for robot pickup.

consideration should be given to the power source provided with the robot. Some robots have provisions for tooling power, especially in part-handling robots, and it is an easy task to tap into this source for actuation of tooling functions. As previously mentioned, many of the robots are provided with a pneumatic power source for tooling actuation and control.

Mechanics

Tooling for robots may be designed with a direct coupling between the actuator and workpiece, as in the case of an air cylinder that moves a drill through a workpiece, or use indirect couplings or linkages to gain mechanical advantage, as in the case of a pivot-type gripping device. A gripper-type hand may also have provisions for mounting interchangeable fingers to conform to various part sizes and configurations. In turn, fingers attached to grippers may have provisions for interchangeable inserts to conform to various part configurations.

Sensors

Sensors are incorporated in tooling to detect various conditions. For safety considerations sensors are normally designed into tooling to detect workpiece or tool retention by the robot during the robot operation. Sensors are also built into tooling to monitor the condition of the workpiece or tool during an operation, as in the case of a torque sensor mounted on a drill to detect when a drill bit is dull or broken. Sensors are also used in tooling to verify that a process is completed satisfactorily, such as wire-feed detectors in arc welding torches and flow meters in dispensing heads. More recently, robots specially designed for assembly tasks contain force sensors (strain gages) and dimensional gaging sensors in the end-of-arm tooling.

37.5. TYPES OF END-OF-ARM TOOLING

There are two functional classifications of end-of-arm tooling for robots: grippers for handling parts and tools for doing work on parts. In general, there is more effort required in developing grippers for handling parts for machine loading, assembly, and parts transfer operations than for handling tools. It is generally more difficult to design special-purpose grippers for part handling than it is to attach a tool to the end of a robot's arm.

The following sections describe the five basic types of end-of-arm tooling, including grippers and tools.

37.5.1. Attachment Devices

These devices are simply mounting plates with brackets for securing tools to the robot tool mounting plate. In some cases attachment devices may also be designed to secure a workpiece to the robot tool mounting plate, as in the case of a robot manipulating a part against a stationary tool where the cycle time is relatively long. In this case, the part is manually secured and removed from the robot tool mounting plate for part retention.

37.5.2. Support and Containment Devices

As shown in Figure 37.6, lifting forks, hooks, scoops, and ladles are typical examples of this type of tooling. Again, no power is required for the tooling, as the robot simply moves to a position beneath a part to be transferred, lifts to support and contain the part or material, and performs its transfer process.

37.5.3. Pneumatic Pickup Devices

The most common example of this type of tooling is a vacuum cup (Figure 37.7) which attaches to parts to be transferred by a suction or vacuum pressure created by a venturi transducer or a vacuum pump. Typically used on parts with a smooth surface finish, vacuum cups are available in a wide range of sizes, shapes, and materials. Parts with nonsmooth surface finishes can still be picked up by a vacuum system if a ring of closed-cell foam rubber is bonded to the surface of the vacuum cup, which conforms to the surface of the part and creates the seal required for vacuum transfer. Venturi vacuum transducers are relatively inexpensive and are used for handling small, lightweight parts where a low vacuum flow is required. Vacuum pumps, quieter and more expensive, generate greater vacuum flow rates and can be used to handle heavier parts. With any vacuum system, the quality of the surface finish of the part being handled is important. If parts are oily or wet, they will tend to slide on the vacuum cups. Therefore some type of containment structure should be used, in addition to the vacuum cups, to enclose the part and prevent it from sliding on the cups. In some applications a vacuum cup with no power source can be utilized. By pressing the cup onto the part and evacuating the air between the cup and part, a suction is created capable of lifting the part. However, a stripping device or valve is required to separate the part from the cup during part release. When a venturi or vacuum pump is used, a positive air blow-off may be used to separate the part from the vacuum cup. Vacuum cups have temperature limitations and cannot be used to pick up relatively hot parts.

Another example of a pneumatic pickup device is a pressurized bladder. These devices are generally specially designed to conform to the shape of the part. A vacuum system is used to evacuate air from the inside of the bladder so that it forms a thin profile for clearance in entering the tooling into a cavity or around the outside surface of a part. When the tooling is in place inside or around the part, pressurized air causes the bladder to expand, contact the part, and conform to the surface of the part with equal pressure exerted on all points of the contacted surface. Pneumatic bladders are particularly useful where irregular or inconsistent parts must be handled by the tooling.

Pressurized fingers, shown in Figure 37.8, are another example of pneumatic pickup devices. Similar to a bladder, pneumatic fingers are more rigidly structured. They contain one straight half, which contacts the part to be handled, one ribbed half, and a cavity for pressurized air between the two halves. Air pressure filling the cavity causes the ribbed half to expand and "wrap" the straight side around a part. With two fingers per gripper, a part can thus be gripped by the two fingers wrapping around the outside of the part. These devices also can conform to various shape parts, and do not require a vacuum source to return to their unpressurized position.

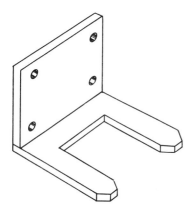

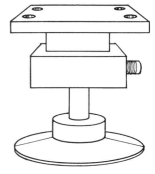

Fig. 37.6. Support and containment device.   Fig. 37.7. Vacuum cup pickup device.

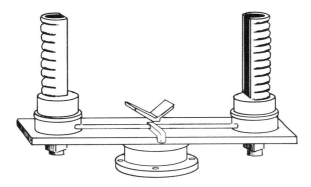

Fig. 37.8. Pneumatic pickup device (pressurized fingers).

37.5.4. Magnetic Pickup Devices

These devices comprise the fourth type of end-of-arm tooling and can be considered when the part to be handled is of ferrous content. Either permanent or electromagnets are used, with permanent magnets requiring a stripping device to separate the part from the magnet during part release. Magnets normally contain a flat part-contact surface but can be adapted with a plate to fit a specific part contour. A recent innovation in magnetic pickup devices uses an electromagnet fitted with a flexible bladder containing iron filings, which conforms to an irregular surface on a part to be picked up. Like vacuum pickup devices, oily or wet part surfaces may cause the part to slide on the magnet during transfer. Therefore containment structures should be used in addition to the magnet to enclose the part and prevent it from slipping. Three other concerns arise in handling parts with magnets. If a metal-removal process is involved in the application, metal chips may also be picked up by the magnet. Provisions must be made to wipe the surface of the magnet in this event. Also, residual magnetism may be imparted to the workpiece during pickup and transfer by the magnetic tooling. A demagnetizing operation may be required after transfer if this is detrimental to the finished part. If an electromagnet is used, a power failure will cause the part to be dropped immediately, which may produce an unsafe condition. Although electromagnets provide easier control and faster pickup and release of parts, permanent magnets can be used in hazardous environments requiring explosion-proof electrical equipment. Normal magnets can handle temperatures up to 60°C (140°F), but magnets can also be designed for service in temperatures up to 150°C (300°F).

37.5.5. Mechanical Grip Devices

Mechanical grip devices, the fifth category, are the most widely used type of tooling in parts-handling applications. Either pneumatic, hydraulic, or electrical actuators are used to generate a holding force which is transferred to the part by linkages and fingers. The most commonly used power source for finger closure actuation is a pneumatic cylinder whose bore and stroke is selected in relation to the available operating air pressure to provide an optimum amount of holding force on the part. The grip force of the fingers may be varied by regulating the pressure entering the tooling actuators. The grip force may be further reduced by mounting soft conforming inserts in the fingers at the point of contact with the part to be handled. Some recent innovations in standard commercially available grippers contain sensors and features for controlling the amount of grip force exerted on the part, in addition to dimensional gaging capabilities incorporated in the gripper. In these hands, the fingers can close on parts of various sizes until a predetermined force is attained and stop at that point. Still other standard grippers are equipped with strain gages in the fingers to detect the position of parts within the gripper and adjust the robot arm accordingly to center the fingers around the part.

The motions of the fingers in a mechanical grip device vary, and an appropriate finger motion can be selected to best suit the part shape and its constraints at the pickup and release station. The simplest finger motion involves a gripper with one stationary or fixed finger and one moving finger, commonly referred to as a "single-action" hand (see Figure 37.9). This hand functions by having the robot move the open hand to a position around the part and the moving finger close on the part, clamping it against the stationary finger. This hand requires that the part be free to move during the pickup process to allow for clearance of the stationary finger around the part. If a part is fixtured tightly during pickup or release, this hand is not the most appropriate choice. By placing a V-block insert in either of the fingers, this single-action hand can center the part during the grip process and locate it accurately.

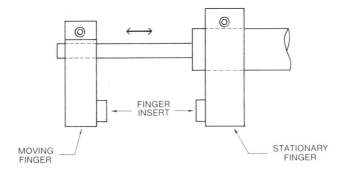

Fig. 37.9.   Single-action pickup hand.

Where parts are tightly fixtured in the pickup or release position, a "double-action" hand can be utilized. This hand contains two moving fingers that close simultaneously to hold a part. The motion of the fingers can be either pivoting or parallel. Pivoting fingers, as shown in Figure 37.10, have greater limitations, as they must be designed to match the part shape at the contact points at a specific angular orientation of the fingers. Since parallel motion double-action hands (Figure 37.11) move in a straight-line motion in closing on a part, they do not have the angular orientation consideration to meet. Thus they can also handle a wide range of part sizes automatically. By placing V-block locators in the fingers as inserts, the double-action hand can center the parts not consistently oriented at the pickup station if they have provisions for movement. On tightly fixtured parts, such as in the unloading of a lathe, both fingers have clearance around the part in the open position and simultaneously close to center on the part.

Some mechanical grip devices contain three moving fingers that simultaneously close to grasp a part or tool. These hands are particularly useful in handling cylindrical-shaped parts, as the three-point contact centers round parts of varying diameters on the centerline of the hand. Machine loading operations, where round parts are loaded into chucks or over mandrels, are best suited for use of a three-fingered centering gripper. Still other mechanical grippers contain four moving fingers that close simultaneously to center a square or rectangular-shaped part on the centerline of the hand.

With most of the hands described, the position of the fingers can be reversed to allow gripping of the internal surfaces of parts, if required.

37.6.   GENERAL DESIGN CRITERIA FOR END-OF-ARM TOOLING

Although robot end-of-arm tooling varies widely in function, complexity, and application area, there are certain design criteria that pertain to almost all robot tooling. These general guidelines follow and are expanded later in this chapter.

First, end-of-arm tooling should be as *lightweight* as possible. This primarily affects the performance of the robot. The rated load-carrying capacity of the robot, or the amount of weight that can be attached to the robot tool mounting plate, includes the weight of the end-of-arm tooling and that of the part being carried. The load that the robot is carrying also affects the speed of its motions. Robots can move at faster rates carrying lighter loads. Therefore, for cycle time considerations, the lighter the tool, the faster the robot is capable of moving. The use of lightweight materials such as aluminum or magnesium for hand components and lightening holes, whenever possible, are common solutions for weight reduction. Figure 37.12 shows a double-action pickup hand with lightening holes.

Second, end-of-arm tooling should be as *small in physical size* as possible. Minimizing the dimensions of the tool helps to minimize the weight also. Minimizing the size of the tooling provides for better clearances in workstations in the system. In relationship to the robot, most load-carrying capacities of robots are based on moment of inertia calculations of the last axis of motion, that is, a given load capacity at a given distance from the tool mounting plate surface. Therefore, by minimizing the size of the tooling and the distance from the tool mounting plate to the center of gravity of the tooling, robot performance is enhanced.

At the same time, it is desirable to handle the *widest range of parts* with the robot tooling. This minimizes changeover requirements and reduces costs for multiple tools. Although minimizing the size of the tooling somewhat limits the range of parts that can be handled, there are techniques for accomplishing both goals. Adjustable motions may be designed into the tooling so that it can be quickly and easily manually changed to accommodate different-sized parts. Interchangeable inserts may be put in tooling to change the hand from one size or shape part to another. The robot may also automatically interchange hands or tools to work on a range of parts, with each set of tooling

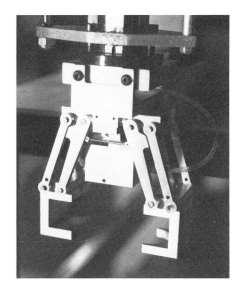

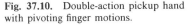

Fig. 37.10. Double-action pickup hand with pivoting finger motions.

Fig. 37.11. Double-action pickup hand with parallel finger motions.

designed to handle a certain portion of the entire range of parts. This addresses weight and size considerations and reduces the total number of tools required. Figure 37.13 shows a standard dual part-handling tool for gripping parts varying within a certain size range.

Maximizing rigidity is another criterion that should be designed into tooling. Again, this relates to the task performance of the robot. Robots have specified repeatabilities and accuracies in handling a part. If the tooling is not rigid, this positioning accuracy will not be as good and, depending on part clearances and tolerances, may cause problems in the application. Excessive vibrations may also be produced by attaching a nonrigid or flimsy tool on the tool mounting plate. Since robots can move the tooling at very high rates of speed, this vibration may cause breakage or damage to the tool. Providing rigid tooling eliminates these vibrations.

The *maximum applied holding force* should be designed into the tooling. This is especially important for safety reasons. Robots are dynamic machines that can move parts at high velocities at the end of the arm, with only the clamp force and frictional force holding it in the hand. Because robots typically rotate the part about a fixed robot base centerline, centrifugal forces are produced. Acceleration and

Fig. 37.12. Double-action pickup hand showing interchangeable fingers with lightening holes and V-block locating features.

Fig. 37.13. Standard tooling for parts-handling applications, with such features as parallel motion fingers, dual part-handling capability, and compliance in hand.

deceleration forces also result when the robot moves from one point to another. The effect of these forces acting on the part makes it critical to design in an applied holding force with a safety factor great enough to ensure that the part is safely retained in the hand during transfer and not thrown as a projectile with the potential of causing injury or death to personnel in the area or damage to periphery equipment. On the other hand, the applied holding force should not be so great that it actually causes damage to a more fragile part being handled. Another important consideration in parts transfer relating to applied holding force is the orientation of the part in the hand during transfer. If the part is transferred with the hand axis parallel to the floor, the part, retained only by the frictional force between the fingers and part, may have a tendency to slip in the hand, especially at programmed stop points. By turning the hand axis perpendicular to the floor during part transfer, the required holding force may be decreased, and the robot may be able to move at higher speed because the hand itself acts as a physical stop for the part.

Maintenance and changeover considerations should be designed into the tooling. Perishable or wear details should be designed to be easily accessible for quick change. Change details such as inserts or fingers should also be easily and quickly interchangeable. The same type of fastener should be used wherever possible in the hand assembly, thereby minimizing the maintenance tools required.

37.7. ESTABLISHING DESIGN CRITERIA FOR END-OF-ARM TOOLING

As previously mentioned, the development of end-of-arm tooling concept and design should proceed only after analysis of the part to be handled or worked on and the process itself has been thoroughly completed. This analysis can be divided into two phases: a preengineering, data collection phase and a design phase. The preengineering phase involves analysis of the workpiece and process with emphasis on productivity considerations, whereas the second phase involves the actual design of the tooling to best meet the criteria developed during the first phase.

37.7.1. Preengineering and Data Collection

Workpiece Analysis

The part being transferred or worked on must be analyzed to determine critical parameters to be designed into the end-of-arm tooling. The *dimensions and tolerances* of the workpiece must be analyzed to determine their effect on tooling design. The dimensions of the workpiece will determine the size and weight of the tooling required to handle the part. It will also determine whether one tool can automatically handle the range of part dimensions required, whether interchangeable fingers or inserts are required, or whether tool change is required. The tolerances of the workpieces will determine the need for compliance in the tooling. Compliance allows for mechanical "float" in the tooling in relation to the robot tool mounting plate to correct misalignment errors encountered when parts are mated during assembly operations or loaded into tight-fitting fixtures or periphery equipment. If the part tolerances vary so that the fit of the part in fixture is less than the repeatability of the robot, a compliance device may have to be designed into the tooling. Passive compliance devices such as springs may be incorporated into the tooling to allow it to float to accommodate very tight tolerances. This reduces the rigidity of the tooling. Other passive compliance devices such as remote center compliance (RCC) units are commercially available. These are mounted between the robot tool mounting plate and the end-of-arm tooling to provide a multiaxis float. RCC devices, primarily designed for assembly tasks, allow robots to assemble parts with mating fits much tighter than the repeatability that the robot can achieve. Active compliance devices with sensory feedback can also be used to accommodate tolerance requirements.

The *material and physical properties* of the workpiece must be analyzed to determine their effect on tooling design. The best method of handling the part, by vacuum, magnetic, or mechanical-grip pickup, can be determined. The maximum permissible grip forces and contact points on the part can be determined, as well as the number of contact points to ensure part retention during transfer. Based on the physical properties of the material, the need for controlling the applied force through sensors can also be resolved.

The *weight and balance* (center of gravity) of the workpiece should be analyzed to determine the number and location of grip contact points to ensure proper part transfer. This will also resolve the need for counterbalance or support points on the part in addition to the grip contact points. The static and dynamic loads and moments of inertia of the part and tooling about the robot tool mounting plate can be analyzed to verify that they are within the safe operating parameters of the robot.

The *surface finish and contour* (shape) of the workpiece should be studied to determine the method and location of part pickup (i.e., vacuum on smooth, flat surfaces, mechanical grippers on round parts, etc.). If the contour of the part is such that two or more independent pickup means must be applied, this can be accomplished by mounting separate pickup devices at different locations on the tool, each gripping or attaching to a different section of the part. This may be a combination of vacuum cups, magnets, and/or mechanical grippers. Special linkages may also be used to tie together two different pickup devices powered by one common actuator.

Part modifications should be analyzed to determine if minor part changes that do not affect the functions of the part can be made to reduce the cost and complexity of the end-of-arm tooling. Often, simple part changes, such as holes or tabs in parts, can significantly reduce the tooling design and build effort in the design of new component parts for automation and assembly by robots.

Part inconsistencies should be analyzed to determine the need for provision of out-of-tolerance sensors or compensating tooling to accommodate these conditions.

In *tool-handling* rather than part-handling applications, the workpiece should be analyzed to determine the characteristics of the tool required. This is especially true for the incorporation of protective sensors in the tooling to deal with part inconsistencies.

Process Analysis

In addition to a thorough analysis of the workpiece, the manufacturing process should be analyzed to determine the optimum parameters for the end-of-arm tooling.

The *process method* itself should be analyzed, especially in terms of manual versus robot operation. In many cases physical limitations dictate that a person perform a task in a certain manner where a robot without these constraints may perform the task in a more efficient but different manner. An example of this involves the alternative of picking up a tool and doing work on a part or instead picking up the part and taking it to the tool. In many cases the size and weight-carrying capability of a person is limited and forces him to handle the smaller and lighter weight of the part or the tool. A robot, with its greater size and payload capabilities, does not have this restriction. Therefore it may be used to take a large part to a stationary tool or to take multiple tools to perform work on a part. This may increase the efficiency of the operation by reducing cycle time, improving quality, and increasing productivity. Therefore, in process analysis, consider the alternative of having the robot take a part to a tool or a tool to a part, and decide which approach is most efficient. When a robot is handling a part, rather than a tool, there is less concern about power-line connections to the tool, which experience less flexure and are less prone to problems when stationary than moving.

Because of its increased payload capability, a robot may also be equipped with multifunctional end-of-arm tooling. This tooling can simultaneously or sequentially perform work on a part that previously required a person to pick up one tool at a time to perform the operation, resulting in lower productivity. For example, the tooling in a die casting machine unloading application may not only unload the part, but also spray a die lubricant on the face of the dies.

The *range and quantity of parts or tools* in the manufacturing process should be analyzed to determine the performance requirements for the tooling. This will dictate the number of grippers or tools that are required. The tooling must be designed to accommodate the range of part sizes either automatically in the tool, through automatic tool change, or through manual changeover. Manual changeover could involve adjusting the tool to handle a different range of parts, or interchanging fingers, inserts, or tools on a common hand. To reduce the manual changeover time, quick disconnect capabilities and positive alignment features such as dowel pins or locating holes should be provided. For automatic tool change applications, mechanical registration provisions, such as tapered pins and bushings, ensure proper alignment of tools. Verification sensors should also be incorporated in automatic tool change applications.

Presentation and disposition of the workpiece within the robot system affects the design of end-of-arm tooling. The position and orientation of the workpiece at either the pickup or release stations will determine the possible contact points on the part, the dimensional clearances required in the tooling to avoid interferences, the manipulative requirements of the tooling, the forces and moments of the tooling and part in relation to the robot tool mounting plate, the need for sensors in the tooling to detect part position or orientation, and the complexity of the tooling.

The sequence of events and cycle time requirements of the process have a direct bearing on tooling design complexity. Establishing the cycle time for the operation will determine how many tools (or hands) are needed to meet the requirements. Multiple parts-handling tools often allow the robot to increase the productivity of the operation by handling more parts per cycle than can be achieved manually. The sequence of events may also dictate the use of multifunctional tooling that must perform several operations during the robot cycle. An example of this is in machine unloading, where the tooling not only grasps the part, but also sprays a lubricant on the molds or dies of the machine. Similarly, robot tooling could also handle a part and perform work on it at the same time, such as automatic gaging and drilling a hole.

The sequence of events in going from one operation to another may cause the design of the tooling to include some extra motions not available in the robot by adding extra axes of motion in the tooling to accommodate the sequence of operations between various system elements.

In-process inspection requirements will affect the design of end-of-arm tooling. The manipulative requirements of the tooling, the design of sensors or gaging into the tooling, and the contact position of the tool on the part are all impacted by the part-inspection requirements. The precision in positioning the workpiece is another consideration for meeting inspection requirements.

The *conditional processing* of the part will determine the need for sensors integrated into the tooling, as well as the need for independent action by multiple-handed grippers.

The *environment* must be considered in designing end-of-arm tooling. The effects of temperature, moisture, airborne contaminants, corrosive or caustic materials, and vibration and shock must be evaluated, as will the material selection, power selection, sensors, mechanics, and the provision for protective devices in the tooling.

37.8. SUMMARY

37.8.1. Design Tips for End-of-Arm-Tooling

The following list presents some tips that are useful in designing end-of-arm tooling (EOAT).

1. Design for quick removal or interchange of tooling by requiring a small number of tools (wrenches, screwdrivers, etc.) to be used. Use the same fasteners wherever possible.

2. Provide locating dowels, key slots, or scribe lines for quick interchange, accuracy registration, and alignment.

3. Break all sharp corners to protect hoses and lines from rubbing and cutting and maintenance personnel from possible injury.

4. Allow for full flexure of lines and hoses to extremes of axes of motion.

5. Use lightweight materials wherever possible, or put lightening holes where appropriate to reduce weight.

6. Hardcoat lightweight materials for wear considerations, and put hardened, threaded inserts in soft materials.

7. Conceptualize and evaluate several alternatives in EOAT.

8. Do not be "penny-wise and pound-foolish" in EOAT; make sure enough effort and cost is spent to produce production-worthy, reliable EOAT and not a prototype.

9. Design in extra motions in the EOAT to assist the robot in its task.

10. Design in sensors to detect part presence during transfer (limit switch, proximity, air jet, etc.).

11. For safety in part-handling applications, consider what effect a loss of power to EOAT will have. Use toggle lock gripper or detented valve to promote safety.

12. Put shear pins or areas in EOAT to protect more expensive components and reduce downtime.

13. When handling tools with robot, build in tool inspection capabilities, either in EOAT or peripheral equipment.

14. Design multiple functions into EOAT.

15. Provide accessibility for maintenance in EOAT design, and quick change of wear parts.

16. Use sealed bearings for EOAT.

17. Provide interchangeable inserts or fingers for part changeover.

18. When handling hot parts, provide heat sink or shield to protect EOAT and robot.

19. Mount actuators and valves for EOAT on robot forearm.

20. Build in compliance in EOAT or fixture where required.

21. Design action sensors in EOAT to detect open/close or other motion conditions.

22. Analyze inertia requirements, center of gravity of payload, centrifugal force, and other dynamic considerations in designing EOAT.

23. Look at motion requirements for gripper in picking up parts (single-action hand must be able to move part during pickup; double-action hand centers part in one direction; three or four fingers center part in more than one direction).

24. When using electromagnetic pickup hand, consider residual magnetism on part and possible chip pickup.

25. When using vacuum cup pickup on oily parts, a positive blow-off must also be used.

26. Look at insertion forces of robot in using EOAT in assembly tasks.

27. Maintain orientation of part in EOAT by force and coefficient of friction or locating features.

37.8.2. Future Considerations

To date, most of the applications of industrial robots have involved a specially designed hand or gripper. Current research is ongoing to develop more flexible general-purpose grippers that can adapt to a variety of sizes and shapes of parts. The state-of-the-art is nowhere near duplicating the complexity of the human hand. However, with increased sensory feedback integrated into robot tooling, more sophisticated tasks are being completed by the robot. This trend will continue in years to come.

REFERENCES

1. Lundstrom, G., Glennie, B., and Rooks, B. W., *Industrial Robots Gripper Review,* IFS Publications, Bedford, England, 1977.

2. Okada, T. and Tsuchiya, S., On a versatile finger system, *Proceedings of the 7th International Symposium on Industrial Robots,* Tokyo, Japan, October 1977, pp. 345–352.

3. Mori, K. and Sugiyama, K., Material handling device for irregularly shaped heavy works, *Proceedings of the 8th International Symposium on Industrial Robots,* Stuttgart, West Germany, May 1978, pp. 504–513.

4. Van der Loos, H., Design of three-fingered robot gripper, *Industrial Robot,* Vol. 5, No. 4, December 1978, pp. 179–182.

5. Chelponov, I. B. and Kolpashnikov, S. N., Mechanical features of gripper in industrial robots, *Proceedings of the 13th International Symposium on Industrial Robots,* Chicago, Illinois, April 1983, pp. 18.77–90.

CHAPTER 38

STRATEGY FOR ROBOT APPLICATIONS

FRED A. CIAMPA

Ford Motor Company
Dearborn, Michigan

38.1. INTRODUCTION

This chapter describes the evolution of robotics in a heavy manufacturing industry. It begins with the *universal transfer device* (UTD) in the early 1960s thru the early 1970s, when the first automobile assembly plant was automated with robot applications. The late 1970s, with the new competitive challenge from off-shore companies and the strategy changes needed by the Ford Motor Company to make substantial gains in both quality and productivity, are also discussed.

The chapter outlines in detail the strategy for the Ford Robotic Application Center to expand the robot knowledge and understanding in a decentralized company and to keep the engineer in tune with the fast-changing high-technology area of robotics. The Center also provides the facilities for the line engineer to develop a complete robotic system, that is, robot, end-of-arm tooling, sensors, conveying devices, and the like. This strategy will correct the major problem experienced on the production floor with a robotic system.

Where the company plans to go by the 1990s, and what the needs are from the robot suppliers to minimize the problems for application development, ease of use, and general maintenance are defined for the maturing robotic industry in this chapter.

38.2. HISTORY

The history of robotics at Ford Motor Company really began with the installation of a Versatran robot for press loading at the Canton Forge Plant in August 1961. This installation was Ford's first, and one of the earliest in the American automotive industry. At this time, Ford referred to these mechanisms as *universal transfer devices* or UTDs. This terminology was used to circumvent industrial relations implications related to the unwillingness of some workers to accept robots as technological tools rather than as mechanical devices intended to displace people.

By the end of 1973 the Ford Motor Company had 66 robots installed worldwide with 33 in the Kansas City, Missouri, Assembly Plant for spotwelding on the Maverick and Comet automobiles. This assembly plant was one of the leaders in the use of robots until the end of the decade. Most of these robots were the Unimate 4000 Series and Versatrans. Some of these applications are shown in Figures 38.1–38.3. At that time most robotic applications were used for spot welding and press loading and unloading (Figures 38.4–38.5).

In March 1973 Unimation Inc. and the Ford Motor Company launched a program to jointly develop a programmable automation system. This concept, the forerunner of the present state of the art for robots, was faster, smaller, and far more precise than the then current robots for repeatability— 0.010 in. (0.0025 mm) instead of 0.050 in. (0.0127 mm). The concept incorporated two robots (Figure 38.6A and 6B) for small assembly, and major changeover would require only a program change and new tooling details. The assembly feasibility of small components requiring multiple robots and wrist motions at previously unheard of accuracies was proven and used in subsequent applications even though this program never got beyond a demonstration unit.

Over the years, it became apparent from seminars and news publications that the terms *robot* or *industrial robot* were being accepted by the industry and the public. The Ford Motor Company also discontinued using the phrase *universal transfer device* (UTD), and referred to them as robots or

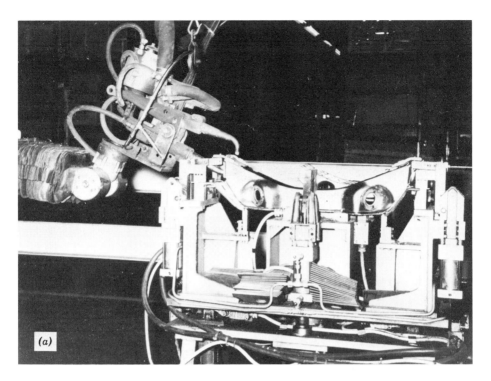

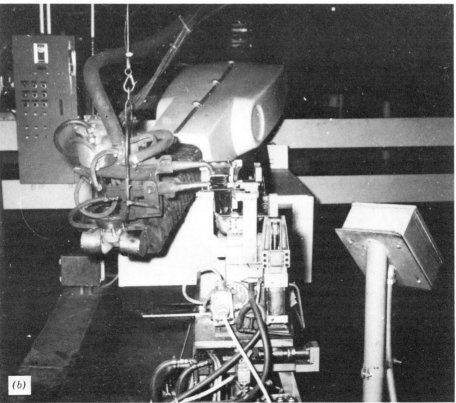

Fig. 38.1. (*a*) A sideview of a robot spot welding an automobile cross member at the Kansas City, Missouri Assembly Plant in 1971. (*b*) A front view of the same robot application.

Fig. 38.2. Four robots on each side of an automobile body weld line at the Kansas City, Missouri assembly line in 1973.

industrial robots in 1980. By 1983 the Ford Motor Company worldwide operations had 1100 robots in use. Their use extended beyond spot welding to include the following:

Arc welding

Palletizing and depalletizing

Machine loading and unloading

Deburring

Performing assembly operations

Paint spraying

Applying sealers and adhesives

Laser bar code reading

Examples of some of the applications are shown in Figures 38.7–38.12. Many of these applications included the use of fiber optics, photo cells, infrared sensor and vision systems for part location and orientation. Even with this progress in robotic application, the Ford Motor Company recognized that further strides had to be taken to compete with the world competition. This dictated that the company make additional effort to automate with more sophisticated systems. It also created a dependency on up-front ease of use by the factory worker, an untrained workforce in technology.

38.3. STRATEGY

The Ford Motor Company is in the midst of a new era for automobile manufacturers, one that offers both tough challenges and exciting opportunities. More than ever, competition is keener and the customers are discerning when they shop for an automobile. To ensure success, Ford's objective is to continue to design and build products using the latest technology available which provides high levels of durability, quality, and reliability for the lowest price possible to the customer.

To meet this competitive challenge of the 1980s, substantial gains in both quality and productivity must be achieved. Historically, incorporation of the latest technology has been a significant contributor to productivity improvements. It is expected that automation and robotic systems, in particular, will

Fig. 38.3. A robot welding automobile rocker panels at the Kansas City, Missouri Assembly Plant in 1973.

be the key technological ingredient that will keep the improvement process properly directed. This necessitates organizing for robotics and automation.

There are several reasons why it was necessary for a large industrial company like the Ford Motor Company to organize for robotics.

Decentralized Operations. Ford is a large multinational corporation with manufacturing facilities and subsidiaries spanning the globe. With this scope, a need was identified to interchange robot technology effectively throughout the highly decentralized operations. It is important to expand this knowledge and understanding concerning successful application developments and thus help to maintain a competitive worldwide market position.

Keeping up with New Developments. It is necessary to keep the manufacturing planners and engineers current with the rapid growth of this relatively new high-technology industry. This growth applies to the number of new pieces of hardware and software being introduced, the complexity, and the increasing number of suppliers.

Keeping up with New Products. A few years ago there were really only a few robot manufacturers and a limited number of models to choose from. But today, there are more than 30 U.S.-based firms, and when Europe and Japan are included, this number approaches 120.

Organized Approach for Rapid Growth of Applications. At present there are 1100 robotic applications within the Ford Motor Company, and a conservative estimate of 2500 robots will be employed in the company's manufacturing operations by 1985 (Figure 38.13). This represents more than a 125% increase in a relatively short time. The task of selecting the proper equipment for this growing number of robotic applications requires an organized approach.

New Emphasis on Technology Orientation. Finally, there is a need to organize to increase the emphasis on the application of new technology for the manufacturing and assembly facilities. Historically, more than 80% of the company's advance engineering and research budget allocations were product oriented.

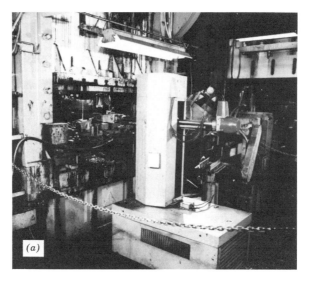

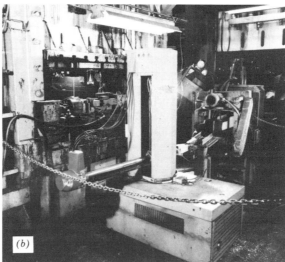

Fig. 38.4. (*a*) Robot loading a stamping press at the frame plant at the Rouge complex in Dearborn, Michigan in 1971. (*b*) The same robot unloading the press.

In December 1982 the Robotics and Automation Applications Consulting Center was established as a means for accomplishing the task of organizing for robotics. This center is comprised of the following functions:

Manufacturing technology applications
Computer-aided manufacturing systems
Training
Application development
Qualification and performance testing

38.3.1. Manufacturing Technology Applications

The increased momentum to implement robotics and other programmable automation, optimizing robot selection and design of the peripheral process systems, end-effector tooling, and software program-

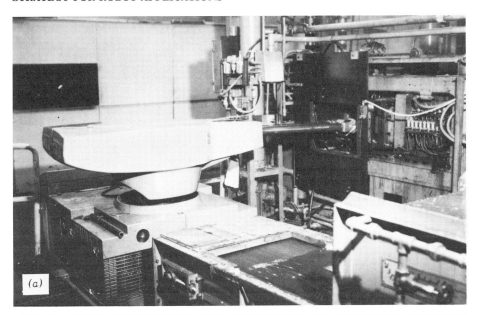

Fig. 38.5.   (*a*) Robot unloading a cluster of parts from a die cast machine at the Rawsonville, Michigan Plant in 1971. (*b*) The same robot loading the parts cluster on a water spray conveyor.

ming—all were placed on the line manufacturing engineers. In general, these engineers have knowledge of robotics and computer technology but collectively have a difficult time keeping up with the latest state of the art.

This prompted the establishment of a new organizational group within the Robotic Center, referred to as Manufacturing Technology Applications Consultants (MTAs). This group is composed of people from the manufacturing and assembly operations having in-depth experience in engineering and production operations. Their primary mission is to identify new technologies and potential opportunities for appropriate use in manufacturing operations, as well as to assist the line manufacturing engineers in the planning and implementation of the manufacturing systems.

38.3.2. Computer-Aided Manufacturing Systems

To complement the MTAs, the computer-aided-manufacturing systems activity that already existed within the company was realigned to form another part of the new organization. This group provides the engineering base for the development and refinement of advanced computer programming, editing techniques, and diagnostic software for troubleshooting robotic systems.

Fig. 38.6. Two views (*a, b*) of an experimental robotic application demonstrating small parts assembly feasibility.

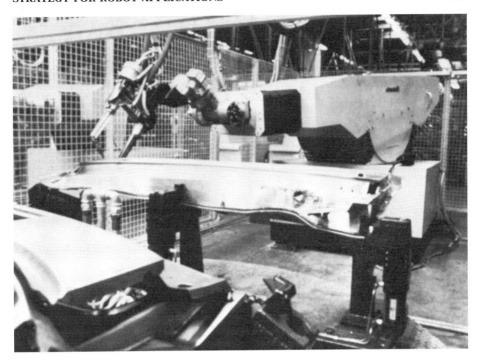

Fig. 38.7. A robot spot welding at the Broadmeadows Assembly Plant (Ford of Australia) in 1979.

38.3.3. Training and Worker Acceptance

The training activity provides necessary education and training for company hourly, salary, and management employees. Involving the hourly worker in the training programs prior to a robotic application in his or her area breaks down the barrier caused by the fear of the unknown, *A Robot*. This has virtually eliminated any worker acceptance problems, which was confined to a minimal number of employees. The training acquaints them with the application, programming, operating, and maintenance of robots and ancillary equipment. The familiarity of the robots obtained by the employees during the training classes promotes the acceptance of the robotic applications by the affected employees. Video tapes, slides, films, and hands-on experience on a variety of robots are used in the presentation of the courses. In addition, commercially available audiovisual training programs are available for the robot users.

38.3.4. Application Development

This activity provides laboratory facilities at the Robotic Center for the line engineers from the manufacturing and assembly plants within the Ford Motor Company (Figure 38.14). This activity will spearhead the strategy for developing the robotic system for particular hard-to-do applications with each manufacturing, assembly, or material-handling engineer. The center and activity support the engineers in the development of specific robot system applications under simulated production conditions. Thus the engineer will be able to refine his or her original robotic system concept to a more simplified, problem-free, workable system, as well as determining cycle time and the robot with the best working envelope before purchasing the equipment for the production application. The development and tryout includes peripheral equipment such as conveyors, end-of-arm tooling and grippers, fixturing and sensing devices, and parts-handling/feeding mechanisms for specific process applications as shown in Figures 38.15 and 38.16.

The development of a robotic application permits the engineer to revise and simplify the original concept to a more problem-free workable system. This precludes the engineer from "wishing I could do it over" when the previous development is done at the robot vendor's floor or in the plant after all of the equipment has been purchased. The engineer can determine the actual cycle time and simplify the robot end-of-arm tooling and the automation's presentation and removal of the workpiece to and from the robot before it is purchased. This reduces the major problems experienced in a manufacturing

Fig. 38.8. (*a*) A robot unloading a transmission case from a floor conveyor in the Livonia, Michigan Transmission Plant in 1979. (*b*) The same robot loading the transmission case on a moving overhead conveyor.

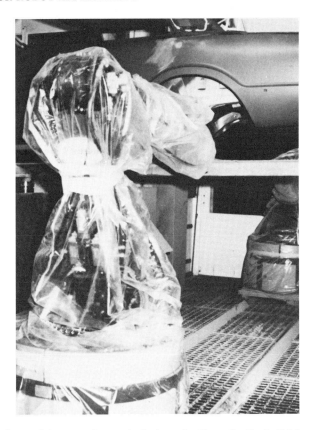

Fig. 38.9. A robot applying a coating on the body underside at the Genk, Belgium Assembly Plant in 1979. The plastic wrapping protects the robot from the coating material overspray.

plant with a robotic application. The problems experienced in production operations at the manufacturing and assembly plants with a robotic system are mainly associated with the end-of-arm tooling and peripheral equipment. The problems are, typically, poor repeatability and reliability of providing workpieces to the robot; poor repeatability and reliability of the end-of-arm tooling in assembling workpieces; poor reliability of a machine fixture for machine loading. Such problems are caused by cycle time constraints of the application, and by the need for unique one-time design of tooling and peripheral equipment required for each application. The application development activity allows line engineers to develop a total system fully by actual cycling of the system to examine performance and durability characteristics. Unsatisfactory performance or durability characteristics that are revealed during the cycle testing can be corrected and reevaluated. This permits developing a system fully prior to installation on the factory floor, where timing constraints preclude optimizing the system.

Approximately 20 applications have been developed at the center in the first eight months of its operation. The line engineers associated with these development applications have exhibited great enthusiasm for the opportunity to work and develop their concepts prior to finalizing the applications. However, the relatively short period that the center has been in operation has not yet allowed for any of the applications to be "tried and proven" in the production environment.

38.3.5. Qualification and Performance Testing

A function of the center is assessing the capabilities of new entries into the robotics market. Additionally, performance testing and measurements, similar to those described in Chapter 10, Performance Testing, are made on products that are currently in use within the Company to provide data relative to specific applications. Examples include coordinate accuracy, repeatability, and power consumption, as well as indicated reliability problems under varying payloads, speeds, and reach.

The testing is necessary to substantiate performance claims on a common baseline and to develop

Fig. 38.10. (*a*) A sideview of four robots performing body inspection at the Wixom, Michigan Assembly Plant. (*b*) A frontal view of the same robotic application.

a performance data base for all robots. These data provide line manufacturing and assembly engineers with the information to assist in the selection of the best equipment for each application.

The qualification and performance testing evaluates the robot for rated accuracy and repeatability as influenced by weight, speed, and reach. Accuracy is defined as the difference between the position the robot tool point automatically goes to and the originally taught position. Repeatability is defined as the envelope of variance of the robot tool point position with repeated cycles. Even though the

Fig. 38.11. A robot removing burrs on the edge of an oil hole on a crankshaft at the Cleveland, Ohio Engine Plant in 1982.

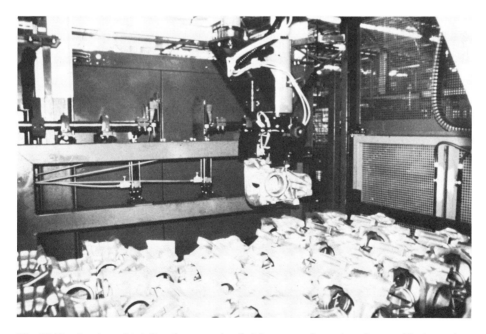

Fig. 38.12. A robot with infrared sensors depalletizing rear axle castings from a shipping rack at the Duram, Germany Plant.

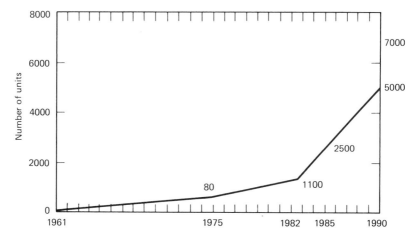

Fig. 38.13. Trends in industrial robot installations at Ford.

test method developed by the Ford Robotic Center measures accuracy and repeatability, the emphasis is on repeatability in order to assess the influence of dynamics. Speed, acceleration, deceleration, and extension and weight of the tooling and workpiece from the end of the robot wrist are only some of the parameters influencing the dynamics of the robots.

Repeatability assessments are performed to rate system capability statistically (Figure 38.17). The basic test equipment is designed to measure positioning of the robot tool point in three planes, as shown in Figures 38.18 and 38.19. This is done by noncontact sensors to prevent errors caused by contact between the gaging block and robot tool point. Repeatability of the robot system is examined in two basic conditions of operation: (1) initial start-up periods and (2) after warm-up or normal operating periods. This enables determination of repeatability extremes during typical operation by

Fig. 38.14. An engineer developing a robotic application for assembling an electric motor at the Ford Robotic Center.

Fig. 38.15. A robotic application for loading and unloading moving overhead conveyor being developed at the Robotic Center.

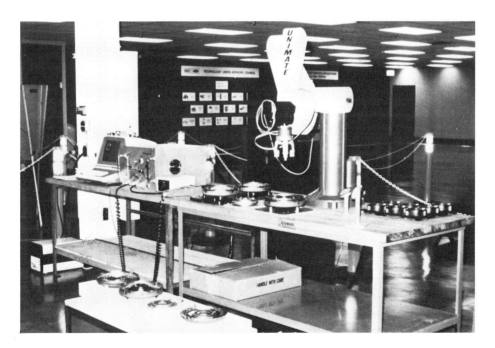

Fig. 38.16. A robotic application for assembling an automatic transmission torque converter being developed at the Robotic Center.

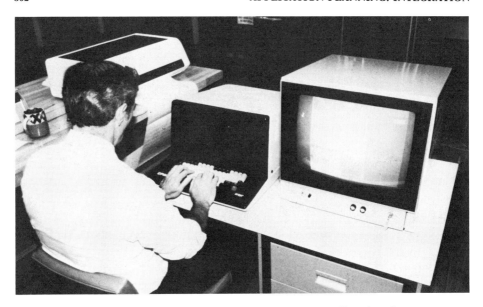

Fig. 38.17.   Computer statistically analyzing the repeatability of a robot.

combining prenormal and normalized operating conditions. (Prenormalized assumes the robot manufacturer's warm-up cycle is not conducted.)

Repeatability is categorized into two parameter groups of testing. The two groups are maximum and median (50%) payload conditions (as rated by the robot manufacturers). The two parameters are tested with two variables—reach and speed. Table 38.1 is a matrix description of the test parameters and variables.

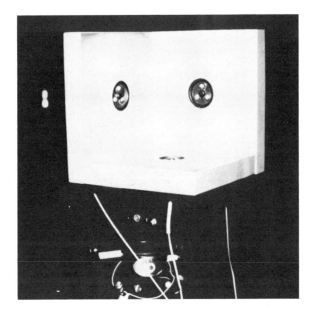

Fig. 38.18.   Noncontact three-plane gaging block for measuring the robot performance.

Fig. 38.19. A robot tool block simulating the tool point approaching the gage block during a test cycle.

The results of the test data are compiled and reported in terms of:

Mean of the repeatability envelope, or $\overline{X}$.
Range of the repeatability envelope, or R.
Upper and lower 3 sigma capability.

38.4. FUTURE OF ROBOTICS IN THE FORD MOTOR COMPANY

As previously discussed, it is estimated that about 2500 robots will be in use in the Ford Motor Company by the end of 1985, an increase of more than 125% from 1983. The 2500 robots are forecasted to double to 5000 by 1990. As more robots are installed and the engineers gain familiarity with robotic technology, these projections will probably be changed to reflect even higher rates of usage—up to 7500 by 1990. Robotic technology is also expected to proliferate to areas such as nonreprogrammable automation and flexible machining cells. This will further improve productivity and quality to meet the Company's goal of making the best automobile at the lowest price. The Ford workers are depending on the industrial robot suppliers and academia for further product innovation to complement this internal development and training effort.

38.5. WHAT DOES INDUSTRY REQUIRE IN ROBOTICS?

Lack of uniformity in construction, product design, and the operating nature of the robots from the commercially available suppliers requires serious attention. The industry offers a wide variety of good

TABLE 38.1

	Variables	
Payload	Reach	Speed
Robot	Max	Max
Manufacturer's	50%	Max
Maximum	50%	50%
Rating	Max	50%
Robot	Max	Max
Manufacturer's	50%	Max
Median (50%)	50%	50%
Rating	Max	50%

products, but variety in itself carries a dilemma to users in many ways. Proliferating the factory floor with unlike devices raises a major concern in the eyes of the robot users in the techniques of programming, characteristic differences in operating one product or another, and variation in components used in the design of the robot systems.

There is a need to rethink market strategies by producers to participate jointly in standardizing systems designs to accommodate major issues confronting users. Standardizing teach methods and symbols and the mounting configuration for the end-of-arm tooling or grippers would not infringe on the robot manufacturer's competitive edge. From the eyes of the user, such unique features as those just described may even prevent the purchase of any robot. The user is first of all interested in whether or not the system application will operate satisfactorily. Second, can the system be economically justified or affordable? And third, are the basic skills for operating and maintaining the system considered within the level of skills available in the plant?

ACKNOWLEDGMENTS

I wish to express my gratitude to everyone at the Ford Motor Company Robotics and Automation Applications Consulting Center for providing me the background for writing this chapter through our close association with our various application programs. My special gratitude goes to Jim Dillon, the Director of the Center, for providing me the opportunity of being involved with the latest technology encountered at the Center. Also, Tom Helzerman, Norbert Michalowicz, Bob Richards, and Gary Sitzman deserve special acknowledgment since they helped provide the photographs and material used for this chapter. John DiPonio deserves special thanks for providing the opportunity for the people at the Center to contribute to this handbook.

PART **8**
FABRICATION AND PROCESSING

CHAPTER 39

AN OVERVIEW OF FABRICATION AND PROCESSING APPLICATIONS

JOHN D. MEYER

Tech Tran Corporation
Naperville, Illinois

39.1. OVERVIEW OF INDUSTRIAL ROBOT APPLICATIONS

39.1.1. Reasons for Using Robots

The first commercial application of an industrial robot took place in 1961, when a robot was installed to load and unload a die casting machine. This was a particularly unpleasant task for human operators. In fact, many early robot applications took place in areas where a high degree of hazard or discomfort to humans existed, such as in welding, painting, and foundry operations. Even though these early robots did not necessarily perform their tasks more economically than humans, the elimination of hazardous and unpleasant manual operations was sufficient justification for their use.

In recent years robots have also been used in many applications where they offer clear economic advantage over human workers. Although human labor rates have continued to escalate, the hourly operating and depreciation costs for robots have remained relatively constant. Thus, in many instances, robots can perform tasks considerably less expensively than humans. Savings of 50–75% in direct labor costs are not uncommon.

Another closely related reason for using industrial robots is increased productivity. Robots are not only less expensive than manual labor, but they also frequently have higher rates of output. Some of this increased productivity is due to the robot's slightly faster work pace, but much is the result of the robot's ability to work almost continually, without lunch breaks and rest periods.

In addition to their economy and their ability to eliminate dangerous tasks and increase productivity, robots are also used in many applications where repeatability is important. Although today's robots do not possess the judgmental capability, flexibility, or dexterity of humans, they do have a distinct advantage of being able to perform repetitive tasks with a high degree of consistency, which in turn leads to improved product quality. This improvement in consistency is important when justifying robots for applications such as spray painting, welding, and inspection.

These four benefits—reduced costs, improved productivity, better quality, and elimination of hazardous tasks—represent the primary reasons for using industrial robots in today's factories. In the future an additional benefit, greater flexibility, is also expected to play a major role in robot justification. As flexible manufacturing systems and the totally automated factory become realities in the future, the robot's ability to adapt to product design changes and variations in product mix will become an increasingly important factor in their use.

39.1.2. Robot Capabilities

In general, robots possess three important capabilities which make them useful in manufacturing operations: transport, manipulation, and sensing.

Transport

One of the basic operations performed on an object as it passes through the manufacturing process is material handling or physical displacement. The object is transported from one location to another to be stored, machined, assembled, or packaged. In these transport operations, the physical characteristics of the object remain unchanged.

The robot's ability to acquire an object, move it through space, and release it makes it an ideal candidate for transport operations. Simple material-handling tasks, such as part transfer from one conveyor to another, may only require one- or two-dimensional movements. These types of operations are often performed by nonservo robots. Other parts-handling operations may be more complicated and require varying degrees of manipulative capability in addition to transport capability. Examples of these more complex tasks include machine loading and unloading, palletizing, part sorting, and packaging. These operations are typically performed by servo-controlled point-to-point robots.

Manipulation

In addition to material handling, another basic operation performed on an object as it is transformed from raw material to a finished product is processing, which generally requires some type of manipulation. That is, workpieces are inserted, oriented, or twisted to be in the proper position for machining, assembly, or some other operation. In many cases, it is the tool that is manipulated rather than the object being processed.

A robot's capability to manipulate both parts and tooling makes it very suitable for processing applications. Examples in this regard include robot-assisted machining, spot and arc welding, and spray painting. More complex operations, such as assembly, also rely on the robot's manipulation capabilities. In many cases the manipulations required in these processing and assembly operations are quite involved, and therefore either a continuous-path or point-to-point robot with a large data storage capacity is required.

Sensing

In addition to transport and manipulation, a robot's ability to react to its environment by means of sensory feedback is also important, particularly in sophisticated applications like assembly and inspection. These sensory inputs may come from a variety of sensor types, including proximity switches, force sensors, and machine vision systems.

State-of-the-art robots have relatively limited sensing capabilities. This is due primarily to the difficulty with which today's robots can be effectively interfaced with sensors and, to a lesser extent, to the availability of suitable low-cost sensing devices. As control capabilities continue to improve and sensor costs decline, the use of sensory feedback in robotics applications will grow dramatically.

In each application one or more of the robot's capabilities of transport, manipulation, or sensing is employed. These capabilities, along with the robot's inherent reliability and endurance, make it ideal for many applications now performed manually, as well as in some applications now performed by traditional automated means.

39.1.3. Types of Applications

By the end of 1983 there were approximately 8000 robots installed in the United States. These installations are usually grouped into the seven application categories shown in Figure 39.1.[1] This figure also shows the major robot capabilities used in each application and the type of benefits obtained. A more detailed list of application examples by type is contained in Figure 39.2.[1] A brief description of each application category is contained in the following paragraphs.

Material Handling

In addition to tending die casting machines, early robots were also used for other material-handling applications. These applications make use of the robot's basic capability to transport objects, with manipulative skills being of less importance. Typically, motion takes place in two or three dimensions, with the robot mounted either stationary on the floor or on slides or rails that enable it to move from one workstation to another. Occasionally, the robot may be mounted overhead, but this is rare. Robots used in purely material-handling operations are typically nonservo, or pick-and-place, robots.

Examples of material-handling applications include: transferring parts from one conveyor to another; transferring parts from a processing line to a conveyor; palletizing parts; and loading bins and fixtures for subsequent processing. A robot unloading glass tubes from a conveyor and placing them on a pallet is depicted in Figure 39.3.

The primary benefits in using robots for material handling are to reduce direct labor costs and remove humans from tasks that may be hazardous, tedious, or exhausting. Also, the use of robots typically results in less damage to parts during handling, a major reason for using robots for moving fragile objects. In many material-handling applications, however, other forms of automation may be more suitable if production volumes are large and no workpiece manipulation is required.

APPLICATION	EXAMPLES	ROBOT CAPABILITIES JUSTIFYING USE			PRIMARY BENEFITS OF USING ROBOTS			
		TRANSPORT	MANIPULATION	SENSING	IMPROVED PRODUCT QUALITY	INCREASED PRODUCTIVITY	REDUCED COSTS	ELIMINATION OF HAZARDOUS/UNPLEASANT WORK
MATERIAL HANDLING	PARTS HANDLING PALLETIZING TRANSPORTING HEAT TREATING	●					●	●
MACHINE LOADING	DIE CAST MACHINES AUTOMATIC PRESSES NC MILLING MACHINES LATHES	●	●			●	●	
SPRAYING	SPRAY PAINTING RESIN APPLICATION		●		●		●	●
WELDING	SPOT WELDING ARC WELDING		●			●	●	●
MACHINING	DRILLING DEBURRING GRINDING ROUTING CUTTING FORMING		●	●		●	●	
ASSEMBLY	MATING PARTS FASTENING		●	●		●	●	
INSPECTION	POSITION CONTROL TOLERANCE			●	●			

Fig. 39.1.   Major categories of robot applications and rationale for use.

Machine Loading and Unloading

In addition to unloading die casting machines, robots are also used extensively for other machine loading and unloading applications. Machine loading and unloading is generally considered to be a more sophisticated robot application than simple material handling. Robots can be used to grasp a workpiece from a conveyor belt, lift it to a machine, orient it correctly, and then insert or place it on the machine. After processing, the robot unloads the workpiece and transfers it to another machine or conveyor. The greatest efficiency is usually achieved when a single robot is used to service several machines. Also, a single robot may be used to perform other operations while the machines are performing their primary functions.

Other examples of machine loading and unloading applications include: loading and unloading of hot billets into forging presses; loading and unloading machine tools, such as lathes and machining centers; stamping press loading and unloading; and tending plastic injection molding machines.

MANUFACTURING OPERATION	SAMPLE ROBOT APPLICATIONS
MATERIAL HANDLING	• Moving parts from warehouse to machines • Depalletizing wheel spindles into conveyors • Transporting explosive devices • Packaging toaster ovens • Stacking engine parts • Transfer of auto parts from machine to overhead conveyor • Transfer of turbine parts from one conveyor to another • Loading transmission cases from roller conveyor to monorail • Transfer of finished auto engines from assembly to hot test • Processing of thermometers • Bottle loading • Transfer of glass from rack to cutting line • Core handling • Shell making
MACHINE LOADING/UNLOADING	• Loading auto parts for grinding • Loading auto components into test machines • Loading gears onto CNC lathes • Orienting/loading transmission parts onto transfer machines • Loading hot form presses • Loading transmission ring gears onto vertical lathes • Loading of electron beam welder • Loading cylinder heads onto transfer machines • Loading a punch press • Loading die cast machine
SPRAY PAINTING	• Painting of aircraft parts on automated line • Painting of truck bed • Painting of underside of agricultural equipment • Application of prime coat to truck cabs • Application of thermal material to rockets • Painting of appliance components
WELDING	• Spot welding of auto bodies • Welding front end loader buckets • Arc welding hinge assemblies on agricultural equipment • Braze alloying of aircraft seams • Arc welding of tractor front weight supports • Arc welding of auto axles
MACHINING	• Drilling aluminum panels on aircraft • Metal flash removal from castings • Sanding missile wings
ASSEMBLY	• Assembly of aircraft parts (used with auto-rivet equipment) • Riveting small assemblies • Drilling and fastening metal panels • Assembling appliance switches • Inserting and fastening screws
OTHER	• Application of two-part urethane gasket to auto part • Application of adhesive • Induction hardening • Inspecting dimensions on parts • Inspection of hole diameter and wall thickness

Fig. 39.2. Examples of robot applications.

Fig. 39.3. Robot unloading glass tubes from conveyor and placing them on pallets.

Although adverse temperatures or atmospheres can make robots advantageous for machine loading and unloading, the primary motivation for their use is to reduce direct labor costs. Overall productivity is also likely to increase because of the longer amount of time the robot can work compared to humans. In machine loading and unloading, it is both the manipulative and transport capabilities that make use of robots feasible.

Spraying

In spraying applications, the robot manipulates a spray gun which is used to apply some material, such as paint, stain, or plastic powder, to either a stationary or moving part. These coatings are applied to a wide variety of parts, including automotive body panels, appliances, and furniture. In those cases where the part being sprayed is on a moving conveyor line, the robot's sequence of spraying motions is coordinated with the motion of the conveyor. A spray-painting robot is shown in Figure 39.4. Relatively new applications for spraying robots include the application of resin and chopped glass fiber to molds for producing glass-reinforced plastic parts and spraying epoxy resin between layers of graphite broadgoods in the production of advanced composites.

The manipulative capability of the robot is of prime importance in spraying applications. A major benefit of their use is higher product quality through more uniform application of material. Another benefit is reduced costs by eliminating human labor and reducing waste coating material. Another major benefit is the reduced exposure of humans to toxic materials.

Welding

The largest single application for robots at present is for spot welding automotive bodies. Spot welding is normally performed by a point-to-point servo robot holding a welding gun. Arc welding can also be performed by robots, as shown in Figure 39.5. However, seam tracking can be a problem in some arc welding applications. A number of companies are developing noncontact seam trackers, which would greatly increase the usage of robots for arc welding.

Robots are used in welding applications to reduce costs by eliminating human labor and to improve product quality through better welds. In addition, since arc welding is extremely hazardous, the use of robots can minimize human exposure to harsh environments.

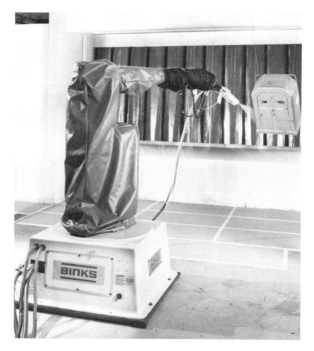

Fig. 39.4. Spray painting robot.

Machining

In machining applications, the robot typically holds a powered spindle and performs drilling, grinding, routing, or other similar operations on the workpiece. An example of a robot deburring parts is shown in Figure 39.6. In machining operations the workpiece can be placed in a fixture by a human, by another robot, or by a second arm of the same robot performing the machining. In some operations the robot moves the workpiece to a stationary powered spindle and tool, such as a buffing wheel.

Robot applications in machining are limited at present because of accuracy requirements, expensive tool designs, and lack of appropriate sensory feedback capabilities. Machining is likely to remain a somewhat limited application until both improved sensing capabilities and better positioning accuracy are achieved.

Assembly

One of the areas of greatest interest today is the development of effective, reasonably priced robots for assembly. Currently available robots can be used to a limited extent for simple assembly operations, such as mating two parts together. However, for more complex assembly operations, robots are subject to the same limitations as in machining operations, namely, difficulties in achieving the required positioning accuracy and sensory feedback.

Examples of current robot assembly operations include the insertion of light bulbs into instrument panels, the assembly of typewriter ribbon cartridges, the insertion of components into printed wiring boards, and the automated assembly of small electric motors.

However, more complex assembly tasks typically cannot be performed by currently available robots. A number of companies are conducting research in sensory feedback, improved positional accuracy, and better programming languages that will permit more advanced assembly applications in the future.

Inspection

A small but rapidly growing number of robot applications is in the area of inspection. In these applications, robots are used in conjunction with sensors, such as a television camera, laser, or ultrasonic detector, to check part locations, identify defects, or recognize parts for sorting. Such robots have been used to inspect valve cover assemblies for automotive engines, sort metal castings, and inspect the dimensional accuracy of openings in automotive bodies.

Fig. 39.5. Arc welding robot.

As in assembly and machining operations, a high degree of accuracy and extensive sensory capabilities are required for inspection applications. In the future this is expected to be one of the high-growth application areas as low-cost sensors and improved positioning accuracy evolve.

39.1.4. User Industries

A breakdown of current robot installations by applications and industry is given in Table 39.1.[1] Not surprisingly, the industry with the most robots installed is the auto industry, with about 40% of the total U.S. robot population. Within the auto industry, welding is the most common robot application, with about 70% of all robots in that industry used for welding.

Although robots are used in almost every industry and type of application, the majority of installations are concentrated in a relatively few plants and types of applications. For example, it is estimated that just 10 plants contain nearly one-third of all robots installations and that the three categories of welding, material handling, and machine loading account for approximately 80% of all current applications. At the same time, it must be remembered that the market penetration of robots has been relatively limited in even the most common applications.

As robot technology begins to diffuse within industry it will affect almost every manufacturer, from furniture producers to pharmaceutical firms. And as robot capabilities continue to improve, new applications will undoubtedly be uncovered.

39.2. ROBOTS IN FABRICATION AND PROCESSING

39.2.1. Definition of Fabrication and Processing

The previous section provided an overview of all major types of industrial applications. However, most of these application types are discussed in detail in subsequent handbook chapters on welding; material handling and machine loading; assembly; inspection, quality control, and repair; and finishing, coating, and painting.

As a matter of convenience, for this chapter, we define *fabrication and processing* as those applications

Fig. 39.6. Robot deburring metal parts.

not covered in later handbook chapters. As used here, fabrication and processing include the types of applications described in the following sections. In some instances, these applications reflect the use of robots for specific manufacturing functions, such as machining or heat treating, whereas in other cases they involve the use of robots in specific industries, such as plastics processing and glassmaking. In these latter cases, there will be some obvious duplication with information presented in subsequent chapters.

39.2.2. Die Casting[2-7]

As mentioned previously, die casting has historically been one of the major application areas for industrial robots. In fact, in 1961 the first commercial robot installation took place, which involved tending a die casting machine. Today, die casting is the second largest category of robot applications in the United States.

  In retrospect, die casting was an ideal area for the initial use of robots, since this application possesses a number of important characteristics that make it amenable to robotization. Die cast parts are produced in relatively large volumes, but it is also important to keep the time required for equipment changeover to a minimum. The parts are precisely oriented when they are removed from the die casting machine, which makes them suitable for robot handling with standard or slightly modified grippers. Because of the maturity of the die casting process, there are few equipment or product design changes that would necessitate retooling. Furthermore, die casting operations are notoriously hot, dirty, and hazardous, making it a particularly unpleasant environment for human workers. And

TABLE 39.1. U.S. ROBOT POPULATION BY APPLICATION AND INDUSTRY (1982)
(NUMBERS SHOW RANK ORDER OF APPLICATION BY INDUSTRY)

U.S. INDUSTRY

APPLICATION	AUTO	FOUNDRY	LIGHT MFG	ELECTRICAL, ELEC- TRONICS	HEAVY EQUIP- MENT	AERO- SPACE	TOTAL
WELDING	1				1		2200 (35%)
MATERIAL HANDLING		1	1	1			1550 (25%)
MACHING LOADING	2	2			2		1250 (20%)
SPRAY PAINTING, FINISHING	3		2	3	3	1	600 (10%)
ASSEMBLY				2		2	200 (3%)
MACHINING							100 (2%)
OTHER							300 (5%)
TOTAL	2500 (40%)	1250 (20%)	1050 (17%)	700 (11%)	600 (10%)	100 (2%)	6200 (100%)

last, die casting is a relatively competitive industry which benefits from both the cost reduction provided by robots and improved product quality resulting from their consistent performance.

Robots can be used to perform a number of functions in die casting. In simple installations, the robot is used to remove the part from the die and place it on a conveyor. In more sophisticated applications, the robot may perform a number of tasks, including part removal, quenching, trim press loading and unloading, and periodic die maintenance. The robot may also be used for insert placement and, in the case of aluminum die casting, loading the cold-shot chamber. In some cases, a single robot can service two die casting machines. The specific functions performed depend on a number of factors, including casting cycle times, physical layout, and robot speed and type.

Although robots have been installed in a number of die casting plants, a number of important points must be considered in planning applications. Overall layout of the installation must be carefully thought out in those cases where the robot performs more than just part unloading or when the robot services more than one die casting machine. Similarly, the interfacing requirements between the robot and other equipment may become complex. Also, additional sensory inputs may be required to insure that all parts have been removed from the die and that the robot maintains its grip on the sprue.

Optimizing the cost-effectiveness of the installation may also be challenging. Although maximizing throughput of the die casting machine is clearly of primary importance, deciding which type of robot should be used and which functions should be performed by the robot is not so obvious. In some installations it is more appropriate to use other automated techniques for such functions as die lubrication and metal ladling. This would permit the use of a less costly, nonservo robot. In other cases, however, the use of continuous-path, servo robots to perform these functions may more than justify the increased cost.

The benefits to be obtained by using robots in die casting have been well established. It is not uncommon to replace as many as two workers on a shift by one robot, and direct labor cost reductions of as much as 80% have been reported. Also, 20% increases in throughput are possible because of consistent cycle times and the elimination of rest periods and lunch breaks required by humans. Significant increases in product quality have also been obtained. This is primarily due to consistent cycle times and constant die temperatures which result in better-quality parts and less scrap. Net yield increases of 15% have been achieved. Other benefits of using robots in die casting include increased die life, reduced floor space for material handling and storage operations, and a significant decrease in the cost of safety equipment since human operators are not required. The added benefit of removing humans from tedious and unpleasant tasks associated with die casting should not be overlooked.

39.2.3. Foundry Operations[2-7]

Although foundries represent one of the most difficult operating environments for human workers in industry today, the use of robots in this area has been relatively slow in materializing. This is probably primarily due to the diversity of castings typically encountered in most foundries and the relatively low-technology approach usually undertaken in such facilities. These factors notwithstanding, a considerable number of robots have been effectively employed in foundries, and additional installations are expected in the future.

Foundry applications of industrial robots have ranged from ladling of molten metal into molds to final cleaning of castings. Robots have been particularly useful in mold preparation, where they have been used for core handling and for spraying and baking of refractory washes on copes and drags. Robots have also been used for traditional material-handling operations, such as removing castings from shakeout conveyors.

Another major use of robots is beginning to emerge in casting cleaning operations. When the casting is first removed from the mold it is still attached to gates and risers and is likely to have a considerable amount of flash which needs removing. Traditionally, removal of these unwanted appendages has been done manually and is an extremely unpleasant and costly task. Robots have met with some success in grinding flash and chipping and cutting away gates and risers.

Another well-established robot application in foundries is mold making for investment casting operations. In this application, wax pattern *trees* are repeatedly coated with a ceramic slurry and stucco sand to build up the mold shell. As many as 12 coats may be required before the mold has reached its desired size. Following these repeated dipping and drying operations, the mold is heated so that the wax patterns are melted out. The mold is then fired and used for investment casting. The mold is destroyed when the casting is removed. A number of these mold-making operations are totally automated through the use of industrial robots.

The primary reasons for using robots in foundry operations are cost reduction and elimination of unpleasant or hazardous tasks for human workers. Also, improvements in product quality which come about through the consistency of robot operations are also a significant benefit in foundries. This latter benefit is particularly true in investment casting where more uniform molds translate into much higher yields of good castings.

The use of robots in foundries also presents a number of challenges, particularly with respect to the operating environment and interfacing the robot with other equipment. The abrasive dust encountered in foundries may require the use of protective covers to prevent damage to the robot. In mold-making operations for investment casting, interfacing the robot with slurry mixtures, fluidized beds, conveyors, and drying ovens is usually a complex task because of the variation in processing requirements that may be reflected in typical product mixes. And last, implementation of casting cleaning operations may be difficult because of the sensory feedback normally required to deflash castings.

39.2.4. Forging[2, 3, 5-7]

Since 1974 a number of robots have been used in forging operations. These applications have ranged from loading and unloading of forging presses to the movement of workpieces from one die station to another. By far, the largest category of applications in forging is material handling. Robots have been used to load furnaces; move heated billets between furnaces and drop hammers or forging presses; and move forged workpieces from presses to drawing benches, trim presses, conveyors, or pallets. Robots have also been used to apply lubricant to both workpieces and dies.

Robots have been used primarily in closed die forging and heading operations, since these are relatively precise processes. To a lesser extent, they have also been used in drop forging, upset forging, roll forging, and swaging, processes which have a relatively large degree of variability between workpieces.

Because of the harsh environment encountered in forging operations, the primary motivation for using robots is to eliminate unpleasant tasks for human workers, which also reduces direct labor costs. Another major reason for using robots in forging is to improve product quality, which can be accomplished through the robot's consistent operation. Furthermore, increases in throughput of up to 30% have been reported because of the robot's ability to work almost continuously.

Forging applications present unique challenges for robots. Handling of the hot billets may require water-cooled grippers or the periodic cooling of tooling by immersion in water baths. Similarly, the heavy shock loads produced by forging may necessitate the use of special grippers for isolation from these forces. Additional problems are created by time-varying changes in the process resulting from warpage and die wear. And last, it must be remembered that forging is still a relatively hazardous process characterized by extremely high forces and frequent abnormalities, such as workpieces not ejecting properly from dies. To overcome these difficulties, robots used in forging operations typically require a high level of interlocking with other equipment and sensory inputs to insure proper functioning of the process.

39.2.5. Heat Treatment[2, 6]

As in applications such as die casting, foundry operations, and forging, the use of robots in heat treating primarily involves material handling and machine loading and unloading tasks. Robots typically are used to load and unload heat-treating furnaces, salt baths, and washing and drying stations.

The motivations for using robots in heat treating include elimination of unpleasant and hazardous tasks, cost reduction, improved product quality, and increased productivity. Generally, medium- and low-technology robots can be employed, and few unusual difficulties are encountered beyond protecting the robot from the high temperatures normally encountered in heat treating.

39.2.6. Forming and Stamping[2, 3, 5-7]

Press work, such as stamping, forming, and trimming, is another area where robots were applied early in their development. Such applications have ranged from feeding presses for stamping small parts to loading and unloading large presses for forming automotive body panels. Again, the applications are primarily machine loading and unloading and typically involve the use of medium- or low-technology pick-and-place robots.

Robots are usually used in press work applications for two reasons. First, press loading is considered to be a very dangerous task, and robots are used primarily for safety and to minimize hazard to human workers. The second benefit is cost reduction through both the elimination of human labor and increased productivity.

Although robots have been successfully applied in press working, there are a number of cases when their use is not appropriate. Many stamping and forming operations are highly automated, particularly for long production runs, and robots have difficulty competing on an economic basis under these circumstances. Similarly, many press work applications are relatively high-speed operations, and robots are simply incapable of achieving the necessary operating speeds. These factors tend to limit robot applications to those situations where production quantities are moderate and manual techniques predominate, or where low-technology robots and special-purpose robot designs, such as two-armed robots, are appropriate.

39.2.7. Machining[2, 4, 6-8]

As mentioned previously, machining applications for robots are somewhat limited at present. However, robots have been used to perform such tasks as drilling, routing, reaming, cutting, countersinking, broaching, and deburring. Many of these applications have been in the aerospace industry, with the exception of deburring, which has seen more widespread use. Since many machining applications tax the capabilities of today's most sophisticated robots, it is likely to be some time before robots are widely used for this function.

The major difficulty in using robots for machining operations is positioning accuracy. Present robots simply have difficulty achieving the necessary precision and repeatability needed to locate tooling accurately. Because of this, most robot machining operations rely on the extensive use of jigs and fixtures, which quickly erases any advantages the robot provides in terms of flexibility and low cost. Considerable research and development is being conducted in an attempt to improve positioning accuracy by means of sensory feedback. Although some progress has been made in this regard, the use of such sensors is rare, and improved solutions to the problem are still being sought.

In some machining operations additional difficulties arise. For example, the robot's operating speed may not be high enough when compared to other automated and manual techniques. Similarly, in tasks such as deburring programming becomes a major problem because of the number of complex motions that must be executed by the robot.

39.2.8. Plastics Processing[2, 3, 5-7, 9]

In plastics processing the most common application of robots by far has been for unloading injection molding machines. Other applications have included unloading transfer molding presses and structural foam molding machines, handling large compression-molded parts, and loading inserts into molds. Robots have also been used for spray painting and applying a variety of resins, as well as performing many secondary operations such as trimming, drilling, buffing, packaging, and palletizing of finished plastic products.

Estimates indicate that nearly 5% of all injection molding machines are now tended by robots, and in Japan this is one of the most common robot applications. The robot may be mounted on either the top or bottom of the molding machine, or it may be a stand-alone unit servicing more than one machine. Increasingly, special-purpose, low-cost robots are being employed for this function.

The major reason for using robots in plastics processing is to achieve cost savings through both the reduction of human labor and increases in throughput. Additionally, product quality may increase significantly because of more uniform processing cycles and consistent handling of delicate workpieces.

The use of robots in plastics processing is no more difficult than other types of robot applications. The same level of interfacing with other equipment, such as conveyors and trim presses, is still required, and overall equipment layout is important. Gripper design may be a problem, however, because of workpiece size, quantity, and limpness.

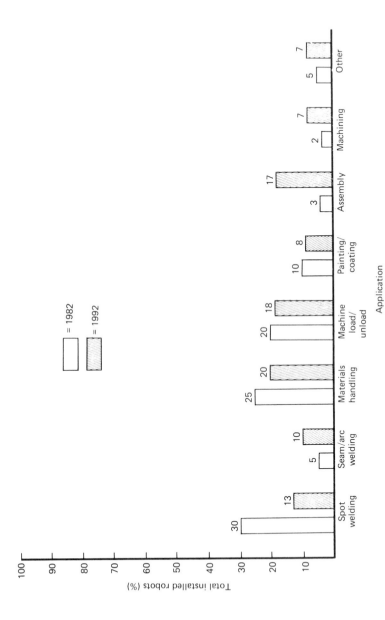

Fig. 39.7. Current and projected U.S. robot usage by application (1982 and 1992).

39.2.9. Other Areas

Robots are being used or considered for use in almost every conceivable industry. A brief sampling of some additional applications which might be considered as part of fabrication and processing is as follows:

1. **Electronics Processing.**   Electronics is starting to emerge as one of the important user industries for robotics. Applications include machine loading and unloading for parts fabrication and presses, component placement for hybrid microcircuit assembly, component insertion for printed circuit board assembly, cable harness fabrication, and robot-assisted test and inspection.

2. **Glassmaking.**   Robots are used in glassmaking because of their ability to withstand high temperatures and handle fragile workpieces, which eliminates unpleasant and hazardous tasks for humans and reduces overall costs. Robots have been used for charging molds with molten glass and for handling both sheet and contoured glass products.

3. **Primary Metals.**   Robots are beginning to be used in the production of primary metals such as steel and aluminum. In addition to traditional material-handling applications, robots have been used for such tasks as charging furnaces with ingots and furnace tapping.

4. **Textiles and Clothing.**   The use of robots in textile and clothing manufacturing presents unique problems because of the limp nature of the workpieces. However, robots are being used for such applications as material-handling in spinning mills and for automatic placement and sewing of clothing items.

5. **Food Processing.**   In addition to material-handling and packaging applications in food processing, robots have also been used for such tasks as decorating chocolates and actual food preparation.

6. **Chemical Processing.**   The chemical processing industry, which is normally thought of as a continuous-flow operation, is starting to use robots for a variety of applications, including material-handling and maintenance activities, such as cleanup of chemical reactors.

39.3. FUTURE TRENDS IN FABRICATION AND PROCESSING

There is almost universal agreement that the use of industrial robots will increase dramatically during the next decade. Anticipated growth, in terms of annual sales increases, is on the order of 40%. At that rate, the U.S.-installed base of industrial robots should reach 100,000 units somewhere in the early 1990s. If each robot displaces two workers, then about 200,000 jobs could be directly affected by robots. However, many new jobs would also be created for such positions as robot programmers, troubleshooters, and maintenance personnel.

Along with this rapid growth in robot sales, a number of important technological and product developments are expected to take place. These include the development of smaller and lighter-weight robots, an increase in payload capacity relative to the weight of the robot, and dramatically improved grippers. In the sensor area, major improvements are expected in machine vision systems, tactile sensing, and low-cost force sensors. Major developments are also anticipated in robot control and programming capabilities, including the use of hierarchical control concepts and off-line programming. At the same time, robot prices are expected to decline as production rates increase.

These trends and anticipated developments can only help to accelerate the use of robots for fabrication and processing applications. Improved capabilities, particularly in control and sensory technology, coupled with declining costs should make many currently difficult applications cost-effective realities in the future.

As robot sales continue to grow, some shift in applications is anticipated. Current and projected robot use by type of application is shown in Figure 39.7.[1] It is anticipated that traditional robot applications such as spot welding, material handling, and painting will decline somewhat in terms of their respective market shares, while other emerging applications, such as assembly and machining, will increase significantly. Since machining, assembly, and similar uses of robots generally require more sophisticated equipment and interfacing, their relative impact on future manufacturing operations will be even more significant.

REFERENCES

1. *Industrial Robots: A Summary and Forecast,* Tech Tran Corporation, Naperville, Illinois, 1983.

2. Engelberger, J. F., *Robotics in Practice,* AMACOM, New York, 1980.

3. Hunt, V. D., *Industrial Robotics Handbook,* Industrial Press, New York, 1983.

4. Hartley, J., *Robots at Work: A Practical Guide for Engineers and Managers,* IFS (Publications), Bedford, England, 1983.

5. Tanner, W. R., *Industrial Robots from A to Z: A Practical Guide to Successful Robot Applications,* The MGI Management Institute, Larchmont, New York, 1983.

6. Tanner, W. R., Ed., *Industrial Robots,* Vols. 1 and 2, Society of Manufacturing Engineers, Dearborn, Michigan, 1981.

7. Warnecke, H. J. and Schraft, R. D., *Industrial Robots: Application Experience,* IFS (Publications), Bedford, England 1982.

8. Molander, T., Routing and Drilling with an Industrial Robot, *Proceedings of the 13th International Symposium on Industrial Robots and Robot 7,* April 1983, Chicago, Illinois, Society of Manufacturing Engineers, Dearborn, Michigan, 1983.

9. Meyer, J. D., Industrial Robots in Plastics Manufacturing—Today and in the Future, *Proceedings of the Regional Technical Conference on Automation, Tooling, and Thermosets,* March 1983, Mississauga, Ontario, Society of Plastics Engineers, Greenwich, Connecticut, 1983.

CHAPTER 40

ROBOTS IN THE AUTOMOBILE INDUSTRY

M. P. KELLY

BL Technology
Cowley, England

MICHAEL E. DUNCAN

Cambridge University
Cambridge, England

40.1. INTRODUCTION

The automobile industry has some well-established needs: productivity, consistency, reliable and untiring operators who are prepared to work at times in hostile environments. These needs can be catered by automation, and, since its inception, the automobile industry has been at the forefront in both developing and implementing automated systems and processes.

There is, however, a further key need, flexibility, that is, the ability to handle product (and production) variants and to react quickly, and with the minimum of investment to facelifts and new models. The robot, because of its programmability, becomes therefore an essential tool in achieving flexible, automated manufacture.

This chapter investigates robotics in the automobile industry by examining certain specific applications, both in terms of positive achievements and of the problems that must be understood and overcome.

40.2. ROBOT TECHNOLOGY

40.2.1. General Observations

Some general remarks are appropriate as an introduction:

Typically, robots are not purpose-built for specific applications; therefore in the majority of cases, existing and readily available robots must be "engineered into" the particular application, which can impose severe constraints. These constraints tend to work against a prime rule for implementing robots—"keep it simple." There may be an evolution into purpose-built programmable equipment.

Currently available robots may be criticized for a low level of self-diagnostics, a lack of perception capability, and widely differing programming systems.

It is difficult to retrofit a robot because it must be interfaced with other facilities. All facility changes, installations, or improvements should be considered as offering an opportunity to install robots in appropriate operations.

Given the state of current robot technology, rules of thumb relevant to robotics applications include: (1) robots prefer working on stationary objects, consequently there is pressure to engineer indexing production lines; (2) avoid destroying component orientation; (3) minimize the number of locations, fixings, and parts; (4) go for commonality and modularity.

Experience has shown that the personnel developing and implementing a process robot need to be supported by an in-depth knowledge of that process (e.g., fusion welding, paint spraying, etc.)

The implementation of robotics for its own sake should be discouraged. There should be a clear rationale for the application and an equally clear rationale in the specification and ultimate selection of a particular robot.

Experience gained in the implementation of robot systems has led to the development of guidelines that provide useful references (e.g., see Teale[1]). See also the list of applications rules at the end of this chapter.

40.2.2. In the Automobile Industry

The theme of rationale brings us to the needs of the automobile industry, which require exploring in a little more detail. This industry is obviously a major robot user; recent figures indicate that more than a third of all robots work in automobile manufacture (see Figure 40.1). Several factors prompt the introduction of robots and robotic systems.

Hostile Operator Equipment

Within the automobile industry there are certain environments that must be considered as hostile, and socially there is an obligation to remove operators from them. These, perhaps, are the priority areas for implementing robotics. Such processes as welding, paint spraying, underseal application, sealing/adhesive bonding, fettling, and handling hot components fall into this category.

Quality

Robots both demand and achieve quality. Also they can be used to assist in the measuring of quality. Consistent panel fit-up is of extreme importance in, for example, fusion welding. To achieve this there is a need for more accurate panels, jigs, and fixings, which in turn will produce in the finished vehicle an improved form as typified by door, trunk (boot), and hood (bonnet) fits, and by the integrity of styling lines and features.

Automated processes will produce improved and consistent quality: paint will be evenly applied, underseal will be directed to the specified areas, adhesive will be continuously applied, arc welds will be of improved quality, resistance welds will be made at specified pitch and location, consistent joint sealing to contain water leaks and seam corrosion will be achieved. Robotics can be used to achieve quality in terms of body preparation for subsequent painting operations, for example, wash-down or blow-down processes.

As an aid to automated inspection, robots also have a role. Examples of this are body scanning and component inspection prior to automatic assembly. (See chapters on inspection in Section 12.)

Productivity

There are clear productivity and economic attractions in using robots as part of an automated system. Equally, the often-stated advantages of mobility and working tirelessly for the whole shift without needing breaks or relief are valid.

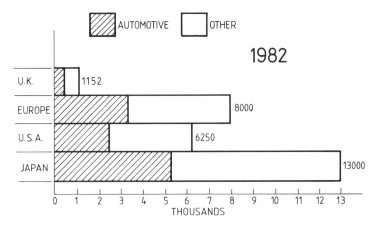

Fig. 40.1. Total and automotive world robot populations in 1982.

Flexibility

This is the prime attraction. As stated previously, the programmable nature of a robot enables it to handle a variety of tasks, which is important in the automobile industry when body variants, or even different models, must be handled on the same line.

Furthermore, once the particular model's life is ended, the robot's life is not. It can be reprogrammed for the next model or relocated to another task. This is a particularly important point to note when preparing an economic justification for robotics.

Management of Automation

Several conclusions may be drawn from experience to date of the successful introduction of robotic systems in the automotive industry. The fundamental lesson already learned in robotic implementation is the need for in-depth experience of the processes and materials to be applied, for example, welding, sealing, bonding. The automotive process engineers have this in abundance. This knowledge, when allied with a relatively short exposure to robotics, creates first-rate applications engineers. These skills must be encouraged and exploited if a positive contribution in robotics implementation is to be achieved.

Equally, management must be made aware of, and sympathetic to, the need for flexible automation. This is not always the case, particularly at middle management levels where robotics can be seen as providing a political platform. Equally, the upper-management stance of legislating robots into production must be avoided. It is clear that robotics must not be looked upon as a discrete technology; some would argue that robotics is not a technology but simply a tool in the application of flexible automation systems. The future of robotics is multidisciplined, involving all aspects of engineering, including the design engineer. Automation in general, and robotics in particular, will become increasingly dependent on the product designer to achieve access, reach, correct materials, accuracy, and so on. To date, the robot applications engineer has managed to live with conventional design criteria, but the inevitable move into assembly will depend upon the designer to the extent that final assembly by robot is almost impossible to envisage without a fundamental rethinking of product design.

While mentioning a multidisciplined approach to robotics, it is worth stressing the essential role of systems engineers in creating totally interfaced systems that are optimized, intercommunicate, and are supported by process control technology, diagnostics, line output data, and the range of software and sensing systems expected in a modern manufacturing unit.

Corporations are realizing that robotics should not be viewed simply against conventional one- or two-year payback periods. Whole-life costing is appropriate, as is consideration of space savings, improved quality, and adaptability. Equally important is the role that robotics will play in protecting the future of the automotive industry. Increasing interest is being shown by many industries in the strategic implications of the implementation of robot and automated systems.[2, 3] Put simply, the future relies upon efficient use of resources—materials, space, plant, people, inventories, money—and robotics has a major contribution to make in meeting the challenges this implies.

Complementing this is the increasing support being made available from governments. This is typified in Japan by support for leasing and loan arrangements, and in the United Kingdom by the DoI robot initiative.[4] In the future, apart from continuing govenment support, there will be an increase in private enterprise support for leasing. There are already signs of this happening in the United States.

The rest of Section 40.2 discusses specific automotive processes in which robots play an important role. Additional details about these applications can be found in the respective chapters of the handbook.

40.2.3. Resistance Welding

Spot, or resistance, welding is the most widespread application of robotics in the automobile industry. The principle requirements are accuracy (± 1.0 mm) and a good weight-handling capability. Both electric and hydraulic powered robots are used, although the majority are hydraulic, and they can be either floor or gantry mounted.

The spot welding process involves squeezing the parts to be joined between two electrodes which produce a 6-mm-diameter "spot" joint. It is a fast, reliable process capable of manual or automated operation. However, the welding gun is heavy and cumbersome, and there are often access problems. The process itself is unpleasant, being noisy and potentially dangerous. Automation not only relieves the operator of a difficult job but also achieves reliable spot spacing and a faster work rate. On the other hand, automation requires accurate panels, subassemblies, and locations.

The trend is toward systems that incorporate a matched set of different robots ranging from three- to six-axis machines. One example of such a system, shown in Figure 40.2, is employed in the assembly of the British Leyland Metro. In the United Kingdom, both Ford and British Leyland (BL) have robot lines for spot welding. While BL uses Unimation equipment in this application, Ford combines

Fig. 40.2. Resistance welding line for the BL Metro.

both Nimak and KUKA robots in lines of 26 machines.[5] To date, spot welding offers the best examples of multirobot systems in the automotive industry.

Anticipated developments in machine specifications include higher wrist load and torque capabilities, increased use of combined transformer/gun units, and a reduction in the cable clutter on the arm. Software development will lead to the solution of such production problems as guns sticking to the component, automatic electrode cleaning, and improved (adaptable) process control techniques. The requirements of a spot welding robot are summarized in Table 40.1.

40.2.4. Arc Welding

Arc, or fusion, welding is considered a major growth area for the application of robotics. The process is very hostile to the operator, generating noise, fumes, and intense light (flash); automation produces high-quality welds with greater consistency and at a faster rate. The robot equipment is typically a five-axis electric powered machine with continuous path and linear interpolation capability. Accuracy of ± 0.2 mm is required.

The process involves feeding a consumable wire through a torch which provides a protective envelope of shielding gas (CO_2). The wire arcs onto the workpiece and melts into a "weld" pool. The process is applied to automobile subassemblies mainly for reasons of strength, low distortion, no slag, high speed, applications where one-sided access only is required, and sealing.

The quality of this process is very operator-dependent, and the robot provides a substitute for the increasingly scarce manual arc welding skills. However, a major problem in robot application relates to panel accuracy and fit-up. This has stimulated considerable research and development work in joint recognition and process adaptive control. An example of such work is the system developed jointly by BL, the Oxford University, and others[6] which uses a solid-state camera to track down the seam which is illuminated by laser. The system is shown in Figure 40.3.

There is some discussion as to whether the complex systems that would evolve from such sophisticated technology might not be out of place in an industrial environment. Meanwhile, there can be no reduction in the efforts to achieve accurate and consistent panel fit by designers and tool makers. The requirements of an arc welding robot are summarized in Table 40.2.

TABLE 40.1.   FEATURES OF RESISTANCE WELDING SYSTEMS

Mechanical and dynamic characteristics	Adequate weight handling capacity—high-load machines Heavy-duty wrist assembly Accuracy (± 1.0 mm) Fast travel for interweld movements
Programmability and control	Tool center point interpolation Interface capability with welding controls Consistent attitude of welding gun to workpiece—less panel damage
System characteristics	Ability to handle awkward guns Integration of robots with conventional multiwelders Integrated, compact welding guns
System requirements	Good access to vehicle interiors to allow gun and robot arm

40.2.5.   Adhesive Bonding

There are a number of instances where consistent application of adhesive is critical for quality reasons. These include containment of leaks or corrosion from joints, achievement of an effective bond, and the like. The use of robotics is fairly novel, but the increased use of adhesives in vehicle body assembly has highlighted the need for automation.

Adhesives are used where, for instance, finish requirements, base materials, or lack of access prevents welding, where there is a requirement for a combined join and seal, load distribution in the join, increased torsional stiffness, sound deadening, or where assembly volumes are low. The adhesive is extruded as a bead or as spots (see Figure 40.4). Typical applications include the build of items such as hood and trunk lids, doors, and the like.

Automation offers consistent performance and releases the operator from a tedious job. Moreover, the industry is moving toward the use of epoxy-based adhesives which carry a potential dermatitis risk. Developments necessary to promote the use of robotics in the application of adhesives include

Fig. 40.3.   Fusion welding using a torch with seam-tracking capability.

TABLE 40.2.   FEATURES OF FUSION WELDING SYSTEMS

Mechanical and dynamic characteristics	Accuracy (±0.2 mm)
Programmability and control	Simple editing and programming Ability to weave, to alter the weave pattern and plane Continuous-path capability Tool center point interpolation Interface capability with welding control parameters and ability to select various programs
System characteristics	Automatic nozzle cleaning Sensing devices/feedback covering wire feed, gas flow, and water flow
System requirements	Good panel fits (essential) Full system approach Good housekeeping

compliant nozzles, joint-following capability, programmable material-flow control, and interchangeable or multifunction heads. It will also be necessary to impose strict material quality control to contain blockages. The requirements of a robot for the application of adhesives are shown in Table 40.3.

40.2.6.   Painting

The use of robots to paint panel sets is fairly well established, although further work is required to integrate the robot within a system including color changes, body identification, material parameter control, and so on. Also paint spraying of automotive bodies is already automated to some extent by the use of conventional roof- and side-coating machines. However, such machines do not coat inside the hood and trunk or internal shut areas. A fully automated spray booth will be achieved when robots can be exploited to paint such internal areas. Two approaches are possible, the use of an automatic door opening device or to paint the body minus doors and other opening panels, these to be painted separately and added to the body later. Whichever is chosen, the fully automated spray booth is inevitable with substantial benefits in improving quality, energy savings (booth extraction), the ability to use paint with potentially harmful constituents, and so on. Some imagination may be necessary in the engineering of installations to achieve the required access as shown in Figure 40.5.

The control of color changes is a potential problem area, and initial developments may be better

Fig. 40.4.   Adhesive application, extruded as spots.

TABLE 40.3. FEATURES OF ADHESIVE APPLICATION SYSTEMS

Mechanical and dynamic characteristics	Accuracy
	Good wrist articulation
	Fast action
	Comparatively low weight capacity (of spot welding)
	Adequate reach without the need for long wrist extensions
Programmability and control	Continuous path capability when applying continuous beads
	Tool center point interpolation
	Control of applicator attitude to panel
	Interface with associated equipment
System characteristics	Adhesive flow control
	Sensing devices/feedback on adhesive flow
System requirements	Accurate parts and accurate parts location

made in a single color area, for example, undercoats. However, technology is becoming available to monitor and control paint parameters automatically. Full exploitation of this will enable adaptive control of the automatic spray booth to match changing circumstances. The requirements for a robotic paint spraying system are summarized in Table 40.4.

40.2.7. Underseal

Correct application of underseal is important if it is to be effective against corrosion. However, the working environment is extremely unpleasant, and consequently it represents an attractive robot application.

Typically the requirement is for a five- or six-axis hydraulic robot capable of continuous-path operation and accuracy of ±1.0 mm. Apparently a straightforward implementation exercise (see Figure 40.6), experience has proved otherwise. Problems have arisen with accessibility line stoppages, material blockages, body stability, and programming. Apart from protecting the operator from a hostile environment, the benefits include improved quality and material savings. Table 40.5 presents a summary of the requirements of a robotic system for undersealing.

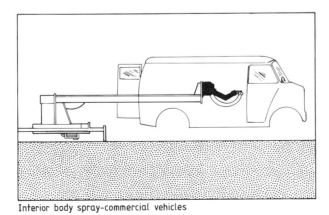

Interior body spray-commercial vehicles

Fig. 40.5. Installation to spray paint inside van bodies. (After GEC, UK illustration).

TABLE 40.4. FEATURES OF PAINT-SPRAYING SYSTEMS

Mechanical and dynamic characteristics	Multiaxis manipulator, 6–7 axis—preferably flexiarm for dexterity
	Hydraulic or pneumatic motive power—safety requirements
	Accuracy/repeatability to better than ±3.0 mm
	Small physical size with a large working envelope
Programmability and control	Numerical control (NC) off-line programming capability
	Program editing
	Diagnostic capability
	Interface with associated equipment and supervisory controls
System characteristics	Gun control capability flow sensors
	Line synchronization capability plus a negative response to track the conveyor when reversed
	Rail system—seventh axis—to allow robot to complete the work cycle if the conveyor stops
System requirement	Consistent conveyor speeds with no irregular movement
	Accurate presentation of the body
	Identification system to enable processing of several models

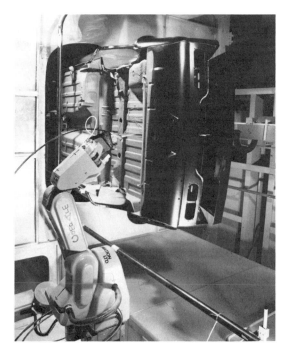

Fig. 40.6. Underseal application.

TABLE 40.5. FEATURES OF UNDERSEAL APPLICATION SYSTEMS

Mechanical and dynamic characteristics	Multiaxis manipulator, 6–7 axis; preferably with flexiarm for dexterity
	Hydraulic or pneumatic motive power—safety requirement
	Accuracy/repeatability better than ±3.0 mm
	Small physical size with large working envelope
Programmability and control	NC off-line programming capability
	Program editing
	Diagnostic capability
System characteristics	Gun control capability
	Flow sensors
	Line synchronization capability, plus negative response to track conveyor when reversed
	Rail system to enable robot to complete work cycle if line is stopped
	Tip blockage detection/correction
System requirements	Consistent conveyor speeds with no irregular movements
	Accurate body presentation, sway within ±5.0 mm
	Static charge grounding of body, grounding of robot and gun
	Controlled airflow in booth to deflect overspray

40.2.8. Seam Sealing

Automotive design requires substantial seam sealing to prevent leaks and noise transmission. The actual process of seam sealing is labor intensive, unpleasant, and an obvious quality-critical application. It is clearly an ideal, if challenging, robot opportunity. It is safe to predict that considerable resources will be devoted to this application over the immediate future.

The major problems will be accessing joints, particularly on moving conveyors. Subsidiary problems remain to be overcome in terms of precise joint location, interference of cables and fluid lines in restricted access areas, and perhaps, above all else, the application method. The method will need to be selected from conventional spray, airless spray, extrusion, or some combination of these. To support the application development, material development will be necessary to provide pumpability, gap filling, and freedom from blockages.

40.2.9. Handling and Assembly

In parallel with arc welding, handling and assembly can be predicted as the major growth area for automotive robotics. Handling activities will range from simple pick-and-place tasks utilizing low-cost three-axis robots to more complex machine loading or transfer tasks. It is unlikely that this level of technology will be particularly demanding in terms of part or component accuracy, gripper design, or perception. The challenge is simply to identify the application, select the correct equipment, and implement.

Component assembly is clearly much more demanding, but achievable by exploiting the application engineers' ingenuity. The final system will be engineered around a purpose-built robot-handling device (possibly multiarmed), material feeders, assembly jigging, and component selection. The element of component selection raises the fundamental point of component orientation; rather than developing vision systems to compensate for lack of orientation, it is obviously more sensible not to destroy orientation in the first place.

The introduction of robotics into the final assembly operation is unlikely in relation to current automotive design. Isolated tasks may be achievable, but the majority of final assembly tasks are complex and extremely demanding of the operators' manual dexterity in reaching awkward areas, handling a variety of materials and components, and progressively building up complicated assemblies. Door and fascia design are good examples of the requirement for large numbers of components that must be assembled prior to the complete assembly being fixed to the car body. In this application the product designer and automation engineer must work together.

40.2.10. Other

Reference has been made to the quality implications of the introduction of robots. Going full circle, robots themselves have been applied to quality-control tasks, including at BL, leak testing. This application,[7] illustrated in Figure 40.7, involves the injection of 0.5 liter of helium gas into the car body under slight pressure. Robots, programmed to follow a given path relative to the car body irrespective of the track speed, manipulate a "sniffer" over some 60 m of the car body. Windows, doors, trunk lid, and seams are covered, yielding 400 items of information on the location and intensity of leaks. An interesting application—it was necessary to reengineer the track to achieve sufficiently accurate location of the car body.

40.3. THE FUTURE

The robot population of BL in 1982 and its distribution is shown in Table 40.6. In the same year Ford (U.K.) employed some 220 of such machines. The future will see a dramatic increase in the number of robots used in the automotive industry, and an estimate of this growth for several major companies is shown in Figure 40.8. Well-established applications, such as resistance welding, will continue to grow in the short term. Longer term, the much wider use of structural adhesives will supplant the resistance welding process with robots applying the adhesives.

Practical perception systems will enhance robot performance in arc welding, grinding, fettling, seam sealing, and assembly operations, leading again to robot growth as vital elements of truly flexible manufacturing systems (FMS). A major robotic impact will be made in automotive paint shops, as the need to conserve energy increases. The development of alternative painting materials offering improved performance will add further impetus.

Robotics of the future will progressively move to a CAD/CAM orientated data base, offering off-

Fig. 40.7. Robot sniffer searching for leaks.

TABLE 40.6. ROBOT APPLICATIONS IN BL, 1982

Application	Number	Type
Resistance welding	36	Unimate
Fusion welding	18	ASEA
Adhesive bonding[a]	3	Unimate/ASEA
Paint spraying[a]	2	GEC
Underseal application[a]	4	Trallfa/GEC
Inspection[a]	2	Trallfa

[a] Indicates rapid growth areas.

line programming capability, which, together with essential inspection elements, will provide the means for totally automatic manufacture.

The possibility exists through the development of advanced high-level languages and the use of artificial intelligence of creating totally intelligent robots. These will be capable of self-teach, will have an awareness of their surroundings, and the possibility of mobility. Equally, the future may see the development of *expert* and *supervisory* robots, as described in Chapter 23, Expert Systems and Robotics.

Major growth can also be expected in handling and assembly tasks. This will range from simple machine-loading tasks to complex assembly of components. Final assembly will continue to remain a major challenge, and successful implementation will depend on a fundamental reappraisal of current design and manufacturing practices.

40.4. APPENDICES

40.4.1. Implementation Rules

As a generalization, these rules apply to all aspects of technology transfer.

1. Do not implement on an ad hoc basis—have a plan and see each implementation as part of that plan—this will avoid incompatible systems, a multitude of robot types and all that means with regard to spares, training, programming, and so forth.
2. Involve from the start all levels of management and the labor force: do not label projects "confidential."

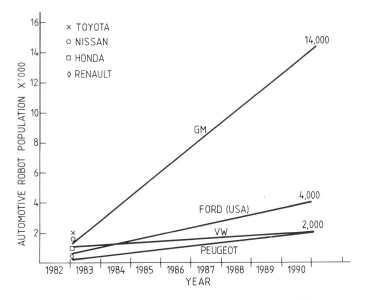

Fig. 40.8. Trends in automotive robot population to 1990.

3. A "protechnology" corporate attitude is essential, including clear policies and programs.
4. An initial feasibility study is important:
   (a) Is the application suitable?
   (b) Analyze product and process specifications.
   (c) System proposal—layouts, costs, processes.
   (d) Robot selection.
   (e) Process and associated equipment.
   (f) Maintenance.
   (g) Safety.
   (h) Systems.
   (i) Turnkey suppliers.
5. Financial justification should include:
   (a) Conventional payback criteria.
   (b) Whole-life robot costing.
   (c) Adaptability.
   (d) Space savings.
   (e) Quality improvements.
   (f) Materials and work-in-progress savings.
   (g) Available grants.
6. Provide training.
7. Create interdisciplinary project teams—including the recipient (make the recipient project leader). "Recipient commitment is vital."
8. Aim for user-friendly systems.
9. A knowledge of robots is not enough, in-depth knowledge of process application is needed.
10. Full system approach is a must.
11. Retrofit is difficult because of constraints of:
    (a) Product.
    (b) Facility.
    (c) Tooling.
12. Understand the limits of robots:
    (a) Deaf
    (b) Dumb
    (c) No smell
    (d) No feel
    (e) No vision

    Beware the misleading analogy with the human hand. "Robots are dumb, the perfect machine in an imperfect world."
13. Involve the product designer.
14. The first time be simple.
15. Do not destroy order or orientation.
16. To install and commission count on six man-months per cell, approximately.
17. Finally:

COMMUNICATE

BE PATIENT

FAILURE IS A MANAGEMENT FAILURE

40.4.2. Things That Can Go Wrong

Some examples:

Implementing technology for the wrong reasons.
Insufficient attention to detail.
Cuts in budgets.
Technical specialists leave.
Product design changes during commissioning.
Projects that end up as robot laboratories or zoos.
Project targets change.

Supplier's equipment does not perform as expected/specified.

Insufficient development effort.

Layout/process shortfalls.

Software bugs.

Lack of continuity in project management.

Suppliers go bankrupt.

Lack of turnkey, total system approach.

REFERENCES

1. Teale, D., Considerations for Implementing Robots, *Proceedings of the 4th British Robot Association Annual Conference,* 1981, pp. 171–178.
2. Stauffer, R. N., Equipment Acquisition for the Automatic Factory, *Robotics Today,* April 1983, pp. 37–40.
3. Van Blois, J. P., Strategic Robot Justification: a Fresh Approach, *Robotics Today,* April 1983, pp. 44, 45, and 48.
4. *Government Support for Industrial Robots,* Department of Industry, United Kingdom, 1982.
5. No End in Sight for Ford's Growing Family of Robots, *The Industrial Robot,* December 1982, pp. 222–227.
6. Clocksin, W. F., Davey, P. G., Morgan, C. G., and Vidler, A. R., Progress in Visual Feedback for Robot Arc Welding of Thin Sheet Steel, *Proceedings of the 2nd International Conference on Robot Vision and Sensory Control,* 1982, pp. 189–200.
7. Robots to Snuff Out Those Water Leaks, *The Industrial Robot,* September 1982, pp. 150–152.

BIBLIOGRAPHY

The following papers are considered to be relevant and useful follow-up reading:

Engelberger, J. F., Robots and automobiles: applications, economics and the future, Society of Automotive Engineers, Paper No. 800377, March 1980.

Kelly, M. P., Robots for the Automotive Industry, *Proceedings of Automan '81,* 1981.

Kelly, M. P., The Future of Robots in the Automotive Industry, *ISATA,* Wolfsburg, September 1982.

Kelly, M. P., Robots in the Paint Shop, *Proceedings of Automan '83,* 1983.

CHAPTER 41
ROBOT APPLICATIONS IN AEROSPACE MANUFACTURING

WALLACE D. DREYFOOS

PAUL F. STREGEVSKY

Lockheed-Georgia Company
Marietta, Georgia

41.1. THE ROLE FOR ROBOTS IN AEROSPACE MANUFACTURING

Most of the shop-floor tasks in an aerospace factory are labor intensive. To curb rising labor costs, the industry has turned to automation. Fixed automation has been cost-effective for high-turnout tasks. But many tasks remain in which the parts worked on are few or complex, but the task itself is comparatively simple. For these tasks, fixed automation may not be feasible, available, or affordable. Instead they are being automated by programmable robots.

Table 41.1 identifies aerospace factory tasks that have been robotized, generally in production runs. For most tasks shown, companies not cited were likewise using or investigating robots. The chief payoff is a cost saving, usually attained by robotizing a procedure that used large sums of labor, for example, drilling, assembly, layup, deburring, sanding, inspection, and wire-harness routing. For other tasks, the robot pares costs by reducing scrap and costly rework. Machining and drilling benefit accordingly, the more so because of the high cost of aerospace materials.

Many of the applications are more task-specific than aerospace-specific, and accordingly are discussed under their respective headings elsewhere in this handbook.

The most extensive use for aerospace robots has been, and will likely continue to be, drilling and riveting. Each year, billions of holes are drilled into sheet-metal parts for aircraft assembly. Through these holes fasteners are then installed. Tolerances are demandingly close. For example, holes in interchangeable panels may have to be round to within +0.075 mm (+0.003 in.)/−0.000. For standard sheet-metal assembly, rivet spacing and location must be repeatable to ±0.76 mm (±0.030 in.) or, for some assemblies, closer.

With few exceptions, meeting these tolerances has been the task of human hands and eyes, guided only by hard or fixed tooling such as templates. The job is tedious, and as a worker fatigues his/her faculties weaken, fostering slowdown and faulty workmanship. The demanding tolerances, awkward quarters, and monotonous repetition of aerospace drilling and riveting make these chores ill-suited to manual labor. Indeed, it has been found that by using a robot to position the drill to each location, productivity is improved by as much as sixfold.[1]

Composite layup, which combines features of materials transfer and assembly, is a labor-intensive task for which aerospace is likewise turning to robots. Northrop has demonstrated an order-of-magnitude saving in cost and time by using a robot to vacuum-lift precut resin-impregnated graphite fabric plies, lay them up into tapered horizontal-stabilizer spars, and transfer the stack of plies to a second work table.[2] The heavy layups sag during transfer; to maintain adequate vacuum against them, the robot uses a compliant foam-faced end effector measuring 0.9 × 1.5 m (3 × 5 ft).

Spray painting of large aircraft assemblies can be hazardous to painters who must work within cavities, such as wheel wells, where they may ingest airborne paint. A versatile cost-saving alternative has been developed by Fairchild-Republic: a mobile paint-spraying robot.[3] The robot, made largely of aluminum, rides on air bearings along a 7.7-m (30-ft) rail, painting fuselage sections and other large parts by following a programmed path. The entire railed system can be moved from one spray booth to another.

Aerospace welding has not been extensively robotized. One reason is that early industrial robots lacked the actuator power needed to provide the high-speed head movement, high welding pressure,

TABLE 41.1.   REPRESENTATIVE ROBOTIZED TASKS IN AEROSPACE MANUFACTURING[a]

Kind of Task	Company	End Product	Description
Joining/Assembly			
Installing small parts	Boeing Aerospace Co.	Cruise missiles	Fastener feeding; installing cover fasteners
Welding	General Electric Company	Gas turbine engines	Fan frame hubs
Riveting	Fairchild-Republic Co.	A-10	Horizontal stabilizer
	Lockheed-Georgia Company	C-130	Bulkheads for floors and wings. Performed with stationary drill/riveter.
	Martin Marietta	Space shuttle	Portable drill/riveter is manipulated by robot arm.
Metal Working			
Chamfering	Pratt & Whitney Aircraft	Gas turbine engines	Vane slots in compressor shrouds
Deburring	Avco Aerostructures Div.	British Aerospace 146	Wing panels Panels up to 18×3 m (60×10 ft) possible using rail-mounted robot.
Drilling	General Dynamics, Ft. Worth Div.	F-16	Pilot holes in graphite/epoxy skin panels
	McDonnell Aircraft Co.	F-18	Canopies and windshield frames. Uses vision and five-axis DNC.
	Grumman Aerospace Corp.	F-14 and EA-6B	Fuselage panels. Rail-mounted robot serves four work stations and needs no templates to guide drill.
Machining	Martin Marietta Aerospace	Space shuttle	Removes 3.6 cm/sec (1.5 in./sec) from external tank ablator-covered parts. Uses tactile sensors to follow aerodynamic contours.
Routing	Grumman Aerospace Corp.	F-14 and EA-6B	Various sheet-metal parts. Rides along 6.1-m (20-ft) rails to serve four work stations.
Sanding	Boeing Aerospace Co.	Cruise missiles	Removes machine marks from wings.
Trimming	Boeing Aerospace Co.	Cruise missiles	Nose caps

TABLE 41.1. (*continued*)

Kind of Task	Company	End Product	Description
Material Manipulation			
Composite layup	Northrop Corp.	F-5 and F-18	Vacuum lifting and stacking of precut plies.
Electronic circuit board assembly	Westinghouse	Avionics	Installs microcircuits onto circuit boards, using two TV cameras.
Dipping	Pratt & Whitney Aircraft	Gas turbine engines	Dips wax patterns into ceramic slurries for investment casting.
Loading	General Electric Company	Gas turbine engines	Titanium slugs
Wire-harness routing	(Several)	Various aircraft	Gantry-mounted robots explored for electrical cables.
Applying Agents			
Applying sealant	Avco Aerostructures Div.	Fuel cells	Mixes and dispenses, allowing quality inspection at point of use.
Brazing	General Electric Co.	Gas turbine engines	Applies braze alloy to turbine nozzle supports.
Flame spraying	Garrett Turbine Engine Co.	Gas turbine engines	Metallic coatings
Plasma spraying	Pratt & Whitney Aircraft	Gas turbine engines	Ceramic and Zirconate protective coatings
Spray painting	Fairchild-Republic Co.	A-10	Mobile, rail-mounted robot; paints forward fuselage, nacelles, landing-gear tub, and other assemblies.
Other			
Inspection	General Electric Company	Gas turbine engines	Uses GE-developed light sensor to inspect 500 blades and vanes per hour.
	Pratt & Whitney Aircraft	Gas turbine engines	Positions turbine blades for radiographic inspection of cooling passages.

[a] Companies not shown are also using or investigating these robotic applications.

and high forge pressure required when spot welding crack-sensitive aerospace alloys. A second reason is that large aerospace parts often have complex curvatures that must be welded in six axes. Two robots would be needed—one for welding, the other for either welding or part rotation. Coordinating them can be difficult.

In place of robot welders, an automated multiaxis welder has been developed at the Lockheed-Georgia Company. The welder can provide cost-effective, high-speed joining of limited-quantity lots of aircraft-alloy parts. An existing five-axis multihead turret welder has been united with a three-axis positioner. The entire system is controlled and integrated by a microprocessor.

41.2. MAJOR ROADBLOCKS AND THEIR SOLUTIONS

Despite their increasing role in automotive assembly lines, robots have been slow to make inroads in aerospace manufacturing. Not until 1979 were production aircraft panels being drilled by robot; not until 1982 were production drilling and riveting robotized as a combined operation. Three chief roadblocks have retarded widespread acceptance: inadequate repeatability, lack of off-line programmability, and cost inefficiencies when using robots to produce small lots.

41.2.1. Inadequate Repeatability

Until the early 1980s, most industrial robots lacked the close repeatability demanded by aerospace tolerances. If a robot had a reach large enough to perform aerospace fabrication—1.2–2.7 m (4–9 ft)—it could lose the rigidity essential for close-tolerance work. This is a serious drawback, because many aerospace tasks require that the robot's arm be fully extended, making it prone to rapid and harsh vibration. Moreover, at long extensions the increased bending moment makes accurate programming difficult, especially when large or heavy components must be handled.

For more precise repeatability, electric robots may be preferred for their servomotors. These motors can be better suited for small translations, as in hole-to-hole drilling, because they are free of the initial surge of many large hydraulic and pneumatic actuators. But until 1982, few large or heavy-duty robots were electric; most were hydraulic.

A hydraulic robot may lose its nominal accuracy when it performs intermittent batchwork common to aerospace. Whenever the robot has been shut down for any length of time, a fast warmup exercise should be run to stabilize the hydraulic components' temperature to prevent program positioning errors.[4] Alternatively, these components can be kept heated when not in use.

For robotic drilling, the robot may be more rigid, and hence accurate, if it holds a lightweight drill rather than a heavy part. One problem caused by lack of rigidity at the end of a robot's arm is the shape of the holes drilled by robots in some tough materials such as graphite/epoxy composite. If an end effector fails to hold a drill steadily perpendicular to the material's surface, the hole becomes bulged or oblong.

In aerospace assembly, the industrial robot must satisfy conflicting demands: it must have powerful actuators to move heavy parts, yet must move subtly from hole to hole. Several techniques on the aerospace floor make large-payload assembly robots more precise and repeatable for close work. One easy technique is to program the robot to pause briefly between movements, thereby allowing its arm to settle.

Several robot manufacturers offer smaller hydraulic actuators that, by reducing speed, can double repeatability.[5] Performance remains high because even at lower speeds most of a drilling robot's time is occupied in the stationary task of drilling, not in translating from hole to hole.

Robots that must lift heavy loads held at long reaches must have extra rigidity built into them. Repeatability of ±0.13 mm (±0.005 in.) is attainable from robots with a reach of 2.5 m (8.2 ft) and a payload of 68 kg (150 lb). Their rigidity comes from features like a cast-iron base, precision ball screws on all their axial drives, and ground, hardened spiral bevel gears in the wrist.

Rigidity can likewise be enhanced in the aerospace factory itself. First, the end effector must be designed to hold the workpiece or tool snugly. Second, brakes can be fitted to the wrist and base, the least stiff axes. The brakes lock out these axes except when they are needed. Using this technique, Grumman Aerospace Corp. eliminated the costly need for templates and bushings ordinarily required to guide hand-held drills and routers.[6]

Third, a stiffer, moderate-sized robot mounted on rails or tracks can be used to vastly exceed the reach of a larger robot. It can also be more productive by serving multiple work stations.[3,6] The rails can be mounted overhead or on the floor.

A compliant end-effector can be used with hard tooling to guide a robotically held drill. Accuracy becomes sufficient for assembling interchangeable panels.[4] This technique eliminates the need for complex feedback but adds costs in tooling and maintenance.

Feedback systems—tactile or visual sensors, mounted on the end-effector—can improve accuracy. Using a sensory feedback loop developed at one aerospace firm, a robot with a quoted accuracy of ±1.3 mm (±0.050 in.) has performed at ±0.13 mm (±0.005 in.).[7] Early two- and three-dimensional vision systems often were not rugged enough to operate in high-vibration and dirty environments.

These environmental shortcomings have largely been corrected. As discussed in Chapter 17, Depth Perception for Robots, three-dimensional vision has become a promising technology for robotic tasks requiring depth perception. Several approaches are under way that are improving resolution, computing speed, low-light capability, and cost.

In 1982 the Air Force Materials Laboratory began funding a program to develop a sensory-based, computer-controlled three-dimensional vision module that would generate robot control commands six times more precise than those driving robots previously.[5,8] The arm-mounted sensor would monitor the location of a robot's end-effector in relation to its target on a workpiece and issue corrective commands to the robot's controller. The goal was a positioning accuracy of ±0.13 mm (±0.005 in.) and a calculation cycle of 100 msec. Accuracy would no longer depend on a robot's load, extension, and arm peculiarities, or on the precise alignment of parts and machines.

41.2.2. Lack of Off-Line Programmability

Widespread aerospace use of robots has been further hampered by the difficulty in programming robots from off-line data bases such as CAD/CAM. Most robots have had to be walked through their paces for each task. This can take weeks, during which time the robot is unavailable for production and workers are idled. For small aerospace lots, new software developed by teach-and-show becomes prohibitively expensive. Limited to its originally programmed task, a robot can be difficult to justify.

Robots produced since 1982 have begun to incorporate features that will eventually allow easy off-line programmability. These machines will be especially useful in flexible batchwork centers, where batches of different parts could be processed in turn by a single full-time robot.

The prerequisite for effective off-line programming is a good CAD/CAM system and the means to tie it into the robots. CAD/CAM can thus integrate robots and vision systems into the total manufacturing process. Special graphic-display systems can greatly ease software development by letting the programmer visualize the steps he/she is commanding. With the development of better software programs, graphic simulation is becoming ever more sophisticated and affordable.

Off-line programming will become easier and universal once the aerospace industry selects a standard robotic programming language expressly for use with CAD/CAM. One attempt at such a language, named MCL or Manufacturing Control Language, has been developed from APT by McDonnell Douglas Corporation for the U.S. Air Force. (See also Chapter 18, Elements of Industrial Robot Software.) The computer code has been made available in written form or on tape from McDonnell Douglas Automation Company. However, MCL's adoption by the industry in its original form is uncertain because it was not written to be compatible with all the robots and design information formats used throughout aerospace.

In 1982 the Air Force began funding development to define the manufacturing requirements for off-line programming, compare these requirements to MCL, and enhance or rewrite MCL to better meet these requirements.[9] A parallel effort was funded to use the improved version to demonstrate off-line programming and other advanced robotic technologies.[10]

In the absence of a standard language, aerospace has increasingly pressed robot manufacturers to furnish the means to develop robotic software off-line. Manufacturers have responded by developing products that are compatible with aerospace computer programming. Among these are interface modules that link robots to particular computers and graphic simulators. By 1983 one such module could provide a direct interface with MCL, the Gerber Manufacturing machine tool code, and a number of popular CAD systems. Such versatility can be important because often the CAD data points must be translated into two separate languages—one for the robots and one for the numerically controlled machines that they operate.

41.2.3. Overcoming Cost Inefficiencies of Batch Production

In aerospace, complex metal parts generally are produced in batch lots of 10–50 units.[11] Spare-part runs can be as small as five units. Products made this way may cost anywhere from 10–30 times more than they would if made in a continuous process, or mass production.[12] But given its low production rates, aerospace cannot use high-volume production methods. If a robot processes no more than one or two different aerospace parts, it will likely be idle much of the time. Without careful planning, robotic batchwork can yield a poor return on investment.

For their initial robot applications, many aerospace firms settle for a low return on investment to gain production experience with robots. Thorough planning, however, can ensure that any robotic application will be as productive as possible.

Several steps can be taken beforehand to assess whether a robot can perform cost-effectively in a low- to medium-volume or batchwork environment. Annual volume, batch size, and task complexity must be considered, together with setup time, setup cost, and downtime.[13] Following this, a more detailed layout of the programmable system should be developed and a new task sequence generated. If the application is still feasible technically and economically, the return on investment is ready to be calculated by traditional methods.[14]

Because aerospace parts are produced in modest quantities, it is rarely cost-effective to redesign or retool the product for easier robotic assembly. Aerospace tooling and machinery are costly, and each aerospace part requires an average of 3½ tools, from fabrication through assembly. Therefore existing tooling, machines, and product designs must be retained.

A mix of human, programmable robot, and fixed automation (such as parts loaders or automatic drill/riveters) can improve the system cost-effectiveness. Strong emphasis must be devoted to the human/machine interface. The manufacturing planner should have a clear understanding of which tasks are to be the robot's and which the human attendant's. Tasks whose demands cause frequent human error and costly scrappage are prime candidates for robots. Humans can best fulfill the tasks requiring judgment, intelligence, or dexterity beyond the easy reach of current-day robots. These tasks may include complex assembly, seeking and finding, performing certain stop/go decisions, and adapting to new physical conditions. (See also Chapter 32, Human Factors in Planning Robotic Systems.)

Every effort should be made to keep the robot busy and productive. If a batchwork robot is employed only part-time, idleness can easily neutralize its higher productivity. By processing two or more different parts in batch series, a single robot may become cost-effective where it otherwise would not be. Programmable or universal fixtures and end-effectors are desirable to minimize changeover time and tooling cost for different parts.

In planning which parts will be processed in the same robotic work cell, group technology should be applied to ensure that different parts are of the same "part family." For the multiapplication robot, substantial savings accrue. By grouping similar parts into part families, based on either their geometric shapes or operation processes, it is possible to reduce costs through the use of fewer robots, more effective design data retrieval, and a reduction in tooling and setup times, in-process inventory, and total throughput time.[15]

Three-dimensional vision will go far to permit robots to cope with unstructured environments such as multipurpose work cells. With three-dimensional depth perception, robots can avoid costly assembly errors, search for out-of-place parts, distinguish between similar parts, and correct for positioning discrepancies. As three-dimensional technology matures, it promises to make robots easier to program and use for small or serial batches, and even for mobile use around the factory. The Air Force has promoted three-dimensional intelligent robots and has sponsored their development for the aerospace factory of the future.[16, 17]

41.3. ROBOTIC ASSEMBLY AT LOCKHEED

In 1982, the Lockheed-Georgia Company became the first aerospace manufacturer to robotize the assembly of sheet-metal parts. The pilot cell was called the Basic Robotic Riveting Cell. It teamed a robot and its controller with a stationary automatic drill/riveter to drill and rivet webs for cargo underfloor bulkheads of the C-130 Hercules transport. A finished assembly measured 0.46 × 0.66 m (1.5 × 2.2 ft) and typically contained seven detail parts and 88 rivets. All underfloor bulkhead web assemblies, representing a variety of configurations, were built in the same two universal fixtures. Larger assemblies, such as wing bulkheads, were later added.

The goal of Lockheed's Basic Robotic Riveting Cell was to demonstrate the feasibility of robotic assembly. A high return on investment was not required, but the pilot robotic application had to be kept simple to reduce risks and improve the confidence of both staff and management. The shop cell emerged from hundreds of industrial and mechanical engineering decisions, most of them aimed at achieving reliability through simplicity.[18, 19]

41.3.1. Task Definition

For simplicity, the scope was limited to assembling conventional, moderate-sized assemblies. The bulkheads webs (Figure 41.1) were chosen as representative assemblies; they were constructed from aluminum sheet with attached aluminum stiffeners and brackets.

The robot would operate with the existing automatic drill/riveter, which would be controlled and monitored by the robot through the robot controller. After manual preassembly, the robot would pick up the fixtured workpiece, position it in the drill/riveter, advance it to each fastener position, and return the completed assembly to the pickup table. Its role would therefore be quite active and its working envelope large. The operator would be retained, but instead of positioning the webs in the drill/riveter by hand, he/she would load and unload the fixture and initiate and monitor the robot.

41.3.2. Design Considerations

Since this cell was a pilot robotic cell, the paramount guideline was to keep equipment and operation simple. The robot would be a popular, proven, hydraulic model, with no vision system or other active feedback device. The automatic drill/riveter, proven in 30 years of use, would perform the actual

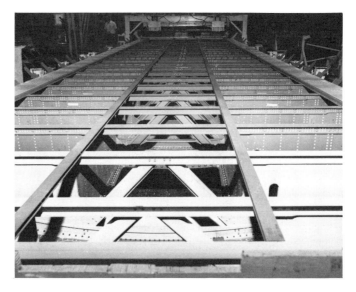

Fig. 41.1. Aircraft underfloor structure, showing robotically assembled bulkhead webs.

fastener installation. The robot would feed the workpiece through the machine with a precision and speed that could not be maintained by a human.

A laboratory work cell was set up that would verify all functions. The robot would be the most critical cell element; all operations and hardware would be designed to work within its limitations of accuracy, programmability, and reach.

Rather than have the robot apply a portable drill, the massive drill/riveting machine would be used. It represented a large investment of capital and could rivet as it drilled. Moreover, it would clamp the parts tightly together during drilling to prevent burrs and chips from working their way between the panel and stiffeners.

The long, fixtured workpieces would weigh a hefty 27 kg (60 lb) and would be held cantilevered from the fully extended arm. To minimize wobble and droop, an end-effector would have to be designed that would weigh little and grip the fixture reliably, firmly, and with high repeatability.

The mechanical abilities of the robot would be complemented by the judgment and dexterity of the operator. He/she would preassemble the detail parts into the fixture, position the fixture onto the pickup table, start and stop the robot, check inventories, and inspect finished work. During the robotic assembly he/she would listen for skipped or misplaced rivets while fixturing the next set of parts. Operation of the robot and drill/riveter would be controlled and integrated by a robot controller furnished by the robot manufacturer. Sequences would be initiated and halted by the operator from a remote console.

In-process inspection was desired because of the high cost of repairing even a single misplaced fastener or deformed panel. But artificial vision or tactile feedback were judged too ambitious for the pilot cell. Instead, all inspection—predrilling, in-process, and postprocess—would be conducted by the operator or a quality inspector. Several inspection criteria were chosen. These included drilling angle, hole roundness, hole size, hole spacing, countersink angle and diameter, and type of rivet. In addition, sheet metal would be checked by the operator to ensure that scratches were acceptably few, shallow, and short.

Off-line programming was not available, so the robot would be programmed by conventional teach methods.

41.3.3. Configuration of the Basic Robotic Riveting Cell

The original Basic Robotic Riveting Cell is shown in Figure 41.2. It consisted of a Cincinnati Milacron six-axis T³ robot, a Lockheed-designed end-effector, a Gemcor G-400 Drivmatic automatic drill/riveter, a Cincinnati Milacron robot controller, left- and right-web workpiece fixtures, a preassembly and pickup table, and an operator.

The T³ was selected because it combined a large work envelope with a high load capacity. It would have to position the workpiece from hole to hole within ±0.8 mm (±0.030 in.). The T³ surpassed

Fig. 41.2. Lockheed's Basic Robotic Riveting Cell. In the background are robot controller and automatic drill/riveter.

this, using an improved-repeatability programming package developed by the manufacturer for aerospace use.

The specially designed end-effector was a rigid aluminum box. On its face were three cone-shaped locators, spaced 120° apart, that extended into matching holes in the workpiece fixture. To tightly grip the fixture, the two fingers of the end effector passed through a hole in the center of the three locator holes, whereupon they spread open, locking the fixture between the fingers and the face of the end-effector.

The drill/riveter clamped the parts together, drilled, applied sealant, inserted a rivet, and upset the rivet, at the rate of 14 rivets a minute. A human averages 11 rivets a minute.

The robot controller was electrically interfaced to the robot and riveter. It relayed signals between the two machines to coordinate their operation. The signals from the robot prompted the riveter to lower its ram and execute a cycle. The signals from the riveter verified that the ram had been lowered and the cycle completed.

41.3.4. Cell Operation

During preassembly, the cell operator positioned the stiffeners onto the web panel using corner pilot holes as guides and one of several retaining methods (clamps, temporary fasteners, or adhesive) to hold the assembly together. Then he placed the fixture onto the pickup table, securing it over the locating pegs. While the robot processed this workpiece, the operator preassembled the next workpiece on the alternate universal fixture.

To initiate robotic assembly, the operator stationed himself/herself at the remote console, where he/she set the end-of-cycle switch on. When the robot swung into place in front of the pickup table, the operator pushed the "continue" button, causing the robot arm to move forward. The operator then signaled the robot whether it was to pick up and assemble a left web or a right one.

Upon receiving a signal to start, the robot picked up the indicated fixture and rotated to the automatic drill/riveter. It then signaled the drill/riveter to lower its ram, and positioned the workpiece below the drill bit where the first fastener was required. Signaling through the controller, the robot instructed the machine to drill the hole and install the rivet, and waited for a cycle completion signal. Upon receiving that signal, it advanced the fixture to the next fastener position. When a row of fasteners was thus completed, the robot controller opened the riveter's ram, and the robot moved the fixture

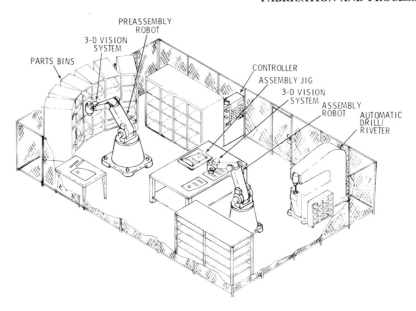

Fig. 41.3. Conceptual design of advanced robotic assembly center.

to the next row. This sequence continued until the web had been completely assembled, whereupon the robot returned the finished assembly to the fixture table, its task complete.

41.3.5. Evaluation of System Benefits and Limitations

The Basic Robotic Riveting Cell met its goal of demonstrating the production feasibility of robotic assembly. The robot reduced the cost of assembling bulkhead webs by 25%. Larger assemblies were later run; their programs were written on-line while the robot was in-between jobs. Given the steady production of the Hercules airplane, the robot would have ample opportunity to recover its purchase cost. Still to be overcome was the slowness and complexity of on-line programming. But with the production experience gained from this pilot cell, Lockheed had the confidence to use robots for more sophisticated assembly duties.

41.3.6. Future Developments in Robotic Assembly at Lockheed

In 1983 Lockheed-Georgia's original robot was relocated to become part of a totally automated Advanced Robotic Assembly Center (Figure 41.3). The new center would use two robots. The first, equipped with a three-dimensional vision sensor, would bin-pick parts and arrange them into a jig; the second robot would pass the jig through a new drill/riveter, possibly inspecting the workpiece during and after assembly. Programs and data would be transferred from off-line sources.

REFERENCES

1. Hohn, R., Application Flexibility of a Computer-Controlled Industrial Robot, SME Paper MR76–603, presented at the First North American Industrial Robot Conference, October 1976; *Industrial Robots—Volume 1: Fundamentals,* 2nd ed., Robotics International of SME; Dearborn, Michigan, 1981, pp. 224–242.

2. Stansbarger, D. L. and Schable, H. G., *Composite Manufacturing Operations Production Integration (Flexible Composite Automation),* final report by Northrop Corporation, Aircraft Division, for Wright-Patterson Air Force Base, Ohio, Contract F33615-78-C-5215, 1983.

3. Barone, P. A., Robotic Paint Spraying at Fairchild-Republic Company, *Robots VI Conference Proceedings,* Robotics International of SME, Dearborn, Michigan, 1982, pp. 321–332.

4. Lockett, J. H., Small Batch Production of Aircraft Access Doors Using an Industrial Robot, SME Paper MS79–783, presented at the Robots II Conference, sponsored by the Society of Manufacturing Engineers and the Robotic Institute of America, Detroit, Michigan, 1977.

5. Lowndes, J. C., USAF Seeks Increased Robot Precision, *Aviation Week & Space Technology,* March 1, 1982, pp. 69–72.

6. Brown, S. F., Automated Sheet Metal Work Cell, *American Metal Market/Metalworking News,* January 24, 1983, pp. 8–9.

7. Robotic Spraying Examined in External Tank Applications, *Aviation Week & Space Technology,* August 2, 1982, pp. 52–53.

8. Advanced Robotic Systems Technology Applications, reports by Robotic Vision Systems, Inc., for Wright-Patterson Air Force Base, Ohio, contract F33615-82-R-5136, 1983+.

9. Advanced Robotic Systems Technologies and Applications, reports by McDonnell Douglas Corporation for Wright-Patterson Air Force Base, Ohio, Contract F33615-82-R-5072, 1983+.

10. Advanced Robotic Systems Technologies and Applications, reports by Fairchild-Republic Co. for Wright-Patterson Air Force Base, Ohio, Contract F33615-82-C-5134, 1983+.

11. Movich, R. C., Robotic Drilling and Riveting Using Computer Vision, presented at Robots V Conference, October 1980; *Industrial Robots—Volume 2: Applications,* 2nd ed., Robotics International of SME, Dearborn, Michigan, 1981, pp. 362–381.

12. Cook, N. C., Computer-Managed Parts Manufacture, *Scientific American,* Vol. 232, No. 2, February 1975, pp. 22–29.

13. Abraham, R. G., Beres, J. H., and Yaroshuk, N., Requirements Analysis and Justification of Intelligent Robots, *Proceedings of the Fifth International Conference on Industrial Robots,* September 1975.

14. Abraham, R. G. and Beres, J. H., Cost-Effective Programmable Assembly Systems, presented at the First North American Industrial Robot Conference, October 1976; *Industrial Robots—Volume 2: Applications,* 2nd ed., Robotics International of SME, Dearborn, Michigan, 1981, pp. 429–451.

15. Ham, I., Group Technology, in Salvendy, G., Ed., *Handbook of Industrial Engineering,* Wiley, 1982, pp. 7.8.1–7.8.19.

16. Intelligent Task Automation, technical reports by Martin Marietta for Wright-Patterson Air Force Base, Ohio, Contract F33615-82-C-5139, 1983+.

17. Intelligent Task Automation, technical reports by Honeywell for Wright-Patterson Air Force Base, Ohio, Contract F33615-82-C-5092, 1983+.

18. Dreyfoos, W. D., Robotic System for Aerospace Batch Manufacturing, Task C—Definition of Robot Assembly Capability, Technical Report AFML-TR-79-4202, Wright-Patterson Air Force Base, Ohio, December 1979.

19. Ooten, G. D. and Plumley, W. J., Assembly and Riveting by Robots, 1980 Design Engineering Show/East and ASME Conference and Seminars, October 28–30, 1980, Paper 4.3.

FURTHER READING

Aerospace Factory of the Future, *Aviation Week & Space Technology,* August 2, 1982, p. 40.

Stansbarger, D. L., and Schable, H. G., Composite Manufacturing Operations Production Integration (Flexible Composite Automation) (Final Report by Northrop Corporation, Aircraft Division), Wright-Patterson Air Force Base, Ohio, Contract F33615-78-C-5215, 1983.

CHAPTER 42

ROBOTS IN CASTING, MOLDING, AND FORGING

WILLIAM E. UHDE

UAS Automation Systems
Bristol, Connecticut

The purpose of this chapter is to analyze the application of robots in casting (Section 42.1), plastic molding (Section 42.2), and forging (Section 42.3). These areas were the first to apply robots and continue to be good candidates for robotization. Additional foundry applications are covered in Chapter 43, Robots in Foundries.

42.1. DIE CASTING

42.1.1. The Process

Most of the high-pressure castings manufactured in the United States are cast in horizontal die casting machines. One die is moved against another by hydraulic force in the horizontal plane. Dies are positioned on plattens which normally are moved along tie bars. Molten metal is delivered under pressure into the die cavity, thereby forming a casting. The most common metals cast by this process are aluminum and zinc, although magnesium and brass are sometimes used. Experimental work has been done with steel.

Robots have been used in a variety of ways to process parts from these machines. A hot chamber machine automatically delivers molten metal into the die to form the part. A cold chamber machine can have a vacuum or mechanical loading system added to it. Occasionally a robot has been used to actually perform the loading operations in cold chamber machines where automatic injection of metal is not available. This is rare and not very cost-effective. Robots are used to remove the parts from the die area, process them through quench tanks (water baths), and deliver them to a trim press where they are separated from sprues, gates, and other scrap. Robots can also be used to deliver steel inserts into dies in which metal, aluminum, or zinc is formed around the steel. In some cases magnetic iron is used as an insert, usually where the resulting casting is used as part of a measuring device or meter.

42.1.2. Practical Die Casting Applications

Die casting applications of robots (see Figures 42.1 and 42.2) include the following typical cases:

1. Robots are used first of all to remove chilled castings from the dies. The robot may be used to process castings from either one or two machines in an alternating unload pattern. A robot with sufficient controls is necessary in order to take care of events that are abnormal in the work cell such as a machine producing a faulty casting. The robot must be able to exclude one machine and continue to operate and remove parts from the other machine which is manufacturing castings of acceptable quality. Air cooling may be required to further cool sensitive aluminum parts.

2. The robot can also perform quenching operations by either using a curtain quench or dipping parts into a quench tank. The quench tank is usually filled with recirculating water to cool the castings sufficiently so that the casting may stabilize and become ready for trimming.

3. Robots have been employed to further process castings from the quench tank to the trim press. Trim presses can be either vertical or horizontal in design. When parts are separated from the sprue and gates, the robot may have to retrieve either the part or the scrap, and place it in storage.

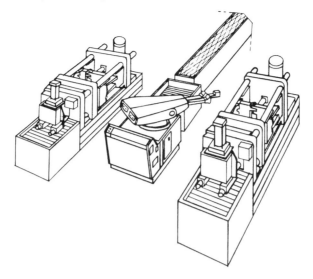

Fig. 42.1. Robot unload of two die casting machines with quench.

4. Where insert loading is contemplated, there is an additional operation to add to the procedure. It is important to place the insert orientation equipment within the reach of the robot. The robot must also have sufficient accuracy to get the inserts into the die.

5. Occasionally parts must be air cooled before trimming, and the robot can hang the part on an outgoing conveyor which will bring the part into the next work area for trimming. This is usually done with sensitive aluminum parts.

Application Considerations

Contract die casting shops (job shops) require maximum utilization from automation equipment. Casting and trim dies may not be compatible for automation, or available at the same time. Lot runs are usually short. The best plan is to assign the robot to two die casting machines and alternately unload the two castings to secondary quenches or outgoing conveyorization. Therefore two die casting machines

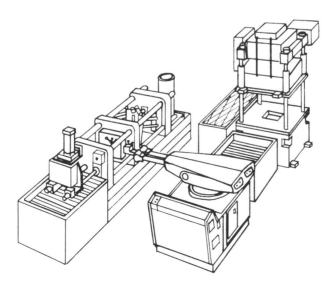

Fig. 42.2. Robot unload, quench, and trimming operation with part removal to conveyor.

are kept operating with a minimum investment in automatic handling, and both pieces of equipment operate with relatively high efficiency.

The trim press, with its lower capital investment, is located in another area, and hand-trimming operations are employed mainly because trimming operations are much faster than the casting operations. Therefore one trim operator can keep up with two or more die casting machines. The penalty is increased process storage between casting and trimming, and the slower casting cycle governs between two die casting machines.

A die casting can be made every 8–60 sec depending on material and size, so care is taken to match casting machine tonnage and casting size. Other factors involve the intricacy of the casting being manufactured and the cooling capability of the die.

42.1.3. Die Care and Part Inspection

Automatic lubrication of the die can be accomplished with a fixed gun and nozzle low-pressure spray system attached to the die casting machine. A dynamic system, with a reciprocating arm that moves in between the die halves, can be employed to spray both die faces where fixed lubrication guns are not adequate. These systems can be attached to the top of the fixed platten.

Sometimes the robot is used to augment the fixed lubrication guns or reciprocating guns with an additional spray gun held in the robot's hand. This is to assure that the areas difficult to reach (usually deep cavities) are properly lubricated and ready for the next shot.

Damage to the die can ensue if some piece of metal is left in the die cavity and another shot is attempted. Inspection devices should be employed to assure complete part removal. The robot would present the part for inspection after removal from the die. A safety problem occurs when the dies do not close properly, resulting in metal escaping from between the die halves during the shot process. The robots can stand the trauma of flying molten metal better than manpower, but effective die care is the best solution.

42.1.4. Electrical Interfacing

Naturally, the die casting machines (DCM) around the robot must be interfaced to the robot electrically to assure that all the operations called for are actually performed (Table 42.1). For instance, the lubrication cycle usually occurs every time a shot is made in a machine, but it does not necessarily have to. The robot with a little bit of intelligence can assign lubricant to be applied at varying intervals, depending on die needs.

42.1.5. Robot Selection

The robot selected should be intelligent enough for the job at hand, and it should be reliable and maintainable. Memory length should be large enough to process parts through trim operations and flexible enough to unload two die casting machines. Random program selection or some subroutine capabilities are required to eliminate malfunctioning equipment from the work cycle automatically.

The robot must be easily programmed so that it can be adapted to the wide range of parts to be run in the die casting machine. This is particularly true in the case of the contract shop which does not know what it will be running six months from now. It must be able to react to its environment by stopping when conditions do not justify continuing the process. Therefore if the part is removed from the machine damaged, the robot would not deliver a signal for a new shot to be made, and also it would ring an alarm for service while it continues to process what it has through the quenching operation and deliver it for inspection by manpower for corrective action.

When unloading two die casting machines alternatively, if a bad casting is made by one of the

TABLE 42.1. ELECTRICAL INTERFACING CIRCUITS

Robot Outgoing Signals	Robot Incoming Signals
1. Start DCM cycle.	1. Die cast machine service signal.
2. Incomplete casting alarm.	2. Casting complete.
3. Start trim operation.	3. Quench temperature in range.
4. Quench unit start.	4. Trim die open and available.
5. Start die lubrication.	5. Output conveyor clear.
6. Start trim die cleaning.	6. Inserts available.
	7. Inserts contained in tool.
	8. Core available for pickup.

machines, the robot should keep the other operational. If the robot does not have this capability, not only is productivity lost, but also die temperatures would be lowered. This would result in bad castings upon start-up, due to chill marks on the castings, until die temperatures move up into operating range. In fact, a machine that is allowed to vary its die temperatures substantially may produce a wide range of bad parts, as well as shortening the operating life of the die.

42.1.6. Automatic Die Casting Versus Manual Operation

Through years of experience and many installations, it has been determined that a die cast machine cycle is far more regular when it is automatically unloaded and parts are processed on a regular time interval. This leads to higher part output and results in a lower part reject rate. There is also a higher uptime for the machine, with less downtime for maintenance and adjustment. Overall, the productivity of a particular die casting machine is substantially increased. In addition, if there is an additional machine in the work cell (DCM or trim press), the labor savings is more than the one person that is associated with one die cast machine. The environment in the die cast machine is unpleasant. The dies are hot, and there is always the danger that the die may not completely close and in turn spray metal around the environment when a shot is made through the partially opened die. The die casting machine is not shut down for coffee breaks or lunch breaks. The machine produces regularly throughout the day.

42.1.7. Automatic Die Casting Economics

Die casting machines are normally kept operating at least two shifts a day, and in many cases are run around the clock. The labor savings become even more dramatic. One person can be used effectively to monitor the activity of six different robots working with up to 12 different casting machines at a time. If a $50,000 robot is used to operate over a three-shift period between two die casting machines, it typically replaces five of the six persons originally required. Five persons replaced at approximately $20,000 a year in direct and indirect costs would have a replacement value of approximately $100,000. A $50,000 robot would need additional peripheral equipment and labor to get it into operation, including a safety fence and electrical interface. A total of $70,000 worth of installed cost would result in a $\frac{3}{4}$-year payback period. If the plant operates on two shifts, the $70,000 investment in robotics would result in a payback of $1\frac{1}{4}$ years. The payback is adversely affected by the number of die changes put into the machine, and by maintenance requirements throughout the year. Users report good justification if a die can be kept operational for 16 hours at a time. These factors can reduce operating time to 75% over the two-shift period. This increases the payback period to 1.8 years. However, a conclusion can be rapidly reached that existing hot chamber die casting machines should in all cases be automated, where die quality allows it.

42.1.8. Part Gripping Techniques

End-of-arm tooling is required to manipulate the parts properly from one location to another. The most convenient grip point is the stub, which is called the *sprue* or *biscuit*. This area, just as in sand casting, connects to the parts themselves. In a pressure casting operation, the sprues or biscuits are stable in diameter or cross section. They vary only according to their length. This length can be controlled normally between 0.75 and 1 inch (1.9–2.5 cm) which is enough to utilize contact-point tooling around the diameter or around the cross-sectional area of the sprue or biscuit. If a diamond head is cast instead of a round area, accurate orientation of the part about the surface can be assured for continuing finishing operations such as trimming. In cases where trimming is not to be considered, the typical round biscuit or sprue can be utilized to process castings to the quench operation and into an output conveyor. Hand tooling for the robot is inexpensive because it is simple in design and easy to maintain.

Usually employed is a toggle action hand, which asserts a firm grip on the surface of the sprue. The fingers on the hands can have metal inserts, pins, or other devices that dig into the surface and hold the castings securely for the removal from the die for inspection and quenching operations.

In some cases, the end user desires to actually grasp the part. If a part is to be trimmed and it is to be recovered from the trim press die by the robot, it then is important for the robot to grasp the part and manipulate the part through various operations leading to the trim operation. The hand can then be designed to fit the part, and after the sprue and gates are cut off the part can be recaptured from the trim press by the robot tooling and placed in a storage container.

When the robot is used to load inserts, the tooling becomes more complicated, and the work environment becomes a little more complicated as well. The robot must be capable of processing parts and at the same time be able to load inserts into a die. A double-pocket hand or a second hand is required to process the inserts. The robot must be accurate enough to place inserts into the die cavity.

Additional pins or guides may be required in the die so that the robot tooling can orient itself on those guides to place the insert properly. This is of particular value where the teaching methods of the robot can lead to variability in the robot's position over a period of time during the placement cycle. Inserts can be oriented on slides, magazines, or bowl feeders that can deliver product to a particular pickup location for the robot. Sensors can be built into those end stops to make sure that a part has been picked up from the nest and also that the product or parts are present so that the robot can call a halt to the procedure if parts are missing from the pickup nest prior to insertion.

Another problem that must be addressed is the ability to remove flash, particularly very thin flash, where it is formed. This flash must be removed by air jets by the application of high-pressure water over the surface of the die to insure that the proper material is removed.

It is important to maintain the dies properly and to keep them in good working condition, otherwise the automatic operation will not be successful. An additional task the robot can perform is the manipulation of floating cores into the die and retrieval from the trimming operation. The processing is similar to that of transferring inserts, but the core is not to be retained in the part so it is removed for reinsertion into the machine within the work cell.

42.2. GRAVITY CASTING APPLICATIONS

42.2.1. More Popular Robot Uses

Robots have been employed for a long time in certain casting-related activities. Greensand mold preparation, including mold spray of compounds to assure casting-surface quality, and the application of flame over the surface of the mold to set the compounds and to force moisture from the mold surface are popular. Pouring of nonferrous metals into permanent gravity molds has also proved successful. In investment casting the robot is employed to build the mold by processing it through slurry and sanding operations.

With the future improvment in the flexibility of robot controls, other tasks such as the cutting of parts from their sprues and gates, the processing of cores, and the processing of shells through casting operations will become more popular.

42.2.2. Investment Casting

This industry has been one of the top five fastest growing in the United States. Although the process, known as "lost wax" casting, dates as far back as man's ability to keep bees, the process did not have great popularity beyond the artistic community because of its cost, but with the advent of robots and the fact that machining is getting expensive, this process has become popular for a variety of industrial products.

First, a mold is made from a combination of waxes in the shape of the final product. Wax sprues and gates are added. The mold components are made in a molding machine with metal dies in a manner very similar to injection molding. The components are joined by persons employing heated irons, which melt the wax surface and allow joining. Premelted wax is used to build up and make the joining easier. Once a mold *tree,* or assembly of parts, has been built, the product is sent to the investing room for shell construction.

The investing room (see Figure 42.3) has strict controls over temperature and humidity. The mold is first etched. A slurry is applied to the surface, followed by a fine sand. Under manual application, manipulation is carefully done to avoid bubbles on the surface of the mold. The fine sands are employed first to get a fine-quality surface on the finished casting so that machining is minimized. The shell is allowed to dry before the second application of slurry and sand. The sands become coarser in repeated applications to build thickness and strength. From 7 to 12 coats are layered on the shell, which finally becomes a hard ceramic shell able to withstand the forces of the casting process.

The completed shell is next sent to the autoclave for removal of the wax mold by melting. Vacuum and steam are used to draw out the residual wax. The wax is salvaged and recirculated to the wax mold area. The shell is then taken to the casting room. After casting, gates and sprues are cut off, and other secondary operations are performed.

42.2.3. The Robot Requirements

The robot has been first applied to the investing room where the shells are built. An intelligent robot is required in order to mix programs as required to provide the proper manipulations in the application of slurry and sand. Both fluidized beds and rain sanders are employed to allow the grains of sand to gradually contact the part with minimal pressure. Although originally lighter-weight-capacity robots were used to emulate manual methods, eventually larger robots were used to handle hundreds of pounds at one time. Each robot now can replace up to 16 workers over a three-shift day because of this increased weight-carrying capacity.

Because the process is not particularly time tolerant, high robot reliability is essential, and manual backup must be provided where possible. When handling weights beyond manual ability, a spare

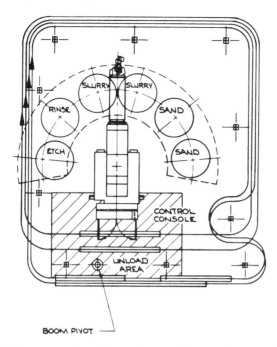

Fig. 42.3. Single cell investment casting dip room layout with one robot performing all dipping operations.

robot is kept in the wings. This means that robots supplied to those facilities also must have established datum points (zeroing abilities) in order to substitute for each other. At least five axes of motion are required for even slurry distribution over the mold surface.

42.2.4. Robot Tooling

End-of-arm tooling consists of a hand which can grip and lift the handle or rod that is attached to the mold at its top. The tooling may require several grip points to process several molds at one time. The rod usually has a cross brace to maintain radial orientation of the mold by its axis of gyration. The molds are individually rotated about those axes during slurry and sanding applications to assure even distribution of material over the surface. Ten to well over 100 rpm are used to remove bubbles from shell surfaces, depending on mold design and materials used. The robot then returns the product to its original orientation for placement on a conveyor or rest stand.

42.2.5. System Design

Most often, a job shop requires one robot to process parts through all of the shell-building operations. Different drying cycles require circulating conveyors of varying length for proper recirculation and the robot must respond with the right program to the mold in hand. It turns off slurry tanks while inserting product and activates fluidized beds and rain sanders for sanding operations. The robot also controls the indexing of conveyors and other fluid tanks within the reach needed to get to all slurry preparation tanks and sanding units required to make a complete shell. Auxiliary process controllers may be required if the robot is performing batch processing under manual direction. Part identification tags such as bar code readers can be used where the process is run without any human intervention.

In a large system the tendency is to design a line where a robot processes a mold through a slurry and sander and then passes the product to the next robot by conveyor. Such lines do exist, but a better idea is to use this method only for the primary dips (first and second) and then to divide the resultant product to multiple robots each performing backup dips utilizing a recirculating conveyor. This can be done because backup slurries and sands are usually the same for subsequent dips. Therefore, a series parallel system design evolves, which would operate with higher reliability. The larger lines utilize process controllers which store the various molds in order, monitor their progress, and direct the robots as to programs to be employed. The resultant system can then run one-piece lots if desired.

42.3. PLASTIC MOLDING

42.3.1. Injection Molding

This plastic molding process is typically used for thermoplastic materials. (See Figure 42.4.) The material to be molded is supplied in a granule form and moved from a hopper to a cylinder from which a plunger forces the granules through a heat chamber into the mold. Then the mold halves open, and the product is withdrawn. Sometimes a three-piece mold is employed to produce product, particularly where complex shapes are to be formed. The plastic material is cooled down from as high as 650°F (340°C), and the product is formed through thousands of pounds per square inch pressure (millions of N/m²).

Many automotive parts are injection molded today, as well as many parts utilized in household appliances and home furnishings. The robot is typically employed to remove the part from the mold either by grasping a sprue and runner assembly or, if the mold is of a sprueless-runnerless type, to grasp the part and remove it and process it to various finishing operations. As in die casting, two machines can be alternately unloaded by the one robot. Usually cycle times are between 15 and 60 sec, and sometimes even longer. This allows time for the robot to perform finishing operations in the work area.

The operators in the typical injection molding shop are employed to separate parts from sprues and runners before allowing the product to continue to the unload station or to the boxing stations. Sometimes the operator in the local area performs the boxing operations on the product as well.

A robot is typically used at an injection molding machine work cell where parts must not be dropped because of fragility or configuration, or where runs are so short that it is not economic to build a totally automatic mold to drop the part through the bottom of the machine.

42.3.2. Use of Automation Versus Manual Methods

There has been resistance in replacing operators at this task simply because of the operator's flexibility in performing secondary operations right at the machine's site, without the need for special machinery. The operator is most easily replaced by a robot where secondary operations are not to be performed on the product. If the mold design is sprueless and runnerless and there are not other secondary operations required, other than boxing for shipment, a robot can be employed. Another condition for utilization of a robot is when a part must be carefully manipulated through the mold or the product cannot be dropped in any way. The manufacture of lenses and other optics is a classic example of this condition.

When secondary operations must be performed at the work site for economic reasons, the parts must be run in large lots, and molds must not be removed often and replaced with different configurations. If they are, there may be more time involved in secondary fixturing than in running the operation. A typical example is where a part that has a gate or runner requires machining operations in the local work cell such as drilling and tapping holes to complete the part.

The robot can then remove the part or parts from the die area and process them in a logical order through mechanical trimming and the various machining operations by handling the part from

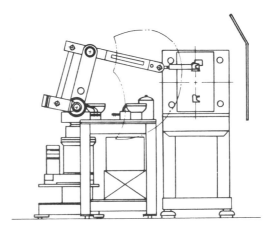

Fig. 42.4. Injection molding machine with a robot unloading parts and insert loading of die from bowl feeders.

station to station. The robot can also be employed to grasp the sprue or gates and insert the product into a press die. The parts are stripped from the gates and runners by the action of the press, then allowed to gather or run down a slide and accumulate in a box in an unoriented fashion. The robot recirculates scrap by lifting it and placing it in a chopper in the work center. Small pins or tacks can be manufactured by utilizing the robot to remove the sprue and the runner by grasping them with the parts attached and then passing the parts through a cutting shear which allows the parts to pop or strip off the runner assembly as the robot passes it through the knives. In this way the product falls into the box and is gathered up in an unoriented fashion at the appropriate time.

Many parts, such as hubs of wheels, require inserts to provide bearing surfaces for high-speed operations. The robot can be used to load inserts into the injection molding machine. These inserts can be loaded in multiples if they are delivered at the proper centerline distances from each other in the mold itself. Typically, the robot enters the mold area of the injection molding machine, removes the sprue, runner, and previously molded parts, and reorients its hand, either by withdrawing from the machine and reentering it in a different attitude or by turning the hand in the machine to place the inserts into the proper cavity. The machine can assist in seating the inserts through vacuum assist, which is part of the molding machines' capabilities. Pins or other guides can be used to make sure that the tooling accurately positions the inserts for loading in the mold area or cavity. The inserts are brought into the work area either in magazines for feeding into the pickup locations or through bowl feeders which deliver the inserts to pickup nests, all of these situated at the proper spacing for placement within the mold.

42.3.3. Robot Tooling

Mechanical grips of the toggle type can be employed when gripping the part by the sprue or the runner. Parts that are rigid at the time of removal can be reliably removed still attached to the sprue. Inspection is required outside of the mold area to determine that all of the parts are present before closing the molds, and then the parts are processed to any secondary operation required.

Mechanical grips can also be used to contact the inserts, pick them up, and process them to the mold area. Individual pancake type actuator cylinders can be used to contact, grip the inserts, and process them. In certain cases it may be more advisable to grasp the surface of the part in a different manner. Vacuum cups can be used where enough surface area is present to be contacted by a vacuum cup and that surface is smooth. The surface of the part must be flat or of such contour that a vacuum cup can make contact over a substantial enough area to remove it and process it. It must be remembered that in certain cases vacuum cups can cosmetically change the appearance of the surface of the part slightly. This must be taken into account when specifying vacuum cups for processing product. The temperature of the part must be cool enough so that there is no material migration from the vacuum cup to the surface of the product.

Specially shaped vacuum cups can be easily obtained from various sources throughout the country. Robots can be used to process parts from injection molding machines where the parts are relatively large and difficult for manpower to handle easily. Garbage cans and large appliance parts can be easily removed by a robot and processed to the output conveyors for manual processing. The robot has the advantage of being able to reorient large parts to manipulate them through the tie bars where dropping the parts through the bottom of the injection molding machine is not practical or impossible. The advantage of using a robot in an injection molding operation is that die care can be accomplished automatically and under strict control. It can also be kept constant by time, volume, and location, thereby avoiding any problems caused by irregular applications by operators. Irregular cycle periods cause scrap to be produced because of the warming up and cooling down of the mold in an unpredictable fashion. The problem is not always seen in the very next shot after an irregular cycle has occurred, but usually several pieces downstream.

42.3.4. Compression Molding

Another operation in which robots have been successfully employed is the processing of thermosetting materials which require compression molding as a means of manufacture into usable products. Electrical circuit and power components are a typical example of a thermoset product.

Thermoset material does not turn plastic when reheated, and a different molding process is used from that used by thermoplastic materials. Thermoset material is usually delivered in extruded bars that are cut into pellets. These pellets are typically fed into a heating chamber or to a delivery point where a robot can pick the product up and deliver it to a heating chamber that raises the temperature of the product to plastic condition. It may take 60 sec or longer to heat the product enough for further processing. The robot then delivers the product to a press mold and upon removal activates the press to start its cycle. Typically, the press is a vertical molding machine. The press squeezes the part into the shape of the product to be formed; and the product is allowed to set in the mold, usually for 60 sec or longer. After this time, the robot picks the product out of the mold and processes it to any finishing operations that may be required, such as surface grinding, routing, or drilling.

Part-sensing equipment is used in the local area to check that the part has been completed and that is is being processed properly. There is always the chance that the part may be on the wrong side of the mold because of improper ejection. There is a need for automatic mold lubrication with this process, as well. Since molding processes are very similar to die casting, Section 42.1 should be referred to for additional information.

42.4. FORGING

Over the years robots have been employed in many different types of forging applications, some of them successful, and some of them not. Robots have been applied to *hammer forge* operations, *upsetter* operations, *roll forges, hot forming presses,* and *draw bench* applications. In some cases the robot acts as the forging machine operator. In other cases it has acted in the role of the forge helper. In many cases teleoperators are used for forging applications, as described in Chapter 9, Teleoperator Arm Design.

42.4.1. Forge Hammers

The most popular forging machine in the United States is the forge hammer. Although some hammers are electrically driven, most forging hammers are either hydraulic, steam hydraulic, or air driven. Smaller forge hammers may rely on gravity, and they are called *board hammers.* But hammers are falling more into disuse as other forging machines with more automatic productivity potential gradually take over.

One-half of the forming die is on the anvil, and the other half of the die is on a ram that moves up and down, either under force by air, steam, or by gravity. Under the control of an operator, the hammer is allowed to strike the part that is lying between the two dies a certain number of times, depending on the observation of the forging operator. The operator determines when to take a part out of one die and move it into the next.

Typically, a three-die operation consists of a buster die to preform the billet and clear it of most of the scale. The part is then moved to a blocker die, which roughly forms the outline of the part; the part is finally completed in the finish die. More dies are used if the part is of an unusually complicated shape. Anything from a few to a dozen strokes may be required to finish a part and process it from the die area. The resulting product is called a platter. The part is formed within the platter, and flash is usually all around the perimeter of the part. The flash must be trimmed from the part. It can be trimmed after the parts cool down, or it may be removed while the part is hot. The resulting part may require coining, as well, to restore flatness, since the flash causes uneven cooling in the platter, resulting in a warped or bent part.

The function of a robot in this application can be to act as a forge helper. When working heavier parts, the robot can be used to load and unload furnaces and process the billets to the forging bed, where the operator can take over and process the parts through the various forming cycles. The robot then can maneuver the finished product to a trim operation, if hot trim is desired, or it can remove the finished product to a bucket or to a bin for cooling.

The forge hammer cannot be automated successfully. Programmable forge hammers have been built. These units can control and monitor the force and the number of blows imparted to the platter. In this case the robot can act as a manipulator processing the billets from die to die. The billets may be turned 90° or turned over between dies, and they must be carefully lined up with the die impressions to prevent forming a double impression in the final or finish die. This requires a highly accurate six-axis robot.

There is no time to perform a trimming operation on the product unless the hot trim die is part of the forging operation. If trimming is to be done on a separate press, the part is usually dropped off the back of the forging machine, collected in a tote, allowed to cool, and taken to the trimming area for cold trim and coining. In some cases the product can be dropped off the back of the programmable hammer into a conveyor direct to the hot trimming operation.

Under test conditions robots have also been applied to the Chambersburg horizontal impacter. This is a forging machine in which both die halves move against the billet or platter horizontally. At the end of the forging cycle the manipulator that moves the product through the impactor delivers the platter to the outstretched hands of the robot. The robot then can process the parts through the final hot trimming operations. In general, the forming machine and dies are far more expensive than the trimming equipment. Therefore it is usual not to allow the impactor to move to an unload position to position parts for pickup by a robot. Usually the product is dropped through the bottom of the forge machine and collected in a carrier. Therefore, in actual practice, robots have found limited application with this type of forging machine.

Where employed, the robot will take the product to the trim press, place the platter in the trim area, allow the part to be pushed through by the trim die, and then remove the scrap from the face of the die and place it into a cargotainer for removal from the area.

42.4.2. Upsetter Operations

The upsetter is also a common forging machine on the U.S. scene. It is used for the forming of round, long slender parts that have a particular shape other than round or a large head at one end. The machine works in the following manner. Two horizontal dies move in to grip the part along the long side of the round area. Then a ram or header comes in from the back of the machine to form one head.

The function of the horizontally moving dies is not only to hold the part. They also form the part where offsets are required along the long dimension. Typically a crank or a wrench handle may be offset along its length. These forging machines can have two to five dies laid out vertically in the horizontal plane of the machine and can have a stroke from 60–80 cycles per minute. Many times the die moving against the side of the part is not enough to keep the part from slipping backward during the ram's stroke. Therefore in many cases a backrest or backstop is used to restrain the part from being shifted backward during the stroke of the ram.

Robot Utilization at an Upsetter

The robot (see Figure 42.5) can be positioned two ways in front of an upsetter. It can be placed on the floor with its arm facing the ram of the upsetter, or it can be placed 90° to that position, which allows the robot the ability to work sideways with regard to the dies in the upsetter.

It has been found that when dealing with small upsetters of 1.5 in. up to 2.5 in. (3.8–6.4 cm) capacity the throat of the upsetter must be ground away to accommodate the hand of the robot. This is particularly true since the robot wrist will be moving to the left and right of the throat area to move parts from die to die.

It is best if induction heating is used in this particular work area. Induction heating can be employed to provide scale-free material in an oriented fashion for the robot to pick up. In the pocket or end of the conveyor it is important to put heat-sensing equipment to make sure that the billet to be worked is within the proper temperature range. Another switch indicates that the product is properly seated for placement into the dies.

The robot will proceed by picking up the part, placing it in the first die, and exercising the upsetter dies. The robot works the part into the die impression, exercises the dies of the machine, and waits for the dies to reopen. As soon as they reopen, the robot takes the rod containing the partially formed part and places it into the next die by moving to the left out of the die area and downward to the next die area, and back into the next impression. The process is repeated until the product is complete. The robot is fastest if it can rotate an offset hand by swivelling the rod downward to the next die impression. It is important to remember that the part be twisted somewhat about its own long axes. This will prevent a large flash buildup in one area of the shafting.

The in-out reference on the part is usually provided by the backstop in each impression location. The upsetter must be single stroked. That means that the machine is only allowed to go through one closing and opening cycle at a time, and the robot must signal the start for every closing. When a press is set up for manual operation, it is usually operated by foot pedal by the operator, and by holding the foot pedal down he makes a continuous rather than a single-stroke process out of the operation. It may require a special kit from the manufacturer of the upsetter to provide the single-stroke capability required.

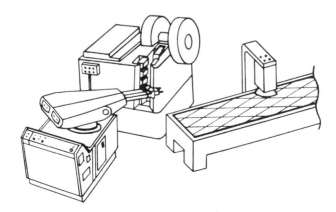

Fig. 42.5. Robot processing tie rods through an upsetter forging machine.

The completed product can be dealt with in two ways. The bottom of the upsetter is open, which allows the part to be dropped through the bottom of the unit to a quench tank or to a conveyor that brings the product out the back end of the upsetter. The other way is for the robot to move the product up, withdraw, and put it down on a conveyor or in a tote next to the machine. Die lubrication and cooling is supplied by continuously falling liquid from the top of the machine.

42.4.3. Press Forging

The robot is placed in front of the press much in the same manner as one would arrange a cold press stamping operation (see Figure 42.6). In this particular operation, the heating equipment can be placed in the back of the press, and the part is brought forward into a pickup nest on one side of the die area. Usually the side picked is nearest the busting die. The robot then progressively moves the parts through the various die impressions, busting, blocking, and finishing, and at the end of the cycle places the part in either an output conveyor for processing to the trim operation or to a cargotainer for delivery to a coin and trim area. The press must be single stroked and must be equipped with automatic die lubrication. Lubrication can be provided by reciprocating guns coming from the back of the press. Occasionally the robot may be required to pick up a lubricating gun and work through several die areas that may be difficult to clean. Air should also be delivered to the die areas to clear off any flash that may be deposited after each stroke.

Occasionally the robot will not be used as the forging mechanism, particularly with regard to parts that require twisting within the die about their own vertical centerline. In this case, the robot is used as a forge helper to deliver billets from gas or electric furnaces to the buster side of the die. This is valuable when parts are very heavy and manpower cannot lift them easily. The robot then

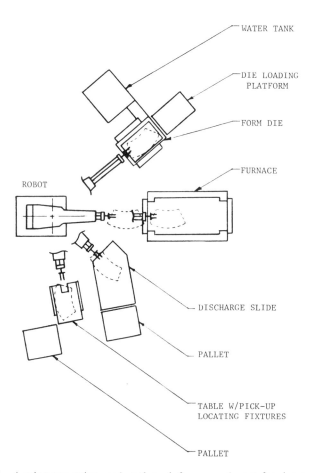

Fig. 42.6. A robot processing product through furnace and press forming operations.

would be responsible for taking the finished product back out of the die and placing it in another location for the finishing operation.

Another subset of this type of operation is high-energy-rate forming machines. These machines are used occasionally to form a completed product in one die and in one die stroke. The robot is used in the same manner as in a typical press operation except productivity is very high. Lubrication and heat dissipation are very important.

42.4.4. Roll Forging

Roll forging is a forming operation rolling against a hot billet, usually forming a compound shape along the length. Two rolls, one situated above the other, turn at a rate of approximately 60 rpm with the billet passing between them. There is an open area in one quadrant to allow for multiple insertions of the billet. The billet goes through several die impressions until formed into the required shape.

A robot is usually placed in front of the machine so that the billet may be inserted linearly into the die. The robot hand is mounted to a sliding member that accommodates any variations in speed between the two machines. Robots have been successfully applied where short lengths are to be worked. Where long billet lengths are to be worked, too much activity is spent putting the robot back on its base after an interface malfunction to make this application worthwhile.

REFERENCES

1. Engelberger, J. F., *Robotics in Practice,* AMACON, New York, 1981.
2. Various documents published by Unimation, Inc., 1976–1983.
3. Engelberger, J. F., Application of robots in die casting, Technical Paper No. 35, Society of Die Casting Engineers, 1964.
4. Barker, A. J., A highly mechanized die casting operation, *Machinery,* Vol. 125, September 1974, pp. 377–381.
5. Laurent, B. P., Use of a robot to obtain a high quality die cast turntable, *Foundry Trade Journal,* Vol. 138, March 1975, pp. 425–428.
6. Canner, J. B., Two arms are better than one, SME Paper No. MS75-253, 1975.
7. Rhea, N., Robots improve a die casting shop, *Tooling and Production,* Vol. 43, No. 12, March 1978, pp. 74–75.
8. Oakland, M. R., Automated aluminum die casting, SME Paper No. MS78-675, 1978.
9. Harris, W. D., Robots in the middle, SME Paper No. MS79-406, 1979.
10. Kellock, B. C., Industrial robot for investment foundry, *Machinery and Production Engineering,* Vol. 129, October 27, 1976, pp. 487–488.
11. Laux, E. G., Automated investment casting shelling process, SME Paper No. MS78-678, 1978.
12. Ostrowski, D., Robots automate investment casting, *Modern Castings,* Vol. 69, No. 6, June 1979, pp. 58–59.
13. Gregory, B., Robot in plastic molding, SME Paper No. MS75-245, 1975.
14. Broderick, W., Part extractors keep injection machine humming, *Plastics Engineering,* Vol. 32, March 1976, pp. 32–33.
15. Campbell, J., Close encounters of the fourth kind, *Industrial Robot,* Vol. 6, No. 3, September 1979, pp. 135–139.
16. Lindbom, T. H., Unimate as a forging hammer operator, *Proceedings of the 3rd International Symposium on Industrial Robots,* Zurich, Switzerland, May 1973, pp. 155–161.
17. Rooks, B. W. et al., Automatic handling in hot forging research, *Proceedings of the 1st Conference on Industrial Robot Technology,* Nottingham, U.K., IFS Publications, March 1973, pp. R8 119–128.
18. Konstantinov, M. and Zakov, Z., Multi-grippers hot purge manipulators, *Industrial Robot,* Vol. 2, No. 2, June 1975, pp. 47–55.
19. Franchetti, I. et al., Automation of forging by means of robots, *Industrial Robot,* Vol. 5, No. 3, September 1978, pp. 121–122.
20. Appleton, E. et al., Open die forging with industrial robots, *Industrial Robot,* Vol. 6, No. 4, December 1979, pp. 191–194.
21. Saladino, J., Upset forging with industrial robots, SME Paper No. MS80-704, 1980.

CHAPTER 43
ROBOTS IN FOUNDRIES

STEN LARSSON

ASEA AB
Västeras, Sweden

Robots have been used in die casting (see Chapter 42) much longer than in gray iron foundries. Die casters have been using them for more than 20 years to unload die casting machines because the robots can easily handle hot castings while humans wait for the castings to cool. In these operations the robots are basically simple manipulators, and pneumatic rather than electronic robots are most often used. They remove the hot castings from the die casting machines and place them on a conveyor or into a container.

In 1979 in the United States alone, 298 robots of all kinds were in use for foundry and die casting work, according to the Society of Manufacturing Engineers (SME). By 1983 more than 600 robots were in use in metal casting. The SME projects that around 900 robots will be used in the United States in metal-casting applications by 1985, a growth rate of 250% in five years.

Recent developments provide new applications of robots in foundries. In this chapter the following are described:

Fettling
Cutting grey-iron castings
Refractory brick handling

43.1. FETTLING

Manual cleaning, fettling, of castings is one of the most arduous and hazardous jobs in industry. It is becoming increasingly difficult to recruit personnel for this work, and the turnover of people in cleaning departments of foundries is substantially higher than in foundries as a whole.

Increasing knowledge of the harmful effects of vibrating hand tools on blood vessels, nerve fibers, and on bones in hands and arms has led to an increasing demand for automatic aids for cleaning castings, all over the world. Instead of lowering the harmful effects of vibration by reducing the size of hand tools, and thereby their efficiency, it is now possible to use industrial robots in place of people in foundry cleaning departments. Additional robotic applications (often for the same robots) are now being found in the same foundries which are turning to robotics for the first time.

43.1.1. Advantages of Robotic Fettling

Robotic cleaning of casting provides economic as well as environmental benefits to foundry managers. The robot operates nonstop and is screened off from the workers, who are thus protected from a job that in the long term carries severe risks of injury. Instead, the manual workers concentrate on supervising the robot, checking the quality of cleaned castings, and, when required, perform robot maintenance.

ASEA's Industrial Robot Division, Västeras, Sweden, has developed a completely automated installation for cleaning gray and ductile iron castings. The system is now being marketed worldwide. The foundry fettling installation went into operation in late 1982 at Volvo Komponenter, Arvika, Sweden. The robotic cleaning installation works together with a handling and hopper system and uses four separate types of tools, all under adaptive computer control.

No specialized computer knowledge is needed for programming the ASEA IRB 60 robot used in the system. An operator familiar with cleaning castings can manually set those points to be searched by the robot, and then the operator enters the point in the computer program. The robot automatically compensates for tool wear. It searches for the condition of tool edges before it starts an operation.

The program also features adaptive control to sense the size and position of risers and external flash, and locates cavities for internal grinding.

When a tool has become so worn that it must be changed, the robot does the job. The tool attachments are designed so that the new tool automatically takes up the same position as the old one. The cleaning installation at Volvo has four tools, as explained later. The complete system is controlled by ASEA's SII electronic robot controller which is based on two Motorola 68000 16-bit microprocessors and floppy disk program storage.

43.1.2. Reducing the Cleaning Cycle

One of the problems initially faced in the Volvo foundry in Arvika was the weight of the gearbox housing castings that had to be handled by the robot. Together with the risers, the gearbox housing weighs more than 60 kg (132 lb), which is the maximum handling capacity of the six-axis ASEA IRB 60 robot.

The problem was solved by making the robot first use a hydraulically operated cutting tool to remove the risers as shown in Figure 43.1. This operation cuts the weight of the casting to a more manageable 55 kg (121 lb).

Castings are fed to the robot by a roller conveyor parts-handling and hopper system. The hopper holds up to 96 parts at a time. Castings are automatically transferred to the robot and returned to the hopper after cleaning.

"The robot and hopper system lets us operate the installation unmanned at night," reports the Project Manager at Volvo Komponenter. "We load the hopper in the evening after the final shift, and the robot works all night. By morning we have a supply of cleaned castings equal to the production of an eight-hour shift."

The robot at Volvo works with four tools: a cutting wheel, a grinding wheel, a chisel, and a rotary file. In the first step of the cleaning operation, risers are cut from the gearbox as the robot presses the hydraulic cutting wheel against the cast part.

During the next three steps, the robot lifts the 55 kg (121 lb) gearbox and holds it up to the stationary tools. Outer edges of the gearbox are ground, and burrs inside are removed, partly with the chisel, and partly with the rotary file (see Figure 43.1b).

The computer program for cleaning each component lasts 7 min—3 min less than required for manual cleaning. The reduction in cleaning time is 30%, and this represents a sizable cost saving.

When a new type of casting is to be cleaned, the robot program is changed. A new program is loaded from a tape cassette into the robot controller, which takes 15 sec. At the same time the robot gripper module is changed to permit handling the new casting.

43.1.3. Consistent Cleaning Quality

"We are going to work hard on automating our foundry," declares the Volvo Project Manager. "Four or five robot lines for cleaning will be needed before the end of this decade to maintain production— and to keep us competitive. Manual cleaning requires six employees per line, and they aren't able to provide the same consistent, high quality as the robot. Equally important, we have totally eliminated personnel hazards such as back injuries from lifting heavy castings, as well as getting our people out of the hot, dusty atmosphere of the cleaning room; and we have solved a high personnel turnover and absentee problem in our foundry."

43.2. PLASMA ARC CUTTING OF GREY IRON CASTINGS

High-speed plasma arc cutting has been applied by researchers at the University of Rhode Island to cleaning gates, visors, and sprue from large gray iron engine block castings (see Figure 43.2). A 20,000°F (11,000°C) plasma arc is guided by a six-axis ASEA IRB 60 electric robot to cut sections of $\frac{1}{8}$ to $\frac{1}{2}$ in. (0.32 to 1.3 cm) thickness, at speeds between 30 and 70 in./min (1.3 and 3 cm/sec). Arc cutting of gray iron was not used until recently because it created a thick, hard layer of white iron carbide. A much thinner layer that can be rapidly removed results, however, under controlled arc parameters.

43.3. ROBOTIC REFRACTORY BRICK HANDLING

Another new area being explored by both foundries and steel mills is the robotic handling of refractory bricks, since the same robots that clean castings are not busy when the primary melters are down for relining. European brick makers already have extensive experience with brick handling robots because the manual handling of brick, like cleaning castings, is not a popular job. A man can wear out three pairs of gloves in a single week handling brick in a brickyard.

In Sweden, the brickmaker Hoganas AB placed a robot on-line in 1980 to stack bricks on pallets. The robot handles 8000 bricks per 8-hr shift. When the job was done by human hands, maximum capacity per shift reached only 5000 bricks.

(a)

(b)

Fig. 43.1. ASEA IRB6 robot pulls die casting from mold and quenches it (*a*), grinds off flash (*b*), drops part in straightener dies (*c*), removes part and performs five other operations before dropping finished die castings on conveyor belt (*d*). Productivity is 500% higher for the die casting operation using a robot than it is for manual operation.

(c)

(d)

Fig. 43.1. (*Continued*)

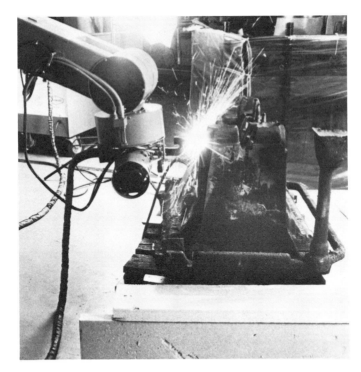

Fig. 43.2. ASEA IRB60 electric robot with gripper-mounted plasma arc cutting torch removes gates, risers, and sprue from gray iron engine block casting.

The brick is handled by an ASEA IRB 60 robot with a capacity of 60 kg (132 lb). Programming for brick handling takes just a few minutes, as the programs are usually quite simple. Once the right moves are in the robot's memory, it performs them with near-perfect repeatability through thousands of working cycles. The unit stacks the bricks to accuracies of a few thousandths of an inch—well beyond the capability of unassisted human hand-eye coordination.

Sensors enable the ASEA robot to detect problems. When it "sees" or "feels" a problem, the robot stops what it is doing immediately and calls for help by an alarm system. When the condition is corrected, the robot resumes work where it left off.

Brick handled by the robot at Hoganas measures 250 × 130 × 130 mm (10 × 5 × 5 in.). Each brick weighs about 9 kg (20 lb). Conveyors carry the brick directly to the palletizing station from a tunnel kiln. On the conveyor, they are stacked 300 mm (12 in.) high in five rows. Approximately 250 mm (10 in.) of air space is left between each row of bricks.

The robot's gripper handles five bricks at a time, using five movable suction cups. When the bricks are picked up, a vacuum sensor checks to see that they actually number five and are securely held. When the bricks are stacked on the pallet and released, an optical laser sensor verifies that all five bricks have, in fact, found their way to the proper location on the pallet.

The brick-handling robot is mounted on a 1-m (3-ft) high pedestal, giving it the working height needed to handle the top bricks in the stacks on both conveyor and pallet. If there had been too little space for floor mounting, the robot could have been inverted and mounted overhead. Robots work just as hard upside down as they do right-side up.

REFERENCES

1. Shimogo, J., et al., A total system using industrial robots for electric arc furnace operation, *Proceedings of the 3rd International Symposium of Industrial Robots,* Zurich, Switzerland, May 1973, pp. 359–374.

2. Synnelius, S., Industrial robots in foundries, *Industrial Robot,* Vol. 1, No. 5, September 1974, pp. 210–212.

3. Gray, W. E., Industrial manipulations in the foundry industry, SME Paper No. MR76-604, 1976.

4. Mori, M., et al., Applications of robot technology for tapping work of carbide electric furnaces,

Proceedings of the 7th International Symposium on Industrial Robots, Tokyo, Japan, October 1977, pp. 293–300.

5. Munson, G., Foundries, robots and productivity, *Proceedings of the 8th International Symposium on Industrial Robots,* Stuttgart, West Germany, May 1978, pp. 303–319.

6. Tomasch, M. B., Materials handling: key to foundry mechanization, *Foundry Management and Technology,* Vol. 106, July 1978, pp. 26–27.

7. Kerr, J., Britain's first robot fettling shows what can be done, *Engineer,* Vol. 248, March 22, 1979, p. 15.

8. Hasegawa, Y., Analysis of complicated operations for robotization, SME Paper No. MS79-287, 1979.

9. Alves, A. I., Thoughts and observations on the application of industrial robots to the production of hot P/M forgings, *Robotics Today,* Spring 1980, pp. 30–31.

10. Ferloni, A., ORDINATORE: A dedicated robot that orients objects in a predetermined direction, *Proceedings of the 10th International Symposium on Industrial Robots,* Milan, Italy, March 1980, pp. 655–658.

CHAPTER 44

FLEXIBLE MACHINING CELL WITH ROBOTS

RALPH L. MAIETTE

UAS Automation Systems
Bristol, Connecticut

44.1. ROBOTS IN FLEXIBLE MACHINING CELLS

The use of an industrial robot at the center of a flexible machining cell goes far beyond the mere utilization of that robot for simple pick-and-place operations. By combining the capabilities of the programmable manipulator with automatic controls provided by numerical control (NC), computer numerical control (CNC), and direct numerical control (DNC) machine tools, the system can provide a high degree of flexibility and a great potential for productivity.

A *flexible machine tool cell* (or machining center) can be defined as a group of machine tools arranged so that they can be tended by one or more industrial robots. Another kind of cell uses a robot loader with a multiple-operation, rotary-table, tandem machine tool incorporating automatic in-process gaging and tool-wear compensation. An example of a flexible machining cell is depicted in Figure 44.1.

Input and output conveyors complete the installation. The entire system draws from established principles of group technology and computer-integrated manufacturing. Flexible machine tool cells can effectively reduce in-process inventory, cut lead times, minimize direct and indirect labor costs, maximize use of capital equipment, and increase the number and quality of parts made per shift.

Two factors must be considered in deciding whether and how to utilize a robot in a cell:

1. Is there a new process or machining operation being implemented? This would require installation of new machine tools, gages, conveyors, and handling devices.
2. Is an existing layout, with existing machine tools and other manually operated equipment, to be rearranged and upgraded into a cell?

In either case, it is important to take into account the capabilities of the robot being considered. Robot manufacturers should be questioned as to the capacity of the robot in terms of programmability, control, weight-handling capacity, speed of operation, accuracy and repeatability, arm reach, and degrees of motion.

Another consideration is the number of robots to be employed. With a single machine tool cell, a single robot can be used. But when the output of one cell is to be transferred to a second or even a third cell, additional robots should be considered.

It is important to have the robot(s) operating as much of the time as possible. When large metal parts are being machined, for example, the workpieces spend a fairly long time at the machine tool. If this time exceeds 20 sec, the robot should not stand idle but should be employed in other activities while the cutting proceeds.

Ample space should be provided to accommodate the reach and movements of the robot while it is operating. Robot tooling and the design of the arm or gripper must also be considered during the planning stage. Such factors as the quantity of parts to be handled simultaneously should also be taken into account.

44.2. PLANNING AN EFFECTIVE FLEXIBLE MACHINE TOOL CELL

In planning a flexible machine tool cell, a designer must consider a number of important functions and capabilities:

1. **Priority Capabilities.** The system must have priority machine selection capabilities. This means the robot should be programmed to accept signals ("service calls") only from machine tools that are not occupied with a job at a particular time. Further, if signals are received from several machines simultaneously, a priority decision must be built into the robot program so it will serve the machines in the proper order.

   If any of the machines is not operating, the robot should be programmed to serve the other machines that are working in the cell. When the down machine is back in service, the robot will then service that machine as required. With this arrangement, no manual intervention is required to keep the robot producing parts only on the normally operating machines.

2. **Multiple Parts Handling.** Where feasible, the robot should handle more than one part at a time. This is particularly important in machine tool loading, where the robot can pick up two parts at a time and load them into the chucks or collets. A robot can perform the operation smoothly and quickly, whereas a human operator might have to load the parts one at a time because of their size or the way they are presented.

   The robot hand can pick up a large part and a small part and load them into their respective fixtures on a machine tool. When running families of parts, it is simple to program the robot to handle a wide range of part sizes.

3. **On-Line Inspection.** An on-line inspection operation should be included in the flexible machine tool cell, and is one of its most important components (see further discussion of this topic in Chapter 45, and in Part 12). This enables parts to remain within the cell and not be brought off the line for inspection, saving time and ensuring better control of the manufacturing process.

   For example, the robot can be programmed to off-load parts onto a gage station that will check the various critical dimensions. If any of these is out of tolerance, information can be fed back to the machine tool so compensation can be made automatically. If the parts are found to be far out of tolerance, the robot can be taught to reject them. The robot's high degree of flexibility in being able to be programmed to make such decisions maintains the flow of good, in-tolerance parts through the system.

4. **Buffer Storage.** Buffer storage (storage between any two consecutive machining cells) should be provided. If one cell is inoperative because of lack of parts, machine tool failure or gaging failure, buffer storage ensures that its destination cell need not be shut down but can continue to operate. The inoperative source cell can then be retooled or maintenance performed to get it back into operation.

5. **Palletizing Finished Parts.** The robot can be used to palletize finished parts after inspection and completion. The considerations here are the size and number of parts the robot can reach

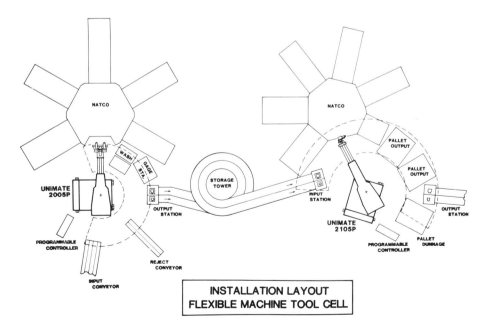

Fig. 44.1. Installation layout of flexible machining cell.

and place on the pallet, and the number of levels of parts. The robot can be programmed to pick up spacers or pieces of dunnage (protective packing material) to fit between each level of parts on the pallet.

6. **Input/Output Devices.** The use of input and output devices should be considered. Parts must be brought into the cell, and the robot must be informed that they are in place so it can pick them up properly. The robot must also be told when a part has been positioned on the output device. Because of the flexibility of the robot in the system, this function can be performed without a human operator.

7. **Quick, Easy Changeover.** Another point to consider in the planning of a flexible machine tool cell is the provision for quick and easy operation changeover. If part size or configuration changes, or if an extra operation must be added, the robot should be readily programmable to accommodate the change.

8. **Tool Changing.** In certain machining cells a robot can also be employed in tool and fixture changing. Discussion of this application is included in Chapter 57.

44.3. EXAMPLE OF A FLEXIBLE MACHINING CELL

Figure 44.1 shows the use of two flexible machining cells, or complexes, with a storage tower between them. The layout was designed for Sperry Vickers, North American Group, in Omaha, Nebraska.

The parts being handled are hydraulic pump cover castings made of ductile cast iron. There are 28 varieties of parts ranging in weight from 3 to 8 lb (1.4–3.6 kg) and ranging in size from $3 \times 3.5 \times 0.75$ in. to $6 \times 6.75 \times 3.5$ m. ($7.6 \times 8.9 \times 1.9$ cm to $15.2 \times 17.1 \times 8.9$ cm). This is the first system known to the author designed to accommodate such a wide range of part sizes and weights.

Each of the two cells is located in a 40-ft (13.3-m) bay that accommodates a multiple-operation rotary-table tandem machining center designed especially for this application. Input, output, and reject conveyors are shown along with a wash and gage station in the first cell and pallet stations in the second cell. At the center of each cell is an industrial robot. These particular robots—the 2105 Unimate and the 2005 Unimate—can handle up to 250 lb (112.5 kg), and each has five axes of freedom enabling them to perform a broad range of motions. The entire machining complex is surrounded by a 5-ft (1.7-m) high chain-link fence as a safety measure. The input conveyor is a two-lane, gravity-fed blue-steel design. The lanes are adjustable to accommodate two parts of the same configuration and size, or one part only in either track.

The first robot uses its dual-cam gripper to pick two parts from the set-down stand in front of the machine tool. The parts are loaded into the five-station, dial-index machine (NATCO) where several machining operations are performed, including milling, rough and finish boring, drilling, and facing.

The robot unloads the machine tool and places the two semifinished castings into a wash station. Next, the robot reloads the machine tool with two raw castings on the set-down stand and starts the machining cycle of the machine tool. Then the robot goes to the wash station, regrasps the two washed castings, and places them on a gage where the critical bore diameters are checked.

While these parts are being gaged, the robot rotates to the dual-lane input, picks up two raw parts, rotates to the set-down stand in front of the first machine tool and deposits the two raw parts. Next the robot rotates to the gage and picks up the two gaged parts. These are either placed on their respective cold-rolled steel pallets on the output conveyors, or, if out of tolerance, on the reject conveyor. If one part is good, it will go through the output conveyors. Only the bad part will be rejected.

The output conveyor is a double-lane pallet-handling system with a circular storage tower that has a buffer storage capacity of 200 pallets, one part per pallet. This conveying mechanism feeds the two parts on the small steel pallets to the input of the second machining cell.

The second robot then loads the castings one at a time into the second machine tool. This is a special, six-way machining center that does horizontal and vertical milling, drilling, chamfering, reaming, spot facing, and tapping. The robot unloads the machine tool by picking the parts out of the machine tool fixture, and again makes a decision to either place them on an outgoing conveyor to a manual operation down the line, or palletize them in containers.

The containers are mounted on automatically leveling lift tables. After a layer of parts is palletized, the robot signals the table to drop the correct distance to allow it to load the next layer. The robot then goes to a dunnage rack, picks up the dunnage (layers of plastic in this case) and places it on top of the previously loaded layer of parts.

Machining time for each part is 45 sec for the first cell, 36 sec for the second cell.

44.4. GRIPPERS FOR MACHINING CELLS

Figure 44.2 shows a typical double-part gripping hand used for machining-cell applications. This is a cam-actuated hand with special polyurethane grippers so that the parts will not be marred. These

Fig. 44.2. Typical double-part, cam-operated gripping hand.

Fig. 44.3. Typical single-part, cam-operated gripping hand with dual purpose.

provide compliance to accommodate the rough casting tolerances of the raw parts. A cam will keep closing until it meets resistance; thus a large part and a small part can be picked up with the same gripper during the same program step.

Figure 44.3 shows a typical single-part gripper hand that has a dual purpose. The hand has special pins that pick parts off the input pallet at the second machine cell. The pins grasp each part in previously drilled clearance holes on its cover casting. When the palletizing program calls for dunnage to be added between part layers, the program allows the hand to use its vacuum gripper to pick up the plastic dunnage and place it into the pallet.

44.5. SYSTEM CONTROL AND PROGRAMMING

The system control for each cell is built around a Westinghouse PC-900 programmable controller. The control has two basic functions: (1) controlling the robot through its external memory address port, and (2) interlocking the operation of all the interconnected equipment in the cell.

The operation of each cell is directed by the programmable controller. The controls on the front panel of the controller cabinet allow an operator to choose either of two operating modes, manual or automatic. In the manual mode the operator can access 11 robot programs for programming the robot or checking robot operation. In automatic mode the operator can select any of eight different output options. The output option selected will determine the operating sequence for the cell.

A teach-control unit is used to program the robots to follow the sequence of movements required for the various machining operations. By "walking" the robot through the cycle of steps in loading and unloading the covers, the sequence is recorded in its memory. The programming of the supervisory controls of the system was developed before the robots were installed. After installation, the robots were taught the actual movements to adapt them precisely to the process.

44.6. EFFECTIVENESS OF THE SAMPLE APPLICATION

The system described is a good working example of an effective flexible machine tool cell, designed to take into account the considerations outlined in Section 44.2 of this chapter. Good planning allows two robots to accomplish work previously done by six operators (three per shift, over two shifts). Productivity is higher because the robotic cells can be relied upon to repeat the same operations at consistent speed throughout the day.

This example is just one illustration of how robots can be used as integral parts of flexible machine tool cells. A well-designed machining center is a proven, efficient, and effective way to increase production, improve product quality, and realize higher profitability in many types of manufacturing.

CHAPTER 45

AN INTEGRATED LASER
PROCESSING ROBOTIC CELL

ALBERT M. SCIAKY

IIT Research Institute
Chicago, Illinois

45.1. INTRODUCTION

This chapter is intended primarily to illustrate the application of robotic principles in a flexible automated manufacturing cell. Considerable work is being done by the U.S. National Bureau of Standards (NBS) at the Automated Manufacturing Research Facility.[1] The facility, which is now in the development stage, will resemble a flexible manufacturing system designed to handle the major part of the part mix now manufactured in the NBS instrument shop for internal use. However, this facility is addressing only the manufacture of individual parts by chip-forming metal removal, and is intended primarily to be a vehicle for providing a realistic test of the NBS concepts of the process of deterministic metrology, the demonstration of certain control architectures, and the generation of standards for the interfaces in computer-integrated manufacturing systems.

At IIT Research Institute, a Flexible Automated Manufacturing Technology Evaluation Center (FAMTEC) has been set up with the cooperation of the Illinois Institute of Technology to develop and demonstrate manufacturing cells in which the other processes such as heat treating, welding, polishing, forming, and bending are integrated into the cell.

The availability of high-energy beams such as lasers and electron beams, when integrated into the manufacturing cell, opens significant opportunities for cost savings and product improvement. The combination of processes can greatly reduce the time parts are in transit as well as the work-in-process inventory. It is believed that the control architecture at FAMTEC, in its essential parts, can be the same as that being developed at the NBS facility.

Interfacing with computer-aided design systems as well as shop scheduling, inventory control, and manufacturing requirements planning (MRP) systems in the future will also be considered.

45.2. MANUFACTURED PART SELECTION

The type of part chosen for the demonstration facility is the plug from a process control valve. Process control valves are used to regulate the flow or pressure of a variety of gases and liquids. This is done by moving the plug in relation to an orifice to obtain a variable flow restriction. They are used extensively throughout the process industries and come in sizes ranging from 0.5 to 4 in. pipe diameter. Approximately 30.6 million valves were manufactured in the United States in 1977.

Process control valve plugs are produced in small batches because each application needs some special features. The nose of the plug is ordered with one of several shapes designed to give a specific flow characteristic (see Figure 45.1). The material from which the plug is machined also varies according to the fluid required. A low-temperature slurry, for example, requires a different material from that used with a high-temperature corrosive gas. In some cases, heat treatment or cladding of specific areas of the valve plug may be required to extend its useful life (Figure 45.2). Table 45.1 gives examples of material choices for various applications.

The demonstration cell has produced the shapes shown in Figure 45.3. A cut-away model of the process control valve and the valve plug can be seen in the assembly.

45.3. CELL CONFIGURATION

The demonstration cell covers a floor area 10 × 10 ft. The laser source is external to the cell, and the beam is transmitted to the work station by a system of mirrors through protective duct work. A

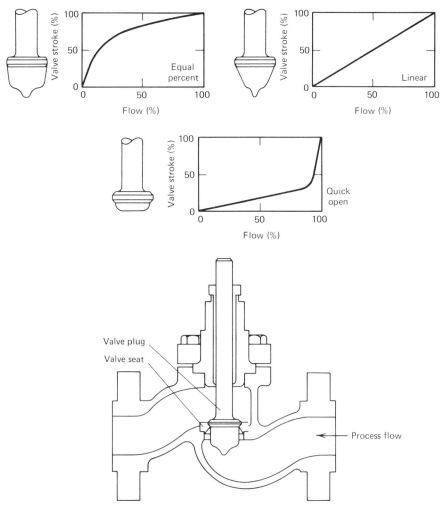

Fig. 45.1. Process control valve and plug flow characteristics.

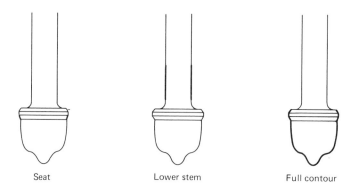

Fig. 45.2. Satellite clad variations of valve plug.

TABLE 45.1. MATERIAL AND TREATMENT VARIATIONS FOR SEVERAL APPLICATIONS

Application	Material							Treatment	
	Carbon Steel	Aluminum	Brass	304SS	310SS	316SS	440SS	Heat Treat	Clad
High temp corrosive					X				
Low temp corrosive				X					X
Low temp corrosive	X								X
High temp abrasive					X				X
Low temp abrasive							X	X	
Low temp abrasive	X							X	
Steam				X					
Oxidizing				X	X	X			
Acidic				X		X			
Cryogenic		X	X			X			
General purpose	X	X		X					

869

 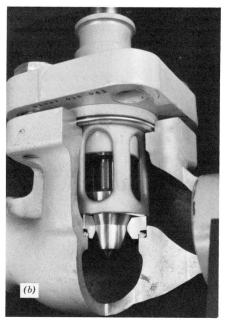

Fig. 45.3. Plug manufactured in demonstration facility. (*a*) Four plug types. (*b*) Plug in position.

Plexiglas enclosure protects personnel from possible scattered laser radiation during operation and encloses the robot work envelopes.

The cell consists of a raw-stock feed station, an industrial robot, a CNC lathe, a laser heat-treating station, a cleaning station, an automatic inspection station, and a set of finished-part bins (Figure 45.4).

Feed Station

This is a simple magazine holding up to 10 blanks which can be fed from outside the cell enclosure. The blanks are $1\frac{5}{8}$-in. diameter bar stock, 4 in. long, and are released one at a time by a pneumatically operated escapement. This provides clearance for the robot fingers to grasp the part. A part presence limit switch is provided in the pickup position. The robot checks this switch before attempting to pick up the part and will not continue if a part is not there.

Robot

This is a GCA/DK P800 industrial robot distributed in the United States by GCA Corporation Industrial Systems Group. It is an accurate, high-speed robot designed for multipurpose industrial applications. The five-axis configuration was found to be compact while providing a large work envelope.

The electric drives provide the necessary speed range to handle the rapid moves as well as the carefully controlled speeds required for laser heat treating. The maximum payload is 66 lb (30 kg), and the maximum reach is 54 in. (135 cm) from the centerline of the robot.

The controller is a microprocessor-based point-to-point control with teach/playback as well as manual data input (MDI) edit programming. Linear interpolation was not supplied on this particular unit but is available and would be desirable. Discrete inputs are provided for program branching, and outputs can also be activated by the robot control program. Internal program-timing functions are also provided. The discrete input and output functions are indispensable to the proper integration of the robot in the manufacturing cell for they serve as a means of communication with the rest of the system.

CNC Lathe

A Kitako Variant 10 CNC lathe is used in the cell. It has a 10-hp variable-speed spindle, a six-position turret, and a hydraulic chuck. It is equipped with a Fanuc 5T controller. The machine was originally designed to be run by a human operator. Integrating the lathe into the cell required that

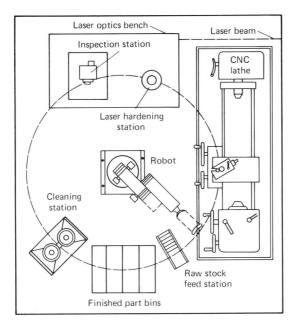

Fig. 45.4. Demonstration cell configuration.

the operator functions of loading and unloading the parts, loading punched tapes, and pushing buttons be performed without operator intervention.

The robot performed the part-loading and unloading functions, and the lathe controller was modified so that NC control programs could be downloaded from the cell control computer. The various push-button operations were also wired to the cell control computer interface.

Laser Heat-Treating Station

A low-vibration modular structure which holds four water-cooled copper mirrors was constructed using Klinger optical frame members and mirror mounts. The parallel beam from the laser source is directed first to a 5-m focusing mirror. The beam then becomes convergent and is directed by the remaining three mirrors to the bottom of the rotary optic fixture where it enters with a diameter of approximately 7.5 mm.

The rotary optics fixture (see Figure 45.5) contains a counterbalanced rotating disk on which are mounted two copper mirrors. The beam enters through the hollow shaft which has an ID of approximately 15 mm. It is reflected upward by the rotating mirrors, and thus the beam sweeps out a cylindrical path which is intercepted by a fixed water-cooled conical mirror at the top of the fixture. The beam is then reflected by the conical mirror and applies a ring of laser energy to the part. The focal point of the beam is approximately 30 mm beyond the center of the part (also the center of rotation of the beam). Therefore the beam is slightly defocused, and the variations in part diameter do not greatly affect the case depth obtained by the heat-treating process.

The primary parameters controlling the heat-treating process can be programmed to accommodate practically any shape of part. They are as follows:

Laser power	1–10 kW
Beam rotation	500–8000 rpm
Axial movement	0–150 mm/sec

One of the advantages of this method is that, since the part is not rotated, powders can be easily applied to the surface when cladding or surface alloying is desired. A coating station could easily be added with the robot taking part in the process.

Cleaning Station

This station is a simple arrangement of two commercially available, slotted-paper polishing wheels mounted with a 25-mm clearance between them. They can be seen in the lower left-hand corner of

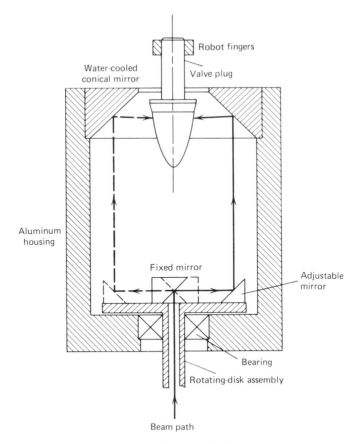

Fig. 45.5. Rotary optics fixture.

Figure 45.6, the overview of the cell. The operation is intended to provide a uniform surface condition to help identify possible surface defects at the inspection station. The robot simply passes the part between the wheels and needs only to rotate the part approximately 180°.

Inspection Station

This unit utilizes a Model VS110 Machine Vision System. A standard Vidicon camera is used to view the part while the robot slowly rotates it under the camera. (Note: A detailed discussion of this type of system is given in Chapter 14.)

Critical part dimensions were checked, and monitoring of possible surface defects was investigated. The dimensional checks were performed with relative ease, but checking for surface defects required delicate threshold adjustments in the system as well as carefully controlled lighting of the part. A simple accept/reject signal was given to the cell control computer, and the robot was signaled to route the part to the appropriate bin.

45.4 CELL CONTROL COMPUTER

This consists of an Apple II plus computer with a 64K memory extension, a parallel interface card, two floppy disk drives, and an "Isaac" interface manufactured by Cyborg Corporation. This interface is designed to operate as an Apple peripheral device and comes with a software package called "Labsoft," which is an extension of Applesoft BASIC. The particular configuration used contains 16 discrete inputs, 16 discrete outputs, 16 analog inputs, 4 analog outputs, and 4 Schmitt trigger inputs. In addition, a real-time clock, a counter, and a timer are built in. Although designed as a general-purpose laboratory multichannel data acquisition and control system, it was found to be relatively easy to use as the cell controller. Simple commands allow the status of the inputs and outputs to be stored as variables in

Fig. 45.6. Overview of cell.

the memory. In this application, the cell sequence logic is written in BASIC and simply tests the status of various inputs to set the status of the outputs according to a prescribed program.

In addition, the cell computer receives requests for the number of parts of each type to be manufactured. Additional parts are automatically made to make up for the rejects, if any. The numerical control part programs are stored on the floppy disk, and the requested program is downloaded to the lathe through the parallel interface card. The lathe has enough memory to hold the program for one complete part.

During the cell operation, the status of the inputs and outputs is displayed on the screen. Also, descriptive messages are scrolled as the operations are performed. This enables the system to run without being constantly watched by an operator. If a malfunction occurs, the last operations are displayed as a guide for the operator or maintenance person to take corrective action.

45.5. SEQUENCE OF OPERATION

In the photograph shown in Figure 45.6 the robot is seen in the center of the cell. In the foreground, from left to right, are the cleaning station, the finished-parts bin, and the raw-stock feed station. The CNC lathe is on the right, and the laser optics station is behind the robot. The inspection station camera mount is also behind the robot, but a portion of it can be seen. The lathe controller and robot controller cabinets can be seen on the left, outside the Plexiglas enclosure. The cell control computer is to the left of the cabinets, beyond the edge of the picture.

During the cell operation, there are two parts in the system. While one part is being machined in the lathe, another part is heat treated, polished, and inspected. The cycle evolves as follows:

When the machining is completed, the robot moves in and unloads the lathe (Figure 45.7).

The robot carries the plug to a holding fixture and deposits it there (Figure 45.8).

The robot moves a blank from the feed station to the lathe (Figures 45.9–45.11), and the lathe begins machining the stem.

Fig. 45.7. Robot removes machined plug from lathe.

The robot then returns to the holding fixture, picks up the part, takes it to the rotating optics fixture (Figure 45.12) where it is laser heat treated, and returns the part to the holding fixture.

By this time, the valve stem machining operation is completed; the robot picks up the part from the lathe, turns it around, and reinserts it for the nose machining operation.

During the machining of the nose, the robot picks up the heat-treated part from the holding fixture, takes it to the polishing station, and then to the inspection station.

After inspection, the part is carried over to the appropriate bin (finished parts or reject) (see Figure 45.13).

By this time, the machining operation is completed, the robot moves to the lathe, and the cycle begins again.

45.6. DISCUSSION

The cell clearly demonstrates the feasibility of process integration in the flexible manufacturing cell. The cell produces one part every 8 min. The important benefits to be gained are that the work-in-process inventory can be drastically reduced when compared to carrying out the operations of machining,

Fig. 45.8. Robot places machined plug in holding fixture.

Fig. 45.9. Robot removes blank from feed station.

Fig. 45.10. Robot carries blank to lathe.

Fig. 45.11. Robot inserts blank in lathe chuck.

Fig. 45.12. Part is heat-treated at laser optics station.

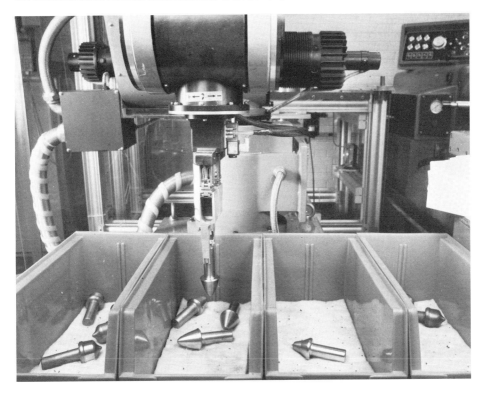

Fig. 45.13. Robot places part in finished parts bin.

heat treating, cleaning, and manual inspection in different parts of the plant. Moreover, since the response time needed to process the first part is only 16 min, the finished-parts inventory can be virtually eliminated.

Future developments should include the evaluation of algorithms for adaptive machining. With the data obtained in the vision system, NC part program changes could be generated automatically for specific repairs to be made on oversize parts.

The flexibility of the laser offers many other possibilities. A welding operation could be integrated into the cell. Laser welding offers high speed and low distortion, filler metal is often not required, and joint efficiencies are very high. Mechanical parts joined by laser welding can be held to close assembly tolerances.

Considerable savings in machining time and in materials can be realized by designing laser-welded mechanical parts instead of machining them from a single piece of raw stock or a casting.

45.7.   CONCLUSIONS

Industrial robotics has come of age. The integration of robotic technology into manufacturing systems is providing a degree of flexibility which, until now, has been obtainable with manual methods only.

The abundance of inexpensive, reliable, and sophisticated microprocessor technology is making flexible automated manufacturing feasible and practical. In the past, the basic processes in manufacturing such as cutting, machining, welding, and heat treating have been considered separately and performed in different locations. The development of new tools such as high-power lasers, electron beams, electric discharge machines, and plasma arc machines provides us with the opportunity of combining processes in flexible manufacturing cells with significant increases in productivity.

ACKNOWLEDGMENTS

The writer wishes to thank Mr. John P. Lamoureux and Mr. Richard R. Ledford for their dedicated cooperation and support in the realization of the demonstration cell.

REFERENCES

1. Simpson, J. A., Hochen, R. J., and Albus, J. A. The Automated Research Facility of the National Bureau of Standards, *Journal of Manufacturing Systems,* 1983.
2. Fisher, E. L., Nof, S. Y., and Seidmann, A., Robot System Analysis: Basic Concepts and Survey of Methods, *Proceedings of the IIE Conference,* Cincinnati, November 1982.
3. Dupont-Gatelmand, C., Survey of Flexible Manufacturing Systems, *Journal of Manufacturing Systems,* Vol. 1, No. 1, 1982, pp. 1–15.

CHAPTER 46
ROBOTS IN THE WOODWORKING INDUSTRY

KEN SUSNJARA

Thermwood Robotics Corporation
Dale, Indiana

46.1. THE WOODWORKING INDUSTRY

The woodworking industry is a broad term that includes a variety of processes and operations. It begins with forestry and lumber operations, includes rough mill, drying and treating, dimensioning or cutting and shaping to size, assembly, painting, finishing or coating, and packaging.

For the purpose of investigating robots and the woodworking industry, we restrict ourselves to *dimensioning, assembly, finishing,* and *packaging* and do not consider forestry, lumber, or drying operations. We also limit ourselves primarily to the furniture industry, and exclude other woodworking areas such as building products.

In broad, general terms, the furniture industry can be separated into two major categories. The first of these is upholstered furniture including chairs, sofas, and seating. The second major category is cabinets, casegoods, and tables. It is the second category that we address in this chapter.

46.1.1. Dimensioning

Typical operation of a casegoods facility begins with incoming material, which is stored in an incoming warehouse and moved primarily by fork truck. This material is then segregated into lots or batches called *cuttings* and processed through the dimensioning portion of the plant. This dimensioning area will cut, size, mold, and sand various components to their final form prior to assembly.

The dimensioning area in the woodworking industry today is characterized by a large number of single-purpose machines. Each of these machines accomplishes a particular task, be it shaping, drilling, molding, or sanding. Each of these single-purpose machines is loaded and unloaded by hand. Material is moved from one pallet or furniture cart to the process and then to another pallet or furniture cart. There is generally in-process storage between the various operations required for a particular component. The cuttings are moved on pallets or wheeled carts from each production machine to the in-process storage area and then to the next production machine as they work their way through the facility. As can be seen by this description, a great deal of human material-handling labor is associated with the production of these various components. A large potential savings exists by simply combining a number of special-purpose machines into multipurpose machines, eliminating the material handling and in-process storage among the various processes. In addition, many of the tasks within the industry are simple enough that a potential exists for automation that uses robots.

46.1.2. Assembly

Assembly techniques for furniture vary from company to company and product to product. Closely mating parts that utilize a variety of adhesives and glues are commonly used. Staples, powered nailing machines, screws, and other mechanical fasteners are also used extensively. One assembly problem unique to the furniture industry concerns the fit of mating parts. The size of lumber components and thus their ability to fit together in an assembly can change with a number of environmental factors. For example, the size of lumber changes proportionately to the amount of atmospheric humidity. Because of these factors, automating the assembly process in the woodworking industry takes on an increased degree of difficulty.

Many pieces of furniture require assembly fixtures to align properly and square the various components with one another. Other pieces of furniture, however, can be assembled freestanding and obtain their alignment primarily from the fit between each of the components.

Short batch runs combined with a high level of variability in the various components to be assembled make the assembly process in furniture manufacturing a fairly difficult area to automate using industrial robots.

46.1.3. Finishing

Finishing the assembled piece of furniture can be accomplished in a number of different ways. Most furniture plants have some type of finishing conveyor lines. The most common conveyor consists of a series of metal pallets onto which is placed the piece of furniture to be finished. These pallets have a pin in the center protruding downward where they fit into a metal slot that runs the length of the conveyor. Two guide rails support the pallet with small wheels or rollers, and a chain located in the guide slot pulls the pallet along by way of the center pin. In this way, the pallet can be moved lengthwise past spray stations. They can be spun 90° and close-packed for travel through ovens and curing stations, reducing the necessary length of those stations. They can also be spun 360° by the operator during spray finishing, allowing access to the entire piece of furniture being finished.

Some furniture plants utilize an overhead conveyor from which is hung a metal platform on a large L-shaped bracket. The piece of furniture to be finished can then be placed on this metal platform which hangs approximately 1 foot from the floor. It is typical in the average furniture plant to place a variety of different pieces of furniture on the line at one time. Chests, dressers, headboards, and nightstands are typically placed on the line at random intervals, relying on the adaptability and flexibility of the individual operator to provide the final finish.

In many finishing rooms, sanding, buffing, and polishing tasks occur in conjunction with the more common spray-finish applications. One example of this type of task is sanding a sealer material on

Fig. 46.1. Robots depalletizing pressed wood boards at Okamura's Takahata Plant, Japan, for loading on a tenoner in a dimensioning department. The suction cup grippers allow tilting heavy panel boards without moving their surfaces.

Fig. 46.2. Machine loading/unloading of wood parts by robot.

the top of a case piece or desk. In this instance, after a sealer has been sprayed on the top and allowed to dry, several people using vibratory sanders sand the top to provide a smooth surface for the next clear coat.

In general, some finishing rooms in the woodworking and furniture industry can be automated using spray-painting robots with little fundamental change to manufacturing procedures, whereas others may require extensive changes to both manufacturing procedures and production lines and processing equipment. The extent to which each facility must change to accommodate industrial robots must be determined on an individual basis.

Before we discuss individual jobs and possible robotic applications, it is important to understand the current level of technology in the woodworking industry.

46.2. ROBOTIZATION POTENTIAL

Woodworking in the United States (not necessarily typical of other countries) is characterized by a series of single-purpose woodworking machines. These single-purpose machines are usually electrically driven and generally contain relatively unsophisticated control systems. Most woodworking plants have mechanical and electrical maintenance capabilities necessary to maintain and operate this type of equipment. Technical developments in the science of cutting and processing wood have occurred at a relatively slow pace, allowing this technology to be absorbed by most of the industry.

Since the late 1970s or so, certain pieces of numerically controlled equipment have gained acceptance by the woodworking industry. This movement began with the availability of numerically controlled routers. Although not yet accepted, a number of machinery supply firms are working with or offering computerized setups to reduce downtime between cuttings. The basic thrust of the industry, however, still remains single-purpose machines with advances being confined to further automation of these devices.

This approach is in contrast to the manufacturing approach taken by the European woodworking industry. European woodworking operates at a much higher level of automation than its counterpart

Fig. 46.3. Programming by leading a robot through a palletizing operation.

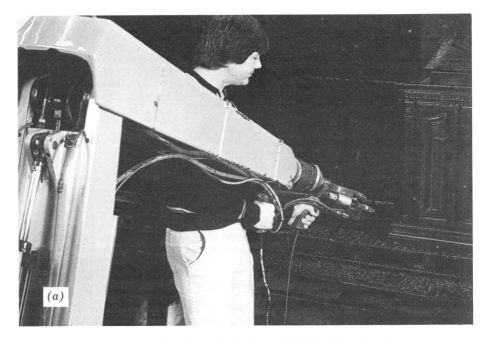

Fig. 46.4. Three examples of robotic painting and finishing of wood products.

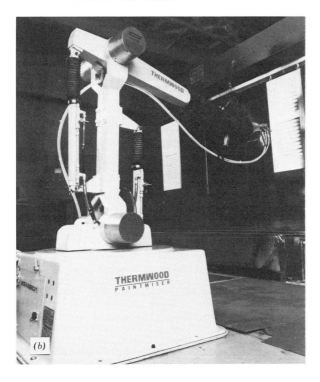

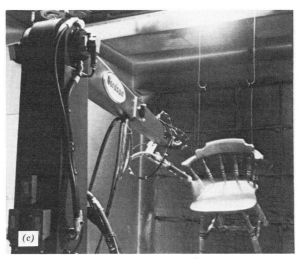

Fig. 46.4. (*Continued*)

in the United States. European automation, however, is primarily confined to what we call "hard automation." In stark contrast to the method of production used in the United States, the Europeans have opted for large, integrated, multifunction machines designed to produce an entire product or subassembly. One example of this might be a door machine where large sheets of pressboard or particleboard are fed into one end of the system, and without being touched by human hands, are processed at high speed into completed kitchen cabinet doors, totally finished with all the pulls and hardware installed. The reason for this sharp contrast in industry production methods seems to lie in the requirement of the marketplace. The methods used by the Europeans to manufacture furniture very productively allow little in the way of end-product design flexibility. The European market, however, is dominated by modern furniture and lends itself to this type of hard automation. On the other hand, the American marketplace is both highly diverse and constantly changing. Because of this, the American manufacturer feels flexibility is much more important to success and survival than productivity. It is important for the furniture industry to be productive and, even more important that, in achieving this productivity, it does not sacrifice the high degree of flexibility it enjoys today.

The furniture industry thus is a perfect candidate for the flexible automation achieved through robotics. As future population trends begin to impact the industry, making it increasingly more difficult to obtain labor, the industry will find itself at a crossroads. The most effective way of dealing with these problems in the future seems to be an extensive use of industrial robots and other programmable devices.

46.3. APPLICATION SELECTION

In examining the use of industrial robots in the woodworking industry, the major criteria must be the selection of jobs that can be accomplished without resorting to complex, expensive sensory systems. Most woodworking plants have the technology and capability in-house to handle the more basic types of robots. However, few, if any, have the necessary expertise to handle vision systems, tactile sensing, and similar sophisticated capabilities. At the same time, the standard woodworking plant is unstructured to the level that few, if any, jobs exist that can be accomplished by very simple robots.

46.3.1. Dimensioning Tasks

In examining the initial use of robots in the furniture industry, we can begin with the dimensioning section of the plant. Closer examination of the special-purpose machines will find a large number of machines, for example, belt sanders and tenoners, that require two people for their operation. The first person removes material from the furniture cart or pallet and places it in the machine. As the material moves through the machine, the necessary operations are conducted, and the second person unloads the component from the exit end. During the normal operation of this type of equipment, continuous adjustment and quality checks must be made to ensure that production is operating properly. Because of this, it is virtually impossible to totally automate the operation. The best solution in these situations is to in some way automate the in-feed end of the machine, allowing the robot to remove parts from the pallet and feed them into the machine, allowing the second individual to unload the equipment, make the necessary adjustments, and perform the quality-check procedures (see Figures 46.1 and 46.2).

Several problems exist in accomplishing this task. First, the stacks of material provided must be reasonably square and straight, and the material must be stacked consistently. This will require a certain level of regimentation that may be difficult to impose on some facilities. The second problem is that a robot program to remove 50, 100, or 200 components from a pallet and feed them into a machine can be quite complex. Since the characteristics of the material feeding through the equipment can vary from day to day, this system will only work if programming can be made very simple and straightforward.

One robot manufacturer has addressed this problem by using a lead-through-teach robot (Figure 46.3). A lead-through-teach robot is one in which the motions are input by physically moving the arm through the desired sequence (see more in Chapter 19, Robot Teaching). It has also supplied special software to allow for stacking. This software moves the robot to the point at which the last part was found, and if it does not find a part there, moves the arm in a straight line toward the floor until it encounters the next part. In this way it is not necessary to program each and every position at which a part can be found. It is only necessary to program the top part while the robot control system allows for the changing height of the stack.

More conventional material-transfer robots can be used to perform this task if the pallets or boards are placed on a lift table that raises the height of the board each time the top part is removed. In this way, each part is presented to the robot in exactly the same location that the preceding part was presented. While this method is not as elegant as the first method, it is a workable approach, especially when working with small, lightweight parts.

46.3.2. Finishing Tasks: Painting

The finishing room in many woodworking operations provides fertile ground for automation using robotics. By batching a product to be run rather than attempting to run completely at random, spray-painting robot applications are possible (see Figure 46.4). Most furniture conveyors will require some type of modification allowing the robot to control an attitude of the pallet carrying the product to be finished. This technology has been demonstrated at industry trade shows and can be accomplished with minor modifications to the conveyor system. Most spray-painting robot controls allow for storage of multiple programs and random selection of those, so that it is not necessary to completely change the randomized product finishing. It is necessary, however, that each product be consistently located on the pallet and that information identifying which product is on the pallet somehow be communicated

Fig. 46.5. Chair assembly experiments at Purdue University.

to the robot. This can be done in several ways, including manually inputting the information when the part is loaded, bar chart readers, and even simple vision systems located just prior to the spray booth.

46.3.3. Finishing Tasks: Sanding

Another area in the finishing room providing a certain level of consistency which holds promise for robot applications is sanding or polishing tops of furniture. Again, using a lead-through-teach robot, it is possible to sand the top of a passing piece of furniture. This is a difficult, tedious, and tiresome job in most plants, and provides a level of consistency that lends itself to operation by an industrial robot. Again, although total automation of the task may not be practical, one person and a number of robots working together could possibly accomplish what five or six people do today. This one individual would primarily concern himself with product placement, checking the integrity of the sanding paper, and small amounts of touch-up. Again, the flexibility of an individual is maintained while removing the largest portion of the undesirable physical labor. A similar situation exists in furniture assembly, where a number of laboratories around the world have performed experiments to overcome the problems of material variability and other difficulties, as discussed in Section 46.1 (see Figure 46.5).

46.4. CONCLUSION

This chapter describes potential robot applications in the woodworking industry and illustrates the method by which robots can be integrated into the industry. To be successful, robots must fit the needs and operating methods of the woodworking industry and must not require major structural changes inconsistent with the market requirements of the individual companies. For the foreseeable future, the approach should be confined to increasing productivity through elimination of manual labor, while maintaining a human presence. Although it might be technically possible for the myriad of decisions that are made each day to be automated, this automation could not occur with equipment that the typical woodworking plant would be able to install, operate, and maintain. It is, therefore, feasible to automate primarily by removing the physical labor, increasing throughput, while maintaining individual human decision making and judgment.

REFERENCES

1. Prak, A. L., Robots in furniture finishing, SME Paper No. MS75-251, 1975.
2. Susnjara, K., *A Manager's Guide to Industrial Robots,* Corinthian Press, Shaker Heights, Ohio, 1982.
3. Robotics in furniture production, *Cabinet Maker & Retail Furnisher,* No. 4324, February 4, 1983, pp. 22–24.
4. Robots! Okamura's programmable workforce, *Wood and Wood Products,* Vol. 88, No. 3, March 1983.
5. Two experts view robotics present and future, *Wood and Wood Products,* Vol. 88, No. 3, March 1983.

CHAPTER 47

ROBOT APPLICATIONS IN EASTERN EUROPE

TIBOR VÁMOS

JÓZSEF MARTON

Hungarian Academy of Sciences
Budapest, Hungary

This chapter illustrates 10 typical applications of industrial robots in Eastern Europe as follows:

1. Welding (Figure 47.1).
2. Assembly (Figures 47.2, 47.3).
3. Material handling in machining (Figures 47.4, 47.5, 47.6, 47.9).
4. Machine loading by mobile robots (Figures 47.7, 47.8).
5. Press tending (Figure 47.10).

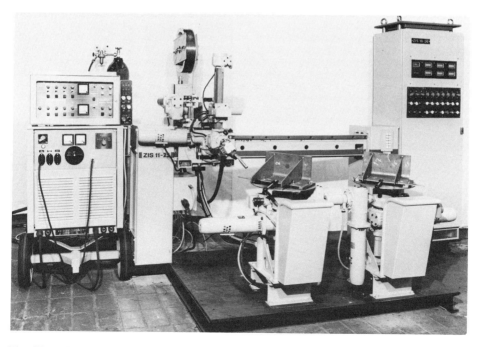

Fig. 47.1. Welding. Robot model ZIS 11 25 made in East Germany (GDR), is set up from the modular elements of the ZIS 995 building-block system. (Courtesy Prof. E. Peisler.)

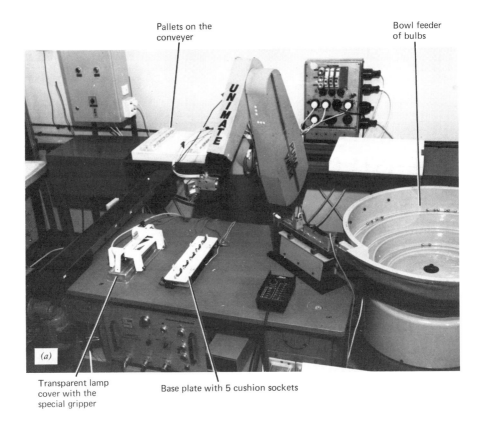

Pallets on the conveyer

Bowl feeder of bulbs

(a)

Transparent lamp cover with the special gripper

Base plate with 5 cushion sockets

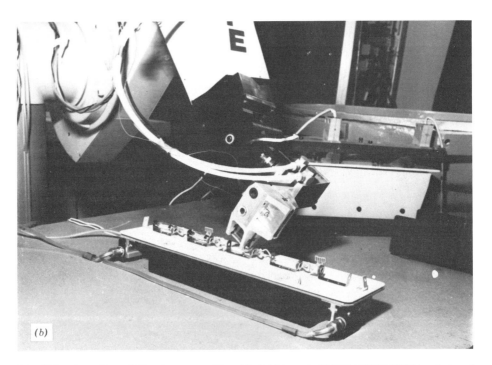

(b)

Fig. 47.2. Assembly. Ceiling lamp assembly with British-made UNIMATION PUMA robot and Hungarian-made Bakony accessories. (a) General view, (b) The robot snaps in the bulb. (Courtesy IKARUS.)

Fig. 47.3. Assembly. Keyboard assembly at Robotron in East Germany (GDR) with stepping motor driven robot PHM4, also made in East Germany.

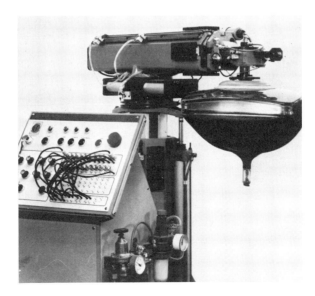

Fig. 47.4. Glass handling. Hungarian pick-and-place robot MTE 55 designed for television picture-tube handling. (Courtesy TUNSGRAM.)

Fig. 47.5. Machine loading. A lathe from Bulgaria with built-on-console part-handling robot ATM 001. The robot is made in Bulgaria by a FANUC licence. (Courtesy BEROE.)

Other popular robot applications in Eastern Europe include the following:

1. Fabrication, for example, deburring, die casting.
2. Injection molding.
3. Foundry applications.
4. Forging.
5. Finishing and painting.

Source material for the illustrated applications and others is listed in the references.

REFERENCES

Bulgaria

1. *Industrial Robots for Automation of Manufacturing Processes* (in Hungarian translation), BEROE Bulgarian Robot Factory, Stava Zagova, 1982.
2. Nachev, G., Adaptive GMA Welder, Now a Reality? *Proceedings of the 1st International Conference on Robots in Automobile Industry,* Birmingham, U.K., April 1982.

Hungary

3. Journal, *Mashinostroenie,* Editor: MSNTI, Moskva, and KG-INFORMATIC, Budapest, Vol. 1, 1983.

Czechoslovakia

4. Dubsky, B., Robotics, A New Branch of Science, Paper No. 1, *Applied Robotics Conference,* CSSR, May 1975, pp. 1/1–54 (in Czech).
5. Kveton, J., Kain, V., and Novotny, P., Further Developments of Robotics Within the ZPA

Fig. 47.6. Machine loading. One version of modular robot IR S2, made in East Germany (GDR), serves one (a) or two (b) NC lathes.

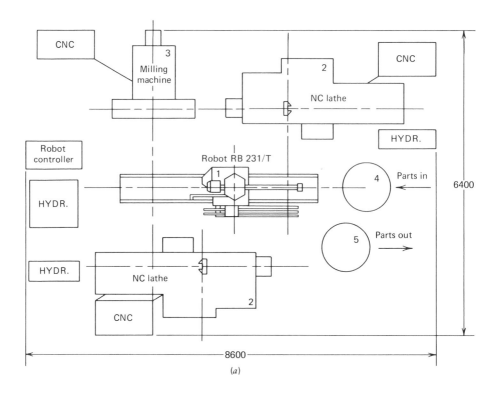

(a)

(b)

Fig. 47.7. Machine loading by a mobile robot. Automatic manufacturing cell ATM-100 (made in Bulgaria) producing shafts, served by a track-mounted robot RB-231/T, which is made in Bulgaria with a VERSATRAN license (Courtesy BEROE). (*a*) Cell layout, (*b*) overview.

Fig. 47.8. **Machine loading by a mobile robot.** One of the most popular robots from the USSR, type Universal-15.01, tends four NC lathes in a machining cell for shafts. (*a*) Overview of cell, (*b*) robot close-up.

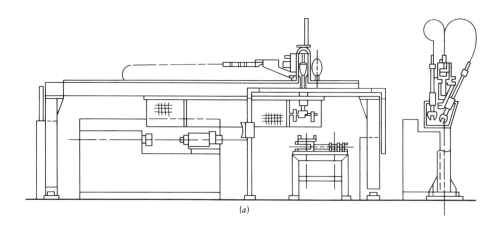

(a)

(b)

Fig. 47.9. Part handling by overhead mobile robot. Machine tool service with a hydraulic overhead-mounted robot, built from the Bulgarian building-block system PIRIN. (Courtesy BEFOE.) (*a*) Front and side view of robot, (*b*) overview of system.

Fig. 47.10. Press tending. Two separate one-armed pick-and-place robots of the same type, ECHD 65, made in E. Germany (GDR), tend a press. (Courtesy Prof. E. Peisler.)

Group of Companies, Paper No. 48, *Applied Robotics Conference,* CSSR, May 1975, pp. 48/a–8 (in Czech).

6. *Proceedings of Applied Robotics 77,* Czechoslovak Scientific and Technical Society, Karlovij Vary, October 1977. (Includes 50 papers, 10 in English, the rest in Russian and Czech.)

7. Umetani, Y., Recent Advances in R&D and Utilization of Robots in Eastern Europe (in Japanese, with some listings in English), *Robot,* Vol. 22, March 1979, pp. 4–13. Describes a 1978 technical tour of Czechoslovakia, Bulgaria, and Poland.

East Germany

8. Journal, *Fetigungstechnik und Betrieb,* Berlin, Vols. 32, 33.

Poland

9. Buc, J., and Morecki, A., Study and Application of the Industrial Robot in Poland, *Proceedings of the 1st International Symposium on Industrial Robots,* Washington, D.C., March 1979, pp. 1–15.

10. Buc, J., Industrial Robots Designed and Manufactured in Institute of Precision Mechanics—development and application in Polish industry, *Proceedings of the 1st International Symposium on Industrial Robots,* Milan, Italy, March 1980, pp. 643–655.

11. Tadeusz, S., Application of Industrial Robots in Construction of Assembled Automats, SME Paper No. MS-19-254, 1979.

12. Morecki, A., Kaczmarczyk, A., and Strzelecki, T., Education in Robotics in Poland, *Proceedings of the 13th International Symposium on Industrial Robots,* Chicago, Illinois, April 1983, pp. 15.33–43.

USSR

13. *Promyshlennye roboty, Katalog* (Industrial robots, catalog, with 54 robots), Yurevich, E. I., Ed., Leningrad Polytechnic Institute, Moskva, 1978, 110 pages.
14. Yurevich, E. I., Second Generation Robots of Leningrad Polytechnic Institute, *Industrial Robot,* Vol. 5, No. 3, September 1978, pp. 127–130.
15. Yurevich, E. I., Industrial Robots in the USSR, *Industrial Robot,* Vol. 6, No. 1, March 1979, pp. 26–30.
16. *Promyshlennye robot, Katalog* (Industrial robots, catalog, with 110 robots) Vasilev, V. C., Ed., Moskva, 1981.
17. Journal, *Mechanizatsija i automatizatsija proizvodstra,* Moskva, Vol. 1980, 1981, 1982, 1983.

Yugoslavia

18. Vukobratovic, M. T., Development of Industrial Robots in Yugoslavia, in Brooks, B., Ed., *Developments in Robotics,* IFS Publications, Bedford, England, 1983.

PART 9

WELDING

CHAPTER 48

ROBOTS IN SPOT WELDING

MARIO SCIAKY

Sciaky S. A.
Vitry-Sur-Seine, France

48.1. INTRODUCTION

One of the main applications of robots has been and still is in automotive spot welding. This is not because it is a simple process; on the contrary, it is a complex one. However, it is a good example of a cost-justified application that also relieves humans from a tedious and difficult job. Furthermore, the automotive industry around the world joined forces with robot makers to improve and advance this application area. J. Engelberger tells in his book[1] how it all happened. Traditionally, large, heavy car body parts had been held by clamping jigs, tacked together by operators using multiple welding guns and then spot welded manually too.

In 1966 the first steps were taken to use a robot to guide the welding guns and to combine the control of the robot with the control of the gun. In 1969 General Motors in the United States installed 26 Unimate robots on a car body spot welding line. Then in 1970 Daimler-Benz in Europe used Unimate robots for body-side spot welding. Since then robotic spot welding has mushroomed, and at the end of 1982 constituted about 25% of all robot applications in Japan and more than 30% of all robot applications in the United States.

Spot welding robots often work three shifts (with the third devoted partly to overall line maintenance), and produce 80 and sometimes well over 150 cars per hour. All major car manufacturers use robots for spot welding because of their speed, accuracy, and reliability. The repeatability and consistent positional accuracy that spot welding robots can achieve provide for a much better quality product than in manual operation. In some cases this can be done with fewer spots welded in locations that can be accurately and precisely controlled.

The objective of this chapter is to explain the spot welding operation, the characteristics of robotic spot welders, and the special considerations in planning robot spot welding applications. Readers are referred to Chapter 34, Planning Robotic Production Systems, which also discusses aspects of spot welding, to illustrate the detail of the production planning process. Chapter 36, Modular Robots Implementation, uses automotive spot welding in several examples. Finally, the following chapters of Part 9, Welding, focus on robot spot welding in production and in operation.

48.2. THE SPOT WELDING OPERATION

Welding is the process of joining metals by fusing them together. This is distinguished from soldering and brazing, which join metals by adhesion. In spot welding, sheet metal sections are joined by a series of joint locations, *spots,* where heat generated by an immense electric current causes fusion to take place (see Figure 48.1).

Three factors are critical to successful spot welding:

1. Pressure between electrodes.
2. Level of current.
3. Welding time.

The pressure exerted by the electrodes on the joined surfaces controls the resistance of the local column of material being welded. With too much pressure the air gap is minimized, resistance is reduced, and higher current is required to generate the heat for fusion. With insufficient pressure the resistance is higher and the spot may burn because of excessive heat. Thus, there is an optimal relationship between the three factors, depending on the material of the workpieces and their thickness.

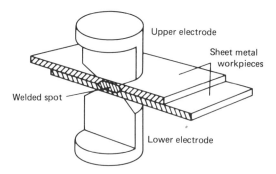

Fig. 48.1. Arrangement of electrodes and workpieces in resistance spot welding.

48.2.1. Welding Sequence

A typical spot welding sequence includes the following four steps:

STEP 1:   Squeeze—hold the two surfaces between the electrodes.
STEP 2:   Weld—turn current on for required duration; heat is generated at the spot.
STEP 3:   Hold—keep the electrodes closed for the duration required to cool the spot; usually, cooling water is circulated through the electrodes.
STEP 4:   Release—release the electrodes' grip and rest.

The spot welding machine is automatically controlled to repeat this sequence accurately and repeatedly, and the control is adjustable for different welding conditions. The electrical components of a spot welding machine are depicted in Figure 48.2.

Materials most appropriate for spot welding are ferrous metals—they are electrical conductors, and they do not have low resistance (such as in aluminum and copper) that requires excessively large current. The most frequent workpieces are made of cold-rolled low-carbon steel, but high-strength steel and galvanized material are also common. Thickness ranges from 0.6 to 1.0 mm. Parts are usually stamped and blanked out in such a way as to facilitate the welding process. The most typical application areas for spot welding are, therefore, the manufacture of automobile bodies, domestic appliances, sheet metal furniture, and other sheet metal fabrications.

48.3.   OVERVIEW OF SPOT WELDING METHODS

Resistance spot welding lines have gone through several transformations since inception. Early assembly lines consisted of a long conveyor on which the components were assembled by spot welding. The

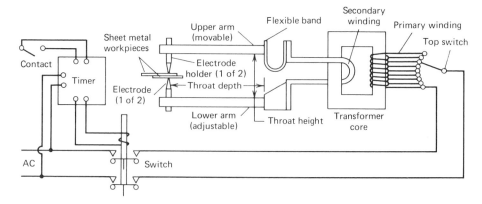

Fig. 48.2. Electrical components of a spot welding tool.

workers operated manual overhead welding sets, each performing his specific task. The system is still used today for low production rates and has the advantages of high flexibility and adaptability; model changes are made possible without undue expense.

The search for improved productivity brought about the use of multiple spot welding machines and transfer lines. The principle consists of moving the parts to be assembled through a series of automatic welding stations, each having one electrode for each spot weld. A line could have as many as several hundred electrodes. These automatic lines were highly inflexible, and a major drawback was the risk that it would be impossible to amortize them if the model for which they were designed did not sell well.

With industrial computers and robots appearing on assembly lines, it is possible to return to the original flexibility offered by use of welding guns, but with the advantages of increased accuracy and production quality. Production facilities are able to accommodate three or four different body styles smoothly and in any order with multilevel computer management control coordinating all stages of production.

48.4. STRUCTURE AND COMPOSITION OF A ROBOTIC SPOT WELDER

A spot welding robot consists of three main parts:

1. A mechanical assembly comprising the body, arm, and wrist of the robot.
2. A welding tool, generally a welding gun.
3. A control unit.

48.4.1. Mechanical Assembly

This is an articulated mechanical structure with the following functions:

To position the operational extremity of the robot—the tool it carries—at any point within its working volume.

To orient this tool in any given direction so that it can perform the appropriate task.

Performance Criteria

When a choice must be made among several robots for a given application, it is necessary to compare their mechanical performances. These include:

1. The number of degrees of freedom or axes.
2. The maximum stroke along each axis and the volume of movement.
3. The resulting maximum speed.
4. The maximum permissible load at maximum speed for a given distance between the center of gravity and the point of application.
5. The usable volume generated by the displacement of the operating point (welding gun electrodes).
6. The positioning accuracy.
7. The repeatability.

Table 48.1 gives numerical values for these and other criteria required of industrial spot welding robots.

TABLE 48.1. TYPICAL RANGES OF MECHANICAL REQUIREMENTS FOR SPOT WELDING ROBOTS

Tool load	40–100 kg
Torque	120–240 Nm
Displacement speed	
Linear	0.5–1.5 m/sec
Angular	60–180 degree/second
Accuracy	±0.5 or ±1 mm
Repeatability	1.5 or 2 mm

48.4.2. The Welding Tool (Gun)

The welding tool is considered to be a resistance welding gun, composed of a transformer, a secondary circuit, and a pressure element.

When the spot weld distribution lines are straight and present no access problem, the attainable rate is 60 spots per minute. Welding two pieces of steel sheet metal each 1 mm thick at this rate requires a 10-cycle current pulse of 10 kA. The welding force to be applied by the electrodes to maintain pressure as explained before should be approximately 3000–3500 N. The power rating of the welding transformer and the secondary voltage will depend on the impedance of the secondary circuit.

These electrical considerations, along with the problem of accessibility, can be met with one of the three following transformer-mounting configurations (see Figure 48.3):

1. Overhead transformers.
2. On-board transformers.
3. Built-in transformers.

Table 48.2 gives numerical indications of the merits and inconveniences of each of these systems, as explained in the following sections.

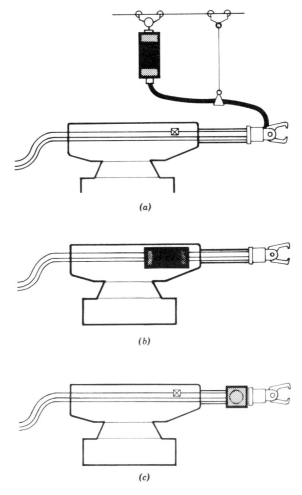

(a)

(b)

(c)

Fig. 48.3. Three different methods of adapting resistance spot welding equipment to robots. (a) Suspended overhead transformer with secondary cable. (b) On-board transformer with a short secondary cable. (c) Transformer integrated into welding gun.

TABLE 48.2. COMPARISON OF VARIOUS SPOT WELDING EQUIPMENT SUITABLE FOR ROBOTS

Characteristic	Transformer Type		
	Suspended	On-board	Incorporated into Gun
Secondary cable			
Length (m)	3.5	1.0	0
Cross section (mm²)	150	150	—
Secondary voltage (V)	22.3	10.0	4.3
Maximum short-circuit current (kVA)	312	140	60
Power at 50% duty cycle (kVA)	160	70	30
Transformer mass (kg)	120	51	18

Welding Guns with Overhead Transformers

The welding transformer is suspended above the robot and is mounted on a track to follow the movements of the robot wrist without undue traction on the cables connecting the welding gun to the transformer. The length of these cables is sufficient to absorb the displacements and rotations of the wrist, and they are usually held by a balancing device, which bears some of the weight. The movements of robots, especially the wrist rotations, make tensile and torsional stresses unavoidable on the welding gun; these stresses are transmitted to the robot wrist. Moreover, during welding, the repeated application of 10 kA of welding current generates intense electrodynamic stresses, which are transferred to the gun and the wrist mechanism.

On-board Transformers

An improvement on the preceding configuration consists of mounting the welding transformer on the robot, as close as possible to the welding gun. The design of the robot must be compatible with this approach; it is generally easy to achieve on Cartesian coordinate robots, and is possible on polar or spherical coordinate robots.

When the transformer can be mounted on the arm, the secondary cable length is significantly reduced. It is even possible to connect the transformer secondary outputs, by means of rigid conductors, close to the wrist. Such a transformer is heavy, and the overload makes necessary a reduction of the robot speed.

With some types of polar coordinate robots with articulated arms, the transformer is used as a counterweight to balance the weight of the gun and cable at the tip of the arm. There also exist certain robot designs where the transformer is housed inside the arm, and the secondary cable is run inside.

Welding Guns with Built-in Transformers

An attractive solution, which cannot be applied in all cases, is to use a welding gun comprising a built-in transformer specifically designed for use with robots. The main advantage is the elimination of the heavy secondary cables, replacing them with primary cables of smaller cross-sectional area. The secondary impedance becomes much lower since it is limited to the gap of the gun, and the size of the transformer is consequently reduced. For example, a 30 kVA transformer measures 325 × 135 × 125 mm and has a mass of 18 kg. The transformer, fully integrated into the welding gun, is directly attached to the active face of the wrist so that its center of gravity is as close as possible to this face to minimize the moment of inertia. Guns of this type are shown in Figure 48.4.

This system has two disadvantages:

1. The welding gun is bulkier. This can lead to penetration difficulties when carrying out certain welds where access is limited, for instance, in wheel wells.
2. A markedly greater mass; the transformer doubles the mass of the gun. A weight of 40 kg is acceptable; at values greater than 50 kg this can become a problem for most robots.

More than the static load, it is the stress generated on the active face of the robot wrist that can become excessive. Indeed, spot welding robots must allow for a torque of at least 120 Nm. This

Fig. 48.4. Modular 33 kVA, 3.5 kN welding guns with integrated transformers. Both have four different mounting surfaces for adaptation to the piece-part geometry and self-equalizing secondary circuits. (*a*) Scissors-type gun. (*b*) C-type gun.

corresponds to a 50-kg gun whose center of gravity is 240 mm from the axis of the wrist. Many robots furnish torques exceeding 200 Nm and, as such, are better suited to handle welding guns with built-in transformers.

48.4.3. Control System

The control unit of a spot welding robot fulfills three essential tasks:

Actuator control.
Programming.
Process control.

Actuator control and part programming are features common to most robotic installations; it is the addition of specific process control and its interlinking into the production process that is particular to robotic spot welding. Provisions are normally made in the control system to provide for flexibility by intervention from upper-level computer control that is coordinated with material purchasing, stock control, and customer orders; the type of models produced and the line speed are accordingly adjusted.

The process control in resistance welding involves an AC phase-shift controller and counting the number of periods of weld time. This can be accomplished by a conventional weld-control timer, but a more effective (and flexible) method is to integrate the process control into a centralized control that includes the robot positioning and the adaptive parameter information. This system makes efficient use of the real-time control system since the weld process takes place when the robot axes are static. Unification of all control functions eliminates the proliferation of interconnections resulting in systems composed of robot controllers, programmable logic controllers, weld timers, and communications links. Interfacing is greatly simplified.

The soundness of the spot welds can be monitored electrically by the dynamic resistance method. Information gathered from this serves two essential purposes. First, if the weld is determined to have insufficient penetration, respot robots further down the line add spots around the faulty point and the centralized control alerts personnel of the possibility of a malfunctioning gun or improper choice of parameters. Second, the dynamic resistance monitoring reveals if there was sufficient "splash" in the weld to cause the electrodes to stick together; the robot head cannot move. Concurrently, the opening and closing positions of the weld gun secondary circuit are monitored to prevent the axis controllers from indexing the gun to the next weld position. Immediate intervention is necessary.

Centralizing the weld process control and the axis control is necessary for adaptive parameter control. As various models come down the line, different weld distributions and sheet metal thicknesses are encountered having various heat-sinking capacities. The weld parameters must be programmed to accommodate for this.

Combining the overall line logic control with the weld control allows for power interlocking of the weld guns. In some cases, power interlocking ensures significant savings in the power distribution system along a line. In other instances, better quality welds are produced because the network is capable of delivering the necessary current to each gun.

Computer-assisted maintenance is essential when many robots are involved. Electrode wear and replacement is a common problem. As the electrodes flatten with use, the current can be programmed to be gradually incremented to maintain a constant current density in the spot weld. Electrode lifetime is thus lengthened; the control system informs maintenance personnel when particular electrodes require replacement. Axis-displacement times are monitored and compared with predetermined values to warn of possible mechanical wear.

48.5. PLANNING ROBOTIC SPOT WELDING LINES

The planning of a robotic assembly spot welding line to meet a manufacturer's specific production needs requires detailed investigations to optimize the proposed solution. This section examines the particular aspect of spot welding by robots and their planning implications.

Design data from the product to be manufactured can be brought together to give a first estimate of the facility requirements. These include the following:

1. The parts to be assembled.
2. The geometrical conformation of these parts and the corresponding number of stations required.
3. The distribution of the spot welds and the number of robots required to weld them.
4. The production rate and the number of lines required to meet the production needs.
5. The desired degree of flexibility.

To complete the design, additional information is needed:

6. The basic principles relating to the transfer and positioning of the assembly.
7. The final selection of the robot, its equipment, and its installation.
8. The environment and the available space.

These decisions are based on the final design of the line and may be heavily influenced by political, social, and economic considerations. Normally, several designs are proposed; the final choice should reflect the best possible compromise between the technology and its cost, as explained in Chapter 33, Evaluation and Economic Justification.

48.5.1. Parts to Assemble

The first step in a project is to undertake a detailed study of the parts to be assembled for purposes of classification. The operational procedure is determined by this classification.

In the automotive field there are numerous examples of various small reinforcing parts welded onto one main part. Such an assembly is shown in Figure 48.5: tunnel reinforcements, seat belt hooks, and cable brackets are welded onto the stamped panel. Extensive loading and small-part-handling equipment must be provided for in this type of assembly. Another example of this type of construction is a suspension arm equipped with cable attachment lugs, spacers, and stiffeners.

A second classification is the assembly of several parts of the same size. These include the front end, composed of baffles, apron, and radiator cross-members, and the rear floor frame made up of longerons and cross-members.

A third classification is the "toy tab" method; approximate geometry is obtained by the mechanical interlocking of the subassemblies. An example is the "body-in-white" composed of the underbody, body sides, roof braces, and the roof.

A similar classification can be used on other fields such as household appliances and metal furniture.

48.5.2. Geometrical conformation

The geometrical references are the significant zones of a part or subassembly that define its theoretical position in the X, Y, and Z dimensions. Compliance with the references of the component elements of a subassembly thus guarantees the geometry of the completed assembly. Reference positioning is therefore an essential function of assembly. Except for built-up assemblies (an assembly or subassembly obtained by layout and preliminary tack welding) where the geometry is an inherent element, it is essential, before any welding assembly design, to perform a detailed analysis of the part reference definition function. Geometric designs of this kind must be conducted in cooperation with the user; either the latter specifies his references and their positions, or the assembly-line manufacturer defines and implements them with the user's approval.

Main references are those that must be maintained throughout the production process, from the stamping phase on. Secondary references are those that are used only when assembling the unit. Figure 48.6 shows a floor panel with its references. The references are maintained by mechanical elements (fixed or movable reference pins, locators, and clamps) for each assembly.

The reference definitions must be preserved in each phase if the entire assembly is not completed in a single operation. The station or stations thus defined are *geometry conforming*. The location of these references makes it difficult for the robots to reach the welding areas. For the assembly of a number of small parts onto a main element (a main part on which are welded various smaller stiffeners), access may prove particularly troublesome.

48.5.3. Weld Distribution

In general, the parameters pertaining to the spot welds, such as quantity, location, and strength, are specified by the product designer. These data are drawn from research, design studies, and tests previously conducted during the product planning and design phase for composition, shapes, and assembly.

Selection of Tack Welding Points

In the geometrical conformation station(s), it is essential to carry out simultaneously designs concerning the number of points required for the assembly geometry and the relative positioning of the robot gun with respect to the reference elements (Figure 48.7).

Operational Procedure

The order in which the parts are loaded along the line must be carefully planned to allow for maximum accessibility of the welding tools until completion of the assembly.

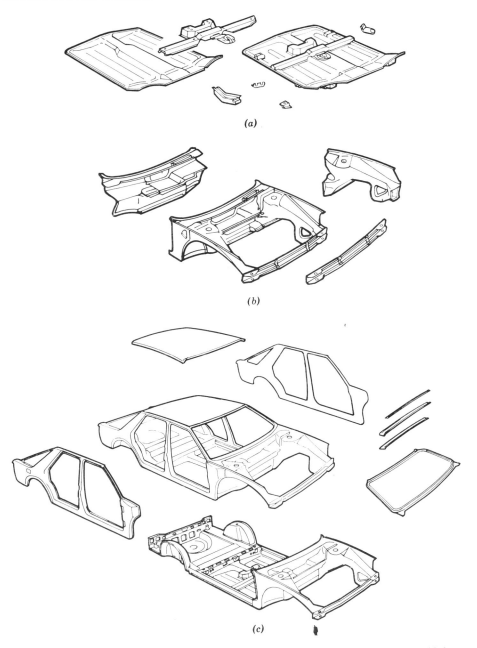

Fig. 48.5. Examples of automobile sheet metal assemblies. (*a*) Stiffeners and accessories added to floor panel. (*b*) Front-end composition. (*c*) Final assembly of underbody, body sides, and roof.

Spot Weld Grouping

At this stage it is necessary to define the grouping of spot welds that can be welded by a single welding element. Similarly, a first approximation must be made of the "base times" required for welding the various weld groups as a function of their position in the assembly and of the existing layout elements.

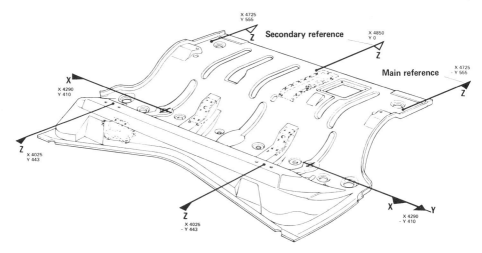

Fig. 48.6. Main and secondary dimensional references for a floor panel.

Welding Gun Planning Phase

On the basis of the foregoing study, the minimum number and the configuration of the various welding elements necessary for these welds or weld groups can be established. If certain points are physically impossible to weld because of the welding method or the assembly itself, this will become apparent at this stage of the study.

48.5.4. Production Rate

Based on the production procedure, the production rate dictates the overall design and the components of an assembly line. In the automotive industry lines are usually moving continuously, and robots must track the lines to perform the spot welding operation "on the go."

Global Line Utilization Factor

This factor is a function of all of the elements entering into the makeup of the line, ranging from the supply of workpieces to be welded to the flow of finished products coming off the line. This factor depends on the design of the welding stations: it decreases if several stations are installed in succession; it increases when intermediate buffer stocks are provided.

Welding Time

The desired rate, taking into account the line utilization factor initially selected, converts into the cycle time per part. After deducting the product-handling and layout time, a preliminary estimate of the time available for the actual welding operation can be made. The importance of minimizing handling and layout time should be stressed. (See Chapter 30, Robot Ergonomics: Optimizing Robot Work.)

Preplanning of the Assembly Line

The welding time thus established provides a preliminary basis for the choice of the type of assembly line to be implemented. In fact, when the calculated time is obviously insufficient (a large number of points to be welded in a minimum period of time), a first option concerns paralleling a number of welding lines for the same assembly.

Minimum Number of Robots

The welding and displacement times, taking into account accelerations and decelerations, make it possible to preplan the number of points that can be welded in one cycle by a single robot. Taking into account the spot weld grouping and the necessary or possible types of welding guns, it is possible to define the minimum number of robots theoretically necessary for the assembly. This figure, of course, must be subsequently verified.

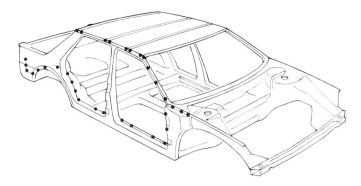

Fig. 48.7. Typical tack weld points for body side.

48.5.5. Flexibility

The flexibility of an assembly line is defined by its degree of adaptability with respect to the various products that are, or may have to be, processed on that line, simultaneously or not. This flexibility is defined by the following options:

Suitability of the tooling for several different products.

Capacity for adaption and the time needed for changeover to another product.

Degrees of Flexibility

The user may request various degrees of flexibility for the lines, which may represent large investment costs.

A line may be initially designed to process only one type of product. Its overall design, however, makes possible partial or complete retooling at the end of the production run, and the replacement of the original product by a product of related design.

A line may be initially equipped for one part type, but its design provides for the accommodation of several variants by adding or adapting the appropriate tools. An alternate is to design the line capable of producing several versions of a single-base product model.

A totally flexible line is designed to produce without preliminary adjustments, in any order, a variety of different products and their variants. Of course, these different products must be similar and involve a comparable production technique.

Replacement Flexibility

This type of flexibility is defined by the ease with which a line can continue to operate at a reduced rate in the event of failure of one of its components (robot or tooling). The line-management programming may be such that the line robots automatically compensate for the missing work of a faulty robot, taking into account problems concerning the specific type of welding gun and the specific capacity of the robot. An alternative is to install respot robots at the end of the line to make welds that were missed.

48.5.6. Part Positioning and Transport

Position of the Part during Processing

Large-scale assemblies, such as floor pans and bodies-in-white, should be handled in their normal or "car" position. Subassemblies like body sides are handled in several positions, depending on the loading restrictions of the smaller parts and the location and accessibility to the weld points. The geometrical conformation tooling must be taken into account.

Part Transfer

In some plants, floor pans and body-in-white assemblies are handled on skids or lorries. These handling devices are kept in all production stages as well as off the assembly line and are removed only after final assembly. These transfer units become an integral part of the assembly whether they are used as a transfer means or remain idle. For high production rates, skids cannot be used for part handling, but they remain fastened to the assembly when alternative transport means are employed.

Carriages are used when two different types of subassemblies are to be processed with geometrical conformation or when the subassembly is already assembled but its geometry must be sustained all along the line.

Automatic guided-vehicle systems are commonly used in flexible shops for the preparation of subassemblies such as doors, hoods, and dashboards.

48.5.7. Environment

Positioning of the Robot

The distribution and scatter of welding points on a given assembly or subassembly can determine the position of the workpiece with respect to the robot when this is physically possible or, if not, make it necessary to install the robot in a certain position. The robot can be installed on the floor, on a base, overhead, or inclined (see Figure 48.8). When the production line must be straddled, it is possible to install two, three, or even four robots on one gantry structure.

Floor Space

The space available around the work station to be equipped may sometime be limited, and the general layout in this case calls for the selection of gantry robots. This type of robot configuration lends itself to more compact installations. A transfer line with gantry robots is typically 4–5 m wide, while the same line equipped with floor-mounted robots occupies 2–3 m, additional, on either side, for a total line width of 8–10 m.

48.6. SELECTION OF ROBOT MECHANICAL CONFIGURATION

Certain generalizations can be made when selecting a robot for a specific spot welding task. Each task is associated with specific geometrical criteria that favor particular mechanical configurations. However, economic factors might modify this choice.

In automotive spot welding five typical spot welding tasks can be identified.

48.6.1. Geometrical Conformation and Tack Welding Stations

Because of its specific design, this case is one of the most difficult to deal with owing to the inherent shape of the body requiring geometrical conformation external to the volume. To weld the appropriate spots, it is thus necessary to pass through the surrounding conformation tooling. This type of station normally requires six robots: four positioned laterally on each side of the body and two at the front and the rear to weld the trunk and front end. (Three such parallel stations are shown in Figure 48.8.)

Regarding the lateral robots, those having cylindrical movements pose certain operational problems because the sweeping movements of the axes during penetration require large openings in the surrounding conformation tooling. For this reason, robots with polar movements and linear penetration are better suited to this type of situation and are thus frequently used. Nongantry Cartesian-coordinate robots are ideal for this application. The only factor limiting their widespread use is the large amount of floor space they require.

As for the two front and rear robots, it is almost impossible to use standard electric robots because of limitations on the positions they may occupy when suspended; they require special balancing adjustments for each configuration. Moreover, this also makes it difficult to install them in the middle of the line.

Hydraulic polar movement robots are the most suitable: they can be installed overhead, and their linear or articulated penetration permits welding at the bottom of the trunk. Their polar axis allows access to lateral points. Cartesian-coordinate gantry robots are also used, but their configuration is less favorable, the access to lateral points being critical. In all applications for this type of machine, it is essential that the robot head be as compact as possible and that the three wrist axes be arranged to minimize their sweep.

48.6.2. Underbody Lines

The main characteristic of this type of assembly is the need to perform all weld points from the outside (with some exceptions). In this kind of application, the robots must be chosen for their capacity to use large, heavy welding guns with long throat depths. The possible choices are thus restricted from the very start.

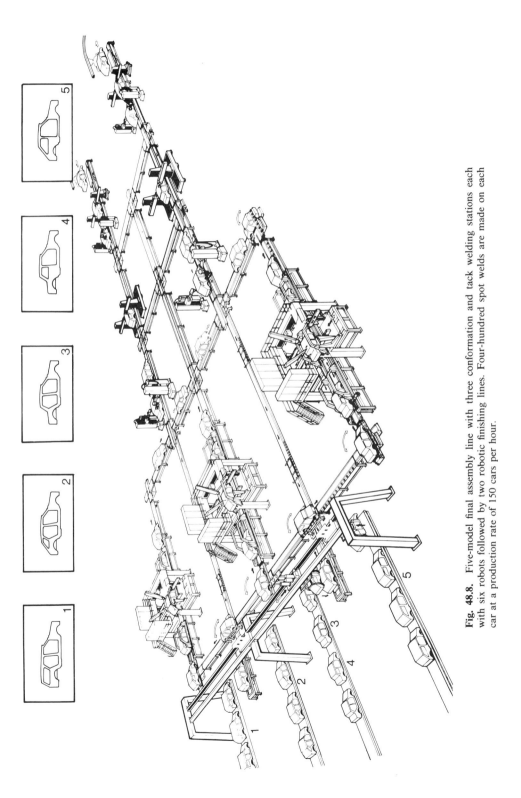

Fig. 48.8. Five-model final assembly line with three conformation and tack welding stations each with six robots followed by two robotic finishing lines. Four-hundred spot welds are made on each car at a production rate of 150 cars per hour.

Cartesian-coordinate robots (Figure 36.4) are the preferred solution to this problem. Indeed, this is one of the prime examples in which their application is economically attractive. In the gantry configuration, as in Figure 36.13, these robots can weld points symmetrically by lateral penetration; in some cases the robots are installed on pivoting arms on the sides of the gantry (Figure 36.14) to increase their flexibility.

Articulated-arm robots and cylindrical-movement robots are possible second choices. Compared with linear-penetration robots, they possess the inherent advantage of not requiring installation on a raised base for accessibility.

48.6.3. Body Side or Floor Frame Lines

These parts, characterized by large openings, permit welding with lateral access as well as from the center. Cartesian-coordinate robots are by far the best choice for this particular application, as shown in Figure 36.11. Polar-movement robots with linear penetration are compatible with this application but have the disadvantage of requiring a raised base.

48.6.4. Lines with Vertical-Plane Welding

When side panels are oriented in their "road" position or slightly inclined, or for final body welding (Figure 36.15), gantry robots are difficult to use. Most other configurations are satisfactory, especially those with linear penetration to reach points inside the body.

48.6.5. Final Assembly Lines

All types of robots can be used for these applications and are selected primarily on the basis of access. In the final assembly line shown in Figure 48.8 several different models are employed that provide access to the vehicle interior and enough flexibility to accommodate all five vehicle models.

REFERENCES

1. Engelberger, J. F., Spot Welding Applications, Chapter 11 in *Robotics in Practice,* American Management Association, 1980.

2. *Robotic Engineering in Car Body Assembly,* Report published by Sciaky S. A., France, 1982.

3. D. Leroux, *Les Robots, Strategies Industrielles,* Hermes, Paris, 1982.

4. *Developments in Mechanized, Automated and Robotic Welding,* Conference Proceedings, IFS Publications, 1981.

CHAPTER 49
Robots in Arc Welding

BRUCE S. SMITH

ASEA, Inc.
Troy, Michigan

49.1. ARC WELDING TECHNOLOGY

Arc welding is a technique by which workpieces are joined with an airtight seal between their surfaces. The principle of the process is the fusion of two metal surfaces by heat generated from an electric arc. During the welding process there is an ongoing electric discharge—thus the sparks between the welding electrode and the work. The resulting high temperatures of over 6000°F (3300°C) melt the metal in the vicinity of the arc. The molten material from the electrode is added to supplement the welding seam. Whereas spot welding is performed with alternating current, arc welding is performed with direct current, usually at 100–200 A at 10–30 V.

Originally, arc welding used carbon rods for electrodes. However, these did not add material to the weld, so metal filler rods had to be added. Modern methods essentially have replaced the carbon arc welding by providing quality solutions to the welding requirements. Some of these methods are listed in Table 49.1. In some methods, to prevent oxidation of the molten metal, electrodes are coated with flux material that melts during the welding process. Inert gas such as helium or oxygen serves also to prevent oxidation.

Automatic, electric arc welding equipment is comprised of a line through which continuous electrode wire is fed. The wire forms a tip that is graded along the desired weld trajectory. The following factors are critical to good welding results:

1. Correct feed rate of electrode wire.
2. Optimal distance between the electrode and workpiece.
3. Correct rate of advance of the electrode.

A typical sequence in automatic arc welding operation is comprised of five steps as follows:

1. Turn on inert gas flow over work area.
2. Start weld cycle: begin wire feed, turn power on.
3. Stop wire feed.

TABLE 49.1. COMMON ARC WELDING METHODS

Method	Comment
TIG —Tungsten insert gas	Welding with a tungsten electrode
MIG —Metal inert gas	Most common, using continuously fed metal electrode from a coil
GMAW—Gas metal arc welding	
GTAW —Hot-wire TIG	Welding processes with high weld-material deposition rate
SAW —Submerged arc welding	
FCAW —Flux cored arc welding	

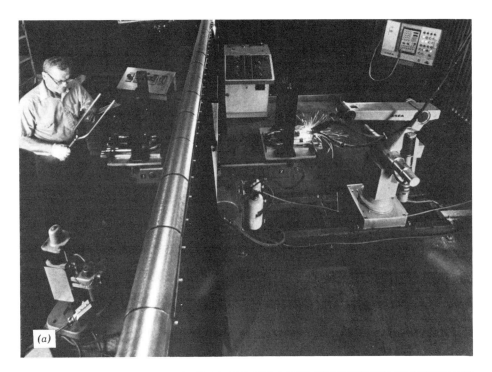

Fig. 49.1. (a) Arc welding robot station. While the robot is welding, the operator prepares to fixture the next part on a second positioner table. (b) Close-up of robot during welding.

4. Turn power off.

5. Turn gas flow off.

49.2. SELECTING A ROBOT FOR ARC WELDING

Robots are attractive for arc welding for several reasons, mainly:

1. Robots replace human operators in the unpleasant, hazardous environment of radiation, smoke, and sparks from the arc welding.

2. Robots relieve human operators from carrying and guiding heavy welding guns, which frequently must be guided in uncomfortable positions.

3. Robots can consistently perform the precise welding motions.

The robot control system can easily communicate with the automatic control of the arc welder to synchronize the necessary robot motions with the welding sequence of steps.

Usually, electrically driven robots are preferred for arc welding. Motion speeds required in arc welding are relatively slow, and the weight of the welding gun is relatively small too. When heavy welding guns, with a water cooling system, are applied, a hydraulic robot could be considered. Except for straight-line welding, continuous-path control is required. Interpolation is required to simplify accurate control of nonlinear welding lines.

49.3. THE ROBOTIC WELDING SYSTEM

A complete robotic welding system includes the robot, its controls, suitable grippers for the work and the welding equipment, one or more compatible welding positioners with controls, and a suitable welding process (with high-productivity filler metal to match). The system installation also requires correct safety barriers and screens, along with a plant materials-handling flow that can get parts into and out of the work station on time (Figure 49.1).

49.3.1. Welding Positioners

The welding positioner is a critical and often overlooked part of that system. Robots are not compatible with perfectly good welding positioners designed for humans. For example, the ASEA IRb 60/2 robot has a position accuracy of ±0.016 in. (±0.40 mm) with up to nine axes of rotation or motion and a 132-lb (60-kg) maximum load capacity. This position accuracy is useless if it is not matched by the positioner handling the weldment. Second, most welding positioners are not designed for operation under microprocessor control. One solution to the control problem is to retrofit an existing positioner with new controls compatible with the robot. That still leaves the problem of positioning accuracy unsolved.

A human operator can tolerate much more gear backlash or small amounts of rational "table creep" during tilting than a robot can. The human will compensate for the variation without thinking about it. The robot expects the weld joint to be where its program thinks it is, not where the positioner actually puts it. A conventional welding positioner can not be retrofitted at a reasonable cost for the precision required by a robot. The machine would have to be rebuilt.

Several major manufacturers of welding positioners already have new models on the market designed specifically for use with robots.

Modern robotic positioners have backlash controlled to within ±0.005 in./in. (±0.001 mm/mm) radius on all gears. A few years ago this would have been unheard-of accuracy for any large-geared machine. Figure 49.2 shows different varieties of part positioners for arc welding robots. The specifications for the illustrated models are as follows:

(a) *ESAB Orbit 500*

Workpiece:
  Max weight   1100 lb (500 kg)
  Max diameter   57 in. (1460 mm)
Indexing: 90°   360°
  Rotation 2.7 sec 7.2 sec
  Tilting   3.3 sec 8.9 sec

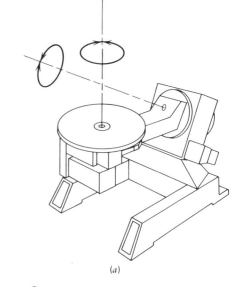

(a)

(b) *ESAB Orbit 160 R*

Workpiece:
  Max weight   352 lb (160 kg)
  Max diameter 45 in. (1150 mm)
Indexing: 90°   360°
  Rotation 2.3 sec   6.8 sec
Station interchange (180°): 4.5 sec

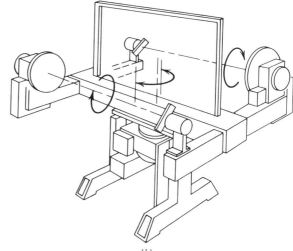

(b)

(c) *ESAB Orbit 160 RR*

Workpiece:
  Max weight   352 lb (160 kg)
  Max diameter   45 in. (1150 mm)
Indexing: 90°   360°
  Rotation 2.3 sec 6.8 sec
  Tilting   2.7 sec 7.2 sec

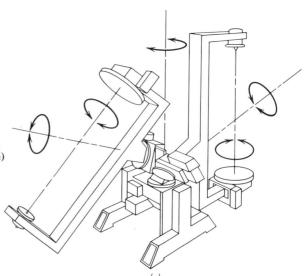

(c)

(d) ESAB Orbit MHS 150

Workpiece:
  Max weight   330 lb (150 kg)
  Max diameter   45 in. (1150 mm)
Indexing: 90°   360°
  Rotation   6.0 sec   12 sec
  Tilting   4.0 sec   —
Station interchange (180°): 7.0 sec

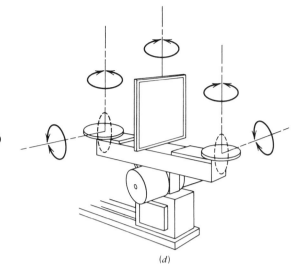

(d)

(e) ESAB Orbit MHS 500

Workpiece:
  Max weight   1100 lb (500 kg)
  Max diameter   Any
Indexing: 90°   360°
  Rotation   11 sec   22 sec
  Tilting   5.0 sec   —

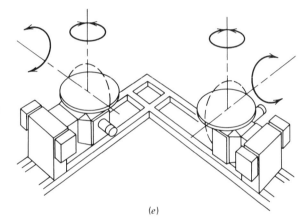

(e)

Fig. 49.2. Five types of part positioners for arc welding robots. (*a*) ESAB Orbit 500; (*b*) ESAB Orbit 160 R; (*c*) ESAB Orbit RR; (*d*) ESAB Orbit MHS 1503; (*e*) ESAB Orbit MHS 500.

All the components of a robotic work station must be interfaced so that each performs its function as initiated by the program. Typically, a robotic positioner will include multiple-stop mechanical or solid-state limit switches on all axes for point-to-point positioning (open-loop) or with DC servodrives for positioning with feedback to the control (closed-loop).

Speed is important in a robotic work station. The full rotation of the worktable positioner on two different axes (while handling a full load) is typically between two to three times faster than a conventional geared positioner. A full rotation in both axes with simultaneous motion takes about 9 sec. A 90° table rotation may take 3 sec.

49.3.2. Robot Control

Programming modes for a welding controller (Figure 49.3) include speed and position settings, arc voltage, wire-feed speed and pulsing data, and other operating parameters. The starting and ending phases of each weld (with special parameters such as pre- and postflow of shielding gas, puddle formation, creep start, crater filling, and afterburn time) are programmed in separately stored routines.

Movements of the welding robot are programmed independently by a separate control. Both control systems coordinate their functions on the handshaking principle (mutual computer affirmatives after each concluded clock cycle). The operator determines when the robot should switch from one workpiece

Fig. 49.3. Complete set of arc welding robot station and the programming unit to program its control.

to the next. The operator also can interrupt and adjust the program at will, even during the welding process.

Often, there is a requirement for the welding robot to move over a long distance to weld large workpieces, for example, in shipbuilding. A robotic linear track system can be used. For example, a track made by ESAB Co. allows the robot to move at operating speeds of 1870 in./min (47.5 m/min) for an ASEA IRB 60 robot, or 2160 in./min (54.9 m/min) for a smaller ASEA IRB-6 robot. The track system can move the entire robot up to 100 ft (30.5 m) or more along a weld seam, or extended workpiece, and stop with a position repeatability accuracy of ±0.004 in. (0.1 mm).

49.4. ECONOMICS OF ARC WELDING

Without analyzing total welding costs, manufacturing methods, present production requirements, and part design, a new robot could become a financial and production disaster. Conversely, with proper planning a user may increase productivity by 50–400%. Many users amortize their investment within one or two years, as found by ASEA in a recent survey of successful welding robot installations.

The total capital cost for a complete robotics welding system including controls, materials handling equipment, accessories, and tools can be double or triple the cost of the robot. The reason is mainly due to the cost of peripheral equipment that is designed for most effective robot operation. For instance, poor fit-up is a problem a skilled welder can usually adjust, but a robot can not solve by itself. At best, poor fit-up requires lower welding speeds. At worst, it may mean incomplete welds.

Accessibility of the joint to the welding gun nozzle and electrode is always a design consideration in welding. Humans can compensate for bad design with exceptional skills. Robots cannot, but if joint design and part tolerances are correct, the work is properly positioned, and the program is well defined, robots can increase weld quality consistently over people.

The cost of bad welds can be excessive. Improving weld quality, alone, can be enough to pay for some robotic welding systems. For example, one weld repair on heavy code welded plate can cost from $400 to $4000 per 30 cm (1 ft) for initial welding, inspection, X-ray, tearing out the bad weld, rewelding, and reinspection. A bad weld can cost much more in money, and sometimes lives, if the weld causes a failure when the part is in service.

Welding positioners designed to work with robots are essential for parts with weld joints the robot cannot reach, and for many joints the robots can get to. Simply reaching the joint is not enough to maximize the productivity of the fusion-welding process. The reason is that both oxy-fuel and arc welding speeds are strongly influenced by the position of the weld metal when it is being deposited.

Downhand welding often can increase weld-metal deposition rates by a factor of 10 or more compared with working out-of-position. What that means, using humans, is that one welding operator working downhand frequently can deposit as much weld metal as 10 operators working overhead. The same thing applies to a robot. Simply turning the work into the downhand position can increase productivity significantly. A robotic welding system can achieve maximum weld-metal deposition rates.

Other advantages of a robotic welding system are:

1. Arc-on time is maximized because the robot does not tire and take breaks.
2. Robots are easy to adapt to new work.

Labor rates for manual welding average 85% of the total welding costs. Even highly automated processes such as submerged arc welding, working with a welding-head positioner, typically require labor rates that are still more than 50% of total welding costs. Furthermore, the United States has a major skilled welder shortage. The situation is expected to get worse through the rest of this century.

49.4.1. Weld Direction and Cost

Downhand welding costs less than vertical or overhead welding because of gravity. In flat downhand welding, gravity allows higher deposition rates and lower labor costs per pound of deposited weld because the molten weld metal is held in position by gravity.

The higher the deposition rate, the greater the chances are that the metal will sag and run out of the weld. Welding out-of-position causes problems whether a human or a robot deposits the weld metal. The preferred solution is to turn the work so that the weld is in the flat downhand position and to use a robot to do the work.

Operators using a welding-head manipulator, multiwire, multipass welding, and a high-production GMAW, FCAW, or subarc system do not need a welding robot, since they already employ a dedicated, automated welding system. Welding robots can be justified if the work falls between intermittent stick-electrode welding and highly automated production welding.

49.4.2. Using Robots and Floor Space

Manufacturing floor space can be the most restrictive operating cost, and robotic welding systems make effective use of available floor space. Space comparisons with other materials-handling systems will generally favor a robotic welding system.

49.4.3. Using Multiple Positioners

The use of two positioners (or one with two tables, as shown in Figure 49.4) and one robot makes extra sense for another important reason. Since positioners are materials-handling devices specifically designed for handling and assembling weldments, the time used to load the work can be cut in half by keeping the workpiece on one positioner after it is welded and using the second positioner to keep the robot welding.

Some applications use work stations with three positioners and one robot. The first positioner is used for parts assembly and tack welding. The robot welds the work on the second positioner. The third positioner is used for parts inspection and nondestructive testing. A robot can be programmed to turn to any of the three positioners that is ready for welding. The labor time saved, alone, between stops for loading and unloading, has changed the entire workflow of some metal-fabricating plants.

49.5. WELDING COST CALCULATIONS

49.5.1. Cost of Filler Metals

No arc welding process is capable of converting all of the filler metal into weld metal. Some of the electrode's weight (unless it is an inert-gas-shielded solid wire) is converted into slag. A surprisingly large amount of most filler metals becomes spatter, and some of the electrode becomes fumes. If stick electrodes are used, a couple of inches of each electrode (typically one-sixth or 17% of it) also is thrown away as a stub. That immediately increases SMAW filler-metal costs by 20%.

Under ideal conditions the gas-metal-arc welding process (GMAW), using a solid wire, can be 99% efficient. At the other extreme, some shops deposit only 30% of the SMAW electrodes that they actually buy. The real cost per pound of filler metal is not that paid for the product in a carton or can, but the amount of electrode that can be converted into useful weld metal.[1] Therefore weld-metal cost is calculated as:

$$\frac{\text{filler metal cost (\$/lb)}}{E_d}$$

Fig. 49.4. A robotic welding positioner with two tables in operation. The curtain in the center of the positioner shields the operator from the welding sparks.

where E_d is the fractional amount of the electrode deposited as weld metal (rather than the amount of the electrode that actually is consumed), that is, the electrode deposition efficiency.

The effects of differences in electrode deposition efficiencies among various electrodes were not as obvious before the advent of welding robots as they are now. Labor cost overwhelmed the total cost of welding, and filler-metal costs accounted for only a few percent (typically less than 5%) of total welding costs. With the increased productivity made possible by the use of a welding robot in a properly designed work station, filler-metal costs become a significantly greater part of the total cost of depositing solid weld metal.

An analysis of filler-metal deposition efficiency and deposition rates can indicate that it may pay to use expensive wire to help the robot lower total costs. This could mean using high-deposition-rate processes such as flux-cored wire welding (FCAW), submerged-arc welding (SAW), or even hot-wire TIG (GTAW).

49.5.2. Deposition Efficiency

The influence of deposition efficiency on costs must not be taken lightly. Figure 49.5 shows how the relative cost of filler metal increases as deposition efficiency drops off. Deposition efficiency has an equivalent effect on labor costs (or robot amortization) discussed in Section 49.5.4.

49.5.3. Gas Costs

Most cored wires and all solid wires are designed to be used with shielding gases during actual welding. The contribution of the shielding gas cost to the overall weld-metal cost cannot be disregarded. Gas cost per pound of deposited weld metal can be calculated as

$$\frac{\text{gas price (\$/ft}^3) \times \text{gas flow rate (ft}^3/\text{h)}}{MR \times E_d}$$

where MR is instantaneous filler metal burn-off rate expressed in pounds per hour, and E_d is the electrode deposition efficiency expressed as a decimal.

Obviously this cost element can be kept to a minimum by selecting less expensive shielding-gas mixtures and setting the lowest possible flow rates, depending on the wire deposition rates and the

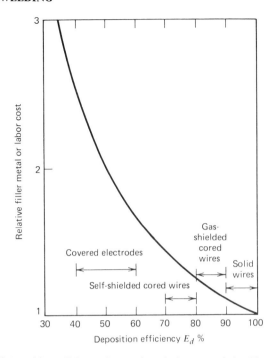

Fig. 49.5. Effect of deposition efficiency E_d on the relative cost of the filler metal or labor. Base labor cost or the cost of the electrode is 1. Covered electrodes have the lowest deposition efficiency, and thus the highest relative cost; solid wires boast the highest E_d, and the lowest relative cost.

arc power levels needed to obtain those rates. The welding equipment can have a significant effect on this cost because reduced flow rates are possible with well-designed nozzles.

On the other hand, lower gas-shielding flow rates too far increases the risk of producing weld porosity and other defects. In some wires, the nature of the shielding gas or gas mixture directly influences the weld-metal transfer mode (for example, at least 85% argon in the shielding gas is required to produce spray-transfer welding instead of dip-transfer).

The operating characteristics of some welding wires are quite sensitive to even small changes in shielding-gas composition. A few percent change in a gas-mixture component can significantly change the shape of weld beds, the wetting characteristics of molten metal, and the depth of penetration into the base metal.

Shielding-gas composition also has an effect on the weld-metal deposition efficiency E_d because the amount of spatter, fume, and slag produced depends on how oxidizing the gas is. Straight CO_2 shielding, although low-cost, produces high spatter. This spatter reduces filler-metal recovery, increases maintenance costs and downtime. The E_d for solid mild-steel GMAW wires is less than pure CO_2 shielding (about 93% deposition efficiency) than with argon-rich gas mixtures such as 98% argon + 2% CO_2, or 95% argon + 5% oxygen, which produces deposition efficiencies of about 99% with properly deoxidized, solid mild-steel wires. The specially deoxidized wire costs a little more, as does the shielding gas, but that cost is more than made up by the high-deposition efficiency of the wire.

49.5.4. Labor Costs

Labor is the most important element in fabricating costs. It must even be included in robot welding, either directly as the cost of the robot tender or programmer, or indirectly as labor cost to load and unload the robot work station. Besides depositing weld metal, welders and other workers must prepare joints for welding, position the components (or program the robotic positioner to do the job), fixture and tack weld parts, remove slag and spatter, and inspect the finished welds.

Shaving costs for welding filler metal, shielding gas, or equipment to save money will quickly be offset by increased labor cost to remove more spatter, repair bad welds, or for additional inspection required by using less than optimum welding conditions.

When determining present cost of labor (in manual system) for depositing a given amount of weld metal, consider the fraction of each welder's day devoted to depositing weld metal (their duty cycle DC) as well as the rates at which the metal is deposited while the worker is welding (the actual deposition rate, $MR \times E_d$, where MR is the melt rate and E_d is the deposition efficiency). The labor cost per pound of deposited weld metal can be calculated as

$$\frac{\text{hourly wages} + \text{overhead}}{MR \times E_d \times DC}$$

The influence of deposition rate and duty cycle on the relative cost of labor and overhead are shown in Figure 49.6. A number of features are revealed here. First, labor costs are very high when deposition rates are low. Relatively small changes in the welding rate cause large changes in cost. Labor costs are relatively low when the welding processes allow high deposition rates to be used, and even small changes in those rates cause significant changes in labor cost. Finally, small changes in the operator's effectiveness as measured by the duty cycle (percent arc-on time) can have a large effect on total cost.

In other words, pay attention to increasing deposition rates ($MR \times E_d$ or melt rate times deposition efficiency) as well as arc-on time, whether humans are doing the welding or robots do it. By increasing deposition rates from 2 to 4 lb/h (0.9–1.8 kg/h), relative costs can be reduced from 50 to 25 units or 50%. The same 2 lb/h (0.9 kg/h) change at a 20 lb/h (9 kg/h) deposition rate still reduces costs from 5 to 4 units, or 20%.

By increasing the duty cycle (arc-on time) from 25 to 50% when depositing weld metal at 4 lb/h (1.8 kg/h), the per-pound cost for labor can be reduced from 25 to 12.5 units. Or if the labor and overhead rate is $15/h for manual welding, the per-pound cost for the weld metal can be reduced by $7.50.

The same relative filler-metal cost reductions apply to robot welding, except that increasing a robot's productivity by 50% will cut in half the time to amortize the robotic work station.

49.5.5. Total Welding Cost

To obtain the total welding costs, the preceding three elements (filler metal cost, gas costs, and labor costs) must be added together. Their relative effects on the overall cost are summarized in Figure

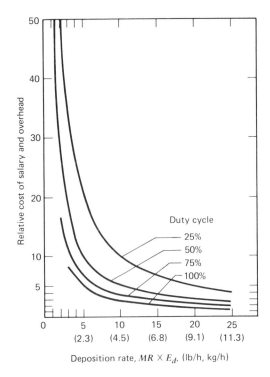

Fig. 49.6. Effect of deposition rate (melting rate times deposition efficiency) and the welder's effectiveness (duty cycle) on labor costs and overhead.

49.7. In this illustration, labor and overhead costs are assumed to be $15/h; 1982 list prices have been used to calculate the filler-metal and gas costs in these examples.

According to the analysis in Figure 49.7, $\frac{1}{8}$-in. (3.2-mm) E7018 SMAW stick electrodes are compared with 0.052-in. (1.3-mm) Verti-Cor® wire costs for depositing welds in the vertical position.

The welder's duty cycle was chosen as 30% for covered stick electrodes and 45% for the cored wire. They appear to represent shop experience with fair accuracy, assuming production-welding conditions. Welding robots have not been in use long enough to give comparable duty cycles for a cored wire, but a complete robotic work center with two or more positioners could probably deliver an arc-on duty cycle in excess of 75–80%. Since no one yet has sufficient experience with robots, a robotic cost analysis is not included in Figure 49.7.

Illustration

Because deposition rates are very low for covered SMAW electrodes compared with a flux-cored wire, small improvements are seen to reduce labor costs significantly. Those for welds made with covered electrodes are calculated to be about $22/lb ($48/kg) in 1982 dollars. With Verti-Cor® wire meeting the E70T-1 standard, however, both the deposition rates and duty cycles are improved, reducing the labor cost to about $8/lb ($18/kg) of weld (using 1982 prices). This $14/lb ($31/kg) difference more than compensates for the higher cost of the Verti-Cor® wire and the need for a shielding gas. The material costs differ by $0.50/lb ($1.10/kg), leaving an overall saving of $7.50/lb ($16.50/kg). Even if the covered electrodes were a gift, it would not pay a shop to use them instead of the flux-cored wire. If a robot welding station were used, the further increased arc-on time probably would be equivalent to getting the cored wire free.

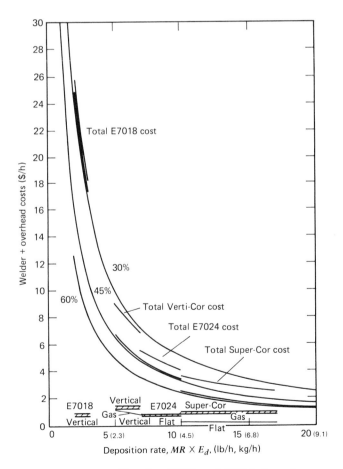

Fig. 49.7. Total costs for Verti-Cor, Flux-Cor, E7018, and E7024 electrodes are compared when gas cost, duty cycle, filler-metal cost, deposition rate, and labor costs are all considered. Here, the labor and overhead costs are assumed to be $15/hr.

Again, the key to these savings is reduced labor costs if human welders are used. If robots are used, most of the labor costs are transferred into capital costs, which can be amortized, whereas labor costs for manual welding remain a constant part of the total welding costs as long as the work is done manually. Labor costs per hour for manual welding increase in proportion to electrode and gas costs. Labor costs per hour for robot welding remain low as long as the work station is designed to keep up with the robot.

Increasing labor efficiency even slightly has a big effect on welding costs when the work is done manually, or even when a semiautomatic welding process is used. Not needing to change stick electrodes helps greatly. But changing electrode wire also is a factor in wire-fed processes. However, a robot work station can include large drum-fed wire containers instead of spooled wire, which substantially reduces wire-changing time. Easier slag removal (or no slag) helps. A more stable arc helps. At deposition rates of 5 lb/h (2.3 kg/h) and with wages and overhead at \$15/h, increasing a manual welder's efficiency from 30 to 35% reduces the weld metal cost by \$1.40/lb (\$3.00/kg). That saving alone could pay for all filler metals and shielding gas.

49.5.6. Joint Design Saves Money

Savings also are possible by reducing the amount of weld metal needed to fabricate a joint. The joint gap is designed to provide room for a welding electrode to reach into the joint, as well as to ensure complete fusion in the weld. Most welds designed for human operators are overwelded to get good average results. Just changing the included angle of a simple V-groove joint from 90 to 60° will reduce weld metal requirements by 40%. Several major filler-metal suppliers provide detailed tables for calculating the volumes of various weld joints. One of the better references is Lincoln Electric's *The Procedure Handbook for Arc Welding*[2] (see the chapter entitled "Determining Welding Costs").

49.6. ARC WELDING CASE STUDIES

Figure 49.8 shows 12 welded parts made by robotic work stations and the productivity increases that were observed in an ESAB Company survey. Productivity was measured in terms of the amount of weld metal deposited. The minimum increase in productivity reported is 60%; the maximum is 860%. The products in the figure range from truck frames to an exercise bicycle. What is not shown is the effort, pain, and frustration experienced by the manufacturing engineers who produced these results. The next example will tell about that. The particulars for each of the parts shown in Figure 49.8 follow.

49.6.1. A Robotic Arc Welding Experience

This example involves a series of products produced by General Electric's Mobile Radio Department at Lynchburg, Virginia. A manual MIG operation was replaced by a GMAW operation with ASEA's IRBG and ESAB MHS 150 positioner. The robotic welding experiences gained by G.E. in the application are summarized in the following list:

Simple parts are better to start with than complex parts. There is a great temptation to begin with complex assemblies that give high paybacks. That is a terrible mistake. Learn first on something simple.

Small assemblies are easier to fixture than large ones. They provide better accessibility and more repeatability. Parts with short welds and short programs require less time to set up and program. Long weld-programming runs should be limited to parts with large yearly volumes and high payback.

Manufacturing processes prior to welding should maintain tolerances that are compatible with the robot's ability. Existing quality-control measures should be reviewed and modified to ensure that these tolerances are met.

Improved operator skills, tooling, and fixturing will be required as a part of parts preparation. In short, make the parts right the first time because trying to compensate with fixturing at the welding station is not only more expensive but it often will not work.

Repeatability goes hand in hand with parts preparation. Getting consistent parts input may require changing preweld manufacturing equipment, changing machine operators, or changing materials vendors.

A good quality-control organization is essential. A thorough quality-control study must be made prior to purchasing the robot to determine repeatability of the assembly methods. This study should include several of the same types of assemblies and should take "stack-up" tolerances into consideration. The results can be combined statistically and compared with the specifications for the robot and positioner to determine compatibility with the robotic welding station.

(a) *Revolving Frame (Oregon): robotic productivity increase—200%*
Dimensions: $44 \times 21.7 \times 5.9$ in. ($1118 \times 551 \times 150$ mm)
Weight: 375 lb (170 kg)
Plate thickness: 0.75–1.25 in. (19–31.7 mm)
Number of welds: 10
Length of weld: 31 in. (787 mm)
Time/unit manual welding: 1.5 hr
Time/unit robot welding: 0.5 hr

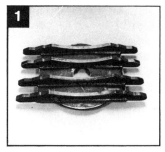

(a)

(b) *Rack (Sweden): robotic productivity increase—200%*
Dimensions: $31.5 \times 24 \times 17.7$ in ($800 \times 610 \times 450$ mm)
Weight: 44 lb (20 kg)
Plate thickness: 0.12 in. (3 mm)
Number of welds: 113
Length of weld: 88.6 in. (2250 mm)
Time/unit manual welding: 30 min
Time/unit robot welding: 10 min

(b)

(c) *Exercise Bicycle Frame (Finland): robotic productivity increase—136%*
Dimensions: $31.5 \times 23.6 \times 8.7$ in. ($800 \times 600 \times 220$ mm)
Weight: 22 lb (10 kg)
Plate thickness: 0.10 in. (2.5 mm)
Number of welds: 28
Length of weld: 53.1 in. (1350 mm)
Time/unit manual welding: 9.2 min
Time/unit robot welding: 3.9 min

(c)

(d) *Engine Mounting (Sweden): robotic productivity increase—250%*
Dimensions: $7.9 \times 13.8 \times 17.7$ in. ($200 \times 350 \times 450$ mm)
Weight: 26 lb (12 kg)
Plate thickness: 0.24–0.47 in. (6–12 mm)
Number of welds: 33
Length of weld: 60.4 in. (1535 mm)
Time/unit manual welding: 47 min
Time/unit robot welding: 13.4 min

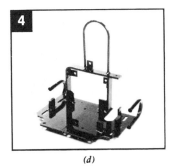

(d)

Fig. 49.8. Twelve products produced by robot welding. (*a*) Revolving frame; (*b*) rack; (*c*) exercise bicycle frame; (*d*) engine mounting; (*e*) gearbox casing; (*f*) truck frame cross member; (*g*) car seat frame; (*h*) container component; (*i*) battery holder; (*j*) trolley tray; (*k*) link part; (*l*) motorcycle frame detail.

(e) **Gearbox Casing (Belgium): *robotic productivity increase—112%***
Dimensions: 28.3 × 15.7 × 16.5 in. (720 × 400 × 410 mm)
Weight: 243 lb (110 kg)
Plate thickness: 0.63 in. (16 mm)
Number of welds: 32
Length of weld: 288 in. (7320 mm)
Time/unit manual welding: 69.4 min
Time/unit robot welding: 33.0 min

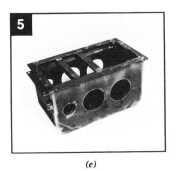

(e)

(f) **Truck Frame Cross Member (Sweden): *robotic productivity increase—100%***
Dimensions: 31.5 × 7.9 × 5.9 in. (800 × 200 × 150 mm)
Weight: 55 lb (25 kg)
Plate thickness: 0.24–0.31 in. (6–8 mm)
Number of welds: 12
Length of weld: 41.3 in. (1050 mm)
Time/unit manual welding: 8 min
Time/unit robot welding: 4 min

(f)

(g) **Car Seat Frame (Sweden): *robotic productivity increase—60%***
Dimensions: 18.5 × 18.5 × 5.9 in. (470 × 470 × 150 mm)
Weight: 8.8 lb (4 kg)
Plate thickness: 0.035–0.12 in. (0.9–3 mm)
Number of welds: 22
Length of weld: 21.7 in. (550 mm)
Time/unit manual welding: 2.4 min
Time/unit robot welding: 1.5 min

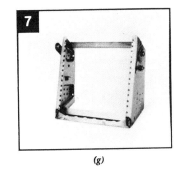

(g)

(h) **Container Component (Belgium): *robotic productivity increase—65%***
Dimensions: 78.7 × 7.9 × 20 in. (2000 × 200 × 50 mm)
Weight: 99 lb (45 kg)
Plate thickness: 0.12–0.60 in. (3–15 mm)
Number of welds: 45
Length of weld: 90.6 in. (2300 mm)
Time/unit manual welding: 14 min
Time/unit robot welding: 8.5 min

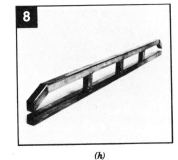

(h)

Fig. 49.8. (*Continued*)

(i) *Battery Holder (Sweden): robotic productivity increase—129%*
Dimensions: 25.0 × 10.6 × 2.8 in. (635 × 270 × 70 mm)
Weight: 2 lb (0.9 kg)
Plate thickness: 0.12 in. (3 mm)
Number of welds: 12
Length of weld: 29.5 in. (750 mm)
Time/unit manual welding: 3.9 min
Time/unit robot welding: 1.7 min

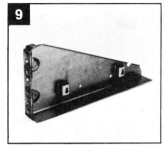

(i)

(j) *Trolley Tray (Holland): robotic productivity increase—140%*
Dimensions: 23.6 × 31.5 × 3.9 in. (600 × 800 × 100 mm)
Weight: 22 lb (10 kg)
Plate thickness: 0.06 in. (1.5 mm)
Number of welds: 44
Length of weld: 63.0 in. (1600 mm)
Time/unit manual welding: 10.8 min
Time/unit robot welding: 4.5 min

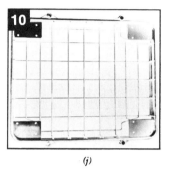

(j)

(k) *Link Part (Oregon): robotic productivity increase—860%*
Dimensions: 8.0 × 3.0 × 0.5 in. (203 × 76.2 × 12.7 mm)
Weight: 3.1 lb (1.4 kg)
Plate thickness: 0.25–0.50 in. (6.4–12.7 mm)
Number of welds: 4
Length of weld: 8.5 in. (215.9 mm)
Time/unit manual welding: 4.5 min
Time/unit robot welding: 0.5 min

(k)

(l) *Motorcycle Frame Detail (Spain): robotic productivity increase—439%*
Dimensions: 7.0 × 18.0 in. (175 × 450 mm)
Weight: 6.6 lb (3 kg)
Plate thickness: 0.08–0.12 in. (2–3 mm)
Number of welds: 20
Length of weld: 15 in. (380 mm)
Time/unit manual welding: 7.0 min
Time/unit robot welding: 1.3 min

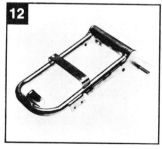

(l)

Fig. 49.8 (*Continued*)

49.6.2. Weld Pattern

For welding cabinets and doors, three types of weld patterns were tested (Figure 49.9): a straight-line weld, three-point weave, and a five-point weave. All bead welds were made on 0.050-in. (1.3-mm) thick steel sheet. GMAW welding parameters were adjusted to provide the best weld bead appearance for each pattern. (The company did not have pulsed-current MIG equipment, which would have greatly simplified the welding of thin-gage metal.)

Next the gap size and location of simulated weld joints were varied while these weld parameters were held constant. Test assemblies were welded, and the resulting joints were examined with the following results:

Straight-line welds are adequate for off-location joints, but are not very good for gaps.

The three-point weave pattern is most effective for dealing with mislocations and gaps. The five-point weave provided good coverage but left undercuts and ridges caused by the rapid traverse speeds required on the thin sheet to prevent burn-through.

More about weaving and seam tracking in arc welding is explained in Chapter 50.

49.6.3. Product Redesign

When faced with major fixturing and preweld manufacturing changes, it proved more economical in the G.E. installation to redesign the parts to be welded. For example, one of the cabinets caused repeatability problems owing to joint fit-up and accessibility. It has been redesigned so that robotic welding occurs prior to spot-welding assembly. In addition, the cabinet components are roll and die formed to give closer tolerances than the press brake could produce. The weld joints are now more accessible. These design changes are expected to result in a substantial cost saving due to reduced weld time and reduced labor in the forming operations.

49.6.4. Fixture Design

Several general conclusions made in the preceding application study are relevant whenever a new weld fixturing is required.

Start with very simple fixtures, or clamp the part directly to the positioner table. Optimize the part location to provide the best accessibility and welding gun angle for the robot work envelope.

Use existing experience in manually welding a particular part to develop the best weld parameters for robotic welding.

When the part positioning problems have been ironed out and several test assemblies have been welded, only then consider going to more permanent fixturing.

To reduce setup and cycle time, multiple-step or full-component welding programs should be avoided. However, in some cases these more complex programs are necessary because of part size or to avoid complex fixturing and manufacturing changes.

Air- or water-cooled copper-backing blocks may be required to reduce burn-through on thin sections. These blocks can be built into the fixture or held in place by clamps.

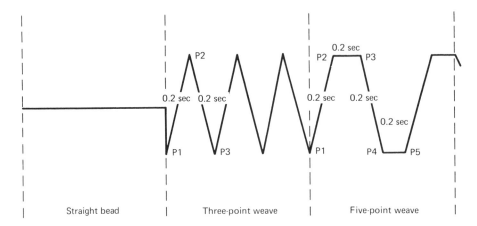

Fig. 49.9. GMAW weave patterns for robotic arc welding.

49.6.5. Workpiece Size and Weight

The positioner and fixtures should be purchased or designed to be compatible with the part size and weight and the robot's tolerances. Close consideration should be given to the part's center of gravity when designing a system.[3] A heavy part or fixture that is not balanced on the holding station can overload a positioner's tilt and rotation functions. Adequate clearance should be allowed for the part to be rotated and tilted in any direction. In conjunction with this requirement, the robot should initially be programmed in a safe location, or by off-line programming, to avoid accidental damage to the positioner while it is in operation.

Part-Locating Positions

When the fixture design has been finalized, make provisions to locate the part accurately and fasten it securely to the positioning station.

Clamping

Start with manual clamping prior to going to hydraulic, pneumatic, or electric systems. Use bolts, welding clamp pliers, or toggle clamps. More sophisticated systems can be considered after testing has established the proper clamp size and location and required clearances for robot and positioner motions.

Setup

Parts loading for a second assembly should take place simultaneously with robot welding on the first assembly. With this type of system, part-loading time is not critical as long as it is less than the robot welding program's cycle time. Provide a second MIG or TIG welding station in the robot enclosure for tack welding and touch-up work.

REFERENCES

1. The Real Cost of Depositing a Pound of Weld Metal, *Metal Progress,* American Society for Metals, April 1982.
2. *The Procedure Handbook for Arc Welding,* Lincoln Electric Company, 22801 St. Claire Avenue, Cleveland, Ohio 44117.
3. The best reference for calculating the center-of-gravity weldments is *The New Handbook of Positioneering,* available from the Aronson Machine Company, Arcade, New York 14009.

BIBLIOGRAPHY

1. Welding, in Tanner, W. R., Ed., *Industrial Robots,* SME, 1981, Ch. 9.
2. Wadsworth, P. K., Thorne, R. G., and Middle, J. E., Feasibility Studies for Small Batch Robotic Arc Welding, *Proceedings of the 13th International Symposium on Industrial Robots,* April 1983, Chicago, Illinois, pp. 6.1–14.
3. Hohn, R. E. and Holmes, J. G., Robotic Arc Welding—Adding Science to the Art, *Proceedings of Robot VI,* March 1982, Detroit, Michigan, pp. 303–320.
4. Sullivan, M. J., Application Considerations for Selecting Industrial Robots for Arc Welding, *Welding Journal,* Vol. 54, No. 4, April 1980, pp. 28–31.
5. Industrial Robots: A Delphi Forecast of Markets and Technology, Society of Manufacturing Engineers, Dearborn, Michigan, 1982.
6. Lagernof, B., Current Industrial Robot Applications for Arc Welding, *Proceedings of the 9th International Symposium on Industrial Robots,* Washington, D.C., March 1979.
7. Keag, R. J., Jr., Equipment Applications for Robots, *Robotics Today,* September 1982, pp. 24–38.
8. Stauffer, R. N., Welding Robots: The Practical Approach, *Robotics Today,* August 1983, pp. 43–44.
9. Soroka, D. P. and Sigman, R. D., Robotic Arc Welding . . . What Makes a System?, *Welding Journal,* Vol. 61, September 1982, pp. 15–21.
10. Welding and Robotics: The New World of Robotic Arc Welding, *Iron Age,* Vol. 226, March 25, 1983, pp. 1–30.
11. Cook, G. E., The Application of Microcomputers in Automated Arc Welding Systems, *IEEE Transactions on Industry Applications,* IA-17, No. 6, November/December 1981, pp. 619–625.

CHAPTER 50

THE OPERATION OF ROBOTIC WELDING

PETER G. JONES
JEAN-LOUIS BARRE
DAN KEDROWSKI

Cybotech
Indianapolis, Indiana

50.1. INTRODUCTION

Chapters 48 and 49 describe planning for robot spot and arc welding applications. This chapter describes the operational details of robotic welding, illustrated in several applications. In each application, the basic robot system is augmented as required to perform automotive spot welding and arc welding of heavy equipment, respectively. Figure 50.1 depicts a diagram for a spot welding application, augmented to operate with user-supplied equipment. The combination of equipment must be tightly integrated so that it is easy to install and to operate.

50.2. SPOT WELDING

The unit of automatic work for the robot is the cycle. In a given automotive application of a robot, each body style requires a specific cycle. Within each cycle, subelements called *trajectories* define lesser but integral elements. Each trajectory is made of points, positions at which the robot either makes a spot weld or simply passes through, depending on programming. In an automotive application, front fender and front door sections might be common to the several body types; therefore, the trajectories for front fender, front window, and front door contours could repeat on two-door, four-door, and station wagon bodies. Trajectories unique to each style of body would then be added to the common trajectories, to establish the specific cycles for each style of automotive body that moves through the robot's work envelope. As the parts are positioned in the robot's work envelope, application equipment signals to the robot system which cycle is required.

The major components of the spot welding system are as follows:

Robot. In this example, a CYBOTECH six-axis robot, combined with welding transformer, welding gun, and hydraulic pressure control assembly, including gun operating pressure control and gun-position monitoring devices.

Power Unit. Standard power unit; receives 460 VAC, and supplies hydraulic and electrical power to the robot system.

Hand Control. Standard hand control (teach pendant), which accommodates user-defined functions. Application software permits multifunction use of several hand control push buttons.

Robot Controller. The robot controller (called RC-6 in the CYBOTECH system) uses a spot welding application software package to effect the required controls and interface with user-supplied programmable controllers.

Input/Output Interface Box (I/O Box). A special I/O box, which adds to the usual I/O functions a weld gun servo controller card, power supply, and AC input. The I/O box retains its normal operational relationship with the robot controller, providing the robot system's interface with the equipment that is required for the application, but is not part of the robot system.

Weld Gun Servo Controller Card. Mounted in the I/O box, the weld gun servo controller card interfaces with the RC-6 controller, the hydraulic pressure control assembly, the I/O box, the

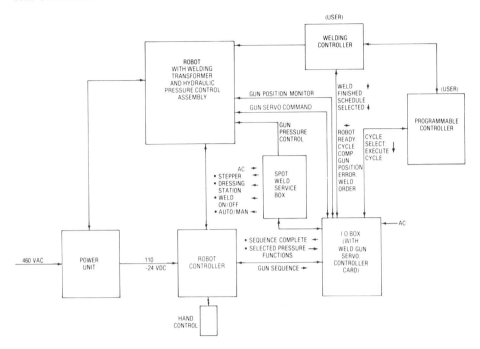

Fig. 50.1. A block diagram for a spot welding system, augmented to operate with user-supplied equipment.

welding controller, and the programmable controller. The weld gun servo controller card is unique to spot welding; it controls the position of the jaws of the welding gun, synchronizing the gun with the rest of the spot welding system.

Spot Weld Service Box.   The spot weld service box is unique to spot welding; it gathers together the manual controls for service and maintenance of the welding gun, and the controls for setting operating pressures for the hydraulically activated gun-control system. Its functions include the following:

1. Provides local manual operation of the welding operation.
2. Provides modification of welding current during operation.
3. Provides routine in-service maintenance on the gun such as dressing the tips or replacing worn electrodes.
4. Provides the means for exerting the correct pressure on the tips during welding.
5. Provides local selection or automatic selection of the gun-tip pressure for each welding point.

Welding Controller.   Controls the specific welding process in terms of squeeze duration and heating required for specific spot welds.

Programmable Controller (PC).   Signals to the RC-6 the specific cycle required, and initiates cycle execution. Also directs the action of the welding controller, coordinating welding parameters delivered at each point.

50.2.1.   Functional Requirements

The largest user of spot welding is the automobile industry. Robots integrated into the process must be synchronized to work within the overall rhythm of the assembly-line process. For each robot station a programmable controller can notify the robot of the arrival/departure of workpieces and coordinate the welding process of that specific robot. A weld controller will provide precise control of the numerous parameters that control the welding process for the various combinations of metals and thicknesses that come through the robot's work envelope.

Within a given line, such as the underbody line, body-side line, or body-framing line, segments composed of multiple work stations (for example, a robot cell, a manual station, and standard automatic equipment) can be managed by larger programmable controllers called sequencers. The entire line can be coordinated by a final programmable controller (sequencer), which integrates operation of subordinate elements, moving parts along the line only when all stations on the line have reported in as finished and ready for the next phase. Information fed into the network of programmable controllers comes from manually operated signals and automatically indicated changes of status from the various equipment, including robots, along the line.

In the automotive application described here, these are the typical parameters of the spot welding process:

Welds per minute: 50 spots/Robot

Duration of welding cycle: $\frac{1}{4}$ to $\frac{1}{3}$ sec (current flow) for each spot

Voltage: 6–8 V

Amperage: 20,000 A

Frequency: 60 cycles/sec

In existing applications robots make from slightly more than 37% to more than 75% of the total spot welds on a given vehicle.

The design of a spot welding production line is described in Chapter 48. Briefly, planners must consider the most efficient way to build up specific parts, the order of flow along a line, the need to allow for variable flow, whether robots or other machines are most desirable for specific functions, and how they must be integrated. Of course, such elements as materials movement to and away from the overall process and coordination between major production segments become vital considerations. For example, if several smaller parts must be held together to form a larger subunit, such as a body side, what is the best way to get them together, and then to do the preliminary welding? Should robots be used to establish fundamental part geometry, or should they be used only after the larger parts are firmly and solidly joined? Can robots be used for transitional functions like loading/unloading conveyors, placing parts into position for welding? In every instance, planners must strike a balance between robots and other machines; between the desire for speed and the need for reliability; between human intervention and the need to automate.

The following description covers spot welding by a single robot, on a single part. The objective is to illustrate how a robot system functions in this application.

Figure 50.2 depicts a spot welding cycle. Trajectory 1 is the front door frame, and trajectory 2 is the rear door frame. Trajectory 1 is comprised of numbers 1–14. Point 1 is the starting, or home, position for the robot. Points 2, 3, and 4 are passing points, through which the robot moves to reach the first working point, 5. Each of the working points, 5–14, must be precisely identified in terms of the welding parameters required. The points are so identified by weld identification numbers, shown on the outside of the trajectory. Each weld identification number carries with it the electrical and mechanical parameters for the weld at the specified point as specified by welding engineers responsible for setting welding parameters for every point to be welded: pressure to be exerted by the electrodes;

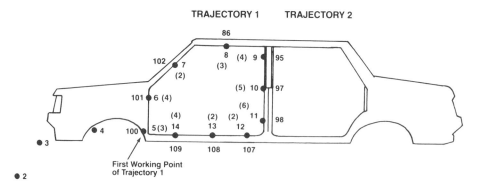

Fig. 50.2. A spot welding cycle example.

duration of the welding impulse; shaping and development of the pulse. The parameters are implemented by the robot system through weld schedule numbers, transmitted by the RC-6 to the spot weld service box and welding controller. The numbers in parentheses are weld schedule numbers. For each point to be welded, the spot weld service box transmits the required control signals to the hydraulic pressure control assembly to achieve the correct pressure and schedule of pressure development at the point during welding. The overall requirements vary according to the metals involved, the thicknesses, and the number of thicknesses; all variables must be predetermined and then programmed into the data for each trajectory that the robot is to execute. The welding controller delivers the required welding current, modulated to produce the appropriate heating cycle in the welded point.

Thus each point on the door frame to be welded has a welding identification number, supplied by the user. Each point is entered into a table by welding identification numbers, including the specified weld schedule. Using the robot controller's keyboard, the operator places this table into the controller's memory. In terms of the robot system's operation, the weld schedule number (1–6) associated with each point for the spot weld service box is the means for establishing the match-up between pressure identification number and required welding parameters. For example, pressure identification number 101 is point 6 on the robot's work trajectory, and is assigned weld schedule number 4. Weld schedule 4 selects the specified pressure, controlled through the spot weld service box, and the welding controller provides preestablished electrical welding parameters associated with weld schedule 4.

Having entered the welding identification number and associated pressure schedule into the table in the controller's memory, the operator then teaches the robot the trajectory. At this time, or later, the operator can use the hand control's multiple-function capabilities to correlate every point on the trajectory with the specified weld schedule, assigning welding identification numbers and weld schedules for each point on the trajectory in accordance with the prescribed schedule.

The operator can also modify the welding identification number, using the editing capabilities integrated into the hand control's operation. The operator can check the exact parameters operative at any point along any of the spot welding trajectories, using status display information available through the robot controller's display screen. For each point the operator can check: trajectory number, point number, robot speed assigned for that number, the welding identification number and weld schedule required for spot welding, and other pertinent data. In the execution phase, the display screen shows a running account of progress through the cycle under way, point-by-point, and trajectory-by-trajectory.

50.2.2. Welding Operational Sequence

The following is the operational sequence, from detection of the part in the robot's work envelope, through completion of the cycle for that specific part (refer to Figure 50.1).

1. Based on the application system's detection of a given part, the programmable controller specifies a cycle number to the robot controller (cycle select). If the cycle agrees with a cycle in the controller's memory, cycle verification occurs, the controller sends ROBOT READY, and the programmable controller sends EXECUTE CYCLE. The robot controller sends the robot from its home position to the first working point of the cycle (point 5 in Figure 50.2).

2. As the robot approaches the first working point, the robot controller sends the weld schedule to the spot weld service box and the weld controller. The spot weld service box sends preestablished pressure-control voltage to the hydraulic pressure-control assembly, based on the pressure schedule transmitted by the robot controller, and the welding controller establishes the other parameters as specified.

3. The robot reaches the first working point:

(a) RC-6 sends gun sequence to the weld gun servo controller card in the I/O box.

    (1) The weld gun servo controller card sends a servo command to the hydraulic pressure-control assembly on the robot and monitors feedback circuitry to determine that the welding gun has closed to the correct position, which corresponds to the pressure specified for that point.

    (2) As the weld gun servo controller card receives confirmation that the gun is closed to the position prescribed in the gun sequence, it transmits the weld order to the programmable controller.

    (3) The programmable controller enables the welding controller, causing the welding controller to generate the correct pulse (heating and duration) to spot weld the type of metal and number of thicknesses between the jaws of the welding gun at the present welding identification number.

    (4) When it has finished welding at the point, the welding controller sends WELD FINISHED to the weld gun servo controller card in the I/O interface box.

    (5) On receipt of WELD FINISHED, the weld gun servo controller opens the welding gun by commands to the hydraulic pressure-control assembly servo valve, and when

the gun opens to a preset distance, as reported by feedback circuitry, the weld gun servo controller card sends SEQUENCE COMPLETE to the robot controller.

(b)   The robot controller sends the robot to the next programmed welding point.

(c)   When the robot's cycle ends, the robot controller sends CYCLE COMPLETE to the programmable controller. The programmable controller is tied to other controllers along the line and to the programmable controller (sequencer) that synchronizes operation of the entire line. When all robot operations, other machine operations, and any required manual operations on the line are in the appropriate condition, the sequencer operates the transfer line, moving parts as required to maintain flow along the line.

4.   If the programmable controller requests a cycle that is not valid for the RC-6 controller, or if the robot controller does not verify a valid cycle and send ROBOT READY, the selection process repeats. The robot does not move until all conditions are satisfied.

5.   Control circuitry in the spot weld service box monitors electrode wear, sending a signal to the programmable controller to stop welding when electrodes are excessively worn (gun position error).

50.3.   ARC WELDING OPERATION: GMAW APPLICATION

This section describes operation of a CYBOTECH robot system in gas metal arc welding (GMAW). The base robot system is augmented by equipment required for continuous arc welding, with seam tracking, and with adaptive feedback for error correction. This application integrates special software, available to the user through the MODE key switch on the robot controller.

Figure 50.3a is a block diagram of the complete arc welding system. The robot controller, hand control, I/O box, and power unit are standard components, with optional elements and features for both the hand control and robot controller. The robot is a standard six-axis robot. It is important to notice that the control is more than merely a device for teaching the robot. In this system the hand control gives the operator the ability to make real-time changes to individual parameters in a given weld table. The operator can also change entire tables during operation. Figure 50.3b depicts a robot, CYBOTECH Model H80, arc welding a part that is mounted on a positioning table. The H80 is driven by hydraulic motors, weighs 1850 kg, and has a load capacity of 80 kg in the three main axes (two rotations at 1.0 rad/sec and one translation at 66 cm/sec; rotations at first and third axes are up to $\pm135°$, and translation at second axis is up to 1.6 m). Its wrist can exert up to 20 kg-m at 3 rad/sec in axes 4 and 5, and up to 10 kg-m in the sixth axis, also at 3 rad/sec. (Rotations of the wrist's fourth, fifth, and sixth axes are, respectively, up to $\pm167°$, $\pm105°$, $\pm172°$.) The position accuracy of the H80 is ±0.5 mm (±0.02 in), and the position repeatability is ±0.2 mm (±0.008 in). Elements of the system are as follows:

Optional Axes (Positioning Table).   Controlled by an optional axis-control board in the robot controller, the positioning table is controlled by the same kinds of signals from the robot controller. In the teaching phase, the operator maneuvers the table and robot as necessary to get the required relationships between the robot's welding torch and the welding seam of the part mounted on the table. Multiple-function capabilities of the hand-control manual control push buttons provide complete control of all eight axes involved. In program execution, the robot controller coordinates all eight axes, presenting the seam to the robot's torch in the location and orientation taught, modified by adaptive feedback or changes inserted by the operator, dynamically, during execution.

Welding Controller.   A subassembly located in the welding power supply, the welding controller sets voltage and amperage as established by robot controller, holding at programmed levels during execution, unless signals are otherwise modified.

Welding Power Supply.   This is the electrical power supply for arc welding. The robot controller controls amperage and voltage by welding controller interface.

Input/Output Interface Box.   A Standard CYBOTECH I/O box, in this application it continuously links the robot controller to controls of cooling water and shielding gas.

Wire Feed Motor.   Installed on robot, controlled by input from robot controller, tachometer feedback to welding controller provides closed-loop control of wire-feed speed, which affects the characteristics of the welding arc. The wire provides filler for the welding seam; the wire chosen for specific applications must be compatible with the metals being welded.

50.3.1.   Process Parameters

In operation, the welding system controls the following parameters, which collectively determine the quality of the welding. Before welding, the operator must enter these welding process variables into weld sequence parameter tables, each with an assigned weld sequence number. Thereafter the weld sequences can be recalled to apply specific combinations required for a given welding task. Also, the tables can be modified, either with the robot controller or with the hand control.

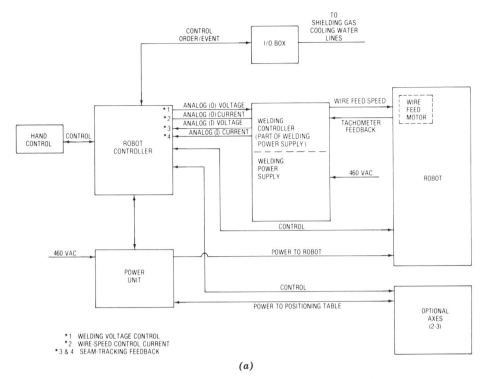

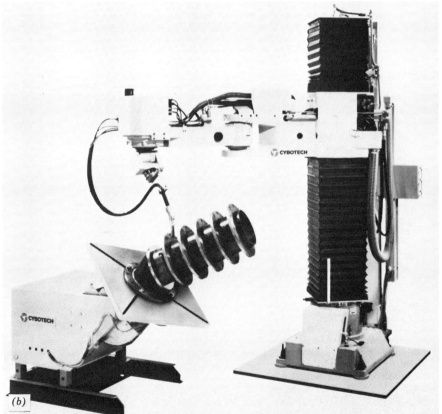

Fig. 50.3. (*a*) A block diagram of an arc welding system. (*b*) Robotic arc welding of a part mounted on a positioning table (CYBOTECH Model H80).

1. PREFLOW: time prior to welding used to establish a flow of shielding gas to protect the arc from atmospheric gases.
2. RUN-IN TIME: time required to establish an arc.
3. RUN-IN VOLTAGE: voltage to establish the arc.
4. RUN-IN FEED: feed rate of wire to establish the arc.
5. WELD VOLTAGE: sustaining welding voltage, after run-in.
6. WELD FEED: sustaining wire-feed rate.
7. CRATER TIME: time required to fill and finish the crater at the end of welding.
8. CRATER VOLTAGE: voltage required during crater finishing.
9. CRATER FEED: wire-feed rate required during crater finishing.
10. BURNBACK TIME: time required to burn the consumable electrode wire free of the welding surface.
11. BURNBACK VOLTAGE: voltage required for burnback.
12. POSTFLOW TIME: provides extra shielding gas to insure coverage of the crater as it cools.
13. WEAVE WIDTH: a lateral vector set at the robot controller, adjustable with hand control; determines the horizontal width of the welding seam.
14. LEFT DWELL: time robot holds torch on left edge of seam during seam tracking.
15. RIGHT DWELL: time robot holds torch on right edge of seam during seam tracking.
16. CROSS TIME: time robot takes to move torch laterally across the welding seam.
17. STICKOUT CURRENT: welding current level, which the RC-6 monitors to track a weld seam at a constant height (stickout).
18. TRAVEL SPEED: speed of the torch along the welding seam.

Variables 1–12 control the arc process; 13–16 are the basic parameters for weaving; 17, STICKOUT CURRENT, is the arc current that the robot controller maintains by moving the robot vertically with respect to the plane of the welding seam, either moving *in* to the seam, or moving *out* away from the seam. Variable 18, TRAVEL SPEED, sets the speed for torch movement along the seam.

50.3.2. Operation

The application sequence section of program software prompts the operator about programming a weld sequence while teaching a trajectory. Each of the preceding variables can be varied in terms of time, speed, voltage, current, or distance, depending on the variable itself and the limits established.

To weld, the operator teaches a trajectory. On a given point he/she can specify start or stop welding, the weld sequence number, and any other required parameters. After teaching the trajectory and linking it to a cycle, the operator can run the welding operation like any other cycle, and can modify the major welding parameters from the hand control dynamically while welding: welding voltage, welding wire feed rate, stickout tracking current, weave width, dwell time, and welding travel speed.

Electronics for seam tracking are integral to the robot controller control, using data from the arc to correct both the cross-seam and stickout directions. The robot controller produces a weaving pattern that has programmable weave width, left-and-right dwell, and crossing time. Oscillation across the weld joint induces a change in the arc current. The seam-tracking electronics samples and analyzes the arc data at each sidewall and makes a proportional correction at the beginning of each weave cycle. Automatic current control maintains a constant stickout height above the workpiece. Figure 50.4, timing of arc welding sequence, displays the interaction of parameters and control signals during welding.

Weaving

During welding, *weaving* widens the welding seam and provides lateral torch motion, required for seam tracking. Weaving is controlled by a weaving vector, taught to the robot controller and inserted into the welding trajectory at the desired point, as are parameters such as speed. Figure 50.5 displays the weaving effect. In each case, weaving moves the torch *across* the welding seam, as the torch moves parallel to the seam centerline. This holds true for curvilinear paths as well. The operator can teach and store weaving vectors for recall.

Selection of WEAVE on the controller's MENU automatically places TOOL GEOMETRY in effect. In TOOL GEOMETRY the reference axes for torch movement are in a coordinate system centered on the tip of the welding torch, shown in Figure 50.6, tool mode operation. The X-axis is the reference for STICKOUT. All of these directions are based on a perspective facing the end of the torch, in the plane of welding. Thus, positive STICKOUT (+X) moves the torch closer to the

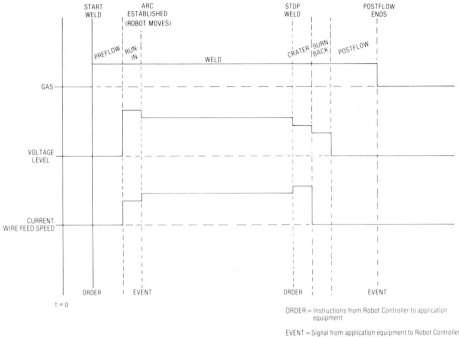

Fig. 50.4. Timing of arc welding sequence.

welder, increasing STICKOUT current; negative STICKOUT (−X) raises the torch from the weld, decreasing STICKOUT current. All commands issued by the robot controller as part of the program or as results of adaptive feedback for path control are referred to the **TOOL GEOMETRY** reference system.

Using a third function of the manual control buttons, the hand control provides real-time modification of parameters during welding. In this application two buttons on the hand control reserved for user-defined functions are marked **WELD CONTROL** and **WELD MODIFY**. Pressing **WELD CONTROL**, the operator can use the six pairs of manual control buttons to increase/decrease welding voltage, wire-feed rate, stickout current, weave width, dwell times, and speed (torch travel speed). By pressing **WELD MODIFY**, the operator inserts the modifications into the currently used welding sequence table.

Adaptive Control

Adaptive feedback enables automatic error correction for stickout and weave. Required information flows from analog inputs from the welding controller through a serial link to the robot controller.

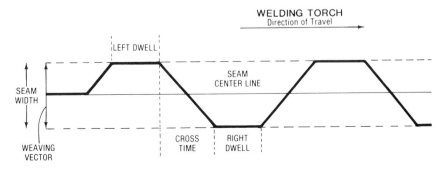

Fig. 50.5. The weaving effect during welding.

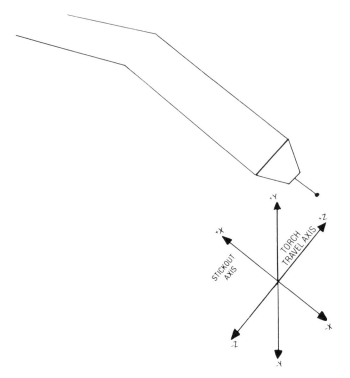

Fig. 50.6. Arc welding tool mode operation.

The system extracts error-correction data for both stickout and weave centerline. TOOL GEOMETRY allows the robot controller to make changes directly into the torch reference system. The origin of the coordinate system lies at the tip of the torch, and the X-axis is coincident with the centerline of the end portion of the torch.

Teaching trajectories and creating cycles with the welding application are essentially the same as with the standard software. Using the hand control and/or the robot controller, the operator starts or stops welding at any given point along a trajectory, modifies parameters and sequences during welding. When the operator stops welding (presses STOP WELD on the robot controller control screen), system software automatically performs a controlled welding shutdown, filling the crater and burning back the electrode wire. Postflow shielding gas protects the crater from atmospheric gases until the welding cools.

A software feature called MULTIPASS allows the operator to rerun the same welding trajectory with multiple welding sequences, so that filling a very large weld seam can be done automatically, using the root pass, memorized by the controller and replayed with lateral bias as required to fill the seam.

50.3.3. Remarks

Application software allows the operator to run a cycle or any portion of a cycle without arc to check the robot's accuracy. Status display screens (on the controller monitor) enable the operator to check the following parameters:

1. Speed of travel.
2. Weaving.
   (a) Width.
   (b) Dwell time.
   (c) Cross time.
3. Weld sequence table number.
   (a) Tracking: cross-seam only.
   (b) Tracking: Stickout only.
   (c) Tracking: both (cross-seam and stickout).

50.4. SUMMARY

Technological advances growing from current research and development will improve all phases of welding. More efficient power supplies will enable robotic welding systems to increase deposition rates (rate at which filler material is deposited) without spatter. Alloys are used more and more in the overall effort to reduce product weight. Welding alloys will require changed welding parameters and the development of new material for welding wire. Other basic welding technologies under development include laser, plasma, and gas tungsten arc welding.

Additionally, robot design will be improved to get a robot ideally formed for arc welding, with more highly specialized control systems. Refined vision systems will be able to track the welding seam more efficiently than current systems and will offer direct, faster-reacting control of weld fill rates.

REFERENCES

Cary, H. B., *Modern Welding Technology,* Prentice-Hall, Englewood Cliffs, New Jersey, 1979.

H80 ROBOT Operation Manual, CYBOTECH Corporation, Indianapolis, Indiana, 1982.

Jefferson, T. B., Ed., *The Welding Encyclopedia,* 14th ed., Welding Engineer Publications, Inc., Morton Grove, Illinois, 1961.

Kearns, W. H., Ed., *Welding Handbook,* 7th ed., Vol. 3, American Welding Society, New York, 1980.

Robotic Spot Welding System, CYBOTECH Corporation, Indianapolis, Indiana, 1982.

Robotic Welding System, CYBOTECH Corporation, Indianapolis, Indiana, 1982.

CHAPTER 51

ARC WELDING OF ALUMINUM PARTS

JOHN A. KALLEVIG

Honeywell, Inc.
Golden Valley, Minnesota

51.1. INTRODUCTION

During 1982 the Honeywell Production Technology Laboratory purchased a robot welding system to allow it to consider welding applications. The application under consideration required the robot to be able to make a continuous 360° weld around a 1.27-cm thick aluminum casting. This is made more difficult since the casting also extends out parallel to and within 5 cm of the tube, which severely limits access to the tube for hand welding. To prevent burn-through of the tube, it is necessary to use a small bead and complete the weld in under 12 sec in one continuous 360° pass. The system also must be capable of general metal inert gas (MIG) welding of both aluminum and steel.

51.2. SYSTEM DESCRIPTION

An Automatix Robot Welding System was purchased. This system uses a Hitachi five-axis robot capable of lifting 10 kg with a repeatability of ±0.2 mm. The robot reach of 1.3 m is sufficient for any foreseeable job. Figure 51.1 is a photograph of the robot welder.

The robot controller is manufactured by Automatix using a Motorola 68000 CPU. It has 512 k bytes of memory and two tape drives for program storage. Programs are written in a higher-order programming language called RAIL which is based on Pascal. It also has circular interpolation which allows the robot to make a circular weld through three points that it is taught. This is particularly valuable for this application because of the small-diameter tube and its tendency to burn through the thin wall. Figure 51.2 is a photograph of the robot controller.

To limit the heat input into the work and to allow excellent control of the weld-seam location, a Miller Pulstar welding power supply was specified. This 450-A machine can be used as a conventional direct current MIG welder, or it can be used as a pulsed power supply. In this instance, the power supply puts out a continuous background current that is not sufficient to melt the wire or to weld. A second current rides piggyback on top of the first in the form of pulses that occur at 60 or 120 Hz. Whenever the pulse occurs, the current becomes high enough to weld.

Molten metal is ejected from the torch in the form of a spray along the axis of the wire. The electromagnetic forces are stronger than the gravity forces so it becomes easy to weld overhead or in other positions with very little splatter. Deposition rates as high as 96% are advertised.

Since very small beads are desirable, 0.76-mm 5056 aluminum wire is used. This wire is very weak and does not push very well. Since the torch is about 4.6 m from the wire feeder, a Cobramatic King Cobra welding torch was specified. This torch has two motors, one at the feeder which pushes the wire, and a second on the torch which pulls the wire. This second motor only must push the wire to the end of the torch, and it does this very well.

The computer controls everything including the wire feed rate, the background voltage, and the peak current. This is good because it is often necessary to change the weld parameters "on the fly." For example, during the weld of the tube to the plate, after the weld has progressed about halfway, it starts to get too hot. Therefore, at that point, the welding voltage and current are reduced, and the weld traverse speed is increased. Figure 51.3 is a photograph of the tube welded to a plate and Figure 51.4 shows a section of the weld. Note the excellent penetration without any burn-through.

An argon gas cover is used over the puddle during the weld process.

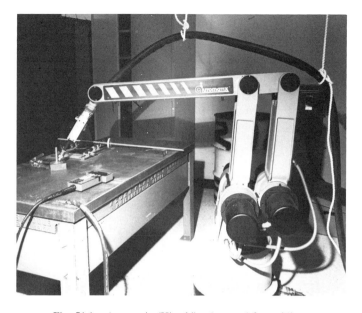

Fig. 51.1. Automatix (Hitachi) robot used for welding.

51.3. ROBOT TASK

Honeywell has set up an automated facility to manufacture canisters for the SLUFAE Project (Surface Launched Unit Fuel Air Explosive). One end of the canister consists of a bulkhead which is 1.27 cm thick. It has a casting protruding from the center. About 5 cm from the casting is a 1.27-cm diameter tube that protrudes 30 cm from the bulkhead. It is necessary to weld the tube to the bulkhead so that fuel will not leak from the canister into the air. The tube has a 0.32-cm wall, and, after welding, a 0.64-cm ball must be able to pass through the tube. If this weld is not continuous, crater cracks occur whenever the operator stops welding, causing leaks. The large thermal mismatch tends to make

Fig. 51.2. Automatix controller.

Fig. 51.3. Aluminum tube welded to an aluminum plate.

the welder burn through or collapse the tube. The length of the tube effectively prevents welding from above the tube. The welding torch must rotate around the tube.

The original plan called for heating the canister to 150°C and TIG welding it by hand. However, to reach the projected production rate would require many ovens and much space for cooling the canisters.

A second method required the purchase of a torch that mounts right on the tube and rotates around it to make the weld. This method was never tested because of contractual limitations.

It has been proven in the lab that the robot can make the weld and perform other welding tasks as well. Owing to the location of the casting, the robot torch must come between the casting and the tube to start the weld. It then backs up and draws the weld around the tube. As it welds, the torch angle is changed so that the torch pushes the weld puddle around the tube back to the starting point. This must occur in one smooth, continuous motion. Any hesitation causes burn-through, and any jerkiness causes a poor bead that may not completely seal the joint.

Since the weld puddle tends to get too hot about halfway around the tube, the weld parameters must be changed at that point. To simplify programming, instead of treating the weld as one 360° weld, it is treated as two 180° semicircles. The first semicircle is taught as three points, and the circular interpolation feature of the robot calculates a semicircular path through the points, complete with correct torch angles. At the end of this path, the second semicircle is taught, using the third point of the first semicircle as the first point in the second semicircle. Owing to the different torch angles required, it is not possible to use the first point in the first semicircle as the last point in the second semicircle.

Table 51.1 gives the weld schedules used. The first column is for the first half of the weld, and the second column is for the remainder of the weld. Note that the wire feed rate is dropped from 14.8 cm/sec for the first half to 14.4 cm/sec for the second half, while the torch speed is increased from 3.6 mm/sec to 5.1 mm/sec. This is necessary to prevent burning through the tube.

Predwell refers to the time that the torch remains stationary after the arc is initiated. Fill delay

Fig. 51.4. Section of weld of aluminum tube welded to an aluminum plate.

TABLE 51.1

		First Semicircle	Second Semicircle
Wire feed	cm/sec	14.8	14.4
Max current	amperes	205	190
Voltage	volts	21.0	20.5
Predwell	seconds	0.3	0.0
Fill delay	seconds	0.0	0.25
Preflow	seconds	0.5	0.0
Postflow	seconds	0.0	1.0
Torch speed	mm/sec	3.6	5.1

is sometimes referred to as crater fill and refers to the time that the torch remains stationary at the end of the weld path until the arc is cut off.

Note also that the gas cover is initiated for 0.5 sec before the arc is turned on, and it is held on for 1.0 sec after the arc is cut off to minimize contamination of the weld puddle.

The robot also is able to weave; but, of course, this feature is not used for this application.

51.4. ROBOT WELDER BENEFITS

Fusion-welding robots offer the following potential benefits:

Lower welding assembly unit cost.
Improved quality and product uniformity.

Fig. 51.5. Close-up of torch welding around the tube.

Reduced costs to train certified welders because the computer controls all of the weld process parameters.

Reduced operator exposure to heat, smoke, and radiation.

Increased operator safety.

Costs to train and certify welders to military requirements have risen to approximately $5000 per operator. It is difficult to hire experienced welders and maintain them on the job.

The actual welding time for the robot is the same as for a human welder. The big time savings occur between the welds. The operator will typically lift a helmet, move to the next position, drop his helmet, and weld. The robot moves very quickly to the next position. A robot welder will work for about $5 per hour including benefits. Human welders cost considerably more.

The robot welder is very consistent. Once it has been taught the weld, it will repeat the same weld over and over. By the same token, if a little flaw occurs, all of the parts will have that flaw. Experience has shown that it is often necessary to reteach the points at the start of each day. The reason for this may be that the robot accuracy is being stretched to the limit.

The joints that allow the robot to rotate, extend, and lift are very strong. However, the joint on the wrist for raising and lowering the torch is much weaker, and the wrist joint for rotating the torch is the weakest of all. (These two joints were appreciably strengthened in the AID 800 A model.) Figure 51.5 shows a close-up of the torch welding around the tube. Note that the tube is positioned at about a 45° angle rather than horizontal. This is desirable because if the weld is made with the canister standing on its end, the weld cables and hoses hit the robot arm. This is enough to deflect joints 4 and 5 so that during the weld cycle, the torch will not track to the points originally taught. By tipping the canister, this interference is minimized.

For most welding applications, 5 DF are probably enough. However, under certain circumstances, 6 DF would be very helpful in positioning the torch in the proper welding position at the proper torch angle.

51.5. FUTURE PLANS

Welding robots show great potential for increased productivity, quality, and cost savings. However, in a factory environment, parts often do not fit together the way that they should. Of course, the robot does not know about the mismatch and will weld where it expects the seam to be, not where it actually is.

Another problem occurs when the parts get too hot. The weld may burn through or just have poor penetration. These two problem areas point out the need for sensors to track the seam and correct the weld path, to monitor the weld puddle temperature and width, and to correct the weld parameters to insure that an excellent weld is made every time on the seam. Honeywell has a project currently to find these sensors and integrate them into the robot system.

REFERENCES

1. Potap'yevskiy, A. G., Lapchinskiy, V. F., et al., *Pulse-Arc Welding of Aluminum Alloys,* FTD–MT–24–23–68, translated by Translation Division, Foreign Technology Division, Wright-Patterson Air Force Base, Ohio, 15 May 1968.

2. *Pulsed Spray Welding Process Manual,* Airco Welding Products, August 1972.

CHAPTER **52**

ARC WELDING
OF AC MOTORS

KENICHI ISODA

KAZUHIKO KOBAYASHI

Hitachi
Tokyo, Japan

52.1. WELDING OF AC MOTOR HOUSING

The frames of small AC induction motors are made of cast iron, but the frames of medium-sized motors are made of steel. We show here the system for assembling and welding automatically the steel frame of a 55–132-kw AC motor (diameter: 345–553 mm; weight: 70–220 kg). It is common to use a welding robot after partial manual welding or welding with a fixture to keep the accurate position of workpieces during the welding operation. This requires a lot of manual work for preparation. Manual work is minimized by the system described here.

   The components for welding are shown in Figure 52.1. For the steel-frame motor, there are seven kinds of workpieces, such as a cylinder, fins, and stays. The total number of the components is 45.

52.2. THE WELDING SYSTEM

The welding station consists of three subsystems (Figure 52.2): (1) two robots; (2) a positioner; (3) conveyors. The two articulated robots (Hitachi Process Robot) are electrically driven, with 5 DF. One is for arc welding, the other is for supplying and positioning of components. The positioner tilts the motor cylinder so that the robot can reach the proper place of the workpiece. Indexing of the tilting by the positioner is at 15°. The conveyer system carries the cylinder from the previous station to the welding station. The hand for gripping the workpiece uses a magnetic chuck and an air-driven clamp.

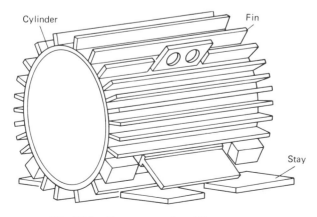

Fig. 52.1. Components of steel-frame motor.

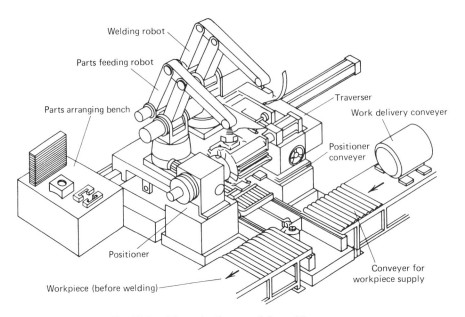

Fig. 52.2. Schematic diagram of the welding system.

Fig. 52.3. An overview of the system.

The welding system is able to position and weld components within 0.5-mm accuracy. In addition to the welding robots themselves, other subsystems are also developed at Hitachi.

52.3. EVALUATION

Since installing the system, 75% of welding, measured in total arc welding length of the motor frame, has been automated. Also, a 20% speedup in production rate has been achieved. A total of 900 location points for parts handling and 1500 location points for welding are taught to the robots. The overview of the system is shown in Figure 52.3.

CHAPTER 53

ARC WELDING AND SPOT WELDING CASES

KENNETH R. HONCHELL

Cincinnati Milacron
Lebanon, Ohio

53.1. ARC WELDING APPLICATIONS

53.1.1. Arc Welding Aluminum Housings

Dahlstrom Manufacturing Company of Jamestown, New York, is a relatively small—400 employees—job shop that provides proof that robots can play an important role in improving productivity, even in smaller plants. Dahlstrom has two Cincinnati Milacron T³-566 robots that they use to arc weld a wide variety of parts. Two of these parts are discussed here.

One of the parts welded by a robot is an aluminum repeater housing, a large, tapered open-bottom box that houses the electronic relays for a residential area.

Dahlstrom did not go into welding aluminum with the robot set on success at any cost. They had three parameters on which they absolutely refused to compromise. First, they insisted on using the same power source they had been using; second, they did not want additional weld clean up; and third, they wanted to use the same wire they had used doing the job manually.

Dahlstrom had no difficulty meeting those requirements. In fact, with the robot they are actually getting a better-quality weld, with a more uniform bead that requires less clean up.

Welding speed is a very high 72 in./min. In addition to the time savings this high speed provides, Dahlstrom has found that although they get the same quality weld they got at slower speeds, they get less heat distortion. No problems were caused with brittleness from laying so much metal so fast.

An interesting application of creative engineering in the case of the repeater housing was Dahlstrom's decision to go from seven aluminum pieces per housing to 17 to avoid difficult welding angles (e.g., inside welds) for the T³ robot. With the penetration they are getting with the robot welding, they have eliminated the need to weld the inside of the seams, but they are still welding twice the footage that they were previously (37 feet versus 19, or 11.3 m vs. 5.8 m) in one-third the time. Floor-to-floor time has been reduced from 2½ hours to 60 min. This change has also eliminated many of the inaccuracies and imperfect fits that are a natural consequence of press brake forming and similar processes that had been necessary with the old seven-piece process.

The welding is done in two stages. First, with the fixturing (designed and built by Dahlstrom) in place, the robot puts down a series of 100 tack welds. These serve to secure the pieces and allow for the removal of the fixturing, which in turn allows for interference-free welding of the entire seams, the second stage of the process.

Beyond the seam welding itself, applying the T³ robot to the repeater housing changed the manufacturing process for that part in virtually every respect. Dahlstrom rethought the entire process—the decision to go from seven component pieces to 17 was just one aspect of this—and the result was a 35–40% reduction in material and labor costs. Cutting sheet aluminum into smaller, simpler shapes resulted in a reduction in material waste. The most significant time savings occurred in press brake time, straightening, grinding, and final assembly, due generally to improved uniformity in the housings resulting from welding with the robot. Shown in Figure 53.1, the Cincinnati Milacron T³-566 Industrial Robot tack welds an aluminum repeater housing, after which the fixturing is removed for welding the entire seams.

Fig. 53.1. Cincinnati Milacron Model T³-566 tack welding an aluminum housing.

53.1.2. Arc Welding Steel Bases

Another part to which Dahlstrom has successfully applied the Milacron robot is a tubular-steel base for computer main frames. This part is a 30 in. × 30 in. (76 × 76 cm) assembly composed of 12 pieces of 1 in. × 2 in. × 0.125 in. (2.5 × 5.1 × 0.3 cm) wall steel tubing. It requires 44 welds, each 2 in. (5 cm) long.

In going from manual to robotic welding of this part, Dahlstrom cut floor-to-floor time from 51 to 16 min. At the same time they made no changes in fixturing or the welding process itself.

Lot sizes for the base assemblies run from 35 to 75, larger than the 10 to 25 typical for the repeater housing, but still small compared to the numbers often associated with robot applications. Figure 53.2 shows the T³-566 Industrial Robot welding a base assembly for a computer main frame. The robot welds 44 2-in. (5-cm) seams from a variety of angles in under 12 min. It took 45 min to weld this assembly manually.

Dahlstrom estimates payback from savings in labor, production time, and material costs at 2–3 years. Equally important, however, is the dramatic improvement in finished part quality and the expansion of the company's capabilities, which have enabled it to attract new work and new customers. These factors have directly affected employee attitudes, too, because they have counteracted the fear of displacement often associated with robots. No Dahlstrom employees have lost their jobs as a result of the change from manual to robotic welding.

Dahlstrom has shown that the robot can be used successfully and cost-effectively in a batch-type production situation.

53.2. SPOT WELDING OF CAR BODIES

Finish spot welding of the Ford Sierra and Fiesta car bodies is carried out by a total of 48 T³-586 robots installed by Cincinnati Milacron at two automated body assembly plants at Dagenham, United

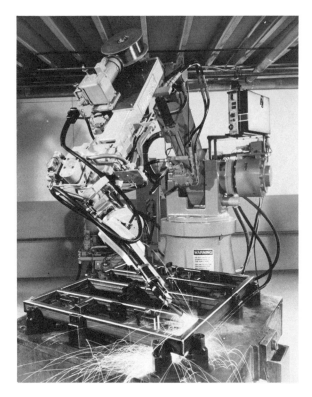

Fig. 53.2.   T³-566 arc welding a steel base frame.

Kingdom, and Genk, Belgium. More than 300 spot welds are applied to each body shell, on average, in a cycle that has bodies leaving each of four identical production lines at a rate of 50/hr—a total production capability of 200 cars every hour.

Each of the four lines—two at Dagenham, two at Genk—is comprised of 12 T³-586 robots. One further robot is provided at each plant for training purposes, as well as for program simulation, maintenance purposes, or for breakdown. (Changeover time to remove a robot from the line and replace it with a second unit is approximately 30 min.) Such contingencies are typical of the production planning process which requires immediate and round-the-clock response to significant changes in the output schedule.

Heavy-duty build skids carry tacked body shells onto the welding lines and into fixtures. A model code identification is also carried by each body/skid through the line. This identification is achieved through electronic signals processed by the line controller and enables each robot's computer control to select the welding program appropriate to that particular style. Body shells of any version can appear in the line at random and are welded in what can be properly described as a flexible manufacturing system.

Prior to entry into the spot welding lines, the various panels and subassemblies that make up what is called the "body-in-white" (roof, both sides, floor assembly, bulkhead, front and rear assemblies, and fenders) are tack welded together. In this condition, the shell is sufficiently stable to be located onto the general-purpose build skid and transported to the final welding lines.

A programmable controller at the head of each line controls the status of each of the stations in the line. (This controller decides which of the two lines can accept the new body shell. It also monitors other functions such as component present/absent; in/out of position; clamped/unclamped; initiate welding cycles; and provide feedback when welding is complete to initiate transfer between stations.) An illuminated panel is also located at the head of each line to indicate visual operating status at each station.

At the first station, the car body is lifted clear of the build skid, and locators and clamps applied. The clamps hold the body in a rigid and accurate configuration, producing just the correct amount of stress during progression through the welding lines.

Transfer between stations is carried out by a shuttle system. The body shell, after location and

Fig. 53.3. T³-586 robots spot welding car bodies.

clamping, is lowered again onto the build skid which is moved mechanically to the next station where it is again lifted clear of the build skid and locked in position by precision locators. Initiation of the transfer between stations normally occurs when all 12 robots have completed their full welding cycles.

Although a continuous throughput rate is planned, contingency arrangements are provided so that each line continues to function if one or more of the stations are empty. In such cases, the robots are inhibited from operating, and station transfer is initiated after the last weld has been made by those robots that are in cycle. Figure 53.3 shows the finish spot welding described.

PART **10**

MATERIAL HANDLING AND MACHINE LOADING

CHAPTER 54

ROBOTS IN MATERIAL HANDLING

JOHN A. WHITE

Georgia Institute of Technology
Atlanta, Georgia

JAMES M. APPLE

Systecon, Inc.
Duluth, Georgia

54.1 INTRODUCTION

To many, the primary function of a robot is to perform material-handling tasks. In support of that belief, we cite the definition of a robot provided by the Robot Industries Association:

A reprogrammable, multifunctional manipulator designed to move material, parts, tools, or special-ized devices through variable programmed motions for the performance of a variety of tasks.

Notice the robot is designed to *move* material, parts, tools, and specialized devices. The distance moved can be large or small, and the material moved can include containers, products, assemblies, parts, supplies, tools, fixtures, packing materials, raw materials, finished goods, in-process materials, machines, and so on. Regardless of the distance or material moved, *the robot is designed to move material.*

Because of the expanding role of the robot in material handling, this chapter focuses on that role and reviews (1) basic material-handling concepts, including material-handling system design, (2) applications of robots in material handling, and (3) the integration of the robot in material-handling systems.

In addition to the role of the robot as a material handler, frequently the robot must interface with other material-handling equipment. In this chapter we consider the robot both as a material handler and as an interface point for other material-handling equipment.

54.2. MATERIAL HANDLING

The robot can be considered to be a material-handling device when it picks up and puts down whatever it carries in each work cycle. Applications such as spot welding, spray painting, and operations in which the robot maintains possession of a tool are considered to be productive operations. Those in which the robot loads a machine, positions a component on an assembly, or simply transfers a part or tool are considered to be material-handling tasks. In the latter, the robot becomes a part of the handling system and, as such, is subject to the same design procedures that govern other components in the system. Since the robot belongs to the "material-handling family," it is important to review briefly some background information related to material handling.

An holistic view of material handling results in the conclusion that *material handling* means much more than simply *handling material.* Material handling can be defined as: Using the right *method* to provide safely the right *amount* of the right *material* in the right *orientation* and the right *condition* to the right *place,* at the right *time* and at the right *cost.* Material-handling systems include the movement, storage, and control of material, with considerable emphasis placed on control.

In designing material-handling systems, the material-handling equation depicted in Figure 54.1 is used. A questioning attitude prevails, with the first question (What?) focusing on the characteristics

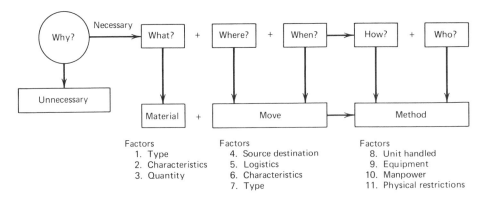

Fig. 54.1. The material-handling equation. (Adapted from Production Handbook, The Ronald Press Co., New York, 1972.)

of the material that will influence the way in which the material is handled, stored, and controlled. After determining the material characteristics, the questions of "Where?" and "When?" are used to establish the requirements for moves or flow.

The combination of material characteristics and flow requirements establishes the *material-flow system requirements definition*. The material-flow system becomes a material-handling system by a methods transformation by answering the questions "How?" and "Who?" Specifically, how the material is to be handled, stored, and controlled and who is to be involved establishes the alternative methods to be evaluated. Even though the thrust of a particular design effort may be the identification of potential robotics applications, the "how" and "who" in the material-handling equation should be used to develop options using other, more conventional, less flexible automated transfer methods. Additionally, mechanically assisted manual handling should not be overlooked. It must be remembered that under certain conditions some handling tasks that *can* be performed by a "robot" also can be more easily performed at a lower cost with less flexible automation.

The evaluation of the alternative material-handling systems is performed by answering the question, "Which?" Economic and noneconomic issues, as well as quantifiable and nonquantifiable factors, are considered in the selection process. For robotics applications, some less quantifiable factors influencing the selection of the most appropriate method include the needs for short-term or long-term flexibility; part-to-part, batch-to-batch, or year-to-year. Also, uninterrupted production may require provision for manual or redundant back up capacity.

Throughout the entire design process the question "Why?" is asked. The importance of eliminating unnecessary materials and moves, as well as infeasible methods, is emphasized. Additionally, opportunities for improving existing methods and combining operations, inspections, and moves are considered throughout the process.

The design of material-handling systems is not unlike the design of other engineered systems. It follows the systematic, six-step engineering design process:

1. Define the problem or opportunity.
2. Analyze the requirements.
3. Generate alternative designs.
4. Evaluate the alternatives.
5. Select the preferred alternative.
6. Implement the system.

54.2.1. Principles and Objectives

In designing material-handling systems, a number of principles should be carefully considered. The following twenty principles of material-handling have been adopted by the College-Industry Council on Material-Handling Education:

1. **Orientation Principle.** Study the problem thoroughly prior to preliminary planning to identify existing methods and problems, physical and economic constraints, and to establish future requirements and goals.

2. **Planning Principle.** Establish a plan to include basic requirements, desirable options and the consideration of contingencies for all material-handling and storage activities.

3. **Systems Principle.** Integrate those handling and storage activities that are economically feasible into a coordinated operating system including receiving, inspection, storage, production, assembly, packaging, warehousing, shipping, and transportation.

4. **Unit Load Principle.** Handle product in as large a unit load as practical.

5. **Space Utilization Principle.** Make effective utilization of all cubic space.

6. **Standardization Principle.** Standardize handling methods and equipment wherever possible.

7. **Ergonomic Principle.** Recognize human capabilities and limitations by designing material-handling equipment and procedures for effective interaction with the people using the system.

8. **Energy Principle.** Include energy consumption of the material-handling systems and material-handling procedures when making comparisons or preparing economic justifications.

9. **Ecology Principle.** Use material-handling equipment and procedures that minimize adverse effects on the environment.

10. **Mechanization Principle.** Mechanize the handling process where feasible to increase efficiency and economy in the handling of materials.

11. **Flexibility Principle.** Use methods and equipment that can perform a variety of tasks under a variety of operating conditions.

12. **Simplification Principle.** Simplify handling by eliminating, reducing, or combining unnecessary movements and/or equipment.

13. **Gravity Principle.** Utilize gravity to move material wherever possible, while respecting limitations concerning safety, product damage, and loss.

14. **Safety Principle.** Provide safe material-handling equipment and methods that follow existing safety codes and regulations in addition to accrued experience.

15. **Computerization Principle.** Consider computerization in material-handling and storage systems when circumstances warrant, for improved material and information control.

16. **System Flow Principle.** Integrate data flow with the physical material flow in handling and storage.

17. **Layout Principle.** Prepare an operation sequence and equipment layout for all feasible system solutions, then select the alternative system that best integrates efficiency and effectiveness.

18. **Cost Principle.** Compare the economic justification of alternate solutions in equipment and methods on the basis of economic effectiveness as measured by expense per unit handled.

19. **Maintenance Principle.** Prepare a plan for preventive maintenance and scheduled repairs on all material-handling equipment.

20. **Obsolescence Principle.** Prepare a long-range and economically sound policy for replacement of obsolete equipment and methods with special consideration to after-tax life-cycle costs.

In designing material-handling systems to support manufacturing the following objectives should be considered:

1. Create an environment that results in the production of high-quality products.

2. Provide planned and orderly flows of material, equipment, people, and information.

3. Design the layout and the material-handling system to accommodate changes in manufacturing technology, processing sequences, product mix, and production volumes.

4. Reduce work-in-process.

5. Provide controlled movement and storage of material.

6. Eliminate manual material handling at and between work stations.

7. Deliver parts to work stations in predetermined quantities and physically positioned or oriented to allow automatic parts feeding, insertion, and assembly.

8. Deliver tooling to machines in a controlled position or orientation to allow automatic loading/unloading and automatic tool change.

9. Integrate processing, assembly, inspection, handling, storage, and control of material.

10. Utilize space most effectively, considering overhead space and impediments to cross traffic.

54.2.2. Handling-System Elements

The selection of material-handling equipment requires a careful consideration of the following functions: material containment, material movement, material storage, and material control. Material movement can be further subdivided into micromovements and macromovements. In specifying the container to be used, a number of parameters must be considered, including the following:

1. Size of the container, cube, and footprint.
2. Weight of the container, empty and loaded.
3. Protection requirements for the product.
4. Modularity of the container, subcontainer, and supercontainers.
5. Standardization of dimensions, designs, and materials for the container.
6. Durability of the container.
7. Stackable/nestable/collapsible requirements.
8. Slave/one-way/returnable container usage.
9. Conveyable requirements.
10. Tracking/identifying requirements.

Many of the best systems are developed without the use of a container, as such, but material containment must still be designed as part of the system. Parts may be handled individually directly from moving belt, trolley, or carousel conveyors.

Micromovement is appropriate for the movement that occurs at a work station, whereas macromovement includes the moves between work stations and between departments. Relatively speaking, macromoves are long moves and micromoves are short moves. The parameters that influence the selection of equipment for performing either micromoves or macromoves include the following:

1. Load size, including consideration of container and/or material.
2. Load weight.
3. Number of loads per move.
4. Mix of load.
5. Frequency of moves.
6. Distance of move.
7. Response time for the move.
8. Obstacles to the flow path.
9. Path becoming an obstacle to other moves.
10. Changing throughput or flow requirements.

Additionally, the following parameters influence the selection of the equipment to be used in performing micromoves at the workplace:

1. Operator, robot, or machine reach.
2. Access to the work station.
3. Interface with the operator, robot, or machine.
4. Cycle time at the work station.
5. Weight of an individual part.
6. Other material characteristics of the part.

Parameters that influence the storage of material (work-in-process, tooling, supplies, fixtures, etc.) include the following:

1. Space required and available.
2. Throughput requirements.
3. Centralized versus decentralized storage considerations.
4. Load size and weight.
5. Load stackability.
6. Special characteristics of the material being stored.
7. Order profile for storages/retrievals.
8. Requirements for selective retrieval.
9. Facility constraints.
   (a) Floor loading.
   (b) Floor condition.
   (c) Column spacing.
   (d) Clean stacking height.
   (e) Sprinklers/codes.

The hardware and software required for material control depends on the following parameters:

1. Requirements for real-time control versus batch control.
2. Requirements for continuous tracking versus interrupted tracking.
3. Transaction rate.
4. Management reporting requirements.
5. Requirements for accurate control.
6. The need to reestablish priorities.
7. Workload schedule.

Among the numerous equipment alternatives available for containing, moving, storing, and controlling material are the following major categories:

1. Containers
    (a) Tote boxes, baskets, and magazines
    (b) Pallets and skids
    (c) Platforms and fixtures
    (d) Skid boxes and racks
2. Industrial trucks
    (a) Hand truck and cart
    (b) Pallet jack
    (c) Walkie stacker
    (d) Pallet truck
    (e) Platform truck
    (f) Tractor-trailer
    (g) Automated guided vehicle
    (h) Counterbalanced lift truck
    (i) Narrow-aisle straddle truck
    (j) Narrow-aisle reach truck
    (k) Narrow-aisle sideloader
    (l) Narrow-aisle turret truck
    (m) Narrow-aisle storage/retrieval truck
    (n) Narrow-aisle order-picker truck
    (o) Straddle carrier
    (p) Mobile crane
3. Conveyors
    (a) Chute conveyor and slide rails
    (b) Flat belt conveyor
    (c) Troughed belt conveyor
    (d) Magnetic belt conveyor
    (e) Roller conveyor
    (f) Wheel conveyor
    (g) Slat conveyor
    (h) Chain conveyor
    (i) Bucket conveyor
    (j) Trolley conveyor
    (k) Power-and-free conveyor
    (l) Tow conveyor
    (m) Car-on-track conveyor
    (n) Screw conveyor
    (o) Vibrating conveyor and vibrating parts feeders
    (p) Pneumatic conveyor
    (q) Lift-and-carry conveyor
    (r) Air film conveyor
4. Balancers, manipulators, and robots
    (a) Work balancer
    (b) Industrial manipulator
    (c) Rectangular coordinate robot
    (d) Cylindrical coordinate robot
    (e) Spherical coordinate robot
    (f) Jointed or articulated arm robot
5. Monorails, hoists, and cranes
    (a) Monorail, unpowered carriers
    (b) Monorail, powered carriers
    (c) Monorail, intelligent powered carriers
    (d) Hoist
    (e) Jib crane
    (f) Bridge crane
    (g) Gantry crane
    (h) Tower crane
    (i) Stacker crane
    (j) Manual stacker
6. Automated storage/retrieval systems
    (a) Unit load AS/RS
    (b) Miniload AS/RS
    (c) Microload AS/RS
    (d) Aisle captive picking vehicle
    (e) Deep lane AS/RS
    (f) Automated item retrieval systems
    (g) AS/R carousels
    (h) AS/R revolving shelving
7. Automatic identification systems
    (a) Optical character recognition
    (b) Hand-held, wand bar code scanner
    (c) Fixed beam, bar code scanner
    (d) Moving beam, laser bar code scanner
    (e) Magnetic code reader
    (f) Microwave transponder
    (g) Pattern and shape recognition
    (h) Reflective code readers
8. Storage drawers and racks
    (a) Storage bins
    (b) Shelving
    (c) Storage drawers
    (d) Selective, one-deep pallet rack
    (e) Double-deep pallet rack
    (f) Portable stacking rack
    (g) Cantilever rack
    (h) Decked cantilever rack
    (i) Drive-in rack
    (j) Drive-through rack
    (k) Edge stacked, plate rack
    (l) Pipe rack
    (m) Pigeon hole rack
    (n) Case flow rack
    (o) Pallet flow rack
    (p) Mobile shelving
    (q) Mobile rack
9. Other handling equipment: dial index table

Descriptions and photographs of many of the equipment alternatives listed can be found in Tompkins and White,[5] as well as Apple,[1] Kulweic,[3] Sims,[4] and issues of *Material Handling Engineering* and *Modern Materials Handling.* The list provided is not comprehensive, since it does not include air film handling, automatic palletizers, automatic depalletizers, ball transfer tables, bowl feeders, and dock boards, lift tables, shrinkwrap/stretchwrap equipment, among others. However, it should provide sufficient coverage to indicate the large variety of equipment alternatives available for consideration in containing, moving, storing, and controlling materials.

54.3. THE ROBOT HANDLING MATERIAL

By including both micromoves and macromoves in the definition of material handling, it is easy to understand why the robot has had a major impact on the field of material handling. In 1983 more than 50% of the robots installed in the United States were performing material-handling tasks. Typical of the material-handling role played by robots are the following applications:

1. Palletizing
2. Depalletizing
3. Case packing
4. Bin picking
5. Kitting
6. Storage/retrieval
7. Machine loading/unloading
8. Conveyor sortation
9. Parts feeding
10. Material delivery

Examples of robots performing material-handling tasks are given in Figures 54.2 through 54.14. Notice, in many of the applications the robot is interfacing with other material-handling equipment. It is very common to find robots performing pick-and-place operations and interfacing with containers,

Fig. 54.2. A robot reserves wrenches from dial index table for broaching operation. (Courtesy Copperweld Robotics.)

Fig. 54.3. A robot picks parts from a carousel parts feeder. (Courtesy GMF Robotics.)

Fig. 54.4. A robot picks cylindrical shapes from a conveyorized carrier and returns them after machining. (Courtesy Cincinnati Milacron.)

Fig. 54.5. A robot palletizes blocks from a conveyor line. (Courtesy ASEA, Inc.)

Fig. 54.6. Pick-and-place robot arms feed and clear a broaching machine at 240 pieces/hour. (Courtesy Apex Broach and Machine Company.)

Fig. 54.7. Handling parts into a press and back to a specially fixtured pallet. (Courtesy Prab Robots, Inc.)

Fig. 54.8. Handling a separator sheet for palletizing cylindrical shapes. (Courtesy Prab Robots, Inc.)

Fig. 54.9. With magnetic tooling, a robot "gang" loads motor frames on a pallet. (Courtesy Prab Robots, Inc.)

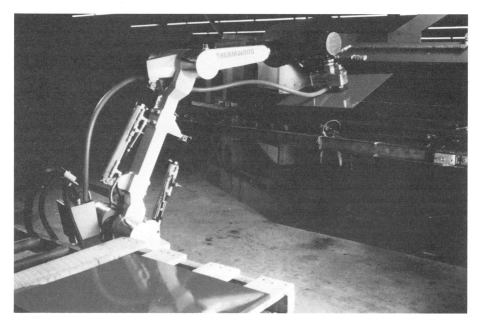

Fig. 54.10. A robot with vacuum tooling transfers plastic sheets into a processing operation. (Courtesy Thermwood Corp.)

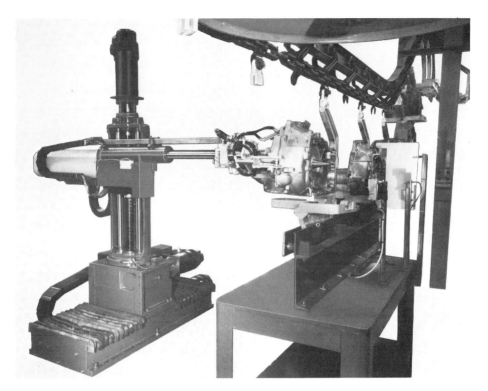

Fig. 54.11. Loading and unloading transmission cases from a moving monorail conveyor. (Courtesy Pickomatic Systems, Inc.)

Fig. 54.12. Gripping on the sides, a robot transfers furniture panels from a prepositioned pallet. (Courtesy Unimation.)

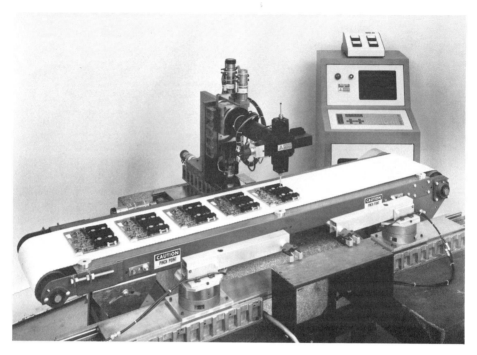

Fig. 54.13. Its travel synchronized with the conveyor, a robot probes components on a printed circuit board. (Courtesy Anorad Corp.)

Fig. 54.14. A pair of plastic tail-lamp lenses is deposited on a slide rail takeaway as the molding sprue is dropped into a collection bin below the rails. (Courtesy Conair, Inc.)

Fig. 54.15. A robot suspended from a bridge selects instrument panels from wire baskets and loads them on an assembly conveyor. (Courtesy GCA Corp.)

conveyors, guided vehicles, monorails, miniload and microload automated storage/retrieval systems, AS/R carousels, and AS/R revolving shelving.

To provide mobility to the robot, it can be mounted on an automated guided vehicle, mounted on a car-on-track conveyor, and hung from a gantry crane, bridge crane, or monorail, as illustrated in Figures 54.15 and 54.16. Power for the robot can be obtained by using festoon cable, power conductor bars in conjunction with the guided vehicle conveyor and monorail, or by providing automatic plug-in capability for the robot at strategically located "power stations."

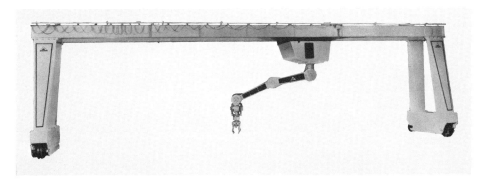

Fig. 54.16. A gantry-mounted robot has a nearly unlimited work envelope. (Courtesy Air Technical Industries.)

Increased usage of monorail and robot combinations is anticipated owing to the ability to utilize overhead space and avoid impediments to cross traffic at floor level.

As robots interact with other material-handling equipment, a number of critical operating parameters must be accommodated by either the robot or the handling system. Table 54.1 lists several key parameters and possible methods for addressing them.

One of the primary benefits gained from the use of robots to perform material-handling tasks is the renewed and increased understanding of material-handling systems design. Specifically, it has been learned that effective use of a robot depends on having well-planned material flow. Experience indicates that people, too, would be more productive if the same degree of attention were given to controlling material flow to human operators as is required for automated operators. The importance of the adage "grab hold and don't turn loose" in moving material through a manufacturing system is more clearly understood in the absence of human operators.

Based on an exposure to robots as material handlers, "lessons learned" are presented in the following:

1. Handling less is best.
2. Grab hold and don't turn loose.
3. Eliminate, combine, and simplify.
4. Moving and/or storing material incurs costs.
5. Preposition materials and tooling.

Handling less is best suggests that handling should be eliminated if possible. It also suggests that the number of times material is picked up and put down and the distance material is moved should be reduced.

Grab hold and don't turn loose, as previously noted, emphasizes the importance of maintaining the physical control of material. Too often parts are dumped into tote boxes or wire baskets. Subsequently, someone must handle each part individually to orient and position it properly for the next operation. Too little attention is paid to the impact downstream of handling decisions made upstream.

Eliminate, combine, and simplify suggests that the principles of work simplification and methods improvement are appropriate in designing in-process handling and storage systems. Handling and storage can frequently be eliminated completely by making changes in processing or production scheduling. Standardizing containers, for example, allows containers to cross departmental boundaries and eliminates the need to transfer the contents of one container into another. Also, handling might be combined with some operations.

Moving and/or storing material incurs costs serves as a reminder that inventory levels should be kept as small as economically feasible. Moving material incurs costs because of the personnel and/or equipment required to perform the move; also, studies have shown that damage to the material is more likely as the frequency of moves is increased. Finally, to move material requires a corridor of space and neither space nor controlled movement is free! Storing material incurs costs because of the

TABLE 54.1. ROBOT/HANDLING SYSTEM INTERFACE

Operating Parameter	Resolved by Handling System	Robot
Speed of moving parts or assemblies	Index conveyor to provide temporarily fixed position	Synchronize robot movement to match conveyor speed
Part orientation	Use fixtured container or fixtures on conveyor Use locating guides to shift part into position Use part-orienting devices	Use internal or external vision system to determine part location and orientation Use tactile sensing to adjust robot to precise position
Queuing parts or assemblies at the work station	Accumulating conveyor Recirculation of parts	Pick parts as they arrive and place in a queuing position for later use
Control and communication	Signal part presence with sensors Use machine readable codes on part or container to indicate proper robot program	Select appropriate program based on specific part recognition with vision system

cost of the space, equipment, and personnel required, as well as the costs of interest, insurance, shrinkage, and inventory taxes. Basically, ownership of material is a costly process.

Preposition materials and tooling has two aspects to be considered. First, parts should be prepositioned to facilitate automatic load/unload, insertion, inspection, assembly, and so on. Second, when material is delivered to a work station and/or machine center, it should be placed in a prespecified location with a designated orientation. Too often, direct labor personnel spend time searching for material that supposedly has been delivered and placed somewhere in the general area.

54.4. INTEGRATED SYSTEMS

The predominate approach used today to introduce factory automation technology into manufacturing is to selectively apply automation and to create *islands of automation.* Such islands must be bridged to form integrated factory systems.

Since 1978 we have used the term *islands of automation* to describe the transition from conventional or mechanized manufacturing to the automatic factory. Interestingly, some appear to use the term as though it were a worthy end objective. On the contrary, the creation of such islands can be a major impediment to achieving an integrated factory.

Manufacturing examples of islands of automation often include numerically controlled machine tools; robots for assembly, inspection, painting, and welding; lasers for cutting, welding, and finishing; sensors for test and inspection; automated storage/retrieval systems for storing work-in-process, tooling, and supplies; smart carts, monorails, and conveyors for moving material from work station to work station; automated assembly equipment and flexible machining systems. Such islands are often purchased one at a time and justified economically by cost reductions.

To integrate the islands of automation it is necessary to link several machines together as a unit. For example, a machine center with robots for parts loading and unloading can best be tied to visual inspection systems for quality. Computer numerical control machine tools can all be controlled by a computer that also schedules, dispatches, and collects data. Selecting which islands to link can be most efficiently pursued on the basis of cost, quality and cycle time benefits.[6]

In some cases the islands of automation will be very small (e.g., an individual machine or work station). In other cases the islands might be department-sized.

As an example of the creation of relatively small islands of automation, consider an appliance manufacturer who installed a number of robots along an existing assembly line. The resulting labor reduction generated a cost savings; the robots were certainly justified economically. However, an opportunity to increase productivity for the total system was missed. Materials were delivered to the robots and removed from the robots using the existing material-handling system. Because the production rates for the robots differed from the manual rates, materials were stacked on the floor around the robots and in the aisles. From a myopic point of view, the robot was impressive, whereas from a systems point of view, an island of automation had been created.

An example of a relatively large island of automation was observed in a gearbox plant for a truck manufacturer. Castings were fed from a magazine to a robot, which subsequently fed the casting to each of three machines and then placed the semifinished part on a conveyor. The conveyor delivered the part to a second robot, which fed the part to each of three additional machines and then placed the semifinished part on a second conveyor. The part was delivered to a third robot, which fed the part to each of three more machines and placed the finished part on a peg-rack dolly for pickup and delivery by tugger to a storage area. Three islands of automation had been linked together to form a much larger island.

Interestingly, the castings were delivered in a wire basket to the first robot by lift truck. The castings had come from the foundry in a nearby building. At one point the castings were on a belt conveyor positioned and spaced in such a way they could have been automatically placed in the delivery container in a controlled fashion. However, they were dumped into the wire basket. Consequently, at the gearbox building someone had to reach into the wire basket, grasp a casting, orient it properly, and place it in the magazine that fed the first robot.

From a systems viewpoint, islands of automation are not necessarily bad, so long as they are considered to be interim objectives in a phased implementation of an automated system. However, to obtain an integrated factory system, the islands of automation must be tied together or synchronized. Systems synchronization frequently occurs by way of the material-handling system; it physically builds bridges that join together the islands of automation. Likewise, information bridges can be provided through the control system.

54.5. SUMMARY

In summary, the robot has had a significant impact on the field of material handling. More specifically, the design of *integrated* material-handling systems for both manufacturing and distribution has been influenced by the expanding availability and capabilities of robots.

It is anticipated that more and more frequently robots will be found to be the right *method* for providing the right *amount* of the right *material* in the right *orientation* and the right *condition* to the right *place,* at the right *time,* and at the right *cost.*

The robot will evolve from being an island of automation to being a key element of the integrated system. They will be physically integrated with other material-handling equipment, and the control system for the robot will be integrated with the overall material-handling control system.

REFERENCES

1. Apple, J. M., *Material Handling Systems Design,* Wiley, New York, 1972.

2. Apple, J. M., Jr. and Strahan, B. A., Proper Planning and Control—The Keys to Effective Storage, *Industrial Engineering,* April 1981.

3. Kulweic, R., *Material Handling Handbook,* Wiley, New York, 1983.

4. Sims, E. R., Jr., *Warehouse Modernization and Layout Planning Guide,* NAVSUP Publication 529, Department of the Navy, Naval Supply Systems Command, Washington, D.C., 1978.

5. Tompkins, J. A. and White, J. A., *Facilities Planning,* Wiley, New York, 1984.

7. White, J. A., Integrated Systems in the Automatic Factory, *Industrial Engineering,* Vol. 14, No. 4, April 1982, pp. 60–68.

8. White, J. A., Robots—A Materials Handling Alternative, *Modern Materials Handling,* Vol. 37, No. 6, May 1982, p. 25.

CHAPTER **55**

WORKPIECE HANDLING AND GRIPPER SELECTION

SEIUEMON INABA

Fanuc Corporation
Yamanashi-Ken, Japan

55.1. INTRODUCTION

This chapter surveys various applications of robots for workpiece handling and provides guidelines for gripper selection. Chapter 56, Robotic Loading of Machine Tools, describes the considerations for selecting among alternative approaches of robotic loading/unloading.

55.2. APPLICATIONS OF ROBOTS FOR HANDLING IN A MACHINING CELL

Robots for machining cells are sorted roughly into the types to be mounted or incorporated on machine tools and the self-standing types (pedestal) to be separated from machine tools.

Separate robots, independent of machine tools, are constructed as cylindrical coordinate types in most cases. (See Figures 55.1 and 55.2.) The maximum load capacity at wrist for the robot shown varies from 47 to 120 kg. Robots, to be mounted or incorporated on machine tools, are constructed as dual cylindrical coordinate types in most cases, as shown in Figures 55.3 and 55.4. The maximum load capacity at wrist for the robot shown varies from 20 to 60 kg.

55.3. EXAMPLES OF ROBOTIC WORKPIECE HANDLING

Figures 55.5 and 55.6 show examples of applications of self-standing robots to lathes. The robot grips workpieces from a rotary workpiece feeder one by one and loads them onto a lathe. The robot also grips machined workpieces from the lathe and puts them in a separate place on the rotary workpiece feeder. As a result, blanks on the rotary workpiece feeder are processed to machined parts after a certain time.

Figures 55.7 through 55.10 depict examples of applications of self-standing robots to machining centers. The robot sequentially picks up workpieces stacked on a table from the top and stacks machined workpieces in regular sequence.

Figures 55.11 and 55.12 show examples of applications of self-standing-type robots to grinding machines. The robot employs a hand suitable for handling cylindrical workpieces. An exclusive pallet is used for stacking workpieces (Figure 55.13).

55.4. WORKPIECE HANDLING BY A ROBOTIC VEHICLE

A machining cell is employed for machining, while an assembly cell is employed for assembling, both as major components of factory automation. In an automated factory, workpieces are carried between an automatic warehouse and such cells by unmanned carriers. In machining cells, a robot then carries the workpieces from the work station to NC machine tools. (See Figures 55.14 and 55.15.)

A pallet carried by an unmanned carrier is transferred to the loading station of the workpiece station. Then, the pallet shifts to the active station. The self-standing robot sequentially grips workpieces on the pallet and sets them one by one onto the mounting jig on the machining center. After machining, the robot stacks machined workpieces onto the pallet on the active station. When all workpieces have been machined, the pallet shifts to the unloading station. The unmanned carrier carries the pallet on the unloading station to a specified position.

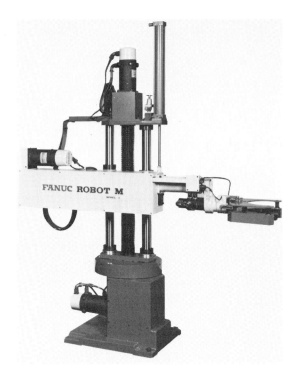

Fig. 55.1. Overview of robot.

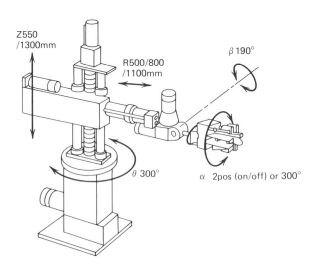

Fig. 55.2. Robot motion.

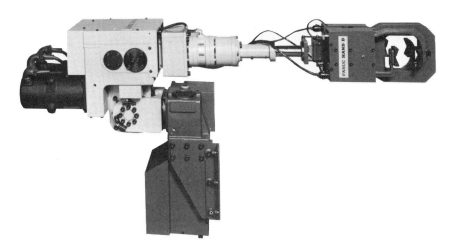

Fig. 55.3. Overview of robot.

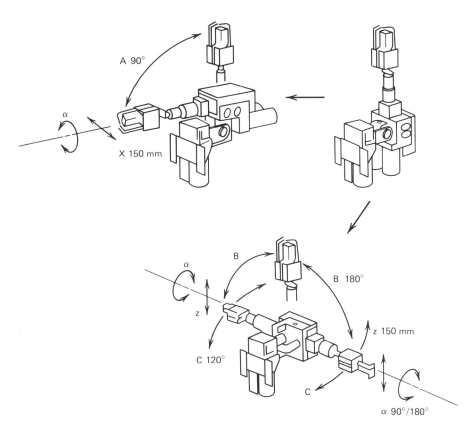

Fig. 55.4. Robot motion.

Fig. 55.5. Overview of an application to a lathe.

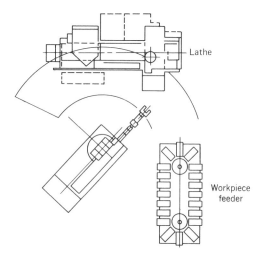

Fig. 55.6. Layout in the example of an application to a lathe.

Fig. 55.7. Overview of an application to a machining center.

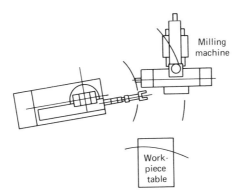

Fig. 55.8. Layout of an application to a machining center.

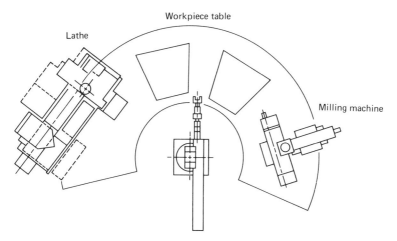

Fig. 55.9. Layout for servicing two machine tools.

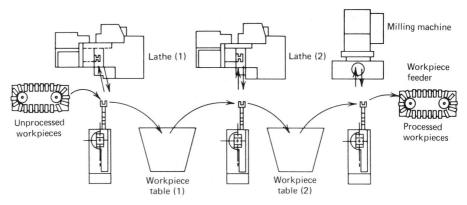

Fig. 55.10. Layout for carrying a workpiece by a robot.

Fig. 55.11. Overview of an application to a grinding machine.

Fig. 55.12. Gripping a cylindrical workpiece.

Fig. 55.13. Pallet for stacking cylindrical workpieces.

55.5. GRIPPER (HAND) SELECTION

Hands with two or three fingers are employed for gripping workpieces (see Figure 55.16).

55.5.1. Hand with Three Fingers

A hand with three fingers can grip a cylindrical or rectangular workpiece.

Gripping of Cylindrical Workpieces

The weight and size of grippable workpieces and the gripping methods using fingers are determined according to the gripping conditions (see Figures 55.17, 55.18, and 55.19).

Gripping of Rectangular Workpieces

Rectangular workpieces are gripped by a hand with two or three fingers. The weight and sizes of grippable workpieces and gripping methods using fingers are determined according to the gripping conditions (see Figures 55.20 and 55.21).

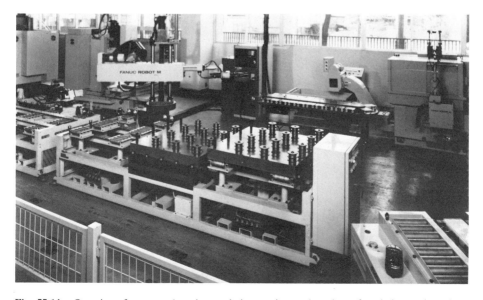

Fig. 55.14. Overview of unmanned carrier, workpiece station, and carriage of workpieces using robots.

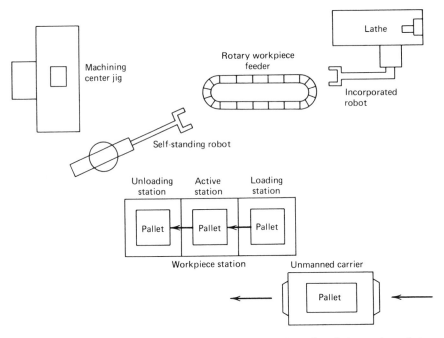

Fig. 55.15. Unmanned carrier, workpiece station, and carriage of workpieces using robots.

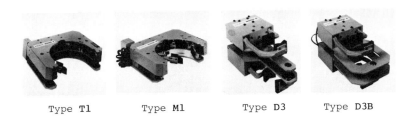

Type T1   Type M1   Type D3   Type D3B

Fig. 55.16. Hands for gripping workpieces.

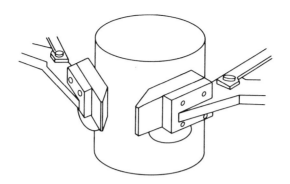

Fig. 55.17. Outer diameter gripping.

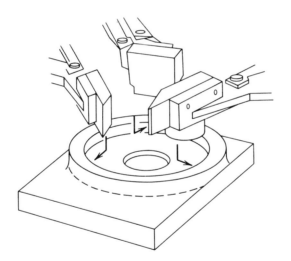

Fig. 55.18. Inner diameter gripping.

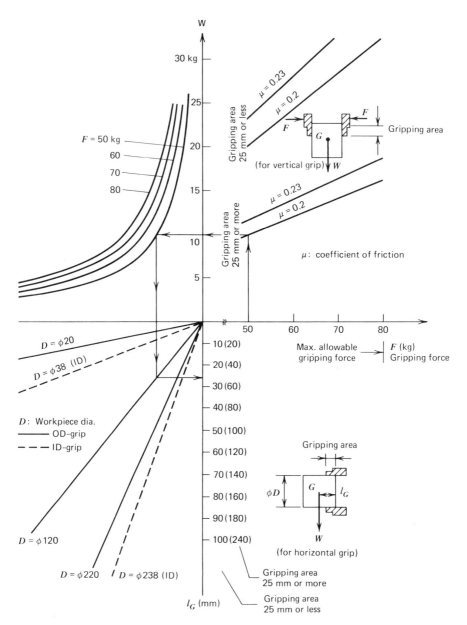

Fig. 55.19. Gripping conditions using a hand with three fingers (cylindrical workpiece). For a vertical grip, when gripping force $F = 50$ kg, and gripping area is 25 mm or less, the maximum weight is 10 kg. For horizontal grip, when workpiece diameter is 120 mm, workpiece gravity center position l_G is 26 mm or less. The flexible push mechanism limits use of Hand T. Usually $\mu = 0.2$.

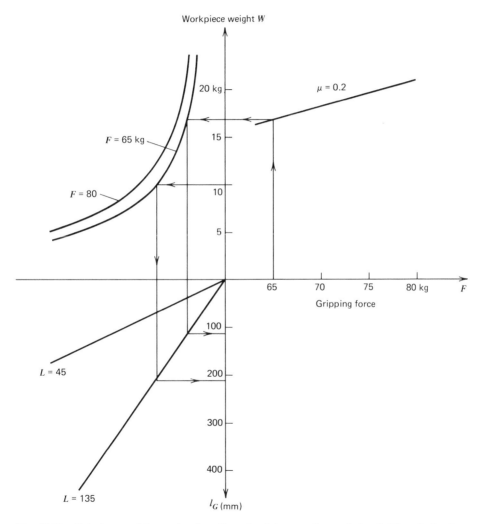

Fig. 55.20. Gripping conditions using three-finger hand (rectangular workpieces). The gripping force when the width across two flats of workpiece, $B = 129$ mm. With the same service pressure, F varies with the workpiece width B. With the gripping force 65 kg, the maximum workpiece weight is 17 kg. When $L = 135$, l_G is 110 mm. When $F = 65$ kg, $W = 10$ kg, and $L = 135$ mm, l_G is 210 mm. The flexible push mechanism limits the use of the Hand T.

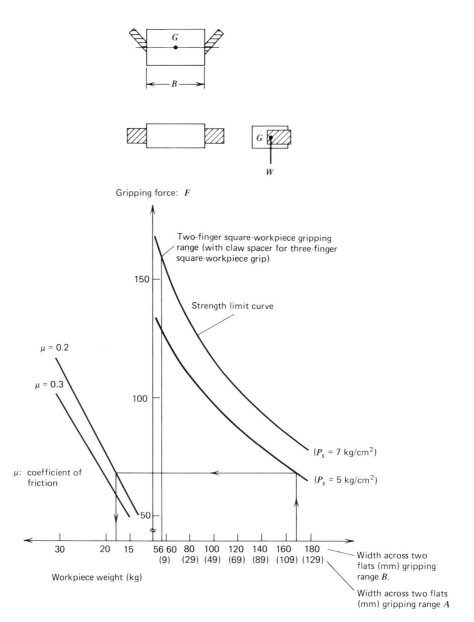

Fig. 55.21. Gripping conditions using two-finger hand (rectangular workpieces). When the width across two flats of workpiece is 170 mm and no claw spaces are provided, the maximum workpiece weight is 18 kg. The hand material strength limits the maximum gripping force and workpiece weight. Always grip the workpiece at the center of gravity so that no moment is applied. The flexible push mechanism limits the use of Hand T.

Insertion of a Workpiece into Lathe Chuck

A workpiece is inserted into the chuck by a hand with the push buffer mechanism, as shown in Figures 55.22 and 55.23. One must be careful with the gripping positions of workpieces if they are particularly long.

55.5.2. Hand with Two Fingers

The hand with two fingers can grip cylindrical workpieces as depicted in Figure 55.24. The weight and size of grippable workpieces and the gripping methods of fingers are determined by the gripping conditions, as shown in Figure 55.25.

REFERENCES

1. Barash, M. M., Integrating machinery systems with workpiece handling, *Industrial Robot,* Vol. 3, No. 2, June 1976, pp. 62–67.

2. Bey, I., Automation of workpiece handling in small batch production, *Proceedings of the 1st International Conference on Flexible Manufacturing Systems,* Brighton, U.K., October 1982.

3. Lian, D., Peterson, S., and Donath, M., A three-fingered, articulated robotic hand, *Proceedings of the 13th I.S.I.R.,* Chicago, Illinois, April 1982, pp. 18/91–101.

4. Lien, T. K., Workpiece handling by robot in a flexible manufacturing cell, *Proceedings of the 8th I.S.I.R.,* Stuttgart, West Germany, June 1978, pp. 242–254.

5. Lundstrom, G., Glemme, B., and Rooks, B. W., *Industrial Robots Gripper Review,* IFS Publications, Bedford, U.K., 1977.

6. Schafer, H. S., and Malstrom, E. M., Evaluating the effectiveness of two-fingered parallel jaw, *Proceedings of the 13th I.S.I.R.,* Chicago, Illinois, April 1983, pp. 18/112–121.

7. Schekulia, K., Workpiece handling on machine tools with industrial robots (in German), *Ind. Fertigung,* Vol. 65, November 1975, pp. 685–686.

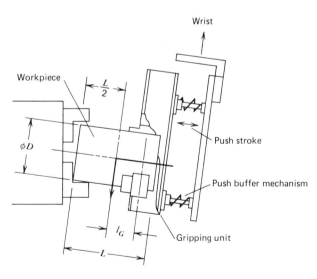

Fig. 55.22. Insertion of workpieces into lathe chuck. *L:* length of workpiece. *D:* diameter of workpiece. l_G: eccentric gravity distance.

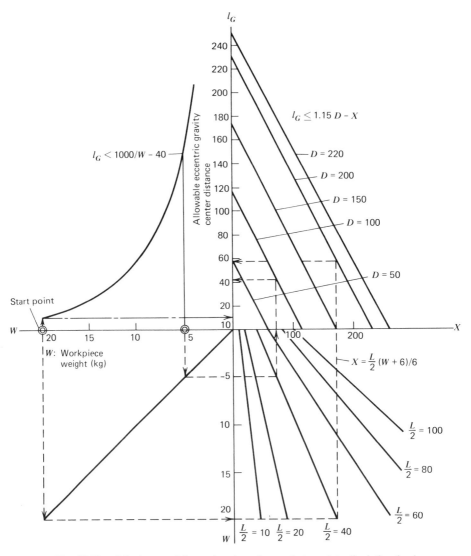

Fig. 55.23. Gripping conditions when inserting workpieces into the lathe chuck.

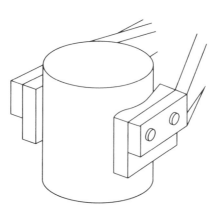

Fig. 55.24. Outer diameter (OD) gripping.

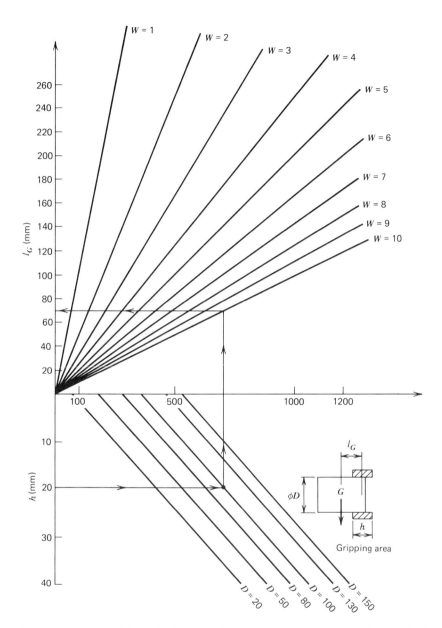

Fig. 55.25. Gripping conditions of a two-finger hand. Example: When the gripping area is 20 mm, outer diameter 100 mm, and workpiece weight 10 kg, allowable eccentric center of gravity in horizontal grip is 70 mm.

CHAPTER 56

ROBOTIC LOADING OF MACHINE TOOLS

THOMAS E. KLOTZ

Mazak Corporation
Florence, Kentucky

56.1. WAYS OF MOUNTING LOADING ROBOTS

There are currently three major ways to mount robots for machine loading: (1) robot carts, or automatic guided vehicles (AGV); (2) floor- or machine-mounted robots; (3) gantry, or overhead robots.

56.1.1. Robot Carts

Robot carts are typically wire- or rail-guided, and can be found in a variety of sizes that can handle part load requirements from several hundred pounds up to 20,000 pounds (100–9000 kg).

Parts are loaded directly on the cart, usually on pallets. Sometimes the pallets can also be utilized as fixtures on NC machines. In a very few applications, robots have been directly mounted on the cart to better handle delicate or odd-shaped parts. More commonly, a local robot will unload parts from the cart, or the loaded pallet will move directly from cart to machine table. The robot cart has found a place in flexible manufacturing facilities, where there is a need to transport raw material, finished machine parts, and in-process parts over relatively great distances (see example in Figure 56.1).

The wire- or rail-guided robot cart can solve a number of production control problems.

The cart has virtually an unlimited traveling range within the confines of the plant. It has the ability to transport raw material or finished parts over great distances without utilizing vital floor space that could later be designated as a manufacturing area in the plant.

Preloaded pallets can be set up in advance so the material, upon demand by the machine tool and its local robot, can be dispatched by the use of the robot cart. The carts can be designed to handle steel pallets or inexpensive wood pallets.

The wire-guided robot cart is adjustable to existing manufacturing floor plans because of the relatively limited amount of floor space required to install it. The robot cart can utilize the pedestrian traffic areas and aisles; for safety reasons, carts are designed with sensors that enable them to recognize on coming pedestrians or obstructions. These sensors will cause the robot cart to slow down or stop until the obstructions are removed from its path.

Unmanned shifts would require tremendous amounts of material to be warehoused at the machine tool site. This warehouse affects the floor space required within the plant. It also prevents efficient placement of machine tools. A robot cart bringing parts and materials from an outside inventory can solve this problem, and machine tools can be placed closer and more efficiently to minimize required floor space. Further details on cart-mounted robots can be found in Chapter 59, Mobile Robot Applications.

56.1.2. Gantry, Overhead Robots

The gantry, or overhead, robot is utilized where there is a requirement for quickly moving workpieces from machine to machine. The gantry robot can move along its X and Y axes traveling over relatively greater distances at higher traverse speeds, while still providing a high degree of accuracy for positioning. These robots can be used for auxiliary tasks beyond just loading parts. For instance, gantry robots

Fig. 56.1. A robot cart (automatic guided vehicle, AGV) for machine tool part loading in flexible manufacturing facilities. A similar cart is used by Mazak to transport and change complete tool magazines.

have been used for cleaning jobs such as removing cutting chips from large workpieces with the use of high-powered vacuum systems in preparing the part for its next process. See also Chapter 60, Gantry Robots and Their Applications.

56.1.3. Floor- and Machine-Mounted Robots

Floor- and machine-mounted robots are preferred for local loading tasks. A floor-mounted, pedestal robot can serve one or more machines. (See examples in Chapter 57, Machine Loading Application Cases.) A machine-mounted robot is usually dedicated to a single machine and can therefore have fewer axes of motion, typically two, three, or four (see Figures 56.2 and 56.3). In 1983 about 15% of all Japanese and 6% of all U.S. CNC machines were equipped with some form of automatic load/unload of parts. Some machine-mounted robots can be used to retrofit existing machinery (e.g., Fanuc Series o).

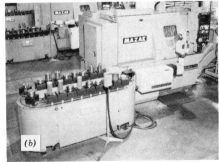

Fig. 56.2. Electric robot arms mounted on machine tools for part loading/unloading in an FMS line. (*a*) Twin-spindle machining center equipped with Mazak Flex II. (*b*) CNC lathe equipped with Mazak Flex I. The teach box shown in the front is used to instruct the robot how to handle up to 60 different workpieces weighing 29 to 40 kg each.

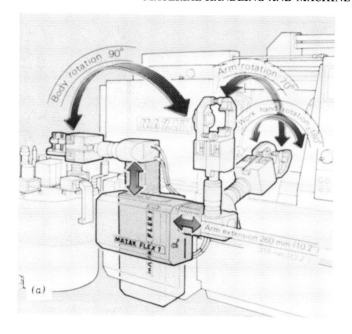

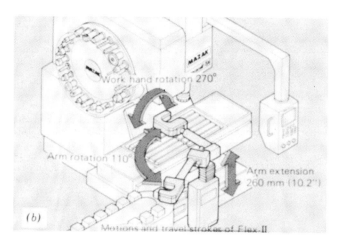

Fig. 56.3.   Motions of machine-mounted robot loaders.

Floor- and machine-mounted robots often use a pallet pick-and-place or a conveyor feeder to feed parts to and from their location. These robots can be programmed to pick or place parts on a geometrically designed surface, putting the parts in rows or palletizing them in a given area (Figure 56.4). This, however, only allows the robot to work unattended for a limited amount of time, and thought should be given to manipulating or moving the palletized parts. A typical operation of a machine-mounted robot is shown in Figure 56.5.

56.1.4.   Robot Drive

With regard to the robot actuation in machine loading, hydraulic or electric servo motors are the main source of power for movement and manipulation. A hydraulic robot has weight-lifting capabilities that usually exceed those of an electric robot. Hydraulic robots are chosen for loading large workpieces that may weigh 300, 400, or 500 lb (135, 180, 225 kg). On the other hand, hydraulic robots do not

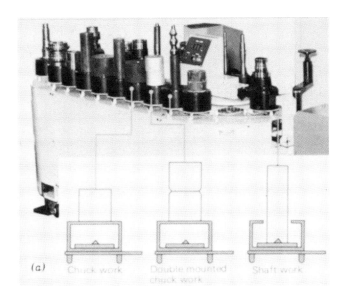

Fig. 56.4. Parts positioned on a pallet conveyor (a) wait for robot loader. A full-capacity conveyor can provide 6 hr of unmanned operation, assuming each workpiece has a cycle time of 6 min. (b) Parts are positioned in jigs for accurate placement.

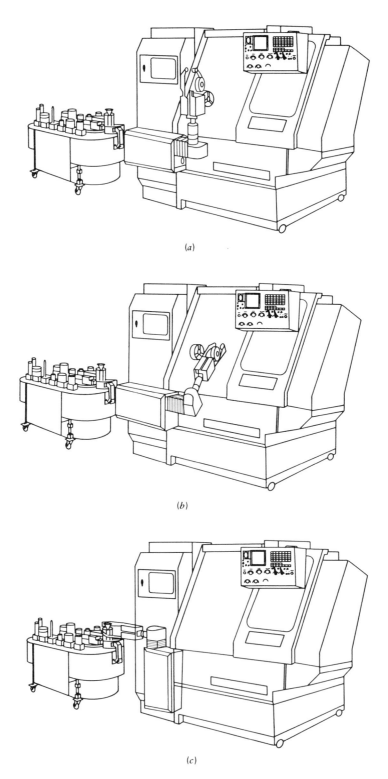

Fig. 56.5. Typical operations of a machine-mounted robot. (*a*) Standby position (body up, arm up). This is the initial position of robot service. It also is the position awaiting completion of the process by the machine. (*b*) Loading/unloading of workpiece on chuck (body up, arm down). This is the position of loading/unloading of workpiece on chuck. Machine is still. (*c*) Loading/unloading of workpiece on conveyor (body down, arm up). This is the position of loading/unloading of workpiece on conveyor.

have the position accuracy of electric robots. Electric robots are frequently used to handle lighter workpieces that require delicate picking and placing and, at the same time, require high accuracy for the placement of the part in its fixture, or in assembly of the products where accuracy is the key to a successful operation.

56.2. PREPARATION OF PARTS FOR ROBOT HANDLING

A robot is normally designed to pick up a part in the same attitude and position each time; therefore proper part preparation for robot handling is required. Jigs and fixtures are commonly supplied to provide accurate part placement for this purpose (see Figure 56.4b). In less common situations, machine tools are loaded with parts by vision-equipped robots, which can locate parts that are placed at less accurate positions. Such capability may be required in many machine tools that have programmable spindle orientation. This orientation is necessary for parts that are arbitrary in shape, and it is required that the arbitrary shape does not interfere with the placement of the part in its proper holding fixture.

Part handling by the robot between operations is usually required also. A machine-mounted robot is designed to swivel its arm to load a part for one machine operation, then set the part down and repick it at a different orientation for the next operation (see Figure 56.6a,b). Fixtures are used to hold and maintain parts positions during these motions. Group technology is often applied to design reusable fixtures for part families. Attempts are also being made to develop programmable fixtures for different tasks; however, such solutions are relatively expensive. With regard to gripper selection for part loading, see Chapter 55, Workpiece Handling and Gripper Selection.

It should be noted that robots can be useful for most machine-loading operations; however, when high-precision processes are required, special care must be taken. Typical positioning accuracy of machine-mounted robots is ±0.019 in. (±0.05 mm). Automatic work-measuring equipment can be installed to compensate the cutting tool for improved accuracy. Robots may not be justified when parts require extremely close tolerances that are beyond the accuracy of the robots. In some cases a robot loader may be useful for the initial stages of the machine tool operation (rough and semifinish), but not in later stages when higher tolerances are specified.

Part unloading and removal from a machine is just as important as supplying raw material and loading new parts. Various alternatives exist for part removal, including removal by square wooden pallets, roller conveyors that are gravity fed or mechanized, or removal of finished or completed items by robot carts. A thorough plan must be thought out so that a proper part flow can be established and bottlenecks and machine blocking are avoided. Techniques for part-flow planning are discussed in Chapter 31, Quantitative Techniques for Robotic System Analysis.

56.3. ROBOT HANDLING OF PART MIX

The following considerations should be taken into account when robots must load a mix of different part types.

1. **Part Identification.** Parts must be identified by auxiliary equipment, such as laser code readers, a vision system, or similar sensory devices. A workpiece recognition unit (Figure 56.7) can distinguish between random parts by automatically measuring their outer diameter. The robot has to follow different programs for different part types, and may require other adjustments as described next.

2. **Lifting Capacity.** The robot may need to adjust its lifting weight capabilities depending on part weight. The accuracy of workpiece placement in a fixture is crucial to the tolerances that are to be maintained by the machine tool. The weight-adjusting feature gives the robot the ability to pick up light or heavy workpieces with consistent placement accuracy.

3. **Gripper.** The robot may need to adjust its gripping capability depending on the part shape and size. In some cases, a robot can change its gripper automatically to handle different part types properly.

56.4. EVALUATION OF ROBOTIC LOADING

Considerations for applying alternative techniques for robotic loading of machine tools have been discussed. With regard to local loading tasks, machine-mounted robots are more economical relative to both pedestal robots and fully automatic, dedicated loaders. However, pedestal robots have the advantage of being able to perform additional nonloading tasks. The advantage of an automatic dedicated loader is that it can load parts even when the spindle is moving.

A machine-mounted robot requires about 18 sec to load a chuck. This is compared to 5 sec by a human loading an 8-lb (about 4 kg) workpiece, or longer for heavier parts. Nevertheless, robotic loaders are preferred for unmanned periods and certainly have the advantage of always being ready, at the machine, when needed. In addition, with heavy parts a human operator must apply mechanical

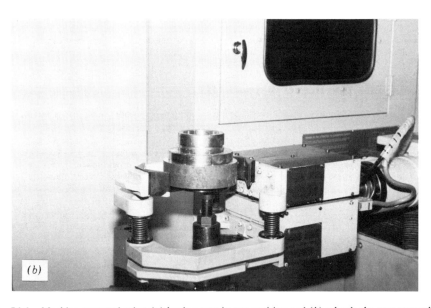

Fig. 56.6. Machine-mounted robot (*a*) loads a part in one position and (*b*) reloads the part at another position.

Fig. 56.7. Workpiece recognition unit to identify parts by their diameter.

assistance anyway, so a robotic arm may be a solution. On the other hand, human operators must load heavy, sagging parts, or when very high precision is necessary.

REFERENCES

1. Machine Loading, in Tanner, W. R., Ed., *Industrial Robots,* Vol. 2, *Applications,* 2nd ed., SME, 1981, Ch. 2.

2. Auer, B. H., Industrial robot feeds numerical controlled machine tools, *Industrial Robot,* Vol. 1, No. 4, June 1974, pp. 174–177.

3. Evans, J. M. et al., Robot feeding of an NC machine tool, *Proceedings of the 13th International Automation Control Conference,* San Francisco, CA, 1977, pp. 720–724.

4. Fritz, W. E., Machine loading with robots, SME Paper No. MS77-739, November 1977.

5. Gandy, T. G., A simple robot system for loading/unloading internal grinders, *Proceedings of the 13th International Symposium on Industrial Robots,* Chicago, Illinois, April 1983, pp. 2/25–31.

6. Holmes, J. G., Integrating robots into a machining system, *Proceedings of the 13th Fall Industrial Engineering Conference,* Houston, Texas, November 1979, pp. 247–256.

7. Inaba, H. and Sakakibara, S., Unmanned machine tool system with an industrial robot, *Proceedings of the 9th ISIR,* Washington, D.C., March 1979, pp. 29–38 (also appeared as SME Paper No. MS79-247).

8. Inaba, H. and Sakakibara, S., Flexible automation of unmanned machining and assembly cells with robots, *Proceedings of the 1st International Conference on Flexible Manufacturing Systems,* Brighton, U.K., October 1982.

9. Kelly, B., Machine loading robots on a planetary pinion machine line, SME Paper No. MS78-676, 1978.

10. Okuda, N. et al., NC machining system with programmable robot, *Proceedings of the 7th ISIR,* Tokyo, Japan, October 1977, pp. 99–106.

CHAPTER 57

MACHINE LOADING APPLICATION CASES

KENNETH R. HONCHELL

Cincinnati Milacron
Lebanon, Ohio

57.1. CRANKSHAFT MAIN BEARING CAPS CELL

At J I Case, Components Division, in Racine, Wisconsin, a Cincinnati Milacron T³-566 robot operates as the central unit in a manufacturing cell dedicated to the machining of crankshaft main bearing caps (see Figure 57.1). In addition to the robot, this cell consists of an in-feed and out-feed conveyor, a part orienter, a six-station dial index machine with two-position fixturing, and a vertical slab broach. All components in the cell are interfaced with the robot through the robot's computer control.

The robot control operates the manufacturing cell, with a skilled machine operator standing by at a host control panel supplied by Cincinnati Milacron. The sequence of operations is as follows:

1. The T³-566 robot takes a raw part from the in-feed conveyor and delivers it to the dial indexing machine.

2. The robot uses its dual gripper to unload a semimachined part, load a raw part, unload a machined part, and reload the semimachined part in the two-position fixture of the dial indexing machine. All drilling and milling operations are performed at the dial index machine. Total cycle time is 42 sec.

3. The robot takes the machined part to the part orienter where the part is indexed 180°.

4. The robot takes the part from the part orienter to the broaching machine.

5. The robot moves the finished part from the broach and loads the next part.

6. The robot takes the finished part from the broach to the conveyor, places the part on the out-feed portion of the conveyor, and obtains another raw part from the in-feed side.

Total cycle time for each finished part is 42 sec, with all operations being completed during the dial index machine's cycle time. At 100% machine capability, this produces a yield of 86 finished parts per hour.

Installation of this manufacturing cell provided a 0.8-year payback through reductions in both direct and indirect labor costs, improvement of in-process work flow, and improved quality.

57.2. HANDLING CONNECTING RODS

Cummins Engine Company, Inc., Columbus, Indiana, uses Cincinnati Milacron T³-566 industrial robots to load and unload steel connecting rods in a machine cell. The work cell includes (see Figure 57.2):

Two Cincinnati T3 robots

Three leading conveyor systems

One Michigan pin bore machine

One Detroit wrap broach machine

57.2.1. Operation Details

1. The incoming rods are presented to the first robot in the proper orientation by the parts conveyor. The first robot picks up two rods, one at a time, then moves to the Michigan pin bore.

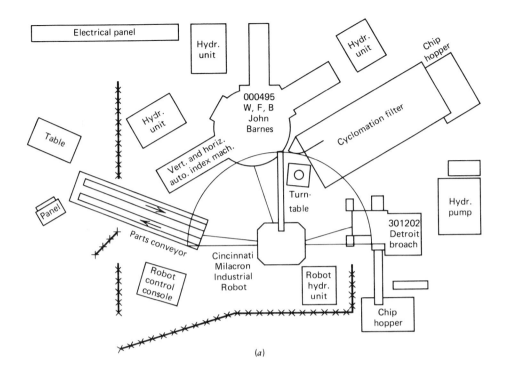

(a)

(b)

Fig. 57.1. (a) Layout diagram of the manufacturing cell with a Cincinnati Milacron T³ 566 robot for material handling. (b) The material-handling robot in operation.

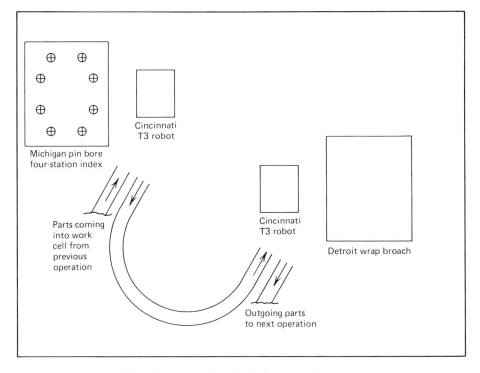

Fig. 57.2. Robotic work cell for connecting rods.

2. At the pin bore the robot loads both rods at the same time. After loading the rods in the machine, the robot signals the machine to clamp the parts. After clamping, the robot signals the machine to cycle.

3. When the pin bore has indexed, a gage comes down and simultaneously gages both parts. If the parts do not check, the robot signals the operator to intervene.

4. If the parts check, the robot enters the machine after the machine has signaled that it has unclamped. It picks up both rods and moves to the outgoing parts conveyor. It deposits the rods in the conveyor one rod at a time.

5. The robot then moves to the incoming conveyor and repeats the cycle. The rods move from the pin bore on the conveyor and are presented to the second robot in the proper orientation to be picked up.

6. After picking up the rod, the robot loads the rod in the wrap broach and signals the machine to clamp and cycle.

7. After the rod has been machined, the robot picks up the rod and moves to the outgoing conveyor. It deposits the rod in the conveyor and moves to the incoming conveyor where it starts the cycle over.

The first robot has a dual gripper and handles two rods at a time. The second robot has a single gripper and handles only one rod at a time.

The hand-coordinated motion of the robots moves the arm through as many as all six of the available axes (X, Y, Z, roll, pitch, and yaw) simultaneously. Because all axes are coordinated, the arm moves in a single straight line from point to point and maintains hand orientation even though as many as six axes are in motion. This straight-line motion is essential to place the "pin end" of the connecting rod on the tapered locator in the machine fixtures, since the clearance between the locator and the connecting rod is only about 0.004 in. (0.1 mm).

Parts come out of the work cell at the rate of 162 rods per hour. This will increase to 200 rods per hour after a cycle change on the broach. One operator monitors the two robots' operation and must periodically intervene when they have size or cycle problems. The system is considered successful. It was found that installation was not very difficult and that the robots run for a long time without problems.

CHAPTER 58
ROBOTIC TOOL CHANGING

RICHARD (BEN) CARTWRIGHT

Unimation, Inc.
Danbury, Connecticut

58.1. INTRODUCTION

Tool changing by robots as an alternative to dedicated, automatic tool changers is becoming attractive owing to flexibility and relative lower cost. A robot equipped with special grippers can handle a large variety of tools, and the tools can be shared quickly by several machines.

This chapter describes a tool-loading application wherein the robot, which is positioned on an oversized 6-ft (180-cm) high riser, loads and unloads two identical vertical milling machines with one of 32 tools that weigh from 15 to 45 lb (7 to 20 kg). All the tools are stored in a rack approximately 9 ft (270 cm) wide in front of the robot, between the two milling machines. The parts being machined are locomotive generator housings, which are brought to the milling machines on auto transfer devices. The entire operation is controlled by a General Electric host computer that directs the robot controller and signals it which part type is coming to the robot and which set of tools to select to load and unload into the milling heads.

58.2. THE OPERATION

The parts being machined in this application are locomotive generator housings, approximately 3 ft (90 cm) square × 5 ft (150 cm) in length. They are cast in a hollow configuration, and the nature of the milling head is such that the tool is fed inside the part, and several final, inside contours are machined utilizing up to eight different tools. There are five different style housings, all with varied tool-selection requirements.

The robot, a 2000 Series Unimate five-axis robot, is positioned between two Giddings and Lewis vertical milling machines. The sequence of operations in this application requires that the Unimate rotate, upon command, to either spindle head and grasp the cutting tool to be removed. The robot then sends an output signal to the machine tool to unclamp the tool. The robot controller looks for an unclamped (negative pressure on pressure switch) condition, which permits the Unimate to continue its program and remove the tool, and to rotate to the rack and deposit the cutting tool in its appropriate keyed tool nest.

The Unimate is elevated on a 6-ft (180-cm) high special riser platform so as to bring the mean horizontal load/unload position in line with the retracted position of both the machine tool spindle heads. The robot faces a multilayered tool rack with eight tools per machine (see Figure 58.1). There are also eight spare tools per machine, and they are used when the machine tool CNC control indicates that the sharpness of the tools has been seriously reduced. This is accomplished by measuring the resistance to the feed of the spindle head. At such time that a new tool is required, the robot receives a command from the host computer, and the robot through its controller now selects the spare identical tool so that the process can continue uninterrupted, with a substantial reduction in the need for manual intervention.

58.3. THE CUTTING TOOLS

The cutting tools are all mounted on a standard number 50 taper lock mandrel (see Figure 58.2). At the base of the cutting tool there is a V-shaped groove similar to a V-belt-pulley configuration, which the special robot hand tooling was designed to grasp. This pickup area was designed to keep things simple. In grasping all cutting tools at this common position, the hand tooling will successfully manipu-

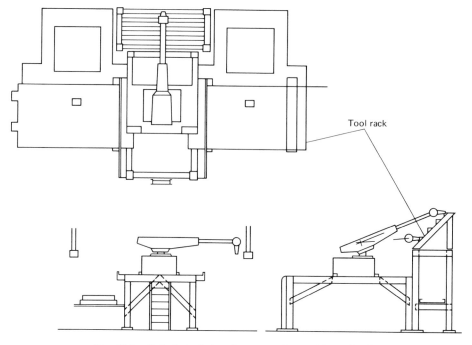

Fig. 58.1. Robotic tool-changing system layout (three views).

late all 32 tools without any addition of spacers, or any adjustment whatsoever. A variety of hand tooling and grippers useful for tool changing are shown in Figure 58.3.

When loading the cutting tools into a spindle head, the flange of the V groove has two female key slots 180° apart that register with two male keys in the spindle head. The key slots in the spindle head are always in the vertical Z axis through spindle orientation. To enable the robot to maintain the cutting tools in this orientation, female keys were attached to the tool nests. Each set of keys in each nest had a different angular location owing to the 32 different locations of the nests in relation to the radial position of the Unimate arm. The nesting approach enabled the Unimate to grasp the cutting tool in a finite, repeatable position every time a tool was loaded.

Within the spindle head of the Giddings and Lewis CNC milling machines, a special hydraulic taper-lock mechanism was designed to clamp and unclamp the number 50 taper tool mount. This mechanism was designed to provide a positive grip on the tool and, at the same time, release the tool (upon an instruction from the robot controller) and break the taper-to-taper bind, so that the Unimate robot can remove the part and replace it with the next scheduled tool. A system safeguard provided by the robot input/output includes limit switches, positioned on either side of the tool rack, that the robot would be programmed to actuate by brushing the tool through a whisker actuator on the limit switch, both in the load and unload programs. This assures the system that (1) the tool has been removed and (2) that the new tool has reached the spindle load position. An additional safeguard to assure tool presence is a limit switch in series with a solenoid on a pressure switch that is part of the hydraulic tool clamp mechanism. The pressure switch reads out a positive or negative pressure, which indicates a clamped or unclamped condition. If a negative pressure is sensed, the robot is not permitted to continue, therefore avoiding any misplacement of tools.

58.4 CONTROL

The Unimate robot is set up with a 32-program memory, to provide a single program for each tool, which in turn simplifies the tool-selection process. The external robot controller is a Modicon programmable controller linked to the robot through its external memory address port.

The function of the controller is to provide the logic required for decision making in control applications. A programmable controller PC is a solid-state device whose logic can be altered without wiring changes, which has been designated as a direct replacement for relays and "hard wired" solid-state electronics. Features of a PC compared to other industrial control devices include the following:

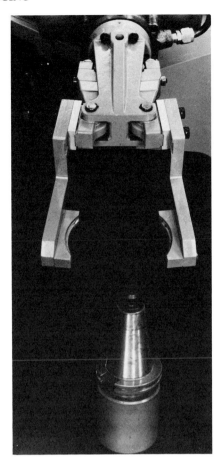

Fig. 58.2. Taper lock mandrel (with groove) to hold tools; appropriate gripper.

Solid-state device for maximum reliability and fast response.

Designed to operate in hostile industrial environment (i.e., heat, electrical transients, and vibration) without fans, air-conditioning, or electrical filtering.

Programmed with a simple ladder diagram language.

Easily reprogrammed with a portable panel if requirements change.

Controller is reusable if equipment is no longer required.

Indicator lights provided at major diagnostic points to simplify troubleshooting.

Maintenance is simple, based upon module replacement, and insures minimum downtime and maximum production.

Programmable controllers were introduced in the late 1960s in the automobile industry to avoid the costly and time-consuming rewiring of control systems at model changeover. They have evolved rapidly with the advent of more sophisticated solid-state devices and are now found in every industry worldwide.

Today, the programmable controller is used as a vital building block in distributed control systems. A computer or other intelligent device is used to monitor and control the activities of many PCs in a plant through a "data highway" called a MODBUS, which is nothing more than a cable similar to a telephone cable. Such a communication scheme is utilized here between the hierarchical General Electric control computer and the Modicon PC.

58.5. EVALUATION

The two key points of this application are: first, the simplicity and, second, the flexibility. Tool changing is not a new technology, but tool changing utilizing robotics is. In addition, there are substantial cost-effective results, as well as the ability to be able to manipulate a whole range of tools. By changing the hand tooling, the robot is able to handle yet another whole family of tools, thus the flexibility.

The comparison of cost to a dedicated, automatic tool changer is quite dramatic. If studied over a period involving a major product change, the cost of the robotic approach is less than half that of a dedicated loader. The savings become apparent when considering the initial capital outlay for both approaches and considering a major product change. The only modification to the Unimate would be the hand tooling. The dedicated loader would require extensive changeover and may not be compatible at all. When one robot changes tools for two or more machines, as in the foregoing example, the savings are even more significant.

Unimation has already implemented many successful machine tool loading work cells, utilizing Unimate robots to manipulate parts through several cells connected by conveyors. These cells, coupled with tool changing robots, can provide total automation and total flexibility.

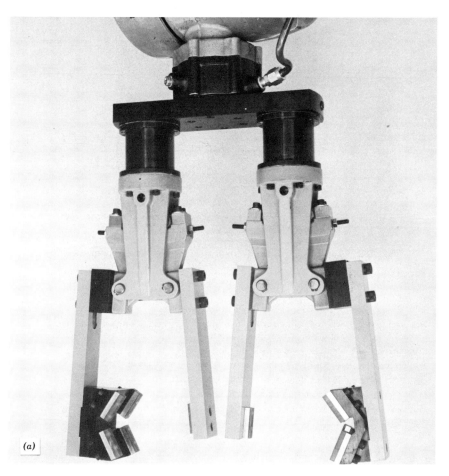

(a)

Fig. 58.3. (a) Typical special dual hand tooling designed to handle round tool shanks, or any cylindrical part within a specified diameter range. (b) A typical special dual hand with cam type actuation that will grasp a wide range of diameters without tool-gripper changing. (c) A typical triple hand for machine tool loading and tool changing.

(b)

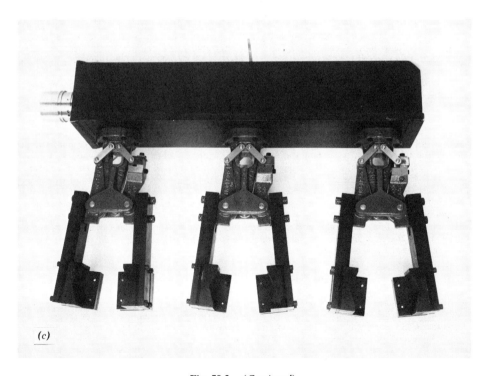

(c)

Fig. 58.3. (*Continued*)

CHAPTER 59

MOBILE ROBOT APPLICATIONS

HANS J. WARNECKE

JOACHIM SCHULER

Fraunhofer Institute for Manufacturing Engineering and Automation
Stuttgart, West Germany

59.1. INTRODUCTION

Industrial robots remain, especially if including their peripheral devices, investment-intensive automation devices. Therefore the economic application of industrial robots with acceptable amortization periods always depends on a good utilization of the industrial robot, with the exception of the application for humane reasons because of difficult working conditions (noise, heat, gases, etc.). The different possibilities offered for such good utilization in the different application areas, and the measures to be taken to increase that utilization, will have a great influence on further distribution of industrial robots.

Stationary industrial robots are used for tool and workpiece handling (see Chapter 56, Robotic Loading of Machine Tools). In tool handling, for example, in coating, spot and arc welding as the most important application areas, the industrial robot usually performs extensive tasks at the workpiece, which leads to good utilization, both technical (operation area) as well as temporal (degree of utilization).

Apart from limitations in load capacity and in sensory ability, it is the size of the operating range that restricts the applicability of industrial robots in many cases. Insufficient utilization of the technical and temporal availability of industrial robots, especially in the area of loading and unloading of machine tools, limits the scope for an economically successful application even if a technical realization might be possible.

In workpiece handling, for example, the supply of machine tools, the relatively long machining times cause frequent stops while the industrial robot "waits" for the following handling operation, which results, consequently, in a low degree of utilization for the robot. It is not always possible to fill this waiting time with additional tasks such as feeding or transferring of machined components to measurement, deburring, or washing stations. Also the multiple operation of a circular arrangement of several machine tools around a stationary robot can seldom be realized satisfactorily. Often disadvantages such as bad access to the machine tool, missing space, the required timing of machining sequences, as well as the need, out of security reasons, to stop the complete manufacturing installation, even with small failures, or with setup changes at one station, do not allow the successful application of this kind of multiple operation.

In contrast, mobile robots offer new possibilities for an increase in technical as well as temporal utilization and therefore for the economic application of industrial robots.[1,2] Mobility is achieved with an automated transport system.

59.1.1. Classification of Mobile Robots

A mobile robot is defined as a freely programmable industrial robot which can be automatically moved, in addition to its usual five or six axes, in another one, two, or three axes along a fixed or programmed path by means of a conveying unit. The robots may be classified according to the number of additional degrees of freedom for the mobility (Figure 59.1):

Linear mobility.
Area mobility.
Space mobility.

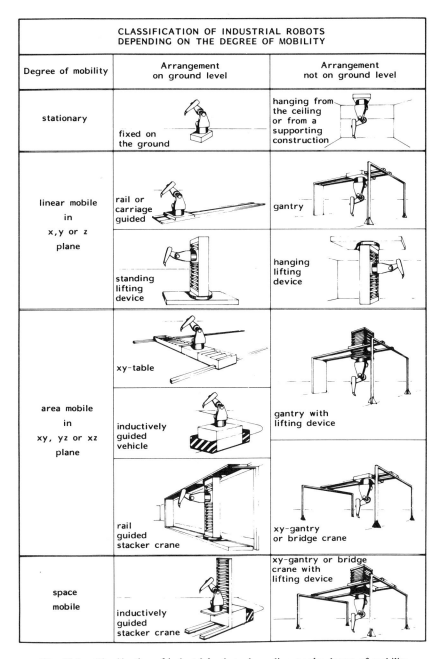

CLASSIFICATION OF INDUSTRIAL ROBOTS DEPENDING ON THE DEGREE OF MOBILITY		
Degree of mobility	Arrangement on ground level	Arrangement not on ground level
stationary	fixed on the ground	hanging from the ceiling or from a supporting construction
linear mobile in x,y or z plane	rail or carriage guided	gantry
	standing lifting device	hanging lifting device
area mobile in xy, yz or xz plane	xy-table	
	inductively guided vehicle	gantry with lifting device
	rail guided stacker crane	xy-gantry or bridge crane
space mobile	inductively guided stacker crane	xy-gantry or bridge crane with lifting device

Fig. 59.1. Classification of industrial robots depending on the degree of mobility.

Linear-mobile robots are characterized by one translatory (usually) horizontal axis. Some applications of this kind are already in service using a linear track.[1-5] Inflexibility in regard to branches and changes of the travel path, as well as interruptions in other functions of the internal material flow, are the disadvantages of this track-bound transport principle. Designs for achieving linear mobility of the robot other than ground level, for example, with a gantry, are also known.[6,7]

Floor-bound mobility is obtained by battery-powered carts which either recognize their environment by means of sensory devices and compare it with a programmed pattern or follow along a wire embedded into the floor.[8-10] The ease of implementing branches into the travel path and the possibility of embedding the guiding wire into the floor like a net allow servicing an area although, in fact, the cart has a linear guidance system.

Only limited-area mobility is offered by industrial robots mounted on an x-y table. Figure 59.2 shows the layout of two robots facing each other, each movable in two directions. Solutions to serve an area with an industrial robot that is not floor-bound can be realized with the aid of a modified loading gantry or a stacker crane. Both cases are so expensive that such solutions are only justifiable under certain circumstances.

Space-mobile applications, with freedom of movement in three axes, can be achieved, for example, by a modified bridge crane. Solutions with floor-bound devices, for example, with a specially designed inductively-guided stacker crane, may also be possible. Considering the present technical development, mobile robots with area mobility may become useful in enlarging the economic application of industrial robots.

59.2. SPECIFICATION OF AREA MOBILE ROBOTS

The specifications for area mobile robots are influenced by their main tasks:

Handling at the place of operation.

Transport of the industrial robot.

Transport and automated exchange of workpieces, components, or auxiliary means for the handling operation.

Because the mobile robot will be applied, for the most part, within highly automated manufacturing systems, it must operate autonomously and as far as possible without manual intervention. Independence in regard to energy supply and control is to be provided. A possible connection with a stationary supply unit must be executed automatically. To limit the costs of the unit it should be designed modularly, using components available on the market. Assuming that most of the applications of mobile robots will be in serving several stations, cost-intensive functions and the "intelligence" of the system should be implemented in the conveying device, which usually exists only once in a system, rather than to integrate them into every station. The conveying device should not interfere with ongoing operations of the material flow. To transport workpieces, tools, and the like a universal connection between the conveying device and goods to be transported must be provided in the form of a pallet which can be loaded and unloaded automatically.

The performance of handling tasks demands sufficient stability from the mobile robot because of

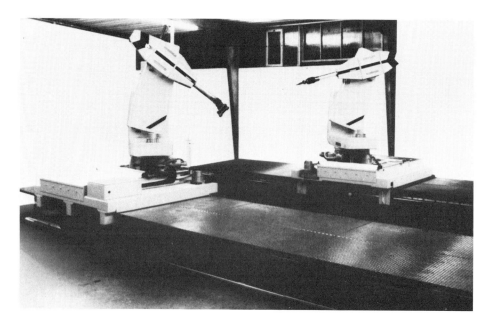

Fig. 59.2. Mobile robots with carriage guide and cross slide.

the displacement of the center of gravity by the cantilever arm and the dynamic forces resulting from the movement of the handling device. The achievable accuracy for positioning the mobile system should be equal to the accuracy of today's stationary robots.

59.3. REALIZATION OF AN INDUCTIVELY GUIDED MOBILE ROBOT

Figure 59.3 shows the design of a mobile robot developed at the Fraunhofer Institute for Manufacturing Engineering and Automation (IPA), Stuttgart, West Germany. It consists of the following components:

An inductively guided vehicle as a "chassis" for the industrial robot.

A freely programmable handling device, in this case, a standard five- or six-axis industrial robot.

A transport platform (rotary pallet table) for holding and transporting of pallets with workpieces, these to be handled by the industrial robot.

A control and energy supply for vehicle and industrial robot as well as positioning aids for the independent, fully automated supply of several work stations.

The inductively guided vehicle used as a chassis for the industrial robot and the pallet system is designed for forward and reverse driving with equal speeds. The position accuracy of approximately 10 mm, achievable at the stopping of the vehicle, is not sufficient for distinct handling operations. To increase the position accuracy of the mobile robot in relation to the stationary reference system, the vehicle is equipped with a lifting platform which carries the industrial robot as well as the pallet system. This lifting platform can be connected by means of four rigid legs with floor-positioning elements at the place of operation. The necessary relative movement for the fine-positioning of the lifting platform is achieved by means of a "floating" support of the lifting platform on the vehicle, which can be locked during movement.

The lifting platform carries, in its forward position in the main driving direction, the pallet system, and behind this the industrial robot is positioned along the long side of the vehicle, which is preferably facing the working or stopping stations. Thus the robot is able to reach the pallet which carries the workpieces or auxiliary devices for the handling operations required. In 90° movements, the sequential covering of the complete pallet area is possible. The automated exchange of pallets by the vehicle is accomplished in respectively lifting and lowering the platform at pallet stations.

The controls of the industrial robot and vehicle are two seperate units connected by an interface. Information and commands coming from the stationary central control are transferred inductively at defined transferring stations to the mobile unit. Numerical working programs for the industrial robot can also be transferred.

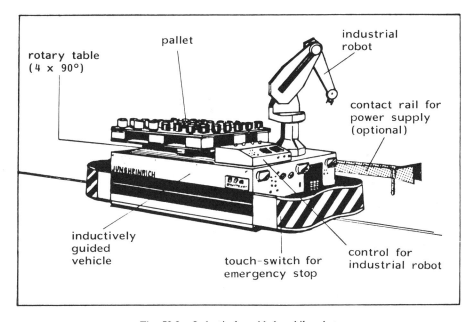

Fig. 59.3. Inductively guided mobile robot.

The industrial robot and the inductively guided vehicle with the auxiliary devices are supplied with energy by one battery each. Thus the mobile industrial robot is able to operate autonomously for a limited time without external energy supply. The connection to the electrical power supply takes place at work stations, where an autonomous operation is not required. The electrical connection is accomplished by means of a sliding contact mounted on the vehicle and a contact rail at some work stations. At the same time the batteries can be recharged.

59.4. AREAS OF APPLICATION FOR MOBILE ROBOTS

The areas of application for mobile robots with inductive guidance are to be found in those places not satisfactorily covered by automation with stationary robots. This applies specifically to cases where stationary robots cannot be used because of insufficient utilization or limited operation space. The connection of both the main material flow functions, transport and handling, with an equally suitable transport and handling unit, opens a wide field of applications for workpiece and tool handling, especially in unattended or almost unattended shifts of flexible automated manufacturing systems.

59.4.1. Workpiece Handling

The supply of workpieces to lathes or of manufacturing cells for rotary components (lathes combined with a workpiece buffer) is used as an example for the application of mobile robots in workpiece handling.

If the workpiece at the machine tool is transferred directly into its machining position as shown in Figure 59.4, the industrial robot performs, after the positioning at its place of operation, one single but complex handling task. This task consists of the removal of a machined component and the supply of an unmachined component brought along on the vehicle. The succeeding stop and go at several work stations is heavily dependent on the machining sequences because the mobile robot must be present at the machine tool at the end of each machining sequence to avoid stoppages. A strong machining-sequence dependency and an extensive planning operation for the disposition of the components through the central computer are characteristic for this application configuration.

If supplying a workpiece buffer storage as depicted in Figure 59.5, the mobile robot drives to the appropriate manufacturing cell after removing a pallet loaded with unmachined workpieces. After positioning the platform, the industrial robot exchanges in sequential-handling operations the unmatched workpieces brought along with the machined ones stored in the buffer of the machine tool. Thus the loading of the workpieces is performed in the main machining time of the machine tool. After the workpiece exchange, the vehicle drives with the machined components either, according to the machining progress, to the next work stations or back to storage. This workpiece-supply concept, largely independent

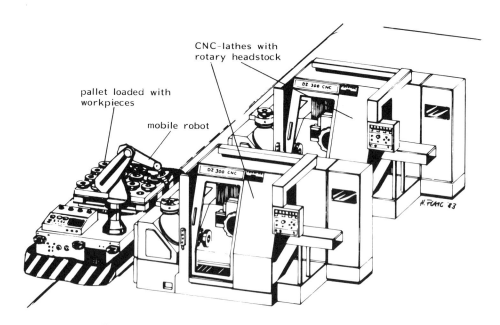

Fig. 59.4.   Loading lathes by an inductively guided mobile robot.

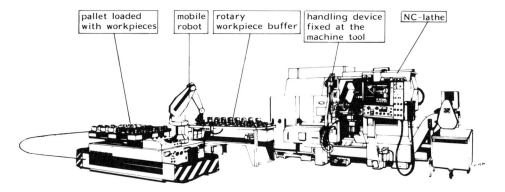

Fig. 59.5. Loading a workpiece buffer by an inductively guided mobile robot.

from the machining sequence, shows the advantages of an integrated transport and handling system, especially when short workpiece machining times are predominant. Figure 59.6 shows the layout of manufacturing cells for rotary components supplied by a mobile robot according to the described principles.

A special case of workpiece or better component handling is illustrated by the example of tool supply to manufacturing cells or machining centers for prismatic components. Recently flexible manufacturing cells for drilling and milling, consisting of a machining center, pallet storage, and automated control and supervision devices, have been showing a rapid increase in applications as well as in the number of suppliers.

The cells are used successfully in a three-shift operation for the "unmanned" machining of mixed

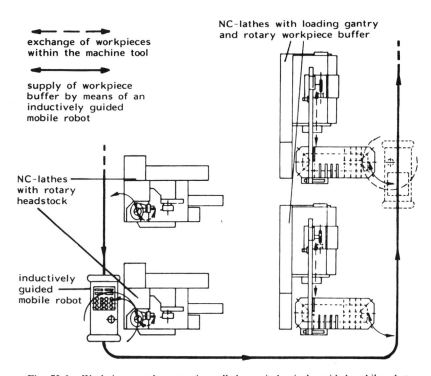

Fig. 59.6. Workpiece supply to turning cells by an inductively guided mobile robot.

components (to a certain degree), buffered in the pallet storage. For manufacturing tasks such cells will be the concept for the coming years. However, they will be limited by two factors:

1. Short machining times per workpiece require a large stock of workpieces on pallets to last for an unattended shift. The high pallet costs are therefore a threat to the economy of a manufacturing cell.

This situation demands a workpiece exchange not dependent on the number of pallets in the pallet buffer. Here the mobile robot appears to be a suitable solution:

2. An extensive workpiece mix in the pallet buffer system, desirable because of small batch sizes or high cost for workpiece-dependent jigs, can be realized at the control side without problems through calling the workpiece-specific machining program. However, this workpiece requires a multitude of different machining operations for the different workpieces. These tool requirements often exceed the capacity of the tool magazine, even if extended.

Figure 59.7 shows the automatic exchange of machining tools at a machining center with pallet buffer storage by a mobile industrial robot with a tool supply pallet. Figure 59.8 presents the layout of the machining center serviced by a mobile robot.

59.4.2. Tool Handling

A future area of application could be tool handling with mobile robots, which requires, however, intensive development efforts. A prerequisite for the actual guiding of the machining tools is that the reaction forces resulting from the machining are not too high, a condition which also applies to stationary installed industrial robots. For the machining operation, the actual tool, for example, a welding gun, is fixed onto the robot's arm, while the auxiliary equipment, for example, the welding transformer or the shielding gas cylinders, is transported on the pallet (Figure 59.9). Thus it is possible to weld or to coat on large stationary components from different sides. If coordinated at the control side, the simultaneous machining would be possible with several independently operating industrial robots.

59.4.3. Drive-Path Layout

Possible layouts can be classified into three different categories (Figure 59.10):

1. **Isolated Application at Single Work Stations.** Supply of several widely spaced machine tools or work stations with workpieces or tools, for example, flexible multiple machine attendance.
2. **Linear Overlapping.** Attendance of several work stations arranged side by side in a line, with or without overlapping of the operation areas for the machining or attendance along a line, for example, welding of long seams by joining several shorter seams by moving the vehicle.

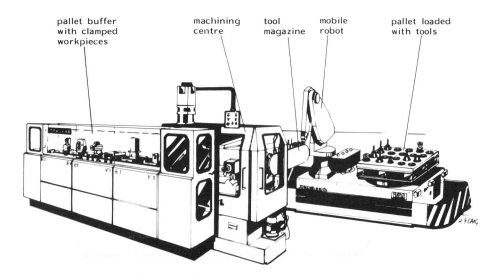

pallet buffer   machining   tool   mobile   pallet loaded
with clamped   centre   magazine   robot   with tools
workpieces

Fig. 59.7.   Loading a tool magazine of a machining center by an inductively guided mobile robot.

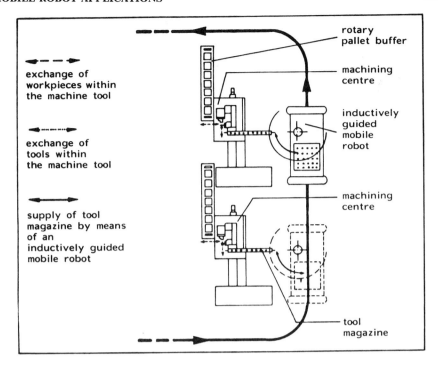

Fig. 59.8. Tool supply of machining centers by an inductively guided mobile robot.

3. **Area Overlapping.** Laying of the drive path, netted to reach the full overlapping of the operation area, for an area covering machining or attendance, for example, for welding, assembly, or coating of large workpieces from different sides. The task may be even simultaneous with several mobile robots.

59.4.4. Security Aspects

When operating industrial robots, no one is permitted to stay in the operation range. Obstacles in the drive path of the vehicle, if touched by the vehicle, must cause its immediate stop. This is achieved

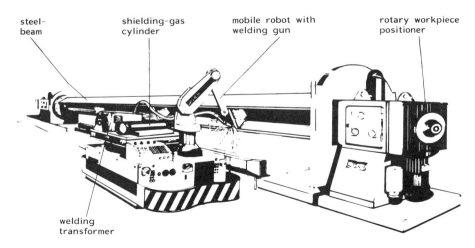

Fig. 59.9. Welding of large components by an inductively guided mobile robot.

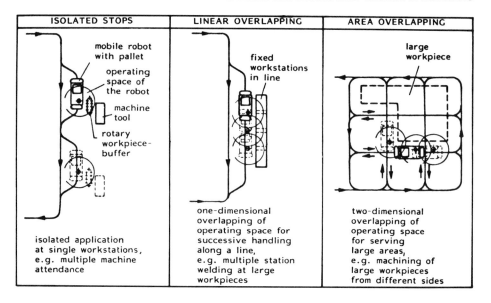

ISOLATED STOPS	LINEAR OVERLAPPING	AREA OVERLAPPING

Fig. 59.10.   Layout variants of inductively guided mobile robots.

by mechanical emergency stop devices, proven on inductively guided vehicles. The securing of the operation area of the industrial robot, however, should be achieved by nontactile methods and without guards, to enable free access to the stations for manual setup, supervision, or maintenance operations if the mobile robot is not in action.

59.5. SUMMARY AND OUTLOOK

The concept of a self-driving mobile robot allows a practically unlimited, freely selectable, and changeable-area mobility, according to the guidewire laid into the floor. The automation of transport and handling tasks in one common technical solution is possible through this combination of the inductively guided vehicle, the industrial robot, and the automatically exchangeable transport pallet.

The discussed applications for the mobile robot in workpiece and tool handling result from the integration of transport and handling functions. From this concept important and promising incentives can be expected, especially on the material flow concepts of flexible automated manufacturing areas.

REFERENCES

1. N. N.: Lichtbogenschweißroboter Technische Rundschau No. 40, 1981, pp. 33–34.
2. Norlin, B., Roboter erweitern Maschinennutzungszeit, *VDI-Nachrichten,* Vol. 36, No. 16, 1982, p. 10.
3. Tsao Kwang-chuan, Hsing-dao Ch., and Hsiao-tsu Ch., Loading a lathe with the aid of a robot, *The Industrial Robot,* Vol. 8, No. 2, 1981, pp. 98–99.
4. N. N., Robot welders attract the smaller firms, *The Industrial Robot,* Vol. 7, No. 4, 1980, p. 260.
5. Engelberger, J. F., The Use of Industrial Robots for Loading Machine Tools, Internationaler Kongreß Metallbearbeitung, Leipzig, 1982.
6. N. N., BMW buys KUKA protal robots, *The Industrial Robot,* Vol. 7, No. 4, 1980, p. 259.
7. Weisel, K. and Katoh, A., Beachheads for robotics, *Proceedings of the 5th International Symposium on Industrial Robots,* 1975, pp. 1–10.
8. Warnecke, H. J. and Schuler, J., Areas of Application for Mobile Robots, *Proceedings of the 2nd European Conference on Automated Manufacturing,* 1983, pp. 261–270.
9. Warnecke, H. J. and Schuler, J., Mobile Robots—A Solution for the Integration of Transport and Handling Functions, *Proceedings of the 2nd International Conference on Automated Guided Vehicle Systems,* 1983, pp. 185–194.
10. Wada, R. and Shima, Y., Höherentwickeltes flexibles Fertigungssystem durch CAD-CAM-Kopplung, *Werkstatt und Betrieb,* Vol. 116, No. 6, 1983, pp. 331–336.

CHAPTER 60
GANTRY ROBOTS AND THEIR APPLICATIONS

JOSEPH P. ZISKOVSKY

GCA/Industrial Systems Group
St. Paul, Minnesota

60.1. GANTRY ROBOTS

A gantry robot can be described as an overhead-mounted, rectilinear robot with a minimum of 3 DF and normally not exceeding 6 DF (Figure 60.1). The robot is controlled by a multimicroprocessor controller allowing it to interact with a multitude of other devices. The bench-mounted assembly robots that have a gantry design are not included in this definition. They fall in the assembly robot category.

Large work envelopes, heavy payloads, mobility, overhead mounting, and the capability and flexibility to do the work of several pedestal-mounted robots are some of the advantages of implementing a gantry robot versus a floor- or pedestal-mounted robot.

Gantry robots have been around for many years in various forms, from refueling systems in the nuclear reactor cell to large material-handling systems in the mining industry. There are also pseudogantry robots which are composed of primarily a pedestal robot mounted in the inverted position and on slides, allowing it to traverse over the work area.

Gantry Robot Terminology

Since gantry robots are somewhat unique, some terms are used that do not pertain to pedestal robots, as follows:

Superstructure: Also called the gantry support structure or box frame. This is the structure upon which the robot will be elevated from the floor. (A in Figure 60.1) It is an integral and essential portion of a gantry robot system.

Runway: The longitudinal or X axis of the gantry robot. It is normally the passive side rails of the superstructure. (B in Figure 60.1)

Bridge: The transversal or Y axis of the gantry robot. The bridge is an active member of the robot riding on the runway rails and supporting the carriage. (C in Figure 60.1)

Carriage: The support structure for the Z axis. Provides the Y axis motion on the bridge. (D in Figure 60.1)

Telescoping tubes/masts: Depending on the robot this is the vertical or Z axis of the gantry robot. In the case of telescoping tubes, they come together, allowing for a lower ceiling. A sliding mast slides along its length up and down, requiring a ceiling height equal to its stroke above the superstructure. (See Figure 60.2.)

Repeatability: The farther the work point is from the centerline of rotation, the greater the linear effect is felt owing to the radian repeatability. For example: the linear repeatability for a representative three-axis gantry robot is ± 0.1 mm (± 0.004 in.). The same robot in the four-axis configuration with a radian repeatability of ± 0.0008 radians and a work point of 25 cm (10 in.) from the centerline would have an effective linear repeatability of ± 0.3 mm (± 0.012 in.) over the entire work envelope at maximum payload. In many cases the calculated repeatability is greater than the actual measured repeatability. Each case must be evaluated individually.

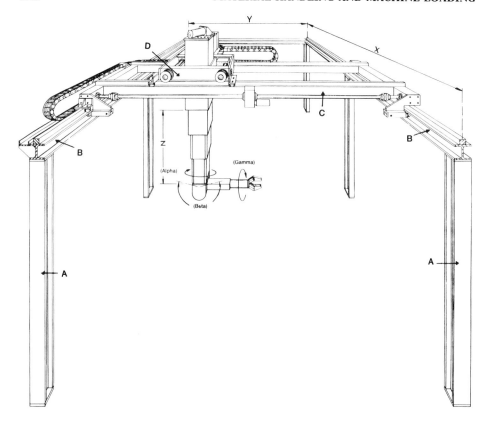

Fig. 60.1. Gantry robot.

Longitudinal axis: X axis.
Transversal axis: Y axis.
Vertical axis: Z axis.

60.2. GENERAL SPECIFICATIONS FOR GANTRY ROBOTS

The gantry robot is comprised of up to six axes depending on the manufacturer and model. They have three linear axes: X, Y, and Z, and three rotational axes: alpha (theta 1), beta (theta 2), and gamma (theta 3). In many cases these robots have very large work envelopes. Tables 60.1, 60.2, and 60.3 show the standard minimum and maximum specifications for today's typical gantry robots. These specifications were obtained from a survey of current product specification sheets from CYBOTECH, NIKO, DURR, and GCA.

One distinct advantage of the gantry robot is its ability to be expanded beyond the standard range. Systems have been built and installed that have axis measurements of 170 m (500 ft) in the X axis, 13 m (40 ft) in the Y axis, 8 m (25 ft) in the Z axis, and handling up to 15 tons as shown in Figures 60.2 and 60.3. This system had two telescoping tubes mounted on one bridge. The one on the left represents the tubes in the extended configuration. It must be noted that with this large weight and range, the repeatability of the robot will be degraded. Table 60.3 shows representative repeatability values.

Tables 60.5 and 60.6 depict how speed and weight relate while attempting to keep repeatability constant. In evaluating the payload-carrying capabilities of gantry robots, the configuration of the robot should be kept in mind. As can be seen in Tables 60.5 and 60.6, when the rotational axes are used, the effective payload in any orientation at the faceplate is dependent on the torque of the specific axes being used. On the other hand, even though the effective payload is reduced because of torque limitations, the dead-lifting capabilities of the linear axes will allow the manipulation of heavy loads.

As an example, using Table 60.5, for a specific five-axis robot with a reach of 214 cm (84 in.) and a speed of 91 cm/sec (36 in./sec), the effective payload is 45 kg (100 lb) in any orientation. The

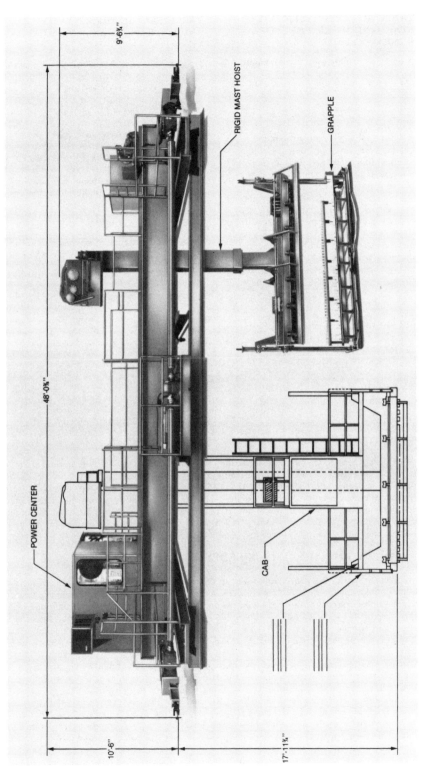

9'-6¾"

RIGID MAST HOIST

GRAPPLE

48'-0½"

POWER CENTER

CAB

10'-6"

17'-11¼"

Fig. 60.2. Large gantry system with dual telescoping tubes.

1013

TABLE 60.1. REPRESENTATIVE STANDARD WORK ENVELOPE (LINEAR AXES)[a,b]

Axis	Minimum	Maximum
X	10	40
Y	10	20
Z	3	10

[a] In feet.

[b] When applying a gantry robot, the required work envelope may necessitate a larger superstructure. In the case of a four-axis gantry robot, for the center of the Z axis to reach a 10 ft × 20 ft work envelope would normally require a superstructure of 12 ft × 22 ft, and on some gantry robots even greater work space (see Figure 60.3). X and Y axes may be extended beyond these nominal distances with modifed power and control systems to account for line drop and feedback sensing over a large distance.

TABLE 60.2. REPRESENTATIVE RANGES OF ROTATIONAL AXES

Axis	Range
Alpha (theta 1)	330°
Beta (theta 2)	210°
Gamma (theta 3)	330°

TABLE 60.3. REPRESENTATIVE REPEATABILITY RANGES AT MAXIMUM PAYLOAD[a]

3 Axes	± 0.004 in. to	± 0.008 in.
4 Axes	± 0.004 in. to ± 0.0008 radians	± 0.010 in.
5 Axes	± 0.007 in. to ± 0.0012 radians	± 0.015 in.
6 Axes	± 0.007 in. to ± 0.0014 radians	± 0.020 in.

[a] Repeatability for robots is the accuracy to which the robot can return to a taught or learned point. For rotational axes the distance to the work point from the centerline of rotation must be considered (see Section 60.3).

TABLE 60.4. REPRESENTATIVE MAXIMUM VELOCITY RANGES BY AXIS[a]

Axis	Minimum	Maximum
X	36 in./sec	39 in./sec
Y	36 in./sec	39 in./sec
Z	36 in./sec	39 in./sec
Alpha (theta 1)	60°/sec	120°/sec
Beta (theta 2)	60°/sec	120°/sec
Gamma (theta 3)	60°/sec	120°/sec

[a] Maximum velocity specification is not a true measure of the capability of a robot. Often the speed must be reduced to handle a large payload, or the distance moved is not large enough to allow the robot to reach maximum velocity.

same robot, using Table 60.6, can dead lift 63 kg (140 lb) as long as all axes are normal to the floor.

The dead-lifting payload can be a great deal larger as the speed and reach (Z axis) is decreased. A five-axis robot with a shorter reach of 142 cm (56 in.) and a speed of 30 cm/sec (12 in./sec) can dead lift four times its rotational axes capability. This unique capability allows the gantry robot the flexibility to handle one payload for orientation or palletizing and then to have the capability of picking up the entire pallet or assembled product. Each specific robot should be examined by the potential user to determine the interaction of weight, speed, repeatability, and so on.

60.3. GANTRY ROBOTS VERSUS PEDESTAL ROBOTS

The evolutionary step to the gantry robot was an obvious one. The normal configuration for a robot was bench or floor mounted. These robots were more or less fixed, dedicated machines, working at one location with a maximum limited work area and in many cases could service only one machine. Attempts were made to resolve this by putting the robots on slides. This required the use of valuable floor space and specific orientation of the machines. The gantry robot, on the other hand, has the capability of servicing many machines with a smaller floor space requirement, as is shown in Figure 60.4.

There are many parameters in which the gantry robot has greater stability than a pedestal robot. Table 60.7 shows a representative comparison of two robots, one a gantry and one a pedestal robot. Both robots have the same reach or travel (limiting the gantry to two linear axes, Y and Z), the same repeatability at centerline (centerline being the center of the base for the pedestal robot and on the centerline of one of the superstructure support legs for the gantry); both robots have the same centerline speed. It is easily seen that for a given load the gantry robot does not derate its load capability as it reaches further out as does the pedestal robot. Likewise, repeatability is not affected. Both robots have inherent vibrations; these increase as the pedestal robot reaches to maximum extension. As is seen in Table 60.8 a deviation occurs in both robots when various factors are applied, but it is again evident that the effects on the gantry robots are reduced because the gantry lifts rather than reaches, allowing it to achieve better results when compared to a pedestal robot.

Robot structure

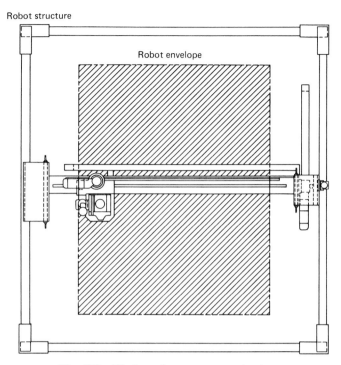

Fig. 60.3. Work envelope versus superstructure.

The repeatability for robots is defined as the accuracy to which the robot can return to a taught or learned point. This is in contrast to accuracy as used in the machine tool industry. A gantry robot is a flexible machine where points are taught or programmed. For the linear axes (X, Y, and Z) these are linear measurements. For the rotational axes (alpha, beta, and gamma) the measurements are best related to radian measurement as shown in Table 60.3; linear repeatability values can be assigned, but the actual repeatability is dependent on two factors:

1. The radian repeatability.
2. The distance the work point (measured point) is from the centerline of the rotation.

60.4. IMPLEMENTATION GUIDELINES FOR GANTRY ROBOTS

The same general guidelines that were discussed in Part 6 of this handbook, Application Planning, apply to gantry robots, but because of their unique nature, additional factors and guidelines should

TABLE 60.5. PAYLOAD AND SPEED TRADE-OFF WHILE HOLDING REPEATABILITY CONSTANT

Z Axis Travel	Speed (in./sec)	Payload (lb)			
		3 Axes	4 Axes	5 Axes[a]	6 Axes[b]
84 in.	36	265	200	100	100
	18	520	750	100	100
56 in.	24	670	610	230	230
	12	945	880	230	230
26 in.	12	1430	1365	230	230
	6	2209	2135	230	230

[a] With standard 30-in. arm.
[b] Within 4 in. of faceplate, on a standard 30-in. arm.

TABLE 60.6.   PAYLOAD AND SPEED TRADE-OFF
(DEAD LIFTING WITH 2 AXIS ONLY)

Z Axis		Load Capacity (lb.) Including Weight of End-Effector (Dead Vertical Lift)			
Tube Travel Maximum (in.)	Tube Speed Maximum (in./sec.)	3 Axis	4 Axis[a]	5 Axis[b]	6 Axis[c]
84	36	265	200	140	100
84	18	520	455	395	355
56	24	535	470	410	370
56	12	940	880	820	780
28	12	1430	1365	1305	1265
28	6	2200	2135	2075	2035

[a] Alpha and beta are limited to 7000 in./lb for one and two moving tubes.

[b] Alpha and beta are limited to 3500 in./lb for three moving tubes.

[c] Gamma is limited to 1000 in./lb independent of tubes.

be considered. As discussed in Section 60.2, the gantry robot can have a wide range of work envelopes. In laying out the application work area, sufficient concern should be given to the following important requirements:

1. The required superstructure will be larger than the work envelope being envisioned.
2. Adequate headroom overhead is necessary for the Z axis movement of the mast gantry robot, not only where there may be any Z axis motion but also within the entire structure.
3. Since the gantry will be working overhead, the robot must be able to go over any machines and the like in its work area or must go around these objects.

Today's gantry robots have the capability of handling very heavy payloads (see Section 60.2). A note of caution: heavier payloads sometimes require stronger and heavier end effectors. End effectors for gantry robots sometimes can become very complex, since they can perform more than one task. These end effectors can become very heavy, reducing the effective payload.

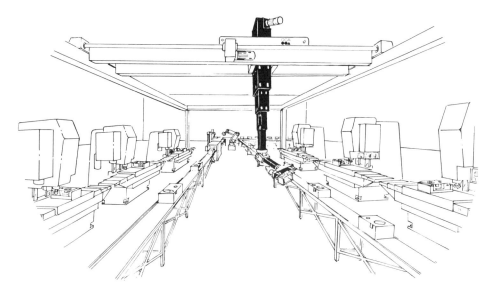

Fig. 60.4.   One robot servicing many machines.

TABLE 60.7. REPRESENTATIVE COMPARISON BETWEEN A GANTRY AND A PEDESTAL ROBOT

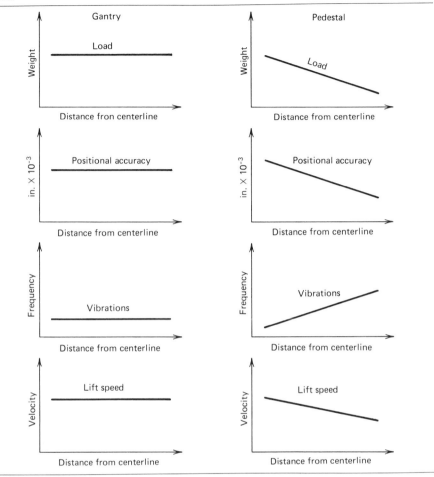

Another factor to be considered is cycle time constraints. Because of the large work areas, consideration must be given to the dynamics involved in moving a great distance at speed and the effect on accuracy. There are three interrelating areas of concern or trade-offs:

1. Speed versus distance.
2. Accuracy versus speed.
3. Accuracy versus distance.

As the work envelopes are made larger, speed must be reduced and finer acceleration/deceleration control must be implemented to maintain the final accuracy. If fast cycle times are required, the distance between end points may have to be shortened. These factors become critical when handling very heavy payloads, which was discussed earlier.

Of course, gantry robots allow the capability of one robot to do the work of many floor-mounted pedestal robots, as discussed in Section 60.3. The layout of what interacts with the gantry robot becomes more of a consideration, and in many cases is easier to implement and allows more efficient utilization of resources.

60.5 SAFETY PLANNING WITH GANTRY ROBOTS

Safety planning, as with all robot applications, is very important. If one keeps in mind what the robot is and can do, and the respect due it, many of the potential problems are solved. Applying the

TABLE 60.8. POSITIONAL ACCURACY VERSUS OTHER PARAMETERS

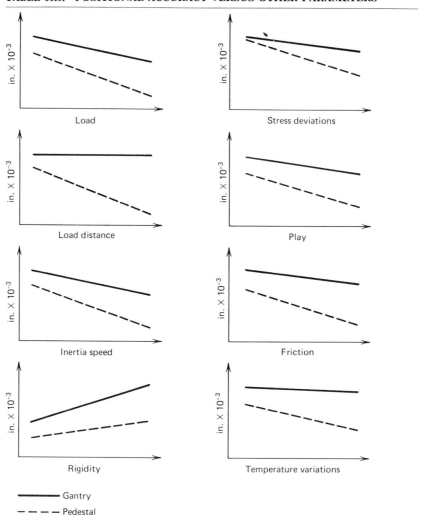

"R³ FACTOR"¹ (*R*obots *R*equire *R*espect) for safety leads one to examine the potential hazards and implement the appropriate safety guards to reduce the risks associated with robot usage. Gantry robots present some problems, from a safety point of view, that floor-mounted pedestal robots do not.

Since the gantry robot is overhead, it is not readily in view. Guards and warnings must be in place to protect the work envelope from the unaware. Interfacing the gantry to many machines can create additional safety concerns. If there is a problem with one machine, the remaining machines must still be serviced. The problem machines should be locked out—yet not inhibit the movement of the robot to the other machines. This can be done in many cases by software, but with proper layout planning, hardware lockouts can be implemented.

Access to the work area except for authorized personnel should be forbidden. As with any robot application, a risk evaluation of the hazards should be done in each of the three modes of operation, automatic, programming, and maintenance. The sources of these hazards should be examined in light of the application.¹ Even with safeguards, training, and proper maintenance, accidents can occur. The goal is to reduce risks. This can be done by applying the R³ factor during all phases of the implementation of a gantry robot. Safety depends on respect for a robot's capabilities and limitations. With respect, safeguards can be incorporated into gantry applications.

Fig. 60.5. (*a*) Depalletization of instrument panels. (*b*) Palletization of instrument panels.

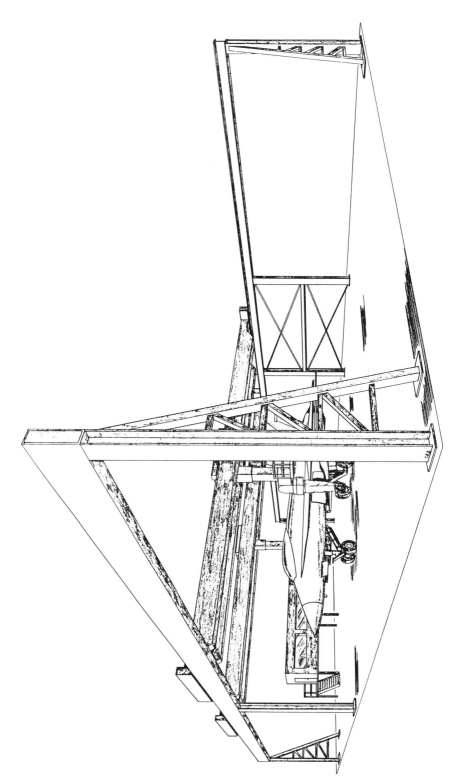

Fig. 60.6. Painting of large aircraft.

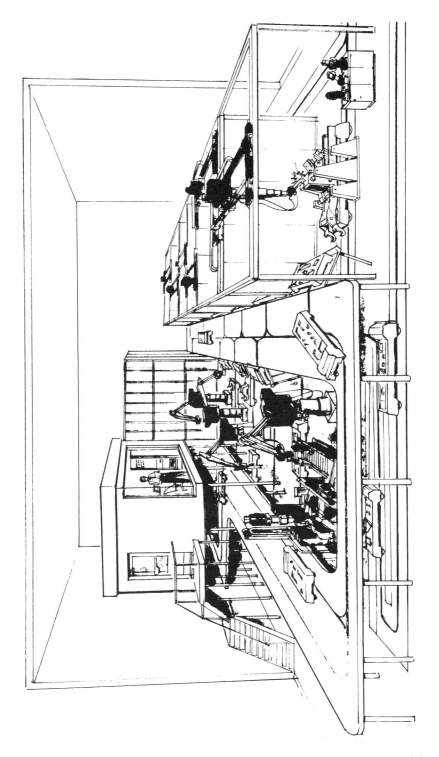

Fig. 60.7. Flexible manufacturing system with pedestal and gantry robots.

60.6. EXAMPLES OF GANTRY ROBOT APPLICATIONS

Material Handling

Advantages	Large work envelopes
	Large payloads
	Flexibility
Constraints	Requires high ceiling, over 3 m (10 ft)
Example	Palletizing, depalletizing of instrument panels (Figure 60.5), furnace unloading, handling large composite sections

Machine Loading

Advantages	Handle many machines
	Handle variety of parts
Constraints	Cycle time requirements
	Payload versus distance

Welding/Cutting

Advantages	Large weldment
	Continuous long welds/cuts
	Use arc, plasma, laser
Constraints	Needs large expensive fixturing
Examples	Plasma cutting and welding of 3-m (10-ft) diameter pipe

Painting

Advantages	Cover large areas
	Remove humans from environment
Constraints	Cycle time requirements
Examples	Painting of aircraft (Figure 60.6)

Deburring, Routing, and Drilling

Advantages	Continuous over large areas
	Can work on flat floor, no vertical fixture required
Constraints	Repeatability requirements may require template

60.7. SUMMARY

The gantry robot is ideal for the emerging factory of the future. It provides flexibility to cover large areas with good repeatability. Gantry robots and pedestal robots can be combined to create a flexible manufacturing system where the proper robot is applied to the proper job (Figure 60.7).

REFERENCE

1. Ziskovsky, J. P., The "R³ Factor" of Industrial Safety, *Proceedings of the 13th I.S.I.R.*, 1983, Chicago, Illinois, pp. 9.1–9.12.

CHAPTER 61

ROTATIONAL WORKPIECE HANDLING IN FMS

HANS J. WARNECKE

ROLF D. SCHRAFT

MARTIN C. WANNER

Fraunhofer Institute for Manufacturing Engineering and Automation
Stuttgart, West Germany

61.1. GENERAL

Flexible manufacturing systems with industrial robots have a long and complex history including serious technical setbacks. One of the first really satisfactory plants in West Germany was developed from 1975 to 1980 by the Zahnradfabrik Friedrichshafen (ZF) in a joint project with several institutes including the IPA, sponsored by the Ministry of Research and Development.[1] The planning procedure and overall layout of the system is presented in this chapter.

61.2. PLANNING PROCEDURE

The aim was to develop a system for the complete machining of a defined family of workpieces used in truck gearboxes. The following criteria for the selection of a suitable workpiece family were considered:

Number of different machining processes: $\leqslant 10$.
Number of manufacturing cells: 8–10.
Machining of different geometrical shapes: 2–3.
Number of shifts: 1.5–2.
Weight of workpieces: > 5 kg.

From different gearbox designs four families of workpieces were selected (see Figure 61.1) according to the aforementioned criteria. The specification for the FMS had been completed by an investigation of the transferability of engineering. Transferability increases if:

The system has a high number of small and independent serviceable components.
The combination of the components leads to a complete FMS.
The system can be kept variable as far as size and kind of tool machines are concerned.
The system is serviceable in a low degree of automation with a possible increase to full automation.

The most important steps in the planning procedure were the following:

Planning the material flow.
Planning the manufacturing cells. Integration of machine tools, robot, workpiece carrier, and the overall control system.
Organization of the whole system.

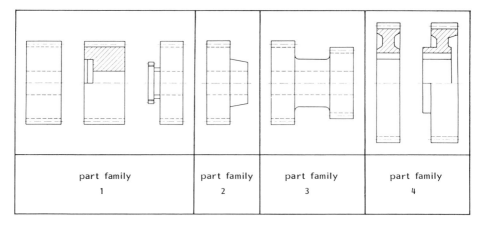

Fig. 61.1. Workpiece families of the ZF-FMS.

61.3. INFLUENCE OF THE FMS ON THE ROBOT DESIGN

During the layout planning of the individual manufacturing cells several robot designs were compared under the following specifications[2]:

Working space (see Figure 61.2), including areas of collision and reachable positions.

Rotatory degree of freedom in the X-Y plane

Dimensions of the workpieces: diameter = 60–280 mm; thickness = 20–80 mm; weight = 0.5–25 kg.

Positioning accuracy $\leq$ 1 mm.

Speed $\geq$ 1.2 m/sec, high acceleration.

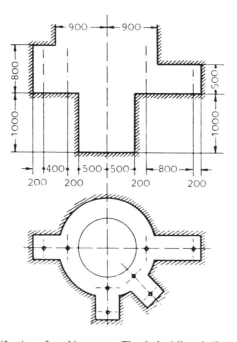

Fig. 61.2. Specification of working space. The dashed lines indicate area of collision.

Controller specifications: PTP-control, teach-in possible, 1000 points in workspace as minimum requirement for one program, batch programs possible, 72 input/output channels.

The most influential parameters were the requirements in workspace and workload. For this application the ZF T III L (described in detail in Chapter 5) was selected. It should be noted here that the original T III L had to be modified to fulfill these requirements. One major modification was the introduction of the telescopic R-axis because of collision problems.

61.4. DESCRIPTION OF THE PLANT

The plant layout is shown in Figures 61.3 and 61.4, and elements of the layout follow:

1. A control buffer for 180 workpiece carriers with a total of 5400–10,800 positions depending on workpiece diameter. Each workpiece carrier has 60 (for 120 mm) to 30 positions (280 mm).

2. A CNC/DNC-controlled gantry robot is used for the handling of the workpiece carriers from the central buffer to decentral buffers at each manufacturing cell. The gantry robot PVF has a translatory Y-axis of 40 m (up to 100 m possible), a translatory X-axis of 4 m, and a vertical Z-axis of 2.3 m. Overall speed limit is 1 m/sec.

3. A DEC/PDP 11 is used for the control of the material flow from the central buffer to decentral buffers, for the DNC-control of the manufacturing cell, calculation, and time schedule. An information system for the manufacturing sequence, NC-programs, tool lists, and overall system control is included.

4. For the manufacturing process: 13 independent cells including 14 tool machines, 14 robots type T III L, 15 decentral buffers, central workpiece input and output station for quality control (see Figure 61.5). In more detail:
   (a) Four cells for turning (CNC/DNC).
   (b) One cell for internal broaching, including tool change for three internal broaches (PC).
   (c) One cell for a drilling unit (PC).
   (d) Three cells for hobbing (six axes) (CNC/DNC).
   (e) One cell for gear grinding (PC).
   (f) Two cells for gear-tooth rounding and chamfering (PC).
   (g) One cell for generating by shaping (PC).

61.5. DESCRIPTION OF A MANUFACTURING CELL

Figure 61.6 shows a manufacturing cell for turning including a CNC-turning machine, a T III L handling robot, and the vertical turning buffer station. The overall flexibility allows economical manufacturing of very small batch numbers.

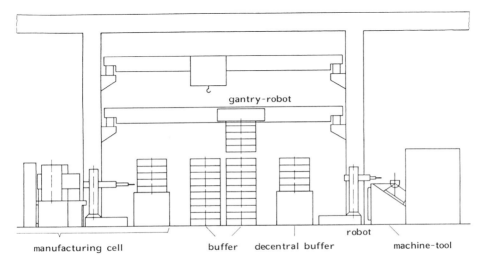

Fig. 61.3. Cross section of the ZF-FMS.

Fig. 61.4. View of the FMS showing (from left to right) robot, decentral buffer, buffer, and gantry robot.

61.6. SUMMARY

The most influential features of this FMS-concept are the following:

Reduction of about 20% in manpower compared to manual systems.

Reduction of about 30% in tool change time achieved by DNC-links from the machine tools to the central processor unit for seven manufacturing cells.

Reduction of time through the system by 25% compared to manual systems.

Optimum batch numbers in the range of 250–400 workpieces.

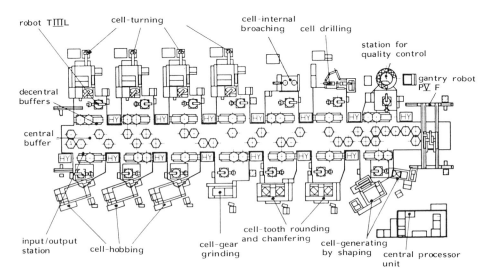

Fig. 61.5. Overall view of the total system.

Fig. 61.6.   Detail of the FMS: manufacturing cell for turning.

Increase in quality due to improved procedures.
High cost share for the buffers.

In practical application the system has shown a remarkable reliability.

61.7.   HANDLING AT CNC-TURNING MACHINES

One of the projects following the described FMS is an application for loading and unloading two CNC turning machines. Designed as an independent manufacturing cell for coupling members with workloads up to 60 kg, the system has the ability to work in a third shift. The overall solution is shown in Figure 61.7, and elements of the system follow.

1. Presentation of the workpieces by two indexing chain conveyors (1) including workpiece carrier.
2. Handling of the workpieces by a T III L cylindrical coordinate robot (4). Programming in teach-in mode.
3. Turning by two CNC double-spindle machines (2,3).
4. Inspection in a quality control station.
5. Finished parts are moved to conveyor (6).
6. Scrap parts are moved into special magazines (7).
7. For the third shift three vertical buffers (5). These buffers enable the system to work for one shift decoupled from the whole system.

Figure 61.8 shows part of the total system during assembly at the manufacturer.

REFERENCES

1. Hörl, A., Flexibles Fertigungssystem für scheibenförmige Rotationsteile, *wt-Z. ind. Fertigung*, January 1982, pp. 9–13.
2. Manogg, H., Industrieroboter konzipiert für die Handhabung von Werkstücken, *Konstruktion*, June 1983, pp. 239–245.
3. Warnecke, H. J., Steinhilper, R., Flexible Manufacturing Systems—New Concepts, EDP-Supported Planning, Application Examples, *Proceedings of the 1st International Conference on Flexible Manufacturing Systems*, Brighton, U.K., 1982.

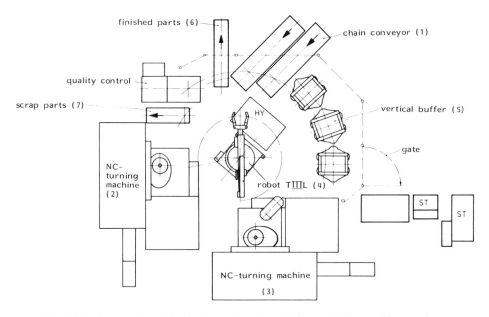

Fig. 61.7. Layout of an individual manufacturing cell for machining coupling members.

Fig. 61.8. Assembly of the manufacturing cell showing parts of the total system: chain-conveyor (1), finished part conveyor (6), scrap parts magazine (7), and robot ZF T III L.

PART 11

ASSEMBLY

CHAPTER 62

PLANNING PROGRAMMABLE ASSEMBLY SYSTEMS

DANIEL E. WHITNEY

C. S. Draper Laboratory, Inc.
Cambridge, Massachusetts

62.1. INTRODUCTION

The basic issues in assembly system design are economic and technical. As products become more complex, manufactured in many models that change design rapidly, it becomes necessary to design flexible assembly systems. People are the most flexible assemblers and the most dextrous, but their performance is variable, difficult to document and hold to a standard. Fixed or special automation is efficient and has uniform performance, but it is too expensive for small production runs because each assembly operation requires its own dedicated station. Robot assembly offers an alternative with some of the flexibility of people and uniform performance of fixed automation. Like people, robots can perform more than one operation: we need not buy one robot for each task. But, like fixed automation, they represent an investment and do not have to be bought again each year the way people must be paid.

In this chapter, we review the current research and applications of robots in assembly at The Charles Stark Draper Laboratory, Inc. The long-range goal of this research is a better understanding of design and operation problems in both conventional and novel assembly systems. The major technical and economic issues involve finding cost-effective ways to perform the assembly operations in space and time sequence, possibly redesigning the product, and selecting the proper assembly technology (i.e., people and machines). The approach is a mixture of research and industrial-application case studies, resulting in a still-evolving research agenda.

The methodology emphasizes taking careful account of technology and economics, and follows these guidelines:

1. Distinguish between global or strategic factors (such as market strategy for the product to be assembled) and local factors (such as process requirements), and ensure that the resulting system is responsive to both.

2. Understand the assembly processes involved, including inspections and tests, and identify needed experiments or developments.

3. Analyze the required operations to find the best organization, without respect to any existing nonautomated system, and, if possible, reexamine the product's design and its subassembly definitions.

4. Examine how to structure the environment of the assembly equipment to make the equipment more efficient or less costly.

5. Use the appropriate technology for each job, not necessarily the most versatile or the most computerized.

6. Apply automated design and evaluation tools wherever possible to synthesize a system's design, study its behavior, and analyze its economics.

The research agenda that follows is also evolving and is shaped and enriched by repeated use of the methodology in actual case studies for industry:

1. Development of economic models of assembly systems to predict unit cost of hypothetical average products at various production volumes using different assembly technologies. A growing data

base of robots and special machines is maintained to provide average cost and performance data. The technologies are modeled parametrically to reflect their average cost to buy, install, and operate, plus their productivity or operating speed. (See Economic Analysis and Unit Cost Models.)

2. Creation of system synthesis tools based on linear and nonlinear programming techniques. These tools can make preliminary designs of specific assembly systems for specific products (as distinguished from average behavior modeling as described previously) by selecting appropriate technology and assigning operations to meet overall throughput requirements at minimum annual costs. (See ADES Synthesis Tool.)

3. Development of product analysis methods to highlight difficult assembly operations, awkward assembly sequences, and inappropriate subassemblies. Operations and sequences that are presumably easy or properly balanced for manual assembly may not be correct for mechanical assembly. Often, we find that the assembly operations to be automated have never been scrutinized before, and large gaps in understanding or unjustified assumptions are found. This usually leads to quite revealing experiments and highly improved operations and organization. (See discussion of product analysis.)

4. Identification of processes lacking adequate engineering models, and development and verification of those models. Processes that have been modeled successfully include rigid-part mating,[1] compliant-part mating[2] (mainly round pegs and holes that are or are not, respectively, rigid compared to the assembly tooling), interference fits accomplished by repeated impacts,[3] square pegs into square holes, screws (both starting and tightening),[4] and staking (see Chapter 64). Processes with less well-developed models, or those not modeled as robotized processes, include welding, grinding, polishing, and deburring. Some of these processes have been robotized, but better models will allow more effective or more economical applications. Extensive literature exists on process models (see, for example, Reference 1 for a bibliography) so this topic is not treated further herein. Also related to this issue is the need for automatic or semiautomatic assembly process planning, possibly applying artificial intelligence techniques, as described in Chapter 23, Expert Systems and Robotics.

5. Extension of process modeling to robot performance modeling and evaluation. This includes conventional kinematic and dynamic models of idealized robots plus performance analysis and calibration methods for real robots; their performance is typically far from ideal. The next few years will find increasing applications of modern control methods to monitor robot performance and to carry out self-calibration on the job. (See References 5 and 6, for example.)

6. Pursuit of case studies to test the methodology and expose the need for additional research and new design tools. (See case studies.)

62.2. TOWARD AN ASSEMBLY SYSTEM DESIGN METHODOLOGY

62.2.1. Global and Local Issues

A piece of manufacturing equipment or a manufacturing system must do its job and must fit into the factory as a whole. That is, the system has both local and global responsibilities. Correspondingly, the product to be fabricated or assembled presents both local and global characteristics that influence the system's design. The local issues are primarily technical, the traditional part-mating or assembly-sequence problems, whereas the global issues are primarily economic and concern management's objectives for the product. Such global issues as its potential volume growth, number of models, frequency of design changes, field repairability, and so on, heavily influence both product and system design. The need for system programmability, for example, must be viewed as one of a broad range of requirements the system must meet.

This morass of requirements has been given preliminary structure according to Table 62.1. It is not a final or complete representation, nor do the items within the blocks all fit comfortably in one and only one place. Traditional assembly system design has focused mostly on meeting local requirements for both product and system, but advanced systems will increasingly require consideration of the impact of global requirements.

The trend toward injecting strategic issues into a traditionally local problem can be seen vividly in Japanese industry. The following quotes are taken from reports by recent visitors to Japan: "Experience has taught the Japanese the value of placing even short-term manufacturing decisions at the service of long-term strategy."[7] And, "Product design [in Japan] is viewed as part of a total product-process system."[8]

62.2.2. Bottom-Up and Top-Down Approaches

The technique originally used by the author and his colleagues to develop a design was to undertake several paper designs and to use some computer-based tools to guide the process. From this start, two distinct tracks have emerged, each with strengths and weaknesses. The tracks have been named "bottom-up" and "top-down," the former an empirical approach and the latter based on analytical management science and the mathematical programming tool ADES.

In brief, the bottom-up approach is characterized by the following:

TABLE 62.1. STRUCTURE OF SYSTEM DESIGN ISSUES

	Global	Local
	Management's objectives:	The parts and assembly operations:
Product	Economics and market	Assembly sequences
	Volume growth	Types of operations
	Design volatility	Geometric constraints
	Quality, reliability, safety	Part size, weight
	Make or buy decisions	Shape, stiffness
	Build to order/stock	Tolerances and clearances
		Tests and inspections
Assembly system	Cost and productivity	System layout
	How it interfaces to the factory	Equipment choice
	Labor support needs	Task assignment
	Failure modes	Part feeding (factory interface issue)
	Space needs	

1. An ad-hoc nature.
2. A bias toward technical issues.
3. Providing a single solution without rigorous comparisons to others.
4. A good chance of being feasible technically but not as good a chance of being optimal economically.

The top-down approach is characterized by the following:

1. A general nature.
2. A bias toward time and cost issues.
3. Implicit consideration of many options.
4. Optimal economically (according to its criterion) but no guarantee of being feasible technically.

62.2.3. Bottom-Up Approach

This approach follows the steps outlined in Table 62.2, which defines a two-phase process: study followed by implementation. The procedure in Table 62.2 leads directly from global to local issues and ensures that both are accounted for. Topic 1 and Topic 2, items A through C, are performed by a multidisciplined team that visits the industrial site and spends several days learning about the problem at many levels. Items D through F typically require another several weeks to several months.

Major Technical Trade-offs

A number of major issues must be resolved during the extended study period. A suitable assembly sequence must be found. Ways to improve the assembleability of the product may be identified. Suitable assembly technology, part-feeding methods, system layouts, and economic justification must also be addressed. These are discussed in turn.

Quite likely, an assembly sequence suitable for mechanized assembly will be different, perhaps totally, from any existing manual sequence, since the latter may contain tasks requiring two hands, flipping an unstable subassembly over, and other operations that are difficult to automate or are time-wasters. First, the individual parts and operations are cataloged on a planning chart (Figure 62.1) which highlights the geometric and dynamic requirements of each operation. An indication of task difficulty and technological options is also provided for each task. The designer must then generate many alternate assembly sequences and critique them for overall feasibility, relevance to available technology, and relation to any other special requirements. This usually requires that he visualize a technique for each operation, including part feeding. Parts trees (Figures 62.2 and 62.3) are used to shorthand these studies.

At the same time, useful subassemblies must be identified. Usefulness can be global or local. For example, any group of parts that is physically stable (can be transported or reoriented without falling apart) is a candidate *local* subassembly. But if it performs an identifiable function in the product and can be tested for that function right after assembly, if it is common to many different product models, or if it comes in a variety of sizes but could be made on one machine using the same operations,

TABLE 62.2. CASE STUDY PROCEDURE

Phase 1

1. Initial analysis of problem which includes detail study of manufacturing facilities.
   A. Briefing by management on their expectations for the study.
   B. Briefing by on-site manufacturing managers of what they feel are their most difficult problems.
   C. Detail tour/study of present manufacturing facilities.
2. Study issues.
   A. Basic plant organization.
      Type of plant (manual, high-level of automation, or hybrid)
      Role of worker (simple monitor, inspector, parts modifier, assembler)
      Attitude of workers toward automation
      Basic economic data
         Annual volume
         Labor rates
         ROI expected
      Local constraints (labor shortages/surpluses, plant area available, locations of sites for demonstrations, institutional problems)
   B. Product structures.
      Clearly defined subassemblies or other ways of modularizing product
      Standardization of subassemblies
      Do subassemblies organize into "islands"? If not, what other organization is suggested?
   C. Initial target(s) list. From the product structure studies, select candidate subassemblies, or portion of final assembly, that appear most feasible:
      Represent/illustrate most generic elements of product problems
      Require minimum design changes
      Many "stack tasks"
   D. Detail study of selected target(s).
      Detail task analysis
         List all tasks (mfg. inspection, material handling or assembly)
         Group tasks
         Identify generic elements if possible
         Part tree analysis
   E. List of applicable/pertinent technology.
   F. First configuration study.
      Identify key tasks/technology issues
      List critical experiments to test proposed options
      Make economic analysis
      Prepare candidate system conceptual designs

Phase 2

1. Perform critical experiments identified under Phase 1.
2. Perform detail configuration studies based on detail drawings, specific tolerances, weights, and so on.
      Perform mock-up studies for part assembly
      Hard mock-up studies using purchased components and simple tooling
3. Detail designs.
4. Validate prototype design on dummy parts or one set of production parts.
      In laboratory
      In customer's R&D center
      On production floor in model shop test area
5. Design/build production system

ASSEMBLY PLANNING CHART

SKETCH OF ASSEMBLY:

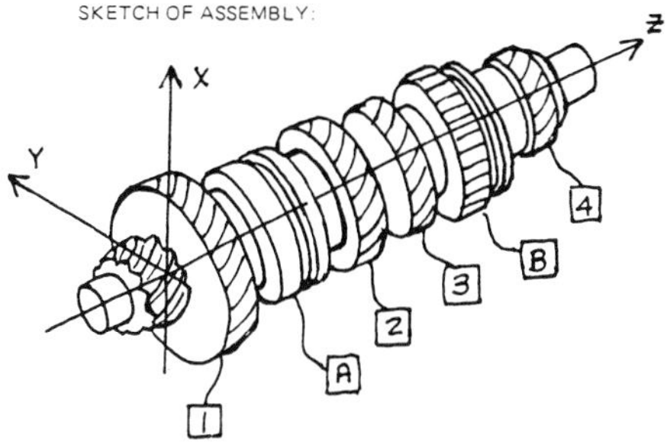

NAME: _SECONDARY SHAFT ASSEMBLY_

DATE: _____

PREPARED BY: _REG_

SHEET _1_ OF _2_

TASK	SEQUENCE	TYPE OF TASK	BASIC ORIENTATION	MOTION(S) REQUIRED	JIGGING/TOOLING REQUIRED	DEGREE OF DIFFICULTY OF TASK	INSPECTION REQUIRED	CYCLE TIME (s)	COMMENTS
Get shaft (temp. nosecone attached)				X-Y		1			
Place shaft in pallet.	1	A	Z		Y	1		3	
Get gear 1,				X-Y		1			
Install on shaft.	2	I	Z			1		3.5	
Get sub-assy. A, *				X-Y	Y	2			
Install.	3	A	Z			2		6	Spline Alignment
Get circlip,				X-Y		1			
Install.	4	S	Z			3		4	3rd Groove
Get splined washer,				X-Y		1			
Install.	5	A	Z			2		3.5	Spline Alignment
Get gear 2				X-Y		1			
Install.	6	I	Z			1		3.5	
Get splined washer,				X-Y		1			
Install.	7	A	Z			2		3.5	Spline Alignment
Get circlip,				X-Y		1			
Install.	8	S	Z			3		3	2nd Groove
Get splined washer,				X-Y		1			
Install.	9	A	Z			2		3.5	Spline Alignment
Get gear 3,				X-Y		1			
Install.	10	I	Z			1		3.5	

ABBREVIATIONS USED:

TASK TYPE

P — PLACE, ORIENT
T — TIGHTEN BOLTS, NUT, etc.
I — INSERT PARTS
M — MEASURE
S — SHAPE MODIFY
A — ALIGN

INSPECTION

B — BOLT TORQUE
G — GAUGE DIMENSION
C — COMPARISON

* Separable sub-assy
A must be flipped
upside-down.

Fig. 62.1. Assembly planning chart.

then its definition and exploitation have *global* implications for the product or for the machine. In fact, it may be beneficial to design and create subassemblies that are impossible in the current design if significant global benefits can be obtained.

This is also the time to study the product, in all its model variations, to identify useful redesigns. These, too, may be divided into global and local. First, global: Concentrate model differences in a few, possibly complex, parts so that the rest of the parts will be common to all models. Reduce part count to keep the assembly system, part feeding and provisioning, and supporting logistics simple. (Do not pursue part-count reduction to the point where the savings are lost in extra part fabrication or materials cost.) If a product line contains subassemblies with similar functions (handles, gear boxes, cords, and switches, for example), consider making these modules identical so that economics of scale

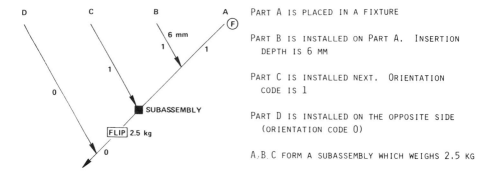

SHORTHAND ASSEMBLY SEQUENCE NOTATION

PART A IS PLACED IN A FIXTURE

PART B IS INSTALLED ON PART A. INSERTION DEPTH IS 6 MM

PART C IS INSTALLED NEXT. ORIENTATION CODE IS 1

PART D IS INSTALLED ON THE OPPOSITE SIDE (ORIENTATION CODE 0)

A,B,C FORM A SUBASSEMBLY WHICH WEIGHS 2.5 KG

Fig. 62.2. Parts tree definitions.

can be gained. Second, local: Design the subassemblies so that most or all assembly operations occur in one direction. Reduce fastener count (a normal consequence of reducing part count) because fasteners are often troublesome to handle and install. Provide external reference surfaces that locate interior points where assembly occurs, and provide generous lead-ins on holes, slots, and grooves.

Figure 62.4 is a good example of a product redesigned with both global and local goals in mind. A smaller part count supports many models; the base acts as the pallet during assembly; final adjustment is made with one motion from the assembly direction.

Each candidate assembly sequence, containing any approved new subassemblies or redesigns, should be critiqued on the basis of its cycle time, number of special tools and fixtures needed, number of nonassembly operations, and overall feasibility. The best one or two should be retained for further study of part feeding, machine layout, technology selection, and economic analysis.

Part feeding has traditionally been accomplished by bowl feeders for small parts, roller conveyors for large parts, or by hand.[9] More recently, techniques such as magazines, stacks, and kits have been used. Table 62.3 shows some of the possibilities recently used and their appropriateness for different part sizes and feeding logistics. For example, Sony assembles its Walkman cassette players using a 7 × 10 in. (17 × 25 cm) manually filled traveling kit containing many stacked parts and two assembly pallets. At IBM, people fill kits with console display components which are fed to RS-1 robots for assembly.[16] Recently Sony has demonstrated automatic pallet loading.

Possible machine layouts have also become more varied, with the basic in-line design being supplemented by lines with gated feeders or station lockout for achieving model mix on a "fixed" machine. Station lockout, a technique used by Hitachi to make cassette tape decks, means turning a station on or off depending on which model is at the station. Many gated feeders are attached to one station, but only one feeder's gate is open while a batch of one model is being made. For a different model, another feeder's gate will be open. The gages in Figure 62.4 are assembled this way. By contrast, the Sony Walkman is made by a machine with several random-access workstations linked by a conveyor system.

The type of technology and its arrangement is also becoming more varied. In the past, a "station" was one piece of fixed technology, a portion of a fixed transfer machine, or a person, working at one worksite. But now it can mean one or more robots at a single worksite, a robot dividing its time among several worksites, a robot serving one or several fixed workheads, or any other feasible combination.

Economic Analyses and Unit-Cost Models

Clearly, advancing technology provides the system designer with more and varied alternatives. Inevitably, economics becomes the tool by which a choice is made. To aid these choices, we have developed or elaborated on three different types of economic models; allowed investment, unit cost, and system synthesis by ADES. The first two are discussed here and more completely in Reference 10.

Allowed investment models are based on the rate of return method. For a candidate automation opportunity, we assume that we can predict the future costs, cost growth-rate, savings and their growth, plus overheads, maintenance, taxes, and depreciation. The intent is to have a rational way of assessing the effects of rising production volume and continuing inflation. In conventional discounted cash flow analyses, these effects usually result in unreasonably large projected future savings (and consequently

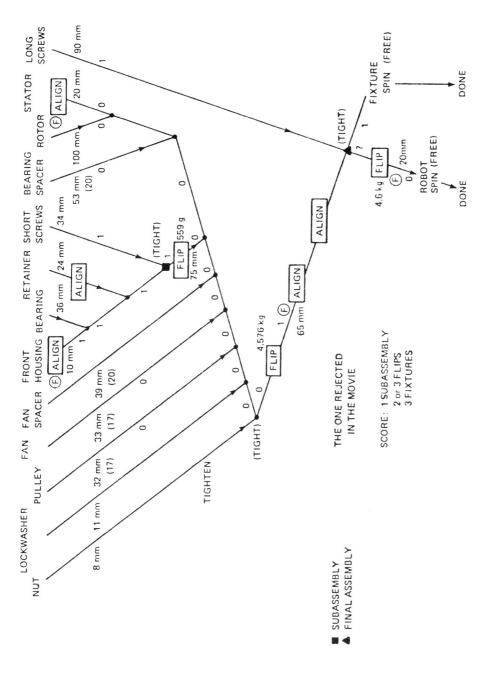

Fig. 62.3. First parts tree for the alternator: a poor choice. (See Figure 62.7 for a picture of the alternator.)

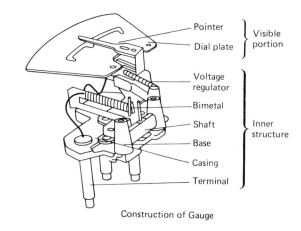

Construction of Gauge

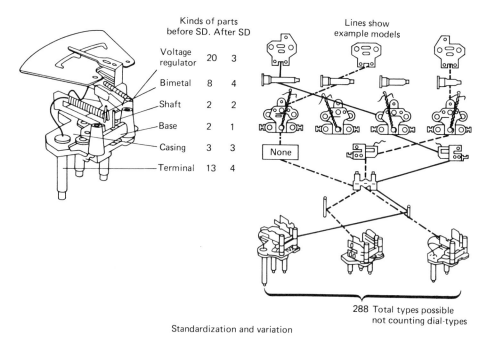

Standardization and variation

Fig. 62.4. Redesign of automobile dash-panel gage to support 288 models from six parts, each of which comes in one to four varieties.

large projected rates of return), so we have introduced the idea of investment horizon. This is the date, usually no more than three or four years in the future, when we agree to stop accumulating discounted savings. Within these limitations and techniques, a curve can be constructed that relates internal rate of return to allowed investment (normalized by first year's savings). Figure 62.5 is an example. This is useful in providing a budget within which the proposed system must be created.

Unit-cost models are based on less complete economic analyses than the foregoing, and they assume some knowledge of the product to be assembled, such as part count, rather than knowledge of future savings. Three types of assembly technology have been modeled: manual, programmable automation, and fixed automation. Each can be assigned an installed cost per "station," a cycle time or throughput per station, and a required rate of return. Different cost and speed assumptions can be compared. For each of a series of required production volumes, the number of needed stations can be calculated, hence the total investment,ʹ and finally the unit assembly cost. Curves like those in Figure 62.6 are the result.

TABLE 62.3. LOGICAL STRUCTURE OF PART FEEDING

	Feed Separately to Each Station	Feed to Many Stations
Small parts <1 in.	Bowl, hopper, magazine	Traveling magazine, pallet, etc. Kit Roving robot
Medium parts <6 in.	Roller conveyor Stack Manual	Kit of the parts for one unit Kit of identical parts Roving robot
Large parts	Conveyor Rack	May not be practical

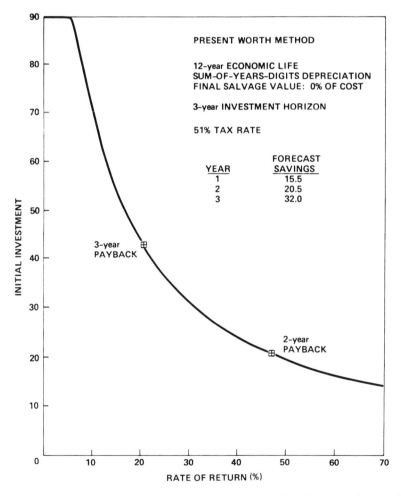

PRESENT WORTH METHOD

12-year ECONOMIC LIFE
SUM-OF-YEARS-DIGITS DEPRECIATION
FINAL SALVAGE VALUE: 0% OF COST

3-year INVESTMENT HORIZON

51% TAX RATE

YEAR	FORECAST SAVINGS
1	15.5
2	20.5
3	32.0

3-year PAYBACK

2-year PAYBACK

Fig. 62.5. Allowed initial investment in automation as a function of required rate of return. Larger rates of return reduce the allowed investment. The return is based on accruing forecast savings over a limited future period. Forecast savings rise because production volumes increase and inflation would raise the cost of manual assembly, the presumed alternative.

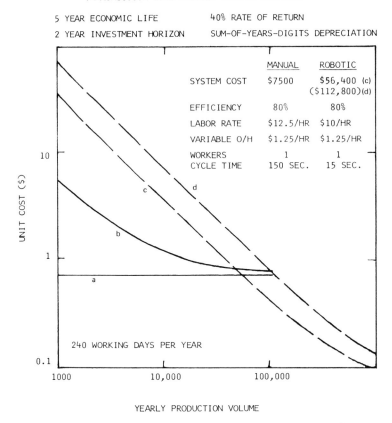

Fig. 62.6. Unit assembly cost as a function of yearly production volume for manual assembly (*a*), manual assembly if required jigs and tools are amortized (*b*), and a robotic assembly system (*c* and *d*).

These curves give a broad-brush view since they use average installed cost estimates; the reader can take his choice. On any one curve, there are one or several stations of only one kind of technology. Hybrid systems are not represented. The product is modeled only as requiring a certain number of presumably identical operations. These models are useful for assessing the relative cost impacts of different assembly technologies and for predicting promising areas for each, assuming that each is substitutable for the other. Despite their basis in average data, the models accurately show the ordinal relationships: manual assembly is the most economical at low-production volumes, fixed automation at high volumes, and programmable assembly at intermediate volumes.

These tools and procedures constitute the bottom-up method. It proceeds from the product through the assembly technology to an economic analysis. System synthesis is based mainly on the designer's intuition, fed by knowledge of available technology and thorough study (and possibly redesign) of the product.

62.2.4. Top-Down Approach

As an alternative to the empirical bottom-up approach, the top-down approach is based on a management science tack. Top-down differs in several significant ways: Economic issues enter right at the beginning, and specific technology, rather than averages, is represented. The product to be assembled is also modeled in detail beyond mere part count. The assembly system is synthesized by a computer program called ADES, rather than by the designer himself. The success of the method still depends heavily on the designer, based on his choice of inputs, inasmuch as the ADES model is necessarily incomplete.

Designer's Decisions and the Role of ADES

ADES is a computer program that aids a designer in configuring an assembly system. As such, it deals mainly with tactical issues. The designer must already have a firm grasp of the strategic issues. We assume, therefore, that the product's design has been adequately studied and modified and that one or more candidate assembly sequences are available. ADES' job is then to select equipment or people from a set of possible "resources" the designer provides and to assign the assembly operations to the selected resources so that the required throughput can be met at minimum annualized cost. Example types of resources include people; fully programmable, tool-changing robots; fixed-stop pick-and-place robots; single fixed workheads; and complete transfer lines. ADES, in its current state, cannot consider geometric constraints such as collisions between robots or contortions of conveyor paths. Thus, the designer must convert ADES' resource selections and task assignments into floor plans. Satisfactory results are not always possible. Examples are given later.

Technical Description and History of ADES

ADES belongs to a class of systems called "resource allocation" programs and comprises several mathematical programming techniques. It has developed through several stages to its present state. Its original version was capable of making very general task assignments[11] in that several resources could share responsibility for each operation. This led to quite efficient resource utilization but complex system floor layouts. Also, only two types of resource versatility could be modeled: totally restricted to one or two operations or totally substitutable and applicable to many operations. Thus totally fixed simple workheads or rather general robots could be modeled, but anything of intermediate versatility was difficult to model realistically. In addition, the resources' time could be allocated to productive work or to dwell times dictated by the need to deliver or remove work or by line imbalance. But time devoted to tool changing, which depends on which operations are assigned to which resources, could not be accounted for.

These limitations were removed in the second generation of ADES.[12] Tool change time is correctly represented, and tool cost is approximately accounted for. Resources of different versatility are modeled by describing which operations are individually feasible plus the maximum number of those (usually fewer than those technically possible) the resource can support simultaneously. To achieve these increases in ADES' fidelity, it was necessary to restrict the generality of task assignment compared to the original ADES. Now, operations cannot be split between several resources. It is still possible, however, to assign several operations to one resource. This, plus tool changing, allows us to model the main economic property of robots or other versatile machines, namely, their ability to do more than one job.

Finally, it should be noted that ADES is not limited to *assembly* system design but generally is applicable to any problem of selecting resources to accomplish a given series of operations. Within the limits of the model, one can design assembly, metal removal, administration, computation, or other types of systems.

To use ADES, the designer must assemble general technological data about feasible resources and specific data about the assembly (or other) operations he needs accomplished, plus the desired sequence of operations. He must then combine these data into an input table that gives the time and variable cost for each resource to perform each operation if the combination (resource × operation) is feasible in his judgement. The needed tools for each operation must be identified along with their variable costs (adjustment, wear) and tool-change time.

The required annualized investment must be given for each resource. This is represented approximately by the purchase price plus a portion of the system's engineering and installation costs, all modified by a capital recovery factor. Finally, each resource's "capacity" must be described. Two factors are involved. One is its availability in minutes per hour. A value of 120, for example, for this datum tells ADES that two identical copies of this resource are available if needed. This gives ADES an additional range of choices. The other factor is the technical versatility, indicated by the maximum number of operations that can be assigned to this resource. Table 62.4 lists some representative assembly and material-handling robots by increasing versatility.[13]

Last, the designer gives the required annual production volume.

For each set of input data, the program returns with its selection of resources plus the operation assignments and corresponding annual operating cost (including amortization of the investment). The percentage utilization of each resource is also given. If none is used 100%, then there is spare capacity. Otherwise, one, sometimes more than one, resource is saturated and thus so is the system. In this case, the annualized cost can be divided by the annual production volume to give a unit cost.

ADES can be used in several ways. A first extension of the basic ADES procedure is to repeat it for several different required production volumes. A set of system designs and unit costs results, similar to that shown in Figure 62.6, but based on a more realistic and particularized model. The set of designs also represents a plan for phased introduction of automation capacity as volume grows. In this case, it is more realistic to run the program in a series with strictly increasing volume. For each

TABLE 62.4. ROBOT PROGRAMMABILITY CLASSIFICATION

Class	Basic Sequence	Sequence Alteration	Location Critical Points	Number Critical Points	Number Of Axes	Example	Approx. Cost (K$)
I	Fixed	Fixed	Fixed	2	1	Single Axis Pneumatic	<5
II	Adjust.	Fixed	Fixed	4–16	2–4	Seiko 100, 200,700	5–10
III	Adjust.	Fixed	Adjust.	4–16	2–4	Seiko 400	10
IV	Adjust.	Boolean	Adjust.	4–32	2–5	Autoplace	7–15
V	Adjust.	Boolean	Adjust.	12–384	3	MHU	25–35
VI	Prog.	Boolean	Prog.	∞	3 + Wrist flip	Fanuc	35
VII	Prog.	Boolean	Prog.	∞	5–6	Unimate	30–45
VIII	Prog.	Algorithm	Prog.	∞	5–6	Puma, T³	35–70

volume, one may include at very low cost the resources already "bought" by all the programs run at lower volumes, simulating reuse of those resources. An automation supplier can use ADES to "test market" a new machine before it is built simply by including its projected cost, versatility, and productivity along with data about existing competitive products. Also, a system designer can test his own selection of resources, letting ADES assign operations. This method can be used to test a system's breakdown resistance, merely by deleting a "broken" resource from the allowed set.

Example Use of ADES

The original version of ADES was tested on assembly of automobile alternators (Figure 62.7). The input data are shown in Table 62.5. The operations are listed along the left side, while the resources and their presumed annualized costs are listed across the top. "PAX" represents a large assembly robot designed by Bendix and used by CSDL to assemble alternators in 1977.[14] "PUMA" represents Unimation's Unimate 600, a medium-size assembly robot. "Autoplace" represents typical fixed-stop pick-and-place robots, which are much less versatile than PAX or PUMA and much less costly. Entries in the table are assumed operation times in seconds, except for "100," which represents the designer's judgment that the operation is not feasible for that resource (parts too heavy in the case of PUMA or lack of tool-changing capacity in the case of Autoplace). These are reasonable but arbitrarily chosen times and capabilities, intended to exercise ADES. They do not represent the true capabilities of any real product, and no endorsement of or comment about any actual product is intended.

The result of a series of runs with increasing production volume is shown in Table 62.6. Over an 8:1 range in volume there is about a 2:1 gain in economy of scale. Note that this is attained by substituting less versatile, less costly resources at higher volumes. More workstations are required to meet these volumes, but each has fewer operations assigned to it; hence, less versatility is needed. The conclusion to be drawn is that between very versatile robots at low production volumes and very restricted fixed automation at high volumes lies a richly varied continuum of hybrid systems composed of stations of intermediate versatility.

Figures 62.8 and 62.9 show two floor layouts created manually from solutions 2 and 7 in Table 62.6. The production volume capacities are greater than the computer specified because the designer knew that tool-change time could be shared over several alternators as long as he provided space for them in front of the tool-changing resource. Note that the system in Figure 62.9 supports shared operation assignments between Autoplace No. 1 and PAX No. 1, and between PAX No. 1 and PAX Nos. 2 and 3. This requires a complex transport sequence and overlapping work regions so that one set of part feeders can be shared. The design may therefore be impractical. The designs are discussed more fully in Reference 15.

Experience working with ADES over the past three years shows that it is a useful tool. It rarely surprises the designer but regularly confirms his expectations and gives him confidence in the designs. As an accompaniment to a manual design procedure, the required data collection and feasibility judgments enforce a useful discipline on the designer, resulting in better designs.

62.3. CASE STUDIES OF ROBOT ASSEMBLY SYSTEM DESIGN

Two specific case studies are described briefly in this section. The first is assembly of a small gunpowder ignitor, the second, assembly of part of a hand-held electric appliance. Finally, some problems of workstation design are treated.

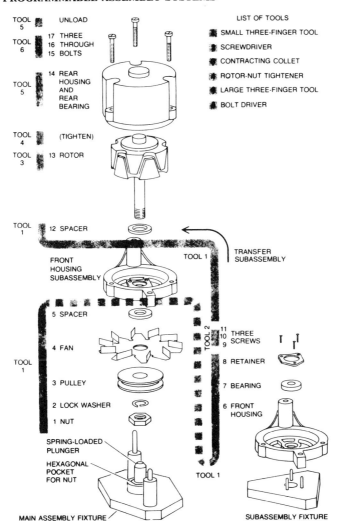

Fig. 62.7. Exploded view of the alternator shows the sequence in which its 17 parts are assembled by the programmable robot and identifies the tools that perform each task. The center rod in the main assembly fixture is a spring-loaded plunger. The collar at the base of the rod contains a hexagonal cavity that firmly holds part *1*, the nut. When the rotor (*13*) is inserted, the plunger is depressed, enabling the rotor's threaded shaft to engage the nut. The rotor is then spun by tool *4* to thread the shaft tightly into the nut. The time now required by entire operation, 2 min 42 sec, could be reduced to 1 min 5 sec if changes were made in design of tools, fixtures and the alternator.

62.3.1. Gunpowder Ignitor

The ignitor (Figure 62.10) is about an inch long and has eight parts, counting the powder charge. The charge ignites when an electric current melts the wire. The product is currently assembled manually with the aid of a press and simple hand tools. While production volume is expected to rise year by year, it is unlikely to become large enough to justify conventional fixed automation. In fact, if robots are used, they must be extremely low cost to compete economically with people, as shown by an analysis like that in Figure 62.5. The challenge therefore is to design a robot assembly system with inexpensive pick-and-place robots having three or four axes that move in simple arcs from one fixed stop to another. This system must not only stack some parts under a press (not too difficult), but must also thread the wire through them (difficult).

TABLE 62.5. INPUT DATA TO SYSTEM DESIGN PROGRAM[a]

Time/Operation τ_{ij} (sec)

Task j	1 PAX $60,000 + $12,000 Tools and Tool Changer	2 2 PAX	3 PUMA $40,000 + $6,000 Tools and Tool Changer	4 2 PUMA	5 Special Screwdriver Station $15,000	6 Autoplace 1, $10,000 + $5,000 for Tools and Rotating Pallet Feeder	7 Autoplace 2, Same as $i = 6$
1 Nut	6		4			3	⟵100⟶
2 Lockwasher	6		4		⟵100⟶	3	
3 Pulley	6		4			3	
4 Fan	6		4			3	
5 Spacer	6		4			3	
6 Front housing	7		6			100	3
7 Bearing	6		4				3
8 Retainer	6		4				3
9 3 screws	8		8		3	⟵100⟶	⟵100⟶
10 Spacer	6		4				
11 Rotor	12		100		⟵100⟶	100	
12 Tighten nut	12		100				
13 Rear housing	12		100		3		
14 3 screws	14		8				

Resource i

[a] Note: Machine 2 consists of two PAX robots each identical to machine 1. A similar interpretation applies to machines 4 and 3.

TABLE 62.6. CHART OF SOLUTIONS TO OPTIMUM DESIGN PROBLEMS BASED ON DATA IN TABLE 62.5 [a]

	Solution Number						
Machines	1	2	3	4	5	6	7
PAX	X	X	X	X			X
2 PAX					X	X	X
PUMA				x		X	
2 PUMA							
SP. SCR. DR.			—		—		—
AUTOPL 1							X
AUTOPL 2							
One shift annual volume	63,000	108,000	168,000	192,000	330,000	360,000	480,000
One shift unit assembly cost based on one year payback	$1.17	0.982	0.809	0.738	0.607	0.652	0.578

[a] A solid line means that a machine is in a solution, while no line means that the machine is not in the solution. Each solution is producing at the maximum production rate and x's indicate the bottleneck machines.

[b] X = unit is fully loaded; x = unit is almost fully loaded

The proposed solution is shown in Figure 62.11. It is based on reversing the manual assembly sequence so that the wire can be attached to the housing and cap in a few simple twists and sideways moves by a fixed-stop robot and a wire-feeding station (Figure 62.12). The final system is shown in Figure 62.13 and contains two such robots with 3 DF each (in-out, up-down, and a 180-degree flip about in-out). The product is built on a small nest that moves from right to left along a conveyor. Robot No. 2 can put the final unit in the "good" or "bad" bin because its controller allows program steps to be skipped based on external sensor signals. The program sends the robot to the "bad" bin first, then to the "good" bin. When a unit tests "good" (most of the time), the trip to the "bad" bin is skipped.

This is not a very flexible system in the sense that it follows one sequence, except at the last station. But it meets the strategic and tactical needs for a low-cost system with the virtues of safety, uniformity of performance, and product quality. It also shows that one can consider simple robots for intricate assembly tasks.

62.3.2. Small Electric Appliance

This product is like many found in the home. It comes in several models of similar size and is typical of several electric products this manufacturer makes. We consider a generic subassembly called the "body," shown in Figure 62.14. Ordinarily, such an item is quite difficult to assemble even manually because of the need to run wires through cramped spaces between the stator and housing to interconnect the stator, brushes, and cord. We assume, and have so advised the manufacturer, that meaningful assembly automation requires redesigning the stator so that its wires terminate on a connector and redesigning the housing so that some electric conductors are molded into it. Thus, wiring is submerged in fabrication. Similar designs are starting to appear in high-volume motors and alternators.

The solutions shown here were generated with the help of ADES. Table 62.7 contains the input data. Note the wide range of versatility and station cost. One resource, the multiple adjustable-stop robot (MASR), is purely mythical. It was made up as a possible compromise technology to see how it would perform against existing equipment. It is presumed to contain real-time programmable stops so that it can stop at many points, but it has no servos or ability to perform controlled trajectories. Thus it is of intermediate versatility and, presumably, intermediate cost.

Two of the solutions are sketched in Figures 62.15 and 62.16, designed to meet 600,000 and 1,300,000 units per shift-year, respectively. Each contains a rotary table for carrying assemblies between workstations. As in the examples in Figures 62.8 and 62.9, higher volumes are met by less versatile but less expensive equipment. Note especially that the MASR plays an important role.

62.4. ROBOT WORKSTATION DESIGN ISSUES

Robot workstation design is a large topic, and only a brief discussion is given here. The emphasis is on analyzing the technical requirements of the stations *before* selecting robots, sensors, computers, and other equipment.

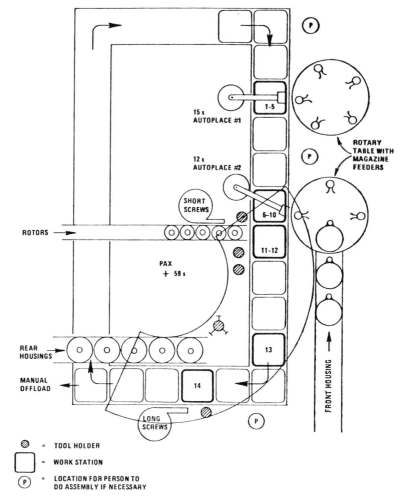

Fig. 62.8. Design for solution 2 of Table 62.6 capable of producing 124,000 assemblies per shift-year. [Computer's solution is capable of only 108,000 assemblies per shift-year.]

Robots perform primarily geometric operations, so the geometry of the station's task must be determined first. This includes the gross size of moves as well as their directions. The tolerances required on all motions, speeds, forces, vibrations, noise, and so on, must also be known. This may bring the design to a temporary halt while the operation is reanalyzed and a better specification, perhaps based on experiments, is created. Vagueness in the specification (e.g., "smooth," "carefully") is common in manual operations, but we have at present no way to convey such terms to a machine. Even to seek to convey them is pointless because they are masks for a lack of understanding of the process. A process that is not understood cannot be automated successfully.

Several design scenarios for carrying out the operation must then be created. Each must be analyzed economically as well as technically to define the requirements of robots, computers, and tooling. Errors or tolerances in the performance of these items must be determined and compared to the task's requirements. This may entail experiments, especially in the case of robots, because the state of knowledge of their performance is generally poor. Manufacturers' data usually are confined to maximum tip speed and repeatability, which are easy to measure but often hard to relate to the needs of the task at hand.

At present, it seems in vogue to sell robots based on their repeatability: the smaller the better. But this also costs money, and the result may far exceed the needs of the task. Some quoted figures appear to be easily swamped by thermal effects. Be that as it may, the task may require something

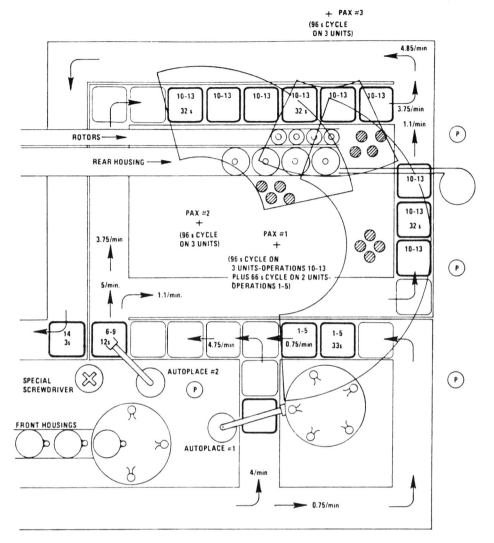

Fig. 62.9. Design for solution 7 of Table 62.6 based on fixed-visit sequence transport configuration and capable of producing 544,000 assemblies/shift-year. [*Note:* the flows shown are the maximum for each location. The system flow is 4.75 units per min.]

totally different. For example, stop-to-stop time is economically much more significant and technically relevant than peak tip speed (which is rarely reached and fleetingly maintained in most assembly tasks). End-point vibrations during stopping, induced by excessive speed, can actually lengthen stop-to-stop time.

Absolute accuracy is rarely quoted but often confused with repeatability. To define it requires that the robot have a controller containing not only the robot's internal coordinates but also its environment's (or world) coordinates, in which the "accurate" moves must occur. Finally, and crucial, the controller must contain a calibration (and a calibration procedure) that relates the two coordinate systems. With these ingredients, the robot is able to go to some X, Y, Z point in its environment for the first time with such and such error. From this capability comes the ability to program the robot off-line using a CAD/CAM data base.

Some operations, such as palletizing, may not require absolute accuracy, but only relative accuracy: from here, go K mm along the X axis. Others, such as inspection, delicate assembly, or other sensor-based operations, will require commandable resolution, the ability to make a given very small move

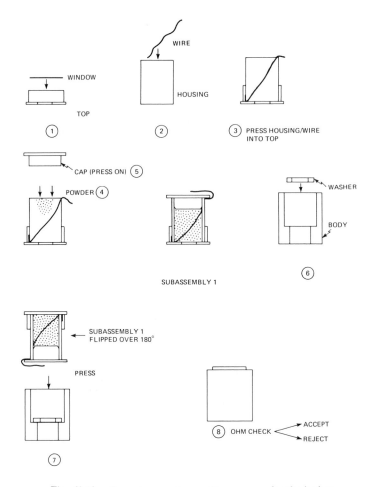

Fig. 62.10. Current manual assembly sequence for the ignitor.

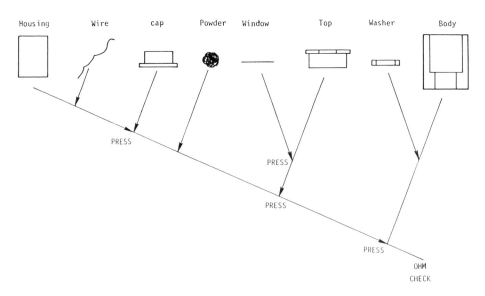

Fig. 62.11. Parts tree for robot assembly of the ignitor.

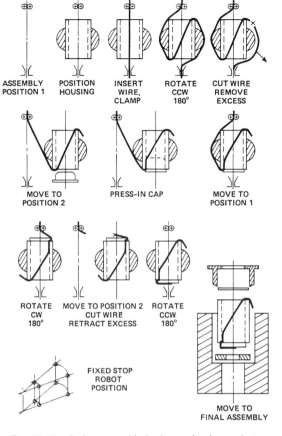

Fig. 62.12. Ignitor assembly by low-technology robot.

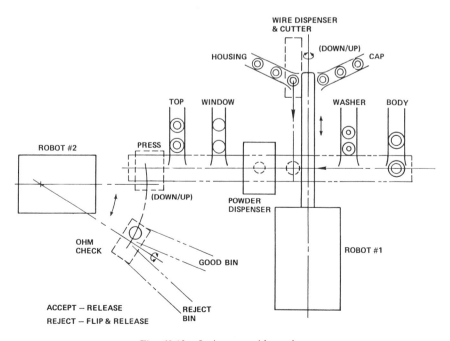

Fig. 62.13. Ignitor assembly station.

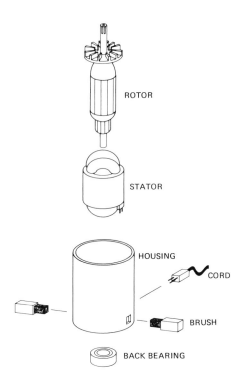

Fig. 62.14. "Body" subassembly.

TABLE 62.7. RESOURCE SUMMARY FOR ADES RUNS ON APPLIANCE BODY

Resource	Number of Tasks	Station Time	Tool Change	Cost
Small pick-and-place (SP&P) robot	1	Task dependent	No	$10,000
Multiple adjustable stop robot (MASR)	4	Task dependent	No	15,000
PUMA	Any	Task dependent	3 sec	45,000
Special press tools	1		No	8,000
Dedicated station	1	1.0 sec	No	30,000
Dedicated station	1	1.5 sec	No	45,000
Dedicated station	1	2.0 sec or more	No	50,000

Body Operations	Operation Time for Each Operation by Each Resource (sec)			
	SP&P	MASR	PUMA	Dedicated Machine
B_1-place back brg.	2.0	2.0	2.5	1.0
B_2-place housing, press	3.0	3.0[a]	2.5[a]	1.5
B_3-place, press stator	4.0	4.0[a]	3.0[a]	1.5
B_4-place rotor	3.0	3.0	2.5	1.0
B_5(1)-place, press brush	2.5	2.5	3.0	1.0
B_6(2)-place, press brush	2.5	2.5	3.0	1.0
B_7-cord	3.0	3.0	3.0	2.0
B_8-remove (unstable)	3.0[b]	3.0[b]	2.5[b]	1.0

[a] Special press tool required for completion of interference fits.

[b] Slight opening of nest may be necessary.

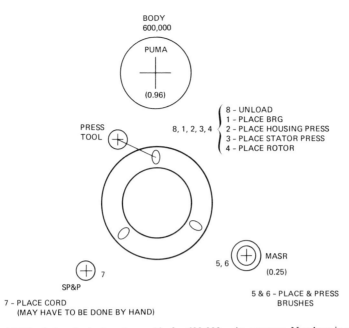

Fig. 62.15. ADES solution for body subassembly for 600,000 units per year. Numbers in parentheses indicate fraction of time the resource is busy.

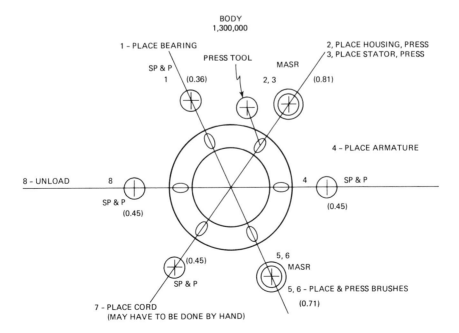

Fig. 62.16. ADES solution for body subassembly for 1,300,000 units per year.

(e.g., 0.01 mm) in a given direction, or the ability to do so at a given update rate. Specifications like these are not quoted for any robot to the author's knowledge, but as CAD/CAM and sensor-based robotics grow in importance, so will the need for specifications of this type.

Sensors are required any time there is a chance of error in parts or process, a need to measure or verify, a need to adapt, or a desire to avoid investment in jigs and fixtures that predetermine the robot's actions. Different sensors are strong and weak in various ways. Touch sensors have little look ahead, but vision sensors cannot see what is causing two parts to jam. An operation that requires keeping the robot's wrist parallel to a surface should not depend on a vision sensor looking perpendicular to that surface. A robot intended to track an undulating and slightly rough contour should not be attached rigidly to its contour follower because the control loop for tracking would need excessive gain and bandwidth to follow the roughness. A compliant wrist would be a better interface. Compliance is also good for absorbing small errors in cases where the robot simply need not follow exactly, or where it just does not have the commandable resolution to do so. The point is that sensors, like other items at the workstation, must be chosen to meet the requirements. Often proper understanding of the task, plus strategic use of compliance, can reduce overall system complexity, lower cost, and improve reliability.

62.5. SUMMARY

This chapter has summarized CSDL experience in applying robots and other technology to assembly. The research approach and results were presented, including the dual theme of economic and technological analyses. Several case studies were described, some designed by the bottom-up method, some using the mathematical programming tool ADES. It was shown that economic assembly systems contain a mix of technology and people and that there is a continuum of system types ranging in versatility as a function of production volume. The utility of simple robots was argued by example, and a mythical robot of intermediate cost and versatility was shown to be worthy of further consideration. Finally, workstation design issues were presented from the point of view that thorough technical analysis must precede equipment selection.

ACKNOWLEDGMENT

The work reported was carried out by a large group of CSDL staff members, MIT and Boston University students, and an MIT professor over several years. It was supported by the National Science Foundation under grant DAR-77-18530 and by several industrial clients.

REFERENCES

1. Whitney, D. E., Quasi-Static Assembly of Compliantly Supported Rigid Parts, *Transactions of ASME Journal of Dynamic System Measurement and Control,* Vol. 104, No. 1, 1982, pp. 65–77.

2. Part Mating Theory for Compliant Parts, CSDL-R-1407, CSDL, Cambridge, Massachusetts, January 1981.

3. Selvage, C. C., Assembly of Interference Fits by Impacts and Constant Force Methods, SM thesis, MIT, Cambridge, Massachusetts, 1979.

4. Blaer, I. L., Reliable Automatic Starting of Threaded Parts, *Russian Engineering Journal,* Vol. 42, No. 12, 1962.

5. Whitney, D. E., Applying Stochastic Control Theory to Robot Sensing, Teaching, and Long Term Control, *Proceedings of the American Control Conference,* 1982.

6. Dubowsky, S. and DesForges, D. T., Application of Model Referenced Adaptive Control to Robotic Manipulators, *Transactions of ASME Journal of Dynamic System Measurement and Control,* 1979.

7. Wheelwright, S. C., Japan—Where Operations Really Are Strategic, *Harvard Business Review,* July–August 1981.

8. Hayes, R. H., Why Japanese Factories Work, *Harvard Business Review,* July–August 1981.

9. Boothroyd, G. and Redford, A. H., *Mechanized Assembly,* McGraw-Hill, New York, 1968.

10. Gustavson, R. E., Engineering Economics Applied to Investments in Automation, *Proceedings of the 2nd International Conference on Assembly Automation,* 1981.

11. Graves, S. C., and Whitney, D. E., A Mathematical Programming Procedure for Equipment Selection and System Evaluation in Programmable Assembly, *Proceedings of the IEEE Decision and Control Conference,* 1979.

12. Graves, S. C. and Lamar, B. W., A Mathematical Programming Procedure for Manufacturing System Design and Evaluation, *Proceedings of the IEEE Conference on Circuits and Computers,* 1980.

13. Seltzer, D. S., Robot Technology, in *Proceedings of the 2nd Annual Seminar on Advanced Assembly Research,* CSDL-P-1447, CSDL, Cambridge, Massachusetts, 1981.

14. Applied Research in Industrial Modular Assembly, CSDL-R-1111, CSDL, Cambridge, Massachusetts, 1978.

15. Design and Control of Adaptable, Programmable Assembly, CSDL-R-1284, CSDL, Cambridge, Massachusetts, 1980.

16. Stauffer, R. N., IBM Advances Robotic Assembly, *Robotics Today,* Vol. 4, No. 5, October 1982, pp. 16–18.

17. Baumann, W., Birner, R., Haeusler, J., and Hartmann, R. P., Operating and Idle Times for Cyclic Multi-Machine Servicing, *Industrial Robot,* Vol. 8, No. 1, March 1981, pp. 44–49.

18. Janjua, M. S., Selection of a Manufacturing Process for Robots, *Industrial Robot,* Vol. 9, No. 2, June 1982, pp. 97–101.

19. Nof, S. Y. and Lechtman, H., Robot Time and Motion System Provides Means of Evaluating Alternate Robot Work Methods, *Industrial Engineering,* April 1982, pp. 38–48.

CHAPTER 63

PLANNING ROBOT APPLICATIONS IN ASSEMBLY

TIBOR CSAKVARY

Cyber Technology, Inc.
Pittsburgh, Pennsylvania

63.1. INTRODUCTION

The assembly process in its definition is very simple. Assembly is the joining and mating of an aggregate of parts. But in execution, it is one of the most complex procedures in industry. The reason is that the word *assembly* is really a collective description of 12 to 17 different types of operation depending on who does the description. For a human operator, provided with the minimum of hand tools, none of these operations is above their capability. They have logic, senses, and two hands with incredible dexterity. By applying these to the tasks at hand, they have assembled products since the beginning of time.

Product assembly technique changed very little from early history to the beginning of the twentieth century. During this period, the many different types of products, from carriages to the first practical automobiles, were assembled by human operators, one product at a time.

The introduction of the assembly-line concept brought a major technological change in the product assembly technique. Although the intelligence and most of the power for the accomplishment of the assembly tasks are still supplied by the human operator, by pacing the assembly on the line and adding mechanical assists, the efficiency of the assembly is greatly increased. With the automation of the assembly operation, the economics of the new technology become even more attractive. Advantageous as it seems today, approximately 70 years after the invention of the assembly line, the application of the assembly-line technique reaches less than 25% of all industrial products assembled in the United States.

If one examines the reasons for the slow adoption of automatic assembly systems, it can be seen that although the incentive is there (the direct labor cost in all manufacturing is highest in assembly and inspection), the overriding factor is that in most cases industrial products are assembled in small batches in many different styles. Economic considerations dictate that automatic assembly lines are built to be dedicated to specific products, assembled in high volume. To design and build dedicated automatic assembly lines for each different product style is economically unacceptable. Dedicated assembly lines could be built to handle a number of different product styles, but the long and frequent changeover time and the cost of additional equipment (parts presenters, pallets, dies, fixtures, tools, etc.) make this concept inefficient and uneconomical. Therefore, currently, the only feasible batch assembly method for industrial product is manual assembly.

Clearly, a new assembly technology is needed for economic and efficient assembly of products manufactured in batches and in many different styles. This new technology is the programmable or flexible automation of the assembly tasks.

63.2. THE PROGRAMMABLE ASSEMBLY SYSTEM

In its concept, programmable assembly combines the most attractive characteristics of the manual and dedicated automatic assembly technique in one system. On one hand, it imitates the dexterity and decision-making capabilities of a human assembly operator; on the other, it approaches the efficiency of a dedicated automation assembly system.

This chapter was written when the author was a Fellow Engineer in the Industry Automation Division, Westinghouse Corporation.

In its actual hardware configuration, a programmable assembly system is a number of assembly stations or assembly centers in a stand-alone or interconnected configuration. They could be connected to each other or to other equipment by a buffered transfer mechanism. Within the assembly station, equipment such as robots, dedicated equipment, programmable feeders, magazines, fixtures, and vision or other sensors are arranged as required by the operations assigned to the station. A suitable computer control system completes the general configuration of the programmable assembly system.

In operation, first the desired program is loaded into the systems computer. The product parts are then loaded in the feeders and magazines of the system. After the system is checked, the computer is initiated, and the operation commences. When the required number of products are assembled, the cycle is repeated for the next product style. In case of a jam or breakdown, the computer, in conjunction with the operator, diagnoses the location and the nature of the problem so that proper action can be taken. The computer also generates the required production data and reports.

63.3. PLANNING PROCEDURE

From the foregoing description, it can be seen that a programmable automatic assembly system is highly flexible and capable of assembling, in batches, a range of product styles within a product family. But its successful development, installation, and operation will depend to a large extent on the procedure used to plan and develop the system. This begins at the point of initiating the project through operational hardware.

Currently, a number of planning procedures are in use. Some are based on highly regimented analytical methods, others are more empirical in nature, still others are little more than "flying by the seat of the pants" method. In a way, all of these procedures are useful, mainly because the user is forced to follow a more or less systematic path toward accomplishing the desired goal. Unfortunately, the success of the applied planning procedure can be judged only at the end of the project when it is evaluated for the results it has accomplished. The planning procedure described here is based on experimental data collected while successfully planning a number of programmable assembly systems.

The following steps are involved in the planning procedure:

1. Project initiation and/or approval.
2. Application survey for candidate products.
3. Preliminary screening of candidate products.
4. Set priority for the candidate products.
5. Selection of the project team.
6. Data collection and documentation of the candidate product.
7. Conceptual system configuration, parallel with initial economic evaluation.
8. Final system configuration and economic evaluation.
9. Detailed design of the selected system configuration.
10. Final design review.

Depending mostly on the availability of resources (i.e., will the project be performed in-house or contracted to an outside supplier?), the sequence of these steps can change somewhat, but the content remains the same.

63.3.1. Project Initiation and/or Approval

In most cases, two sources are available for project initiation. It could originate from upper management or be requested by manufacturing. In both cases, the basic driving force behind the initiation is either economics or social conditions. Marketing pressures brought on by competition, expected or existing production increase, quality improvement, and productivity improvement are some of the major economic factors. Work environment, better use of capabilities, and stress reduction are some of the social factors. In many instances these factors interact in such a way that the establishment of one brings forth another.

Whatever is the initial source of the project, the absolute primary requirement is total support from all persons affecting the outcome of the project. This is a natural requirement for any interactive endeavor in an industrial environment, but for a programmable assembly project it is mandatory. This is particularly important because:

The technology is new.
The available relevant data are limited.
The project requires a system approach with a number of different disciplines involved.

In-house expertise in the field is limited to theoretical or new knowledge.

The relative effect and relations of the variables influencing the projects are not clearly understood.

With proper support and involvement these difficulties can be resolved.

If the project was initiated by upper management, a reasonably fast start is already assured mainly because one elementary phase, the overall project approval, is accomplished. This was not necessarily so when the project was initiated from manufacturing. Anyone who has gone through this type of procedure knows the obstacles involved in collecting signatures for approval. However, proper presentation of clearly defined objectives and supporting data, fashioned in an easily understandable format, will do wonders toward acquiring approval. After all, one of the key functions of management is to evaluate and determine the value of a proposal and approve those that benefit the company the most.

63.3.2. Application Survey for Candidate Products

Programmable assembly is a new, just maturing, technology. For a person with limited or no experience in the field it can be a disheartening experience to try to select and evaluate a suitable product for programmable assembly for application. At the same time, by keeping an open mind it can be a real challenge.

An application survey starts with the preparation of a simple form which contains only a few easily attainable data. It is important to keep the data collection to a minimum because at this point the intent is to consider the widest possible range of products so no possible candidate will be left behind. If there is a rule for product candidate selection at this stage, it is that "if doubtful, list it." The following data are required from each product considered for programmable assembly:

Name of product.

Yearly volume assembled (including number of rejects).

Number of product styles within a product family.

Assembly time per product (actual, including efficiency factor, weighted average if more than 10% assembly time difference exists from style to style).

Table 63.1 lists data to be collected during the preliminary application survey. Although the list shows more information than noted, at present only the data so noted are required. However, if the other data are readily available, it is useful to record it for future use.

When recording assembly time, it is important to record the actual time practiced on the shop floor. Experience shows that frequently the assembly time values documented in manufacturing instructions or operator instruction are quite remote from actual time values. The actual values could be higher or lower than the documented values, but if the time values are not comparable later on, when calculating economic justification, unpleasant surprises could arise. The same is true for the

**TABLE 63.1. APPLICATION SURVEY
DATA SHEET**

Date: Area or Shop No:
Name of product[a]
Volume of product assembled yearly[a]
Number of product styles[a]
Assembly time per product style[a]
Volume in each style
Number of assembly changeovers per style per month
Number of shifts assembled
Number of operators assembling the product
Number of parts in the assembly
Major product dimensions W: L: H:
Weight of the heaviest part in the assembly
Presence of compliant parts (wires, tubes, etc.) in the assembly
Sketch of the product

[a] *Note:* pertinent information.

volume. If the number of products assembled does not include the rejects and reworks, difficulties could develop when calculating equipment capacity.

63.3.2. Preliminary Screening of Candidate Products

Preliminary screening of candidate products for application survey, no matter how limited it is in scope, costs time and money. Therefore anyone who conducts a survey must do it with the best efficiency. This means to uncover the greatest possible number of products that could be candidates for automatic assembly within the shortest period of time.

The screening and selection of product for assembly on a dedicated automatic assembly system is a well-developed quantitative procedure. The single overriding factor affecting the final decision in this procedure is the annual production volume. Many unfavorable product characteristics can be resolved if the required production volume is there. Product selection for programmable automatic assembly is more complex for the following reasons:

The selection must be made among three (manual, dedicated, or programmable) assembly methods.

No key factor (such as volume) is available that dominates, and therefore simplifies, the selection process.

A large number of variables influence the selection.

Uncertainty exists in the determination of variables influencing the selection.

Considering that even a relatively small manufacturing plant could yield a half-dozen or more product candidates for automatic assembly, and that each would take many hours to evaluate, a detailed evaluation of each product would be quite expensive. Since this is a common problem, a simplified procedure has been developed which requires only the data mentioned above and, when applied to a specific product, can answer the following questions:

Is the product suitable for the automatic assembly method?

Should the assembly method be dedicated or programmable?

What is the economic advantage of one method over the other in percents for a specific product?

Knowing the answers to these questions for each specific candidate product, one can quickly screen the products and concentrate on those that look the most promising. The suggested product screening and selection procedure is based on two charts. The first one is shown in Figure 63.1. On the chart, the assembly time in hours is placed against the product volume assembled.

Inside the diagram, the area is separated by a boundary line to two parts. The area inside the chart indicates products that are more advantageous to assemble using manual assembly techniques. The area outside the boundary line indicates products that are candidates for automatic assembly methods. Since there is some uncertainty* in the position of the boundary line, if the plotted point lies close to the manual assembly side of the boundary, the product is still, and should be, a candidate for automatic assembly.

The second chart, shown in Figure 63.2, is somewhat more complex. To develop the chart, trend analysis was used, based on a large number of variables affecting manual, dedicated, and programmable assembly. The chart shows an equilateral triangle with logarithmic scales on all three sides. The base of the triangle is the volume line, representing annual production volume. The scale on the right side of the triangle represents the number of product styles in a family of products. The left side of the triangle shows three logarithmic scales. The first scale, forming the third side of the triangle, represents the relative cost of the manual assembly (MA) for a part. The second scale, parallel with the first one, is the same for dedicated assembly (DA). The third scale is for programmable assembly (PA).

From the chart, it can be noted that the position of the three lines on the left side of the triangles (representing the cost of assembly per part assembled) are shifted relative to each other. This shift is virtual, the lengths of the lines were purposely and arbitrarily cut off where the advantage of one method over the other is obvious. It should be noted that the manual assembly method cost scale on this chart has no significance from the product selection point of view; it is there only for reference.

---

* The uncertainty arises from the fact that labor and overhead rates are different at different plant locations. The reason for showing relative cost of assembly in percentage rather than specific cost is, as pointed out previously, that labor rates are different for different locations and applications; therefore, scores of charts would be needed to cover a range of rates. If it is of interest, the relative cost can be easily converted to specific cost just by taking the existing manual assembly cost per part assembled and inserting it on the manual assembly cost scale in the position indicated by the annual volume and the number of product styles. The specific cost values on the three parallel scales can then be factored according to the inserted value.

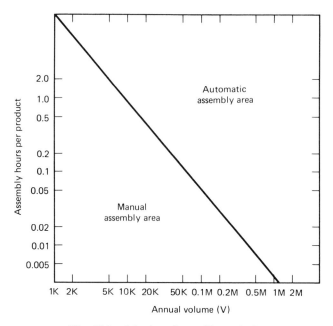

Fig. 63.1. Selection of assembly method.

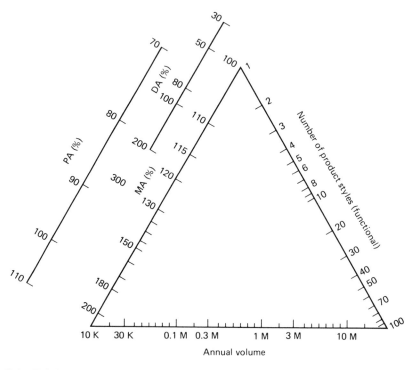

Fig. 63.2. Relative labor cost per part assembled. MA: manual assembly; DA: dedicated assembly; PA: programmable assembly.

The use of the product screening procedure is simple. Once the data listed are available, first determine that the product is a candidate for automatic assembly. This can be done by plotting the annual product volume assembled against the assembly time per product assembled on the chart in Figure 63.1. If the point on the chart lies clearly in the manual assembly zone, the procedure is over, and the product assembly should stay with manual assembly technique. However, if the point lies close to either side of the chart or in the automatic assembly range, it is a candidate for automatic assembly. Exceptions to this are products that are considered for automation for other than economic reasons.

The next step is to find out what type of automatic assembly method is most advantageous to use for assembling the product. First, plot the annual volume on the baseline of the triangular chart in Figure 63.2 and then draw a perpendicular line at this point to the baseline. Next, plot the number of product styles on the right side of the triangle and draw a perpendicular line to this line (to the right side of the triangle). The two lines, one perpendicular to the base, the other perpendicular to the right side of the triangle, intersect each other at some point. This point could be inside or outside of the triangular chart. Find this point and from it draw a perpendicular line to the left side of the triangle.

The line will mark a point on the scales representing the relative cost of assembly per part assembled. Where one assembly method dominates strongly over the other, only two or possibly one scale will be marked. Since, in the first step of the screening procedure, it was determined that the product in question was a candidate for automatic assembly, the scale representing the cost of manual assembly is meaningless here. It is there for two purposes: to provide a conversion scale from relative to specific cost, and to provide a reference line so that, once the selection between the dedicated and programmable assembly method is made, the economics of the selected method can be compared to the present manual assembly method.

63.3.4. Set Priority for the Candidate Product

Once the screening procedure is completed and the results point to one or more candidate products for automatic assembly, priority must be established among the products. Very few manufacturing plants could afford to run a number of automation projects simultaneously, either because of economic restraints or available, suitable manpower. The following factors could affect priority:

Economics

Technical

Environmental

Social

Political

In most cases the decisive factor is economics. It is easy to see why if one keeps in mind the fact that the cost of an automatic assembly station is generally in the range of $25,000 to $40,000 per part assembled at the station. This does not include the cost of the controller, the robot, or a dedicated automatic work head. Also, because of economics, the product that is a candidate for dedicated assembly will have first priority for automatic assembly. But, products for dedicated automatic assembly seldom come through the selection procedure for flexible assembly, mainly because of the high volume usually involved. As such, this type of product is normally destined for automatic assembly at the product conception, design, or pilot production stage. The usual exception is where the volume of the product, during a period of time, increases slowly to the level where it becomes feasible for automation. Therefore the candidate product that promises the shortest payback period is usually the one that will be assigned highest priority.

For technical consideration, the product characteristics should be examined. These characteristics should conform to the following limitations:

1. The major dimensions of the largest parts in the assembly should fit in a 250 mm × 500 mm × 800 mm envelope. Exceptions would be components such as large panels where hardware is assembled at a small area on the panel. Components larger than the specified dimensions would be difficult to fit in most assembly equipment work envelopes.

2. The heaviest part in the assembly should be less than 5 kg. Again, most assembly equipment could handle up to 10 kg, but usually some additional tooling is involved to handle the part, thereby increasing the total weight close to the capacity of the equipment.

3. Flexible wire, tubes, or soft fabric parts should be absent from the assembly. These types of components are not only difficult to handle, but also are even more difficult to locate. If they are present, the product should not be rejected, but it should be determined if they could be replaced with stiff or preformed components, eliminated through redesign, or assembled manually outside of the automatic assembly system.

4. The total number of parts in the candidate product should be less than 20–25 pieces. If the number of components is higher, the product should be broken down to subassemblies and handled as an individual product. Otherwise the automatic assembly system designed and built for a complex product will be prone to excessive breakdown. For a programmable assembly system where different styles of the same family of products are assembled, keeping the number of parts down to a manageable number is even more important, since each part of the product could have a number of variations. Because of this, even a relatively simple product, consisting of only 8 or 10 parts, could require a complex system able to handle dozens of components.

Environmental conditions will also affect product priority. In fact, undesirable environmental conditions have opened up new opportunities for programmable assembly and robotics technology applications. Present-day components of robotics technology can be safely packaged, installed, and reliably operated for extended periods of time in a polluted environment where human operators would not desire or be able to work. High temperature, excessive noise level, contaminated atmospheric conditions, and radiation are the main factors to be considered. Keeping workers away from these environments is not only consistent with government regulations, but also eliminates low-productivity areas which are invariably the result of poor environmental conditions.

Socially undesirable conditions could also affect priority. If a job is undesirable because it is boring, heavy, dangerous, repetitive, and places unreasonably high stress on the operators, it is a prime candidate for automation. These types of jobs are easy to recognize. It is usually characterized by high absenteeism, high employee turnover, and difficulty in filling the position. In a way, these undesirable factors place limits on human tolerance and, together with the increased desire for the improvements in the quality of life, act as an additional driving force toward the use of robotics technology and automation.

Finally, company politics and policy could also affect product selection. It is not uncommon that higher management decides that a time is just right for the application of new technology and within a specified time robotics or other high technology must be applied to the factory floor. The merit of this method is debatable, but if the good intention is channeled in the right direction, it could have a favorable affect on the progress and profitability of the company.

At this point of the project, a decision is required from the upper level of company management. This is to introduce the plans for automation to the locally affected trade union and/or to the operators working on the assembly of the product. Cooperative effort is required to work out problems affecting operation personnel. A direct and honest approach will go a long way to eliminate concerns and channel potentially limiting forces into creativity. Personnel affected by the project will find out about the plan through the grapevine anyway, and probably in a distorted form. It is a well-known fact that few other factors can kill a project or make it unmaintainable faster than resentment from the persons involved in making it work.

63.3.5. Select The Project Team

Whatever determining factor is used to decide which product will be the first one to be assembled on a programmable automatic assembly system, every effort should be made to justify objectively the selection of the product. The product screening method described previously is a good tool to uncover promising candidate products with a minimum of time and effort; it is not and cannot be a substitute for a thorough evaluation procedure. At this point, application engineering and detailed planning should take over.

Planning for programmable assembly, up to this point, could be an in-house effort even for small manufacturing plants; but now an important question must be answered: "Is there enough in-house expertise for a thorough application engineering evaluation to develop the necessary comprehensive plans, or must it be sought from an outside firm specialized in the field?" Since this type of practical and proven expertise is presently a rare commodity, the answer will be independent from the physical size of the plant or the quality of the people working there. Often, the right action is to engage an outside firm for the task.

A logical way to develop your own in-house expertise is to assign a person or a small team to work closely together with the experts engaged to implement the project. This also assures an open and continuous line of communication between the supplier and the customer, or between any two groups working on the venture—a primary requirement for successful implementation of the project.

63.3.6. Data Collection and Documentation of the Selected Product

Product documentation is the next step in the implementation of the project. This is the process where all the pertinent information related to the product is collected. A list of this information is long, the time it takes to collect it is longer.

Also, since this phase of the work is usually the least interesting and the most bothersome (a number of busy people must be disturbed in their daily routines to provide the data) the temptation is great to reduce the scope of the work to the minimum. The documentation will be the foundation of the project, and the soundness of the foundation will affect the final result.

The following list gives the data and documents required to clearly understand, justify, and implement a programmable automatic assembly project:

Monthly and yearly production volume. These data were collected previously, but at this point it would be a good time to check their accuracy.

Number of styles in the product family. (Same as above.)

Engineering product drawings and specifications. Drawings should be available on the completely assembled product, on all the different styles, and on all the product parts. Material and product operation specifications should also be collected. Information derived from the drawings and specifications such as tolerances, material characteristics, product-part surface finishes and conditions, and so on, will have a major affect on the final configuration and cost of the system.

Engineering drawings and specifications on the equipment, tooling, and fixturing used at present to assemble the product. Periodically it turns out that currently used equipment and/or tooling and fixturing can be incorporated into the new system as is, or with minor redesign. If drawings and specifications are available, this could save valuable engineering time.

Task or operation instructions on the present assembly sequence and method. When it is not available or not updated to present date, a commentated videotape of the operation is a good substitute.

Operation time values from individual stations or operator and total cycle time. Normally two types of time values are available: standard time values, based on measurements, tables, and calculations for each station and for the total line; actual time values, based on the operators' and operation efficiency. The additional time value required is the setup time for each currently existing station involved in the assembly of the product.

Average batch size assembled for each product style and the number of batches run per month.

Part defects. These should be listed as to the type of defects and the quality. These are very important data to collect. They will affect the cycle time and the justification of the new automated system. If the data are not available, the next best thing is to ask the line supervisor to collect all the rejected parts, subassemblies, and final assemblies for four or five days in a row, determine the reasons for rejects, then tabulate the results.

Testing and inspection specification during and after assembly. Again, two types of data are needed: quality inspection and test; functional inspection and test. The area of testing and inspection is one of the most important areas where programmable assembly systems can contribute heavily to the quality improvement of a product. The sensory and data-handling capabilities built into the system can provide the base for the desired checkpoints and inspections, and selective assembly (matching parts according to dimensions within the range of the tolerance) will improve the quality of the product if the requirements are clearly documented at the beginning of the project.

Number of persons involved in the assembly of the product such as assembly operators, testing and inspection operators (during assembly), and material-handling operators.

Present material-handling and feeding methods. The following data are needed on the status of the orientation of the parts delivered to the assembly area: types of containers; degree of orientation (zero-, one-, two-, or three-axis orientation); type of feeding devices (bins, cardboard containers, etc.).

Incoming parts inspection methods.

Shop floor area layout and dimensions for the present assembly area.

Market forecast for the product over the next three years. Since the usual payback requirement for a programmable assembly system is three years (for a complex system it could be up to five years), the projected sales volume should be forecast with reasonable accuracy.

Warranty cost for at least one year.

Possibility of design changes in the product. Although at this point it is difficult to predict the type of design change desired, if any, a good indication can be gained from the operator assembling the products. Chamfers, asymmetry, tolerances, material, and function of the parts are some of the factors that manual-assembly operators could point out as candidates for design changes important for successful automatic assembly. In fact, the ideal situation would be to conduct a "Design for Assembly" evaluation on the product prior to implementing the programmable assembly system. For this an excellent and practical procedure was worked out by Dr. G. Boothroyd at the University of Massachusetts and is highly recommended.[1] In many cases significant cost reductions can be achieved in material and labor costs of an assembly when using the Boothroyd procedure for design evaluation.

Present production planning method, data collection method, and data processing equipment. Frequently it is a requirement to provide an interface for communications between the programmable assembly systems computer and the host computer in the plant. Scheduling in-process inventory control, production status reports, statistical data collection, and other reports could be some of the reasons for the request. Therefore selection of the equipment for the system will be affected by these requirements.

Environment. Need to determine the range of operating temperature at the assembly area; the range of humidity at the assembly area; atmospheric conditions such as vapors, dust, and smoke; vibration and noise levels; type of light and light levels.

The previous list contains the data and documentation normally required to develop and implement a programmable automatic assembly system. The depth and breadth of the data required indicate why it is necessary to have a product-selection method that can be used to prescreen the candidates and save all the time and effort involved in fully documenting a product that is not feasible for automatic assembly.

During the data collection and documentation work, it will turn out that some of the required data are not available. New products or a quite matured one are where most of the difficulties will be encountered when attempting to collect the necessary data. In the first case the data and drawings may not be mature or prepared yet; in the second it could be obsolete, not updated, and/or misplaced. Effort should be made to make a note of it and since this type of data is needed for normal production control, it should be developed as time permits. In the meantime, substitute or supporting data should be collected on a best-effort base. This could be a set of product parts and a completely assembled product and descriptive information collected from the assembly operators.

Although actual parts and assembled products could assist in preliminary system configuration, it should not be used too extensively. The danger in using actual product parts for system configuration is that one or two parts cannot define the total requirements, such as tolerances and orientation, but it is easy to assume that they do. Product parts are a help in visualizing the setup requirements, but when the documentation is available assistance in three-dimensional visualization should be the only use of the parts.

63.3.7. Conceptual System Configuration, Parallel with Economic Evaluation

System configuration is an iterative process. It starts with the evaluation of the factors affecting the product and the assembly tasks, proceeds with the selection of a concept dependent on these factors, and evaluates the performance of the selected concept. Then, based on the results, it returns to the starting point to either discard the system concept or refine the configuration and optimize its performance. The method used for conceptual system configuration could be analytical or empirical. The latter method is used here in a systematic form for two reasons: first, at present analytical methods are still under development, and second, at some points the analytical methods also incorporate empirical steps in the system configuration process. Regardless of the method used, a systematic approach will produce the best end result for a given set of data. The major steps involved in the regimented empirical procedure, used successfully for conceptual system configuration, are discussed in detail in the following sections.

Review Product Documentation

The purpose of the review is to become familiar with the product. Its geometry, function, operation, and peculiar characteristics should be evaluated and thoroughly understood. Familiarity with the product will lead to a clear definition of the problem at hand. Seemingly a simple task, the accurate description of the problem is frequently the most difficult one, leading to misunderstandings and friction between project team members or between customer and supplier. This is the time to ask questions, clarify responsibilities, discuss environmental and safety conditions, request additional data if necessary, and consider the general capabilities and restrictions.

Develop Process Flow Chart

The process flow chart is the graphic representation of the assembly tasks in sequence. It can be considered as the logic diagram for the conceptual system configuration work. The main advantage of developing a process flow chart is that it summarizes and presents all the assembly and inspection tasks in one easily viewable, simple format.

In the chart, the principal direction of the process flow represents the sequential assembly tasks. From the principal flow line, side lines branch out. On the left side, the branches represent tasks that involve components and subassemblies always present in the product. On the right side, the branches represent tasks involving components and subassemblies appearing only in certain product styles in the product family. In the chart, assembly tasks are represented by circles, inspection tasks by squares. The task symbols are interconnected by lines to integrate the flow chart. Arrows on the line point in the direction of the flow.

To each task symbol two numbers are attached. One three-digit number represents the operation number. This number refers to the name of the operation contained in the lists prepared parallel with the process flow chart. The second number, two digits, represents the style number of the product. If the part is common to more than one style, all the style numbers should be marked at the operations concerned with the part. If only one product style is planned to be assembled on the system, only the operation numbers should be present at the task symbol. Figure 63.3 shows the general layout of a flow chart for a family of products with three different product styles.

Developing and refining a process flow chart for programmable automatic assembly is one of the most creative phases of the conceptual system configuration. An inventive approach is in order when starting with the layout of the flow chart. Although each product can be assembled a number of different ways, a logical and efficient process usually surfaces shortly after the initial layout is completed. Attention must be paid to the following when developing a process flow chart:

During manual assembly, the operators pick up the part in random orientation, inspect and orient the part, then assemble and inspect again. All these operations (some of them are not even specified in the operator instructions) must be disclosed and accounted for, then included in the process flow chart.

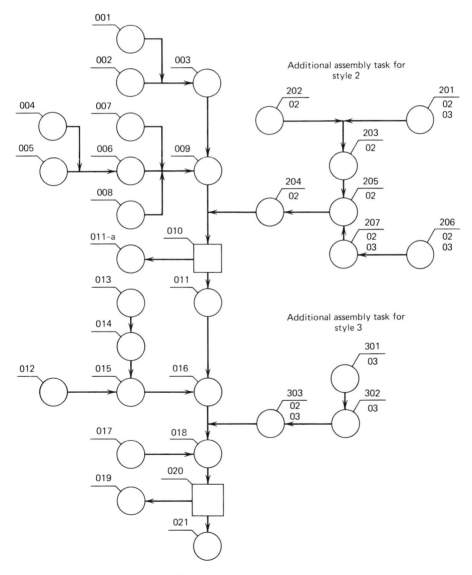

Fig. 63.3. Process flow chart.

Operations, wherever possible, should be combined to reduce cycle time. This is also an excellent opportunity for suggesting product part redesign for ease of assembly and for reducing the number of parts in the assembly by combining the functions of the different parts. Although product design or redesign for assembly is a sure way to reduce product and system cost and improve quality, it should not be counted on to have serious effect on the present system configuration. By the time a suggestion related to product redesign is approved and integrated into the product, the system is usually up and running. Still, it will provide an opportunity for future product cost and quality improvement.

The product should be divided into logical subassemblies. This is most important when the product is complex and contains a large number of parts. The subassemblies then can be treated as individual branches of the process flow chart.

When the process flow chart is completed, it will show all the assembly and inspection tasks, the sequence of the tasks, and indirectly list all the product parts and subassemblies in one comprehensive figure. It will be a solid base for conceptual system configuration. Although minor changes can be expected to occur as the system configuration progresses, experience shows that the overall process flow remains the same.

Develop Cycle Time Chart

In manufacturing, the single most significant factor affecting production is the cycle time. All system design parameters ultimately can be related to this factor. Therefore it is desirable to define operation, or in this case assembly cycle time, as early in the system configuration work as possible. Naturally, the data defined in the early phase of the work will not have the reliability required for the final system design, but they will be accurate enough to permit the development of concepts suitable to start the configuration procedure.

As was the case in the process flow development, graphic representation of the assembly operation time value and the total cycle time provides a very handy tool to assist in system configuration. A cycle time chart is easy to work with since it is similar to the popular Gantt chart. The one drawback of using this chart is that it is cumbersome to update if one time value is changed. This problem is easily eliminated by using a personal computer to operate this chart. But even without a computer, the advantages in the use of the chart far outweigh the drawbacks.

In starting to develop the cycle time chart, the first question asked is how to determine the operation time values when the equipment performing the operation is unknown. Two sources are available. The one is the existing or standard time values and the other is estimated practiced time values. Since the operations are already defined in the flow chart diagram, standard time values can be assigned to the following group of operations:

Drive-threaded components
Rivet
Solder, weld
Deposit
Form (cut, punch, crimp, bend)
Press
Inject
Gage

If no standard time value is available, a 6.0-sec time value for most cases is in the right range. Estimated time value of 4.0 sec can be assigned for the following groups of operation:

Orient
Slide
Twist
Insert
Pull
Remove

In both groups of operation time values, a pick-and-place task is included. One more operation must be accounted for, and this is a transfer. Either a robot arm or a transfer mechanism could do this task. Although the time value depends on the distance of the transfer and the type of equipment, an initial value of 2.5 sec is useful. These are all the operations involved in assembly. At first glance a task might not seem to fit in either group, but closer examination will reveal that it does, and then the appropriate time value can be selected. Figure 63.4 shows the general format of the cycle time chart. Note that some tasks can be performed parallel with the assembly operation, thereby significantly reducing the overall cycle time.

Once all the cycle time charts are completed for the different product styles and summarized, the first approach on the assembly cycle time will result in a value. The accuracy of the value is not known, but since the foregoing figures are on the conservative side, if anything, the determined cycle time value will be on the safe side. Using the previously determined cycle time value and additional data collected during the product documentation phase, the following calculations can be made:

The available time per year per shift is: 8 hr $\times$ [(365 days) $-$ (2 $\times$ 52 weekend days $+$ 10 vacation days)] $=$ 2008 hr/year.

Using 90% figures for factory efficiency: 2008 $\times$ 0.9 $=$ 1807 hr/year/shift.

The required time per year per shift is: Yearly product $\times$ cycle time $-$ A hr/year

Using 90% figure for the assembly system uptime

A hr/year $\times$ 0.9 $=$ required hours/year/shift.

Now, comparing the value of the available hours to the required hours, the following conclusions can be made:

1. If the required time is equal to or less than the available time, a one-shift operation with one programmable assembly system will produce the scheduled production volume. If the required time is significantly more (20% or more) than the available time, two- or three-shift operations are required. Normally, if the difference is less than 20%, clever system configuration can increase the capacity of the system to the point where a one-shift operation is sufficient.

2. If the required volume cannot be produced even with a three-shift operation, double tooling or a multiple system will be the answer.

Further information can be gained on the required system by considering the number of parts in the product. The efficiency of an assembly system is significantly affected by the number of operations it performs. The higher the number of operations, the lower the efficiency, simply because there is more chance for jamming or breakdown. Practical experience dictates that the number of parts assembled

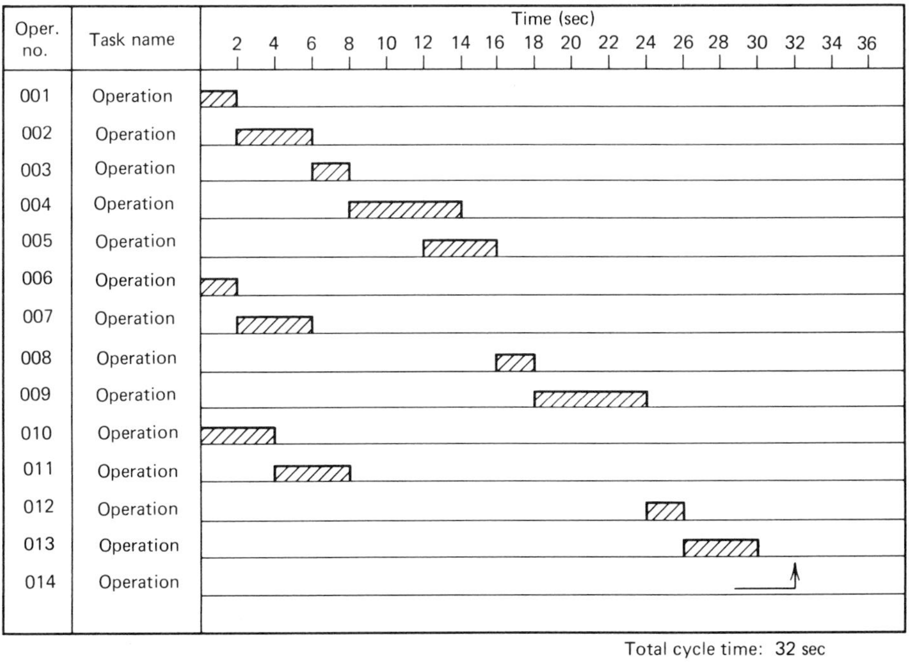

Total cycle time: 32 sec

Fig. 63.4. Cycle time chart.

at one assembly station or with one assembly arm (with some exception such as PC board assembly) should be kept between two and five. Not only the reliability of the system suffers when exceeding this number, but the available space around the equipment will be cluttered and jammed to the point where maintenance will be very difficult. Knowing the total number of parts in the product and assuming an average of four parts assembled per assembly station, the number of stations required will be one-fourth of the total number of parts in the product. At the final system configuration the number of stations could change somewhat because of load balancing.

Determine System Requirement

For a programmable assembly system two basic concepts can be considered. One is the assembly center and the other is the assembly-line concept. Although sometimes it is difficult to draw a sharp dividing line between the two concepts, each has its own individual characteristics.

Programmable assembly centers are concentrated groups of equipment dedicated to complete all operations at one geographic location. Although a number of assembly stations could exist within the assembly center, the product, during the process of assembly, will see only a minimum number of transfers, or none. In its concept, the assembly center consists of one to four robot arms, assembly fixtures, feeders, and tooling. A computer and a control terminal complete the system. In operation, one or two robot arms complete the assembly of the product while, if needed, other arms simultaneously preassemble the next product and/or prepare subassemblies. Figure 63.5 shows a typical conceptual layout of an assembly center.

Programmable assembly lines are, in layout, similar to dedicated or fixed assembly lines. The product is transferred from station to station, and progressively assembled until it is complete. In its concept the programmable assembly line consists of a number of assembly stations interconnected with a buffered

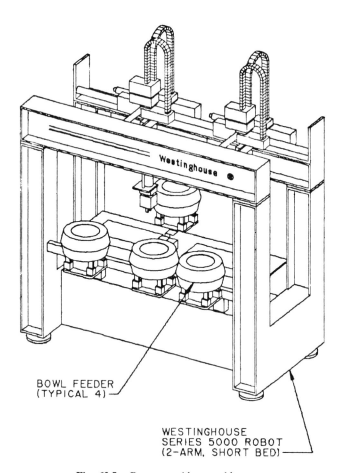

BOWL FEEDER
(TYPICAL 4)

WESTINGHOUSE
SERIES 5000 ROBOT
(2-ARM, SHORT BED)

Fig. 63.5. Programmable assembly center.

transfer system. A multilevel computer system controls the assembly line. The individual stations can be programmable, dedicated, or manual assembly stations.

As the programmable assembly technology advances, the number of the dedicated and manual assembly stations will diminish, but at present they still play a significant role in assembly automation. A well-conceived programmable assembly line provides for technology advance by adding extra stations to the line at the points where, owing to the complexity of the operation, manual assembly is used. The geometrical layout of the system could be closed-loop branched or in-line; the choice depends on local conditions or preference. Figures 63.6 and 63.7 show two conceptual layouts for a programmable assembly line. Figure 63.6 is a closed loop, Figure 63.7 is an in-line arrangement.

Selecting between the two system configurations, that is, assembly center or assembly line, is empirical, although a few basic rules help in the selection. In most cases the choice will fall on the assembly center concept. There is a definite trend toward this concept in the manufacturing industry for the following reasons: multiarm robots used at assembly centers are very efficient. The assembly center can accomplish complex tasks. The system is compact and takes a relatively small area on the factory floor. The robot's controller is used for most of the control task. Task load balancing is simple. It can assemble products using the selective assembly principle. Transfer time is reduced or eliminated.

A programmable assembly line concept should be selected when manual operations are involved at some point in the assembly cycle or when more than a few product styles are planned to be assembled. Since some assembly operations, such as wire handling and soldering, are still too complex to be reliably executed by robots, it is frequently necessary to incorporate manual assembly stations into an automatic system. This is very easily done in a programmable automatic assembly line but not in an assembly center. Also, assembling a number of different product styles on the same system means that additional product parts must be accommodated within the system's boundary. For this, only a limited area is available within the envelope of an assembly center, but there is much less limitation existing for an assembly line.

Conceptual System Configuration

At this point the following information is available for system configuration:

The type of assembly and inspection task.

The preliminary sequence of the assembly and inspection tasks.

An estimate of the operation time values and total cycle time.

An estimate of the number of shifts required to assemble the desired number of products.

The estimated number of assembly stations.

The number of product styles planned to be assembled on the system.

Environmental conditions.

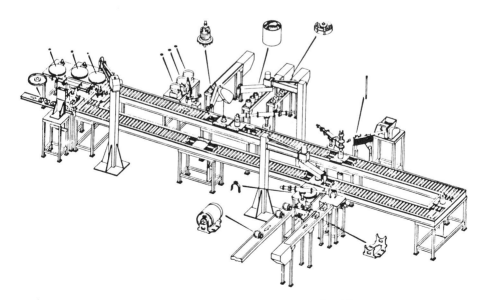

Fig. 63.6. Programmable assembly system, closed loop.

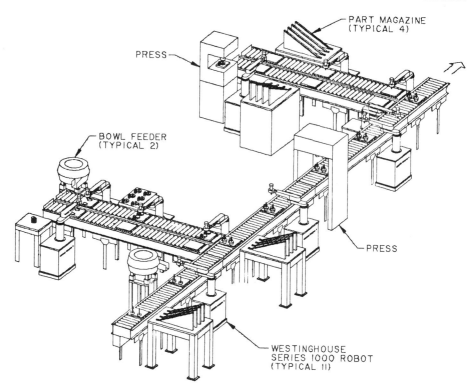

Fig. 63.7. Programmable assembly system, in-line.

Based on this information, the concept for the programmable assembly system can be selected. Once this is done, the next task is to develop a scaled layout of the selected concept. A list of the equipment and its overall dimensions must be determined either for an assembly center or an assembly line. Since each operation requires some type of equipment to perform the operation, the list of the required equipment can be derived from the process flow chart. Dimensions of the equipment can be found from industrial catalogs and from suppliers of the equipment. Once this is available, the outline of the major components of the system is drawn and cut out to form templates. These templates are then arranged, using the process flow chart and the cycle time value charts as a guide, to form the first outline of the conceptual programmable assembly system. If a CAD system is available, the making of the templates can be eliminated, and the layout can be simulated on the CRT. This process is iterative in nature. Once the conceptual configuration of the system is available, it must be checked against the process flow chart to see if all the specified tasks can be accomplished and in the desired sequence. Then, the cycle time chart must be revised to see if the task can be completed within the specified time value.

It is very likely that the first approach will not yield the desired results and that the process will have to be repeated a number of times until the system is optimized. The goal is to complete the assembly and inspection of the product with a minimum of time and investment. The refining of the concept could go on for a long time to achieve this goal, but after the third or fourth iteration, the process arrives at a point where the invested time will not be proportional with the return, and the process should be terminated.

Probably most of the problems in selecting system components for the configuration will arise in connection with the selection of the robot. What simplifies the procedure here is that there are only two kinds of multiarm assembly robots and little more than a half-dozen high-performance single-arm assembly robots to choose from. Work envelope, the type of drive (electric, pneumatic, hydraulic), rigidity of construction, reliability, repeatability, and ease of programming should be the major factors affecting the final choice when specifying a robot for assembly.

Speed, accuracy, degrees of freedom, and so on, although important parameters, depend heavily on the arrangement and type of tooling and fixturing within the system and therefore should be considered in this framework (see Section 63.3.9 in this chapter).

With the selection of the robot or robots, the associated tooling, and the list of equipment previously defined with the aid of the process flow chart, the first estimate on the economics is in order. Two sets of figures are sought here. One is the possible savings expected, the other is the investment required to develop and install the equipment. To estimate the savings, the number of assembly operators saved by the proposed system must be determined.

The savings is the number of operations required for the manual assembly minus the number of operators required by the proposed system. The required number of years for payback and the number of shifts used for manual assembly will modify the savings. When the number of operators saved is multiplied by \$25,000 to \$35,000, the estimated savings can be calculated.

The investment can be estimated by multiplying the number of stations in the assembly system by an average assembly station cost and adding the cost of robot(s) and system control to it. Another way to estimate the investment is to secure cost estimates for the list of equipment determined from the system configuration and add to it the appropriate engineering cost (one to two times the equipment cost). Neither way is very accurate, but it is close enough to determine a budgetary estimate without spending excessive time on collecting cost figures.

By comparing the two sets of estimated figures (savings and investment) a reasonably good picture can be obtained for the feasibility of the concept. If the savings are more than the investment, or the two figures are close to each other, the proposed programmable assembly system is feasible and work can start on refining the concept. If the investment is significantly higher than the savings, the system concept should be reviewed and reconfigured. If the developed new concepts still point to unfavorable economic results, the project still could be justified on environmental, quality, or technical reasons, or discontinued in favor of another product. Rejection of a product is very unlikely at this point, because the preliminary screening at the product selection procedure eliminated most of the unsuitable products.

63.3.8. Finalize System Configuration and Economic Evaluation

Assuming the economics and other factors have justified the feasibility of the developed concept, the rough cutouts and sketches used for concepting must be converted to drawings depicting the final layout, detailed outline, and major dimensions of the system and system components. Up to now, the main effort of the concept development and system configuration was concentrated on organization determination with a minimum amount of time spent on hardware specification. Almost any part of this work can be substituted with available or currently developed analytical methods. To finalize system configuration and lay down the base for system design, the work now must concentrate on technological determination and use the developed organization as a guideline for the design of a cost-effective, reliable, programmable assembly. For this task, no analytical procedure is available; it must come from the creative capabilities of the designer.

The work starts with the review of the operations, the assembled list of equipment required to accomplish these operations, and the concept developed and justified for the assembly system. Starting with a scaled plane-view layout of the robot work envelope or the dedicated workhead, the first station is laid out. Fitted to the outside of the work envelope are the part feeders, transfer mechanisms, other dedicated equipment and storage devices. Inside the work envelope are the work stage, assembly fixture, feeder tracks, grippers, special tooling and fixturing, and test or inspection devices. The type and position of sensory devices are listed, then marked on the layout. When an assembly center or assembly line is being laid out, attention must be paid to provide enough space around the equipment for troubleshooting and maintenance. Any control cabinet or console must be located in a position that, when the operator uses the control, the equipment to be controlled is clearly visible.

If additional stations are required, the layout follows a similar procedure. If the system is an assembly line, buffers should be provided between the stations. Statistical data prove that most breakdowns in an assembly system are caused by part jamming, and that most jams can be cleared within 30–60 sec. Therefore, buffer capacity should be such as to provide enough work for the stations that the jammed station can be stopped, the jam cleared and the station restarted without interfering with the rest of the system.

If at all possible, pallets should not be used in programmable automatic assembly systems as a means of transferring in-process product or as buffer storage devices. Pallets and the associated pallet transfer equipment are expensive, take a large amount of factory floor, and do not contribute anything to the assembly process. For an assembly line, rotary tables, spiral silos, pick-and-place, walking beams, and similar transfer devices can provide buffer storage and the means to transfer the part without losing orientation. For a simple assembly center, the feeders can provide the buffer. For a multiarm assembly center, internal buffers can be designed into the work station.

If manual assembly is required at some point in the operation, two stations should be provided within the system. One station could be a spare for possible future automation, the other is the station for manual assembly. Effort must be made to assure the safety and comfort (noise, light, etc.) of the operator. Free access to and from the system also must be provided so the finished product can be removed and new product components can be brought in.

Once the plane view of the layout is completed, side views should be drawn for the more complex equipment in the system. The side view will assist in visualizing the system and help to discover interferences and other unforeseen problem areas. It also helps in defining the type and dimensions of the major system components.

After completing the system configuration layout, a detailed economic evaluation is in order. Determining the economics of a planned manufacturing system is normally a straightforward procedure. For a programmable assembly system, the evaluation is more complicated: first, because a number of factors additional to the usual ones must be considered, and second, because these factors, although representing real savings, are difficult to quantify. Experience shows that the total quality improvement, shop floor area reduction, yield improvement, warranty cost reduction, in-process inventory cost reduction, and other provide an additional 10% or more savings.

Since the procedure for detailed economic evaluation is well documented, we do not consider it here. But it is believed that a product and a concept that pass through the two previously described screening points and are found feasible will not be difficult to justify economically.

With the completion of the system configuration layout and detailed economical evaluation, the plans should be presented for final review to all persons involved with the project. Within the framework of the basic concept, this is the time for suggesting changes, criticizing decisions, and asking questions. If product redesign changes were considered previously, a final decision must be made at this point. Interfacing to other equipment or production areas also must be finalized. Environmental conditions must be reviewed against the proposed equipment to see if they are capable of operating reliably and safely. During the review, the comments should be noted and saved for future reference. Prior to final approval, a Gantt chart (or bar chart) should be developed showing the scheduling of the major phases and the completion date of the project.

Final approval should signal the end for changes in system configuration. Minor design changes could occur throughout the design and fabrication phase of the programmable assembly system, but after approval, any change in system configuration could represent major cost increases and time delay in completion.

63.3.9. Detailed Design of the Selected System Configuration

The two branches of mechanics, kinetics and statics, provide well-developed tools and methods for designers to develop sound and reliable mechanical structures. And modern control capability based on microprocessors and other LSI (large scale integrated) circuits enable designers to provide the movement and flexibility for these structures to perform the desired tasks. Therefore we examine and discuss the programmable assembly system design procedure, not as a design task, but more in terms of the peculiarities associated with this type of system. To organize the discussion, the system is divided into the following components:

Work stations
Robots
End effectors
Feeders
Transfer mechanisms
Assembly fixtures
Sensors
Final design review

In some instances, this division is somewhat artificial, but it contains all the major components associated with a programmable assembly system in a conveniently usable form.

Work Stations

The function of a work station is to provide a stable, well-defined place for the implementation of the assembly and inspection tasks. Its major components are the station substructure, tooling plate or platform, and locating devices to interface with other equipment. Depending on system configuration, any or all of these components could be missing from the station. For example, in the concept shown in Figure 63.8, only a substructure is required. In the arrangement, a robot, acting as a transfer mechanism, carries the product from station to station. At each station, a component is assembled or inspection is performed. Then, at the end of the cycle, the completed product is disposed of. This type of configuration can be used successfully to replace a pallet carrier transfer mechanism and eliminate the tooling plate or assembly platform.

Most of the complications associated with the work station design for a programmable assembly center are connected to providing utilities to the components of the station. Frequently, space must

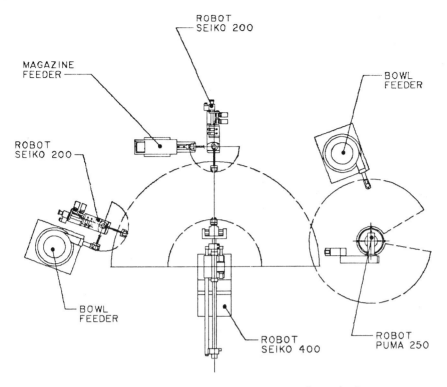

Fig. 63.8. Assembly center with robot as a transfer mechanism.

be provided for literally dozens of electrical signal and power lines, compressed air and other gas lines, hydraulic lines, and vacuum lines. And even more frequently, these lines are laid down and attached to the station randomly because, during the design phase, adequate thought was not given to accommodate them properly. The result can be a type of equipment only a porcupine could love. Careful consideration of positioning utility lines during design not only aides the aesthetics, which reflect the quality of the equipment, but also increases maintainability.

Another problem area in the work station design is the availability of usable space within the work envelope of the station. This is especially common when designing an assembly center. The work station platform or tooling plate must have enough space to accommodate all of the part feeder track ends, orienting devices, assembly fixtures, transfer mechanisms, inspection and test equipment, and sensors necessary for completing the assembly and inspection tasks. A rotary table, installed on the tooling plate, is a practical way to increase the available area. If it is installed correctly, it could serve as a transfer mechanism to interface with other equipment outside the work envelope of the station. As an extra benefit, and frequently the main reason for using a rotary table at assembly centers, the cycle time of the assembly operation can be reduced significantly.

An interchangeable tooling plate is another way to increase the available work space at the work station or assembly center and to increase the efficiency of a programmable assembly system as well. This type of arrangement can be used very advantageously when a number of different subassemblies or a family of products must be assembled, but none of them has enough volume for economic justification of an assembly system. In this case, an individual tooling plate, containing all the fixturing and tooling necessary to perform the desired tasks, could be designed for each subassembly or product style. The tooling plate is then attached to the work station through locating pins and clamps as production schedules require it. When the tooling plate is not in use, it can be stored at a secured location so individual fixtures cannot be removed from the plate or moved out of position. Interchangeable tooling plates must be designed with quick-connect utility connectors.

Another variation of this concept is the multistation assembly line with a transferable robot. Again, when not enough volume is available within one product or a group of subassemblies, an assembly station can be designed for each, but only one robot is used to accomplish all the assembly and inspection tasks. The robot is then moved from station to station to produce assemblies as the production schedule requires it. In this case, sufficient locating and clamping mechanisms must be provided so the robot can be reliably aligned and clamped to the specific work station. Putting the robot on rails facilitates the transfer of the robot to the different assembly stations. Figure 63.9 shows this type of arrangement.

Robots

At present, the use of robots and robotic technology in assembly application, compared to the total number of robots in industrial applications, is small. Although valuable theoretical work has touched almost every facet of robotics in this field, and although simultaneous experimental applications have proven many of the theories, these efforts have not been successful yet in accelerating the use of these technologies. But, as the robots and the associated technology itself are maturing, an increasing number of industrial installations are proving what the theory and experimental applications have indicated, that is, that robots in assembly are cost-effective, reliable, and excellent tools to increase productivity. To a large extent, the success of these and future applications depends on the robot selected for the system and its performance. Selection of the right robot for a specific application is getting to be increasingly difficult, mainly because of the proliferation of the equipment in the field. Additionally, without prior experience in the field, it is difficult for the system designers to select and evaluate the relevant data from the long list presented with each robot. To ease the predicament, a list of the types of primary data needed for the selection of an assembly robot follows:

Work Envelope. This is probably the most important parameter affecting the selection of the robot for a programmable assembly system. It seems as if one can always use a little more space than is available during the design phase of the system. Therefore, it is good practice to select a robot that comfortably accommodates all the necessary equipment required for the assembly tasks.

Repeatability. Since assembly, even in small batches, is a repeat of the same operations, it is important that the tool on the end of the arm arrives back at the intended point within the range of the product part tolerances. But repeatability should not be overemphasized. A tooling aid, such as a remote center compliance (RCC) device, or even the natural compliance built into the robot, permits it to perform significantly better than the published data would indicate. A robot with a mediocre ± 0.10-mm repeatability (full load, full extension, full speed) usually has a capacity of performing the most demanding assembly operations.

Accuracy. Treated similarly as repeatability.

Payload. More than 85% of the parts in a large piece of earth-moving equipment weigh less than 3.0 kg. The ratio is similar to many other types of equipment. Therefore, for most assembly applications, the load-carrying capacity of a robot should be in the 5–10 kg range. If occasionally a single, unusually heavy part is involved in the assembly, a suitable tooling concept should be developed to assist in part handling before considering a robot with larger load-carrying capacity.

Speed. This is probably the most misunderstood parameter in specifying a robot because of the following reason. To optimize the assembly cycle time, the travel distance of the robot arm must be kept to the minimum. This can be done by positioning the part pickup points close to the assembly point, let us say within 500 mm. Assume that the robot arm accelerates/decelerates at 5 m/sec^2 (0.5 g), and the linear speed of the robot arm is 1.0 m/sec. Figure 63.10 shows one cycle of movement for such an arm. From the diagram, it can be seen that the total time for acceleration and deceleration is 1.2 sec, and during this time the arm traveled 600 mm. But the arm needs to travel only 500 mm; therefore, the robot arm, at least in this case, will never reach its top speed. The foregoing numbers are somewhat on the high side; usually the product parts pickup points are positioned closer to the assembly point than 500 mm, and normal acceleration-deceleration is closer to 0.3 g rather than 0.5 g. The example illustrates the fact that there is small advantage in specifying robot arm speed above the 0.6–0.8 m/sec value.

Degree of Freedom and Drive. The assembly task, specified to be performed at the assembly center or station, usually clearly defines the number of degrees of freedom and the type of drive (servo or on-off) required from the robot. Using the minimum number of servo axes and specifying a number of on-off–driven axis movements as required by the assembly tasks is a good practice. For an assembly center, the drive for the three basic axes (X-Y-Z or R-Θ-Z or Θ_1-Θ_2-Z) should be closed-loop servo-driven. Frequently, to facilitate part orientation, a fourth axis, a rotation around the Z axis, is also required. For an anthropomorphic robot, a minimum of five servo-driven axes need to be specified. For assembly stations, one servo-driven axis is usually sufficient, in addition to the required number of on-off drives. What must be remembered when specifying degrees of freedom for a robot is that it will operate within the confinement of a programmable assembly system, and as such, it must satisfy a multitude of requirements. Once a robot is built into the system, adding more degrees of freedom is a major undertaking, even if it is an on-off type.

Control. Specifying the robot control is not much of a dilemma for the simple reason that, by specifying the preceding parameters for the robot, the control is automatically defined. Present

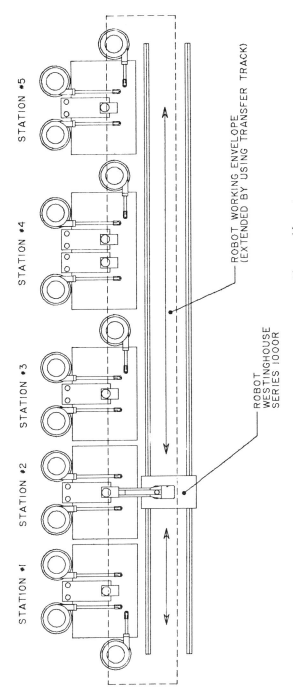

Fig. 63.9. Multistation–single-robot programmable assembly system.

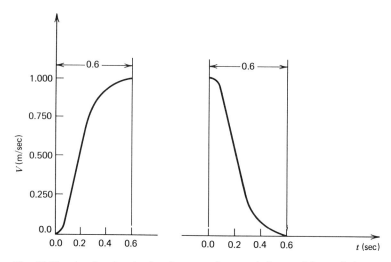

Fig. 63.10. Acceleration-deceleration curve for a typical servo-driven robot arm.

robot controls feature so many capabilities that the user is faced with the very pleasant problem of determining how to utilize all the available control functions to the best advantage of the system. The usual concerns are not the control but the method of interfacing the controller to other computers and the method of external memory storage.

Foundation. Robots with an integrated base need only the customary industrial machinery foundation and anchoring. The situation is different with those types of robots that require a fabricated base for the following reason: Depending on the type, the natural frequency of vibration of a robot is between 3 and 8 Hz. During operation, additional higher-frequency harmonics are also generated. To dampen this low-frequency vibration, a very solid base must be designed and fabricated to which the robot can be securely anchored. A poorly designed and fabricated base will result in undampened vibration and unstable operation of the robot arm, which in turn affects the cycle time of the assembly operation. In addition, the excess vibration will shorten the operating life of the equipment.

End-Effector (Gripper)

Designing an end-effector or gripper for a robot, for the most part, is a task similar to designing tooling for many other types of machines. First, the effect of the workpiece on the design (in this case the product part) must be determined, then the task to be performed must be evaluated, and finally the environment in which the gripper will be used must be defined. Normally, once these three groups of factors are known, the design can proceed. What makes this type of tooling design different from any other is that the operation efficiency designed into the gripper is a major contributor to the overall efficiency of the system.

In any type of assembly station (manual, dedicated, or programmable) to accomplish a specific assembly task, the product part first must be transferred from the parts presentors to the assembly point. Each transfer task takes significant time to accomplish, and any action directed toward reducing or eliminating this time value will reduce the overall cycle time. For a programmable assembly system, where a number of parts are handled at each station, the most effective way to reduce the transfer time (in addition to moving the part pickup points as close to the assembly point as possible) is to handle all parts within the station with one gripper. The application of a multipurpose special gripper provides the following advantages:

Eliminates tool-changing time and the mechanism and interfacing design work associated with the tool changer.

Reduces or eliminates the individual part transfer time.

The disadvantage is that the multipurpose special gripper is by nature more difficult to design and more complex in its operation. But, in almost every case, the advantages of using a multipurpose special gripper outweigh the disadvantages.

Figure 63.11 shows a sample of a multipurpose special gripper. It was designed to handle two different product parts: a die cast disk and a steel shaft. In operation, the gripper sequentially picks

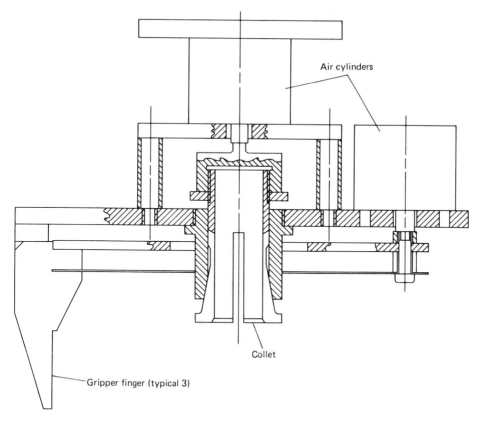

Air cylinders

Collet

Gripper finger (typical 3)

Fig. 63.11.   Multipurpose special gripper.

up the two parts, then delivers the parts to the assembly point where it inserts the parts into the product.

If, for some reason, it is required to change the gripper, a small rotary table can be vertically mounted on the end of the arm. The different grippers then can be attached to this turret and brought into position as the task requires. While the part-delivery time portion of the cycle time will not be reduced in this case, the tool-changing time is still eliminated or significantly reduced. When handling a number of parts with similar material and geometric characteristics, the use of a multipurpose general type gripper is frequently sufficient to accomplish the desired tasks. These types of grippers, either parallel or scissor, are commercially available items and should be used whenever possible to eliminate expensive design work. Figure 63.11 shows a multipurpose general gripper.

The gripper is the most abused component in an assembly system. During setup and programming, it is not unusual to specify the direction of the movement of the robot arm erroneously. If this is not discovered and corrected in time, the resulting disaster generates the most dreaded sound in robotics technology—the sound of a gripper crashing into a fixture or some other part of the system. In addition, when the part pickup and insertion points are not specified accurately, the gripper will be deformed every time it operates because of the positioning error. Although the deformation is small and is mostly elastic, the functional performance of the tool will suffer in the long run. To prevent serious damage to the tool or to facilitate quick repair and realignment, the designer should consider some of the following factors during the design of a gripper:

Adjustment for realignment in the X and Y direction.

Easily removable fingers.

Mechanical fusing (shear pins, etc.).

Locating surfaces at the gripper-arm interface.

Spring loading in the Z (vertical) direction.

Specifying spare gripper fingers.

Product Part Feeder

The function of the product part feeder within the confinement of a programmable assembly system is to provide, separate, orient, and present the individual parts to the assembly mechanism of the system. In addition to the most frequently used vibratory bowl feeder, a number of different feeding devices, such as elevators, silos, and centrifugal bowls, could serve this purpose. The common limitation of all of these devices is that with a given set of tooling they are capable of feeding only one specific product part. This means that every time the assembly system is reprogrammed to assemble a different style of product within the product family either a new feeder, already built into the system, is switched into operation or some of the feeders are replaced with a different one. While replacing feeders is not a particularly difficult procedure when proper locating surfaces are provided, it is still time-consuming and expensive.

To overcome this difficulty, new types of feeders, programmable feeders, are being developed. Some of these developments are based on the widely used vibratory bowl feeder, adding new technology components, such as optical recognition, to determine the type and orientation of the part. Others are completely new developments; the most promising are the belt feeders with programmable tooling or rotary bowl feeders with programmable tooling. These types of feeders can deliver a wide range of product parts with known or desired orientation and without any replacement or retooling when switching to a different product part. The main drawback preventing these feeders from widespread use is that their operational reliability is not yet up to the level of the standard feeders. Figures 63.12 and 63.13 show two types of programmable feeders. Figure 63.12 is a belt feeder. Two belts, moving opposite to each other, orient and deliver the product part to the part pickup point. Figure 63.13 is a rotary bowl feeder. In this case, a rotary ring and adjustable vanes orient and deliver parts to the pickup point.

Selection of a feeder is dependent on the product part geometry, part material, and the number of parts required per unit of time, a reasonable simple decision. Designing the tooling for the feeder, however, requires the skill and experience of a competent designer well versed in the field. Experience shows that most breakdowns in an assembly system are due to part jamming, either because of a misoriented part or a foreign object that has migrated in the part-delivery track. Most of the cause of jamming can be prevented with proper design, but some undesirable object can find its way into the best-designed tooling, causing a jamming condition. Therefore it is important to design the tooling so that it is accessible and open, and quick action is possible to clear jamming without endangering either the operator or the equipment.

When designing feeder tooling for an automatic assembly system, invariably good opportunities are presented to further reduce assembly cycle time in the form of preassembling components. The

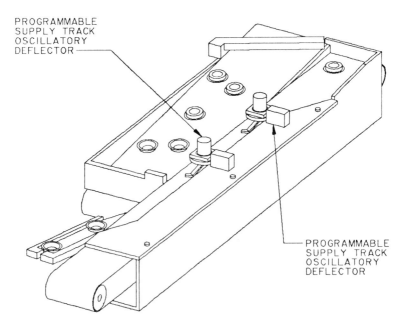

PROGRAMMABLE
SUPPLY TRACK
OSCILLATORY
DEFLECTOR

PROGRAMMABLE
SUPPLY TRACK
OSCILLATORY
DEFLECTOR

Fig. 63.12. Programmable belt feeder.

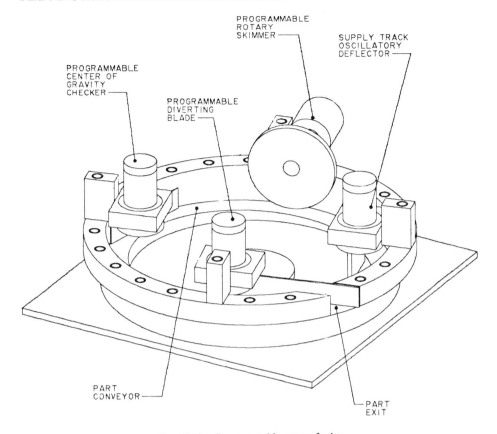

Fig. 63.13. Programmable rotary feeder.

feed tracks can be designed and arranged so that two or three tracks meet at a common point where, using a simple cam or air cylinder actuator, two or more product parts can be preassembled while the assembly robot arm is busy elsewhere. The preassembled components then can be presented to the grippers as one aggregate, thereby saving the time that would have been spent picking up the parts individually.

Magazines and palletized parts should be used only when automatic part feeding is not possible or is impractical. The loading of magazines and palletizing parts is, in most cases, a manual operation and therefore expensive. The possibility should be investigated of product parts being loaded into magazines or palletized either automatically or at the source of manufacturing where it might not cause any extra expense.

Transfer Mechanism

During assembly, either within an assembly center or from station to station with an assembly line, subassemblies or semifinished assemblies are transferred to complete the desired assembly tasks. While in transfer (except in high-speed on-the-fly assembly), no useful work is asserted or no value is added to the product. In fact, the product can be damaged or destroyed, and orientation can be lost due to jamming, vibration, or other causes. In addition, transfer requires design, equipment, drive mechanisms, sensors, and controls, all costing money. These are some of the reasons why product transfer, whenever it is associated with assembly, should be kept to the minimum. Since assembly centers require little or no station-to-station transfer, the problems and expenses associated with the transfer are largely eliminated. This fact has contributed in no small amount to the tendency toward the development and increasing acceptance of the assembly center concept, dedicated or programmable.

The redeeming quality of an assembly system transfer mechanism is that, when it is properly designed or selected, it can serve also as a buffer storage area between stations for a semifinished product. If the use of a transfer mechanism is unavoidable, it should be a totally asynchronous transfer system whose function is to provide enough buffer storage that if one station is temporarily incapacitated, there is enough product to keep the system running until the problem is corrected.

Assembly Fixture

Like almost any other fixtures, the function of an assembly fixture is to locate and support the in-process product or product component. The design and fabrication requirements, therefore, are also similar. But assembly fixtures, working in the confinement of a programmable assembly system, have an expanded function, that is, they must be capable of locating and supporting a variety of product styles within the product family. To satisfy this requirement, a number of approaches can be considered:

> One approach is to replace the assembly fixture with a new one every time the assembly system is set up to assemble a new product style with a geometry different from the previous one. This is an expensive and time-consuming proposition, but when the geometry of the different product styles changes drastically, this might be a reasonable solution.
>
> Interchangeable locating and clamping pads within the assembly fixture are a more practical way to expand the capabilities of an assembly fixture. The pads can be replaced either manually or automatically, depending on the complexity of the fixture.
>
> An excellent way to design an assembly fixture is to look for well-defined surfaces on the different product styles that are common to all. Frequently, these surfaces are not readily recognized (hidden inside the part or existing on different elevations). But, if they exist, their use can greatly simplify the design and reduce the cost of an assembly fixture.
>
> If the assembly system is an assembly center, it is possible to eliminate the assembly fixture altogether. Since assembly centers are usually based on multiarm robots, the concept of cooperative assembly can be used where one robot arm holds and locates the product, while the other arm or arms complete the assembly. In this case, the function of the assembly fixture is transferred to the gripper and/or the tooling or base plate. Multipurpose general grippers are quite adaptable and capable of handling a wide variety of parts, frequently for a fraction of the cost of an assembly fixture.

Sensors

Part recognition, orientation, inspection, the presence or absence of a part, or other types of sensory functions are performed during manual assembly with great efficiency and without external aid; not so during automatic assembly. System efficiency and the quality of assembly largely depend on sensors, strategically located throughout the assembly system. A good picture can be gained of the importance of sensor application just by examining a simple pick-and-place operation, using a pneumatically operated arm:

Operation	Sensory Function
Gripper open	Detect opening
Arm move to pickup point	Detect presence of arm
Grip product part	Detect presence of part at part presenter
	Detect closing of gripper
	Detect presence of part in gripper
Arm moves to place point	Detect presence of arm
Gripper opens	Detect gripper opening
Arm moves to home position	Detect presence of arm
	Detect presence of part in assembly

It can be seen that to reliably verify the accomplishment of the part pickup and placement operation, using a simple pick-and-place arm, nine sensory responses are required. In addition to monitoring the accomplishment of the assembly process, timely responses from the sensors assure that the assembly events proceed in a predetermined manner and deviation from it can be detected in time.

To monitor every possible action in an automatic assembly system is impractical, mainly because of the expense and space limitations. The guideline for a designer to specify or not to specify a sensor is to evaluate the possible consequences of not specifying a sensor. If the probability of an undesirable occurrence is low and the penalty is a short time loss, the sensor might not be needed. But, if the penalty is a possible injury to a person or damage to the system, specifying a sensor will be in order.

Sensor application for a programmable assembly system using a servo-driven robot arm is greatly simplified. Many of the hardware sensory functions can be replaced with the system I/O signals, which are then incorporated into the operating software.

Final Design Review

Upon completion of the system design, a review meeting should be scheduled, inviting all the persons involved in the project. Starting with the overall system, the design is reviewed. It must be understood by the attendees that the purpose of the review is to (constructively) evaluate and criticize the details of the design, and not the concept. The concept was defined and agreed upon previously, and it serves no useful purpose at this point to explore new ideas. However, it is proper and desirable to evaluate the design thoroughly, criticize its shortcomings, and suggest suitable alternative solutions. What must be remembered here is that there is no such thing as a *best design*. But at the same time, the question is, Will the suggested change in the design improve it to a degree where the additional investment in time and money is proportional to the gain that the suggested improvement provides? It is surprising how many times the answer is a definite "no."

Once the design is reviewed and agreement is reached on the proposed change, the design drawings should be corrected and signed off. Fabrication and purchasing of system components then can commence.

63.4 CASE STUDIES

63.4.1. Case Study 1 (See Figure 63.14)

Product. Clutch-bearing assembly

Yearly volume: 720,000 units

Manual assembly time: 38 sec

Number of product styles: 6

Three different types of bearing subassemblies

Two different types of clutch subassemblies

Number of parts: 6 (two subassemblies, four components)

Number of shifts worked: 1

Number of batches per month: 14

Average batch size: 4000 units

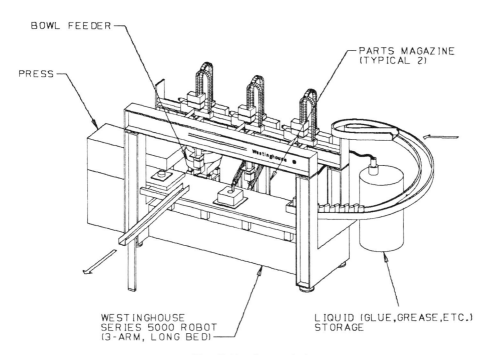

Fig. 63.14. Case study 1.

Average yield: 94%
Weight of the assembly: 0.55 kg

Reason for Automation. Product cost reduction.

System Description. Programmable assembly center, based on the Westinghouse Series 5000 three-arm assembly robot. Included in the system are one spiral storage feeder, one vibratory bowl feeder, two single and one dual magazine, fixtures, grippers, sensors, and control.

System Operation:

ROBOT ARM 1: Pick up a bearing housing from the spiral storage feeder and place it in the assembly fixture. Pick up a shaft and insert in the housing.

ROBOT ARM 2: Pick up subassembly from position 1 and place in assembly fixture at position 2. Pick up clutch assembly and insert it in bearing housing.

ROBOT ARM 3: Dedicated equipment slides two brackets in the fixture in the press. Robot arm 3 picks up subassembly from position 2 and places its fixture in press. Press presses components. Robot arm 3 removes assembly from press.

System Cycle Time. Eight seconds at position 1, 8 sec at position 2, and 9 sec at position 3. Overall cycle time: 9 sec.

Required number of operators. One

63.4.2. Case Study 2 (See Figure 63.15)

Product. Ratchet assembly
Yearly volume: 720,000 units
Manual assembly time: 34 sec
Number of product styles: 4 (four different types of gears)
Number of parts: 6
Number of shifts worked: 2
Number of batches per month: 5
Average batch size: 12,000
Average yield: 98%
Weight of the assembly: 0.12 kg

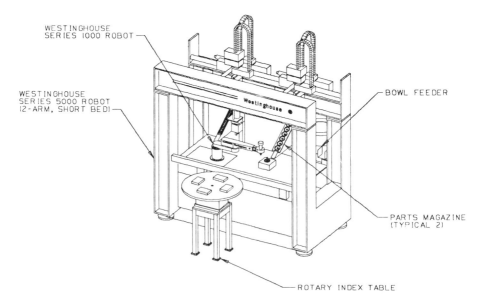

Fig. 63.15. Case study 2.

Reason for Automation. Quality improvement, cost reduction.

System Description. Programmable assembly center, based on the Westinghouse Series 5000 two-arm assembly robot. Included in the system are one vibratory bowl feeder, three magazines, one Series 1000 robot, fixtures, grippers, sensors, and control.

System Operation:

ROBOT ARM 1: Pick up shaft and place it in the assembly fixture. Pick up plastic gear and press the gear on the shaft. Pick up a ratchet gear and press it on the shaft.

ROBOT ARM 2: Pick up the subassembly from position 1 and transfer to position 2. Pick up lever and position it in the assembly fixture. Insert rivet in lever and rivet lever to gear.

ROBOT ARM 3: (Series 1000): Transfer assembly to rotary table. Inspect position and operation of lever.

System Cycle Time. 18 sec

Required Number of Operators. 0.5

63.4.3. Case Study 3 (See Figure 63.16)

Product. Gearbox assembly

Yearly volume: 140,000

Manual assembly time: 92 sec

Number of product styles: 17 (depends on gear arrangements)

Number of parts: 14 (two subassemblies, 12 parts)

Number of shifts worked: 2

Number of batches per month: approximately 60

Average batch size: 200 units

Average yield: 88%

Weight of assembly: 1.2 kg

Reason for Automation. Quality improvement, cost reduction.

System Description. Programmable assembly line based on the Westinghouse Series 5000 and Series 1000R robots. Included in the system are a power-and-free conveyor guided through the two Series 5000 robots, grease press, feeders, tooling and fixturing, screwdriver, sensors, and control.

System Operation:

STATION 101: Pick up and place housing base subassembly on the pallet. Pick up gear part and place in housing base. Pick up shaft and insert in gear in housing. Pick up gear and place it on the shaft. Transfer to next station.

STATION 102: Pick up spacer and place on shaft in the housing base. Pick up shaft and insert it in the housing base. Pick up gear and place it on the shaft. Inject grease in the housing base. Transfer to next station.

STATION 103: Pick up pin and insert it in housing base. Pick up second pin and insert it in housing base. Pick up housing top subassembly and place it on housing base. Transfer to next station.

STATION 104: Drive four screws in housing top.

System Cycle Time. 19 sec

Required Number of Operators. One per shift

63.4.4. Case Study 4 (See Figure 63.17)

Product. Angle bracket

Yearly volume: 3,300,000

Manual assembly time: 13 sec

Number of product styles: 1

Number of parts: 3 and 1 inspection task

Number of shifts worked: 2

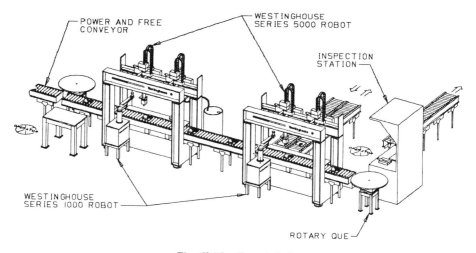

Fig. 63.16.   Case study 3.

Average yield: 96%

Weight of the assembly: 0.20 kg

Reason for Automation.   Volume increase, cost reduction.

System Description.   Double-tooled, high-speed programmable assembly center, based on the Westinghouse Series 5000 two-arm, long-bed, assembly robot. Included in the system are two sets of identical tooling and fixturing mounted on two rotary tables, two spiral silo conveyor feeders, four vibratory bowl feeders, three pick-and-place transfer arms, sensors, grippers, and control.

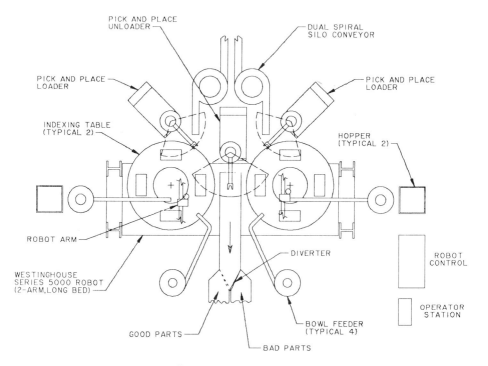

Fig. 63.17.   Case study 4.

System Operation: The equipment assembles two products simultaneously on two sets of tooling. A pick-and-place loader places a bracket in the assembly fixture mounted on a rotary table. The table indexes, the robot arm picks up a lever, rotates it, and positions it inside the bracket. On the next index, an air-cylinder-driven mechanism inserts a pin through the bracket and the lever. During the next indexing of the table the lever movement is inspected on the fly. After indexing, the center pick-and-place arm removes the assembly and transfers it to the delivery chute.

System Cycle Time.   5.5 sec for two products
2.75 sec for one product

Required Number of Operators.   One per shift

ACKNOWLEDGMENT

The author wishes to express his appreciation to Messrs. R. L. Eshleman, D. L. Wolfe, and V. P. Valeri, Managers of the Westinghouse Industry Automation Division for their support of this work.

REFERENCES

1. Boothroyd, G., Poli, C., and Munch, L. E., *Automatic Assembly,* Dekker, New York, 1982.

2. Boothroyd, G. and Dewhurst, P., *Design for Assembly,* Department of Mechanical Engineering, University of Massachusetts, Amherst, Massachusetts, 1983.

3. Csakvary, T., Product selection procedure for programmable automatic assembly technique, *Proceedings of the 2nd International Conference on Assembly Automation,* Brighton, U.K., May 1981, pp. 201–210.

4. Warnecke, H. J., *Proceedings of the 3rd International Conference on Assembly Automation,* Stuttgart, West Germany, May 1982, pp. 1–14.

5. Eversheim, W. and Müller, W., Assembly oriented design, *Proceedings of the 3rd International Conference on Assembly Automation,* Stuttgart, West Germany, May 1982, pp. 177–190.

6. Captor, N. et al., Adaptable-programmable Assembly Research Technology Transfer to Industry, Final Report, Westinghouse Industry Automation Division, Pittsburgh, Pennsylvania, January 1983.

CHAPTER 64
PART MATING IN ASSEMBLY

DANIEL E. WHITNEY

C. S. Draper Laboratory Inc.
Cambridge, Massachusetts

64.1. INTRODUCTION

Robots are being applied to a wide variety of industrial processes: welding, painting, machine loading, grinding, and assembly, to name a few. All these processes traditionally have been accomplished by people, often using hand-held tools. For some operations, merely attaching those tools to the end of a robot has been sufficient. But it has been shown many times in many fields that true automation of a previously manual operation requires considerable study to understand the process itself. People and robots have different, sometimes complementary, strengths and weaknesses. This means that truly intelligent and efficient robot automation requires that the target process be understood well enough to make the best use of robots.

This is particularly true in the case of assembly. The situation can be deceptive because it is so often assumed that robots can do what people do, or soon will be able to. And assembly looks so easy! But it is not known just what people do when they fit two parts together, or is it necessary to find out to create robot assembly. On the contrary, what people do may be irrelevant. The important thing is to consider what the parts need in order to fit together easily and reliably, and then to consider how to equip a robot to satisfy those needs. This chapter approaches the problem from this point of view.

The goal of this chapter is to describe rigid part mating, that is, the assembly of parts that do not deform substantially during assembly. The chapter is divided into two main parts, theoretical and practical. A reader who wishes to get the flavor of the theory but concentrate on practical aspects should read Sections 64.1, 64.3, 64.4, and 64.8. Compliant part mating, where the parts do deform by design during assembly, has also been studied. See References 35, 36, and 37.

Practical assembly system implementations require attention to many things other than part mating, including appropriate product design, part feeding, material handling, inspection, and economic analysis. These important issues are discussed in Chapter 62, Planning Programmable Assembly Systems, as well as in Chapter 63, Planning Robot Applications in Assembly.

Assembly is a geometric problem, and if parts were identical, perfectly made, and perfectly positioned, assembly would always be successful and free of excessive mating force. Practical constraints such as cost and technical limitations cause parts to differ, machines and jigs to wear and, consequently, parts to be misplaced or misaligned at the moment of assembly. The theory and experiments described delineate the events that occur and forces that arise during error-corrupted assembly. From this we can obtain techniques for increasing the likelihood of successful assembly without recourse to expensive methods like eliminating the errors in advance or sensing and correcting them on the fly.

This chapter brings together new work plus several years' prior work by the author and his colleagues. A large part consists of reformulations of work by Simunovic.[16, 17] Other portions appeared in References 1–6. References 15, 23–30, 33, and 34 also treat aspects of the problem. Reference 33 analyzes the assembly of a rigidly supported rigid peg into a hole in a compliantly supported worktable. Reference 34 contains a complete kinematic and force analysis of rigid peg-hole assembly in the one- and two-point contact phases only, utilizing an arbitrary insertion force vector.

The key point of this chapter is that part geometry, the stiffness of jigs or grippers supporting the parts, and friction between the parts are the major factors in rigid part mating. Equations for mating force versus insertion distance are derived, along with limitations on allowed error and recommendations for compliant gripper or jig design which help avoid unsuccessful assembly. All equations are worked out for the case of round pegs inserted into round chamfered holes (modeled as flat tabs and slots) with constant and identical coefficient of friction acting at all contact points. Small angle

1084

approximations are used so that explicit solutions can be obtained and several interesting properties derived and studied. Gravity and inertia are ignored. Experimental verifications utilize three-dimensional parts. Assembly of holes onto pegs has not been extensively studied. However, the results given here can be applied to good approximation.

During assembly, parts must be supported by jigs, fixtures, hands, grippers, and so on. These supports have some compliance, either by design or accident. Rigid part-mating theory can then be said to apply to parts that are rigid compared to the supports. Modeling the supports is an important aspect of part-mating theory. In this chapter, a unified and fairly general method of modeling supports is used, and the equations for mating forces contain the main characteristics of the modeled supports in parametric form so that the influence of different values of support parameters can be studied.

The importance and desirable features of a properly designed support are major results of part-mating theory. Supports with especially good characteristics are called remote center compliances (RCCs).[6, 15] They are discussed in Sections 64.4, 64.5, and 64.8.

The chapter is organized as follows: Section 64.2 contains general descriptions and definitions. Section 64.3 analyzes the geometric conditions of round peg-hole assembly, while Section 64.4 derives the contact forces between parts and concludes with sufficient conditions to avoid "wedging" and "jamming." Section 64.5 derives the insertion force equations and, using a simplified support model, shows how the RCC helps avoid jamming and reduces mating force. Section 64.6 presents complete mating force equations and experimental verifications. Section 64.7 is an analysis of wedging. Section 64.8 describes practical uses for RCCs.

64.2 DEFINITIONS

Consideration of typical part-mating geometry shows that a mating event has these stages (see Figure 64.1):

Approach
Chamfer crossing
One-point contact
Two-point contact

(Under some circumstances, two-point contact does not occur. Also, one-point contact or line contact can recur after a period of two-point contact.)

In general, the part rotates and translates during mating as initial lateral and angular errors between the parts are corrected. Compliant supports must therefore provide both lateral and angular compliance for at least one of the mating parts. If certain symmetry conditions are satisfied, then any compliant support can be represented mathematically by a *compliance center* plus the support's stiffness *laterally at* (K_x), and *angularly around* (K_θ), this center (Figure 64.2). That is, the compliance matrix of the support is diagonal in coordinates whose origin is at the compliance center. The support is assumed mathematically to attach to the part at this point, marked with a black and white circle in Figure 64.2. The distinction between the mathematical attachment point and the point where the part is actually gripped is important, and is discussed in Section 64.5. The forces and moments applied by the support are reexpressed in peg tip coordinates in terms of F_x, F_z, and M. This method of representing supports was introduced by Simunovic[17] and is a powerful tool for analyzing part mating because so many kinds of support can be represented this way.

A part-mating event can then be represented by the path of the supported part (constrained by its shape and the shape of the other mating part), the path of the support (constrained by the robot or machine doing the assembly), the forces and moments applied to the part by the compliances of

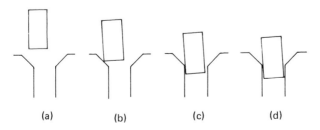

(a)			(b)			(c)			(d)

Fig. 64.1. Four stages of assembly: (*a*) approach, (*b*) chamfer crossing, (*c*) one-point contact, (*d*) two-point contact.

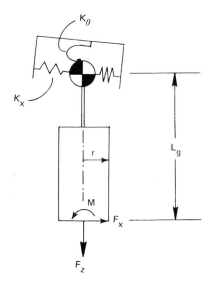

Fig. 64.2. Rigid peg supported compliantly by lateral spring K_x and angular spring K_θ at distance L_g from peg's tip. The black and white circle is the compliance center.

the support as these paths deviate, and the forces applied by the contact and friction forces between the two parts.

Simunovic showed that successful part mating depends on maintaining certain relationships between the applied forces and moments during two-point contact. This guarantees avoidance of jamming, in which the forces applied to the part point in the wrong direction. A similar phenomenon, called "wedging," can occur if two-point contact occurs too early in the mating. Avoidance of wedging depends on control of initial condition errors between the parts.

64.3. GEOMETRY OF PART MATING

We begin by developing the equations that describe the path of the compliance center while the part traverses the phases of chamfer crossing, one-point contact, and two-point contact. The compliance center's location is a basic support design parameter. It will be placed at an arbitrary point on the peg's axis for generality, and the optimum location will be derived later. We assume throughout that the angle θ between peg and hole axes is small and that the assembly device carries the support on a path parallel to the hole's axis while the compliance center moves with the peg. (A more complete analysis would allow a general angle between the path of the support and the hole's axis.) Until the peg touches the chamfer or hole, the compliances are relaxed. In this state the initial lateral and angular errors ϵ_0 and θ_0 (Figure 64.3) combine to place the compliance center a distance U_0 to one side of the hole's axis. Once contact occurs the compliances will deform. These deformations will be accounted for by deviations $U_0 - U$ and $\theta - \theta_0$ from the rest position.

64.3.1. Chamfer Crossing

The geometry of peg and hole during chamfer crossing is shown in Figure 64.4a and the forces are shown in Figure 64.4b. The chamfer is described by its angle α and width w. The geometry is described by

$$U_0 = \epsilon_0 + L_g \theta_0 \tag{64.1}$$

and

$$U = L_g \theta - \frac{z}{\tan \alpha} + \epsilon_0 \tag{64.2}$$

where z ranges from zero at the initial contact to $\epsilon_0 \tan \alpha$ when the tip of the peg reaches the bottom of the chamfer and one-point contact begins. We define

$$\epsilon_0' = \epsilon_0 - cR \tag{64.3}$$

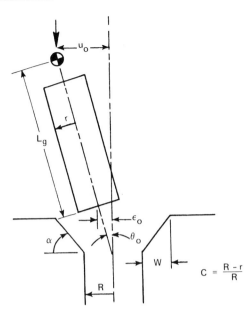

Fig. 64.3. Definition of terms for geometric analysis of part mating. The arrow indicates direction of motion of the support.

where

$$c = \frac{R - r}{R} \tag{64.4}$$

R and r are hole and peg radius, respectively, and c is called the clearance ratio.

A force balance yields

where

$$\left.\begin{aligned}
f_1 &= f_N \mathbf{B} \\
f_2 &= f_N \mathbf{A} \\
\mathbf{A} &= \cos \alpha + \mu \sin \alpha \\
\mathbf{B} &= \sin \alpha - \mu \cos \alpha
\end{aligned}\right\} \tag{64.5}$$

and μ is the coefficient of friction. The contact and applied spring support forces can be expressed in coordinates attached to the peg's tip by

$$\left.\begin{aligned}
F_x &= -f_1 \\
F_z &= f_2 \\
M &= f_2 r
\end{aligned}\right\} \text{ contact forces} \tag{64.6a}$$

$$\left.\begin{aligned}
F_x &= -K_x(U_0 - U) \\
M &= K_x L_g(U_0 - U) - K_\theta(\theta - \theta_0)
\end{aligned}\right\} \text{ support forces} \tag{64.6b}$$

Combining Eqs. (64.1) through (64.6) yields expressions for θ and U during chamfer crossing

$$\theta = \theta_0 + \frac{K_x(z/\tan \alpha)(L_g \mathbf{B} - r\mathbf{A})}{(K_x L_g^2 + K_\theta)\mathbf{B} - K_x L_g r\mathbf{A}} \tag{64.7}$$

and

$$U = U_0 - \frac{K_\theta(z/\tan \alpha)\mathbf{B}}{(K_x L_g^2 + K_\theta)\mathbf{B} - K_x L_g r\mathbf{A}} \tag{64.8}$$

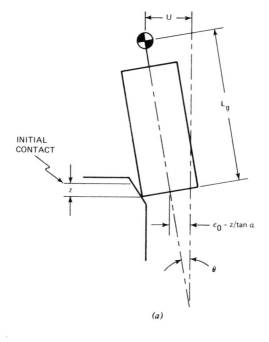

(a)

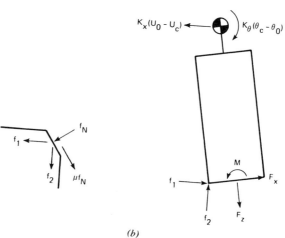

(b)

Fig. 64.4. Geometry *(a)* and forces *(b)* during chamfer crossing.

64.3.2. One-Point Contact

The forces acting during one-point contact are shown in Figure 64.5. A derivation analogous to the above, beginning with geometric constraint

$$U = cR + L_g\theta - l\theta \tag{64.9}$$

yields the following expressions for θ and U during one-point contact

$$\theta = \frac{C(\epsilon'_0 + L_g\theta_0) + K_\theta\theta_0}{C(L_g - l) + K_\theta} \tag{64.10}$$

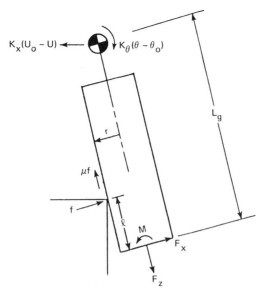

Fig. 64.5. Forces acting during one-point contact.

where

$$C = K_x(L_g - l - \mu r)$$

and

$$U = U_0 - \frac{K_\theta(\epsilon_0' + l\theta_0)}{C(L_g - l) + K_\theta} \tag{64.11}$$

Insertion depth l is defined to be zero just as the tip of the peg reaches the bottom of the chamfer. Setting $l = 0$ in Eq. (64.10) yields θ_1, the value of θ just as one-point contact begins

$$\theta_1 = \theta_0 + \frac{K_x(L_g - \mu r)\epsilon_0'}{K_x L_g(L_g - \mu r) + K_\theta} \tag{64.12}$$

64.3.3. Two-Point Contact

Two-point contact will be analyzed for only one of the four possibilities of initial error ($\pm\epsilon_0$, $\pm\theta_0$). The derivations for the other cases are similar.

Geometric compatibility between peg and hole during two-point contact is governed by

$$R = \left(\frac{l}{2}\right)\tan\theta + r\cos\theta \tag{64.13}$$

which, for small θ, becomes

$$l\theta = 2cR = cD, \quad \text{constant} \tag{64.14}$$

It is easy to show that θ must be less than θ_m, defined by

$$\theta_m \triangleq \sqrt{2c} \tag{64.15}$$

during two-point contact. A smaller upper limit on θ in two-point contact, based on friction considerations, is discussed in the next section. Equations (64.13) and (64.15) are plotted in Figure 64.6.

To determine when two-point contact begins, we note first that during one-point contact, the geometric constraint is

$$U_0 - U = \epsilon_0' + L_g(\theta_0 - \theta) + l\theta \tag{64.16}$$

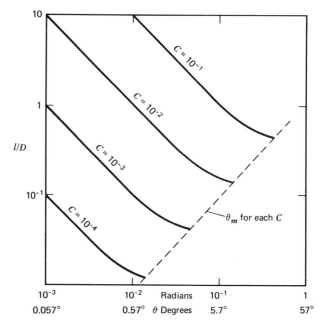

Fig. 64.6. Wobble angle versus insertion depth.

Substituting Eqs. (64.14) into (64.16) and calling θ equal to θ_2 where two-point contact begins, we have

$$U_0 - U_2 = \epsilon_0'' + L_g(\theta_0 - \theta_2) \tag{64.17}$$

where

$$\epsilon_0'' = \epsilon_0 + cR \tag{64.18}$$

Equation (64.17) relates $\theta = \theta_2$ and $U = U_2$ just as two-point contact starts. Using Eq. (64.17) in a force balance analysis similar to the above (see Figure 64.7) yields for U and θ at onset of two-point contact

$$\theta_2 = \theta_0 + \frac{K_x \epsilon_0''(L_g - l_2 - \mu r)}{K_x L_g^2 + K_\theta - K_x L_g(l_2 + \mu r)} \tag{64.19}$$

and

$$U_2 = U_0 - \frac{K_\theta \epsilon_0''}{K_x L_g^2 + K_\theta - K_x L_g(l_2 + \mu r)} \tag{64.20}$$

To find l_2, the insertion depth at which two-point contact begins, substitute Eq. (64.14) into (64.19). This yields a quadratic for l_2

$$\alpha l_2^2 - \beta l_2 + \gamma = 0 \tag{64.21}$$

where

$$\left.\begin{array}{l} \alpha = K_x(\epsilon_0'' + L_g \theta_0) \\ \beta = (L_g - \mu r)\alpha + K_x L_g cD + K_\theta \theta_0 \\ \gamma = cD(K_x L_g^2 + K_\theta - K_x L_g \mu r) \end{array}\right\} \tag{64.22}$$

so that

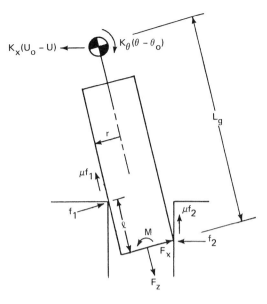

Fig. 64.7. Forces acting during two-point contact.

$$l_2 = \frac{(\beta - \sqrt{\beta^2 - 4\alpha\gamma})}{2\alpha} \qquad (64.23)$$

and

$$l_2' = \frac{(\beta + \sqrt{\beta^2 - 4\alpha\gamma})}{2\alpha} \qquad (64.24)$$

The existence of two solutions shows that there is not only a value l_2 at which two-point contact begins but also a value $l_2' > l_2$ at which two-point contact ends and one-point contact resumes. Once l_2 is known, we can use Eq. (64.13) or (64.14) to find θ_2.

If we assume $K_\theta >> K_x L_g^2$ and $K_\theta \theta_0 >> \mu K_x \epsilon_0'' r$, we obtain the approximations

$$l_2 \cong \frac{cD}{\theta_0} \qquad (64.25)$$

and

$$l_2' \cong \frac{K_\theta \theta_0}{K_x \epsilon_0''} - l_2 \qquad (64.26)$$

These assumptions are, coincidentally, satisfied by all current designs of remote center compliances and a wide range of values of θ_0, ϵ_0, μ, r, and L_g (see Section 64.5). Also, if L_g and θ_0 are small enough, there will be no solution to Eq. (64.21), implying that two-point contact cannot occur.

64.3.4. Discussion

Figure 64.8 summarizes the foregoing results as a crossplot of l versus θ. Figure 64.8 is called a life cycle plot of a peg-hole insertion. The terms l^* and θ_w are defined in later sections, where experimental verification is presented.

Dimensioning practice was surveyed in typical machined parts[31] with the result that clearance ratio c is well (though not perfectly) correlated with part type and use. Figure 64.9 shows the survey results. The combination of Figures 64.6 and 64.9 indicates how difficult an insertion of a given type part might be, based on the geometric parameters. A complete discussion of this point is deferred to the next section. Note that the assumption of small angles is borne out for the part types surveyed.

Finally, note that most part-mating difficulties occur during two-point contact. Measures taken to

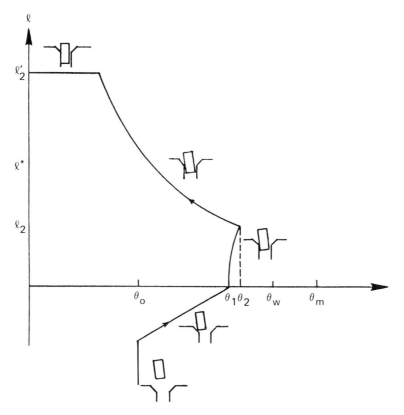

Fig. 64.8. Schematic life cycle plot of insertion.

increase the value of l_2 or to prevent two-point contact will result in more reliable part mating. The most effective of these measures is making L_g as small as possible. This is a major function of the RCC.

64.4. WEDGING AND JAMMING, AND CONDITIONS FOR SUCCESSFUL ASSEMBLY

64.4.1. Wedging and Jamming

To determine whether an assembly will succeed or fail, it is necessary to consider the forces acting on the parts in addition to the geometric compatibilities treated previously. Two phenomena were defined by Simunovic, *wedging* and *jamming,* to describe situations in which the peg seems to stick in the hole during two-point contact.

Jamming is a condition in which the peg will not move because the forces and moments applied to the peg through the support are in the wrong proportions. The correct proportions were originally presented by Simunovic.[16] A simpler derivation and more complete exploration of the problem are presented here.

We follow Simunovic's approach and express the applied insertion force in terms of F_x, F_z, and M at or about the peg's tip, as shown in Figure 64.10. We wish to find equilibrium-sliding-in conditions between the applied forces and the reactions f_1 and f_2.

Simunovic's results are based on ignoring the angle of tilt of the peg with respect to the hole. The equilibrium equations that describe the peg sliding in during two point contact (Figure 64.10a) are then given by

$$F_z = \mu(f_1 + f_2) \tag{64.27}$$

$$F_x = f_2 - f_1 \tag{64.28}$$

$$M = f_1 l - \mu r(f_2 - f_1) \tag{64.29}$$

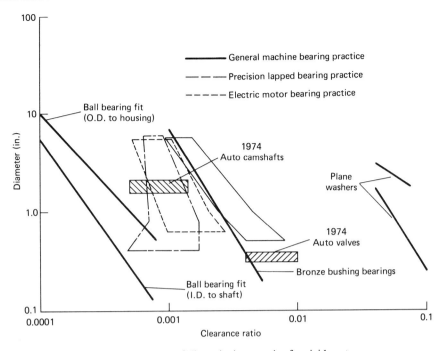

Fig. 64.9. Survey of dimensioning practice for rigid parts.

Combining these equations yields

$$\frac{M}{rF_z} = \frac{l}{2r\mu} - \frac{F_x}{F_z}\left(\frac{l}{2r} + \mu\right) \tag{64.30}$$

Define

$$\lambda = \frac{l}{2r\mu} \tag{64.31}$$

Then Eq. (64.30) can be expressed as

$$y = mx + b \tag{64.32}$$

where

$$y = \frac{M}{rF_z} \tag{64.33}$$

$$x = \frac{F_x}{F_z} \tag{64.34}$$

$$m = -\mu(1 + \lambda) \tag{64.35}$$

$$b = \lambda \tag{64.36}$$

If one draws the peg leaning the other way, one obtains the same equation, except

$$b = -\lambda \tag{64.37}$$

To finish the derivation we must consider the four possible one-point contacts, of which two will suffice for illustration (see Figure 64.10b and c).

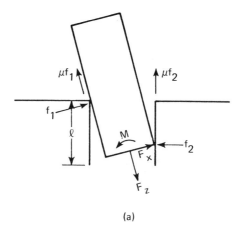

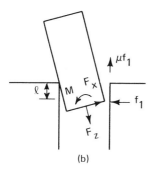
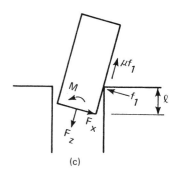

Fig. 64.10. Applied forces and reaction forces. (*a*) Peg in two-point contact. (*b*, *c*) The two right-side one-point contacts.

For Figure 64.10*b*, the equilibrium equations are

$$M + \mu r F_x = 0 \tag{64.38}$$

$$F_z - \mu F_x = 0 \tag{64.39}$$

or

$$\frac{F_x}{F_z} = \frac{1}{\mu} \tag{64.40}$$

$$\frac{M}{r F_z} = \frac{-\mu F_x}{F_z} = -1 \tag{64.41}$$

For Figure 64.10*c*, we have

$$M + l F_x + \mu r F_x = 0 \tag{64.42}$$

$$F_z - \mu F_x = 0 \tag{64.43}$$

or

$$\frac{F_x}{F_z} = \frac{1}{\mu} \tag{64.44}$$

$$\frac{M}{r F_z} = -(2\lambda + 1) \tag{64.45}$$

For the other two one-point contacts, one obtains

$$\frac{F_x}{F_z} = \frac{-1}{\mu} \tag{64.46}$$

$$\frac{M}{rF_z} = 1 \text{ or } (2\lambda + 1) \tag{64.47}$$

It is easy to show that these four points (equations 64.40–41, 64.44–45, and 64.46–47) lie on the two lines which obey Eq. (64.32), restated as

$$\frac{M}{rF_z} = \pm\lambda - \frac{F_x}{F_z}\mu(1 + \lambda) \tag{64.48}$$

In fact, they are the end points of these lines because the lines represent just sliding in, that is,

$$F_z \geq \mu F_x \tag{64.49}$$

or

$$\frac{F_x}{F_z} \leq \frac{1}{\mu} \tag{64.50}$$

for the right-side one-point contacts, and similarly with a minus sign for the left-side ones. Larger F_x/F_z results in one-point contact jams. All of these conditions can be summarized in Figure 64.11. The vertical dotted lines in the diagram describe a line contact. Figure 64.11 may be interpreted as follows. Combinations of F_x, F_z, and M falling on the parallelogram's edges describe equilibrium sliding in. Outside the parallelogram lie combinations that jam the peg, either in one- or two-point contact. Inside, the peg is in disequilibrium sliding or falling in.

Note that as $\lambda \to 0$ (peg shallow in the hole) the parallelogram collapses to a line running between $(1/\mu, -1)$ on the right to $(-1/\mu, 1)$ on the left, showing that the no-jam region is quite small. When $\lambda \to \infty$ (peg deep in the hole) the parallelogram expands to a vertical strip lying between $-1/\mu$ and $1/\mu$, implying that the line and one-point contact jams are still possible, but two-point contact jams are quite difficult to achieve. The F_x/F_z axis intersection of the upper side of the parallelogram is at $F_x/F_z = \lambda/(\lambda + 1)\mu$. A little algebra shows that this must be less than $1/\mu$ for all $\lambda \geq 0$, so Figure 64.11 shows the general case.

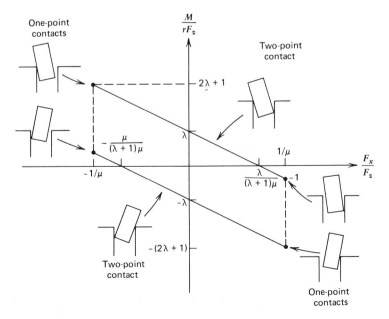

Fig. 64.11. The jamming diagram.

Wedging is also a condition in which the peg appears stuck in the hole, but unlike jamming, the cause is geometric rather than ill-proportioned applied forces. Indeed, wedges can be so severe that no reproportioning of the applied forces can cause assembly to proceed except by deforming and possibly damaging the parts at their contact points.

To model wedging we must assume that at least one of the parts is elastic, although it is still stiff compared to K_x and K_θ. In wedging, the two contact forces f_1 and f_2 can point directly toward each other, storing energy in the elastic part. This is possible if two-point contact occurs when l_2 is small, allowing the two friction cones at the contact points to intersect. Figure 64.12 shows one possible situation, in which l_2 is as large as possible and still allows wedging. The right-side contact force does not point along an extreme of the friction cone, indicating that relative motion between the parts on the right side has stopped. The left-side contact force points along the lower extreme of the friction cone, indicating that the left side of the peg is attempting to move *out* of the hole. This could occur if the peg has been pushed counterclockwise, elastically deformed at the contact points, and released.

Consideration of the geometry in Figure 64.12 allows us to write

$$l_w = \mu d \tag{64.51}$$

where l_w is the largest l_2 at which wedging could still occur. The derivation assumes θ and c are both small. Using Eq. (64.14) in Eq. (64.51) allows us to define θ_w as the smallest θ_2 at which wedging could occur

$$\theta_w = \frac{c}{\mu} \tag{64.52}$$

θ_w is noted on Figure 64.8.

64.4.2. Conditions for Successful Assembly

We are now in a position to complete the conditions for successful assembly.

1. To cross the chamfer we need

$$|\epsilon_0| < w \tag{64.53}$$

where w is the width of the chamfer (Figure 64.3).

2. To avoid wedging, two-point contact must occur at a value of θ_2 obeying

$$|\theta_2| < \frac{c}{\mu} \tag{64.54}$$

Using Eq. (64.19) and assuming l_2 and r are small, we can write

$$\theta_0 + s\epsilon_0 < \pm \frac{c}{\mu} \tag{64.55}$$

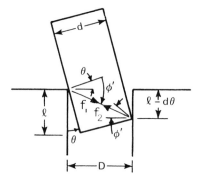

Fig. 64.12. Geometry of wedging condition showing intersection of left- and right-side friction cones; cone half angle $= \phi' = \tan^{-1} \mu$. Figure is drawn for the case of largest l for which jamming can occur.

where

$$s = \frac{L_g}{L_g^2 + K_\theta/K_x} \tag{64.56}$$

These relations are plotted in Figure 64.13. Note that when $L_g = 0$ there is no interaction between the requirements on ϵ_0 and those on θ_0.

3. To avoid jamming, we must maintain

$$\left| \frac{M}{rF_z} + \frac{\mu(1+\lambda)F_x}{F_z} \right| < \lambda \tag{64.57}$$

and

$$\left| \frac{F_x}{F_z} \right| < \frac{1}{\mu} \tag{64.58}$$

Condition 1 is purely geometric and provides a limit on the amount of permissible lateral error in relation to the width of the chamfer. If both peg and hole have chamfers, the widths of both chamfers may be added to obtain the value of w to use in Eq. (64.53). Lateral error ϵ_0 has several sources: location of tooling, tolerance, slop, or wear in parts fixtures, and fabrication tolerances in part dimensions that relate tooling or fixturing points on the parts to center locations of holes where part mating occurs. Since the surfaces on which parts are fixtured can sometimes be quite rough, a careful tolerance study should be made to determine, for example, a 95% confidence limit on ϵ_0. This will indicate whether parts or tooling must be improved, or will allow a suitable value for w to be determined.

Condition 2 is partly geometric but is also related to tooling stiffness. Unless L_g is small, ϵ_0 and θ_0 cannot be determined independently of each other. The reason is that, if there is some lateral error ϵ_0, the peg will tend to tilt while crossing the chamfer. The resulting angle is proportional to L_g and ϵ_0, and inversely proportional to K_θ/K_x. It must be added to any direct error θ_0. θ_0 has several sources: tool approach axis misalignment, part jigging misalignment, peg grip misalignment, part or peg fabrication error, and so on. A tolerance study of these error sources will indicate whether condition 2 is in danger of being violated. As a first approximation, s can be set to zero in Eq. (64.55).

Condition 3 involves the mating forces and is the least likely to be considered in traditional tooling design. The traditional approach is to build solid tooling that seeks to minimize ϵ_0 and θ_0 outright, with little thought given to the forces that might arise during assembly. K_x and K_θ are implicitly assumed to be quite large, and rarely is an attempt made to calculate them. This technique is expensive and, in any case, is extremely difficult to use in robot assembly because of the lower accuracy and

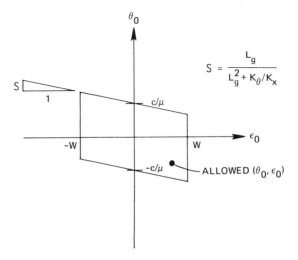

Fig. 64.13. Geometry constraints on lateral and angular error to cross chamfer and avoid wedging.

stiffness of robots compared to traditional assembly machines. This forces us to consider a different approach, called *compliant assembly,* that permits us to satisfy the three conditions relatively easily.

The concept of compliant assembly is relatively new, although lateral compliance (sometimes called "float") has been used in the past. The philosophy of compliant assembly is based on two observations: first, *there will be* lateral and angular errors in assembly despite expensive efforts to eliminate them, and these errors will be larger in robot assembly of low volume or model-mixed items than in traditional mechanized assembly. Second, when errors occur, it is compliance that allows the parts to go together, undocumented and probably low compliance, given the rigid tooling usually employed—thus mating forces will be large and the likelihood of meeting the foregoing conditions will be small. Compliant assembly consists of deliberately placing a known, engineered, and relatively large compliance into the tooling. This "soft tooling" then permits easy or easier satisfaction of all three of the conditions for successful assembly.

The *remote center compliance* (RCC) is such an engineered compliance. The theoretical reasons why it works are explained in Sections 64.5 and 64.6 in terms of the engineering mechanics of part mating. An intuitive discussion, together with several practical examples, is given in Section 64.8. Commercial versions are available with a wide range of known K_x, K_θ, and L_g.

Naturally, remote center compliance can improve the performance or reduce the cost of traditional assembly machines as well. In fact, at present, the majority of RCC applications have been to traditional assembly machines and fixed single-axis work stations.

64.5. INSERTION FORCES, JAMMING AVOIDANCE, AND THE REMOTE CENTER COMPLIANCE

The RCC is a passive device that supports parts and aids their assembly. It can be modeled by Figure 64.2 to good accuracy, except that, by design, the RCC succeeds in placing the support point or compliance center at or near the tip of the peg. That is, L_g is nearly zero. The reasons why this is desirable are made clear shortly. If L_g is to be approximately zero, then the point where the part is physically gripped must be far from the compliance center, where the attachment is mathematically assumed to be. The mathematical point is the important one and, in practical terms, it must be outside the tooling. The ability of the RCC to project its compliance center outside of itself is its main feature. It is the source of its particular abilities to aid assembly and the reason for the word *remote* in its name.

In this section, the RCC is described by a simplified model having $K_\theta = 0$. This makes it easier to show how the RCC avoids jamming, how relatively unimportant it is to make L_g exactly zero, and how insertion force varies if L_g is not zero.

64.5.1. Model of the RCC as Lateral Stiffness Alone

Refer to Figure 64.14, where the simplest compliant suspension is shown, one with only lateral stiffness.

The situation shown in the figure can occur if the peg is initially in error to the left of the hole, where the springs are relaxed. The peg tilts, as shown, while crossing the chamfer, and rocks clockwise during two-point contact. This stretches and squeezes the springs as indicated.

The force and moment applied by the lateral springs can be reexpressed in peg tip coordinates as

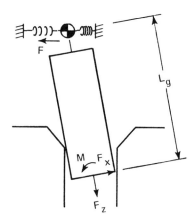

Fig. 64.14. Peg suspended by lateral spring.

$$F_x = -F \qquad (64.59)$$

$$M = L_g F = -L_g F_x \qquad (64.60)$$

Dividing Eq. (64.60) by rF_z, one obtains

$$\frac{M}{rF_z} = -\frac{L_g}{r}\left(\frac{F_x}{F_z}\right) \qquad (64.61)$$

which means that M, F_x and F_z will lie on a line in Figure 64.11 having slope $-L_g/r$ and passing through the origin. This line will intersect the parallelogram at

$$\frac{F_x}{F_z} = \pm \frac{\lambda}{L_g/r - \mu(1 + \lambda)} \qquad (64.62)$$

or at

$$\frac{F_x}{F_z} = \pm \frac{1}{\mu} \qquad (64.63)$$

if the intersection is on the vertical sides of the parallelogram. Since the amount of lateral error and K_x determines F_x, we see that insertion can occur only if F_z is big enough to satisfy Eqs. (64.62) or (64.63).

64.5.2. Relation Between the Value of L_g and the Jamming Diagram

Let us study the effect of different values of L_g. The intersection points in Eq. (64.62) move farther horizontally left and right as L_g/r is made smaller. When we have

$$\frac{L_g}{r} = \frac{l}{r + \mu} \qquad (64.64)$$

we obtain the condition of Eq. (64.63), that is, we just maintain equilibrium in one-point contact, as shown for the upper left corner of the parallelogram by Figure 64.15. Note that to maintain this equilibrium, the point where the spring attaches, marked by the black and white circle, must remain fixed in space with respect to the *hole*. The necessary F_z for given F_x is smallest for this value of L_g. Also, F_x/F_z and M/rF_z can be permitted wider variations without threatening an occurrence of jamming. For these reasons, the RCC attempts to locate the support point close to the tip of the peg or mouth of the hole, where Eq. (64.64) is satisfied approximately.

If L_g gets a bit smaller, the peg will suddenly snap over clockwise into a line contact, and maintain this equilibrium until L_g/r reaches

$$\frac{L_g}{r} = \mu \qquad (64.65)$$

at which point the peg will be in a precarious equilibrium in the other left-side one-point contact. Further lowering L_g will cause the peg to snap into two-point contact of the opposite sense from that shown in Figure 64.15. For stable equilibrium in this condition (without changing the sign of ϵ_0) we need $L_g \leq 0$.

Note finally that, with the spring pulling to the left, the right-side one-point contacts cannot occur.

A more complex analysis in Section 64.6. takes into account suspensions having torsional stiffness as well. We can predict in advance that the line of slope L_g/r will not pass through the origin of Figure 64.11.

64.5.3. Model of a Peg Pulled by a String or Pushed by a Slender Rod

This case represents a pure force (no moment) applied to a point on the peg, corresponding in many ways to Figure 64.14 (see Figure 64.16). It is easy to show that

$$\frac{M}{rF_z} = -\frac{F_x L_g}{F_z r} = -\tan\phi \cdot \frac{L_g}{r} \qquad (64.66)$$

which is the same as Eq. (64.61) and is diagrammed in Figure 64.17.

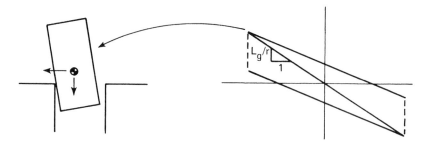

Fig. 64.15. Combination of Eq. (64.59) and Figure 64.11 for the case of peg compliantly held by lateral spring at point $L_g = l + r\mu$ from tip of peg.

Although the two cases (spring and string) have the same equation, there is a difference. When the peg is pulled by the string, F_x and F_z are proportional to each other, whereas in the case of lateral spring support F_x comes from the spring and F_z is an independent pushing force. Therefore in the spring support case one can make F_x/F_z and M/rF_z smaller by making F_z bigger, so that a jammed peg can be unjammed merely by being pushed harder. But Eq. (64.66) shows that M/rF_z depends only on L_g/r and the angle of the string. If the latter are too large, M/rF_z will be too large regardless of $|F_z|$, and the peg, if jammed, will stay jammed.

The reason for pursuing this exercise is that the string/rod model represents a common error-absorbing technique called *float*, in which a part is allowed motion in the x direction by a sliding bearing and, possibly, centering springs. Bearing friction will contribute an x force proportional to F_z, possibly resulting in failure to avoid jamming. Therefore, in the presence of angular error, float can be an unreliable technique.

64.5.4.   Relating the Support Point Location L_g and the Pulling Error Angle ϕ to Jamming

This issue applies to both the lateral spring support and the string-pulling situations because, if the peg is sliding,

$$\tan \phi = \frac{F_x}{F_z} \tag{64.67}$$

in both cases. To relate ϕ to sliding equilibrium, put Eq. (64.67) into Eq. (64.62) to get

$$\tan \phi = \frac{\lambda}{L_g/r - \mu(1 + \lambda)} \tag{64.68}$$

Abbreviate

$$\frac{L_g}{r} = \beta \tag{64.69}$$

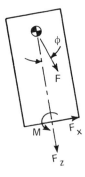

Fig. 64.16.   Peg pulled along F by a string.

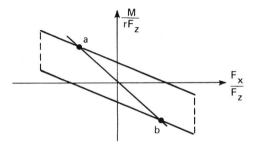

Fig. 64.17. Equilibrium sliding conditions for peg pulled by string: point a corresponds to the sliding solution when $\phi < 0$; point b corresponds to $\phi > 0$.

Let us study two-point contacts in their worst case, namely, when $\lambda \approx 1$ where wedging can barely happen. Two-point contact means

$$\beta \geq \frac{l}{r} + \mu \tag{64.70}$$

and this plus $\lambda = 1$ reduces Eq. (64.68) to

$$\tan \phi = \frac{1}{\beta - 2\mu} \tag{64.71}$$

Using values of β such as

$$\beta = \frac{nl}{r} + \mu \qquad n = 1, 2, 3, \ldots$$

we can graph $\tan \phi$ using Eq. (64.71) (see Figure 64.18).

As a numerical example, let us take $\mu = 0.2$. With $\beta = 10l/r + \mu$, $\tan\phi = 5/19$ or $\phi = 14.74°$. Note that, for a clearance ratio of 0.005, the wedging angle is $\tan^{-1}(c/\mu) = 1.43°$. For smaller β, ϕ will be much larger, and is maximum in the RCC where $\beta \cong l/r + \mu$.

Recall that we must point the insertion-force vector within $\pm\phi$ of the peg's axis to avoid jamming, and must point the peg's axis within $\pm c/\mu$ of the hole's axis to avoid wedging. In this example, using quite typical numbers, we see that in terms of error angles, jamming is 10 times easier to

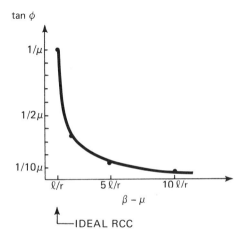

Fig. 64.18. Worst case ϕ to maintain sliding when wedging can barely occur ($\lambda = 1$).

avoid than wedging, even when the center of support or pulling is 10 times farther away from the tip of the peg than recommended by Eq. (64.64).

The previous discussion shows that the compliance center can be quite far from the tip of the peg and still provide jamming avoidance. It is important to realize that the analysis ignores the fact that the peg is tilted so that an insertion force along the peg is not along the hole axis and vice versa. This means that the allowable ϕ region is not symmetric about the peg's axis. When l/d is small, this asymmetry can be significant unless ϕ is several times the wedging angle.

In addition, the value of β affects the amount of insertion force needed. We examine this next.

64.5.5. How Much Insertion Force is Needed and How This is Affected by Support Point Location

Suppose the peg is supported laterally by a spring of stiffness K_x as in Figure 64.14. Repeating Eq. (64.48)

$$\frac{M}{r} + \mu(1 + \lambda)F_x = \lambda F_z \tag{64.48}$$

and describing the compression of the spring by U, we have

$$F_x = -K_x U \tag{64.72}$$

$$M = -L_g F_x$$
$$= K_x L_g U \tag{64.73}$$

Putting Eq. (64.73) into Eq. (64.48), and using

$$\lambda = \frac{l}{2r\mu}$$

yields

$$F_z = K_x U \mu \left[2\frac{L_g}{l} - \frac{1}{\lambda} - 1 \right] \tag{64.74}$$

which says that F_z depends linearly on K_x, U, μ, and L_g. The RCC attempts to put

$$L_g = l + r\mu$$

to achieve a one-point contact equilibrium. This reduces Eq. (64.74) to

$$F_z = K_x U \mu \tag{64.75}$$

the force created by a one-point contact. For given K_x, μ, and U this is the smallest F_z obtainable, save for no contact force at all.

64.6.   INSERTION FORCE WHEN PEG IS SUPPORTED BY BOTH LATERAL AND ANGULAR COMPLIANCES

We present here models for the insertion force during chamfer crossing, one-point contact, and two-point contact when the peg is supported by both lateral and angular springs. To obtain F_z during chamfer crossing, we substitute Eqs. (64.7) and (64.8) into (64.6) to yield

$$F_z = \frac{K_x K_\theta A(z/\tan \alpha)}{(K_x L_g^2 + K_\theta)B - K_x L_g r A} \tag{64.76}$$

To obtain F_z during one-point contact, substitute Eqs. (64.10) and (64.11) into the force-balance equations for one-point contact to yield

$$F_z = \frac{\mu K_x K_\theta(\epsilon_0' + l\theta_0)}{C(L_g - l) + K_\theta} \tag{64.77}$$

To obtain F_z during two-point contact, substitute Eqs. (64.14) and (64.17) into (64.68) to yield

$$F_x = -K_x L_g \left(\theta_0 - \frac{cD}{l} \right) - K_x \epsilon_0'' \tag{64.78}$$

$$M = (K_x L_g^2 + K_\theta) \left(\theta_0 - \frac{cD}{l} \right) + K_x L_g \epsilon_0'' \tag{64.79}$$

Putting these into Eq. (64.48) yields

$$F_z = \frac{2\mu}{l} \left[\left(\frac{\mathbf{D}(\theta_0 - cD)}{l} \right) + \mathbf{E} \right]$$
$$+ \mu \left(\frac{1 + \mu d}{l} \right) \left[\mathbf{F} \left(\frac{\theta_0 - cD}{l} \right) - \frac{\mathbf{E}}{L_g} \right] \tag{64.80}$$

where

$$\mathbf{D} = K_x L_g^2 + K_\theta \tag{64.81}$$
$$\mathbf{E} = K_x L_g \epsilon_0' \tag{64.82}$$
$$\mathbf{F} = -K_x L_g \tag{64.83}$$

Differentiating Eq. (64.80) with respect to l and setting the result equal to zero yields l, the value of l where F_z and the contact forces f_1 and f_2 are maximum

$$l* = \frac{(4\mathbf{D} + 2\mathbf{F}\mu d)cD}{2\mathbf{D}\theta_0 + \mathbf{E}(1 - \mu d/L_g) - \mathbf{F}(\theta_0 \mu d - cD)} \tag{64.84}$$

When $L_g \approx \mu r$ this reduces to

$$l* = \frac{4K_\theta cD}{2K_\theta \theta_0 - K_x \mu r \epsilon_0'} \tag{64.85}$$

The stiffnesses and initial errors influence the result in each instance but it is often true that $K_\theta \theta_0 > K_x \mu r \epsilon_0'$. In this case

$$l* = \frac{2cD}{\theta_0} \simeq 2l_2 \tag{64.86}$$

That is, insertion and contact forces are maximum at an insertion depth that is about twice the depth at which two-point contact first occurs. Substitution of Eq. (64.84) into Eq. (64.80) gives the maximum value of F_z. The peak contact force is approximately this F_z divided by 2μ.

Experiments were designed to test Eqs. (64.76) through (64.84). The parameters were as follows:

Support (a Draper Lab Model 4B RCC):
    $K_x = 7\text{N/mm}$
    $K_\theta = 53000 \text{ N-mm/rad}$

Peg and hole (steel, hardened and ground):
    Hole diameter $= 0.5002$ in. (12.705 mm)
    Peg diameter $= 0.4989$ in. (12.672 mm)
    $c = 0.0026$

Initial errors:
    $\epsilon = 1.4$ mm and 0.85 mm
    $\theta_0 = 0$

Location of support:
    $L_g = 45$ mm and 1 mm (achieved by using two pegs of different lengths)
Coefficient of friction ≈ 0.1 (determined empirically from one-point contact data)

Figure 64.19 is a photo of the apparatus.
Figure 64.20 gives the results of insertion force versus l for $L_g = 45$ mm and $\epsilon_0 = 1.4$ mm, while Figures 64.21 and 64.22 show net lateral force F_x and moment M. Figure 64.23 shows insertion

Fig. 64.19. Photo of the apparatus used to verify Eqs. (64.74), (64.75), (64.78), and geometric relations during part mating. A hardened and ground steel peg is supported by an RCC whose deflections can be measured using integral optical sensors. In this way, ϵ and θ of the peg can be recorded during insertion. The RCC is in turn held by a six-axis force-torque sensor with ten-gram threshold. This allows F_x, F_z, and M to be recorded as well. Finally, the sensor is held in the quill of a milling machine whose motion is measured by an LVDT, allowing l to be recorded.

force for $L_g = 45$ mm and $\epsilon_0 = 0.85$. Finally, Figure 64.24 shows the insertion and lateral force when $L_g = 1$ mm. As predicted, two-point contact does not occur. Figure 64.25 compares the peg's angle θ versus l with the theory of Section 64.3. Angle data were obtained using an instrumented RCC (IRCC). See References 19 and 20 for descriptions and other uses of the IRCC.

In all cases, theory and experiment agree as to general trends and compare fairly well as to absolute magnitudes. Since geometry, K_x, K_θ, and μ cannot be predicted exactly, one can get better "agreement" between theory and experiment by searching for "better" values of these parameters. Such a search would only improve our knowledge of these values, however, and would not increase our understanding of the problem.

64.7. ANALYSIS OF WEDGING

The other phenomenon that can occur in addition to jamming is called *wedging*. Analysis of this case must assume some deformation in the parts. The forces acting are shown in Figure 64.12, where the contact forces point toward one another. If an insertion force is applied, and the parts are allowed to deform, then if the jamming conditions are satisfied, the contact forces will move to the upper boundaries of the friction cone and insertion will occur. At some point the peg, perhaps damaged, will be deep enough into the hole that wedging can no longer occur. The contact forces now can exert a couple about the tip of the peg and turn it parallel to the hole. Prior to this point there was little or no moment, misalignment was not corrected, and high insertion and contact forces occurred. We analyze these forces, using numerous assumptions. A similar analysis appears in Reference 6.

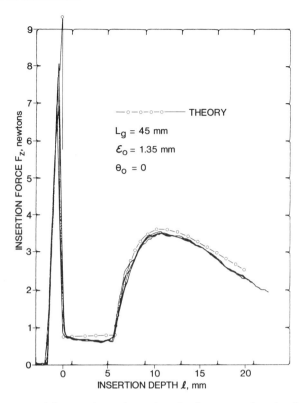

Fig. 64.20. Comparison of theory and experiment: insertion force versus insertion depth with support point 45 mm from tip of peg.

The situation shown in Figure 68.26a is that of a wedged peg which, when pushed by F_z, does not turn clockwise but instead compresses. This is approximately equivalent to the case shown in Figure 64.26b. We intend to use Hertz stress analysis (not totally appropriate) to analyze this case. To do so, we split and image the peg axially, so that the contact point can be represented as an edge of small radius of curvature. Figure 64.26b can be analyzed to yield

$$F_z = 2f\left(\mu + \frac{\theta}{2}\right) \tag{64.85}$$

and Figure 64.26c can be analyzed to yield

$$Q = f_1 \tag{64.88}$$

On the assumption that, for most machined parts, $\theta < \mu$, we have

$$Q \cong \frac{F_z}{2\mu} \tag{64.89}$$

A Hertz stress analysis is used to find the deflection δ at each contact point. The geometric relation between δ and incremental insertion motion Δ

$$\Delta = \frac{\delta}{\theta} \tag{64.90}$$

is used to see how far the peg can move into the hole for given δ. The result is probably an underestimate since the Hertz analysis ignores shear stress, which is large in our case.

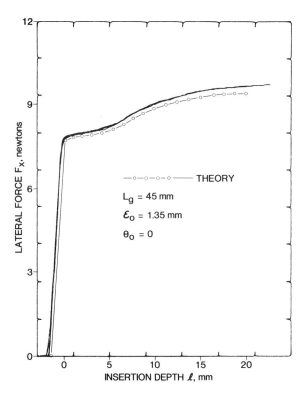

Fig. 64.21. Comparison of theory and experiment: lateral force versus insertion depth with support point 45 mm from tip of peg.

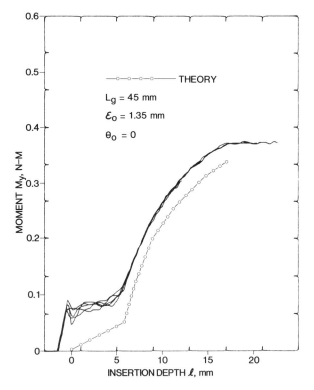

Fig. 64.22. Comparison of theory and experiment: moment normal to insertion axis versus insertion depth with support point 45 mm from tip of peg.

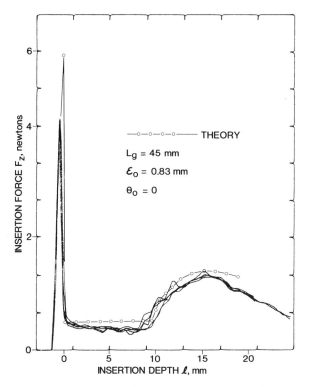

Fig. 64.23. Comparison of theory and experiment: insertion force as in Figure 64.20 except lateral error is smaller.

The Hertz analysis models the hole as a concave body with curvatures

$$\rho_1 = -\frac{1}{R} \tag{64.91}$$

and

$$\rho_2 = 0 \tag{64.92}$$

while the peg is modeled as an ellipsoid with

$$\rho_1 = \frac{1}{R} \tag{64.93}$$

and

$$\rho_2 = \frac{1}{\beta r} \qquad \text{with } \beta \ll 1 \tag{64.94}$$

to indicate a "sharp" corner.

Following the procedure in Harris[32] and using

$$\mu = 0.2$$
$$r = 6.35 \text{ mm}$$
$$R = 6.35635 \text{ mm (corresponding to } c = 10^{-3})$$
$$\beta = 10^{-2}$$
$$\theta = 5 \times 10^{-3} = \frac{c}{\mu}$$

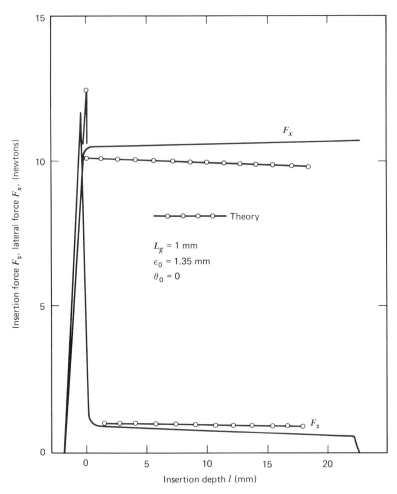

Fig. 64.24. Comparison of theory and experiment: insertion force and lateral force as in Figures 64.20 and 64.21 except support point is almost exactly at tip of peg.

we obtain the results shown in Table 64.1, assuming steel peg and hole. Even with 222 Newtons insertion force the peg moves more than $r/10$ into the hole, indicating that many wedges apparently can be relieved if sufficient force is used and galling or other damage can be tolerated. However, quite high stresses may occur.

These results may explain the apparent fact that wedging does not seem to occur in machine-aided assembly, although it often occurs in manual assembly. Probably, wedging does occasionally occur, but the assembly apparatus is strong enough to force the parts past the wedging region.

64.8. REMOTE CENTER COMPLIANCE

The remote center compliance (RCC), as analyzed earlier, is a unique device for aiding assembly insertion operations. It is entirely mechanical, deriving its properties from its geometry and the elasticity of its parts. Its major function is to act as a multiaxis "float," allowing positional and angular misalignments between parts to be accommodated. Easy matings can be accomplished between two parts, a tool and a part, a part and a fixture, a tool and a tool holder, and many other mating pairs. To show why the RCC is useful and how it works, we first discuss errors in assembly and the role of compliance between parts. The RCC is then described, and many examples of its use are listed, along with the range of equipment it can be attached to.

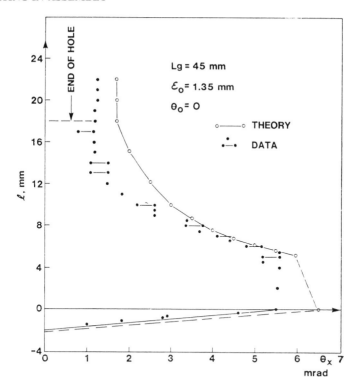

Fig. 64.25. Comparison of theory and experiment: conditions as in Figure 64.20. Plot of inclination angle versus insertion depth, as shown schematically in Figure 64.8.

64.8.1. What the RCC Does, and How It Does It

The RCC is designed to hold a workpiece so that the piece can rotate about its tip, that is, about the point where it engages a mating part. This allows the workpiece to respond to contact forces during insertion, which tend to realign the piece to the insertion axis. If the initial angular error is less than θ_2 in Eq. (64.54), successful insertion will result in spite of the effects of friction and regardless of the initial lateral error (within some limits.)

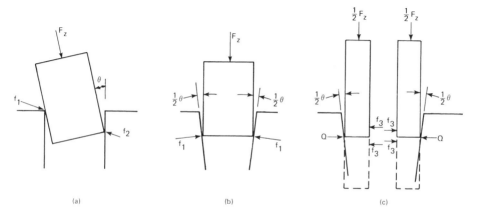

Fig. 64.26. Forces acting during wedging.

TABLE 64.1. RESULTS OF WEDGING ANALYSIS

F_z, N	44.5	222	445	2222	4450
Q, N	111.25	556.25	1112.5	5562.5	11125
δ, mm	1.6×10^{-3}	4.57×10^{-3}	7.36×10^{-3}	2.16×10^{-2}	3.56×10^{-2}
Δ, mm	0.32	0.934	1.47	4.34	6.91

A crucial feature of the RCC is that lateral error and angular error are absorbed independently. Its design permits lateral motion in response to laterally directed contact forces (such as those experienced during engagement) without any accompanying angular motion. A workpiece that engages a chamfer owing to lateral error will then slide down the chamfer toward the mouth of the hole. Thus the part behaves as if suspended compliantly from its *tip*. As long as Eq. (64.54) is obeyed and lateral error stackup can be controlled so that the chamfers engage, successful engagement and mating will occur. The size of chamfers can obviously be chosen to help achieve this, a much less costly approach than attempting to eliminate the errors themselves.

The figures show how the RCC accomplishes these motions. One part of the device holds the piece so that its angular motion is forced to occur about a point in space (the *remote center*) (Figure 64.27). The other part of the device allows the first part to translate (Figure 64.28). During a typical assembly the lateral part does the work during chamfer crossing while the angular part takes over during insertion. Figure 64.29 shows these functions in a commercial version of the RCC.

A typical installation of the RCC places it in a workhead just behind the tool or gripper. The combined length of the tool and gripped part should be such as to put the part's tip at or near the remote center, whose location is fixed with respect to the workhead. Exact coincidence of the tip and the center is not necessary because tests have shown that axial deviations of about 10–15% of the RCC-center distance do not degrade performance significantly.

Designers contemplating using the RCC should bear in mind the following limitations. It is not designed to cope with the case where error is so large that chamfers do not meet. At present, changing the center location during operations is not possible, although methods for accomplishing this exist. If the RCC is to be used to perform insertions along a horizontal axis, some counterbalancing may be necessary. Also, there are limits to the amount of lateral and angular error that can be absorbed by a unit of any one size because of the need to keep stresses in the RCC below elastic limits. Finally, theory has shown that Eq. (64.54) must be satisfied in order that the parts not jam and deform along a line between the contact points.

Within these limits, the RCC can be thought of as a general error absorber with the special ability to perform close clearance insertions. The size of errors it can absorb is large enough (typically 1–2 mm and one to two degrees) that system designers can relax many of their design constraints, such as feeder alignments, pallet uniformity, and part tolerances. This relaxation can be used to reduce

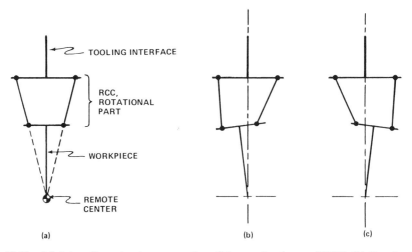

Fig. 64.27. (*a*) A two-dimensional representation of the rotational part of RCC. (*b*) Rotational part of RCC allowing workpiece to rotate counter-clockwise. (*c*) Workpiece rotating clockwise.

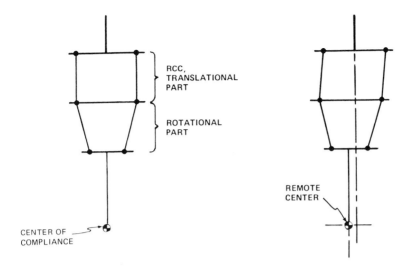

Fig. 64.28. (Left) Planar representation of RCC showing rotational and translational parts. (Right) Translational part of RCC allowing workpiece to translate to the left without rotating.

Fig. 64.29. Commercially available RCC.

cost, reduce time to set up and debug a system, or increase the complexity of a system without increasing its downtime.

64.8.2. Examples of Use of RCCs

The following list is intended to illustrate, but not limit, the possible applications of the RCC. Examples are grouped into clearance fits, interference fits, tooling interfaces, and some speculative possibilities. Examples that have actually been verified in our laboratory are marked with an asterisk (*).

Clearance Fits

Bearings into Housings.*   The bearing can be laterally displaced 2 mm (0.08 in.) and still successfully enter a 40 mm (1.6 in.) diameter hole with 0.01 mm (0.0004 in.) clearance. One mm (0.04 in.) of this lateral error is due to tolerance stackings between the location of the hole and location of jig points. During powered mating, engagement and insertion occur in 0.2 sec.

Shafts into Bearings.*   The shaft easily enters the bearing despite the existence of threads on the end of the shaft.

Gear onto Spline Shaft.*   Here the RCC's ability to comply about the insertion axis is utilized.

Rivets into Holes.   Here the holes may be imprecisely located, or the thin material containing them is variable in shape.

Screws into Threaded Holes.*   This is similar to plain peg and hole except that torsional stops on the RCC allow torque to be exerted on the screw.

Cover onto Box.   The RCC can mate rectangular cross sections as well as circular. The amount of angular misalignment about the insertion axis must be controlled so that chamfers on cover and box engage.

Rivet Tool over Rivet Head.   The RCC allows mating to a convex part too. Similarly, donut onto peg can be performed.

Forging into Die.   Rough forgings are of unpredictable shape. Often they can be dropped into the die and will land correctly. If they land incorrectly the die may be destroyed. If one depends on hidden float the forging may not seat firmly.

Unmachined Side of Casting onto Pallet Locating Pins.*   The casting may be grasped on a machined surface, but the pins will mate with unmachined surfaces whose location is uncertain.

Precise Assembly of Delicate Parts.*   The errors will be quite small here, but so are the clearances. The parts are fragile and have easily spoiled surfaces. The RCC serves to guide them together while protecting them from jamming and large contact forces. Figure 64.30 shows a manually operated inserter based on this idea.

Interference Fits

Nozzle into Housing.*   The nozzle is brass, about 3 mm (0.12 in.) diameter by 5 mm (0.2 in.) long. The housing is aluminum. The interference is about 0.02 mm (0.0008 in.). The holes in the housing can be improperly located by as much as 0.5 mm (0.02 in.), and rigid tooling tends to use the nozzle as a broach during insertion, ruining the assembly. Mounting the tooling on an RCC solved the problem with no increase in insertion time (about 0.1 sec).

Bushing into Housing.   Same idea as nozzle, but larger.

Shaft into Laminate Stack.   Essentially this is a force fit of a round peg into a round hole. The laminations must be supported to avoid peeling them apart. The RCC does not solve this part of the problem.

Thread Rolling Screws into Untapped Holes.*   This is really no different from screws into tapped holes.

Snap Fit of Sheet Metal Parts.   Here it is important that the parts not be too compliant compared to the RCC or else they, rather than the RCC, will bend out of shape in response to the contact force.

Tooling Interfaces

Tool into Machine Tool Changer, or Tool Storage Socket.*   Here the RCC functions as in any clearance, light press or snap fit. Once the tool has been grasped by the machine, however, the RCC may have to be locked tight for the tool to function. This is true of cutting tools, for example.

Tool onto Guide Pins.   The pins may be on the fixture which holds the workpiece. The tool is located to the pins and then is activated. The RCC helps the location process and assures that no large hidden stresses build up in the tool, pins or fixture. This helps reduce wear.

Fig. 64.30. Manually operated inserter for precision assembly. The inserter contains an RCC, 6-axis load cell, and brake. Electronics activates brake if force exceeds preset level.

Drill into Drill Bushing. Here the ability of the RCC to guide the drill into the bushing with little side force greatly reduces bushing wear. It also allows relatively imprecise drilling equipment to drill very precisely located holes.

Tool onto Part Held in Escapement.* The escapement may be at the end of a feeder track, from which the tool must pick it up. The RCC allows for error in the angular and lateral location of the feeder, wear in the escapement, and part-to-part differences.

Force Sensor into Contact with Object.* Here the goal may be to perform force feedback assembly, or to test for presence of a part, or other operation where amount of force exerted must be measured. If there is uncertainty as to the object's location or if the position resolution of the device carrying the sensor is coarse, forces larger than the sensor's operating limit may be encountered, and no meaningful readings will be obtained. This is especially likely if small forces are being sought and a sensitive sensor is in use. The RCC acts as a multiaxis cushion, providing small forces in response to large displacements, keeping the exerted forces small and allowing them to be measured.

Possible, but Speculative Uses

Alignment of Press, Stamp or Mold Dies. Alignment is provided by guide pins or rods, and the RCC holds half of the die as it engages the pins. This makes construction of the machine much easier since initial alignment of the separated die halves is not so critical.

Universal Joint for Laterally and Angularly Misaligned Shafts. Torsional stops would, of course, be necessary.

Coupling Space Manipulators to Precessing, Rotating Target Satellites to Aid Retrieving Them.

Related Concept

"Projected Elastic Center" Technique for Shock Mounting Machinery and Aircraft Engines. This concept has been a major product of Lord Corporation, Erie, Pennsylvania, for many years.

64.8.3. Equipment the RCC Can Be Used with

The RCC can be used for assembly and material handling in conjunction with conventional powered workheads, manually operated workheads, industrial robots, and remote manipulators.

The alignment, setup, and maintenance problems in standard transfer machines can be eased if the workheads are equipped with RCCs. Manually operated workheads can be augmented with RCCs, too, allowing assemblers to perform difficult insertions (fans onto motor shafts, for example) more easily and quickly.

Robot assembly is a most fruitful area for RCCs because the robot must reach to so many different locations. The ability to absorb the errors at all of these places with the same RCC will probably be essential in making such a concept successful.

Remote manipulators are common in underwater and nuclear radiation environments. The inability to line up the manipulator's end effector with the task objects is one reason why manipulation is slow and consequently costly. Previous approaches to this problem involve providing force sensing and feedback to the operator. An alternative is provision of an RCC in the manipulator's wrist. Approximate alignment of the manipulator is then sufficient, saving time, cost, and operator fatigue.

64.8.4. Some Misconceptions

"The RCC works only if it is held vertically so that the insertion axis is vertical." No—the RCC retains its essential properties regardless of which direction it faces. A deadweight load may indeed deflect the RCC, but often this can be counteracted by a position adjustment or a counterweight.

"The RCC is only for close clearance insertions." No—while the RCC is most spectacular in this application, its real significance to a system designer is as a general error absorber.

64.8.5. Summary

The RCC is a controlled, documented, reproducible source of multiaxis compliance or float that allows easy interfacing of mechanical mating parts in spite of initial lateral and angular misalignments. Its greatest potentials lie in accomplishing difficult insertions and in providing a valuable margin of error in constructing and maintaining many kinds of machines.

64.9. CONCLUSIONS

Geometric and force-deformation analyses for rigid part mating have been presented, covering the main geometric phases of assembly plus the phenomena of wedging and jamming. The ability of the RCC to reduce mating forces and the chance of jamming have been explained. The models have been verified by experiments. These results allow calculation of allowed lateral and angular error limits plus tolerance on the ideal location of the compliance center. If maximum allowed values of the contact forces are known, then this plus maximum allowed errors permit calculation of good values for K_x and K_θ.

ACKNOWLEDGMENTS

This work summarizes, reformulates, and extends a large base of work performed by the author and his colleagues over several years. The major contributors are Sergio N. Simunovic, Samuel H. Drake, Paul C. Watson, James L. Nevins, Anthony S. Kondoleon, Donald S. Seltzer, Daniel R. Killoran, Albert E. Woodin, Richard E. Gustavson, and Thomas L. DeFazio. Valuable discussions and corrections were made by Michael P. Hennessey.

The work was supported by the National Science Foundation under Grant No. DAR-79-10341 and predecessor grants, and by the Ecole Polytechnique Federale de Lausanne, where the author spent a sabbatical quarter. All findings and conclusions are the responsibility of the author and do not necessarily reflect the views of the National Science Foundation or the Ecole Polytechnique Federale de Lausanne.

REFERENCES

1. Nevins, J. S., et al., Exploratory Research in Industrial Modular Assembly, C. S. Draper Laboratory Report No. R-800, March 1974.
2. Ibid., Report No. R-850, December 1974.
3. Ibid., Report No. R-921, October 1975.
4. Ibid., Report No. R-996, August 1976.
5. Ibid., Report No. R-1111, August 1977.
6. Drake, S. H., The Use of Compliance in a Robot Assembly System, presented at the IFAC Symposium on Information and Control Problems in Manufacturing Technology, Tokyo, October 1977, and Using Compliance in lieu of Sensory Feedback for Automatic Assembly, Ph.D. thesis, M.I.T. Mechanical Engineering Department, Cambridge, Massachusetts, September 1977.

7. Nevins, J. L. and Whitney, D. E., Computer-Controlled Assembly, *Scientific American,* Vol. 238, No. 2, February 1978, pp. 62–74.

8. Lynch, P. M., Economic-Technological Modeling and Design Criteria for Programmable Assembly Machines, DL T-625, Ph.D. thesis, M.I.T. Mechanical Engineering Department, Cambridge, Massachusetts, June 1976.

9. Watson, P. C. and Drake, S. H., Pedestal and Wrist Force Sensors for Automatic Assembly, CSDL Report No. P-176, June 1975, presented at the 5th International Symposium on Industrial Robots in Chicago, September 1975.

10. U.S. Patent No. 4,094,192, Watson, P. C. and Drake, S. H., Methods and Apparatus for Six Degree of Freedom Force Sensing, June 13, 1978.

11. Whitney, D. E., The Mathematics of Coordinated Control of Prosthetic Arms and Manipulators, *ASME Journal of Dynamic Systems, Measurement and Control,* December 1972, pp. 303–309.

12. Whitney, D. E., Force Feedback Control of Manipulator Fine Motions, *ASME Journal of Dynamic Systems, Measurement and Control,* June 1977, pp. 91–97.

13. U.S. Patent No. 4,098,001, Watson, P. C., Remote Center Compliance System, July 4, 1978, and U.S. Patent No. 4,155,169, Drake, S. H. and Simunovic, S. N., Compliant Assembly Device, May 22, 1979.

14. Whitney, D. E., Nevins, J. L., and CSDL staff, What is the Remote Center Compliance and What Can It Do?, CSDL Report P-728, November 1978, presented at the 9th International Symposium on Industrial Robots, Washington, D.C., 1979.

15. Watson, P. C., A Multidimensional System Analysis of the Assembly Process as Performed by a Manipulator, presented at the 1st North American Robot Conference, Chicago, October 1976.

16. Simunovic, S., Force Information in Assembly Processes, presented at the 5th International Symposium on Industrial Robots, Chicago, September 1975, proceedings published by Society of Manufacturing Engineers.

17. Simunovic, S. N., An Information Approach to Part Mating, Ph.D. thesis, M.I.T., Cambridge, Massachusetts, April 1979.

18. U.S. Patent No. 4,156,835, Whitney, D. E. and Nevins, J. L., Servo-Controlled Mobility Device, May 29, 1979.

19. De Fazio, T. L., Displacement-State Monitoring for the Remote Center Compliance (RCC)— Realizations and Applications, *Proceedings, 10th International Symposium on Industrial Robots,* Milan, March 1980.

20. Seltzer, D. S., Use of Sensory Information for Improved Robot Learning, SME Paper MS79-799, presented at Autofact, Detroit, November 1979.

21. *Product System Productivity Research,* Vol. 2, *Productivity, Technology and PSPR,* collaborative study by C. S. Draper Laboratory and M.I.T. Center for Policy Alternatives, January 1976.

22. Nevins, J. L., et al., Exploratory Research in Industrial Assembly Part Mating, C. S. Draper Laboratory Report R-1276, March 1980.

23. McCallion, H. and Wong, P. C., Some Thoughts on the Automatic Assembly of a Peg and a Hole, *Industrial Robot,* Vol. 2, No. 4, 1975, pp. 141–146.

24. Savischenko, V. M. and Bespalov, V. G., Orientation of Components for Automatic Assembly, *Russian Engineering Journal,* Vol. 45, No. 5, 1965, p. 50.

25. Andreev, G. Y. and Laktionev, N. M., Problems in Assembly of Large Parts, *Russian Engineering Journal,* Vol. 46, No. 1, 1966, p. 60.

26. Laktionev, N. M. and Andreev, G. Y., Automatic Assembly of Parts, *Russian Engineering Journal,* Vol. 46, No. 8, 1966, p. 40.

27. Karelin, M. M. and Girel, A. M., Accurate Alignment of Parts for Automatic Assembly, *Russian Engineering Journal,* Vol. 47, No. 9, 1967, p. 73.

28. Gusev, A. S., Automatic Assembly of Cylindrically Shaped Parts, *Russian Engineering Journal,* Vol. 49, No. 11, 1969, p. 53.

29. Andreev, G. Y. and Laktionev, N. M., Contact Stresses During Automatic Assembly, *Russian Engineering Journal,* Vol. 49, No. 11, 1969, p. 57.

30. Andreev, G. Y., Assembling Cylindrical Press Fit Joints, *Russian Engineering Journal,* Vol. 52, No. 7, 1972, p. 54.

31. Kondoleon, A. S., Application of Technology—Economic Model of Assembly Techniques to Programmable Assembly Machine Configuration, S.M. thesis, M.I.T. Mechanical Engineering Department, Cambridge, Massachusetts, May 1976.

32. Harris, T. A., *Roller Bearing Analysis,* Wiley, New York, 1966.

33. Arai, T. and Kinoshita, N., The Part Mating Forces that Arise When Using a Worktable with Compliance, *Assembly Automation,* Vol. 1, No. 4, August 1981, pp. 204–210.

34. Ohwovoriole, M. S., An Extension of Screw Theory and Its Application to the Automation of Industrial Assemblies, Ph.D. thesis, Stanford University Department of Computer Science, April 1980.

35. Whitney, D. E., Gustavson, R. E., DeFazio, T. L., Kaks, D., and Seltzer, D. S., Part Mating Theory for Compliant Parts, C. S. Draper Laboratory Report R-1407, October 1980.

36. Whitney, D. E., Gustavson, R. E.., and Hennessey, M. P., Part Mating Theory for Compliant Parts, C. S. Draper Laboratory Report R-1537, August 1982.

37. Hennessey, M. P., Compliant Part Mating and Minimum Energy Chamfer Design, S.M. thesis, MIT Mechanical Engineering Department, Cambridge, Massachusetts, June 1982.

38. Whitney, D. E., Quasi-static assembly of compliantly supported rigid parts, *ASME Journal of Dynamic Systems, Measurement, and Control,* March 1982, Vol. 104, pp. 65–77.

CHAPTER **65**

APPLICATIONS OF ROBOTS
IN ASSEMBLY CELLS

SEIUEMON INABA

Fanuc Corporation
Yamanashi-Ken, Japan

65.1. ROBOTS FOR ASSEMBLY CELL

The robots destined for work in the assembly cell are constructed to be suitable for that purpose. They can operate with four- or five-axis control and, also, they can be provided with an automatic hand changer for convenient assembly. Examples are illustrated in Figures 65.1, 65.2, 65.3, and 65.4. The motion range differs according to robot size, as illustrated in Table 65.1 for some Fanuc assembly robots.

65.2. ASSEMBLY USING AHC (AUTOMATIC HAND CHANGER)

The assembly cell robot changes its hand by itself, so that it can assemble parts having different profiles. This is illustrated in Figures 65.5–65.9.

65.3. ASSEMBLY OF PUMPS USING ROBOTS

Three assembly cell robots sequentially grip workpieces on the table for workpieces, or rotary workpiece feeder, to assemble them and carry finished pumps to the rotary workpiece feeder (Figures 65.10, 65.11). Assembly work is done according to the procedure shown in Figure 65.12.

65.4. ASSEMBLY OF MOTORS USING ROBOTS

Three robots for assembly cell A0 and one feeder robot M1 are used to sequentially grip parts from a rotary workpiece feeder or a parts stocker, and carry finished products to a rotary workpiece feeder (Figures 65.13, 65.14).

This assembly cell comprises the work tables for LOCKTIGHT coating, magnetization, pressure fit of oil seal, pressure fit of bearings, assembly of rotor/shell/end bell, screw-in of tie-bolts, and tightening of washers and nuts. Each task is executed by assembly cell robots and automatic machines. The workpieces on each work table are sequentially fed by the feeder robot. Each robot executes the work as depicted in Figure 65.15.

65.5. ASSEMBLY OF ELECTRONIC UNITS BY ROBOTS

This system consists of assembly cell A using robots A1, A0; assembly cell B using robot A1; and feeder robot M (four robots in total), plus parts carry-in /carry-out conveyors. It sequentially assembles carried-in parts, and carries out finished products, as shown in Figures 65.16. 65.17.

Assembly Cell A

Conveyors are installed to feed one kind of base and five kinds of units.

Robot A1 is equipped with an AHC, and it employs hands for gripping printed circuit boards (PCB), rectangular workpieces (OD), thin plates, and other workpieces.

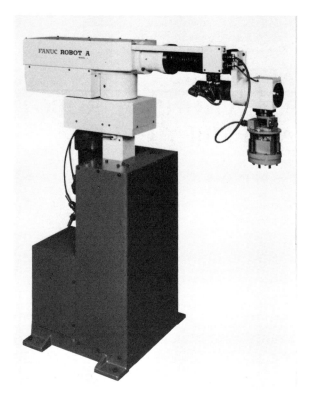

Fig. 65.1. Overview of robot for assembly cell.

Assembly Cell B

Conveyors are installed to feed one kind of base and two kinds of units.

Robot A1 is equipped with an AHC, and it employs hands for gripping thin plates, gripping keyboards, protruding pins, and other work.

Insertion of PCB in assembly cell A is depicted in Figure 65.18. Insertion of various units in assembly cell B is depicted in Figure 65.19.

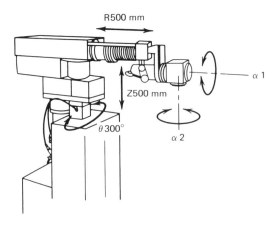

Fig. 65.2. Four-axes control.

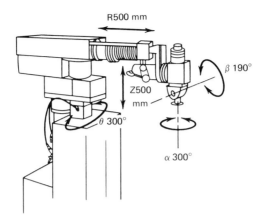

Fig. 65.3. Five-axes control.

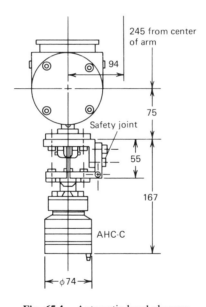

Fig. 65.4. Automatic hand changer.

TABLE 65.1. MOTION RANGE ACCORDING TO ROBOT SIZES

Type	R (mm)	Z (mm)	$\alpha 1$	$\alpha 2$	β
00	300	150	Not provided	Provided	Not provided
0	300	300	Provided	Provided	Provided
1	500	500	Provided	Provided	Provided

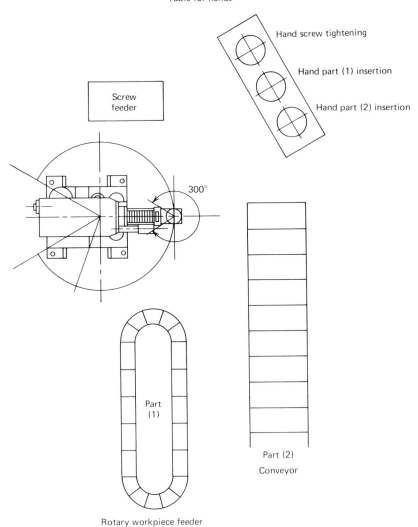

Table for hands

Hand screw tightening

Hand part (1) insertion

Hand part (2) insertion

Screw
feeder

300°

Part
(1)

Part (2)
Conveyor

Rotary workpiece feeder

Fig. 65.5. Robot for assembly cell.

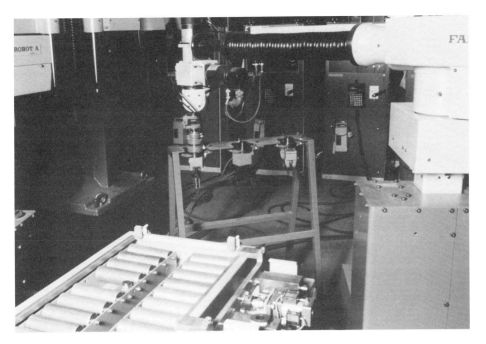

Fig. 65.6. Overview of table for hands.

Fig. 65.7. Assembling hand (1).

Fig. 65.8. Assembling hand (2).

Fig. 65.9. Assembling hand (3).

Fig. 65.10. Overview of assembly cell.

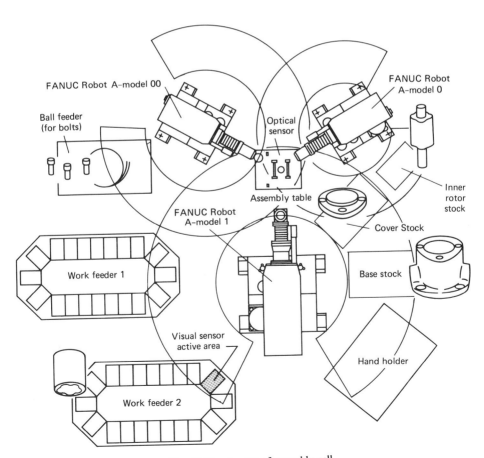

Fig. 65.11. Layout of assembly cell.

Bolt	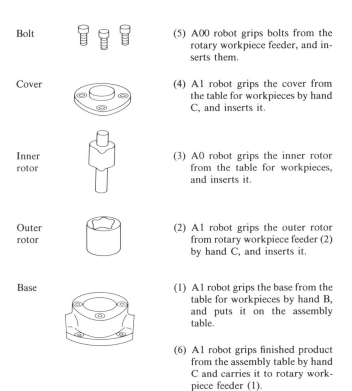	(5) A00 robot grips bolts from the rotary workpiece feeder, and inserts them.
Cover		(4) A1 robot grips the cover from the table for workpieces by hand C, and inserts it.
Inner rotor		(3) A0 robot grips the inner rotor from the table for workpieces, and inserts it.
Outer rotor		(2) A1 robot grips the outer rotor from rotary workpiece feeder (2) by hand C, and inserts it.
Base		(1) A1 robot grips the base from the table for workpieces by hand B, and puts it on the assembly table.
		(6) A1 robot grips finished product from the assembly table by hand C and carries it to rotary workpiece feeder (1).

Fig. 65.12. Pump assembly procedure.

Fig. 65.13. Overview of assembly cell.

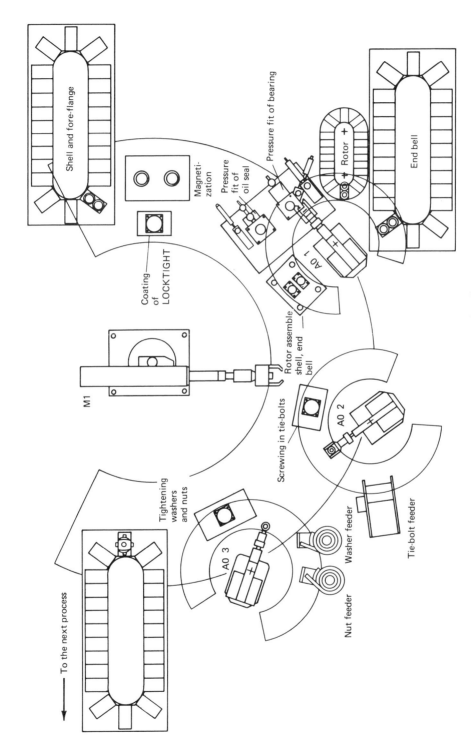

Fig. 65.14. Layout of assembly cell.

Shell and fore-flange

Magnetization

Pressure fit of oil seal

Pressure fit of bearing

Rotor

End bell

Coating of LOCKTIGHT

M1

Rotor assemble shell, end bell

AO 1

Screwing in tie-bolts

AO 2

Tightening washers and nuts

AO 3

Washer feeder

Tie-bolt feeder

Nut feeder

To the next process

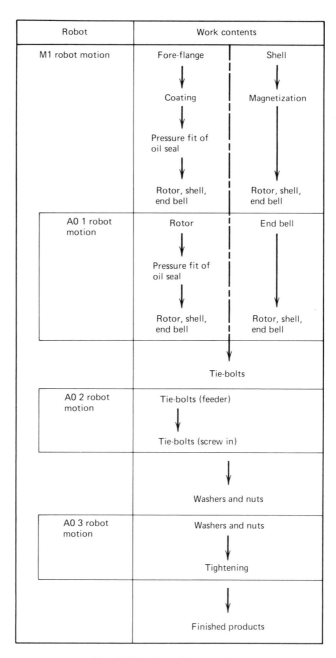

Fig. 65.15. Assembly procedure.

Fig. 65.16. Overview of assembly cell.

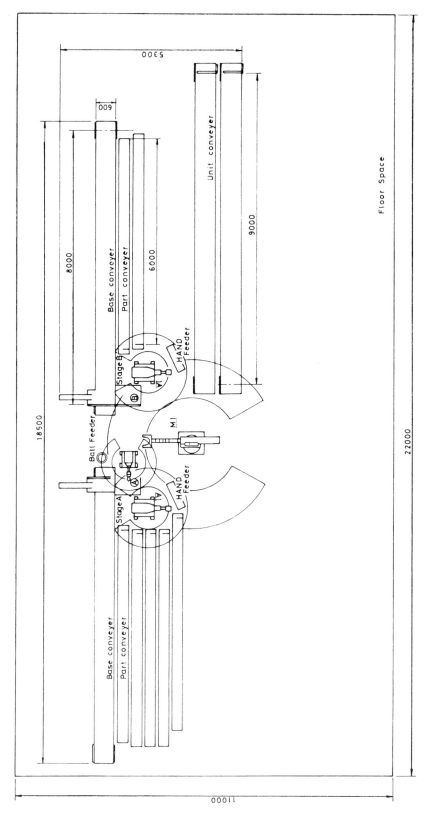

Fig. 65.17. Layout of assembly cell.

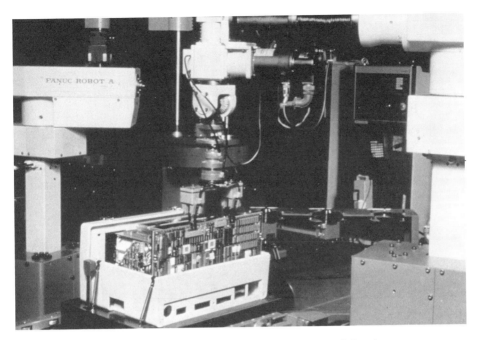

Fig. 65.18. Insertion and assembly of printed circuit board.

Fig. 65.19. Insertion and assembly of units.

CHAPTER 66

ASSEMBLY CASES IN PRODUCTION

KENICHI ISODA

MICHIO TAKAHASHI

Hitachi
Tokyo, Japan

This chapter describes four examples of robot applications from production lines of Hitachi, Ltd. They are all self-designed and developed by the company as well as used within its factories: video tape recorder (VTR) assembly, assembly of rotary compressors, pressure gage assembly, and gearbox assembly.

66.1. SMALL PARTS ASSEMBLY FOR VTR MECHANISM

66.1.1. Outline of VTR Mechanism Assembly Process

The video tape recorder (VTR) mechanism is comprised of pressed parts, like chassis and levers; plastic molded parts; electrical parts, like motors and magnetic heads; soft or flexible parts like rubber belts, and so on (Figure 66.1). Assembly producibility of the product was evaluated at the time of introduction of the robots, according to the Assemblability Evaluation Method, developed by Hitachi, and the following improvements were made:

1. Improved construction, that is, easy-assembly design.
2. Reduction in the number of parts through providing several functions to the parts.
3. Breakdown of complicated parts into functional subassemblies to simplify assembly operations in the assembly line.
4. Connectorized electrical parts.

Consequently, it has been possible to simplify the assembly process and to perform automatic assembly in a short time cycle.

66.1.2. Outline of Automation System

The system has base machines for workpiece conveyance locating, parts supply units, and assembling units including robots as shown in Figure 66.2. Since the base machines are designed to be independently usable, a nonsynchronous assembly line of arbitrary length can be composed by combining these base machines. Some parts are fed from a vibratory bowl feeder, others are placed in magazines; subassembled parts are also arranged in magazines, and then these magazines are distributed from the warehouse to the assembly stations by self-driven vehicles.

The automatic assembly units include small-size assembly robots, pick-and-place units, and single-purpose machines such as screwing machines, oil-applying machines, spring-fitting machines, and rubber belt fitting machines. The robot station is composed of a base machine, a robot, and a magazine supply unit. First, a magazine is supplied and positioned at the predetermined location. Then, the robot grips the parts one by one, starting from the end of the magazine, and fits them onto the chassis positioned on the base machine. Figure 66.3 shows an external view of the robot station.

Small-size assembly robots developed by Hitachi are designed for economical mass production

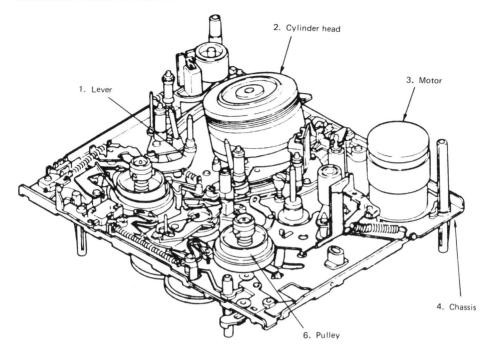

Fig. 66.1. VTR Mechanism.

assembly lines. As a result, improved flexibility and space efficiency are achieved to meet the production requirements for a variety of products.

66.2. ASSEMBLY OF ROTARY COMPRESSORS

66.2.1. Outline of Assembly Process

The T-shaped rotary compressors are mounted on home air conditioners. The rotary compressor consists of a rotary pump, a motor, and a case as shown in Figure 66.4. For the rotary compressor assembly, several elemental assembly functions are required, such as parts mating, centering, screwing, pressed insertion. The sections of the parts come in many forms ranging from circles to rectangles, and most of them require accurate parts mating. Therefore peripheral machines are placed along the rail track. They perform special operations, such as screwing and pressed insertion, that are not able to be performed by the assembly robot. The assembly robot performs accurate mating, loading and unloading of the peripheral machines, and transporting between processes.

66.2.2. Outline of Automation System

Figure 66.5 shows an external view of the rotary pump assembly line. The system consists of three traveling assembly robots, a circular rail track, peripheral machines such as screwing devices, and so on, and a host microprocessor. The traveling assembly robot transfers parts to and from the peripheral machines and completes the assembly after one trip around the circular rail track. The features of the components are as follows:

1. The robot arm has the electrically driven articulated structures, with 5 degrees of freedom (DOF). The arm is suspended by an arm supporter and it enables a wide range of motion capability.

2. The straddle-type drive unit rides on two rails and has a bogie truck construction to enable smooth traveling on the curved rail track. The wheels are covered by a urethane gum layer to prevent noise and vibration.

3. The robot controller adopts a pulse width modulation (PWM) servo amplifier using field effect transistor (MOS FET) and is driven in connection with the robot arm.

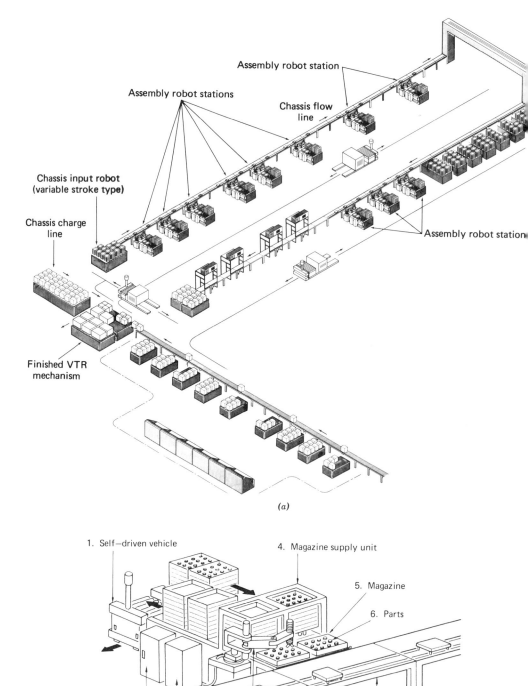

Fig. 66.2. VTR mechanism automatic assembly line. The assembly line (a) is U-shaped, and the total length is 150 m. The base machines (b) adopt a nonsynchronous direct-feed system, which does not use platens.

Fig. 66.3. Overview of the robot station. Eleven units of the assembly robot are introduced into this assembly line.

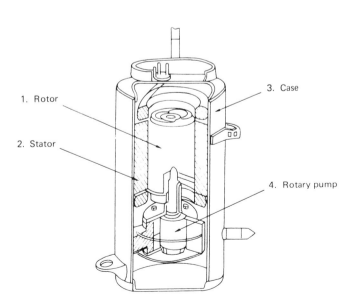

1. Rotor

2. Stator

3. Case

4. Rotary pump

Fig. 66.4. Construction of rotary type compressor. The compressor is mounted on home air conditioners and is composed of a rotary pump, a motor, and a case.

Fig. 66.5. Overview of a rotary pump assembly line. The automatic pump assembly line is composed of three traveling assembly robots, peripheral machines, and a host microprocessor.

4. Hand trajectory generation. A new algorithm for determining hand trajectory has developed. To shorten the moving time of the arm from a given point A to point B by point C, a parabolic interpolation curve technique has been applied—whereas the conventional point-to-point control allows only straight-line interpolation.

5. A mating table that helps the robot to perform a close peg-in-hole mating has been developed. The mating table consists of links and coil springs, and it is able to modify the position and orientation of the workpieces according to the reflecting forces between the mating workpieces.

66.3. ASSEMBLY OF PRESSURE GAGE

66.3.1. Outline of Assembly Process

The pressure gage is made to detect the pressure of gas and liquid and is used in chemical plants, steel factories, and power plants. The process, automated by the following assembly system, is to put stainless steel flanges onto both sides of the pressure-detecting subassembly and to fasten them by four through-bolts and nuts.

In Figure 66.6 the flange assembly is shown. The pressure-detecting element is sandwiched between two flanges, in which gaskets are buried, and held together by nuts and bolts.

66.3.2. Outline of Automation System

Figure 66.7 shows a diagram of the automation assembly system. The system consists of an assembly robot, an automatic screwing machine, and parts supply/discharge devices. The assembly robot grips parts supplied from parts-supply devices one by one, fits them into fixture for assembly, and picks the completed products up after the multiaxis automatic screwing machine has fastened the bolts. Figure 66.8 shows an external view of the system, and Table 66.1 gives the specifications of the system.

An articulated, 6-DF small-sized assembly robot has been adopted. This is because the objects are small, working spaces are limited, and the 6-DF robot is needed owing to complicated parts handling. The weight of a single part is about 3 kgf at maximum, but after the assembly, the weight of product

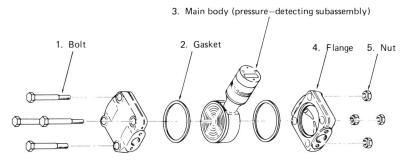

Fig. 66.6.   Components of a pressure gage.

is about 10 kgf. The load capacity of the robot has been increased to 15 kgf by hanging the robot wrist with a spring balancer for reduction of motor load.

A double-handed AHC system has been adopted because the robot must handle various kinds of workpieces. The system has two kinds of hands, a bolt/nut hand and a flange/main body hand. The flange/main body hand is set on the arm tip of the robot. If necessary, it grips the bolt/nut hand and uses it. A remote center compliance device for position slippage compensation of assembled parts is added to the hand tip to increase flexibility.

66.4.   ASSEMBLY OF A GEARBOX

66.4.1.   Outline of Assembly Process

The gearbox is a subassembly device of an elevator system. It is used to drive elevator cargo vertically. As shown in Figure 66.9, the main components of the gearbox are a driving motor, an electromagnetic brake, and a worm reduction gear.

The weight range of the components spreads from a few grams for parts such as washers to 290 kg for a motor. The worm shaft and the sheave shaft are assembled on the subassembly line. The assembled workpieces are inserted into the machine body on the main conveyor. The sheave and brake drum are also mounted at the same station.

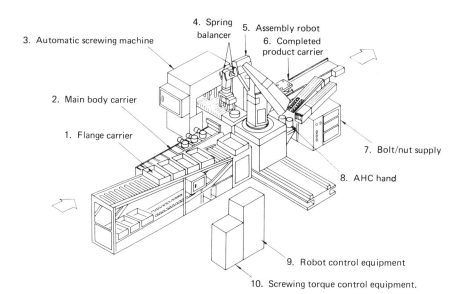

Fig. 66.7.   Schematic diagram of the assembly system. Supply, fitting, and conveyers are arranged around the robot.

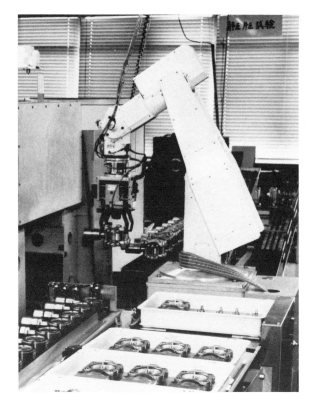

Fig. 66.8. Overview of the system. The robot is unloading fastened parts.

66.4.2. Outline of Automation System

The idea of the system is to combine an inexpensive light-load-capacity robot with an electrical hoist (balancer). The system enables a large load capacity with small cost. The total assembly system consists of the described balancer and single-purpose machines such as a pressing machine for press fitting of ball bearings and a screwing machine for fastening of nuts.

At present, the system performs only the subassembly of a worm shaft that is a part of the whole assembly process of the gearbox.

TABLE 66.1. SPECIFICATIONS OF THE SYSTEM

Workpiece	Pressure Gage
Number of the components assembled	11
Cycle time of the system	3 min
Arm:	
Degrees of freedom of motion	6
Maximum tip speed	1.0 m/sec
Repeatability	± 0.1 mm

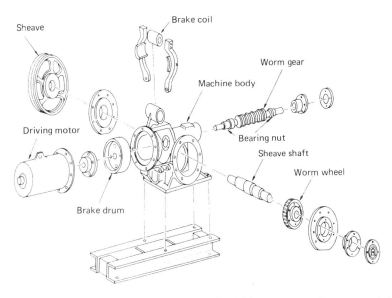

Fig. 66.9. Traction machine for elevator, composed of a driving motor, an electromagnetic brake, a worm reduction gear, and so on. Automatic assembly equipment is applied to the assembly process of the work shaft.

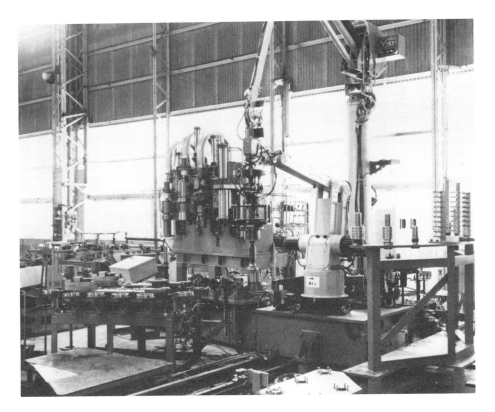

Fig. 66.10. Automatic assembly system of traction machine for elevator. The assembly system composed of a balancer-assisted robot and peripheral machines is able to handle light to heavy parts efficiently.

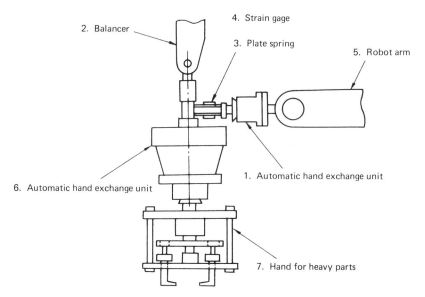

Fig. 66.11.   Robot and balancer cooperation system. The speed of the balancer is controlled by the robot. A force sensor at the wrist of the robot gives the controlling signals during the cooperation movement.

The features of the assembly system are as follows:

1. For light parts, assembly is performed by the robot alone. For heavy parts, if the weight exceeds the load capacity of the robot, the robot fetches the balancer arm and lets it carry the heavy parts.
2. The assembly of various parts can be performed by the single robot-balancer system through the automatic exchange of various types of hands.
3. The robot-balancer system is mounted on a track to enlarge the working area.

The external view of the system is shown in Figure 66.10. In Figure 66.11, at the wrist of the robot, a force sensor using strain gages is located. It measures the force that arises between the robot and the balancer. The robot-balancer coordinated motion is controlled using this measured force.

CHAPTER 67
ROBOTIC ASSEMBLY OF COMPUTER COMPONENTS

KEITH L. KERSTETTER

IBM Corporation
Boca Raton, Florida

67.1. INTRODUCTION: THE ROBOT SYSTEM

Complex, precision assembly and testing represent the smallest application area for today's industrial robotic systems, but it is the area with the greatest potential for growth. Further implementation of robotic assembly applications will depend on many factors, including the economic and social environments. However, from a technical standpoint, the major considerations for an assembly robot are (1) intelligence or computer control, (2) sensory feedback, and (3) design for automation.

Because it does have high-level computer control and sensory feedback, the IBM 7565 Manufacturing System has been able to handle a variety of assembly and test applications both in IBM and at customer locations on products not designed specifically for automated assembly. This chapter describes several IBM applications with the intent of showing the broad spectrum of applications an intelligent robot can address.

The 7565 system consists of a hydraulically powered, rectangular box-frame manipulator controlled by a modified IBM Series/1 computer (Figure 67.1). The system operates under the control of A Manufacturing Language (AML), a sophisticated robotic control language that combines manipulator control with data processing and data communications capabilities. For more information on AML, see Chapter 21, A Structured Programming Robot Language.

The manipulator can be configured for 3, 4, or 6 DF, with a seventh motion available in the servo-controlled, parallel opening and closing of the gripper fingers. The gripper can be equipped with both tactile and optical sensing. Tactile sensors—strain gages in the fingers—can detect forces in the sides, pinch, and tip, allowing the system to "feel" components. Optical sensing in the form of a LED beam between the fingers allows the system to detect the presence or absence of a part or tool. Optical sensing also allows the system to calibrate itself to base or reference points within the work area. This dynamic recalibration capability helps to eliminate the problems of drift caused by machine wear, ambient temperatures, changes in electrical currents or hydraulic fluids, and other variables in the manufacturing environment.

Under the control of AML, the computer monitors all seven axes every 20 msec, or 50 times a second. Furthermore, the control language starts and stops each axis simultaneously, regardless of the individual differences in distance to be traveled. This provides the smooth, fluid motion required for complex precision assembly applications.

Using sensory and positional feedback, sophisticated error recovery routines can be "nested" in the application program, allowing the system to identify and adapt to changes in the work environment. This error-recovery capability is illustrated by the applications.

AML also features data communications capabilities, enabling the robotic system to communicate with a host computer through an RS232C interface; data processing capabilities to keep records, generate reports, and control activity at the work station; and 64 digital inputs and 64 digital outputs to detect error conditions and control peripheral devices around the manipulator.

The rigid box-frame structure allows the robotic arm to travel at speeds to 40 in./sec (102 cm/s) with uniform repeatability throughout the $58 \times 18 \times 18$ in. ($147 \times 46 \times 46$ cm) work envelope. Positional repeatability of the three linear axes is ± 0.005 in. (± 0.13 mm). Maximum distance of arm travel is ± 8.9 in. (± 22.6 cm) for the X-axis; ± 29.4 in. (± 74.7 cm) for the Y-axis; and ± 8.8 in. (± 22.3 cm) for the Z-axis. Roll and yaw motions are each $\pm 135°$, while pitch is $\pm 90°$. Rotary motions have a velocity of $360°$/sec.

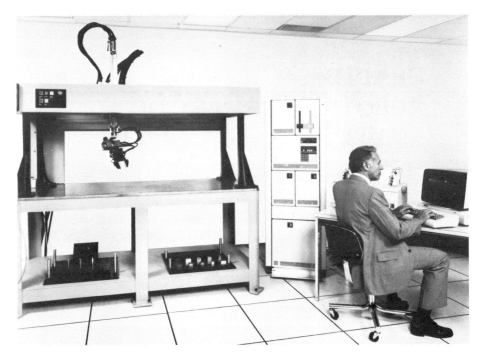

Fig. 67.1. The IBM 7565 Manufacturing System is a hydraulically powered, rectangular box-frame manipulator controlled by an IBM Series/1 computer.

The 7565 system and AML are based on more than 10 years of research and development and internal application of versions of the technology. The three applications described use the RS 1, a prototype.

67.2. AUTOMATED CARTRIDGE RIBBON ASSEMBLY

When IBM introduced its 535 data processing cartridge ribbon in 1980, the System Supplies Division in Dayton, N.J., began manufacturing the replacement cartridges for general-purpose printers by manual assembly. But, because of the high volumes involved, stringent quality control requirements, and the tedious nature of the assembly process, the goal was to implement an automated manufacturing process as rapidly as possible. Today, an almost totally automated system takes the cartridges from shell to packaged product over two 8-hr shifts, with minimal operator intervention. The automated assembly line has yielded a three-time increase in productivity and has virtually eliminated scrap in the manufacturing process.

The success of the automated line is the result of the creative blending of several hard-automation devices with computer-controlled flexible automation. At the heart of the line is a robotic manufacturing system with a 3-DF manipulator. The computer controls and coordinates peripheral hard-automation devices through digital input/digital output.

The cartridge ribbon assembly line is a nonsynchronous, continuous-loop, and constantly moving conveyor belt on a rectangular frame (roughly 40 × 16 ft, or about 12 × 5 m). Pallets traveling on the conveyor are raised above the moving belt by pneumatic lifting mechanisms at several work stations along the line where a variety of tasks are performed. After work at one station has been completed, the pallet is lowered and continues along the conveyor until the finished and tested cartridge is picked off the line by a cam-driven arm that places it on another conveyor for packaging. The empty pallet continues along the main conveyor line, ready for the assembly process to start again.

Placement of the cartridge base on the empty pallet is the first step in the assembly process. This is done by a hard-automation device that picks off the base from a vibratory feeder track and moves it in proper orientation to the pallet. As it grabs the base, the device also sprays a lubricant onto the journals into which gears will be inserted in a subsequent process.

The pallet then moves into the working area of the robotic system's manipulator. Here two gears are inserted, ribbon is loaded into the cartridge, and the top of the cartridge is placed onto the base.

The Series/1 control unit monitors the position and operation of the robotic arm to respond to changes in the work environment such as missing or damaged components and variations in position of feeders and fixtures.

In the cartridge ribbon application, the rotary motors of the manipulator have been replaced with a specially designed eight-position turret containing the tooling for the various assembly processes (Figure 67.2). Currently, only six of the positions are being used, leaving two positions available to accommodate new operations or changes in product design.

The computer also controls fixed automation devices through digital input/digital output to a programmable controller, providing a coordinated flow of pallets through the various work stations. This marriage of fixed and flexible automation illustrates an important point about computer-integrated manufacturing: most often the right manufacturing solution is a proper balance between hard and flexible machines.

The communication between computer, manipulator, and fixed automation provides the Dayton assembly line with an important quality control feature: Each pallet has a memory pin that determines whether or not work should be performed on the cartridge. If any errors occur during the assembly process—a ribbon snapping, for example—a pneumatic punch will depress the memory pin on the pallet, and it will travel around the line with no other work done on that particular cartridge. Error conditions are detected by external sensors, and the information is fed back to the computer.

67.2.1. Robotic Assembly

The first step in the robotic assembly process is to pick up four gears with the turret. The gears—two male and two female—are oriented by vibratory bowl feeders. The arm inserts one male and one female into the prelubricated journals of each cartridge using a zig-zagging motion to snap them in place.

To minimize the number of movements the arm must make and to maximize throughput, the manipulator works sequentially on two pallets at a time for each step of the robotic assembly process. The sequence of events during assembly is (1) pick up four gears, (2) thread ribbon, and (3) insert gears.

To thread the ribbon into the cartridge, the turret spins until a tweezerlike tool is positioned directly above the conveyor. Using this tool, the arm grips the loose end of ribbon from a dereeler. The dereeler is a staging area for the ribbon that will be fed into the cartridge. It consists of a spool

Fig. 67.2. Specially designed eight-position turret mounted to the end of Z-axis contains tooling required to insert gears and thread ribbon into the cartridge.

of ribbon and "stuffing box"—a plastic container into which 13 yards (11.9 m) of ribbon is loosely fed from the spool. This device maintains the proper spool tension to prevent snapping or overfeeding of ribbon during loading. Each pallet is serviced by two dereelers. So, as one spool of ribbon is depleted, an external sensor signals the computer through the programmable controller, which automatically activates the second dereeler.

Holding the loose end of the ribbon with the tool, the robotic arm moves through the ribbon feed track and brings the ribbon through the two gears or drive wheels (Figure 67.3). Next, a fixed automation device winds the 13 yards (11.9 m) of ribbon into the cartridge as another device positions the top of the case. After the ribbon is loaded, a hot wire cuts the loose end from the dereeler. The same procedure is repeated on the second pallet. The arm then returns to insert the gears on two incoming pallets while the forward pallets, with cartridges about 75% complete, proceed through the assembly line.

At subsequent stations, the ribbon is cut and bonded into a continuous loop, the cartridge top is welded to the base, and the completed cartridge is tested at high speed to make sure no twisting or binding has occurred. The completed and tested cartridge moves to the final station, where a cam-driven, hard-automated arm lifts the cartridge from the pallet and places it on a smaller conveyor parallel to the fourth side of the assembly line. This conveyor delivers the cartridge to an automated bagger where it is inserted into a plastic bag and sealed, ready for shipment.

The success of the automated cartridge ribbon line is evident in a few statistics: Compared with manual assembly, the automated line has increased production by about 300%, with a remarkably low scrap rate of less than 0.4%. In just over two years, the line has produced more than a million cartridge ribbons.

67.3. WORD PROCESSOR ASSEMBLY

Three RS 1 robotic manufacturing systems are used for on-line assembly of IBM Displaywriter word processing systems at the company's Communication Products Division plant in Austin, Texas. A fourth robot is used in a laboratory environment for building subassemblies and for training and application development work.

The Austin Process Development Lab (PDL), the group responsible for application development,

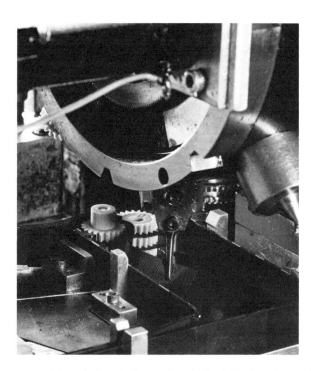

Fig. 67.3. Using a tweezerlike tool, the robotic arm threads the ribbon into the cartridge and through the two drive gears. Sensors detect if the ribbon snaps or is improperly installed, and the turret will automatically rotate to a punch tool to depress the memory pin on the pallet.

adapted the manipulator to Austin's particular environment and came up with a number of innovations that reduce cost and increase throughput and productivity using the flexibility inherent to intelligent robotic systems. The robots on the final assembly line straddle the line so that the conveyor moves through half of the work space under the arm.

The rectangular box frame has been tilted approximately 30° relative from front to back. This serves two purposes. First, it allows use of simple, inexpensive gravity parts feeders without losing the perpendicular orientation of the arm to the parts being fed. And it also allowed the PDL team to develop the cart concept, a highly efficient approach to robotic assembly in a batch-manufacturing environment.

Mobile carts containing all the parts, tools, feeders, and fixtures for a subassembly—a gear plate or base plate, for example—can be positioned in the robot's work space next to the Displaywriter assembly line. If for some reason the robot cannot work on Displaywriter kits on the main assembly line, the system automatically will move to the cart and perform subassembly work until it can resume operations on the main assembly line. This cart concept allows for optimum utilization of the robotic system. The subassembly operations performed on these carts are described after a review of the applications performed by the three on-line robotic systems.

The Displaywriter word processing system is composed of five major elements: keyboard, display, media box (disk drive), system electronics box (SEB), and printer. Robotic assembly is now used for the media box and the SEB.

Figure 67.4 shows the layout of the Displaywriter production line and the relative position of the three RS 1 systems. Robot A performs assembly operations on the SEB, while Robots B and C work together in a serial loop arrangement on the media box. Both the SEB and the media box arrive at the assembly stations in kits which are manually assembled at different stations and then sent to the robotic assembly stations.

Media box kits are delivered to the robotic work station by an elevator and roller conveyor. Kits are routed first to Robot B, then to Robot C, and finally onto the main conveyor for final manual assembly. SEB kits travel only to Robot A for assembly operations and then directly to the main conveyor.

67.3.1. SEB Assembly

Robot A performs two major assembly operations on the SEB. First, it secures the power supply to the cover. In the other operation, it builds a motherboard subassembly, a structural support for the Displaywriter system's logic cards.

As the SEB kit comes into the robotic work area on a conveyor, a digital output-controlled air cylinder secures it in the fixture. The arm pushes the power supply into the fixture and lines up the holes in the power supply with screw inserts in the cover. The arm then moves to pick up a special, multifunction tool which it uses to pick up four screws, driving them one at a time through the power supply into the cover.

The motherboard subassembly consists of a board, two card support frames, two metal brackets, and four screws. Boards are picked off a stack by air-cylinder-driven pickoff knives and placed in a fixture. Using the same multipurpose tool it used to drive the screws, the robot picks up a card support frame and uses the frame to pick up two screws for a double-headed screw dispenser (Figure 67.5). The screw dispenser was designed so that screw positions match the holes on the frame, allowing the robot to move the frame under the dispenser to lift the screws in place. The arm places the support frame on the motherboard, where it is automatically latched in place, then drives the screws in sequence.

The same procedure is repeated for the second support frame, and then all clamps are automatically released. The arm lifts the completed assembly by one of the support frames and places it on top of the power supply in the SEB kit. The kit is released from the fixture, and the robot pushes the pallet back onto the main conveyor line.

67.3.2. Media Box Assembly

Robot B drives eight screws—three different types—during its assembly work on the media box. Using a general-purpose, multifunction air driver tool, the robot picks up the screws fed to it by vibratory bowl feeders and rotary escapement devices (Figure 67.6).

An air cylinder pulls the kit into the work envelope where it is clamped into a fixture. The base plate and the two disk drives are locked in place for the assembly operation. Again, the air cylinder is controlled by digital input/digital output to and from the robotic system's computer.

Holding the general-purpose driver between its fingers, the robot moves to pick up a magnetic driver bit and proceeds to drive two large screws to secure the base plate to the front disk drive. Next, the manipulator changes drive bits, picking up a smaller one to drive one small screw into the front of the disk drive and three small screws into the second disk drive. The arm uses the same

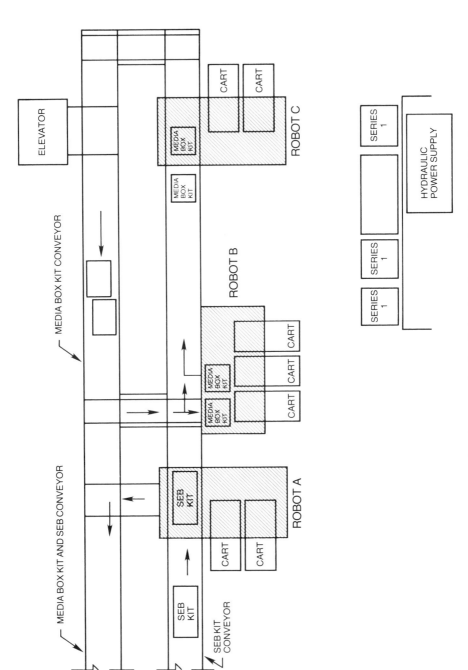

Fig. 67.4. Layout of Displaywriter production line. Robot A performs work on the SEB, while robots B and C work in serial loop arrangement on the media box.

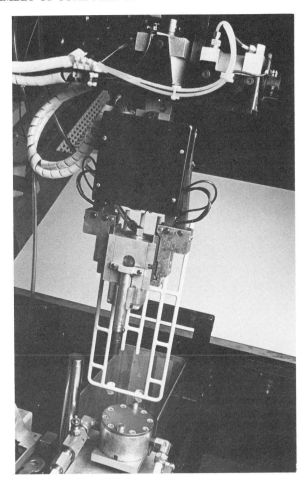

Fig. 67.5. Holding a multipurpose tool between the gripper fingers, the robotic arm picks up a card support frame. The frame is used to pick up screws from a double-headed screw dispenser.

magnetic drive bit to pick up two more screws and prestarts them in the base plate, saving the manual operator time later in the manual assembly area.

When all eight screws are driven, the media box moves to the work space of Robot C, where a plastic strain relief is inserted and secured and four rubber feet are attached to the base plate. After the base plate is clamped into the fixture, the arm picks up the strain relief and places it in the fixture (Figure 67.7). The robot uses sensory feedback to locate the strain relief, which is packaged loosely in the kit. After insertion, Robot C picks up and drives three screws—one at a time—to fasten the strain relief to the base plate.

For the next assembly operation, Robot C picks up four rubber feet, again one at a time, and fastens them to the base plate. Screws are inserted into the feet by a bowl feeder prior to the robotic assembly operation.

The final assembly operation on Robot C is to start two screws on the bottom of the base plate, again to enhance productivity in the manual assembly operation.

67.3.3. Error Recovery

One of the keys to the Austin robotic applications—and to any intelligent robotic applications—is the system's ability to detect a variety of errors in the manufacturing environment and, through programming, to take corrective action to resolve the problem and complete the manufacturing process.

Any number of error conditions can occur during robotic assembly on the Displaywriter. For example, screws may have damaged threads or may be missing the slot on the head. Holes may

Fig. 67.6. The same multipurpose tool is used to drive different types of screws, fed to the system by vibratory bowl feeders and rotary escapement devices.

Fig. 67.7. Robot uses sensory feedback to locate loosely packed strain relief and inserts it in fixture on media box. Three screws are driven one at a time to secure strain relief.

have been tapped improperly. Parts feeders can empty or jam. The robotics systems automatically detect these conditions through sensory feedback in the gripper.

If the arm attempts to drive a screw with no slot, the screw will return with the driver. Tactile, or force, feedback tells the system the screw is still in the magnetic bit when the arm moves to pick up another screw. The faulty screw is rejected and another is retrieved and installed. If a hole were to be improperly tapped or missing, the system would sense the condition and reject the component. Typically, the system will be programmed to try an operation three times. If it is unsuccessful after the third try, a signal is given for operator intervention.

Parts are usually fed into the work area through multiple feeders, so if one were to run out of parts or jam, the system is programmed to move automatically to the next available feeder. Again, if it cannot recover from the error condition, an operator is alerted.

67.3.4. Cart Concept

The cart concept developed by the PDL aptly illustrates the RS 1's error recovery capability, particularly a gear plate assembly application performed by an off-line system. To assemble a gear plate used in the paper-feed mechanism of the Displaywriter printer, the system first moves the manipulator to a reference point or "find" post on the cart which is wheeled into the robot's work space. Using optical sensing, the system calibrates itself to the post to determine the orientation of the parts and tools.

It should be pointed out that this capability eliminates the need for elaborate devices to ensure that components line up exactly in the same place each time the cart is changed. The computer simply adjusts the application steps to correspond to the data it receives during the calibration routine.

Using a general-purpose tool, the arm picks up a metal plate from a gravity feeder and places it on a fixture. With the same tool it picks up two gears and places them on posts on the plate, then returns and installs a third gear (Figure 67.8). The arm then exchanges the general-purpose tool for a C-clip applicator and inserts clips on each post to secure the gears and, with the opposite end of the tool, places a drop of oil on each post to lubricate the completed subassembly.

During assembly, gears may not mesh, C-clips may not fit because the post may be too short or

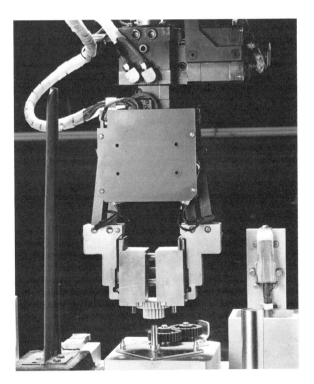

Fig. 67.8. To assemble a gear plate subassembly, robot uses tactile feedback to mesh gears while seating. After installing three gears, the arm exchanges tools and uses a C-clip adaptor to insert clips securing the gears on the posts.

the groove machined improperly, or parts feeders may be empty. If after placing two gears on their posts the system feels resistance as it tries to place the third gear, it has been programmed to move to the gear not seating—determined through tactile feedback—and will tap it from above. If that fails, the arm uses the third gear to turn the others so that the teeth mesh. Similarly, the system detects errors when applying C-clips and automatically discards damaged parts.

67.4. MAINFRAME COMPUTER TESTING

At the Data Systems Division plant in Poughkeepsie, N.Y., the company builds its largest mainframe computers, including the IBM 3081, 3082, and 3083 computers. A major component of the central processor of these systems is the thermal conduction module (TCM) board, which includes thousands of printed circuit connections. Each computer has four to nine TCM boards, depending on the model. A unique robotic tester—a two-armed version of the RS 1—provides 100% reliability testing of the printed circuits in one-twentieth the time it would take manually.

The robotic tester at Poughkeepsie has two arms, each with three linear axes, equipped with specially designed probes at the end of the Z-axis to perform the required test of the electronic network within the TCM board. Another modification is the extension of both Z-axes by about 4 in. (10 cm) to cover the height of the board.

Each of the 11 robotic testers is connected through a communications link to a host computer. Resident within the host is Poughkeepsie's Computer-Aided Repair and Rework System, which controls and manages the testing procedure for the entire system. When an operator keys in the serial number of a board to be tested, the host computer checks the serial number for accuracy and board status (Figure 67.9). The host then sends the appropriate test data to the Series/1 work station controller.

After the board has been tested, the controller sends a summary report and status of the test results to the host system, which automatically generates a list of repair actions for all verified defects.

67.4.1. Two-Armed Testing

The test sequence begins when the operator loads a TCM board into the robotic system's workspace. The Series/1 work station controller automatically sends a signal to a board-lifting fixture to lift the

Fig. 67.9. TCM boards are tested for reliability by a two-armed robotic tester. When operator keys in serial number of a board, a host computer checks number for accuracy and board status, then sends appropriate test data to Series/1 work station controller.

board into a vertical position. Alignment between the robotic arms and the board is done automatically through a program that keys off three reference points on the board. From these three calibration points, the board panel origin is calculated by a coordinate transformation routine, and directional vectors are generated for each arm. The test points are calculated by simple vector additions.

The testing begins at a signal from the operator. Each pair of test points is calculated based on test data sent from the host computer, and the two probes are sent to their respective destinations and engage C-springs on the TCM board (Figure 67.10). The Series/1 computer uses its digital output to signal a meter, and a reading is taken to determine whether that particular segment is good or bad. If a segment fails, it is retried several times at slightly offset locations. If a good test cannot be achieved, the failing segment is recorded and printed for later off-line analysis and rework. After each segment is tested, the Series/1 work station controller prints a summary of the test and signals the fixture to unload the board.

During the testing process, events are constantly monitored by the Series/1 work station controller. For example, the system will immediately discontinue robotic motion if someone were to step onto the safety mat, open the safety doors, or break the safety light curtain. This utilization of sensory feedback helps maintain a safe work environment. The system will automatically resume the test when the problem is corrected.

Use of the robotic tester allows a board to be completely tested in 2.5–5.5 hr, depending on the size and type of board. Manually, it would take more than 100 hr to complete the same test.

67.5. CONCLUSION

The applications in this chapter show some of the creative ways intelligent robotic systems can be used to lower manufacturing costs and increase productivity. Adaptive, flexible automation offers new tools to a wide range of industries to help meet the challenges of batch manufacturing and rapidly changing product designs. Because they can be reprogrammed and retooled to perform radically different operations in a relatively short time and at a lower cost than hard automation, robotic manufacturing systems will be an integral part of the computer-integrated manufacturing plant of the future.

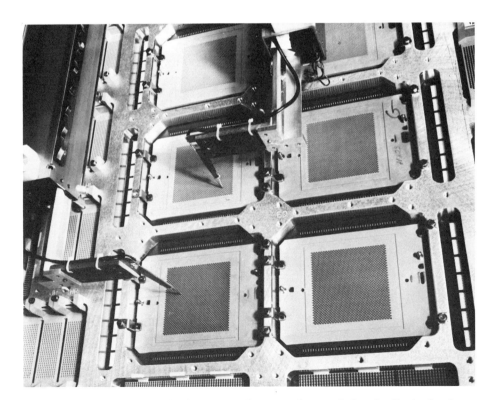

Fig. 67.10. Directed by the Series/1 computer, the two probes attached to the Z-axis of each arm engage C-springs on the TCM board. DI/DO is used to signal a meter reading to determine board status. If a segment fails, it is automatically retried several times at slightly offset locations.

REFERENCES

A list of publications follows in which application articles on IBM's robotic assembly have appeared.

1. Robot Improves IBM Cartridge Ribbon Assembly, *Assembly Engineering,* February 1983.
2. Robots on the Line—An Overview, *Electronic Packaging & Production,* April 1983.
3. Hydraulic Robotic System Automates Assembly of Printer Cartridge Ribbons, *Hydraulics & Pneumatics,* March 1983.
4. IBM Advances Robotic Assembly in Building a Word Processor, *Robotics Today,* October 1982.
5. IBM 7565 Robot Speeds Precision Assembly, *Robotics World,* March 1983.

CHAPTER **68**

MODULAR PROGRAMMABLE ASSEMBLY RESEARCH

RANDALL C. SMITH

DAVID NITZAN

SRI International
Menlo Park, California

68.1. INTRODUCTION

68.1.1. General

The automation of discrete-parts manufacturing, spurred in its practical development by both economic and social considerations, is being implemented worldwide on an ever-increasing scale. Advanced automation expands the productivity of labor, combats inflation, and makes it possible to compete effectively in world markets; it improves the everyday working conditions of the labor force and raises the standard of living of the population as a whole. Automation is especially important for batch manufacturing, which is extremely labor-intensive and accounts for the largest portion of the total cost of discrete-product manufacturing.[1] Unlike the application of hard automation to mass production, where the expense of acquiring special-purpose equipment can be justified by the resulting high volume of production, automation of batch manufacturing must be programmable. Programmable industrial automation is characterized by three salient features[2]:

1. **Flexibility.** The capability of a machine system to perform different actions for a variety of tasks.
2. **Ease of Training.** The facility with which a person can efficiently program a machine system to execute a desired task.
3. **Artificial Intelligence.** The ability of a machine system to perceive new conditions (whether anticipated or not), decide what actions must be performed under those conditions, and plan these actions accordingly.

Although suitable primarily for increasing the productivity of batch manufacturing, programmable automation may also be applicable to mass production for the following purposes:

Reduction of the setup time for manufacturing short-lived products in a competitive world market.

Lowering the cost of production equipment by using components, such as robots, sensors, and computers, that are commercially available and recyclable.

Producing a sufficiently large sample of new products to enable technical and market performance to be tested before investing in a costly hard-automation system that would mass-produce them.

The factory of tomorrow will be characterized by integration of manual, hard-automation, and programmable-automation activities in proportions designed to minimize the total cost of manufacturing and servicing. The ultimate goal of programmable industrial automation is an automated factory based on the advanced technology of programmable computer-aided manufacturing (CAM). Eight CAM functions are distinguished: product design, part fabrication (including metalcutting by numerically controlled machines), part storage and transportation, logistics, materials handling, assembly, inspection, and process planning. At present only the first four functions may be found in a factory employing a flexible manufacturing system (FMS). The other four are still objects of study and experimentation in research centers worldwide.

Among these latter four research topics, programmable assembly is the most challenging. There are two reasons for this. First, programmable assembly is critically important; it may replace manual assembly, which constitutes the largest portion (approximately 22%) of the total labor cost for all durable goods.[3] Second, programmable assembly is complex; it includes materials handling and in-process inspection as well as programmable part presentation, trajectory planning, collision avoidance, arm control guided by multisensory feedback, part mating, and multimanipulator cooperation.

68.1.2. Programmable Assembly Research Issues

The complexity of programmable assembly brings about a variety of research issues. Discussing these issues in detail and describing the activities and accomplishments of worldwide research centers that investigate them are beyond the scope of this chapter. We present, however, a list of some major assembly research issues that face the factory of the future, which is an extension of a similar list that was generated in Nitzan and Whitney.[4]

Technoeconomic Analysis

1. Cost-effectiveness of manual, hard-automation, and programmable assembly as a function of product volume, quality, marketability and other factors.
2. Optimal mix of manual, hard-automation, and programmable assembly, given the foregoing product factors.

System Configuration

1. Modularity of hardware and software components.
2. Layout of programmable assembly stations, transfer mechanisms, and station modules.
3. Communication among the system, station, and module controllers.
4. Factors affecting how general the equipment components (feeders, fixtures, jigs, tools, arms, end effectors, sensors, computers, etc.) should be.

Sensing

1. Efficient utilization of a variety of noncontact (e.g., vision, range, proximity, and olfactory) and contact (touch and force/torque) sensors.
2. Improved transducers of light intensity, range, acoustic, tactile, and temperature signals (point, one- and two-dimensional arrays).
3. Very small noncontact and contact sensors to be mounted with their microprocessors on a manipulator end effector.
4. Fast signal processing in hardware, software, or both.
5. Integration of partial information from different sensors under different time constraints.
6. Sensor modeling, selection, and planning for different tasks and other constraints.

Manipulation

1. Computer control of manipulators (arms, end effectors, tables, actuators, and the like) to achieve task-dependent fast, smooth, and accurate motion (with or without sensors).
2. Multisensory guidance of manipulators in unstructured environments.
3. Collision detection, collision avoidance, and collision-damage prevention among multiple manipulators, other work station equipment, and workpieces.
4. Cooperation between two or more manipulators handling and mating parts that are rigid, semirigid, or limp.

End Effectors

1. Light, dexterous, and semicompliant fingers ("bones and flesh") with high-resolution surface touch sensors ("skin") under local microprocessor control ("smart hand").
2. Programmable application of different tools, each indexed by a multifinger hand, a wrist latch/unlatch lock, a carousel, or any other means.

Part Presentation

1. Programmable feeders of classes of parts as in group technology.
2. Logistics of part storage and retrieval.

Part Mating

1. Application of passive compliance (e.g., using a remote center compliance device), active compliance (e.g., nulling force/torque), or both to mating rigid, semirigid, or limp parts.
2. A programmable fixture with five degrees of freedom to allow vertical part mating.

In-Process Inspection

1. Monitoring three-dimensional parts, equipment, and assembly actions at production speed.

Modeling and Representation

1. Modeling three-dimensional parts and equipment components for design and manufacturing (fabrication, handling, inspection, and assembly) of parts and assemblies.
2. Representation of the foregoing models for automated planning and execution of manufacturing processes.

Product Design

1. Design rules for programmable assembly based on both the limitations and capabilities of programmable automation technology.
2. An "expert system," based on the foregoing rules, in a CAD system to facilitate interactive design of parts and to determine their attachment sequence.

Assembly Programming

1. Use of a robot programming language versus a general-purpose high-level language in programming a given assembly task.
2. Part representation by "showing" on-line versus using CAD data base off-line.
3. Determination and compensation for location (position and orientation) uncertainty of parts, equipment, and sensors.
4. Collision detection and avoidance among work station manipulators, other equipment, and workpieces.
5. Simulation of assembly operations off-line for detection of possible errors.
6. Use of multiple noncontact and contact sensors for guiding manipulators and monitoring execution of assembly steps.

Assembly Planning

1. Programmable assembly "rules," and an "expert system" based on these rules.
2. Sensor-based conditional branching (fan out) and merging (fan in) in a plan for an assembly process to be performed in a partially unstructured environment.
3. Representation of asynchronous time-dependent operations (actions and action monitoring).
4. Allocation of resources and control of parallel operations by the execution controller.
5. Dividing the responsibility for sensor-based operations between an off-line planner and a real-time execution controller.
6. Automated replanning, in case of an unexpected execution failure, for fast error recovery.

68.2. PAST ASSEMBLY RESEARCH AT SRI

Different methods have been proposed for implementing programmable assembly. One of these methods, developed at SRI International (henceforth referred to as SRI) in 1978, uses visual servoing and passive accommodation to align an *x-y* table supporting a subassembly with a part-mating end effector. We summarize this method as follows, before describing the recent developments in programmable assembly at SRI, which is the main concern of this chapter.

We have been exploring innovative configurations and techniques for cost-effective flexible and adaptable assembly with minimum jigging. Two versions of such an assembly process have been demonstrated in our laboratory; in both versions a compressor cover is fastened to its housing by eight bolts.[5]

The assembly station in the first version (see Figure 68.1) consists of the following components: (1) a Unimate arm under direct control of an LSI-11 microcomputer, constrained to move between eight fixed positions, thereby simulating a less costly four-joint limited sequence arm with servoed wrist rotation; (2) a gripper, mounted with a linear potentiometer on a pneumatic-cylinder piston to provide force control and displacement sensing along the vertical direction; (3) a programmable x-y

ASSEMBLY STATION

Fig. 68.1. First version of the assembly station.

table whose frame is driven by two DC servomotors (with 5-mil accuracy) and whose top is either rigidly attached to the frame or free to move relative to it (within ±6.3 mm in either x or y direction) to allow passive accommodation; (4) a GE Model Z7891 solid-state television camera with 100 × 100 elements above the x-y table; (5) auxiliary items, consisting of a pneumatic impact wrench held by the Unimate gripper, a bolt feeder, and two fixtures—one for the impact wrench and the other for the compressor cover; (6) a DEC PDP-11/40 minicomputer, which supervises the Unimate LSI-11 microcomputer and controls the remaining components of the system.

Following a one-time calibration procedure establishing the parameters of transformation between the image coordinates of the television camera and the axes of the x-y table, the compressor housing is placed randomly on the x-y table, and the assembly operation proceeds as follows. The television camera takes a picture of the compressor housing, and the minicomputer analyzes its image, computes its position and orientation, and commands the x-y table to move the housing to a fixed stacking position at the center of the camera's field of view. The Unimate then picks up the cover and places it on the compressor housing while the x-y table is freed to passively accommodate any minor cover misalignment (see Figure 68.2). The minicomputer analyzes the cover image seen by the television camera, computes the position of the first bolt hole, and commands the x-y table, with its top rigidly locked, to move and align the bolt hole at the central stacking position. The Unimate acquires the impact wrench, picks up the first bolt, and inserts it in the cover hole. As the x-y table is now freed to passively accommodate any minor bolt misalignment, the Unimate drives the bolt into the housing by spinning the impact wrench and exerting a constant vertical force downward, and then moves away. Performing in-process inspection, the minicomputer analyzes the cover image seen by the television camera and verifies that the hole image has disappeared (if not, a second trial is executed before aborting the operation). The last three steps are repeated for the remaining seven bolts, after which the Unimate returns the impact wrench to its fixture.

In the second version of the assembly, the following improvements were introduced: (1) The bolting operation is performed by a pneumatic four-joint limited sequence arm, the Auto-Place Series 50, whose rotary joint is driven by a DC servomotor and whose end effector is an impact wrench (see Figure 68.3); (2) the SRI Vision Module[6] is used as a self-contained vision subsystem to sense and

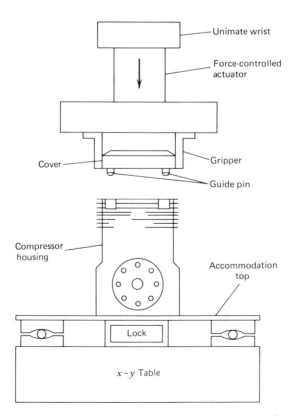

Fig. 68.2. Parts-mating system using passive accommodation.

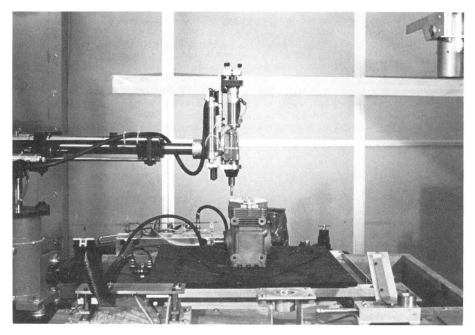

Fig. 68.3.   Limited-sequence arm bolting compressor cover into housing.

process visual images in response to top-level commands from the supervisory PDP-11/40 minicomputer; (3) the Unimate is used to present parts with the aid of visual feedback in lieu of jigging, namely, picking up the compressor housing from a moving conveyor belt and picking up the cover from a tote-box containing semiorderly packed covers.

In the remainder of this chapter we describe recent research on programmable assembly conducted at SRI. Two major topics are covered:

1.   The characteristics of an assembly system we are developing, including hierarchical organization, modularity, distributed processing, and robustness.
2.   Implementation of an assembly station with the foregoing characteristics and its use in demonstration of a programmable assembly task.

68.3.   ASSEMBLY SYSTEM CHARACTERISTICS

The characteristics of the assembly system we are developing are as follows:

Hierarchical Organization.   To simplify system software development by partitioning the overall control problem into manageable subsets that can be dealt with separately and simultaneously.[7]

Modularity.   For rapid reconfigurability of system components and easier modeling of those components because their interconnections and capabilities are well defined.

Distributed Processing.   To take advantage of the opportunities for parallel computation by means of multiple processors at the points of sensing, control, and demand.

Robustness.   To minimize the need for manual intervention in recovering from various fault conditions.

Each of these characteristics, as well as some instances of their implementation, is discussed in the following sections.

68.3.1.   Hierarchical Organization

An assembly *system* computer will control the activities of a number of assembly *stations* and a *transfer mechanism* carrying subassemblies between them. An initial design criterion is that every station must be available for experimentation, training, calibration, setup, and debugging. For such purposes, each station will be operated independently of the rest of the system. In contrast thereto, when the entire system performs a task, some coordination of the individual stations and the transfer mechanism will be necessary. Station-level integration will be accomplished by the system computer.

We next divide each assembly station into its major functional components, called *modules*—for example, manipulators, vision modules, part presenters, and support tables. A station can be composed of these modules in different ways and configurations, thus providing flexibility for batch assembly of a variety of products. Each module is encapsulated to make it self-contained, controllable by its own computer, and easily configured for coordinated operation with other modules.

A module consists of *devices,* each of which may be controlled by a device computer or processor. For instance, a manipulation module may consist of a robot arm with its control microcomputer, an end effector with its microprocessor, and sensors with their microprocessors. If the control of a device is very simple, assigning a special computer to control that device may not be justified; hence, the device will be controlled solely by the module computer.

The outlined four-level hierarchy of system–station–module–device computers is shown in Figure 68.4. This computer system should be able to support a fairly large-scale assembly operation (e.g., a factory with as many as 50 modules working simultaneously).

68.3.2. Modularity

A module will have the following general characteristics:

Perform a generic operation (e.g., manipulation, image acquisition and processing, or part feeding) and control any devices used in that operation.

Use its computer to provide an external interface to these generic operations at a high level.

Use auxiliary sensors to verify expected local conditions; for example, a manipulator will have a sensor to verify that it is "holding an object."

Execute reflex actions, under predefined conditions, that can be detected by the local sensors.

Operate independently of other modules.

Figure 68.5 shows the basic components of a general module with the above-listed attributes. These components are discussed in the subsequent sections.

Module Types

The module computer contains a processor, network interface cards, memory, and input/output interface cards for the analog and digital signals from or to the auxiliary sensors and devices of a module.

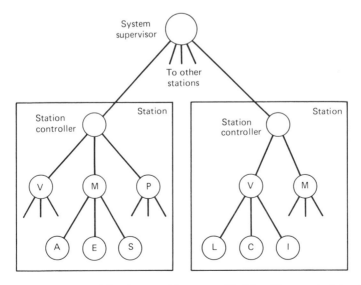

Fig. 68.4. Hierarchy of control for the assembly system. Modules: M = manipulator, V = vision system, P = part presenter. Devices: A = arm, E = end effector, S = sensor, L = lighting, C = camera, I = image processing.

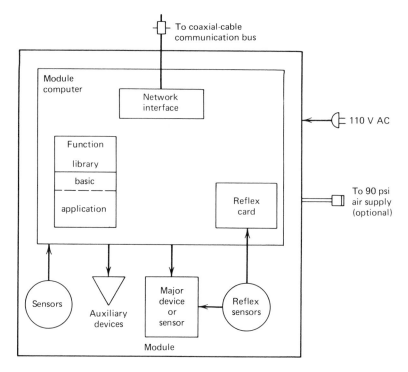

Fig. 68.5.   Basic module organization.

Some specific module types are described in the following sections, along with examples of the basic functions they provide. In general, these functions entail intermodular communication of high-level information, rather than large amounts of raw data.

Manipulator Module

We have developed two manipulator modules, each consisting of a Unimation PUMA 560 robot and an end effector. The end effector of one arm consists of an electrically actuated two-fingered hand, a remote center compliance (RCC) device,[8] and a remote-head video camera. The end effector on our second PUMA 560 consists of a six-axis force/torque sensor mounted on the wrist, and a pneumatic two-fingered hand. The module computer and the PUMA controller for each manipulator are mounted under the arm's supporting stand. Figure 68.6 shows the hardware control and sensing functions associated with the manipulator module, including planned extensions for proximity and touch sensors on the end effector. The module computer, in this case, will provide a means for controlling the hand, reading sensors in the hand and wrist, and moving the arm (indirectly, by communicating with the PUMA controller).

A few examples of the manipulator module functions are as follows:

Where(Result)
> Returns the location (position and orientation) of the manipulator's end effector.

MoveTo(Location)
> Move to the specified location.

Grasp(GraspLocation)
> Approach the given grasp location, open the hand, move to that location, close the fingers, and depart along the approach vector.

StopOnForceZ(Frame, StopForce, Overshoot)
> Moving along the z-axis of a specified coordinate frame, stopping if a contact force along the z-axis exceeds StopForce or if the manipulator travels farther than Overshoot along the z-axis.

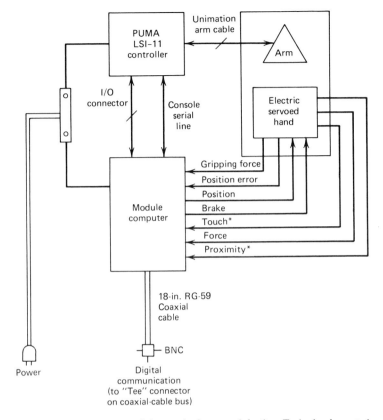

Fig. 68.6. Schematic of the manipulator module. * = To be implemented.

Binary Vision Module

The binary vision module includes an SRI vision module[6] with up to four 128 × 128-element solid-state cameras attached, a module computer, and future illumination and camera control (see Figure 68.7).

In addition to invoking SRI vision module functions, the module computer could be used, as needed, as follows:

Store the application-dependent routines developed around the basic functions of prototype training, recognition, and feature extraction.

Control such auxiliary devices as zoom lenses or lens turrets, and implement control of focus, directional lighting (front/back), and lighting intensity.

Measure illumination uniformity with work surface sensors, and track intensity changes. This measurement would ensure that the lighting conditions during object recognition and object training are identical.

The electronic components of the SRI vision module and the module computer are stored inside a stand like the one on which each manipulator is mounted. The top of the stand forms a work surface and can hold the aforementioned intensity meters and/or a light table for backlighting operations. Cameras are attached to long cables and can be mounted anywhere within the assembly station. One camera is mounted on the hand of one of the manipulators.

Some examples of binary vision module functions are as follows:

Picture(BlobCnt)
    Take a picture, perform connectivity analysis of the image, and return the number of connected regions (blobs), BlobCnt, that meet a certain blob criterion, such as minimum size and roundness.

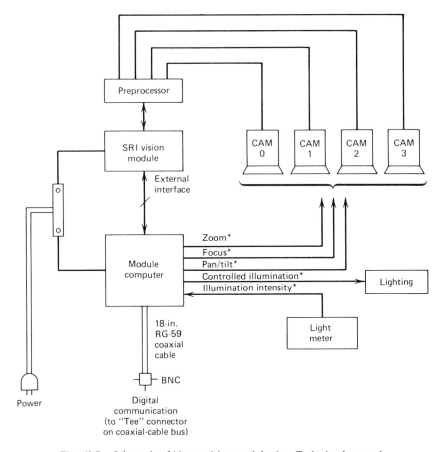

Fig. 68.7.   Schematic of binary vision module. * = To be implemented.

GetFeature(BlobN,FeatN,Result)
>   Return the value of a blob feature (indicated by the index FeatN) of a selected blob (indexed
>   according to decreasing size by BlobN). Features are, for example, blob area, perimeter
>   length, and moments.

Recognize(BlobN,BestDist,NxtDist,ProtoName)
>   Compare a given blob, BlobN, with a set of prototypes and return the smallest distance
>   (in feature space), BestDist, from the blob to any prototype, the second smallest distance,
>   NxtDist, and the name of the prototype, ProtoName, which is the best match.

Find(PartType,Result)
>   Take a picture and look for a prototypical blob, PartType, by comparing each blob in the
>   picture with a pretrained set of prototypes. If a good match is found, use information stored
>   with the prototype to determine a unique set of three-dimensional axes in the part on the
>   basis of its two-dimensional features, and return the part location.

Limited-Sequence Manipulator Module

We have a limited-sequence manipulator module that consists of an Auto-Place Model 50 manipulator
mounted on a servo-controlled rotary table.

A limited-sequence manipulator module may be used effectively to transfer parts to and from a
work surface or an inspection surface. The limited-sequence arm is less costly than servoed manipulators,
is usually faster (because it operates between fixed mechanical stops), and is normally rugged enough
to carry a greater payload. The limited-sequence arm, however, can operate only on parts in a fixed

location (position and orientation) relative to the arm. Two examples of functions for the limited-sequence manipulator module are as follows:

AutoCmnd(RelayStates)
>Set up the AutoPlace relay(s) to the requested state(s). This command actuates the pneumatics, but not the rotary table.

APTMove(Theta)
>Turn the rotary table holding the Auto-Place manipulator to absolute position given by theta radians.

Other Modules

We have been developing or planning other modules, such as the following:

1. **X-Y-Theta Table.** An x-y table whose movable surface can also be rotated about the z-axis. The table is equipped with a translucent top and a row of fluorescent lamps underneath so we can backlight objects resting on the surface.

2. **Part Presenters.** Programmable part presenters previously developed at SRI will be incorporated into the assembly system and extensions of their capabilities explored. One example is the SRI "Eye Bowl"—a standard bowl feeder that utilizes vision rather than mechanical blades for part sorting and feeding.[9]

3. **Programmable Jig.** We wish to explore the concept of a multipurpose, computer-controlled jig that is capable of accepting and rigidly holding parts of many different shapes, which it would then present, upon request, to a manipulator or vision system in a specified orientation. This might take the form of a rugged hand and 3-DF wrist.

68.3.3. Distributed Processing

A local area network is a natural organization of a system in which processing is distributed among numerous computers, which are often separated from one another by, say, 1 m or more. Hardware supporting various network topologies is available commercially. We use a system in which a coaxial cable forms a communication bus connecting all the communicating computers. The bus-network organization serves as follows:

>Allows direct communication between any two computers connected to the coaxial-cable bus.

>Promotes modularity of system components by requiring only a standard network interface for systemwide communication.

>Facilitates reconfigurability of components by permitting them to be connected to the network at any point on the coaxial-cable bus.

>Permits the sharing of expensive system resources, such as printers, graphics devices, and file-storage units.

Each computer connected to the communication bus contains a network interface with a unique name (numerically symbolized) assigned to it. Names are used to identify both the source and the destination of a message. When a message is transmitted on the bus, every network interface compares its name with the message destination and accepts receipt of the message only if there is a match. A special type of broadcast message can be addressed so that all the computers (except the sender) will receive it. This type of message is useful when a module needs help from the system, but does not know the name of the unit that can furnish such help.

A communication software package has been written to provide flexible communication capabilities through the network interface. The selected protocol and the characteristics of the network interface are described in detail in Nitzan et al.[9, 10] Briefly, the package supports a "random-access" protocol whereby any computer may send a message to any other computer or computers if the communication bus is idle. The message is usually one of the following:

1. A command to a module to perform one of its functions with the parameters given in the message. The command may be to supply information, request information, or to perform a specified activity.

2. A reply to a command, containing any results or requested information. The reply also serves to confirm completion of a commanded activity, which may otherwise not have returned any results.

The communication package performs the following functions:

Message creation, retrieval, buffering, and deletion.

Message receipt, acknowledgment, and transmission.

Automatic retransmission of unacknowledged messages.

Detection of any special broadcast messages for later action.

A module will never have more than one buffered message to transmit at any time. However, it can receive and buffer numerous messages and either attend the oldest one or search in its buffer for an expected message of a certain type, from a certain source, or both.

68.3.4. Robustness

We have recently implemented three levels of processing in each module computer:

Reflex Level. To detect hardware or software faults and set the module hardware devices to predetermined states.

Bootstrap Level. To set the module-computer program at a predetermined state in response to reflex activation, and to notify the rest of the system about this event.

Program Level. To implement the main functions of the module. A sensor-monitor routine will be implemented at this level in the future to detect conditions beyond the capabilities of the reflex level.

Reflex and Bootstrap Levels

At the lowest processing level the modules (particularly those incorporating manipulators) need self-protective mechanisms that act as reflexes—hardware responses to a set of external or internal fault conditions. Such conditions include loss of operating power, loss of program control, and human intrusion into the assembly area. If necessary, special sensors will be assigned to detect these conditions. Once enabled, a reflex device will watch for the fault condition it guards against and be triggered if that condition arises. We have designed a reflex card to implement this function and fulfill the following responsibilities:

To provide a mechanism that forces the module's device or sensors into "safe" default states when a reflex is triggered.

To protect the module from power loss.

To protect the module from loss of program control.

To provide sensor-triggered hardware reflexes.

To provide programmable-condition reflexes.

In addition, activation of a reflex may optionally force the program control in the module computer to transfer to a simple bootstrap program resident in a nonvolatile memory on the reflex card. The module subsequently executes a simple program at the bootstrap level that will do the following:

Initialize the network interface for communication.

Use the network to broadcast a message notifying other module computers of this module's current state.

Provide capabilities for loading the program of this module through the network interface when commanded by another module computer.

Supply tools for remote diagnosis of this module through the network interface.

Figure 68.8 depicts the relationship between the reflex and bootstrap control levels. More information about reflexes and the bootstrap functions may be found in Reference 9. The responsibilities at the reflex level are described in the next sections.

Default State Conditioning

Each device or sensor connected to the module computer should be set to a safe default state whenever a reflex is triggered. It is a common practice to provide a computer bus line that carries an INITIALIZE signal to all the computer interfaces. The reflex card triggers generation of this signal when a reflex condition occurs. The INITIALIZE signal, for example, may halt any moving device if the signal is applied directly as an override or a shutdown signal to that device.

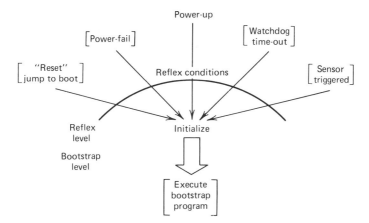

Fig. 68.8. Reflex level.

Power Loss Protection

Like many computers, the module computer has the capability of detecting imminent power loss through a sensing circuit in its power supply. A signal indicating this event is supplied to the reflex card from the computer power supply. The reflex card triggers an INITIALIZE signal on the bus, thus initiating a command to halt moving devices—for instance, prior to the power loss in the module computer. In a "power-up" sequence, the INITIALIZE signal is generated again, and program control is transferred to the bootstrap code.

Protection from Loss of Program Control

The module computer's program may not always be running correctly; for example, it may become deadlocked, halted, or contain errors. For these contingencies, an independent timer, called a "watchdog timer," is included on the reflex card. The watchdog timer must be reset periodically by a properly executing program in the module computer. If it is not reset, the watchdog timer reaches a "time-out" state. This state is a reflex condition that, like any other, will cause generation of an INITIALIZE signal by the reflex card and transfer the module computer's program control to the bootstrap code.

Provision for Sensor-Triggered Reflexes

Certain events detected by sensors connected to the module computer may induce an emergency response from the module. The reflex card provides such a response capability by accepting binary signals from these sensors. The binary signals could, for example, indicate the state of contact/noncontact or proximity/nonproximity of objects with respect to a manipulator's end-effector. One reflexive response might be to halt a moving manipulator any time an intruder is detected in the work space. As another example, a proximity sensor on a manipulator's hand may be used to trigger a "stop-arm" reflex to prevent collision with unexpected obstacles; however, sometimes this reflex must be disabled to permit the hand to reach a target object. The computer program may disable any reflex circuit on the reflex card by transmitting a special code word. This encryption reduces the possibility that the reflexes may be disabled accidentally. Reflex devices will always be placed in the "reflex disabled" state following the INITIALIZE signal, so that the module will react to the fault condition once rather than repeatedly. The triggering condition should be determined by reading the reflex status and removing it before the reflex is enabled again. When the module program begins, it enables those reflex sensors that should be active at that time.

Provision for Programmable-Condition Reflexes

The reflexes just described are directly triggered by simple binary sensor signals. Certain more complex conditions detected by the module-computer program may also warrant initiation of an orderly shutdown. Loss of communication with other devices is one instance of a potential shutdown condition; another example is detection of anomalous conditions computed from local sensor values and internal program states. A RESET instruction should be supplied by the module computer to activate the INITIALIZE bus signal from software. Utilization of this provision will be based on the estimated urgency of the situation. After it carries out the RESET command, the program should transfer execution to the bootstrap program.

Sensor-State Monitoring at Program Level

The reflex and bootstrap levels are concerned with initializing the module, getting it loaded and running, and providing a uniform method for detecting and reacting to local anomalous conditions. The main functions of the module are performed at the program level, that is, controlling the module's main device or sensor. The program level has the following responsibilities:

Reset the watchdog timer periodically.

Provide full, flexible intermodule communication by the network interface.

Implement the generic functions defined for the module type to control the module's devices or sensors.

Provide periodic sensor-state monitoring.

Examples of module functions and a description of module intercommunication capabilities have been given in Sections 68.3.2 and 68.3.3.

Sensor-State Monitoring

A background process for monitoring the numerous sensors associated with a module is under development; it is presented as a principal component of the program level of each module. The monitoring process is intended to read the values of local sensors periodically and to compare these values with their expected range, which is given in a table. When an actual sensor value is found outside the expected range, a programmable-condition reflex may be executed or, less drastically, a message reporting the anomaly may be broadcast. Range entries in the sensor table will be made in one of two ways:

Explicitly, through a MONITOR command given to the module, indicating which sensor to use and the expected range values of that sensor in the next interval.

Implicitly, by execution of a generic function that imposes known constraints upon a sensor's values.

A MONITOR command can be used, for example, to instruct a manipulator module to monitor the forces and torques acting on its end effector and to assure that they remain within a specified range for a given application. Monitoring will then proceed independently until the MONITOR command is cancelled.

In some cases the effect of a generic function on a set of sensors is known a priori, and sensor monitoring can be initiated implicitly. For example, let us consider two functions for a manipulator hand: GRASP and RELEASE. The value of a binary touch sensor on the hand's fingers after execution of the GRASP command is expected to be ON. The GRASP routine itself will verify this condition and enter the tolerance range for the sensor value (unnecessary in this instance) in the sensor range table. Execution of the RELEASE routine will generate an entry in the table corresponding to the OFF value for the contact sensors (if, indeed, they were off). The sensor value corresponding to the ON or OFF state can be checked repeatedly by a sensor monitor routine. Thus, an external application program directing the manipulator module need not verify continuously that an object in the hand is still there, but will instead be notified immediately if the object is dropped.

Other operations similarly impose anticipated constraints on associated sensors. Should a discrepancy occur between the sensor table range and the actual value of a sensor, the safest approach will be to execute a "programmable-condition" shutdown reflex and enter the bootstrap level. Since the reflex does not destroy the resident program, state information can be retrieved from the module by means of the diagnostic routines available in the module's bootstrap program. Such information may be used in future work to determine why a module failed and to direct the recovery of that module or the entire station accordingly.

68.4. ASSEMBLY STATION

68.4.1. Station Configuration

Figure 68.9 shows the current configuration of our assembly station. It consists of a station controller, a binary vision module with three video cameras, and two manipulator modules whose black and white arms are called Arm 1 and Arm 2, respectively. The end effector of Arm 1 consists of a plastic remote center compliance device, the front end of a video camera (the back end is mounted on the manipulator arm), and a two-fingered hand. The end effector of Arm 2 consists of a six-axis wrist

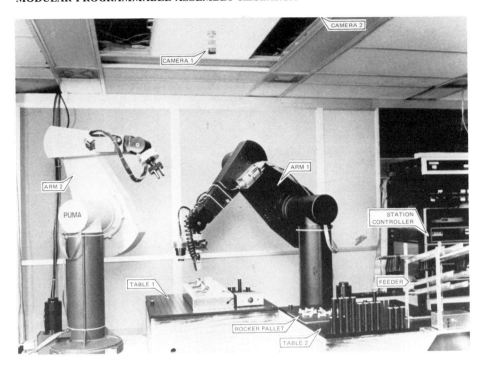

Fig. 68.9. Assembly station.

force/torque sensor and a two-fingered hand. Not included in this configuration are the limited-sequence-manipulator module and the other modules described in Section 68.3.2. These modules may be incorporated into the assembly station as needed during the performance of other assembly tasks.

This station configuration includes two support surfaces. The first, located between the manipulators, is the assembly area (the binary vision module, excluding its cameras, is mounted beneath it). The second surface, located near Arm 1, supports general-purpose part feeders and pallets within reach of that arm. Both surfaces have fixed cameras mounted above them, and both have a grid of holes, which are used to provide mechanical support components and to aid in calibrating the coordinate frames of the manipulators, the support surfaces, and the cameras relative to one another.

The station controller, a small computer, is used to control the sequential and parallel operations of the modules by means of commands on the communication network. The station controller also controls a printer, a speech output device, and two disks. Messages from the modules to the operator are printed or spoken; files stored on the disks are sent to the module computers upon request. In addition, the station controller has menus of module commands that can be executed interactively by the user, as well as a cross-network debugger that allows him to examine, alter, and set breakpoints in the programs of the module computers.

68.4.2. Calibration of Coordinate Frames

Communication of object locations (position and orientation) among modules is facilitated if there is a common coordinate system, or *reference frame*, in which to define these locations. For instance, a camera may be used to determine the location of a part; that information, expressed in terms of the coordinates of the reference frame, will enable any manipulator within reach to access the part. Each manipulator module will need to know only its relationship with the reference frame rather than all its pairwise relationships with other coordinate frames in the assembly station. We therefore calibrate all the station modules to a station reference frame.

In Figure 68.10 we define the base coordinate frames of two manipulators, R1 for Arm 1 and R2 for Arm 2, the coordinate frames of their respective end effectors, E1 and E2, and two coordinate frames, T1 and T2, on Table 1 and Table 2, respectively. Camera 1 overlooks Table 1 and Camera 2 overlooks Table 2. Frame T1 is chosen to be the reference frame, and the other frames must be related to it, as indicated by the dashed arrows, by means of transforms—4 × 4 homogeneous coordinate-transformation matrices.[11]

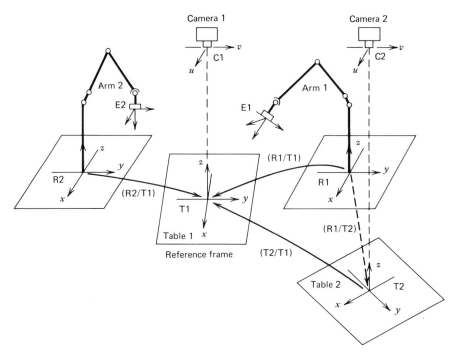

Fig. 68.10.   Station coordinate frames and transforms.

We use the following general notation. The position vector $(x, y, z, 1)$ of a point P in an arbitrary homogeneous coordinate frame F is denoted by P(F). The same point may be described by P(F1) or P(F2), where F1 and F2 are two different frames. We denote the transform from Frame F1 to Frame F2 by [F1/F2], where P(F1) = [F1/F2] * P(F2). Using this notation, note that P(F2) = [F2/F1] * P(F1), where [F2/F1] is the inverse of [F1/F2]. We depict [F1/F2] by an arrow pointing from the origin of Frame F1 to that of Frame F2 (see the examples in Figure 68.10); note that the inverse of [F1/F2] would be depicted by an arrow in the opposite direction.

Using the preceding notation, we obtain P(R1) = [R1/T1] * P(T1). The elements of transform [R1/T1] can be determined by mounting a pencillike tool on the wrist of Robot 1 and leading that robot so that its tool tip touches three points in Frame T1—its origin, a point on its $+x$ axis, and a point on its $+y$ axis. Reading the robot positions in Frame R1, the coordinates corresponding to the origin yield the translation vector of transform [R1/T1], while those of the other two points yield its rotation matrix. A similar sequence using Arm 2 and the same calibration points in T1 may be performed to derive [R2/T1]. At this stage, the position of a part given in Frame T1 can be transformed into the frames of both arms.

Next, points in the two-dimensional image plane of each camera are related to a three-dimensional space. Consider first Camera 1, which is calibrated directly to the Table 1 grid it views. Pegs of varying lengths are inserted into the grid holes, which are easily identifiable integer coordinates in the grid framework. The white tops of the pegs can be seen as bright spots by the camera. A set of known peg-top (x, y, z) positions in Frame T1 and the corresponding set of (u, v) image coordinates in a two-dimensional Frame C1 are utilized, using a least-mean-squares fitting method, to produce the camera calibration matrix.[10] The camera calibration matrix can be used to compute the image coordinates (u, v) of a given point (x, y, z) in the calibration frame. Generally, however, we do the reverse—the x and y coordinates of a point in the calibration frame are obtained as a function of the camera calibration matrix, the image point (u, v), and the known z-coordinate of that point. For Camera 1, the calibration frame is equivalent to the reference frame (Frame T1).

Calibration of Camera 2 is similar to that of Camera 1, except that the relation between the calibration frame (Frame T2) and the reference frame (Frame T1) must be derived. The relation [T2/T1] is determined by making the tool tip of Arm 1 touch points on the origin, the $+x$ axis, and the $+y$ axis of Frame T2 to derive [R1/T2], and computing [T2/T1] = [T2/R1] * [R1/T1].

The hand-held camera is calibrated similarly to the other cameras, except that only one peg is used and the camera is moved to view it from different locations. The camera is rigidly attached to the tool-mounting flange on the wrist of Arm 1; hence, points seen by the camera will be initially referenced to a frame FL fixed in that flange. Assume that the peg is placed in a grid hole on Table 1 and that its position in Frame T1 is known. After each time the arm is moved, we compute the peg-top position in Frame FL, using the relation Peg(FL) = [FL/R1] * [R1/T1] * Peg(T1). From the resulting list of (x, y, z) positions in Frame FL and the corresponding (u, v) image coordinates, the camera calibration matrix can be computed as for Camera 1 or Camera 2. Relations between the hand-held camera and other frames are illustrated in the next section.

68.4.3. Example of Intermodular Communication

Most of the communication that occurs in the context of our assembly station is between the station controller and the modules it commands or queries. However, one important example of direct communication among the modules themselves involves the use of a camera mounted on a manipulator's hand.

A stationary camera can supply information about the location of an object it views in a fixed coordinate frame, such as Camera 1 over Table 1 in Figure 68.9, because the relationship between the camera and the table is constant. A mobile camera, on the other hand, can supply information about the position of a part relative only to the camera's viewing location. If the part location is desired with respect to a fixed frame, such as T1, then the location of the viewing camera must be known. Figure 68.11 shows schematically the transforms between frames associated with finding a part PARTX1 in a pallet on Table 1 by means of a camera attached to a flange FL that holds the end effector of Arm 1. The following sequence of commands illustrates a direct communication between a binary vision module using a hand-held camera and the manipulator module.

MoveTo (ARM1, AbovePallet). The station controller commands Arm 1 to move to Location AbovePallet above a pallet with a desired part. Since AbovePallet is described with respect to T1, Arm 1 converts AbovePallet to a location in its own frame, using the relation [R1/AbovePallet] = [R1/T1] * [T1/AbovePallet], and moves to that location. After the move, Frames E1 and AbovePallet will coincide.

SetCamera(VM1, HandCamera). The station controller commands vision module VM1 to select the hand-held camera for input.

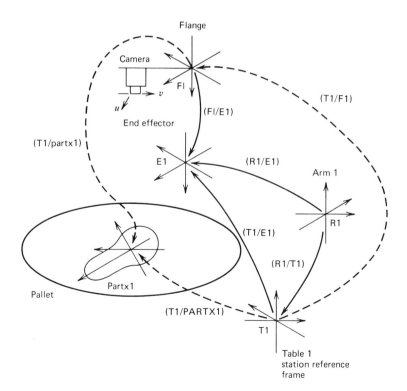

Fig. 68.11. Using the hand-held camera.

Find(VM1,PARTX, Result).   The station controller commands VM1 to take a picture with the selected camera, compute the location of a part of Type PARTX in reference coordinates, and return this result.

The task of VM1 is thus to compute transform [T1/PARTX1]. Since [T1/PARTX1] = [T1/FL] * [FL/PARTX1], VM1 will first ask and obtain from Arm 1 the value of [T1/FL]; it will then find PARTX1 and compute its transform, [T1/PARTX1], as follows:

WhereFL(ARM1, Result).   The vision module asks Arm 1 for the location in T1 of its tool-mounting flange FL to which the camera is attached. The camera has been previously calibrated so that points it views will be referenced to a coordinate system fixed to the flange.

Reply(VM1, Result).   Reading its current value of [R1/E1], Arm 1 computes [T1/FL], using the expression [T1/FL] = [T1/R1] * [R1/E1] * [E1/FL], and gives its value to the vision module. The latter then takes a picture with the hand-held camera, recognizes an instance (PARTX1) of PARTX, if present, and determines [FL/PARTX1] on the basis of two-dimensional image features and additional prototype information.

Reply(StationController, Result).   The vision module tells the station controller where PARTX1 is in T1, using the relation [T1/PARTX1] = [T1/FL] * [FL/PARTX1].

68.4.4.   Assembly Demonstration

A few demonstrations of the assembly station have been performed, the most recent one involving the assembly of part of a DEC LA-34 printer carriage.[10] That assembly contained four part types (see Figure 68.12):

1.   A friction shaft with four plastic rocker arms, each containing a hole.
2.   Four plastic rockers, each of which snaps into the hole of one of the rocker arms.
3.   Two small rollers, each of which snaps into the front end of two adjacent rockers.
4.   Two large rollers, each of which snaps into the back end of two adjacent rockers.

All the parts are acquired by Arm 1 from their locations on Table 2 (see Figure 68.9). The shaft and the two rollers slide into fixed pickup locations on three feeders, each consisting of two inclined

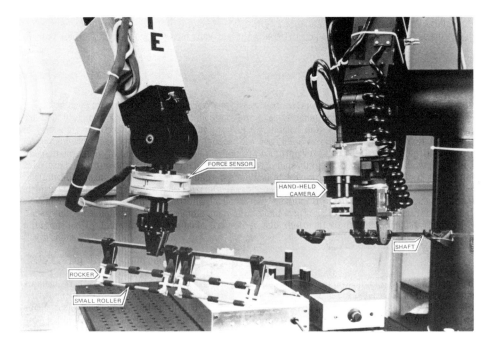

Fig. 68.12.   Simultaneous performance of initial and final assembly steps.

rails. The rockers are presented on a sticky pallet under Camera 2 in one stable state (upright), but at arbitrary locations.

Table 1 supports a simple fixture with "V" notches for aligning cylindrical parts, such as the friction shaft and the rollers. The fixture is attached to a light table that backlights it and a small surrounding area. Using Camera 1, the location of the fixture is determined usually from above by recognizing and locating a reference target on it; the locations of a few fixture components, such as the holding support for the friction shaft, are computed according to their relative locations with respect to the target. Meanwhile, Arm 1 acquires a friction shaft from its feeder and places the shaft on its support (after the latter has been located). After Camera 1 locates the fixture, Camera 2 is used to locate a rocker on the pallet. Subsequently Arm 1 acquires the rocker and places it beside an empty rocker arm whose hole has been located by the hand-held camera on the arm. That camera then determines the precise location of the rocker.

Part mating with force feedback is performed next by Arm 2 because Arm 1 has no wrist force sensor. Given the locations of each rocker and its destination hole, Arm 2 grasps the rocker and places it in the hole while making small corrective movements based on information from its force sensor. That information is used to determine when the rocker should be set into place as well as to protect the assembly from excessive forces that Arm 2 might accidentally exert. An increase in the applied force followed by a sudden drop in that force provides confirmation that the two parts have been snapped together successfully. While the rocker is being inserted by Arm 2, Arm 1 acquires one of the four rollers needed and places it behind the shaft support on the fixture.

The preceding cycle is repeated until all four rockers have been inserted in their respective rocker arms and all four rollers have been placed in their fixture. Subsequently, Arm 1 turns the friction shaft subassembly over onto the fixtured rollers and holds the shaft in place, as shown in Figure 68.13. Arm 2 then presses every rocker down until it is snapped onto the corresponding front and back rollers; force sensing is used again to verify this operation. Finally, Arm 1 releases the friction shaft, and Arm 2 removes the completed assembly from the assembly area.

68.4.5. Conclusions

The assembly station and assembly demonstration described in this chapter have exhibited the following capabilities:

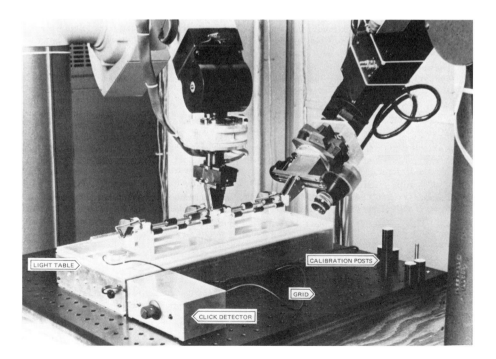

Fig. 68.13. Pressing rockers onto rollers.

No special-purpose fixtures—some parts are presented to the manipulator on a pallet, others are on a slide. The assembly is assisted by a simple fixture with "V" notches for centering cylindrical workpieces.

Utilization of medium-level module commands, such as MOVETO, FIND, and GRASP.

Binary vision to locate and identify parts on a pallet for acquisition. Three cameras are used during the assembly—two fixed in the ceiling and one attached to the end effector of one of the manipulators.

Force feedback to actively control compliance and insertion forces in snap-together, part-mating operations.

A stationwide calibration for two manipulators and multiple cameras, enabling part locations to be given with respect to a common reference frame.

Concurrent operation of two manipulators in a central assembly area, while the binary vision system recognizes and locates parts elsewhere.

Synchronization of manipulator motions to avoid collisions in the central assembly area.

Simple sensors to verify operation, including the use of finger separation to ascertain the presence of a part in the hand; a click detector, employing a microphone mechanically coupled to the assembly area, to verify that semirigid parts have snapped together successfully, or to detect a drop of a part.

These capabilities are important for the development of not only programmable assembly, but also of programmable automation in its broadest potential range of applications.

REFERENCES

1. Cook, N. H., Computer-Managed Part Manufacture, *Scientific American,* Vol. 232, February 1975, pp. 86–93.

2. Nitzan, D. and Rosen, C. A., Programmable Industrial Automation, *IEEE Transactions on Computers,* Vol. C-25, December 1976, pp. 1259–1270.

3. Nevins, J. et al., Exploratory Research in Industrial Modular Assembly, NSF Grants GI-39432X and ATA74-18173A01, Reports 1–3, Draper Laboratories, Cambridge, Massachusetts, June 1, 1973, to August 31, 1975.

4. Nitzan, D. and Whitney, D., Research in Automated Assembly and Related Areas—Report of the ARC Programmable Manufacturing Systems Group, *Proceedings of the Joint Automatic Control Conference,* Vol. 4, Philadelphia, Pennsylvania, October 1978, pp. 21–28.

5. Rosen, C. et al., Machine Intelligence Research Applied to Industrial Automation, Eighth Report, NSF Grant APR75-13074, SRI Project 4391, SRI International, Menlo Park, California, September 1978.

6. Gleason, G. J. and Agin, G. J., A Modular Vision System for Sensor-Controlled Manipulation and Inspection, *Proceedings of the 9th International Symposium and Exposition on Industrial Robots,* Washington, D.C., March 1979.

7. Albus, J. S., *Brains, Behavior, and Robotics,* BYTE Books, Subsidiary of McGraw Hill, Peterborough, New Hampshire, 1981.

8. Whitney, D. et al., Short- and Long-Term Robot Feedback: Multiaxis Sensing, Control, and Updating, *Proceedings of the Tenth NSF Conference on Production Research and Technology,* Detroit, Michigan, February 28–March 2, 1983, pp. 119–122.

9. Nitzan, D. et al., Machine Intelligence Research Applied to Industrial Automation, Eleventh Report, NSF Grant DAR80-23130, SRI International, Menlo Park, California, January 1982.

10. Nitzan, D. et al., Machine Intelligence Research Applied to Industrial Automation, Twelfth Report, NSF Grant DAR80-23130, SRI International, Menlo Park, California, January 1983.

11. Paul, R. P., *Robot Manipulators,* MIT Press, Cambridge, Massachusetts, 1981.

PART 12

INSPECTION, QUALITY CONTROL, AND REPAIR

CHAPTER 69
APPLYING ROBOTIC INSPECTION IN INDUSTRY

JERRY KIRSCH

KERRY E. KIRSCH

Kirsch Technologies
St. Clair, Michigan

69.1. QUALITY, INSPECTION, AND TESTING

The concept of product quality can be defined as the totality of features and characteristics of the product that bear on its ability to satisfy a given function.* In setting quality standards, an industrial organization must strike a balance between the costs of achieving a given quality level and the benefits accrued from the quality product. The benefits are usually the profits arising from increased sales of the final product. The benefits can also be viewed as improved production effectiveness when parts are processed within a given quality range.

To assure quality, parts and products must be inspected and tested. Inspection is defined as a process of careful search for nonconformities or errors, that is, falling outside the specified quality range. Inspection in industry has three primary functions:

1. Preventing nonconforming parts or materials from proceeding further in the production process. The purpose is to avoid production of poor-quality products.

2. Collecting data on specific characteristics of parts or materials for use in decisions regarding overall quality. Here the purpose is to identify whether imperfections are severe enough to be considered nonconforming.

3. Collecting data on specific characteristics or parts or materials to provide feedback for the manufacturing process. The purpose is to correct the process before more poor-quality parts are produced.

Inspection usually implies measurement of certain part properties, for example, geometric dimensions, surface finish, position accuracy, assembly integrity, and so on for quality control. *Testing,* on the other hand, implies some active examination of specific operational functions for quality control.

69.2. NEED FOR INDUSTRIAL INSPECTION EQUIPMENT

Industrial robots have been accepted for a variety of reasons which vary from task to task. By far the greatest single area of acceptance has been in spot welding. This has come about as a result of the improved consistency of welds made by robots over those made by humans. Spot welding guns are rather heavy and bulky, making this operation difficult and demeaning for humans and resulting in high costs for these operations. Spot welding robots have therefore been accepted for four reasons: flexibility, reduced cost, social aspects of the job, and improved quality of the finished product. Arc welding and spray painting are similar applications which have the added incentive to automate of relatively high skilled-labor cost. Although machine tending robots may or may not improve quality, they often are highly cost-effective in these dehumanizing jobs.

This chapter was written when both authors were with Copperweld Robotics.

* European Organization for Quality Control, *Glossary of Terms Used in Quality Control,* 4th ed., July 1976.

Applying this same thinking to inspection, it is clear that there is an incentive to automate. By allowing machines to perform tasks traditionally done by humans, there is a loss of the operator's intelligence and decisions which normally go without notice. Even in the most mundane of jobs the operator is required to inspect, think, and make decisions. Someone whose job it is to insert a screw in a hole must be sure that the screw has threads and that there is a hole to put it in. This person also has the wherewithal to throw a screw in the trash if it has no threads or put a workpiece aside if it has no hole. Although this example seems trivial to the onlooker, it can be disastrous for an automated system without these capabilities.

Human inspectors also are subjective in their duties. We can say, in general, that there is no straight good or bad, only varying degrees of badness with respect to the ideal. Quality is often defined as within specifications. This is clear when looking at almost any blueprint that states a dimension as a specific number qualified with a tolerable variance. This means that if a component meets all of its specifications it is acceptable and can be considered a quality part. Being subjective, humans therefore make a judgment call based on some reference. These references can range from a single scale, or go-no-go gage, to more sophisticated tools like optical comparators. Experience is also another method humans employ in place of mechanical references. Although experience is valuable to humans, it is subjective and its reliability is highly variable.

Machines, however, are highly objective in their operation. They apply logical judgment to any decisions required of them. The removal of any doubt, which is often injected when inspection is performed by the human inspectors, is always desirable. The consistency of these automated forms of inspection may be the single largest reason for their acceptance. As with all forms of automation there must be economic justification to motivate acceptance. Not only is there substantial labor savings in automating inspection, but there is also the cost of quality as compared with the lack of it, which often alone justifies inspection automation.[1]

69.3. AUTOMATED INSPECTION

Automated inspection has been with us for some time in various forms. Simple mechanical probes have been, and continue to be, used to determine presence or absence in many applications. Currently researchers and industry are approaching the inspection field using more sophisticated techniques. The trend in automated inspection today is toward the emulation of human vision.[2] There are currently several suppliers of vision systems which incorporate two-dimensional matrix array cameras to perform gaging and inspection tasks. Early forms of three-dimensional inspection are also beginning to appear utilizing stereo techniques, structured light, and other methods. (See Chapter 13, Sensor Based Robots, Chapter 14, Vision Systems, and Chapter 16, Depth Perception for Robots).

There are three major components to a vision system, sensors, electronic hardware, and software. Sensors come in many varieties. Currently the most prevalent are solid-state devices. They can be configured in one dimension, typically called a line scan camera, which is a single line of picture elements (pixels or pels) with lengths of up to 2048 elements or more. More common are two-dimensional cameras, which may be typically configured as 128×128 or 256×256 elements. These sizes are convenient to deal with as they are natural binary numbers; however, many cameras offer nonsquare formats. The imaging device is a solid-state array with individual picture elements that output a voltage corresponding to the light intensity received by the element. Electronics within the sensor scan the array in much the same way you are reading this page, left to right from top to bottom. Analog voltage signals are then output corresponding to the light intensity received by each pixel. To allow this information to be sorted out by an external device, some reference must accompany the video information. For this reason markers are inserted in the video indicating the end of each line as well as the end of each page. When these markers (syncs) are included in the video it is called composite video. Some manufacturers of sensors also provide a separate output of digital pulses called a pixel clock to allow one to discriminate accurately the timing between each pixel element.

The standard time to output a single frame of video is one-sixtieth of a second. This means that a 256×256 matrix camera which has a total of 65,536 individual elements outputs one element in approximately 200 nsec, or at a rate of nearly 4 MHz. This is a great deal of information acquired in a very short time. The trend in sensor technology is toward increasing the size of the array to improve its resolution. By doubling the size of a 256×256 array to 512×512, the number of elements increases by a factor of 4 to 262,144, and the scan rate is increased to nearly 16 MHz.

There are three basic approaches to vision that are being used in industry currently:

Image buffering
Edge detection
Windowing

Image buffering (frame grabbing) is a system that digitizes an entire image and stores it in memory, where it may be analyzed by a computer. The gray levels of each picture element are stored as binary numbers in words typically 6 or 8 bits in size. Once an image has been stored a computer can then read and write to the data. The image buffer approach depends heavily on software to perform a task. For this reason it is limited in speed only by the processor speed and complexity of the program.

Generally most of the information that is supplied from the camera is of little or no interest. What is of interest can often be described in terms of numbers as features. There are many useful features that can be quantified such as the object's area, centroid location, orientation and perimeter, among many others. These features can often be employed in identification, gaging, and inspection tasks.

Edge detection is a method of obtaining information about a scene without acquiring an entire image. Edge-detection-based systems record the locations of transition from black to white and white to black and store these locations in memory. A computer can then connect these points through a process called connectivity. Once the edges are connected the objects within the field of view can be separated into "blobs" which can be analyzed for their respective features.

In a further effort to reduce data and reduce program complexity a windowing approach can be taken. Windowing is a method where only selected areas of the image are analyzed. These areas, or windows, may surround a hole or some other aspect of a part in the field of view. Within a window a simple analysis may occur, such as a counting of the light or dark elements to determine a hole size. Such other operations can also be performed as finding the vertical or horizontal location of a transition between light or dark, or the total number of lines containing light or dark elements.

The artificial vision industry in the early 1980s can be compared to that of the robot industry in the mid- to late 1970s. There is a great deal of interest in vision and much is being said about the potential, yet the current market is quite small. Although sales are being made, the enormous potential has yet to be realized. It is estimated that the artificial vision market will surpass the one billion dollar figure by the early 1990s and this figure may even be low. New companies are beginning to appear on a regular basis in much the same way that robot companies sprang up in the late 1970s and early 1980s. The artificial vision community today is in many ways closely aligned with the robot industry. Many of the vision companies display their products and make technical presentations at the robot trade shows. Vision systems are applied to robotic inspection, as described in the rest of this chapter and in the following chapters.

69.4. ROCKER COVER INSPECTION

An automotive plant that manufactures engine rocker arm covers had a requirement to increase levels of quality in their process. The covers manufactured are made for two models of engines, each type having both a right- and left-hand part, four part styles. They are assembled manually on two large-diameter dial index tables. Operators stationed around the table insert brackets, clips, and weld nuts and a baffle into fixtures that retain them in location prior to a spot welding operation that affixes the details to the cover.

The spot welding takes place on each of the two dial index tables followed by the automatic removal by a sliding arm having a part gripper jaw. The assembled covers are then visually inspected by an operator viewing the assembly as it moves by on a belt conveyor. This inspection was found to be difficult, especially when running production with the assembly having seven different parts welded to the cover. In many cases the inspector could not detect that parts were missing or that wrong parts were present. In some cases wrong parts would fit into the fixtures on the dial table.

The automatic welder also would occasionally create pinholes at points of spot welds in the assembly. These holes resulted in oil leaks and could not be detected visually by the inspector.

The defective parts then flowed through a washer and dryer, followed by painting and shipping to engine assembly plants. These plants, upon discovering the defects, would then return the entire shipment of the covers for reinspection and repair, followed by reshipment back to the assembly plant—to say the least, a costly quality problem.

69.4.1. The Application Scope and Task Performed

A dial index table was installed in a position where two conveyors transport assembled parts from each of the two assembly welding machines. The finished assembled covers are then manually deposited, two at a time, on positioning fixtures on the dial table. The operator depresses two palm buttons advancing the index table. Proximity sensors positioned over the table at one point detect whether the operator, in fact, loaded both fixtures; this information is also stored in memory. At the following station an overhead ram moves downward and depresses both covers to seal them as they would be found on the engine. Seals are mounted on the dial table around each of the positioning fixtures. Once the ram is in its down position a leak pressure test is performed. The test cavities are charged with 0.73 KPa (5 psi), and pressure readings are individually compared to acceptable limits of pressure decay, which is a result of leakage. Accept or reject decisions are then made for each part and stored in memory. Spot welds having small holes are rejected. The previous method in this testing was to pressurize the part under water and manually inspect for bubbles. This is obviously a costly and time-consuming process which was performed only on an audit basis.

Moving to the next station on the index table (Figure 69.1), the two cover subassemblies are visually inspected to insure that proper parts were assembled and that all parts are present. Opto-Sense with four matrix array cameras was programmed to inspect all of the part models. A model select switch is positioned on the operator's control panel. This inspection task requires the resolution of four cameras, and they are aligned to overlap a portion of their field of view. Two cameras view each cover.

Incandescent front lighting was found to be most suitable in this application. The windowing technique was also found to be most suitable in performing the inspection task. Both accept and reject for visual reasons are stored in memory. It is important to note that both the leak pressure test and visual tasks were proven and tested before commencing with the design of this system.

At the next station the valve engine covers are unloaded by a point-to-point robot. The robot arm, having two gripper hands, simultaneously grips both parts and removes them from the fixtures on the table. Empty fixtures are then indexed toward the operator's load position.

If a part is not in the fixture, as detected earlier, no inspection tasks are performed, and the robot is signaled as to which gripper hand is not to be activated to insure that it does not attempt to grip and lift an empty position fixture bolted to the dial table. The robot proceeds loading accepted parts on a take-away conveyor. Once the robot moves the rejected parts, it places the rejects in one of three reject unload positions, sorting by reason of reject—visual, leak, or both.

The results of this application can be summarized as follows:

Elimination of two quality problems.

Elimination of shipping and return and reshipping parts to car assembly plants.

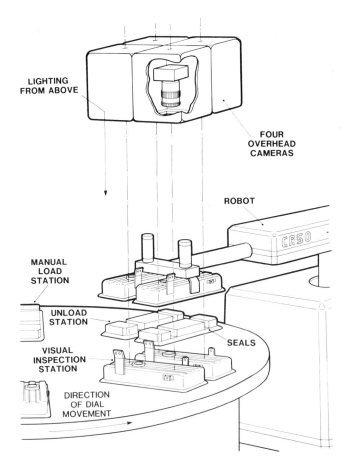

Fig. 69.1. Rocker cover inspection system.

Elimination of oil leak warranty costs.

Increase in production, with one person loading 600 indexes per hour for an inspection rate of 1200 covers per hour.

Efficient space utilization, as the completed machine is mounted on a 9-ft square unitized base.

Reduction in labor to reinspect because robot sorts rejects by reason.

69.5. SPRAY PAINTING APPLICATIONS

A manufacturer of appliances has a group of four spray painters on one shift and three on another. Their job was to spray paint the interior of five different models of rectangular metal containers with powdered paint. Any one of the five models can be found on an overhead moving conveyor. The conveyor's speed was changed before each shift to accommodate the change in numbers of spray painters. Any one of the five models of containers can be loaded onto unevenly spaced conveyor hooks that accommodate all of the models and transport them to the spray-painting booth. This spray painting task was found to be tedious and perhaps dangerous, as the material applied contained toxic chemicals.

69.5.1. The Application Scope and Tasks Performed

Two DeVilbiss spray painting robots were installed in the same area in which the spray painting takes place. They were programmed to work together and cover the areas to be painted on all five models. A Copperweld Opto-Sense Camera System was installed approximately 45 ft (15m) upstream of the spray painting booth. Opto-Sense was programmed to identify any of the models as well as to perform a check determining whether or not any of the containers had previously been painted and not removed from the conveyor line.

In Figure 69.2, Camera 1 identifies the model of container which is transported between the camera and a translucent plastic screen. This plastic screen diffuses incandescent lights which provide back light directed toward Camera 1.

Opto-Sense was programmed in this application to use a windowing technique discerning the model by correlating the overall length and width with the different openings found in the various models. A hole was provided in the plastic light-diffusing material to provide an opening for Camera 2 so that it could view the interior of each container through it. The same lighting arrangement that provides the back light for Camera 1 provides front lighting for Camera 2, which illuminates the interior of the container. Camera 2, looking for previously painted containers, signals the robots not to repaint containers found already painted.

An encoder was placed on the conveyor's drive motor, and it pulses position information to Opto-Sense's control. With the distance varying between conveyor hooks, a switch is made by the hook at the Opto-Sense station. This switch triggers Opto-Sense to perform its visual inspection, after which the results are stored in its computer's memory. These results are then shifted to Opto-Sense's memory, which, with the pulses of the optical encoder, knows when to provide information as to which task the Trallfa robots must perform. This information is provided by Opto-Sense before the hooks contact switches at each robot station, which triggers each robot to commence with the correct program for the model on the hook.

The results of this application can be summarized as follows:

Elimination of seven spray painters performing a tedious and possibly dangerous task.

Improvement in quality of spray paint coverage with a reduction in material sprayed.

Removed possibility of respray of a previously painted part.

Opto-Sense vision eliminated the requirement for an operator who would have been required to identify models and manually feed this information to the Trallfa robots.

It is obvious that the user in this case is enjoying a handsome return on his investment.

69.6. APPLICATION OF SEALANT

The use of sealants and adhesives in assembly is found to be growing throughout many industries. One application in an automotive plant requires a person to apply beads of material in various zones on a sheet metal subassembly. These subassemblies are positioned in front of the operator who then applies the sealant. The sealant material is automatically dispensed from 55-gal drums. As the operator triggers a hand-held gun to dispense the material, he manually moves the gun through various paths of motion in the various zones.

While going through these motions the speed at which the operator works must be controlled to insure that no less than $\frac{1}{16}$ in. (1.6 mm) diameter bead of material is applied on the prescribed paths.

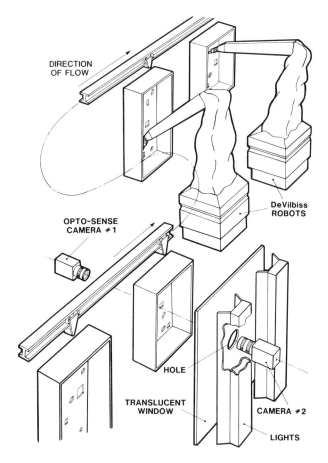

DIRECTION
OF FLOW

DeVilbiss
ROBOTS

OPTO-SENSE
CAMERA #1

HOLE

TRANSLUCENT
WINDOW

CAMERA #2

LIGHTS

Fig. 69.2.   Robotic inspection in spray painting.

It was found that, on occasion, gaps would appear on the paths. These gaps were created by the presence of air pockets in the sealant material within the 55-gal drum. Where gaps appeared, the operator filled them to insure a continuous path to meet the job's requirements.

69.6.1.   Application Scope and Tasks Performed

An ASEA robot (see Figure 69.3) was positioned at this work station and programmed to follow the paths in the various zones. The sealant gun was positioned at the end of the robot's arm in a relationship with an Opto-Sense camera. Fiber optics direct light onto the path in which material has been dispensed by the robot. Opto-Sense accesses visual information from its matrix array camera, analyzes the width of the bead, and searches for gaps. Small beads and gaps, when found, are stored in Opto-Sense's memory by the zones in which they are found. Once the robot has completed its motions in the various zones, it receives information from Opto-Sense as to the zones where it must repeat the job, where dispensing errors were visually found.

After the retracing in these zones has reapplied the adhesive material, Opto-Sense rechecks the way in which the sealant has been applied.

Shortly after this application, with its ongoing inspection abilities, it was discovered that each 55-gal drum contained voids created by air pockets in the range of 25% of the volume. This problem was resolved quickly and the robot's need to reapply additional material was reduced. The air pocket problems were not detected while this task was manually performed.

The results of this application can be summarized as follows:

Labor and material savings with consistent robot speed and accurate repeated motion.

Quality check with vision mounted at the end of the robot arm.

Material savings by reduction of air pockets in 55-gal drum: 450 gal per hour.

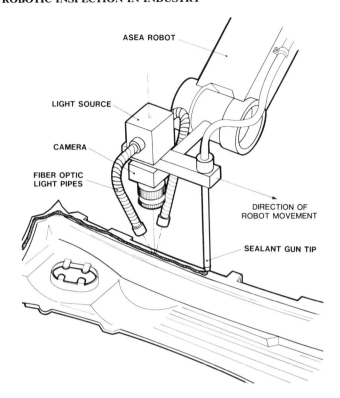

ASEA ROBOT

LIGHT SOURCE

CAMERA

FIBER OPTIC
LIGHT PIPES

DIRECTION OF
ROBOT MOVEMENT

SEALANT GUN TIP

Fig. 69.3. Robotic inspection in sealant application.

69.7. SUMMARY

The industrial applications described here are successful robotic inspection examples of what can be accomplished using robots and vision today. The success was not obtained until the difficulties in the application were understood and overcome. First-time applications today and in the future will result in some suffering by both the user and suppliers. The robot industry will lose some of its glamour in the near future. This loss has already begun as users become increasingly aware of the necessity for robots and the seriousness of their industries' requirements. But future robots will drive the technology to levels far beyond our imaginations. These future trends will follow those of the computer industry. Those who have waited too long have not been able to gain the experience and benefits as the industry grows.

REFERENCES

1. Crosby, P. B., *Quality is Free,* McGraw-Hill, New York, 1979.
2. Glorioso, R., Osorio, M., and Colon, C., *Engineering Intelligent Systems,* Digital Press, Bedford, Massachusetts, 1980.

FURTHER READING

Robotic Inspection

Agin, G. J., Vision systems for inspection and manipulator control, *Proceedings of the 1977 International Automatic Control Conference,* San Francisco, California, June 1977, pp. 132–138.

Agin, G. J., Servoing With Visual Feedback, *Proceedings of the 7th International Symposium On Industrial Robots,* October, 1977.

Albrecht, M. et al., Automated visual inspection for components manufacture and assembly in the automotive industry, *Proceedings of the 5th International Conference on Automatic Inspection and Product Control,* Stuttgart, Germany, June 1980, pp. 51–66.

Artley, J. W., Automated visual inspection systems can boost quality control affordability, *Industrial Engineering,* December 1982, pp. 28–30.

Atkinson, B. M. and Heywood, P. W., Automated inspection—a challenge of vision, *Proceedings of the 6th International Conference on Automated Inspection and Product Control,* Birmingham, U.K., 1982.

Batchelor, B. G., Automatic visual inspection in industry, *Industrial Robot,* Vol. 5, No. 4, December 1978, pp. 174–175.

Bretchi, J., *Automated Inspection Systems for Industry,* IFS Publications, 1982.

Engelberger, J. F., *Robots in Practice,* Amacom, New York, 1980.

Gauging: Practical Design and Application, IFS Publications, 1981.

Gleason, G. J. and Agin, G. J., A modular vision system for sensor-controlled manipulation and inspection, *Proceedings of the 9th International Symposium on Industrial Robots,* Washington, D.C., March 1979, pp. 57–70.

Hill, A. G., A general applied economic study of automated inspection and product control, *Proceedings of the 6th International Conference on Automated Inspection and Product Control,* Birmingham, U.K., 1982.

Kelly, R., Birk, J. and Wilson, L., Algorithms to visually acquire workpieces, *Proceedings of the 7th International Symposium on Industrial Robots,* October 1977.

Kochhar, A. K. and Burns, N. D., Microcomputer Based Automatic Testing and Quality Control Systems, in *Microprocessors and their Manufacturing Applications,* Arnold, London, 1983.

Macri, G. C. and Calengor, C. S., Automatic dimensional inspection utilizing robots, SME Paper No. MS79–795, 1979.

Macri, G. C. and Calengor, C. S., Robots combine speed and accuracy in dimensional checks of automotive bodies, *Robotics Today,* Summer 1980, pp. 16–19.

Makhlin, A. G., Westinghouse visual inspection and industrial robot control system. *Proceedings of the 1st International Conference on Robot Vision and Sensory Control,* Stratford, U.K., 1981.

Makhlin, A. G., Robot control and inspection by multiple camera vision system, *Proceedings of the 11th International Symposium on Industrial Robots,* Tokyo, 1981.

March, A. A. C., Optomation—GE inspection and measurement systems based on CID solid state array cameras, *Proceedings of the 6th International Conference on Automated Inspection and Product Control,* Birmingham, U.K., 1982.

Ouda, H. and Ohashi, Y., Introduction of visual equipment to inspection, *Industrial Robot,* Vol. 6, No. 3, September 1980, pp. 131–135.

Rushlow, G. O., Robot operated body inspection system, *Proceedings of Autofact 5,* Detroit, Michigan, November 1983.

Schmid, D., Industrial robot vision with video camera for detection of material defects, *Proceedings of the 2nd International Conference on Robot Vision and Sensory Control,* Stuttgart, W. Germany, 1982.

Scott, A. J., Why should I use automatic inspection?, *Proceedings of the 6th International Conference on Automated Inspection and Product Control,* Birmingham, U.K., 1982.

Toutant, R. T., Robotic testing and inspection applications, *Proceedings of Autofact 5,* Detroit, Michigan, November 1983.

Unser, M. and de Coulon, F., Detection of defects by texture monitoring in automatic visual inspection, *Proceedings of the 2nd International Conference on Robot Vision and Sensory Control,* Stuttgart, W. Germany, 1982.

Von Hipple, E., Users and innovators, *Technology Review,* January 1978.

Von Hipple, E., Successful industrial products from customer ideas, *Journal of Marketing,* 1978.

Quality Control and Inspection

American Society for Quality Control, *Glossary and Tables for Statistical Quality Control,* ASQC, Milwaukee, Wisconsin, 1973.

American Society for Quality Control (ASQC), *Guide for Reducing Quality Costs,* Milwaukee, Wisconsin, 1977.

Drury, C. G., Improving Inspection Performance, in Salvendy, G., Ed., *Handbook of Industrial Engineering,* Wiley, 1982, Chapter 8.4.

Duncan, A. J., *Quality Control and Industrial Statistics,* 4th ed., Irwin, 1974.

European Organization for Quality Control, *Glossary of Terms Used in Quality Control,* 4th ed., July 1976.

Juran, J. M. et al., *Quality Control Handbook,* 3rd ed., McGraw-Hill, New York, 1974.

Kennedy, C. W. and Andrews, D. E., *Inspection and Gaging,* 5th ed., Industrial Press, New York, 1977.

Society of Manufacturing Engineers, *Handbook of Industrial Metrology,* Prentice-Hall, Englewood Cliffs, New Jersey, 1967.

Teel, K. S. et al., Assembly and inspection in microelectronic systems, *Human Factors,* Vol. 10, 1968, pp. 217–224.

CHAPTER 70

ADVANCED ROBOTIC INSPECTION APPLICATIONS

KENICHI ISODA

YASUO NAKAGAWA

Hitachi
Tokyo, Japan

In factories, inspection is as important as manufacturing itself. Visual inspection by a human inspector supports product quality to a great degree because functional testing can not always inspect everything. Much of the production of microelectronics technology requires very fine visual inspection and this is very difficult to accomplish with human eyes. The following are examples of Hitachi's robotic inspection systems developed to solve the problems.

70.1. AUTOMATIC VISUAL INSPECTION OF SHADOW-MASK MASTER PATTERNS

70.1.1. Pattern Defects and Criteria of Inspection

Master Pattern

A color picture tube (CPT) has a shadow-mask at the back of its front panel. It is made of thin iron plate, on the surface of which there are many small rectangular holes arrayed in rows; the total number of these holes is several hundred thousand per shadow-mask.

These rectangular holes are formed by contact-exposure printing and chemical etching. To obtain high-quality shadow-masks, it is essential to keep the quality high for the master plate used in contact-exposure printing. The patterns on the master plate are damaged during production or when used for contact-exposure printing many times. So the master plate should be inspected and retouched for repetitive use.

A master pattern for shadow-masks is a dry plate, the size of which is about 610 × 800 mm, and on its surface many stripe patterns are photo-printed within a frame line of a CPT panel. Figure 70.1 shows examples of the stripe patterns. Two kinds of stripe patterns, "thick stripes" and "fine stripes," are used to make one shadow-mask. The sizes of these stripe patterns are a little different in different sorts of CPTs. Table 70.1 shows an example of the sizes, where an arrow in the width W of "fine stripes" means that the width is decreased from the center to the edge of a CPT smoothly, that is to say "graded."

A designed stripe pattern is an exact rectangular pattern, but in real patterns there are some distortions, something like roundness of the corner, a tiny unevenness of the edge line, and other irregularities. Figure 70.2 shows the magnified photograph of the real pattern. In automatic inspection these tiny irregularities should be tolerated.

Criteria of Inspection

Table 70.2 shows the criteria for inspection, that is, the kinds of defects and the minimum sizes to be detected. The defects can be classified into 10 kinds. These defects are caused by pinholes or alien substances in the process of pattern making, by scratches during use of master plates, and by mistakes in retouching defect patterns.

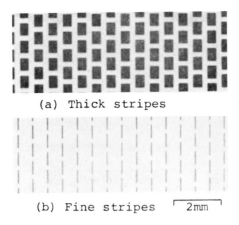

(a) Thick stripes

(b) Fine stripes ⌐ 2mm ⌐

Fig. 70.1. Shadow-mask master patterns.

70.1.2. Basic Concepts of Pattern Inspection

It is advantageous to use comparative inspection between two neighboring patterns, because the difference of sizes between neighboring patterns is small, and in fine stripes the sizes of patterns change gradually along with inspection sequence. On the other hand it is essential to extract tiny defects by individual inspection of patterns because the limit size of defects is small and is only a little larger than the detectable minimum size.

With these considerations, we designed the algorithm of interpretation, as shown in Figure 70.3. After the video signal is thresholded, the obtained binary signal of patterns is divided by *pattern extraction* into *effective patterns* and *invalid patterns*. Figure 70.4 shows this process schematically. In Figure 70.4 the left-side neighboring patterns are neglected by pattern extraction which functions just when the bright, transparent area has appeared during a scan. The effective patterns obtained are interpreted by six methods. Among these methods, *area comparison, bridge width comparison, maximum width comparison* and *center position comparison* are comparative inspections between a pattern just below the inspection area and a correct pattern inspected just previously. *Width change* and *center change* examine just the pattern below the inspection. If any defects are detected, the error signal is transmitted to a defect marker.

TABLE 70.1. SIZES OF STRIPE PATTERNS

	THICK STRIPES	FINE STRIPES
W	0.38	0.10 → 0.07
L	0.67	
B	0.13	
Px	0.60	
Py	0.80	

unit:mm

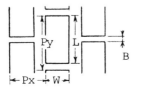

(a) Thick stripes

(b) Fine stripes ⌈0.2mm⌉

Fig. 70.2. Details of patterns.

TABLE 70.2. CRITERIA OF PATTERN INSPECTION

No.	Name	Feature	Limit size
1	chip-1		10 μm
2	chip-2		10 μm
3	projection		10 μm
4	pinhole		10 μm
5	missed		10 μm
6	too large		10 μm
7	too small		10 μm
8	dot		10 μm
9	bridge width		10 μm
10	position error		10 μm

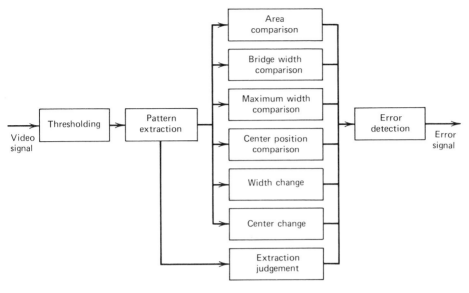

Fig. 70.3. Structure of interpretation algorithm.

70.1.3. System Configuration

Figure 70.5 shows a schematic structure of the inspection machine. It consists of detecting heads which include microscopes and image sensors on the real image planes, a sliding table on which a master plate is set with the surface upward, driving motors, and circuits for judgment and control. Five microscopes are arrayed transversely against the table sliding direction. Longitudinal scanning is done by moving the table to and fro. During this scanning each of the microscopes detects a row of stripe patterns respectively. Every time the longitudinal scan is completed, the gang of microscopes is moved transverse, step by step, so that each of the microscopes can inspect a new row. Signals from the image sensors are led to circuits for interpretation, and if a defect is detected, a marker is actuated to print a mark just at the back of the defect. By using a solid-state linear image sensor, it is possible to detect 200 stripes per second per detecting head in the resolution of 7 μm. The algorithm, the *extraction interpretation,* can detect 10 kinds of limit-size defects at a high detection rate. As the conditions of interpretation are relaxed, the results of automatic inspection are coincident with the results of visual inspection by skilled operators.

Figure 70.6 shows the automatic inspection machine developed for real use.

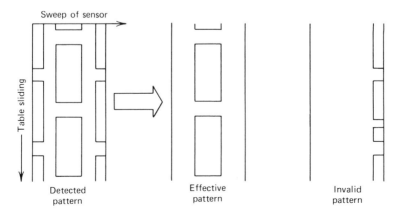

Fig. 70.4. Pattern extraction.

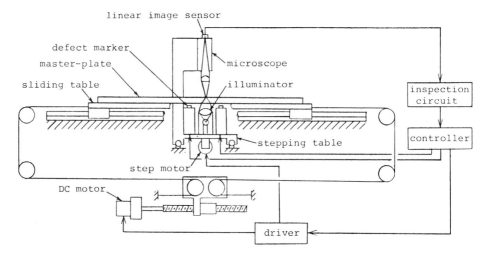

Fig. 70.5. Structure of automatic inspection machine.

70.2. AUTOMATIC INSPECTION OF DEFECTS ON CONTACT PARTS

70.2.1. Contact Points of Relay Switch and Their Defects

As shown in Figure 70.7, 16 pairs, or a total number of 32 contact points are welded on a stainless spring sheet. The dimensions of each pair are also shown in Figure 70.7. A contact point is made of a Cu chip gilded with Au and Pd.

Figure 70.8 shows an example of defects in appearance that will have bad effects on electrical contact when the switch is turned on, and on insulation when the switch is turned off. These defects are caused while contact points are welded on the spring, and they are roughly classified as follows:

Defects in the welding position of contact points (Figure 70.8*a*).
Defects in the shape and surface of contact points (Figure 70.8*b, c, d, f*).
Defects in the surrounding of contact points (Figure 70.8*e, f*).

Visual inspection criteria for these defects have been as follows:

1. Displacement from regular position:
Displacement should be less than ±0.1 mm each direction along x and y axis.

Fig. 70.6. Automatic inspection machine for real use.

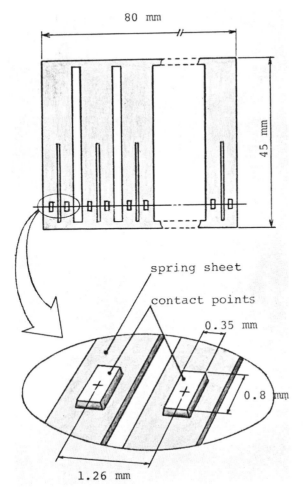

Fig. 70.7. A contact and relay sheet for a relay switch.

2. Deformation:
   The contact size should be larger than 0.32 × 0.8 mm.

3. Scratch:
   A scratch should be less than 10 μm in width as well as less than 50 μm in length.

4. Adherence of alien substance:
   Adherence of alien substance should be less than 50 μmϕ.

5. Welding splash:
   A welding splash should be less than 20 μm in width at 0.2 mm from the side of contact points.

6. Collapse:
   A collapse should be less than 0.2 mm in length from the side of contact points.

70.2.2. Basic Concepts of Inspection Devices

Optical System of Detection and Its Characteristics

Figure 70.9 shows the optical system and its formation. The system has a television camera with a 0.5 in. vidicon tube for detecting defects. Since the resolution of the camera is 650 lines per image window width, it is appropriate to reflect two contact points in one image window, judging from the size of the defects to be detected. An objective lens of 5× magnification was used to reflect the images of these two contact points on the effective surface of the image tube.

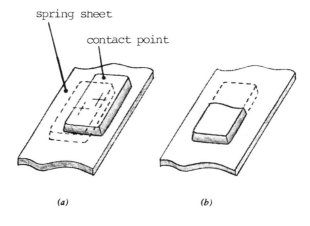

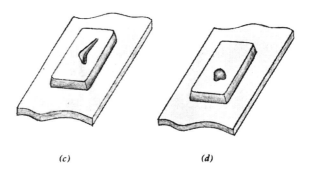

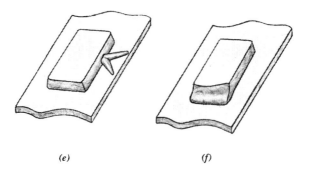

Fig. 70.8. Illustration of contact point defects to be detected. (*a*) Displacement; (*b*) deformation; (*c*) scratch; (*d*) adherence of alien substance; (*e*) welding splash; (*f*) collapse.

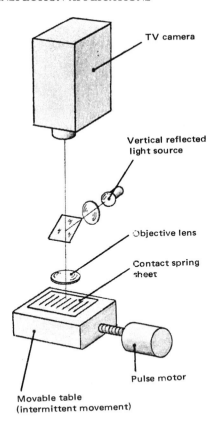

TV camera

Vertical reflected
light source

Objective lens

Contact spring
sheet

Pulse motor

Movable table
(intermittent movement)

Fig. 70.9. Optical system for detecting defects of the contact points.

The vertical-reflected illumination method was adopted. When lighted from above, as shown in Figure 70.10, the surface of contact points is the brightest and the spring surface is second brightest; on the other hand, the outline of the contact point is dark, and the scratch, welding splash, and so on, which should be detected as defects, are also dark. Therefore we can discriminate them with sufficient contrast.

A pulse motor intermittently shifts a movable plate on which the contact spring is placed. To simplify the shifting mechanism and to save shifting time, a pair of television cameras is used. In this way, two different pairs of contact points are reflected on each television screen simultaneously.

Defects Detection Algorithm

Figure 70.11 shows the defects detection algorithm.

First Field. The first field is to determine the threshold level to convert the video signal into the binary symbol of 1 or 0.

Second Field. By converting the video signal into the binary value of 1 or 0 based on a pertinent threshold level, decided in the first field, we search the location of the center of contact point from its outline and inspect the displacement. Figure 70.12 shows how to locate the center point.

Third Field. We limit the inspection range by setting the frames and set a different new threshold level to detect the defects other than displacement.

In the first place, we set two frames (as shown in Figure 70.13) centered on the contact point center determined in the second field. The size of the inside frame is 0.32 × 0.8 mm and that of the outside frame is 0.72 × 1.2 mm.

Taking the size of defects to be detected into consideration, as shown in Figure 70.13, we scan all over the inside frame with a 7 × 7 pixel window (54 × 54 square μm). A continuous sequence of

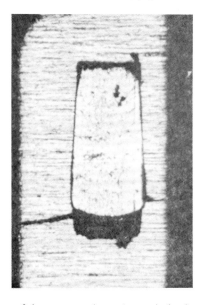

Fig. 70.10. Image of the contact point under vertical reflected illumination.

1's extending across the entire window represents a feature large enough to be judged a defect. If any such sequence is found, we decide the contact is defective.

Still taking the inspection criteria into consideration, we scan with a 3×3 pixel window (23×23 square μm) along the perimeter of the outside frame, as shown in Figure 70.13. If for any window, all of the nine pixels indicate the symbol 1, the contact is considered defective.

70.2.3. System Configuration

Figure 70.14 shows the appearance of the device installed in the production line for practical use. As mentioned before, there are two sets of television cameras, but the signal-processing circuit is single and processes the signals coming from both camera sets. The time necessary for the inspection of 32 contact points is 5.4 seconds, and its breakdown is as follows:

Inspection and judgment (32 contact points)	2.1 sec
Shifting (10 times of intermittent motions)	2.9 sec
Elimination of afterimage and time lag for receiving signal	0.4 sec
Total time	5.4 sec

70.3. AUTOMATIC VISUAL INSPECTION OF SOLDER JOINTS ON PRINTED CIRCUIT BOARDS

70.3.1. Solder Joint Defects

Examples of solder joint defects are shown in Figure 70.15. This figure illustrates the sectional shape of solder joints, where the soldered surface is on top and the loaded electric parts are underneath. Although defects can be classified in detail into more than 10 categories, there are five fundamental types of defects, four shown in this figure. The defect not shown is a solder bridge, which can be detected by an electric tester and is thus excluded from those defects to be detected by automatic visual inspection. No-solder in Figure 70.15 can also be detected by electric tester if there is no contact, but in general the lead of the part is touching the land of the board, and electric contact exists. So, no-solder cannot be detected by the tester in many cases.

In detection of solder joints, it is necessary to detect the shape of the joint correctly without being influenced by the gloss and blur of the solder surface.

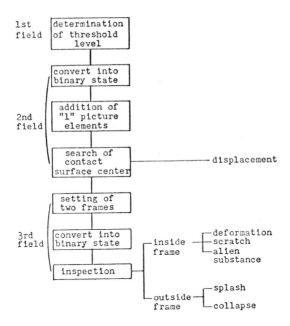

Fig. 70.11. Format of defects-detecting algorithm.

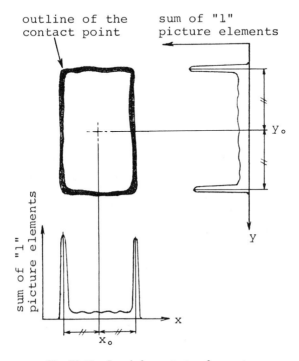

Fig. 70.12. Search for contact surface center.

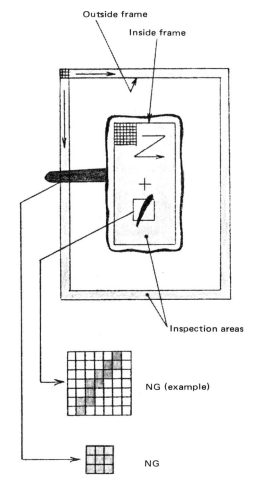

Fig. 70.13. Inspection range-scanning matrixes.

Fig. 70.14. Device installed in production line.

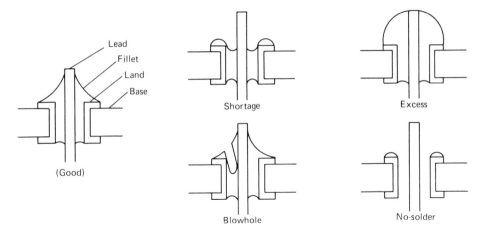

Fig. 70.15. Solder joint defects.

70.3.2. Basic Ideas of Automatic Inspection

Some basic configurations of the light-section method are illustrated in Figure 70.16. In Figure 70.16a a slit projector and an image detector face each other so that their optical axes form a 90° angle, and a slit projected from the oblique direction is detected from the opposite oblique direction. In Figure 70.16b a slit is projected on the object from a vertical direction, and its image detected from a plane parallel to the object surface. In the former the detector detects the direction reflection of the slit image, and in the latter it detects the diffused reflection part. But in either case the projected slit on the target is in the objective plane of the image detector. Thus these methods have an advantage in focusing where the target shape does not influence the defocusing, and are applied widely as basic configurations. These methods are effective for the object whose sectional shape does not change quickly in the plane including the optical axes of the projector and detector. In the case of a quick shape change, the object may obstruct the detection of the projected slit, or an unnatural sectional shape may be obtained for the oblique directional cutting in Figure 70.16a. These shortcomings cannot be neglected in the case of solder joints. For this, we adopt the configuration of Figure 70.16c, that is, the slit is projected from the vertical direction and is detected from the oblique direction. Using this method, the solder shape can be detected stably, not influenced by neighboring joints. And defocusing is not so difficult when the image is focused on the middle of the solder surface.

The biggest merit of the application of the light-section method to three-dimensional shape inspection is that the detection process is essentially sequential. In general, three-dimensional shape detection requires a greater amount of input information than does a plane image. In the light-section method it is possible to compress the input information to obtain condensed information sequentially. The process model of shape inspection with the light-section method is shown in Figure 70.17. At first, a light-section waveform is extracted from a gray-scale image obtained by an image detector. Next, the waveform is analyzed, and the data are transformed into more meaningful information such as, "This waveform has some irregularity." Moving the XY table stepwise in the Y direction, gray-scale images are detected and processed in the same way repeatedly. Finally, the meaningful information is summarized and interpreted.

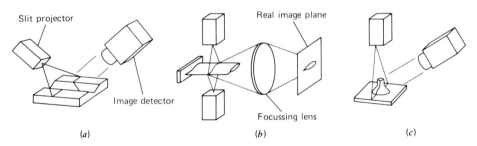

Fig. 70.16. Light section method.

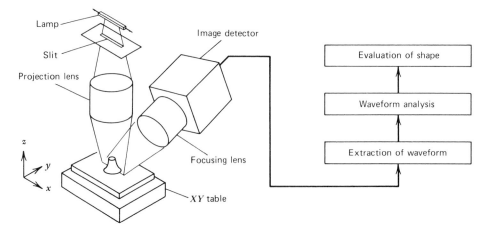

Fig. 70.17.   Automatic inspection system using light-section method.

70.3.3.   System Configuration

Figure 70.18 shows the organization of a prototype inspection system, and Table 70.3 shows the principal specifications of the system. The prototype system is shown in Figure 70.19; television cameras watch a projected slit at a 35° angle from the horizon. The judgment algorithm is described in the following paragraphs.

Extraction of Light-Section Waveform

As shown in Figure 70.20a, a light-section image obtained by the television camera is a gray-scale image where only the projected slit is bright. A waveform is extracted from the image by picking the brightest point in each vertical sweep as shown in Figure 70.20b and c.

   If the brightness at x is less than V1 for all heights z, the output waveform has zero height at x. If there is a brightness peak between V1 and V2, at height $z1$, the waveform has height z at x. If the brightness peaks above V2 at z, the height at x is the average of two heights, above z and below z, where the brightness equals V2.

Evaluation of Waveform

A waveform is evaluated by checking a number of parameters. In counting these parameters, particular regions for each parameter are set up, and the counting is done only in these regions.

   Figure 70.21 shows four parameters, that is, area S_l, S_h, S_b, and broken length W, which correspond to the four fundamental defect types previously mentioned. By this method all defects can be detected without overlooking any. An inspection speed of 10 points per second is possible for dual in-line IC joints.

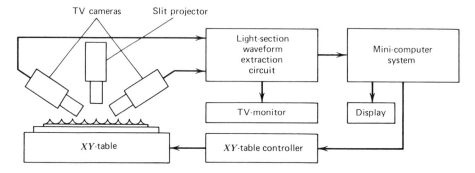

Fig. 70.18.   System organization of a prototype inspection system.

TABLE 70.3. PRINCIPAL SPECIFICATIONS

Slit Projector

Light source	:	halogen lamp, 300 W
Slit	:	0.2 mm W, 10 mm L
Projection lens:		macrolens, $f = 90$ mm, mag. 1X

Image Detector

TV camera	:	$\frac{2}{3}$ in. vidicon
Focusing lens	:	$f = 95$ mm, mag. 0.5X
Size of pixel	:	20 μm/pixel (x)
		18 μm/pixel (y)

Light-Section Waveform Extraction Circuit

Data point	:	240 pixels

70.4. AUTOMATIC INSPECTION SYSTEM FOR PRINTED CIRCUIT BOARDS

70.4.1. Introduction

The printed circuit boards used in modern electronic computers must be of high reliability. Figure 70.22 shows an example of a printed circuit board having a minimum pattern width of 0.14 mm. Figure 70.23 shows an example of a defect in a printed circuit pattern. If shorts or breaks are present in the pattern they will be detected by electrical conduction tests. However, if defects are of the type shown in Figure 70.23, they cannot be detected by this method.

The reliability of printed circuit boards has, up to the present time, been guaranteed by means of visual inspection for defects. The quality of this type of inspection is influenced by the degree of experience of the inspector and may even be influenced by the inspector's mood at the time.

The continuing automation of the printed circuit board production process has naturally resulted in demands for automatic rather than visual inspection of these products.

Fig. 70.19. Prototype inspection system.

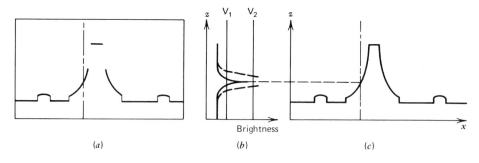

Fig. 70.20. Waveform extraction. (*a*) Original image; (*b*) video signal for one vertical sweep; (*c*) waveform.

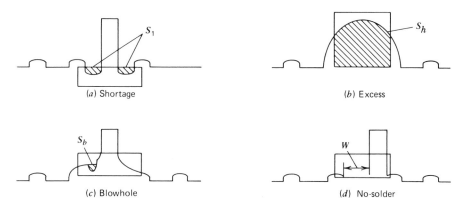

(*a*) Shortage

(*b*) Excess

(*c*) Blowhole

(*d*) No-solder

Fig. 70.21. Parameters for interpretation.

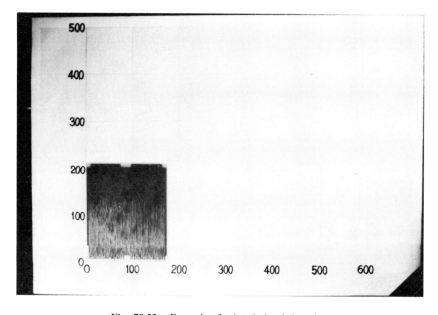

Fig. 70.22. Example of printed circuit board.

Fig. 70.23. Example of defect in a printed circuit board.

70.4.2. Automatic Inspection Technology

Illumination

Figure 70.24 illustrates illumination of the pattern for accurate detection of defects. The pattern is vertically illuminated through a half-mirror as well as from eight slant directions by means of optical fiber units. The image of the pattern is focused on the detector by the lens above the half-mirror. When the pattern is illuminated at a slant, a ringlike pattern is obtained as shown in Figure 70.25a. When illuminated both at a slant and from above, an almost perfect ring pattern is obtained as shown in Figure 70.25b.

Since the area around the through-holes is raised, this method of illumination is valid even when the cross section of the pattern is flat.

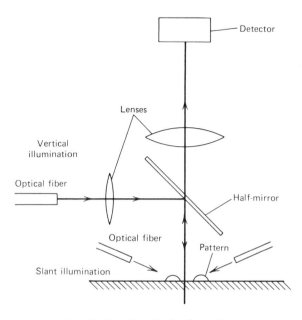

Fig. 70.24. New illumination system.

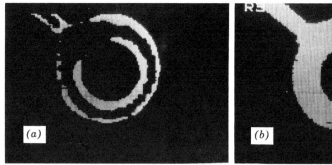

Fig. 70.25. Effect of new illumination system. (*a*) Detected pattern, slant illumination; (*b*) Detected pattern, vertical and slant illumination.

Pattern Detector and Binarization of Detected Signals

A linear image sensor is used as the pattern detector. The linear image sensor commonly consists of either 1024 or 2048 minute ($10 \sim 30$ μm) photodetectors arranged in-line.

Complete illumination is normally difficult when the pattern is illuminated by reflected light. Therefore binarizing with a uniform threshold level of Vth may cause the pattern width to appear different and so be detected, depending on its position (refer to Figure 70.26). Simultaneously, the latitude allowed for noise will be reduced.

These factors resulted in the development of the *light intensity optimization binarization method* to permit setting of binary levels in response to light intensity distribution. This method involves the initial storage in memory of a light intensity distribution $V1(x)$ produced by the detection of a uniform reflecting surface, the binary threshold levels $Vth(x)$ for all locations on the circuit board then being calculated and set automatically with the following equation.

$$Vth(x) = V_0 + k \cdot V1(x)$$

where V_0 = Offset
       k = Constant determining level height

$$0 < k < 1$$

$Vth(x)$ changes in accordance with the light intensity distribution shown by the dotted line in Figure 70.26 and therefore permits the accurate detection of patterns irrespective of their position.

Defect Detection Method

The defect inspection method[4] developed is based on the pattern comparison method. The principle used is shown in Figure 70.27, in which two patterns f and g are aligned for comparison. As the

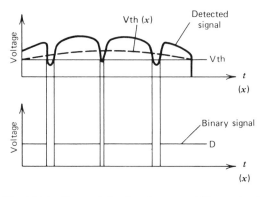

Fig. 70.26. Detected signal and converted binary signal.

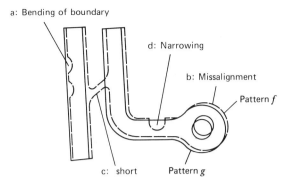

Fig. 70.27. Overlapped image of patterns f and g with small missalignment.

two are not aligned accurately and the patterns themselves include relative dimensional differences resulting from the production process, they cannot be aligned perfectly.

Any defects of a size less than the alignment inaccuracy must of course be recognized as defects. This required the development of a method of comparison in which the local features of the pattern are extracted and used as a basis for recognizing these defects.

All images are binary (either 1 or 0) and take the form of small squares termed pixels, which are stored in the memory.

Extraction of Boundary Lines

Comparatively wide defects are detected by extraction of boundaries running in a direction that differs from that of the boundaries of the reference pattern. This method may be used for detection of all defects of a width greater than a fixed value, and for detection of isolated defects.

A practical method for the extraction of boundaries, given as follows, is illustrated in Figure 70.28. A boundary is determined when the following conditions are established, where (P) is the value of pixel P.

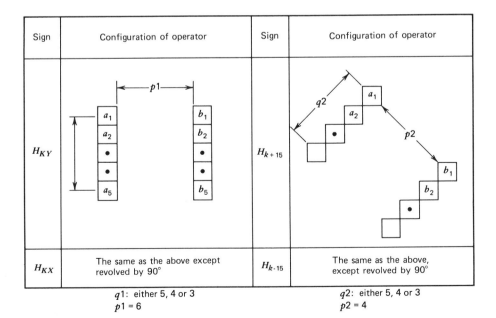

Fig. 70.28. Extraction method of boundary. For example, a boundary is determined when $(a1) = \cdots = (a5)$, $(b1) = \cdots = (b5)$, $(a1) \neq (b1)$.

$$(a1) = \ldots = (am)$$
$$(b1) = \ldots = (bm)$$
$$(al) \neq (bl)$$

Boundary line extraction is performed in four directions ($0°$, $90°$, $+45°$, $-45°$). Selection of q1, p1, q2, p2 sets the limits of the length and inclination for possible boundary lines.

Extraction of Small Line-Width Patterns

This method is used to detect narrow defects not extracted as boundary lines. These defects include fine wiring and whisker wires which, although extremely small, may cause electrical short circuits.

A practical method for the extraction of extremely fine patterns is illustrated in Figure 70.29. Such patterns are determined when the following conditions are established.

$$(a1) = \ldots (a8) = (a9) = \ldots (a16)$$

and any one of (b1) . . . (b4) is not equal to (a1).

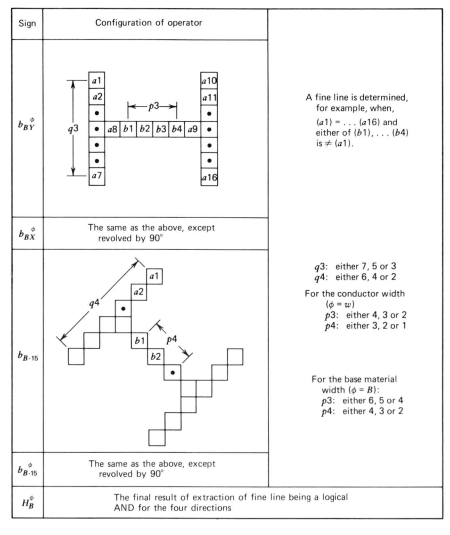

Sign	Configuration of operator	
b_{BY}^{ϕ}		A fine line is determined, for example, when, $(a1) = \ldots (a16)$ and either of $(b1), \ldots (b4)$ is $\neq (a1)$.
b_{BX}^{ϕ}	The same as the above, except revolved by 90°	
$b_{B\text{-}15}$		$q3$: either 7, 5 or 3 $q4$: either 6, 4 or 2 For the conductor width $(\phi = w)$ $p3$: either 4, 3 or 2 $p4$: either 3, 2 or 1 For the base material width $(\phi = B)$: $p3$: either 6, 5 or 4 $p4$: either 4, 3 or 2
$b_{B\text{-}15}^{\phi}$	The same as the above, except revolved by 90°	
H_{B}^{ϕ}	The final result of extraction of fine line being a logical AND for the four directions	

Fig. 70.29. Extraction method of fine line.

The extraction of fine patterns is performed for four directions (0°, 90°, +45°, −45°), the results of this extraction being a logical AND for the four directions. This differs from the case of the boundary lines in that portions including normally undetected patterns are extracted. The production of the AND therefore precludes any undesirable influence on defect detection performance and permits reductions in circuit size. Selection of q3, p3, q4, p4 regulates the detection criteria for the distance of defects from the patterns and boundary lines inspected for fine defects.

Different inspection criteria may be set for the conductor portion ((bi) = 1) and the base material portion ((bi) = 0) by altering p3 and p4.

Comparison of Extracted Features

Refer to Figure 70.30 for an explanation of extraction of pattern features and defect recognition: (1) Figure 70.30 shows the two patterns (f and g) being compared, (2) the images Fk and Gk extracted from the boundary lines in the Y direction, and (3) the images Fb and Gb extracted as fine patterns. Defect recognition involves the comparison of Fk and Gk (or Fb and Gb). When corresponding points on the reference pattern and printed circuit board pattern exhibit the same features, the pattern is determined as having no defects. If this relationship is not established, a defect is determined. In (3) an electrical short defect (narrow in this case) is recognized.

Image Processor for Defect Detection

The following provides an explanation of a method of implementing the preceding inspection with the use of a hard-wired circuit (refer to Figure 70.31). This circuit binarizes the image signals obtained from a pair of linear image sensors and includes three components. First is a noise filter circuit to remove fine, nondefect, isolated patterns (pinholes, excess conductors) from the binary signal. A feature extraction circuit extracts local features of the pattern image after noise filtering, and a feature comparison circuit compares the extracted features within a fixed area.

The feature comparison circuit compares each window of $n \times n$ pixels to determine whether or not the same features exist and therefore permits scanning even if the two patterns are not perfectly aligned.

The pi and qi (Figures 70.28 and 70.29) are set to small values relative to the sizes under comparison to remove undesirable influences caused by quantization error arising when the two patterns are binarized.

Figure 70.32 shows examples of detected defects. In the actual inspection system three images are combined and displayed on a color television monitor.

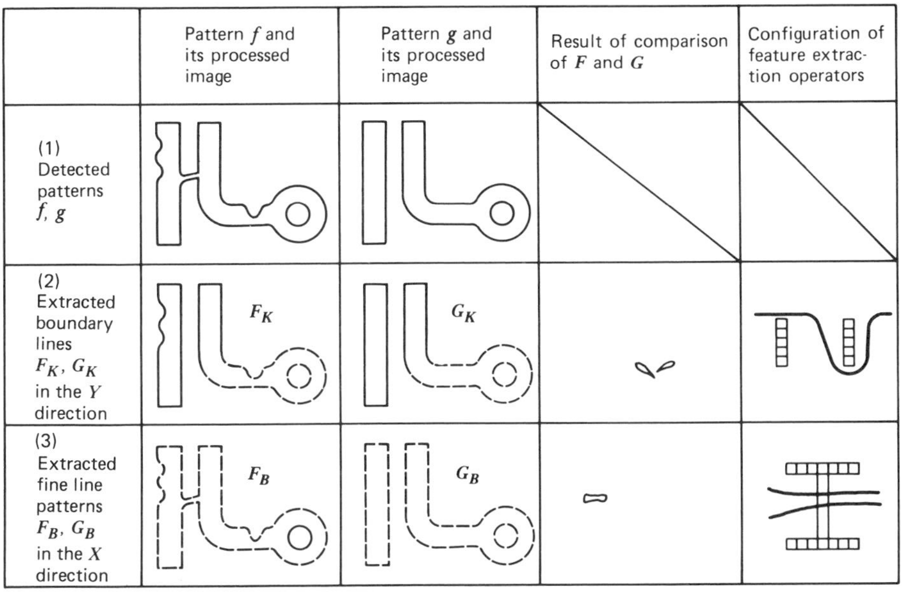

	Pattern f and its processed image	Pattern g and its processed image	Result of comparison of F and G	Configuration of feature extraction operators
(1) Detected patterns f, g				
(2) Extracted boundary lines F_K, G_K in the Y direction	F_K	G_K		
(3) Extracted fine line patterns F_B, G_B in the X direction	F_B	G_B		

Fig. 70.30. Extraction of features and comparison of the extracted feature patterns.

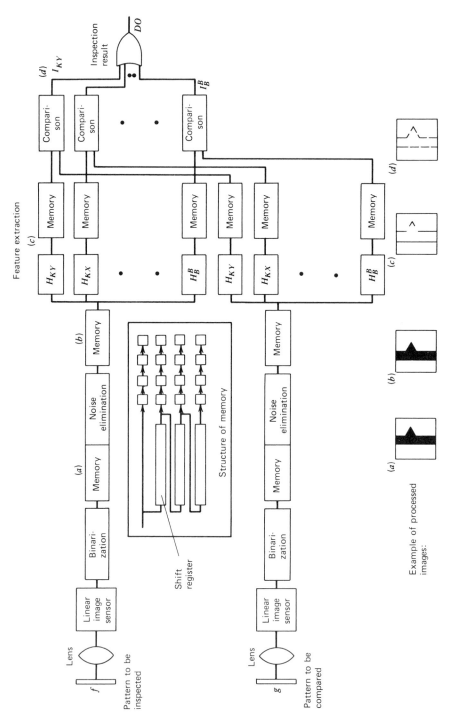

Fig. 70.31. Real-time hard-wired logic circuits for the defect detection method.

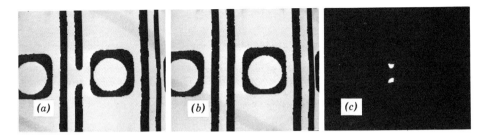

Fig. 70.32. Example of detected defects. (*a*) Pattern to be inspected; (*b*) pattern comparison; (*c*) inspection result.

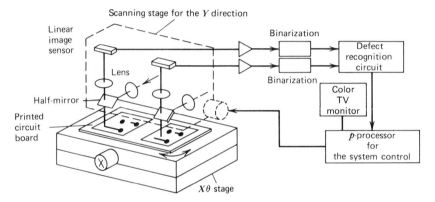

Fig. 70.33. Schematic structure of the inspection system.

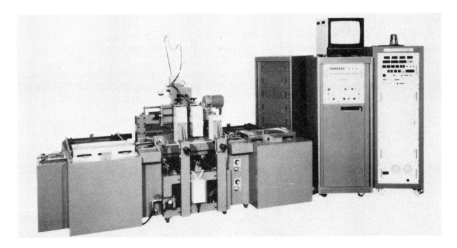

Fig. 70.34. Automatic inspection system for printed circuit boards.

70.4.3. Inspection System

An inspection system using the technologies described has been developed. The scheme of the system is shown in Figure 70.33. The printed circuit board is mounted on the $x\theta$ stage, a number of the same circuit patterns being printed repeatedly above this circuit board. This board is illuminated and the corresponding portions of the circuit pattern are projected onto a pair of linear image sensors. The image signals obtained from these sensors then are fed to the defect recognition circuit.

This system is such that pixel size Δp may be altered electrically, and an isolated pattern (size Dn) eliminated as noise, patterns (sizes q1, q2) extracted as border lines or (sizes p3, p4, q3, q4) as fine patterns, and the allowable alignment error r may be set as desired.

Figure 70.34 shows a general view of the system. The inspection system itself is located in the center, the loader and unloader at either side, and the electrical circuits at right. Handling of circuit boards is completely automated.

The printed circuit boards are stored in a magazine and delivered one at a time to the $x\theta$ stage. The system includes a detector table on which two detectors are mounted, this table being capable of movement in the Y direction so that the distance between the detectors may be altered.

The printed circuit board is fed to the inspection stage, and the misalignment of the two patterns in the Y direction is detected by the two detectors, this misalignment being corrected by fine adjustment on the θ axis. Following the completion of alignment, the inspection stage is moved in the X and Y directions and the board inspected. If a defect is detected, its coordinate is temporarily stored in the memory and it is marked with ink later when the defect is moved under the marking head. Following the completion of inspection, the board is fed to the unloader on the side of the system opposite to the loader. Table 70.4 lists the specifications of this system.

TABLE 70.4. SPECIFICATIONS[a]

Minimum detectable defect size	40 μm
Minimum line width of pattern	140 μm
Detector	CCD linear image sensor
Binarization method of detected image signals	Light intensity optimization method
Detector scanning speed	5 MHz
Inspection speed	20 cm²/sec
Handling of boards	Full automatic
Recording of detected defect information	Marking red ink onto detected defects
Stroke of scanning stages	500 × 600 mm

[a] Pixel size 20 μm.

REFERENCES

1. Ejiri, M., Mese, M., and Ikeda, S., A process for detecting defects in complicated patterns, *Computer Graphics and Image Processing,* Vol. 2, No. 4, December 1973, p. 326.

2. Sterling, W. M., Automatic nonreference inspection of printed wiring boards, *Proceedings of Pattern Recognition and Image Processing,* August 1979, p. 93.

3. Jarvis, J. F., A method for automating the visual inspection of printed wiring boards, *Proceedings of the SITEL-ULG Seminar on Pattern Recognition,* Belgium, November 19, 1977, p. 9.1.1.

4. Hara, Y. et al, Automatic visual inspection of LSI photomasks, *Proceedings of the 5th International Conference on Pattern Recognition,* December 1980, p. 273.

CHAPTER 71

ROBOT-OPERATED BODY INSPECTION SYSTEM

JAMES A. KAISER

General Motors Corporation
Warren, Michigan

71.1. DIMENSIONAL CHECK METHODS

The traditional methods of dimensionally checking automobile body openings during the body assembly stage of manufacturing included the use of both "apply" fixtures and off-line precision-machined "hard" fixtures. Apply fixtures were either stored on a rack next to the line or hung over the line on a counterbalance. They were used when required by placing them in an opening (such as windshield, door, or trunk opening) located from key points, to make the measurement. The off-line hard fixtures were fixed to a surface plate. To use these fixtures it was necessary to remove an in-process body from the assembly line and place it in the fixture. The checks were then made manually using feeler gages and the fixture units mounted on a side frame. This method has many drawbacks, the most severe of which is the limitation on the number of bodies that can be checked per shift. A need was recognized to gather more data more quickly in order to bring the manufacturing process under statistical control. One of the most successful systems developed to meet this need was the ROBI (Robot-Operated Body Inspection) system.

71.2. SYSTEM DESCRIPTION

The ROBI system was developed to check automobile bodies automatically without removing them from the assembly line. The system consists of two six-axis servo-driven robots mounted on DC electrically driven, multiposition tracks. (See Figure 71.1.) Laser probes are mounted on each robot wrist to take the dimensional readings from the bodies.

The system occupies one station on a stop-and-go shuttle line. The body is shuttled into the station on a pallet which is then locked into place by a shot pin. Lifters then raise the body about 5.0 mm, and the robot performs the check. On the continuously moving line, three stations are required for a power and free shuttle to provide an idle check station. The body moves into the first station where the line chain is disengaged. The body is then rapidly advanced into the second station on precision rails, where it stops. A positioner locks into the front gate lock on the build truck. After the check is performed, the body is rapidly advanced into the third station to catch up with the moving line, and the chain is reengaged.

71.3. SYSTEM OBJECTIVES

The ROBI system objectives were set as follows:

1. To check the same points and obtain the same readings as the master checking fixture.
2. To be adaptable enough to check various styles and sizes of bodies.
3. To be flexible enough through programming to be able to selectively check specific points when required (i.e., to zero in and keep detail checking specific problem areas, or to completely check a body).
4. To provide software powerful enough to generate useful information for quality and process control.
5. To check some portion of every body manufactured.

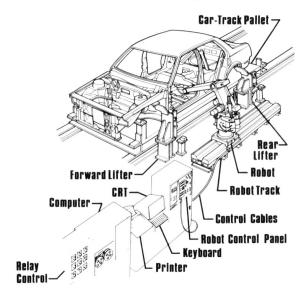

Fig. 71.1. The ROBI system.

71.4. SYSTEM COMPONENTS

To meet these objectives, several factors had to be considered and several alternatives examined for each of the four main system components: the probe, the robots, the tracks, and the computer.

The probe used for the ROBI System is the Optocator produced by the Swedish firm of Selcom AB, basically a light-source reading device. The light source is a low-power infrared laser diode. The beam is reflected off the sheet metal surface and read by a camera which calculates the measurement through simple triangulation. (See Figure 71.2.) Dual probes are used on each robot to obtain two-directional readings with minimal robot movement. Each probe includes a light source, a camera unit with lens and detector, and probe processing electronics. The output signal is linear in a digital serial format of the distance to the surface to be measured. This particular probe was selected because of its accuracy and speed in obtaining data. The reading resolution is better than 0.05 mm, and the reading range is 32 mm.

In the search for the best robot to meet the goals of the ROBI system, many were considered. At the time of the search, only a few electric robots were available. Today, many more are on the

SELCOM LASER PROBE
PRINCIPLES OF OPERATION

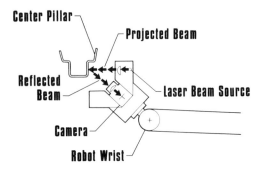

Fig. 71.2. Laser probe principle of operation.

market. The all-electric ASEA IRb-60 with six axes of freedom was selected, as it met the requirements for speed, repeatability, reach, low power usage, and compatibility with external equipment. These robots are able to obtain high repeatability without warm-up time or hydraulic temperature fluctuations. Since no settling time was needed, more points could be checked with this robot than with the others considered in the required span of time.

The working envelope required to accommodate the larger-style bodies is significantly greater than the envelope of a single robot. To expand the system envelope, programmable electric DC servo-driven tracks were added. This allowed each robot to move between six locations to optimize its position along the length of the body and to check various points at the front, middle, and rear of each body. The robot can move along the track during the checking of a body if necessary. Covers had to be added to protect the precision ways from grit and damage, and riser blocks of varying heights between the robots and the tracks were used to vertically adjust the working envelope to fit the conditions in each plant.

The system is operated by a DEC PDP 11/44 computer with two disk drives. Commands are entered by way of a CRT keyboard. The results of inspection cycles are displayed on the CRT screen or on a hard copy printer. The computer is interfaced with the Selcom Probes, the lifter or the building truck positioner, the respot line or power-and-free conveyor, the ASEA robots, and the tracks.

71.5. OPERATION DESCRIPTION

To place the system in operation the tool engineer uses the CRT keyboard to select the body opening(s) he or she wants checked.

The robot will continue to check that opening or sequence of openings on each body until the command is changed. The car line and style information is passed on to the ROBI computer from the style read station. Upon receiving the style information and the "Ok to download" signal at the end of the last check cycle, the computer will download the appropriate programs to the robots. The robots will then make a floor track position move if it is required. When checking the same opening as the last body checked, no track move is made.

The next step is for the robots to position the probes at verification blocks to make readings. This may seem unnecessary, but it will guarantee that the track has indexed properly and nothing has been moved. If everything is still reading within the set limit of programmed position, then the robot moves on; if not, it will abort to home position.

If the verification is within limits, the robot will move to a "ready-to-check" position that is still clear of the moving line. The system waits for a signal that the body is properly positioned in the station. If there is not enough time left to complete the check cycle, the robots will abort to the home position, and the body will index out of station. If enough time remains, the robots will then move into the body opening, taking the probes from point to point to read the opening. When checking is complete, the robots will go to home position and give a "clear" signal, the body will unlock and index out. At each check point a reading is taken, and it is compared to master data in the computer. The difference or error is printed out for each check point. Each robot can read up to 20 points per body at a line rate of 60 bodies per hour. All check data is stored in the computer for one week so that various summaries can be generated. Summaries can be requested by opening, body style, sideframe gate, or building truck.

The ROBI system is considered to be highly successful in having met all its goals. However, some problems were encountered when installing the system. All programs had to be adjusted and fine-tuned at each plant. This took a considerable amount of time. Availability of plant personnel to work with the group installing the system was limited, thus making it necessary for the general office personnel to do more of the programming and setup work. The usual problems with the robots, tracks, hardware, software, and line interfacing were encountered during start-up. All of the problems were solved and corrected. It took about three months to debug all the systems. The system started with a PDP 11/34 computer and had to be upgraded to a PDP 11/44 computer to get reports and be able to use the CRT for system changes at the same time that the robots were checking a body. For optimum usage of and confidence in the system, it had to be correlated to the hard, off-line checking fixture. The ROBI system is mastered by taking master data from a body in the ROBI station and then checking the same body in the hard fixture. The deviations found by the hard fixture are entered into the ROBI computer as offsets. This system of robot, track, and probe has proven to be repeatable within ±0.2 mm (±0.008 in.).

The ROBI system has a few constraints that should be noted. At present, the system cannot check holes for size and location; however, it has been used to check pin locations. A perfect one-to-one correlation to the off-line hard-checking fixture cannot be obtained owing to the multiplicity of building trucks.

The ROBI system has proven to be a useful tool in helping to monitor quality and to collect data for process control. The system has been installed on-line in eight assembly plants, and more installations are being planned. The installation and body-programming time in the plants for all eight systems was about 36 man-months completed over a 10-month period. The system and mechanical development time, before installation began, took two men about one year. The control system design and development and computer programming took about 60 man-months.

A spin-off of this system is a robot-operated panel inspection system (ROPI) which checks individual body panels and subassemblies in component plants. This is a prototype system utilizing a single ASEA IRB-6 industrial robot with five axes of freedom, and equipped with a double-laser-beam opticator probe. The robot positions the noncontact probe at each check point, inspecting for dimensional accuracy of surface and trim edges and verifying the presence of critical holes. One of these systems is currently installed and being evaluated at present.

CHAPTER 72

INSPECTION AND REPAIR OF NUCLEAR PLANT

DUNCAN B. LOWE

Taylor Hitec Ltd.
Chorley, Lancashire, United Kingdom

72.1. PREAMBLE

Hostile environments were the first to welcome robots, and the inside of a nuclear reactor is certainly hostile. However, the particularly sensitive and complex requirements of a nuclear reactor call for human supervisory control of any robotic devices employed. Therefore teleoperators can provide a very useful solution.

The development of remote manipulators, or teleoperators, discussed in Section 3.3 and in Chapters 9, Teleoperator Arm Design, and 17, Control of Remote Manipulators, is the focus of this chapter.

The nuclear power industry is under steadily increasing obligations to provide evidence that plant and equipment is consistently satisfying safety and integrity requirements, imposed by various authorities. Nowhere is such evidence more mandatory than from regions where critical systems and components are inaccessible to man or deny his involvement by their hostile state.

At the same time there are obvious incentives for the industry to prolong the operational life of both existing and future power plants, again creating a demand for evidence by safety and licensing organizations.

A consequence common to these two influences is the need for adequate survey and sampling equipment, the most specialized and technically advanced of which is that required for remote operation in hostile environments. If the evidence procured by these means indicates a need for adjustment or repair, then the scope and dexterity of the implements must be further increased.

Such are the motivations for the design and development of remotely controlled manipulators for working in hitherto inaccessible regions of reactor interiors, and there are already firm indications that similar devices will be called for in the processing of radioactive wastes and the decommissioning of nuclear plants.

The predominant constraint upon the design and function of in-reactor tools is the access route to the worksite. In existing operational reactors few or no specific routes were incorporated in their design, hence access must utilize whatever route can be used with safety (usually a fuel or control channel) to insert and position a manipulative device at the worksite. Once there, the manipulator responds to transmitted commands, direct or programmed, to employ the inspection device, tool, and so on in a predetermined task.

The foregoing describes a *deployed robot,* which is the essential nature of all the equipment to be discussed in this chapter.

72.2. DESIGN CONSTRAINTS

In addition to the restrictions on cross-sectional dimensions imposed by the access routes, there are further factors affecting the design approach, created by the interior environment of a nuclear reactor, as follows:

1. **Temperature.** Inspection and maintenance operations employing remote equipment are carried out with the reactor "shut down." However, the ambient temperature at the worksite can approach 150°C.
2. **Radiation.** Residual radiation levels of up to 30 rads/hour in gas-cooled reactors can prevail in the operational zone.

3. Radioactive Contamination.   The access route is a penetration into a contaminated environment. Radioactive particulate matter is deposited on the inserted equipment during its residence within the hostile worksite.

72.2.1. Fail Safe

Once entered into a hostile containment, it is essential that the position in space of the device be precisely known at all times; otherwise collision, breakage, and worse still, lodging of the device can occur. The last is of particular concern to the designer, for not only is an irretrievable device an embarrassment and expensive, but it can also render inoperative the plant concerned for an extensive period or even permanently. In a nuclear reactor such a situation could create safety hazards in addition to loss of use and consequently is quite unacceptable. The device and its deployment gear must therefore be designed to fail in a manner that will ensure that it can be extricated from whatever confines it may encounter.

Should an accident or malfunction prevent withdrawal of the equipment by normal means, the components must be so designed as to collapse sufficiently to allow extraction when extra force is applied. The manner of collapse is critical. No component must break free to be left behind, and the forces required to cause collapse must not inflict significant damage to the reactor internals. In other words, a fail-safe mechanism must be ensured to avoid the quite unacceptable risk of its becoming inextricably lodged in a reactor.

72.3. FOLDING ARM MANIPULATORS

The evolution of the deployed robot for in-reactor work began some 15 years ago and in its course has drawn on the advantages of the rapid advances in technology, particularly in remote control, to meet the constantly increasing demands on its scope, accuracy, and dependability. The first devices that could be regarded as deployed robots were the folding arm types, which have progressed through several generations and still have a role to play.

Figure 72.1 portrays a typical folding-arm manipulator which has entered the reactor core region by a vacated fuel channel. An automatically actuated seal unit closes the top of the fuel channel to prevent ingress of air to the CO_2 atmosphere within the reactor containment or, conversely, escape of that CO_2 to the "clean" zone above the reactor.

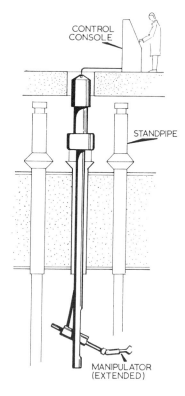

Fig. 72.1.   A typical folding arm manipulator.

Command of the operations originates at a console in the safe region. An umbilical linking the control center to the implement carries command signals, telemetry, and services. Visual feedback from the work zone employs closed circuit television through a remotely controlled camera on the manipulator, feeding a Visual Display Unit (VDU) in the control console. Essential illumination is provided by high-intensity lamps, also carried on the manipulator.

The actuating systems are electropneumatic, and to facilitate fail-safe retrieval, all relevant motions can be recovered in an emergency situation. The "wrist" section of the manipulator can be forcibly back-driven through a slipping clutch by applying a measured excess withdrawal load, and the pivoting section has a supplementary hand wind to return the motion.

Deleterious effects of radiation on the equipment cannot be prevented but are minimized by the following:

1. The use of radiation-resistant insulators.
2. Stabilized glass in optical components.
3. Approved materials for seals and the like.
4. Inclusion of shielding where practicable.

Cooling gas purges the entire manipulator assembly constantly during operation to prevent overheating of vulnerable components.

Figure 72.2 illustrates the most recent of a series of folding-arm manipulators produced for in-reactor use. It is 13 m long with a mast diameter of 219 mm and an all-up weight of 2.2 tonnes. The mast is designed to be split for transportation and to reduce the headroom necessary over the reactor. It has 7 DF, each driven by a pneumatic motor, in preference to hydraulics, to avoid the risk of oil entering the reactor.

Each motion is fitted with a safety device to limit the load that can be applied to the reactor and a positional feedback system giving an accuracy of $\pm0.5°$ for angular movements and ±1.27 mm for linear movements. The maximum payload is 30 kg at a reach of 1.7 m, although greater reaches can be achieved through design modifications.

At the end of the arm is a dovetail feature which incorporates two pneumatic and four electrical supplies to power a wide variety of quickly interchangeable tooling.

The manipulator has its own built-in high-resolution mono television camera which will remotely pan over 45° and tilt up to 120°. This camera is fitted with remote-focus iris and zoom functions giving a magnification of up to 6×. A twin-bulb quartz iodine 250 W lamp unit is fitted below the camera. This lamp is variable in intensity and mechanically linked to the camera, ensuring that the light is always correctly positioned.

The manipulator can work in an air or CO_2 atmosphere up to 75°C and requires two supplies of compressed air (or gas) at 600 kPa and 1400 kPa (kPa = kilo-Pascal where 1 Pascal = 1 Newton per square meter). Special attention has been paid to safety, and all movements have been designed to allow the manipulator to be retrieved from the reactor in the event of drive failure.

Similarly, maintenance problems have been given careful consideration, and 50% of the mast's surface is in the form of removable fitted cover plates, thus allowing excellent access for maintenance.

BASIC SPECIFICATION

Degrees of freedom	7
Tube outside diameter	219 mm
Overall length	13 m
Weight	2168 kg

MOVEMENTS

Vertical	2.2 m	at	150 mm per min
Azimuth	±360°	at	1 rpm
Swing	140°	at	0.5 rpm
Extension	609 mm	at	150 mm per min
Wrist	±180°	at	1 rpm
Knuckle	90°	at	1 rpm
Tool rotate	±180°	at	1 rpm
Camera tilt	120°		
Camera pan	+22½°		
Payload at max extension	30 Kg		
Max working temperature	75°C		
Max reach	1676 mm		

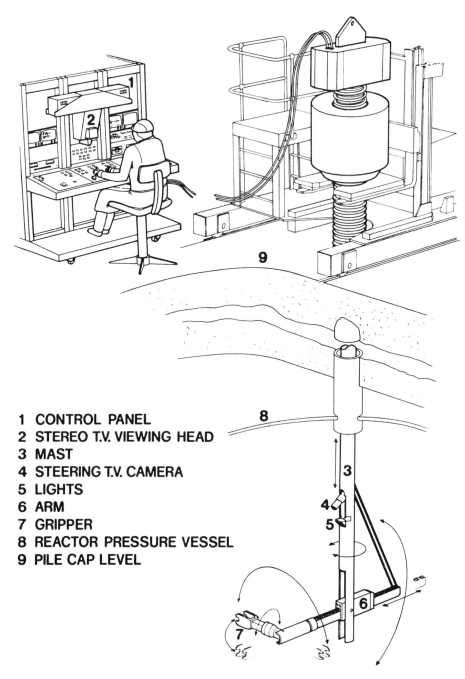

1 CONTROL PANEL
2 STEREO T.V. VIEWING HEAD
3 MAST
4 STEERING T.V. CAMERA
5 LIGHTS
6 ARM
7 GRIPPER
8 REACTOR PRESSURE VESSEL
9 PILE CAP LEVEL

IN CORE POWER MANIPULATOR

Fig. 72.2. A recently commissioned folding-arm manipulator for remote ultrasonic inspection of a reactor vessel from within.

72.3.1. The Control System

The control panel incorporates the following three levels of operation:

1. Backup control
2. Manual control
3. Teach and repeat

and harnesses the power and scope of microprocessors to achieve them.

Backup Control

In the event of computer or system failure while operating in the reactor, it is a necessity that some basic level of control be provided to enable the manipulator to be retrieved. The facility provided for this feature is a set of momentary pushbuttons, each of which activates a solenoid valve in the pneumatic system.

Manual Control

Manual control of the manipulator for all normal driving operations is achieved using two joysticks at the control console. All the manipulator functions are split into logical pairs, and utilizing a set of pushbuttons, any pair of movements can be switched to any joystick. Feedback to the operator is through a pictorial representation presented on the visual display unit in the console.

Teach and Repeat

The main feature of the automated control system is route memorizing (teach and repeat system). This enables an operator to record in a rehearsal mode a complete sequence of manipulator joint movements which are required to perform subsequently a specific task within the reactor. Whenever the recorded task must be repeated, after initialization, the control system automatically reproduces the manipulator movements, eliminating all the operator thinking time. Recording manipulator trajectories can be carried out within the reactor or in a mimic test facility prior to the manipulator's being deployed into the reactor.

The following list outlines all the routines that are made available in the system:

1. Learn sequence.
2. Replay sequence.
3. Replay sequence in reverse.
4. Replay sequence and learn new sequence.

72.3.2. Other Control Features

Software Position Limits

To improve operational safety of the manipulator, the power of the microcomputer is used to create settable software limit stops for each driven function.

Information Display

The information displayed on the VDU includes positional feedback for each joint, the settable position limits, and which movements are at their limit of movement.

72.4. MANIPULATORS WITH MONOARTICULATE CHAINS

The current requirements of the nuclear inspection authorities are already outstripping the capabilities of manipulators being commissioned at present. To achieve the reach, dexterity, and accuracy now called for requires a totally new design approach for the deployment mechanism, and this was eventually realized in the use of a monoarticulate chain.

Figure 72.3 illustrates the basic principles of this system. The "chain links" are of box section, joined by pivots at their upper corners, allowing the chain to articulate in one direction, but because the lower corners of the links abut, articulation in the opposite direction is prevented. These properties create a chain that can be reeved or coiled but that forms a rigid element when extended. Precision machining of the links and minimal clearances at the pivots produce a relatively light but stiff structure which is, in effect, a modular beam of variable length, thus greatly increasing lateral reach and reducing the headroom required for the equipment.

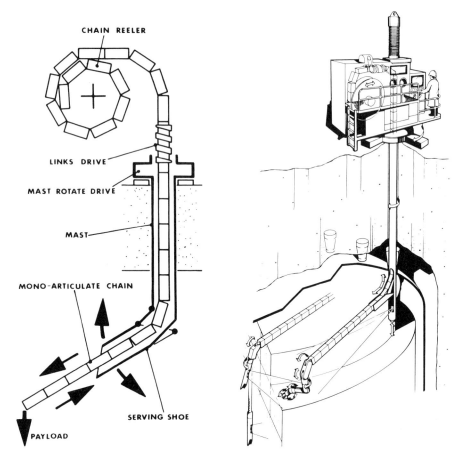

Fig. 72.3. Basic principles of the monoarticulated chain deployment system. (Acknowledgment to CEGB for design contribution.)

Fig. 72.4. An application of the deployment chain system for reactor inspection. (Acknowledgment to CEGB for design contribution.)

To utilize this implement to the best effect it is fed vertically down through a hollow mast and subsequently through a serving shoe. The latter is adjustable in its angular relation to the mast, thus dictating the trajectory of the chain as it extends outward in the rigid mode.

The raising and lowering of the chain is driven by means of a rotary scroll gear engaging with protruding rollers at the link pivots.

Provision of a rotary drive to the mast completes the system as a deployment manipulator. With the chain fully reeled in and the serving shoe retracted, the manipulator is compacted within the envelope of the mast, allowing the whole device to be cleanly entered into a reactor access route. Once installed, the serving shoe adjustment, chain feed-out, and mast rotation drive work in concert to project the payload at the chain's end to the worksite.

72.4.1. The Next Phase in Development

The diagram in Figure 72.4 shows an entire system employing the chain manipulator to place and mobilize a television inspection head within a reactor vessel. The mast drive and chain-reeling gear are housed in a portable "drive cabinet" which locates and rests upon the charge hall floor. This cabinet is sealed to the top of the access tube, thus forming a containment with a controllable interior atmosphere. With such facility, the payload can be retrieved without withdrawal of the manipulator.

Control and actuation of the drives is totally remote, effected from a console linked to the drive cabinet by an umbilical. A second umbilical links the payload to the drive cabinet, passing through the hollow chain links and thus well protected from damage or snagging.

The basic specification is as follows:

Payload: 30 kg basic (+25 kg snagging overload)

Reach (horizontal): 7.3 m

Reach (vertical): 14.0 m

Positional repeatability: ±1 mm

Working environment: CO_2 or air at 150°C

72.5. SPECIALIZED CHAIN ROBOT

The chain deploys the payload, which can be any one of several inspection or task-performing workheads. Of the latter the most comprehensive is a robot arm. Commercially available robots have excessive bulk for this application and additionally incorporate materials and features unacceptable for in-reactor use.

A specially designed robot arm has recently been produced in the United Kingdom which achieves the power, strength, and accuracy required within the size parameters imposed by the deploying manipulator, and the prototype is in manufacture. There were unusual constraints on the design of the arm to facilitate its deployment by the chain manipulator. Its cross-sectional profile could not exceed that of the chain links, for it must pass along the chain-serving route, yet within it had to be housed the prime movers and gearing systems, services, feedback, and other electronic units. The configuration and functional abilities of the robot arm are outlined in Figure 72.5.

As a mechanism, the arm must fail safe, that is, in the event of drive failure or snagging, the joints are designed to back-drive under excess withdrawal load. Thus, the arm is "straightened" for rescue recovery.

The wrist unit has coincident axes, differential gearing, producing 3 DF, as follows:

Wrist roll: ±180°.

Wrist pitch: 90° from straight-out mode.

Tool rotate: ±180° from straight-out mode (or, alternatively, continuous).

Through these motions, the wrist can place a tool at any point on a hemisphere of radius equal to the distance between the coincident axes and the tool extremity.

A further 3 DF are reproduced through the following:

Elbow pitch: 115° from straight.

Shoulder pitch: 115° from straight.

Shoulder rotate: ±180°.

The robot arm thus formed has the ability to visit virtually any locus in a hemisphere space of 1 m in radius.

Technology has now produced small, high-powered DC servo motors to give the designer a feasible alternative to hydraulics, and these were chosen as the prime movers. The drives are swept by a

Fig. 72.5. An advanced manipulator designed and developed for in-reactor use.

continuous flow of cooling gas, both to permit their use at elevated temperatures and to lift their rated performance for long-term operation.

To position the manipulator joints accurately and repeatably, with fine resolution, high-performance servo drives are required. All drives are operated under closed-loop control, that is, the desired joint angle is compared with the actual joint angle measured by a feedback transducer, and any error between the two amplified and used to provide a correcting signal to the joint drive motor. The position demand signal for each joint is generated by the control computer in response to operator commands or a trajectory repeat program. Tachogenerators are provided on each joint to give velocity signals. Orientation in space of the robot arm is automatically interpolated from positional feedback signals from the drives of azimuthal rotation, chain feed-out, and serving shoe elevation.

Another utilization of the chain-deployed robot arm manipulator is at present under consideration for the examination of active liquor tanks housed in concrete cells. (See Figure 72.6.) The enhanced reach coupled with the robot arm dexterity almost doubles the coverage attainable with manipulators of earlier design.

In the United Kingdom development is continuing on the deployed robot in the conviction that such devices will be increasingly applied as operators, maintenance organizations, and inspectorates become increasingly aware of their potential.

72.6. ROBOTS IN DECOMMISSIONING

The problems involved in decommissioning a nuclear plant give rise to needs for equipment similar to that already described for maintenance use. Again, the deployed robot has a role to play in the majority of schemes, where it is necessary to work upon, extricate, and process components by remote means, due to their irradiated or contaminated state.

Recently design studies have been applied to proposals for the decommissioning of nuclear reactors at the end of their operational life, also for the retrieval, treatment, and safe disposal of radioactive wastes.

One example of an application in reactor dismantling is shown in Figure 72.7 where an adapted industrial robot is the "working hand" of a large special-purpose machine. The task for which this equipment is designed is the dismantling of the graphite core structure and its associated facilities, followed by the progressive cutting up and removal of the core containment vessel. All of the materials involved are radioactive to varying levels and prohibit direct human involvement.

The basic element of the equipment is a rotating turret, braced against the circular vessel through three legs, each of which is suspended from a winding unit. Beneath the turret is mounted a hinged beam carrying the robot. Power and services are fed to the machine and the robot through umbilicals

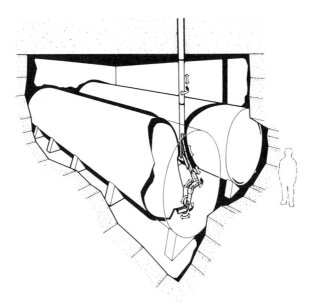

Fig. 72.6. A chain-deployed manipulator devised for remote ultrasonic weld inspection.

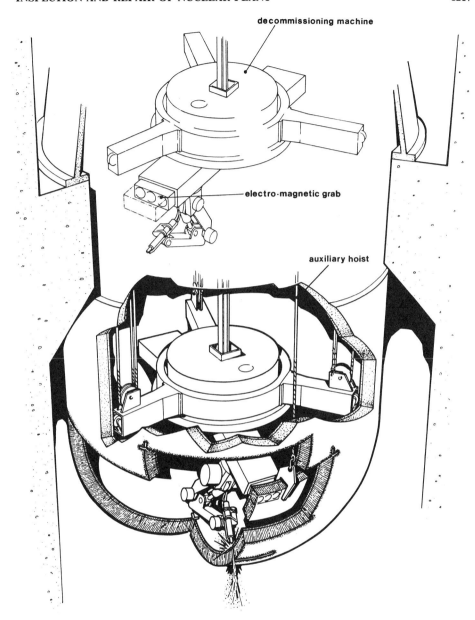

Fig. 72.7. A reactor decommissioning machine with an on-board robot.

from the plant sited in the clean zone above the reactor. The robot arm has a variety of tasks and an appropriate set of interchangeable tools with which to accomplish them.

In the early stages of the work the robot arm holds and directs a specialized flame cutter and, working to a programmed pattern, cuts up a large inner vessel (hot box) into slabs sized for subsequent treatment and disposal. When the inner vessel has been removed, the robot arm then works in partnership with a hoist-and-grab system in the disassembly of the graphite block core structure. In this work the robot is under the direct control of the operator, who uses visual information from television cameras surveying the work to guide and command the movements and manipulations required.

When the core has been completely removed, the robot resumes its more complex role of remote cutting, first to slice up the massive core support grid and subsequently to sectionalize the vessel.

This final task involves comprehensive preprogramming of the robot to achieve the progressive cutting up of the plate and its backing of heavy mineral insulation, in sections suitably sized for disposal. Many obstacles must be negotiated to achieve this in the hemispherical base of the vessel, where compound curves must be followed by the cutting torch in addition to the execution of the severence pattern to form the sized slabs.

The cylindrical body of the vessel presents fewer problems and allows the cutting to progress on a tier-by-tier principle, winching up the machine for each successive tier and rotating the turret beam to take the robot through a full circle of programmed cutting. An electromagnetic grab assists by holding each severed plate during and after severance cutting.

A second example is a specially designed, large manipulator arm, intended for the retrieval of radioactive debris from storage vaults. Figure 72.8 shows a typical unit.

In this case the arm must be lowered through a contrived access lock some 7 m before it can be articulated for the tasks of retrieval. Such duties are arduous and require robust engineering in the mechanisms and limbs, but, at the same time, the control must be quite precise.

Hydraulics are employed as the prime movers for heavy robots of this type, giving high power

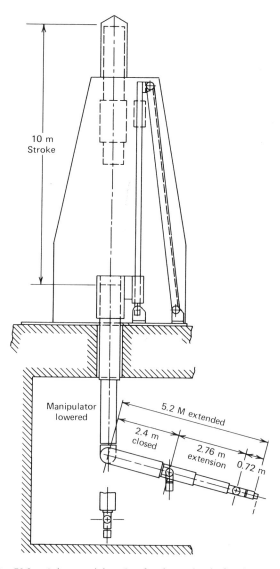

Fig. 72.8.   A heavyweight robot for the retrieval of active waste.

from small-volume drives and having adequate accuracy of control when servo systems are introduced in the command networks. Preprogrammed operation is rarely required for these heavyweight implements owing to the constantly changing nature of operations, although there are no technical barriers to programmed control.

72.7. SUMMARY

Concerning the status of the machines, they are as yet confined to the United Kingdom Nuclear Power Industry: one is in the final stages of works commissioning, a second is in the final manufacturing phase, and a third has recently entered the detail design and manufacturing stage. Prototype chain units with basic drive systems have existed for several years as development tools.

The current cost to the U.K. customer for an entire system is in the region of 1.6 million pounds sterling.

The foregoing discussion gives only a glimpse of the applications and technology involved in employing remote manipulators for the nuclear world. Developments are frequent to meet increasing sophistication in demand, calling strongly upon innovative design and technology to hold the pace.

REFERENCES

1. Constant, J. A. and Hill, K. J., Teleoperators in the nuclear industry, *Proceedings of the 6th I.S.I.R.,* Nottingham, U.K., 1976, pp. H1/1–12.
2. Okamoto, H. et al., Remotely operated systems for the maintenance work of control rod drives in boiling water reactors, *Proceedings of the 7th I.S.I.R.,* Tokyo, Japan, 1977, pp. 323–328.
3. Zook, C. R., Remote maintenance development, SME Paper No. NS79-786, 1979.

CHAPTER 73
A VISION-GUIDED *X-Y* TABLE FOR AUTOMATIC INSPECTION

MICHAEL J. W. CHEN

Machine Intelligence Corporation
Sunnyvale, California

73.1. OVERVIEW

This chapter demonstrates an example of utilization of machine intelligence in automation. A system was developed to perform precision part inspection and automated workpiece handling. This system consists of a robot which is utilized to perform a simple pick-and-place function, a Machine Intelligence VS-100 Machine Vision System which provides a vision library with the functions of masking and programmable image overlay, and an *X-Y-*Θ table. This setup illustrates a simplified approach to machine vision and automation. In this complete sensorimotor system, BASIC was the programming language used to develop and integrate the control software for the inspection process by using the MI DS-100 Machine Vision Development System. By calling vision, *X-Y-*Θ, and robot commands, the task of parts inspection under high- and low-resolution cameras, sorting, as well as disposition, is shown to be easy to conceptualize and implement. This robot system can perform tasks without the necessity of prealigning or jigging workpieces. Numerous other applications can be accomplished by adopting a similar methodology. Readers are referred to Chapter 14, Vision Systems, for explanation of the terminology used here.

73.2. A DYNAMIC VISUAL INSPECTION SYSTEM

Some of the general principles and methodologies used in implementing an application-specific automated system are discussed in this chapter through the introduction of the design and integration of an inspection system. The functional details of this setup, together with the design strategies are described.

Let us first visualize the behavior of a worker who inspects, classifies, and disposes workpieces. The prototypes of the workpieces will first be shown to the person. After memorizing the shape, size, and the features of the prototypes, the person will be able to start the inspection process. Typically, the scene will be "scanned" in general to capture recognizable parts and to determine their approximate location and orientation. The line of gaze, often in saccadic movements, will be projected on one part under oculomotor control. This brings the field of view under the high-resolving power of the area centralis (fovea). A comparison will be made between this part with the stored prototypes in the person's memory. A classification will then be made, and the part will be deposited by hand in an appropriate location. During the inspection process, if the dimensions of the parts are small or the defects are not distinct enough for the foveal vision to discern the details, some kind of eye-hand coordination (under brain control) will be necessary to bring the workpiece closer to the worker's eyes for the workpiece to occupy a larger area in the field of view. Under certain circumstances, magnified or microscopic views will be necessary to gain better resolution before any sorting decision about the part can be made.

It is known in human visual physiology that there exists peripheral as well as foveal vision. The former monitors a larger field of view than the latter. Human beings mainly use peripheral vision to capture visual targets and sudden events in their environments. From a physiological point of view, peripheral vision in general has only crude resolution and discrimination capabilities. Through the

An earlier version of this chapter was presented at the 26th International Symposium on Sensor Robot Technology, 1982.

functions of eye movements, targets that require fine-grain pattern recognition are brought into the foveal region which anatomically has denser cell distribution. Such arrangement by nature provides the economy of visual processing by concentrating upon the important features of the images that are more relevant to the species. Often manual assistance will also be necessary to bring the objects to be inspected closer to the retina.

Unlike human vision, machine vision systems in general have even distribution of light-sensing elements such as CCD arrays (without foveal/peripheral distinction). To emulate human inspection behavior using state-of-the-art technology, a dynamic machine-vision inspection system has been developed using artificial components with automated workpiece transportation, shown in Figure 73.1. A dynamic system is defined as one that can automatically transport and dispose of parts and can adapt to changing working environment to a certain degree. A static system, on the contrary, requires jigging, fixturing, and alignment of the workpieces. This diagram consists of four major subsystems: the control system, the sensor system, the image-processing system, and the workpiece-handling system. This artificial system resembles the human eye-brain-hand system in performing automatic inspection, and also for vision-guided material handling and assembly tasks. The workpiece-handling system may consist of *X-Y-Θ* tables, limited-sequence arms, robots, or other positioning devices to emulate the oculomotor or manual functions of human workers. The sensor system may consist of single or multiple visual sensors for partial/overall viewing of scenes, and/or coarse/fine inspection (to emulate human peripheral/foveal or far/close vision).

73.3. MACHINE VISION

The VS-100 vision system[13] (see Table 73.1) is a commercial binary vision system. It receives a video gray-scale image from a solid-state or vidicon camera and thresholds it into a binary (black/white) image that is run-length encoded (see Chapter 14, Vision Systems) for data compression and subsequent processing. Computer algorithms perform a connectivity analysis of the encoded images, building data structures that represent essential features of each contiguous region. The vision system characterizes blobs on the basis of distinguishing features such as area, perimeter, minimum and maximum radii,

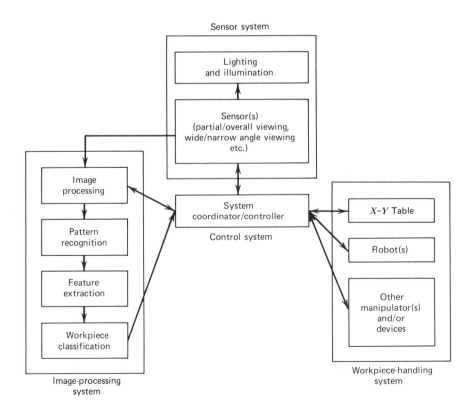

Fig. 73.1. The block diagram of a dynamic visual inspection system with automated workpiece transportation and positioning.

TABLE 73.1.   MODEL VS-100 MACHINE VISION SYSTEM

1. Processors

LSI-11/23 CPU with 256K bytes parity-checked RAM

Custom image data compression logic

Two frame buffers for up to 256 × 256 pixel arrays (binary)

1 display buffer (256 × 256) for graphics and text overlay

Eight programmable image overlay buffers for up to 256 × 256 pixel arrays (binary) (standard for VS-100P, optional for VS-100)

One 12-in. TV display monitor for displaying binary image data, processed image data, and analog images

2. Input/output

Four-port serial RS-232C, or RS-423 interface (DLV11-J) (three ports dedicated to TU-58 tape drive, console terminal and printer; one port available for other devices)

Bidirectional 16-bit parallel TTL interface (DRV11)

Six quad-slots or 12 dual-slots available on LSI-11 backplane for customized applications

3. Cameras

Standard camera is the MI-830A, a 256 × 240 solid-state CCD array with 55 mm, $f2.8$ lens

Optional camera is the MI-820V, a 256 × 256 silicon vidicon with 55 mm, $f2.8$ lens

and number of holes, to name a few. The system can be trained to analyze new objects simply by showing them to the system. Object recognition is performed using a nearest neighborhood classifier (see Chapter 14, Vision Systems) operating on a user-selectable set of the features. Interactions with the system are menu driven, using light pen or keyboard input. Menus allow various system choices such as selection of the threshold value, window size, operating options, and parameters for specific applications. Calibration, training by showing, storing and loading of prototype data can all be accomplished readily. The vision system can also be operated as a satellite processor from an external computer by way of a 16-bit parallel interface. One example of this application is the Univision/PUMA system[5, 14] which runs under the VAL robot control language designed specifically for use with Unimation Inc. industrial robots.

73.4.   MACHINE VISION DEVELOPMENT SYSTEM

Since machine vision is still a general-purpose tool rather than a set of well-defined specifically oriented products, machine vision systems (such as the MI VS-100) are designed to have more capabilities than any single user can employ for a given application. It is therefore, under certain situations, necessary to program the systems using high-level programming language for optimizing the solutions of complicated tasks. Eventually, as generic vision applications arise, user programmability will change from a linguistic task to an operational task akin to robot and NC control. Therefore a main reason to implement a software development system for integrating computer vision with manipulators and other sensors is to support and encourage the intermediate stage during which systems houses and development laboratories identify and create the eventual turnkey packages. This chapter in effect describes an example of creating a turnkey application system by using the DS-100 as the development aid.

The MI DS-100 Machine Vision Development System[7, 14] (Figure 73.2 and Table 73.2) is a commercial BASIC language programming system that is used to develop and test programs for the VS-100 Vision System—the final target system. The development system enables users to use the disjoint silhouettes of objects as the conceptual framework in vision programming, instead of the encoded pixel structure, by providing users with a library of blob-processing-based vision routines, although the pixel-level processing is still accessible to the users. This simplifies user vision-application in programming, conceptualizing, and implementing. The DS-100 system consists of an LSI-11/23 microcomputer with floating-point processor, 256K bytes of memory, serial lines, 16-bit bidirectional parallel interface module, disk drives, CRT terminal, line printer, and other components such as solid-state camera, monitor, light pen, and dual TU-58 cassette tape drives. The DS-100 supports a version of BASIC language that has been extended to include calls to image-processing and pattern-recognition functions, running under the RSX-11M multitasking operating system. Programs are created and modified under the system's text-editing facilities. The programs are compiled, linked with a library of vision routines, and then executed. For the application described in this paper, X-Y-Θ table control routines and robot commands were also created and linked. After the program is debugged, it can be transferred to a cartridge tape for final checkout and subsequent execution on a VS-100 target system.

Fig. 73.2. The MI DS100 Machine Vision Development system used to develop and integrate vision, X-Y-Θ motion, and robot motion control software for this application.

The image-processing and pattern-recognition routines in the DS-100 development system permit the rapid development of application-specific programs. The commands are invoked as BASIC subroutines. According to their functions, they can be divided into six categories: (1) system control, (2) image analysis, (3) measurement, (4) training and recognition, (5) graphics, and (6) factory control. The functional capabilities of this system were discussed in a previous paper by the author.[7]

In addition to the foregoing, a subroutine library was created to operate an X-Y-Θ table. Although this library is not considered as part of the DS-100 development system, it can be used as an example of system integration of vision systems with motion actuators. The X-Y-Θ table that was put under control of BASIC-callable routines was the 8138 Taskmaster manufactured by Tri-Sigma Corporation. This was made possible by utilizing the external-command serial-line input capability of the system. Command strings were sent to the table processor through the control of a DLV-11E serial interface (Figure 73.3). The table routines perform such functions as system initialization, incremental and absolute motions in table or vision coordinates, position retrieval, and parameter setting such as coordinate origin, dwell time, and feed rate. The interrelationship between the table coordinate and the camera coordinate systems is expressed as coordinate-transformation matrices.

73.5. VISION-GUIDED AUTOMATIC INSPECTION SYSTEM

Consider the following problem. We are to sort four kinds of parts: four- and six-sided parts with and without defects (Figure 73.4). These parts are approximately 1.5×2 in. in dimension. The parts are considered as rejects if there are missing drill-holes or if any of the edges is defective. The parts will be singly or multiply presented to the inspection system without prealigning and jigging. The system should perform the inspection, transportation, and disposition of parts automatically. Using the general guideline for designing a generic, dynamic, and automatic visual inspection system as described before, the final design goal is reduced to integrating a vision system with high- and low-resolution cameras for sensing, an X-Y-Θ table for part positioning, and a robot interface for part disposition, with machine vision being the source of control information for the application. The final setup is shown in Figure 73.5.

The vision sensors (two solid-state television cameras) are mounted in the optical tower above the X-Y-Θ table (Figure 73.6). One camera is equipped with a wide-angle lens for identifying and orienting the parts to be inspected; the other camera has a narrow-angle lens for accurate automatic part alignment and detailed inspection. Applications using more than one camera in a single task have been reported in previous studies. Two such examples are the research by Mese et al.[15] of Hitachi

TABLE 73.2. FEATURES OF MODEL DS-100 DEVELOPMENT SYSTEM

1. System Input/Output

The following input-output connections are provided for the system:

  2 each MI-800A (240 × 240 solid-state array cameras)

  2 each MI-820V cameras (256 × 240 Silicon Vidicon)

  1 each MI-810L camera (Linear Array)

  1 each belt trigger signal

Provisions for the following outputs:

  2 each MI-810L camera threshold

  2 each strobe lamp triggers

  1 each video monitor

  1 each horizontal video drive

  1 each vertical video drive

  1 each composite video drive

  1 each belt ready signal

Provision for the following I/O:

  4 each RS-232C serial ports

  2 each 16-bit I/O ports

2. Computer/Manipulator Communications Interfaces

Standard RS-232C or RS-423 serial communication ports for host and/or control computer interface

Data rate jumper—selectable from 150 to 9600 baud

Bidirectional 16-bit parallel TTL interface may be used for manipulator control applications

3. Cameras

The DS-100 System supports 1, 2 or optionally up to 4 cameras singly or in combination from the following types:

MI-800A, 240 × 240 solid-state array (standard)

MI-810L, 256 × 1 solid-state linear array

MI-820V, 256 × 240 silicon vidicon.

4. Monitor

12″ TV display monitor for displaying binary image data, processed image data and analog images

5. User Console

The standard system console and program development terminal is the DEC VT100 terminal

6. System Printer

The DS-100 is provided with a 200 character per second dot-matrix printer. This is a bidirectional 132-column device.

7. Software Features

Operating Systems

RSX-11M

RSX-11S

Language Compiler—DEC BASIC-PLUS-2

Utilities

Editor—DEC EDT and EDI Programming—A full set of development utilities are available

Diagnostics for both the DEC LSI-11 and VS-100 vision system are available.

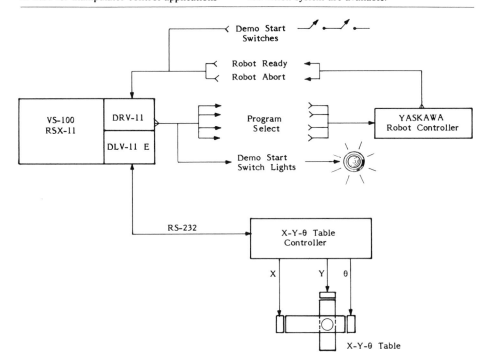

Fig. 73.3. The hardware interface diagram between vision, robot, and *X-Y-Θ* table subsystems.

Fig. 73.4. Top view of some of the parts used in the inspection process.

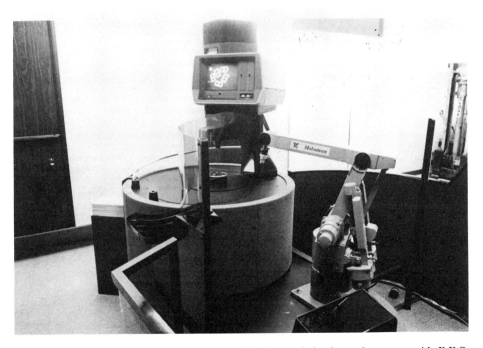

Fig. 73.5. The general work station setup of the high/low resolution inspection system with X-Y-Θ table for workpiece positioning and robot for part sorting.

Fig. 73.6. The computer-controlled wide- and narrow-angle cameras used for overall viewing and fine inspection, with the robot and the X-Y-Θ table operating in the background.

and Kawata et al.[12] of Mitsubishi Electric. They used two television cameras to view two different portions of a chip to determine the position and orientation of the chip from bonding pads of the two images. In our particular setup, the high- and low-resolution cameras have fields of views that overlap, instead of two cameras viewing different portions of the single object.

The X-Y-Θ table serves as an intelligent workpiece conveyor/presenter which loads, transports, positions, and orients parts. On the rotary stage of the table, special backlighting is prepared to create high-contrast images for binary vision. This backlighting is under on/off control signal from the computer (Figure 73.7). In some sense, the X-Y-Θ table performs functionally the mechanical counterpart of human oculomotor system or eye-hand coordination. The low-resolution (wide-angle) system in essence emulates the human far vision. The high-resolution (narrow-angle) system primarily emulates human foveal vision. Such simulation is one good example of cybernetics and artificial intelligence. Through their applications, intelligent humanlike machine behaviors can be achieved.

The robot that is used to perform pick-and-place functions in this setup is a Yaskawa Motoman L-3 robot which has 5 DF (swivel, lower arm motion, upper arm motion, wrist turning, and wrist bending). A two-finger gripper is installed on the robot wrist to perform gripping function. The parts are picked up by inserting the two fingers of the gripper in the central two large holes of the workpieces. This robot is programmed to execute the part pickup, transport, and disposition sequence by using a hand-control teach box and by guiding it through sequences of positions that are recorded in memory. When the teaching operation is completed, the control system is switched to the playback mode. Under control signals from the interface, the robot then repeats the selected sequence of operations. The control signals include the signal to start the pickup operation and also the signal informing the robot to dispose the part in one of the four chutes: four-sided accept, reject, six-sided accept, and reject (Figure 73.3).

To initiate the inspection, the user loads the table with single or multiple parts on the rotary stage of the table and pushes two buttons to start the process. The table then brings the parts within the field of view of a wide-angle camera for part identification. The system selects from the scene one recognizable part for further inspection. If no recognizable part is found in the scene, the stage will return to the loading position waiting for next push-button signal. If a recognizable part is found in the field of view, the position and orientation of this part in camera coordinates will be computed. These coordinate values are then transformed into table coordinates and are passed on to the table controller to move the table accurately under the narrow-angle camera for fine, high-resolution inspection. The software first aligns the part using a line that the system computes between the centers of the two large holes in the part. The selected part is then imaged, positioned, and oriented to precisely align with the coordinate system of a previously stored "good" part image.

Once the part is precisely positioned, a comparison is made using the programmable image-overlay feature. Any logical "difference" or mismatch is compared and displayed. If the mismatch exceeds the tolerable error, then the part is considered a reject. The processor then directs the *X-Y-Θ* table to move the part to a specified pickup location and passes the inspection decision to the Yaskawa Motoman robot which subsequently deposits it in the appropriate chute. The programmable image overlay system provides high-speed dimensional comparison of the parts to be inspected. This feature permits the incoming image to be logically combined with one or more prestored images at video rates, thereby providing a very fast two-dimensional inspection tool for fixed position objects.

The software performs coordinate-transformations between the table coordinate and the two camera coordinate systems. This forms the theoretical basis of using vision to guide the run-time behavior of the table. The transformations between the camera coordinates and the table coordinate can be represented as matrix operations. Since only two-dimensional information processing is involved (the vision system perceives two-dimensional scenes, and the *X-Y-Θ* table performs two-dimensional translational and rotational planar motions), the third dimension (depth) can be neglected in the computations. Instead of using a more general 4×4 matrix operation for three-dimensional robot processing[16] 3×3 matrix operations can be used since the depth is maintained fixed. Furthermore, if table rotation and translation are considered separately, the matrices can be reduced to 2×2.

In the calibration phase (Figure 73.8), automatic scaling is done to obtain precise scaling factors between the table and the vision systems. The calibration procedure is written to allow fine adjustment of loading-point coarse/fine inspection and robot-pickup position and orientation. This procedure also determines the angular inclinations among the three coordinate systems.

The accuracy of the visual inspection is to approximately 0.5% (limited by the 256×256-pixel digitizing) of the field of view of the high-resolution camera, which in the demonstration provides a visual resolution of approximately 0.008 in. (0.203 mm). Much finer inspection can be achieved if greater magnification and higher-resolution processing are used in combination with the higher accuracy of table motion.

By using the preceding setup, the robot is given the equivalent functions of a seeing robot. The robot appears to be intelligent and possesses some of the capabilities of a vision-guided one, although,

Fig. 73.7. The robot's gripper is picking up an identified part at a fixed position from the stage mounted on the rotary of the *X-Y-Θ* table.

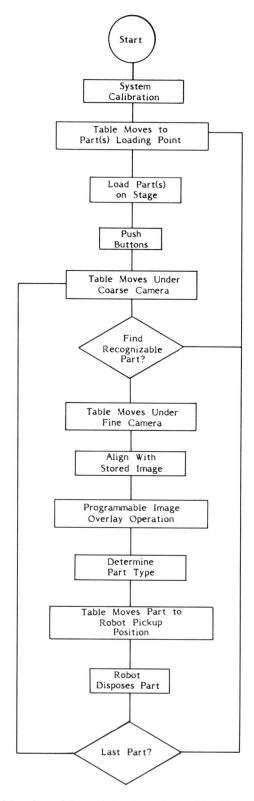

Fig. 73.8. The general flowchart of the workpiece inspection and handling system with X-Y-Θ table under BASIC program control for precision part(s) alignment and presentation.

in effect, only the *X-Y-*Θ table is guided by vision. The significance of this is that instead of using a more expensive system such as the PUMA/Univision combination, we can interface inexpensive components such as *X-Y-*Θ tables and limited-sequence arms to perform the task and to gain equivalent functional capabilities. This will be more cost-effective and suitable for some industrial applications.

Solving an automation problem as defined requires multiple levels of hardware and software integrations. The problem itself is rather complicated. All the system components, the vision system, the *X-Y* table, and the robot, must function properly individually and collectively. For the system to be sufficiently intelligent to respond properly to varying situations, different operating conditions and contingencies must be taken into consideration. Situations such as parts touching one another, stacked on purpose, or taken away during inspection operation are all accounted for and countermeasures taken. To the application programmer and system integrator, the technical details are made invisible. The task is mainly to sequence vision commands logically with table and robot commands, and to consider contingent situations. The development system has reduced the complicated programming problem from coping with all the technical details to the more conceptual layout of logical flow by taking advantage of the semantics of the system and by modularizing the functional components of the software system. The automation task is thereby simplified and made to appear more intelligent and user-friendly.

73.6. CONCLUSION

This chapter demonstrates that machine intelligence can be applied to industrial automation environments for parts inspection and handling. An *X-Y-*Θ table was used for parts transfer, positioning for coarse- and fine-grain pattern recognition, locating, and orienting for pickup by robots. The robot can be a simple pick-and-place device performing a monotonous task. The capabilities of the software enable the system to perform calibration, coordinate transformations among the two camera-coordinate and the table-coordinate systems, as well as image processing and pattern recognition. No prepositioning of the parts is required for inspection. The significance of the application described is summarized in this section.

1. The analogy of this system to the human physiological visual system was discussed. The wide- and narrow-angle cameras in essence emulate peripheral/foveal and far/close vision. The *X-Y-*Θ table partially emulates the oculomotor functions and eye-hand coordination.

2. The task of system integration was simplified through well-defined semantics of vision, *X-Y-*Θ table, and robot commands. The semantics are embodied in the library of subroutines for each component. The concept of hierarchical control systems[2,3,11] in robotics can be more easily applied using this method.

3. Most current robots require costly fixturing and jigging of the workpieces. By expanding the sensory capability of the robot by adding vision, the robot can now work in a less structured environment. One example of the earlier attempts is the combination of the MI VS-100 Machine Vision System with Unimation PUMA robots.[6,17] This system has the ability to acquire randomly located workpieces to reduce fixturing, jigging, and part-presentation requirements. The robot receives information on the location and orientation of the real world from the vision system. Internal transformations between the world coordinates and the joint coordinates are performed. A common Cartesian frame of reference is established during system calibration. In the setup described, coordinate transformations are performed among the two camera systems and the table. Since the robot is trained to pick up parts from fixed position and orientation, no coordinate transformations need be established between the robot and cameras or the table. It is demonstrated here that to attain the goal of unstructured parts presentation and acquisition, we can either implement a more expensive vision-guided robot system with real-time robot joints-world transformations or utilize less expensive components such as a mobile *X-Y-*Θ table and limited-sequence arms interfaced with vision system.

4. The ease of utilizing the MI Development System in conceptualization and implementation of an application is exemplified in this setup. The BASIC language makes robotics and automation programming more user-friendly and approachable. In addition, the application programmers are also provided a tool to cope with situations limited by connectivity analysis. A strategy similar to that described by Bolles and Cain[4,5] has been utilized to circumvent the touching-parts problem. It is demonstrated that the main restriction of binary vision to isolated parts can be bypassed through proper selection of key feature variables or utilization of the local-feature-focus method. In other words, the users can introduce their subjective discretion in selecting certain attributes of the parts for discriminatory purposes. They can define their own special feature variables as aids to cope with application-specific problems. This often shortens recognition-processing time and makes vision operations modifiable without modifying the internal programs of the vision systems. This is considered an important feature that a "future" machine vision system should possess. The need by the automation industry to have a versatile and flexible vision system has been mentioned in different publications (such as Agin[1] and Chin[9]). This application demonstrated that the DS-100 Development System provides a powerful solution to this demand although intrinsic requirements and limitations of binary vision

(e.g., controlled lighting and lack of three-dimensional surface descriptions such as texture) are unavoidable. Despite the limitations, there are many applications in manufacturing that are applicable to this approach. Furthermore, binary processing of filtered color scenes[8] has been attempted and has proved to be a feasible approach to augment the capabilities of binary vision systems, yet retaining the advantage of high-speed processing.

In the world of industrial automation, an important consideration is not only what can be done, but also what can be easily integrated into turnkey systems. The cost-effectiveness and the man-hours required in system setup, maintenance, and improvement must be taken into consideration. It is demonstrated through setting up this inspection station that a turnkey system can be integrated with shortened engineering time without requiring great expertise in computer engineering. The DS-100 system provides users with the tool and the environment to implement such systems with minimal effort.

ACKNOWLEDGMENTS

The author is grateful to Mr. Tye Shultz for sharing the software-development effort in setting up this inspection station. Also, thanks are due to Dr. David Milgram, Dr. Earl Sacerdoti, Dr. Charles Rosen, Mr. Gerald Gleason, Mr. Ted Panofsky, Mr. Dennis McGhie, and Mr. Rick Held for their valuable suggestions, and to Mr. John Baxter and Ms. Jane Ferrier for assisting in the preparation of the manuscripts for publication.

REFERENCES

1. Agin, G. J., Computer Vision System for Industrial Inspection and Assembly, *IEEE Computer Society Magazine,* May 1980.
2. Albus, J., *Brains, Behavior, and Robotics,* McGraw-Hill, New York, 1981.
3. Albus, J., Barbera, A., and Fitzgerald, M., Programming a Hierarchical Robot Control System, *Proceedings of the 12th International Symposium on Industrial Robots, 6th Conference on Industrial Robot Technology,* 1982.
4. Bolles, R. and Cain, R., Recognizing and Locating Partially Visible Workpieces, *Proceedings of PRIP (Pattern Recognition and Image Processing),* IEEE Computer Society, 1982.
5. Bolles, R. and Cain, R., Recognizing and Locating Partially Visible Objects: The Local-Feature-Focus Method, *The International Journal of Robotics Research.* Vol. 1, No. 3, Fall 1982.
6. Carlisle, B., Roth, S., Gleason, J., and McGhie, D., The PUMA/VS-100 Robot Vision System, *1st International Conference on Robot Vision and Sensory Control,* Stratford-upon-Avon, U.K., April 1981.
7. Chen, M. and Milgram, D., A Development System for Machine Vision, *Proceedings of PRIP (Pattern Recognition and Image Processing),* IEEE Computer Society, 1982.
8. Chen, M. and Milgram, D., Binary Color Vision, *Proceedings of the Second International Conference on Robot Vision and Sensory Control,* Stuttgart, Germany, 1982.
9. Chin, R., Machine Vision for Discrete Part Handling in Industry, A Survey, *IEEE Computer Society Conference Record, Workshop on Industrial Applications of Machine Vision,* 1982.
10. Chin, R. and Harlow, C., Automated Visual Inspection: A Survey, *IEEE Transactions on Pattern Analysis and Machine Intelligence,* Vol. PAMI-4, No. 6, November 1982.
11. Graupe, D. and Saridis, G., Principles of Intelligent Controls for Robotics, Prosthetics, Orthotics, *Workshop on the Research Needed to Advance the State of Knowledge in Robotics,* April 1980.
12. Kawata, S. and Hirata, Y., Automatic IC Wire Bonding System with TV Cameras, *SME Assembly VI Conference,* 1979.
13. Machine Intelligence Corporation, *VS-100 Machine Vision Reference Manual,* 1981.
14. Machine Intelligence Corporation, *DS-100 Machine Vision Development System Technical Manual,* 1982.
15. Mese, M., Yamazaki, I., and Hamada, T., An Automatic Position Recognition Technique for LSI Assembly, *Proceedings of the 5th International Joint Conference on Artificial Intelligence,* August 1977, pp. 685–693.
16. Paul, R., Manipulator Path Control, *Proceedings of the 1975 International Conference on Cybernetics and Society,* September 1975.
17. *Univision Supplement—User's Guide to VAL, A Robot Programming and Control System,* Version 13(VSN), 2nd ed., July 1981.

CHAPTER 74

ROBOTIC INSPECTION OF CIRCUIT BOARD SOLDER

WILLIAM E. McINTOSH

Honeywell Inc.
Golden Valley, Minnesota

74.1. INSPECTION TASK BACKGROUND

One step in the manufacture of circuit boards is the wave solder operation in which component leads are automatically soldered to circuit board lands. To ensure high-quality boards, two types of inspections are often done on the finished boards.

One inspection is an electrical inspection. In terms of solder quality, this type of inspection can locate solder bridges between solder joints. These are often caused by component leads that are bent over rather than sheared off. An electrical test will not detect solder joints with insufficient or even missing solder so long as the lead is in temporary or partial contact with the land.

The second common type of inspection is a visual inspection. A visual inspection would be to locate such solder defects as solder insufficiencies, excessive solder, missing solder, and bent-over leads that could form a bridge after packaging and shipping of the boards.

Visual inspection of solder joints can be an inefficient operation when performed entirely by humans. Operator inconsistency takes its toll. Often solder defects are missed, and often solder joints are touched up that do not need to be touched up.

The Honeywell Production Technology Laboratory has developed a system to automate the first-pass inspection of solder joints on a particular Honeywell product line. The inspection goals are as follows:

1. Produce a better-quality inspection by standardizing inspection criteria.
2. Eliminate effects of operator inconsistency.
3. Increase operator productivity by screening the solder joints so that only those joints with a high probability of being defective will be inspected and touched up by the operator.

DESIGN CRITERIA AND OBJECTIVES

Criterion	Objective
Flaws detected (%)	Greater than 99% of all flaws
False flaws (%)	Less than 5% of all joints flagged as defective
Speed	40 msec. per solder joint
Component availability	Off-the-shelf
Cost	Less than $75,000 total
Return on investment	In two to three years by savings from less inspector labor and fewer field failures
Development time	Four man-months engineering time
Board handling	Flexible, adaptable
User level	Shop floor operator, menu-driven software

74.1.1. System Overview

The inspection of the solder joints is performed by a computer image processor operating on a two-dimensional, black and white image from a television camera of a portion of the board. The camera field of view on the board is approximately 30 mm × 20 mm. For processing, the scene is digitized into an array of pixels 235 pixels wide by 216 pixels high. Assuming a maximum resolution of 1 pixel, a solder bridge of 0.1-mm diameter can be detected.

A robot with a board-handling tool is used to index the board in front of the camera. The robot provides flexible board handling and repeatability to 0.1 mm in positioning the board in front of the stationary camera.

The robot positions the circuit board as directed by the image processor. When the move to a board location is completed, the image processor takes a "snapshot" of the board area, inspects each solder joint in the scene individually, and accumulates data on potential solder flaws. When the board is finished, the robot sets the board down while the image processor downloads the flaw data to a microprocessor. The microprocessor postprocesses the data, stores production statistics, and prints out the inspection results for the board as the image processor inspects the next board. The inspection results are used by the operator to locate quickly those solder joints that need to be touched up. A block diagram of the system is shown in Figure 74.1.

74.1.2. Inspection Strategy

Each inspection scene consists of up to 40 solder joints, depending on the board location. The solder joints are connected to traces which are connected to other solder joints in the scene or which terminate at the boundary of the scene. To inspect the solder joints the image processor must be able to distinguish the joints from the traces and background board. This is accomplished through special scene illumination and camera filtering.

The circuit boards for which this system was developed have solder masking over the traces and background board. This solder mask is a saturated, dark green color. When yellow-orange light is used to illuminate the board, and an orange-red filter is placed in front of the camera lens, everything green in the scene is made to look dark. In this way the traces are suppressed, and the image processor is presented with a scene in which the traces have disappeared, leaving only the solder joints.

Figure 74.2 shows how the solder joints would appear to the image processor without the special illumination and filtering, and Figure 74.3 shows how the solder joints appear with the special illumination and filtering. The "tails" on the solder joints are component leads and trace exits.

Figures 74.2 and 74.3 also show three solder flaws in the top row of solder joints. A bridge between

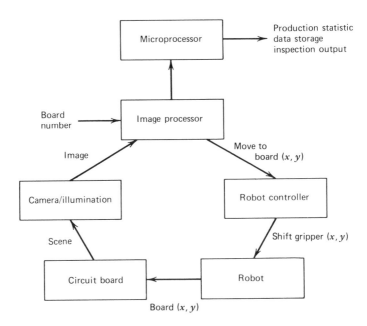

Fig. 74.1. Inspection system.

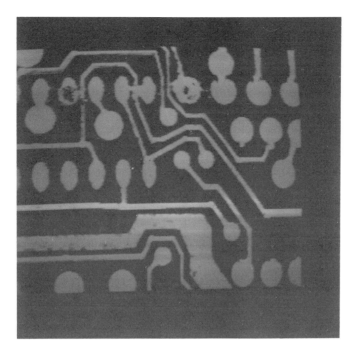

Fig. 74.2. Solder joints with traces visible.

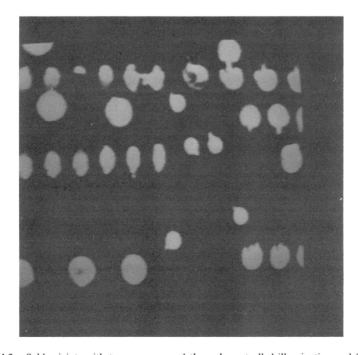

Fig. 74.3. Solder joints with traces suppressed through controlled illumination and filtering.

two IC pads (and a trace) is shown, as well as a hole and a solder insufficiency. There is also a double pad in the scene that is not a flaw.

The inspection performed by the image processor is much simpler and faster when the solder joints are isolated in the input scene than if the image processor had to separate the joints from the traces before the inspection. With isolated solder joints, the image processor can be programmed in a straightforward way to step from each joint to the next, measuring features of each joint and comparing their values to predetermined values and tolerances.

74.2. SYSTEM COMPONENTS

74.2.1. Illumination

The illumination task is to illuminate the solder side of the circuit board with 360° of bright, yet soft and diffused orange light. The solder joints must be evenly illuminated from the sides as well as from the top. Specular reflections from anywhere in the scene must be kept to a minimum. Low initial and operating cost is desirable, as well as the capability for easy change of bulbs. The light fixture should be rigid but lightweight.

The light fixture is shown in Figure 74.4, along with the robot, circuit board, camera, and triangular calibration frame. The camera looks through the light fixture. The light fixture frame consists of two concentric rings of 1.59-mm sheet metal joined by three support members in the shape of a truncated cone. Illumination is supplied by two concentrically mounted ring fluorescent tubes.

Two theater gels are placed inside the fluorescent tubes so that the light from the tubes must shine through them to reach the board. One gel is a frosty diffusion gel and the other is an orange gel. The light frame is ordinarily covered with a highly reflective material to keep the light inside the viewing area.

74.2.2. Camera and Optics

The lens and camera task is to focus and sense the light in the field of view on the circuit board and to transmit the light-level data to the image processor for digitization and processing. The components used are a Panasonic BS-170 solid-state television camera with a 50-mm Cosmicar C-mount CCTV lens with 11 mm of lens extension.

The camera is black and white, is internally synched, and produces an RS-170 composite video output. The photosensor is a CCD array of 256 × 404 discrete elements, the output of which is latched to produce the composite video output.

74.2.3. Robot Board Gripping Tool

The task of the gripping tool is to pick up and hold circuit boards firmly and repeatably. The tool is a two-position pneumatic gripper. When the gripper closes, the board is forced into grooved slots so

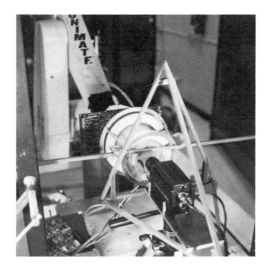

Fig. 74.4.   Inspection station components.

that each board will always be in the same position in the tool. A clamping force of 8.8 N is adequate to hold the board firmly. The tool always picks up a board in the same way from a nest. The board is held in the gripper in such a way that the tool approach axis is normal to the board.

74.3. SYSTEM SETUP AND CALIBRATION

This task is to ensure an accurate, repeatable mapping of flaw locations and sizes from the camera frame of reference to the circuit board frame of reference as the board is indexed in an arbitrary plane in the robot work volume. The setup and calibration sequence must be done with each new placement of the camera in the robot work volume. The setup and calibration is performed manually.

The result of the setup procedure is the definition of the plane in the robot work volume in which the robot indexes the circuit board. The result of the calibration procedure is the determination of camera coordinate correction factors necessary to map accurately camera coordinates to board coordinates.

The camera setup procedure is as follows: The camera is mounted in a triangular calibration frame, Figure 74.4. Three points of the frame define a plane parallel to the plane in which the inspection will occur. The camera is mounted within the frame so that the camera axis is always normal to the plane regardless of the plane's position and orientation.

A calibration pointer is placed in the board-gripping tool. The pointer is aligned with the tool approach vector, which is normal to the plane in which the circuit board is held by the gripper. With the pointer in its hand, the robot is taught the locations in space of the three corner points of the calibration frame. The pointer diameter "just fits" into the corner hole diameters, and the depth of the corner point holes ensures that the pointer tip goes in straight.

These restrictions on how the pointer meets the corner holes ensure that the tool approach vector is normal to the calibration plane and parallel with the camera axis at each point. Small rotations of the gripper about the approach vector are compensated for in the calibration procedure.

These three points define a reference coordinate frame or a transformation with respect to the robot world coordinate frame. The origin of this frame is chosen as one of the points on the base of the calibration frame. The X axis passes through the other point on the base of the calibration frame. The third point, at the top of the calibration frame, defines the plane of the Y axis. The Z axis is parallel with the camera axis, and its positive direction is toward the camera.

The actual plane in which the robot will move is defined by shifting the reference coordinate frame in the newly defined negative Z direction. This is a shift in a straight line along the camera axis, away from the camera. This defines a plane that is parallel and unrotated with respect to the previously defined plane. The purpose of this shift is to focus and size the camera field of view, and to make room for the light fixture. The exact distance to be shifted can be determined by trial and error.

At this point the camera coordinates to board coordinates calibration procedure can begin. The calibration parameters to be obtained are the scale factor relating pixel units to millimeters, and the X and Y correction factors to be added to the computed, nominal values of board coordinates to make the board coordinates exact. These correction factors are necessary for three reasons. One reason is to accommodate the repeatable inaccuracies in the robot movement. Another reason is to accommodate a possible small rotation of the camera sensor element about the optical axis of the camera. The final and most important reason is that even if the camera sensor element were aligned perfectly with the calibration frame it would be difficult to exactly align the board edges with the edges of the camera field of view. There is always some rotation of the board relative to the camera scene over the full range of travel by the board in front of the camera. This results from small unavoidable rotations of the board tool about its approach axis during the setup phase.

The circuit board is 275 mm long and 140 mm high measured from center to center of corner tooling holes. The origin of the board is defined to be the center of the upper-left tooling hole as the camera sees the board. The board X axis is defined to be parallel to the length of the board, extending from the origin to the upper-right tooling hole. The Y axis extends downward from the origin and passes through the lower-left tooling hole. These X and Y axes are nominally parallel with the defined robot X and Y axes.

The board is divided into a grid pattern of fields of view 8 rows deep and 11 columns wide. Each row is 20 mm high and each column is 25 mm wide. The camera field of view is a little larger than this, but these dimensions allow some overlap from scene to scene so that no joints on the edges of the field of view are missed during the inspection. The robot follows the board grid pattern as it moves during inspection. The robot shifts in the board and robot X direction 25 mm at a time, and 20 mm at a time in the Y direction. The nominal coordinates of a point on the board are simply the camera coordinates in millimeters added to the shifts made so far by the robot.

The pixel-to-millimeter scale factor is computed using a 10-mm calibration square in the camera field of view. The width and height of the square in pixel units is measured, and the number of pixels per millimeter is used as the scale factor.

The positional correction factors are computed by placing a circuit board in front of the camera as the camera repeatedly takes pictures and draws a target at a particular place on the screen. The board is shifted and rotated manually until the upper-left tooling hole is over the target and the lower-left tooling hole is over the target after a simple Y direction shift by the robot. That is, the circuit board is aligned so that its left edge remains exactly parallel with the left edge of the camera field of view as the robot travels over its full range in the Y direction.

The board is then returned to the position in which the upper-left tooling hole is on the target. This position is taught to the robot as its starting position. The board is then translated in the robot X direction 250 mm (10 shifts) until the upper-right tooling hole is in the camera scene. The position of the center of the upper-right tooling hole in camera coordinates is recorded.

The board is then translated in the robot Y direction 140 mm (seven shifts) until the lower-right tooling hole is in the camera field of view. The position of the lower-right tooling hole in camera coordinates is recorded.

The coordinate correction factors can now be computed. There are two correction factors for the X position of a point on the circuit board, and two for the Y position. One correction for each position coordinate is a function of the nominal X coordinate of the point, and one correction is a function of the nominal Y coordinate of the point. The difference in millimeters between the measured X coordinates of the upper- and lower-right tooling holes, divided by the distance between them (140 mm), becomes an X correction factor which is a function of the nominal Y coordinate of a point. The difference between the nominal X position of the lower-right tooling hole and its exact X position (275 mm), divided by the length of the board (275 mm), becomes the second X correction factor.

Analogous corrections are computed for the Y coordinate. The difference in millimeters between the Y coordinates of the upper-left and upper-right tooling holes, divided by the distance between them (275 mm), becomes the Y correction factor dependent on the X coordinate of a point. The difference between the nominal Y position of the lower-right tooling hole and its exact Y position (140 mm), divided by the width of the board (140 mm), is the second Y correction factor, this one dependent on the nominal Y coordinate of a point.

For example, the actual X coordinate of a point on the board is its nominal X coordinate plus the nominal X coordinate multiplied by the correction factor dependent on X travel, plus the nominal Y value multiplied by the X correction factor dependent on the Y travel.

74.4. IMAGE PROCESSOR

The image processor task is to direct the robot as to when and where its moves should be made, to do the inspection of the solder joints when the move is completed, to accumulate the solder flaw data for the board, and to download the data to the postprocessor.

The equipment used is the Automatix Inc. Autovision II image processor. The image processor receives the composite video data from the camera and digitizes it into a rectangular array 235 pixels wide by 216 high. Each pixel in this array is digitized with six bits so that there are 64 possible gray levels. The threshold gray level separating objects (solder joints) from background (traces and boards) is computed automatically for each board as a first step before the inspection. The threshold is computed by locating the gray level midway between the two peaks of the gray level histogram in a particular scene on the board. One peak corresponds to objects and one peak corresponds to background.

The image processor automatically implements any one or all 45 object-recognition algorithms commonly referred to as SRI and URI algorithms (many of which were developed or refined at the Stanford Research Institute and the University of Rhode Island in the 1970s—see Chapter 14, Vision Systems). Object-recognition features include such object characteristics as area, perimeter, bounding box size, and moments of area.

The following object features are used in the solder inspection:

1. Presence of hole in solder joint.
2. Size of solder joint, measured by size of a bounding box which "just fits" around the solder joint as well as anything else solder colored attached to it.
3. Solder joint peround. Peround is a measure of shape equal to the area of an object divided by its perimeter squared, and normalized so that for a circle peround equals 1.0. Anything more irregularly shaped than a circle would have a peround of less than 1.0.

The image processor is programmed in a high-level, PASCAL-like language to go from solder joint to solder joint in a scene, measuring these features for each joint and comparing their values to preset values with tolerances. The presence of a hole in a solder joint is always an error condition. If a solder blob has too large a bounding box there may be a bridge between two joints. A box too small indicates missing solder or a spot of spurious solder. An irregularly shaped solder joint indicates missing solder or a solder bridge.

If the image processor detects a joint with either a hole or an irregular shape it gives the joint a second chance to pass the inspection. The flaw may be a shadow or other "optical flaw" rather than a true flaw. The image processor sets a local processing window around the joint, reduces the threshold gray level by ten levels to simulate more light on the joint, and inspects it again. If the joint fails again it is considered a true flaw.

When a flaw is detected its XY location on the board, its size, and flaw type are stored by the image processor until the end of the cycle when the data are downloaded to the microprocessor.

74.5. CONTROL AND INTERFACING

74.5.1. Robot and Robot Controller

The robot and robot controller task is to receive and decode commands from the image processor, and to move a circuit board accordingly in the defined inspection plane within the robot work space. The equipment used is a Unimate PUMA 560 manipulator and controller with the VAL operating system.

The inspection plane in which the robot is to move is divided into "high-order" and "low-order" shift increments in both X and Y directions. All robot movements in the inspection plane are SHIFT commands with respect to the initial starting position defined in the setup phase.

The number of millimeters associated with any shift is arbitrary within the reach capability of the robot. Based on the size of the board to be inspected and the nature of the communication interface, the high-order shifts in each direction are set equal to 32 mm, and the low-order shifts in each direction are set equal to 16 mm.

74.5.2. Image Processor–Robot Interface

This interface task is to provide communication between the image processor and the robot controller so that the image processor can direct the robot motions. The interface is implemented over the software-accessible parallel ports of each system, using six AC input lines and six AC output lines for each system. AC is used because each component comes equipped with AC opto-isolated I/O modules.

The six lines consist of the following:

1. A data-valid line.
2. An acknowledge line.
3. Four lines for communication of four-bit numbers.

Handshaking is done in software. The protocol is as follows:

Sender:

1. Set up four-bit data lines.
2. Wait until acknowledge (in) line goes low.
3. Set data-valid line (out) high.
4. Wait until acknowledge (in) line goes high.
5. Set data-valid line (out) low.

Receiver:

1. Wait until data-valid line (in) goes high.
2. Latch in data on four-bit lines.
3. Set acknowledge (out) line high.
4. Wait until data-valid line (in) goes low.
5. Set acknowledge (out) line low.

The data passed from the image processor to the robot are commands numbered from 0 to 15. Each command received by the robot tells the robot to execute a specific subroutine that performs a specific task. Three specific tasks used in this inspection are:

1. Pick up a board from the nest and move to the start position in front of the camera.
2. Set board down at end of cycle, move to rest position and await further instructions.
3. Move to a board position.

For task number 3 the image processor computes the number of high- and low-order shifts necessary for the robot to move from its starting position to the desired position. The command to the robot is a particular number which signals the robot to expect four more numbers corresponding to the number of shifts of each type in each direction the robot is to make. The robot signals the image processor with a particular number when its move is completed.

74.5.3. Microprocessor

The task of the microprocessor is to receive the inspection data from the image processor, postprocess the data, format and print out the inspection results, and maintain statistical data on the production run. The processor is an Apple II Plus microcomputer with the Apple Pascal operating system.

When the image processor is finished with a board, the inspection data are downloaded to the Apple over a one-way RS-232 serial line, one line at a time. The data are stored in memory until all the data are received. The data are then written to a file on a floppy disk. When this is finished the postprocessing program is executed.

The purpose of postprocessing the data is to eliminate from the flaw category those oversized or irregularly shaped solder joints (without holes) that are supposed to be on the board. These are double and triple joints, and joints with connected test points. There are about 80 such joints on the board.

The postprocessing is accomplished by comparing data from the inspection file to data in a stored reference file. The reference file is generated in advance from a known good board, and consists of the XY board locations and sizes of all solder joints found to be oversized or of an odd shape on the good board. The comparison is accomplished by looking in the reference file for solder joint coordinates matching the coordinates in the inspection file of a joint that is oversized or out of shape and contains no hole. If a match-up between the coordinates and size of the joint is found, the joint in the inspection file is not included in the flaw file.

The inspection file data are accumulated in the same order as the reference file data. This means that the search through the reference file for a coordinate match-up can be localized to a particular area in the reference file, saving time.

After the postprocessing the final flaw data are printed out as a table and as a flaw type code letter on a graphical representation of the board.

74.5.4. Image Processor–Microprocessor Interface

The task of this interface is to allow one-way communication from the image processor to the microprocessor for flaw data download.

The image processor has a software switch that allows everything written to its terminal screen to be dumped out over one of two RS-232 communication ports as strings of ASCII characters. The microprocessor has an asynchronous RS-232 I/O card which receives the ASCII characters one at a time. A PASCAL program calls an assembly language procedure which takes the ASCII characters from the I/O buffer, puts them into a string and returns the string to the calling program. The program then writes the information to a text file on floppy disk in a format suitable for postprocessing.

74.6. SYSTEM EVALUATION

This system for automatic inspection of solder joints offers a relatively inexpensive, flexible system for automating the first-pass inspection of solder on circuit boards with solder masking. Improvements to the system are being made to bring the inspection speed up to the design goal by upgrading the image processor and by optimizing the implementation of the automatic thresholding, the postprocessing, and the communication links. Improved techniques for scene illumination will soon be implemented. Eventually, the system will be upgraded to where solder touch-up will be done automatically by a laser using the flaw coordinate information.

Work in solder inspection continues at the Production Technology Laboratory in several areas. One area is the inspection of solder on boards with no solder masking, that is, boards on which the traces form part of the scene. This approach requires the use of a board CAD file or the generation of a pseudo-CAD file. The purpose of the CAD file is to use prior knowledge of what is expected on a board at a certain location. The CAD file becomes the reference file for comparison to inspection data.

Another area of laboratory activity is three-dimensional mapping of solder-joint topography using structured laser light. This is the approach taken by Nakagawa at Hitachi.[1] This approach yields a three-dimensional profile of the solder joint, but is very slow (about 1 sec per joint). Much higher scanning rates will have to become available before this approach becomes feasible.

The approach to solder inspection described in this chapter is a two-dimensional approach. This approach cannot yield information about solder defects that are only visible from an oblique viewing angle. One example of such a defect is insufficient solder, where the entire land is covered but the solder does not extend far enough up the lead. Another example is a solder joint containing a peak or globule in the fillet that does not extend beyond the perimeter of the fillet. Two other examples are excessive solder on the joint in which the solder extends too far along the lead ("icicle"), and insufficient wetting of the solder joint.

These solder defects are important, but not as important as solder bridges or insufficient solder around the lead. These two solder flaws account for 90% of all solder flaws occurring in one Honeywell manufacturing facility. These are the defects for which this system was designed. However, a three-dimensional approach will have to be implemented to do a complete solder inspection to military specifications.

Other areas of research include inspection of solder using the thermal signature of a solder joint heated by a laser pulse,[5] solder inspection by gray-scale topography mapping (shape from shading), and inspection of solder on the component side of circuit boards using a robot-held camera.

REFERENCES

1. Nakagawa, Y., Automatic Visual Inspection of Solder Joints on Printed Circuit Boards, *Proceedings of SPIE,* Vol. 336, *Robot Vision,* May 1982, pp. 121–127.

2. Naval Air Systems Command, Department of the Navy Process Specification, Procedures and Requirements for Preparation and Soldering of Electrical Connections, Publication No. WS-6536D, March 14, 1980.

3. Newsletter of the Production Technology Laboratory, Honeywell Corporate Production Technology Laboratory, 815 Zane Avenue North, Minneapolis, Minnesota, published quarterly.

4. Publications of the IPC Task Force on Automatic Solder Joint Inspection, The Institute for Interconnecting and Packaging Electronic Circuits, 3451 Church Street, Evanston, Illinois.

5. Vanzetti, R. and Traub, A. C., Automatic Solder Joint Inspection in Depth, *Proceedings of the 7th Annual Seminar on Soldering Technology and Product Assurance,* February 1983, Naval Weapons Center, China Lake, California.

CHAPTER 75

TESTING AND SORTING OF PRINTED CIRCUIT BOARDS

RONALD D. McCLEARY

UAS Automation Systems
Bristol, Connecticut

75.1. INTRODUCTION

This chapter presents some of the difficulties encountered while attempting to automate, by the use of robotics, the process of testing fully loaded printed circuit boards. It is an example of the use of a robot in a material-handling application specific to testing. Although the application seems relatively simple, it illustrates an area of a very large potential for robots, and clearly shows that a significant system engineering effort is required before operation, including the careful preparation of specialized software.

The chapter concentrates on the specific testing process utilizing bed-of-nails printed circuit board testers. Solutions are suggested to some of the difficulties based on a system that the author installed while employed at Unimation Inc. Alternative solutions, in some cases providing better productivity, are also discussed.

75.2. THE MANUAL TESTING METHOD

In the manual method (see Figure 75.1), which is widely used in the electronics industry today, an operator sits in front of a bed-of-nails tester, with a tote box full of untested circuit boards on one side and a tote box full of tested and passed circuit boards on the other side. Somewhere in between, the operator has a tote box for boards that have failed the bed-of-nails test. Attached to these rejected boards, usually by operator-applied masking tape, is a printout from the bed-of-nails tester indicating the nature of the faults on that particular board.

When the operator has either emptied an incoming tote box or filled one of the outgoing tote boxes, the operator then gets up, replaces that tote box with another one, and continues the operation.

There are many variations on this theme, some involve conveyorization for the part input and output, others where a repair technician works alongside the test operator repairing any rejected boards for further testing.

75.2.1. Disadvantages of Manual System

The disadvantages of the manual testing system are as follows:

1. Poor material handling on input and output of work center.
2. Lack of in-process buffer storage.
3. Typically, only single-station test fixtures are used.
4. Operator tagging of rejected boards is inefficient both in time and accuracy.
5. Operators do not work at a constant rate, and as the day goes on their efficiency drops off. Further loss of production is due to coffee breaks, lunch breaks, and other personal breaks.

75.2.2. Possible Manual System Improvements

Before automating the system by robotics, several improvements can be effected in an existing manual system to increase productivity. Some examples follow in the next sections.

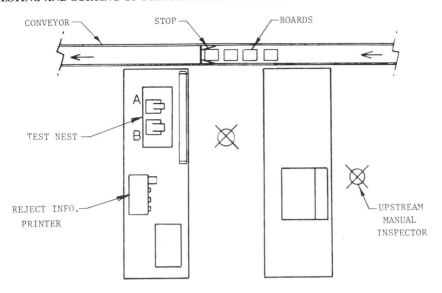

Fig. 75.1. Layout of manual testing system.

Improved Input/Output Material Handling

Although presenting circuit boards directly to the operator on conveyors can be very cumbersome, it is not much less effective and still quite productive to have fully automatic transporter systems that carry containers of untested circuit boards to the operator with a small buffer storage spur at the work station. In the same way the finished boards, having been tested, may be removed from the work center, whether good or bad, by additional conveyorized transporter systems.

Use of Data Base Systems for Repair Information

Some companies with bed-of-nails test systems have been using computerized data base systems which automatically pick up a serial number from the circuit board, then store the results of the test. This information in the computerized data base system may then be recalled by a repair technician at the work station. This approach removes the need for the operator to tape a computer printout to the circuit board to identify the nature of the fault, thus increasing the efficiency of the operator and improving the reliability of failed part information. As an added benefit of the data base, the manufacturer may develop accurate statistics of the performance of the printed circuit board manufacturing facility.

Use of Dual-Station Test Fixtures

Another possibility for increasing the productivity of manual, automatic testing of printed circuit boards is the use of two-station test nests. With a single-station test nest, the operator must wait while the tester checks the circuit board (typically about 30 sec) before unloading and reloading another part to be tested. With a two-station test nest, the operator can improve efficiency by approximately 10% by being able to load a second test nest while the first one is being tested. Once the test is complete, the test machine will automatically begin testing the board in the second nest, and so on.

75.3. FLEXIBLE AUTOMATION ADVANTAGES

In the preceding section we discussed how the efficiency of the human operator may be significantly improved by altering the procedures currently in use in a typical testing work center. There are, however, still many areas in which the overall productivity of the testing work center can be greatly improved through the use of flexible automation, in this case, **robotics.**

Advantages of robotics have been discussed in previous chapters (see Chapter 33, Evaluation and Economic Justification). With regard to testing, these advantages can be summarized as follows:

1. **Consistency.** The introduction of an industrial robot to a process greatly improves the consistency of the process. Once programmed, the robot performs its task in a very repeatable manner,

both in physical positioning and in performance correctness. This consistency will result in productivity increases because of not passing boards by error that should have been rejected, or rejecting boards that should have been passed, which then causes them to go through a repair area and eventually be retested.

2. **Flexibility.** The intelligent industrial robot allows the manufacturer to have a flexible manufacturing system similar to when the human operator was present. This is a big advantage in companies where parts are produced in small to medium lots.

   The robot, being a reprogrammable unit, allows for major installation alterations without obsolescence. Small, minor lot changes can automatically be compensated for by the robot. If one type of board is replaced by a similar board requiring different test characteristics, the robot can sense the change and instruct the tester so that the proper test can be performed.

3. **High Utilization.** The robot gives the manufacturer the maximum utilization of the expensive test equipment. When coupled to a buffer storage system, the robot does not stop for breaks or shift changes, giving maximum utilization of the costly test equipment.

The greatest increase in testing productivity can be achieved when the technology of the flexible automation, robotics, is combined with the productivity improvements discussed in Section 75.2. The result is a flexible manufacturing system (FMS) which can accommodate varieties of product in a very efficient manner.

75.4. ROBOTIC TESTING INSTALLATION

While employed by Unimation Inc. (Systems Division) the author installed a system (see Figures 75.2 and 75.3) utilizing a UNIMATE model 560 PUMA industrial robot. This robot has a weight capacity of approximately 5 lb (2.25 kg), repeatability ±0.004 in. (0.1 mm) and six fully coordinated degrees of motion.

The system, as it existed prior to the introduction of the PUMA robot, implemented many of the items mentioned in Section 75.2.

In the manual operation, the operator has parts entering the workcell on a conveyor transporting individual parts. Good parts, after being successfully tested, were placed on another conveyor which took the parts to the next testing operation. Rejected parts, with the computer printout taped to them by the operator, were placed in racks which had to be manually emptied after approximately 2 hr. The work station already utilized a two-station test fixture which allowed the operator to work more efficiently.

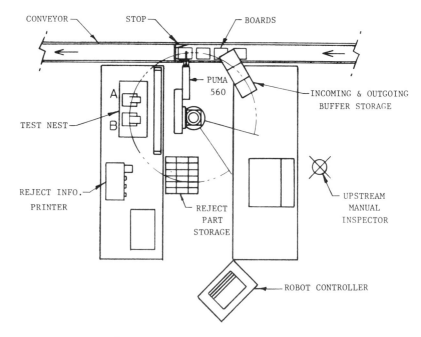

Fig. 75.2. Layout of robotic testing system.

Fig. 75.3. General view of robotic testing station.

The disadvantages of the system with a manual operator are as follows: Parts entering the system on the input conveyor would become backed up when the operator was not operating at maximum efficiency, causing operations preceding the testing operation to be stopped since their output (the test station's input conveyor) was backed up.

Additionally, if the test station output conveyor became backed up due to a slowdown further along the conveyor line, the testing operations in the work center had to be stopped since there was no place to put the acceptable parts after testing.

Also, it was necessary for the operator on occasion to get up and replenish full storage trays for the output of rejected parts, thus decreasing the operator's overall efficiency.

Taking into account these areas of inefficiency, as well as the fact that the operator would occasionally break for coffee, lunch, and so on, the overall average cycle time for the work center was approximately 23 sec per part whereas the actual testing time per part was only approximately 20 sec. With the dual-station test fixture, had the operator been working at 100% efficiency, the cycle time per part would have been approximately 20 sec.

To flexibly automate this system using the UNIMATE Puma 560 industrial robot required very few modifications to the existing equipment. Those modifications that were required are as follows:

1. An accurate and repeatable stop had to be installed on the incoming conveyor with a limit switch capable of detecting both part presence and orientation.

2. An interface had to be developed for the robot input/output to communicate with the FAULT-FINDER test unit.

3. Accurate and repeatable buffer storage locations (existing board carriers) were utilized for storage of overflow from the input conveyor and for overflow going to the output conveyor.

4. An accurate and repeatable system had to be developed for rejected boards and the computer printout of reject information.

5. Photo switches and their interface had to be utilized to identify overload conditions on the conveyors as well as empty locations for part set-down.

6. A robot hand-tooling gripper had to be developed and fabricated for picking up the circuit boards and loading the tester, throwing a small switch on some circuit boards during test, and for picking up, tearing off, and loading the computer-printed-out reject information into the tote prior to loading the reject board into the tote.

7. A Val™ software program had to be developed, written, and debugged for the PUMA robot to handle all operations.

75.5. SEQUENCE OF OPERATIONS

In the final system configuration, the robot waited at a home position and would constantly scan a variety of input signals monitoring the status of the workcell. A typical sequence of operations would be as follows:

1. Robot receives signal from test station A that the test is complete.
2. Test automatically begins immediately at test station B, which was previously loaded.
3. The robot checks to see if the board has passed or failed. If the board passes, the robot would move to station A (the tester) and remove the part.
4. The robot checks to see if the output conveyor load station is empty. If "yes" the robot places the part on the output conveyor station and retracts. In the event that the output conveyor load station is occupied, the robot waits 1 sec, then checks the station again. If this station is still occupied after 1 sec, the robot then proceeds to the output conveyor buffer storage and places the part into the output conveyor buffer.
5. If the robot receives a signal that the part in station A was a reject part, it then goes to the computer printout, tears it off, and places it in the next available compartment in the reject part tote bin.
6. The robot moves to test station A and removes the part and places it into the same reject tote bin locations as the label, placing the part on top of the label.
7. Once the robot has completed the foregoing, it will check the input conveyor station to see if a part is present and properly oriented. If there is a part present and oriented at this location, the robot then picks up a part at the input conveyor and loads test station A.
8. If the robot finds that no part is present, or that a part is present but not properly oriented in the input conveyor, it will indicate this as a fault condition, then proceed to the input conveyor buffer storage tray and remove a part from this tray and load it into test nest A.

The robot then returns to its home position and scans through the signals, again waiting for a signal indicating that his services are required again. Should no signal be received, the robot will attempt to do some housecleaning chores.

As an example, if no signals demanding its operations are received, the robot will check to see if the output conveyor load station is now empty. If so, the robot will check to see if any parts are present in the output conveyor buffer storage tray. Again, if this condition is met, the robot will unload the next circuit board in the tray and load it onto the output conveyor load station. This will continue until either the output conveyor becomes backed up, the output conveyor buffer becomes empty, or until the PUMA robot receives a signal from the automatic test set indicating that its services are required.

The services required of the robot will be of the following nature: either a test station has completed the test and the board has passed or failed; there is a circuit board in the test station that requires an operation be performed on it as part of the test (flipping a switch); or a board was immediately rejected by the tester, which would indicate that it was not properly seated in the test nest, in which case the robot would pick it up and reload it.

As previously indicated, the system, as automated by the use of the robot, now contains the following elements which make up a good, flexible machining work center:

1. Parts are presented to the work center by conveyor.
2. Buffer storage exists for incoming parts/outgoing parts that have been rejected.
3. A two-station test nest has been implemented as in the manual case, which allows the robot to maintain cycle times of 20 sec per board, as no time is lost due to loading and unloading of circuit boards.

75.6. CONCLUSION

Two major improvements could be made to the system as it now exists.

1. The robot still must place a computer printout with the boards that are rejected. There is still some room for error with this system. A better solution would be to implement the computer data base system described earlier.
2. It is still necessary for an operator to remove the full trays of rejected parts. A better solution would be a small section of conveyor that would transport the rejects to the repair area. This may be hard to justify on a cost-only basis as an addition. It would be less costly and more effective to implement this in the early planning stage.

The testing system described in this chapter was installed at a leading manufacturer of printed circuit boards. As mentioned, the manufacturer had already implemented some of the productivity improvements prior to the installation of the robot. The company has realized a productivity improvement since the robot is able to work through lunch and coffee breaks. Justification of the system was on a three-shift basis with a payback of approximately one year.

REFERENCES

1. Chin, R. et al., Automation inspection technique (of printed circuit boards), SME Paper No. AD77-729, November 1977.

2. Kuln, G. and Pavel, G., Testing equipment in an automatic assembly line for seat belts, *Proceedings of the 5th International Conference on Automatic Inspection and Product Control*, Stuttgart, Germany, June 1980, pp. 415–424.

3. The automation of testing, *Proceedings of IEEE Conference No. 91*, September 1972.

4. Ciarcia, S., Computerized testing, *Byte*, Vol. 5, No. 12, 1980, pp. 44–70.

5. Healey, J. T., *Automatic Testing and Evaluation of Digital Integrated Circuits*, Reston Publishing Co., 1981.

PART **13**

FINISHING, COATING, AND PAINTING

CHAPTER 76
ROBOT APPLICATIONS IN FINISHING AND PAINTING

TIMOTHY J. BUBLICK

The DeVilbiss Company
Toledo, Ohio

76.1. INTRODUCTION

Industrial finishing robots, regardless of their manufacturer, are more often than not quite similar in appearance, componentry, and purpose. All spray finishing robots, as an example, are comprised of three basic components: a control center, manipulator, and hydraulic unit (Figure 76.1).

The most common reasons for using spray finishing robots are to reduce labor requirements, remove operators from potentially hazardous environments, reduce coating material and energy consumption, reduce the number of rejects, improve quality and consistency, and assure repeatability.

Finishing areas of manufacturing facilities are normally the least up-to-date areas in the plant. Finishing in some industries is still considered a function of man and a skill or art rather than a science. When consideration is given to the implementation of a robot, it is advantageous not to rely on the methods of the past to be the only guidelines for the future.

Finishing robots in themselves should not be viewed as complete solvers of all painting problems. Rather, they should be recognized as integral parts of an automated system. And, like other forms of automation, finishing robots should possess the capabilities to complete their tasks and complement the total production operations.

Today's painting robots can consistently duplicate the best work performed by skilled production spray painters, providing sufficient time is allowed to adjust and refine the programming. Once the robot is programmed, it will repeat the exact motions of the sprayer/programmer and provide quality results whether the application is a final top coat, primer, sealer, mold release, or almost any other material.

The computer memory of the robot also provides the capability to interface with other equipment supportive to the finishing system. These include color changers, turntables, conveyors, lift and transfer tables, and other host computers. In fact, using the robot's memory capabilities can alleviate many of the design and operating concerns associated with present-day production finishing systems.

Today in the finishing industry, robots are applying automotive exterior top coat and underbody primer, stains on wood furniture, sound deadeners on appliances, porcelain coating on kitchen and bathroom fixtures and appliances, enamel on lighting fixtures, and even the exterior coating on the booster rockets that propelled the Space Shuttle into orbit (Figure 76.2).

These proven applications speak well for the versatility of painting robots and indicate that a great majority of production painting operations can at present, or in the future, be done economically and efficiently by robots. The success of converting to automation depends primarily on a well-planned program that begins with a thorough examination of robotics based on application objectives set for the particular situation.

76.1.1. Application Objectives

Establishment of parameters is essential to the selection and successful implementation of a spray finishing robot, and the process should begin by answering the following questions: Why select a finishing robot, When to select a finishing robot, Where could a finishing robot be used, How to select a finishing robot.

Fig. 76.1. The main components of a spray finishing robot include a manipulator, control center, and hydraulic unit. (Photo courtesy of DeVilbiss.)

Fig. 76.2. Applying a primer or top coat to auto bodies is just one application of a spray finishing robot. (Photo courtesy of Devilbiss.)

Why Select a Finishing Robot

The answer to the first question is relatively simple. All the reasons summed up previously reflect one basic goal—to reduce operating costs and increase return on investment. Unfortunately, finishing robots have been installed in some plants simply because of a management desire to possess the latest technology. Underutilization of a programmable robot is as unprofitable as adapting limited automation to a task requiring a high degree of flexibility.

When to Select a Finishing Robot

The "when" of selecting a finishing robot depends entirely upon the individual manufacturer. Factors include status in the market, type of product manufactured, condition of the plant and production line, labor costs, applicability of government health and safety regulations, material and energy costs, and many others particular to the specific manufacturer. Following a complete investigation, the findings may indicate that considerable modifications are required in other areas of the plant. For example, it would be inefficient and not very cost-effective to install a painting robot capable of handling 20% more production on a line where the other components can handle no more than a 5% increase.

Where Could a Finishing Robot Be Used

Although the answer to this question could be "anywhere," the most appropriate application must be determined on an individual basis. Once again, depending upon the specific conditions, it may be that another form of automated finishing, such as conventional spray guns, electrostatic spray guns, or rotational atomizers, would be the wise decision in the long run. Again, it is essential to conduct a coordinated investigation because of the many factors that can affect the decision to install a finishing robot. Representatives of the company's engineering, production, finance, and finishing departments, along with capable robot vendors, must be part of the process to draw meaningful conclusions.

How to Select a Finishing Robot

Without a doubt, this question can be the most difficult to answer. The actual selection of the finishing robot requires the coordination of data with information about robots and robot manufacturers. The most effective method of obtaining robotic information is to allow one or more robot manufacturers to prepare system proposals based on their examination of the company's product, processes, and finishing requirements. Most robot manufacturers can duplicate or nearly duplicate the finishing environment of the requesting company and conduct extensive tests with the actual products at application labs or, in some cases, in the potential customer's plant.

76.2. JUSTIFICATION

The justification of installing a robotic finishing system includes two major areas: compatibility with the total production environment, which includes specific aspects and features of the robot itself, and cost.

76.2.1. Application Compatability

In general, when considering a robot, the first thought that must be kept in mind is that the robot is part of a finishing system. Finishing robots have not been successful when implemented as "stand alone" robots. Also, when investigating finishing robots for a particular application, consider only those applications that had a high level of success.

The interfaces between the robot and the conveyor, the robot and the part or product, the robot and the spray booth, and the robot and the total environment are essential for system success. A robot can be made into a movable item by mounting it on a turntable, lift table, or transverse table. Parts or products can be rotated or indexed, and the conveyor can be continuous, intermittent, or indexing. The robot pays for itself when it is applying paint to a product, not while it is sitting idle waiting for a product to come into position.

For that reason, during the evaluation a great deal of thought must be given to the interface between the conveyor and the robot. The robot must be synchronized with the line and be compatible with the type of conveyor in use (Figure 76.3). If there is no conveyor, but plans include the installation of a conveyor system, the system design must fit the physical layout and also be compatible with the product or parts and the robot. It is essential that the robot velocity vary with conveyor speed changes to maintain the correct gun-to-target position for a quality finish. Two features must also allow for conveyor stops and starts, which means consideration must be given to avoid light spots or sags when conveyor stopping or starting occurs.

Product identification is another important consideration when installation of a robot is proposed

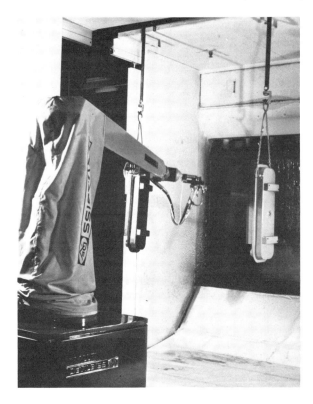

Fig. 76.3.   Spray finishing robot painting products as they move on a conveyor line. (Photo courtesy of DeVilbiss.)

on an existing line or new line. Rarely is a finishing system used to coat a single product or part configuration. Human painters can identify different parts and products, but the robot relies on mechanical or photoelectric sensing devices to indicate a part or product change. This same consideration must also be given to color changes.

Modifications to the spray booth may also be necessary. The booth may need to be enlarged to accommodate the robot. However, there are instances where a single robot has been positioned to alternately service two spray booths (Figure 76.4). Although the production advantages of this type of installation are obvious, the modifications can be extensive. Because of the absence of a human painter inside the booth, air supply and exhaust volumes can be reduced to levels required for control of volatile materials. Recirculated exhaust air, air staging, and lower operating temperature also become more practical because of the absence of humans.

Although humans may be absent from the booth during the painting sequence, that does not mean they are not needed. Since robots are more sophisticated than other finishing systems, training programs must be developed for the company's maintenance personnel to handle routine servicing of the robot's mechanical and electrical components.

In addition to evaluation of the relationship of the robot to the production and finishing processes as a whole, there are a number of specifics regarding the relationship of the robot, its components, and its capabilities to the application.

The spray tool is frequently the most overlooked component of the system. Surprising as it may seem, this is one area in which unique possibilities may aid the design engineer and provide the user with maximum coating efficiency.

Seldom does a human spray finisher minimize pattern size or fluid flow to achieve uniform coating efficiency on a product. Also, when an electrostatic system is added, a deterrent is the Faraday cage effect* in tight areas where deep penetration of the coating is required. Less efficient spray caps or

---

* The Faraday cage effect occurs in electrostatic applications where negatively charged paint particles are attracted to one side of a grounded object being painted. This results in paint buildup on the corners or edges of the object rather than in an even coating.

Fig. 76.4. Two spray booth areas are serviced simultaneously by a single finishing robot. (Photo courtesy of DeVilbiss.)

spray guns have been incorporated into systems due to problems with overspray contamination and dirty spray caps. Some automated systems have built-in factors for scheduled downtime for relief operators. Valuable production time must be sacrificed to allow crews to clean spray caps and maintain the system and the quality of the finish. Innovative engineering and the programmable robot can help to solve these problems.

76.2.2. The Robot Manipulator

Workers play a key role during the design of a robotic finishing system. Therefore close attention must be given to the interaction between the programmer, the robot, and the product. The programmer plays a vital role, and both the physical limitations of the programmer and the robot must be considered.

Manual spray finishing permits coating of remote and hard-to-reach areas because of the dexterity of the painter. Unless prior consideration is given to the size of the robot's arm and the spray-finishing tool, the coating capability of the robot will be limited. Most spray finishing robots are equipped with a pitch-and-yaw axis and a rotation axis at the worktool. However, this intricate painting capability cannot be achieved by a robot with this type of arrangement. Limitations posed by the axially related pitch-and-yaw, coupled with the size and shape of the actuators, can seriously restrict the flexibility of the unit.

Ideally, the robot arc should have at least as much freedom as the human wrist. This degree of freedom was not achieved until the introduction of the flexi-arm, manufactured by The DeVilbiss Company (Figure 76.5). This design provides full arching without regard to the pitch-and-yaw axis. It also minimizes the arm size and eliminates the need for electrical wiring, hydraulic hoses, and actuators on the end of the arm. The arm can be easily fitted with paint and air hosing and covered to ease cleanup and protect the equipment from overspray. This type of arm can be fitted with a seventh axis to further increase gun mobility and permit spraying the backside and underside of parts and hard-to-reach areas. When the robot is outfitted in this manner it has better reach and access than the human spray finisher.

Equipping the spray gun with a multiple-rotation capability permits rolling of the gun to coat

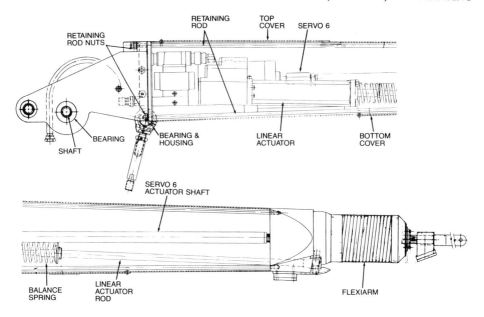

Fig. 76.5. The flexi-arm, first manufactured by DeVilbiss, allows greater freedom of mobility during the coating process.

interior surfaces. This is ideal for coating engine compartments, frame structures, undercarriages, and similar wall-type structures.

76.2.3. Programmable Positioning

The robot system provides more versatility and flexibility than manual or automated systems. However, because robots are limited to their work envelopes, it is sometimes necessary to increase their spatial coverage. This is accomplished using positioning axes, which can be programmed to reposition the robot manipulator to gain the optimum working position during the coating process. By repositioning the manipulator as necessary, it is possible to reach areas that would normally be outside the work envelope. In many cases, one robot mounted on an X, Y, and/or Z axis, or any combination thereof, can do the work of several robots or human spray finishers.

Positioning axes can be designed to elevate the robot manipulator to increase its vertical reach or to enable tracking of a moving conveyor. They can also reposition the manipulator closer to or farther away from an object. It is possible to supplement the vertical reach of a robot from the typical 7 ft to 10 or 12 ft (3.3–4.0 m) to gain access to tall objects. At the same time, it is possible to track a moving conveyor and maintain full synchronization. Varying degrees of sophistication exist with these devices. Their controls vary from simple end-stop, nonservo devices to servo-controlled units that provide virtually infinite control of position and speed (Figure 76.6).

Programming of the axes is controlled by the operator from the teach device. The spray programmer positions the axes with fingertip-controlled switches during the teach cycle. During production, the axes will position the manipulator and maintain full synchronization.

76.2.4. Fluid and Air Control

Control of fluid and air pressure is a major concern in any finishing system. The control of these functions is the key to obtaining a quality finish and efficient use of coating materials.

There are a number of methods available to achieve this control. Most automatic and manual spray guns are equipped with manual adjustments to perform these functions. However, these controls are normally preset by the spray finisher to coat the large surface areas of the product. They also are usually set to a level that requires the least amount of effort and skill by the worker. With feathering techniques, an experienced worker may elect to minimize fluid flow for certain areas, but under most conditions this is not the case. On automatic systems, feathering features usually do not exist. Consequently, the flow and pattern are set for maximum area coverage.

By taking advantage of the programmable output functions of the robot controller, it is possible

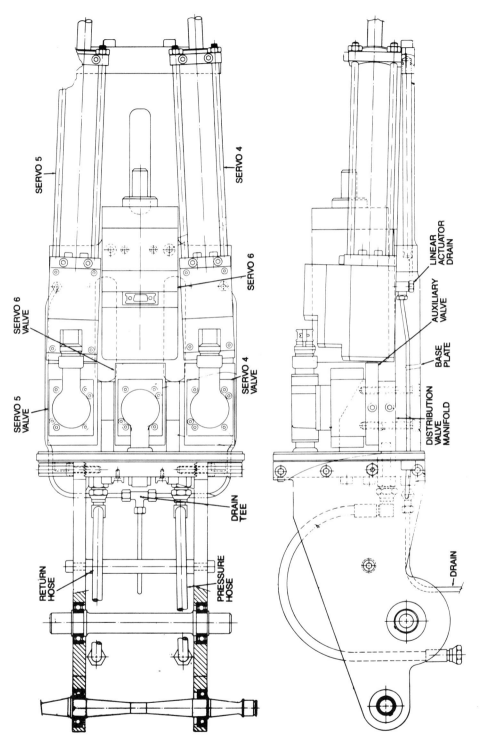

Fig. 76.6. Positioning axes, controlled by servo devices, give a finishing robot greater spacial coverage.

to have fully automatic control over fluid flow, atomization, and pattern size. A standard automatic-type spray gun is modified to permit remote and separate control over atomization and pattern-forming air. Through a series of air-piloted and shuttle valves, remote-control regulators monitor fluid and air pressure to the spray gun. By using only two functions, it is possible to achieve three levels of controls, more than adequate for most applications.

The functions are programmed by the robot trainer during the teach cycle (Figure 76.7). The actuation of a miniature, three-position toggle switch conveniently located on the teaching handle performs the programming and permits the trainer to vary pressure and spray pattern to achieve the desired results and compensate for variation in products and materials.

With this technique it is necessary to preset the desired levels of control prior to programming. Should this not be desirable, or should three levels of control be inadequate, analog control can provide the trainer with virtually infinite fingertip control at the teach handle.

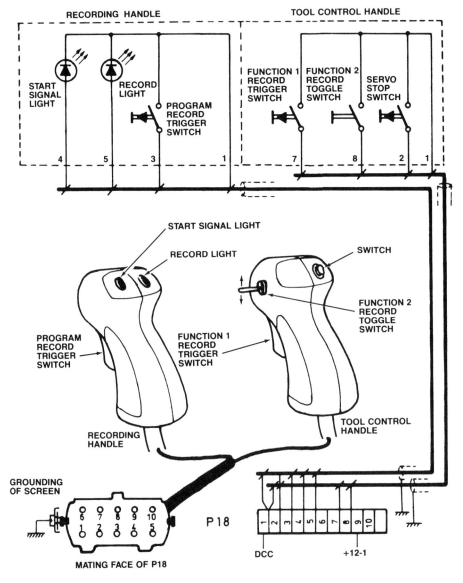

Fig. 76.7. Wiring diagram of robot teach handles. Functions are programmed by a robot trainer during the teach cycle.

76.2.5. Cleaning Spray Caps

A problem for most automated systems is the requirement to clean spray guns to maintain the quality of the finish. This results in the additional cost of relief painters and cleanup crews in production downtime. Again, the programmable output functions of the robot can solve this problem.

In most cases, the cleanup required during production of an automated system or even on a hand gun consists of cleaning or changing the retaining ring and spray cap of the gun. If the gun is mounted on an automatic machine, the system may have to be shut down. But not with the robot. A spray-cleaning nozzle is mounted at a convenient location, and the robot is programmed to move the spray gun to the cleaning device and position the gun in front of the nozzle. At this time the programmable output function controls the spray-cleaning jets for the proper duration. Using a counter or timer and appropriate controls, it is possible to operate this cleaning program to meet the requirements of production and maintain clean air caps throughout the day. The cleaning cycle is determined by the frequency set on the counter or timer. Using proper solvents and nozzle design, the cleaning cycle can be completed in approximately 2–3 sec.

76.2.6. Electrostatic Control

The electrostatic application of coatings is well known for its benefit of materials savings. However, on many products the electrostatic attraction, ionization, and Faraday cage effect can pose problems. These effects can be compounded by a product design that requires deep-coating penetration or includes sharp corners, by insulated components within an assembly or other material, or by a problem with the substrate.

With human sprayers and conventional automation, any one of these problems could prohibit electrostatic application. With robotic applications, programmable output functions can solve the problem. The solution is in programming one of the output functions to control the electrostatic power supply and cycle the voltage as required. This control can be built into the teach handle of the robot, enabling the programmer to control the electrostatics as required throughout the coating process. Thus the desired penetration can be achieved in recessed areas by eliminating the Faraday cage effect created by the high voltage. It is also possible to achieve the benefits of electrostatics while, at the same time, selectively eliminating the detrimental effects associated with electrostatic coating.

In a spray application, more so than in a pick-and-place or machine loading/unloading situation, ambient environmental conditions can play a significant role in the success or failure of a robot installation. It is essential that the characteristics of the coating material remain consistent in terms of viscosity, specific gravity, temperature, or pressure. If these items vary to any great degree, the finish will also vary. As a result, even though the robot may be performing flawlessly, lack of control over these ambient conditions can greatly affect finish quality.

There are many instances where a robot is installed in a finishing operation where the success is dependent on factors other than the robot. When initial failures occur, many times the robot is viewed as the cause when, in fact, problems have arisen with the material being sprayed, part orientation, or environmental factors in the booth.

76.2.7. Cost Evaluation

Perhaps to many industrial executives, the cost evaluation of a finishing robot is at least as important, if not more important, than the application compatibility studies. As a result, cost justification usually is the responsibility of those involved in the overall investigation of finishing robots.

To explain the cost evaluation process for a finishing robot, it is useful to examine a typical hand spray-painting operation that is under consideration for conversion to a robot application system. The example consists of a baking enamel applied by conventional spray handguns to household appliance components. The preliminary data that must be collected include the number of painters now applying the coating material, the amount of wages and benefits paid to them, the cost of the coating material, the amount used, and the reject rate. Also to be considered are health and safety regulations for the painters. In addition to these factors, other specific data related to the product and finish are required. For the purpose of this example, the input data is contained in Figure 76.8.

Manual Spray Application

The operation used in this example has two painters coating products during two 8-hr shifts—one painter per shift. The flow rate has been set at 1 qt/min to meet production requirements. The reject rate from finishing errors averages approximately 15%.

It is assumed that the spray booth in which the finish is being applied is suitable. Also, the paint-circulating system supplying the coating material to the spray gun used on the robot is compatible with the robot installation.

Substrate area $= 1.6$ ft^2 (average per hanger)
Conveyor speed $= 21$ ft/min
Part centers $= 3$ ft 0 in.
Flow rate $= 32$ oz/min
Manual spray deposition efficiency $= 40\%$
Robot spray deposition efficiency $= 40\% + 15\% = 55\%$
Spray gun "on time" $= 50\%$
Specified film build $= 1$ mil
Material cost $= \$5.00$/gal

Fig. 76.8.   Data on material savings calculation.

Material Savings

The proposition in this example is to replace one painter per shift with one robot. In essence, a direct comparison can be made between manual versus robotic application in terms of material savings. Figure 76.9a and b contains the material savings cost calculations for the manual spray and robotic spray based upon the data given in Figure 76.8. Note that in Figure 76.8, deposition efficiency for a manual sprayer is given at 40%. Based upon reports from robot users, a properly programmed robot can improve deposition efficiency 15–20%. The more conservative 15% figure is used in this example. For both the manual and robot applications, the spray gun "on time" factor has been set at 50%.

Another consideration in this application is that the specifications call for one mil coating coverage of the product. The majority of hand spray finishers will normally apply a heavier coating to assure proper coverage of the product. A properly programmed robot utilizing correct fluid flow rates and atomization air pressures can achieve and control the specified film deposition, which can save a considerable amount of coating material.

Using a cost figure of $5/gal, the annual cost of overspray alone for the manual spray application is $89,640. For the robot application, the cost is $67,230 per year, a savings in material cost of $22,410, or approximately $11,200 per shift. Again, this figure represents coating material savings with the robot that result from a reduction of overspray. Not included are the material savings that can occur through reduced rejects, which robot users have placed as high as 13%.

Material deposited manually on substrate:
32 oz/min $\times$ 0.5 "on time" $\times$ 0.4 deposition efficiency $= 6.4$ oz/min

Overspray:
(32 oz/min $\times$ 0.5 "on time") $- 6.4$ oz/min $= 9.6$ oz/min

Yearly overspray (two shifts):
9.6 oz/min $\div$ 128 oz/gal $\times$ 60 min/hr $\times$ 16 hr/day $\times$ 249 days/yr $= 17,928$ gal/yr

Yearly cost of overspray:
17,928 gal/hr $\times$ \$5.00/gal $= \$89,640.00$ per year

(*a*)

Material deposited by robot on substrate:
32 oz/min $\times$ 0.5 "on time" $\times$ (0.4 $\div$ 0.15) deposition efficiency $= 8.8$ oz/min

Overspray:
(32 oz/min $\times$ 0.5 "on time") $- 8.8$ oz/min $= 7.2$ oz/min

Yearly overspray (two shifts):
7.2 oz/min $\div$ 128 oz/gal $\times$ 60 min/hr $\times$ 16 hr/day $\times$ 249 day/yr $= 13,446$ gal/yr

Yearly cost of overspray:
13,446 gal/yr $\times$ \$5.00/gal $= \$67,230.00$ per year

(\$89,640.00 manual overspray cost) $-$ (\$67,230.00 robot overspray cost) $=$
                                    \$22,410.00 paint saving per year with robot

(*b*)

Fig. 76.9.   Data on spray calculations. (*a*) Manual spray, (*b*) robot spray.

Payback and Return on Investment

To establish the payback time for an industrial finishing robot, the following equation is used with the values assigned according to Figure 76.10:

$$Y = \frac{C}{WS + RS + AS + DE(R) - RM - RO}$$

Y is the number of years to realize a payback on the original investment or total cost of the robot C, which has been set at $80,000 and includes the robot and controls. Accessories for spraying and interfacing with allied equipment are not included. This cost will vary depending upon the simplicity or complexity of the overall system.

The wages and benefits WS have been estimated for one operator at $22,000 per year, or $44,000 per year for a two-shift operation. Annual robot savings (see previous section) has been calculated at a very conservative $22,410. AS, annual savings for operating and maintenance (spray booth), is estimated at $2000. This figure is indicative of reduced maintenance due to higher transfer efficiency and less overspray accumulation in the booth. The robot depreciation DE is the total initial cost, $80,000, over a 7-year period, or $11,428 per year on a straight-line basis. The tax rate R is a relatively standard 50%. For annual robot maintenance RM, $3000 has been allowed and consists of $2500 in parts and $500 in labor. The figure represents both preventive maintenance and maintenance due to failure. Robot reliability will vary with the severity of the application and the in-service time required. Here, the operating time has been established at 99% uptime with a mean time between failures of approximately 850 hr. Robot operating cost RO has been estimated at $3000 and represents the energy required to operate the unit and costs incurred during programming time. These values are inserted into the payback formula to produce the following equation:

$$Y = \frac{80,000}{44,000 + 22,410 + 2000 + 11,428(0.50) - 3000 - 3000} = 1.17 \text{ years}$$

Many manufacturers will wish to establish a return on investment (ROI) table in addition to calculating the payback for an industrial finishing robot. For this example, the same values assigned in Figure 76.10 for payback are used for the ROI table, Figure 76.11. (The ROI figures require the aid of a computer programmed for such calculations.) As noted in the table, 6% annual rate of inflation has been used over a 7-year period. The resultant ROI for this example is 121%.

The calculations for this particular example show economic justification for the installation of an industrial spray painting robot. However, if these figures had not been so favorable, perhaps another method of automation or even continued manual spray application would be the most cost-effective

FORMULATING PAYBACK

Payback time in years = cost/yearly savings

$$Y = \frac{C}{WS + RS + AS + DE(R) - RM - RO}$$

Y = Years payback		
C = Total first cost (robot, accessories, control)	=	80,000
WS = Wages & benefits — 2 workers — one year	=	44,000
RS = Yearly robot savings (material, quality, energy, OSHA)	=	22,400
AS = Yearly savings (operating and maintenance)	=	2,000
DE = Robot depreciation expense 80,000/7	=	11,428
R = Tax rate	=	50%
RM = Yearly robot maintenance	=	3,000
RO = Operation and programming costs	=	3,000

$$Y = \frac{80,000}{44,000 + 22,400 + 2,000 + (11428)(.50) - 3,000 - 3,000}$$

Y = 1.17 years payback time

NOTE: ALL FIGURES CALCULATED ON A TWO SHIFT BASIS.

Fig. 76.10. Calculations on formulating payback time.

RETURN ON INVESTMENT
Cash Flow For Robot Investment

	YEAR 1	YEAR 2	YEAR 3	YEAR 4	YEAR 5	YEAR 6	YEAR 7
Initial investment	− 80,000						
Investment tax credit	8,000						
Start-up costs	− 5,000						
Start-up tax credits	2,500						
Depreciation rate (.50)	11,429	8,163	5,831	3,645	3,645	3,645	3,645
Wages & benefits of replaced workers*	44,000	46,640	49,438	52,405	55,549	58,882	62,415
Tax cost*	− 22,000	− 23,320	− 24,719	− 26,202	27,774	− 29,441	− 31,207
Maintenance cost of robot*	− 3,000	− 3,180	− 3,371	− 3,573	− 3,787	− 4,015	− 4,256
Tax credit for maintenance*	1,500	1,590	1,686	1,787	1,894	2,007	2,128
Material savings*	22,400	23,744	25,169	26,678	28,279	29,976	31,775
Tax cost for savings*	− 4,500	− 4,770	− 5,056	− 5,359	− 5,681	− 6,022	− 6,383
Operating and programming expenses*	− 3,000	− 3,180	− 3,371	− 3,573	− 3,787	− 4,015	− 4,256
Tax credit for operating & programming*	1,500	1,590	1,685	1,786	1,893	2,007	2,128
TOTALS	− 26,171	47,277	47,292	47,594	50,231	53,024	55,989

*6% inflation added per year
Return on investment = 121%

Fig. 76.11.   Calculations on formulating return on investment.

for this particular application. On the other hand, there are going to be some applications where it will be required by law to remove the hand sprayer from exceptionally hostile spray booth environments. These applications will still require a painter's flexibility and agility to complete the finished product. But because of serious health hazards and the toxicity level of some materials, painters will not be permitted in the exposed atmosphere. There are numerous applications where film buildup is extremely critical, such as in the aerospace industry where weight and uniformity are closely controlled. It may not be possible to meet these rigid specifications consistently, eight hours per day, five days per week. No doubt there are other aspects that may be specific to a plant or operation and may in fact assist in the justification of a robot. For the most part, however, the foregoing calculations should serve as a general guideline for the evaluation of the cost-effectiveness of a robotic finishing system.

76.3.   ROBOT SYSTEM DESIGN

Having examined the rationale for selecting an industrial finishing robot, an examination of the robot itself is in order. It was noted in the introduction that industrial finishing robots are similar in appearance and consist of three basic components, a control center, a manipulator, and an hydraulic or electric unit.

The control center acts as the brain of the robot system, providing data storage of position data, memory, outside function control, and control interface with other programmable controllers or host computers. The memory medium for finishing robots can be solid-state, floppy disk, tape cassette, or any other available type of mass memory.

The control center is normally designed to interface through input-output with other controls, such as programmable controllers, and other mechanisms, like transfer tables, lift tables, and spray gun triggering. It is necessary for this type of interface to provide a synchronous program between the robot movements and other components.

The second component, the robot manipulator, is the working end of the robot. This component is the actual spray finisher and must duplicate the movements of the human body. Hip movement, arm movement, and most important, wrist movement must be duplicated for the robot to achieve a quality finish like the spray painter. A robot-finished job is only as good as the program. If the program taught to the robot is questionable, the robot-repeated work will be unacceptable.

The third item of the basic robot system is the hydraulic power pack, which will vary in size depending on the equipment manufacturer. However, the purpose is the same, to supply the hydraulic power for the manipulator. In most cases, the hydraulic power pack can be quite a distance from the manipulator without causing any operational problems. Most equipment manufacturers supply the hydraulic power pack with both supply and return filters, along with high-temperature and low-oil-level protection.

76.3.1.   Feature Capabilities

Most finishing robots are of the continuous-path variety, capable of multiple-program storage with random access. These two features allow for the smooth, intricate movements duplicating the human

Fig. 76.12. The teaching handle, attached to the manipulator arm, is used during programming of a lead-through-teach robot. (Photo courtesy of DeVilbiss.)

wrist which is needed to perform high-speed, efficient spray finishing. The random access of multiple programs is a requirement normally needed on production lines that paint a variety of different style parts. It is seldom that a particular production line will paint the same style and color of parts.

The successful robot finishing installations have utilized lead-through teach-type robots which are taught by leading the robot manipulator, or teach arm that simulates the manipulator, through movements and physically spraying the part. This allows for the hand-eye coordination necessary to accomplish adequate spray finishing.

Programming the robot in this manner is simple. The operator attaches a teaching handle to the manipulator arm (Figure 76.12) and plugs it into a receptacle on the robot base. He then leads the arm through the designed program sequence to define the path and relative velocity of the arm and spray gun. After programming, the operator switches the control from "programming" to "repeat" and puts the robot into automatic mode. Robot applications have been used with a teach control pendant, but this type of teach method is usually successful only when the part is stationary.

Finishing robots require some additional features that are not normally required for other types of robots. In most cases, finishing robots must be equipped with noise filters to prevent interference of electrical noise from electrostatic spraying devices located on the end of the robot or in the near vicinity. Also usually required is an explosion-proof remote control operator's panel in the spray area, so the operator can safely turn the unit on or off, select the operation of the spray gun, or have the robot repeat a single or multiple cycle.

Other features that enhance the robot finishing system are gun and cap cleaners and a cleaning receptacle, into which the robot can submerge the spray gun so the gun's exterior can be automatically cleaned after prolonged use or after color changes.

76.3.2. System Operation

Finishing robots must be designed and built for reliable operation in a dirty, solvent-filled atmosphere. In most cases, the control center is in an area susceptible to paint, solvent, heat, and other elements harmful to computer technology. The control center must be designed to operate easily for paint shop personnel who are not normally trained in sophisticated computer control machinery.

The equipment used in a robot finishing system must be properly maintained, especially since the manipulator end of the robot is subjected to solvent, mist, and accumulation of overspray. Robots will operate satisfactorily within a spray atmosphere for long periods; however, unmaintained equipment will eventually begin to cause production downtime and loss of productivity.

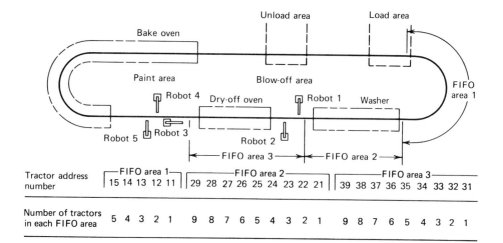

Fig. 76.13. Diagram of robot finishing system at John Deere and Company's Tractor Works facility.

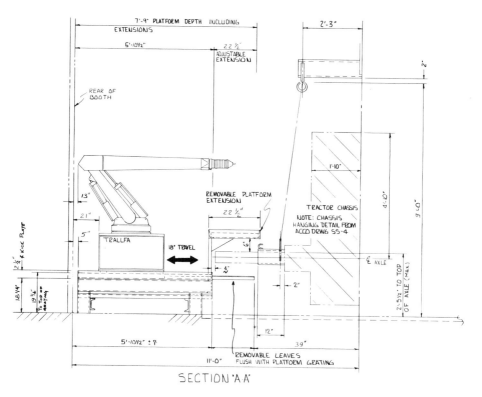

SECTION 'A A'

Fig. 76.14. Diagram of robot positioning when coating tractor chassis.

76.3.3. System Design Application

The first continuous-path robotic finishing system at John Deere and Company was installed as part of the chassis plant operation at the company's Tractor Works plant in Waterloo, Iowa. The criterion was to provide capacity to paint tractors at 92% efficiency in an 880-min day. In addition, there were eight basic tractor chassis models with a total of 36 paint programs to handle all variations. The system's goal was to paint 95% of two-wheel drive tractors and 90% of four-wheel drive tractors. The first design consideration was the number of robots needed to store information about the different styles and variations of chassis and paint programs.

After initial testing, it was determined that three robots could paint the tractor chassis to specifications. Figures 76.13 and 76.14 show the general layout and the cross section of the robots as they were installed in the spray booth relative to the tractor chassis. One robot was located on each side of the conveyor line, and the third was installed in a pit beneath the tractor chassis.

The robot in the pit sprayed the underside of the moving chassis and axles, as well as the chassis sides. The chassis then proceeded into the next robot station where the two robots sprayed the sides, top, and remaining areas. Owing to the size of the tractor, the system was designed so the robot moved in and out 18 in. perpendicular to conveyor travel. This allowed the robot to paint the end of the axle and the middle of the tractor chassis.

Robot programming was kept to a minimum to ensure a satisfactory start-up time for the new system. To reduce the number of programs, a feature termed *program linking* was used. Program linking incorporates the composite of three individual programs, one for the front end of the chassis, one for the center, and one for the rear. By linking these three segments as required for each chassis model, the paint code was determined.

The benefits of program linking were easier and faster programming during installation and reduced memory storage. The latter benefit was illustrated with the two front ends, which required painting variations on two-wheel drive chassis in addition to the option of rockshaft or no rockshaft. Without program linking, each of the six variations would have required 4.5 min per program, or 27 total minutes, to finish that model. Through program linking, each front end required only 1.5 min, or 4.5 total minutes. The middle section was the same for all chassis of the model, requiring 1.5 min, and the rockshaft/no rockshaft added another 3 min, for a total of 9 min, or two-thirds less time than if each model had a continuous program.

76.3.4. System Monitoring

Various safety interlocks and monitoring points have been incorporated into the system. The system monitors robots in the "home" position, so the conveyor will not start if the robot is not in its proper position. A limited switch failure utilizes two limit switches during the paint cycle, each acting as a backup for the other. If either fails to function, an alarm is triggered from the main control panel. However, the system will continue to function with either limit switch operative.

Another monitor built into the system is a manipulator safeguard. If the robot should contact or become entangled with a solid object, the hydraulic pressure of the robot arm will be reduced, allowing the arm to be pushed out of the way. In addition, high hydraulic temperature and low hydraulic oil pressure are monitored to assure that no damage occurs to the hydraulic unit of the robot.

This finishing system has been in operation for some time and has proven to be a reliable production operation. Benefits such as paint and energy savings have reached levels that were estimated, or have exceeded these levels. Other benefits, which are difficult to evaluate, have been the reduction of touch-up and the removal of workers from an unpleasant and possibly unhealthy atmosphere.

76.4. ROBOT PERFORMANCES

The robot is truly the ideal spray finisher. The fact that robots lack human senses makes them almost immune to the environment. They can work in conditions unfit for humans—they are not bothered by irritating odors, high noise levels, darkness or light, or extreme temperatures.

With robots, air supply and exhaust volumes in spray booths can be decreased to levels necessary only for control of volatile materials. It is possible to save energy by lowering operating temperatures in spray booths. These advantages open up many avenues for increasing productivity and conserving energy.

Robots have proven to be a reliable and efficient means of finishing and are the painters of the future. And, if implemented properly, they can begin to save money a short time after installation.

CHAPTER 77

ROBOTS FOR SEALING AND ADHESIVE APPLICATIONS

PATRICK J. BOWLES
L. WAYNE GARRETT

General Electric Company
Louisville, Kentucky

77.1. ROBOTIC DISPENSING SYSTEMS

Industrial robots are increasingly utilized for material dispensing, especially in application of sealants and adhesives. In this context, the example of cake decorating by robots (see Chapter 1) is also appropriate. Four major factors motivate this application area[1]:

1. **Applying Just the Right Amount of Adhesive, Sealant, or Other Materials.** Unlike manual application, where operators often dispense too much material that may cause poor quality, robots can accurately control the amount and flow of material dispensed. In addition to the better quality, material cost savings can be very significant over long periods (up to about 30%).

2. **Consistent, Uniform Material Dispensing.** Robots can maintain, with high repeatability, a consistent bead of material while laying it along accurate trajectories. Furthermore, where two or more components must be mixed while they are dispensed, as is the case, for example, in certain adhesives, robots can provide better control.

3. **Process Flexibility.** As in other application areas, the robot can be used to dispense materials according to different programs, depending on the particular operation that is required. It can be easily reprogrammed when design changes occur.

4. **Improved Safety.** The use of robots reduces the health hazards to workers from dispensed materials, including allergic reactions to epoxy resins and other substances, and potential long-term problems.

These four general factors are similar in many ways to those discussed in Chapter 76. Similar to painting and finishing, the major applications of sealing and adhesive dispensing are in the automotive, appliance, aerospace, and furniture industries. In general, any process that requires joining of component parts in a variety of production is a potential candidate for robotic dispensing. In the area of adhesive application, one can certainly state that adhesive bonding is as important to the assembly of plastic parts as welding has always been to metal joining.

Typical examples of robotic dispensing are the following:

Sealer to car underbody wheelhouse components.

Silicone on truck axle housing.

Urethane bead on windshield periphery before installation.

Two-component polyurethane or epoxy adhesive to automobile hoods made of sheet molding compound (SMC) between outer and inner shells.

Sealant application in the appliance industry, as described in Section 77.3.

An earlier version of this chapter appeared in *Adhesive Age,* April 1983.

77.2. COMPONENTS OF ROBOTIC DISPENSING SYSTEMS

The general structure of a robotic dispensing system is shown in Figure 77.1.

The Robot

Typically, a robot for accurate material dispensing is electrically actuated to achieve smooth motion and high repeatability. Five- to six-axis robots are required for flexible motions. Although speeds are usually on the order of 10–15 in. (25–38 cm) of bead per second, the robot must be able to move at different speeds at different segments of the bead path. However, a key requirement is the ability to maintain a constant bead size along all corners and part contours. Robots can be floor or overhead mounted, depending on the particular dispensing orientation. A robot can be mounted on a slide when material must be dispensed onto large parts, such as aircraft wings. Additional general considerations in selecting the robot can be found in Chapter 5, and in Parts 6 and 7 of the *Handbook*.

Type of Dispensed Material

In general, each material will have unique properties that must be considered in the application planning. A good summary of adhesives used in assembly and their properties can be found in Reference 2. Two-part adhesives usually require static or dynamic mixing devices. When a mixer is attached to the robot manipulator, motion vibrations may cause inconsistencies in the bead. Hot-melt materials do not require mixing, but the hot temperatures (300–400°F or 150–200°C are typical ranges for sealants and adhesives) may present a problem to some robots.

Container, Pump, and Regulators

Pump selection depends on the properties of the material, the container size, and the dispensing rate. The dispensing system must track the level of material in the container and stop the automatic operation when the material is depleted. Another issue is the timing control. Certain materials dry out, harden, or solidify if not mixed or if left unused for a period of time. Some adhesives harden within a few minutes after mixing and must be controlled very carefully. In such cases, the system should have the capability to automatically purge spoiled material and clean the container and lines. Regulators of filters and line pressure are also essential accessories that are required in the dispensing system.

Programmable Controller

A programmable controller is used to supervise the overall dispensing option and communicate between the robot, the container, and the dispenser ("gun"). Typically, it is responsible for the on/off activation of the dispenser in coordination with the robot motion, and for the control of the material level.

Dispenser ("Gun")

Usually an automatic dispenser is attached to the robot wrist, together with an inlet hose for material and one for cleaning solvent. When a mixing unit must be used, as in two-component adhesives, the combined weight of dispenser, mixer, and three hoses necessitates mounting on the robot arm. In this case, an additional tube is attached to the wrist and leads the mixed material from the mixer to the workpiece.

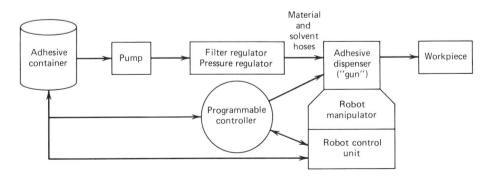

Fig. 77.1. General structure of a robotic adhesive-dispensing system.

An important issue is the control of bead integrity. An air jet sensor located on the dispenser nozzle can be applied to detect any break in bead application. This type of sensor must be cleaned frequently to prevent any clogging. In more sophisticated application, a vision system is used to monitor the bead integrity. When a break is detected, the robot is instructed to return and dispense material to the missing areas.

77.3. ROBOTIC DISPENSING OF FOAMED, HOT-MELT SEALANT

In this section, a particular robotic dispensing system of foamed, hot-melt sealant is examined. This system was developed and installed for sealing perforations in multiple refrigerator cases at General Electric's Decatur, Alabama, Refrigerator Plant. It combines the GE P5 Robot with Nordson's Foam-Melt System.* Production results have been excellent and follow the list of potential benefits mentioned in Section 77.1.

The major appliance industry has witnessed a problem in the manufacturing of refrigerators that until recently has only found moderately acceptable solutions. The refrigerator outer case contains holes and gaps necessary for fabrication and assembly. To prevent foam leaks and moisture migration, these perforations must be sealed prior to the injection of urethane foam insulation.

Historically, these perforations were manually sealed with tape or "gum putty." Recently, General Electric's Advanced Technology Section, working with Advanced Manufacturing Engineers, successfully developed a system to automatically solve this problem.

77.3.1. The Sealing Process

The process of sealing perforations in refrigerator cases with an adhesive appeared initially to be relatively straightforward. The manufacturing parameters in GE's Decatur plant varied, however, and an extensive evaluation was required. These parameters were as follows:

Two refrigerator models were to be sealed with provisions for future new models.

These models included two different case sizes, 11 and 14 ft.³

The potential sealant application points varied from 15 to 20 points.

Various hole sizes, contours, and locations were mandated by design in each case.

Production rates varied from 15 to 18 sec.

Cases were moving on a conveyor at the point of manufacturing, lying horizontally with the open side up.

The substrates for application of sealant were both prepainted and galvanized steel.

Extensive planning and evaluation were required to demonstrate successfully the elements of automation proposed for case sealing. Studies were performed to evaluate (1) foamed hot-melt adhesives as opposed to conventional sealants, and the suitability of those materials, (2) the capabilities of robotic systems versus hard automation, and (3) the automatic material handling and fixturing.

77.4. AUTOMATION COMPONENTS OF THE SEALANT ROBOTIC DISPENSING SYSTEM

The system includes the following components:

1. Sealant.
2. Robotic system.
3. Material-handling system.

77.4.1. Foamed Hot-Melt Adhesive Sealant

In 1981 Nordson introduced a commercially available foamed hot-melt adhesive system. Early studies of sealing refrigerator cases with conventional hot melt were unsuccessful because of the sagging and running of the material. *Foamed* hot-melt adhesive was selected for trial in this application for two reasons. First, a foamed adhesive material is more cohesive than the same conventional hot-melt adhesive. This reduces material sag and run and enables holes to be bridged easily without penetration. Second, material savings are approximately 50% greater (for a material with 100% expansion) with a foamed rather than unfoamed adhesive.

Extensive development work was conducted in which sample refrigerators were manually sealed

---

* Registered trademark of Nordson Corporation.

with a foamed adhesive. Holes were successfully bridged with the foamed sealant by raising one end of the case, applying the sealant above a hole, and allowing gravity to cause the material to flow over the holes. The sample refrigerators were then processed through a urethane foam injection system. The urethane foam provides the thermal insulation and structural rigidity between the liner and case and is produced by an exothermic reaction of two particular chemicals. Therefore the foamed hot-melt sealant must be able to withstand the urethane foam temperature and pressure. No urethane foam leaks were detected from any perforations manually sealed with the foamed adhesive sealant.

The adhesive application equipment for the final production installation consisted of Nordson's (1) FM103A FoamMelt Unit, (2) H-20 Gun with a 12-in. (30-cm) heated extension, (3) two 16-ft (5.3-m) hoses, and (4) specific accessories. This system proved to be successful in actual production.

In selecting the production sealant, several materials were tested. The results showed that a particular amorphous polypropylene-based sealant met the requirements of the application. This material was chosen for the following reasons:

The material foams with nitrogen and has excellent foamability with an expansion of more than 100%.

Adhesion is excellent to galvanized and prepainted steel.

Material setup is fast on steel substrates.

It can withstand the temperatures and pressures encountered in the urethane foam injection process.

The material foams and applies at moderate temperatures, which reduces energy consumption.

It passes General Electric's odor and taste tests for major appliances.

Meets the Federal Food and Drug Administration (FDA) approved standards.

No filler is added that might cause equipment wear.

Previous studies demonstrated suitability for this application.

Fig. 77.2. Robot automatically dispenses foamed hot-melt adhesive extruded from a dispensing gun for sealing holes in refrigeration cases.

Fig. 77.3. General view of robotic sealant dispensing cell. Hydraulic device (right) lifts and presents refrigerator cases properly fixtured and oriented to the robot. Each case is presented at 30° to enhance sealant flow coverage. Robot controller with programmable controller and foam container/extruder are shown left. The robot is mounted on a roller track for off-line programming and maintenance.

77.4.2. Robotic System

Sealing perforations in refrigerator cases with foamed hot-melt adhesives could be accomplished with either hard automation or a robotic system. Based on the manufacturing parameters of varied case sizes, hole sizes, contours and locations, hard automation proved to be impractical. Criteria for a robotic system meeting the manufacturing parameters were established. Detailed feasibility studies and capability demonstrations were conducted involving several robot manufacturers.

The General Electric Model P5 Process Robot was demonstrated and proved to be the most advantageous for this particular application. Tests confirmed the required cycle time and working envelope. Although selection of the P5 Robot involved many technical characteristics, three were essential: First, the varied hole locations and contours demonstrated the need for point-to-point programming with a teach pendant. (See Chapter 19, Robot Teaching.) Second, the manufacturing repeatability of hole locations and fixturing tolerance (±0.03 in., or ±0.76 mm) demonstrated the need for highly repeatable robot arm positioning; the P5 Robot is specified with a repeatability of ±0.008 in. (±0.203 mm). Third, the application required a six-axis robot. It is particularly interesting to note that although the P5 Robot maneuvers with five axes, a pseudoaxis of the arm is available by offsetting the gun centerline with the robot wrist centerline.

77.4.3. Automated Material Handling

The production process is limited without the benefits of automated material handling. In the GE Decatur plant, four constraints dictated the type of material-handling system needed for case sealing production:

The cases moved continuously side-by-side along a horizontal conveyor with the open side up.

The required maximum production cycle time was 15 sec.

Two different size cases were processed on the line.

Fig. 77.4. Robot is shown dispensing sealant along its programmed point-to-point path.

Fig. 77.5. Close-up of sealant application. Gun tip must be positioned at proper orientation and distance from holes to provide successful sealing.

A hydraulic tip-up station was required to raise one end of the case to allow the foamed adhesive to flow by gravity over the holes.

With these constraints in mind, a material-handling system was conceived and designed (Figures 77.2–77.5). A hydraulic walking-beam type transfer system was incorporated as the basis for handling. This was necessary because the time required to seal a case is about equal to the designated production cycle time; little time was left for a completed case to leave the sealing station and another to enter. To account for different size cases, adjustable pickup pads were designed and built.

Finally, a hydraulic tip-up station was provided to allow the foamed sealant to flow by gravity over the holes in the case. This tip-up proved to be critical for the flow concept developed in early investigations and thus was incorporated into this equipment.

77.5. SYSTEM INTEGRATION

Manufacturers today are producing many individual "cells" of high technology. Often, the marriage of two or more technologies is overlooked or simply not possible. The sealing of refrigerator cases discussed here is an example of where two high-technology components are combined with the latest in automation techniques. The GE Advanced Technology Section engineered the entire system for electrical control integration and interaction with mechanical components.

The electrical system was controlled by a GE programmable controller. The programmable controller integrates the master control panel with the P5 Robot, FoamMelt Unit, hydraulic system, material-handling system, and input-output devices. Through ladder-diagram programming the controller commands the sequence of operations both in the manual and automatic modes. For example, the controller instructs the robot when to move through its motions. The robot then instructs the controller when to dispense adhesive. The controller further instructs the FoamMelt Unit to dispense a timed quantity of adhesive. Similarly, the programmable controller sequences the operation of other system devices.

Although many mechanical design features were incorporated into the system, a select few were subtle, yet very important. First, the dispensing gun was attached at an offset to the P5 Robot arm. This enabled a five-axis robot to simulate a six-axis robot. Second, the suspension of the adhesive hoses removed restrictional forces from the robot arm. This proved necessary for consistent repeatability. Finally, the robot was placed on a roller track for off-line programming and maintenance. This allowed the walking beam to continue the transfer of parts while the robot was being maintained or reprogrammed with an auxiliary case. These features provided some additional support for avoiding problems.

The automatic sealing of refrigerator cases using foamed hot-melt adhesive and a robot has demonstrated a process of potential benefit to many industries today. Sealant applications are plentiful and possibly limited only by the adhesive chemistry and robotic characteristics. For example, robots can be used to seal body panels in the automotive industry or provide a means for in-place gasketing in the appliance, automotive, or other industries. Manufacturers again have another tool for productivity and quality improvement.

REFERENCES

1. Dueweke, N., Robotics and Adhesives—An Overview, *Adhesive Age,* April 1983 (the April 1983 issue of *Adhesive Age* is devoted to robotics and adhesives).

2. Larson, M., Update on Adhesives, *Assembly Engineering,* June 1983, pp. 9–12.

FURTHER READING

DeFrayne, G., High Performance Adhesive Bonding, *SME Paper,* 1983.

Satriana, M. J., *Hot Melt Adhesives,* Noyes Daya Corp., 1974.

Shields, J., *Adhesive Handbook,* CRC Press, 1970.

Young, J. D., The Use of Robots in the Spraying of Coating and Adhesives, *Industrial Robot,* Vol. 7, No. 1, March 1980, pp. 45–46.

ROBOTICS TERMINOLOGY

JOSEPH JABLONOWSKI

American Machinist
New York, New York

JACK W. POSEY

Purdue University
West Lafayette, Indiana

INTRODUCTION

What Is a Robot?

The notion of what constitutes a "robot" has become more complex as the machines themselves have proliferated. The early concept of an industrial robot, pioneered in the mid-1950s by George C. Devol and Joseph F. Engelberger, developers of *Unimate* (for universal assistant), called for a servo-controlled manipulator built essentially for replacing workers in loading machines. Since then, the term has taken on many connotations: at one extreme are the broad-brush notions that label anything automatic as robotic, and at the other are those who instill a personification and look for human-like qualities in order to call the machine a robot.

Clouding the discussion is the fact that each manufacturer has its own tailored notion. Furthermore, additional complications come from various international comparisons, wherein different national groups categorize automated manipulators according to different criteria. Two current robot definitions follow.

The *R*obotic *I*ndustries *A*ssociation (RIA, previously known as Robot Institute of America) adopted this definition:

A robot is a reprogrammable, multifunctional manipulator designed to move material, parts, tools, or specialized devices through variable programmed motions for the performance of a variety of tasks.

The definition proposed—but not yet formally adopted—by the *I*nternational *O*rganization for *S*tandardization (ISO) is only slightly different:

A robot is an automatic, position-controlled, reprogrammable, multifunctional manipulator having several axes, capable of handling materials, parts, tools, or specialized devices through variable programmed operations for the performance of a variety of tasks.

The ISO adds a further comment:

A robot often has the appearance of one or more arms ending in a wrist; its control unit uses a memorizing device, and sometimes it can use sensing and adaption appliances that take account of the environment and circumstances. These multipurpose machines are generally designed to carry out repetitive functions and can be adapted to other functions without permanent alteration of the equipment.

The term "robot" comes from the Czech word "robota," which means compulsory labor. The Czech playwright, Karel Capek, first used the word in his 1921 theatrical drama, R.U.R. (*R*ossum's

1271

Universal *R*obots). In the play the central character, Rossum, developed drudges that would work for him. The term gained considerable stature with the 1940s science-fiction classic, *I, Robot,* by Isaac Asimov (see Foreword of this Handbook). It has continued to be a part of popular literature right up to the android heroes of the *Star Wars* motion pictures.

Objective and Sources

This terminology section provides a comprehensive list of terms important to the understanding of the field of industrial robotics. The definitions of the terms reflect their use in the field. Some computer terminology of special importance to robotics has also been included.

All of the authors have contributed terminology to this list by gleaning terms and definitions from their respective chapters. Almost all of the terms and definitions come from the material of the Handbook. To ensure consistency some definitions from the *RIA Glossary* were used.

This terminology is not implied to be an exhaustive, final word on the vocabulary of the field of robotics, but rather is a reflection of the importance the authors have placed on various theories, practices, and equipment in the field.

Cross References

Each word listed in this section is also listed in the index. Synonyms are handled in the following manner: The terms are listed and then followed, if necessary, by the synonyms after "Also known as . . . ." Each of the synonyms is listed as a separate entry in the index, and one of the page numbers for the entry is the page number for the terminology definition. For example, the definition for cable drive would appear as follows:

> *Cable Drive: Also known as tendon drive. The transmission of mechanical power from an actuator to a remote mechanism via flexible cable and pulleys.*

In the index, the page with this definition is listed with the heading for cable drive as well as the heading for tendon drive.

Legend

(   )   Indicates a qualification of a word or term,
           for example,

                    Finishing (Robotic)

,   Indicates a specification of a word or term,
           for example,

                    Gripper, Soft
                    Gripper, Swing
                    Gripper, Translational

ROBOTIC TERMS

A

Access Route: A path of entry into an otherwise sealed or prohibited zone, such as in a nuclear reactor, used to insert and position a manipulative device.

Accommodation, Active: A control technique that integrates information from sensors with the robot's motion in order to respond to felt forces. Used to stop a robot when forces reach set levels, in guiding tasks like edge tracing and insertion, and to provide the robot with a capability to compensate for errors in the positioning and orientation of a workpiece.

Accommodation, Passive: The capability of a manipulator to correct for residual positioning errors through the sensing of reaction forces and torques as sensed by a compliant wrist. No sensors, actuators, or controls are involved.

Accuracy:

1. The quality, state, or degree of conformance to a recognized standard or specification.
2. The ability of a robot to position its end effector at a programmed location in space. Accuracy is characterized by the difference between the position the robot tool-point automatically goes to and the originally taught position, particularly at nominal load and normal operating temperature.

Compare with *repeatability*.

Actuator: A motor or transducer that converts electrical, hydraulic, or pneumatic energy into power for motion or action.

Ada: A high-level programming language used by the U.S. Department of Defense for its versatility.

Adaptive Control: A control method used to achieve near-optimum performance by continuously and automatically adjusting control parameters in response to measured process variables. Its operation is in the conventional manner of a machine tool or robot with two additional components:

1. At least one sensor which is able to measure working conditions, and
2. A computer routine which processes the sensor information and sends suitable signals to correct the operation of the conventional system.

Adaptive Robot: A robot equipped with one or more external sensors, interfaced with other machines, and communicating with other computers. This type of robot could exhibit aspects of intelligent behavior, considerably beyond unaided human capabilities, by detecting, measuring, and analyzing data about its environment, and using both passive and active means for interaction. Sensory data could include signals from the electromagnetic spectrum, acoustic signals, measurements of temperature, pressure, and humidity, measurements of physical and chemical properties of materials, detection of contaminants, and electrical signals.

Adaptive System: A robot system with adaptive control capability having:

1. At least one sensor which is able to measure the working conditions, and
2. A computer routine which processes the sensor information and sends suitable signals to correct the operation of the conventional system.

ADES (Assembly Design and Equipment Selection): A computer program developed at C. S. Draper Laboratory, Cambridge, Massachusetts, to aid a designer in configuring an assembly system. Used in system synthesis for assembly system economic analysis, ADES selects equipment, including robots, or people from a set of possible *resources* the designer provides, and assigns the assembly operations to the selected resources so the required throughput can be met at minimum annualized cost.

AGV (Automatically Guided Vehicle): Also known as robot cart. Wire- or rail-guided carts used to transport raw materials, tools, finished parts, and in-process parts over relatively great distances and through variable, programmable routes. The parts are usually loaded on pallets.

Air Jet Sensor: A sensor, located on the dispenser nozzle of a dispensing system, that uses an air jet to detect any break in bead application.

Air Motor: A device that converts pneumatic pressure and flow into continuous rotary or reciprocating motion.

AL: A research-oriented robot motion language developed by the Stanford University Artificial Intelligence Laboratory, and designed primarily to deal with robot positions and velocities. AL's features include joint interpolated motion, force sensing and application, changeable tools, and vision verification. It is based on the ALGOL language.

ALGOL (*Algorithmic Language*): A scientific-oriented computer language resembling Fortran and PL/I that expresses arithmetic and logical statements concisely.

Algorithm: A set of specific rules for the solution of a problem in a finite number of steps.

AML (*A Manufacturing Language*): An interactive, structured robot programming language developed by IBM and capable of handling many operations including interfacing and data processing that go beyond the programming of robot motions. Command categories include robot motion and sensor commands. AML supports joint interpolated motion and force sensing.

Analog Control: Control signals that are processed through analog means. Analog control can be electronic, hydraulic, or pneumatic.

Android: A robot that resembles a human being.

Anthropomorphic Robot: Also known as a jointed-arm robot. A robot with all rotary joints and motions similar to a person's arm.

APL (A Programming Language): This high-level programming language was developed in 1962 by Harvard University and IBM and is primarily for mathematical applications. APL is especially suitable for mathematical modeling.

APT (Automatically Programmed Tool): A high-level programming language, originally developed at Massachusetts Institute of Technology in 1959, under sponsorship of the Air Force, for numerical control of multiple-axis machine tools.

Arm: An interconnected set of links and powered joints comprising a manipulator which support or move a wrist, hand, or end effector.

Articulated Robot: A robot arm which contains at least two consecutive revolute joints acting around parallel axes resembling human arm motion. The work envelope is formed by partial cylinders or spheres. The two basic types of articulated robots, vertical and horizontal, are sometimes called anthropomorphic because of the resemblance to the motions of the human arm.

Artificial Intelligence: The ability of a machine system to perceive anticipated or unanticipated new conditions, decide what actions must be performed under the conditions, and plan the actions accordingly. The main areas of application are expert systems and computer vision.

Assembly (Robotic): Robot manipulation of components resulting in a finished assembled product. Presently available robots can be used to a limited extent for simple assembly operations, such as mating two parts together. However, for more complex assembly operations, robots are subject to difficulties in achieving the required positioning accuracy and sensory feedback. Sample applications include the insertion of light bulbs into instrument panels, the assembly of typewriter ribbon cartridges, the insertion of components into printed wiring boards, and the automated assembly of small electric motors.

Assembly Center (Robotic): Also known as assembly cell, or assembly station. A concentrated group of equipment, such as manipulators, vision modules, part presenters, and support tables, dedicated to complete assembly operations at one physical location. A computer and a control terminal complete the system. In operation, one or two robot arms complete the assembly of the product while, if needed, other arms simultaneously preassemble the next product and/or prepare subassemblies.

Assembly Cycle Time: The time required for a robot to accomplish a group of tasks for the assembly of a specific product.

Assembly Language: Computer language composed of brief expressions in mnemonic codes that are later translated into machine-level language for execution by the computer.

Assembly Planner: A system for automating robot programming. The assembly planner examines a computer-assisted design database to produce a task-level program. A task planner then creates a robot-level program from the task-level program.

Assembly Process: The joining and mating of an aggregate of parts.

Asynchronous Program Control: A type of robot program control structure that allows the execution of event-driven steps of a program. Typical events can be hardware errors, program function key interrupts, or sensors exceeding specified ranges.

Automated Inspection: The use of any one of several techniques to determine the presence or absence of features. The techniques include simple mechanical probes and vision systems. Some vision systems incorporate two-dimensional matrix array cameras while others utilize stereo techniques, structured light, and other methods for three-dimensional inspection.

Automatic Operation: The operation mode wherein the robot performs its programmed tasks through continuous program execution without human intervention.

Automatic Visual Inspection: The use of a robot, camera, image processor, and microprocessor for detection of flaws, compiling of statistics, and inspection of data output. A robot can be used to pick up and/or hold precisely in place, and in the required orientation, the object to be inspected. The camera transmits light level data to the image processor for digitization and processing. The image processor's task is to direct the robot as to when and where its moves should be made, to carry out the inspection by analyzing the image received from the camera, and to accumulate inspection data for downloading to the microprocessor. The microprocessor receives the inspection data from the image processor, post-processes the data, formats and prints out inspection results, and maintains statistical data on the production run.

Automation: Automatically controlled operation of an apparatus, process, or system by mechanical or electronic devices that replace human observation, effort, and decision.

AUTOPASS: An experimental task-oriented robot programming language developed by IBM, by which a manufacturing task to be accomplished is described. In a task-oriented language such as AUTOPASS, the robot must recognize various high-level, real-world terms. AUTOPASS supports parallel processing, can interface with complex sensory systems, and permits the use of multiple manipulators. It is a PL/I-based language.

Availability: The probability that at any point in time a piece of equipment will be ready to operate at a specified level of performance. It is a measurement of how often the equipment is ready when needed.

Axis: A traveled path in space, usually referred to as a linear direction of travel in any three dimensions. In Cartesian coordinate systems, labels of X, Y, and Z are commonly used to depict axis directions relative to Earth. X refers to a directional plane or line parallel to Earth. Y refers to a directional plane or line that is parallel to Earth and perpendicular to X. Z refers to a directional plane or line that is vertical to and perpendicular to the Earth's surface.

Axis, Prismatic: Also known as translational axis. An assembly between two rigid members in a mechanism enabling one to have a linear motion relative to and in contact with the other.

Axis, Rotational: Also known as rotatory axis. An assembly connecting two rigid members in a mechanism which enables one to rotate in relation to the other around a fixed axis.

B

Backlash: Free play in a power transmission system, such as a gear train, resulting in a characteristic form of hysteresis.

Bang-Bang Control: A binary control system which rapidly changes from one mode or state to the other. In motion systems, this applies to direction only. Bang-bang control is often mechanically actuated, hence the name.

Bang-Bang Robot: A robot in which motions are controlled by driving each axis or degree of freedom against a mechanical limit stop.

Base: The platform or structure to which a robot arm is attached; the end of a kinematic chain of arm links and joints opposite to that which grasps or processes external objects.

BASIC (Beginners All-purpose Symbolic Instruction Code): A high-level computer language developed in 1964 at Dartmouth College as an easy-to-learn language for students.

Batching: The operation of a robot system such that tasks of one family of operations, called a batch, are performed together. Batching is required when a system cannot easily perform a variety of tasks in random order. Different setups may be required for different families of tasks. Batching reduces the number of setups by performing one family of tasks, followed by another batch from another family, and so on.

Batch Manufacturing: The production of parts or material in discrete runs, or batches, interspersed with other production operations or runs of other parts or materials.

Baud: The unit of measure of signalling speed in data communications. Baud is equal to the number of bits of signal events per second.

Beam Arm Structure: Also known as monocoque arm structure. A robot arm constructed of beams in which the covering absorbs most of the stresses the arm is subjected to.

Bilateral Manipulator: A master-slave manipulator with symmetric force reflection. Both master and slave arms have sensors and actuators; positional error between the two arms results in equal and opposing forces to each.

Bilateral Manual Control: A type of remote teleoperator or robot operation permitting the operator to feel the forces and torques acting on the arm/hand while manually controlling the motion of the arm. The operator is kinematically and dynamically coupled with the remote arm and can command the robot with *feel* and control with a *sense of touch.*

Binary Picture: A digitized image in which the brightness of the pixels can have only two different values, such as white or black.

Bit (Binary digit): In computer terms, a single number equivalent to one or zero. Microprocessors used in robot controls can commonly process data in groups (bytes) of eight, sixteen, or thirty-two bits at a time. The memory capacity of a computer is usually measured in kilobytes (Kbytes, or simply K).

Blob: A cluster of adjacent pixels of the same nature (gray level, or color in binary image, etc.) which usually represents an object or region in the field of view.

Branching: The transfer of control-program execution to an instruction other than the next sequential instruction. In unconditional branching, the next instruction is predetermined. In conditional branching, the next instruction selected is based on some sort of test (for example, in BASIC: IF x condition is true, THEN go to instruction number n).

Brightness Resolution: The number of discrete values a pixel may assume in an image. Pixels in a binary image may have only one of two values. A non-binary image is called a gray scale image.

C

Cable Drive: Also known as tendon drive. The transmission of mechanical power from an actuator to a remote mechanism via a flexible cable and pulleys.

CAD/CAM: An acronym for Computer Aided Design and Computer Aided Manufacturing.

Cartesian Path Tracking: The travel of the hand or end effector of a manipulator in a path described by Cartesian coordinates. The manipulator joint displacements can be determined by means of an inverse Jacobian transformation.

Cell: A manufacturing unit consisting of two or more work stations or machines, and the material transport mechanisms and storage buffers that interconnect them.

Center: A manufacturing unit consisting of two or more cells and the material transport and storage buffers that interconnect them.

Center of Gravity: That point in a rigid body where the entire mass of the body could be concentrated and produce the same gravity resultant as for the body itself.

Chain Drive: The transmission of power via chain and mating-toothed sprocket wheels.

Chain Robot Arm: A robot arm designed especially for use as a mono-articulate chain manipulator. Special design considerations include a cross-sectional profile no larger than that of the chain links, a fail-safe return to a straightened position, high-powered DC or AC servo motors, and continuous gas cooling of the drives. Actual and planned applications include repair and inspection of nuclear reactors and active liquor tanks housed in concrete cells.

Circular Interpolation: A robot programming feature which allows a robot to follow a circular path through three points it is taught.

Closed-Loop Control: The use of a feedback loop to measure and compare actual system performance with desired performance. This allows the robot control to make any necessary adjustments.

Comparative Inspection: An inspection technique in which an item to be inspected is compared with a newly inspected acceptable item.

Compartmentation: Dividing workspace for robots so that no interference zones exist.

Compiler: A computer program that converts another program written in a high-level language into binary coded instructions the machine can interpret.

Complex Joint: An assembly between two closely related rigid members enabling one member to rotate in relation to the other around a mobile axis.

Compliance: A feature of a robot which allows for mechanical float in the tooling in relation to the robot tool mounting plate. This enables the correction of misalignment errors encountered when parts are mated during assembly operations or loaded into tight-fitting fixture or periphery equipment.

Compliant Assembly: The deliberate placement of a known, engineered, and relatively large compliance into tooling in order to avoid wedging and jamming during rigid part assembly.

Compliant Motion: The motion required of a robot when in contact with a surface because of uncertainty in the world model and the inherent inaccuracy of a robot.

Compliant Support: In rigid part assembly, compliant support provides both lateral and angular compliance for at least one of the mating parts.

Computed Path Control: Also known as Cartesian motion. A control scheme, such as an acceleration limit, a minimum time, and so on, wherein the path of the manipulator endpoint is computed to achieve a desired result in conformance to a given criterion.

Computer-Aided Design (CAD): The use of an interactive-terminal workstation, usually with graphics capability, to automate the design of products. CAD includes functions like drafting and parts-fitup.

Computer-Aided Manufacturing (CAM): Working from a product design likely to exist in a CAD database, CAM encompasses the computer-based technologies that physically produce the product, including part-program preparation, process planning, tool design, process analysis, and part processing by numerically controlled machines.

Computer-Integrated Manufacturing (CIM): The philosophy dictating that all functions within a manufacturing operation be database-driven and that information from within any single database be shared by other functional groups. CIM includes major functions of operations management (purchasing, inventory management, order entry), design and manufacturing engineering (CAD, NC programming, CAM), manufacturing (scheduling, fabrication, robotics, assembly, inspection, materials handling), and storage and retrieval (inventories, incoming inspection, shipping, vendor parts).

Computerized Numerical Control (CNC): A numerical control system with a dedicated mini- or microcomputer which performs the functions of data processing and control.

Computer Vision: Also known as machine vision. The use of computers or other electronic hardware to acquire, interpret, and process visual information. It involves the use of visual sensors to create an electronic or numerical analog of a visual scene, and computer processing to extract intelligence from this representation. Examples of applications have been in inspection, measurement of critical dimensions, parts sorting and presentation, visual servoing for manipulator motions, and automatic assembly.

Configuration (of robot system): The description and specification of joint configuration, including the kinematic and/or structural features, the number of degrees of freedom, the joint travel range, and the type of drive for the robot.

Configuration Design (of robotic system): The determination of equipment location, material handling equipment, buffer storage, and accessories based on a preliminary set of operations and workstations comprising a robotic system. The result is a detailed system design for a robotic system.

Connectivity Analysis: An image analysis technique used to determine whether adjacent pixels of the same color constitute a blob, for which various user-specified features can be calculated. Used to separate multiple objects in a scene from each other and from the background.

Contact Sensor: A grouping of sensors consisting of tactile, touch, and force/torque sensors. A contact sensor is used to detect contact of the robot hand with external objects.

Continuous Path Control: A type of robot control in which the robot moves according to a replay of closely spaced points programmed on a constant time base during teaching. The points are first recorded as the robot is guided along a desired path, and the position of each axis is recorded by the control unit on a constant time basis by scanning axis encoders during the robot motion. The replay algorithm attempts to duplicate that motion. Alternatively, a continuous path control can be accomplished by interpolation of a desired path curve between a few taught points.

Continuous Path System: A type of robot movement in which the tool performs the task while the axes of motion are moving. All axes of motion may move simultaneously, each at a different velocity, in order to trace a required path or trajectory.

Control: The process of making a variable or system of variables conform to what is desired.

1. A device to achieve such conformance automatically.
2. A device by which a person may communicate commands to a machine.

Control Hierarchy: A system in which higher-level control elements are used to control lower-level ones and the results of lower-level elements are utilized as inputs by higher-level elements. Embodied in the phrase *distributed intelligence,* in which processing units (usually microprocessors) are dispersed to control individual axes of a robot.

Controlled Path Motion: A type of robot control involving the coordinated control of all joint motions simultaneously to achieve a desired path between two programmed points. Each axis moves smoothly and proportionally to provide a predictable, controlled path motion.

Control Loops: Segments of control of a robot. Each control loop corresponds to a drive unit which actuates one axis of motion of the manipulator.

Control System: The system which implements the designed control scheme, including sensors, manual input and mode selection elements, interlocking and decision-making circuitry, and output elements to the operating mechanism.

Contour Following: A feature of a robot control program permitting the robot to move along a desired surface that is not defined completely. The robot gripper along with associated sensors is positioned at the beginning of the contour to be tracked. As the gripper is moved along the contour, the sensors feed data back to the control unit to ensure that constant contact is maintained with the contour.

Conveyor Tracking Robot: A robot synchronized to the movement of a conveyor. Frequent updating of the input signal of the desired position on the conveyor is required.

Coordinated Joint Motion: Also known as coordinated axis control. Control wherein the axes of the robot arrive at their respective endpoints simultaneously, giving a smooth appearance to the motion.

Coordinate Transformation: In robotics, a 4×4 matrix used to describe the positions and orientations of coordinate frames in space. It is a suitable data structure for the description of the relative position and orientation between objects. Matrix multiplication of the transformations establishes the overall relationship between objects.

Correspondence Problem: In stereo depth imaging, determination of the pixel pairs in the two images that correspond to the same object point.

Cybernetics: The theoretical study of control and information processes in complex electronic, mechanical, and biological systems, considered by some as the theory of robots.

Cycle (Program):   The unit of automatic work for a robot. Within a cycle, sub-elements called trajectories define lesser but integral elements. Each trajectory is made up of points where the robot performs an operation or passes through depending on the programming.

Cycle Time: The period of time from starting to finishing an operation. Cycle time is used to determine the nominal production rate of a robotic system.

Cylindrical Robot: Also known as cylindrical coordinate robot, or columnar robot. A robot, built around a column, that moves according to a cylindrical coordinate system in which the position of any point is defined in terms of an angular dimension, a radial dimension, and a height from a reference plane. The outline of a cylinder is formed by the work envelope. Motions usually include rotation and arm extension.

D

Dedicated Assembly System: The equipment used to assemble a single product, or several products with only minor design variations. Such a system is usually comprised of hard tooling.

Degrees of Freedom: The number of independent ways the end effector can move. It is defined by the number of rotational or translational axes through which motion can be obtained. Every variable representing a degree of freedom must be specified if the physical state of the manipulator is to be completely defined.

Deployed Robot: A manipulator that enters an otherwise sealed and prohibited zone through an authorized path of entry in order to employ an inspection device, tool, and so on, in a predetermined task under direct or programmed control.

Derivative Control: A control scheme whereby the actuator drive signal is proportional to the time derivative of the difference between the input (desired output) and the measured actual output.

Design Process (of robots):   A multistep process beginning with a description of the range of tasks to be performed. Several viable alternative configurations are then determined, followed by an evaluation of the configurations with respect to the sizing of components and dynamic system performance. Based on appropriate technical and economic criteria, a configuration is then selected. If no configuration meets the criteria, the process may be repeated in an iterative manner until a configuration is selected.

Direct Dynamics: Also known as forward dynamics, or integral dynamics. The determination of the trajectory of the manipulator from known input torques at the joints.

Direct Kinematics: Also known as forward kinematics. The determination of the position of the end effector from known joint displacements.

Direct Matching: A class of techniques in image analysis which matches new images or portions of images directly with other images. Images are usually matched with a model or template which was memorized earlier.

Disparity, Convergent: Also known as crossed disparity. In stereo image depth analysis, convergent disparity is a condition of an object point located in front of the fixation point. The optic axes of two cameras will have to cross or converge in order to fixate at the object point.

Disparity, Divergent: Also known as uncrossed disparity. In stereo image depth analysis, divergent disparity is a condition of an object point located behind the fixation point. The optic axes of two cameras will have to diverge or uncross in order to fixate at the object point.

Dispensing System (Robotic): Several components combined to apply a desired amount of adhesive, sealant, or other material in a consistent, uniform bead along accurate trajectories. The components are: a five- or six-axis robot; containers, pumps, and regulators; a programmable controller; and a dispenser or gun. Examples of robotic dispensing include: sealer to car underbody wheelhouse components, urethane bead on windshield periphery before installation, and sealant applications on appliances.

Distributed Numerical Control (DNC): The use of a computer for inputting data to several physically remote numerically controlled machine tools.

Distal: The direction away from a robot base toward the end effector of the arm.

Drift: The tendency of a system to gradually move away from a desired response.

Drive Power: The source or means of supplying energy to the robot actuators to produce motion.

Drum Sequencer: The mechanically programmed rotating device that uses limit switches to control a robot or other machine.

Duty Cycle: The fraction of time during which a device or system will be active, or at full power.

Dynamic Accuracy:

1. Degree of conformance to the true value when relevant variables are changing with time.
2. Degree to which actual motion corresponds to desired or commanded motion.

E

Edge Detection: An image analysis technique in which information about a scene is obtained without acquiring an entire image. Locations of transition from black to white and white to black are recorded, stored, and connected through a process called connectivity to separate objects in the image into blobs. The blobs can then be analyzed and recognized for their respective features.

Edge Finding: A type of gray-scale image analysis used to locate boundaries between regions of an image. There are two steps in edge finding: locating pixels in the image that are likely to be an edge, and linking candidate edge points together into a coherent edge.

Elbow: The joint which connects the upper arm and forearm on a robot.

Emergency Stop: A method using hardware-based components that overrides all other robot controls and removes drive power from the robot actuators to bring all moving parts to a stop.

EMILY: A high-level, functional robot programming language developed by IBM using a relatively simple processor. Considered to be a primitive motion level language that incorporates simple and straight-line motion but not continuous path motion. EMILY, which has gripper operation commands, is based on assembly-level language.

Encoder: A transducer used to convert linear or rotary position to digital data.

End Effector: Also known as end-of-arm tooling or, more simply, hand. The subsystem of an industrial robot system that links the mechanical portion of the robot (manipulator) to the part being handled or worked on, and gives the robot the ability to pick up and transfer parts and/or handle a multitude of differing tools to perform work on parts. It is commonly made up of four distinct elements: a method of attachment of the hand or tool to the robot tool mounting plate, power for actuation of tooling machines, mechanical linkages, and sensors integrated into the tooling. Examples include grippers, paint spraying nozzles, welding guns, and laser gauging devices.

End Effector, Turret: A number of end effectors, usually small, that are mounted on a turret for quick automatic change of end effectors during operation.

End-of-Axis Control: Controlling the delivery of tooling through a path or to a point by driving each axis of a robot in sequence. The joints arrive at their preprogrammed positions on a given axis before the next joint sequence is actuated.

Endpoint Control: Control wherein the motions of the axes are such that the endpoint moves along a prespecified type of path line (straight line, circle, etc.).

Endpoint Rigidity: The resistance of the hand, tool, or endpoint of a manipulator arm to motion under applied force.

Error Recovery: An ability in intelligent robotic systems to detect a variety of errors and, through programming, take corrective action to resolve the problem and complete the desired process.

Exoskeleton: An articulated teleoperator mechanism whose joints correspond to those of a human arm and, when attached to the arm of a human operator, will move in correspondence to his/her arm. Exoskeletal devices are sometimes instrumented and used for master-slave control of manipulators.

Expert System: A computer program, usually based on artificial intelligence techniques, that performs decision functions which are similar to those of a human expert and, on demand, can justify to the user its line of reasoning. Typical applications in the field of robotics are high-level robot programming, planning and control of assembly, and processing and recovery of errors.

Extension: A linear motion in the direction of travel of a sliding motion mechanism, or an equivalent linear motion produced by two or more angular displacements of a linkage mechanism.

External Sensor: A feedback device that is outside the inherent makeup of a robot system, or a device used to effect the actions of a robot system that are used to source a signal independent of the robot's internal design.

Eye-In-Hand System: A robot vision system in which the camera is mounted on, or near, the robot gripper. This eases the calculation of object location and orientation, and eliminates blind-spot problems encountered when using a static overhead camera.

F

Fail Safe: Failure of a device without danger to personnel or damage to product or plant facilities.

Fail Soft: Failure in performance of some component part of a system without immediate major interruption or failure of performance of the system as a whole and/or sacrifice in quality of the product.

Feature Extractor: A program used in image analysis to compute the values of attributes (features) considered by the user to be possibly useful in distinguishing between different shapes of interest.

Feature Set (Image): A group of standard feature values gleaned from a known image of a prototype and used to recognize an object based on its image. The feature set is computed for an unknown image and compared with the feature sets of prototypes. The unknown object is recognized as identical to the prototype whose feature values vary the least from its own.

Feature Values: Numerical values used to describe prototypes in a vision system. The feature values that are independent of position and orientation can be used as a basis of comparison between two images. Examples of single feature values are the maximum length and width, centroid, and area.

Feature Vector: A data representation for a part. The data can include area, number of holes, minimum and maximum diameter, perimeter, and so on. The vision system usually generates the feature vector for a part during teach-in. The feature vector is then used in image analysis.

Feedback: The signal or data sent to the control system from a controlled machine or process to denote its response to the command signal.

Finishing (Robotic): A finishing robot is usually of the continuous path type and capable of multiple program storage with random access. These two features allow for the smooth, intricate movements duplicating the human wrist which are needed to perform high-speed, efficient finishing tasks such as spray painting. Usually, lead-through programming is used to teach finishing movements. Finishing robots usually require some additional features not normally required for other types of robots: noise filters to prevent interference of electrical noise from electrostatic spraying devices, an explosion-proof remote control operator's panel in the spray area, gun and cap cleaners, and a cleaning receptacle for cleaning the gun after prolonged use or color changes.

First Generation Robot Systems: Robots with little, if any, computer power. Their only *intelligent* functions consist of *learning* a sequence of manipulative actions, choreographed by a human operator using a teach-box. The factory world around them has to be pre-arranged to accommodate their actions. Necessary constraints include precise workpiece positioning, care in specifying spatial relationships with other machines, and safety for nearby humans and equipment.

Fixed Assembly System: A group of equipment dedicated to assemble one product type only.

Fixed Coordinate System: A coordinate system fixed in time and space.

Fixed-Stop Robot: Also known as nonservo robot or open loop robot. A robot with stop-point control but no trajectory control. Each of its axes has a fixed mechanical limit at each end of its stroke and can stop only at one or the other of these limits. See *bang-bang robot.*

Flexi-Arm: A manipulator arm designed to have as much freedom as the human wrist and especially suited to spray finishing applications. The design provides full arching without regard to the pitch-and-yaw axis, minimizes the arm size, and eliminates the need for electrical wiring, hydraulic hoses, and actuators on the end of the arm. The arm can be fitted with a seventh axis to further increase gun mobility so the arm has better reach and access than a human spray finisher.

Flexibility (Gripper): The ability of a gripper to conform to parts that have irregular shapes and to adapt to parts that are inaccurately oriented with respect to the gripper.

Flexibility, Mechanical: Pliable or capable of bending. In robot mechanisms this may be due to joints, links, or transmission elements. Flexibility allows the endpoint of the robot to sag or deflect under a load, and vibrate as a result of acceleration or deceleration.

Flexibility, Operational: Multipurpose robots that are adaptable and capable of being redirected, trained, or used for new purposes. Refers to the reprogrammability or multi-task capability of robots.

Flexibility-Efficiency Trade-off: The trade-off between retaining a capability for rapid redesign or reconfiguration of the product to produce a range of different products, and being efficient enough to produce a large number of products at high levels of production and low unit cost.

Flexible Arm: A robot arm with mechanical flexibility.

Flexible Manufacturing System (FMS): An arrangement of machine tools that is capable of standing alone, interconnected by a workpiece transport system, and controlled by a central computer. The transport sub-system, possibly including one or more robots, carries work to the machines on pallets or other interface units so that accurate registration is rapid and automatic. FMS may have a variety of parts being processed at one time.

Flexion: Orientation or motion toward a position where the joint angle between two connected bodies is small.

Float: In rigid part assembly a common error-absorbing technique in which a part is allowed motion by a sliding bearing and, possibly, centering springs. Bearing friction, however, may result in failure to avoid jamming; therefore float can be an unreliable technique for error absorption.

Floor-Mounted Robot: Also known as pedestal robot. A robot with its base permanently or semi-permanently attached to the floor or a bench. Such a robot is working at one location with a maximum limited work area and in many cases servicing only one machine. Floor-mounted robots often use a pallet pick-and-place or a conveyor feeder to feed parts to and from their location.

Floor-to-Floor Time: The total time elapsed for picking up a part, loading it into a machine, carrying out operations and unloading it (back to the floor, bin, or pallet, etc.). This time measurement generally applies to batch production.

Folding Arm Manipulator: A manipulator designed to enter an enclosed area, such as the interior of a nuclear reactor, through a narrow opening. A control console outside the enclosed area is connected to the manipulator by an umbilical cord which carries command signals, telemetry, and services. Once inside the enclosed area, the manipulator unfolds and/or extends into a working position. Visual feedback from the work zone employs closed-circuit television through a remotely controlled camera on the manipulator feeding a visual display unit in the control console. High-intensity lamps on the manipulator provide illumination. The manipulator is designed to allow all relevant motions to be recovered in an emergency situation.

Force Reflection: Also known as bilateral master-slave control. A category of teleoperator control incorporating the features of simple master-slave control and also providing the operator with resistance to motions of the master unit which corresponds to the resistance experienced by the slave unit.

Force-Torque Sensors: The sensors that measure the amount of force and torque exerted by the mechanical hand along three hand-referenced orthogonal directions and applied around a point ahead and away from the sensors.

Forearm: That portion of a jointed arm which is connected to the wrist and elbow.

Fortran: A high-level computer language developed at IBM in 1954. Fortran is the acronym for *For*mula *Tran*slator, and is applicable in scientific work. Some dedicated robot-programming languages are based on Fortran.

Forward Dynamics: The computation of a trajectory resulting from an applied torque.

Forward Kinematics: The computation of the position or motion of each link as a function of the joint variables.

FREDDY: A pioneering robot system developed at Edinburgh University, Scotland, in the mid-1970s that used television cameras, a touch-sensitive manipulator, and a motor-controlled mobile viewing platform to study the acquisition of perceptual descriptions.

FUNKY: Robot software developed by IBM for advanced motion guiding that produces robot programs through the use of a function keyboard and manual guiding device. Considered a point-to-point level language that is inherently unstructured, FUNKY has support for gripper commands, tool operations, touch sensor commands, and interaction with external devices.

G

Gantry Robot: An overhead-mounted, rectilinear robot with a minimum of three degrees of freedom and normally not exceeding six. Bench-mounted assembly robots that have a gantry design are not included in this definition. A gantry robot can move along its x and y axes traveling over relatively greater distances than a pedestal-mounted robot at high traverse speeds while still providing a high degree of accuracy for positioning. Features of a gantry robot include large work envelopes, heavy payloads, mobile overhead mounting, and the capability and flexibility to operate over the work area of several pedestal-mounted robots.

Gantry Robot Coordinate System: The x, y, and z axes of a gantry robot consist of the following components:

x axis: Runway. The longitudinal axis, normally the passive side rails of the superstructure of the gantry robot.

y axis: Bridge. The transverse axis, an active member of the robot riding on the runway rails and supporting the carriage of the gantry robot.

z axis: Telescoping tubes or masts. The vertical axis, supported by the carriage.

Geometric Dexterity: The ability of the robot to achieve a wide range of orientations of the hand with the tool center point in a specified position.

Geometric Modeller: A component of an off-line programming system which generates a world model from geometric data. The world model allows objects to be referenced during programming.

Graceful Failure: Failure in performance of some component part without immediate major interruption or failure of the system as a whole.

Grasp Planning: A capability of a robot programming language to determine where to grasp objects in order to avoid collisions during grasping or moving. The grasp configuration is chosen so that objects are stable in the gripper.

Gray-Scale Picture: A digitized image in which the brightness of the pixels can have more than two values which are typically 128 or 256. A gray-scale picture requires more storage space and more sophisticated image processing than a binary image.

Gripper: The grasping hand of the robot which manipulates objects and tools to fulfill a given task.

Gripper, External: A type of mechanical gripper used to grasp the exterior surface of an object with closed fingers.

Gripper, Internal: A type of mechanical gripper used to grip the internal surface of an object with open fingers.

Gripper, Soft: A type of mechanical gripper which provides the capability of conforming to part of the periphery of an object of any shape.

Gripper, Swing Type: A type of mechanical gripper which can move its fingers in a swinging motion.

Gripper, Translational: A type of mechanical gripper which can move its own fingers, keeping them parallel.

Gripper, Universal: A gripper capable of handling and manipulating many different objects of varying weights, shapes, and materials.

Gripper Design Factors: Factors considered during the design of a gripper in order to prevent serious damage to the tool or facilitate quick repair and alignment. The factors include: parts' or tools' shape, dimension, weight, and material; adjustment for realignment in the x and y direction; easy-to-remove fingers; mechanical fusing (shear pins, etc.); locating surface at the gripper-arm interface; spring loading in the z (vertical) direction; and specification of spare gripper fingers.

Gripping Surfaces: The surfaces, such as the inside of the fingers, on the robot gripper or hand that are used for grasping.

Gross Volume of Work Envelope: The volume of the work envelope determined by shoulder and elbow joints.

Group Technology: A technique for grouping parts to gain design and operational advantages. For example, in robotics group technology is used to ensure that different parts are of the same *part family* when planning part processing for a work cell, or to design widely useable fixtures for part families. Part grouping may be based on geometric shapes, operation processes, or both.

Growing (Image): Transformation from an input binary image to an output binary image. Growing increases the number of one type of pixel for purposes of smoothing, noise elimination, and detection of blobs based on approximate size.

Guarded Motions: The motion required of a robot when approaching a surface. This motion is required because of uncertainty in the world model and the inherent inaccuracy of a robot. The goal of the guarded motion is in achieving a desired manipulator configuration on an actual surface while avoiding excessive forces.

H

Hand: A fingered gripper sometimes distinguished from a regular gripper by having more than three fingers, and more dexterous finger motions resembling the human hand.

Handchanger: A mechanism analogous to a toolchanger on a machining center or other machine tool, that permits a single robot arm to equip itself with a series of task-specific hands or grippers.

Hand Coordinate System: A robot coordinate system based on the last axis of the robot manipulator.

Hard Automation: Also known as fixed automation, or hard tooling. A non-programmable, fixed tooling which is designed and dedicated for specific operations that are not easily changeable. It may be reconfigured mechanically and is cost effective for a high production rate.

HELP (High Level Procedural Language): A robot programming language, based on PASCAL/Fortran, developed at the DEA Corporation in Turin, Italy. HELP supports structured program design for robot operation and features flexibility to multiple arms, support of continuous path motion, force feedback and touch sensor commands, interaction with external devices, and gripper operation commands.

Heuristic Problem Solving: In computer logic, the ability to plan and direct actions to steer toward higher-level goals. This is the opposite of algorithmic problem solving.

Hierarchical Control: A distributed control technique in which the controlling processes are arranged in a hierarchy and distributed physically.

High-Level Language: A programming language that generates machine code from function-oriented statements that approach English.

High-Level Robot Programming: The control of a robot with a high-level language that contains commands that perform computations of numerous elementary operations in order to simplify complicated robot operations.

Hold: A stopping of all movement of the robot during its sequence in which some power is maintained on the robot; for example, on hydraulically driven robots, power is shut off to the servo valves but is present in the main electrical and hydraulic systems.

Home Robots: Small mobile vehicles fitted with a relatively slow-moving arm and hand, and visual and force/tactile sensors, controlled by joysticks and speech, with a number of accessories specialized for carrying objects, cleaning, and other manipulative tasks.

Homogeneous Transform: A 4×4 matrix which represents the rotation and translation of vectors in the joint coordinate systems. It is used to compute the position and orientation of any coordinate system with respect to any other coordinate system.

HRL (*H*igh *R*obot *L*anguage): Robot motion software, based on LISP and Fortran, developed at the University of Tokyo. HRL is used to describe manipulator motions for mechanical assemblies and disassemblies. Its features include language extensions, world models, and orbit calculation commands.

Hybrid Teleoperator/Robot Systems: A partially controlled robot for performing service tasks. Most of the intelligence is supplied by a human operator interfaced in a user-friendly manner to control switches, joysticks, and voice input devices to control the physical motion and manipulation of the robot.

Hydraulic Motor: An actuator consisting of interconnected valves and pistons or vanes which converts high-pressure hydraulic or pneumatic fluid into mechanical shaft translation or rotation.

I

Image Analysis: The interpretation of data received from an imaging device. For the three basic analysis approaches that exist see image buffering, edge detection, and windowing.

Image Buffering: An image analysis technique in which an entire image is digitized and stored in computer memory. Computer software uses the image data to detect features, such as an object's area, centroid location, orientation, perimeter, and others.

Imaging: The analysis of an image to derive the identity, position, orientation, or condition of objects in the scene. Dimensional measurements may also be performed.

Induction Motor: An alternating-current motor wherein torque is produced by the reaction between a varying or rotating magnetic field that is generated in stationary-field magnets and the current that is induced in the coils of the rotor.

Inductosyn: Trademark for Farrand Controls resolver, in which an output signal is produced by inductive coupling between metallic patterns, versus glass-scale position resolvers that use Moire-fringe patterns.

Industrial Robot: See the introduction to this terminology.

Inspection (Robotic): Robot manipulation and sensory feedback to check the compliance of a part or assembly with specifications. In such applications, robots are used in conjunction with sensors, such as a television camera, laser, or ultrasonic detector, to check part locations, identify defects, or recognize parts for sorting. Application examples include inspection of printed circuit boards, valve cover assemblies for automotive engines, sorting of metal castings, and inspection of the dimensional accuracy of openings in automotive bodies.

Integral Control: A control scheme whereby the signal driving the actuator equals the time integral of the error signal.

Intelligent Robot: A robot that can be programmed to execute performance choices contingent on sensory inputs.

Interactive Manual-Automatic Control: A type of remote robot operation. Data from sensors integrated with the remote robot are used to adapt the real-time control actions to changes or variances in task conditions automatically through computer control algorithms.

Interface: A shared boundary which might be a mechanical or electrical connection between two devices; it might be a portion of computer storage accessed by two or more programs; or it might be a device for communication with a human operator.

Interface Box (Input/Output): This provides the robot system's interface with equipment required for an application, but not part of the robot system. For example, in spotwelding, an interface box can be used to control cooling water, shielding gas, a weld gun servo controller card, power supply, and AC input.

Interfacing (Robot With Vision): Calculating the relative orientation between the camera coordinate frame and robot coordinate system so that objects detected by the camera can be manipulated by the robot.

Interference Zone: Space contained in the work envelopes of more than one robot.

Interlock: To arrange the control of machines or devices so that their operation is interdependent in order to assure their proper coordination and synchronization.

Interpolator: A program in a system computer of a numerically controlled machine or robot that determines the calculated motion path (e.g., linear, circular, elliptic, etc.), between given end points.

Inverse Dynamics: The determination of torques to be exerted at the joints to move the manipulator along a desired trajectory, and to exert the desired force at the end effector.

Inverse Kinematics: The determination of joint displacements required to move the end effector to a desired position and orientation.

Islands of Automation: An approach used to introduce factory automation technology into manufacturing by selective application of automation. Examples include numerically controlled machine tools; robots for assembly, inspection, painting, or welding; automated assembly equipment; and flexible machining systems. Islands of automation should not be viewed as ends in themselves but as a means of forming integrated factory systems. They may range in size from an individual machine or work station to entire departments.

J

Jamming: In part assembly, jamming is a condition where forces applied to the part for part mating point in the wrong direction. As a result, the part to be inserted will not move.

Job and Skill Analysis: A method for analyzing robot work methods. Job analysis focuses on what to do while skills analysis focuses on how.

Job Shop: A discrete parts manufacturing facility characterized by a mix of products of relatively low volume production in batch lots.

Joint: A rotary or linear articulation or axis of rotational or translational (sliding) motion in a manipulator system.

Joint Coordinate: The position of a joint.

Joint Coordinate System: The set of all joint position values. In non-Cartesian robots, actually not a coordinate system.

Joint Level Control: A level of robot control which requires the programming of each individual joint of the robot structure to achieve the required overall positions.

Joint Rate Control: A category of teleoperator control which requires the operator to specify the velocity of each separate joint.

Joint Space: The space defined by a vector whose components are the angular or translational displacement of each joint of a multi-degree-of-freedom linkage relative to a reference displacement for each such joint.

K

Kinematic Chain: The combination of rotary and/or translational joints, or axes of motion.

Kinematic Model: A mathematical model used to define the position, velocity, and acceleration of each link coordinate and the end effector, excluding consideration of mass and force.

Kinematics (of Robot, Manipulator): The study of the mapping of joint coordinates to link coordinates in motion, and the inverse mapping of link coordinates to joint coordinates in motion.

L

LAMA (*Language for Mechanical Assembly*): Robot programming software that is part of a system capable of transforming mechanical assembly descriptions into robot programs. LAMA allows a programmer to define assembly strategies. Force feedback is accomplished in the system by force sensors on the wrist that are capable of resolving X, Y, and Z components of the force and torque acting on the wrist.

LAMA-S: An APL-based manipulator-level language developed at Project Spartacus, Iria, France, and designed as a research tool for robot programming. This supports parallel tasks at the execution level and incorporates force feedback by reading gripper motor currents.

Laser Scanner: A laser device used in the three available distance measurement techniques: phase shift, time of flight, and triangulation.

Laser Sensor: A range-measuring device which illuminates an object with a collimated beam. The backscattered light, approximately coaxial with the transmitted beam, is picked up by a receiver. The range, or distance, is determined by either:

1. Measuring the time delay for a pulse of light to travel from the sensor to the object and back.
2. Modulating the beam and measuring the phase difference between the modulations on the backscattered and the transmitted signals.

Lateral Resolution: The ability of a sensor, such as an ultrasonic sensor, to distinguish between details in the direction of a scan. In simple ultrasonic sensors, lateral resolution is poor but can be improved by using the concept of back-propagation.

Lead-Through Programming: Also known as lead-through teaching, programming by guiding, or manual programming. A technique for programming robot motion, usually following a continuous path motion, but sometimes referring also to teaching point-to-point motions by using a teach box. For continuous path motions, the operator grasps a handle which is secured to the arm and guides the robot through the desired task or motions while the robot controller records movement information and any activation signals for external equipment. This contrasts with off-line programming, which can be accomplished away from the manipulator.

Learning Control: A control scheme whereby experience is automatically used to change control parameters or algorithms.

Level of Automation: The degree to which a process has been made automatic. Relevant to the level of automation are questions of automatic failure recovery, the variety of situations which will be automatically handled, and the conditions under which manual intervention or action by human beings is required.

Light-Section Inspection: The use of a slit projector to project a slit of light on an object to be inspected, and an image detector to interpret the slit image of the object. Depending on the specific application, the projector and image detector may be oriented to provide a direct reflection or diffused reflection. A feature of light-section inspection is that the detection process is essentially sequential, thereby allowing relatively easier image analysis compared to other three-dimensional inspection techniques.

Limit Detecting Hardware: A device for stopping robot motion independently from control logic.

Limited-Degree-of-Freedom Robot: A robot able to position and orient its end effector in fewer than six degrees of freedom.

Limited Sequence Manipulator: A non-servo manipulator that operates between fixed mechanical stops. Such a manipulator can operate only on parts in a fixed location (position and orientation) relative to the arm.

Limit Switch: An electrical switch positioned to be switched when a motion limit occurs, thereby deactivating the actuator that causes the motion.

Linear Interpolation: A computer function automatically performed in the control that defines the continuum of points in a straight line based on only two taught coordinate positions. All calculated points are automatically inserted between the taught coordinate positions upon playback.

Link: A basic member of a robot arm that transmits motion between joints, and to the end effector.

Link Coordinate: A coordinate system attached to a link of a manipulator.

LISP: A high-level computer language implemented at the Massachusetts Institute of Technology in 1958. LISP, the acronym for *list p*rocessing, is useful in artificial intelligence applications.

Load Capacity: The maximum total weight that can be applied to the end of the robot arm without sacrifice of any of the applicable published specifications of the robot.

Load Deflection:

1. The difference in position of some point on a body between a nonloaded and an externally loaded condition.
2. The difference in position of a manipulator hand or tool, usually with the arm extended between a nonloaded condition (other than gravity) and an externally loaded condition. Either or both static and dynamic (inertial) loads may be considered.

Location Analysis: The first step of refining a robotic system design. The details of the placement of workstations, buffers, material handling equipment, and so on, are worked out. Location analysis may also be applied to location of an automated facility with respect to existing facilities and location of workpieces, accessories, and tools within a workstation. Typical factors to be considered are: transport time or costs, workstation dimensions, transporter reach, fixed costs, and capacity limits.

Logic: A means of solving complex problems through the repeated use of simple functions which define basic concepts. Three basic logic functions are *AND, OR,* and *NOT.*

Long-Term Behavior: The characteristic time required to achieve thermal stability.

M

Machine Language: The lowest-level language used directly by a machine.

Machine Loading/Unloading (Robotic):   The use of the robot's manipulative and transport capabilities in ways generally more sophisticated than simple material handling. Robots can be used to grasp a workpiece from a conveyor belt, lift it to a machine, orient it correctly, and then insert or place it on the machine. After processing, the robot unloads the workpiece and transfers it to another machine or conveyor. Some applications include loading and unloading of: hot billets into forging presses; machine tools such as lathes and machining centers; and stamping presses. Another application is the tending of plastic injection molding machines. The primary motivation for robotic machine loading/unloading is the reduction of direct labor cost. Overall productivity is also increased. The greatest efficiency is usually achieved when a single robot is used to service several machines. A single robot may also be used to perform other operations while the machines are performing their primary functions.

Machine-Mounted Robot: A robot usually dedicated to the machine it is mounted on. The robot is designed to swivel its arm to load a part for one machine operation and then set the part down and repick it at a different orientation for the next operation. Fixtures are used to hold and maintain part position during these motions.

Machine Vision Inspection System: The combination of a control system, sensor system, image processing system, and workpiece handling system to automatically inspect, transport, and handle the disposition of objects. The system also can adapt to changing working environments to a certain degree. The workpiece handling system may consist of x-y-θ tables, limited sequence arms, robots, or other positioning devices to emulate the oculomotor or manual functions of human workers. The sensor system may consist of single or multiple visual sensors for partial/overall viewing of scenes and/or coarse/fine inspection (to emulate human peripheral/foveal or far/close vision).

Machining (Robotic):   Robot manipulation of a powered spindle to perform drilling, grinding, routing, or other similar operations on the workpiece. Sensory feedback may also be used. The workpiece can be placed in a fixture by a human, another robot, or a second arm of the same robot. In some operations, the robot moves the workpiece to a stationary powered spindle and tool, such as a buffing wheel. Because of accuracy requirements, expensive tool designs, and a lack of appropriate sensory feedback capabilities, robot applications in machining are limited at present and are likely to remain so until both improved sensing capabilities and better positioning accuracy are achieved.

Machining Center: A numerically controlled metalcutting machine tool that uses tools like drills or milling cutters equipped with an automatic tool-changing device to exchange those tools for different and/or fresh ones. In some machining centers, programmable pallets for part fixturing are also available.

Magnetic Pickup Devices: A type of end-of-arm tooling that can be used to handle parts with ferrous content. Either permanent magnets or electromagnets are used, with permanent magnets requiring a stripping device to separate the part from the magnet during part release.

Main Reference: A geometric reference which must be maintained throughout a production process, for example, spot welding. The compliance with the references of the component elements of a sub-assembly guarantees the geometry of the complete assembly.

Major Axes (Motions):   These axes may be described as the independent directions an arm can move the attached wrist and end effector relative to a point of origin of the manipulator such as the base. The number of robot arm axes required to reach world coordinate points is dependent on the robot configuration.

MAL (*Multipurpose Assembly Language*):   This is a Fortran-based robot programming software developed by the Milan Polytechnic Institute of Italy, primarily for the programming of assembly tasks. Multiple robot arms are supported by MAL.

Manipulation (Robotic):   The handling of objects by moving, inserting, orienting, twisting, and so on, to be in the proper position for machining, assembling, or some other operation. In many cases it is the tool that is being manipulated rather than the object being processed.

Manipulator: A mechanism, usually consisting of a series of segments, or links, jointed or sliding relative to one another, for grasping and moving objects, usually in several degrees of freedom. It is remotely controlled by a human (manual manipulator) or a computer (programmable manipulator). A manipulator refers mainly to the mechanical aspect of a robot.

Manipulator Level Control: A level of robot control which involves specifying the robot movements in terms of world positions of the manipulator structure. Mathematical techniques are used to determine the individual joint values for these positions.

Manual Manipulator: A manipulator operated and controlled by a human operator. See *teleoperator*.

Manual Teaching: A method of robot programming in which an operator using a control box (called a teach pendant) moves each axis of the robot manually until the combination of all axial positions yields the desired position of the robot. The position coordinates are then stored in the computer memory. The process is repeated for each required position until the task program is completed. Manual teaching is usually used for point-to-point motions, as opposed to *lead-through programming*.

MAPLE: A PL/I-based robot language developed by IBM and used for computations with several extensions for directing a robot to execute complex tasks. MAPLE supports force feedback and proximity sensory commands, gripper commands, coordinate transformations, and simple and straight-line motion.

Mass Production: The large-scale production of parts or material in a virtually continuous process uninterrupted by the production of other parts or material.

Master-Slave Control: A teleoperator control that allows an operator to specify the end position of the slave (remote) end effector by specifying the position of a master unit, sometimes with a change of scale in displacement or force.

Material Handling (Robotic): The use of the robot's basic capability to transport objects. Typically, motion takes place in two or three dimensions with the robot mounted stationary on the floor, on slides or rails that enable it to move from one workstation to another, or overhead. Robots used in purely material handling operations are typically non-servo or pick-and-place robots. Some application examples include: transferring parts from one conveyor to another; transferring parts from a processing line to a conveyor; palletizing parts; and loading bins and fixtures for subsequent processing. The primary benefits in using robots for material handling are reduction of direct labor costs, removal of humans from tasks that may be hazardous, tedious, or exhausting, and less damage to parts during handling. It is common to find robots performing material handling tasks and interfacing with other material handling equipment such as containers, conveyors, guided vehicles, monorails, automated storage/retrieval systems, and carousels.

MCL (Manufacturing Control Language): A high-level programming language developed by the McDonnell Douglas Aircraft Company and designed for off-line programming of work cells that may include a robot. MCL is structured with major and minor words that are combined to form a geometric entity or a description of desired motion. It supports more than one type of robot and peripheral devices, as well as simple and complex touch and vision sensors. Robot gripper commands are included. The language is a combination of Fortran and assembly based on a modification to IBM 360 APT.

Mechanical Grip Devices: The most widely used type of end-of-arm tooling in parts handling applications. Pneumatic, hydraulic, or electrical actuators are used to generate a holding force which is transferred to the part via linkages and fingers. Some devices are able to sense and vary the grip force and grip opening.

Memory: A computer device that accepts data, holds it, and permits it to be retrieved.

Microprocessor: The basic element of a central processing unit that is constructed as a single integrated circuit.

Minor Axes (Motions): The independent attitudes relative to the mounting point of the wrist assembly on the arm by which the wrist can orient the attached end effector.

Mobile Robot: A freely moving programmable industrial robot which can be automatically moved, in addition to its usual five or six axes, in another one, two, or three axes along a fixed or programmed path by means of a conveying unit. The additional degrees of freedom distinguish between linear mobility, area mobility, and space mobility. Mobile robots can be applied to tasks requiring workpiece handling, tool handling, or both.

Mobile Robot, Area: A mobile robot whose mobility is characterized by translation in two axes. Area mobility can be achieved by an X-Y table, inductively guided vehicle, gantry with lifting device, rail-guided stacker crane, or X-Y-gantry or bridge crane.

Mobile Robot, Linear: A mobile robot whose mobility is characterized by one translational axis, usually horizontal. Linear mobility can be achieved by rail or carriage guides, a gantry, or standing or hanging lifting devices.

Mobile Robot, Space: A mobile robot whose mobility is characterized by translation in three axes. Space mobility can be achieved by an inductively guided stacker crane, X-Y-gantry, or bridge crane with lifting device.

Modal Analysis: A ground vibration test to determine experimentally the natural frequencies, modeshapes, and associated damping factors of a structure.

Modular Robots: Robots that are built of standard independent building blocks, such as joints, arms, wrists, grippers, controls, and utility lines, and are controlled by one general control system. Each modular mechanism has its own drive unit and power and communication links. Different modules can be combined by standard interface to provide a variety of kinematic structures designed to best solve a given application requirement. "Mobot," a contraction for *mo*dular ro*bot,* is a tradename of the Mobot Corporation of San Diego.

Modulator/Demodulator (MODEM): An electronic device that decodes for reception and is used to send and receive digital data over voice telecommunication lines. The digital signals are used to modulate or encode carrier signals that travel over the communication line for sending.

Monitoring: The comparison of the actual performance of a robotics system with management goals, and ascertaining the economic (and other) returns from the automated operations.

Mono-Articulate Chain Manipulator: A manipulator at the end of a special type of chain used to enter an enclosed area through a narrow opening. The chain is constructed of box section links in such a manner as to allow the chain to articulate in one direction but not another. This results in a chain which can be reeled or coiled but which forms a rigid element when extended.

Motion Economy Principles: These are principles that guide the development, troubleshooting, and improvement of work methods and work places, adapted for robot work.

Motion Hold: A means of externally interrupting continuance of motion of the robot from any further sequence or action steps without dissipating stored energy.

Motion Planning, Fine: Dealing with uncertainty in the world model by using guarded motions when approaching a surface and compliant motions when in contact with a surface.

Motion Planning, Gross: Planning robot motions that are transfer movements for which the only constraint is that the robot and whatever it is carrying should not collide with objects in the environment.

Motion-Velocity Graphs: Graphs which show regions of maximum movement and velocity combinations for common arm and wrist motions. Such charts are used to ascertain the applicability of a robot for a particular task.

Mounting Plate: The means of attaching end-of-arm tooling to an industrial robot. It is located at the end of the last axis of motion on the robot. The mounting plate is sometimes used with an adaptor plate to enable the use of a wide range of tools and tool power sources.

Multigripper System: A robot system with several grippers mounted on a turret-like wrist, or capable of automatically exchanging its gripper with alternative grippers, or having a gripper for multiple parts. A type of mechanical gripper enabling effective simultaneous execution of two or more different jobs effectively.

Multiple Stage Joint: A linear motion joint consisting of sets of nested single-stage joints.

N

Nearest Neighbor Classifier: A method of object classification by statistical comparison of computed image features from an unknown object with the features known from prototype objects. The statistical distance between object and prototype is computed in the multidimensional feature space.

Noncontact Sensor: A type of sensor, including proximity and vision sensors, that functions without any direct contact with objects.

Nulling Time: The time required to reduce to zero, or close to zero, the difference between the actual and the programmed position of every joint.

Numerical Control: A method for the control of machine tool systems. A part program containing all the information, in symbolic "numerical" form, needed for processing a workpiece is stored on a medium such as paper or magnetic tape. The information is read into a computer controller which translates the part program instructions to machine operations on the workpiece. Also see *computerized numerical control.*

O

Object Level Control: A type of robot control where the task is specified in the most general form. A comprehensive database containing a world model and knowledge of application techniques is required. "Intelligent" algorithms are required to interpret instructions and apply them to the knowledge base to automatically produce optimized, collision-free robot programs.

Off-Line: Devices not under the direct control of the present computer operating system. The processor operates independently of peripheral equipment which is off-line.

Off-Line Programming: Developing robot programs partially or completely without requiring the use of the robot itself. The program is loaded into the robot's controller for subsequent automatic action of the manipulator. An off-line programming system typically has three main components: geometric modeller, robot modeller, and programming method. The advantages of off-line programming are: reduction of robot downtime, removal of programmer from potentially hazardous environments, a single programming system for a variety of robots, integration with existing computer-aided design/computer-assisted manufacturing systems, simplification of complex tasks, and verification of robot programs.

One-Dimensional Scanning: The processing of an image one scan line at a time independent of all other scan lines. This simplifies processing but provides limited information. It is useful for inspection of products such as paper, textiles, and glass.

On-Off Control: A type of teleoperator control in which joint actuators can be turned on or off in each direction at a fixed velocity.

Open-Loop Control: Control of a manipulator in which preprogrammed signals are delivered to the actuators without measuring the actual response at the actuators. This is the opposite of *closed-loop control.*

Operating System: A software that controls the execution of computer programs and one that may provide scheduling, allocation, debugging, data management, and other functions.

Optimal Control: A control scheme whereby the system response to a commanded input, given the dynamics of the process to be controlled and the constraints on measuring, is optimal according to a specified objective function or criterion of performance.

Orientation: Also known as positioning. The consistent movement or manipulation of an object into a controlled position and attitude in space.

Orientation Finding: The use of a vision system to locate objects so they can be grasped by the manipulator or mated with other parts.

Overshoot: The degree to which a system response to a step change in reference input goes beyond the desired value.

P

PAL: A robot motion software developed at Purdue University in Indiana, by which robot tasks are represented in terms of structured Cartesian coordinates. PAL incorporates coordinate transformation, gripper, tool, and sensor-controlled operation commands.

Palletizing/Depalletizing: A term used for loading/unloading a carton, container, or pallet with parts in organized rows and possibly in multiple layers.

Pan:

1. Orientation of a view, as with a video camera, in azimuth.
2. Motion in the azimuth direction.

Parallel Program Control: A robot program control structure which allows the parallel execution of independent programs.

Part Classification: A coding scheme, typically involving four or more digits, which specifies a discrete product as belonging to a part family according to group technology.

Part Orientation: See *orientation.*

Part Program: A collection of instructions and data used in numerically controlled machine tool systems to produce a workpiece.

PASLA (Programmable Assembly Robot Language): A robot programming language developed at Nippon Electric Company, Ltd. (NEC), of Japan, that incorporates coordinate guidance and sequencer functions. It is a motion-directed language that consists of twenty basic instructions.

Path Accuracy: For a path-controlled robot, this is the level of accuracy at which programmed path curves can be followed at nominal load.

Path Measuring System: A part of the mechanical construction of each axis which provides the position coordinate for the axis. Typically, for translational axes, potentiometers or ultrasound are used for path measuring systems. But for rotational axes, resolvers, absolute optical encoders, or incremental encoders are used. A path measuring system may be located directly on a robot axis or included with the drive system.

Pattern Comparison Inspection: The comparison of an image of an object with a reference pattern. The image and pattern are aligned with one another to detect deviations of the object being inspected. The features extracted from the object image are compared with the features of the reference pattern. If both have the same features, the object is presumed to have no defects.

Pattern Recognition: A field of artificial intelligence, in which image analysis is used to determine whether a particular object or data set corresponds to one of several alternatives or to none at all. The analysis system is provided in advance with the characteristics of several prototype objects so that it can classify an unknown object by comparing it with each of the different prototypes.

Payload: The maximum weight that a robot can handle satisfactorily during its normal operations and extensions.

Performance Specifications: The specification of various important parameters or capabilities in the robot design and operation. Performance is defined in terms of:

1. The quality of behavior.
2. The degree to which a specified result is achieved.
3. A quantitative index of behavior or achievement, such as speed, power, or accuracy.

Peripheral Equipment: The equipment used in conjunction with the robot for a complete robotic system. This includes grippers, conveyors, part positioners, and part or material feeders that are needed with the robot.

Perspective Transform: The mathematical relationship between the points in the object space and the corresponding points in a camera image. The perspective transform is a function of the location of the camera in a fixed coordinate system, its orientation as determined by its pan and tilt angles, and its focal length.

Pick-and-Place: A grasp and release task, usually involving a positioning task.

Pick-and-Place Robot: Also known as bang-bang robot. A simple robot, often with only two or three degrees of freedom, that transfers items from a source to a destination via point-to-point moves.

Pinch Point: Any point where it is possible for a part of the body to be injured between the moving or stationary parts of a robot and the moving or stationary parts of associated equipment, or between the material and moving parts of the robot or associated equipment.

Pitch: Also known as bend. The angular rotation of a moving body about an axis that is perpendicular to its direction of motion and in the same plane as its top side.

Pixel: Also known as photo-element or photosite. This is a digital picture or sensor element. Pixel is short for *pic*ture-c*ell*.

PLACE: A simulator software based on Fortran and developed by McDonnell Douglas Automation Company, St. Louis, Missouri. It was designed to create, analyze, and edit cell descriptions which were collections of *CAD* information describing the robot, tooling, parts, end effectors, NC equipment, and other components of a robotic application. Output of the simulation included computer graphic wire frame figures of cells.

Playback Accuracy: The difference between a position command taught, programmed, or recorded in an automatic control system, and the position actually produced at a later time when the recorded position is used to execute motion control.

Pneumatic Pickup Device: The end-of-arm tooling such as vacuum cups, pressurized bladders, and pressurized fingers.

Point-To-Point Control: A robot motion control in which the robot can be programmed by a user to move from one position to the next. The intermediate paths between these points cannot be specified.

Point-To-Point System: The robot movement in which the robot moves to a numerically defined position and stops, performs an operation, moves to another numerically defined position and stops, and so on. The path and velocity while traveling from one point to the next generally have no significance.

Position Control: A control by a system in which the input command is the desired position of a body.

Position Finding: The use of a vision system to locate objects so they can be grasped by a manipulator or mated with other parts.

Positioning Table: Also known as positioner. A device used in robotic arc welding to hold and position pieces to be welded. The movable axes of the table are sometimes considered additional robot axes. The robot controller controls all axes in order to present the seam to be welded to the robot's torch in the location and orientation taught or modified by adaptive feedback or changes inserted by the operator, dynamically, during execution.

Presence Sensing: The use of a device designed, constructed, and installed to create a sensing field or area around a robot(s) and one that will detect an intrusion into such field or area by a person, another robot, and so on.

Pressurized Bladder: A pneumatic pickup device which is generally designed especially to conform to the shape of the part. The deflated bladder is placed in or around the part. Pressurized air causes the bladder to expand, contact the part, and conform to the surface of the part, applying equal pressure to all points of the contacted surface.

Pressurized Fingers: A pneumatic pickup device that has one straight half which contacts the part to be handled, one ribbed half, and a cavity for pressurized air between the two halves. Air pressure filling the cavity causes the ribbed half to expand and "wrap" the straight side around a part.

Priority Capability: A feature of a robot program used to control a robot at the center of a flexible machining center. If signals (service calls) are received from several machines simultaneously, the priority capability determines the order the machines will be served by the robot.

Prismatic Motion: The straight-line motion of an arm link relative to the link connected to it.

Process Control: The control of the product and associated variables of processes (such as oil refining, chemical manufacture, water supply, and electrical power generation) which are continuous in time.

Programmable: A robot capable of being instructed, under computer control, to operate in a specific manner or capable of accepting points or other commands from a remote source.

Programmable Assembly Line: A number of assembly stations interconnected with a buffered transfer system. A multilevel computer system controls the assembly line. The individual stations can be programmable, dedicated, or manual assembly.

Programmable Assembly System: A number of assembly stations or assembly centers in a stand-alone or interconnected configuration, possibly connected to one another or other equipment by a buffered transfer mechanism. Within the assembly station, equipment such as robots, dedicated equipment, programmable feeders, magazines, fixtures, and vision or other sensors are arranged as required by the operators assigned to the station. A suitable computer control system completes the configuration.

Programmable Automation: Automation for discrete parts manufacturing characterized by the features of flexibility to perform different actions for a variety of tasks, ease of programming to execute a desired task, and artificial intelligence to perceive new conditions, decide what actions must be performed under those conditions, and plan the actions accordingly. Although this is suitable primarily for increasing the productivity of batch manufacturing, programmable automation may also be applicable to mass production for the following reasons:

1. Reduction of the setup time for manufacturing short-lived products in a competitive world market.
2. Lowering the cost of production equipment by reusing components, such as robots, sensors, and computers, that are commercially available and recyclable.
3. Producing a sufficiently large sample of new products to enable testing their technical and market performance.

Programmable Controller: Also known as PC. A solid-state device which has been designated as a direct replacement for relays and "hard wired" solid-state electronics. The logic can be altered without wiring changes. Its features include: high reliability and fast response; operation in hostile industrial environments without fans, air conditioning, or electrical filtering; programmed with a simple ladder diagram language; easily reprogrammed with a portable panel if requirements change; controller is reusable if equipment is no longer required; indicator lights provided at major diagnostic points to simplify troubleshooting.

Programmable Feeder: A part feeder that can deliver a wide range of product parts with known or desired orientation and without any replacement or retooling when switching to a different product part.

Programmable Fixture: A multipurpose, computer-controlled fixture that is capable of accepting and rigidly holding parts of different shapes which it then presents, upon request, to a manipulator or vision system in a specified orientation. This might take the form of a rugged hand and three-degree-of-freedom wrist.

Programmable Manipulator: A mechanism which is capable of manipulating objects by executing a program stored in its control computer (as opposed to manual manipulator, which is controlled by a human).

Programming (Robot): The act of providing the control instructions required for a robot to perform its intended task.

Pronation: The orientation or motion toward a position with the back or protected side facing up or exposed. See *supination.*

Proportional Control: A control scheme whereby the signal which drives the actuator is monotonically related to the difference between the input command (desired output) and the measured actual output.

Proportional-Integral-Derivative (*PID*) Control: A control scheme whereby the signal which drives the actuator equals a weighted sum of the difference, time integral of the difference, and time derivative of the difference between the input (desired output) and the measured actual output. Used especially in process control.

Prosthetic Robot: A programmable manipulator or device that substitutes for lost functions of human limbs.

Protocol: A defined means of establishing criteria for receiving and transmitting data through communication channels.

Proximal: The area on a robot close to the base but away from the end effector of the arm.

Proximity Sensor: A device which senses that an object is only a short distance (e.g., a few inches or feet) away, and/or measures how far away it is. Proximity sensors typically work on the principles of triangulation of reflected light, elapsed time for reflected sound, intensity-induced eddy currents, magnetic fields, back pressure from air jets, and others.

Pseudo-Gantry Robot: A pedestal robot installed in the inverted position and mounted on slides which allows it to traverse over a work area.

PUMA: An acronym for *P*rogrammable *U*niversal *M*achine for *A*ssembly, which originated as a 1975 developmental project at General Motors Corporation and resulted in the specification for a human-arm-size, articulated, electrically driven robot that was later commercialized by Unimation, Inc.

R

Radius Statistics: These statistics are shape features computed from the set of radius vectors of a blob. They are used in image analysis for determining orientation, distinguishing shapes, and counting corners.

RAIL: A generalized robot programming language based on PASCAL, and developed by Automatix, Inc., Billerica, Massachusetts, for control of vision and arm manipulation with many constructs to support inspection.

RAPT (*Robot APT*): An APT-based task-level robot programming language developed at the University of Edinburgh in Scotland. RAPT describes objects in terms of their features (cylinders, holes, and faces, etc.) and has means for describing relationships between objects in a subassembly.

Rate Control: A control system where the input is the desired velocity of the controlled object.

RCCL (*Robot Control "C" Library*): A robot programming system based on the "C" language and developed at Purdue University of Indiana, to specify tasks by a set of primitive system calls suitable for robot control. RCCL features include manipulator task description, sensor integration, updatable world representation, manipulator independence, tracking, force control, and Cartesian path programming.

RCL: An assembly-level, command-oriented language developed at Rensselaer Polytechnic Institute of New York to program a sequence of steps needed to accomplish a robot task. RCL supports simple and straight-line motion but has no continuous path motion. The force feedback, gripper, and touch sensor commands are included in the language.

Recognition Unit: A sensory device used to distinguish between random parts (e.g., by measuring their outer diameter).

Record-Playback Robot: A manipulator for which the critical points along desired trajectories are stored in sequence by recording the actual values of joint position encoders as the robot is moved under manual control. To repeat the performance of a recorded trajectory, the stored points are played back to the robot servo system.

Rectangular Robot: Also known as rectangular coordinate robot, Cartesian coordinate robot, Cartesian robot, rectilinear coordinate robot, or rectilinear robot. A robot that moves in straight lines up and down, and in and out. The degrees of freedom of the manipulator arm are defined primarily by a Cartesian coordinate axis system consisting of three intersecting perpendicular straight lines with origin at the intersection. This robot may lack control logic for coordinated joint motion.

Relational Image Analysis Methods: Methods for image analysis that depend on local features rather than global feature values.

Relative Coordinate System: Also known as tool coordinate system. A coordinate system whose origin moves relative to world or fixed coordinates.

Remote Center Compliance Device (RCC): In rigid part assembly, a passive support device that aids part insertion operations. The RCC can be used with assembly robots as well as with traditional assembly machines and fixed single-axis workstations. The main feature of the RCC is its ability to project its compliance center outside itself, hence the source of its particular abilities to aid assembly and the reason for the word *remote* in its name. Its major function is to act as a multi-axis "float," allowing positional and angular misalignments between parts to be accommodated. The RCC allows a gripped part to rotate about its tip or to translate without rotating when pushed laterally at its tip. It enables successful mating between two parts, a tool and a part, a part and a fixture, a tool and a tool holder, and many other mating pairs.

Repeatability: The envelope of variance of the robot tool point position for repeated cycles under the same conditions. It is obtained from the deviation between the positions and orientations reached at the end of several similar cycles. Contrast with *accuracy*.

Replacement Flexibility: The ease with which a production line can continue to operate at a reduced rate in the event of failure of one of its components (robot or tooling). One way to attain replacement flexibility is by controlling other line robots to automatically compensate for the missing work of a faulty robot. Another way is to install robots at the end of a line to perform operations that were missed.

Replica Master: A control device which duplicates a manipulator arm in shape, and serves for precise manipulator teaching. Control is achieved by servoing each joint of the manipulator to the corresponding joint of the replica master.

Resolution: The smallest incremental motion which can be produced by the manipulator. Serves as one indication of the manipulator accuracy.

Resolved Motion Rate Control:

1. A control scheme whereby the desired velocity vector of the endpoint of a manipulator arm is commanded and from it the computer determines the joint angular velocities to achieve the desired result.

2. Coordination of a robot's axes so that the velocity vector of the endpoint is under direct control. Motion in the coordinate system of the endpoint along specified directions or trajectories (line, circle, etc.) is possible. This is used in manual control of manipulators and as a computational method for achieving programmed coordinate axis control in robots.

Resolver: A rotary or linear feedback device that converts mechanical motion to analog electrical signals that represent motion or position.

Reversal Error: The deviation between the positions and orientations reached at the ends of several repeated paths.

Rigidity: The property of a robot to retain its stiffness under loading and movement. Rigidity can be improved by features such as a cast-iron base, precision ball screws on all axial drives, ground and hardened spiral bevel gears in the wrist, brakes on the least stiff axes, and end effector design that permits a workpiece or tool to be held snugly.

ROBEX: An off-line programming system developed at the Machine Tool Laboratory in the Federal Republic of Germany for the control of a robotic arm. ROBEX supports collision detection and evasion.

Robot: See the definition of "Industrial robot" in the introduction to this section.

Robotic Assembly: The combination of robots, people, and other technologies for the purpose of assembly in a technologically and economically feasible manner. Robot assembly offers an alternative with some of the flexibility of people and the uniform performance of fixed automation.

Robot Calibration (for vision): The act of determining the relative orientation of the camera coordinate system with respect to the robot coordinate system.

Robot Ergonomics: The study and analysis of relevant aspects of robots in working environments. It is used to provide tools for the purpose of optimizing overall performance of the work system, including analysis of work characteristics, work methods analysis, workplace design, performance measurement, and integrated human and robot ergonomics. Robot work should be optimized to:

1. Minimize the time/unit and cost of work produced.
2. Minimize the amount of effort and energy expanded by the operator (robot and/or human).
3. Minimize the amount of waste, scrap, and rework.
4. Maximize quality of work produced.
5. Maximize safety.

Robotics: The science of designing, building, and applying robots.

Robotic Welding System: This is currently the largest application area for industrial robots, including mainly spot and arc welding. A robotic welding system comprises one or more robots, controls, suitable grippers for the work and the welding equipment, one or more compatible welding positioners with controls, and a suitable welding process with high-productivity filler. The correct safety barriers and screens along with a suitable material handling system are also required.

Robot Learning: The improvement of robot performance by experience. Robots learn by three different methods:

1. Being taught by an operator via a teach box
2. Being taught via off-line geometric database and programs
3. Learning from on-line experience.

Robot learning is found in three main areas in industry:

1. Robots learning about their operation
2. Human operators learning to accept and work with robots
3. Organizations learning to introduce, integrate, and effectively utilize robots.

Robot-Level Language: A robot system computer programming language with commands to access sensors and specify robot motions.

Robot-Man Charts: Detailed relative lists of characteristics and skills of industrial robots and humans that were developed at Purdue University. These charts are used to aid engineers in determining whether a robot can perform a job, or as a guideline and reference for robot specifications. The charts contain three main types of characteristics: physical, mental and communicative, and energy.

Robot Modeller: A component of an off-line programming system which describes the properties of jointed mechanisms. The joint structure, constraints, and velocity data are stored in the system to give a kinematic representation of the robot. Robot modellers are classified as kinematic, path control, or generalized types.

Robot Operated Body Inspection **(ROBI) System:** A system developed at General Motors to automatically check gross dimensional accuracy of automobile bodies without removing them from the assembly line. The system consists of two six-axis servo-driven robots mounted on DC electrically driven, multiposition tracks. Noncontact optical probes mounted on each robot wrist take dimensional readings from the bodies.

Robot Operated Panel Inspection **(ROPI) System:** A system developed at General Motors to check individual body panels and sub-assemblies in automobile component plants. A robot with five degrees of freedom and equipped with a noncontact optical probe that inspects for dimensional accuracy of surface and trim edges, and verifies the presence of critical holes.

Robot Selection Data: The following data are typically needed for the selection of a robot for a given task: work envelope, repeatability, accuracy, payload, speed, degrees of freedom, drive, control, and foundation type.

Robot Systems: A "Robot System" includes the robot(s) (hardware and software) consisting of the manipulator, power supply, and controller; the end effector(s); any equipment, devices, and sensors required for the robot to perform its task; and any communications interface that is operating and monitoring the robot, equipment, and sensors. (This definition excludes the rest of the operating system hardware and software.)

Robot (Robotics) Technician: An individual with the training or experience to test, program, install, troubleshoot, or maintain industrial robots. Robot technicians are generally trained in a two-year college program.

Robot Time and Motion (RTM) **Method:** A technique developed at Purdue University to estimate the cycle time for given robot work methods without having to first implement the work method and measure its performance. RTM is analogous to the Methods Time Measurement technique for human work analysis. This system is made up of three major components: RTM elements, robot performance models, and an RTM analyzer.

Robot Workstation Design: The use of geometric requirements of a workstation, including gross size of moves as well as their directions, and design scenarios for carrying out operations in conjunction with an economic analysis to select robots, computers, and tooling for an operation.

Roll: Also known as twist. The rotational displacement of a joint around the principle axis of its motion, particularly at the wrist.

ROSS: An object-oriented language suitable for use in a multi-robot distributed system. All processing in ROSS is done in terms of message passing among a collection of "actors" or "objects."

RPL (*Robot Programming Language*): A robot motion software developed at SRI International. RPL was designed for the development, testing, and debugging of control algorithms for manufacturing systems consisting of manipulators, sensors, and auxiliary equipment. Continuous path motion is supported by this language. Features of RPL include a manual teach mode and commands for object recognition, gripper operation, feedback, touch sensors, and vision. The programming is based on LISP cast in a Fortran-like syntax.

RS-232C, RS-422, RS-423, RS-449: The common standard electrical interfaces for connecting peripheral devices to computers. These are based on standards from IEEE and ANSI.

Run-Length Encoding: A method of storing compressed data of a binary image. Each horizontal line of the image is represented by a run-length representation containing only the column numbers at which a transition from 0 to 1, or vice versa, takes place. Run-length encoding can provide considerable efficiency in image processing compared to pixel-by-pixel analyses.

S

Seam Tracking: A control function to assure the quality of the weld seam. Various parameters of a robotic arc weld are monitored and the data is then used to correct both the cross-seam and stickout directions. Lateral welding torch motion is required and provided by a weaving motion for seam tracking.

Search Routine: A robot function that searches for a precise location when it is not known exactly. An axis or axes move slowly in one direction until terminated by an external signal. It is used in stacking and unstacking of parts, locating workpieces, or inserting parts in holes.

Secondary Reference: These are geometric references used when assembling a main assembly component. The compliance with the references of the components guarantees the correct geometry of the completed assembly.

Second Generation Robot Systems: A robot with a computer processor added to the robot controller. This addition makes it possible to perform, in real time, the calculations required to control the motions of each degree-of-freedom in a cooperative manner to effect smooth motions of the end effector along predetermined paths. It also becomes possible to integrate some simple sensors, such as force, torque, and proximity, into the robot system, providing some degree of adaptability to the robot's environment.

Selective Compliance Robotic Arm for Assembly (SCARA): A horizontal-revolute configuration robot designed at Japan's Yamamachi University. The tabletop-size arm sweeps across a fixtured area and is especially suited for small-parts insertion tasks.

Semi-Autonomous Control: A method for controlling a robot whereby an operator sets up the robot system for some repetitive task and a subroutine then takes over to complete the assigned task.

Sensing: The feedback from the environment of the robot which enables the robot to react to its environment. Sensory inputs may come from a variety of sensor types including proximity switches, force sensors, and machine vision systems.

Sensor: A device such as a transducer that detects a physical phenomenon and relays information to a control device.

Sensor Coordinate System: A coordinate system mounted above the working space of the robot, assigned to a sensor.

Sensor System: The components of a robot system which monitor and interpret events in the environment. Internal measurement devices, also considered sensors, are a part of closed axis-control loops and monitor joint position, velocity, acceleration, wrist force, and gripper force. External sensors update the robot model and are used for approximation, touch, geometry, vision, and safety. A data acquisition system uses data from sensors to generate patterns. A data processing system then identifies the patterns, and generates frames for the dynamic world-model processor.

Sensor-Triggered Reflex: An emergency response of a manipulator induced by a sensor upon detection of certain events such as an intruder being detected in the workspace or an impending collision with unexpected obstacles.

Sensory Control: The control of a robot based on sensor readings. Several types can be employed:

1. Sensors used in threshold tests to terminate robot activity or branch to other activity.
2. Sensors used in a continuous way to guide or direct changes in robot motions.
3. Sensors used to monitor robot progress and to check for task completion or unsafe conditions.
4. Sensors used to retrospectively update robot motion plans prior to the next cycle.

Sensory-Controlled Robot: Also known as intelligent robot. A robot whose program sequence can be modified as a function of information sensed from its environment. The robot can be servoed or non-servoed.

Sequencer: A controller which operates an application through a fixed sequence of events.

Sequential Program Control: A robot program control structure which allows the execution of a program in ordered steps. Standard structured programming language constructs for sequential control are branches, IF tests, and WHILE loops.

Servo-Controlled Robot: A robot driven by servo mechanisms, that is, motors or actuators whose driving signal is a function of the difference between command position and/or rate, and measured actual position and/or rate. Such a robot is capable of stopping at, or moving through, a practically unlimited number of points in executing a programmed trajectory.

Servo Control Level: The lowest level of the control hierarchy. At this level, drive signals for the actuators are generated to move the joints. The input signals are joint trajectories in joint coordinates.

Servomechanism: An automatic control mechanism consisting of a motor or actuator driven by a signal which is a function of the difference between commanded position and/or rate, and measured actual position and/or rate.

Servo-System: A control system for the robot, in which the control computer issues motion commands to the actuators, internal measurement devices measure the motion and signal the results back to the computer. This process continues until the arm reaches the desired position.

Servovalve: A transducer whose input is a low-energy signal, and output is a higher-energy fluid flow which is proportional to the low-energy signal.

Shake: The uncontrollable vibration of a robot's arm during, or at the end of, a movement.

SHAKEY: A pioneering man-sized mobile robot on wheels equipped with a range-finding device, camera, and other sensors. The SHAKEY project was developed in the late 1960s at Stanford Research Institute to study robot plan formation.

Shell Arm Structure: A robot arm structure designed to yield lower weight or higher strength to weight ratios. Such design is typically more expensive and generally more difficult to manufacture. Cast, extruded, or machined hollow beam-based structures are often applied, and though not as structurally efficient as pure monocoque designs, they can be more cost effective.

Shoulder: The manipulator arm linkage joint that is attached to the base.

Shrinking (Image): The transformation from an input binary image to an output binary image. This decreases the number of one type of pixel for purposes of smoothing, eliminating noise, and detecting blobs based on their approximate size.

SIGLA: An assembly-level manipulator model language developed by the Olivetti Corporation in Italy, in which the focus is on the end effector's motion through space. SIGLA supports multiple arms, gripper operation, touch sensors, force feedback, parallel processing, tool operations, interaction with external devices, relative or absolute motion, and an anticollision command.

Single-Point Control of Motion: A safeguarding method for certain maintenance operations in which it is necessary to enter the restricted work envelope of the robot. A single-point control of the robot motion is used such that it cannot be overridden at any location, in a manner which would adversely affect the safety of the persons performing the maintenance function. Before the robot system can be returned to its regular operation, it requires a deliberate separate action, by the person responsible for it, to release the single-point control.

Single Stage Joint: A linear motion joint made up of a moving surface which slides linearly along a fixed surface.

Skeletonizing: The transformation from an input binary image to an output binary image. This is similar to shrinking in that blobs are guaranteed to never shrink so far that they entirely disappear.

Skill Twist (Robot-Caused):   The change in level of skill requirements of jobs eliminated and jobs created by industrial robotics. The performance requirement of jobs eliminated are mostly at the semi-skilled or unskilled level, while those created are generally at a higher technical level.

Slew Rate: The maximum velocity at which a manipulator joint can move; a rate imposed by saturation in the servo loop controlling that joint.

Slip Sensors: Those sensors that measure the distribution and amount of contact area pressure between hand and objects positioned tangentially to the hand. They may be single-point, multiple-point (array), simple binary (yes-no), or proportional sensors.

Space Robot: A robot used for manipulation or inspection in an earth orbit or deep space environment.

Spatial Resolution: A value describing the dimensions of an image by the number of available pixels, for example, 512 × 512. A relatively larger number of pixels implies a relatively higher resolution.

Speed Control Function: A feature of the robot control program used to adjust the velocity of the robot as it moves along a given path.

Spherical Robot: Also known as a spherical coordinate robot or a polar robot. A robot operating in a spherical work envelope, or a robot arm capable of moving with rotation, arm inclination, and arm extension. A cylindrical-coordinate robot with the addition of pitch.

Splined Joint Trajectory: A technique for following a path described in Cartesian coordinates. Several points are selected from the Cartesian path and transformed into angular displacements of the robot's joints that are then controlled to move along straight-line segments in joint coordinates. These may or may not correspond precisely to the straight line specified in Cartesian coordinates, resulting in an error between the two paths that can be reduced by adding intermediate points.

Spotwelding Robot: A robot used for spotwelding and consisting of three main parts: a mechanical structure comprising the body, arm and wrist; a welding tool; and a control unit. The mechanical structure serves to position the welding tool at any point within the working volume and orient the tool in any given direction so that it can perform the appropriate task.

Spraying (Robotic):   Robot manipulation of a spray gun to apply some material, such as paint, stain, or plastic powder, to either a stationary or moving part. These coatings are applied to a wide variety of parts, including automotive body panels, appliances, and furniture. Other uses include the application of resin and chopped glass fiber to molds for producing glass-reinforced plastic parts, and spraying epoxy resin between layers of graphite broadgoods in the production of advanced composites. The benefits of robotic spraying are: higher product quality through more uniform application of material, reduced costs by eliminating human labor and reducing waste coating material, and reduced exposure of humans to toxic materials.

Springback: The deflection of a manipulator arm when the external load is removed.

SRI Vision Module: A self-contained vision subsystem developed at SRI International to sense and process visual images in response to top-level commands from a supervisory computer.

Static Deflection: Also known as static behavior or droop. Deformation of a robot structure considering only static loads and excluding inertial loads. Sometimes static deflection is meant to include the effects of gravity loads.

Stepping Motor: A bi-directional, permanent-magnet motor which turns through one angular increment for each pulse applied to it.

Stereo Analysis, Area-Based: A stereo image depth analysis in which a succession of windows in the left image are matched to corresponding windows in the right image by cross-correlation.

Stereo Analysis, Edge-Based: A stereo image depth analysis characterized by candidate points for matching which represent changes in image intensity.

Stereo Imaging: The use of two or more cameras to pinpoint the location of an object point in a three-dimensional space.

Stiffness: The amount of applied force per unit of displacement of a compliant body.

Stop (Mechanical): A mechanical constraint or limit on some motion. It can be set to stop the motion at a desired point.

Structured Light: A method of illumination for machine vision designed so that the three-dimensional pattern of light viewed causes visible patterns to appear on the surface of objects being viewed. The patterns are then easily determined.

Supervisory Control: The overall control and coordination of the robot system including both the internal sections of the system and synchronization with the external environment comprising the operator, associated manufacturing devices, and possibly a higher-level control computer. This enables a higher level of robot decision making. Lower-level controllers perform the control task continuously in real time, and usually communicate back with the supervisory control.

Supination: An orientation or motion toward a position with the front, or unprotected side, face up or exposed. See *pronation*.

Support and Containment Device: A type of end-of-arm tooling requiring no power such as lifting forks, hooks, scoops, and ladles. The robot moves to a position beneath a part to be transferred, lifts to support and contain the part or material, and performs the transfer process.

System Configuration: An iterative design process consisting of evaluation of the factors affecting the product and the production tasks, selection of a design concept based on these factors, and evaluation of the performance of the selected concept. Following the results of the evaluation, a system can be refined and reevaluated, or discarded.

System Planning: An early phase of robotic system selection during which the concern is to establish feasibility of the project and determine the initial estimates for system size, cost, return on investment, and other variables of interest to upper management.

T

T3: A robot programming language developed by Cincinnati Milacron, Inc., for the T3 industrial robot to program by teaching robot tasks. The language is based on programmed motions to the required points. T3 supports straight-line motion taught with a joystick, continuous path motion, and interaction with external devices, sensors, and tool operations.

Tactile Sensing: The detection by a robot through contact of touch, force, pattern slip, and movement. Tactile sensing allows for the determination of local shape, orientation, and feedback forces of a grasped workpiece.

Task Interpreter: A part of the robot control program controlling the step-by-step execution of the robot task and responsible for fetching, analyzing, and initiating each step that is performed. The task interpreter monitors real-time events and uses the information to direct task execution, communicate with the operator, and collect production statistics.

Task-Level Language: The computer programming languages for robot programming which require specifying task goals for the positions of objects and operations. This goal specification is intended to be completely robot-independent; no positions or paths dependent on the robot geometry or kinematics are specified by the user.

Task Planner: A part of a task-level programming language which transforms task-level specifications into robot-level specifications. The output of the task planner is a robot program for a specific robot to achieve a desired final state when executed in a specified initial state.

TEACH: A robot programming software developed by the California Institute of Technology and the Jet Propulsion Laboratory to provide commands for vision, force, and other sensors and process synchronization commands which provide for concurrency in a systematic manner. This software supports multiple manipulators and other peripheral devices simultaneously.

Teaching By Doing: A method of providing detailed points of a path, particularly a continuous path, by recording points while the robot hand is led manually by an operator through the points.

Teach Pendant: Also known as teach box. A portable, hand-held programming device connected to the robot controller containing a number of buttons, switches, or programming keys used to direct the controller in positioning the robot and interfacing with auxiliary equipment. It is used for teach pendant programming.

Teach Pendant Programming: The use of a teach pendant for teach programming which directs the controller in positioning the robot and interacting with auxiliary equipment. It is normally used for point-to-point motion and controlled path motion robots, and can be used in conjunction with off-line programming to provide accurate trajectory data.

Teach Programming: Also known as teaching. A method of entering a desired control program into the robot controller. The robot is manually moved by a teach pendant, or led through a desired sequence of motions by an operator. The movement information as well as other necessary data are recorded by the robot controller as the robot is guided through the desired path.

Teach Restrict: A facility whereby the speed of movement of a robot, which during normal operation would be considered dangerous, is restricted to a safe speed during teaching.

Teleoperation: The use of robotic devices which have mobility, manipulative and some sensing capabilities, and are remotely controlled by a human operator.

Teleoperator: A device with sensors and actuators for mobility and/or manipulation, remotely controlled by a human operator and used in remote or hazardous environments.

Telescoping Joint: A linear motion joint consisting of sets of nested single-stage joints.

Template Matching: The comparison of sample object image against a stored pattern, or template. This is used for inspection by machine vision.

Third Generation Robot Systems: Robot systems characterized by the incorporation of multiple computer processors, each operating asynchronously to perform specific functions. A typical third generation robot system includes a separate low-level processor for each degree of freedom and a master computer supervising and coordinating these processors, as well as providing higher-level functions.

Three Roll Wrist: A wrist with three interference-free axes of rotational movement (pitch, yaw, and roll) that intersect at one point to permit a continuous or reversible tool rotation which simplifies the required end effector design by its extensive reachability. It was originally designed by Cincinnati Milacron, Inc.

Thresholding: A procedure of binarization of an image by segmenting it to black and white regions (represented by ones and zeroes). The gray level of each pixel is compared to a threshold value and then set to 0 or 1 so that binary image analysis can then be performed.

Tilt: The orientation of a view, as with a video camera, in elevation.

Time Effectiveness Ratio: A measure of performance for teleoperators which, when multiplied by the task time for the unencumbered hand, yields the task time for the teleoperator.

Tool Center Point (TCP): A tool-related reference point that lies along the last wrist axis at a user-specified distance from the wrist.

Tool Changing (Robotic): An alternative to dedicated, automatic tool changers that may be attractive because of an increased flexibility and a relatively lower cost. A robot equipped with special grippers can handle a large variety of tools which can be shared quickly and economically by several machines.

Tool Coordinate System: A coordinate system assigned to the end effector.

Tool-Coordinate Programming: Programming the motion of each robot axis so that the tool held by the robot gripper is always held normal to the work surface.

Torque Control: A method of control of the motions of a robot driven by electric motors. The torque produced by the motor is treated as an input to the robot joint. The torque value is controlled by the motor current.

Torque/Force Controller: A control system capable of sensing forces and torques encountered during assembly or movement of objects, and/or generating forces on joint torques by the manipulator, which are controlled to reach desired levels.

Touch Sensors: Those sensors that measure the distribution and amount of contact area pressure between hand and objects perpendicular to the hand. Touch sensors may be single-point, multiple-point (array), simple binary (yes-no), proportional sensors, or may appear in the form of artificial skin.

Tracking: A continuous position-control response to continuously changing input requirements.

Tracking (Line): The ability of a robot to work with continuously moving production lines and conveyors. Moving-base line tracking and stationary-base line tracking are the two methods of line tracking.

Tracking, Moving-Base Line: Line tracking that requires the robot to be mounted on some form of transport system, such as rails, which will allow the robot to move parallel to the production line at a synchronized speed.

Tracking, Stationary-Base Line: Line tracking whereby the control unit of a stationary unit uses sensor data to dynamically modify the robot motion commands to permit work to be performed on objects moving on the line.

Training By Showing: The use of a vision system to view actual examples of prototype objects in order to acquire their visual characteristics. The vision system can then classify unknown objects by comparison with the stored prototype data.

Trajectory: A sub-element of a cycle that defines lesser but integral elements of the cycle. A trajectory is made up of points at which the robot performs or passes through an operation, depending on the programming.

Transducer: A device that converts one form of energy into another.

Translation: A movement such that all axes remain parallel to what they were (i.e., without rotation).

Transport (Robotic): The acquisition, movement through space, and release of an object by a robot. Simple material handling tasks requiring one- or two-dimensional movements are often performed by non-servo robots. More complicated operations, such as machine loading and unloading, palletizing, part sorting, and packaging, are typically performed by servo-controlled, point-to-point robots.

Triangulation Ranging: Those range-mapping techniques that combine direction calculations from a single camera and the previous known direction of projected light beams.

U

Ultrasonic Sensor: A range-measuring device which transmits a narrow-band pulse of sound towards an object. A receiver senses the reflected sound when it returns. The time it takes for the pulse to travel to the object and back is proportional to the range.

Unit Task Times: A method of predicting teleoperator performance in which the time required for a specific task is based on completion of unit tasks or component subtasks.

Universal Fixture: A fixture designed to handle a large variety of objects. See *programmable fixture*.

Universal Transfer Device (UTD): A term first applied to a Versatran robot (Versatran was one of the pioneering robot manufacturers that was later acquired by Prab Company of Michigan) used for press loading at the Canton Forge Plant of the Ford Motor Company, and later to other robots at Ford plants. The use of the term was discontinued in 1980.

Upper Arm: That portion of a jointed arm which is connected to the shoulder.

V

Vacuum Cups: A type of pneumatic pickup device which attaches to parts being transferred via a suction or vacuum pressure created by a venturi transducer or a vacuum pump. They are typically used on parts with a smooth surface finish, but can be used on some parts with non-smooth surface finishes by adding a ring of closed-cell foam rubber to the cup.

VAL (Versatile Assembly Language): An assembly-level robot programming language developed by Unimation, Inc., in the late 1970s and an outgrowth of work done at California's Stanford University that provides the ability to define the task a robot is to perform. VAL's features include continuous path motion and matrix transformation.

VAL-II: An enhanced and expanded assembly-level robot control and programming system based on VAL and developed by Unimation, Inc. VAL-II includes the capabilities of VAL as well as a capability for communication with external computer systems at various levels, trajectory modifications in response to real-time data, standard interfaces to external sensors, computed or sensor-based trajectories, and facilities for making complex decisions.

Velocity Control: A method of control of the motions of a robot driven by electric motors. The robot arm is treated as a load disturbance acting on the motor's shaft. The velocity of the robot arm is controlled by manipulating the motor voltage.

Vision, Three-Dimensional: The means of providing a robot with depth perception. With three-dimensional vision, robots can avoid assembly errors, search for out-of-place parts, distinguish between similar parts, and correct positioning discrepancies.

Vision, Two-Dimensional: The processing of two-dimensional images by a computer vision system to derive the identity, position, orientation, or condition of objects in the scene. It is useful in industrial applications, such as inspecting, locating, counting, measuring, and controlling industrial robots.

Vision System: A system interfaced with a robot which locates a part, identifies it, directs the gripper to a suitable grasping position, picks up the part, and brings the part to the work area. A coordinate transformation between the camera and the robot must be carried out to enable proper operation of the system.

W

Warmup: A procedure used to stabilize the temperature of a robot's hydraulic components. A warmup usually consists of a limited period of movement and motion actions and is carried out to prevent program positioning errors whenever the robot has been shut down for any length of time.

Weaving: In robotic arc welding, this is a motion pattern of the welding tool to provide a higher-quality weld. The robot controller produces a weaving pattern by controlling weave-width, left-and-right dwell, and crossing time.

Wedging: In rigid part assembly, a condition where two-point contact occurs too early in part mating, leading to the part that is supposed to be inserted appearing to be stuck in the hole. Unlike jamming, the cause is geometric rather than ill-proportioned applied forces.

Welding (Robotic): Robot manipulation of a welding tool for spot or arc welding. Robots are used in welding applications to reduce costs by eliminating human labor, improve product quality through better welds, and, particularly in arc welding, to minimize human exposure to harsh environments. Spot welding automotive bodies, normally performed by a point-to-point servo robot, is currently the largest single application for robots. In such applications, robots make from about 40 to over 75 percent of the total spot welds on a given vehicle.

Welding Positioner: The equipment, sometimes programmable, used to place parts to be welded in a precise position the robot can reach, thus maximizing the productivity of the welding process by positioning the parts for the most appropriate welding.

Windowing (Image): An image analysis technique in which only selected areas of the image are analyzed. The area, or windows, may surround a hole or some other relevant aspect of a part in the field of view. Various techniques can be used to study features of the object in the window.

Windup: A colloquial term describing the twisting of a shaft under torsional load that may cause a positioning error; the twist usually unwinds when the load is removed.

Work Envelope: Also known as the robot operating envelope. The set of points representing the maximum extent or reach of the robot tool in all directions.

Working Range:

1. The volume of space which can be reached by maximum extensions of the robot's axis.
2. The range of any variable within which the system normally operates.

Worksite Analysis: A procedure to analyze existing manual or automated worksites in order to prepare performance specifications for the robot system.

Workspace: The envelope reached by the center of the interface between the wrist and the tool using all available axis motions.

Workstation: A location providing a stable, well-defined place for the implementation of related production tasks. Major components may include a station substructure or platform, tool and material storage, and locating devices to interface with other equipment. The workstation is traditionally defined as one segment of fixed technology, a portion of a fixed transfer machine, or a person working at one worksite. Now station can mean one or more robots at a single worksite, a robot dividing its time among several worksites, a robot serving one or several fixed workheads, or any other useful combination.

World-Coordinate Programming: Programming the motion of each robot axis such that the tool center point is the center of the path with no regard to tool pose.

World-Coordinate System: A Cartesian coordinate system with the origin at the manipulator base. The X and Y axes are perpendicular and on a plane parallel to the ground, and the Z axis is perpendicular to both X and Y. It is used to reference a workpiece, jig, or fixture.

World Model: A model of the robot's environment containing: geometric and physical descriptions of objects; kinematic descriptions of linkages; descriptions of the robot system characteristics; and explicit specification of the amount of uncertainty there is in model parameters. This is useful in task-level programming.

Wrist: A set of joints, usually rotational, between the arm and the hand or end effector, which allow the hand or end effector to be oriented relative to the workpiece.

Wrist Force Sensor: A structure with some compliant sections and transducers that serve as force sensors by measuring the deflections of the compliant sections. The types of transducers used are strain-gauge, piezoelectric, magnetostrictive, and magnetic.

X

X-Y-θ Table: This is used primarily for positioning parts by translational and rotational planar motions. It can be integrated into a vision system and serve as an intelligent workpiece conveyor/presenter which loads, transports, positions, and orients parts.

Y

Yaw: The angular displacement of a moving joint about an axis which is perpendicular to the line of motion and the top side of the body.

APPENDIX

INDUSTRIAL ROBOTICS AROUND THE WORLD

The intent of this appendix is to summarize the current market for industrial robots, project some near-term directions of development, and present an overview of the characteristics of selected robots from around the world. It is divided into five sections:

A.1. Industrial Robot Market Characteristics

A.2. Specifications of Selected Industrial Robots

A.3. Addresses of Industrial Robot Organizations

A.4. Addresses of Industrial Robot Manufacturers

A.5. Robot Journals

The following is a list of tables and figures which appear:

Table A.1. World Robot Population, End of 1982

Table A.2. Robot Density by Country, End of 1981

Table A.3. Forecasted Growth of Annual Robot Production, Worldwide

Table A.4. Cumulative Installed Robots in U.S. Industry

Table A.5. Japanese Estimates of High- and Low-Grade Robot Share of Annual Production

Table A.6. Installed Operating Industrial Robots by Application and by Industry

Table A.7. Robot Specifications

Figure A.1. World Market Shares by Country—1980, 1985, and 1990

Figure A.2. Robot Arm Joint Configurations

Figure A.3. Common Robot Work Envelopes

A.1. INDUSTRIAL ROBOT MARKET CHARACTERISTICS

Worldwide robotic sales passed the 1 billion dollar mark in 1981 and could near 10 billion dollars by the year 1990, according to U.S. Department of Commerce figures.[4] Output is expected to climb from 15,000 to near 135,000 robots per year over the same period. (See Figure A.1.)

Table A.1 gives the world robot population at the end of 1982. Figures are given for each major robot-producing country in actual numbers and as a percentage of world robot population. Japan with 51% and the United States with 18% are clear leaders in this category.

Another interesting indicator of robot use is robot density, defined as the ratio of the number of robots to the active industrial work force. Table A.2 gives figures for robot density by country, with Sweden and Japan far outranking other countries in this indicator of robot use.

Projected patterns for world robotics development are summarized in Table A.3, Forecasted Growth of Annual Production, and Figure A.1, World Market Shares by Country—1980, 1985, and 1990.

Growth in robotics both for new applications and for replacement will be shared by all of the advanced industrial countries, with Japan continuing to produce and employ the majority of the world's output for most of this decade. At the same time, Western Europe—especially West Germany, the United Kingdom, Sweden, France, and Italy—will make up yet another market of roughly the same magnitude or larger than the United States. Conservative figures imply that the U.S. market will steadily increase in relative importance, comprising 20% of world demand by the end of the decade (versus its present 13% value share).

The growth dynamics of a particular country can be seen in the U.S. market, where more than

---

Data for this appendix were collected by E. L. Fisher from Purdue University.

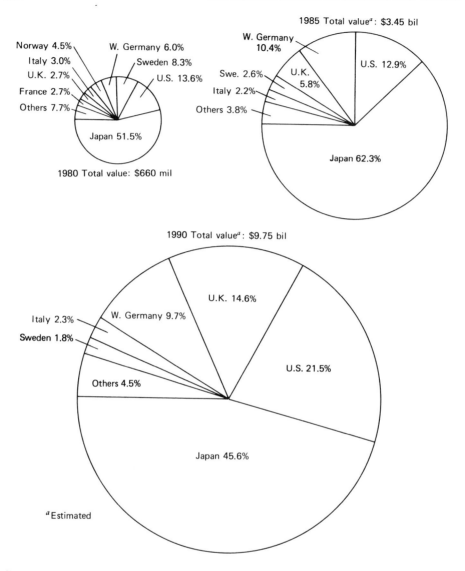

Fig. A.1. World market shares by country. *Note:* the value for the United Kingdom for 1990 may be exaggerated, but it is based on substantial current government support. (Based on Reference 4.)

50 firms are currently manufacturing industrial robots. Table A.4 summarizes both present and forecasted cumulative robots installed in the United States through 1990.

The robot production can be split into two groups: (1) simple, task-repetitive robots and (2) more sophisticated robots, having enhanced sensory as well as other intelligent capabilities. The Japanese have attempted to quantify this trend as shown in Table A.5, and their estimates should be representative of patterns observed in the United States and worldwide.

The world's industrial robots are busily working in a number of different application areas. We conclude this section with a breakdown of robot population by application area for the two largest robot producers, Japan and the United States. This summary appears in Table A.6.

A.2. SPECIFICATIONS OF ROBOTS

The purpose of this section is to review specific characteristics of some current robot models found worldwide. Industrial robots are commonly classified in a number of different ways, for example, by

TABLE A.1. WORLD ROBOT POPULATION, END OF 1982

	Number	Percent
Japan	18,000	51
United States	6,200	18
Western Europe:		
West Germany	2,800	8
Sweden	1,600	5
United Kingdom	800	2
France	700	2
Italy	500	1
Norway	400	1
Other	400	1
Total	7,200	20
U.S.S.R.	3,000	9
Eastern Europe	600	2
Total world population	35,000	100

Source. Reference 2.

TABLE A.2. ROBOT DENSITY BY COUNTRY

Country	Robots per 10,000 Employed in Manufacturing			
	1974	1978	1980	1981
Sweden	1.3	13.2	18.7	29.9
Japan	1.9	4.2	8.3	13.0
W. Germany	0.4	0.9	2.3	4.6
United States	0.8	2.1	3.1	4.0
France	0.1	0.2	1.1	1.9
United Kingdom	0.1	0.2	0.6	1.2

Source. Reference 5.

TABLE A.3. FORECASTED GROWTH OF ANNUAL ROBOT PRODUCTION, WORLDWIDE

	1980		1985		1990	
	Units	Value	Units	Value	Units	Value
World	7500–8500	$660 mil	52,000–56,000	$3.4–3.5 bil	130,000–140,000	$9.5–10.0 bil
Japan			31,000	$2150 mil	57,500	$4450 mil
United States			7,700	$ 445 mil	31,300	$2100 mil
West Germany			5,000	$ 360 mil	12,000	$ 950 mil
United Kingdom			3,000	$ 200 mil	21,500[a]	$1420 mil
Sweden			2,300	$ 90 mil	5,000	$ 180 mil
Italy			1,250	$ 75 mil	3,500	$ 225 mil
Norway			1,000	$ 50 mil	2,000	$ 103 mil
France			1,000	$ 50 mil	2,800	$ 150 mil

Source. Reference 4.
[a] May be exaggerated but is based on substantial current government support in the United Kingdom.

TABLE A.4. CUMULATIVE INSTALLED ROBOTS IN U.S. INDUSTRY

	1980	1981	1982	1983	1985[a]	1990[a]
Cumulative installed (year end)	3100	4500	6200	8200	14400	30000

Source. Revised from Reference 2.
[a] Forecast.

TABLE A.5. JAPANESE ESTIMATES OF HIGH- AND LOW-GRADE ROBOT SHARE OF ANNUAL PRODUCTION

	1980		1985		1990	
	Units (%)	Value (%)	Units (%)	Value (%)	Units (%)	Value (%)
High-grade robots (having instruction retrieval, sensory, and reader functions)	7.4	30.5	18.5	44.2	24.8	55.2
Low-grade robots (simple task-repetition capabilities only)	92.6	69.5	81.5	55.8	75.2	44.8

Source. Reference 4.

size, geometry, control type, or application. For general purpose there is no one preferred way. Regardless of their type, it can be said that industrial robots generally have three major components:

1. **Mechanical System.** The robot's body, arm, wrist, and end effector. The latter can be in the form of grippers, robotic "hands," or similar special-purpose devices (welding or painting mechanisms, for example).
2. **Servo-System and Sensors.** Precisely controls and positions the robot's mechanical components.
3. **Computer-Control System.** Contains specific programming tasks and sequences that direct and control the robot operations.

One of the key components of a robot's mechanical system is its "arm," which allows it to achieve a position in *x-y-z* space. Several joint configurations of robot arms are available, each producing a distinct working geometry. These are (see Figure A.2):

Rectangular
Cylindrical
Spherical
Articulated (or jointed) arm

Figure A.3 illustrates work envelope shapes produced by these joint configurations. Another component of the mechanical system is the end effector. This component must often be designed to fit a specific application. A number of examples which are generally available are given in Chapter 37, End-of-Arm Tooling.

Several pertinent characteristics for industrial robots are now defined, with a summary and specific values given for a number of robot models in Table A.7.

Important Note:

An attempt was made to include selected models from a number of countries in this table. It should be emphasized that Table A.7 includes a sample of robots, and in *no way* includes a complete list of current robots. The models that are included were chosen to demonstrate specific detail. Another

TABLE A.6. INSTALLED OPERATING INDUSTRIAL ROBOTS [a]

(1) By Application

	Japan (1982)	United States (1982)	United States (1990 [b])
Welding	25%	35%	23%
Material handling	20%	20%	12%
Assembly	20%	2%	12%
Machine loading	8%	15%	19%
Painting	3%	10%	6%
Foundry	2%	15%	11%
Other	22%	3%	17%
	100%	100%	100%

(2) By Industry

SIC [c]	Industry	United States
33	Primary Metals	18%
34	Fabricated Metals	15%
35	Machinery, Nonelectrical	11%
36	Electrical Machinery	3%
36	Electronics	3%
37	Automotive	43%
37	Aerospace	0.5%
	All Others	6.5%
		100.0%

[a] Based on reports by the Japanese Industrial Robot Association (JIRA) and the Robotic Industries Association (RIA): Japan—approximately 18,000 robots installed (U.S. definition); United States—approximately 6,200 robots installed.
[b] Forecast.
[c] Standard industrial classification.

caution: robot specifications change rapidly, and manufacturers must be consulted directly for accurate, up-to-date data. The following is a legend for the entries of Table A.7.

Velocity

Velocity at the end effector is given as available and is a maximal value in millimeters per second unless noted otherwise.

Actuator Type

The actuator type given is one of three categories: electric, hydraulic or pneumatic. Some models may employ more than one type, or in some cases a particular model may be available in more than one acuator type.

Repeatability

The repeatability of a machine indicates the proximity of a repeated movement, under the same precise conditions, to the same location. It is given in Table A.7 as $\pm x$, where x is in millimeters.

Payload

The payload given in Table A.7 is the maximal rated lift capacity for each robot model in kilograms.

RECTANGULAR

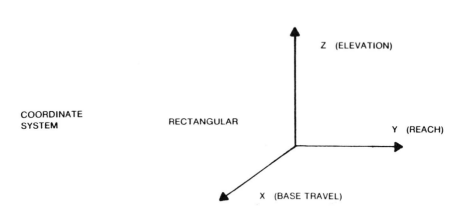

COORDINATE
SYSTEM

RECTANGULAR

Z (ELEVATION)

Y (REACH)

X (BASE TRAVEL)

TYPICAL
ROBOT
DESIGN

ELEVATION

BASE
TRAVEL

REACH

Fig. A.2. Robot arm joint configurations.

CYLINDRICAL

COORDINATE
SYSTEM

CYLINDRICAL

Z (ELEVATION)

R (REACH)

Θ (BASE ROTATION)

TYPICAL
ROBOT
DESIGN

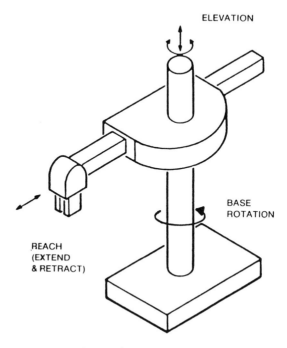

ELEVATION

BASE
ROTATION

REACH
(EXTEND
& RETRACT)

Fig. A.2. *Continued*

SPHERICAL

COORDINATE
SYSTEM

SPHERICAL

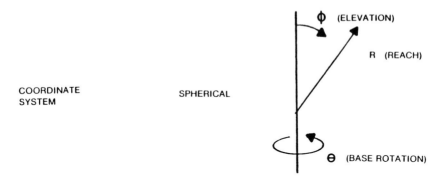

TYPICAL
ROBOT
DESIGN

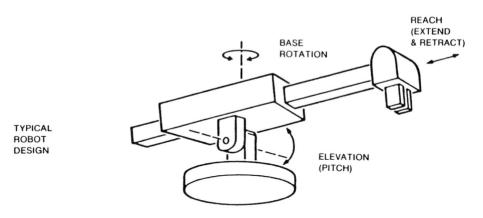

Fig. A.2. *Continued*

ARTICULATED ARM

COORDINATE
SYSTEM   JOINTED

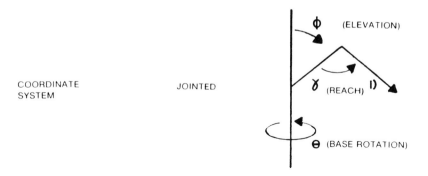

TYPICAL
ROBOT
DESIGN

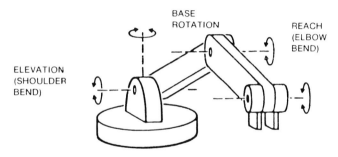

Fig. A.2.   *Continued*

Cost

The value given for cost is an average figure specified by the manufacturer. In general, the typical
price categories for industrial robots are[4]:

Materials handling	Under $20,000/unit
Welding, spraying, drilling, and similar single-purpose industrial robots	$20,000–$50,000/unit
Multifunctional	Over $50,000/unit

Where an average was not stated, but a range was available, the latter is given.

Application

Using categories as included in References 1 and 3, the following applications and corresponding
abbreviations were used in Table A.7:

Die casting	DC
Forging	FO
Investment casting	IC
Machine tool loading/unloading	ML
Part transfer	PT
Spray painting	SP

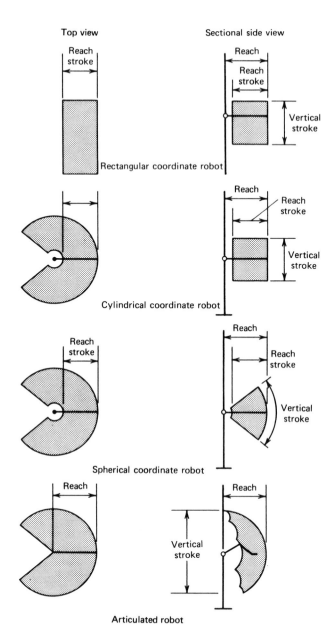

Fig. A.3. Common robot work envelopes.

TABLE A.7 ROBOT SPECIFICATION SURVEY

Company Name	Model	Velocity (mm/s)	Actuator Type	Repeatability (±mm)	Payload (kg)	Typical Cost (U.S. $)	Applications
Advanced Robotics Corp. (U.S.)	Cyro 750	250	E	0.20	15	135,000	WE
Air Technical Industries (U.S.)	Rapid Transfer Robot	3000	P	1.50	5	12,500	DC, IC, ML, PT, FI, PM, IN
Alop Kremlin Robotique, S.A. (France)	AKR 3000	2000	H	2.00	15	90,000	SP
ASEA, Inc. (Sweden, U.S.)	IRb 6	1500	E	0.20	6	60,000	DC, FO, IC, ML, WE, IN
	IRb 60	2900	E	0.40	60	90,000	DC, FO, IC, ML, WE, IN
Automatix, Inc. [a] (U.S.)	Cybervision I	1000	E, P	0.08	28	95,000	PT, SA, WE, EA, IN
	Robovision II	1000	E	0.20	30	85,000	WE
Bendix (U.S.)	AA-620	2500	E	0.05	20	70,000	DC, FO, IC, ML, PT, SA, FI, PM, WE, MA, IN
	ML-670	2500	E	0.13	70	95,000	DC, FO, IC, ML, PT, SA, FI, PM, WE, MA, IN
Cincinnati Milacron (U.S.)	T3-566	1250	H	1.25	45	80,000	FO, IC, ML, PT
	T3-586	900	H	1.25	100	85,000	FO, IC, ML, PT, FI, PM, WE, MA, IN
	T3-776	650	E	0.25	70	90,000	DC, FO, IC, ML, PE, FI, PM, WE, MA, IN
Copperweld Robotics, Inc. [b] (U.S.)	CR-5	N/S[j]	P	0.08	1	15,000	DC, FO, IC, ML, PT, SA, FI, PM, EA
	CR-50	N/S	P	0.60	11	40,000	DC, FO, IC, ML, PT, SA, FI, PM, EA
	CR-100	N/S	P	0.05	5	45,000	ML, PT, SA, PM, EA
Cybotech, Inc. [c] (U.S.)	G-80	1000	H	0.20	80	182,500	PT, WE, MA, IN, IC
	H-80	1 r/s[d]	H	0.20	80	110,000	IC, FO, PT, WE, MA, IN
	P-15	2000	H	5.00	15	141,000	PT, WE, MA, IN
Dainichi Kiko, Ltd. (Japan)	Robosky SD350-71	N/S	H, E	0.50	350	N/S	PE, TP, SA, WE, SP
	BA-4700	400	P, E	1.00	350	N/S	SP, FI
GCA Co. (U.S.)	XR 100 Series	900	E	0.50	45–1000	150,000	PT, WE, ML, PR, SA
General Electric Corp. [e] (U.S.)	A12-Allegro	670	E	0.025	10	110,000	FI, EA, IN, SA
	P5 Process	1000	E	0.20	40	50,000	WE, SA
GMF Robotics, Inc. [f] (U.S., Japan)	A-00	1200	E	0.05	10	17,000	ML, PT, SP, MA, EA, IN
	M-0	500	E, P	0.50	20	21,500	NL, PT, EA, IN
	S-0	500	E	0.50	10	40,000	DC, FO, IC, ML, PT, SA, FI, PM, WE, MA, EA, IN
Hall Automation (U.K.)	Camparm	1900	E	2.00	N/S	90,000	SP, FI
Hitachi, Ltd. (Japan)	Process Robot PW-10 II	1100	E	0.20	10	N/S	WE
	"Mr. Aros" JP	100	H	1.00	5	N/S	WE

TABLE A.7 (*Continued*)

Company Name	Model	Velocity (mm/s)	Actuator Type	Repeatability (±mm)	Payload (kg)	Typical Cost (U.S. $)	Applications
IBM Corp. (U.S.)	7535	1450	E	0.05	6	28,500	ML, PT, SA, EA, IN, ED
Ivo Lola-Ribar (Yugoslavia)	ILROT 5	1500	E	0.40	5	N/S	ML, PT, SA, IN
Kuka (W. Germany)	IR 200	1200	E	1.00	60	125,000	ML, PT, SA, WE, MA
Matsushita Electric Co. Ltd. (Japan)	Pana Robot AW 1400	170	E	0.20	10	N/S	WE
Microbo S.A. (Switzerland)	MR-02	2700	E, H	0.02	1	50,000	PT, EA, MA
Microbot Inc. (U.S.)	Alpha	500	P	0.50	2	10,500	IC, ML, ED
Mitsubishi Electric Co. Ltd. (Japan)	Melfa RW-211	1400	E	0.20	10	74,000	WE
Mitsubishi Heavy Industries Ltd. (Japan)	Robitus RC-RH	800	E, H	1.00	45	N/S	WE
Nordson Corp. (U.S.)	Robot System	120	H	0.80	15	110,000	SP, FI
Okamura Corp., Ltd. (Japan)	RC-07 AR	1500	E, H	5.00	110	N/S	PT, ML, PR
Prab Robots, Inc. (U.S.)	Model FA	900	E, H	1.25	115	72,000	DC, FO, IC, ML, PT, PM, MA
	4200	N/S	H	0.20	56	32,000	DC, FO, IC, ML, PM, PT, MA, PR
Prva Patoletka (Yugoslavia)	UMS-3	1000	H	1.20	30	N/S	ML, PT, SA
Reis Machines, Inc. (W. Germany)	RR 625	N/S	E	0.4	25	65,000	DC, FO, IC, ML, PT, PM, WE, IN
Rhino Robots, Inc. (U.S.)	Rhino Charger	760	E	0.75	45	27,000	DC, FO, IC, ML, PT, PM, IN, ED
Satt Automation AB (Sweden)	A3	700	H	0.10	50	42,500	DC, FO, ML, PT, PM, EA
Seiko Instruments, Inc. (Japan, U.S.)	M-100	1000	P	0.01	2	5,700	ML, PT, SA, EA, IN
	M-700	312	P	0.03	1	9,600	ML, PT, SA, EA, IN
Star Seiki Co., Ltd. (Japan)	T-600	1000	P	0.50	5	N/S	PM
Taiyo Ltd. (Japan)	TCN 25	1500	E	0.50	25	N/S	MA, PR
Thermwood Corp. (U.S.)	Taskhandler 25	750	H	3.20	12	39,500	DC, ML, PT
Tokico Ltd. (Japan)	705 RCS	800	E, H	2.00	5	75,000	SP
Toshiba Seiki Co., Ltd.	TOSMAN IX-15SN	700	E, H	1.00	20	30,000	WE, PT

Trallfa[g] (Norway)	TR-3500	H	180	1.00	N/S	N/S	PT, SP
United Technologies Corp. (U.S.)	Niko 150	E	1500	0.15	15	58,000	DC, FO, IC, ML, PT, SA, FI, PM, WE, MA, IN
Westinghouse/Unimation Inc.[h] (U.S.)	Unimate 2000[i]	H	0.6 r/s[d]	1.25	135	60,000	DC, FO, IC, ML, PM, PT, WE, PR
	Unimate 4000	H	0.36 r/s[d]	2.00	205	70,000	DC, FO, IC, ML, PT, WE, PM, PR
	Puma 260	E	1000	0.05	1	42,000	PT, SA, FI, EA, IN
	Puma 550	E	1000	0.10	3	47,000	ML, PT, SA, FI, WE, EA, IN, ED
Yaskawa Electric Ltd. (Japan)	Motoman L3	E	1180	0.10	3	42,000	WE, IN, SA, MS
Broye, Bulgaria	RB-232	H	914	1.25	23	N/S	MA
Vukov Institute, Preshow, Czechoslovakia	PR-16P	P	1000	0.20	16	N/S	ML
Agricultural Machinery Corp., E. Germany	IR5E	E	1 r/s[d]	0.20	5	N/S	WE, SP, FI
"	IR60E	E	1 r/s[d]	0.80	60	N/S	WE, SP, FI
(Hungary)	MTE-55	P	400	0.50	25	N/S	ML, PT
Tekoma, Poland	MP-25	P	150	0.10	2	N/S	SA, EA
(USSR)	MAH-63S	H	300	3.00	63	N/S	FO, MA, ML
"	PR-10I	H	800	0.30	10	N/S	ML
"	Universal 5.02	E	340	1.00	5	N/S	ML, PT, MA
"	Universal 15.01	H	450	3.00	15	N/S	ML, PT, MA
"	Universal 15.04	H	675	1.00	15	N/S	ML, PT, MA

[a] Cybervision and Robovision employ enhanced versions of Hitachi's process robot model.
[b] Copperweld Robotics was recently acquired by Rimroc.
[c] Cybotech is a joint venture between Ransburg Corp. of the United States and Renault Corp. of France.
[d] Radians per second.
[e] Allegro model is based on Italian Olivetti model; the P-5 model is based on Hitachi process robot model.
[f] GMF Robotics is a joint venture between General Motors of the United States and Fanuc, Ltd. of Japan.
[g] Trallfa data taken from DeVilbiss Co. TR-3500S specifications–DeVilbiss markets the Trallfa robot line in the United States.
[h] Unimation was recently acquired by Westinghouse, and their robot product lines have been merged.
[i] This is the most sold robot model to date.
[j] N/S–not specified.

Small part assembly	SA
Finishing	FI
Plastic molding	PM
Welding	WE
Machining	MA
Electronic assembly	EA
Inspection	IN
Processing	PR
Education	ED
Measurement	MS

REFERENCES

1. Japanese Industrial Robot Association, *The Specifications and Applications of Industrial Robots in Japan,* 1984.
2. Tech Tran Co., *Industrial Robots—A Summary and a Forecast,* 2nd ed., Technical Report, Naperville, Illinois, 1983.
3. Technical Data Base Corp., *1984 Robotics Industry Directory,* 1984.
4. U.S. Department of Commerce, *The Robotics Industry,* Technical Report, April 1983.
5. Robots: The Users and the Makers, *The OECD Observer,* No. 123, July 1983, Organization for Economic Cooperation and Development, Paris, France, pp. 11–17.

A.3. ADDRESSES OF INDUSTRIAL ROBOT ORGANIZATIONS

Belgium

Belgian Industrial Robot Association (BIRA)
c/o Chef de Service/DEI
Fabrique Nationale—Brugge
Ten Briele 2
B-8200 Brugge
BELGIUM

Denmark

Danish Robot Association
c/o Technological Institute
Division of Industrial Automation
Gregersensvej, 2630 Taastrup
DENMARK

Finland

Robotics Society
Oy Nokia Ab Robotics
Pursimiehenkatu 29–31
SF-00150 Helsinki 15,
FINLAND

France

Association Francaise de Robotique Industrieele
(AFRI)
89 Rue Falgueire
75015 Paris
FRANCE

Italy

Societa Italiana Robotica Industriale (SIRI)
Instituto di Electrotechnica ed Electronics

Politechnico di Milano
Piazza Leonardo da Vinci 32
20133 Milano
ITALY

Japan

Japan Industrial Robot Association (JIRA)
Kikai Shinko Kaikan Building
3–5–8 Shiba-kown
Minato-ku
Tokyo 105
JAPAN

Netherlands

Contactgroep Insutriele Robots (CIR)
Landbergstratt 3
2628 CE Delft
NETHERLANDS

Singapore

Singapore Robotic Association
5, Portsdown Road
Off Ayer Rajah Road
Singapore 0513
Republic of Singapore

Spain

Comite Espanol de Robots Industriales
c/o Instituto de Automatica Industrial (C.S.I.C.)
CRA Valencia KM.22800—la Poveda
Arganda de Rey (Madrid)
SPAIN

Sweden

Swedish Industrial Robot Association (SWIRA)
Box 5506
Storgatan 19
S-114 85 Stockholm
SWEDEN

United Kingdom

British Robot Association (BRA)
35–39 High Street
Kempston Bedford MK42 7BT
UNITED KINGDOM

United States

Robotic Industries Association (RIA)
One SME Drive

P.O. Box 1366
Dearborn, MI 48121
U.S.A.

West Germany

Fraunhofer Institute of Manufacturing,
Engineering and Automation
Nobelstrasse 12
7000 Stuttgart 80
WEST GERMANY

Yugoslavia

Robotics Dept.
Mihailo Pupin Institute
POB 906
11000 Beograd SFR,
YUGOSLAVIA

A.4 ROBOT MANUFACTURERS

Austria

IGM Industrie Zentrum
Mo-sud Strasse
2a Halle M8
A-2351 Wiener Neudorf
AUSTRIA

Voest Alpine AG
Postfach 2
A 4010 Linz
AUSTRIA
(*Tel: 0732 585 1*)

Belgium

Distribel
33 rue Godwin, Ensival
B-4850 Verviers
BELGIUM
(*Tel: 087 33 11 56*)

FN Robotics SA
Shell Building
60 rue Ravenstein
B-1000 Brussels
BELGIUM
(*Tel: 02 511 2500*)

L.V.D.
Nijrerheidslaan
2–8630 Gullegem
BELGIUM

Tecnomatix
Herentalsebaan
71–2100 Deurne
BELGIUM

Woit & Cotrico
rue du Compas
19–1070 Bruxelles
BELGIUM

Canada

Can Engineering Sales
P.O. Box 428
6800 Montrose Road
Niagara Falls,
Ontario
CANADA

Diffracto Ltd.
2775 Kew Drive
Windsor, Ontario
CANADA NVT 159

Pavesi International
Burlington
Ontario
CANADA
(*Tel: 416 631 6909*)

Wexford Robotics Ltd.
2118 Queen
Regina, Sask.
CANADA S4T4C3
(*Tel: (306) 522 7469*)

Finland

Nokia Robotics AB
Matinkatu 22
02230 Espoo 23
FINLAND
(*Tel: 358 0 8035610*)

Oy Nokia Ab Robotics
Pursimiehenkatu 29–31
SF-00150 Helsinki 15
FINLAND

Oy W Rosenlaw AB
PB 51
SF-28101 Pori 10
FINLAND

France

ACMA-Cribier (Renault)
3–5 rue Denis Papin
95250
FRANCE
(*Tel: 3/4135490*)

Afma Robots
BP 315, St. Avertin
37173 Chambray les Tours Cedex
FRANCE
(*Tel: 47/276066*)

AKR (Aoip Kremlin Robotique)
6 rue Maryse Bastie
ZI de Sait-Guenault
91031 Evry Cedex
FRANCE
(*Tel: 6/077 9615*)

Bertin & Cie
B.P. 3
78370 Plaisir
FRANCE
(*Tel: 3/056 25 00*)

C.G.M.S.
98 rue D'Ambert
B.P. 1825
45008 Orleans Cedex
FRANCE
(*Tel: 38/86 25 14*)

Citroen
133 quai Andre Citroen
75747 Paris, Cedex 15
FRANCE
(*Tel: 578 61 61*)
Commercy
55200 Commercy
FRANCE
(*Tel: (29) 010104*)

Continental Parker
51 rue Pierre
92110 Clichy
FRANCE
(*Tel: 793 3330*)

Dimenco PSP
16 rue Gay Lussac
25000 Besancon
FRANCE
(*Tel: 81/53 81 32*)

DOGA
avenue Gutenburg
B.P. 53, 78311 Maurepas Cedex
FRANCE
(*Tel: 3/062 41 41*)

Dubus
40 rue Marceau
93100 Montreuil

FRANCE
(*Tel: 859 51 84*)

Duffour et Igon
rue de l'Oasis
31300 Toulouse
FRANCE
(*Tel: 61/42 35 36*)

H. Ernault-Somua
32 avenue de l'Europe
78140 Velizy-Villacoublay
FRANCE
(*Tel: 946 96 40*)

France Euromatic
8 rue du Commerce
68400 Reidisheim
FRANCE
(*Tel: 89/64 15 33*)

Holbronn Freres
4 rue Jeane Moulin
94130 Nogent sur Marne
FRANCE
(*Tel: (1) 873 6945*)

Industria
28 avenue Clara
94420 Le Plessis Trevise
FRANCE
(*Tel: 01/576 53 78*)

Kasto-France
6 rue Pierre et Marie Curie
94430 Chennevieres sur Marne
FRANCE
(*Tel: 576 20 13*)

Lanquepin
8 rue Proudhon
93214 La Plaine St Denis
FRANCE
(*Tel: (1) 203 0381*)

Pharemme
Les Nouhauts B.P. 1
87370 Saint Sulpice Lauriere
FRANCE
(*Tel: 55/71 44 11*)

SCEMI
61 rue de Funas
38300 Bourgoin-Jallieu
FRANCE
(*Tel: 74/93 20 04*)

Sciaky
119 quai Jules Guesde
94400 Vitry/Seine
FRANCE
(*Tel: 680 85 07*)

Sietam
38–48 avenue du President Kennedy

91170 Vitry-Chatillon
FRANCE
(*Tel: 6/996 91 80*)

Sodimat
Sapignies 62121 A
62121 Achiet-le Grand, Cedex 6
FRANCE
(*Tel: 6/903 78 79*)

Sormel SA
rue Becquerel
Z.I. Chateaufarine, B.P. 1565
25009 Besancon Cedex, France
(*Tel: 8181 4245*)

Italy

AISA
via Roma 20
26020 Cumigano
ITALY

Ansaldo SpA
viale Sarca 336
20126 Milan
ITALY

Basfer SpA
via Iseo 60
20052 Monza
ITALY

Camel Robot srl
piazza Addolorata 5
20030 Palazzolo Milanese
Milan
ITALY

Comau SpA
via Rivalta 30
10095 Grugliasco (Turin)
ITALY

DEA SpA
Corso Torino 70
10024 Moncalieri (Turin)
ITALY

Duplomatic SpA
via Alba 18
21052 Busto Arsizio (Varese)
ITALY

FATA-Bisiach & Carru
Strada Statele 24, 12km
10044 Pianezza (Turin)
ITALY

Gaiotto Impianti SpA
Statele Milano-Crema km 27
26100 Vaiano Cremasco (Cremona)
ITALY

Jobs SpA
via Marcolini 11

29100 Piacenza
ITALY

Norda SpA
via Vallecamonica 14/F
25100 Brescia
ITALY

Olivetti OCN SpA
Stradele Torino
10090 S
Barnardo D'Ivrea (Turin)
ITALY

Prima Progetti SpA
Strada Carignano 48/2
10024 Moncalieri (Turin)
ITALY

Robox Elettronica Industriale
36 via Sempione-Strada Privata Mainini
28053 Castelletto Ticino Novara
ITALY
(*Tel: 0331 922086*)

Japan

Aida Engineering Ltd.
Automatic Machine Dept.
No 2–10 Oyama-cho, Sagamihara-Shi
Kanagawa-Ken 229
JAPAN
(*Tel: 0427 (72) 5231*)

Citizen Watch Co., Ltd.
840 Shimotomi takeno Tokorozawa City 359
Saitama Pref
JAPAN
(*Tel: (0429) 42-6271*)

Daido Steel Co. Ltd.
7–13 Nishi Shinbashi 1-chome
Minato-ku, Tokyo
JAPAN
(*Tel: 03/501 5261*)

Daikin Kogyo Co. Ltd.
700–1 Hitotsuya
Settsu City, Osaka 564
JAPAN
(*Tel: 06/349 7361*)

Dainichi Kiko Co. Ltd.
Kosiacho Nakakoma-gun
Yamanashi Prefecture
JAPAN
(*Tel: 05528/2 5581*)

Fanuc Ltd.
Engineering Administration Dept.
5–1 Asahiugaoka 3-chome
Hino City, Tokyo
JAPAN
(*Tel: 0425/84 1111*)

Fuji Electric Co. Ltd.
12–1 Yuraku-cho 1 chome
Chiyoda-ku, Tokyo 100
JAPAN
(*Tel: 03/211 7111*)

Fujitsu Ltd.
1015 Kamiodanaka Nakahara-ku
Kawasaki-City Kana 211
gawa-Pref.
JAPAN
(*Tel: (044) 777-1111*)

Harmo Japan
7621–10 Fujizuka
Nisha-minoco, Ina-City
Nagano Pref
JAPAN
(*Tel: 399-45*)

Hirata Industrial Machines
5–4 Myotaiji-machi
Kumamoto 860
JAPAN

Hitachi Ltd.
Industrial Components & Equipment Div.
4–1 Hammatsu-cho 2-chome
Minato-ku, Tokyo
JAPAN
(*Tel: 03/435 4272*)

Hikawa Industry Co., Ltd.
22–1 Futamuradai 1-chome
Toyoake-City 470–11
Aichi-Pref.
JAPAN
(*Tel: (05613) 4-1611*)

Ichikoh Engineering Co. Ltd.
1297–3 Ninomiya-cho
Maebashi City
JAPAN
(*Tel: 2072/68 2131*)

Ikegai Iron Works
1–21 Shiba 4-chome
Minato-ku
Tokyo
JAPAN 108
(*Tel: (03) 452-8111*)

Ishikawajima-Harima Heavy
Industries Co., Ltd.
Shin-Ohtemachi Bldg., 2–1, Ohtemachi
2-chome, Chiyoda-ku, Tokyo 100
JAPAN
(*Tel: (03) 244-6496*)

Kayaba Industry Co. Ltd.
Engineering Administration Dept.
4–1 Hammatsu-cho 2-chome
Minato-ku, Tokyo
JAPAN
(*Tel: 03/435 3511*)

Kawasaki Heavy Industries Ltd.
Hydraulic Machinery Div.
4–1 Hammatsu-cho 2-chome
Minato-ku, Tokyo
JAPAN
(*Tel: 03/435 6853*)

Kitamura
1870 Toide-cho
Takaoka City, Toyama Pref., 939–11
JAPAN
Tel: (0766/63-11000)

Kobe Steel Ltd.
Machinery & Engineering Div.
8–2 Marunouchi 1-chome
Chiyoda-ku, Tokyo
JAPAN
(*Tel: 03/218 7553*)

Komatsu Ltd.
3–6 Akasaka 2-chome
Minato-ku, Tokyo 107
JAPAN
(*Tel: 03/584 7111*)

Koyo Automation Systems Co. Ltd.
26–3 Toyocho 1-chome
Koto-ku, Tokyo 135
JAPAN
(*Tel: 03/615 2611*)

Kurogane Crane Co. Ltd.
60 Shibacho, Minami-ku
Nagoya City 457
JAPAN
(*Tel: 052/822 3211*)

Kyoritsu Engineering Co. Ltd.
Miyado Building, 6–19 Hacchobori
Naka-ku Hiroshima Pref
JAPAN
(*Tel: 0822/28 9747*)

Kyoshin Electric Co. Ltd.
20–7 Ikegami 6-chome
Ota-ku, Tokyo 146
JAPAN
(*Tel: 03/751 2131*)

Marol Company Ltd.
1.34 2-chome, Ohashi-cho
Nagata-ku, Kobe
JAPAN
(*Tel: (078) 611 2151*)

Matsushita Industrial Equipment Co. Ltd.
2–7 Matsuba-Cho
Kadoma City, Osaka 571
JAPAN
(*Tel: 06 901-1171*)

Meidensha Electric Mfg. Co. Ltd.
Mechatronics Business Div.
2-1-17 Osaki Shinagawa-ku,

Tokyo
JAPAN 104
(*Tel: 492/1111*)

Mitsubishi Electric Corporation
Engineering Dept.
2–3 Marunouchi 2-chome
Chiyoda-ku, Tokyo
JAPAN 100
(*Tel: 03/218 2111*)

Mitsubishi Heavy Industries Co. Ltd.
Precision & Machinery Division
5–1 Marunouchi 2-chome
Chiyoda-ku, Tokyo
JAPAN 100
(*Tel: 03/212 3111*)

Mizano Iron Works
Kanimachi Kanigun
Gifu Pref 509 02
JAPAN

Motoda Electronics Co. Ltd.
Kamikitazawa 4-chome
Setagaya-ku, Tokyo 156
JAPAN
(*Tel: 03/303 8491*)

Murata Machinery Ltd.
3 Minamiochiai-cho Kishoin Minami-ku,
Kyoto City 100
JAPAN
(*Tel: (075) 681-9141*)

Nachi-Fujikochi Corporation
Machine Tool Division, World Trade Centre
4–1 Hammatsucho, 2-chome, Minato-ku
Tokyo
JAPAN
(*Tel: 03/435 5111*)

Nagoya Kiko
38 Mori Koshi, Shinden-cho
Toyoake City
Aichi Pref
JAPAN
(*Tel: (0562) 92 7111*)

Nippon Electric Co. Ltd.
Production Facilities Development Div.
1–17 Shibaura 2-chome
Minato-ku, Tokyo
JAPAN
(*Tel: 03/451 5131*)

Nippon Robot Machine Co. Ltd.
73 Yoge Nihongi-cho
Anjo City, Aichi Pref. 446
JAPAN
(*Tel: 0566/74 1101*)

Nitto Seiko Co., Ltd.
Umegahata 20, Inokura-cho, Ayabe City
Kyoto 623

JAPAN
Tel: ((0773) 42-3111)

Okamura Corporation
2944 Urazato 5-chome
Yokosuka City, Kanagawa Pref. 237
JAPAN
(*Tel: 0468/65 8201*)

Oki Electric Industries Co., Ltd.
7–12, Toranomon 1-chome, Minato-ku,
Tokyo 125
JAPAN
(*Tel: (03) 501-3111*)

Osaka Denki Co., Ltd.
3-31, 4-Chome, Nishimikuni Yodogawaku
Osaka 532
JAPAN
(*Tel: (06) 394-1191*)

ORII Corp.
6 Suzuhawa, Isehara City,
Kanagawa Pref. 259–11
JAPAN
(*Tel: 0463 93-0811*)

Osaka Transformer Co. Ltd.
1–11 Tagawa 2-chome
Yodogawa-ku, Osaka 532
JAPAN
(*Tel: 06/301 1212*)

Pental Co. Ltd.
1–8 Yoshi-cho, 4-chome
Soka City
Saitoma Pref. 340
JAPAN
(*Tel: 0489 22:1111*)

Sanki Engineering Co. Ltd.
Sanshin Building
4–1 Yurakucho 1-chome Chiyoda-ku, Tokyo,
JAPAN 100
(*Tel: 03/502 6111*)

Sankyo Seiki Mfg. Co. Ltd.
17–2 Shinbashi 1-chome
Minato-ku, Tokyo
JAPAN 105
(*Tel: 03/508 1156*)

Shawa Kuatuski
3–19 Kanda-Sakumacho
Chiyada-ku, Tokyo
JAPAN

Shinko Electric Co. Ltd.
3–12 2 Nihonbashi
Chuo-ku, Tokyo 103
JAPAN
(*Tel: 03/274 1111*)

Shinmeiwa Industry Co. Ltd.
1–1 Shinmeiwa-cho

Takarazuka City, Hyogo Pref.
JAPAN 665
(*Tel: 0798/52 1234*)

Shoku Corporation
1010 Minorudai, Matsudo City
Chiba Pref.
JAPAN
(*Tel: 0473/64 1211*)

Star Seiki Co. Ltd.
252 Kawachiya Shirden
Komaki City, Aichi Pref. 485
JAPAN
(*Tel: 0568/75 5211*)

Sumitomo Heavy Industry Ltd.
Shumisho Building, 1 Mitoshirocho
Chiyoda-ku, Tokyo
JAPAN
(*Tel: 03/296 5183*)

Taiyo Ltd.
48 Kitaguchi-cho
Higashiyodogawa-ku
Osaka
JAPAN 533
(*Tel: 06/340 1111*)

The Japan Steel Works Ltd.
Hibiya Mitsui Bldg. 1–2 Yurakucho 1-chome
Chiyoda-ku, Tokyo 100
JAPAN
(*Tel: (03) 501-6111*)

Tokico Ltd.
6–10 Uchikanda 1-chome
Chiyoda-ku, Tokyo,
JAPAN 101
(*Tel: 03/292 8111*)

Tokyo Keiki Co. Ltd.
2–16 Minami Kamata
Ohta-ku, Tokyo 144
JAPAN
(*Tel: 03/732 2111*)

Tokyo Shibaura Electric Co., Ltd.
1–6, Uchi-Saiwaich 1-chome
Chiyoda-ku, Tokyo
JAPAN

Toshiba Seiki Co. Ltd.
14–33 Higashi Kashiwagaya 5-chome
Ebina City, Kanagawa Pref.
JAPAN 243
(*Tel: 0462/31 8111*)

Toyoda Machine Works Ltd.
1 Asahi-cho, 1-chome
Karia City Aichi Pref. 448
JAPAN
(*Tel: 0566/22 2211*)

Yaskawa Electric Mfg. Co., Ltd.
Chiyoda-ku

6–1 Ohtemachi 1-chome
Tokyo
JAPAN 100
(*Tel: 03/217 4111*)

Yasui Sangyo Co. Ltd.
3711 Mannohara-Shinden
Fujinomiya-Shi
Shizuoka-ken,
JAPAN
(*Tel: 05442 62124*)

Norway

Øglaend
4301 Sandes P.B. 115
NORWAY
(*Tel: 04 605000*)

Trallfa
P.O. Box 113
N 4341 Bryne
NORWAY
(*Tel: 04 48 1800*)

Spain

Campania Anomina de Electrodos
Infanta Carlota 56
Barcelona
SPAIN
(*Tel: (93) 666 50152*)

Inser SA
José Ortega y Gasset 62
Madrid
SPAIN

Oficina Tecnica Comercial (OTC)
Padilla, 382.5°
Barcelona
SPAIN
(*Tel: (93) 309 6462*)

Sweden

ASEA AB
Industrial Robot Division
S 72183 Vasteras
SWEDEN
(*Tel: 021 100000*)

ASEA AB (Previously Electrolux)
Industrial Robot Division Stockholm
Fagelviksvagen 3
S-145 53 Norsborg
SWEDEN
(*Tel: 046 753 89100*)

Atlas Copco
S-105 23
Stockholm
SWEDEN

ESAB AB
Box 8004

S-40277 Gothenburg
SWEDEN

Satt-Kaufeldt AB
P.O. Box 32 006
S-12611 Stockholm
SWEDEN
(*Tel: 08 810100*)

Spine Robotics AB
Flojelbergsgaten 14
S 43137 Molindal
SWEDEN
(*Tel: 031 870710*)

Torsteknik AB
Box 130
S-385 00 Torsas
SWEDEN

Switzerland

Automelec S.A.
Case postale 8
137, rue des Pondireres
CH-2006 Neuchâtel
SWITZERLAND

Cod Inter Techniques S.A.
16, rue Albert-Gos
CH-1206 Geneve
SWITZERLAND

Ebosa
Kapellstrasse 26
CH 2540 Grenchen
SWITZERLAND

Microbo
3 avenue Beauregard
CH 2035 Corcelles
SWITZERLAND
(*Tel: 1941 3831 5731*)

Schweissindustrie Oerlikon Buhrle AG
Birchstrasse 230
8050 Zurich
SWITZERLAND
(*Tel: 01 301 2121*)

United Kingdom

Airstead Industrial Systems Ltd.
New England House
New England Street
Brighton BN1 4GH, UK
Tel: 0273 689793

Ajax Machine Tool Ltd.
Knighton Heath Estate
847/855 Ringwood Road
Bournemouth BH1 UK

ATM Engineering Ltd.
Unit 9, Earls Way
Church Hill Ind Est

Thurmaston, Leicester LE4 8DH UK
(*Tel: 0533 693396/7*)

British Federal Welder and Machine Co. Ltd.
Castle Mill Works
Birmingham New Road
Dudley, West Midlands DY1 4DA, UK
(*Tel: 0384 54701*)

Cirrus Equipment Ltd.
Heming Road
Redditch
Worcs B98 0DN UK
(*Tel: 0527 27882*)

Fairey Automation Ltd.
Techno Trading Estate
Bramble Road, Swindon
Wilts SN2 6HB, UK
(*Tel: 0793 481161*)

Frazer Nash
Vine House
143 London Road
Kingston-upon-Thames
Surrey KT2 6NW UK

H.H. Freudenberg Automation
Cobden House, Cobden Street
Leicester, LE1 2LB UK

GEC Robot Systems Ltd.
Boughton Road
Rugby CV21 1BD UK
(*Tel: 0788 2144*)

George Kuikka Ltd.
Hill Farm Avenue
Leavesden, Watford
Herts, UK
(*Tel: 09273 70611*)

Haynes & Fordham
Unit 4, Moorfield Ind Est
Yeadon
Leeds LS19 7BM, UK
(*Tel: 0532 507090*)

INA Automation Ltd.
Forge Lane, Minworth
Sutton Coldfield
West Midlands B76 8AP, UK
(*Tel: 021 351 4047*)

Lamberton Robotics Ltd.
26 Gartsherrie Road
Coatbridge
Strathclyde ML5 2DL, UK
(*Tel: 0236 26177*)

Lansing Industrial Robots
Kingsclere Road
Basingstoke
Hants, UK
(*Tel: 0256 3131*)

Lincoln Electric Ltd.
Welwyn Garden City
Herts AL7 1QA UK
(*Tel: 070732 24581*)

Martonair Ltd.
St. Margarets Road
Twickenham TW1 1RJ UK
(*Tel: 01 892 4411*)

Marwin Production Machines
Waddons Brook
Wednesfield
Wolverhampton WV11 3AA, UK
(*Tel: 0902 65363*)

Modular Robotic Systems Ltd.
30/31 St. George's Square
Worcester WR1 1HX UK
(*Tel: 0905 612881*)

Pendar Robotics Ltd.
Unit 10, Rassau Industrial Estate
Ebbw Vale
Gwent NP3 5SD, UK
(*Tel: 0495 307070*)

Remek Micro Electronics, Ltd.
35, Barton Road, Water Eaton
Industrial Estate
Bletchley, Milton Keynes MK2 3HY
UK

Ringway Power Systems Ltd.
Churchill House, Talbot Road
Old Trafford
Manchester M16 0PD, UK
(*Tel: 061 872 6829*)

Taylor Hitec Ltd.
77 Lyons Lane
Chorley, Lancs PR6 0PB, UK
(*Tel: 02572 65825*)

Wickman Automated Assembly Ltd.
Herald Way, Brandon Road
Binley
Coventry CV3 2NY, UK
(*Tel: 0203 45080*)

WRA Ltd.
Units 2/3, Wulfrun Trading Estate
Stafford Road
Wolverhampton, UK
(*Tel: 0902 711201*)

United States

Accumatic Machinery Corp.
3537 Hill Avenue
Toledo
Ohio 43607, USA
(*Tel: 419 535 7997*)

Acrobe Positioning Systems Inc.
3219 Dolittle Drive
Northbrook
Illinois 60062 USA

Action Machinery Co.
PO Box 3068
Portland
Oregon 97208 USA

Admiral Equipment Co. Ltd.
305 West North Street
Akron
Ohio 44303, USA
(*Tel: 216 253 1353*)

Advanced Robotics Corp.
Route 79
Newark Industrial Park
Hebron, Ohio 43025, USA
(*Tel: 614 929 1065*)

Ameco Corporation
PO Box 385
Menomonee
Wisconsin 53051 USA

American Robot Corp.
201 Miller Street
Winston-Salem
North Carolina 27103, USA
(*Tel: 919 748 8761*)

Anorad
110 Oser Avenue
Hauppauge
New York 11788, USA
(*Tel: 516 231 1990*)

Armax Robotics Inc.
38700 Grand River Avenue
Farmington Hills
Michigan 48018, USA
(*Tel: 313 478 9330*)

ASEA Inc
1176 E. Big Beaver
Troy, Michigan 48084, USA
(*Tel: 313 528 3630*)

Automation Corporation
23996 Freeway Park Drive
Farmington Hills
Michigan 48024, USA

Automatix Inc.
1000 Tech Park Drive
Billerica
Massachusetts 01821, USA
(*Tel: 617 667 7900*)

Barrington Automation Ltd.
1002 South Road
Fox River Grove
Illinois 60021 USA

Binks Corp.
9201 W. Belmont Ave
Franklin Park, Illinois 60131, USA
(*Tel: 312 671 3000*)

Ceeris International Inc
1055 Thomas Jefferson St. NW
Ste 414
Washington DC 20007, USA
(*Tel: 202 342 5400*)

Cincinnati Milacron
Industrial Robot Division
215 S. West Street, Lebanon
Ohio 45036, USA
(*Tel: 513 932 4400*)

Comet Welding Systems
900 Nicholas Road
Elk Grove Village
Illinois 60007, USA
(*Tel: 312 956 0126*)

Control Automation Inc.
PO Box 2304
Princeton
New Jersey 08540 USA

Cybotech Corp
P.O. Box 88514
Indianapolis
Indiana 46208, USA
(*Tel: 317 298 5136*)

Cyclomatic Inc.
7520 Corvey Court
San Diego, California 92111 USA
(*Tel: 714 292 7440*)

Dependable-Fordath Inc.
400 SE Willimette St.
Sherwood
Oregon 97140 USA

DeVilbiss Company
837 Airport Boulevard
Ann Arbor,
Michigan 48104, USA
(*Tel: 313 668 6765*)

Dynamcac Inc.
410 Forest Street
Marlboro
Massachusetts 01752 USA

Everett/Charles
Automation Modules Inc.
9645 Arrow Route, Suite A
Rancho Cucamonga, California 91730, USA
(*Tel: 714 980 1525*)

EWAB America
292 W. Palatine Road
Wheeling
Illinois 60090 USA

Fared Robotic Systems Inc.
3860 Revere Street, Suite D
P.O. Box 39268
Denver, Colorado 80239, USA
(*Tel: 303 371 5868*)

Fleximation Systems Corporation
53 Second Avenue
Burlington
Massachusetts 01803 USA

GCA/PAR Systems
3460 Lexington Avenue, North
St Paul
Minnesota 55112, USA
(*Tel: 612 484 7261*)

General Electric
Automation Systems
1285 Boston Avenue, Bridgeport
Connecticut 06602, USA
(*Tel: 203 382 2876*)

General Numeric Corp.
390 Kent Avenue
Elk Grove Village
Illinois 60007, USA
(*Tel: 312 640 1595*)

GMF Robotics Corp.
5600 New King Street
Troy
Michigan 48098 USA

Graco Robotics Inc.
12899 Westmore Ave
Livonia, Michigan 48150, USA
(*Tel: 313 261 3270*)

Hellstar Corporation
1600 N. Chestnut
Wahoo
Nebraska 68066 USA

Hobart Brothers Co.
600 W. Main
Troy
Ohio 45373 USA

Hodges Robotics International Corp.
3710 North Grand River Avenue
Lansing
Michigan 48906, USA
(*Tel: 517 323 7427*)

IBM Advanced Manufacturing Systems
1000 NW 51st Street
Boca Raton
Florida 33432, USA
(*Tel: (305) 998-2000*)

Industrial Automates Inc.
6123 W. Mitchell St.
Milwaukee

Wisconsin 53214, USA
(Tel: (414) 327 5656)

Intarm
P.O. Box 53
Dayton, Ohio 45409, USA
(Tel: (518) 294 0834)

Intelledex Inc.
33840 Eastgate Circle
Corvallis
Oregon 97333 USA

International Robotmation Intelligence
2281 Las Palmas Drive
Carlsbad
California 92008, USA
(Tel: 714 438 4424)

ISI Manufacturing Inc.
31915 Groesbeck Highway
Fraser
Michigan 48026, USA
(Tel: 313 294 9500)

Keller Technology Corporation
Robotics Automation Systems
2320 Military Road
Tonawanda
New York 14150, USA

Lamson Corp.
P.O. Box 4857
Syracuse
New York 13221, USA
(Tel: 315 432 5467)

Livernois Automation Co.
25315 Kean
Dearborn
Michigan 48124, USA
(Tel: (312) 278 0201)

Lynch Machinery Corp.
2300 Crystal St.
P.O. Box 2477
Anderson, Indiana 46018, USA
(Tel: (317) 643 6671)

Machine Intelligence Corp.
330 Potrero Ave.
Sunnyvale, California 94086, USA
(Tel: 408 737 7960)

Mack Corp.
3695 East Industrial Drive
Flagstaff
Arizona 86001, USA
(Tel: 602 526 1120)

Manca Inc.
Link Drive
Rockleigh
New Jersey 07647, USA
(Tel: (201) 767 7227)

Microbot Inc.
453-H Ravendale Drive
Mountain View
California 94043, USA
(Tel: 415 968 8911)

Mobot Corp.
980 Buenos Avenue
San Diego
California 92110, USA
(Tel: 714 275 4300)

Pickomatic Systems Inc.
37950 Commerce Drive
Sterling Heights
Michigan 48077, USA
(Tel: 313 939 9320)

Positech Corporation
114 Rush Lake Road
Laurens
Iowa 50554 USA

Prab Robots Inc.
5944 E. Kilgore Road
Kalamazoo
Michigan 49003, USA
(Tel: 616 349 8761)

Precision Robots Inc.
6 Carmel Circle
Lexington
Massachusetts 02173, USA
(Tel: 617 862 1124)

Reeves Robotics Inc.
Box S
Issaquah
Washington 98027, USA
(Tel: 206 392 1447)

Ron-Con Ltd./Bra-Con Industries
12001 Globe Road
Livonia
Michigan 48154 USA

Robotic Sciences International Inc.
2709 South Halladay
Santa Ana
California 92705, USA
(Tel: 714 979 6831)

Robotiks Inc.
507 Prudential Road
Horsham
Pennsylvania 19044, USA
(Tel: 215 674 2800)

Sandhu Rhino Robots
308 S. State Street
Champaign
Illinois 61820, USA
(Tel: 217 352 8485)

Schrader-Bellows
US Rt. 1 N

Wake Forest
North Carolina 27587, USA
(*Tel: 919 556 4031*)

Seiko Instruments USA Inc.
2990 W. Lomita Boulevard
Torrance
California 90505, USA
(*Tel: 213 530 8777*)

Sigma Sales Inc.
6505C Serrano Avenue
Anaheim Hills
California 92807, USA
(*Tel: 714 974 0166*)

Sterling Detroit Company
261 E. Goldengate
Detroit
Michigan 48203, USA
(*Tel: 313 366 3500*)

Sterltech
PO Box 23421
Milwaukee, Wisconsin 53223, USA
(*Tel: 414 354 0493*)

TecQuipment Inc.
PO Box 1074, Acton
Massachusetts 01720, USA
(*Tel: (617) 263 1767*

Thermwood Corporation
P.O. Box 436
Dale
Indiana 47523, USA
(*Tel: 812 937 4476*)

Unimation Inc.
Shelter Rock Lane
Danbury
Connecticut, 06810, USA
(*Tel: 203 744 1800*)

United States Robots Inc.
1000 Conshohocken Road
Conshohocken
Pennsylvania 19428 USA

United Technologies Automotive Group
5200 Auto Club Drive
Dearborn
Michigan 48126 USA

VSI Automation Assembly Inc.
165 Park Street
Troy
Michigan 48084 USA

Westinghouse Electric Corporation
See Unimation Inc.
400 Media Drive
Pittsburgh
Pennsylvania 15205, USA
(*Tel: 412 778 4347*)

West Germany

Carl Cloos Schweisstechnik GmbH
D-6342 Haiger
WEST GERMANY
(*Tel: 02773850*)

Fibro GmbH
Postfach 1120
D 6954 Hassmersheim
WEST GERMANY

Gebr. Felss
7535 Konigsbach Stein 2
Gutensbergstr. 4
WEST GERMANY

G.D.A.
5 Am Bahnhof
D-8915 Fuchstal
WEST GERMANY
(*Tel: 08243 2012*)

Jungheinrich KG
Friedrich-Ebert-Damm 184
2000 Hamburg 70
WEST GERMANY
(*Tel: 040 66 43 50*)

KUKA Schweissanlagen + Roboter
Zugspitzstrasse 140
D-8900 Augsburg 43
WEST GERMANY
(*Tel: 0821 7971*)

Mantec GmbH
Postfach 2620
D-8520 Erklangen
WEST GERMANY
(*Tel: 09131/16200*)

Messer Griesheim GmbH
Landsbergerstrasse 432
D-8000 Munich 60
WEST GERMANY

Nimak–MAG
Postfach 192
D-5248 Wissen
WEST GERMANY
(*Tel: 02742 4024/4025*)

Ottensener Eisenwerk GmbH
Steinwerder
D-2000 Hamburg 11
WEST GERMANY
(*Tel: 040 306859*)

Pfaff
Postfach 3020/3040
D-6750 Kaiserslautern
WEST GERMANY
(*Tel: 0631 881*)

Produtec GmbH
Heilbronnerstrasse 67

D-7000
Stuttgart 1
WEST GERMANY

Robert Bosch GmbH
Geschaftsbereich Industrieausrustung
7000 Stuttgart 30, Kruppstrasse 1
Postfach 300220
WEST GERMANY
(*Tel: 0711 811 5225*)

Siemens AG
Rupert-Mayer-Strasse 44
8000 Munchen 70, Postfach 70 00 75
WEST GERMANY
(*Tel: 089 722 26126*)

Union Carbide Deutschland GmbH
(*See also Nimak–MAG*)
Postfach 133, D-5248 Wissen
WEST GERMANY
(*Tel: 02742 751*)

VFW Fokker GmbH
Hunefeldstrasse 1–5
D-2800 Bremen 1
WEST GERMANY

Volkswagen AG
Industrial Robot Division
3180 Wolfsburg
WEST GERMANY

Walter Reis GmbH & Co
D-8753 Obernburg
WEST GERMANY
(*Tel: 04 188 113*)

ZF
Postfach 25 20
D-7990 Friedrichshafen 1
WEST GERMANY
(*Tel: 07541 701-1*)

A.5. ROBOT JOURNALS

Australia

Australian Machinery and Production
    Engineering
Australian Welding Journal
Electrical Engineer
Journal of the Institution of Engineers, Australia
Metals Australia

Austria

Schweisstechnik
Diagramm

Belgium

Manutention mécanique et automation
Revue M (Mécanique)

Bulgaria

Mashinostroene
Problemi na tekhnicheskata
    kibernetika i robotikata
Teoretichna i prilozhna
    mekhanika

Canada

Canadian Machinery and Metalworking

Czechoslovakia

Slévárenstvi
Strojirenska vyroba
Technická práce

East Germany

Feingerätetechnik
Fertigungstechnik und Betrieb

Hebezeuge und Fördermittel
Maschinenbautechnik
Messen—Steuern—Regeln
Metallverarbeitung
Schweisstechnik
Seewirtschaft
Sozialistische Rationalizierung in der
    Elektrotechnik/Elektronik
Technische Gemeinschaft
VEM—Elektro—Anlagenbahn
Die Wirtschaft
Wissenschaftlich—Technische Informationen
Wissenschaftliche Zeitschrift der Technischen
    Hochschule Ilmenau
Zeitschrift für angewandte Mathematik und
    Mechanik
ZIS—Mitteilungen

Finland

Konepajamies

France

Energie fluide. Hydraulique, pneumatique
    asservissements, lubrification
Fondeur aujourd'hui
Machine moderne
Machine—outil
Manutention
Métaux déformation
Nouvel automatisme
Soudage et techniques connexes
L'Usine nouvelle

Hungary

Automatizálás
Bányászati és kohászati lapok öntöde az országos
    magyar bányászati és kohászáti egyesület
    lapja

Gépgyartástechnológia
Ipargazdaság
Mérés és automatika

Italy

La tecnica professionale. Collegio ingegneri
ferroviari italiani
Macchine. Rassegna tecnika dell'industria
metalmeccanica
Rivista de meccanica
Transport industriali

Japan

Automation
Chemical Engineering
Hydraulics and Pneumatics
Japan Economic Journal
Japan Light Metal Welding
Journal of the Instrumentation Control
Association, Japan
Journal of the Japan Welding Society
Mechnical Automation
Mechanical Design
Mechanical Engineering
Mitsubishi denki giho
Mitsubishi juko giho
Promoting Machine Industry in Japan
Robot
Robotpia (Mechatronic, Science & Society in
Robot Age)
Science of Machine
Technology Reports of the Kyushu University
The Japan Robot News
The Journal of the Institute of Electrical
Engineers of Japan
Toshiba Review
Transactions of the Institute of Electronics and
Communication Engineering, Japan
Transactions of the Society of Instruments and
Control Engineers
Welding Technique

Netherlands

Ingenieur
Iron Age Metalworking International
Metalbewerking
Polytechnisch tijdschrift. Uitgave A—
Werkuigbouwkunde en elektrotechniek

New Zealand

Automation and control

Poland

Automatcka kolejowa
Biuletyn informacyjny Institutu maszyn
matematycznych
Mechanik
Przeglad mechaniczny
Przeglad spawalnictwa
Wiadomosci elektrotechniczne

Zeszyty naukowe. Akademia gorniczohutnicza
Zeszyty naukowe. Politechnika Slaska

Rumania

Constructia de masini

Spain

Regulación y mando automatico

Switzerland

C. I. R. P. Annals (International Institution for
Production Engineering Research)
Elektroniker
Management Zeitschrift Ind. Organis.
Schweißtechnik Soudure
Schweizerische technische Zeitschrift
Schweizer Maschinenmarkt
Technica. Illustrierte technische Zeitschrift
Technische Rundschau
Zeitschrift Schweisstechnik

United Kingdom

Assembly Automation
Assembly Engineering
Automation
Automotive Engineering
British Foundryman
Control and Instrumentation
Design Engineering Materials and Components
Electrical Review
Engineer
Foundry Trade Journal
Hydraulic Pneumatic Mechanical Power
Industrial Robot
International Journal of Man-Machine Studies
Machinery and Production Engineering
Manufacturing Engineering
Materials Handling News
Mechanism and Machine Theory
Metals and Materials
Metalworking Production
New Electronica
New Scientist
Pattern Recognition
Plastics in Engineering
Robot News International
Sensor Review
Sheet Metals Industries
Welding and Metal Fabrication

United States

American Machinist
American Metalmarket
Artificial Intelligence
ASME Transactions on Dynamics,
Measurement and Control
Assembly Engineering
Automatic Machining
Compressed Air

Computer Graphic and Image Processing
Design News
Electronics
Futurist
Hydraulics and Pneumatics
IEEE Spectrum
IEEE Transactions on Automatic Control
IEEE Transactions on Industrial Electronics and
   Control Instrumentation
IEEE Transactions on Power Apparatus and
   Systems
IEEE Transactions on Systems, Man and
   Cybernetics
IIE Transactions
Industrial Engineering
Industrial Robots International (USA)
Information and Control
International Journal of Computer and
   Information Science
International Journal of Robotics Research
Iron Age
Journal of Manufacturing Systems
Journal of Robotic Systems
Machine and Tool Blue Book
Manufacturing Engineer (USA)
Material Handling Engineering
Mechanical Engineering
Modern Material Handling
Plating and Surface Finishing
Popular Science
Product Engineering
Production (USA)
Robotics. An International Journal.
Robotics Age
Robotics Today
Robotics World
Tooling and Production
Welding Design and Fabrication
Welding Engineer
Welding Journal

USSR

Avtomatika i telemekhanika
Avtomatizatsiya proizvodstvennykh protsessov
   v mashinostroenii i priborostroenii
Avtomatizatsiya tekhnologicheskikh protsessov
Avtomatizirovannyy elekroprivod
Vestnik mashinostroeniya
Voprosy dinamiki i prochnosti
Izvestiya Akademii nauk SSSR. Tekhnicheskaya
   kibernetika
Izvestiya vysshikh uchebnykh zavedeniy.
   Mashinostroeniye
Izvestiya vysshikh uchebnykh zavedenii:
   Elektromekhanika
Izvestiya Leningradskogo
   elektrotekhnicheskogo instituta
Kuznechno-stampovochnoye proizvodstvo
Liteynoye proizvodstvo
Mashinovedeniye
Mashinostroitel'

Mekhanizatsiya i avtomatizatsiya proizvodstva
Mekhanizatsiya i elektrifikatsiya sel'skogo
   khozyaystva
Priborostroeniye
Promyshlennyy transport
Svarochnoye proizvodstvo
Stanki i instrument
Stroitel'nyye i dorozhnyye mashiny
Sudostroeniye
Trudy Vsesoyuznogo nauchno-
   issledovatel'skogo i proektno-
   tekhnologicheskogo instituta ugol'nogo
   mashinostroeniya
Trudy Leningradskogo politekhnicheskogo
   instituta
Trudy Moskovskogo energeticheskogo instituta
Elektrotekhnika
Elektrotekhnicheskaya promyshlennost':
   Elektroprivod.

West Germany

Biological Cybernetics
Blech Rohre. Profile
Der Plastverarbeiter mit Sonderdruck «Der
   Kunststoffmarkt»
Die Maschine
Die Computerzeitung
DVS—Berichte
Elektronik
Elektronikindustrie
Elektrotechnische Zeitschrift
Feinwerktechnik und Messtechnik
Fördern und Heben
Industrie—Anzeiger
Kunststoffe
Lecture Notes in Computer Science
Maschinen—Anlagen + Ferfahren
Maschinenmarkt + Europa Industrie Revue
Maschine und Werkzeug
Praktiker
Produktion
Regelungstechnik
Regelungstechnische Praxis
Schweissen und Schneiden
Siemens Energietechnik
Technische Zentralblatt für praktische
   Metallbearbeitung
TU. Sicherheit Zuverfässigkeit in Wirtschaft—
   Betrieb—Verkehr
Verbindungstechnik
VDI—Berichte. Düsseldorf
VDI—Nachrichten
VDI—Z. Zeitschrift für die Gesamte Technik
Werkstatt und Betrieb
Werkstattstechnik
ZwF. Zeitschrift für wirtschaftliche Fertigung

Yugoslavia

Automatika

INDEX